2024
INTERNATIONAL
BUILDING CODE® ILLUSTRATED
HANDBOOK

About the International Code Council®

The International Code Council is the leading global source of model codes and standards and building safety solutions that include product evaluation, accreditation, technology, codification, consulting, training and certification. The International Code Council's codes, standards and solutions are used to ensure safe, affordable and sustainable communities and buildings worldwide.

The International Code Council family of solutions includes the ICC Evaluation Service (ICC ES), S. K. Ghosh Associates, the International Accreditation Service (IAS), General Code, ICC NTA, ICC Community Development Solutions, Alliance for National & Community Resilience (ANCR) and American Legal Publishing.

Office Locations:
Headquarters:
200 Massachusetts Avenue, NW, Suite 250
Washington, DC 20001
888-ICC-SAFE (888-422-7233)
www.iccsafe.org

Eastern Regional Office
900 Montclair Road
Birmingham, AL 35213

Central Regional Office
4051 Flossmoor Road
Country Club Hills, IL 60478

Western Regional Office
3060 Saturn Street, Suite 100
Brea, CA 92821

MENA Regional Office
Dubai Association Centre Office, One Central
Building 2, Office 8, Dubai World Trade Centre Complex
PO Box 9292, Dubai, UAE

OCEANIA Regional Office
Level 9, Nishi Building
2 Phillip Law Street
Canberra ACT 2601 Australia

2024 INTERNATIONAL BUILDING CODE® ILLUSTRATED HANDBOOK

Douglas W. Thornburg, AIA, CBO
John "Buddy" Showalter, PE

Library of Congress Control Number: 2024951028

McGraw Hill books are available at special quantity discounts to use as premiums and sales promotions or for use in corporate training programs. To contact a representative, please visit the Contact Us page at www.mhprofessional.com.

2024 International Building Code® Illustrated Handbook

Copyright © 2025 by the International Code Council. All rights reserved. Printed in the United States of America. Except as permitted under the United States Copyright Act of 1976, no part of this publication may be reproduced or distributed in any form or by any means, or stored in a data base or retrieval system, without the prior written permission of the publisher.

1 2 3 4 5 LBC 28 27 26 25

ISBN 978-1-266-20578-1
MHID 1-266-20578-0

This book is printed on acid-free paper.

Sponsoring Editor
Lara Zoble,
McGraw Hill (MH)

Production Supervisor
Richard Ruzycka, MH

Acquisitions Coordinator
Olivia Higgins, MH

Project Manager
Radhika Jolly,
KnowledgeWorks
Global Ltd. (KGL)

Copy Editor
Girish Sharma, KGL

Proofreader
Manish Tiwari, KGL

Art Director, Cover
Anthony Landi, MH

Composition
KnowledgeWorks Global Ltd.

ICC Staff

Executive VP and Director of Business Development
Mark Johnson

Senior VP, Product Development
Hamid Naderi

VP and Technical Director, Products and Services
Doug Thornburg

Information contained in this work has been obtained by McGraw Hill from sources believed to be reliable. However, neither McGraw Hill nor its authors guarantee the accuracy or completeness of any information published herein, and neither McGraw Hill nor its authors shall be responsible for any errors, omissions, or damages arising out of use of this information. This work is published with the understanding that McGraw Hill and its authors are supplying information but are not attempting to render engineering or other professional services. If such services are required, the assistance of an appropriate professional should be sought.

About the Authors

Douglas W. Thornburg, AIA, CBO, is currently Vice President and Technical Director of Products and Services for the International Code Council (ICC), where he provides administrative and technical leadership for the ICC product development activities. Prior to employment with ICC in 2004, he spent 9 years as a code consultant and educator on building codes. Doug also spent 10 years with the International Conference of Building Officials (ICBO) where he served as Vice President/Education. He began his career with the City of Wichita, Kansas, where he had inspection, plan review, and building administration responsibilities.

In his current role, Doug also continues to create and present building code seminars nationally and has developed numerous educational texts and resource materials. He was presented with the ICC's inaugural Educator of the Year Award in 2008, recognizing his outstanding contributions in education and training.

A graduate of Kansas State University and a registered architect, Doug has over 44 years of experience in building code education and administration. He has authored a variety of other code-related support publications, including the *Significant Changes to the International Building Code.*

John "Buddy" Showalter, PE, is a senior staff engineer with the International Code Council's (ICC) product development group. He develops technical resources and provides education in support of the structural provisions of the *International Building Code* and *International Residential Code.* He is the project lead in the development of *Mass Timber Buildings and the IBC* and associated education programs, jointly developed by ICC and the American Wood Council (AWC).

A graduate of Virginia Tech, Showalter has also been a member of the editorial board for *STRUCTURE* magazine for nearly 20 years. Before joining ICC, Showalter spent 26 years with AWC where he led its technology transfer program with oversight responsibility for publications, website, helpdesk, education, and other technical media. He has over 35 years of experience in the development and support of building codes and standards, authoring or coauthoring more than 70 publications and technical articles for industry-related trade journals.

Contents

Foreword xv

Preface xvii

Acknowledgments xix

Chapter 1
Scope and Administration 1

☐ Section 101	Scope and General Requirements 2	
☐ Section 102	Applicability................ 4	
☐ Section 103	Code Compliance Agency 6	
☐ Section 104	Duties and Powers of Building Official 6	
☐ Section 105	Permits 9	
☐ Section 107	Construction Documents 11	
☐ Section 108	Temporary Structures, Equipment, and Systems..... 12	
☐ Section 109	Fees 12	
☐ Section 110	Inspections 13	
☐ Section 111	Certificate of Occupancy 14	
☐ Section 112	Service Utilities 16	
☐ Section 113	Means of Appeals 16	
☐ Section 114	Violations.................. 16	
☐ Section 115	Stop Work Order 16	
☐ Section 116	Unsafe Structures and Equipment................. 18	
☐ KEY POINTS 18		

Chapter 2
Definitions 19

☐ Section 201   General 20
☐ Section 202   Definitions.................. 20
☐ KEY POINTS 50

Chapter 3
Occupancy Classification and Use 51

☐ Section 302   Occupancy Classification and Use Designation 52
☐ Section 303   Assembly Group A.......... 54
☐ Section 304   Business Group B........... 59
☐ Section 305   Educational Group E 60
☐ Section 306   Factory Group F............ 62
☐ Section 307   High-Hazard Group H 64
☐ Section 308   Institutional Group I......... 71
☐ Section 309   Mercantile Group M 74
☐ Section 310   Residential Group R......... 74
☐ Section 311   Storage Group S............ 77
☐ Section 312   Utility and Miscellaneous Group U................... 78
☐ KEY POINTS 80

Chapter 4
Special Detailed Requirements Based on Occupancy and Use 81

☐ Section 402   Covered Mall and Open Mall Buildings 82
☐ Section 403   High-Rise Buildings......... 92
☐ Section 404   Atriums 98
☐ Section 405   Underground Buildings..... 101
☐ Section 406   Motor-Vehicle-Related Occupancies 103
☐ Section 407   Group I-2................. 108
☐ Section 408   Group I-3 113
☐ Section 409   Motion-Picture Projection Rooms 114
☐ Section 410   Stages, Platforms, and Technical Production Areas 114
☐ Section 411   Special Amusement Areas 117
☐ Section 412   Aircraft-Related Occupancies 118
☐ Section 413   Combustible Storage 120
☐ Section 414   Hazardous Materials 120
☐ Section 415   Groups H-1, H-2, H-3, H-4, and H-5.................... 124
☐ Section 416   Spray Application of Flammable Finishes........ 128
☐ Section 417   Drying Rooms.............. 129
☐ Section 418   Organic Coatings 129
☐ Section 420   Groups I-1, R-1, R-2, R-3, and R-4.................... 129

Section 422	Ambulatory Care Facilities 131
Section 423	Storm Shelters 133
Section 424	Play Structures.......... 133
Section 427	Medical Gas Systems...... 134
Section 428	Higher Education Laboratories 134
KEY POINTS 136

Chapter 5
General Building Heights and Areas 139

Section 502	Building Address 140
Section 503	General Building Height and Area Limitations 140
Section 504	Building Height and Number of Stories 143
Section 505	Mezzanines and Equipment Platforms....... 144
Section 506	Building Area 148
Section 507	Unlimited-Area Buildings ... 161
Section 508	Mixed Use and Occupancy................ 169
Section 509	Incidental Uses........... 183
Section 510	Special Provisions......... 186
KEY POINTS 191

Chapter 6
Types of Construction 193

| Section 602 | Construction Classification ... 194 |
| Section 603 | Combustible Material in Type I and II Construction............. 203 |
| KEY POINTS .... 205 |

Chapter 7
Fire and Smoke Protection Features......................... 207

Section 702	Multiple-Use Fire Assemblies 208
Section 703	Fire-Resistance Ratings and Fire Tests........... 208
Section 704	Fire-Resistance Rating of Structural Members....... 215
Section 705	Exterior Walls............ 220

Section 706	Fire Walls 240
Section 707	Fire Barriers............. 251
Section 708	Fire Partitions............ 253
Section 709	Smoke Barriers........... 256
Section 710	Smoke Partitions 257
Section 711	Floor and Roof Assemblies 257
Section 712	Vertical Openings......... 258
Section 713	Shaft Enclosures.......... 261
Section 714	Penetrations 265
Section 715	Joints and Voids 273
Section 716	Opening Protectives....... 276
Section 717	Ducts and Air Transfer Openings............... 282
Section 718	Concealed Spaces......... 288
Section 719	Fire-Resistance Requirements for Plaster.... 294
Section 720	Thermal- and Sound-Insulating Materials....... 294
Section 721	Prescriptive Fire Resistance 294
Section 722	Calculated Fire Resistance ... 299
KEY POINTS 300

Chapter 8
Interior Finishes..................... 303

Section 802	General 304
Section 803	Wall and Ceiling Finishes ... 305
Section 804	Interior Floor Finish........ 308
Section 805	Combustible Materials in Types I and II Construction............. 309
KEY POINTS 310

Chapter 9
Fire Protection and Life-Safety Systems.......................... 311

Section 901	General 312
Section 902	Fire Pump and Riser Room Size 314
Section 903	Automatic Sprinkler Systems 315
Section 904	Alternative Automatic Fire-Extinguishing Systems 341

☐ Section 905	Standpipe Systems	341
☐ Section 907	Fire Alarm and Detection Systems	348
☐ Section 909	Smoke Control Systems	358
☐ Section 910	Smoke and Heat Removal	362
☐ Section 911	Fire Command Center	365
☐ Section 912	Fire Department Connections	365
☐ Section 913	Fire Pumps	366
☐ Section 914	Emergency Responder Safety Features	366
☐ Section 915	Carbon Monoxide Detection	367
☐ Section 916	Gas Detection Systems	368
☐ Section 917	Mass Notification Systems	369
☐ KEY POINTS		370

Chapter 10
Means of Egress 371

☐ Section 1001	Administration	373
☐ Section 1002	Maintenance and Plans	373
☐ Section 1003	General Means of Egress	374
☐ Section 1004	Occupant Load	379
☐ Section 1005	Means of Egress Sizing	392
☐ Section 1006	Number of Exits and Exit Access Doorways	400
☐ Section 1007	Exit and Exit Access Doorway Configuration	409
☐ Section 1008	Means of Egress Illumination	413
☐ Section 1009	Accessible Means of Egress	415
☐ Section 1010	Doors, Gates, and Turnstiles	421
☐ Section 1011	Stairways	441
☐ Section 1012	Ramps	451
☐ Section 1013	Exit Signs	454
☐ Section 1014	Handrails	457
☐ Section 1015	Guards	465
☐ Section 1016	Exit Access	471
☐ Section 1017	Exit Access Travel Distance	473
☐ Section 1018	Aisles	476
☐ Section 1019	Exit Access Stairways and Ramps	477
☐ Section 1020	Corridors	478
☐ Section 1021	Egress Balconies	483
☐ Section 1022	Exits	484
☐ Section 1023	Interior Exit Stairways and Ramps	485
☐ Section 1024	Exit Passageways	490
☐ Section 1025	Luminous Egress Path Markings	491
☐ Section 1026	Horizontal Exits	492
☐ Section 1027	Exterior Exit Stairways and Ramps	496
☐ Section 1028	Exit Discharge	499
☐ Section 1029	Egress Courts	501
☐ Section 1030	Assembly	503
☐ Section 1031	Emergency Escape and Rescue	518
☐ KEY POINTS		522

Chapter 11
Accessibility 525

☐ Section 1101	General	528
☐ Section 1102	Compliance	529
☐ Section 1103	Scoping Requirements	529
☐ Section 1104	Accessible Route	531
☐ Section 1105	Accessible Entrances	533
☐ Section 1106	Parking and Passenger Loading Facilities	534
☐ Section 1107	Motor-Vehicle-Related Facilities	536
☐ Section 1108	Dwelling Units and Sleeping Units	536
☐ Section 1109	Special Occupancies	540
☐ Section 1110	Other Features and Facilities	541
☐ Section 1111	Recreational Facilities	546
☐ Section 1112	Signage	546
☐ KEY POINTS		547

Chapter 12
Interior Environment 549

☐ Section 1202	Ventilation	550
☐ Section 1203	Temperature Control	554

Section 1204	Lighting 555
Section 1205	Yards or Courts 556
Section 1206	Sound Transmission 557
Section 1207	Enhanced Classroom Acoustics 557
Section 1208	Interior Space Dimensions 558
Section 1209	Access to Unoccupied Spaces 560
Section 1210	Toilet and Bathroom Requirements 560
KEY POINTS	 561

Chapter 13
Energy Efficiency 563

Chapter 14
Exterior Walls 565

Section 1402	Performance Requirements 566
Section 1403	Materials 567
Section 1404	Installation of Wall Coverings 568
Section 1405	Combustible Materials on the Exterior Side of Exterior Walls 570
Section 1406	Metal Composite Material (MCM) 570
Section 1407	Exterior Insulation and Finish Systems 571
Section 1409	Insulated Metal Panels 572
Section 1412	Soffits and Fascias at Roof Overhangs 573
KEY POINTS	 574

Chapter 15
Roof Assemblies and Rooftop Structures 575

Section 1504	Performance Requirements 576
Section 1505	Fire Classification 577
Section 1506	Materials 578
Section 1507	Requirements for Roof Coverings 579
Section 1511	Rooftop Structures 580
Section 1512	Reroofing 582
KEY POINTS	 583

Chapter 16
Structural Design 585

Section 1601	General 586
Section 1602	Notations 587
Section 1603	Construction Documents ... 587
Section 1604	General Design Requirements 587
Section 1605	Load Combinations 600
Section 1606	Dead Loads 602
Section 1607	Live Loads 603
Section 1608	Snow Loads 614
Section 1609	Wind Loads 620
Section 1610	Soil Loads and Hydrostatic Pressure 625
Section 1611	Rain Loads 627
Section 1612	Flood Loads 628
Section 1613	Earthquake Loads 630
Section 1614	Atmospheric Ice Loads 633
Section 1615	Tsunami Loads 634
Section 1616	Structural Integrity 634
KEY POINTS	 635

Chapter 17
Special Inspections and Tests 637

Section 1701	General 638
Section 1702	New Materials 640
Section 1703	Approvals 641
Section 1704	Special Inspections and Tests, Contractor Responsibility, and Structural Observation 647
Section 1705	Required Special Inspections and Tests 652
Section 1706	Design Strengths of Materials 672
Section 1707	Alternate Test Procedure ... 672
Section 1708	In-Situ Load Tests 672
Section 1709	Preconstruction Load Tests 673
KEY POINTS	 681

Chapter 18
Soils and Foundations **683**

☐ Section 1801	General 684
☐ Section 1802	Design Basis 684
☐ Section 1803	Geotechnical Investigations 685
☐ Section 1804	Excavation, Grading, and Fill 689
☐ Section 1805	Dampproofing and Waterproofing 690
☐ Section 1806	Presumptive Load-Bearing Values of Soils 697
☐ Section 1807	Foundation Walls, Retaining Walls, and Embedded Posts and Poles 698
☐ Section 1808	Foundations 703
☐ Section 1809	Shallow Foundations 709
☐ Section 1810	Deep Foundations 714
☐ KEY POINTS	 743

Chapter 19
Concrete **745**

☐ Section 1901	General 746
☐ Section 1902	Coordination of Terminology 747
☐ Section 1903	Specifications for Tests and Materials 747
☐ Section 1904	Durability Requirements ... 748
☐ Section 1905	Seismic Requirements 749
☐ Section 1906	Footings for Light-Frame Construction 750
☐ Section 1907	Slabs-on-Ground 751
☐ Section 1908	Shotcrete 751
☐ KEY POINTS	 752

Chapter 20
Aluminum **753**

☐ Section 2002	Materials 754
☐ KEY POINTS	 754

Chapter 21
Masonry **755**

☐ Section 2101	General 756
☐ Section 2102	Notations 756
☐ Section 2103	Masonry Construction Materials 757
☐ Section 2104	Construction 758
☐ Section 2105	Quality Assurance 759
☐ Section 2106	Seismic Design 760
☐ Section 2107	Allowable Stress Design ... 764
☐ Section 2108	Strength Design of Masonry 765
☐ Section 2109	Empirical Design of Adobe Masonry 765
☐ Section 2110	Glass Unit Masonry 766
☐ Section 2111	Masonry Fireplaces 766
☐ Section 2112	Masonry Heaters 767
☐ Section 2113	Masonry Chimneys 767
☐ Section 2114	Dry-Stack Masonry 767
☐ KEY POINTS	 768

Chapter 22
Steel **769**

☐ Section 2201	General 770
☐ Section 2202	Structural Steel and Composite Structural Steel and Concrete 771
☐ Section 2203	Structural Stainless Steel ... 773
☐ Section 2204	Cold-Formed Steel 773
☐ Section 2205	Cold-Formed Stainless Steel 774
☐ Section 2206	Cold-Formed Steel Light-Frame Construction ... 774
☐ Section 2207	Steel Joists 775
☐ Section 2208	Steel Deck 776
☐ Section 2209	Steel Storage Racks 777
☐ Section 2210	Metal Building Systems 778
☐ Section 2211	Industrial Boltless Steel Shelving 778
☐ Section 2212	Industrial Steel Work Platforms 779
☐ Section 2213	Stairs, Ladders and Guarding for Steel Storage Racks and Industrial Steel Work Platforms 779
☐ Section 2214	Steel Cable Structures 779
☐ KEY POINTS	 780

Chapter 23
Wood 781

☐ Section 2301	General	782
☐ Section 202	Definitions	782
☐ Section 2302	Design Requirements	783
☐ Section 2303	Minimum Standards and Quality	783
☐ Section 2304	General Construction Requirements	795
☐ Section 2305	General Design Requirements for Lateral Force-Resisting Systems	808
☐ Section 2306	Allowable Stress Design	809
☐ Section 2307	Load and Resistance Factor Design	815
☐ Section 2308	Conventional Light-Frame Construction	817
☐ Section 2309	Wood Frame Construction Manual	857
☐ KEY POINTS		858

Chapter 24
Glass and Glazing 861

☐ Section 2402	Glazing Replacement	862
☐ Section 2403	General Requirements for Glass	862
☐ Section 2404	Wind, Snow, Seismic, and Dead Loads on Glass	863
☐ Section 2405	Sloped Glazing and Skylights	865
☐ Section 2406	Safety Glazing	868
☐ Section 2407	Glass in Handrails and Guards	881
☐ Section 2408	Glazing in Athletic Facilities	882
☐ Section 2409	Glass in Walkways, Elevator Hoistways, and Elevator Cars	882
☐ KEY POINTS		883

Chapter 25
Gypsum Panel Products and Plaster 885

☐ Section 2501	General	886
☐ Section 2508	Gypsum Construction	886
☐ Section 2510	Lathing and Furring for Cement Plaster (Stucco)	889
☐ Section 2511	Interior Plaster	890
☐ Section 2512	Exterior Plaster	891
☐ KEY POINTS		893

Chapter 26
Plastic 895

☐ Section 2603	Foam Plastic Insulation	896
☐ Section 2605	Plastic Veneer	902
☐ Section 2606	Light-Transmitting Plastics	902
☐ Section 2608	Light-Transmitting Plastic Glazing	903
☐ Section 2609	Light-Transmitting Plastic Roof Panels	903
☐ Section 2610	Light-Transmitting Plastic Skylight Glazing	903
☐ KEY POINTS		904

Chapter 27
Electrical 905

☐ Section 2702	Emergency and Standby Power Systems	906
☐ Section 2703	Lightning Protection Systems	907
☐ KEY POINTS		907

Chapter 28
Mechanical 909

Chapter 29
Plumbing 911

☐ Section 2902	Minimum Plumbing Facilities	912
☐ Section 2903	Installation of Fixtures	917
☐ KEY POINTS		918

Chapter 30
Elevators and Conveying Systems ... 919

☐ Section 3001	General	920
☐ Section 3002	Hoistway Enclosures	920

☐ Section 3003	Emergency Operations 924		☐ Section 3309	Fire Extinguishers 953
☐ Section 3006	Elevator Lobbies and Hoistway Door Protection . . . 924		☐ Section 3310	Means of Egress 954
☐ Section 3007	Fire Service Access Elevator 926		☐ Section 3311	Standpipes 954
			☐ Section 3313	Water Supply for Fire Protection 954
☐ Section 3008	Occupant Evacuation Elevators 928		☐ Section 3314	Fire Watch During Construction 954
☐ Section 3009	Private Residence Elevators 928		☐ KEY POINTS	 955

☐ KEY POINTS 929

Chapter 31
Special Construction 931

Chapter 34
Reserved 957

☐ Section 3102	Membrane Structures 932
☐ Section 3103	Temporary Structures 933

Chapter 35
Referenced Standards 959

☐ Section 3104	Pedestrian Walkways and Tunnels 936
☐ Section 3106	Marquees 938
☐ Section 3111	Solar Energy Systems 938
☐ Section 3113	Relocatable Buildings 939
☐ Section 3114	Intermodal Shipping Containers 939

Appendices 961

☐ Appendix A	Employee Qualifications 962
☐ Appendix B	Board of Appeals 962
☐ Appendix C	Group U Agricultural Buildings 963
☐ Appendix D	Fire Districts 963
☐ Appendix E	Supplementary Accessibility Requirements 965
☐ Appendix F	Rodentproofing 965
☐ Appendix G	Flood-Resistant Construction 965
☐ Appendix H	Signs 966
☐ Appendix I	Patio Covers 966
☐ Appendix J	Grading 966
☐ Appendix K	Administrative Provisions . . . 967
☐ Appendix L	Earthquake Recording Instrumentation 967
☐ Appendix M	Tsunami-Generated Flood Hazard 967
☐ Appendix N	Replicable Buildings 967
☐ Appendix O	Performance-Based Application 968
☐ Appendix P	Sleeping Lofts 969

☐ KEY POINTS 940

Chapter 32
Encroachments into the Public Right-of-Way 941

☐ Section 3201	General 942
☐ Section 3202	Encroachments 942

☐ KEY POINTS 946

Chapter 33
Safeguards During Construction 947

☐ Section 3301	General 948
☐ Section 3302	Owner's Responsibility for Fire Protection 948
☐ Section 3303	Demolition 948
☐ Section 3304	Site Work 949
☐ Section 3306	Protection of Pedestrians . . . 950
☐ Section 3307	Protection of Adjacent Property 953
☐ Section 3308	Temporary Use of Streets, Alleys, and Public Property 953

Metric Conversion Table 971

☐ Metric Units, System International (SI) 972
☐ Soft Metrication 972
☐ Hard Metrication 972

Foreword

How often have you heard these questions when discussing building codes: "What is the intent of this section?" or, "How do I apply this provision?" This publication offers the code user a resource that addresses much of the intent and application principles of the major provisions of the *2024 International Building Code®* (IBC®).

It is impossible for building codes and similar regulatory documents to contain enough information, both prescriptive and explanatory narrative, to remove all doubt as to the intent of the various provisions. If such a document were possible, it would be so voluminous that it would be virtually useless.

Because the IBC must be reasonably brief and concise in its provisions, the user must have knowledge of the intent and background of these provisions to apply them appropriately. The IBC places great reliance on the judgment of the building official and design professional for the specific application of its provisions. Where the designer and official have knowledge of the rationale behind the provisions, the design of the building and enforcement of the code will be based on informed judgment rather than arbitrariness or rote procedure.

The information that this handbook provides, coupled with the design professional's and building official's experience and education, will result in better use of the IBC and more uniformity in its application. As lengthy as this document may seem, it still cannot provide all of the answers to questions of code intent; that is why the background, training, and experience of the reader must also be called on to properly apply, interpret, and enforce the code provisions.

The preparation of a document of this nature requires consulting a large number of publications, organizations, and individuals. Even so, the intent of many code provisions is not completely documented. Sometimes the discussion is subjective; therefore, individuals may disagree with the conclusions presented. It is, however, important to note that the explanatory narratives are based on many decades of experience by the authors and the other contributors to the manuscript.

Preface

Internationally, code officials and design professionals recognize the need for a modern, up-to-date building code addressing the design and installation of building systems through requirements emphasizing performance. The *International Building Code*® (IBC®) meets this need by providing model code regulations that safeguard the public health and safety in all communities, large and small. The *2024 IBC Illustrated Handbook* is a valuable resource for those who design, plan, review, inspect, or construct buildings or other structures regulated by the 2024 IBC.

The IBC is one of a family of codes published by the International Code Council® (ICC®) that establishes comprehensive minimum regulations for building systems using prescriptive and performance-related provisions. It is founded on broad-based principles that use new materials and new building designs. Additionally, the IBC is compatible with the entire family of *International Codes*® published by the ICC.

There are three major subdivisions to the IBC:

1. The text of the IBC
2. The referenced standards listed in Chapter 35
3. The appendices

The first 34 chapters of the IBC contain both prescriptive and performance provisions that are to be applied. Chapter 35 contains those referenced standards that, although promulgated and published by separate organizations, are considered part of the IBC as applicable. The provisions of the appendices do not apply unless specifically included in the adoption ordinance of the jurisdiction enforcing the code.

The *2024 IBC Illustrated Handbook* is designed to present commentary only for those portions of the code for which commentary is helpful in furthering the understanding of the provision and its intent. This handbook uses many drawings and figures to help clarify the application and intent of many code provisions.

The handbook examines the intent and application of many provisions for both the non-structural- and structural-related aspects of the IBC. It addresses in detail many requirements that are considered as "fire- and life-safety" provisions of the code. Found in IBC Chapters 3 through 10, these provisions focus on the important considerations of occupancy and type of construction classification, allowable building size, fire and smoke protection features, fire protection systems, interior finishes, and means of egress.

The discussion of the structural provisions in this handbook is intended to help code users understand and properly apply the requirements in Chapters 16 through 23 of the 2024 IBC. Although useful to a broad range of individuals, the discussion of the structural provisions was written primarily so that building officials, plans reviewers, architects, and engineers can get a general understanding of the IBC's structural requirements and gain some insight into their underlying basis and intent. To that end, the numerous figures, tables, and examples are intended to illustrate and help clarify the proper application of many structural provisions of the IBC.

Because the IBC adopts many national standards by reference rather than transcribing the structural provisions of the standards into the code itself, in some cases the discussion in this handbook pertains to the provisions found in a referenced standard such as ASCE 7 rather than the IBC. The structural provisions addressed focus on the general design requirements related to structural load effects; special inspection and verification, structural testing, and structural observation; foundations and soils; and specific structural materials design requirements for concrete, masonry, steel, and wood.

Questions or comments concerning this handbook are encouraged. Please direct any correspondence to *handbook@iccsafe.org*.

Participation in ICC Code Development Process. Architects, engineers, designers, builders, and other professionals can freely participate in the ICC Code Development Process by submitting proposed code changes, collaborating with colleagues in developing code language and submitting changes, participating in giving testimony, and becoming ICC members to have voting opportunities in person or online at Committee Action Hearings. The Code Development Process is conducted via the ICC's state-of-the-art, cloud-based cdpACCESS system. Committee Action Hearings and Public Comment Hearings are broadcast live so anyone can follow the testimonies and actions taken. All building design and construction professionals are encouraged to participate in the ICC Code Development Process and have a say in the outcome of future editions of the *International Building Code* or any other of the ICC *International Codes*. Because architects, engineers, and other design and construction professionals apply the code to actual buildings and experience first-hand the effectiveness of code provisions, it is very critical for them to participate in the Code Development Process and improve the code each cycle.

Acknowledgments

The publication of this handbook is based on many decades of experience by the authors and the other contributors. Since its initial publication, the handbook has become a living document subject to changes and refinements as newer code regulations are released. This latest edition reflects extensive modifications based on the requirements found in the *2024 International Building Code®* (IBC®).

The initial handbook, on which the nonstructural portions of this document are based, was published in 1988 and based on the *Uniform Building Code*. It was authored by Vincent R. Bush. In developing the discussions of intent, Mr. Bush drew heavily on his 25 years of experience in building safety regulation. Mr. Bush, a structural engineer, was intimately involved in code development work for many years.

In addition to the expertise of Mr. Bush, major contributions were made by John F. Behrens. Mr. Behrens's qualifications were as impressive as the original author's. He had vast experience as a building official, code consultant, and seminar instructor. Mr. Behrens provided the original manuscript of the means of egress chapter and assisted in the preparation of many other chapters.

Revisions to the handbook occurred regularly over the years, with nonstructural content based on the provisions of the *International Building Code* authored by Doug Thornburg, AIA, CBO. Mr. Thornburg, a certified building official and registered architect, has over 44 years of experience in the building regulatory profession. Previously a building inspector, plans reviewer, building code administrator, seminar instructor, and code consultant, he currently develops and reviews technical publications, reference books, resource materials, and educational programs relating to the *International Codes®*. Mr. Thornburg was the recipient of the International Code Council Educator of the Year Award in 2008, recognizing his outstanding contributions in education and training.

The basis of the discussion on the structural provisions in the *2024 IBC Illustrated Handbook* is the *2000 IBC Handbook—Structural Provisions*, authored by S. K. Ghosh, Ph.D., and Robert Chittenden, SE. Dr. Ghosh initially authored Chapters 16 and 19, and Mr. Chittenden authored Chapters 17, 18, 20, 21, 22, and 23. John Henry, PE, former ICC principal staff engineer, then authored the commentary addressing IBC Chapters 16 through 23 for several editions. Mr. Henry was presented with the ICC's John Nosse Award for Technical Excellence in 2011, recognizing his outstanding contributions and technical expertise. William Bracken, PE, a registered civil and structural engineer with over 30 years of related experience, authored a portion of the commentary in the 2018 edition.

The structural provisions in relation to Chapters 16 through 23 in the 2021 edition were updated by Chris Kimball, SE. Mr. Kimball is a licensed structural engineer, civil engineer, and ICC master code professional. He is also certified by the ICC in many other disciplines, including building official and fire code official. Mr. Kimball earned a master's degree with an emphasis in structural engineering and currently serves as the Vice President of West Coast Code Consultants, Inc. (WC^3), a third-party plan review and inspection company. He has performed plan reviews for thousands of complex projects throughout the United States, is an ICC-approved instructor, and has authored several publications, including the *2018 International Existing Building Code Handbook*. Mr. Kimball has provided code training classes to building official, design professional, and contractor organizations throughout the United States.

The structural provisions in relation to Chapters 16 through 23 in this edition have been updated by John "Buddy" Showalter, PE. Mr. Showalter is a licensed civil engineer and is a senior staff engineer with the International Code Council's (ICC) product development group. He develops technical resources and provides education in support of the structural provisions of the *International Building Code* and *International Residential Code*. He is the project lead in the development of *Mass Timber Buildings and the IBC* and associated education programs, jointly developed by ICC and the American Wood Council (AWC). A graduate of Virginia Tech, Showalter has also been a member of the editorial board for *STRUCTURE* magazine for nearly 20 years. Before joining ICC, Showalter spent 26 years with AWC where he led its technology transfer program with oversight responsibility for publications, website, helpdesk, education, and other technical media. He has over 35 years of experience in the development and support of building codes and standards, authoring or co-authoring more than 70 publications and technical articles for industry-related trade journals.

Great appreciation is due a number of individuals who contributed their vast expertise and experience to the development of this publication. Special recognition goes to Sandra Hyde, PE, Managing Director of ICC Product Development; Jay Woodward, ICC Senior Staff Architect; and Kevin Scott, President of KH Scott and Associates, for their valued review and input.

The information and opinions expressed in this handbook are those of the present and past authors, as well as the many contributors, and do not necessarily represent the official position of the International Code Council. Additionally, the opinions may not represent the viewpoint of any enforcing agency. Opinions expressed in this handbook are only intended to be a resource in the application of the IBC, and the building official is not obligated to accept such opinions. The building official is the final authority in rendering interpretations of the code.

2024
INTERNATIONAL
BUILDING CODE®
ILLUSTRATED
HANDBOOK

CHAPTER 1

SCOPE AND ADMINISTRATION

Section 101 Scope and General Requirements
Section 102 Applicability
Section 103 Code Compliance Agency
Section 104 Duties and Powers of Building Official
Section 105 Permits
Section 107 Construction Documents
Section 108 Temporary Structures, Equipment, and Systems
Section 109 Fees
Section 110 Inspections
Section 111 Certificate of Occupancy
Section 112 Service Utilities
Section 113 Means of Appeals
Section 114 Violations
Section 115 Stop Work Order
Section 116 Unsafe Structures and Equipment
Key Points

> In addition to the code's scope, this chapter covers general subjects such as the purpose of the code, the duties and powers of the building official, performance provisions relating to alternative methods and materials of construction, applicability of the provisions, and creation of the code compliance agency. This chapter also contains requirements for the issuance of permits, subsequent inspections, and certificates of occupancy. The provisions in this chapter are of such a general nature as to apply to the entire *International Building Code®* (IBC®).

Section 101 *Scope and General Requirements*

101.2 Scope. The scope of the code as outlined in this section is that the IBC applies to virtually anything that is built or constructed. The definitions of "Building" and "Structure" in Chapter 2 are so inclusive that the code intends that any work of any kind that is accomplished on any building or structure comes within its scope. Thus, the code would apply to a major high-rise office building as well as to a retaining wall creating a significant elevation change on a building site. However, certain types of work are exempt from the permit process as indicated in the discussion of required permits in this chapter.

Whereas initially the IBC appears to address all construction-related activities, the design and construction of most detached one- and two-family dwellings and townhouses, as well as their accompanying accessory structures, are intended to be regulated under the *International Residential Code®* (IRC®). However, in order for such structures to fall under the authority of the IRC, two limiting factors have been established. First, each such building is limited to a maximum height of three stories above grade plane as established by the definition of "Story above grade plane" in Section 202. In broad terms, where a floor level is located predominantly above the adjoining exterior ground level, it would be considered in the total number of stories above grade plane for evaluation of its regulation by the IRC. It is quite possible that a residential unit with four floor levels will be regulated by the IRC, provided that the bottom floor level is established far enough below the exterior grade that it would not qualify as a story above grade plane, but rather as a basement. Additional occupiable floor levels may also be permitted under the allowances in the IRC for habitable attics and mezzanines. For further discussion on the determination of a story above grade plane as similarly regulated in the IBC, see the commentary on Section 202. Second, each dwelling unit of a two-family dwelling or townhouse must be provided with a separate means of egress. Although the definition of an IBC means of egress would require travel extending to the public way, for the purpose of this requirement, it is acceptable to provide individual and isolated egress only until reaching the exterior of the dwelling at grade level. Once reaching the exterior at grade, building occupants could conceivably share a stairway, sidewalk, or similar pathway to the public way. The IRC does not regulate egress beyond the structure itself; thus, any exit discharge conditions would only be applicable to IBC structures.

Townhouse design and construction is also regulated by the IRC. Section 202 defines a townhouse as a grouping of three or more townhouse units in the same structure. The townhouse units must each extend individually from the ground to the sky, with open space provided on at least two sides of each dwelling unit. The effect of such limitations maintains the concept of "multiple single-family dwellings."

The requirement for open space on a minimum of two sides of each townhouse unit allows for interpretation regarding the degree of openness. Although not specific in language, the provision intends that each townhouse be provided with a moderate degree of exterior wall, thus allowing for adequate fire department access to each individual unit. A townhouse condition occurs where the four established criteria are met. The presence of a lot line or property line between attached dwelling units, or the lack of such a line, has no bearing on its designation as a townhouse.

Structures such as garages, carports, and storage sheds are also regulated by the IRC where they are considered accessory to the residential buildings previously mentioned. Such accessory buildings are limited in height under the same limitations applied to dwellings, a maximum of three stories above grade plane. The maximum floor areas of dwellings and accessory structures constructed under the IRC are unlimited; however, it is probable that local zoning ordinances will provide some degree of regulation.

Even though the IRC may use the IBC as a reference for certain design procedures, the intent is to use only the IRC for the design and construction of one- and two-family dwellings, multiple townhouse units (townhouses), and their accessory structures. This does not preclude the use of the IBC by a design professional for the design of the types of residential buildings specified. However, unless specifically directed to the IBC by provisions of the IRC, it is not the intent of the IRC to utilize the IBC for provisions not specifically addressed. For example, the maximum allowable floor area of a residence based on the building's type of construction is not addressed in the IRC. Therefore, there is no limit to the floor area permitted in the dwelling unit. It would not be appropriate to use the IBC to limit the residence's floor area based on construction type.

101.2.1 Appendices. A number of subjects are addressed in Appendices A through P. The topics range from detailed information on the creation of a board of appeals to more general provisions for grading, excavation, and earthwork. Although the code clearly indicates that the appendices are not considered a part of the IBC unless they are specifically adopted by the jurisdiction, this does not mean they are of any less worth than those set forth in the body of the code. Although there are several reasons why a set of code requirements is positioned in the Appendix, the most common are that the provisions are limited to a small geographic location or are of interest to only a small number of jurisdictions.

Jurisdictions have the ability to adopt any or all of the appendices based on their own needs. However, just because an appendix has not been adopted does not lessen its value as a resource. In making decisions of interpretation of the code, as well as in evaluating alternate materials and methods, the provisions of an appendix may serve as a valuable tool in making an appropriate decision. Even in those cases where a specific appendix is not in force, the information it contains may help in administering the IBC.

101.3 Purpose. Various factors are regulated that contribute to the performance of a building in regard to the health, safety, and welfare of the public. The IBC identifies several of these major factors as those addressing structural strength, egress capabilities, sanitation and other environmental issues, fire- and life-safety concerns, energy conservation, fire/explosion hazards, and other dangerous conditions. In addition, the safety of fire fighters and emergency personnel responding to an emergency situation is an important consideration. The primary goal of the IBC is to address any and all hazards that are attributed to the presence and use of a jurisdiction's buildings and structures, and to safeguard the public from such hazards.

The purpose of the code is more inclusive than most people realize. A careful reading will note that in addition to providing for life safety and safeguarding property, the code also intends that its provisions consider the general welfare of the public. This latter item, *general welfare*, is not so often thought of as being part of the purpose of a building code. However, in the case of the IBC, safeguarding the public's general welfare is a part of its intent, which is accomplished, for example, by provisions that ameliorate the conditions found in substandard or dangerous buildings. Moreover, upon the adoption of a modern building code such as the IBC, the general level of building safety and quality is raised. This in turn contributes to the public welfare by increasing the tax base and livability. Additionally, substandard conditions are reduced, and the subsequent reduction of unsanitary conditions contributes to safeguarding the public welfare. For example, the maintenance provisions of the *International Property Maintenance Code*® (IPMC®) and the continued enforcement of the IBC slow the development of substandard conditions. A rigorous enforcement of the IPMC will actually reduce the conditions that contribute to the deterioration of the existing building stock. Thus, public welfare is enhanced by the increased benefits that inure to the general public of the jurisdiction as a result of the code provisions.

The concept of "minimum" requirements is the established basis for the technical provisions set forth in the IBC. The requirements are intended to identify the appropriate level of regulation to achieve a balanced approach to the design and construction of buildings. On the one hand, it is critical that an appropriate degree of safety be established in order to protect the general public. Conversely, it is also important that the economic impact of the regulations be considered, as well as a building's efficiency of use. It is this balance of concerns that provides for the necessary degree of public health, safety, and welfare within appropriate economic limits. The establishment of multiple occupancy classifications with varying requirements for each is a basic example of this philosophy.

101.4 Referenced codes. A number of other codes are promulgated by the International Code Council® (ICC®) in order to provide a full set of coordinated construction codes. Seven of those companion codes are identified in this section, as they are specifically referenced in one or more provisions of the IBC. The adoption of the IBC does not automatically include the full adoption of the referenced codes, but rather only those portions specifically referenced by the IBC.

For example, Section 903.3.5 requires that water supplies for automatic sprinkler systems be protected against backflow in accordance with the *International Plumbing Code*® (IPC®). As a result, when the IBC is adopted, so are the backflow provisions of the IPC for those buildings protected with an automatic sprinkler system. The extent of the reference is backflow protection; therefore, that is the only portion of the IPC that is applicable based on this reference. Broader references are also provided, such as many of the references to the *International Fire Code*® (IFC®). Section 307.2 requires that hazardous materials in any quantity conform to the requirements of the IFC. Although the entire IFC may not be adopted by the jurisdiction, the provisions applicable to hazardous materials are in force with the adoption of the IBC.

The *International Existing Building Code*® (IEBC®) is referenced in its entirety for any building undergoing a repair, alteration, addition, relocation, or change of occupancy.

Section 102 *Applicability*

102.1 General. Where there is a conflict between two or more provisions found in the code as they relate to differences of materials, methods of construction, or other requirements, the most restrictive provision will govern. Typically, the code will identify how the conflicting requirements should be applied. For example, the occupant load along with the appropriate factor from Section 1005.3.1 is used to calculate the total capacity required for egress stairways—often referred to as the *calculated* width. Section 1011.2 also addresses the minimum required width for a stairway based on the absolute width necessary for the use of a stairway under any condition, deemed to be the *component* width. When determining the proper minimum width required by the code, the more restrictive, or wider, stairway width would be used. See Application Example 102-1.

In addition, where a conflict occurs between a specific requirement and a general requirement, the more specific provision shall apply. Again, the IBC provisions typically clarify the appropriate requirement that is to be applied without the need to determine the appropriate general/specific relationship. As an example, Section 1011.5.2 limits the height of stair risers to

> **GIVEN:** An occupant load of 130 assigned to each of two stairways in a nonsprinklered office building.
> **DETERMINE:** The required minimum width of each stairway.
>   1. Based on Section 1005.3.1, the minimum calculated width would be:
>   0.3 inches/occupant × 130 occupants = 39 inches
>   2. Based on Section 1011.2, the minimum required width would be 44 inches.
>
> **SOLUTION:** Therefore, the more restrictive condition, 44 inches, would apply.
> For SI: 1 inch = 25.4 mm.
>
> **CONFLICTING REQUIREMENTS**

Application Example 102-1

7 inches (178 mm) as a general requirement for stairways. However, Section 1030.14.2.2 allows for a maximum riser height of 8 inches (203 mm) for stepped aisles serving assembly seating areas. Because the greater riser height is only permitted for a specific stair condition, rather than for all stairways in general, it is intended to apply where those special means of egress provisions established in Section 1030 are applicable.

Occasionally it is difficult, during the comparison of two different code provisions, to determine which is the general requirement and which is the specific requirement. In some cases, both requirements are specific, but one is more specific than the other. It is important that the intent of this section be applied in reviewing the proper application of the code. Where it can be determined that one provision is more specific in its scope than the other provision, the more specific requirement shall apply, regardless of whether it is more or less restrictive in application. The provisions of Chapter 4 provide multiple examples of the general/specific relationship. There are many special detailed requirements in Chapter 4 for various uses and occupancies that are intended to supersede general provisions found elsewhere in the code.

102.4 Referenced codes and standards. Differences between the code and the various standards it references are to be expected. Unlike the companion *International Codes*® where consistency between them can be provided, there is not necessarily a conscious effort to see that the standard and code publications are completely compatible with each other. In fact, such compatibility would be impossible to achieve due to a variety of factors. As a result, it is critical that the code indicates that its provisions are to be applied over those of a referenced standard where such differences exist. For example, the provisions of National Fire Protection Association (NFPA) 13R addressing sprinkler systems in residential occupancies allow for the omission of sprinklers at specified exterior locations, including decks and balconies. However, the provisions of IBC Section 903.3.1.2.1 mandate sprinkler protection for such areas where specific conditions exist. In this case, the provisions of the IBC for sprinkler protection would apply regardless of the allowances contained in NFPA 13R.

There are also times when the standard being referenced includes subject matter that falls within the scope of the IBC or the other *International Codes*. It is intended that the requirements of a referenced standard supplement the IBC provisions in those areas not already addressed by the code. In those areas where parallel or conflicting requirements occur, the IBC provisions are always to be applied. For example, IBC Section 415.9.3 mandates that "the construction and installation of dry cleaning plants shall be in accordance with the requirements of the IBC, the *International Mechanical Code*, the *International Plumbing Code*, and NFPA 32." Although NFPA 32 addresses construction and installation criteria for dry cleaning plants, only those portions of the standard that are not addressed within the IBC, IMC, and IPC are applicable.

Section 103 *Code Compliance Agency*

This section recognizes the creation of a jurisdictional enforcement agency charged with implementing, administering, and enforcing the IBC. The term *building official*, as used in the IBC, represents the individual appointed by the jurisdiction to head the code compliance agency. Although many jurisdictions utilize the title *Building Official* to recognize the individual in charge of this agency, there are many other titles that are used. These include *Chief Building Inspector, Superintendent of Central Inspection, Director of Code Enforcement*, and various other designations. Regardless of the title selected for use by the individual jurisdiction, the IBC views all of these as equivalent to the term *building official*.

The building official, in turn, appoints personnel as necessary to carry out the duties and responsibilities of the code compliance agency. Such staff members (deputies), including inspectors, plan examiners, and other employees, are empowered by the building official to carry out those functions set forth by the jurisdiction. Where the IBC references the building official in any capacity, the code reference also includes any deputies who have been granted enforcement authority by the building official. Where an inspector or plan reviewer makes a decision of interpretation, they are assuming the role of building official in arriving at that decision. There is an expectation on behalf of the jurisdiction that such employees possess the knowledge and experience to take on this responsibility. The failure to grant appropriate authority will often result in both ineffective and inefficient results.

For those jurisdictions desiring guidelines within the text of the code for the selection of agency personnel, Appendix A addresses minimum employee qualifications for various positions. Experience and certification criteria for building officials, chief inspectors, inspectors, and plans examiners are set forth in this Appendix chapter.

Section 104 *Duties and Powers of Building Official*

104.2 Determination of compliance. The IBC is designed to regulate in both a prescriptive and performance manner. An extensive number of provisions have been intentionally established to allow for jurisdictional interpretation based on the specifics of the situation. This section establishes the building official's authority to render such interpretations of the IBC. In addition, the building official may adopt policies and procedures that will help clarify the application of the code. Although having no authority to provide variances or waivers to the code requirements, the building official is charged with interpreting and clarifying the provisions found in the IBC, provided that such decisions are in conformance with the intent and purpose of the code.

The authority to interpret the intended application of the IBC is a powerful tool available to the building official. With such authority comes a great degree of responsibility. Such interpretations must be consistent with the intent and purpose of the code. It is therefore necessary that all reasonable efforts be made to determine the code's intent in order to develop an appropriate interpretation. Various sources should be consulted to provide a broad background from which to make a decision. These could include discussions with peers, as well as information found in various educational texts and technical guides. However, it must be stressed that the ultimate responsibility for determining the appropriateness of an interpretation lies with the jurisdictional building official, and all opinions from others, both verbal and written, are just that, opinions. The building official must never relinquish their authority to others in the administration of these very important interpretive powers.

104.2.1 Listed compliance. The IBC allows for the adoption of new technologies in materials and building construction that currently are not covered by the code, thereby providing additional performance options. The IBC thus encourages state-of-the-art concepts in design, construction, and materials as long as they meet the performance intended by the code. Where a product listing is appropriate, then the fact that the product is listed and installed in accordance with the listing specifications and the manufacturer's instructions becomes the approval of the product.

104.2.2 Technical assistance. Although not specifically addressed, the informal utilization of technical assistance by the building official has historically been a method for determining code compliance. In a more formal sense, this assistance in determining compliance can be provided in the form of a technical report and opinion. The allowance for a technical report gives jurisdictions the benefit of expert opinions provided by knowledgeable persons on the item under consideration. Technical reports are required to be prepared by an individual, firm, or corporation acceptable to the jurisdiction, and must be provided without charge to the jurisdiction.

104.2.3 Alternative materials, design, and methods of construction and equipment. This section of the IBC may be one of the most important. It allows for the adoption of new technologies in materials and building construction that currently are not covered by the code. Furthermore, it gives the code even more of a performance character. The IBC thus encourages state-of-the-art concepts in design, construction, and materials as long as they meet the performance intended by the code. When evaluating the alternative methods under consideration, the building official must review for equivalency in quality, strength, effectiveness, fire safety, durability, and general safety. It is expected that all alternatives, once presented to the building official for review and approval, be thoroughly evaluated by the code compliance agency for compliance with this section. If such compliance can be established, the alternatives are deemed to be acceptable. An allowance is made for the use of the ICC Performance Code (ICCPC) provided it does not deal with a structural material or design. This recognition could apply where a single material, element, or system cannot conform to the prescriptive requirements that are otherwise met for the project.

The provisions of this section, similar to those of Sections 104.2 and 104.2.4, reference the *intent* of the code. It is mandated that the building official, when evaluating a proposed alternative to the code, only approve its use where it can be determined that it complies with the intent of the specific code requirements. Thus, it is the responsibility of the building official to utilize those resources necessary to understand the intended result of the code provisions. Only then can the code be properly applied and enforced.

Similar to the approach taken where modifications are requested under the criteria of Section 104.2.4, the request for acceptance under this section should be made in writing to the building official. At a minimum, the submittal should include: (1) the specific code section and requirement, (2) an analysis of the perceived intent of the provision under review, (3) the special reasons as to why strict compliance with the code provision is not possible, (4) the proposed alternative, (5) an explanation of how the alternative meets or exceeds the intended level of compliance, and (6) a request for acceptance of the alternative material, design, or method of construction. Where approval of the request is not granted, it is mandated that the building official state in writing the reasons for denial. Such an action provides for a clear understanding as to why the alternative was not accepted, thereby providing guidance for future submittals.

Evaluation reports, can address and delineate a review of the appropriate testing procedures to support the alternative as code compliant. The use of an evaluation report may be helpful in reducing additional testing or documentation that is necessary to indicate compliance. It is important that the building official evaluate not only the information contained within the report, but also the technical expertise of the individual or firm issuing the report. It must be noted, however, that an evaluation report is simply a resource to the building official to assist in the decision-making process. The report itself does not grant approval, as acceptance is still under the sole authority of the building official.

One of the most commonly utilized evaluation reports is the ICC Evaluation Service (ICC-ES) Evaluation Report. ICC-ES is a nonprofit, limited liability company that does technical evaluations of building products, components, methods, and materials. If it is found that the subject of an evaluation complies with code requirements, then ICC-ES publishes a report to that effect and makes the report available to the public. However, ICC-ES Evaluation Reports. like all other evaluation reports, are only advisory. The authority having jurisdiction is always the final decision maker with respect to acceptance of the product, material, or method in question.

104.2.4 Modifications. The provisions of this section allow the building official to grant modifications to the requirements of the code under certain specified circumstances. The building official may modify requirements if it is determined that strict application of the code is impractical and, furthermore, that the modification is in conformity with the intent and purpose of the code. Without this provision in the IBC, the building official has very little discretionary enforcement authority and, therefore, would have to enforce the specific wording in the code, no matter how unreasonable the application would be.

The code does not intend to allow the building official to issue a variance to the provisions of the code to permit, for example, the use of only a single exit where two are required. This is clearly not in conformity with the intent and purpose of the code, no matter how difficult it may be to meet the requirements. In fact, the code is very specific that any modification cannot reduce health, accessibility, structural, and fire- and life-safety requirements.

Where the building official grants a modification under this section, the details of such an action shall be recorded. This document must then be entered into the files of the code compliance agency. By providing a written record of the action taken and maintaining a copy of that action in the agency files, the building official always has access to the decision-making process and final determination of his or her action should there be a need to review the decision.

Although it is expected that a permanent record be available for future reference when a modification is accepted, there is perhaps an even more important reason for the recording and filing of details of the approving action. The willingness to document and archive the modification action indicates the confidence of the building official in the decision that was made. A reluctance to maintain a record of the action taken typically indicates a lack of commitment to the action taken.

104.4 Right of entry. This section has been compatible with Supreme Court decisions since the 1960s regarding acts of inspection personnel seeking entry into buildings for the purpose of making inspections. Under present case law, an inspection may not be made of a property, whether it be a private residence or a business establishment, without first having secured permission from the owner or person in charge of the premises. If entry is refused by the person having control of the property, the building official must obtain an inspection warrant from a court having jurisdiction in order to secure entry. The important feature of the law regarding right of entry is that entry must be made only by permission of the person having control of the property. Lacking this permission, entry may be gained only through the use of an inspection warrant.

If entry is again refused after an inspection warrant has been obtained, the jurisdiction has recourse through the courts to remedy this situation. One avenue is to obtain a civil injunction in which the court directs the person having control of the property to allow inspection. Alternatively, the jurisdiction can initiate proceedings in criminal court for punishment of the person having control of the property. It cannot be repeated too strongly that criminal court proceedings should never be initiated against an owner or other person having control of the property if an inspection warrant has not been obtained. Because the consequences of not following proper procedures can be so devastating to a jurisdiction if a suit is brought against it, the jurisdiction's legal officer should always be consulted in these matters.

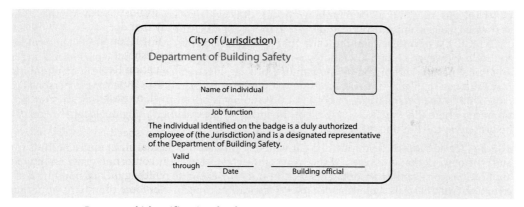

Figure 104-1   Personnel identification badge.

104.5 Identification. For the benefit of all individuals involved, inspection personnel of the code compliance agency are mandated to carry proper identification. The display of an identification card or badge, an example of which is shown in Figure 104-1, signifies the function and authority of the individual performing the inspection.

104.8 Liability. It is the intent of the IBC that the building official not become personally liable for any damage that occurs to persons or property as a result of the building official's acts so long as they act in good faith and without malice or fraud. This protection is also extended to any member of the Board of Appeals, as well as any jurisdictional employee charged with enforcement of the IBC. Nevertheless, legal action is occasionally undertaken in an effort to hold civil officers personally liable for their acts. This section requires that the jurisdiction defend the building official or other protected party if a suit is brought against them. Furthermore, the code requires any judgment resulting from a suit to be assumed by the jurisdiction.

Case law regarding tort liability of building officials is constantly in a state of flux, and past doctrines may not now be applicable. Therefore, the legal officer of the jurisdiction should always be consulted when there is any question about liability.

Section 105   *Permits*

This section covers those requirements related to the activities of the code compliance agency with respect to the issuance of permits. The issuance of permits, plan review, and inspection of construction for which permits have been issued constitute the bulk of the agency's duties. It is for this reason that the code goes into detail regarding the permit-issuance process. Additionally, the code provides detailed requirements for the inspection process in order to help ensure that the construction for which the inspections are made complies with the code in all respects.

105.1 Required. Prior to obtaining a permit, the owner of the property under consideration, or the owner's authorized agent, must apply to the building official for any necessary permits that are required by the jurisdiction. One or more permits may be required to cover the various types of work being accomplished. In addition to building permits that address new construction, alterations, additions, repairs, moving of structures, demolition, or change in occupancy, trade permits are required to erect, install, enlarge, alter, repair, remove, convert, or replace any electrical, gas, mechanical, or plumbing systems. It is evident that almost any work, other than cosmetic changes, must be done under the authority of a permit.

Typically, a permit is required each time a distinct activity occurs that is regulated under the code. However, certain alterations to previously approved systems can be performed under an annual permit authorized by the building official. Electrical, gas, mechanical, or plumbing installations are eligible for such consideration when one or more qualified trade persons are employed by the person, firm, or corporation who owns or operates the building, structure, or premises where the work is to take place. In addition, the qualified individuals must regularly be present at the building or site.

105.2 Work exempt from permit. It would seem that the IBC should require permits for any type of work that is covered by the scope of the code. However, this section provides limited applications for exempted work. This section not only exempts certain types of building construction from permits, but also addresses electrical, gas, mechanical, and plumbing work that is of such a minor nature that permits are not necessary.

It is further the intent of the IBC that even though work may be exempted from a permit, such work done on a building or structure must still comply with the provisions of the code. As indicated in Section 101.2, the scope of the IBC is virtually all-inclusive. This may seem to be a superfluous requirement where a permit is not required. However, this type of provision is necessary to provide that the owner, as well as any design professional or contractor involved, be responsible for the proper and safe construction of all work being done.

A common example of exempted work is a small, one-story detached accessory structure such as a storage shed. Although the code does not require a permit for an accessory building not exceeding 120 square feet (11 m^2) in floor area, all provisions in the code related to a Group U occupancy must still be followed regardless of size.

105.3 Application for permit. In this section, the IBC directs that a permit must be applied for, and describes the information required on the permit application. The permit-issuance process, as envisioned by the IBC, is intended to provide records within the code compliance agency of all construction activities that take place within the jurisdiction and to provide orderly controls of the construction process. Thus, the application for permit is intended to describe in detail the work to be done. In this section, the building official is directed to review the application for permit. This review is not a discretionary procedure, but is mandated by the code.

The code also charges the building official with the issuance of the permit when it has been determined that the information filed with the application shows compliance with the IBC and other laws and ordinances applicable to the building at its location in the jurisdiction. The building official may not withhold the issuance of a permit if these conditions are met. As an example, the building official would be in violation in withholding the issuance of a building permit for a swimming pool because an adjacent cabana was previously constructed without a permit.

105.4 Validity of permit. The code intends that the issuance of a permit should not be construed as permitting a violation of the code or any other law or ordinance applicable to the building. In fact, the IBC authorizes the building official to require corrections if there were errors in the approved plans or permit application at the time the permit was issued. The building official is further authorized to require corrections of the actual construction if it is in violation of the code, although in accordance with the plans. Moreover, the building official is further authorized to invalidate the permit if it is found that the permit was issued in error or in violation of any regulation or provision of the code.

Although it may be poor public relations to invalidate a permit or to require corrections of the plans after they have been approved, it is clearly the intent of the code that the approval of plans or the issuance of a permit may not be done in violation of the code or of other pertinent laws or ordinances. As the old saying goes, "Two wrongs do not make a right."

105.5 Expiration. The IBC anticipates that once a permit has been issued, construction will soon follow and proceed expeditiously until completion. However, this ideal procedure is not always the case and, therefore, the code makes provisions for those cases where work has not started, or alternatively where the work, after being started, has been suspended for a period of time. In these cases, the IBC allows a period of 180 days to transpire before the permit becomes void. The code then requires that a new permit be obtained. It is assumed by the code that the code compliance agency will have expended some effort in follow-up inspections of the work, etc., and, therefore, the original permit fee may be retained in order to compensate the agency for the work. The allowance for full or partial refunds should be addressed by policy under the authorization granted by Section 109.6. The building official has the authority to grant one or more extensions of time, provided the permit holder can demonstrate a justifiable cause as to why the permit should not be invalidated. The time period for such extensions cannot exceed 180 days; however, additional extensions may be granted if approved by the building official.

There are several reasons why it is important to establish a limitation on the validity of a permit, such as purging the agency files of inactive permits. Additionally, it keeps the project on track with the code edition in effect at the time of permit issuance.

Section 107 *Construction Documents*

Plans, specifications, and other construction documents, along with other applicable data, must be filed with the permit application. Such submittal documents are intended to graphically depict the construction work to be done. The IBC sets forth the necessary information that must be provided to the building official at the time of plans submittal, as well as procedures for deferred submittals.

107.1 General. In this section, the IBC directs that at the time of application for a permit, construction documents and other essential data on the project be submitted. The code requires that plans, engineering calculations, and any other information necessary to describe the work to be done be filed along with the application for a permit. Where geotechnical investigations are required, a written report shall be submitted at the time of permit application. A statement of special inspections that may be required by Section 1704.2.3 is also to be submitted with the application for a permit. Based on the statutes of the jurisdiction, the plans, specifications, and other documents may need to be prepared by a registered design professional. Under special circumstances, the building official may also require that the submittal documents be prepared by a registered design professional even when not mandated by jurisdictional statute. The building official is permitted to waive the requirement for the filing of plans and other data, where not required by statute to be prepared by a registered design professional, provided the building official ensures that the work for which the permit is applied is of such a nature that plans or any other data are not necessary in order to indicate and obtain compliance with the code.

107.3 Examination of documents. In this section, the building official is directed to review the plans, specifications, and other submittal documents filed with the permit. The building official is not at liberty to check only a portion of the plans. On this basis, the structural drawings, as well as the engineering calculations, must be checked in order for the building official to provide a full examination.

107.3.4 Design professional in responsible charge. The building official has the authority to require the designation of a design professional to act in responsible charge of the work being performed. The function of such an individual is to review and coordinate submittal documents prepared by others and, if necessary, coordinate any deferred or phased submittal items. In some cases, the design of portions of the building has been completed and may be dependent on the manufacturer of proposed prefabricated elements, such as for truss drawings. Therefore, the code specifically allows deferring the submittal of portions of plans and specifications.

Section 108 *Temporary Structures, Equipment, and Systems*

This section authorizes the issuance of permits for the erection of temporary buildings and structures such as those erected at construction sites. The regulation of temporary viewing stands and other miscellaneous, temporary structures would also fall under this section. Reference is made to Section 3103 for the special requirements applicable to temporary buildings. For applying the applicable code provisions, "temporary" is considered as being erected for a period of less than 180 days. The *International Fire Code* (IFC) also has a significant number of provisions applicable to such structures. Additionally, the IBC identifies key criteria that must be applied to temporary structures, including conformity with structural strength, fire safety, means of egress, accessibility, light, ventilation, and sanitary requirements. Permits are typically required, as are construction documents. The scope of Section 3103 includes special event structures, tents, umbrella structures, and other membrane structures.

Section 109 *Fees*

Permits required by the jurisdiction are not valid until the appropriate fees have been paid. The IBC provides for each individual jurisdiction to establish its own schedules for permit fees. The fees collected by the code compliance agency are typically set at a level to adequately cover the cost to the agency for services rendered.

109.3 Permit valuations. The code uses the concept of valuation to establish the permit fee. This concept is based on the proposition that the valuation of a project is related to the amount of work to be expended in the various aspects of administering the permit. Also, there should be some excess in the permit fee to cover agency overhead. Essentially, the valuation is considered as the cost of replacing the building. The valuation also includes any electrical, plumbing, and mechanical work, even though separate permits may be required for the mechanical, plumbing, and electrical trades.

To provide some uniformity in the determination of valuation so that there is a consistent base for the assignment of fees, the IBC directs the building official to determine the value of the building. To assist in obtaining uniformity, "Building Valuation Data (BVD)" can be accessed on ICC's website. Thus, building officials may utilize a common base in their determination of the value of buildings. However, ICC strongly recommends that all jurisdictions and other interested parties actively evaluate and assess the impact of the BVD table before utilizing it in their current code enforcement activities. As an option, any other appropriate method may also be used by the jurisdiction as the basis for determining the proper valuation of the work.

109.4 Work commencing before permit issuance. When work requiring a permit is started without such a permit, the IBC allows the building official to assess an additional fee above and beyond the required permit fees. This fee is intended to compensate the code compliance agency for any additional time and effort necessary in evaluating the work initiated without a permit. It may often be necessary for the building official to cause an investigation to be made of the work already done. The intent of the investigation is to determine to what extent the work completed complies with the code, and to describe with as much detail as possible the work that has been completed.

109.6 Refunds. This section authorizes the building official to establish a policy for partial or complete refunds of fees paid to the department. Although not specified in the code, there are a variety of reasons why some level of a fee refund would be appropriate. One instance would be where the permit fee is collected in error. Another reason for authorizing a refund of the fees paid would be that circumstances beyond the control of the applicant caused delays and the eventual expiration of the building permit. Typically, the building official will withhold from the refund any monies expended by the agency for related administrative activities that have taken place. It should be noted that the building official, when approving a fee refund, is authorizing the disbursement of public funds. Therefore, the building official must be sure that there is good cause for the refund to the applicant.

Section 110 *Inspections*

The inspection function is arguably the most critical aspect of code compliance agency operations. An important concept views that, as with the issuance of permits, inspections that presume to give authority to violations of the code are invalid. In general, the IBC charges the permit applicant with the responsibility of ensuring that the work to be inspected remain accessible and exposed until it is approved. Any expense incurred in the process of providing for an accessible and exposed inspection, such as the removal of gypsum board or insulation, is not to be borne by the building official or jurisdiction. Therefore, it is critical that work should not proceed beyond the point where an inspection is required.

110.3 Required inspections. The code mandates a variety of inspections during the progress of construction. Several such inspections are common to almost all types of buildings, including:

- Footing and foundation inspection
- Concrete slab and/or under-floor inspection
- Frame inspection
- Gypsum panel products and/or lath inspection
- Energy efficiency inspection
- Final inspection

These inspections are critical in ensuring the construction is compliant with the provisions of the code as in many cases such verification would be difficult to achieve.

Fire-resistance-rated and/or smoke-resistant elements are often constructed with the inspection of penetration and joint protectives necessary. Where balconies and other elevated surfaces are considered as weather-exposed and the structural framing is protected by an impervious moisture barrier, elements of the moisture barrier system must be inspected. The inspection of wood cover providing fire-resistance ratings in Type IV-A, IV-B, and IV-C buildings is also mandated. Special inspections as addressed in Section 1705 require special expertise to ensure compliance. Another type of inspection is required in flood hazard areas. After the lowest floor elevation or the elevation of dry floodproofing, where applicable, is established, but prior to additional vertical construction, the required elevation certification shall be submitted to the building official.

The code also gives the building official the authority to require and make other inspections where necessary to determine compliance with the code and other laws and regulations enforced by the code compliance agency. There are times where unique concerns of the jurisdiction make it necessary to provide some degree of inspection. Where such review is necessary, the inspections should be handled in a manner consistent with those specifically identified in the IBC.

In each case for the required inspections identified in Section 110.3, the IBC is very specific as to how far the construction must have progressed prior to a request for an inspection. If it is necessary to make a re-inspection because work has not progressed to the point where it is ready for inspection, the building official may charge an additional fee under the general provisions of Section 109. A re-inspection fee is not specifically mandated, however, and normally would not be assessed unless the person doing the work continually calls for inspections before the work is ready for inspection. Such actions would typically cause increased costs to the jurisdiction not covered by the original building permit.

110.5 Inspection requests. In general, the IBC charges the individual holding the building permit with notifying the code compliance agency when it is time to make an inspection. The permit holder may also authorize one or more agents for this responsibility. The code places a duty upon the permit holder or their authorized agent to have the work to be inspected accessible and exposed, so it can be evaluated as to code compliance.

110.6 Approval required. The code intends that no work shall be done beyond the point where an inspection is required until the work requiring inspection has been approved. Moreover, it is intended that work requiring inspection not be covered until it has been inspected and approved.

Section 111 *Certificate of Occupancy*

The tool that the building official utilizes to control the uses and occupancies of the various buildings and structures within the jurisdiction is the certificate of occupancy. The IBC makes it unlawful to use or occupy a building or structure unless a certificate of occupancy has been issued for that use. Furthermore, the code imposes the duty of issuing a certificate of occupancy upon the building official when they are satisfied that the building, or portion thereof, complies with the code for the intended use and occupancy.

Prior to use or occupancy of the building, the building official shall perform a final inspection as addressed in Section 110.3.12. If no violations of the code and other laws enforced by the code compliance agency are found, the building official is required to issue a certificate of occupancy. Figure 111-1 illustrates the information that must be provided on the certificate of occupancy.

Where a portion of a building is intended to be occupied prior to occupancy of the entire structure, the building official may issue a temporary certificate of occupancy. This situation would occur where partial occupancy is requested prior to completion of all work authorized by the building permit. Prior to issuance of a temporary certificate of occupancy, it is critical that the building official ascertain that those portions to be occupied provide the minimum levels of safety required by the code. In addition, the building official shall establish a definitive length of time for the temporary certificate of occupancy to be valid.

In essence, the certificate of occupancy certifies that the described building, or portion thereof, complies with the requirements of the code for the intended use and occupancy, except as provided for existing buildings undergoing a change of occupancy as regulated by the *International Existing Building Code* (IEBC). However, any certificate of occupancy may be suspended or revoked by the building official under one of three conditions: (1) where the certificate is issued in error, (2) where incorrect information is supplied to the building official, or (3) where it is determined that the building or a portion of the building is in violation of the code or any other ordinance or regulation of the jurisdiction. The most common reason for suspending or revoking a certificate of occupancy is that the building is being used for a purpose other than that intended when approval for occupancy was granted. When a permit is issued, plans reviewed, inspections made, and a certificate of occupancy given, there is an expectation that the building will be used for specific activities. As such, the hazards associated with such

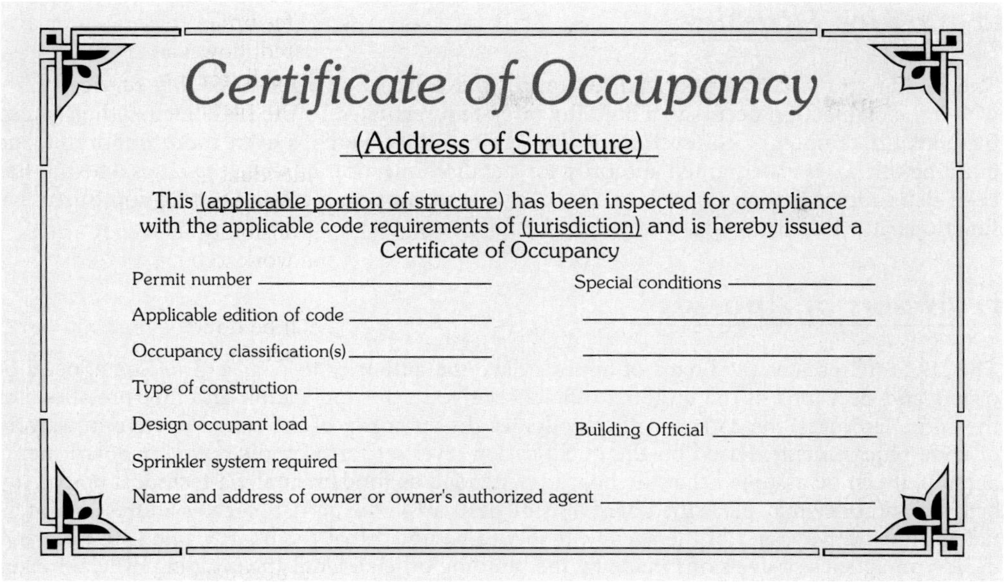

Certificate of Occupancy

(Address of Structure)

This (applicable portion of structure) has been inspected for compliance with the applicable code requirements of (jurisdiction) and is hereby issued a Certificate of Occupancy

Permit number _____

Applicable edition of code _____

Occupancy classification(s) _____

Type of construction _____

Design occupant load _____

Sprinkler system required _____

Name and address of owner or owner's authorized agent _____

Special conditions _____

_____

_____

_____

Building Official _____

Figure 111-1 **Certificate of occupancy.**

activities can be addressed. However, where an unanticipated use occurs, there is the potential that the necessary safeguards are not in place to address the related hazards. An unauthorized use can be intentional or unintentional; however, it is no less a concern until the use is discontinued or the necessary remedies are in place.

A key distinction in the application of the IBC concerns the terms "use" and "occupancy." Most buildings have multiple uses, but often contain only a single occupancy. For example, an office building may have business areas, small storage areas, and small assembly areas, as well as support areas such as restrooms and mechanical equipment rooms. Although multiple uses occur in the building, it only contains a single occupancy, Group B. It is anticipated that all of the hazards that are anticipated as part of the building's function can be effectively addressed through classification as a single occupancy group.

The definition's reference to a change in purpose or level of activity is not intended to address minor changes in use that are typical when new ownership or tenancy occurs. Rather, the need for a new certificate of occupancy only occurs where change results in an increase in hazard level. As an example, where a Group B business office building undergoes a change of use to a Group B ambulatory care facility, additional fire- and life-safety safeguards are necessary in order to address the increased hazard due to the possible presence of health-care recipients who will be temporarily incapable of self-preservation. Although both types of facilities are classified as Group B occupancies, there is a distinct difference in their uses, and, as such, a new certificate of occupancy must be issued to address the change. Conversely, the provisions are not intended to require a new certificate of occupancy where the change in purpose or activity creates little, if any, change in hazard level. For example, a change from a Group M grocery store to a Group M electronics store would not require issuance of a new certificate of occupancy. In an evaluation of whether a change of occupancy occurs, the relationship of the new use to that of the previous use must undergo an evaluation conducted in accordance with the definition of "change of occupancy" in Chapter 2.

Section 112 *Service Utilities*

The building official is authorized to control the connection release for any service utility where the connection occurs to a building or system regulated by the IBC. Such authority also includes the temporary connection of the service utility. Perhaps even more important, the building official is also granted authority for the disconnection of service utilities where it has been determined that an immediate hazard to life or property exists. As in all administrative functions, it is important that due process be followed.

Section 113 *Means of Appeals*

The IBC intends that the board of appeals have the authority to hear and decide appeals of orders and decisions of the building official relative to the application and interpretations of the code. However, the code specifically denies the authority of the board relative to waivers of code requirements. Based on the qualification level set forth by the code for board membership, it can be assumed that the board is intended as fundamentally a technical body. Any broader authority may place the board outside of its area of expertise, such as addressing internal administrative issues of the agency. It should be noted that the board's role is to hear and decide on appeals of decisions made by the building official. Until the building official makes their determination, the issue is not subject to board review.

The importance of a board of appeals should not be taken lightly. Its role is equivalent to that of the building official when it comes to technical questions placed before it, thus the need for highly qualified board members. The board is not merely advisory in its actions, but rather is granted the authority to overturn the decisions of the building official where the determination is within the scope as described by the code. An example would be an appeal requesting the use of an alternative method of construction previously denied by the building official. The alternative method would be permitted if found by the board to be equivalent to or better than that set forth in the code. Where the board of appeals is desired to take on a more expansive role, such as code adoption review or contractor licensing oversight, those duties should be specifically granted by the jurisdiction. More detailed information on the qualifications and duties of a board of appeals is found in IBC Appendix B.

Section 114 *Violations*

The provisions of this section establish that violations of the code are considered unlawful and such violations shall be abated. Where necessary, a notice of violation may be served by the building official to the individual responsible for the work that is in violation of the code. Further action may be taken where there is a lack of compliance with the notice. The jurisdiction shall establish penalties based on the various specified violations.

Section 115 *Stop Work Order*

The stop work order is a tool authorized by the IBC to enable the building official to demand that work on a building or structure be temporarily suspended. Intended to be utilized only under rare circumstances, this order may be issued where the work being performed is dangerous, unsafe, or significantly contrary to the provisions of the code. A stop work order is often a building official's final method in obtaining compliance on issues of extreme importance. All other reasonable avenues should be considered prior to the issuance of such an order.

Stop Work Order

The stop work order shall be a written document indicating the reason or reasons for the work to be suspended. It shall also identify those conditions where compliance is necessary before the cited work is allowed to resume. All work addressed by the order shall immediately cease upon issuance. It is important that the stop work order be presented to the owner of the subject property, the agent of the owner, or the individual doing the work. Only after all issues have been satisfied may work continue. Because of the potential consequences involved with *shutting down* a construction project, it is critical that the appropriate procedures be followed. Although a stop work order can be issued at any time during the construction process, the most common application of the provision is where work has commenced before or without the issuance of a building permit. The order is given to notify those affected parties that all work be stopped until a valid permit is obtained. An example of a stop work order is shown in Figure 115-1.

STOP WORK

DEPARTMENT OF BUILDING SAFETY
NOTICE
This building has been inspected and
❑ Footing/foundation
❑ Concrete slab/underfloor
❑ Framing
❑ Lath/gypsum board
❑ Energy efficiency
❑ Rough electrical
❑ Rough gas
❑ Rough plumbing
❑ Rough mechanical
❑ Final inspection
IS NOT ACCEPTABLE
Please call _____
before any further work is done.

_____
Date   Inspector
Do not remove this notice.

- -
DETACH and bring this portion of card with you.
LOCATION: _____

_____
Date

_____
Building Official

_____
Phone

Figure 115-1 **Stop work order.**

Section 116 *Unsafe Structures and Equipment*

The provisions of this section are intended to define what constitutes an unsafe building and unsafe use of a building. Unsafe buildings and structures are considered public nuisances and require repair or abatement. The abatement procedures are indicated in this section, including the creation of a report on the nature of the unsafe condition. Written notice shall be provided and the method of service of such notice is specified in Section 116.4. Where restoration or repair of the building is desired, the provisions of the IEBC for existing buildings are appropriate.

> **KEY POINTS**
> - The IBC provides minimum standards to safeguard the health, safety, and welfare of the public.
> - Rendering interpretations of specific code provisions is the responsibility of the building official, to be based on the purpose and intent of the code.
> - Modifications to specific IBC provisions may be acceptable where strict application of the code is impractical.
> - Alternative designs, materials, and methods of construction to those detailed in the code are to be evaluated by the building official based on an equivalency to the prescribed regulations.
> - Tests may be mandated by the building official in order to verify compliance with the code.
> - The issuance of permits, plan review, and inspection of construction for which permits have been issued constitutes the bulk of the duties of the typical code compliance agency.
> - The code intends that the issuance of a permit should not be construed as allowing a violation of the code or any other local ordinance applicable to the building.
> - Submittal documents must be appropriately submitted to the building official for review, and approved prior to receiving a building permit.
> - The building official has the authority to require the designation of a registered design professional to review and coordinate submittal documents prepared by others.
> - A certificate of occupancy, granted by the building official, indicates that the structure is lawfully permitted to be occupied.
> - The board of appeals provides a mechanism for individuals to challenge the interpretation of the building official on technical issues in the IBC.

CHAPTER 2

DEFINITIONS

Section 201 General
Section 202 Definitions
Key Points

> A number of definitions are applicable specifically to the *International Building Code®* (IBC®) and may not have an appropriate definition for code purposes in the dictionary. Therefore, over 700 words and terms are defined in this chapter to assist the user in the proper application of the requirements. Those words and terms specifically defined in the code are italicized where they occur throughout the text in order to identify for the user that a specific definition can be found in this chapter. Although infrequent, the definitions of some terms are contained within the text of the requirement. For example, the definition of day care is implied in the description of Group E occupancies. Other frequently used and significant terms are undefined (i.e., 1-hour fire-resistance-rated construction), and their meaning can be discerned only from their context. There are numerous definitions in this chapter, but only selected definitions are included in this commentary.

Section 201 *General*

An important feature of this section is the requirement that ordinarily accepted meanings be used for definitions that are not provided in the code. Such meanings are based on the context in which the term or terms appear. The code defines terms that have specific intents and meanings insofar as the code is concerned, and leaves it up to the user to apply all undefined terms in the manner in which they are ordinarily used. Where there is any question as to the meaning or application of a particular definition, the building official shall make the determination under the interpretive authority granted in Section 104.2 based on the context and intent of the term's usage in the code.

Section 202 *Definitions*

ACCESSIBLE MEANS OF EGRESS. In concert with efforts to make buildings accessible and usable for persons with disabilities, it is necessary that safe egress for individuals with disabilities also be provided. Therefore, accessible egress paths of continuous and unobstructed travel are required. Generally consistent with the provisions of other exiting systems, an accessible means of egress shall begin at any accessible point within the building and continue until reaching the public way. The primary provisions regulating accessible means of egress are located in Section 1009.

AISLE. Where furniture, fixtures, and equipment limit the potential travel paths within the means of egress system, aisles and aisle accessways are created. Typically, aisles accept the contribution of occupant travel from adjoining aisle accessways. At times, multiple aisles may converge into a main aisle, which then may lead to exit access doorways or exits. Aisles are common throughout most buildings and are considered portions of the exit access. Aisles in most applications are addressed in Section 1018. Those aisles, including stepped aisles, that are present in assembly seating areas are regulated by Section 1030.

AISLE ACCESSWAY. The path of travel from an occupiable point in a building to an aisle is considered to be an aisle accessway. Often called a "row" in everyday language when adjacent to seats or fixtures, an aisle accessway is typically used by small numbers of people prior to converging with other persons at an aisle. An example of aisle accessways regulated by Section 1018 is shown in Figure 202-1. Where aisle accessways occur in assembly seating areas, such as those found in theaters and arenas, they must comply with the provisions of Section 1030.

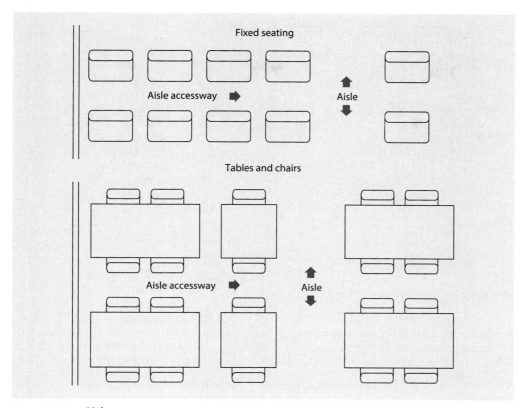

Figure 202-1 **Aisle accessways.**

ALTERNATING TREAD DEVICE. By appearance more of a ladder than a stairway, an alternating tread device is shown in Figure 202-2. Typically, steps are supported by a center rail placed at a severe angle from the floor. The key difference between this device and other forms of stairways addressed in the IBC is that the user, by nature of the design of the device, can never have both feet on the same level at the same time. The use of an alternating tread device is selectively permitted in certain areas not occupiable by the public, some of which are identified in Section 1011.14.

AMBULATORY CARE FACILITY. Facilities where individuals are provided with medical care on less than a 24-hour basis are classified as Group B occupancies. However, two separate definitions highlight the fact that there are two unique types of persons that occupy such facilities. The important difference involves the self-preservation capabilities of the individuals. Where the occupants are capable of self-preservation (the ability to respond to emergency situations without assistance from others) throughout the period of time in which they are at the facility, the building is considered an "outpatient clinic" as defined in this chapter. If such self-preservation cannot be accomplished due to the application of sedation or similar procedures, or where the care recipients are already incapable of self-preservation upon arrival, then the facility is by definition an "ambulatory care facility." The need for separate definitions is due to special provisions in Section 422 that specifically regulate those types of facilities where individuals are temporarily or permanently incapable of self-preservation.

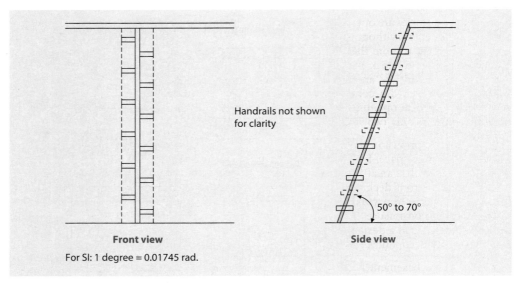

Figure 202-2 **Alternating tread device.**

Ambulatory care facilities are more highly regulated than outpatient clinics in regard to smoke compartmentation, incidental uses, automatic sprinkler system protection, and fire alarm system requirements.

ANNULAR SPACE. The open space created around the outside of a pipe, conduit, or similar penetrating item where it passes through a vertical or horizontal assembly is considered the annular space. The code addresses methods to maintain the integrity of a fire-resistance-rated assembly, including methods to protect any annular space around a penetration.

APPROVED. Throughout the code, the term *approved* is used to describe a specific component, material, or method of construction, such as an approved barrier in interior exit stairways addressed in Section 1023.8. Where *approved* is used, it merely means that such a design, material, or method of construction is acceptable to the building official (or other authority having jurisdiction), based on the intent of the code. It would seem appropriate that the building official base their decision of approval on the result of investigations or tests, if applicable, or by reason of accepted principles.

AREA OF REFUGE. It is common for an area of refuge to be included as a portion of the accessible means of egress in a nonsprinklered multistory building. The intent of the refuge area is to provide a location where individuals unable to use stairways can gather to await assistance or instructions during an emergency evacuation of the building. The size and construction requirements for areas of refuge are provided in Section 1009.6.

ATRIUM. The recognition of an atrium in the IBC as more than just a design feature is due to the openings between stories that pose a significant hazard under fire conditions. In order to address this unique approach to vertical openness, special provisions were established in the 1980s to provide alternative protection through concepts applicable to both open courts and shaft enclosures. Requirements specific to atriums are set forth in Section 404.

The definition of atrium is limited to addressing two conditions: (1) an atrium is a vertical space enclosed at the top, and (2) an atrium typically connects three or more stories. Where no more than two stories are open to each other, there are multiple allowances in Section 712.1 that allow such openness without consideration or regulation as an atrium.

It is important to note that there may be limited circumstances where an open two-story condition exists and one or more of the requirements of Section 712.1.9 cannot be met. In such situations, by definition, an atrium condition cannot be considered due to the "three-story" threshold. However, if a complying atrium can be considered as an acceptable approach to addressing a vertical opening of three or more stories, it should also be acceptable where the opening connects only two stories.

ATTIC. Several provisions apply to the attic area of a building, such as those relating to ventilation of the attic space. In order to fully clarify the portion of a building defined as an attic, this chapter identifies an attic as the space between the ceiling framing at the top story and the underside of the roof deck, slab, or sheathing. An attic designation is appropriate only if the area is not considered occupiable. Where this area has a floor, it would typically be defined as a story. A common misuse of IBC terminology is the designation of a space as a *habitable* or *occupiable* attic. Such a designation is inappropriate insofar as once such a space is utilized for some degree of occupancy, it is no longer deemed an attic.

BASEMENT. A basement is considered to be any story that does not meet the definition of "story above grade plane." There are limited provisions in the code that are specifically applicable to basements. One such significant requirement is established in Section 903.2.11.1.3 mandating the sprinklering of basements where adequate exterior openings are not provided. In short, the code regulates a below-ground story based on its qualification as a story above grade plane. See the commentary in this chapter on Story Above Grade Plane.

BLEACHERS. Structures designed for seating purposes containing tiered or stepped seating two or more rows high are considered bleachers. The definition of "grandstand" is identical to that of bleachers, recognizing that the terms are interchangeable. Bleachers may, or may not, be provided with backrests. In addition, bleachers may be located either inside or outside a structure. The specific provisions for bleachers are found in ICC 300, *Bleachers, Folding and Telescopic Seating, and Grandstands*. Similar seating areas that are considered as building elements would not be defined as bleachers and are not regulated by ICC 300. A building element, also defined in Section 202, is deemed to be a fundamental component of the building's construction as listed in Table 601.

BUILDING AREA. The term *building area* describes that portion of the building's floor area to be utilized in the determination of whether or not a structure complies with the provisions of Chapter 5 for allowable building size. It is not to be confused with the term *floor area*, which is the basis for occupant-load determination in Chapter 10 for means of egress evaluation, nor the term *fire area* as used in the application of automatic sprinkler requirements in Chapter 9.

The definition of building area is the area included within the surrounding exterior walls of the building, and the definition further states that the floor area of a building or portion thereof not provided with surrounding exterior walls shall be the usable area under the horizontal projection of the roof or floor above. The intent of this latter provision is to address where a structure may not have exterior walls or may have one or more sides open without an enclosing exterior wall. Examples would include a canopy covering pump islands at a service station, or the covered drive-through area of a fast food restaurant. Where a column line establishes the outer perimeter of the usable space under the roof, it is also typically the extent of building area. Beyond the column line, the overhead cover is simply viewed as a projection. See Figure 202-3. If all of the area beneath the roof above can be considered usable space, then the building area is measured to the leading edge of the roof above. See Figure 202-4.

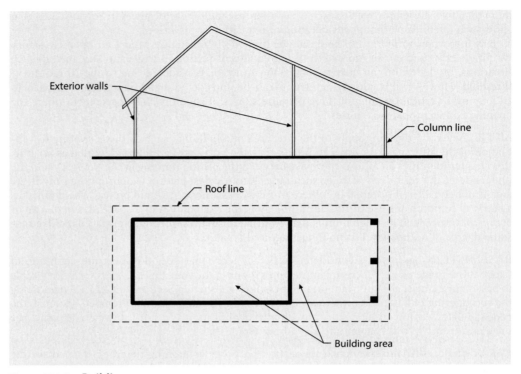

Figure 202-3 **Building area.**

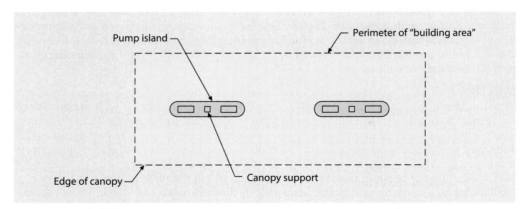

Figure 202-4 **Building area.**

BUILDING ELEMENT. Primary structural frame members, bearing walls, nonbearing walls and partitions, floor construction including secondary members, and roof construction including secondary members are considered to be building elements for the purposes of the IBC. Such elements are primarily regulated based on two criteria: fire resistance and combustibility. In determining a building's type of construction, the building elements are evaluated based on the criteria previously mentioned.

BUILDING OFFICIAL. Regardless of title, the individual who is designated by the jurisdiction as the person who administers and enforces the IBC is considered by the code to be the building official. In addition, all other individuals who have been given a similar enforcement authority, such as plans examiners and inspectors, are also considered building officials to a limited degree under the IBC. A further discussion of the duties and responsibilities of the building official is found in the commentary on Section 104.

CARE SUITE. Special means of egress provisions are provided in Section 407.4.4 for care suites in Group I-2 occupancies. The definition of "care suite" establishes the scope of such special provisions. The concept of suites recognizes those arrangements where staff must have more supervision of care recipients in specific treatment and sleeping rooms. Therefore, the general means of egress requirements are typically not appropriate under such conditions. The special allowances for care suites are not intended to apply to day rooms or business functions of the health-care facility. Intensive-care and critical-care units are typically configured as care suites.

CEILING RADIATION DAMPER. Designed to protect air openings that occur in fire-resistance-rated roof/ceiling or floor/ceiling assemblies, ceiling radiation dampers are listed devices that automatically limit the radiative heat transfer from a room or space into the cavity above the ceiling. The damper, in conjunction with the fire-resistive ceiling membrane, protects the structural system within the floor/ceiling or roof/ceiling assembly from failure that is due to excessive heat.

CHANGE OF OCCUPANCY. Every building, or portion of a building, must be assigned an occupancy classification with respect to its use by placing it into one of the specific occupancy groups identified in Chapter 3. These groups are used throughout the code to address everything from allowable building size to required fire protection features. The occupancy classification process represents identifying a distinct hazard, or more typically a group of hazards, that must be specifically addressed throughout the applicable code requirements. For example, the classification of Group A recognizes those assembly-related concerns associated with theaters, night clubs, places of worship, and other facilities where large numbers of people congregate in a concentrated manner. Where there is a change in the use of the building that modifies the type or extent of the hazards created by the new use, a change of occupancy occurs.

Where no change in occupancy classification occurs, a change of occupancy exists only in those buildings where (1) there is a change in a building's purpose or level of activity, (2) that functional change is such that the current IBC requires a greater degree of regulation than presently exists in the current building, and (3) the greater degree of regulation required by the current IBC occurs only in the areas of accessibility, structural strength, fire protection, means of egress, ventilation, or sanitation. The intent is to limit the application of a change of occupancy where there is no change in classification to only those new uses that present a higher risk to the life safety or welfare of the occupants than was created by the previous use.

It is generally accepted that where a change from one occupancy group to another group, such as from a Group B to a Group M, as well as a change of divisions within an individual group, such as from a Group F-2 to a Group F-1, constitutes a change of occupancy. However, due to the scope of the definition, there is a question as to how to address a building that undergoes a change in occupancy classification where no greater degree of safety, accessibility, structural strength, fire protection, means of egress, ventilation, or sanitation is required by the code. An example is the remodeling of a Group A-2 assembly space in a manner where it would be reclassified as Group B. Where the new use poses an equal or lesser hazard level, but such use is classified as a different occupancy, it would seem appropriate to also consider it as a change of occupancy even though no modifications would be required.

COMMON PATH OF EGRESS TRAVEL. It is important to limit the distance of egress travel that occurs where only a single path or single point of egress is available to the user of the means of egress. For this reason, such limited travel is regulated as a common path of egress travel.

Very similar to the concept addressed in the limitations of dead-end conditions in corridors, the intent of regulations regarding a common path of egress travel is based on the lack of at least two separate and distinct paths of egress travel toward two or more remote exits. Every room, space, or area that only has access to one exit access doorway or exit must be regulated for common path limitations. Included as the initial part of the permitted travel distance, a common path of egress travel only occurs within the exit access portion of the means of egress. See Figure 202-5.

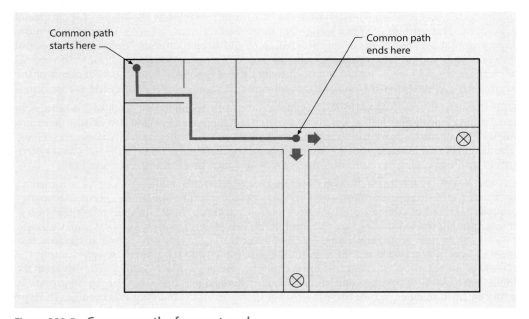

Figure 202-5 **Common path of egress travel.**

CORRIDOR. A corridor is considered a component of the exit access portion of the means of egress that provides an enclosed and directed route of egress travel along the path to an exit. Regulated by Section 1020, corridors may or may not be required to be fire-resistance rated, but in all cases, an enclosure of some sort is anticipated. The determination as to whether or not a design element is to be regulated as a corridor is to be made by the building official. Many factors may be considered prior to making this determination, which could possibly include length, degree of enclosure, length-to-width ratio, adjacent spaces, and other considerations. However, because of the reliance upon corridors as an important egress element, it is critical that they be appropriately regulated. A corridor's primary purpose is for the movement of occupants, both as a part of the building's circulation and its use as a means of egress. Although some spaces may have one or more of the characteristics of a corridor as previously mentioned (length, degree of enclosure, length-to-width ratio, etc.), their primary function is that of rooms, and they should not be considered corridors. A classic example can be found in many observation buildings in zoos, where a very long, narrow element is used as a means for occupants to view displays of wildlife. Although in plan view it may appear to be a corridor, it actually functions as a room and should be regulated as such. It should be noted that the placement of a few pieces of furniture or equipment within a corridor in an effort to consider the space a room rather than a corridor is not appropriate. It should be noted that corridors are never mandated by the IBC, but rather are utilized as design elements. Where provided, however, such components must comply with the code.

Definitions 27

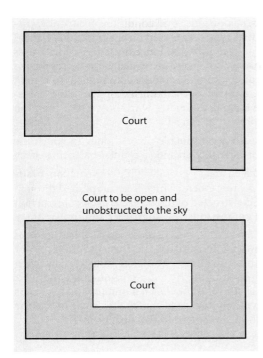

Figure 202-6 **Definition of courts.**

COURT. Open and unobstructed to the sky above, an exterior area is considered a court where it is enclosed on at least three sides by exterior walls of the building or other enclosing elements, such as a screen wall. Regulations for courts, including those used for egress purposes, are found throughout the code. Examples of courts are shown in Figure 202-6. Although the IBC does not mandate a minimum depth for consideration as a court, it is expected that certain design and structural features of the building that create minor exterior wall offsets would not require designation as a court. The determination of the presence of a court under such conditions is subject to the building official's discretion.

COVERED MALL BUILDING. A covered mall building consists of various tenants and occupancies, as well as the common pedestrian areas that provide access to the tenant spaces. Although the most common tenants of covered mall buildings are retail stores, various other uses are also commonly found in such a structure. Restaurants, drinking establishments, entertainment and amusement facilities, passenger transportation terminals such as airports and bus stations, offices, and other similar uses are often located in a covered mall building. There are a limited number of access points to a covered mall building; therefore, the tenants share a few major entrances into the mall. Although an anchor building is not to be considered a part of the covered mall building, it is possible to design an exterior perimeter building as merely another of the tenants within the covered mall building. It has become increasingly popular to create large-scale projects resembling covered mall buildings without roofs over the pedestrian circulation areas. Various "tenant space" buildings and "anchor buildings" are situated around unroofed pedestrian ways (open malls) in a manner very similar to that for covered mall buildings. The inclusion of open mall buildings in the IBC recognizes that the same benefits should be available as for enclosed structures, provided the appropriate measures are taken. Where the mall area is open to the sky, equivalent or better life safety and property protection is provided.

When one thinks of a covered mall building, its use is typically associated with retail sales and related activities. However, the provisions of Section 402 may also lend themselves to other occupancies such as office uses and transportation facilities. Recently, more attention has been given to the application of the covered mall building concept to large educational buildings. Many of the characteristics of an exciting and efficient school environment are consistent with those of a covered mall building, including spacious areas open to each other, vertical interaction between floor levels, ease of supervision, and limited points of entry. Educational facilities even have their own *anchor buildings* such as the auditorium, gymnasium, media center, and other spaces. However, because the criteria of Section 402 are more traditional in their approach, some of the provisions cannot be directly applied to an educational occupancy. It will be necessary to utilize the alternate methods and design allowance in Section 104.2.3 to more specifically address those issues.

CUSTODIAL CARE. When classifying buildings, or portions of buildings, by occupancy, custodial care activities are anticipated where residential-type care is provided. Conditions 1 and 2 of both Groups I-1 and R-4, as well as Group I-4 all recognize the hazards associated with uses where such care is provided. Individuals receiving custodial care are expected to be capable of self-preservation, although many such persons require limited or physical assistance from staff. This assistance often consists of helping someone transfer to a walking aid or providing necessary instructions. Other aspects of custodial care include assistance with those tasks common to daily activities, such as taking medications or preparing a meal.

DOOR, BALANCED. Balanced doors are a special type of double-pivoted door in which the pivot point is located some distance in from the door edge, thus creating a counter-balancing effect.

DRAFTSTOP. Required by the code only in concealed areas of combustible construction, draftstops are utilized in large concealed spaces to limit air movement, which is accomplished through the subdivision of such spaces. Draftstops are to be constructed of those materials or construction identified by the IBC that effectively create smaller compartments within attics and similar areas.

DWELLING UNIT AND DWELLING. A *dwelling unit* is considered a single unit that provides living facilities for one or more persons. Dwelling units include permanent provisions for living, sleeping, eating, cooking, and sanitation, thus providing a complete independent living arrangement. A dwelling unit, while typically addressed in the IBC as a portion of a Group R-2 occupancy, may also be classified as Group I-1, R-1, or R-3. A *dwelling* is a building that contains either one or two dwelling units. Dwellings are typically regulated under the provisions of the *International Residential Code®* (IRC®), as noted in the exception to IBC Section 101.2.

EGRESS COURT. That portion of the exit discharge at ground level that extends from a required exit to the public way is considered an egress court. An egress court may be a yard, a court, or a combination of a yard and a court that extends from the end point of an exit, typically an exterior door at grade level, until it reaches a public way. An example is illustrated in Figure 202-7.

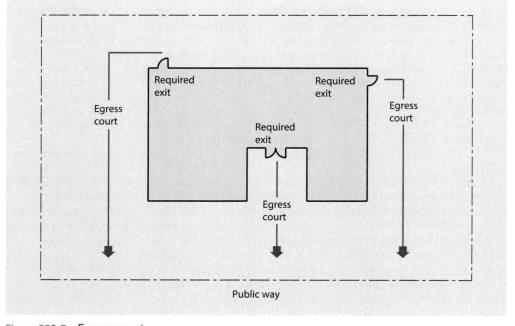

Figure 202-7 **Egress court.**

EMERGENCY ESCAPE AND RESCUE OPENING. Required in sleeping rooms and basements in limited residential occupancies, emergency escape and rescue openings are intended to allow for a secondary means of escape or rescue in the event of an emergency. Typically an operable window or door, such an exterior opening is regulated by Section 1031 for minimum size, maximum opening height from the floor, and operational constraints. Although the provisions are found in Chapter 10 of the code, the intent of the opening is for emergency escape or rescue access, and it is not intended to be considered an element of a complying means of egress.

EXIT. An exit is the first portion of egress travel where the code assumes that the occupant has obtained an adequate level of safety so that travel distance limitations are no longer a concern. In addition, an exit typically provides only single-direction egress travel. The adequate level of safety is provided by building elements that completely separate the means of egress from other interior spaces in the building. Inside the building, fire-resistance-rated construction and opening protectives are utilized to provide a protective path of egress travel between the exit access and exit discharge portions of the means of egress. Building elements that are considered exits include exterior exit doors at the level of exit discharge, interior exit stairways and ramps, exit passageways, exterior exit stairways, exterior exit ramps, and horizontal exits.

EXIT ACCESS. The exit access is identified as the initial component of the means of egress system, the portion between any occupied point in a building and the exit. Leading to one or more of the defined exit components, the exit access makes up the vast majority of any building's floor area. Because the exit access begins at any point that may potentially be occupied, it is probable that only those concealed areas, such as attics and under-floor spaces that are typically unoccupied, fall outside the scope of the means of egress.

EXIT ACCESS DOORWAY. The term "exit access doorway" is commonly used in IBC Chapter 10 to establish a reference point within the exit access for applying various means of egress provisions, including those addressing arrangement, number, separation, opening protection, and exit sign placement. Although one would expect the term "doorway" to be limited to those situations where an actual door opening, either with or without a door, is present, the IBC definition expands this traditional meaning by including certain access points that do not necessarily include doorways, such as at unenclosed exit access stairs and ramps. In fact, any point at which the exit access is narrowed so as to create a single point of travel could potentially be considered as an exit access doorway, as shown in Figure 202-8 where a narrow, directed path of egress travel is provided.

EXIT ACCESS STAIRWAY. Stairways intended to serve as required exit components within the means of egress are considered as interior exit stairways. All other stairways are regulated as exit access stairways. As expected, an exit access stairway is considered an exit access component of the egress system. Exit access stairways, like most other exit access components, are not intended to provide any type of fire-resistive protection for the building occupants as they travel through the means of egress. Thus, most exit access stairways are permitted to be unenclosed under the allowances of Section 1019.3. However, since stairways create vertical openings between stories, enclosure may be required under some conditions to restrict the vertical spread of fire, smoke, and gases. Exit access stairways may selectively serve as a portion of the required means of egress, while non-required stairways (convenience stairs) are also considered as exit access stairways.

EXIT DISCHARGE. Exit discharge is the final portion of the three-part means of egress system and is that portion between the point where occupants leave an exit and the point where they reach a public way. For conceptual ease, all exterior egress travel at ground level is considered a part of the exit discharge. An egress court is the primary component of exit discharge. Regulated in Section 1028.2, limited interior exit discharge is also permitted under specified conditions.

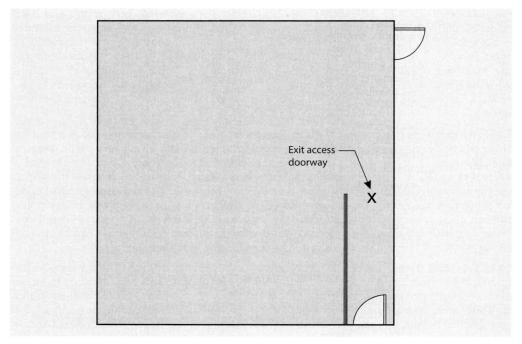

Figure 202-8 **Exit access doorway.**

EXIT DISCHARGE, LEVEL OF. The story within a building where an exit terminates and the exit discharge begins creates a condition defined as the level of exit discharge. Because exit discharge occurs substantially at ground level, the level of exit discharge typically provides a horizontal path of travel toward the public way. At times, the code uses this term to define another floor level's relationship to egress at grade level. For example, in most buildings, the first story above the level of exit discharge is typically the building's second story above grade plane, as shown in Figure 202-9. Where a mezzanine or similar floor level change occurs within the first story above grade plane, it is not considered as located above the level of exit discharge. Rather, its location is viewed as *within* the level of exit discharge.

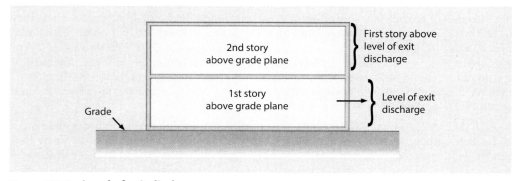

Figure 202-9 **Level of exit discharge.**

EXIT, HORIZONTAL. A horizontal exit is considered a component of the exit portion of the means of egress. The concept of a horizontal exit is to provide a refuge compartment adequately separated from fire, smoke, and gases generated in the area of a fire incident. As the occupants pass through the horizontal exit, they enter an area intended to afford safety from the fire and smoke of the area from which they departed. The concept of a horizontal exit is much different from the other exit elements set forth in the IBC, as its primary function is a refuge area rather than a path of egress travel.

EXIT PASSAGEWAY. Much like an interior exit stairway or ramp in providing smoke and fire protection, an exit passageway is an exit component that is separated from the remainder of the building by fire-resistance-rated construction and opening protectives. The horizontal path of protected travel afforded by an exit passageway often extends to the exit discharge or the public way. An exit passageway is often utilized as the horizontal extension of egress travel connecting an interior exit stairway or ramp to the exterior of the building. Its function is to maintain a level of occupant protection equivalent to that of the interior exit stairway or ramp to which it is connected. An exit passageway is also to be used for protected horizontal travel where the continuity of an interior exit stairway or ramp enclosure must extend to another enclosure. Occasionally, an exit passageway is utilized by the designer as a means to provide compliance with travel distance, exit separation, or other means of egress provision.

EXTERIOR EXIT STAIRWAY. To be classified as an exterior exit stairway, it must be open on at least one side. The open side must then adjoin an open area such as a yard, egress court, or public way. By preventing full enclosure of the stairway, an exterior exit stairway will be sufficiently open to the exterior to prevent the accumulation of smoke and toxic gases. Additional criteria for defining an exterior exit stairway used as a means of egress are found in Section 1027.3.

F RATING. Through-penetration firestop systems, perimeter fire containment systems, and continuity head-of-wall systems are provided with F ratings to indicate the time periods in which they limit the spread of fire through the penetration or void. Tested to the requirements of ASTM E814 or UL 1479, penetration firestop systems for fire-resistance-rated assemblies will typically have an F rating no less than the required rating of the assembly penetrated. Perimeter fire containment systems are generally required to be tested in accordance with ASTM E2307 requirements. Continuity head-of-wall systems are to be tested in accordance with ASTM E2837.

FIRE AREA. Many of the provisions of Section 903 requiring the installation of automatic sprinkler systems utilize the fire area concept. A fire area is considered a compartment that will contain a fire such that the maximum fire size will be limited to the size of the compartment. An in-depth review of fire areas is found in the discussion of Section 901.7.

FIRE BARRIER. Fire-resistance-rated walls are considered fire barriers if constructed under the provisions of Section 707. The purpose of such assemblies is to create a barrier that will restrict fire spread to and from other portions of the building. All openings within a fire barrier must be protected with a fire-protective assembly. Fire barriers are often utilized to separate incompatible uses within the building, to create fire areas, or to provide for egress through a protected exit system. Fire barriers may selectively be fire-resistance-rated at 1, 2, 3, or 4 hours.

FIRE DAMPER. Regulated by test standard UL 555, fire dampers are devices located in ducts and air-transfer openings to restrict the passage of flames. Fire dampers close automatically upon the detection of heat, maintaining the integrity of the fire-resistance-rated assembly that is penetrated. The actuation of fire dampers creates some restriction to airflow from migrating throughout the duct system or through transfer openings, although not to the level of that required for a smoke damper.

FIRE DOOR ASSEMBLY. Where openings occur in fire-resistance-rated assemblies, fire door assemblies are permitted as a method to protect the openings. Addressed in Section 716, fire door assemblies include not only the door but also the door frame, the door hardware, and any other components needed to provide the necessary fire-protective rating required for the specific application.

FIRE PARTITION. A fire partition is a wall or similar vertical element that is utilized by the code to provide fire-resistive protection in limited applications. As a general rule, fire partitions are required to be of 1-hour fire-resistance-rated construction and mandated for use in eight specified locations. The most common applications for fire partitions among those identified in Section 708 are for the construction of fire-resistance-rated corridors and for walls separating dwelling units in apartment buildings.

FIRE-PROTECTION RATING. Opening protectives, such as fire door and fire window assemblies, are assigned a fire-protection rating in order to identify the time period in which the protective is expected to confine a fire by resisting or preventing the movement of excessive flames through the opening protective. The specific rating requirement, in either hours or minutes, varies based on the details of the fire-resistance-rated assembly in which the opening protective is located.

FIRE-RATED GLAZING. Fire separation elements such as fire partitions, fire barriers, and fire walls often include glazing in some forms, such as glazed wall assemblies, fire windows, and/or fire doors with vision panels. The definition of "fire-rated glazing" encompasses both types of such glazing addressed by the code, fire-resistance-rated glazing and fire-protection-rated glazing. Fire-resistance-rated glazing, addressed in Sections 703.4 and 716.1.2.3, must be tested in accordance with ASTM E119 or UL 263 as a wall assembly. Fire-protection-rated glazing, established for use by Sections 716.2.5 and 716.3, is to be tested in accordance with NFPA 257 or UL 9 as an opening protective. Both types of glazing are collectively referred to as fire-rated glazing.

FIRE-RESISTANCE RATING. Identified by a specific time period, a fire-resistance rating is assigned to a tested component or assembly based on its ability to perform under fire conditions. A fire-resistance-rated component or assembly is intended to restrict the spread of fire from a specified area or provide the necessary protection for the continued performance of a structural member. This performance is based on fire resistance, defined as the property of materials or assemblies that prevents or retards the passage of excessive heat, hot gases, or flames.

FIRE-RESISTANT JOINT SYSTEM. Where a linear opening is placed in or between adjacent fire-resistance-rated assemblies to allow independent movement of the assemblies, it is considered by the IBC as a joint. One or more joints may be provided to address movement caused by thermal, seismic, wind, or any other similar loading method. Designed to protect a potential breach in the integrity of a fire-resistance-rated horizontal or vertical assembly, a fire-resistant joint system is a tested assembly of specific materials designed to restrict the passage of fire through joints. The fire-resistance ratings required for the joint systems, as well as other requirements, are addressed in Section 715.

FIRE SEPARATION DISTANCE. The fire separation distance describes the distance between the exterior surface of a building and one of three locations—the nearest interior lot line; the centerline of a street, alley, or other public way; or an imaginary line placed between two buildings on the same lot. The method of measurement is based on the distance as measured perpendicular to the face of the building. See Figure 202-10. The fire separation distance is important in the determination of exterior wall and opening protection based on a building's location on the lot. See the discussion of Table 705.5 for a further analysis of this subject.

FIRE WALL. Scoped by Sections 414.2.3 and 503.1, and fully described by Section 706, one or more fire walls are building elements used to divide a single building into two or more separate

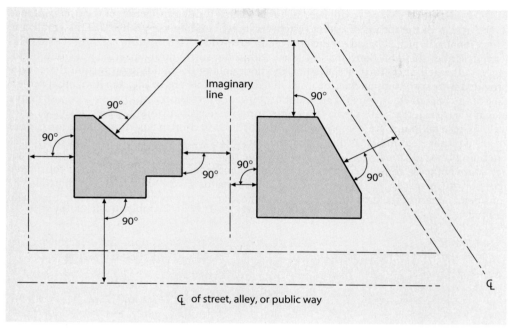

Figure 202-10 **Fire separation distance.**

buildings for the purpose of applying the allowable height, allowable area, type of construction, and control area requirements of the IBC. In addition, a fire wall is selectively identified in the code as a means to serve a variety of purposes where separate buildings are not created, including the use of horizontal exit separation as addressed in Section 1026.2. Starting at the foundation and continuing vertically to or through the roof, a fire wall is intended to fully restrict the spread of fire from one side of the wall to the other. Fire walls are higher level fire-resistance-rated elements than both fire barriers and fire partitions. Because the primary concept of fire walls is to create smaller buildings within one larger structure, with the code selectively regulating each small building individually rather than collectively, it is critical that a fire wall be capable of maintaining structural stability under fire conditions. If construction on either side of a fire wall should collapse, such a failure should not cause the fire wall to collapse for the prescribed time period of the rating of the wall.

FIREBLOCKING. Fireblocking is mandated by the code to address the spread of fire through concealed spaces of combustible construction. Experience has shown that the greatest damage to conventional wood-frame buildings during a fire occurs when the fire travels unimpeded through concealed draft openings. Materials identified by the IBC as effective in resisting fire spread through concealed spaces include 2-inch (51-mm) nominal lumber, gypsum board, and glass-fiber batts.

FLIGHT. The use of the term "flight" is specifically defined in order to establish its use within the code. The definition addresses two separate issues. First, a flight is made up of the treads and risers that occur between landings. As an example, a stairway connecting two stories that includes an intermediate landing consists of two flights. Second, the inclusion of winders within a stairway does not create multiple flights. Winders are simply treads within a flight and are often combined with rectangular treads within the same flight.

FLOOR AREA, GROSS. As evidenced in Table 1004.5, the determination of the occupant load in the design of the means of egress system for most building uses, other than assembly and educational, is typically based on the gross floor area where fixed seating is not provided. This term describes the total floor area included within the surrounding exterior walls of a building, and the definition further states that the floor area of the building or portion thereof not provided with surrounding exterior walls shall be the usable area under the horizontal projection of the roof or floor above. The intent of this latter provision is to cover where a structure may not have exterior walls or may have one or more sides open without an enclosing exterior wall. Where buildings are composed of both enclosed and unenclosed areas, the gross floor area is typically determined as illustrated in Figure 202-11. Projections, such as eave overhangs, extending beyond an exterior wall or column line that are not intended to create usable space below are not to be considered in the determination of gross floor area. Areas often considered accessory-type spaces, such as closets, corridors, elevator shafts, and stairways, must also be considered a part of the gross floor area, as are those areas, occupied by furniture, fixtures, and equipment.

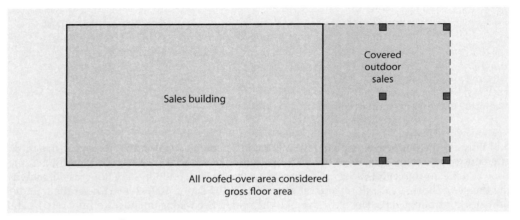

Figure 202-11 **Gross floor area.**

FLOOR AREA, NET. The net floor area is considered the portion of the gross floor area that is typically occupied. Normally unoccupied accessory areas such as corridors, stairways, closets, toilet rooms, equipment rooms, and similar spaces are not to be included in the calculation of net floor area. In addition, the measurements are based on clear floor space, allowing for the deduction of building construction features such as interior walls and columns, as well as elevator shafts and plumbing chases. The use of net floor area in the calculation of design occupant load is typically permitted only in assembly and educational uses as set forth in Table 1004.5. It is important to note that in calculating net floor area, as well as gross floor area, the floor space occupied by furniture, fixtures, and equipment is not to be excluded in the calculation. Within an individual room or space, the floor area utilized for aisles and aisle accessways should also be included when calculating the net floor area. The floor-area-per-occupant factor established in Table 1004.5 includes any such anticipated furnishings and circulation paths in the establishment of an appropriate density estimate. The application of "net floor area" is not intended within an individual room.

FLOOR FIRE DOOR ASSEMBLY. Although a fire door is typically viewed as an element protecting an opening in a vertical building element such as a wall, it is possible that such doors can be effective if installed horizontally for the protection of an opening in a fire-resistance-rated floor.

The floor fire door assembly, like other fire door assemblies, includes the door, frame, hardware, and other accessories that make up the assembly, and provides a specified level of fire protection for the opening.

FOLDING AND TELESCOPING SEATING. Folding and telescoping seats are structures that provide tiered seating, which can be reduced in size and moved without dismantling. Utilized quite often in school gymnasiums, such seating presents the same concerns and risks as permanently installed bleacher seating when occupied. Such seating is regulated by ICC 300, *Standard for Bleachers, Folding and Telescopic Seating, and Grandstands.*

GRADE PLANE. The code indicates that the grade plane is a reference plane representing the average of the finished ground level adjoining the building at its exterior walls. Under conditions where the finished ground level slopes significantly away from the exterior walls, that reference plane is established by the lowest points of elevation of the finished surface of the ground within an area between the building and lot line, or where the lot line is more than 6 feet (1,829 mm) from the building, between the building and a line 6 feet (1,829 mm) from the building. Where the slope away from the building is minimal (typically provided only to drain water away from the exterior wall), the elevation at the exterior wall provides an adequate reference point.

The method for calculating grade plane can vary based on the site conditions. Where the slope is generally consistent as it passes across the building site, it may only require the averaging of a few points along the exterior wall of a rectangular-shaped building, as illustrated in Figure 202-12. Where the slope is inconsistent or retaining walls are utilized, or where the building footprint is complex, the determination of grade plane can be more complicated. In such cases, a more exacting method for calculating the grade plane must be utilized. In addition, where fire walls are present, the elevation points should be taken at the intersections of the fire wall and the exterior walls.

This definition is extremely important in determining a building's number of stories above grade plane as well as its height in feet. In some cases, the finished surface of the ground may be artificially raised with imported fill to create a higher grade plane around a building so as to decrease the number of stories above grade plane or height in feet. The code does not prohibit this practice, provided the building meets the code definition and restriction for height or number of stories above grade plane. See Figure 202-13.

It is important to note that for the vast majority of buildings, it is not necessary to precisely calculate the grade plane. In such buildings, a general approximation of grade plane is sufficient to appropriately apply the code. A detailed calculation is only necessary in those limited situations where it is not obvious how the building is to be viewed in relationship to the surrounding ground level.

GRANDSTAND. The definition of grandstand is also applicable to bleachers. Further information is provided in the discussion of the definition of bleachers. Grandstands are to be regulated by ICC 300, *Standard for Bleachers, Folding and Telescopic Seating, and Grandstands.*

GUARD. A component or system of components whose function is the minimization of falls from an elevated area is considered a guard. Placed adjacent to the elevation change, a guard must be of adequate height, strength, and configuration to prevent someone from falling over or passing through the guard. Outside of the code, this element is more commonly referred to as a guardrail.

HABITABLE SPACE. An area within a building, typically a residential occupancy, used for living, sleeping, eating, or cooking purposes would be considered habitable space. Those areas not considered to meet this definition include bathrooms, closets, hallways, laundry rooms, storage rooms, and utility spaces. Obviously, habitable spaces as defined in this section are those areas usually occupied, and as such are more highly regulated than their support areas.

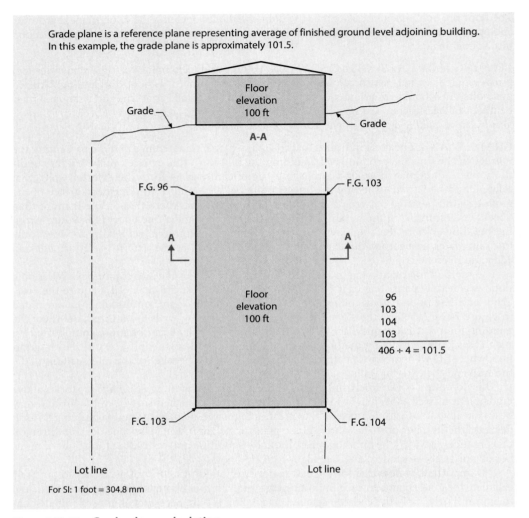

Figure 202-12 **Grade plane calculation.**

Although typical, it is not necessary that a room or area be finished in order to be considered habitable space. It is not uncommon for a residential unit to have a large basement that is not completely finished-out. Nevertheless, the basement may be used as living space, particularly for children who use it as a playroom. Such a basement would be considered habitable space, as the definition is simply based on the use of the room or area.

HANDRAIL. Typically used in conjunction with a ramp or stairway, a handrail is intended to provide support for the user along the travel path. A handrail may also be used as a guide to direct the user in a specified direction.

HEIGHT, BUILDING. Once the elevation of the grade plane has been calculated, it is possible to determine the building's height. This height is measured vertically from the grade plane to the average height of the highest roof surface. Examples of this measurement are shown in Figure 202-14.

Definitions 37

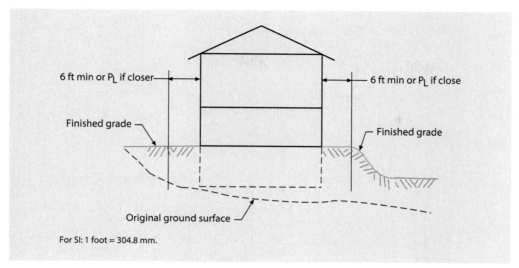

Figure 202-13 **Use of built-up soil to raise finished grade.**

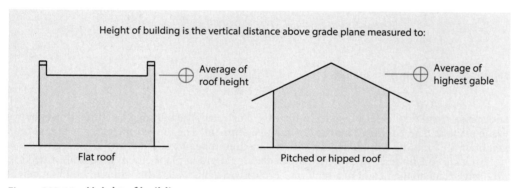

Figure 202-14 **Height of building.**

Where the building is stepped or terraced, it is logical that its height is the maximum height of any segment of the building. It may be appropriate under certain circumstances that the number of stories in a building be determined in the same manner. Because of the varying requirements of the code that are related to the number of stories, such as means of egress, type of construction, and fire resistance of shaft enclosures, each case should be judged individually based on the characteristics of the site and construction. In addition to those factors better related to the number of stories, other items to consider are fire department access, location of exterior exit doors, routes of exit travel, and types of separation between segments.

Figure 202-15 illustrates one example in which the height of the building and number of stories are determined for a stepped or terraced building. In the case of a stepped or terraced building, the language *total perimeter* is used to define the situation separating the first story above grade plane from a basement and is intended to include the entire perimeter of the segment of the building. Therefore, in the cross section of Figure 202-16, the total perimeter of the down-slope segment would be bounded by the retaining wall, the down-slope exterior wall,

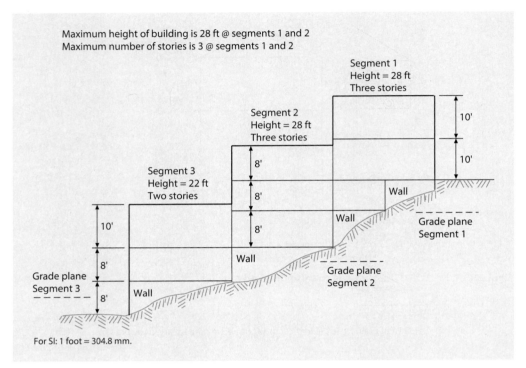

Figure 202-15 **Terraced building.**

and the east and west exterior walls. In the case illustrated, the building has three stories above grade plane and no basement for the down-slope segment. The measurement for the maximum height of the building would be based on the maximum height of the down-slope segment.

Similar to an unnecessarily detailed calculation of grade plane, there is seldom a need to precisely calculate the height of a building. Typically, a general determination of building height is adequate to ensure compliance with the code. For example, it is not necessary to go into great detail evaluating the average roof elevation of a built-up roof that has a low degree of slope for drainage purposes. The need for a more exacting determination of roof height is directly related to any uncertainty that may occur in reviewing for code compliance.

HIGH-RISE BUILDING. A high-rise building has long been defined as a building having one or more floor levels used for human occupancy located more than 75 feet (22,860 mm) above the lowest level of fire department vehicle access, as illustrated in Figure 202-17. Most moderately large and larger cities have apparatus that can fight fires up to about 75 feet (22,860 mm); thus, the fire can be fought from the exterior. Any fires above this height will require that they be fought internally. Also, in some circles, 75 feet (22,860 mm) is considered to be about the maximum height for a building that could be completely evacuated within a reasonable period of time. Thus, the fire department's capability plus the time for evacuation of the occupants constitute the criteria used by the IBC for defining a high-rise building.

The determining measurement is to be taken from the lowest point at which the fire department will locate their fire apparatus to the highest floor level that is viewed as occupiable. This would include a mezzanine floor level that may occur within the highest story, as well as an occupiable roof (such as a tennis court, swimming pool, or sun deck on the roof of a condominium building).

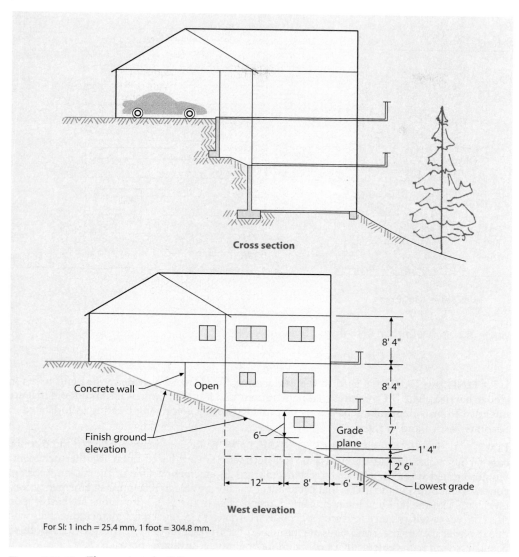

Figure 202-16 **Three-story building.**

HORIZONTAL ASSEMBLY. A horizontal assembly is for the most part the horizontal equivalent of a fire barrier. It is utilized to restrict vertical fire spread through an established degree of fire resistance as mandated through various provisions in the IBC. The specifics for horizontal assemblies, which include both floor and roof assemblies, are found in Section 711.

INTERIOR EXIT STAIRWAY. One of several exit components established in the code, an interior exit stairway provides a fire-resistance-rated enclosure for vertical egress in buildings. Section 1023 identifies the requirements for an interior exit stairway, including its construction and termination; permissible openings, penetrations, and ventilation; and necessary signage. One or more interior exit stairways are typically required in multistory buildings in order to satisfy the various means of egress requirements in Chapter 10.

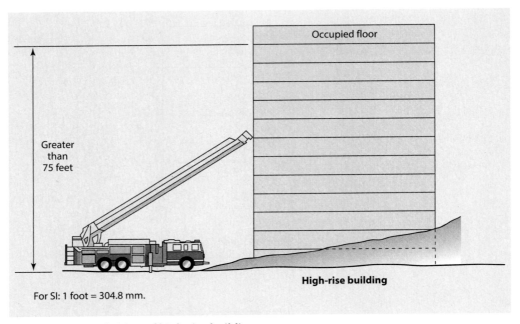

Figure 202-17 Definition of high-rise building.

L RATING. An L rating is used in the evaluation of air leakage at penetrations and joints in smoke barriers. The IBC requires that penetrations and joints in those walls and floors that are intended to provide smoke separations be provided with a complying L rating as indicated in Sections 714.5.4 and 715.9.

LIMITED PHYSICAL AND VERBAL ASSISTANCE. Requirements applicable to custodial care facilities classified as Group I-1 or R-4 occupancies vary based upon whether occupants require limited verbal or physical assistance during an emergency. One of the most important concepts embodied in the IBC is that occupants typically are able to recognize an emergency condition and exit the building in an efficient manner. Most of the code's provisions are based upon this concept of self-reliance for egress purposes. However, there are a great many individuals who due to physical or mental conditions are incapable of self-preservation. The IBC provides for these occupants with requirements that take on more of a protect-in-place approach.

This definition describes a middle-ground-type of occupant capabilities between those individuals able to evacuate independently and those incapable of self-preservation. The Condition 2 occupancy classifications for Groups I-1 and R-4 include facilities where residents, because of frailness, cognitive impairment, or other conditions, need limited assistance while exiting the building. Staff will be required to assist care-recipients during evacuation who may use mobility devices or can self-propel in a wheelchair, help with transferring out of bed, assist with balance while walking stairways, or provide some other type of physical assistance. It may also be necessary that staff provide extra prompting or repeat instructions to residents with dementia in order to complete the evacuation process.

MASS TIMBER. Mass timber consists of Type IV structural elements that are primarily of solid, built-up, panelized, or engineered wood products that meet minimum Type IV cross-section dimensions. Mass timber represents both the large wood building elements historically recognized as heavy timber (Type IV-HT) construction and the three newly established construction types (Types IV-A, IV-B, and IV-C).

MEANS OF EGRESS. The means of egress describes the entire travel path a person encounters under exiting conditions, beginning from any occupiable point in a building and not ending until the public way is reached. Often encompassing both horizontal and vertical travel, the means of egress should be direct, obvious, continuous, undiminished, and unobstructed. It includes all elements of the exiting system that might intervene between the most remote occupiable portion of the building and the eventual place of safety—typically the public way. Therefore, the means of egress includes all intervening components such as aisle accessways, aisles, doors, corridors, stairways, and egress courts, as well as any other component that might be in the path of travel, as depicted in Figure 202-18. There are three distinct and separate portions of a means of egress—the exit access, the exit, and the exit discharge.

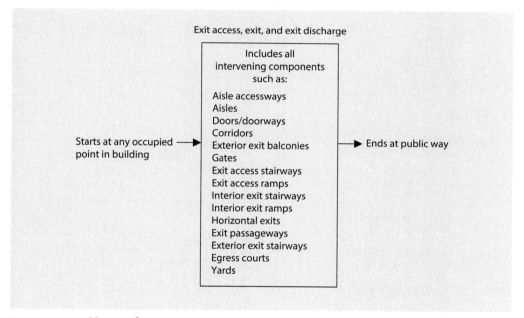

Figure 202-18 **Means of egress.**

MEMBRANE PENETRATION. Similar to a through penetration in its performance requirements, a membrane penetration is an opening through only one membrane of a wall, floor/ceiling, or roof/ceiling assembly. An example of a very common membrane penetration is an electrical box.

MEMBRANE-PENETRATION FIRESTOP. Where a penetrating item such as a pipe or conduit passes through a single membrane of a fire-resistance-rated wall, floor/ceiling, or roof/ceiling assembly, a membrane-penetration firestop may be required in order to adequately protect the penetration. Such a firestop consists of a device or construction that would effectively resist the passage of flame and heat through the opening in the membrane created by the penetrating item. A fire-resistance rating is assigned to a membrane-penetration firestop to indicate the time period for which the firestop is listed.

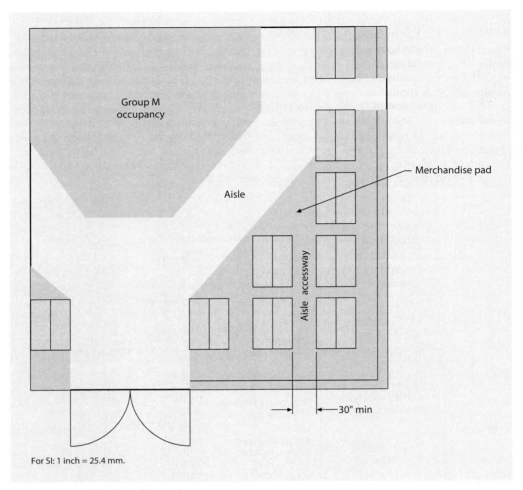

Figure 202-19 **Merchandise pad.**

MERCHANDISE PAD. Typically found in a mercantile use, a merchandise pad is created where racks, displays, shelving units, and similar fixtures are grouped into a specific area. Aisle accessways are provided within the merchandise pad to allow for customer circulation and are utilized as exit access elements. Such aisle accessways connect to the aisles that partially or entirely surround the merchandise pad. An example is shown in Figure 202-19.

MEZZANINE. A mezzanine is merely a code term for an intermediate floor level placed within a room between the floor and ceiling of a story. Typically limited in floor area to 33⅓ percent of the area of the room or space into which it is located, a mezzanine is regulated under the provisions of Section 505. A floor level fully complying with the provisions of Section 505 is permitted to be considered a mezzanine in order to utilize those special allowances applicable to such a condition. The use of the mezzanine provisions is a design option, as an elevated floor level complying with the scoping provisions of Section 505 is always permitted to be considered as an additional story rather than a mezzanine.

OCCUPANT LOAD. Viewed as the basis for the design of the means of egress system, the occupant load is the number of persons to be considered for means of egress purposes both within and from any space, area, room, or building. Variations in code requirements are often caused based on the anticipated occupant load served by the specific building egress element. Although the establishment of an occupant load is critical in the application of many means of egress provisions located in Chapter 10, it is also often necessary when addressing minimum requirements for fire alarm systems, sprinkler systems, and plumbing fixture counts. Section 1004 establishes the appropriate methods for the determination of occupant loads.

OCCUPIABLE SPACE. A number of provisions in the code apply only to those spaces, rooms, or areas typically occupied during the course of a building's use. This definition clarifies that an occupiable space is intended for human occupancy, and as such is provided with means of egress, as well as light and ventilation facilities. Occasionally the code refers to the term *normally occupied* space. For example, Section 1023.4 limits openings in interior exit stairways to those necessary for exit access from *normally occupied spaces*. Although not defined in the code, these spaces are generally occupied for extended periods of time during the building's use. There is an expectation that a fire or other hazardous condition would be quickly identified and addressed, rather than go unnoticed for an extended time. Examples of those spaces that would not be considered *normally occupied* include storage rooms, mechanical equipment rooms, and toilet rooms.

PANIC HARDWARE. The term *panic hardware* describes an unlatching device that will operate even during panic situations, so that the force of individuals against the egress door will cause the door to unlatch without manual manipulation of the device.

PENTHOUSE. Regulated by Section 1511.2, and further defined by Section 1511.2.2, penthouses are structures placed on the roofs of buildings to shelter various types of machinery and equipment, as well as stairway and vertical shaft openings. The definition clarifies that a penthouse is intended to be unoccupied. Rooftop structures, such as tanks, towers, spires, and domes, are not considered penthouses, but are regulated to some degree by other provisions in Section 1511.

PHOTOVOLTAIC (PV) SUPPORT STRUCTURE, ELEVATED. Often referred to as "solar shade structures," overhead PV support structures are most commonly constructed over vehicle parking spaces of surface parking lots and on the uppermost level of parking garages. In addition, such structures are built in a variety of other locations with or without vehicles parked beneath. These independent PV panel support structures are designed with usable space underneath with minimum clear height of 7 feet 6 inches (2,286 mm), intended for secondary use such as providing shade or parking of motor vehicles. Per Section 3111.3.5, elevated PV support structures with PV panels installed over open-grid framing or over a noncombustible deck shall have PV panels tested, listed and labeled with a fire type rating. PV panels marked "not fire rated" shall not be installed on elevated PV support structures. Elevated PV support structures with a PV panel system installed over a roof assembly shall have a fire classification in accordance with Section 1505.9. Roof structures that provide support for PV panel systems shall be designed in accordance with Section 1607.14.3.

PLATFORM. The distinctions between the definitions of a stage and a platform are very important because of the requirements applicable to each element. A platform differs from a stage due to the lack of overhead hanging curtains, drops, scenery, and other effects that a stage contains. The amount of fire loading associated with a platform is typically much less than for a stage. Thus, the fire-severity potential is much lower. In order to be regulated as a platform, a raised performance area cannot have overhanging curtains; however, horizontal sliding curtains are permitted.

PRIMARY STRUCTURAL FRAME. Any structural member or component that is essential to the vertical stability of the building under gravity loads is to be classified and protected as part of the primary structural frame. For the primary structural frame, the following components qualify:

- Columns
- Beams and girder trusses—directly supported by columns
- Roof and floor panels and slabs—monolithic concrete and timber construction
- Braced frames and moment frames—transfer gravity loads to the foundation
- Bearing walls—support gravity loads and transfer them to the foundation

PRIVATE GARAGE. There are fundamentally two types of parking garages regulated by the IBC, private garages and public garages. There is no specific definition in the code for public garage, but Section 406.4 indicates that a public garage is any vehicle parking garage that does conform to the requirements for a private garage. Such public parking garages are further classified as either open parking garages or enclosed parking garages based upon specific characteristics.

Private garages are limited to those facilities that comply with not only the definition criteria in this chapter, but also the limitations set forth in Section 406.3. A private garage, in addition to its restriction on floor area in Section 406.3.1, is also regulated as to its use. Motor vehicles stored in a private garage are limited to only those vehicles used by the owner and/or tenants of the building or buildings on the same premises as the garage. In addition, the repair and/or serving of vehicles for business purposes cannot occur within a private garage.

Although most private garages are associated with residential occupancies, there is no limit as to what occupancy classifications can be served by a private garage. For example, a small parking facility serving only those tenants in an office building could be considered as a private garage provided the requirements of Section 406.3 are met.

PUBLIC WAY. The code defines a public way essentially as a street, alley, or any parcel of land that is permanently appropriated to the public for public use. Therefore, the public's right to use such a parcel of land is guaranteed. The building occupants, having reached a public way, are literally free to go wherever they might choose. They are certainly free to go so far as to escape any fire threats in any building that they might have been occupying. There is an expectation that the public way be a continuation of the egress path, providing for continued exit travel away from the building until the necessary safety level has been achieved.

PUBLIC-OCCUPANCY TEMPORARY STRUCTURE. Regulated by Chapter 31, public-occupancy temporary structures are new buildings or structures that are used by the public, or that support public events, where the public expects similar levels of reliability and safety as offered by permanent construction. Public-occupancy temporary structures are often assembled with reusable components and designed for a particular purpose and limited period, which is defined as a temporary event when the period is less than 1 year. Public-occupancy temporary structures in service for a period that exceeds 1 year must meet IBC requirements for new buildings.

RAMP. Where the slope of a travel path exceeds 1 unit vertical to 20 units horizontal (1:20), it is considered a ramp. Where the travel path has a slope of 5 percent or less (less than or equal to 1:20), it is considered merely a walking surface.

REPAIR GARAGE. Where a building, or portion of a building, is used for painting operations, body and fender work, engine overhauling, or other major repair of motor vehicles, it is considered to be a repair garage. Other activities included in the scope of those occurring in repair garages include service and maintenance work such as tire replacement, oil changes, and brake work. The designation as a repair garage is important in the occupancy classification of such activities as Group S-1.

Where the work being performed falls outside of the scope of service or repair, such as vehicle detailing, window tinting, or the installation of audio or security components, its consideration as a repair garage may not be appropriate. The activities may be considered more of a business or sales use rather than that of repair or maintenance.

SCISSOR STAIRWAY. A unique design element, the scissor stair allows for two independent paths of travel within a single interior exit stairway. A scissor stair is considered a single means of egress because the exit paths are not isolated from each other with the required level of fire resistance.

SECONDARY STRUCTURAL MEMBERS. Secondary structural members are components that are built into an assembly that supports a portion of a floor or a roof, or only their own self-weight. The following members are considered as secondary:

- Roof trusses connected to a girder truss
- Roof purlins and subpurlins connected to beams
- Floor joists and trusses
- Nonbearing walls
- Bracing in the roof, floor, or walls that is specifically designed to resist wind or seismic loads and is redundant for gravity systems.

SELF-CLOSING. In order to eliminate a portion of the human element in maintaining the integrity of fire-resistance-rated wall assemblies, doors in such assemblies are typically required to be provided with self-closing devices that will ensure closing of the doors after having been opened. Occasionally, automatic-closing fire doors are installed in specific locations on account due to the nature of the situation. Under such conditions, the automatic-closing fire assemblies are to be regulated by NFPA 80.

SHAFT. A shaft is considered the enclosed space that extends through one or more stories of a building. Its function is to connect vertical openings in successive floors that have been created to accommodate elevators, dumbwaiters, mechanical equipment, or similar devices, as well as for the transmission of natural light or ventilation air.

SHAFT ENCLOSURE. A shaft enclosure is the building element defined by the boundaries of a shaft, which typically includes its surrounding walls and other forms of construction. Regulated by Section 713, it is required to be of fire-resistance-rated construction. A shaft enclosure is the most traditional application listed in Section 712.1 for addressing vertical openings in buildings.

SLEEPING UNIT. The single required characteristic of a sleeping unit is that it is used as the primary location for sleeping purposes. The room or space that has sleeping facilities may also provide for eating and living activities. It could have a bathroom or a kitchen but not both, as this would qualify it as a dwelling unit. Guestrooms of Group R-1 hotels and motels would typically be considered sleeping units. Sleeping units are also commonly found in congregate living facilities, such as dormitories, sorority houses, and fraternity houses, and are regulated as Group R-2 occupancies.

Group R occupancies are not the only types of uses where sleeping units are located. Several of the varied uses classified as Group I occupancies also contain resident or patient sleeping units. The proper designation of these spaces as sleeping units is important in the application of Section 420 mandating the separation of sleeping units in Group R and I-1 occupancies, as well as addressing the appropriate accessibility provisions of Chapter 11.

SMOKE BARRIER. Required under various circumstances identified by the code, smoke barriers are either vertical or horizontal membranes, or a combination of both, intended to restrict the movement of smoke. Walls, floors, and ceiling assemblies may be considered smoke barriers where they are designed and constructed in accordance with the provisions of Section 709. Smoke barriers are typically required where it is anticipated that building occupants will not be able to evacuate the building quickly under emergency conditions. Refuge areas created through the use of smoke barriers allow occupants to be isolated from fire and smoke conditions within the building until such time other action is required.

SMOKE COMPARTMENT. Where smoke barriers totally enclose a portion of a building, the enclosed area is considered a smoke compartment. By completely isolating the compartment from the remainder of the building by interior walls, floors, and similar elements, smoke can be either contained within the originating area or prevented from entering other areas of the building. The use of smoke compartments is predominant in Group I-2 and I-3 occupancies.

SMOKE DAMPER. Test standard UL 555S states that leakage-rated dampers (smoke dampers in the terminology of the IBC) are intended to restrict the spread of smoke in heating, ventilating, and air-conditioning (HVAC) systems that are designed to automatically shut down in the event of a fire, or to control the movement of smoke within a building when the HVAC system is operational in engineered smoke-control systems. The IBC simply identifies smoke dampers as listed devices designed to resist the passage of air and smoke through ducts and air-transfer openings. Smoke dampers must operate automatically unless manual control is desired from a remote command station.

SMOKE-PROTECTED ASSEMBLY SEATING. Where the means of egress for assembly seating areas is designed to be relatively free of the accumulation of smoke, the seating is considered to be smoke protected. Protection may be achieved in one of the two ways, either through the design of a passive system or through ventilation by mechanical means. In order to qualify as smoke protected, the seating area and its exiting system must comply with the provisions of Section 1030.6.2, which addresses the methods of smoke control, the minimum roof height, and the possible installation of sprinklers in adjacent enclosed spaces. Exterior seating facilities such as stadiums or amphitheaters are commonly considered as open-air assembly seating due to the natural ventilation that is available. Where smoke-protected or open-air assembly seating is provided, selected means of egress requirements are modified recognizing the reduced potential for smoke accumulation in the seating area.

STAIR. Where one or more risers are provided to address a change in elevation, a stair is created. A stair may simply be a slight change in height from one floor level to another, commonly referred to as a step, or may be a series of treads and risers connecting one floor level or landing to another. Also described in the code as a flight of stairs, a stair does not include the landings and floor levels that interrupt stairway travel.

STAIRWAY. Where one or more flights of stairs occur, including any intermediate landings that connect the stair flights, a stairway is created. The term *stairway* describes the entire vertical travel element that is made up of stairs, landings, and platforms.

STAIRWAY, SPIRAL. A spiral stairway is a stairway configuration where the treads radiate from a central pole. The treads are uniform in shape, with a tread length that varies significantly from the inside of the tread to the outside. The dimensional characteristics of a spiral stairway cause it to be limited in its application.

STORY. Although seemingly quite obvious, the definition of a story is that portion of a building from a floor surface to the floor surface or roof above. In the case of the topmost story, the height of a story is measured from the floor surface to the top of the ceiling joists, or to the top of the roof rafters where a ceiling is not present. The critical part of the definition of a story involves the definition of *story above grade plane*.

It is not uncommon for a roof level to be used for purposes other than weather protection or mechanical equipment. A roof patio, garden, or sports area is sometimes provided in order to utilize as much of the building as possible. Although an occupiable roof does not meet the definition of "story," there are certain provisions in the IBC that would be applicable due to the fact occupants can be present. For example, an occupiable roof must be provided with a complying means of egress designed for the anticipated occupant load of the roof level. Full compliance with applicable accessibility requirements must also be provided. Required fire alarm system protection should be extended to such occupiable roofs. However, the roof level would not be considered part of the building area for allowable area purposes. In addition, its consideration in the evaluation of a building's allowable number of stories varies based upon the provisions of Section 503.1.4. A careful analysis should be made when determining which provisions are applicable to an occupiable roof.

Although stories are typically identified as above-ground floor levels, basements are also considered as stories. For example, a building with four stories above grade plane and one level of basement would be considered as a five-story building. The provisions specific to stories are typically addressing concerns within a building, while those related more toward stories above grade plane and basements are focused on exterior issues.

STORY ABOVE GRADE PLANE. Throughout the code, the number of qualifying stories in a building is a contributing factor to the proper application of the provisions. As an example, a building's allowable types of construction are based partly on the limits on height in stories placed on various occupancy groups. In this case, the code is limiting construction type based on the number of stories *above grade plane*. The code defines a story above grade plane as any story having its finished floor surface entirely above grade plane. However, floor levels partially below the grade at the building's exterior may also fall under this terminology. The critical part of the definition involves whether or not a floor level located partially below grade is to be considered a story above grade plane. There are two additional criteria that are important to the determination if a given floor level is to be considered a story above grade plane:

1. If the finished floor level above the level under consideration is more than 6 feet (1,829 mm) above the grade plane as defined in Section 202, the level under consideration is a story above grade plane, or

2. If the finished floor level above the level under consideration is more than 12 feet (3,658 mm) above the finished ground level at any point, the floor level under consideration shall be considered a story above grade plane.

Where either one of these two conditions exists, the level under consideration is also considered a story above grade plane.

Conversely, if the finished floor level above the level under consideration is 6 feet (1,829 mm) or less above the grade plane, and does not exceed 12 feet (3,658 mm) at any point, the floor level under consideration is not considered a story above grade plane. By definition, it is regulated as a basement. Figures 202-20 and 202-21 illustrate the definitions of "story," "basement," and "story above grade plane."

Although the criteria for establishing the first story above grade plane in Item 2 indicates that such a condition occurs where the 12-foot (3,658-mm) limitation is exceeded, the application of this provision is not that simple. It is not the intent of the code to classify a story that is completely below grade except for a small entrance ramp or loading dock as a *story above grade plane*, provided there is no adverse effect on fire department access and staging. An analysis of the impact of such limited elevation differences is necessary to more appropriately apply the code's intended result.

T RATING. The T rating is defined as the time required for a specific temperature rise on the unexposed side of a penetration firestop system or continuity head-of-wall system. More specifically, the penetration firestop system as well as the penetrating item must provide for a maximum

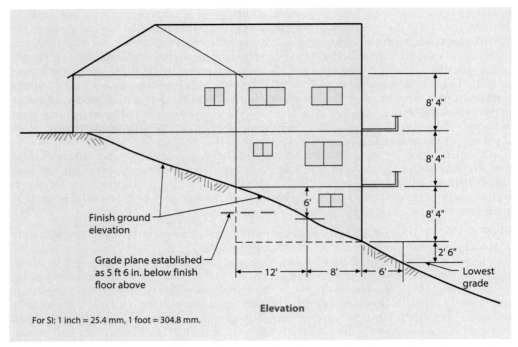

Figure 202-20 Three-story building with two stories above grade plane and one basement level.

increase in temperature of 325°F (181°C) above its initial temperature for the time period reflected in the fire-resistance rating. The maximum temperature rise is also limited through the void on the non-fire side on a continuity head-of-wall system. The establishment of the rating of a through-penetration firestop is determined by tests in accordance with ASTM E814 or UL 1479, while a continuity head-of-wall system is to be tested to ASTM E2837. A more detailed discussion of the subject is found in Chapter 7.

TECHNICAL PRODUCTION AREA. Many auditoriums, arenas, and performance halls, as well as other types of entertainment and sport venues, are provided with elevated technical support areas used for lighting, sound, scenery, and other performance effects. The code regulates these as technical production areas. The areas may or may not be associated with a stage, but are typically an integral part of the production. These spaces are generally limited in floor area, and access is always restricted to authorized personnel. The term "technical production areas" is intended to encompass all technical support areas, regardless of their traditional name. Special requirements for these types of areas are addressed in Section 410.

THROUGH PENETRATION. A through penetration is considered an opening that passes through an entire assembly, accommodating various penetrating items such as cables, conduit, and piping. Where membrane construction is provided, such as gypsum board applied to both sides of a stud wall, a through penetration would pass entirely through both membranes and the cavity of the wall.

THROUGH-PENETRATION FIRESTOP SYSTEM. In order to adequately protect the penetration of a fire-resistance-rated assembly by conduit, tubing, piping, and similar items, a through-penetration firestop system is sometimes required. Such a system may selectively

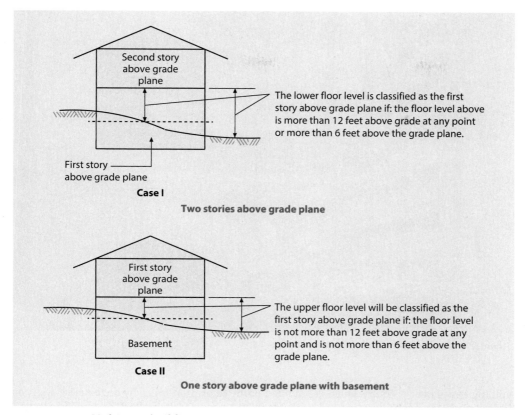

Figure 202-21 **Multistory building.**

include various materials or products that have been designed and tested to resist the spread of fire through the penetration. Through-penetration firestop systems are fire-resistance rated based on the criteria of ASTM E814 or UL 1479, and are provided with an hourly rating for both fire spread (F rating) and temperature rise (T rating).

WALKWAY, PEDESTRIAN. Described as a walkway used exclusively as a pedestrian trafficway, a pedestrian walkway provides a connection between buildings. A pedestrian walkway may be located at grade, as well as above ground level (bridge) or below grade (tunnel). The provisions addressing pedestrian walkways are optional in nature and utilized primarily to allow for the consideration of the connected buildings as separate structures. Regulations for pedestrian walkways and tunnels are found in Section 3104. An example of a pedestrian walkway is shown in Figure 202-22.

WINDER. A winder, or winder tread, is a type of tread that is used to provide for a gradual change in direction of stairway travel. Although the total directional change created by winders is typically 90 degrees (1.57 rad), other configurations are also acceptable. Owing to a reduced level of safe stairway travel, winders may typically only be used as a required means of egress stairway when located within a dwelling unit.

YARD. Used throughout the code to describe an open space at the exterior of a building, a yard must be unobstructed from the ground to the sky and located on the same lot on which the

Figure 202-22 **Pedestrian walkway.**

building is situated. A court, which is bounded on three or more sides by the exterior walls of the building, is not considered a yard. Both a yard and a court are expected to provide adequate openness and natural ventilation so that the accumulation of smoke and toxic gases will not occur.

It is not intended that exterior areas devoted to parking, landscaping, or signage be prohibited to qualify as a yard, provided access to and from the building is available and maintained for both the occupants and fire department personnel. It is also important to recognize that the code provisions sometimes require a *yard* and at other times an *open space*, as well as references to *fire separation distance*. Although the differences may appear to be subtle, each term is applied somewhat differently.

KEY POINTS

- The IBC defines terms that have specific intents and meanings insofar as the code is concerned.
- All terms specifically defined in the IBC are listed in Section 202 along with their definitions.
- Words and terms specifically defined in this chapter are italicized wherever they appear within the code.
- Ordinarily accepted meanings are to be used for definitions that are not provided in the code, based on the context in which the term or terms appear.

CHAPTER 3

OCCUPANCY CLASSIFICATION AND USE

Section 302 Occupancy Classification and Use Designation
Section 303 Assembly Group A
Section 304 Business Group B
Section 305 Educational Group E
Section 306 Factory Group F
Section 307 High-Hazard Group H
Section 308 Institutional Group I
Section 309 Mercantile Group M
Section 310 Residential Group R
Section 311 Storage Group S
Section 312 Utility and Miscellaneous Group U
Key Points

> This chapter of the code is simply an extension of Chapter 2 in that there are no technical requirements provided but rather descriptions of various uses and their assignment into specific classifications. These "occupancy" classifications are used as the basis for many of the provisions established elsewhere in the code. Potential uses are assigned an occupancy classification based on the specific occupant and content hazards that are anticipated to exist. Uses having similar hazard characteristics are all grouped into the same occupancy classification. Such characteristics may include the number, density, mobility, and awareness of the probable building occupants, as well as the combustibility, quantity, and environment of the building's contents. Proper occupancy classification is a major factor in achieving the code's intended purpose of establishing "minimum requirements to provide a reasonable level of safety, health, and general welfare" as set forth in Section 101.3.

Section 302 *Occupancy Classification and Use Designation*

302.1 Occupancy classification. Every structure, or portion of a structure, must be classified with respect to its use by placing it into 1 of the 26 specific occupancy groups identified in the code. These groups are used throughout the *International Building Code®* (IBC®) to address everything from building size to fire-protection features. The occupancy groups are organized into 10 categories of a more general nature, representing the following types of uses: assembly, business, educational, factory/industrial, high hazard, institutional, mercantile, residential, storage, and utility/miscellaneous.

The provisions of Section 302.1 provide direction to:

1. Classify all buildings into one or more of the 26 groups identified in the IBC.

The occupancy classification is typically established by the design professional during the code analysis phase. Most of the time, the designer's determination will be consistent with that of the building department. However, where there is disagreement as to the proper classification of the various uses within the building, it is the building official's responsibility to make the final decision. This authority is granted in Section 104.2 dealing with the interpretive powers of the building official. Although the IBC lists in some detail the uses allowed within a specific occupancy classification, the building official will at times also be called upon to judge whether or not a selected classification is appropriate under specific conditions. Assigning occupancy classification often depends not only on the use, but also on the extent and intensity of that use. The use may be so incidental to the overall occupancy that its effect on fire and life safety is negligible. As an example, the administrative office area in an elementary school performs a business-type function, but such a use is so incidental to the general operation and activities of the school that assigning it to a separate occupancy group would quite probably be unproductive. Therefore, the building official's judgment will often be relied upon to classify occupancies that could potentially fall into more than one group.

Several of the occupancy groups are further subdivided to more precisely recognize the difference in hazards that need to be addressed. Within the Group I-2 institutional group, Condition 1 occupancies consist primarily of nursing homes, while Condition 2 represents hospitals and similar facilities. Condition 1 and 2 classifications are also assigned to Group I-1 and R-4 occupancies to better apply the necessary custodial care requirements based on whether the occupants require limited verbal or physical assistance. Along with other specific variations in application, it is apparent that there are, in fact, even more occupancy classifications than the 26 initially established. It is intended that appropriate code requirements can be more accurately applied where even slight differences in hazard levels can be identified.

Occupiable roofs are to be assigned one or more occupancy classifications in a manner consistent with the classification of uses inside a building, based upon the fire and life safety hazards posed by the rooftop activities. The IBC is considered as an "occupancy-based" code, where the primary difference in requirements between buildings is due to the varying uses that are anticipated. As such, it is critical that an occupancy classification be assigned to any occupiable portion of a building in order that the appropriate fire and life safety criteria are applied. For example, where a rooftop contains a restaurant having dining seating for 50 or more persons, the occupiable roof would be classified as a Group A-2 occupancy in order to address those hazards associated with such an assembly use.

2. Address any room or space within the building that will be occupied at different times for different purposes.

Although an uncommon occurrence, a building space may at times be used for an activity that is considerably different from its typical use. It is important that the hazards associated with those different uses also be addressed. The code is basically asking that such a space be assigned multiple occupancy classifications, with the requirements of each assigned occupancy group to be applied. For practical purposes, the classification of the space should probably be based on the more restrictive of the occupancies involved. This would account for most of the requirements that would be in place. Any additional requirements that would be applicable because of the other occupancy classification would then be layered on top of the other provisions. Another method would be to establish the occupancy classification of the major use, which has the greatest occurrence, with a layering of the other occupancy requirements on top. Whatever the procedure, it is important that all anticipated uses, and the hazards these uses pose to the building's occupants, be taken into account.

A common example of a space used for various functions is a high school gymnasium. The space takes on various occupancy classifications based on the varied activities that occur, including Group E (physical education classroom), Group A-3 (community activities), Group A-4 (spectator gymnasium), and Group M (weekend craft shows). The classification of the space, which will result in the necessary safeguards being put in place, requires a comprehensive review of the anticipated activities and the hazards involved. A seasonal change in occupancy is another occurrence that must be considered. The creation of a *haunted house* for Halloween activities in a space typically used for other purposes is not uncommon. Regardless of the occupancy classification assigned, it is important that all of the anticipated uses be identified in order to apply the necessary code requirements.

3. Regulate buildings having two or more distinct occupancy classifications under the provisions of Section 508.

Many buildings cannot simply be classified under a single designation. A hotel, considered a Group R-1 occupancy, typically includes assembly spaces classified as Group A. In addition, Group M and B occupancies may be present. Each distinct occupancy will be regulated based on the specific hazards that the individual uses create. The relationship between one occupancy and another is also very important. Where multiple occupancy conditions exist, the provisions of Section 508 are applicable.

There are two basic approaches to assigning occupancy classifications where buildings have multiple uses. One approach is to evaluate the building as individual areas, assigning classifications specific to the use that is under evaluation. Once this process is complete, a re-evaluation should occur to determine which classifications can be revised to reflect that of the major use. The other option is to initially classify the building as a single occupancy. Then each anticipated use that cannot be adequately addressed under the major classification will be assigned its own classification. Whatever approach is used, the goal is to make sure that the code provisions that are intended to address the anticipated uses of the space, and their potential hazards, are put in place.

There is an expectation that small support and circulation elements be included within the occupancy classification of the area in which they are located. This would include toilet rooms, storage closets, mechanical equipment rooms, and corridors. These spaces do not take on a unique classification unless they pose a unique hazard that can only be addressed with a different classification. A small reception area/office space of 360 square feet (33.45 m^2) would most probably not be considered a Group B occupancy where it is a portion of a Group R-1 hotel. On the other hand, a 7,200-square-foot (669-m^2) office/accounting/reservations center could hardly be considered as merely an extension of the Group R-1 classification. At some point, the use of a space and its relationship to other spaces in the building provide for a need to assign a separate classification.

 4. Classify a use into the group that the occupancy most nearly resembles, based on life and fire hazard, when the use is not described specifically in the code.

The code intends to divide the many uses possible in buildings and structures into 10 separate general groupings where each group by itself represents similar hazards. The perils contemplated by the occupancy groupings are of the fire- and life-safety types and are broadly divided into two general categories: those related to people and those related to property. The people-related hazards are divided further by activity, number of occupants, their ages, their capability of self-preservation, and the individual's control over the conditions to which they subjected. The property-related hazards are divided further by the quantity of combustible, flammable, or explosive materials and whether such materials are in use or in storage.

The uses to which a building may be put are obviously manifold, and as a result the building official will, on more than one occasion, either find or be presented with a use that will not conveniently fit into one of the occupancy classifications outlined in the code. As indicated previously in this commentary, under these circumstances, the IBC directs the building official to place the use in that classification delineated in the code that it most nearly resembles based on its life and fire risk. This requirement gives the building official broad authority to use judgment in the determination of the hazard of the affected group and, as a result of this evaluation, determine the occupancy classification that the hazards of the use most nearly resemble.

Occasionally, there may be a question as to which classification is to be assigned to a specific use. The owner of a building and the building official may have a difference of opinion as to the proper occupancy classification, or the building official may face a use that appears to fit into one of the code-described groups, but after further analysis it is determined that the hazards representative of the code-defined group are not present in the use proposed. In such situations, the building official should use their authority to place the use in the occupancy classification that it most nearly resembles based on its life- and fire-hazard characteristics. It must be remembered that the purpose of occupancy classification is solely to have the ability to properly regulate for the hazards associated with the building's expected use(s).

Section 303 *Assembly Group A*

The first item to be considered in this section is the description of an assembly occupancy as "the use of a building or structure, or portion thereof, for the gathering of persons for purposes such as civic, social or religious functions, recreation, food or drink consumption, or awaiting transportation." The description of an assembly occupancy is further defined by the numerous examples of Group A uses listed in Section 303.

The concerns unique to assembly uses are based primarily on two factors: the sizable occupant loads and the concentration of those occupant loads. Both conditions must exist to warrant a Group A classification. For example, a Group M big-box store has the potential for a large occupant load; however, the anticipated density of occupants within the store is not as concentrated as those densities are regulated under the Group A provisions. In order to be classified as a Group A occupancy, it is expected that the occupants are concentrated in a manner that will cause some inefficiency in egress under emergency conditions. Classification as an assembly use, and subsequently a Group A occupancy, requires that the occupants be present in a density that poses unique hazards due to the high concentration of people.

Assembly uses are further divided based on other factors unique to the activities that take place. A review of the uses specifically listed as Group A occupancies indicates that although some are general in nature, such as libraries and arenas, others are specific to a room or area within a building (courtroom, lecture hall, and waiting areas in transportation terminals). This concept of identifying uses is not limited to Group A occupancies, but rather is consistent throughout all of the occupancy group descriptions. It is critical that the building official thoroughly evaluate the uses that are anticipated and assign occupancy classifications based on the hazards that have been identified.

303.1.1 Small buildings and tenant spaces. Classification of a small assembly building or tenant space housing limited occupants as a Group A occupancy is considered unnecessary due to the limited hazards that are expected to be present. An assembly use with an occupant load of fewer than 50 persons and not accessory to another use, such as a small free-standing chapel or a café located in a strip shopping center, is to be classified as Group B. As a result, a Group A occupancy classification is typically only assigned to a stand-alone assembly use containing at least 50 occupants.

Where a Group B café or similar establishment has an associated kitchen, the kitchen is also to be classified as a portion of the Group B occupancy. This approach is consistent with the classification of commercial kitchens associated with Group A-2 dining facilities as indicated in Section 303.3.

In the calculation of occupant load for a café or similar establishment with customer seating, only the occupant load of the seating area should be considered when determining the occupancy classification. For example, a café with 48 occupants in the dining area and 4 occupants in the kitchen area would be classified as a Group B occupancy. This methodology is based upon the recognition that less than 50 occupants are in an assembly condition.

303.1.2 Small assembly spaces. Where the area used for assembly purposes is accessory to another occupancy in the building and contains an occupant load of fewer than 50 persons, the room or space can merely be considered an extension of the other occupancy. As an example, a break room with an occupant load of 30 in a large manufacturing facility could simply be considered a portion of the Group F occupancy. See Figure 303-1. As an option to the designer, a Group B classification can be assigned to the break room.

These two classification options are also available where the occupant load exceeds 49 but the floor area does not exceed 749 square feet (70 m^2). For example, a 715-square-foot (66.4 m^2) casino gaming area in a Group M retail sales establishment could be classified as Group B or a portion of Group M even though the occupant load of the gaming area would be 65.

Another example as shown in Figure 303-2 illustrates how a small accessory assembly space with an occupant load of 50 or more need not be classified as a Group A occupancy. In the example, a small 480-square-foot (44-m^2) chapel, having an occupant load of 68 and located within an assisted living facility classified as a Group I-1 occupancy, can simply be classified as an extension of the Group I-1 classification. As a design alternative, a classification of Group B is also permitted to be assigned to the chapel.

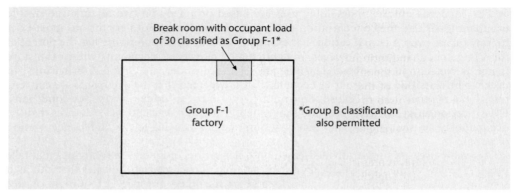

Figure 303-1 **Accessory space classification.**

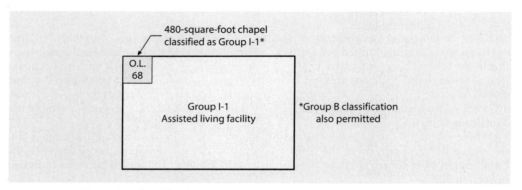

Figure 303-2 **Group I-1 classification.**

303.1.3 Associated with Group E occupancies. The classification of those assembly areas associated with a Group E occupancy as part of the Group E is permitted; however, the application of the Group E classification for the entire educational facility is based on the assembly areas being subsidiary to the school function. This would seem to indicate that the users of the associated assembly spaces are limited to students, teachers, administrators, and others directly involved in educational activities. A typical example of an associated assembly space is a library or media center that is used almost exclusively by students of the school. On the other hand, gymnasiums and auditoriums located in high school buildings are often used for community functions and other outside activities such as sports tournaments, craft shows, and community theater productions that have no relationship to normal educational uses. In such cases, a classification of Group A is often viewed as more appropriate based on these unrelated uses. It is up to the building official to identify the various activities intended for these assembly spaces and apply the proper occupancy classification based upon the potential hazards due to such activities. See the discussion of Section 302.1 regarding rooms or spaces intended to be occupied at different times for different purposes. Even in those situations where a Group E classification is appropriate for associated assembly uses, it is important that the assembly accessibility provisions of Chapter 11 be applied. In addition, the assembly means of egress provisions set forth in

Section 1030 should also be used in designing the means of egress system for the assembly areas, regardless of occupancy classification.

303.1.4 Accessory to places of religious worship. Small and moderately sized educational rooms and auditoriums accessory to places of religious worship are permitted to be classified as a portion of the major occupancy rather than individually. This would result in their classification as part of the overall Group A-3 occupancy. The allowance is simply a design option that can be used to eliminate or reduce any potential mixed occupancy conditions. If the designer wishes to classify such spaces individually, such as using a Group E occupancy classification for any religious educational rooms, such a classification is also permissible.

303.2 Assembly Group A-1. A factor involving human behavior in theaters classified as Group A-1 assembly rooms is the fact that in many cases the occupants are not familiar with their surroundings and the lighting level is usually low. Thus, when an emergency arises, the occupants may perceive the danger to be greater than presented, and panic may occur because of the fear of not being able to reach an exit for escape. In addition, the concentration of occupants in such uses is quite dense. The presence of a stage and its distinctive hazards that occur in some Group A-1 occupancies cause unique concerns, addressed by the special provisions of Section 410. Although not required for classification as a Group A-1, the presence of fixed seating is common in such occupancies. Similar multipurpose spaces where seating is not permanent are typically classified as Group A-3 occupancies due to the potential for other types of assembly activities.

303.3 Assembly Group A-2. Group A-2 occupancies include uses primarily intended for the consumption of food or drink, and include dining rooms, cafeterias, restaurants, cafes, nightclubs, taverns, and bars. The fire record in occupancies of this type is not very good, based in part on the delay in occupant response to a fire or other emergency incident. Because of the common presence of loose tables and chairs, aisles are often difficult to maintain, resulting in obstructions to egress travel. Overcrowding conditions, low-lighting levels, and the consumption of alcoholic beverages also increase the risks associated with many of these types of occupancies. The gaming floor areas of casinos are also classified as Group A-2 based in great part due to the congestion and distractions often encountered.

Included in the uses classified as Group A-2 occupancies are those commercial kitchens directly associated with Group A-2 restaurants, cafeterias, and similar dining facilities. Although a commercial kitchen does not pose the same types of hazards as an assembly use, the allowance for a similar classification has traditionally been viewed as appropriate.

Where the primary function of an assembly space or building is not food or drink consumption but drinking and dining activities do occur, it is not always appropriate to assign a Group A-2 classification. For example, an arena or stadium used for sports events and concerts typically contains eating and drinking as a part of the available activities. And yet, such buildings would be classified as Group A-4 and A-5 occupancies, respectively, due to their primary usage being sporting events with spectator seating. Classification as a Group A-2 occupancy should be limited to those situations where the primary use and expected hazards are related to food and drink consumption.

303.4 Assembly Group A-3. Occupancies classified as Group A-3 have varying degrees of occupant density, numerous types and numbers of furnishings and equipment, and fire loading that can vary from low to high. The hazards for uses in this category are similar to most of those found in Group A-1 and A-2 occupancies. Where a use does not conveniently fit into one of the other four Group A classifications, a Group A-3 designation is typically appropriate. The classification of an assembly occupancy as a Group A-3 is also common where varying assembly uses are likely to occur at different times within the same space. For example, a meeting room at a hotel is typically used at differing times for various functions, including seminar presentations, dining activities, trade shows, and wedding receptions. Although these functions may have different Group A designations when viewed individually, as a group they pose a hazard level that can be appropriately addressed with the Group A-3 classification. Therefore, most multipurpose rooms are simply classified as Group A-3 occupancies.

The mere presence of recreational activities does not necessarily warrant the classification of a building or space as a Group A-3 occupancy. Rather, it is the concentration of occupants participating or viewing the activity that should be used to determine the appropriate classification. The example illustrated in Figure 303-3 is based on a golf instruction academy with various "stations" at which students work on their skills individually or with an instructor. Support facilities such as restrooms, locker rooms, merchandise sales, and a snack bar are also provided. Although the occupant load of the building exceeds 49, there is little or no concentration of the occupants in a single area. Therefore, classification as Group A occupancy would not be warranted. A more appropriate classification of Group B would better describe the hazards and conditions found within a use of this type.

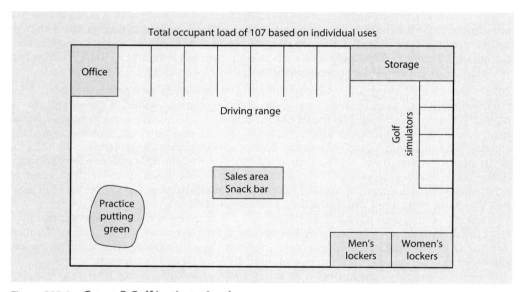

Figure 303-3   **Group B Golf Institute Academy.**

303.5 Assembly Group A-4. The combination of spectator seating and sporting events creates a condition within a building that warrants a specific occupancy classification within the Group A classification. A Group A-4 facility contains those occupant-related hazards found in other assembly occupancies, namely high occupant loads in concentrated areas, along with large areas having limited occupants and little, if any, fire loading conditions. The focus of the Group A-4 designation is that there is a significant number of spectators present in a relatively concentrated environment.

303.6 Assembly Group A-5. Uses classified as Group A-5 are similar in nature to Group A-4 occupancies, with the controlling difference being that Group A-5 occupancies are structures related to outdoor activities. Therefore, the fire hazard for Group A-5 occupancies is typically less than that for those classified as Group A-4 occupancies, and significantly lower than the other assembly Group A occupancies. The controlling expectation recognizes there will be little to no smoke accumulation under fire conditions in the assembly areas of a Group A-5 occupancy.

However, there still exist the hazards of crowding a large number of occupants within a relatively small space. The hazard of panic is assumed to be a large portion of the overall concern for Group A-5 structures. Generally, associated spaces such as concession stands, locker rooms, storage areas, press boxes, and toilet rooms are included as a portion of the Group A-5 classification. However, where uses within the building create conditions more hazardous than anticipated by the Group A-5 designation, such uses must be classified according to their individual characteristics. For example, an enclosed 400-seat restaurant within an open-air sports stadium should be appropriately classified as a Group A-2 occupancy. With all things considered, the primary consideration in classifying a use as Group A-5 is the lack of smoke accumulation. Where there is doubt as to the openness of the assembly seating area, an analysis must be undertaken to determine whether a Group A-5 classification is appropriate.

Section 304 *Business Group B*

The most common use classified as a Group B occupancy is an office building, or a portion of a building containing office tenants or office suites. Those areas of such business occupancies where records and accounts are stored are also considered part of the Group B classification. Examples of uses involving office, professional, or service-type transactions are listed in the code. Group B occupancies are among the least regulated uses in the IBC due to the limited hazards presented. Low to moderate occupant loads and fire loading, along with the typical presence of ambulatory occupants, create environments that do not require a significant amount of fire- and life-safety regulation.

Airport traffic control towers are considered Group B occupancies and are regulated under the special-use provisions of Section 412.2. Car wash structures and motor vehicle showrooms are also considered business occupancies. Since a car wash facility or vehicle showroom contains a limited number of vehicles that are present in a very controlled condition, it is anticipated that the fire risk is limited, and classification as a Group B occupancy is appropriate.

Medical offices, including both outpatient clinics and ambulatory care facilities, are classified as Group B occupancies. By definition, an outpatient clinic is not expected to serve patients who will be temporarily rendered incapable of self-preservation due to their treatment. On the other hand, an ambulatory care facility is expected to have one or more individuals present who are temporarily rendered incapable of self-preservation due to the application of nerve blocks, sedation, or anesthesia. Such facilities may also include individuals who are already incapable of self-preservation upon arrival. Although both types of facilities would be classified as Group B occupancies, the unique concerns applicable to ambulatory care facilities are further addressed in the special provisions of Section 422. Several fire- and life-safety requirements applicable to ambulatory care facilities are not mandated for other types of Group B occupancies.

Educational occupancies above the 12th grade, including college classrooms and training rooms for adult education, are considered Group B occupancies. This designation is not specified as incumbent on the number of students (occupants) in the room; however, a lecture hall is specifically listed in Section 303.4 as a Group A-3 occupancy. Therefore, it should be assumed that a college classroom having an occupant load of 50 or more is to be classified as Group A-3 rather than Group B. The hazards associated with a large classroom are consistent with those for other assembly spaces and should be regulated accordingly. Recognition of educational occupancies serving students above the 12th grade is deemed necessary so as to differentiate from educational uses serving lower grade levels classified as Group E. Where the Group B higher education facilities include laboratories used for testing, analysis, research, teaching, or development activities, such laboratories would be considered as an extension of the Group B classification if in compliance with Section 428.

Food processing establishments and commercial kitchens of limited size are considered as Group B occupancies in those cases where they do not serve dining facilities. Small catering operations, bakeries, and similar facilities are viewed as limited in hazard level where they do not exceed 2,500 square feet (232 m^2) in floor area. More commonly, the Group B classification is applied to small retail food and beverage uses where cooking activities occur and the public is present for a limited time period, but no dining area is provided, such as a pizza carry-out business. These small, quick-transaction food sales tenants do not pose concerns to the degree of restaurants or large food processing operations; therefore, a classification of Group B is considered appropriate where the floor area is limited in size.

Where not exceeding the maximum quantities of hazardous materials allowed by the code in Section 307, testing and research laboratories may be considered Group B occupancies. Where the allowable quantities are exceeded, such laboratories would be classified as Group H. A review of Section 428 is necessary for laboratories located in university and college buildings.

Facilities specifically used for testing, research, and development activities related to lithium-ion or lithium metal batteries are also to be considered as Group B occupancies. The fire potential presented by lithium-ion and lithium metal batteries has been highlighted during code discussions of energy storage systems (ESS). An ESS installed in a building having an occupancy other than Group F-1 is considered as a part of that other occupancy. The occupancy classification for the portion of the building containing the ESS does not change due to its presence as appropriate safety levels for such installations can be maintained through compliance with *International Fire Code* (IFC) Section 1207. These IFC provisions require extensive protection features for such installations, including detection, suppression, fire separation, and explosion control, along with large-scale testing to document the effectiveness of chosen protection levels. Based on this understanding of how facilities with lithium-ion and lithium metal batteries are regulated, the classification of Group B is deemed appropriate for related laboratory-type functions.

The Group B classification of training and skill development programs outside of a school or academic program includes uses such as tutoring programs and instrumental music training. Where the occupant load and concentration of occupants is such that an assembly use is created, a classification of Group A is typically required where an occupant load of 50 or more is established. Although it is specified that there are no occupant age criteria associated with classification as a Group B occupancy, it may be necessary to apply a higher level classification where the hazards anticipated based on the specific use fall outside of those typically encountered within a Group B occupancy. For example, skill development activities conducted for infants or toddlers should more appropriately be considered as a Group E or I-4 occupancy if the potential hazard level is consistent with that of a day-care facility. As with any occupancy designation, the intent of the classification is to be able to correctly apply the appropriate minimum standards of health, safety, and public welfare.

Classification of a space or building as Group B is also addressed in Sections 303.1.1 and 303.1.2. Assembly buildings and tenant spaces with occupant loads of less than 50 are to be considered as Group B occupancies. In addition, those assembly spaces accessory to other occupancies may also be considered as Group B where the occupant load does not exceed 50 or the floor area is less than 750 square feet (70 m^2). Although the use of such buildings and spaces is assembly in nature, the reduced hazard due to the limited occupant load or floor area allows for a classification as Group B.

Section 305 *Educational Group E*

The Group E classification is assigned to schools, including primary, middle, and high schools, as well as day-care facilities. All Group E occupancies have three features in common: they are limited to the education, supervision, or personal care of persons at an educational level no higher than the 12th grade; the occupants are only in the facility for a limited time each day; and there are at least six persons being educated, supervised, or cared for at the same time.

305.1 Educational Group E. The classification of school classrooms as Group E occupancies is typically a straightforward decision. It is also common to classify the administrative offices within a school building as an extension of the Group E function. Even the media center and lunchroom are generally viewed as just additional areas of the Group E occupancy, but what about other assembly spaces such as the gymnasium and auditorium? There is a unique feature involved in educational occupancies—the use of school buildings for assembly purposes outside the scope of the educational use. For example, many school auditoriums are used for community theater and other productions to which the public at large is invited. Also, the school gymnasium in many cases is used for neighborhood recreation activities or sporting events where all ages of occupants are present. It is not uncommon for school auditoriums and gymnasiums to be rented out to groups for special functions. When these additional uses are anticipated, there is adequate reason that they be classified as Group A occupancies. Therefore, on account of these multipurpose uses in many school buildings, it is necessary that the code requirements applicable to all expected uses be enforced in order to satisfy the safety requirements for each use. See the additional discussion of Section 303.1.3 on multipurpose areas within educational facilities.

305.1.1 Accessory to places of religious worship. It is common in buildings of religious worship that support spaces are provided in addition to the main worship hall. Such spaces typically include rooms for educational activities for persons of all ages, including children. Such educational areas are permitted to be classified as an extension of the Group A-3 worship hall rather than Group E, thus eliminating a mixed-occupancy condition. The limitation of the provision to rooms and auditoriums with occupant loads of less than 100 has limited, if any, application, as larger assembly spaces would typically be considered Group A-3 occupancies as well. This allowance is also addressed in Section 303.1.4.

305.2 Group E day-care facilities. Day-care facilities considered as Group E occupancies are limited to those facilities where children are provided with educational, supervision, or personal care services for periods of less than 24 hours per day. It is anticipated that the children will be able to egress efficiently with limited assistance from staff. Facilities that provide custodial care services in a day-care environment are more appropriately assigned a Group I-4 classification. In addition, full-time care facilities cannot be considered Group E occupancies. The number of children housed in a day-care operation classified as Group E is not limited; however, where the number of children is five or fewer, the use is to be classified as part of the primary occupancy.

The provisions of Section 305.2 only address care facilities for children over the age of 2½ years. However, the Group E classification is also applicable to facilities that provide infant/toddler care (2½ years of age or younger) where the conditions established in Section 308.5.1 are met. In such cases, the rooms housing the infants and/or toddlers must be located on the level of exit discharge, and each of such rooms must have an exit door directly to the exterior of the building. See Figure 305-1. Although the term "directly" as applied in this provision is subject to interpretation, it is expected that there is no intervening corridor, room, or other area that would contribute additional occupants in the egress path.

305.2.1 Within places of religious worship. It is common for child care to be available at places of religious worship during worship services and other activities. Nurseries, "cry rooms," and other child care spaces are often occupied during the same limited time period as the other activities within the worship facility. The occupancy classification of such spaces is permitted to be consistent with that of the main occupancy so that a mixed-occupancy condition need not be created. Although this allowance is limited to only those child-care facilities where the children are older than 2½ years of age, a similar allowance for facilities caring for younger children and adults is found in Section 308.5.2.

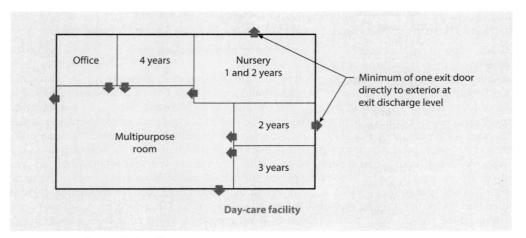

Figure 305-1 **Group E classification.**

305.2.2 Five or fewer children. Where five or fewer children are receiving day care within a building housing another use that can be considered the primary occupancy, the day-care operation is simply to be considered a portion of that occupancy. For example, a small day-care activity within an office environment would be classified as a portion of the Group B office use. Where there is no use in the building other than a day-care facility serving five or fewer children, it is assumed that a Group R-3 classification is to be applied. One of the listed Group R-3 occupancies in Section 310.4 is "care facilities that provide accommodations for five or fewer persons receiving care."

305.2.3 Five or fewer children in a dwelling unit. Where care is being provided to no more than five children and such care occurs within a dwelling unit, the occupancy classification of the dwelling unit is not to be modified due to the presence of the child care activities. For example, child care for five or fewer children in a dwelling unit complying with the scope of the *International Residential Code*® (IRC®) is permitted with no additional requirements applicable to the day-care operation. An occupancy classification of Group R-2 would continue to be appropriate for such child-care activities in a dwelling unit classified as a Group R-2 occupancy. Although the code text specifically calls out classification as a Group R-3 occupancy, it would be more appropriate to maintain the same classification as the other dwelling units within the building. It is intended that the presence of a child-care use in a dwelling unit where the number of children receiving care does not exceed five will have no effect on the occupancy classification determination or the applicable code that is to be enforced.

Section 306 *Factory Group F*

Although the potential hazard and fire severity of the multiple uses in the Group F occupancy classification is quite varied, these uses share common elements. The occupants are adults who are awake and generally have enough familiarity with the premises to be able to exit the building with reasonable efficiency. Public occupancy is usually quite limited, if at all, and most occupants are aware of the potential hazards the use creates. Group F occupancies are generally regarded as factory and industrial uses. The degree of hazard between the uses is very broad, and therefore the occupancy is divided into two categories.

Many manufacturing and industrial uses contain some degree of hazardous material as a necessary part of the manufacturing process. However, where the amount of hazardous material does not exceed the maximum allowable quantities set forth in Table 307.1(1) or 307.1(2), a classification as Group H is not appropriate. Because of the similarity between the different types of uses in Group F-1 occupancies and those in Group H occupancies, care must be exercised when determining the appropriate classification, and operators of Group F-1 occupancies should be apprised of the limitations on the quantities of hazardous materials that are allowed.

Some of the activities specifically listed as Group F occupancies also occur in a limited sense as accessory functions, and as such are not to be classified as Group F. For example, *food processing* is identified as a Group F-1 occupancy, but this is not to say that a kitchen serving a restaurant or cafe should be classified as such. Kitchens are considered to be classified as a portion of the major occupancy that they serve, typically either a Group A-2 or a Group B occupancy. The food processing operations designated as Group F-1 occupancies primarily include factories that produce canned or packaged items in bulk, as well as commercial kitchens that support catering operations or similar activities. Smaller facilities, those under 2,500 square feet (232 m^2), are to be classified as Group B as addressed in the discussion of Section 304.1.

Electrical energy storage systems provide standby or emergency power, an uninterruptable power supply, load shedding, load sharing, and similar capabilities. Such systems are extensively regulated by the IFC, allowing for recognition as a moderate-hazard condition in the determination of occupancy classification. The Group F-1 designation is to be applied where the use of the energy storage system and equipment contains lithium-ion or lithium metal batteries. Facilities used for the manufacturing of lithium-ion batteries or vehicles powered by lithium-ion or lithium metal batteries are also specifically included in the Group F-1 category.

The hazard from uses in Group F-2 occupancies is very low; in fact, the activities are deemed as among the lowest hazard groups in the code. It is assumed that the processing, fabrication or manufacturing of noncombustible materials will pose little, if any, fire risk to the building or its occupants. Foundries would be considered Group F-2 occupancies, as would facilities used for steel fabrication or assembly. Manufacturing operations producing ceramic, glass, or gypsum products are also included in this classification. Although very high temperatures are critical to the processes and operations of these types of uses, such heat is controlled and not a concern due to any combustibles or other fire loading that may be present.

Distilleries and similar alcohol beverage manufacturing facilities are classified within the Group F category based upon the alcohol content of beverages being produced. A designation as a Group F-1 occupancy is based on an alcohol content of more than 20 percent. Accordingly, where the alcohol content of beverages in a Group F occupancy does not exceed 20 percent, a classification of Group F-2 should be assigned. Beverages with an alcohol content greater than 20 percent are considered to present a higher level of hazard and therefore are placed into the higher manufacturing classification of Group F-1. Where above the 20 percent threshold, the alcohol is considered an ignitable liquid and required to be more highly regulated. Conversely, where the alcohol content does not exceed 20 percent, the fire hazard is considered minimal. IFC Chapter 32, regulating high-piled combustible storage, lists beverages in glass or ceramic containers with up to 20 percent alcohol content as Class I commodities. The alcohol content does not raise the flammability of the liquid to an extent where additional levels of protection are necessary, and for the most part can be considered nonflammable or noncombustible. As a result, the manufacturing and packaging of beverages with an alcohol content not exceeding 20 percent is to be classified as a Group F-2 occupancy.

Section 307 *High-Hazard Group H*

High-hazard Group H occupancies are characterized by an unusually high degree of explosion, fire, or health hazard as compared to typical commercial and industrial uses. The identification of hazardous occupancies is provided in this section.

There is one common feature about Group H occupancies—they are designated as Group H based on excessive quantities of hazardous materials contained therein. Where the quantities of hazardous material stored or used in a building exceed those set forth in Section 307, a Group H classification is warranted. On the other hand, where such quantities are not exceeded, a Group H classification is not appropriate.

Because of the technical nature of the operations and materials found in Group H occupancies, a number of specific terms are defined by the IBC in Chapter 2. The definitions are intended to assist the code user in applying the provisions of this chapter, as well as other portions of the code relating to high-hazard uses.

Group H-1 occupancies are those buildings containing high-explosion hazard materials. Materials that have the potential for detonation must be housed in buildings regulated in a very special manner, and designed and constructed unlike any other occupancies described in the code. Examples of detonable materials include explosives and Class 4 oxidizers.

Group H-2 generally includes those occupancies that contain materials with hazards of accelerated burning or moderate explosion potential, including materials with deflagration hazards. Common occupancies included in this category are those operations where flammable or combustible liquids are being used, mixed, or dispensed. The potential for a hazardous incident is increased because of the materials' exposure to the surrounding area. Occupancies containing combustible dusts may also be considered as Group H-2, as dusts in suspension, or capable of being put into suspension, in the atmosphere are a deflagration hazard. In the determination of occupancy classification for a facility where combustible dusts are anticipated, a technical report and opinion must be provided to the building official that provides all necessary information for a qualified decision as to the potential hazards due to the combustible dusts.

Buildings containing materials that present high-fire or heat-release hazards are classified as Group H-3 occupancies. Where flammable or combustible liquids are present in such occupancies, they must be stored in normally closed containers or used in low-pressure systems. Because of the enclosed nature of these liquids, the hazard level is not nearly as severe as it is for Group H-2 occupancies. Other hazardous materials such as organic peroxides and oxidizers, based on their hazard classification, may also be used or stored in Group H-3 occupancies.

Group H-4 occupancies are those containing health-hazard materials such as corrosives and toxics. Section 202 defines "Health hazards" as those "chemicals for which there is statistical significant evidence that acute or chronic health effects are capable of occurring in exposed persons." Quite often, a material considered a health hazard also possesses the characteristics of a physical hazard. It is important that all hazards of materials be addressed.

Occupancies classified as Group H-5 are those uses containing semiconductor fabrication activities, including the ancillary research and development areas. The Group H-5 category was created in order to address the explosive and highly toxic materials used in semiconductor fabrication by providing specific requirements for the particular operations conducted, while at the same time providing a level that allows reasonable transaction of the fabrication process.

A more complete commentary on hazardous materials is provided under the discussion of Section 414.

307.1 High-Hazard Group H. The concept of maximum allowable quantities of hazardous materials as the basis for occupancy classification is further extended through the use of control areas as regulated by Section 414.2.

Maximum allowable quantities. Occupancy classifications of buildings containing hazardous materials are based on the *maximum allowable quantities* concept. Tables 307.1(1) and 307.1(2), together with their appropriate footnotes, identify the maximum amounts of hazardous materials that may be stored or used in a control area before the area must be designated as a Group H occupancy. The maximum quantities of hazardous materials permitted in non–Group H occupancies vary for different states of materials (solid, liquid, or gas) and for different situations (storage or use). The allowable quantities are also varied based on protection that is provided, such as automatic sprinkler systems and storage cabinets.

Control areas. Areas in a building that contain hazardous materials in amounts that do not exceed the maximum allowable quantities and that are properly separated from other areas containing hazardous materials are called *control areas*. Any combination of hazardous materials, up to the maximum allowable quantities, is permitted in a control area. A control area may be an entire building or only a portion of the building. It can be part of a story, an entire story, or even include multiple stories. A control area is classified, based on it use, as other than Group H as its sole purpose is to avoid a Group H classification. As such, control areas are not applicable to a Group H condition. The detailed requirements regarding control areas are set forth in Section 414.

The control-area method is based on the concept of fire-resistance-rated compartmentation. It regulates quantities of hazardous materials per compartment (control area), rather than per building. The limit for the entire building, using control areas, is then established by limiting the total number of control areas allowed per story and the quantities of hazardous materials that are located in each control area. The control-area concept was introduced in an effort to regulate buildings of different sizes in a consistent manner. It is based on a premise that the storage and use of limited quantities of hazardous materials (not exceeding maximum allowable quantities) in areas that are separated from each other by fire-resistance-rated separations do not substantially increase the risk to the occupants or change the character of the building to that of a hazardous occupancy, subject to a limitation on the number of control areas. The fire-resistive separations are relied on minimizing the risk of having multiple control areas involved simultaneously during an emergency.

There is no special occupancy designation for a control area. For example, a control area in a manufacturing occupancy is merely part of the Group F occupancy. Further discussion of control areas is addressed in the commentary of Section 414.2. In addition, the use of laboratory suites in college and university laboratory settings as discussed in Section 428 is based on the control area concept.

Increased quantities. Given this basic understanding of maximum allowable quantities and control areas, the various options in the code for increasing the quantities of hazardous materials within a building are as follows:

1. Buildings are generally allowed to have up to the basic maximum allowable quantities of hazardous materials without restriction with respect to separations or protection. In this case, the entire building is designated as a control area. The boundaries of the control area are the boundaries of the building (i.e., exterior walls, roof, and foundation). See Figure 307-1.
2. Using the footnotes to Tables 307.1(1) and 307.1(2), the maximum allowable quantities of many materials can be increased by providing automatic sprinkler protection throughout the building and/or by using approved storage cabinets, safety cans, or other code-approved enclosures to protect the hazardous materials. It is important that the increases identified in the footnotes only be used where applicable.

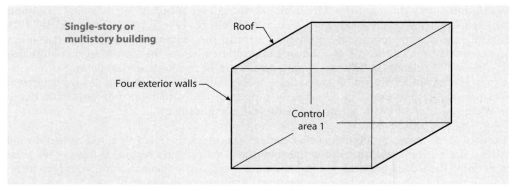

Figure 307-1 **Control area boundaries for one control area.**

3. Four other options are available to further increase the quantities of hazardous materials in any building:

   3.1. Provide additional control areas as limited by Table 414.2.2,

   3.2. Provide one or more fire walls in conformance with Section 706,

   3.3 Apply the exemptions for materials and conditions listed in Table 307.1.1, or

   3.4 Construct the building as required for a Group H occupancy.

Assuming additional control areas are used, each control area must be separated from all other control areas by minimum 1-hour fire barriers, or 2-hour fire barriers if required by Section 414.2.4. Vertical isolation of control areas must be accomplished by floors having a minimum 2-hour fire-resistance rating. Under limited conditions, the floor construction of the control area separation may be reduced to 1 hour. Its application is limited to fully sprinklered two- or three-story buildings of Type IIA, IIIA, IV, or VA construction. In all cases, construction of supporting such floors shall have an equivalent fire-resistance rating. A designated percentage of the maximum allowable quantities of hazardous materials is allowed in each control area per Table 414.2.2. See Figure 307-2.

The permitted number of control areas decreases vertically through the building, as does the quantities of hazardous materials per control area. As hazardous materials are located higher and higher above ground level, they become more difficult to address under emergency conditions. As with many other conditions regulated by the code, a key factor is the ability of the fire department to access the incident area. The higher the hazardous materials are located in the building, the more restrictive the provisions become, owing to the limitations on fire department access and operations.

The number of control areas permitted within a single structure may also be increased by dividing the structure into two or more buildings with fire walls as specifically permitted by Section 414.2.3. The limits on control areas can be applied independently to each portion of a structure separated by fire walls complying with Section 706. As a result, an increased amount of hazardous materials may now be present without classification as a Group H occupancy where fire walls divide a structure into separate buildings.

Storage and use. One other fundamental concept involved in applying the maximum allowable quantities is *situation of material*. The maximum allowable quantities in the code are based on three potential situations: storage, use-closed, and use-open.

Though not defined by the code, the term *storage* is generally considered to include materials that are idle and not immediately available for entering a process. The term *not immediately*

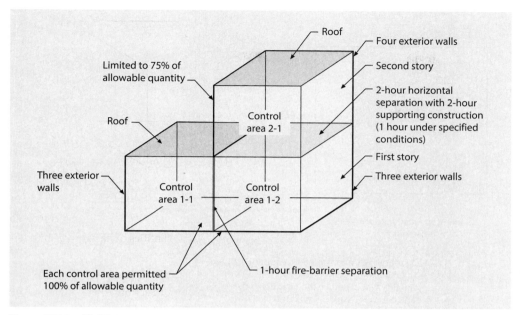

Figure 307-2 **Multistory areas.**

available can be thought of as requiring direct human intervention to allow a material to enter a process or, alternatively, as using approved supervised valving systems that separate stored material from a process. In the case of liquids and gases, storage is generally considered to be limited to materials in closed vessels (not open to the atmosphere). For example, materials kept in closed containers such as drums or cans are in storage because deliberate action (opening the drum or can) would be required to use the material. However, when a container or tank is connected to a process, the question arises whether the material in the container or tank is in storage or in use.

In general, the quantity of material that would be considered to be in use is the quantity that could normally be expected to be involved in a process, or that could reasonably be expected to be released or involved in an incident as a result of a process-related emergency. Consider, for example, a process having hazardous materials that are piped from an underground storage tank outside of a building to a dispensing outlet within a building. Because the tank is connected to a process within the building, it could be argued that the contents of the tank are available for use in the building [see definition of "Use" (Material) in Section 202] and that the amount should be counted toward the maximum allowable quantities. However, if an approved, reliable arrangement of valving is provided between the supply and the point where the material is dispensed, it would be reasonable to conclude that the quantity on the supply side of such valving that is outside of the building would be unlikely to impact incidents occurring within the building and, therefore, need not be counted toward the maximum allowable quantities. This reliable arrangement of valving can be considered an interruption of the connection between the confined material (storage) and the point where material is placed into action or made available for service. See Figure 307-3.

The difference between use-closed and use-open is basically whether the hazardous material in question is exposed to the atmosphere during a process, with the exception that gases are defined as always being in closed systems when used insofar as they would be immediately dispersed (unless immediately consumed) if exposed to the atmosphere without some means of containment.

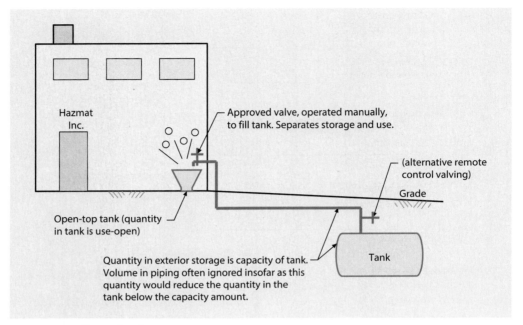

Figure 307-3 **Example of storage versus use.**

Table 307.1(1)—Maximum Allowable Quantity per Control Area of Hazardous Materials Posing a Physical Hazard. This table sets forth maximum allowable quantities for physical hazard materials. All three situations (storage, use-closed, and use-open) are considered. For the specific case of gases, maximum allowable amounts are all listed under storage and use-closed because the definition of "Use" includes all gases.

With two exceptions, any combination of materials or situations listed in this table is allowed in each control area. These two exceptions are (1) as provided by Footnote h, the aggregate of IA, IB, and IC flammable liquids, and (2) as provided by Footnote b, which requires that aggregate quantities of materials in both use and storage must not exceed the allowable quantity for storage.

Specific footnotes to the table provide the following information:

Footnote a. This footnote references Section 414.2 for the use of control areas. For additional information, see the discussion of control areas in Sections 307.1 and 414.2.

Footnote b. This footnote requires quantities of materials that are in use to be counted as both storage and use when comparing quantities to those permitted. For example, a single control area in a manufacturing facility would be permitted up to 30 gallons (116.25 L) of a Class II combustible liquid in use and a maximum of 120 gallons (454.2 L) in storage without being considered a Group H occupancy. However, a total of 150 gallons (567.75 L) would be prohibited. The total of both use and storage is limited to 120 gallons (454.2 L), with no more than 30 gallons (116.25 L) permitted in use.

Footnote c. This footnote provides a cross-reference to Section 428 regulating college and university laboratories classified as Group B occupancies. Where such laboratories contain hazardous materials, their classification and regulation based on the quantities of such materials is not subject to the limitations of Table 307.1(1). Rather, the provisions of Section 428 and IFC Chapter 38 are to be applied.

Footnote d. This footnote allows certain materials to have exempt amounts doubled when stored or used in sprinklered buildings. Compounding with the increases provided by Footnote e is allowed when both footnotes are applicable. Materials and situations referencing both Footnotes d and e can receive four times the listed maximum allowable quantity when both footnotes are applied. See Application Example 307-1.

> **GIVEN:** A fully sprinklered Group F-1 storage building housing Class II combustible liquids. The Class II liquids are all stored in approved safety cans. The entire building is a single control area.
>
> **DETERMINE:** The maximum allowable quantity of the Class II liquids in storage in order to maintain the Group F-1 classification.
>
> **SOLUTION:**
>
> | | |
> |---|---|
> | Basic MAQs per Table 307.1(1) | 120 gallons |
> | Sprinkler increase per Footnote d (100%) | + 120 gallons |
> | | 240 gallons |
> | Safety can increase per Footnote e (100%) | + 240 gallons |
> | Total of maximum permitted for Group F-1 classification | 480 gallons |

Application Example 307-1

Footnote e. This footnote allows exempt amounts for certain materials in storage to be doubled when approved storage cabinets, gas cabinets, safety cans, etc., as applicable, are employed. Also, see Footnote d and Application Example 307-1.

Footnote f. This footnote allows certain materials to be stored or used in unlimited quantities in sprinklered buildings. When the building is fully sprinklered, it has the effect of classifying the building as other than a Group H occupancy. Application of this footnote is limited to the storage or use of Class IIIB combustible liquids or Class I oxidizers. An unlimited quantity of each of these materials is permitted in a fully sprinklered building without requiring a designation of Group H.

Footnote g. This footnote limits storage and use of certain materials to sprinklered buildings. Where the quantity does not exceed the maximum allowable, the building is required to be sprinklered throughout if a non-Group H classification is desired.

Footnote h. Where flammable liquids are concerned, the maximum allowable quantities are regulated both individually and cumulatively. To be considered an occupancy other than Group H, the control area must not contain more than the maximum allowable quantities for each type of flammable liquid (Classes IA, IB, and IC), as well as the combination of such limits established by Table 307.1(1).

Footnote j. Substantial increases in the maximum allowable quantities are permitted for Class 3 oxidizers used for maintenance and sanitation purposes. A common application of the increased quantities occurs in health-care facilities.

Footnote k. If the net weight of the pyrotechnic composition of fireworks is known, that weight is used in applying the limitations of the table. Otherwise, 25 percent of the gross weight of the fireworks is to be used, including the packaging materials.

Footnote o. The listing of combustible dusts in Table 307.1(1) is unique in that it is the only material listed where a maximum allowable quantity has not been established in order to determine occupancy classification. The classification of Group H-2 is to be based solely on a determination by a qualified person, firm, or corporation, along with concurrence of the building official. Reference is made to Section 414.1.3 where the general information that is required to make such a determination is set forth.

Table 307.1(2)—Maximum Allowable Quantity per Control Area of Hazardous Materials Posing a Health Hazard. This table is similar in nature to Table 307.1(1), except that the maximum allowable quantities listed in Table 307.1(2) are for health hazard materials. For discussions of specific footnotes, see the above discussion of Table 307.1(1).

307.1.1 Occupancy exemptions. Prior to an analysis of MAQs and control areas in the determination of occupancy classification where hazardous materials are present, it would be wise to review the varied conditions that are specified in Table 307.1.1 under which a Group H classification is not to be assigned in spite of the fact that the quantities of materials exceed the amounts set forth in Table 307.1(1) or 307.1(2). Where one or more exemptions apply to hazardous materials used, handled, or stored, the occupancy classification for that specific material and application is to be determined based upon which occupancy group it most nearly resembles. A key requirement in the exemption from a Group H classification is the mandate that compliance with all applicable provisions in the IFC related to the storage, use, and handling of hazardous materials must be achieved. For example, the storage of aerosol products is to be designated a Group S-1 classification provided the storage facility is in compliance with the IFC. It is important that the code user review all of the exemptions listed in Table 307.1.1 in order to properly classify those buildings, or portions of buildings, where hazardous materials are present.

The application of flammable finishes is not considered a Group H condition provided the requirements of Section 416 are applied, as well as all applicable provisions of IFC Chapter 24. Detailed requirements set forth in the IFC for spray booths, spraying spaces, and spray rooms, as well as for spaces where dipping operations and powder coating occur, provide a degree of safety that allows for spray finishing and similar activities to be considered as non–Group H facilities.

Where flammable or combustible liquids or gases are contained within a closed piping system and utilized solely for operational functions, such as where hydraulic fluid is utilized within heavy equipment and machinery, the amount of such fluids is not considered in the classification process. Refrigeration systems do not contribute to Group H consideration due to their regulation under the IFC and the *International Mechanical Code*®. The use and/or storage of hazardous materials for agricultural purposes does not create a Group H occupancy designation provided the materials are limited to storage and use as needed on the premises.

Distilling, brewing, and the storage of alcohol beverages, regardless of alcohol content and the quantity of liquid, are not considered as Group H occupancies where in compliance with the IFC. This allowance in large part takes into consideration the automatic sprinkler requirements for the manufacture of distilled spirits, or the bulk storage of distilled spirits or wine. Where the alcoholic beverages being manufactured or stored exceed 20-percent alcohol content, regardless of the liquid quantity, a sprinkler system shall be provided throughout the fire area containing the Group F-1 or S-1 occupancy. Due to the fire protection afforded by an automatic sprinkler system, along with other safeguards established in the IFC, the manufacture and storage of alcoholic beverages in any quantity is not to be considered as a Group H occupancy. Manufacturing activities are classified as Group F-1 or Group F-2 based on the alcohol content, while storage occupancy provisions address the concerns associated with classification as Group S-1 or Group S-2.

Additional allowances are also provided within the provisions such that it is not appropriate, due to the safeguards and limitations imposed, that a Group H classification be assigned.

307.8 Multiple hazards. As previously noted, most hazardous materials possess the characteristics of more than one hazard. This section requires that all hazards of materials must be addressed. For example, if a material possesses the hazard characteristics of a Class 2 oxidizer and a corrosive, the material would be regulated under the provisions for both Group H-3 and H-4 occupancies. The more restrictive provisions of each occupancy must be satisfied.

Section 308 *Institutional Group I*

Group I occupancies are institutional uses and in the IBC are considered to be basically of two broad types. The first are those facilities where individuals are under supervision and care because of physical limitations of health or age. The second category includes those facilities in which the personal liberties of the occupants are restricted.

In both types, the occupants either are restricted in their movements or require supervision in an emergency, such as a fire, to escape the hazard by proceeding along an exit route to safety. There is actually a third category in which the occupants enjoy mobility and are reasonably free of constraints but do require a measure of professional care and are asleep for a portion of the day.

As Group I occupancies are people-related occupancies, the primary hazard is from the occupants' lack of free mobility or mental capacity needed to extricate themselves from a hazardous situation. On the other hand, the hazard from combustible contents is typically very low and, as a result, the occupancy requirements for Group I occupancies are essentially based on the physical or mental limitations of the occupants. Also, the occupants of most Group I occupancies are usually institutionalized for 24 hours or longer and therefore are asleep at some point during their stay. Thus, the protection requirements of the code are more comprehensive than in almost any other people-related occupancy. It should be noted that institutional occupancies described as Groups I-1, I-2, and I-4 may be classified as Group R-3 where care is provided for five or fewer persons. As an option for such small institutional uses, the structure need only comply with the IRC. The code typically recognizes that such small occupant loads in educational or institutional environments can be adequately addressed for fire and life safety through the provisions for residences.

Although the Group I-3 classification is only applicable where six or more individuals are restrained or secured, there is no indication as to the proper classification where five or fewer persons are involved. Where the intended detention or restraint occurs as an ancillary use within some other occupancy, classification would be based on that of the major occupancy. For example, up to five individuals could be restricted in areas such as interrogation rooms for alleged shoplifters in a covered mall building, jewelry viewing rooms for customers of a retail store, and time-out rooms in a school, without classifying that portion of the building a Group I-3 occupancy. Further discussion on allowances for locking devices for such spaces can be found in the commentary on Section 1010.2.4, Item 1.

308.2 Institutional Group I-1. The occupants housed in buildings classified as Group I-1 occupancies live in a supervised environment where custodial care services are provided, such as assistance with cooking, bathing, and other daily tasks. Types of uses included in this category are halfway houses, alcohol and drug rehabilitation facilities, assisted living facilities, social rehabilitation facilities, and group homes. It is possible that a listed use may rather be considered a Group I-2 or I-3 occupancy if the residents are incapable of self-preservation because of injury, illness, or incarceration. For example, an alcohol treatment center may provide lockdown for a number of persons under care. Where this number exceeds five, a Group I-3 classification would be more appropriate than a Group I-1.

The naming, classification, and regulation of these uses by individual state licensing agencies may not exactly correlate with the IBC list of uses. The actual "condition" of such uses also varies between states. It is for these reasons that the Group I-1 list of uses is included under the general occupancy classification and not under the specific condition. The Group I-1 classification should correlate how the specific state licensing regulations relate to the code's occupancy, care type, number of persons receiving care, and condition of evacuation capability. A similar review and analysis is warranted for Group R-4 occupancies where the only differing factor from Group I-1 is the number of persons receiving care.

The specific classification is based on the number of residents. In this case, a Group I-1 is the proper classification where more than 16 occupants are receiving custodial care. The threshold of 17 or more is based on the number of supervised individuals who reside in the facility and does not include associated staff members. Where the number of residents is between 6 and 16 inclusive, a Group R-4 classification is appropriate. Those supervised residential facilities that provide custodial care services for five or fewer occupants can be either considered a Group R-3 occupancy or designed and constructed under the provisions of the IRC.

The care uses recognized as Group I-1 occupancies are further divided into two classifications, Condition 1 and Condition 2. Condition 1 includes those facilities where the care recipients are capable of self-preservation and no assistance is required for building evacuation or relocation under emergency conditions. Condition 2 recognizes a higher degree of hazard in that the persons receiving care require some limited amount of verbal or physical assistance in order to evacuate or relocate in response to an emergency situation. Provisions related to number of stories, smoke barriers, sprinkler systems, and smoke detection are more stringent in Condition 2 occupancies as opposed to those applicable to Condition 1 facilities. It is anticipated that most assisted living, memory care, and residential board and care facilities will appropriately be classified as Condition 2 facilities.

Group I-1 custodial care recipients and their conditions are different from those in Group I-2 facilities, establishing different emergency preparedness concepts per the IFC. Group I-2 medical care recipients typically have higher acuity levels than Group I-1 custodial care recipients. Individuals in custodial care still participate in fire drills with or without assistance from others, versus medical care which implements defend-in-place strategies during emergencies. Group I-1, Condition 2 implements essentially a staged evacuation concept by utilizing smoke compartments per the IFC.

"Limited verbal or physical assistance" with evacuation, as defined in Chapter 2, assumes differences in abilities between medical care and custodial care recipients, with assumed limits placed on custodial care. Such custodial care recipients are conscious and not on life support systems, therefore bed movement is not required under emergency egress situations as is typically required in Group I-2 facilities. A custodial care recipient requiring assistance with emergency evacuation may require help getting out of bed into a wheelchair or walker, but then can proceed with or without assistance from others during evacuation. Custodial care recipients receiving assistance during evacuation also assumes varying mental limitations requiring verbal assistance, such as a person with mental disabilities, dementia, or Alzheimer's disease.

308.3 Institutional Group I-2. The primary feature that distinguishes the Group I-2 occupancy from the others is that it is a medical care facility in which the patients are, in general, nonambulatory and incapable of self-preservation. This classification includes hospitals, detoxification facilities, psychiatric hospitals, and nursing homes. The nursing homes included in this category are deemed to provide intermediate care or skilled nursing care. Foster care facilities for the full-time care of infants (under the age of 2½ years) are included in this classification, as the code assumes that the very young require the same protection as is provided for those individuals whose capability of self-preservation is severely restricted. Where care is provided for a limited time period, such as at an outpatient health-care clinic or ambulatory care facility, a classification of Group I-2 is not appropriate. In such cases, a Group B classification is warranted, even in those cases where some of the patients are incapable of self-preservation.

Due to the diversification of how medical care is provided in the five characteristic uses as established in the code, the Group I-2 classification is divided into two basic categories:

Condition 1, long-term care (nursing homes), and Condition 2, short-term care (hospitals). Changes in how care is delivered have recently included a general increase in the ratio of floor area per patient in hospitals due to the increase in diagnostic equipment and the movement toward single-occupant patient rooms, as well as the trend to provide more residential-type arrangements in nursing homes, such as group suite living and cooking facilities. Although most applicable code requirements apply equally to both medical care conditions classified as Group I-2, the division of uses does allow for varying provisions based upon the type of care being provided. An example of differences in the requirements is addressed in the discussion of Section 407.5.1 regarding smoke compartments.

308.4 Institutional Group I-3. The uses of Group I-3 occupancies encompass jails, prisons, reformatories, detention centers, and other buildings where the personal liberties of the residents are similarly restricted. For guidance on the classification of detention facilities having occupant loads of five or less, see the general discussion of Section 308. The classification of Group I-3 buildings shall also include one of five occupancy conditions. Several provisions specific to Group I-3 occupancies vary based on which condition is anticipated, such as the manner of subdividing resident housing areas. The conditions are described as follows:

1. The highest level of freedom assigned to a Group I-3 occupancy is considered Condition 1. Free movement is permitted throughout the sleeping areas and the common areas, including access to the exterior for egress purposes. A facility classified as Condition 1 is permitted to be constructed as a Group R occupancy, most likely a Group R-2. There may also be cases where it is more appropriate to classify the use as some occupancy other than Group R. For example, an industrial building included within a Condition 1 facility would most probably be classified as a Group F-1 occupancy.

2. Condition 2 buildings permit free movement between smoke compartments; however, access to the exterior for egress purposes is restricted because of locked exits.

3. Access between smoke compartments is not permitted in Condition 3 occupancies, except for the remote-controlled release of locked doors for necessary egress travel. Movement within each individual smoke compartment is permitted, including access to individual sleeping rooms and group activity spaces.

4. Condition 4 buildings restrict the movement of occupants to their own space, with no freedom to travel to other sleeping areas or common areas. Movement to other sleeping rooms, activity spaces, and other compartments is controlled through a remote release system.

5. The lack of freedom provided in Condition 5 facilities is consistent with that of Condition 4. However, staff-controlled manual release is necessary to permit movement throughout other portions of the building.

308.5 Institutional Group I-4, day-care facilities. Where custodial care is provided for persons for periods less than 24 hours at a time, an occupancy classification of Group I-4 is appropriate. The code further restricts this category by limiting the care to individuals other than parents, guardians, or relatives. This occupancy classification is appropriate for day-care facilities with children no older than 2½ years of age, as well as older persons who are deemed to require assistance if evacuation or relocation is necessary due to an emergency situation. The need for supervision and custodial care services is the primary factor that contributes to this type of use being classified as an institutional occupancy.

Day-care facilities for children above the age of 30 months are considered educational occupancies (Group E). Where certain conditions are met, such a facility caring for infants/toddlers (30 months or less in age) may also be classified as Group E. See the discussion on Section 305.2.

Where the care activities are for adults who can physically respond to an emergency situation without physical or verbal assistance, an institutional classification may not be appropriate. Rather, the facility could more likely be classified as Group B.

Section 309 *Mercantile Group M*

The mercantile uses listed in this section are mostly self-explanatory. For the most part, occupants of this type of use are ambulatory individuals, with any young children supervised by parents or other adults. Although the occupancies may contain a variety of combustible goods, the possibility of ignition is limited. High-hazard materials may be present in small quantities, but not enough of the material is present to be considered a Group H occupancy. For the limitations on hazardous materials in a Group M occupancy, see the discussion on Section 414.

As a sales operation, a service station whose primary function is the fueling of motor vehicles is considered a Group M occupancy. The Group M designation applies to any building or kiosk used to support the function of vehicle fueling. This classification would also apply to a canopy constructed over the pump islands. By assigning the structures associated with the fueling of motor vehicles a classification the same as that for other sales operations, such as convenience stores, there is no question as to the proper application of the code. Through the design and construction of a motor-vehicle service station in conformance with Section 406.7 and the IFC, there is no distinct uncontrolled hazard that would cause separate and unique occupancies to be assigned. Conversely, where service or repair activities are involved, a Group S-1 classification is warranted. This would include operations such as the exchange of parts, such as tire and muffler shops, as well as service-oriented activities (oil change and lubrication work).

Section 310 *Residential Group R*

Group R occupancies are residential occupancies and are characterized by:

1. Use by people for living and sleeping purposes.
2. Relatively low potential fire severity.
3. The worst fire record of all structure fires.

The basic premise of the provisions in this section is that the occupants of residential buildings will be spending about one-third of the day asleep and that the potential for a fire getting out of control before the occupants are awake is quite probable. Furthermore, once awakened, the occupants will be somewhat confused and disoriented, particularly in hotels.

310.1 Residential Group R. The four unique residential classifications are based on occupant load and density as well as the permanency of the occupants. Therefore, hotels, motels, and similar uses in which the occupants are essentially transient in nature are distinct in classification from apartment houses. The reason for this distinction is the occupants' lack of familiarity with their surroundings. This in turn leads to confusion and disorientation when a fire occurs while the occupants are asleep. Because of this key difference, hotels and motels are considered Group R-1 occupancies, whereas apartment houses are designated as Group R-2. The uses classified as Group R-3 and R-4 include both transient- and nontransient-type facilities housing limited numbers of occupants.

310.2 Residential Group R-1. Transient residential uses, classified as Group R-1 occupancies, typically include buildings containing sleeping units or dwelling units. In addition to hotels and motels for transient guests, the Group R-1 classification also includes boarding houses and congregate living facilities where the number of transient occupants exceeds 10 persons. Such facilities are permitted to be classified as Group R-3 rather than Group R-1 where the occupant load does not exceed 10; however, they are not permitted to be constructed in accordance with the IRC. Consider a transient lodging operation consisting of a large number of single-family cabins containing living, cooking, sleeping, and sanitation facilities. The cabins could be classified in whole as a Group R-1 occupancy, or more probably, each individual cabin would be considered as a Group R-3 structure, provided the occupant load of the cabin does not exceed 10 persons. However, it would not be appropriate to apply the provisions of the IRC to this type of transient use.

310.3 Residential Group R-2. Included in the Group R-2 occupancy classification along with apartment buildings are dormitories, fraternity and sorority houses, convents, monasteries, and living quarters for emergency services personnel such as fire fighters, paramedics, and other emergency responders. These latter types of uses are considered congregate living facilities. By definition, they contain one or more sleeping units where the residents share bathroom and/or kitchen facilities. Another type of nontransient congregate living facility is a boarding house. Common in residential areas adjacent to college and university campuses, boarding houses are also often located in urban areas of cities. An occupancy classification of Group R-2 is only appropriate for congregate living facilities where the occupant load of the facility is 17 or more persons. A lesser number will result in a Group R-3 classification. On the other hand, apartment buildings containing three or more dwelling units are always considered Group R-2 occupancies regardless of the building's occupant load. The Group R-2 classification is also appropriate for live/work units as regulated by Section 508.5. Live/work units include those dwelling units or sleeping units where a significant portion of the space includes a nonresidential use operated by the tenant.

310.4 Residential Group R-3. Group R-3 occupancies are generally limited to small congregate living facilities and lodging houses with limited occupant loads. Other Group R-3 occupancies include mixed-occupancy buildings containing only one or two dwelling units, as well as those small facilities used for various types of care. It is expected that the occupant load of a Group R-3 occupancy will be quite low. Typically, dwellings would not be classified as Group R-3 occupancies, as they will be regulated by the IRC. Only where the dwelling falls outside the scope of the IRC, will the Group R-3 classification for such structures be appropriate. For example, a four-story-above-grade-plane dwelling would be regulated as an R-3 occupancy, as would a single dwelling unit located above a small retail store. As previously indicated, boarding houses and congregate living facilities, both transient and nontransient in nature, may be classified as Group R-3 where the occupant load is relatively low.

Where the use of a single-family dwelling classified as a Group R-3 occupancy consists of care activities, the IRC may be used, provided the building falls under the scoping provisions of the IRC. It should be noted that Group R-3 uses permitted to be constructed under the provisions of the IRC are required either to be protected by an NFPA 13D automatic sprinkler system or to comply with the IRC residential sprinkler system provisions of Section P2904.

Where a structure is used as a lodging house having no more than five guest rooms, such as a small bed-and-breakfast facility, it is initially considered as a Group R-3 occupancy. However, where the lodging house is also owner-occupied, the provisions of the IRC may be applied for its design, construction, and use. This allowance recognizes the single-family dwellings with limited guests create little, if any, additional hazard beyond those structures regulated under the IRC. In both cases, where the number of guest rooms in a lodging house exceeds five, a classification of Group R-1 is appropriate.

310.5 Residential Group R-4. Assisted living facilities, halfway houses (Figure 310-1), and other 24-hour custodial care facilities are to be classified as Group R-4 occupancies, provided the number of residents under custodial care does not exceed 16. Where the number of

Figure 310-1 **Group R-4 halfway house.**

residents is five or less, the use is considered a Group R-3 occupancy. The list of Group R-4 uses is fully consistent with those designated as Group I-1 (more than 16 persons receiving care), as the only difference between the two classifications is the number of persons who receive custodial care. Two conditions of use are established in a manner consistent with that for Group I-1 occupancies. The discussion of Section 308.2 addresses these conditions in significant detail.

The Group R-4 classification applied to supervised custodial care facilities differentiates them from Group R-3 care facilities based solely on the number of persons receiving care. For the most part, there is no difference between the hazards that are anticipated at such facilities. Therefore, the code mandates that those uses classified as Group R-4 comply with the requirements for Group R-3 occupancies except for those requirements specific to Group R-4 occupancies.

As an example, assume an assisted living facility has accommodations for up to 12 persons who receive custodial care. The facility would be classified as a Group R-4 occupancy since the number of care recipients is more than 5 but less than 17. However, most of the applicable code requirements would be based on a Group R-3 classification. Stair riser heights would be limited to 7¾ inches with minimum tread runs of 10 inches required, as set forth in Exception 3 to Section 1011.5.2. A reduction in the required 42-inch height of guards would be permitted per Section 1015.3, Exceptions 1 through 4. Unless the code specifically identifies a requirement as applicable to a Group R-4 occupancy, the provisions for Group R-3 shall be met.

There are, however, a number of provisions specifically established for Group R-4 occupancies. Using the example in the preceding paragraph, Table 506.2 is to be used as the basis for allowable building areas based on the Group R-4 classification, with an allowable area factor of 7,000 square feet (650.3 m^2) if the building is of Type VB construction and provided with an NFPA 13R sprinkler system. Smoke alarms are generally required under the provisions of Section 907.2.11.2, and an Accessible unit would be mandated per Section 1108.6.4.

Section 311 *Storage Group S*

In general, the Group S designation includes storage occupancies that are not highly hazardous as well as uses related to the storage, servicing, or repair of motor vehicles. Such storage uses are classified into two divisions based on the hazard level involved. Group S-1 describes those buildings used for moderate-hazard storage purposes, whereas low-hazard uses make up the Group S-2 classification.

Before addressing the two different types of storage uses, consideration should be given to the classification of borderline uses. An exclusionary rule is used to assist in determining those moderate-hazard storage uses that are to be classified as Group S-1. For example, a Group S-1 occupancy is used for storage uses not classified as a Group S-2 or H occupancy. The building official will often be called upon to decide which classification is most appropriate when a use can fall within the two Group S occupancy classifications. As guidance in making this decision, it is usually more appropriate to choose the most restrictive occupancy, which is the Group S-1. Classifying the use into the more restrictive category would allow the building to be protected at a higher level and address the worst-case situations that might occur. By classifying the use into the least restrictive category, it would typically reduce the required controls, causing a potential problem where the building operator chooses to store combustible materials within the building.

311.1.1 Accessory storage spaces. The classification approach to storage rooms is similar to that of other support areas within a building. Where the hazard level within the storage area is such that the provisions for the general building use do not adequately address the risks posed by the storage use, the storage room is to be classified as a Group S occupancy. There has been a historic consensus that small storage areas pose only a limited concern above that anticipated due to the major use of the building, thus allowing such small storage spaces to be included as part of the major classification.

However, the classification of larger storage rooms has historically been one of the most elusive issues in the IBC. Assigning a Group S occupancy classification to a warehouse, or other significant storage area, has never been questioned. However, where the room or space poses little, if any, hazard above that created by the occupancy to which the storage use is accessory, it is not necessary that a Group S classification be applied. The current approach to classifying a storage area is formally addressed such that storage rooms or spaces that are accessory to other uses are to be classified as part of the occupancy to which they are accessory, regardless of the size of the storage area.

The classification of storage spaces does not vary based upon the size of the storage space. There is no square footage or percentage threshold, such as 100 square feet or 10 percent, over which the Group S classification will be applied. Where the storage use is considered as accessory to the other uses in the building, it is to be classified in accordance with those other uses. The key point is the hazard level that storage brings to the building. It is assumed that accessory storage uses pose little additional hazard above the occupancies which they serve. Where storage activities pose a significantly higher hazard than the other uses in the building, they would typically not be considered accessory and therefore classified as a Group S occupancy.

311.2 Moderate-hazard storage, Group S-1. Group S-1 occupancies are typically used for the storage of combustible commodities. A complete list of all products allowed in this use would be very lengthy; however, many of the more common storage items are identified by the code. In general, buildings classified in this manner would be used for the storage of commodities that are manufactured within buildings classified as Group F-1 occupancies. Commodities that constitute a high physical or health hazard, and exceed the maximum allowable quantities set forth in Section 307, would be stored in the appropriate Group H occupancy. Repair garages for motor vehicles are also considered Group S-1 occupancies. See the discussion in Section 202 regarding the definition of a repair garage.

In a manner consistent with their occupancy classification in the manufacturing process, the storage of lithium-ion or lithium metal batteries is considered a moderate hazard condition. As such, a classification of Group S-1 is appropriate. In addition, facilities that service or repair vehicles powered by lithium-ion or lithium metal batteries are classified no differently than general vehicle repair garages. The presence of such batteries does not increase the hazard level above that anticipated in a Group S-1 occupancy.

Facilities intended for the storage of alcoholic beverages are to be classified as Group S occupancies regardless of quantities. This allowance is based in part upon the requirements of IFC Chapter 40, Storage of Distilled Spirits and Wines. Such storage facilities are further classified within the Group S category based upon the alcohol content of the beverages being stored. Designation of a Group S-1 occupancy is based on an alcohol content of more than 20 percent. As a result, where the alcohol content of beverages in a storage condition does not exceed 20 percent, a classification of Group S-2 is appropriate. Additional information is provided in the discussion of changes to the Group F occupancy classifications.

311.3 Low-hazard storage, Group S-2. Group S-2 occupancies include the storage of noncombustible commodities, as well as open or enclosed parking garages. Buildings in which noncombustible goods are packaged in film or paper wrappings or cardboard cartons, or stored on wooden pallets are still considered Group S-2 occupancies. This also includes any products that have minor amounts of plastics, such as knobs, handles, or similar trim items. It is important, however, that the commodities being stored are essentially noncombustible, insofar as the provisions that regulate Group S-2 occupancies are based on an anticipated minimal fire load.

The classification of a warehouse or similar storage building as a Group S-2 occupancy is incumbent upon the expectation that combustible materials will not be stored within the building. This expectation is almost always difficult to achieve due to the transient nature of the materials being stored. It is not uncommon for the types of materials being stored to vary significantly from one month to the next, particularly for those storage activities not directly related to a specific manufacturing function. Therefore, the most common type of Group S-2 storage condition would be an extension of a Group F-2 manufacturing operation, where the types of materials being stored are consistent with those being manufactured or produced.

Section 312 *Utility and Miscellaneous Group U*

This section covers those utility occupancies that are not normally occupied by people, such as sheds and other accessory buildings, carports, small garages, fences, tanks and towers, and agricultural buildings. The fire load in these structures and uses varies considerably but is usually not excessive. Because they are normally not occupied, the concern for fire load is not very great, and as a group these uses constitute a low hazard. It is also important to note that a Group U occupancy is not expected to have any public use.

Group U occupancies can generally be divided into two areas. The first includes those buildings that are accessory to other major-use structures. Although these accessory-use buildings will at times be occupied, the time period for occupancy is typically limited to short intervals. The second type of Group U occupancy is those miscellaneous structures that cannot be properly classified into any other listed occupancy. The structures are not intended to be occupied, but must be classified in order to regulate any hazards they may pose to property or adjoining structures and persons.

The uses classified as Group U have been deemed to pose little, if any, risk to persons who may be present. Where hazards exist that are not typical of those represented by a Group U classification, it is important that another occupancy classification be assigned. For example, an agricultural building used as an arena for horse shows or livestock auctions should typically be regulated as a Group A assembly occupancy rather than a Group U agricultural structure.

A greenhouse best illustrates the approach established in Section 312.1 for classifying structures as Group U occupancies only where they cannot be appropriately classified as any other occupancy. Where greenhouses are used for assembly, sales, education, or other activities that are more extensive in scope than addressed in the definition, their classification as Group U is not appropriate. Only where the characteristics and hazards do not cause classification as some other occupancy should a greenhouse be considered as a Group U occupancy.

If the jurisdiction has adopted Appendix C, then it will govern the design and construction of agricultural buildings that come under its purview; however, many urban jurisdictions do not adopt this appendix chapter. In this case, should an occasional agricultural building be constructed, it would be regulated by Section 312.

KEY POINTS

- Proper occupancy classification is a critical decision in determining code compliance.
- Uses are classified by the code into categories of like hazards, based on the risk to occupants of the building as well as the probability of property loss.
- Group A occupancies, typically include rooms and buildings with an occupant load of 50 or more, used for the gathering together of persons for civic, social, or religious functions; recreation, food, or drink consumption; or similar activities.
- The hazards unique to Group A occupancies are based primarily on the sizable occupant loads and the concentration of occupants into very small areas.
- Business uses, such as offices, are classified as Group B occupancies and are considered moderate-hazard occupancies.
- Group E occupancies are limited to schools for students through 12th grade and most day-care operations.
- Manufacturing occupancies, classified as Group F, are classified based on whether the materials being produced are combustible or noncombustible.
- Group H occupancies are heavily regulated because of the quantities of hazardous materials present in use or storage.
- Where amounts of hazardous materials are limited in control areas to below the maximum allowable quantities, the occupancy need not be considered a Group H.
- Both physical hazards and health hazards are addressed under the requirements for Group H occupancies.
- Institutional occupancies, classified as Group I, are facilities where individuals are under supervision and care because of physical limitations of health or age, or that house individuals whose personal liberties are restricted.
- Group M occupancies include both sales rooms and motor fuel–dispensing facilities.
- Residential Group R occupancies are partially regulated based on occupant load or number of units, as well as the occupants' familiarity with their surroundings.
- Group S occupancies for storage are viewed in a manner consistent with Group F manufacturing uses.
- Group U occupancies are utilitarian in nature and are seldom, if ever, occupied.

CHAPTER 4

SPECIAL DETAILED REQUIREMENTS BASED ON OCCUPANCY AND USE

Section 402 Covered Mall and Open Mall Buildings
Section 403 High-Rise Buildings
Section 404 Atriums
Section 405 Underground Buildings
Section 406 Motor-Vehicle-Related Occupancies
Section 407 Group I-2
Section 408 Group I-3
Section 409 Motion-Picture Projection Rooms
Section 410 Stages, Platforms, and Technical Production Areas
Section 411 Special Amusement Areas
Section 412 Aircraft-Related Occupancies
Section 413 Combustible Storage
Section 414 Hazardous Materials
Section 415 Groups H-1, H-2, H-3, H-4, and H-5
Section 416 Spray Application of Flammable Finishes
Section 417 Drying Rooms
Section 418 Organic Coatings
Section 420 Groups I-1, R-1, R-2, R-3, and R-4
Section 422 Ambulatory Care Facilities
Section 423 Storm Shelters
Section 424 Play Structures
Section 427 Medical Gas Systems
Section 428 Higher Education Laboratories
Key Points

This chapter provides specific detailed regulations for those types of buildings, features and uses that have very unique characteristics and hazards. The uses in this chapter, though encompassing only a very small fraction of those commonly encountered, require special consideration. Some of the provisions address conditions that could occur in various occupancy classifications such as covered and open mall buildings, high-rise buildings, and underground buildings. Concerns associated with motor-vehicle-related uses, hazardous occupancies, and institutional uses are specifically addressed. Special elements within a building, such as stages, platforms, motion picture projection rooms, and atriums, are also regulated by this chapter. In addition, a variety of special concerns regarding hazardous materials, processes, and activities are addressed through specific requirements.

In all cases it should be remembered that the provisions found in this chapter deal in a more detailed manner with uses and occupancies also addressed elsewhere in the code. Some of the provisions in this chapter may be more restrictive than the general requirements of the code, whereas others may be less restrictive. The general rules found in other areas of the *International Building Code*® (IBC®) will govern unless modifications from this chapter are applicable.

Section 402 *Covered Mall and Open Mall Buildings*

Provisions for covered mall and open mall buildings included in the IBC set forth specific code requirements for a specific building type. Provisions in this section only apply to covered and open mall buildings having a height of not more than three levels at any one point and not more than three stories above grade plane. Furthermore, the provisions are only those that are considered to be unique to covered and open mall buildings. For those features that are not unique, the general provisions of the code apply. Covered and open mall buildings that comply in all respects with other provisions of the code are not required to comply with these provisions. It should be noted that foyers and lobbies of office buildings, hotels, and apartment buildings are not required to comply with the special provisions for covered or open mall buildings. An example of an open mall and associated buildings is shown in Figure 402-1.

402.1.1 Open mall building perimeter line. It is necessary that a perimeter line for an open mall building be established in order to apply various requirements in Section 402. The line creates the boundary of the tenant spaces, service areas, pedestrian paths, and similar spaces that make up the open mall building, but does not include any anchor buildings or parking garages. An example is shown in Figure 402-2.

402.2 Open space. The special provisions of Section 402 are applicable only where the entire building, including the anchor buildings and attached parking structures, are surrounded by permanent open space at least 60 feet (18,288 mm) in width. See Figure 402-3. The allowance provided in the exception to Section 402.2 for reduced open space surrounding covered mall buildings, including their associated parking garages and anchor stores, is consistent with that for other unlimited area buildings as permitted by Section 507 since a covered mall building contains similar characteristics of those buildings. The reduction in required open space is not permitted where the covered mall building or anchor stores include Group E, H, I, or R occupancies. Where a building is considered an open mall building, the

Covered Mall and Open Mall Buildings

Figure 402-1 **Open mall.**

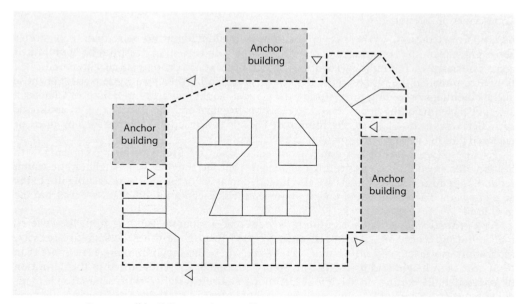

Figure 402-2 **Open mall building perimeter line.**

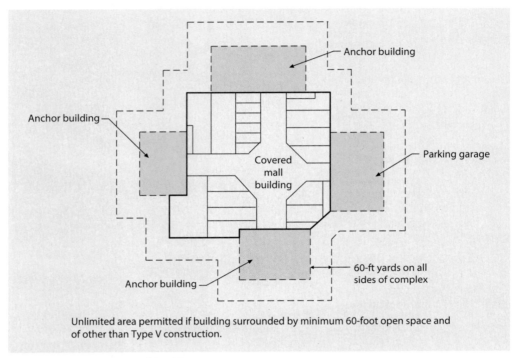

Figure 402-3 **Covered mall building.**

required permanent open space is regulated based on the open mall building perimeter line as described in Section 402.1.1.

402.4 Construction. Where covered and open mall buildings no more than three stories above grade plane are constructed of other than Type V construction, they may be of unlimited area. Associated anchor buildings of Type I, II, III, or IV construction may also have unlimited floor area provided they are no more than three stories in height above grade plane. For those anchor buildings exceeding three stories, the general height and area limitations of Chapter 5 are applicable, including appropriate increases for frontage and the presence of an automatic sprinkler system. In all cases, the minimum type of construction required for any open or enclosed parking garage is that mandated by Section 406.

Tenant spaces must be separated from each other by fire partitions complying with Section 708. In addition to protecting one tenant from the activities of a neighbor, the tenant separation requirements for malls are also intended to assist in the goal of restricting fire to the area of origin. There is no requirement, however, for the separation of tenant spaces from the mall itself.

As a general rule, an anchor building is viewed as a separate building from the covered mall building. Therefore, a fire wall must be used to provide the necessary fire-resistive separation. However, only a fire barrier is required where the anchor building is no more than three stories in height and its use is consistent with one of those identified in the definition of "covered mall building" in Section 202. Although a fire-resistance-rated separation is mandated between an anchor building and the mall, openings in such a separation typically need no fire-protection rating. Anchor buildings, other than those containing Group R-1 sleeping units, constructed of Type I or II noncombustible construction may have unprotected openings into the mall.

Under the concept of covered and open mall buildings, there is no requirement for a fire separation between tenant spaces and the mall. Similarly, the food court needs no separation between adjacent tenant spaces and the mall. The hazards presented by an attached parking garage, however, must be addressed through the separation provided by a minimum 2-hour fire-resistance-rated fire barrier.

402.5 Automatic sprinkler system. An automatic fire-sprinkler system is the primary means of fire protection for a covered mall buildings. Because of the reliance placed on the sprinkler system, this section requires the following additional safeguards:

1. The code requires that the sprinkler system be complete and operative throughout all of the covered mall building before occupancy of any of the tenant spaces. In those areas that are unoccupied, an alternative protection method may be approved by the building official.

2. The mall, anchor buildings, and tenant spaces shall be protected by separate sprinkler systems, except that the code will permit tenant spaces to be supplied by the same system as the mall provided they can be independently controlled.

Sprinkler protection is also required for open mall buildings. This protection must extend to beneath any exterior circulation balconies located adjacent to the open mall.

402.6.2 Kiosks. Kiosks and similar structures, both temporary and permanent in nature, are regulated for construction materials and fire protection owing to their presence in an established egress path. Such structures shall be noncombustible or constructed of fire-retardant-treated wood, complying foam plastics, or complying aluminum composite materials. Active fire protection is provided by required fire suppression and detection devices. Kiosks are also limited in size, and their relationship to other kiosks is regulated. Multiple kiosks can be grouped together, provided their total area does not exceed 300 square feet (28 m^2). At that point, a separation of at least 20 feet (6,096 mm) is required from another kiosk or grouping of kiosks.

402.8 Means of egress. One of the significant areas in which the provisions for covered mall and open mall buildings differ from the general provisions applied to the majority of buildings is the means of egress. Issues such as occupant load determination and travel distance are modified specifically in this section owing to the unique features of a covered or open mall building. It is important to remember that where this section conflicts with the general requirements of the code in Chapter 10, the provisions of this section are applicable.

402.8.1 Mall width. With its added life-safety systems, the mall may be considered a corridor without meeting the means of egress size requirements of Section 1005.1 when the mall complies with the conditions of this section as depicted in Figure 402-4. In this case, the code requires that the minimum mall width be 20 feet (6,096 mm), and this typical cross section shows that the minimum required width may be divided so that a clear width of 10 feet (3,048 mm) is provided separately on each side of any kiosks, vending machines, benches, displays, etc., contained in the mall. In addition, food court seating in the mall would have to be located so as not to encroach upon any required mall width. Understandably, the mall width shall also accommodate the occupant load immediately tributary thereto.

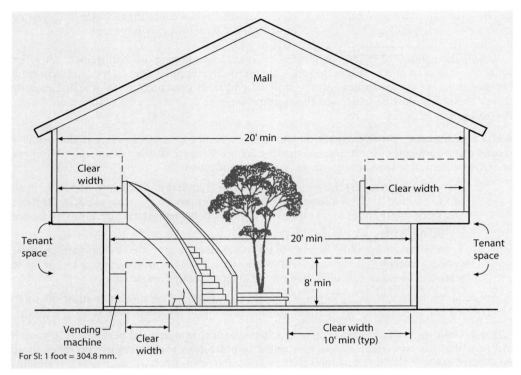

Figure 402-4 **Mall width requirements.**

402.8.2 Determination of occupant load. The determination of the occupant load and minimum required number of exits can be divided into two areas:

1. Tenant spaces.
2. The covered or open mall building.

The maximum occupant load of any individual tenant space is determined in a manner consistent with its use as regulated by Chapter 10. Means of egress requirements for individual tenant spaces are to be based on the occupant load as typically determined. Figure 402-5 depicts the method for determination of the occupant load in a tenant space having a retail area, a storage room, an office, and a bathroom. Although the total tenant space contains 1,500 square feet (139.3 m^2) of floor area, each individual use has a designated occupant load based on the appropriate factor from Table 1004.5. In this example, the occupant load would be calculated at 23.

Not only must the occupant load be determined for each individual tenant space, but the occupant load for the entire covered or open mall building must also be determined. It is highly unlikely that all tenant spaces will be fully occupied at the same time in a covered or open mall building. However, the mall will itself have a significant occupant load independent of the tenant spaces. Therefore, a different method is used to determine the number of occupants from which to base the means of egress from the mall itself. This occupant load is to be determined based on the gross leasable area of the covered or open mall building, excluding

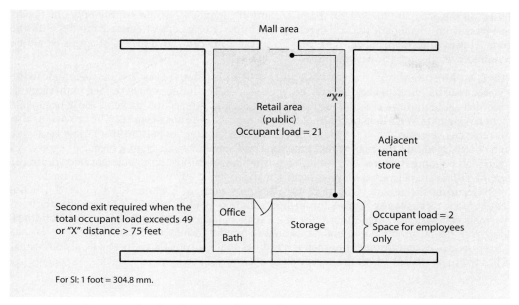

Figure 402-5 **Occupant load and means of egress from tenant space.**

any anchor buildings and those tenant spaces having an independent means of egress, with the occupant-load factor determined by the following formula:

$$\text{Occupant-load factor} = (0.00007)(\text{gross leasable area}) + 25$$

As a result, the net effect is that the total occupant load computed for the covered or open mall building will be something different than the summation of the occupant loads determined for each individual tenant space plus any occupant load that may be assigned to the mall itself.

The occupant-load factor used for egress purposes shall not exceed 50, nor is it ever required to be less than 30. Where there is a food court provided within the covered or open mall building, the occupant load of the food court is to be added to the occupant load of the covered or open mall building as previously calculated in order to determine the total occupant load.

In utilizing several examples, assume a building contains 600,000 square feet (55,740 m²) of gross leasable area. The occupant-load factor, when calculated, would be 67. However, a factor of 50 would be used in determining an occupant load of 12,000. Should a food court be present that seats 600 occupants, the occupant load of 12,000 would be increased accordingly. Where a covered mall building contains 100,000 square feet (9,290 m²) of gross leasable area, the occupant-load factor would be 32. A factor of 32 would then be used to calculate the occupant load of the covered mall building, which would be 3,125 occupants. As the provision is applied to a smaller covered mall building, an occupant-load factor of 30 will be used when the gross leasable area of the covered mall building is less than 71,500 (6,642.3 m²) square feet. As a final note, because anchor buildings are not considered a part of a covered mall building, their occupant load shall not be included in computing the total number of occupants for the mall.

402.8.3 Number of means of egress. Figure 402-5 also depicts the requirements of Section 402.8.3 for the determination of the number of means of egress from the tenant space.

Based on the occupant load of 23, the provisions of Chapter 10 would require only one means of egress from this tenant space. Therefore, the number of means of egress complies with the code. However, if the distance, x, exceeds 75 feet (22,860 mm), two means of egress would be required even though the occupant load is less than 50.

402.8.4 Arrangements of means of egress. The provisions of this section are unique to the covered mall building and are depicted in Figure 402-6. The limitations described in this section encompass a high number and density of occupants, and this section prevents those occupants from having to traverse long portions of the mall to reach a means of egress. The provisions also prevent the overcrowding of the mall, such as if a large number of patrons from these uses were to be discharged into the mall at the same time and some distance from a means of egress. An open mall building is addressed a bit differently in that assembly occupancies located within the perimeter line are permitted to have their main exit open at any point into the open mall.

In securing the intent of Section 402, the provisions of Section 402.8.4.1 establish the requirement that means of egress for anchor buildings shall be provided independently from the mall exit system. Furthermore, the mall shall not egress through the anchor buildings. Moreover, the termination of a mall at an anchor building where no other means of egress has been provided except through the anchor buildings shall be considered to be a dead end, which is limited in accordance with the exception to Section 402.8.6.

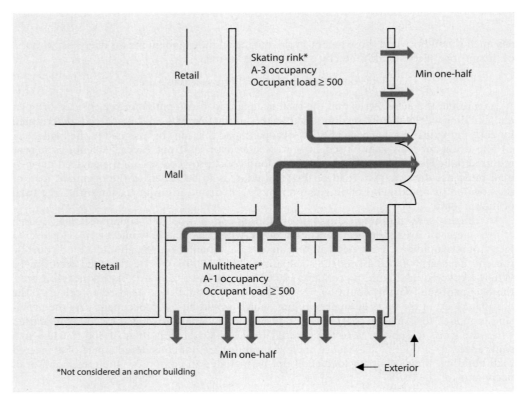

Figure 402-6 **Arrangement of exits.**

402.8.5 Distance to exits. Figure 402-7 depicts the multifaceted provisions of this section as it relates to a covered mall building as follows:

1. The first case is illustrated for travel within the tenant space and includes the provisions applicable to tenant spaces A and B. For tenant space A, the diagram depicts the application of the code for a tenant space with a closed front with only a swinging exit door to the mall. The entrance to the mall would be the point at which occupants from the tenant space pass through the egress door from the tenant space to the mall.

    Tenant space B represents the condition for an open storefront using a security grille instead of a standard egress door. The entrance to the mall in this case is the point at which occupants of the tenant space pass by an imaginary plane that is common to both the tenant space and the pedestrian mall. The location of the assumed required clear exit width along the open front of tenant space B may be placed at any point along the front, and its location would depend only on that which would render the least-restrictive application of the provisions.

    For either tenant space A or B, the code permits the travel distance within the tenant space to the entrance to the mall to be a maximum of 200 feet (60,960 mm).

2. After the occupants exit from a tenant space into the mall, the code permits another 200 feet (60,960 mm) of exit access travel distance to one of the exit elements described in Section 202. This travel limitation also applies to all other locations in the mall where occupants may be located when an exiting condition occurs.

    It can be seen from this discussion, plus the perusal of Figure 402-7, that the travel distances permitted for a covered mall building are generally more liberal than those permitted by Section 1017.2. This liberalized and increased travel distance is based on the rationale that travel within a mall will be within an area where special fire protection features are provided.

3. Another limitation of these provisions regarding travel distance within covered mall buildings is also illustrated by Figure 402-7. In this instance, if the path of travel is through a secondary means of egress door from tenant space B to an exit (in this case an exit passageway), travel distance is not limited once the exit is reached.

4. In the case of exiting via the corridor as depicted for tenant space A, the total travel distance is limited to 200 feet (60,960 mm). This limitation is based on the consideration that a corridor does not offer as much protection as either an exit (such as an exit passageway) or the mall.

    For an open mall building, travel from an individual tenant space to the mall is addressed in the same manner as for a covered mall building. Once the open mall is reached, travel to the perimeter line or to an exit is then regulated.

402.8.6 Access to exits. This section uses the same approach as does Chapter 10 of requiring that the means of egress be arranged so that the occupants may go in either direction to a separate exit. However, in this section the dead end is measured in a manner similar to that of Exception 3 of Section 1020.5. Figure 402-8 shows the manner in which dead-end conditions are regulated.

Regardless of the occupant load served, the minimum width of an exit passageway or corridor from a mall is to be 66 inches (1,676 mm). The exit passageway shown in Figure 402-9 must be at least 66 inches (1,676 mm) wide. The main entrances shown in the same figure are not subject to this requirement insofar as the required capacity of Section 1005.3 will most often require a greater exit width.

Another exiting provision that is unique to a covered or open mall building regards exit passageway enclosures. Section 402.8.7 allows mechanical and electrical equipment rooms,

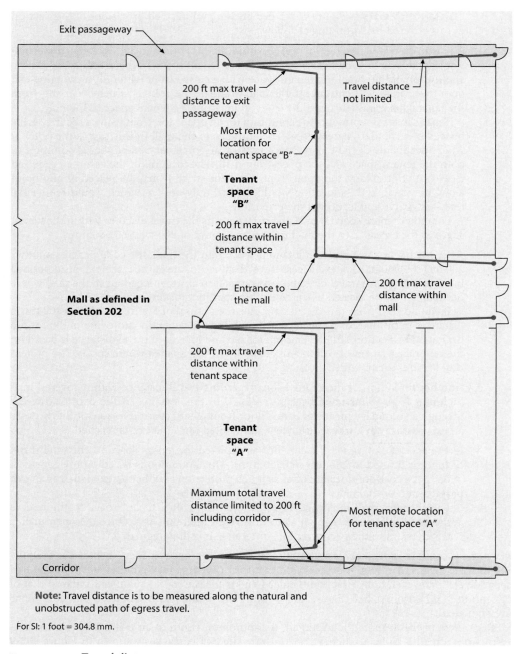

Figure 402-7 **Travel distance.**

Covered Mall and Open Mall Buildings

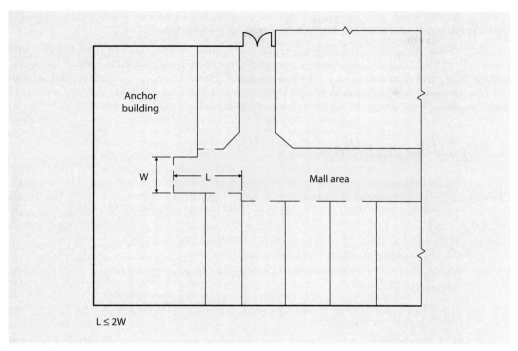

Figure 402-8 **Dead-end mall criteria.**

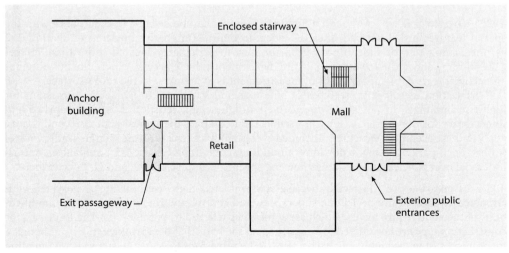

Figure 402-9 **Exit passageway width and capacity.**

building service areas, and service elevators to open directly into exit passageways, provided the minimum 1-hour fire-resistance-rated separation is maintained.

402.8.8 Security grilles and doors. Quite often, mall tenants wish to have the dividing plane between the mall and the tenant space completely open during business hours. Horizontal sliding or vertical security grilles or doors are usually placed across this opening. This section permits their use, provided they do not detract from safe exiting from the tenant space into the mall. To secure that intent, the code requires four limitations outlined in this section.

Section 403 *High-Rise Buildings*

This section encompasses special life-safety requirements for high-rise buildings. The comparatively good fire record notwithstanding, particularly in office buildings, fires in high-rise buildings have prompted government at all levels to develop special regulations concerning life safety in high-rise buildings. The potential for disaster that is due to the large number of occupants in high-rise buildings has resulted in the provisions included in this section.

The high-rise building is characterized by several features:

1. It is impractical, if not impossible, to completely evacuate the building within a reasonable period of time.
2. Prompt rescue will be difficult, and the probability of fighting a fire in upper stories from the exterior will be low.
3. High-rise buildings are occupied by large numbers of people, and in certain occupancies the occupants may be asleep during an emergency.
4. A potential exists for stack effect. The stack effect can result in the distribution of smoke and other products of combustion throughout the height of a high-rise building during a fire.

The provisions in this section are designed to account for the features described above.

403.1 Applicability. Although a high-rise building can be defined in accordance with the special features just described, the IBC elects to define a high-rise building in Section 202 as one having one or more floors used for human occupancy located more than 75 feet (22,860 mm) above the lowest level of fire department vehicle access. Where the building has an occupiable roof, the upper point of measurement is to be taken to the roof's surface.

This section identifies those types of buildings and structures to which the provisions for high-rise buildings do not apply due to their unique characteristics. It is inappropriate and unnecessary for the special safeguards established to address the typical high-rise condition to apply to these exempted structures. Included in this group of structures are aircraft-traffic control towers, open parking garages, portions of buildings containing Group A-5 occupancies, special industrial occupancies, and buildings housing specific Group H-1, H-2, and H-3 occupancies.

403.2 Construction. Primarily because a sprinklered high-rise building is provided with an increased level of fire protection supervision and control, the IBC permits certain modifications of the code requirements, which are sometimes referred to as trade-offs. The trade-offs for construction type are considered to be justified on the basis that the sprinkler system, although a mechanical system, is highly reliable because of the provisions that require supervisory initiating devices and water-flow initiating devices for every floor. In addition, a secondary on-site supply of water is mandated for those high-rise buildings subject to a moderate to high level of seismic risk.

This section permits some degree of reduction in the required fire-resistance ratings of building elements required to be protected on account of type of construction. In the evaluation of the maximum allowable height and area permitted for the building, the original construction type would remain applicable. For example, a building classified as Type IB and regulated as Type IB throughout the code would only need to meet the fire-resistive requirements of Table 601 for a building of Type IIA construction (in other than Group F-1, H-2, H-3, H-5, M, and S-1 occupancies).

In addition to the reductions permitted for building elements identified in Table 601 based on the building's type of construction, the fire-resistance rating of shaft enclosures may be reduced to 1 hour where sprinklers are installed within the shafts at the top and at alternative floor levels.

The reduction in fire-resistance ratings is not applicable in all cases. In all high-rise buildings classified as Type IA, the 3-hour rating for structural columns supporting floors must be maintained. The critical role of columns in the structural integrity of a high-rise structure during fire conditions mandates that their fire resistance not be lessened. For buildings that exceed 420 feet (128 m) in height, no reduction to the required fire resistance of any building elements is permitted, and the required vertical shaft protection cannot be reduced from the general 2-hour requirement. The increased risk of catastrophic damage associated with these very tall buildings requires an increased level of fire resistance.

403.3 Automatic sprinkler system. The automatic fire-sprinkler system required by Section 403.3 must be completely reliable, as must the other life-safety systems. As part of that reliability effort, a high-rise building must be provided with a secondary water supply where required by Section 403.3.3. In Seismic Design Category C, D, E, or F, an on-site automatic secondary water supply shall be provided, with a supply of water equal to the hydraulically calculated sprinkler design demand, including the hose stream requirement, for a duration of at least 30 minutes. As fires can (and do) break out as a result of earthquake damage to the various mechanical systems within a building, it is imperative that the reliability of the sprinkler system be such that any resulting fires can be automatically extinguished.

403.4 Emergency systems. Among the more important life-safety features required by the code are the alarm and communications systems required by this section. Where it is expected that people will be unable to evacuate the building, it is imperative that they be informed as to the nature of any emergency that may break out, as well as the proper action to take to exit to a safe place of refuge. Furthermore, a system is necessary in most cases to provide for communication between the fire officer in charge at the scene and the fire fighters throughout the building. Section 911 provides details of the required fire command center utilized by the fire department to coordinate fire suppression and rescue operations.

In order to provide for efficient and reliable communications among fire fighters, police officers, medical personnel, and other emergency responders, an in-building two-way emergency responder communications system must be installed in all high-rise buildings. The details for such systems are established in Section 510 of the *International Fire Code*® (IFC®). IFC Section 510.4.1 specifies that acceptable radio coverage is satisfied when 95 percent of all areas on each floor of the building meet the signal strength requirements in Sections 510.4.1.1 through 510.4.1.3 for signals transmitted into and out of a building. Additionally, at least 99 percent of all areas designated as critical areas by the fire code official must meet these requirements.

In addition, one of the fire department's duties during a fire event is to expel the smoke after the fire has occurred. Three methods for smoke exhaust are available: through natural means, through the use of mechanical air-handling equipment, or through an equivalency approach. Where the method of natural ventilation is used for the removal of smoke, openable windows or panels are required to be distributed around the perimeter of each floor level. The fire departments can open the appropriate windows as necessary and provide pressurization through the use of fans. The use of fixed tempered glass panels is also acceptable if they are not coated in a manner that will modify the natural breaking characteristics of the glass. Where mechanical air-handling equipment is used for smoke removal purposes, the building's HVAC system is equipped with appropriate dampers at each floor that are arranged in a manner that will stop the recirculation of air through the use of 100 percent fresh air intake and outside exhaust. The panel for controlling this system is to be located in the building's fire command center. In addition, the building official has the authority to approve any other means of smoke removal provided it accomplishes the intended goal of the prescriptive mechanical or natural ventilation approaches described by the code. See Figure 403-1. It must be noted that this smoke exhaust system is for fire department use only and is not intended to be a part of the occupant-related life-safety systems placed in high-rise buildings.

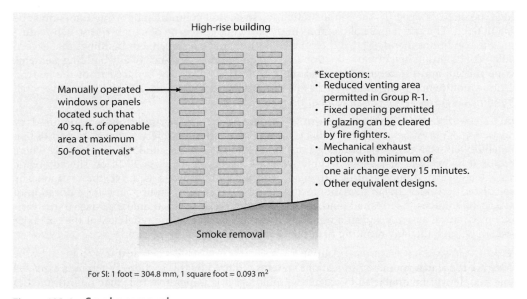

Figure 403-1 **Smoke removal.**

Furthering the intent of the IBC that life-safety systems in high-rise buildings be highly reliable, the code requires that the power supply to the life-safety systems be regulated by the appropriate provisions of NFPA 70, more specifically, Articles 700 and 701. The basis of the reliability is that the building's power be automatically transferable to a standby or emergency power system in the event of the failure of the normal power supply. Those standby power loads required by the code include power and lighting for the fire command center, ventilation and automatic fire detection equipment for smokeproof enclosures, and elevators.

The code further requires that lighting for exit signs, means of egress illumination, and elevator car lighting be automatically transferable to an emergency power system capable of operation within 10 seconds of the failure of the normal power supply. Additionally, all emergency voice/alarm communications systems, automatic fire detection systems, fire alarm systems, and electrically powered fire pumps are to be provided with emergency power.

403.5.1 Remoteness of interior exit stairways. The general requirements for separation of exit or exit access doorways as established in Sections 1007.1.1 and 1007.1.2 are supplemented in this section by adding a minimum required separation distance between the enclosures for interior exit stairways. In addition to maintaining a minimum separation between the doors to the enclosures of one-third the length of the overall diagonal dimension of the area served, the

High-Rise Buildings   95

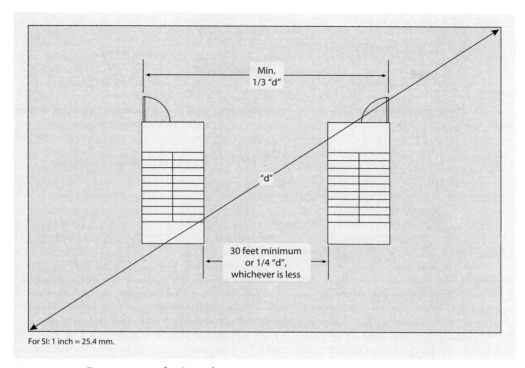

Figure 403-2 **Remoteness of exit enclosures.**

interior exit stairways must be located at least 30 feet (9,144 mm) apart or not less than one-fourth the diagonal dimension, whichever is less. See Figure 403-2. If three or more interior exit stairway enclosures are mandated, at least two of the enclosures must be separated as indicated.

403.5.2 Additional exit stairway. During a fire that requires a full evacuation of a building of extensive height, the fire-fighting operations will reduce the capacity of the egress system. The extended period of time needed to fully evacuate a very tall building means that people will still be evacuating while full fire-fighting operations are taking place. Sound high-rise fire-fighting doctrine provides that the fire department take control of one stair, the one most appropriate to the circumstances of the given fire condition. This can result in a significant reduction in egress capacity of the stairway system. For example, in order to conduct suppression activities in a building with two required stairs of the same width, one-half of the exit capacity is unavailable while the building is still being evacuated. An additional stair is required so that egress capacity will be maintained through the time that full evacuation is complete.

It is important to note that this additional stair is not required to be a dedicated fire department stair. The fire department should be able to choose the stair that is most appropriate for the actual fire event. As a result, it will be necessary for emergency responders to manage evacuation flow to the available stairs.

The application of this requirement is limited to only those buildings over 420 feet (128 m) in height. In addition, it does not apply to Group R-2 occupancies and their ancillary spaces due to the limited occupant load of such uses. In determining if the required egress capacity is provided by the stairway system, it must be assumed that the widest of the stairways is the one that is unavailable for means of egress travel. The remaining stairways must be sized to accommodate the total required egress capacity. The additional stairway's sole purpose is to provide additional egress capacity. Therefore, other means of egress design issues, such as travel distance and exit separation, are not regulated.

The additional exit stairway is not required where occupant evacuation elevators are provided in accordance with Section 3008. The availability of elevators for evacuation purposes provides for a reasonable alternative to an additional stairway.

403.5.3 Stairway door operation. In those cases where it is impractical to totally evacuate the occupants from the building through the stairway system, it must be possible to move the occupants to different floors of the building that are safe by way of the stairway system. For this to happen, the doors to the interior exit stairway enclosures must either be unlocked or be designed for individually or simultaneously unlocking from the fire command station. Upon activation of a fire alarm signal in the area served by the stairway, simultaneous unlocking shall also occur. In addition, failure of the power supply to the door's lock or locking system shall cause the stairway door to unlock.

The IBC further requires that a telephone or other two-way communication system (such as a two-way system with speaker and microphone) be located at every fifth floor in each required stair enclosure for those cases where the stair enclosure doors are to be locked. Moreover, the code requires that this communication system be connected to an approved station that is constantly attended. Thus, anyone trapped in the stairway during a nonfire emergency may call for help without traversing more than two levels. In the case of office buildings and apartment houses, the attended station may be considered the office of the building as long as the office has continuous attendance by responsible individuals who are familiar with the life-safety systems. For hotel buildings, the most likely choice for the attendance station will probably be the hotel telephone operators, and, again, they must be trained to assist the persons trapped within the stair enclosure.

403.5.4 Smokeproof enclosures. Those interior exit stairway enclosures in a high-rise building that serve floor levels located more than 75 feet (22,860 mm) above the lowest level of fire department vehicle access must be designed as smokeproof enclosures. Exit stairways that do not serve floors above the height indicated are not regulated by this section. Section 1023.12 regulates the access, extension, and termination relating to the utilization of smokeproof enclosures and pressurized stairways as a part of the means of egress system. Section 909.20 provides the construction and ventilation criteria for smokeproof enclosures, as well as establishing stair pressurization as an acceptable alternative.

403.5.5 Luminous egress path markings. In high-rise buildings, increased visibility for travel on stairways and through exit passageways is important due to the extreme conditions that may be encountered under emergency conditions. The use of photoluminescent or self-illuminating materials to delineate the exit path is required in high-rise buildings housing Group A, B, E, I-1, M, and R-1 occupancies. Specific requirements related to these egress path markings are set forth in Section 1025 and include the regulation of striping on steps, landings, and handrails; perimeter demarcation lines on the floor and walls, including their transition; acceptable materials; and illumination periods. An example of such markings is shown in Figure 403-3.

403.6 Elevators. In order to facilitate the rapid deployment of fire fighters, at least two fire service access elevators are required in high-rise buildings that have an occupied floor more than 120 feet (36,576 mm) above the lowest level of fire department vehicle access. Usable by fire fighters and other emergency responders, the specific requirements for the elevators are set forth in Section 3007. There are a number of key features that allow fire fighters to use the elevator for safely accessing an area of a building that may be involved in a fire. A complying lobby is required adjacent to the elevator hoistway opening, creating a protected area from which to stage operations. Access to standpipe hose valves is required, as are two-way communication

Figure 403-3 **Luminous egress path markings.**

features. A single fire service access elevator is permitted in the unlikely situation where a building regulated by this section has only one elevator.

The mandate for multiple fire service access elevators is based on information that indicates at least two elevators are necessary for fire-fighting activities in high-rise buildings. In addition, past experience has shown that on many occasions elevators are not available due to shutdowns for various reasons, including problems in operation, routine maintenance, modernization programs, and EMS operations in the building prior to firefighter arrival. A minimum of two fire service elevators provided with all of the benefits afforded to such elevators better ensures that there will be a fire service access elevator available for the fire fighters' use in the performance of their duties.

A more comprehensive discussion of the requirements for fire service access elevators is found in the analysis of Section 3007.

The use of elevators as an evacuation element for occupants of a high-rise building is possible provided the elevators are in compliance with the requirements established in Section 3008. The controls and safeguards provided in Section 3008 create a suitable environment to allow complying elevators to be used for occupant self-evacuation purposes. The presence of such elevators does not reduce the general means of egress requirements established in Chapter 10; however, the additional exit stairway mandated for very tall high-rise buildings by Section 403.5.2 is no longer required. It is important to note that the installation of occupant evacuation elevators in high-rise buildings is not mandated by the code; however, such elevators are permitted for use for occupant self-evacuation and may be utilized as an alternative to the additional stairway requirement.

Section 404 *Atriums*

This section was developed to fill a need for code provisions applicable to the trends in the architectural design of buildings where the designer makes use of an atrium. Prior to the early 1980s, building codes did not provide for atriums, and, moreover, atriums were prohibited because of the requirements for protection of vertical openings. They were, however, permitted on an individual basis, usually under the provisions in the administrative sections of the code permitting alternative designs and alternative methods of construction. The general concept of alternative protection is to provide for both the equivalence of an open court and at the same time provide protection somewhat equivalent to shaft protection to prevent products of combustion from being spread throughout the building via the atrium.

Unprotected vertical openings are often identified as the factors responsible for fire spread in incidents involving fire fatalities or extensive property damage. Section 404 addresses one method for protection of these specific building features in lieu of providing a complete floor separation between stories. Where a conforming atrium is provided, it is considered as an equivalent degree of resistance to fire and smoke spread vertically from story to story. An atrium, specifically defined in Section 202, is a vertical space in a building that is enclosed at the top. Furthermore, the application of Section 404 requires a minimum of three stories to be connected by the opening for all occupancies other than Groups I-2 and I-3. Due to the required smoke compartmentation of such institutional occupancies, in those cases an atrium only requires the connection of two or more stories. The use of an atrium to protect vertical openings between stories is permitted in all occupancies except for Group H.

The atrium allowance is one of the numerous options for addressing vertical openings as set forth in Sections 712.1.1 through 712.1.16. Atriums are identified in Section 712.7 and are considered acceptable when in compliance with the criteria of Section 404. Complying atriums are not to be considered unprotected vertical openings; rather, the vertical openings are protected by means other than enclosure by a shaft or a complete floor assembly. The provisions of Section 404 are not intended to apply to conditions that are addressed by Section 712.1 based on the itemized options, other than when Section 712.1.7 is the chosen method of compliance. In other words, if compliance with Section 712.1 is achieved by applying one of the options other than an atrium as permitted by Section 712.1.7, then the provisions of Section 404 are not applicable. Conversely, if the provisions of Section 404 have been complied with, then the other options listed in Section 712 do not need to be applied for the atrium condition.

404.3 Automatic sprinkler protection. One of the basic requirements for atriums is that the building be provided with an automatic sprinkler system throughout. Two exceptions modify this general requirement. Those areas of the building adjacent to or above the atrium are not required to be sprinklered if appropriately separated from the atrium. This separation must consist of minimum 2-hour fire barriers, horizontal assemblies, or both. In addition, sprinkler protection is not required at an atrium ceiling located more than 55 feet (16,764 mm) above the atrium floor.

404.5 Smoke control. Another major component of the life-safety system for a building containing an atrium is the required smoke-control system. The design of the smoke-control system is to be in accordance with Section 909. Although the exhaust method is typically used as the means of accomplishing smoke control, the code would not prohibit the use of the airflow or pressurization methods where shown to be suitable. One of these methods is often used where the ceiling height makes it difficult to maintain the smoke layer at least 6 feet (1,829 mm) above the floor of the means of egress.

Applicable in other than Group I-2 and Group I-1, Condition 2 occupancies, Exception 1 eliminates the requirement for smoke control in those atriums that connect only two stories. However, as established in the atrium definition, where only two stories are open to each other in a Group I-1 occupancy the condition is not regulated under the provisions of Section 404. Therefore, the application of Exception 1 is limited solely to Group I-3 occupancies, recognizing that atriums in such buildings need not be provided with smoke control where only two stories are connected.

Exception 2 allows an opening connecting multiple stories without the protection afforded by a smoke control system where two conditions are met. First, only the two lowest stories are permitted to be open to the atrium. Second, those stories located above the two lowest stories are required to be separated from the atrium with shaft enclosures having a minimum fire-resistance rating in compliance with Section 713.4. It is recognized that a combination vertical opening condition consisting of both an atrium and a shaft enclosure provides the necessary degree of separation expected between multiple stories, eliminating the need for a smoke control system. See Figure 404-1.

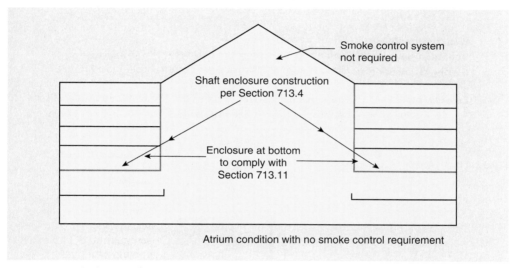

Figure 404-1   **Atrium enclosure.**

404.6 Enclosure of atriums. With some exceptions, an enclosure separation is required between the atrium and the remainder of the building. See Figure 404-2. The basic requirement is for a 1-hour fire-resistance-rated fire barrier with openings protected in accordance with Tables 716.1(2) and 716.1(3). This degree of enclosure, in addition to the other special conditions of Section 404, is intended to provide protection somewhat equivalent to the otherwise mandated shaft protection. Two alternative methods of atrium separation are described in the exception. The special sprinkler-wetted glass enclosure as depicted in Figure 404-3 provides a prescriptive method of achieving equivalency. In addition, the separation may consist of a ¾-hour-rated glass-block wall assembly.

The separation between adjacent spaces and the atrium may be omitted on a maximum of any three floor levels, provided the remaining floor levels are separated as provided in this section. Additional limitations apply to Group I-2 and Group I-1, Condition 2 occupancies. In computing the atrium volume for the design of the smoke-control system, the volume of such open spaces shall be included. Where a smoke control system is not required, it is also unnecessary in other than Group I-2 and Group I-1, Condition 2 occupancies to provide the fire-resistive separation between the atrium and the adjoining spaces.

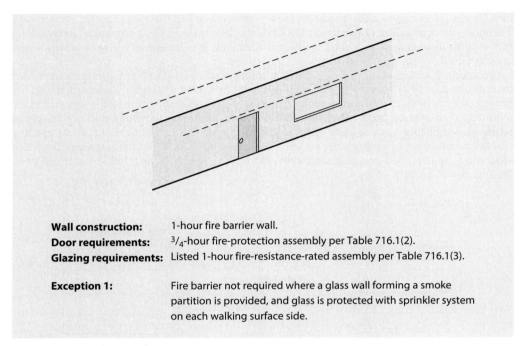

Figure 404-2 **Atrium enclosure**

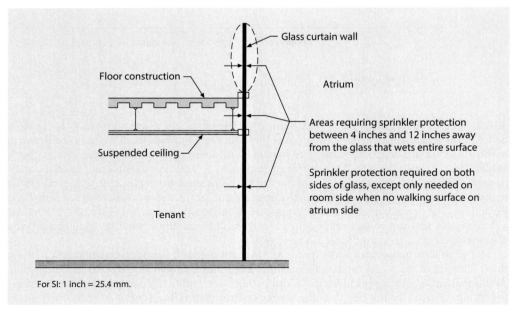

Figure 404-3 **Glass protection.**

Some form of a boundary is required to assist the required smoke control system in containing smoke to only the atrium area. Exceptions 6 and 7 to the enclosure requirements address conditions where escalators or exit access stairways penetrate a required horizontal assembly. These exceptions recognize that no horizontal assembly separation of the atrium is required at those locations where openings are created for such escalators or stairways provided in compliance with Section 712.1.3.1 or Item 4 of Section 1019.3, respectively, is achieved. See Figure 404-4. It is anticipated that the presence of the draft curtains and sprinklers limit the potential of smoke spread through the opening; thus, such areas that only communicate via these specific openings should not be considered as an extension of the atrium.

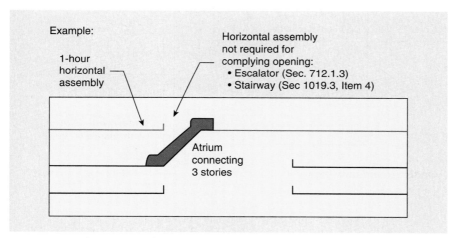

Figure 404-4   **Openings in horizontal assemblies.**

Section 405 *Underground Buildings*

Structures that have occupied floor levels well below ground level, and thus significantly below the level of fire personnel access from the exterior and building occupant discharge to the exterior, present special hazards that are specifically addressed in this section. Much like high-rise buildings, underground buildings can create difficult egress conditions as well as pose many problems for the fire department in their rescue and suppression activities. Fundamental to the protection features of this type of building are the requirements for Type I noncombustible construction and the installation of an automatic sprinkler system. A standpipe system is also required. For clarification, only the underground portion of the structure needs to be of Type I construction, and only those floor levels at the highest discharge level and below need to be sprinklered. See Figure 405-1.

The basic criterion for consideration as an underground building is that a floor level used for human occupancy be located more than 30 feet (9,144 mm) below the finished floor of the lowest level of exit discharge. Exempted from the requirements of Section 405 are many of those buildings and structures that are commonly constructed significantly below grade, including: sprinklered dwellings; parking garages having sprinkler protection; fixed guideway transit systems such as subways; stadiums, arenas, and similar assembly structures; those buildings where the lowest story is the only story that mandates compliance with Section 405 and the

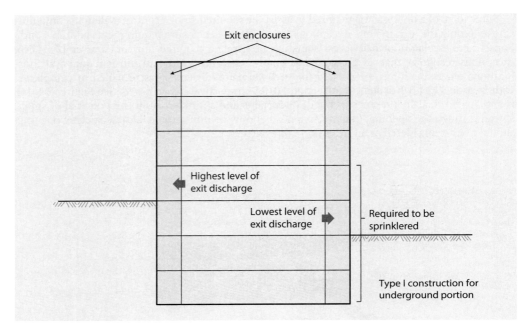

Figure 405-1 **Underground buildings.**

story's floor area and occupant load are very limited; and mechanical spaces that are typically unoccupied. As a result, the application of Section 405 is quite uncommon. However, where applicable, the provisions address a unique hazard by imposing a variety of mandated safeguards.

A valuable concept in fire protection is utilized in the provisions for underground buildings that extend even deeper into the ground. Where an occupied floor level is located more than 60 feet (18,288 mm) below the finished floor of the lowest level of exit discharge, at least two compartments of approximately equal size must be created. The compartmentation must extend throughout the underground portion of the structure, up to and including the highest level of exit discharge. The separation between the two areas is intended to allow for horizontal egress travel to a refuge area if necessary, while also permitting the use of the compartment as a staging area for fire-suppression activities. A smoke barrier is required as the separation element, with door openings also protected in a manner to restrict smoke leakage. Other openings and penetrations are strictly limited. Air supply and exhaust systems, where provided, must be independent of the other compartments. Where the underground portion of the building is served by elevators, each compartment must have access to at least one elevator. An elevator lobby, enclosed by a smoke barrier, may be used to allow a single elevator to serve more than one compartment. See Figure 405-2.

Smoke control is also an important part of the overall fire-protection package. By limiting the spread of smoke to only the originating area of the fire, the remainder of the underground building should be provided with acceptable egress paths. Each compartment shall be provided with its own smoke-control system. A manual fire-alarm system and complying communications system is also an integral part of the fire- and life-safety concept for underground buildings.

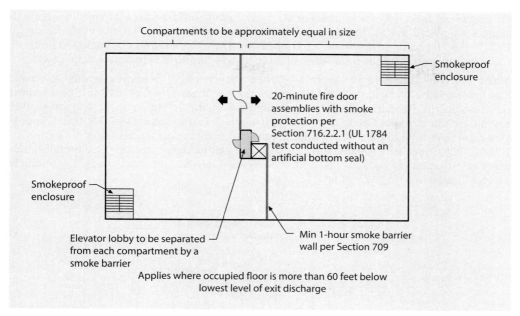

Figure 405-2 **Compartmentation of underground buildings.**

Stairways serving the floor levels of an underground building that are more than 30 feet (9,144 mm) below the discharge level are to be smokeproof enclosures, with at least two means of egress from each floor level. If multiple compartments are formed, each compartment must have at least one exit, with a second egress path available into an adjoining compartment. It is mandatory that multiple exits be provided within enclosures designed to resist the penetration of smoke.

Both standby and emergency power shall be provided to specific loads identified by this section. See the discussion of Section 403.4.

Section 406 *Motor-Vehicle-Related Occupancies*

Although uncommon, fire hazards related to motor vehicles are a concern, particularly when associated with other occupancies. The code regulates occupancies containing motor vehicles, whether they be parked, under repair, or being fueled. The hazards are primarily related to the plastic components and fuel in the vehicles, as the overall fire loading related to vehicle occupancies is typically low.

406.2 Design. There are a number of provisions that are applicable to carports and the various types of garages, regardless of their classification. To allow access for the parking of other than high-profile vehicles, the clear height of each floor level is to be at least 7 feet (2,134 mm). Note that the minimum height of the means of egress system is 7 feet, 6 inches (2,286 mm), based on the general provisions of Section 1003.2. However, Exception 7 to Section 1003.2 allows the clear height to be reduced to 7 feet (2,134 mm) in those areas of parking garages used for vehicular and pedestrian traffic.

Where the parking garage is connected directly to a room containing a fuel-fired appliance, a vestibule is mandated to separate the two spaces. This provides at least two doorways to isolate the equipment from the vehicle area. The vestibule is not necessary where the appliance ignition sources are placed at least 18 inches (457 mm) above the floor. The *International Mechanical Code*® (IMC®) and/or *International Fuel Gas Code*® (IFGC®) should also be consulted for those requirements regulating the installation of mechanical equipment within parking garages.

406.3 Private garages and carports. Where the amount of floor area devoted to vehicle parking is relatively small, the code recognizes the low hazard level by establishing a limited set of safeguards. In addition to the scope established by the definition of "private garage," the maximum floor area per Group U garage is established at 1,000 square feet. Multiple Group U garages are permitted in the same building provided such garages are separated by minimum 1-hour fire barriers, with a maximum aggregate area per building based on the allowable building area provisions of Sections 503 and 506. Figure 406-1 illustrates an example of multiple garages located in a single building.

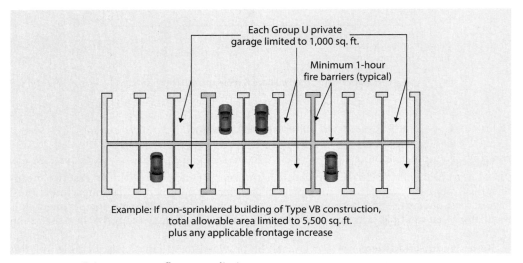

Figure 406-1 **Private garage floor area limits.**

406.4 Public parking garages. There are fundamentally two types of parking garages regulated by the IBC—private garages and public garages. Although there is no specific definition for public garage, the basis for both classifications is Section 406.3 addressing private garages and carports. Those parking structures that fall outside of the scope of Section 406.3 or the private garage definition in Chapter 2 are considered as public parking garages. The primary difference between private and public garages is typically the size of the facility, rather than the use. Strictly limited in permissible area per garage compartment, private parking garages serve only a specific tenant or building. It is important to note that there is no implication that public parking garages must be open to the public, as they are only considered public in comparison to private garages. A public parking garage is then further characterized as one of two types, either an enclosed parking garage or an open parking garage, and regulated accordingly. It is also permissible for a garage that is defined as a private garage to be regulated under the provisions for a public garage.

Guards must be provided in accordance with the general provisions of Section 1015. In addition, all parking areas more than 12 inches (305 mm) above adjacent levels shall be provided with vehicle barriers at the ends of parking spaces and drive lanes. The height of the vehicle barriers cannot be less than 2 feet, 9 inches (835 mm).

406.5 Open parking garages. Studies and tests of fires in open parking garages have shown that, in addition to limited fire loading, the potential for a large fire is uncommon. Based on this data, the IBC establishes special provisions for open parking garages in this section, which in general are less restrictive than those for enclosed parking structures addressed in Section 406.6. The key is that the open parking garage is well ventilated naturally, and as a result, the products of combustion dissipate rapidly and do not contribute to the spread of fire.

To secure the proper amount of openness, the code, as illustrated in Figure 406-2, specifies the following:

1. The building must have openings on at least two sides.
2. The openings must be uniformly distributed along each side.
3. The area of openings in the exterior walls on any given tier must be at least equal to 20 percent of the wall area of the total perimeter of each tier.
4. Unless the required openings are uniformly distributed over two *opposing* sides of the building, the aggregate length of openings considered to provide natural ventilation shall constitute a minimum of 40 percent of the wall length of the perimeter of that tier.
5. The area of openings in the interior walls must be at least 20 percent of the area of the interior walls with openings uniformly distributed.

There are situations where the required openings of open parking garages are located below the surrounding grade. Section 406.5.2.1 mandates that a clear horizontal space be provided adjacent to the garage's exterior openings that allows for adequate air movement through the opening. The dimensional requirements are based on the provisions of Section 1202.5.1.2, which addresses openings below grade when such openings are used for the required natural

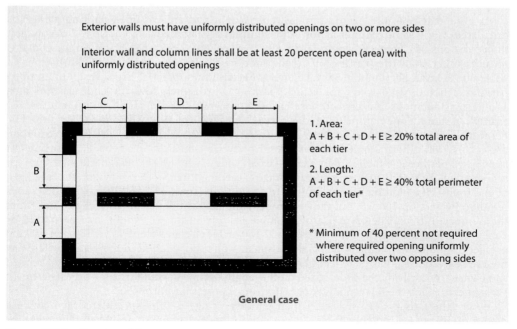

Figure 406-2 **Open parking garages.**

ventilation of a building's occupied spaces. Where openings in the exterior wall of an open parking garage are located below grade level, some degree of clear space must be provided at the exterior of the openings. As the distance of the openings below the adjoining ground increases, the minimum required exterior clear space also increases proportionately. The horizontal clear space dimension, measured perpendicular to the exterior wall opening, must be at least one and one-half times the distance between the bottom of the opening and the average adjoining ground level above. The extent of the required clear space allows for adequate exterior open space to meet the intent and dynamics of natural ventilation requirements for open parking garages. See Figure 406-3.

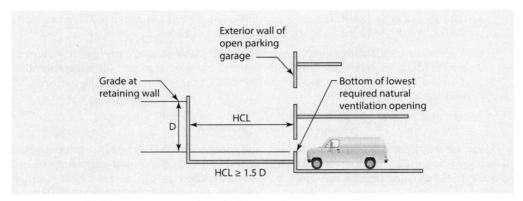

Figure 406-3   **Parking garage openings below grade.**

As a general rule, the maximum allowable height and area of open parking garages is calculated in the same manner as for other buildings. Classified as a Group S-2 occupancy, the provisions of Chapter 5 would apply. However, where the open parking garage contains no uses other than the parking of private motor vehicles, the specific size limitations of Table 406.5.4 and Section 406.5.5 take effect. Because the potential fire severity of an open parking structure used solely for vehicle parking is extremely low, the code permits area and height limitations in excess of those for other nonsprinklered Group S-2 occupancies. For example, a stand-alone nonsprinklered open parking garage of Type IIB construction would be permitted a tabular floor area of up to 26,000 square feet (2,415 m^2) per story if only based on Table 506.2, with a height limit of three stories. However, a tabular floor area of 50,000 square feet (4,645 m^2) per tier is established in Table 406.5.4, with a height limit of eight tiers for a ramp-access garage. For such an open parking garage exceeding three stories in height, the total area of the multistory building is not limited to three times that for a one-story building, as is required by Section 506.2.3, but rather can be computed as the permitted area per tier times the number of tiers. Therefore, in the example just given, the total floor area permitted by Table 406.5.4 would be 400,000 square feet (37,160 m^2) for a stand-alone Type IIB open parking garage. It should be noted that a sprinkler system would be required in the garage where any fire area exceeds 48,000 square feet (4,460 m^2). The maximum height in tiers has been limited somewhat arbitrarily by the code, based on the length of time it would take for fire-department personnel to reach the top of the structure for fire-suppression purposes.

Where the Group S-2 single-use open parking garage is fully sprinklered, it may be a design disadvantage to apply Table 406.5.4 due to the lack of an allowable area increase for sprinkler protection. Because the use of Table 406.5.4 is voluntary, it may be advantageous from a design perspective to apply Table 506.2 for a sprinklered condition rather than Table 406.5.4 when the garage is limited to a single-use condition.

The area and height increases above the tabular limits listed in Table 406.5.4 for single-use open parking garages are those outlined in Section 406.5.5, and are basically keyed to the provision of more natural ventilation area than the minimum required by the code. For unlimited-area buildings permitted by this section, see Figure 406-4.

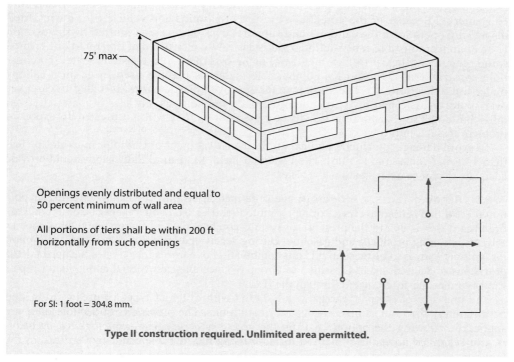

Figure 406-4 **Unlimited area open parking garages.**

In the classification of a Group S-2 parking structure as an open parking garage, the code identifies the following prohibitions:

1. There shall be no automobile repair work performed in the building.
2. There shall be no parking of buses, trucks, or similar vehicles.
3. There shall be no partial or complete closing of the required exterior wall openings by tarpaulins or by any other means.
4. There shall be no dispensing of fuel.

The intent of these limitations is to further ensure low fire loading, low possibility of fire spread, and natural cross ventilation.

406.6 Enclosed parking garages. Any vehicle parking garage that does not meet the criteria of an open parking garage or a Group U private garage is to be regulated under the general allowable height and area provisions for a Group S-2 occupancy. Tables 504.3, 504.4, and 506.2, along with any applicable height and area increases, will limit the height and floor area of an enclosed parking garage, with an allowance for use of the roof for parking purposes. A key requirement is that a mechanical ventilation system and exhaust system must be provided in accordance with the IMC.

406.7 Motor-fuel-dispensing facilities. Because most of the hazards involved with a fuel-dispensing operation are due to the storage and dispensing of flammable liquids, the majority of regulations are addressed by the IFC. The primary provisions of this section apply to canopies that are placed over the fueling areas for the purpose of customer convenience. Because of the potential exposure of gasoline and vehicle fires during fuel-dispensing operations, the canopies and supports over pumps are required by this section to be of noncombustible construction or, alternatively, constructed of fire-retardant-treated wood, complying heavy-timber members, or be of 1-hour fire-resistance-rated combustible construction. Occasionally, combustible materials may be used in or on a canopy under limited conditions. The allowance for approved plastic panels installed in canopies over motor-vehicle pumps is intended to isolate the combustible plastic materials from other buildings so that if the materials become ignited, they will not present an exposure problem to such buildings.

To avoid damage to vehicles and canopies, the clear height of canopies must not be less than 13 feet, 6 inches (4,115 mm). The 13-foot, 6-inch (4,115-mm) dimension should provide adequate clearance for recreational vehicles.

406.8 Repair garages. A repair garage is defined in Section 202 as any building or portion thereof that is used to service or repair motor vehicles. The potential exists for a moderate fire hazard that is due to the presence of various combustible and flammable liquids such as solvents, cleaning products, and gasoline. During repair operations, it is also not uncommon for ignition sources to be present. It is this combination of hazards that creates the need for the provisions of Section 406.8. Classified as Group S-1 occupancies, special concerns for repair garages are primarily regulated through the IFC.

The presence of a repair garage in a building with different types of uses is addressed no differently than other mixed-occupancy conditions. The provisions of Section 508.1 are applicable, allowing the option of using the *accessory occupancies, nonseparated occupancies*, or *separated occupancies* method for addressing the multiple occupancy groups in the building.

Section 407 *Group I-2*

In institutional occupancies, particularly those classified as Group I-2, it is important to balance the fire-safety concerns with the functional concerns of the health-care operations. This section modifies the general code provisions in an effort to achieve such a balance.

407.2 Corridors continuity and separation. Corridors are intended to provide a direct egress path adequately separated from hazards in adjoining spaces. However, in hospitals, nursing homes, and other Group I-2 occupancies, a number of necessary modifications are provided to facilitate the primary functioning of these types of health-care facilities. These modifications recognize the special needs of these occupancies to provide the most efficient and effective health-care services. See Figure 407-1.

In order to provide appropriate waiting spaces for visitors, Section 407.2.1 allows such waiting spaces to be unseparated from the corridors. One reason for this is to permit the waiting areas and similar spaces to be so located as to permit direct visual supervision by health-care facility staff. In exchange for the elimination of the corridor separation, certain

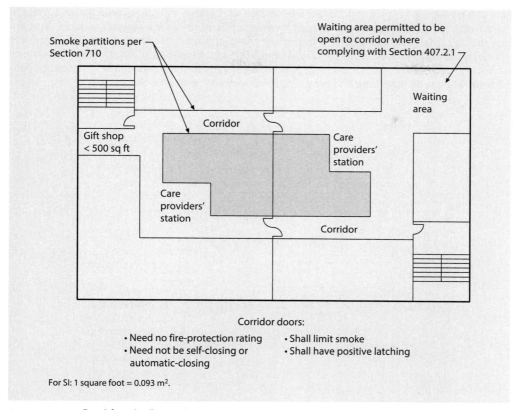

Figure 407-1 Corridors in Group I-2 occupancies.

conditions are imposed on the location of such waiting spaces. Although the scoping language only includes waiting areas and similar spaces, the primary criteria limiting those spaces that can be open to a corridor would seem to be identified in Item 1. Health-care facilities will often create alcoves adjacent to the corridor for the temporary storage of medical supplies, linen carts, food carts, etc. that are necessary to the daily functions of the facility. Without the alcoves, the corridors would be obstructed by these uses. Therefore, the code makes an allowance for such spaces. Allowances are also made for areas associated with the treatment of mental-health patients. Provided the areas are under continuous supervision by facility staff, they may be open to the corridor where six conditions are met as established in Section 407.2.3.

Similarly, Section 407.2.2 makes provisions for the location of clerical stations and other spaces necessary for doctors' and nurses' charting and communications in positions that need not be separated from the corridors. Essentially, these special-use areas are permitted to be located in the corridor. When this arrangement occurs, however, it is necessary that the construction surrounding the clerical station be as that required for corridors.

In nursing home environments, residents are encouraged to spend time outside of their rooms. By providing a variety of shared living spaces open to the circulation/means-of-egress system, socialization and interaction are encouraged. Further, being able to preview activities that are occurring helps to encourage interaction and allows reluctant participants to join at their own pace. Finally, a more open plan allows staff to more easily monitor residents throughout the day. For these reasons, it is important that the physical separation of shared resident spaces from corridors be eliminated. The conditions established in Section 407.2.5, some of which are represented in Figure 407-2, provide the means for recognizing the benefits of openness while also maintaining an appropriate level of egress protection.

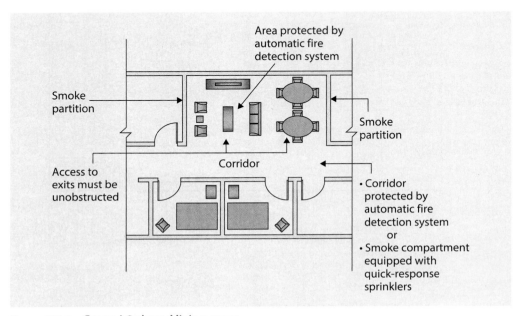

Figure 407-2 **Group I-2 shared living space.**

Another allowance for eliminating the separation between a corridor and surrounding spaces addresses areas with cooking facilities. A part of the desired group environment is also a functioning kitchen that can also serve as the hearth of the nursing home. Instead of a large, centralized, institutional kitchen where all meals are prepared and delivered to a central dining room of the resident's room, the new "household model" nursing home uses decentralized kitchens and small dining areas to create and focus the feeling of home. Allowing kitchens that serve a small, defined group of residents to be open to common spaces and corridors is viewed as critically important to enhancing the feeling and memories of home for older adults. Figure 407-3 identifies several of the six conditions set forth in Section 407.2.6 that must be met where this alternative approach is utilized. Condition 6 references Section 407.2.7 that establishes requirements specific to the installation of the domestic cooking appliances installed in the facility.

407.3 Corridor wall construction. Walls enclosing corridors and other spaces permitted by Section 407.2 to be open into corridors are intended to provide a relatively smoke-free environment during the relocation of patients during a fire emergency. Therefore, such walls must be constructed in accordance with the provisions of Section 710 as smoke partitions. The walls may extend either tight to the floor or roof deck above, or extend tight to the ceiling, provided the ceiling is also constructed to limit smoke transfer.

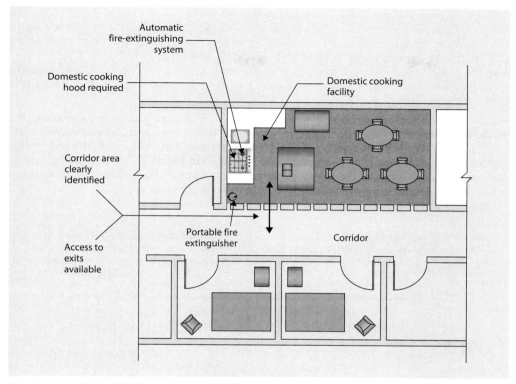

Figure 407-3 **Group I-2 domestic cooking facilities.**

Corridor doors protecting those spaces adjacent to the corridor are not required to have a fire-protection rating, nor are they required to be self-closing assemblies. They must, however, be able to limit the transfer of smoke through the opening but need not be tested for air leakage under UL 1784. One of the most controversial issues relative to the arrangement of health-care facilities such as hospitals and nursing homes is the matter of the installation of door closers on doors to patient sleeping rooms. The health-care industry has long believed it is more important to the proper delivery of health-care services that the doors to patient rooms not be self-closing and therefore constantly closed. In recognition of this special need, self-closing or automatic-closing devices are not required on corridor doors. Positive latching is required, however, and roller latches are not considered acceptable latching hardware. Where positive latching is not desired, typically where sliding doors are installed at patient or treatment rooms, the common corridor arrangement cannot be utilized. In such instances, the spaces could be designed as care suites under the provisions of Section 407.4.4. Corridor-type configurations within such suites are not subject to the requirements of Section 407.3.

Locking devices may be arranged so that they are readily operable from the patient-room side and are readily operable by the facility staff from the opposite side. This special arrangement permits keys or other limited access methods to be utilized for the care recipient rooms. However, egress from the care recipient rooms shall be unrestricted unless such rooms are in mental-health facilities or regulated as controlled egress doors per Section 1010.2.14.

407.4.4 Group I-2 care suites. Special means of egress provisions are provided for care suites in Group I-2 occupancies. The definition of "care suite" in Section 202 identifies the scope of such special provisions. The concept of suites recognizes those arrangements where staff must have more supervision of patients in specific treatment and sleeping rooms. Therefore, the general means of egress requirements are not appropriate under such conditions. The special allowances for suites are not intended to apply to day rooms or business functions of the health-care facility.

407.5 Smoke barriers. Evacuation of a building such as a hospital or nursing home is a virtual impossibility in the event of a fire, particularly in multistory structures. Horizontal evacuation, on the other hand, is possible with a properly trained staff. As a result, the code makes provisions for horizontal compartmentation as illustrated in Figure 407-4, so that if necessary, care recipients can be moved from one compartment to another. This intent is secured by this

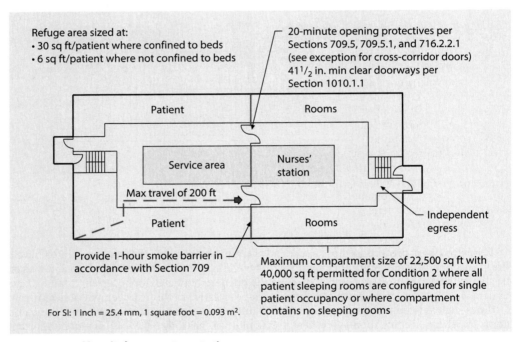

Figure 407-4 **Hospital compartmentation.**

section wherein, under most conditions, each story of a Group I-2 occupancy is required to be divided into at least two smoke compartments by a smoke barrier constructed in accordance with Section 709. Limited by floor area and travel distance, each compartment shall be sized to permit the housing of patients from adjoining smoke compartments. It is expected that in multistory buildings, the floor construction also provides for smoke compartmentation vertically. As such, the concept of smoke resistance must be considered relative to vertical openings and penetrations, including interior exit stairways and shaft enclosures.

407.10 Secured yards. It is not uncommon that a secured exterior area or yard be provided for Group I-2 occupancies, particularly where the facility specializes in the treatment of mental disabilities such as Alzheimer's disease. Where such fencing and locked gates prohibit the continuation of the exit discharge to the public way, the use of safe dispersal areas is acceptable. To adequately provide for temporary refuge, the safe dispersal area must be sized to accommodate the occupant load of the egress system it serves. In all cases, the entire dispersal area must be located at least 50 feet (15,240 mm) from the building.

Section 408 Group I-3

The concerns for both security and fire safety must be balanced when it comes to Group I-3 detention facilities. Special consideration must be given to the secured areas without sacrificing an unreasonable degree of fire and life safety for the occupants. This section addresses the unique conditions that occur in these types of buildings.

Section 408.3 modifies the general requirements for the means of egress found in Chapter 10. A major difference is the allowance for glazing in the doors and walls of interior exit stairways, provided a number of conditions are met. As would be expected, the most dramatic variation from the general requirements has to do with the locking hardware. The requirements vary based on the nature of the detention occupancy. Reference must occasionally be made to the occupancy conditions of Section 308.4 to determine the appropriate egress criteria.

Similar to the provisions of Section 407 for Group I-2 occupancies, smoke compartments must be created where the occupant load per story is 50 or more. Additionally, regardless of occupant load, floor levels utilized as sleeping areas must be divided into a minimum of two compartments. More than two smoke compartments may be necessary on any floor level where the dictated travel distances cannot be provided or where the occupant load of the compartment is excessive. No more than 200 occupants can be assigned to a single compartment. The refuge area must be sized to accommodate the total number of residents that may be contained within the compartment. Independent egress is needed from each compartment so that it is not necessary to travel back into the compartment where travel originated.

An important feature of the Group I-3 provisions is the allowance for multiple floor levels of residential housing to be open to each other without an enclosure. Through the safeguards provided, it is possible to provide increased security by opening up the multiple housing areas to a single common area where visual supervision is more easily accomplished. It is important that independent egress to an exit be provided from each level. The limit of 23 feet (7,010 mm) between the lowest and highest floor levels, as well as the required egress directly out of each story, provides additional qualifications that must be met in order to eliminate the required vertical enclosure protection.

As an additional allowance for security purposes, the fire-protection rating is not required for security glazing installed in 1-hour fire barriers, fire partitions, and smoke barriers that may be present. Rather, equivalent protection is provided through compliance with four specific conditions addressing the glazing and its frame. The use of security glazing is necessary in such facilities to track and contain inmate movement for the protection of other inmates and administrative personnel. Three of the most common types of fire separations are addressed: fire barriers, fire partitions, and smoke barriers. The allowance is not applicable to fire walls, nor is it permissible where the fire separation wall has a required fire-resistance rating of more than 1 hour. The conditions imposed on the security glazing limit the area of each individual glazed panel, mandate sprinkler protection that will wet the entire glazing surface on both sides, regulate the gasketed frame for deflection, and prohibit the installation of obstructions between the sprinklers and the glazing.

Section 409 *Motion-Picture Projection Rooms*

Prior to the 1970s, building codes addressed the subject of motion picture projection rooms based on the hazard of the cellulose nitrate film being used at that time. Actually, production of cellulose nitrate film ceased around 1950, although its use continued thereafter. In fact, even today, some cellulose nitrate film is used at film festivals and special occasions requiring the projection of historically significant films that are still imprinted on cellulose nitrate film. Where this type of film continues to be utilized or stored, it will be regulated under the provisions of NFPA 40. Although the provisions in the codes since 1970 are based on the use of safety film, some of the protection requirements for cellulose nitrate film have been retained in the present requirements, such as ventilation requirements for the projection room.

The intent of the current provisions regulating motion picture projection rooms is to provide safety to the occupants of a theater from the hazards consequent on the light source where electric arc, xenon, or other light-source projection equipment is used. Although not used to any extent today, electric-arc projection lamps emit hazardous radiation. Xenon lamps, which have been highly prevalent as projection lamps, emit ozone. As a result, the provisions of Section 409 are based on the lamps used for projection of the film rather than the type of film to be used, as long as the film is not nitrate based.

The provisions intend to isolate the projection room so that it does not present a danger to the theater audience. As the room is designed for the projection of safety film, there is no intent to provide a special fire-resistive enclosure, and fire protection of openings between the projection room and the auditorium is not required. However, due to the projection lamps, it is the intent of the code to provide an emission-tight separation so that any opening should be sealed with glass or other approved material such that emissions from the projection lamps will not contaminate the auditorium.

Section 410 *Stages, Platforms, and Technical Production Areas*

The provisions in Section 410 are continuously reviewed in an attempt to bring the code requirements in line with the present methods and technologies regarding the use of stages and platforms, as well as related accessory and support areas. Although the basic provisions for life safety have remained essentially unchanged over the years, occasional modifications have been made that are due to the need to accommodate state-of-the-art performances.

It is critical that the definitions found in Chapter 2 related to Section 410 are reviewed in order to properly apply the provisions. There are terms unique to the performing arts that are not generally understood, such as fly gallery, gridiron, and pinrail, which fall under the general term "technical production area." The distinctions between the definitions of a stage and a platform are also very important because of the specific requirements for each. The primary difference between a stage and a platform is the presence of overhead hanging curtains, drops, scenery, and other stage effects. The amount of combustible materials associated with a stage is typically considerably greater than that for a platform. Thus, the fire-severity potential is much higher.

410.2 Stages. An assembly occupancy considered among the most hazardous is a Group A-1 containing a large concentrated occupant load in very close proximity to a performance stage.

The hazard created by the stage is the presence of combustibles in the form of hanging curtains, drops, leg drops, scenery, etc., which in the past have been the source of ignition for disastrous fires in theaters. Modern stages also have an increased hazard from special effects such as pyrotechnics, utilized in so-called *spectaculars*.

Where the stage height exceeds 50 feet (15,240 mm), the fire hazard is even greater because the fly area that is usually above the stage is a large blind space containing combustible materials that have a fuel load considerably greater than that normally associated with an assembly occupancy. Many of the construction requirements for stages are depicted in Figure 410-1.

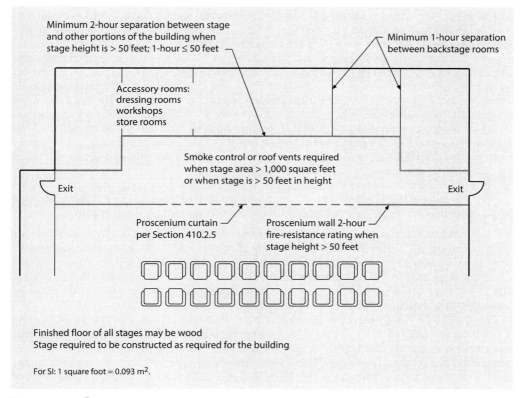

Figure 410-1 **Stages.**

410.2.1 Stage construction. In addition to the features shown in Figure 410-1, any stage may have a finished floor of wood, provided construction of the stage floor or deck is in compliance with this section. As the area above and at the sides of stages can be filled with combustible materials that can be moved both vertically and horizontally, such as curtains, drops, leg drops, scenery, and other stage effects, the code requires that such stages be constructed of the same materials as required for floors for the type of construction of the building and separated from the balance of the building.

410.2.4 Proscenium wall. Where the stage height exceeds 50 feet (15,240 mm), measured from the lowest point on the stage floor to the highest point of the underside of the roof or floor deck above, a proscenium wall must be provided. The proscenium wall is intended by the IBC to provide a complete fire separation between the stage and the auditorium. Extending from the foundation continuously to the roof, the wall is to have a minimum fire-resistance rating of 2 hours. The extent of the separation in a Type I building can be reduced to where it is only continuous between a minimum 2-hour floor slab and a minimum 2-hour floor or roof assembly.

410.2.5 Proscenium curtain. Because the opening in the proscenium wall described in Section 410.2.4 is too large to protect with any usual type of fire assembly, the code requires that it be protected with a fire-resistive fire curtain or water curtain. Where a fire curtain is installed, it must comply with the provisions for fire-safety curtains set forth in NFPA 80 *Fire Doors and Other Opening Protectives*. A fire curtain or water curtain is not required where a complying smoke-control system or natural ventilation is provided. Horizontal sliding doors may also be used to provide the required separation provided they have a minimum 1-hour fire protection rating.

The purpose of the proscenium curtain protection is to provide occupants with additional time to exit the assembly seating area if there is a fire in the stage area. With the benefits afforded by an engineered smoke-control system or natural ventilation, the occupants should be equally or better protected from the hazards of fire than with a proscenium curtain or water curtain. By providing a performance-based alternative to a proscenium curtain, more design options are available where the use of fire-safety curtains is considered impractical or causes obstructions of the production. It is important to note that the elimination of the proscenium curtain is not permitted if the smoke-protected assembly seating provisions of Section 1030.6.2 are being utilized, for example, a decrease in the required egress widths of the assembly seating area.

The requirement for a complying fire curtain is triggered solely by the proscenium wall provisions of Section 410.2.4. Where a proscenium wall is fire-resistance rated for solely a different purpose, such as a bearing wall in a Type IB building, the fire curtain is not required.

410.2.7 Stage ventilation. The Iroquois Theater fire in 1903 was directly responsible for the requirement for automatic vents in the roofs of theater stages. Because of the presence of large amounts of combustible materials, excessive quantities of smoke will accumulate in and above the stage area unless it is automatically vented or removed by a smoke-control system. The removal of smoke is necessary for fire fighting as well as the prevention of panic by drawing off the smoke so that it will not infiltrate the theater auditorium.

The maximum floor area of stages that is permitted without the installation of venting is 1,000 square feet (93 m^2). The stage area to be considered includes the performance area and adjacent backstage and support areas not separated from the performance area by fire-resistance-rated construction. In addition, stages must be equipped with smoke-removal equipment or roof vents where they are greater than 50 feet (15,240 mm) in height. If either of these two conditions exist, stage ventilation is required. The detailed requirements for smoke vents in the IBC are intended to provide reliability and a reasonable assurance that after many years of operation the vents will operate when needed.

410.3 Platform construction. Materials used in the construction of permanent platforms must be consistent with those materials permitted based on the building's type of construction. Therefore, in noncombustible buildings, the platforms must be of noncombustible construction. However, in buildings of Type I, II, and IV construction, the use of fire-retardant-treated wood is permitted where all of the following conditions are met:

1. The platform is limited in height to 30 inches (762 mm) above the floor.
2. The floor area of the platform does not exceed one-third of the floor area of the room in which it is located.
3. The platform does not exceed 3,000 square feet (279 m^2) in floor area.

In those situations where the concealed area below the platform is to be used for storage or any purpose other than equipment, wiring, or plumbing, the floor construction of the platform is to be fire-resistance rated for a minimum of 1 hour. Otherwise, no protection of the platform floor is necessary.

As it is often impractical to construct temporary platforms of fire-resistive materials, the code permits temporary platforms to be constructed of any materials, but restricts the use below the platform to that of electrical wiring or plumbing connected to platform equipment. Therefore, no storage of any kind is permitted beneath temporary platforms, primarily because of the potential for a fire to start and spread undetected.

410.4 Dressing and appurtenant rooms. Not only must a stage exceeding 50 feet (15,240 mm) in height be separated from the adjoining seating area by a minimum 2-hour fire-resistance-rated proscenium wall, but such a separation is also required between the stage and all other portions of the building, including all related backstage areas. Dressing rooms, property rooms, workshops, storage rooms, and all other areas must be separated from the stage with minimum 2-hour fire-resistance-rated fire barriers and/or horizontal assemblies, and all openings must be appropriately protected. A minimum 1-hour fire-resistance-rated separation is required where the stage height does not exceed 50 feet (15,240 mm).

In addition to their required fire separation from the stage, dressing rooms and all other related backstage areas must be separated from each other. One-hour fire-resistance-rated fire barriers and/or horizontal assemblies, along with opening protectives, satisfy the minimum requirements. The hazards caused by the significant fire loading that occurs in conjunction with stages are greatly reduced through the use of compartments.

410.6 Automatic sprinkler system. One of the special areas mentioned in Table 903.2.11.6 that requires a suppression system is stages. The general requirement mandates the sprinklering of not only the stage area but also all support and backstage areas serving the stage. An automatic sprinkler system is an effective tool in limiting the exposure of a fire to the area of origin. Sprinklers are not required for a stage having both a small floor area and a low roof height. Under such conditions, the amount of combustibles in the stage area is typically very limited.

Section 411 *Special Amusement Areas*

The provisions of Section 411 are only applicable to those areas and buildings that are specifically identified in Section 202 as *special amusement areas*. Typically walk-through and ride-through attractions at amusement parks, haunted houses, and escape rooms, such amusement areas are usually classified as Group A occupancies but should be classified as Group B where the occupant load is less than 50. The major factors contributing to the loss of life in fires in buildings with special amusement areas have been the failure to detect and extinguish the fire in its incipient stage, the ignition of synthetic foam materials and subsequent fire and smoke spread, and the difficulty of escape. Provisions for the detection of fires, the illumination of the exit path, and the sprinklering of the structures are required to protect the occupants in such structures. However, special amusement areas without walls or a roof are not required to comply with this section, provided they are designed to prevent smoke from accumulating in such areas. Approved smoke-detection and alarm systems are also required in buildings with special amusement areas. A provision of Section 411.4 is that on the activation of the system as described, an approved directional exit-marking system shall activate in those areas where the configuration of the space is such as to disguise the path and make the egress route not readily apparent.

Exceptions to Section 411.1 identify two conditions where the special provisions for special amusement areas do not apply. In all cases, a special amusement area designed and constructed to prevent the accumulation of smoke within all occupiable spaces would not be regulated by Section 411. The lack of a roof or walls that could contain smoke provides for an environment where the multiple safeguards established in this section are unnecessary. Additionally, and only applicable to those special amusement areas defined as puzzle rooms, compliance with Section 411 is not required where the means of egress system meets some basic performance criteria. These basic egress requirements recognize the limited occupant loads anticipated in a puzzle room condition.

Section 412 *Aircraft-Related Occupancies*

Because of the unique nature of occupancies related to aircraft manufacture, repair, storage, and even flight control, provisions have been developed to address the special conditions that may exist. Although the various uses fall into different occupancy classifications, they all have one thing in common—they are related to aircraft. Additional requirements related directly to aviation facilities are found in Chapter 20 of the IFC.

412.2 Airport-traffic control towers. These provisions are intended to reconcile the differences between the life-safety needs of air-traffic control towers and the life-safety requirements in the body of the code. In developing these provisions, consideration was given to the inherent qualities of the use, which makes the general requirements of the IBC inappropriate. For example, air-traffic control personnel are required to undergo medical examinations to ensure they are of sound body and mind. Recognition was also given to the life-safety record of these uses and specific limitations, which are imposed on the allowable size, type of construction, etc. The provisions also require automatic smoke-detection systems.

412.3 Aircraft hangars. Aircraft hangars are intended to be classified as Group S-1 occupancies. All aircraft hangars are to be located at least 30 feet (9,144 mm) from any public way or lot line, providing adequate spatial separation for neighboring areas. It should be noted that the measurement is taken to the nearest point of any public way, not to the centerline as typically applied. Otherwise, their exterior walls must have a minimum 2-hour fire-resistance rating. The criteria of Section 412.3.1 are not applicable in the regulation of two or more hangars on the same lot as there is no reference to imaginary lines placed between hangars. Where such conditions occur, the general provisions of Section 705.3 and Table 705.5 are to be applied. Because of the concerns about below-grade spaces under any facility where flammable and combustible liquids are commonly present, the code requires the hangar floor over a basement to be liquid and air tight with absolutely no openings. Floor surfaces must also be sloped to allow for drainage of any liquid spills.

412.3.6 Fire suppression. In order to minimize the fire hazards associated with aircraft hangars, fire suppression is required based on the criteria of Table 412.3.6. The table determines the hangar classification (Group I, II, or III) to which the fire suppression must be designed in accordance with NFPA 409, *Aircraft Hangars*. The classification is based on the hangar's type of construction and fire area size. Fire area size is based on the aggregate floor area bounded by minimum 2-hour fire-resistance-rated fire walls. For the purposes of hangar classification, ancillary uses located within the fire area are not required to be included in the fire area size provided they are separated from the aircraft serving area by minimum 1-hour fire barriers. See Figure 412-1.

412.5 Aircraft paint hangars. The hazards involved with the application of flammable paint or other liquids cause aircraft painting operations to be highly regulated. Where the

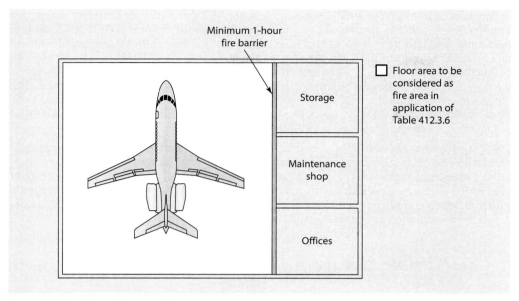

Figure 412-1   **Aircraft hanger fire area.**

quantities of flammable liquids exceed the exempt quantities listed in Table 307.1(1), such hangars are classified as Group H-2 occupancies. They must be built of noncombustible construction, provided with fire suppression per NFPA 409, and ventilated in the manner prescribed by the IMC. Where the amount of flammable liquids within the hangar does not exceed the maximum allowable quantities set forth in Table 307.1(1), the classification is most appropriately a Group S-1 occupancy, and the provisions of this section do not apply.

412.6 Aircraft manufacturing facilities. The traditional travel distance limitations have always been considered somewhat problematic for expansive aircraft manufacturing facilities without incorporating exit passageways or horizontal exits into the building's means of egress system. Due to the compartmentalized nature of horizontal exits, they do not lend themselves to aircraft production processes or movement of the completely assembled aircraft. For similar reasons, exit passageways are generally installed below the floor of the manufacturing level. The use of underground passageways during a fire event or other emergency in such a large, high-volume space is widely viewed as generally contrary to human nature. Once aware of an event, employees typically evacuate the building instinctively at the level with which they are most familiar. It is also relatively common for occupants to want to move away from the point of origin of a fire due to a person's sensory awareness within the entire open space. Given the fact that occupants sense safety as they move away from a fire incident, it is counterintuitive to enter an underground area unless as a final resort.

In spite of these observations, it is important that it can be demonstrated that such large-volume spaces are able to provide a tenable environment for the evacuation or relocation of building occupants. The increased travel allowances established in Table 412.6 are supported by smoke and temperature fire modeling conducted using the National Institute of Standards and Technology Fire Dynamics Simulator computer program. Results of the fire modeling activity, based on conservative assumptions, were used to establish the maximum travel distances provided.

The travel distance allowances for aircraft manufacturing facilities are therefore based on a combination of building features: the minimum height from the finished floor to the bottom of the ceiling, roof slab, or roof deck above; and the contiguous floor area of the aircraft manufacturing facility having the indicated height.

Ancillary spaces within or adjacent to the manufacturing area are permitted to egress through the manufacturing area having a minimum height as established by Table 412.6. The portion of travel within the ancillary spaces is limited to the general travel distances set forth in Table 1017.2 based upon the occupancy classification of the ancillary space. The overall travel distance cannot exceed the distance indicated in Table 412.6.

412.7 Heliports and helistops. Helistops are differentiated from heliports by the presence of refueling facilities, maintenance operations, and repair and storage of the helicopters; thus, helistops pose similar hazards to those posed by aircraft repair hangars. The minimum size of a helicopter landing area is addressed, as are requirements for construction features and egress. Where heliports and helistops are constructed in compliance with the provisions of this section, they may be erected on buildings regulated by this code.

Section 413 *Combustible Storage*

Any occupancy group containing high-piled stock or rack storage is subject to the provisions of the IFC as well as the IBC. Chapter 32 of the IFC regulates combustible storage based on a variety of conditions, including the type of commodities stored, as well as the height and method of storage and the size of the storage area.

This section also specifically addresses any concealed spaces within buildings, including attics and under-floor spaces, that are used for the storage of combustible material. Where combustible storage occurs in areas typically considered unoccupiable, the storage areas are to be separated from the remainder of the building by 1-hour fire-resistance-rated construction on the storage side. The protective membrane need only be applied on the storage side insofar as the location of the hazard has been identified as the storage area only. Openings are to be protected with self-closing door assemblies that are either of noncombustible construction or are a minimum 1¾-inch (45-mm) solid wood. This separation is not necessary in Group R-3 and U occupancies. In addition, those combustible storage areas protected with sprinkler systems need not be separated. The provisions are not intended to apply to those storage rooms that are constructed and regulated as usable spaces within the building, such as closets and portions of basements where combustible products or materials are stored.

Section 414 *Hazardous Materials*

Although the *International Fire Code* (IFC) sets forth most of the requirements applicable to hazardous materials, including their use, storage, manufacturing, processing, and dispensing within buildings, there are a significant number of hazardous material provisions in the IBC. Section 414 establishes a variety of requirements that are applicable to the presence of hazardous materials in buildings regardless of occupancy classification. Conversely, Section 415 only contains provisions that apply to those uses classified as Group H. It is important to note, however, that the exemptions established for hazardous materials in Table 307.1.1 are not subject to the requirements of Section 414. In addition, Table 307.1.1 also exempts the materials listed from classification as a Group H occupancy when in compliance with the specific conditions established in the table. For further guidance, see the discussion of Section 307.1.1.

Hazardous Materials 121

Figure 414-1 outlines the process for determining the code requirements that are a function of the quantities of hazardous materials stored or used. The outline is useful for both design and review. To begin, one must determine the hazardous processes and materials involved in a given occupancy and gain a thorough understanding of the operations taking place. Once the hazardous processes and materials have been identified, it is necessary to classify the materials based on the categories used by the code.

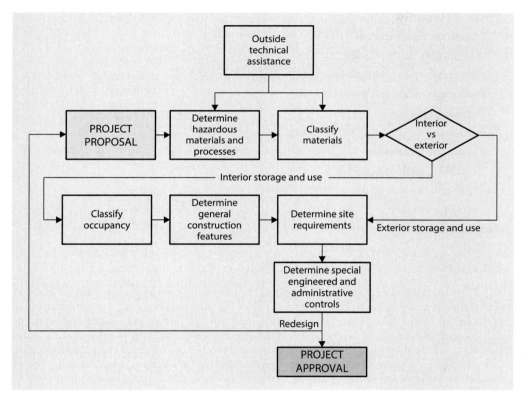

Figure 414-1 **Code approach to hazardous materials.**

Section 414.1.3 provides the means for the building official to acquire outside technical assistance to assist in the review of a project. Such assistance is often critical in assuring that appropriate decisions are made. Classifying materials is a subjective science, requiring judgment decisions by an expert familiar with the characteristics of a particular material to categorize it within the categories used by the IBC and IFC. Accordingly, material classifications must be determined by qualified individuals, such as industrial hygienists, chemists, or fire-protection engineers. Though some jurisdictions employ individuals qualified to make these determinations, most jurisdictions rely on outside experts acceptable to the jurisdiction to submit a report detailing classifications compatible with the system used by the code.

Often, a permit applicant will attempt to submit a cadre of Material Safety Data Sheets (MSDSs) as a means of identifying material classifications. Though these may contain the information necessary to determine the proper classification, they do not normally contain a complete designation of classifications that is compatible with the system used by the IBC.

Therefore, MSDSs are not normally acceptable as a sole means of providing material classifications to a jurisdiction. The building official should understand that it is not the responsibility of the jurisdiction to provide classifications for hazardous materials. Rather, it is the responsibility of the permit applicant to provide material classification information.

In the classification system used by the *International Codes*®, hazardous materials are generally divided into two major categories, physical and health hazards, and 12 subcategories, as follows:

Physical Hazards

    Explosives and fireworks

    Combustible dusts and fibers

    Flammable and combustible liquids

    Flammable solids and gases

    Organic peroxides

    Oxidizers

    Pyrophoric materials

    Unstable (reactive) materials

    Water-reactive materials

    Cryogenic liquids

Health Hazards

    Highly toxic and toxic materials

    Corrosives

414.1.3 Information required. A report is required to allow the building department to evaluate the presence of hazardous materials within the proposed building based on the criteria established by the IBC. Since Tables 307.1(1) and 307.1(2) are critical in the evaluation of buildings containing hazardous materials, information is needed in order to properly utilize the tables. Such information must include the maximum expected quantities of each material in use and/or storage conditions, those fire-protection features that are to be in place, and any use of control areas for isolation of the materials. The submission of a technical report is necessary to allow the jurisdiction to perform a code compliance evaluation. The requirement for a technical report gives jurisdictions the benefit of expert opinions provided by knowledgeable persons in the particular hazard field of concern. Technical reports are required to be prepared by an individual, firm, or corporation acceptable to the jurisdiction, and must be provided without charge to the jurisdiction. Where the quantities of hazardous materials are such that a Group H occupancy is warranted, floor plans must be submitted to the building official identifying the locations of hazardous contents and processes.

414.2 Control areas. As addressed previously in the discussion of Section 307, areas in a building that are designated to contain the maximum allowable quantities of hazardous materials in use, storage, dispensing, or handling are considered control areas. At a minimum, 1-hour fire barriers shall be used to separate control areas from each other. Where required by Table 414.2.2 for the fourth story above grade plane and all stories above, a minimum 2-hour fire-resistance rating is required for such fire barriers. Openings in fire barriers are to be protected in accordance with Section 716. As a general rule, all floor construction that forms the boundaries of control areas is to have a minimum 2-hour fire-resistance rating. Building elements structurally

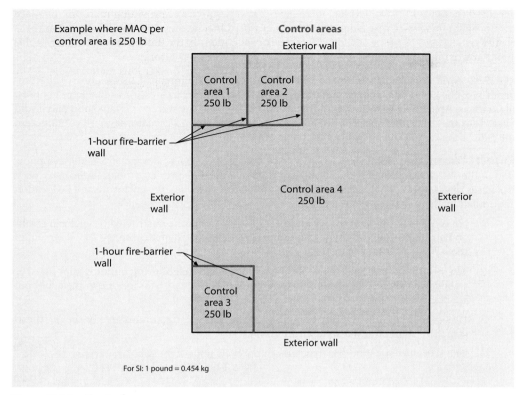

Figure 414-2   **Control areas.**

supporting the 2-hour floor construction shall have an equivalent fire-resistance rating. There is an allowance for those two-story and three-story sprinklered buildings that are primarily of 1-hour fire-resistive construction (Types IIA, IIIA, IV, and VA), which permits 1-hour floor construction of the control area and the supporting construction. It is apparent that a considerable level of fire separation must be achieved in order to increase the quantity of hazardous materials in non–Group H buildings. An example of this provision is illustrated in Figure 414-2.

In addition to the fire-resistive separations enclosing control areas, the quantities of hazardous materials and the number of control areas are limited on a story-by-story basis. The number of control areas, per story, within a structure may be increased through the use of fire walls designed and constructed in accordance with Section 706. For example, in a one-story structure where Table 414.2.2 limits the number of control areas in the building to four, a single fire wall would permit a total of eight control areas (four on each side of the fire wall).

The control area concept is extended to the regulation of laboratories in colleges and universities through the special provisions of Section 428 rather than those found in Section 414.2. Identified as "laboratory suites" rather than control areas, these compartments housing the use and/or storage of hazardous materials in a higher education setting are regulated a bit differently as found in the discussion of Section 428.

414.5.1   Explosion control.   Table 414.5.1 indicates, based on material, the explosion control methods that must be provided where hazardous materials exceed the allowable quantities specified in Table 307.1(1). Explosion control is also required in any structure, room, or space occupied for purposes involving explosion hazards. Once some type of explosion control is required, Section 911 of the IFC must be referenced to identify the details for controlling explosion hazards.

414.5.3 Spill control, drainage, and containment. The intent of this section is the prevention of the accidental spread of hazardous material releases to locations outside of containment areas. Applicable to rooms, buildings, or areas used for the storage of both solid and liquid hazardous materials, the specifics for spill control, drainage, and containment are contained in the IFC.

414.6.1 Weather protection. In order to be considered outside storage or use in the application of the IFC, hazardous material storage or use areas must be primarily open to the exterior. If it is necessary to shelter such areas for weather protection purposes, the enclosure and its location are limited by the following requirements:

1. No more than one side of the perimeter of the area may be obstructed by enclosing walls and structural supports unless the total obstructed perimeter is limited to 25 percent of the structure's total perimeter.
2. The minimum clearance between the structure and neighboring buildings, lot lines, or public ways shall be equivalent to that required for outside storage or use areas without weather protection.
3. Unless increased by the provisions of Section 506, the maximum area of the overhead structure shall be 1,500 square feet (140 m^2).
4. The structure must be constructed of approved noncombustible materials.

Section 415 *Groups H-1, H-2, H-3, H-4, and H-5*

The provisions of this section apply to those buildings and structures where hazardous materials are stored or used in amounts exceeding the maximum allowable quantities identified in Section 307. Applied in concert with the IFC, the requirements address the concerns presented by the high level of hazard as compared to other uses. For a further discussion, see the commentary on Section 414.

415.6 Fire separation distance. This section provides regulations that limit the locations on a lot for Group H occupancies and establish minimum percentages of perimeter walls of Group H occupancies required to be located on the building exterior. Based on the specific Group H occupancy involved, the building must be set back a minimum distance from lot lines, as shown in Figure 415-1. As illustrated in Figure 415-2, the distance is measured from the walls enclosing the high-hazard occupancy to the lot lines, including those on a public way. An exception to this method of measurement occurs where multiple buildings are on the same site and an assumed imaginary line is placed between them under the provisions of Section 705.3. In such a situation, the assumed line is to be ignored in the application of this section for those buildings that manufacture or use explosives and the IFC is utilized to provide adequate spatial separation. Only in such limited cases are assumed lot lines not to be applied. The specific provisions in this section also require that Group H-2 and H-3 occupancies included in mixed-use buildings have 25 percent of the perimeter wall of the Group H occupancy on the exterior of the building. The access capability for fire personnel is greatly enhanced where the hazardous conditions are located in such a manner that allows for exterior fire-fighting operations.

Groups H-1, H-2, H-3, H-4, and H-5

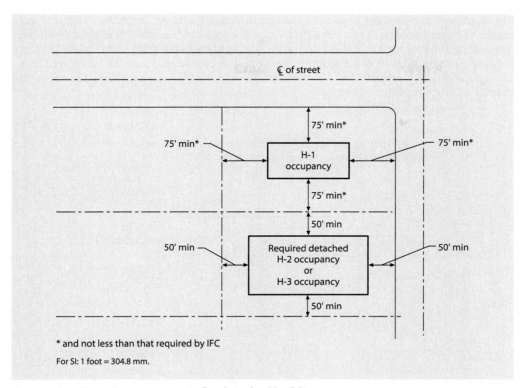

Figure 415-1 **Location on property for detached buildings.**

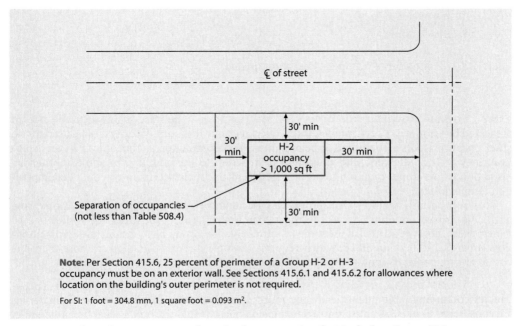

Figure 415-2 **Location on property for mixed occupancies that include a Group H-2 occupancy.**

Allowances are provided for smaller, liquid use, dispensing, and mixing rooms; liquid storage rooms; and spray booths. See Figure 415-3. It should be noted that where a detached building, required by Table 415.6.5, is located on the lot in accordance with this section, wall construction and opening protection are not regulated based on the location on the lot. A minimum fire separation distance of 50 feet (15,240 mm) is required for such buildings. Therefore, the exterior wall and opening requirements of Table 602 have no application.

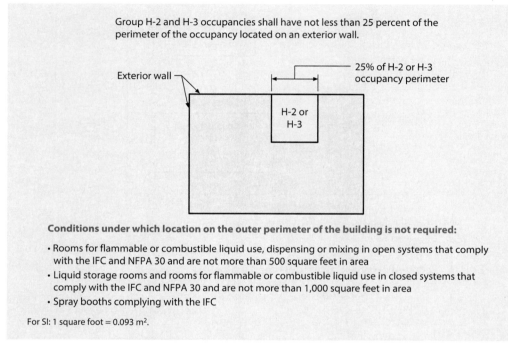

Figure 415-3 **Perimeter of occupancy on exterior wall.**

415.7 Special provisions for Group H-1 occupancies. Because of the extreme hazard presented by Group H-1 occupancies, this section requires that such occupancies be used for no other purpose. Roofs are required to be of lightweight construction so that, in case of an explosion, they will rapidly vent with minimum destruction to the building. In addition, thermal insulation is sometimes required to prevent heat-sensitive materials from reaching decomposition temperatures.

This section also requires that Group H-1 occupancies that contain materials possessing health hazards in amounts exceeding the maximum allowable quantities for health-hazard materials in Table 307.1(2) also meet the requirements for Group H-4 occupancies. This provision is parallel to Section 307.8, which requires multiple hazards classified in more than one Group H occupancy to conform to the code for each of the occupancies classified.

415.8 Special provisions for Group H-2 and H-3 occupancies. Group H-2 and H-3 occupancies containing large quantities of the more dangerous types of physical hazard materials are considered to present unusual fire or explosion hazards that warrant a separate and distinct occupancy in a detached building used for no other purpose, similar to the requirements for a Group H-1 occupancy. The threshold quantities for requiring detached Group H-2 and H-3 occupancies are set forth in Table 415.6.5.

This section also requires water-reactive materials to be protected from water penetration or liquid leakage. Fire-protection piping is allowed in such areas in recognition of both the integrity of fire-protection system installations and the need to protect water-reactive materials from exposure fires.

415.9   Group H-2.   Both this section and the IFC are to be used in the regulation of buildings containing the following hazardous materials operations:

1. Flammable and combustible liquids
2. Liquefied petroleum gas distribution facilities
3. Dry cleaning plants

The hazards presented in these operations, through the storage, use, handling, processing, or transporting of hazardous materials, are unique enough to require special provisions, both in this section and in the IFC. The regulations of buildings in which materials that produce combustible dusts are stored or handled are further established in Section 426.

415.10   Groups H-3 and H-4.   This section identifies several specific issues in Group H-3 and H-4 occupancies. Group H gas rooms shall be isolated from other areas of the building by minimum 1-hour fire barriers and/or horizontal assemblies. Highly toxic solids and liquids must also be separated from other hazardous material storage by fire barriers and/or horizontal assemblies having a minimum 1-hour fire-resistance rating, unless the highly toxic materials are stored in approved hazardous material storage cabinets. A related provision requires liquid-tight noncombustible floor construction in areas used for the storage of corrosive liquids and highly toxic or toxic materials.

415.11   Group H-5.   The Group H-5 occupancy category was created to standardize regulations for semiconductor manufacturing facilities. This section provides the specific regulations for these occupancies. The Group H-5 category requires engineering and fire-safety controls that reduce the overall hazard of the occupancy to a level thought to be equivalent to a moderate-hazard Group B occupancy. Accordingly, the allowable floor areas established for Group H-5 occupancies are the same as those for Group B occupancies.

The code requires that special ventilation systems be installed in fabrication areas that will prevent explosive fuel-to-air mixtures from developing. The ventilation system must be connected to an emergency power system. Furthermore, buildings containing Group H-5 occupancies are required to be protected throughout by an automatic fire-sprinkler system and fire and emergency alarm systems. Fire and emergency alarm systems are intended to be separate and distinct systems, with the emergency alarm system providing a signal for emergencies other than fire. This section also provides requirements for piping and tubing that transport hazardous materials that allow piping to be located in exit corridors and above other occupancies subject to numerous, stringent protection criteria. The provisions for Group H-5 occupancies are correlated with companion provisions in Chapter 27 of the IFC.

Table 415.6.5—Detached Building Required.   This table establishes the threshold quantities of hazardous materials requiring detached buildings. Once the quantities listed in the table are exceeded, a detached building is required. The limitations placed on detached buildings required by Table 415.6.5 and classified as Group H-2 or H-3 are essentially the same as those applied to H-1 occupancies. The applicability of the table is based solely on the total quantity of material present. The material can be in use, in storage, or both. See Application Example 415-1.

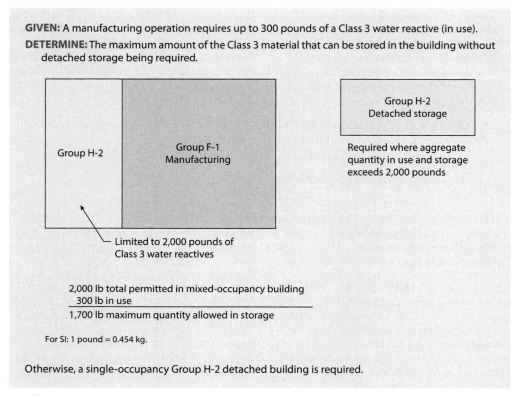

Application Example 415-1

Section 416 *Spray Application of Flammable Finishes*

This section applies to those buildings used for the spraying of flammable finishes such as paints, varnishes, and lacquers. In addition, Chapter 24 of the IFC contains extensive requirements for these types of operations. The IFC addresses a variety of spraying arrangements, each of which is specifically defined and regulated. These include spray rooms, spray booths, spraying spaces, and limited spraying spaces. The IBC provides limited construction provisions only for those arrangements determined to be spray rooms, as well as ventilation and surfacing requirements for all spraying spaces. An automatic fire-extinguishing system is mandated for all areas where the application of flammable finishes occurs, including all spray rooms and spray booths.

The occupancy classification of buildings, rooms, and spaces utilized for flammable finish application is not specifically addressed. Table 307.1.1 indicates that classification as a Group H occupancy is not applicable where such buildings or areas meet the requirements of Section 416 and, by reference, the IFC. Therefore, the occupancy classification analysis would be no different from that for other types of uses. In a manufacturing building, a Group F-1 occupancy classification would be appropriate. Spraying operations within a vehicle repair garage would most likely be considered part of the Group S-1 classification. In a Group B adult vocational school, an area containing spraying activities would be considered a portion of the Group B occupancy.

A spray room, designed and constructed to house the spraying of flammable finishes, must be adequately separated from the remainder of the building. The enclosure must consist of fire barriers, horizontal assemblies, or both, each having a minimum fire-resistance rating of 1 hour. In addition, the walls and ceilings are to be of noncombustible construction or the surfaces must have noncombustible coverings. Spraying rooms must be frequently cleaned; thus, all of the interior surfaces must be smooth and easily maintained. The smooth surfaces also allow for the free passage of air in order to maintain efficient ventilation. The room construction must also be tight in order to eliminate the passage of residues from the room, which should be easily accomplished because of the fire separation required. Spraying spaces not separately enclosed shall be provided with noncombustible spray curtains to restrict the spread of vapors.

Section 417 *Drying Rooms*

Where the manufacturing process requires the use of a drying room or dry kiln, the room or kiln containing the drying operations must be of noncombustible construction. It must also be constructed in conformance with the specific and general provisions of the code as they relate to the special type of operations, processes, and materials that are involved. Clearance between combustible contents that are placed in the dryer and any overhead heating pipes must be at least 2 inches (51 mm). In addition, methods are addressed to insulate high-temperature dryers from adjacent combustible materials.

Section 418 *Organic Coatings*

Defined in Section 202 of the IFC, organic coatings are those compounds that are applied for the purpose of obtaining a finish that is protective, durable, and decorative. Used to protect structures, equipment, and similar items, organic coatings provide a surface finish that resists the effects of harsh weather. The concern for occupancies where organic coatings are manufactured or stored is based primarily on the presence of flammable vapors. As such, this type of use is highly controlled, by both the IBC and the IFC.

The manufacturing of organic coatings creates a high probability that flammable vapors will be present. Therefore, buildings where such materials are manufactured shall be without basements or pits because of the heavier-than-air nature of the vapors. In addition, no other occupancies are permitted in buildings used for the manufacture of organic coatings. The processing of flammable or heat-sensitive material must be done in a noncombustible or detached structure. Tank storage of flammable and combustible liquids inside a building must also be located above grade. In order to isolate the various hazard areas, the storage tank area must be separated from the remainder of the processing areas by minimum 2-hour fire barriers and/or horizontal assemblies. Because of the extreme hazards involved with nitrocellulose storage, it must also be separated by 2-hour fire-resistance-rated fire barriers and/or horizontal assemblies, or preferably located on a detached pad or in a separate structure.

Section 420 *Groups I-1, R-1, R-2, R-3, and R-4*

In residential-type uses, it is important that any fire conditions created in one of the dwelling units or sleeping units do not spread quickly to any of the other units. As residential fires are the most common of fire incidents, it is critical that neighboring units be isolated from the unit of fire origin. The need for an adequate level of fire resistance is enhanced because of the lack of immediate awareness of fire conditions when the building's occupants are sleeping. The provisions are applicable to hotels and other Group R-1 occupancies, apartment buildings, dormitories, fraternity and sorority houses, and other types of Group R-2 occupancies, and between dwelling units of a Group R-3 two-family dwelling. Sleeping units and dwelling units of a supervised residential care facility classified as Groups R-4 and I-1 must also be provided with such fire-resistive separations.

Where dwelling units or sleeping units are adjacent to each other horizontally, the minimum required separation is a fire partition. The wall serving as a fire partition is regulated by Section 708 and typically must have a minimum 1-hour fire-resistance rating. A reduction to a ½-hour fire partition is permitted under the special conditions set forth in Exception 2 of Section 708.3. Where dwelling or sleeping units are located on multiple floors of a building, they must be separated from each other with minimum 1-hour fire-resistance-rated horizontal assemblies as described in Section 711. An allowance for a ½-hour reduction, similar to that permitted for fire partitions, is also available under specified conditions.

In addition to the required separation between adjoining units, dwelling units and sleeping units must also be separated with complying fire partitions and/or horizontal assemblies from other adjacent occupancies. Applicable in multiple-use buildings, this requirement takes precedence over the allowances in Sections 508.2 and 508.3 for accessory occupancies and nonseparated occupancies, respectively. Even in those cases where the multiple-occupancy provisions of Section 508.2 or 508.3 are applied, the separation requirements of Section 420 must be followed. In those cases where the separated occupancy provisions of Section 508.4 are utilized, the more restrictive fire-resistive rating is applied.

It should be noted that the separation of dwelling units and sleeping units from other types of spaces in the building appears to only apply if those spaces are of a different occupancy from that of the residential units. For example, a separation is not required between a Group R-1 sleeping unit in a hotel and the adjacent hotel lobby if the lobby is classified as a portion of the Group R-1 occupancy. However, it would seem that the intent of this section suggests that a separation be provided in order to isolate each individual dwelling unit or sleeping unit, regardless of the classification of the adjacent space. See Figure 420-1.

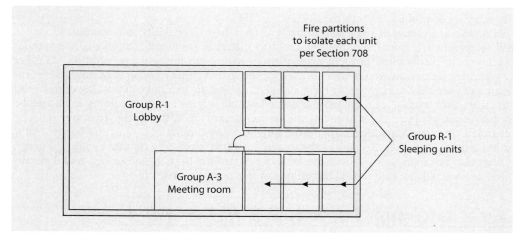

Figure 420-1 **Separation of dwelling units and sleeping units.**

Studies have shown that cooking appliances are the leading cause of fires in residential settings. Electric ranges are by far the leading cause of home cooking appliance fires. Unattended cooking is a factor in the majority of home electric range fires. Physical conditions such as falling asleep or impairment by alcohol or drugs are other contributing factors. Distractions that pull the cook outside of the kitchen (doorbell, social interactions) are another. Due to these concerns, the installation and use of domestic cooking appliances are regulated in both common areas and sleeping rooms of Group R-2 college dormitories. The scope of the provisions is limited to those appliances intended to be used by residents of Group R-2 college dormitories. It does not apply to residential dwelling units on college campuses that are not classified as dormitories. In addition to a limit on the types of appliances and their locations, cooktops and ranges are further regulated due to their increased hazard.

Section 422 *Ambulatory Care Facilities*

Ambulatory care facilities, often referred to as ambulatory surgery centers or day surgery centers, are defined in Chapter 2 as a building or portion of a building "used to provide medical, surgical, psychiatric, nursing or similar care on a less than 24-hour basis to individuals who are rendered incapable of self-preservation by the services provided or staff has accepted responsibility for care recipients already incapable." Classified as Group B occupancies, such facilities are generally regarded as moderate in hazard level due to their office-like conditions. However, additional hazards are typically of concern due to the presence of individuals who are temporarily rendered incapable of self-preservation due to the application of nerve blocks, sedation, or anesthesia. While the occupants may walk in and walk out the same day with a quick recovery time after surgery, there is a period of time where a potentially sizable number of people could require physical assistance in case of an emergency that would require evacuation or relocation.

Although classified as a Group B occupancy in the same manner as an outpatient clinic or other health-care office, an ambulatory care facility poses distinctly different hazards to life and fire safety, such as:

- Patients incapable of self-preservation require rescue by other occupants or emergency responders.
- Medical staff must stabilize the patient prior to evacuation, possibly resulting in delayed staff evacuation.
- Use of oxidizing medical gases such as oxygen and nitrous oxide.
- Potential for surgical fires.

As a result of the increased hazard level, additional safeguards have been put in place. Smoke compartments must be provided in larger facilities, and the installation of fire-protection systems is typically mandated. See Figure 422-1.

Any story containing ambulatory care facilities having more than 10,000 square feet (929 m^2) of floor area must be subdivided into at least two smoke compartments by smoke barriers in accordance with Section 709. The limit on compartment size of 22,500 square feet (2,092 m^2) may require that three or more smoke compartments be provided. Additional compartments may also be required due to travel distance limitations. Any point within a smoke compartment must be no more than 200 feet (60,960 mm) in travel distance from a smoke barrier door. Each smoke compartment must be large enough to allow for 30 square feet (2.8 m^2) of refuge area for each nonambulatory patient. In addition, at least one means of egress must be available from each smoke compartment without the need to return back through the original compartment. Fire-safety evacuation plans must be provided where smoke compartments are mandated. The plans are intended to address the defend-in-place philosophy for emergency response as defined in Section 202 and regulated by Sections 403 and 404 of the IFC.

132 Chapter 4 ■ Special Detailed Requirements Based on Occupancy and Use

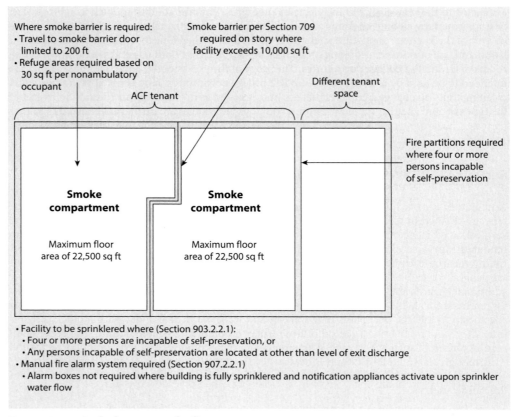

Figure 422-1 **Ambulatory care facility.**

As a general rule, Group B occupancies do not require a sprinkler system based solely on their occupancy classification. However, Section 903.2.2 mandates that a Group B ambulatory care facility be provided with an automatic sprinkler system when either of the following conditions exist at any time:

- Four or more care recipients are incapable of self-preservation, or
- One or more care recipients who are incapable of self-preservation are located at other than the level of exit discharge.

The extent of the sprinkler protection is detailed in Section 903.2.2. In addition, the fire alarm requirements are more stringent than those of other Group B occupancies. Section 907.2.2.1 requires the installation of a manual fire alarm system in all Group B fire areas containing an ambulatory health-care facility. The manual fire alarm boxes are not required if the building is fully sprinklered and the occupant notification appliances activate upon sprinkler water flow. The required automatic smoke detection system is also not required where the same conditions occur.

Although not very common, there are times where physical therapy areas of ambulatory care facilities contain kitchens to educate patients in the safe operation of cooking equipment. Such training may be helpful to the patients for when they are cooking at home, or it may be a portion of nutrition counseling. In addition, domestic cooking facilities are often used for other purposes, such as where the cooking appliances are installed in an employee break room. Due to the potential hazards involved, five fundamental conditions of compliance must be met to allow the installation of domestic cooking appliances in ambulatory care facilities.

Section 423 *Storm Shelters*

ICC-500, *ICC/NSSA Standard on the Design and Construction of Storm Shelters*, establishes minimum requirements for structures and spaces designated as hurricane, tornado, or combination shelters. The standard addresses the design and construction of such shelters from the perspective of the structural requirements for high wind conditions, and addresses minimum requirements for the interior environment during a storm event.

Scoping provisions mandate the construction of complying storm shelters in critical emergency operations facilities where such facilities are located in geographical areas where the shelter design wind speed is at its highest: 250 miles per hour. This geographical area, which includes most of the Interior Plains region of the United States, is identified in detail in ICC 500. Critical facilities, such as emergency operations centers, fire and police stations, and buildings with similar functions, are essential for the delivery of vital services or the protection of a community. It is important to protect the occupants of such critical facilities struck by tornadoes, as well as to maintain continuity of operations for those facilities. Emergency operations facilities, as well as police and fire rescue facilities, are crucial to disaster response because an interruption in their operation as a result of building equipment failure may prevent rescue operations, evacuation, assistance delivery, or general maintenance of law and order, which can have serious consequences for the community after a storm event.

The minimum required size of a storm shelter is based upon the design occupant capacity. For critical emergency operations facilities, this capacity is set as the total occupant load of the offices plus the total number of beds. The required occupant capacity for Group E school facilities is the total occupant load of the classrooms, vocational rooms and offices. This approach recognizes the primary use of the building(s) as an educational facility. Three exceptions modify the base scoping provisions for the determination of the design occupant capacities of required storm shelters, applicable to both Group E occupancies and emergency operations facilities.

The proximity of the shelters to the occupants of the building is a critical factor. Storm shelters are required to be located in the same building being served by the shelters or, if in a separate building, within 1,000 feet (305 m) of the building served by the shelter. The 1,000-foot (305 m) measurement is to be taken from an exterior door of the building being served by the shelter to the door of the shelter. Due to the time-sensitive nature of the potential hazard, it is important that the distance individuals must travel to reach a protected space is limited.

Section 424 *Play Structures*

Play structures for various recreational activities were regulated for some time by the IBC only where such structures were located within covered mall buildings and intended for use by children. The primary concern, consistent with that of other structures located within a covered mall building, was the combustibility of such play structures. Due to the potential fire hazards associated with play structures, the regulations are now applicable where such structures are located within any building regulated by the IBC, regardless of occupancy classification. Because concerns relating to potential fire hazards due to the combustibility and flammability of the structure and its contents are present regardless of the ages of the participants, the provisions are also now applicable to all such structures. Only those play structures that exceed 10 feet (3,048 mm) in height or 150 square feet (14 m^2) in area are to be regulated by the IBC.

Play structures must be constructed of noncombustible materials or, as an option if combustible, must comply with the appropriate criteria established in Section 424. Such alternative methods include the use of fire-retardant-treated wood, textiles complying with the designated flame propagation performance criteria, and plastics exhibiting an established maximum peak rate of heat release.

Where the play structure exceeds 600 square feet (56 m^2) in area or 10 feet (3,048 mm) in height, the interior finishes are to be regulated for flame spread index and it must be designed in conformance with Chapter 16. In addition, play structures exceeding 600 square feet (56 m^2) require a special investigation acceptable to the building official to demonstrate adequate fire safety. Conformance with Chapter 16 is also required where the play structure exceeds 10 feet (3,048 mm) in height, regardless of area.

Section 427 *Medical Gas Systems*

Hospitals and most other health-care facilities typically require the use of medical gases as a critical component of their functions. Oxygen, nitrous oxide, and a variety of other compressed gases are piped into treatment rooms from medical gas storage rooms. These rooms are regulated for fire-resistive separation and ventilation purposes. Provisions address both exterior rooms, which must be located on an exterior wall, and interior rooms, where an exterior wall location cannot be provided. In addition, gas cabinet construction criteria are set forth. These requirements, historically set forth in the IFC, are also located in this chapter of the IBC in a manner consistent with other IFC provisions that have been replicated in the IBC in order to provide a more comprehensive and efficient set of construction regulations.

Section 428 *Higher Education Laboratories*

Colleges and universities often have chemistry, biology, medical, engineering, and other types of laboratories where significant amounts of hazardous materials are stored and used. These teaching and research laboratories are regulated in a specific manner, rather than under the general hazardous materials provisions which often are not appropriate for specialized academic laboratory settings.

There are a number of conditions typically present in higher education laboratories that make them unique, thus requiring unique solutions.

- Lower hazardous materials density in individual laboratory spaces. In academic environments, there are typically a large number of laboratories, but each laboratory only contains a small amount of hazardous material. Individually the quantities in use and storage are relatively low; however, the total quantity of hazardous materials on the story can be significant, even to the point of exceeding the maximum allowable quantities per control area. The lower density condition is considered as a means of mitigating the overall risk.

- Considerable staff oversight of activities utilizing hazardous materials. Most academic laboratories are well staffed with faculty members and support personnel who are well acquainted with hazardous material safety. They are an integral part of the preparation and review of laboratory safety documentations and safety audits.

- Mixed-occupancy buildings. Higher education laboratories are often found in campus buildings that also contain storage, business, and assembly uses. The traditional limits on the permissible amount of hazardous materials on upper floor levels are extremely restrictive, often requiring that lecture halls and classrooms be placed on upper floors so that the lower stories can be utilized for laboratories. This scenario places significant occupant loads on upper floors above those areas devoted to the use and storage of hazardous materials.

There are three primary considerations in regard to the special provisions for higher education laboratories. Specifically, Section 428 addresses the following primary needs:

- Increased general laboratory safety. Requirements and allowances specific to higher education laboratory uses are established in the IBC, along with a reference to NFPA 45, *Standard on Fire Protection for Laboratories Using Chemicals*. In addition, the IFC contains provisions addressing the unique issues of these types of laboratories in a manner even more comprehensive that the IBC.

- Control area limitations. In an increasingly number of cases, the buildings housing higher education laboratories are built taller and/or larger than in the past. In response to this reality, greater numbers of control areas and larger percentages of maximum allowable quantities are necessary. An alternate design approach is provided for such scenarios where traditional control area limitations or construction as a Group H occupancy are not feasible. The "laboratory suite" concept provides an option to allow increased flexibility in the storage and use of hazardous materials while continuing to maintain a Group B occupancy classification.

- Allowances for existing nonsprinklered buildings. Although not regulated due to the scope of the IBC, an approach to regulating existing academic laboratory buildings without fire sprinkler protection is established in the IFC. Limited to very small quantities of hazardous materials, the allowance recognizes the thousands of buildings built decades ago where retrofitting with sprinklers is not practical.

The IBC recognizes that these higher education laboratories can be considered as a part of the Group B occupancy classification afforded to other portions of college and university buildings provided such laboratories comply with Section 428. It is important that the IFC be consulted for any additional requirements related to the storage and use of hazardous materials. In addition, the activities are limited to the testing, analysis, teaching, research, and development on a nonproduction basis.

A key aspect of the academic laboratory provisions is the creation of laboratory suites. Similar in concept to control areas, such suites are fully enclosed by fire-resistance-rated construction in order to provide containment areas. The number of laboratory suites permitted on a story, as well as the amount of hazardous materials that can be used and/or stored within a laboratory suite, is significantly greater than that allowed in control areas.

KEY POINTS

- Special buildings, features, and uses such as covered and open mall buildings, atriums, high-rise buildings, underground buildings, and parking garages are so unique in the type of hazards presented that specialized regulations are provided in the IBC.
- A covered mall building or open mall building consists of various tenants and occupants, as well as the common pedestrian area that provides access to the tenant spaces.
- For those features that are not unique to a covered or open mall building, the general provisions of the code apply.
- The means of egress provisions for a covered or open mall building vary significantly from those for other buildings.
- High-rise buildings are characterized by the difficulty of evacuation or rescue of the building occupants, the difficulty of fire-fighting operations from the exterior, high occupant loads, and potential for stack effect.
- There are a number of provisions for high-rise buildings that are less restrictive than the general requirements, including the reduction in fire resistance for certain building elements.
- The special allowance for a reduction in construction type is not applicable to any high-rise building exceeding 420 feet (128 m) in height.
- Smoke detection, alarm systems, and communications systems are important characteristics of a high-rise building.
- Occupant egress and evacuation, as well as fire department access, are addressed in high-rise buildings through provisions for stairway enclosure remoteness, an additional stairway, luminous egress path markings, fire service access elevators, and occupant evacuation elevators.
- The use of the atrium provisions is typically limited to those multistory applications where compliance with the other vertical opening applications established in Section 712.1 is not possible.
- Buildings containing atriums, high-rise buildings, covered mall buildings, and open mall buildings must be provided with automatic sprinkler systems throughout.
- Another component of the life-safety system for a building containing an atrium is a required smoke-control system.
- An underground building is regulated in a manner similar to that for a high-rise building, as the means of egress and fire department access concerns are similarly extensive.
- Fundamental to the protection features for an underground building are the requirements for Type I construction for the underground portion and the installation of an automatic sprinkler system.
- Private garages and carports are regulated to a limited degree based on the hazards associated with the parking of motor vehicles.
- Special provisions for open parking garages are typically less restrictive than those for enclosed parking structures because of the natural ventilation that is available.
- In Group I-2 and I-3 occupancies, the functional concerns of the health-care operations must be balanced with the fire-safety concerns.

- Stages exceeding 50 feet (15,240 mm) in height present additional risks that are due to the expected presence of high combustible loading such as curtains, scenery, and other stage effects.
- Under-floor areas and attic spaces used for the storage of combustible materials must be isolated from other portions of the building with fire-resistance-rated construction.
- Hazardous materials that are used or stored in any quantity are subject to regulation by Sections 307 and 414, and the IFC.
- An increase in the maximum allowable quantities of hazardous materials in a building not classified as Group H is permitted through the proper use of control areas.
- Group H occupancies are highly regulated because of the hazardous processes and materials involved in such occupancies.
- Special provisions are applicable to the construction, installation, and use of buildings for the spraying of flammable finishes in painting, varnishing, and staining operations.
- Dwelling units and sleeping units must be separated from each other and from other portions of the building through the use of fire partitions and/or horizontal assemblies.
- Health-care offices where care is provided to individuals who are rendered incapable of self-preservation are considered to be ambulatory care facilities.
- Storm shelters, constructed in accordance with ICC-500, are required in critical emergency operations facilities and most Group E occupancies.
- Where combustible dust hazards are present in grain elevators, grinding rooms, and tire rebuilding facilities, special safeguards are established.
- Medical gas systems used in health care-related facilities are regulated as to their fire-resistive separation from the remainder of the building.
- Laboratories in a university or college building can be constructed as suites which provide for expanded use of the control area concept in addressing the use and storage of hazardous materials without creating a Group H classification.

CHAPTER 5

GENERAL BUILDING HEIGHTS AND AREAS

Section 502 Building Address
Section 503 General Building Height and Area Limitations
Section 504 Building Height and Number of Stories
Section 505 Mezzanines and Equipment Platforms
Section 506 Building Area
Section 507 Unlimited-Area Buildings
Section 508 Mixed Use and Occupancy
Section 509 Incidental Uses
Section 510 Special Provisions
Key Points

Chapter 5 provides general provisions that are applicable to all buildings. These include requirements for allowable floor area, including permitted increases for open frontage unlimited-area buildings; and allowable height of buildings. Buildings containing multiple uses and occupancies are regulated through the provisions for incidental uses, accessory occupancies, nonseparated occupancies, and separated occupancies. Miscellaneous topics addressed in Chapter 5 include premises identification and mezzanines.

In addition to the general provisions set forth in Chapter 5, there are several special conditions under which the specific requirements of Chapter 5 can be modified or exempted, including the horizontal building separation allowance and unique provisions for buildings containing a parking garage.

Section 502 *Building Address*

502.1 Address identification. In this section, the *International Building Code®* (IBC®) intends that buildings be provided with plainly visible and legible address numbers posted on the building or in such a place on the property that the building may be identified by emergency services such as fire, medical, and police. The primary concern is that responding emergency forces may locate the building without going through a lengthy search procedure. In furthering the concept, the code intends that the approved street numbers be placed in a location readily visible from the street or roadway fronting the property if a sign on the building would not be visible from the street. Address numbers may be required in multiple locations to help eliminate any confusion or delay in identifying the location of the emergency. The fire code official can require, and must approve, additional address identification locations. Regardless of the sign's location, the minimum height of letters or numbers used in the address is to be at least 4 inches (102 mm) and contrast to background itself. Additional criteria are provided to provide consistency and clarity of the identifying numbers.

Section 503 *General Building Height and Area Limitations*

The IBC regulates the size of buildings in order to limit to a reasonable level the magnitude of a fire that potentially may develop. The size of a building is controlled by its floor area and height, and both are typically limited by the IBC. Whereas floor-area limitations are concerned primarily with property damage, life safety is enhanced as well by the fact that in the larger building there are typically more people at risk during a fire. Height restrictions are imposed to also address egress concerns and fire department access limitations.

The essential ingredients in the determination of allowable areas are:

1. The amount of combustibles attributable to the use that determines the potential fire severity.

2. The amount of combustibles in the construction of the building, which contributes to the potential fire severity.

In addition to the two factors just itemized, there may be other features of the building that have an effect on area limitations. These include the presence of built-in fire protection (an automatic fire-sprinkler system), which tends to prevent the spread of fire, and open space

General Building Height and Area Limitations

(frontage) adjoining a sizable portion of the building's perimeter, which decreases exposure from adjoining properties and provides better fire department access.

A desirable goal of floor-area limitations in a building code is to provide a relatively uniform level of hazard for all occupancies and types of construction. A glance through Table 506.2 of the IBC will reveal that, in general, the higher hazard occupancies have lower permissible areas for equivalent types of construction and, in addition, the less fire-resistant and more combustible types of construction have more restrictive area limitations.

The IBC also limits the maximum building height and number of stories above grade plane based on similar reasons discussed for area limitations. In addition, the higher the building becomes, the more difficult access for firefighting becomes. Furthermore, the time required for the evacuation of the occupants increases; therefore, the fire resistance of the building should also be increased.

The code presumes that when the height of the highest floor or roof level used for human occupancy exceeds 75 feet (22,860 mm), the life-safety hazard becomes even greater because most fire departments are unable to adequately fight a fire above this elevation from the exterior. Furthermore, the evacuation of occupants from the building is often not feasible. Thus, Section 403 prescribes special provisions for these high-rise buildings. Similar concerns for buildings with occupied floors well below the level of exit discharge are addressed in Section 405 for underground buildings.

Coming back to this section, the code specifies in Section 504 both the maximum allowable height in feet (mm) and the maximum number of stories above grade plane. The maximum allowable height in feet is regulated by Table 504.3 primarily by the building's construction type, with limited regard for the occupancy or multiple occupancies located in the building. However, the maximum allowable height in stories above grade plane can vary significantly based on the occupancy group involved as set forth in Table 504.4. Where multiple occupancies are located in the same building, and the provisions of Section 508.4 for separated occupancies are utilized, each individual occupancy can be located no higher than set forth in the table. See Figure 503-1.

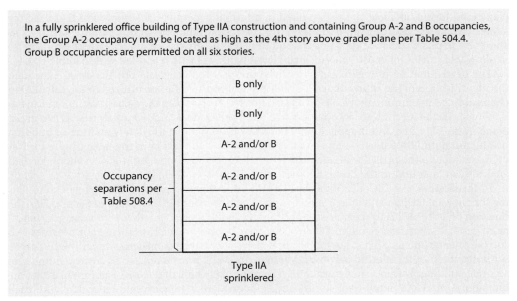

Figure 503-1 **Height limitations—separated occupancies example.**

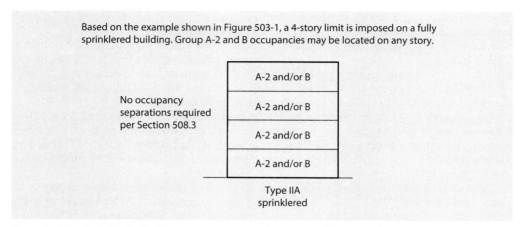

Figure 503-2 **Height limitations—nonseparated occupancies example.**

Where the nonseparated-occupancies provisions of Section 508.3 are applied, the most restrictive height limitations of the nonseparated occupancies involved will limit the number of stories above grade plane in the entire building. See Figure 503-2. In general, the greater the potential fire- and life-safety hazard, the lower the permitted overall height in feet (mm), as well as the fewer the number of permitted stories above grade plane.

503.1 General. Once the occupancy group(s) and type of construction have been determined, the limitations in Tables 504.3, 504.4, and 506.2 must be consulted individually. Allowable building height, as set forth in Table 504.3, is based on height in feet above grade plane. Both "building height" and "grade plane" are defined in Chapter 2 to ensure consistent application of the provisions. The table includes the increased height allowances that are available for various types of sprinklered buildings. The allowable number of stories above grade plane is addressed in Table 504.4, presented in a format generally consistent with that used for Tables 504.3 and 506.2. Again, definitions are very important, and in this case the term "story above grade plane" must be fully understood to apply the provisions appropriately. An increase in allowable stories for most types of sprinklered buildings is also applied within Table 504.4.

The determination of maximum allowable building area is initiated in Table 506.2 through the identification of the appropriate allowable area factor. This factor, utilized in the calculation of a building's maximum allowable area, varies based upon occupancy classification, construction type, installation of an automatic sprinkler system, and number of stories above grade plane in the building, and is applied on a per story basis. Unlike the determination of building height and number of stories above grade plane, the area limitation in the table can be further increased due to the presence of adequate frontage at the building's perimeter. A more detailed analysis can be found in the discussion of Section 506.

In this section, the IBC also indicates that fire walls create separate buildings when evaluating for allowable height and area. Defined and regulated under the provisions of Section 706, the function of a fire wall is to separate one portion of a building from another with a fire-resistance-rated vertical separation element. Where a fully complying fire wall is provided, it provides two compartments, one on each side of the wall, which may each be considered under the IBC to be separate buildings for specific purposes. Multiple fire walls may be utilized to create a number of separate buildings within a single structure. The resulting benefit of the use of a fire wall is that the limitations on height, number of stories, and floor area are then addressed individually for each separate building created by fire walls within the structure, rather than for the structure as a whole.

503.1.1 Special industrial occupancies. This special provision exempts certain types of buildings from both the height limitations and the area limitations found in Sections 504 and 506. Thus, the type of construction is not limited, regardless of building height or area. It is also not necessary to comply with the provisions of Section 507 for unlimited-area buildings to utilize this provision. Applicable to structures housing low-hazard and moderate-hazard industrial processes that often require quite large areas and heights, the relaxation of the general provisions recognizes the limited fire severity, as well as the need for expansive buildings to house operations such as rolling mills, structural metal-fabrication shops, foundries, and power distribution. It is not the intent that buildings classified as Group H occupancies be addressed under the allowances of Section 503.1.1. as evidenced by the specific types of uses listed in the provisions.

503.1.2 Buildings on the same lot. Where two or more buildings are located on the same lot, they may be regulated as separate buildings in a manner generally consistent with buildings situated on separate parcels of land.

As an option, multiple buildings on a single site may be considered one building, provided the limitations of height, number of stories, and floor area based on Sections 504 and 506 are met. The height and number of stories above grade plane of each building and the aggregate area of all buildings are to be considered in the determination. Under this method, the provisions of the code applicable to the aggregate building shall also apply to each building individually. Further regulations for buildings on the same lot are discussed in the commentary for Section 705.3.

503.1.4 Occupiable roofs. A building's maximum allowable height in regard to number of stories above grade plane is determined based upon the building's type of construction and the occupancy classification of the occupancies involved. A story, by definition, is considered as that portion of a building between the upper surface of a floor and the upper surface of the floor or roof next above. Because a roof deck has no floor or roof above it, an occupiable roof does not qualify as a story. However, the presence of occupants and fire loading on an occupiable roof requires such a condition to be specifically addressed.

As a general rule, an occupancy is permitted to be located on the roof only if the occupancy is permitted by Table 504.4 to be located on the story directly below the roof. However, such a limitation does not apply where two conditions exist: (1) the building is fully sprinklered, and (2) occupant notification per Section 907 is provided for the occupied portion(s) of the roof. Where both conditions exist, the rooftop occupancies have no effect on the building's allowable height in stories above grade plane.

In addition to the allowance that an occupied roof is not considered as a story for purposes of applying the IBC, it is also not considered as building area in the regulation of allowable floor area. Addressed in much the same manner as penthouses and other roof structures, the roof area is not a factor in determining the permissible building area. Where penthouses and other enclosed rooftop structures are in compliance with Section 1511, such structures also do not contribute to the building's floor area, height, and number of stories above grade plane.

Section 504 *Building Height and Number of Stories*

As previously addressed, a building's limiting height in feet and number of stories above grade plane is based upon the type of construction, occupancy classification(s), and presence of an automatic sprinkler system. Because automatic fire-sprinkler systems have exhibited an excellent record of in-place fire suppression over the years, Tables 504.3 and 504.4 generally allow greater heights where an automatic fire-sprinkler system is installed throughout the building. The tables typically reflect an increase of one story in the number of stories above grade plane, and 20 feet (6,096 mm) in building height, where the building is provided with an automatic fire-sprinkler system throughout.

There are basically five variations to the general requirements for height and story increases:

1. Extended height and number of stories sprinkler allowances are not permitted for Group I-2 occupancies of Type IIB, III, IV, or V construction, or for Groups H-1, H-2, H-3, and H-5 occupancies of any construction type. These occupancies present unusual hazards that limit their heights even where a sprinkler system is present.

2. One-story aircraft manufacturing buildings and hangars may be of unlimited height when sprinklered and surrounded by adequate open space. Such uses require very large structures and through the safeguards provided, should be adequately protected.

3. For Group R buildings provided with an NFPA 13R sprinkler system, the tables reflect that the increases in height and number of stories above grade plane apply only up to a maximum of 60 feet (18,288 mm) and four stories, respectively. The limitation of four stories above grade plane and 60 feet for buildings sprinklered with a 13R system cannot be exceeded under any circumstances. In those residential buildings where an NFPA 13, rather than an NFPA 13R system, is installed, the limitations of 60 feet (18,288 mm) and four stories above grade plane are not applicable.

4. Roof structures such as towers and steeples may be of unlimited height when constructed of noncombustible materials, whereas combustible roof structures are limited in height to 20 feet (6,096 mm) above that permitted by Table 504.3. In all cases, such roof structures are to be constructed of materials based on the building's type of construction. These requirements are not based on the presence of the sprinkler system. Additional requirements for roof structures can be found in Section 1511.

5. The allowable height in feet and number of stories above grade plane for Type IV-A, IV-B, and IV-C mass timber buildings are based upon a variety of factors due to the unique features of these three construction types. In addition, the expectation is that such mass timber buildings be fully sprinklered. As a result, the typical allowances that the maximum permissible height and number of stories above grade plane for nonsprinklered (NS) buildings be established at 20 feet (6,096 mm) and one story, respectively, below that permitted for sprinklered buildings are not recognized. NS values are nevertheless provided in the table to allow for the calculation of area increases for open perimeter conditions, which are based on NS building areas.

Section 505 *Mezzanines and Equipment Platforms*

A mezzanine is defined in Chapter 2 as an intermediate level or levels between the floor and ceiling of any story. As such, a mezzanine is not considered as a story but is rather deemed to be located "within" a story. An intermediate floor level of limited size and without enclosure has historically been deemed to cause no significant safety hazard. The occupants of the mezzanine by means of sight, smell, or hearing will be able to determine if there is some emergency or fire that takes place either on the mezzanine or in the room in which the mezzanine is located. However, once portions of or all of the mezzanine is enclosed, or the mezzanine exceeds one-third the area of the room in which it is located, life-safety problems such as occupants not being aware of an emergency or finding a safe exit route from the mezzanine become important. Therefore, the code places the restrictions encompassed in this section on mezzanines to ameliorate the life safety that can be created.

505.2 Mezzanines. By virtue of the conditions placed on mezzanines in Section 505, a complying mezzanine is not considered to create additional building area or an additional story for the purpose of limiting building size. In the determination of a building story's floor area for allowable area purposes, the floor area of a complying mezzanine need not be added to the area of the floor below. As previously mentioned, complying mezzanines also do not contribute to the actual number of stories in relationship to the allowable number of stories above grade plane permitted by Table 504.4. The limitations imposed on mezzanines are deemed sufficient to permit such benefits.

In contrast to the above allowances, the floor area of mezzanines must be included as a part of the aggregate floor area in determining the fire area. Because the size of a fire area is based on a perceived level of fire loading present within the building, the contribution of a mezzanine's fire load to the fire loading in the room in which the mezzanine is located cannot be overlooked. Figure 505-1 depicts the proper use of these provisions. The clear height above and below the floor of the mezzanine is also regulated at a minimum height of 7 feet (2,134 mm).

505.2.1 Area limitation. There is no limit on the number of mezzanines that may be placed within a room; however, the total floor area of all mezzanines must typically not exceed one-third of the floor area of the room in which they are located. See Figure 505-2. As illustrated in Figure 505-3, any enclosed areas of the room in which the mezzanine is located are not to be utilized in the calculations for determining compliance with the one-third rule.

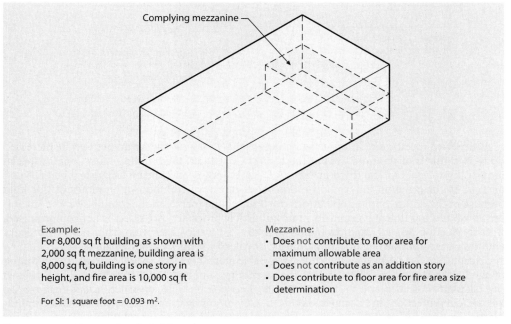

Example:
For 8,000 sq ft building as shown with 2,000 sq ft mezzanine, building area is 8,000 sq ft, building is one story in height, and fire area is 10,000 sq ft

For SI: 1 square foot = 0.093 m².

Mezzanine:
• Does not contribute to floor area for maximum allowable area
• Does not contribute as an addition story
• Does contribute to floor area for fire area size determination

Figure 505-1 **Mezzanine height and area.**

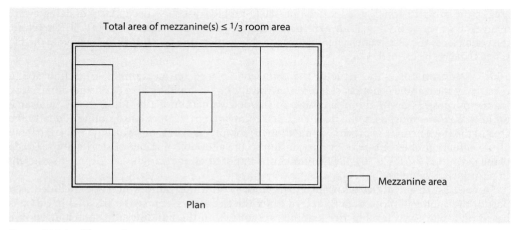

Figure 505-2 **Mezzanine area.**

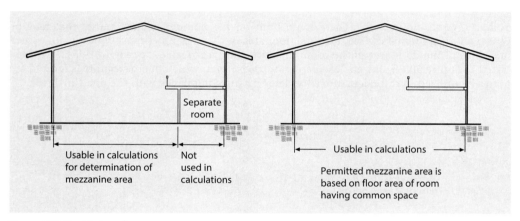

Figure 505-3 **Mezzanine area.**

Where two specific conditions exist, the aggregate floor area of mezzanines may be increased up to two-thirds of the floor area of the room below. First, the building must contain special industrial processes as identified in Section 503.1.1, and second, the building shall be of Type I or Type II construction. Intermediate floor levels are very common in buildings of this kind because of the nature of their operations. By limiting the increased mezzanine size to noncombustible buildings housing primarily noncombustible processes, fire safety is not compromised. Because Section 503.1.1 exempts buildings addressed in this exception from number of story and floor area limits, there is seldom, if ever, any need to apply Exception 1.

Exception 2 also permits an increase in allowable mezzanine size, in this case up to a maximum of one-half of the area or room in which the mezzanine is located. See Figure 505-4. The increased size takes into consideration the enhancements of noncombustible construction, automatic sprinkler system protection, and occupant notification by means of an emergency voice/alarm communication system. By limiting construction to Type I or II, there is no contribution to the fire hazard that is due to the materials of construction. The automatic sprinkler protection will increase the

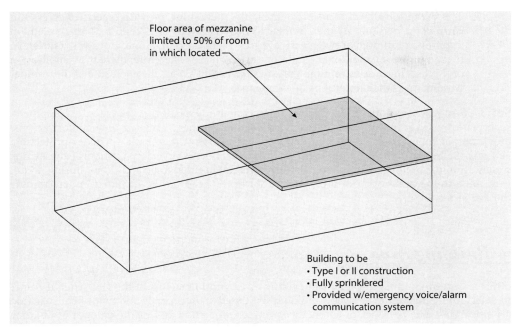

Figure 505-4 **Maximum floor area of mezzanine.**

potential for limiting fire spread. In addition, the occupant notification system increases occupant awareness of a fire condition and allows for evacuation during the early fire stages.

505.2.3 Openness. As a general rule, a mezzanine must be open to the room in which it is located. Any side that adjoins the room will be considered open if it is unobstructed, other than by walls or railings not more than 42 inches (1,067 mm) in height, or columns and posts. There are, however, five exceptions to the requirement for openness that result in most mezzanines being permitted to be partially or completely enclosed. If in compliance with any one of the five exceptions, the mezzanine need not be enclosed.

1. The sole, criterion is that the enclosed area contains a maximum occupant load of 10 persons. The limitation on occupant load is based on the aggregate area of the enclosed space. This exception is consistent with other provisions of the code that relax the requirements where the occupant load is expected to be relatively small. This exception may result in the enclosure of an entire mezzanine or just a portion of it.

2. A mezzanine may also be fully or partially enclosed if it is provided with two means of egress. No further conditions are established under this option.

3. Up to 10 percent of the mezzanine area may be enclosed, typically applicable to toilet rooms, closets, utility rooms, and other similar uses that must, of necessity, be within an enclosure. As long as the aggregate area does not exceed 10 percent of the area of the mezzanine, the enclosure is permitted by the code.

4. Those mezzanines used for control equipment in industrial buildings may be glazed on all sides. This exception is necessary because of the delicate nature of much of today's control equipment and the fact that it may require a dust-free environment.

5. This exception allows enclosure of the entire mezzanine, provided specified egress and sprinkler provisions are met. Similar to the conditions of Exception 2, this exception requires a minimum of two means of egress from the mezzanine. However, it differs in that a number of additional conditions are required. Therefore, there is no application of this exception as Exception 2 satisfies the condition for the enclosure of mezzanines without any other conditions of compliance.

505.3 Equipment platforms. In buildings containing platforms that house equipment, such platforms need not be considered stories or mezzanines, provided they conform to the provisions of this section regulating platform size, extent of automatic sprinkler protection, and guards. In addition, the equipment platforms cannot serve as any portion of the exiting system from the building. Complying platforms are not deemed to be additional stories, do not contribute to the building floor area, and furthermore, need not be included in determining the size of the fire area.

Section 506 *Building Area*

506.1 General. Building area is one of the three limiting factors in the IBC when it comes to determining maximum allowable building size based on occupancy and type of construction classifications. The presence of automatic sprinkler protection and/or the amount of building frontage also contribute to this determination. The process is initiated in Table 506.2 where the allowable area factor is identified, and continued through the application of the frontage modifier. A general overview of the concept regarding allowable building area can be found in the discussion of Section 503.

There are two additional approaches to allowable area compliance that modify or exempt compliance with Section 506. For those buildings constructed under the provisions of Section 507, there is no limit on floor area, provided specified conditions are met. In addition, modifications and exemptions to height and area limitations of Chapter 5 are set forth in Section 510. Although there is no requirement to apply Section 507 or 510 for allowable area determination, the allowances provided in these sections provide alternative methods for compliance.

The code intends that a basement that does not exceed the allowable floor area permitted for a one-story building need not be included in determining the total allowable area of the building. This provision is a holdover from many years back when basements were most commonly used for the service of the building. Today, it is not uncommon to find basements occupied for the same uses as the upper floors. There apparently has been no adverse experience for cases of this type (most likely because of the fire-sprinkler protection required in most basements), and therefore the code provision appears to be satisfactory. The code does not address how to handle a basement that exceeds the area permitted for a one-story building. However, as the code does not permit any story to exceed that permitted for a one-story building, it would seem logical that a basement should also be limited to the same area. If anything, a fire in a basement is more difficult to fight than one in an aboveground story. Thus, it does not seem reasonable or appropriate to permit a single basement with an area exceeding that allowed for a one-story building. Multiple basements are also exempt from inclusion in the building area calculations, but only where the aggregate area of such basements does not exceed the area permitted for a single-story building. Additional requirements for basements extending well below grade level may be found in Section 405 for underground buildings.

506.2.1 Single-occupancy, buildings. The most straightforward determination of allowable area is applied to those buildings no more than three stories above grade plane that contain only one occupancy classification. The formula for the calculation of the maximum allowable area for these types of buildings (in square feet), Equation 5-1, is additive. It is based on the sum of the tabular allowable area factor shown in Table 506.2 plus any increase that is due to building frontage per Section 506.3. Equation 5-1 provides for the maximum allowable floor area for each story of a building having one, two, or three stories above grade plane. An example is shown in Application Example 506-1.

GIVEN: Two-story above grade plane office building
- Type IIB construction
- Fully sprinklered
- Yards and streets as shown

DETERMINE: Maximum allowable area per story above grade plane (A_a)

A_a = $A_t + (NS \times I_f)$ (Equation 5-1)
A_t = 69,000 sq ft (Table 506.2) (Type IIB, Group B, SM*)
I_f = 0.50 (Table 506.3.3, based on 69% of building perimeter open with 30 feet as the smallest open space of 20 feet or more; see discussion of Section 506.3)

A_a = 69,000 + (23,000 × 0.50)
  = 69,000 + 11,500
  = 80,500 sq ft per story

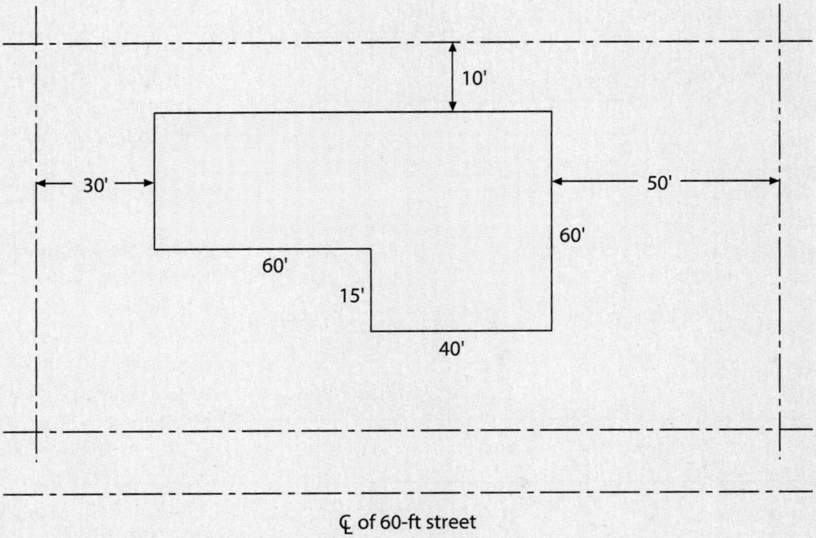

℄ of 60-ft street

For SI: 1 foot = 304.8 mm, 1 square foot = 0.093 m².
*Although not listed in Equation 5-1, the SM factor is the only "applicable" factor for a multi-story condition.

EXAMPLE OF ALLOWABLE AREA DETERMINATION FOR SINGLE-OCCUPANCY BUILDING WITH NO MORE THAN THREE STORIES ABOVE GRADE PLANE

Application Example 506-1

Where the building has four or more stories above grade plane, the total building area based on the sum of the floor areas of all stories above grade plane is further limited. The calculation of the allowable building area is based on Equation 5-2 limiting the total building area to three times the allowable area as calculated in Equation 5-1. It is important to note that any individual story cannot exceed the allowable area as determined by Equation 5-1. See Application Example 506-2.

GIVEN: Four-story above grade plane office building
- Type IIB construction
- Fully sprinklered
- Yards and streets as shown

DETERMINE: Maximum allowable area for the building

$A_a = A_t + (NS \times I_f) \times S_a$ (Equation 506-2)

$A_t = 69{,}000$ sq ft (Table 506.2) (Type IIB, Group B, SM)

$I_f = 0.50$ (Table 506.3.3, based on 69% of building perimeter open with 30 feet as the smallest open space of 20 feet or more; see discussion of Section 506.3)

$A_a = [69{,}000 + (23{,}000 \times 0.50)] \times 3$

$= (69{,}000 + 11{,}500) \times 3$

$= 241{,}500$ sq ft total of all stories above grade plane, with no single story exceeding 80,500 sq ft*

For SI: 1 foot = 304.8 mm, 1 square foot = 0.093 m².
*If all stories have same footprint, the maximum allowable area per story would be 60,375 sq ft.

EXAMPLE OF ALLOWABLE AREA DETERMINATION FOR SINGLE-OCCUPANCY BUILDING WITH FOUR OR MORE STORIES ABOVE GRADE PLANE

Application Example 506-2

The exception to this methodology is applicable to only those four-story above grade plane buildings equipped with an NFPA 13R residential sprinkler system. As the installation of an NFPA 13R sprinkler system in a Group R occupancy does not provide for an increase in allowable area per story as indicated for S13R sprinkler protection in Table 506.2, and the use of an NFPA 13R system is limited to buildings no more than four stories in height, it is considered appropriate to permit the maximum allowable area per story for each of the stories in the residential building. This would include the four-story structure as depicted in Figure 506-1.

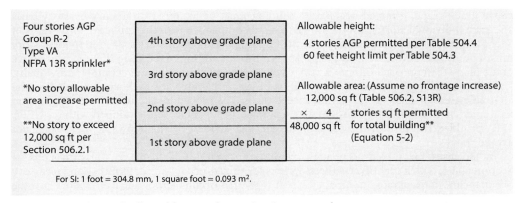

Figure 506-1   **Group R allowable area determination example.**

How does the installation of an automatic sprinkler system relate to a building's allowable floor area? The tabular allowable area factors established in Table 506.2 include any increases that may be appropriate due to the installation of an automatic sprinkler system. Because of the excellent record of automatic sprinkler systems for the early detection and suppression of fires, the IBC allows quite large floor-area increases where an automatic fire sprinkler system is installed throughout the building. As evidenced in Table 506.2, the maximum allowable area of a fully sprinklered (S1) one-story building is typically four times the amount for a NS building; and for a building of two or more stories in height above grade plane, the tabular allowable area is typically three times the floor area permitted for a NS condition. The lesser sprinkler benefit for multistory buildings is based on the assumption by the code that fire department suppression activities are still going to be required even where an automatic fire-sprinkler system is installed. Therefore, a multistory building presents more problems to the fire department than a one-story building, and a smaller floor area allowance is permitted.

Table 506.2 reflects increased floor area allowances for buildings protected throughout with an NFPA 13 sprinkler system, but not those buildings protected with an NFPA 13R system. In addition, the increased floor areas are not applicable to Group H-1, H-2, or H-3 occupancies. The code considers that the conditions requiring the installation of sprinkler systems in these three high-hazard occupancies are such that the sprinkler system should not also be used to provide additional allowable area. In a mixed-occupancy building, this restriction is only applicable to the Group H-2 and H-3 portions. Other occupancies within the building are permitted to utilize the allowable areas established for building protected by an automatic sprinkler system as indicated in Section 506.2.2.1.

506.2.2   Mixed-occupancy, buildings.   For allowable area purposes, the introduction of multiple occupancies into the building requires an additional level of scrutiny. The methods for determining allowable area compliance in mixed-occupancy buildings differ based on the manner in which the various occupancies are addressed. Where the accessory occupancy method is utilized, the allowable area of the main occupancy will govern. Where occupancies are regulated under the nonseparated occupancy method, the allowable area is limited to that of the most restrictive occupancy involved. In the case of separated occupancies, the unity formula must be used to determine if the allowable area is exceeded. Each individual story of a mixed-occupancy building must be evaluated for allowable area compliance, based on the process as addressed in Section 508.1. Further discussion can be found in the narratives of Section 508 for each of the three mixed-occupancy methodologies.

Similar to the approach described in Section 506.2.1 for buildings of three stories or less above grade plane, a mixed-occupancy building must be analyzed on a story-by-story basis and each individual story must be shown to be in compliance with the applicable provisions of Section 508.1. In all cases, regardless of the building's number of stories, each individual story must comply for allowable area purposes. For those buildings exceeding three stories above grade plane the approach differs. Again, each story in the building must comply individually. In addition, the aggregate sum of the ratios of the actual areas of each story divided by the allowable areas of such stories cannot exceed three. The procedure is illustrated in Application Example 506-3. In buildings designed as separated occupancies and equipped with an NFPA 13R sprinkler, the sum of the ratios of actual area divided by the allowable areas shall not exceed four.

In all cases for mixed-occupancy conditions, reference is made to the provisions of Section 508.1, which identify the three options for mixed-occupancy determination (accessory occupancies, nonseparated occupancies, and separated occupancies). Since each story is to be evaluated independently in a multistory condition, it is possible for the designer to utilize different mixed-occupancy options on various stories within the building.

506.3 Frontage increase. The initial requirement of the code, insofar as a frontage increase is concerned, is that it adjoin or have access to a public way. Thus, the structure could extend completely between side lot lines and to the rear lot line, and be provided with access from only the front of the building, and still potentially be eligible for a small frontage increase. Therefore, it follows that if a building is provided with frontage consisting of public ways and/or open space for an increased portion of the perimeter of the building, some benefit should accrue based on better access for the fire department. Also, if the yards or public ways are wide enough, there will be a benefit that is due to the decreased exposure from adjoining properties or possibly other buildings on the same lot.

Because of the beneficial aspects of open space adjacent to a building, the IBC permits increases in the tabular areas established from Table 506.2 based on the amount of open perimeter and width of the open space and public ways surrounding the building. For any open space to be effective for use by the fire department, it is mandated that it be accessed from a public way or a fire lane so that the fire department will have access to that portion of the perimeter of the building that is adjacent to open space. See Figure 506-2.

Open space and public ways—what can and cannot be used. In addition to allowances for public ways, the IBC uses the term *open space* where related to frontage increases in the determination of allowable floor areas. Although the term *open space* is not specifically defined in the IBC, the definition of a *yard* is an open space unobstructed from the ground to the sky that is located on the lot on which the building is situated. It is logical that this definition is consistent with the intended description of open space. This definition seems to preclude the storage of pallets, lumber, manufactured goods, home improvement materials, or any other objects that similarly obstruct the open space. However, it would seem reasonable to permit automobile parking, low-profile landscaping, fire hydrants, light standards, and similar features to occupy the open space. These types of obstructions can be found within the public way, so their allowance within the open space provides for consistency. Because a yard must be unobstructed from

GIVEN: A fully sprinklered, four-story, Type IIA hotel containing a Group A-2 restaurant, Group A-3 meeting rooms, and Group M retail stores. The floor areas of each occupancy are as shown. Inadequate frontage provides for no area increase.

DETERMINE: If the building complies with the allowable area provisions of Chapter 5 if the occupancies are regulated under the provisions of Section 508.4 (separated occupancies).

A-2 8,000 sq ft	R-1 38,000 sq ft	
	R-1 46,000 sq ft	
	R-1 46,000 sq ft	
A-3 24,000 sq ft	R-1 8,000 sq ft	M 14,000 sq ft

SOLUTION FOR TOTAL BUILDING AREA:

Allowable area per occupancy
- A-2: 46,500
- A-3: 46,500
- M: 64,500
- R-1: 72,000

Sum of ratios calculation per story

1st story $\dfrac{24{,}000}{46{,}500} + \dfrac{8{,}000}{72{,}000} + \dfrac{14{,}000}{64{,}500} = 0.52 + 0.11 + 0.22 = 0.85 \leq 1.0$

2nd story $\dfrac{46{,}000}{72{,}000} = 0.64 \leq 1.0$

3rd story $\dfrac{46{,}000}{72{,}000} = 0.64 \leq 1.0$

4th story $\dfrac{8{,}000}{46{,}500} + \dfrac{38{,}000}{72{,}000} = 0.17 + 0.53 = 0.70 \leq 1.0$

Each story complies

Sum of ratios calculation for building
$0.85 + 0.64 + 0.64 + 0.70 = 2.83 \leq 3.0$

Entire building also complies

ALLOWABLE AREA OF MULTISTORY MIXED-OCCUPANCY BUILDING

For SI: 1 square foot = 0.093 m^2.

Application Example 506-3

the ground to the sky, open space widths should be measured from the edge of roof overhangs or other projections, as shown in Figure 506-3.

Regarding the use of public ways for providing frontage increases, the portion of public way that should be used for determining area increases seems to cause confusion. Should the full width of the public way or only the distance to the centerline be used? The confusion evolves from the definition of fire separation distance as established in Chapter 2, which states that

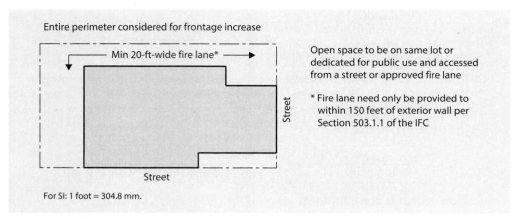

Figure 506-2 **Open space access.**

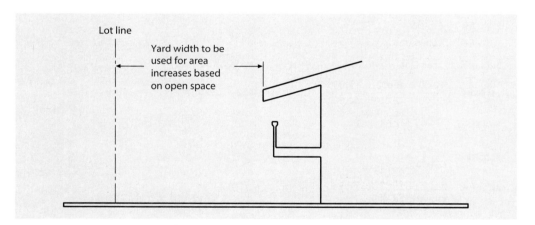

Figure 506-3 **Measurement of open space.**

fire separation distance is measured from the building face to the centerline of a street, alley, or public way. However, the requirement to use the centerline is limited to fire separation distance and is not applicable to Section 506.3. For determining frontage increases for open space, the full width of the public way may be used by buildings located on both sides of the public way as expressed in Item 2 of Section 506.3.2.

The following type of question is also sometimes asked: "Why cannot I use the big open field next door for an area increase?" Section 506.3.1 specifically mandates that open space used for a frontage increase must be on the same lot as the building under consideration, or alternatively, dedicated for public use. There is a good reason for this limitation, insofar as the owner of one parcel lacks control over a parcel owned by another and, thus, the open space can disappear when the owner of "the big open field" decides to build on it. One method by which some jurisdictions have allowed such large open spaces to be used is by accepting joint use of shared yards. It is typically necessary that a recorded restrictive covenant be executed to ensure that the shared space will remain open and unoccupied as long as it is required by the code. The creation of a no-build zone does not seem unreasonable insofar as the aim is to maintain open spaces between buildings. Any covenant should be reviewed by legal counsel to be sure it will accomplish what is intended. In addition, it should clearly describe the reason and applicable code section so that any future revisions or deletions may be considered if the owners wish to terminate such an agreement. In such an event, each building should be brought into current code compliance, or the agreement would be required to remain in effect.

Whereas use of a public way as open space is permitted by the IBC, other publicly owned property is generally not, because the building official usually has no control over the long-range use of publicly owned property, and there is little assurance that such property will be available as open space for the life of the building. Remember that what is today's publicly owned open parking lot could become tomorrow's new city hall, and the open space used to justify area increases would no longer exist. Whereas Section 506.3.1 allows publicly owned property to be considered open space, the intent is such that the property be permanently dedicated for public use and maintained as unobstructed. The term *public way* was used in place of streets because its definition allows the use of a broader range of publicly owned open space while still allowing the building official some discretion as to the acceptability of a particular parcel. *Public way* usually conjures up visions of streets and alleys, but how about other open spaces such as power line right-of-ways, flood-control channels, or railroad rights-of-way? Many such open spaces are generally acceptable, provided there is a good probability that they will remain as open space during the life of the building for which they will serve. Power lines and flood-control channels are usually good bets for longevity, but railroad routes are often abandoned and, therefore, may not be as good a bet. There is also an expectation that the public way is maintained in an unobstructed condition to allow for fire department access, which potentially would disallow the use of waterways and similar features. If the public way does not provide for fire department access, its use for a frontage increase is prohibited. It should be noted that the definition for public way requires any such public parcel of land, other than a street or alley, to lead to a street. Figure 506-4 provides a visual summary of open space and public ways that could be used for open-space area increases.

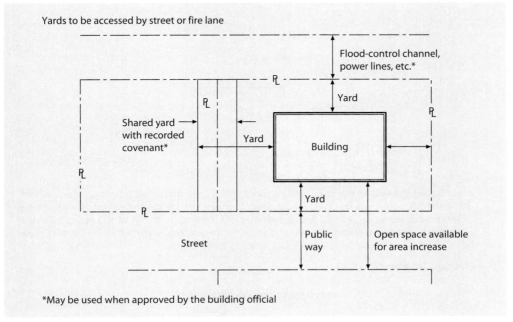

Figure 506-4 **Yards and public ways available for area increases.**

How much increase? Where public ways or open space having a minimum width of 20 feet (6,096 mm) adjoin at least 25 percent of the building's perimeter, the code permits an increase in the building's allowable area per story as established in Table 506.2. The area increase that is due to frontage (I_f) can be determined through the application of Table 506.3.3. The increase is based on two criteria: (1) the percentage of the building's perimeter that has open space or public ways of at least 20 feet (6,096 mm) in width, and (2) the smallest of such open space or public ways that is 20 feet (6,096 mm) or more wide.

Based on the table, the maximum area increase permitted is 75 percent (I_f of 0.75) of the NS tabular area factor set forth in Table 506.2 as shown in Application Example 506-4. In order to receive the maximum increase permitted by the IBC, at least 75 percent of the entire perimeter of the building must adjoin a public way or open space having a width of at least 30 feet (9,144 mm). Where a lesser amount of the perimeter has adequate open area, or where some portions of the complying open space are less than 30 feet (9,144 mm) in width, or both, the area increase for frontage will be reduced as illustrated in Application Examples 506-5 and 506-6.

Whereas 75 percent is generally the largest allowable frontage increase, a greater area increase is permitted for those buildings that comply with all of the requirements for unlimited-area

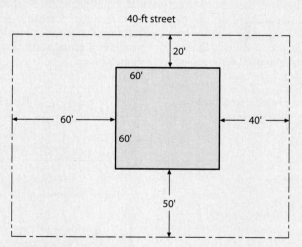

GIVEN: Building width, yards, and street as shown
DETERMINE: Frontage increase for area modification

For SI: 1 foot = 304.8 mm.

In evaluating the allowable area increase for open space surrounding the building, at least 30 feet (9,144 mm) of frontage is provided for the entire perimeter. In the determination of the frontage increase (I_f) based on Table 506.3.3, the percentage of building perimeter of 100 percent and the open space of at least 30 feet (9,144 mm) for the entire perimeter results in a frontage increase factor of 0.75.

Frontage increase (I_f) = 0.75

MAXIMUM FRONTAGE INCREASE

Application Example 506-4

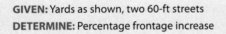

GIVEN: Yards as shown, two 60-ft streets
DETERMINE: Percentage frontage increase

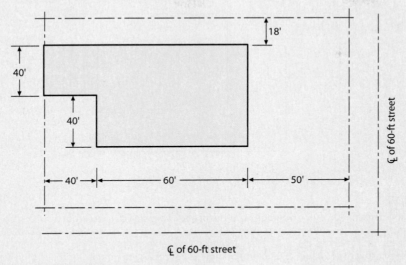

For SI: 1 foot = 304.8 mm.

In evaluating the allowable area increase for open space surrounding the building, it is necessary to determine the total perimeter as well as the portion of the perimeter that is at least 20 feet (6,096 mm) in width. In this example, the total perimeter is 360 feet (109,728 mm), with 220 feet (67,056 mm) of perimeter adjoining a minimum 20-foot (6,096-mm) open space. This results in a percentage of open building perimeter of 61 percent (220/360). All open space that is at least 20 feet (6,096) in width is also 30 feet (9,144 mm) or more. Therefore, in the determination of the frontage increase (I_f) based on Table 506.3.3, the percentage of building perimeter of 61 percent and the open space of at least 30 feet (9,144 mm) for those portions of the exterior wall with a frontage of at least 20 feet (6,096 mm) results in a frontage increase factor of 0.50.

Frontage increase (I_f) = 0.50

FRONTAGE INCREASE

Application Example 506-5

buildings as described in Section 507, other than compliance with the 60-foot (18,288-mm) open space or public way requirement. Where such a condition exists, Table 506.3.3.1 is to be applied.

A maximum frontage increase of 150 percent can be achieved based on at least 75 percent of the perimeter being open with a minimum width of 55 feet (16,764 mm). Once 60 feet (18,288 mm) of accessible open space and public ways is obtained for 100 percent of the building's perimeter, the provisions of Section 507 are applicable and Table 506.3.3.1 need not be used. An example of the increased frontage increase is shown in Application Example 506-7

Must all qualifying open space be considered? When applying Table 506.3.3, it may be advantageous for the designer to not consider portions of the open space that are 20 feet (6,096 mm) or more in width. By electing not to take credit for a qualifying yard, there are

GIVEN: A building fronted by 60-foot street and three yards as shown

DETERMINE: The quantity W to be used in the calculation of I_f (area increase due to frontage)

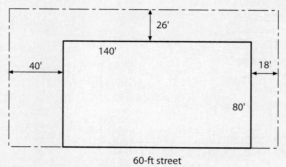

60-ft street

For SI: 1 foot = 304.8 mm.

SOLUTION:

In the determination of the frontage increase for allowable area purposes, both the percentage of perimeter and the width of the smallest public way and/or open space that is 20 feet (6,096 mm) or greater must be considered. In this example, the available perimeter is 82 percent (360/440) and the smallest yard of 20 feet (6,096 mm) or more is 26 feet (7,925 mm) in width. In applying Table 506.3.3, the frontage increase factor is 0.63.

Frontage increase (I_f) = 0.63

CALCULATION OF FRONTAGE INCREASE

Application Example 506-6

situations where the frontage increase (I_f) will be greater than if all yards with a width of at least 20 feet (6,096 mm) are considered. It is acceptable to ignore any open space where its inclusion in the application of Tables 506.3.3 will result in a smaller frontage increase.

The same approach may be taken when Table 506.3.3.1 is applied. It may be advantageous for the designer to not consider portions of the open space that are 30 feet (9,144 mm) or more in width. An example is illustrated in Application Example 506-7.

How much access be provided to an open space? The IBC provides no details as to the degree of fire department access required in order to consider open space for an allowable area increase; it only mandates that access be provided from a street or approved fire lane. It is clearly not the intent of the provisions to mandate a street or fire lane in the complying open space around a building in order to acquire the maximum frontage increase. However, fire personnel access from such streets or fire lanes is necessary. Although it is not a requirement to provide access around a building for fire department apparatus, other than that required by IFC Section 503.1.1, the frontage increase is based on the ability of fire personnel to physically approach the building's exterior under reasonable conditions. For example, where the space adjacent to the building is heavily forested or steeply sloped, the frontage increase addressed in Section 506.3 is not permitted. The presence of a lake or similar water feature next to a building would also reduce or prohibit an allowable area increase. The evaluation of each individual building and its site conditions is necessary to properly apply the code for fire department access.

Building Area 159

GIVEN: Fully sprinklered, one-story retail sales building, yards as shown, 40-ft street
DETERMINE: Percentage frontage increase for area purpose (I_f)

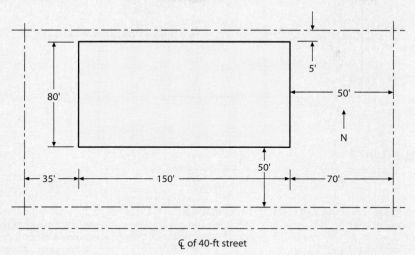

For SI: 1 foot = 304.8 mm.

Section 507.4 will permit the building shown to be unlimited in area provided it meets four conditions. In the example, all conditions are met except the building is not fully surrounded by a minimum 60-foot (18,288-mm) open space and/or yards. As such, Table 506.3.3.1 can be used to determine the frontage increase factor based on (1) a percentage of perimeter of 67 percent (310/460), based on those yards and/or streets having a width of at least 30 feet (9,144 mm), and (2) an open space of 35 feet (10,668 mm), based on the smallest yard or street with a width of at least 30 feet (9,144 mm). In applying Table 506.3.3.1, a frontage increase factor of 0.67 is available.

Frontage increase (I_f) = 0.67*

*A larger increase can sometimes be obtained by not considering some degree of complying open space when applying the table. In this case, by not considering the open space adjacent to the west wall, the frontage increase factor is 1.00 based on (1) a percentage of perimeter of 50 percent (230/460) and (2) an open space of 70 feet (21,336 mm) [smallest yard of 30 feet (9,144 mm) or more].

Frontage increase (I_f) = 1.00

CALCULATION OF FRONTAGE INCREASE

Application Example 506-7

What other conditions affect the determination of any frontage increase? As expected, there are other situations that can arise involving the determination of the frontage increase for a building. It should be noted that, as illustrated in Figure 506-5, the introduction of a fire wall in a large-area building will result in the loss of a portion of the open space at the building perimeter. Figure 506-6 illustrates the permitted frontage increase for an open space shared by two buildings on the same lot.

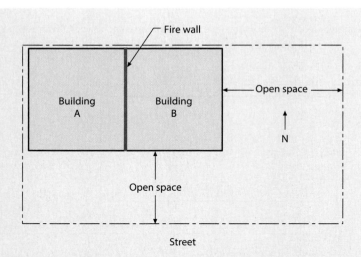

Where a fire wall divides a structure into separate buildings as shown, each building may be evaluated individually for allowable area purposes. The presence of a fire wall potentially eliminates open space available for a frontage increase. As illustrated, the fire wall eliminates the ability for Building A to utilize the open space to the east of Building B. Conversely, any open space to the west of Building A would not be available for use by Building B.

Figure 506-5 **Area increases with fire wall.**

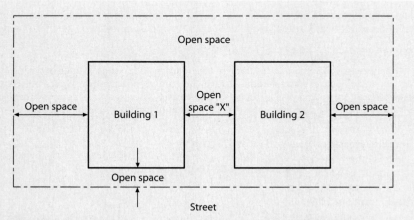

Where two buildings are adjacent on the same lot, the open space "X" between such buildings may be fully applied to both buildings for the purpose of determining an allowable area increase for frontage.

Figure 506-6 **Area increases for open spaces.**

Section 507 *Unlimited-Area Buildings*

There are many cases where very large undivided floor areas are required for efficient operation in such facilities as warehouses and industrial plants. Through the use of adequate safeguards, the IBC recognizes this necessity and allows unlimited areas for these uses under various circumstances. Large open floor space is also desirable for other applications; therefore, such allowances are also permitted for business and mercantile occupancies, as well as specific assembly and educational uses. The use of this section is typically intended to eliminate fire-resistive construction of the building that would be mandated based on the area limitations of Section 506. Contrary to the general philosophy that as a building increases in floor area the allowable types of construction become more restrictive, many of the unlimited-area uses permit the use of any construction type.

It is important to remember that Section 507 is not the only means under which a building can be unlimited in floor area. Those buildings containing occupancies other than Groups H-1 and H-2 may be unlimited in area where built of Type IA construction as set forth in Table 506.2. In addition, unlimited area Type IB buildings are permitted for a significant number of moderate- and low-hazard occupancies. Section 503.1.1 also permits unlimited floor areas in special industrial occupancies that require large areas to accommodate equipment and processes.

507.1 General. Historically, structures constructed under the provisions for unlimited area have performed quite well in regard to fire and life safety. A number of occupancy groups, particularly those relating to institutional, residential, and high-hazard occupancies, are excluded from the benefits derived from the provisions for unlimited-area buildings. Such occupancies pose unacceptable risks that are due to their unique characteristics. As a general rule, only those occupancy classifications specifically identified in this section are permitted to be housed in buildings allowed to be unlimited in area by Section 507. For example, Group I occupancies are not specifically permitted by any of the provisions addressing unlimited-area buildings. Therefore, it would appear no amount of Group I is permitted in such structures. However, Section 507.1.1 identifies the method that allows for a limited degree of such prohibited occupancies. Section 508.2.3 indicates that the allowable area for an accessory occupancy is to be based on the allowable area of the main occupancy. If the main occupancy is permitted by Section 507 to be in an unlimited-area building, a complying accessory occupancy also enjoys the same benefit. Figure 507-1 illustrates this condition.

It is also not uncommon for two or more occupancies regulated under the provisions of Section 507 to be located within the same building. For example, assume a one-story building contains a Group M furniture store and its associated Group S-1 warehouse. Because both Group M and Group S-1 occupancies are permitted in an unlimited-area building complying with Section 507.4, both occupancies are permitted to be located in the same unlimited-area building. See Figure 507-2. Any fire-resistive separation requirement would be based on the applicable mixed-occupancy method of Section 508 applied to the building.

All buildings regulated under the unlimited-area building provisions of Section 507 are required to be surrounded by public ways and yards of substantial width. This continuous open space provides a means for the fire service to access the building as necessary from the exterior, while at the same time maintains a sizable separation from any other structures on the site. The required open space is to be measured from all points along the building's exterior wall in all directions, ensuring that the full perimeter of the building is provided with continuous open space. See Figure 507-3. This differs somewhat from the right-angle method established for an allowable area frontage increase as set forth in Section 506.3 where such continuity of open space is not required at the building corners. Consistent with the measurement method of Section 506.3 to gain a frontage increase, the open space width adjacent to those exterior walls fronting on a public way is permitted to include the entire width of the public way.

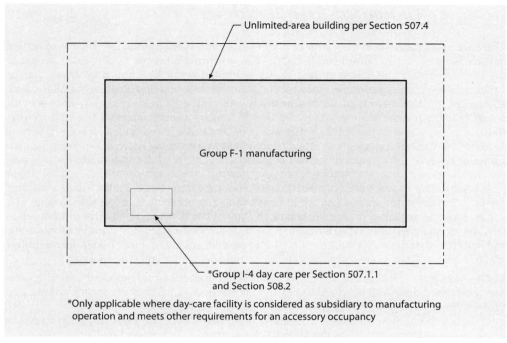

Figure 507-1 **Accessory occupancy in an unlimited-area building.**

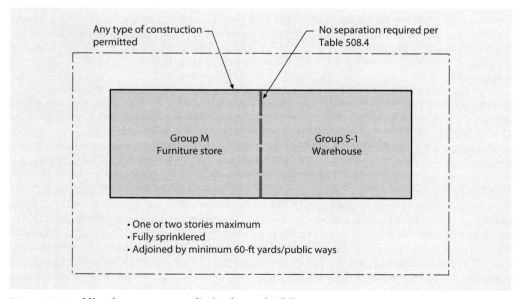

Figure 507-2 **Mixed-occupancy unlimited-area buildings.**

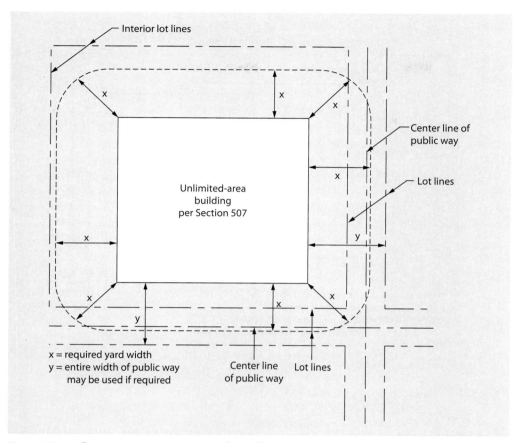

Figure 507-3 **Open space measurement for unlimited-area building.**

507.2.1 Reduced open space. There may be situations where the full 60 feet (18,288 mm) of open space or public ways surrounding an unlimited-area building cannot be obtained or its use is undesirable. The IBC permits a reduction in the required open space under very specific conditions, as illustrated in Figure 507-4. In no case may the permanent open space be reduced to less than 40 feet (12,192 mm) in width. By limiting the amount of reduced open space, requiring a high degree of exterior wall fire resistance, and mandating opening protectives in all openings in the exterior wall facing the reduced open space, the code provides protection equivalent to full 60-foot (18,288-mm) yards or public ways.

507.3 Nonsprinklered, one-story buildings. This section addresses a Group F-2 or S-2 occupancy in a one-story building of any type of construction. Both Group F-2 and S-2 occupancies by definition are low-hazard manufacturing or storage uses, which the code considers to be low fire risks. Fire risk is further reduced by requiring that the building be surrounded by yards or streets with a minimum width of 60 feet (18,288 mm). The relatively low fire loading expected in such occupancies is why the code does not require the installation of an automatic fire-sprinkler system for this application. The use of this provision is applicable to buildings of all construction types; however, it is anticipated that the structure will mirror the contents in the absence of combustible elements.

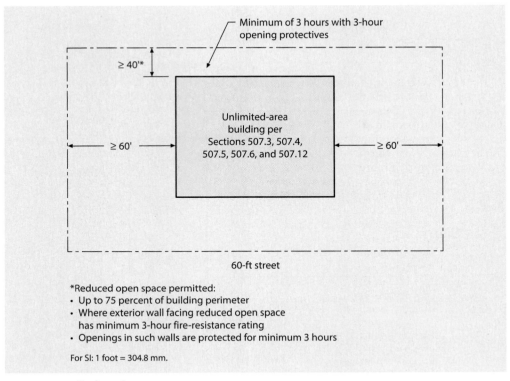

Figure 507-4 **Reduced open space.**

507.4 Sprinklered, one story buildings. Specific moderate-hazard occupancies, limited to Groups B, F, M, and S, are permitted single-story buildings of unlimited area where the building is completely surrounded by streets or yards not less than 60 feet (18,288 mm) in width and the entire structure is protected by an automatic fire-sprinkler system. The limitation of one story does not apply where the building is utilized for rack storage, provided the building is of noncombustible construction, and is not intended for public access. This unlimited-area storage facility, required to conform with Chapter 32 of the IFC regulating high-piled combustible storage, is permitted to be of any height.

In most applications, the use of the unlimited-area provisions simply means that the type of construction is not regulated, regardless of the size of the building's floor area. The code assumes that the amount of combustibles and, consequently, the potential fire severity are relatively moderate. In addition, the protection provided by the automatic fire-sprinkler system plus the fire-department access furnished by the 60-foot (18,288-mm) yards or streets surrounding the building reduce the potential fire severity to such a level that unlimited area is reasonable.

One-story Group A-4 occupancies are also permitted to be of unlimited area where a sprinkler system is provided throughout and a minimum 60-foot (18,288-mm) open space surrounds the building. However, because of the increased risk posed by the anticipated high number and concentration of occupants in such a structure, the construction type of the building is limited to Type I, II, III, or IV. The automatic sprinkler system required in an unlimited-area building housing a Group A-4 occupancy may be omitted in those specific areas occupied by indoor participant sports, including tennis, skating, swimming, and equestrian activities.

Such an omission mandates that exit doors from the participant sports areas lead directly to the outside, and the installation of a fire-alarm system with manual fire-alarm boxes is required. It is anticipated that such sports areas will have little, if any, combustible loading if the uses are limited to those described in the code. If there is a reasonable expectation that other types of uses could occur, it would be inappropriate to omit the sprinkler system in such areas.

Group A-1 and A-2 occupancies are permitted in a mixed-occupancy building when in compliance with the general limitations of Section 507.4 plus additional criteria as established in Section 507.4.1. This allowance does not grant the designated Group A occupancies unlimited area, but rather allows such assembly occupancies to be located within a Group B, F, M, S, or A-4 unlimited-area building under specified conditions. The building must be classified as Type I, II, III, or IV construction. In addition, the Group A-1 and A-2 assembly occupancies must be separated with fire barriers from other occupancies within the building, in accordance with the separated occupancy provisions of Section 508.4.4. For example, in an unlimited-area retail sales building, a Group A-2 restaurant would be required to be separated from a Group M sales tenant by a minimum 2-hour fire-resistance-rated fire barrier. No reduction in the minimum required fire-resistance rating of the fire barrier is permitted for the presence of an automatic sprinkler system. Each individual Group A tenant would also be limited in area by the provisions of Section 506. No additional height increases would apply to the Group A occupancies because of the specific Section 507.4 limitation of one story for the entire building. An additional requirement mandates that all required means of egress from the assembly spaces exit directly to the exterior of the building. See Figure 507-5. Application of this provision does not require the assembly occupancy to be accessory to the major use of the building, nor is the assembly floor area limited to 10 percent of the floor area. As an additional note, it would seem appropriate that the provision also extend to Group A-3 occupancies, as such uses are typically considered equal or lesser in hazard level to the Group A-1 and A-2 classifications.

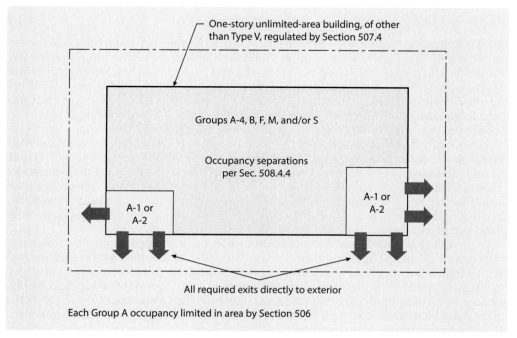

Figure 507-5   **Group A occupancies in unlimited-area buildings.**

507.5 Two-story buildings. In Groups B, F, M, and S, the unlimited-area provisions also apply to structures that are two stories in height. Minimum 60-foot (18,288-mm) open space or public ways must surround the building, and an automatic fire-sprinkler system is required throughout the structure.

507.6 Group A-3 buildings of Type II construction. Although most assembly occupancies are viewed as relatively high hazard because of the concerns associated with the number and concentration of the occupants, certain types of uses classified as Group A-3 occupancies are considered as moderate hazard. Therefore, it is possible to utilize the unlimited-area provisions for specific Group A-3 uses where the specified criteria are met.

Such buildings allowed to be of unlimited area are limited to one story in height, must be of Type II noncombustible construction, and contain only those specific types of assembly uses listed in the code. By limiting the types of buildings to places of religious worship, gymnasiums (without spectator seating), lecture halls, and similar uses, it is anticipated that the fire loading is relatively low. Buildings such as libraries, museums, and similar uses pose a higher risk that is due to the large amount of combustibles expected to be present. The potential for combustible loading is further reduced by the prohibition of a stage as a part of the use, although a platform is acceptable. Installation of an automatic sprinkler system is mandated, as is the presence of a minimum 60-foot (18,288-mm) open space around the building.

507.7 Group A-3 buildings of Types III and IV construction. Buildings of Type III or IV construction are granted unlimited-area status when housing specified Group A-4 occupancies in a manner similar to those Type II buildings identified in Section 507.6. The same requirements apply for sprinkler protection and adequate open space. In addition, the assembly use must be located relatively close to the exterior ground level to expedite the exiting process. As a part of this requirement, any elevation change from the building to the grade level must be accomplished by ramps rather than stairs. This further provides for an efficient means of egress from the assembly building. The additional limitation regarding floor-level location is mandated due to the combustible nature of the building's construction.

507.8 Group H occupancies. Because many large industrial operations, both manufacturing and warehousing, have a need to utilize a limited quantity of high-hazard materials in some manner, it is necessary that Group H-2, H-3, and H-4 occupancies be permitted to a small degree in Group F and S unlimited-area buildings. Because of the allowances given to buildings of unlimited area, it is critical that the high-hazard occupancies be strictly limited in floor area and adequately separated from the remainder of the building.

There are four factors that limit the allowable floor area of the permitted Group H occupancies: the specific classification of the Group H occupancy, the building's type of construction, its floor area, and the location of the high-hazard uses in relationship to the building's exterior wall. Where high-hazard occupancies are located on the perimeter of the building, fire-department access is enhanced, and exposure to interior areas is reduced. Accordingly, the permitted floor area of the Group H occupancies located on the building's perimeter is considerably greater than that allowed for such high-hazard uses completely surrounded by other interior portions of the unlimited-area building.

Where Group H-2, H-3, and H-4 occupancies are located at the perimeter of the unlimited-area Group F or Group S building, their size is restricted by the area limitations of Section 506. In a condition where the high-hazard occupancy is totally enclosed by the unlimited-area building, the size of the Group H occupancy is limited to only 25 percent of the area limitations specified in Section 506. Both of these conditions are shown in Application Example 507-1. The example also illustrates that multiple Group H occupancies that are not located at the perimeter of the building are limited in size based on the aggregate floor area of such occupancies. Similarly, where multiple Group H occupancies do occur on the building's perimeter, the maximum permitted size is also based on the total of all such occupancies. In all situations, the appropriate

GIVEN: A 130,000 square-foot factory building of Type IIB construction regulated under the unlimited area provisions of Section 507.4. One Group H-3 storage room is located on the building's perimeter. Multiple Group H-3 storage rooms are located such that they are not located along an exterior wall.

DETERMINE: The maximum allowable floor areas for the Group H-3 storage rooms.

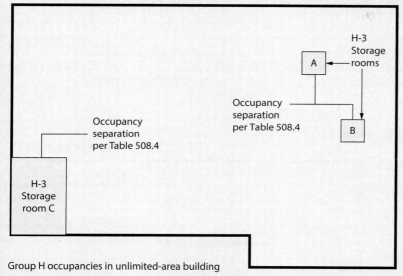

Group H occupancies in unlimited-area building

Aggregate of Group H: (rooms A, B, and C)	10 percent of 130,000 13,000 sf max, *nor*
	Table 506.2 with frontage 14,000 + 7,000 = 21,000 sf
	∴ Aggregate limit of 13,000 sf
Located within building: (rooms A and B)	25 percent of 14,000 ∴ Aggregate limit of 3,500 sf
Located on perimeter: (room C)	13,000 – (area of rooms A and B) ∴ 9,500 sf to 13,000 sf (depending on actual area of rooms A and B)

UNLIMITED AREA GROUP F OR S BUILDING WITH GROUP H OCCUPANCIES

For SI: 1 square foot = 0.093 m².

Application Example 507-1

fire-barrier assemblies mandated by Table 508.4 must be provided. Regardless of the location of the applicable Group H occupancies, in no case may such aggregate floor areas exceed 10 percent of the area of the entire building. The provisions addressing liquid use, mixing and dispensing rooms, liquid storage rooms, and spray-paint booths simply replicate those of Sections 415.6.1, 415.6.2, and 415.6.3, in order to remind code users of the limitations and allowances placed on such rooms and areas.

507.9 Unlimited mixed-occupancy buildings with Group H-5. The Group H-5 classification was created some time ago to standardize regulations for semiconductor manufacturing facilities. The Group H-5 category requires engineering and fire-safety controls that reduce the overall hazard of the occupancy to a level regarded to be equivalent to that of a moderate-hazard Group B occupancy. The mitigating provisions of Section 415.11 have effectively equalized the two occupancies in terms of relative hazard such that, in concert with the three additional criteria established in this section, the allowance for Group H-5 unlimited-area buildings is deemed to be appropriate.

A unique condition not applied to other unlimited-area buildings regulated by Section 507 is the need to limit the unseparated floor area of Group H-5 use. Although an unlimited amount of Group H-5 floor area is permitted within a complying unlimited-area building, it may be necessary to subdivide the floor area using minimum 2-hour separations due to the limits of Item 3. An application of the provisions and the required fire barrier separations is illustrated in Figure 507-6.

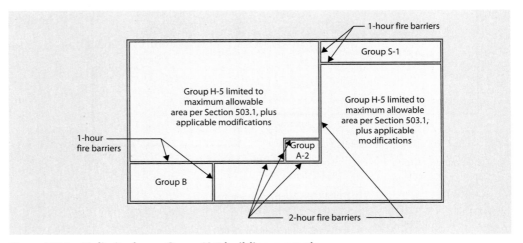

Figure 507-6 **Unlimited area Group H-5 building example.**

507.10 Aircraft paint hangar. The provisions of Section 412.5 address aircraft painting operations where the amount of flammable liquids in use exceeds those maximum allowable quantities listed in Table 307.1(1). Classified as Group H-2, such aircraft paint hangars must be fully suppressed in accordance with NFPA 409 and of noncombustible construction. One-story hangars may be unlimited in floor area where complying with Section 412.5, provided they are surrounded by public ways or yards having a width of at least one and one-half times the height of the building.

507.11 Group E buildings. Because of the various fire- and life-safety concerns associated with educational occupancies, buildings housing uses classified as Group E are typically not eligible for consideration as unlimited-area buildings. Only when the following six criteria are met does the IBC permits the area of a Group E educational building to be unlimited based on Section 507.

1. The building is limited to one story above grade plane.
2. The building is of Type II, IIIA, or IV construction.

3. Two or more means of egress are provided from each classroom.
4. At least one means of egress from each classroom is a direct exit to the exterior of the building.
5. An automatic sprinkler system is provided throughout the building.
6. The building is surrounded by open space at least 60 feet (18,288 mm) in width.

507.12 Motion-picture theaters. Because of their limited combustible loading, motion-picture theaters are granted unlimited floor areas in a manner relatively consistent with other moderate-hazard uses. This specific allowance is not extended to the other uses classified as Group A-1, such as performance theaters, because of their higher fire-severity potential.

In order to address the concerns related to the high-density, high-volume occupant loads often encountered in motion-picture theaters, unlimited area is not permitted where the building is of Type III, IV, and V combustible construction. This restriction further limits the fire load contained within the building construction. In concert with Section 507.4, a fire-sprinkler system must be installed throughout, and minimum 60-foot (18,288-mm) open areas must completely surround the building.

The application of this provision differs from the allowance granted in Section 507.4.1. That provision permits any Group A-1 occupancy, including motion-picture theaters, to be located in an unlimited-area building complying with Section 507.4, provided the limitations of the exception are met. However, it does not allow Group A-1 occupancies themselves to be unlimited in area. On the other hand, this section permits a Group A-1 theater complex to be unlimited in area, provided it is fully sprinklered, of Type II construction, and surrounded by adequate open space.

507.13 Covered and open mall buildings and anchor buildings. The provisions of Section 402 for unlimited-area covered and open mall buildings are referenced for convenience purposes. Note that although the reduction in open space permitted by Section 507.2.1 is not applicable to covered or open mall buildings, a similar reduction is permitted by the exception to Section 402.2.

Section 508 *Mixed Use and Occupancy*

Multiple uses commonly occur within a single building. Each use creates its own distinct hazards, many of which are addressed by the code. However, many of the hazards are similar in nature, which allows the varied uses to be grouped into categories that recognize the common concerns. These categories are identified in Chapter 3 as occupancy groups. Where two or more occupancy groups share a single building, it is necessary to evaluate their relationship to each other as a mixed-occupancy condition. This section provides various methods to address such relationships in regard to occupancy classification, allowable height and area, fire protection systems, and fire-resistance-rated separation.

It should be noted that live/work uses as addressed under the provisions of Section 508.5 are not to be regulated as mixed-occupancy conditions. Section 508.5.2 indicates that live/work units are to be classified as a single occupancy, a Group R-2. As such, the Section 508.5 essentially becomes an exception to applying the mixed occupancy requirements of Section 508.

508.1 General. A mixed-occupancy condition exists where two or more distinct occupancy groups are determined to exist within the same building. In fact, it is quite common for a building to contain more than one occupancy group. For example, hotel buildings of various sizes not only house the residential sleeping areas, but may contain administrative offices, retail and service-oriented spaces, parking garages, and, in many cases, restaurants, conference rooms, and other assembly areas. Each of these uses typically constitutes a distinct and separate occupancy as far as Chapter 3 of the IBC is concerned. Because this situation is not uncommon, the code specifies requirements for buildings of mixed occupancies. Under such circumstances, the designer has available several methodologies (accessory occupancies, nonseparated occupancies, and separated occupancies) to address the mixed-occupancy concerns. The methods that have been established represent a hierarchy of design prerogatives that may be utilized at the discretion of the design professional. Although compliance is required with only one of the three mixed-occupancy methods, it is acceptable to utilize two or even all three methods in the same building, as shown in Figure 508-1. The common format utilized in presenting the requirements for each of the methods allows for a comparison of the provisions. This should assist in determining the most appropriate method, or methods, for the building under consideration. A simple comparison of the three mixed-occupancies methods is shown in Figure 508-2.

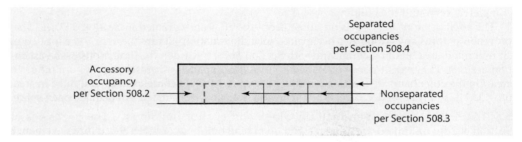

Figure 508-1 **Combination of methods.**

	Accessory Occupancies Section 508.2	**Nonseparated Occupancies Section 508.3**	**Separated Occupancies Section 508.4**
Occupancy Classification	Individually classified	Individually classified	Individually classified
Allowable Area	Based on allowable area of main occupancy	Based on most restrictive of occupancies within building	Determined such that sum of the ratios cannot exceed 1.0
Allowable Height	Based on allowable height of main occupancy	Based on most restrictive of occupancies within building	Based on general provisions of Section 504
Separation	No separation required	No separation required	Separation as required by Table 508.4
Special Conditions	1. Subsidiary to main occupancy 2. Aggregate area ≤ 10 percent of story 3. Aggregate area ≤ value in Table 506.2 for nonsprinklered buildings 4. Not applicable to Groups H-2, H-3, H-4, and H-5	1. Most restrictive provisions of Ch. 9 apply to entire building 2. Not applicable to Groups H-2, H-3, H-4, and H-5	

Figure 508-2 **Summary of mixed-occupancy methods.**

It is important to recognize that there is no relationship between the mixed-occupancy provisions of Section 508.4 and the fire area concept utilized in Section 903.2 for automatic sprinkler systems. As set forth in Section 508.4.1, compliance with any of the three mixed-occupancy methods does not relieve the responsibility to comply with Section 901.7 and Table 707.3.10 regarding the proper separation of fire areas. An example is shown in Figure 508-3.

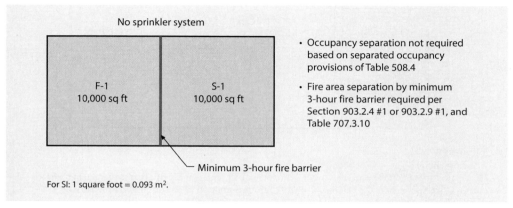

Figure 508-3 Occupancy separation versus fire area separation example.

508.2 Accessory occupancies. Those minor uses in a building that are not considered consistent with the major occupancy designation can potentially be considered accessory occupancies. They often are necessary or complimentary to the function of the building's major use, but have few characteristics of the major occupancy in regard to fire hazards and other concerns. Therefore, accessory occupancies must each be assigned to an occupancy group established in Chapter 3 based on their own unique characteristics. While maintaining the philosophy of a mixed-occupancy building, this section permits such relatively small accessory occupancies to be considered merely a portion of the major occupancy for fire separation purposes. A good example would be a lunchroom seating 120 persons and located in a large manufacturing facility. Whereas the individuals using the lunchroom are generally the same individuals who work elsewhere in the factory, the hazards encountered while they are occupying the lunchroom are quite different from those created in the Group F-1 manufacturing environment. Therefore, the lunchroom must be appropriately classified as a Group A-2 occupancy, creating a mixed-occupancy condition. However, through compliance with the accessory occupancy provisions of this section, the need for a fire-resistance-rated separation between the lunchroom and manufacturing area is eliminated. In fact, no physical separation of any kind would be mandated. It is important to note, however, that in spite of the absence of a fire separation, the two areas would maintain their unique occupancy classifications. They would continue to be classified as Groups A-2 and F-1, respectively, and the building would be considered a mixed occupancy.

As previously indicated, consideration as an accessory occupancy is only possible where the occupancy is subsidiary to the main occupancy of the building. There are several additional criteria that must also be met in order to utilize the accessory occupancy method. The occupancy under consideration cannot exceed 10 percent of the floor area of the story on which the accessory occupancy is located, nor more than that permitted by Table 506.2 for NS (nonsprinklered) buildings. See Figure 508-4. It is specified that the 10 percent limitation is based on the

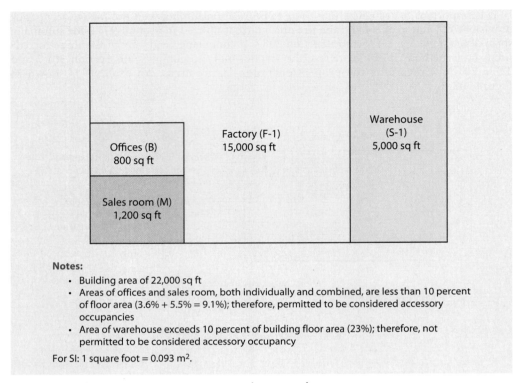

Figure 508-4 Aggregate accessory occupancies example.

aggregate floor areas of all accessory use areas, not individually. Multiple minor occupancies that cumulatively make up more than 10 percent of the total floor area could pose a hazard that the code does not anticipate. There are unique situations—such as where minor occupancies are adequately separated spatially or are of such different types of uses that their aggregate area is not relevant—where the regulation of minor uses as individual areas could potentially be considered. See Figure 508-5. The application of these limits is subject to the interpretation of the building official, based on conditions unique to each building under consideration. The general limitations applied to accessory occupancies are shown in Application Example 508-1. There is limited guidance in the code as to the determination of the allowable location of an accessory occupancy in a multistory building. The provisions address the building as a whole, with a reference to Section 504. In applying Section 504.2 addressing all mixed occupancies, it follows that the accessory occupancy cannot be located higher than that permitted by Table 504.4 for the main occupancy as illustrated in Figure 508-6.

Although the primary allowance provided by the accessory occupancy provisions is the lack of a required fire separation between the accessory occupancy and the remainder of the building, there is also a potential benefit that is due to the manner in which allowable area is regulated. The code calls for the allowable area of the accessory occupancy to be based on the main occupancy of the building. This approach typically allows for a greater allowable area, as well as allowable height, than would be permitted under the conditions for both non-separated occupancies and separated occupancies. Another important benefit provides for occupancies not normally permitted in unlimited-area buildings, as regulated by Section 507,

Mixed Use and Occupancy   173

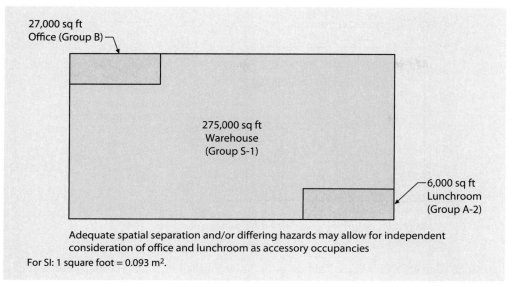

Adequate spatial separation and/or differing hazards may allow for independent consideration of office and lunchroom as accessory occupancies

For SI: 1 square foot = 0.093 m².

Figure 508-5   Individual accessory occupancies example.

General criteria for consideration as accessory occupancy:

√   1. **Accessory to major use.** Conference room's primary function limited to use by employees of factory.

√   2. **Does not exceed 10 percent of floor area.** Conference room does not exceed 7,225 square feet, 10 percent of total floor area.

√   3. **Does not exceed tabular nonsprinklered allowable area.** Conference room does not exceed 9,500 square feet allowed by Table 506.2 for (nonsprinklered) Group A-3 occupancy of Type IIB construction.

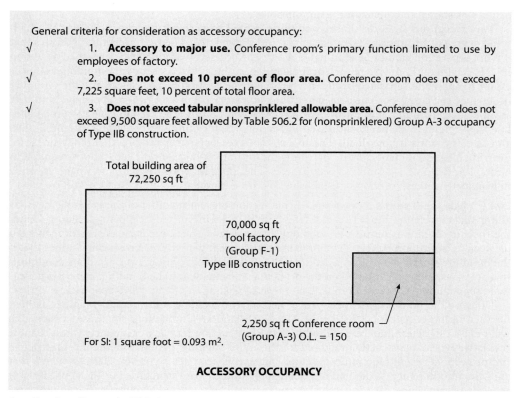

ACCESSORY OCCUPANCY

Application Example 508-1

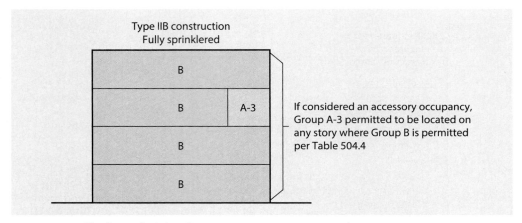

Figure 508-6 **Maximum height of accessory occupancy example.**

to be located in such buildings. This allowance is established in Section 507.1.1. For example, a Group A-2 lunchroom considered an accessory occupancy may be located in a two-story Group B unlimited-area building complying with Section 507.5 with no separation required between the two occupancies. This allowance, along with the limitations for accessory occupancies, is shown in Application Example 508-2.

Two exceptions indicate those conditions related to accessory occupancies under which some degree of fire-resistance-rated separation is mandated. The first exception indicates that the fire separation cannot be eliminated where the accessory occupancy is classified as Group H-2, H-3, H-4, or H-5. Where such occupancies occur in a mixed-occupancy building, it is necessary to apply the separated occupancy provisions of Section 508.4.4. Note that the Group H occupancy can continue to be considered as an accessory occupancy, where applicable, and take advantage of the other allowances provided by Section 508.2. The second exception mandates that the separation elements addressed in Section 420 for buildings containing dwelling units and sleeping units be provided. The use of the accessory occupancy methodology does not override the requirements for fire partition and/or horizontal assembly separations required in Group I-1, R-1, R-2, R-3, and R-4 occupancies. The application to Group R-4 occupancies, while not set forth in the exception, is established by the second paragraph of Section 310.5

508.3 Nonseparated occupancies. This section presents another of the available methods addressing the relationship between different occupancies in a mixed-occupancy building. Under the specific conditions of this methodology, fire-resistance-rated separations are not mandated between adjacent occupancies. The fundamental concept behind this provision assumes that if the building is designed in part to address the most restrictive and most hazardous conditions that are expected to occur based on the occupancies contained in the building, a fire-resistance-rated separation is not needed. In fact, no physical separation of any type is required.

Utilizing the nonseparated-occupancy method, the building must be individually classified for each unique occupancy that exists. The height and area limitations for those occupancies will be used to determine the required type of construction for each occupancy, with the most restrictive type of construction required for the entire building. In addition, the most restrictive fire-protection system requirements (automatic sprinkler systems and fire alarm systems) that apply to an occupancy in the building shall apply to the entire structure. For the application of other code provisions, each individual occupancy will be regulated by only the specific requirements related to that occupancy. A special condition mandates that if a high-rise building is

GIVEN: A 150,000-square-foot two-story office building (75,000 sf/story) housing a 3,750-square-foot employee lunchroom with an occupant load of 250 persons. The building is fully sprinklered, is of Type VB construction, and qualifies for unlimited area under the provisions of Section 507.5.

DETERMINE: Application of the accessory occupancy provisions of Section 508.2 for the lunchroom.

1. Is the accessory occupancy subsidiary to the building's major occupancy?
   Yes, the lunchroom is intended to serve the employees of the office space.
2. Is the accessory occupancy no more than 10 percent of the floor area of the story?
   Yes, 10 percent of 75,000 sq ft = 7,500 sq ft maximum; the lunchroom is 3,750 sq ft.

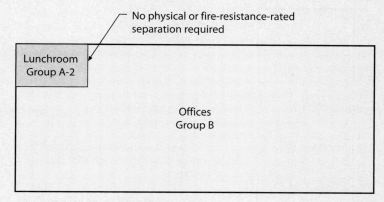

For SI: 1 square foot = 0.093m².

3. Is the accessory occupancy no larger than the tabular values in Table 506.2 for nonsprinklered buildings?
   Yes, the tabular value for a nonsprinklered Group A-2 of VB construction is 6,000 sq ft; the lunchroom is 3,750 sq ft.
4. What is the occupancy classification of the lunchroom?
   Group A-2, based on the individual classification of the use.
5. How are the other requirements of the IBC applied?
   The provisions for each occupancy are applied only to that specific occupancy.
6. What is the allowable height and area of the building?
   The building's allowable height and area are based on the major occupancy involved; in this case it is Group B. Based on the criteria of Section 507.5, the building is permitted to be unlimited in area and limited to two stories in height.
7. What is the allowable height location of the lunchroom?
   Based on Table 504.4 for Group B in a sprinklered Type VB building, the lunchroom is permitted to be located on either the first story or the second story.
8. What is the minimum required separation between the lunchroom and the remainder of the building?
   There is no fire-restrictive or physical separation required, owing to compliance with the provisions of Section 508.2 for accessory occupancies.

Application Example 508-2

regulated under the nonseparated-occupancy provisions, even those portions of the building that are not considered high rise must comply with the high-rise requirements. See Figure 508-7. Another important limitation is that the use of the nonseparated-occupancy method is significantly limited for Group H-2, H-3, H-4, and H-5 occupancies. The key benefit of this mixed-occupancy method is not available because occupancy separations as established in Table 508.4 must be provided where applicable. In addition, the nonseparated-occupancy provisions cannot be utilized to eliminate any required dwelling unit or sleeping unit separation required under the special provisions of Section 420.

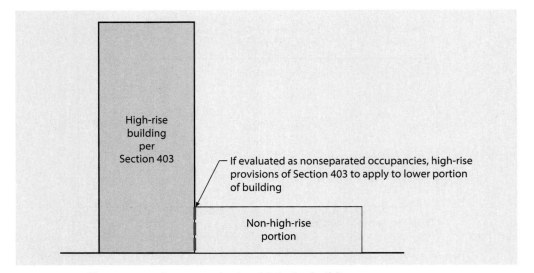

Figure 508-7 **Nonseparated occupancies in a high-rise building.**

The approach to understanding the rationale for the nonseparated-occupancy method may be better understood by viewing the structure as multiple single-use facilities. In evaluating the building described in Application Example 508-3, assume that the building is entirely a Group B occupancy. Based on that assumption, determine its maximum allowable height and area, as well as any required fire protection features. Now evaluate the building as if it were entirely a Group E occupancy, again addressing the maximum allowable height and area, along with the requirements for fire protection systems. By applying the most restrictive height, area, and fire protection provisions of both Group B and E occupancies to the entire building, the conditions for nonseparated occupancies can be determined.

Where no fire separation is provided between a Group I-2, Condition 2 occupancy, and other occupancies in the building due to the application of the nonseparated occupancy provisions, it is important that some critical fire protection features be extended beyond the healthcare portion of the building. Many of these restrictions directly support the defend-in-place concept that hospitals rely on. Within an individual fire area containing a Group I-2, Condition 2 occupancy, the more restrictive provisions of Sections 407, 509, and 712 apply to all occupancies within the fire area. In addition, the more restrictive means of egress provisions apply to the entire path of egress from the Group I-2, Condition 2 occupancy until arriving at the public way.

Section 407 contains provisions that are specific to Group I-2 occupancies that may not necessarily apply to the entire building. Within the fire area that contains the Group I-2, Condition 2

GIVEN: A mixed-occupancy building of Type VB construction, housing both Group B and Group E day-care occupancies. The design occupant loads have been established as 35 for the Group B and 60 for the Group E. Assume there is no automatic sprinkler system and no frontage increase is available.

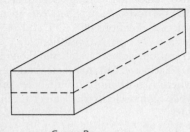

Group B
- 9,000 sq ft/story max
- 2 story max
- No fire protection systems required

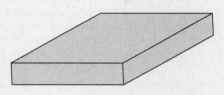

Group E
- 9,500 sq ft/story max
- 1 story max
- Manual fire alarm system required

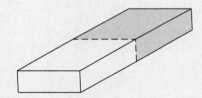

Group B/E as nonseparated occupancy
- 9,000 sq ft/story max
- 1 story max
- Manual fire alarm system required throughout building

For SI: 1 square foot = 0.093 m².

DETERMINE: The limitations that apply if the occupancies are to be considered nonseparated. As the more restrictive requirements for each occupancy must apply to the entire building, the following limitations are imposed in order to eliminate any form of occupancy separation:

SOLUTION:

Maximum allowable area:	9,000 sq ft based on Group B
Maximum allowable height:	1 story based on Group E
Fire protection features:	Manual fire alarm system required throughout building, based on Group E

NONSEPARATED OCCUPANCIES

Application Example 508-3

occupancy, safeguards including those for corridor construction, smoke compartmentation and hospital-specific egress must be maintained in order to support the defend-in-place concept. Separation and/or protection requirements for incidental uses that are specific to Group I-2 occupancies as established in Section 509 must also be provided where such uses occur within other occupancies in the same fire area as the hospital use. The vertical opening limitations set forth in Section 712 for Group I-2 occupancies must also be applied to other occupancies in the same fire area, addressing the concern of unprotected vertical openings between adjacent stories. In all cases, the most restrictive applicable provisions of Sections 407, 509, and 712 are to be applied where one of the nonseparated occupancies is a Group I-2, Condition 2.

Means of egress concepts such as the minimum width appropriate for stretcher and bed traffic must be applied from the hospital area through the exit discharge. This mandate is addressed in a general sense by Section 1004.4 indicating that where two or more occupancies utilize portions of the same means of egress system, those egress components shall meet the more stringent requirements of all occupancies that are served.

508.4 Separated occupancies. Using this method of addressing mixed-occupancy buildings, the code directs that each portion of the building housing a separate occupancy be individually classified and comply with the requirements for that specific occupancy. Furthermore, the code intends that each pair of adjacent occupancies be evaluated through Table 508.4 as to the relationship of the hazards involved, often mandating a fire-resistance-rated separation between them. The allowable number of stories above grade plane for each occupancy is independently regulated based on Table 504.4 based on the type of construction of the building. See Figure 508-8.

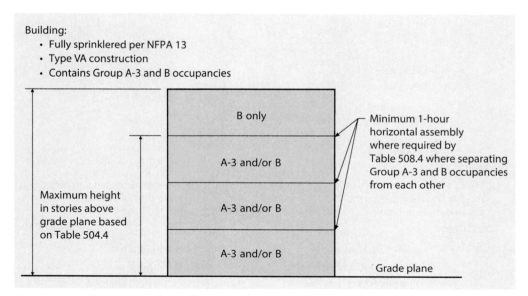

Figure 508-8 **Maximum height of separated occupancies example.**

For allowable floor-area considerations in a building having multiple occupancies, the code uses a formula that is very similar to the interaction formula used in structural engineering where two different types of stress are imposed on a member at the same time.

In the case of a mixed-occupancy building, the code uses this type of formula for the calculation of the allowable building area for each floor. For example, if there are three different occupancies in a building, the formula is as follows:

$$a1/A1 + a2/A2 + a3/A3 \leq 1.0$$

WHERE: $a1$, $a2$, and $a3$ represent the actual areas for the three separate occupancies, and $A1$, $A2$, and $A3$ represent the allowable areas for the three separate occupancies.

See Application Examples 508-4 and 508-5 for examples of this computation.

This formula essentially prorates the areas of the various occupancies so that the sum of percentages must not exceed 100 percent.

It is also appropriate to utilize a variation of this formula for the determination of the allowable area for the total building as evidenced by the provisions of Section 506.2.4. See Application Example 506-3. For all practical purposes, the need to evaluate the building as a whole is only necessary in buildings of four or more stories above grade plane. For two- and three-story buildings, if each story is compliant for allowable area purposes, the entire building will always comply.

Of special note is how the fire protection provisions of Chapter 9 are to be applied under the separated-occupancy method. Any fire sprinkler and fire alarm requirements applicable to an occupancy regulated as a separated occupancy are to be extended to all occupancies where no fire-resistance-rated separation is mandated by Table 508.4. As addressed in the specific

---

GIVEN: A one-story nonsprinklered building housing day care classified as Group E, Group B offices, and a Group A-3 conference room. The building is of Type VA construction. Insufficient open space is available for area increase purposes. Floor areas are as follows:

    Office (B)   4,500 square feet
    Assembly (A-3)   1,000 square feet
    Day care (E)   6,000 square feet

DETERMINE: If the building area is within the allowable area under the "separated occupancies" provisions.

SOLUTION: In accordance with Section 508.4.2:

$$\frac{\text{Actual area of office}}{\text{Allowable area of office}} + \frac{\text{Actual area of assembly}}{\text{Allowable area of assembly}} + \frac{\text{Actual area of E}}{\text{Allowable area of E}} \leq 1$$

$$\frac{4,500}{18,000} + \frac{1,000}{11,500} + \frac{6,000}{18,500} \overset{?}{\leq} 1$$

$$0.25 + 0.09 + 0.32 \overset{?}{\leq} 1$$

$$0.66 \leq 1, \text{ therefore OK}$$

Building is within the allowable area.
For SI: 1 square foot = 0.093 m².

DETERMINING ALLOWABLE AREAS FOR A MIXED-OCCUPANCY BUILDING REGULATED AS SEPARATED OCCUPANCIES

Application Example 508-4

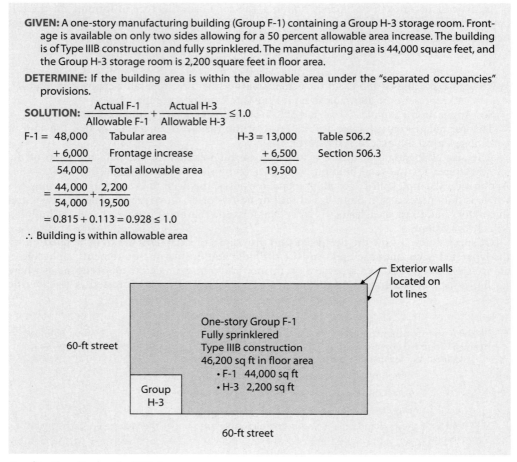

Application Example 508-5

discussion of Table 504.8, there are multiple combinations of occupancies where the table does not require a fire separation due to the similarity of the occupancies under consideration.

Table 508.4 Required Separation of Occupancies. The code has established several alternative methods for addressing mixed-occupancy buildings regarding fire separations between the various occupancies involved. Both Sections 508.2.4 and 508.3.3 for accessory occupancies and nonseparated occupancies, respectively, do not require any fire-resistance-rated separation between occupancies. However, Table 508.4, establishing fire separations under the separated occupancy method of Section 508.4, indicates varying degrees of separation. Fire-resistive separations of 1, 2, 3, and 4 hours are selectively required based on the occupancies involved. In some cases, however, no fire separation is mandated. The intent of the table is to provide for relative separation requirements based primarily on dissimilar risk. The fire-resistance ratings, including the lack of such required ratings in many circumstances, appropriately recognize the degree of dissimilarity between the various occupancies.

In a general sense, the following logic was utilized in formulating the table. High-hazard (Group H) occupancies are required to be separated from each other and from all other occupancies. Ordinary or moderate-hazard commercial/industrial (Groups B, F-1, M, and S-1) occupancies require no separation from each other; however, they are required to be separated from all other occupancies. People-intensive (Groups A and E) occupancies also require no occupancy separation between each other but must be separated from all other occupancies except for fully sprinklered Group F-2 and S-2 occupancies. Group R occupancies require no separation from other Group R occupancies, but such separations are mandated between all other occupancy classifications. Similar criteria apply to the Group I occupancies with a modification for Group I-2. The philosophy of the provisions set forth in the table dictates increased fire-resistance ratings on the basis of greater inherent dissimilar risk. Where Table 508.4 mandates some degree of occupancy separation, an occupancy shall be physically separated from the other occupancy through the use of fire barriers, horizontal assemblies, or a combination of both vertical and horizontal fire-resistance-rated assemblies.

In some cases, the rationale behind the use of fire-resistance-rated separations between incompatible occupancies concerns itself with the amount of combustibles encompassed in the adjoining occupancies and is termed in fire-protection circles as *fire loading*. Thus, if the amount of combustibles or fire loading in one occupancy is quite high while there is a limited fire load anticipated in the other occupancy, some degree of fire separation between the two distinct occupancies is necessary. However, the relationship of fire loads is not the only factor in determining an appropriate occupancy separation. In some cases, the separation is specified to be of 1 hour or more in duration mainly because of what the code implies to be incompatibility between the activities that occur within the two occupancies. For example, the code requires a separation of 1 hour between a Group I-2 occupancy and a Group S-2 occupancy in a fully sprinklered building. The limited amount of combustibles in either occupancy does not justify a fire-resistive separation; however, because of the presumed incompatibility between the two occupancies, the 1-hour separation is considered to be justified.

As a general rule, a reduction of the fire-resistance ratings in Table 508.4 by 1 hour is permitted in buildings equipped throughout with an automatic sprinkler system. The potential of a sizable fire spreading throughout a building is minimal under fully sprinklered conditions.

The fire-resistance ratings set forth by the table are further modified by Footnote b. The required separation for storage occupancies may be reduced by 1 hour where the storage is limited to the parking of private or pleasure vehicles, but may never be less than 1 hour. It would seem the footnote would not apply where no separation is initially required. For example, where the two occupancies involved are Groups S-2 and U parking facilities, a fire-resistive separation between them would not be warranted.

It is necessary to again emphasize that there is no relationship between the separated occupancy provisions of Section 508.4 and the fire area concept utilized in Section 903.2 for automatic sprinkler systems. Compliance with Table 508.4 does not relieve the responsibility to comply with Section 901.7 and Table 707.3.10 regarding the proper separation of fire areas. An example is shown in Figure 508-3.

508.5 Live/work units. A unique concept of building use combines a residential unit with a small business activity. Residential live/work units typically include a dwelling unit along with some public service business, such as an artist's studio, coffee shop, or chiropractor's office. There may be a number of employees working within the residence and the public is able to enter the work area of the unit to acquire service. Live/work units are a throwback to 1900-era community planning where residents could walk to all of the needed services within their neighborhood. These types of units began to re-emerge in the 1990s through a development style known as "Traditional Neighborhood Design." More recently, adaptive reuse of many older urban structures in city centers incorporated the same live/work tools to provide a variety of residential unit types. Provisions specifically addressing live/work units recognize the uniqueness of this type of use.

By definition, a "live/work unit" is primarily residential in nature but has a sizable portion of the space devoted to nonresidential activities. Often service-related in nature, the nonresidential portion is limited in several respects. The unit itself, including both the residential and nonresidential portions, is limited to 3,000 square feet (279 m²) in total floor area. In addition, the nonresidential activities cannot take up more than 50 percent of the unit's total floor area. The portion dedicated to nonresidential use must be located on the first floor of the unit, or where applicable, on the unit's main floor level. In addition to the unit's residents, a limit of five workers or employees is permitted at any one time. An overview of the limitations is shown in Figure 508-9.

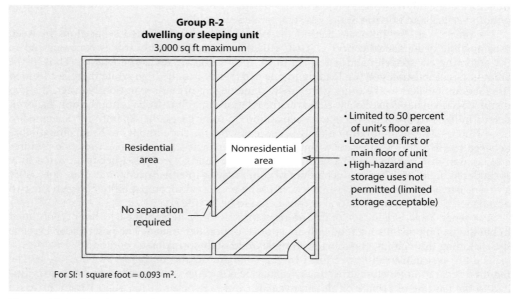

Figure 508-9 **Live/work unit.**

The occupancy classification of a live/work unit is Group R-2 based upon the primary use of the unit. Although differing uses are typically classified based on the characteristics of the varying uses involved and considered as mixed-occupancy conditions, in this case a single classification is considered acceptable. The potential hazards created due to the nonresidential uses are addressed through the special requirements of Section 508.5 that are to be applied in addition to those required due to the Group R-2 classification. Since live/work units are regulated as single-occupancy conditions, the provisions of Section 508 for mixed-occupancy buildings do not apply. Therefore, the application of Section 508.5 should be viewed as an exception to the mixed-occupancy requirements. In addition to the other limitations on use of a live/work unit, significant storage uses and those activities involving hazardous materials are prohibited. The increased fire load found in many storage uses is not considered in the live/work provisions, nor is the potential physical or health hazard that is due to the use or storage of hazardous materials. A very small amount of storage is permitted if it is deemed to be accessory to the nonresidential use.

Even though a live/work unit is classified as a Group R-2 occupancy, there are several issues where the residential and nonresidential portions are regulated independently. Structural floor loading conditions, accessibility features, and ventilation rates are all to be based on the individual function of each space within the unit, as are the design of the means of egress and the determination of required plumbing facilities. In all other cases, including the installation of fire protection features, the provisions applicable to a Group R-2 occupancy are to be applied to the entire live/work unit.

Section 509 *Incidental Uses*

There are times where the hazards associated with a particular use do not rise to the level of requiring a different occupancy classification; however, such hazards must still be addressed because of their impact on the remainder of the building. These functions are identified as incidental uses, which are regulated independently of the mixed-occupancy provisions. Incidental uses are uniquely addressed through the use of fire-resistance-rated separations or automatic sprinkler system protection.

What is an incidental use? There are occasionally one or more rooms or areas in a building that pose risks not typically addressed by the provisions for the general occupancy group under which the building is classified. However, such rooms or areas may functionally be an extension of the primary use. These types of spaces are considered in the IBC to be incidental uses and are regulated according to their hazard level. These areas are not ever intended to be considered different occupancies, creating a mixed-occupancy condition, but rather are classified in accordance with the main occupancy of the portion of the building where the incidental use is located. There is no specific definition for an incidental use, as it is simply described as any of those rooms or areas listed in Table 509.1. If it is not listed, it is not considered an incidental use for code purposes.

The designation and regulation of incidental uses do not apply to those areas within and serving a dwelling unit. Otherwise, the special hazards that may be found in buildings of various uses and occupancies are addressed through the construction of a fire barrier and/or horizontal assembly separating the incidental use from the remainder of the building, the installation of an automatic sprinkler system in the incidental use space, or, in special cases, both the fire separation and automatic sprinkler system.

Incidental uses are listed in Table 509.1. Most of the rooms or areas identified in the table are regulated where located in any of the occupancy groups established by the code, other than dwelling units as previously noted. A few of the incidental uses are to be regulated only where located within a specific occupancy or a limited number of occupancies.

It is common for many of the listed incidental uses to be unoccupied for extended periods of time, creating the potential for a fire to grow unnoticed. Oftentimes, combustible or hazardous materials are present in such areas. Because of the potentially high fuel load and lack of constant supervision, spaces such as furnace rooms, machinery rooms, laundry rooms, and waste collection rooms are selectively considered incidental uses. Other uses, such as paint shops, laboratories, and vocational shops, may cause concern to the point where they too must be protected or separated from other areas of the building. A significant number of incidental uses are located in Group I-2 occupancies and ambulatory care facilities due to the concerns related to occupants who are incapable of self-preservation.

How are incidental uses regulated? The fire-resistance-rated separations required by Table 509.1 are to be fire barriers and/or horizontal assemblies, typically having a minimum fire-resistance rating of 1 hour. For incinerator rooms and paint shops, the minimum required rating is greater. Where an automatic sprinkler system provides the necessary protection, it need only be installed within the incidental use under consideration. Examples of the requirements are illustrated in Figure 509-1.

There is a variety of combinations regarding the separation and protection methods identified in Table 509.1. It should be noted that where an automatic sprinkler system is utilized

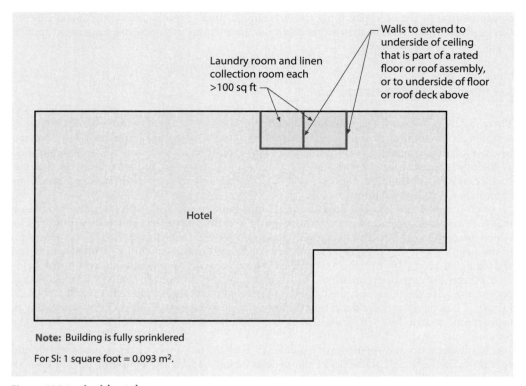

Figure 509-1 **Incidental use.**

without a fire barrier as the protective element, the incidental use must still be separated from the remainder of the building. However, this separation need only consist of construction capable of resisting the passage of smoke. As this requirement is performance in nature rather than prescriptive, the separation is not required to be constructed as a smoke partition or smoke barrier. Although not required to have a fire-resistance rating, walls must either extend to the underside of the floor or roof deck above, or to the underside of a fire-resistance-rated floor/ceiling or roof/ceiling assembly. Doors are to be self-closing or automatic-closing upon detection of smoke, with no air-transfer openings or excessive undercuts. See Figure 509-2. The room must be tightly enclosed, providing for the containment of smoke while assisting in the heat increase necessary to activate the automatic sprinkler system.

Table 509.1 Incidental Uses. Incidental uses are limited to those rooms or spaces listed in this table. The listed rooms have been selected for inclusion because of the increased hazard they present to the other areas of the building. However, it is recognized that the degree of hazard is such that a separate occupancy classification is not warranted. In fact, such a classification may be overly restrictive. The intent of the fire separation and fire protection requirements is to provide safeguards because of the increased hazard level presented by the incidental use.

Furnace rooms and boiler rooms. The hazard potential for fuel-fired heating equipment is addressed once the thresholds established in the table have been exceeded. It should be noted that the limitations are based on individual pieces of equipment, rather than the aggregate amounts from all equipment within the space. For example, the requirements of Table 509.1 are not applicable where there are two furnaces within the furnace room, each with an input rating of 300,000 Btu

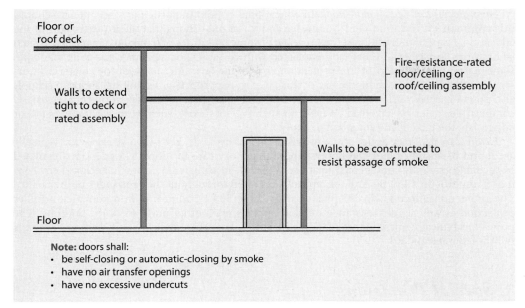

Figure 509-2 **Incidental use smoke separations.**

per hour (87,900 watts). Because no furnace exceeds the 400,000 Btu/hr (117,200 watts) threshold, there is no fire barrier separation or automatic sprinkler system protection required.

The regulated furnace or boiler is anticipated to be located within a room isolated from the remainder of the building. Where such an enclosed space is provided, the furnace room or boiler room must either be separated from other portions of the building with a complying fire barrier and/or horizontal assembly, or isolated by construction capable of resisting the passage of smoke and provided with an automatic sprinkler system. The intent of the requirement is to address the hazards associated with specified boilers and furnaces, along with those hazards created where a concealed accessible space with limited occupancy is provided.

Hydrogen fuel gas rooms. Hydrogen fuel gas rooms are defined in Chapter 2 as rooms or spaces that are intended exclusively to house a gaseous hydrogen system. Special requirements applicable to such rooms are set forth in Section 421. Where the quantities of materials would cause a hydrogen fuel gas room to be classified as a Group H occupancy, the separation requirements of Section 508.4 for separated occupancies will apply rather than those of Table 509.1.

The reference in the table exempting those rooms classified as Group H is not strictly limited to hydrogen fuel gas rooms. In fact, where the quantities of hazardous materials in any of the rooms or areas designated by Table 509.1 exceed those permitted by Section 307.1 causing classification as a Group H occupancy, the use of the table is not appropriate. In a mixed-occupancy building, all spaces with a Group H classification must be separated from the remainder of the building in accordance with Section 508.4 for separated occupancies.

Paint shops. The provisions of Section 416 control the construction, installation, and use of rooms for spraying paints, varnishes, or other flammable materials used for painting, varnishing, staining, or similar purposes. The paint shops regulated by Table 509.1 are those rooms where the same types of spraying operations occur. As a general rule, a minimum 1-hour fire-resistive enclosure is mandated by Section 416.2 to isolate spray rooms from the remainder of the building. However, in all but Group F occupancies, Table 509.1 mandates a higher degree of protection.

Laboratories and vocational shops. In educational buildings, particularly secondary schools, it is common to find multiple laboratories and vocational shops that are an extension of the educational function. Because of the presence of some quantities of hazardous materials in such laboratories, as well as in those labs associated with Group I-2 occupancies, the code mandates some degree of separation and/or protection. Where the quantities of hazardous materials warrant a Group H classification, the use of this table is inappropriate. Vocational shops in schools also pose a hazard to the remainder of the building that is due to the hazardous processes and combustible materials involved. Where no such hazards exist, such as in a computer lab or design lab, the table is not intended to apply.

Health care facilities. Many of the rooms regulated by Table 509.1 involve areas that typically are found in hospitals, nursing homes, and ambulatory care facilities. Those rooms regulated by the table generally contain significant combustible loading within a small, enclosed environment, resulting in a fire hazard not typically expected throughout the remainder of the facility. The protection afforded by compliance with Table 509.1 includes both automatic sprinkler protection as well as fire-resistance-rated separation from other portions of the building. The provisions typically mirror those mandated by the Centers for Medicare and Medicaid Services (CMS) federal standard.

Section 510 *Special Provisions*

The provisions of this section allow for modifications or exceptions to the general provisions for building heights and areas as regulated by Chapter 5 of the IBC. These special provisions are viewed as specific in nature and, based on Section 102.1, take precedence over any general provisions that may apply. Because this section permits, rather than requires, the use of the special conditions established in Section 510, the provisions are optional. Much like the application of the mezzanine provisions of Section 505, only where the designer elects to utilize the special allowances does this section apply.

It is evident that several of the provisions overlap in their scope. For example, Sections 510.2, 510.4, and 510.7 all address a potential condition where an open parking garage is located below a Group R occupancy. It is the choice of the designer which of the three methods to use where such a condition exists, based on the benefits and consequences of each method. Or, as stated above, none of the methods need to be applied. The requirements could simply be based on the general provisions of Chapter 5 for allowable building height and area.

510.2 Horizontal building separation allowance. This section is one of several that contain provisions that might be considered the only exceptions to the principle that a fire wall is the only code-established approach to dividing a single building into two or more separate buildings. See the discussion of Section 503.1 for the application the fire wall concept. In Item 1, the code makes provisions for the use of a minimum 3-hour fire-resistance-rated horizontal assembly, including any horizontal offsets that are provided, as an equivalent construction feature, in certain aspects, to a fire wall. This methodology is often referred to as "podium" or "pedestal" buildings.

The provisions create, in effect, an exception that allows those stories below the horizontal fire separation to be considered portions of a separate building for specific purposes, provided no Group H occupancy is involved. The stories above the horizontal separation must only house Group A, B, M, R, and/or S occupancies. As these occupancies are quite common in multi-story construction, the typical application is to grant the maximum number of stories for the selected type of construction without the penalty of sacrificing one or more stories for a parking garage or other permitted uses located on the level(s) below the 3-hour horizontal assembly. Distinct buildings are also created for use of differing construction types, area limitations, and fire wall continuity.

If the lower levels are classified as stories above grade plane, they would typically be included in the number of total stories permitted by the code. However, as depicted in Figure 510-1, those stories would not be included where the building below the 3-hour horizontal separation is of Type IA construction.

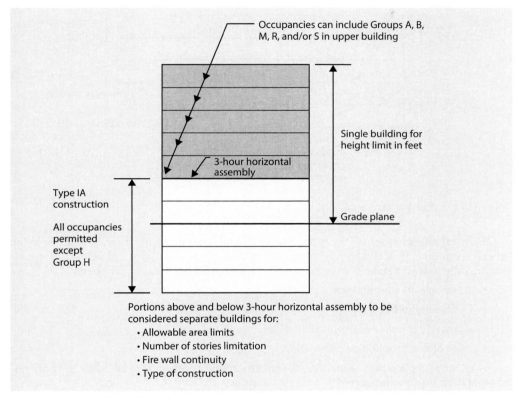

Figure 510-1   Horizontal building separation.

The code lists conditions that must be met in order to take advantage of these provisions. The first condition regulates the fire-resistance rating of the horizontal assembly. In addition, it allows for the construction of vertical offsets within the horizontal assembly provided they are minimum 3-hour fire barriers. Second, the minimum type of construction for portions below the horizontal assembly is established. The third condition addresses the methods for protection of openings that will occur in the fire-resistance-rated horizontal assembly. An allowance for combustible stairways in the Type IA building is established in Condition 4 as shown in Figure 510-2. Fifth, only Groups A, B, M, R, and S are permitted to be located above the horizontal assembly. The sixth condition limits the use of the building below the horizontal assembly to those that are not classified as Group H and requires the lower building to be sprinklered throughout. An often-overlooked provision is the seventh condition, requiring that the overall height in feet (mm) of both buildings not exceed the height limits set forth in Section 504.3 for the least type of construction that occurs in the building.

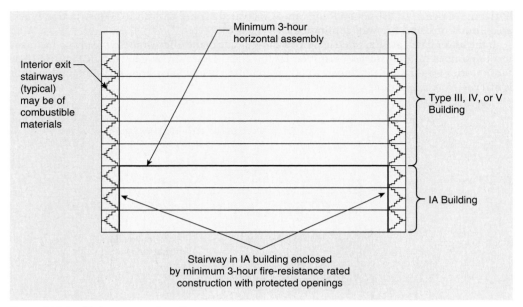

Figure 510-2 **Stairway construction.**

Of critical importance is the scope of the horizontal separation's use as an alternative to the general requirements of Chapter 5. The buildings that are created above and below the separation are only to be considered as separate and distinct buildings for four purposes:

- Allowable area limitations
- Continuity of fire walls
- Allowable number of stories limitation
- Type of construction determination

For all other applications of the IBC, the entire structure both above and below the horizontal separation is to be considered as a single building.

510.3 Group S-2 enclosed parking garage with Group S-2 open parking garage above. The provisions of this section are similar in nature to those of Section 510.2, insofar as the two different parking uses in a single structure, one located above the other, may be considered two separate and distinct buildings for the purpose of determining the type of construction. Five specific conditions must be met in order for an open parking garage, located above an enclosed parking garage, to be regulated on its own for construction type. Details of this special situation are shown in Figure 510-3.

510.4 Parking beneath Group R. Consistent with the concepts expressed by Sections 510.2 and 510.3, the provisions of this section address residential uses located above a first story used as a parking garage. Where parking is limited to the first story, the number of stories used in the determination of the minimum type of construction may be measured from the floor above the garage. The construction type of the parking garage and the floor assembly between the residential area and the garage are further regulated. See Figure 510-4.

510.7 Open parking garage beneath Groups A, I, B, M, and R. The excellent fire-safety record for open parking garages is the basis for this modification in the general provisions for

Special Provisions   **189**

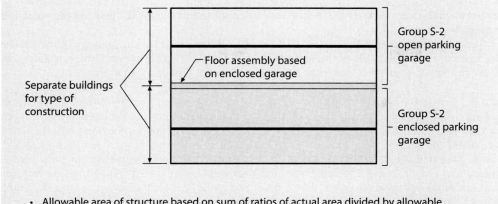

- Allowable area of structure based on sum of ratios of actual area divided by allowable area, which cannot exceed 1
- Enclosed parking garage of Type I or II construction and at least equivalent to open parking garage construction type
- Height and number of tiers in open parking garage limited per Table 406.5.4
- Floor separating open and enclosed garages based on enclosed garage requirement
- Enclosed parking garage limited to parking and small associated uses

Figure 510-3   **Open and enclosed parking structure.**

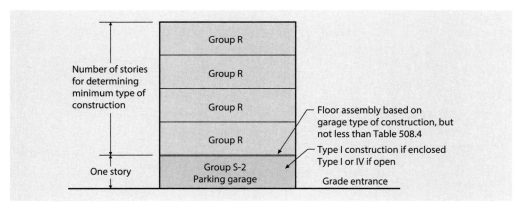

Figure 510-4   **Parking beneath Group R occupancy.**

allowable floor area and allowable height. Where located below assembly, institutional, business, mercantile, or residential occupancies, an open parking garage is regulated for height and area by Section 406.5. Those permitted occupancies located above the parking garage are independently regulated by Section 503.1 for height and area. The only exception requires that the height of the portion of the building above the open parking garage, both in feet (mm) and stories, be measured from the grade plane.

The details of construction type are applicable to each of the occupancies involved; however, the structural-frame members shall be of fire-resistance-rated construction according to the most restrictive fire-resistive assemblies of the occupancy groups involved. Egress from the areas above the parking garage shall be isolated from the garage, with the level of protection at least 2 hours.

Because the provisions of Section 510.2 can also address open parking garages below similar occupancies, the application of this section may be limited. There are several minor differences between the two provisions; however, the general concept remains consistent.

510.8 Group B or M buildings with Group S-2 open parking garage above. A desirable feature in high-density areas is to have offices and/or retail stores on the lower levels of open parking structures. This provision allows for the type of construction for the portion of the structure below the parking garage to be evaluated separately from that of the open parking garage above, provided the uses are properly separated and the means of egress from the garage is independent from that of the first floor (and basement if applicable). This provision reverses the conditions addressed by other provisions of Section 510 where the parking garage is located below other occupancies. The resulting benefit provides for a potential reduction in the type of construction by permitting the evaluation of allowable floor areas independently for the open parking garage and the Group B and/or M occupancies.

510.9 Multiple buildings above a horizontal assembly. Where the varying provisions of Section 510 are utilized to create separate buildings above and below a complying horizontal separation, it is acceptable for two or more buildings to be located above the separation while only one building (such as a parking garage) is located below. For example, a condominium building is permitted to be regulated as a separate building from an adjacent office building even though both are located above a single parking facility designed under the special provisions of Sections 510.2, 510.3, or 510.8, as applicable. An example is shown in Figure 510-5.

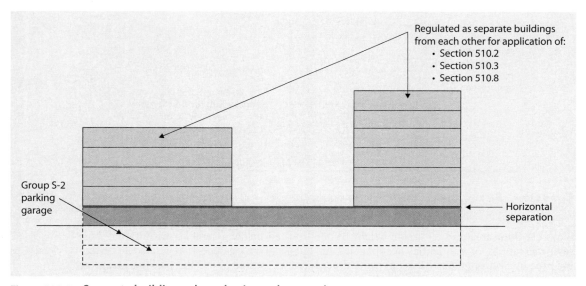

Figure 510-5 **Separate buildings above horizontal separation.**

KEY POINTS

- Buildings are regulated for maximum size based upon the occupancy or occupancies involved, the building's construction type, and special features such as sprinkler protection and open frontage.
- Table 506.2 provides the allowable area factor per story for all occupancies and types of construction.
- The maximum height of a building above grade plane is regulated by Tables 504.3 and 504.4 for the total height in feet and the maximum height in stories above grade plane, respectively.
- Buildings of three or more stories in height are permitted to be three times the allowable area permitted for a single story.
- Stepped or terraced buildings should be reviewed carefully to determine the height, as often each segment can be viewed independently of the others.
- Mezzanines are defined by the code in a very specific manner.
- The area of the basement typically does not need to be included in the total allowable area of the building.
- The installation of automatic sprinkler systems throughout a building typically provides for a sizable increase in the allowable area and building height.
- Sufficient open yards and public ways may be used to increase the allowable area.
- Through the use of adequate safeguards, the IBC allows certain types of buildings and occupancies to have unlimited floor areas.
- In mixed-occupancy buildings, three different methods (accessory occupancies, nonseparated occupancies, and separated occupancies) are available for addressing occupancy classification, allowable height and area, and separation.
- Incidental uses, classified as a part of the building where the use is located, shall be separated, protected, or both.
- The special provisions of Section 510 provide for alternative approaches to the specific height and area requirements of Chapter 5.

CHAPTER 6

TYPES OF CONSTRUCTION

Section 602 Construction Classification
Section 603 Combustible Material in
             Type I and II Construction
Key Points

As its title implies, this chapter develops requirements for the classification of buildings by type of construction. Based on a combination of two general factors, the combustibility potential of the materials of construction and the degree of fire-resistance protecting such materials, there are 12 unique construction types. The use of combustible materials in otherwise noncombustible buildings is also addressed.

Section 602 *Construction Classification*

For the most part, the concepts and applications related to type of construction are focused on one simple mandate: In order to build a structure bigger, it must be built better. Where a building is desired to be greater in height and floor area, the materials of construction and fire-resistive protection of building elements are regulated in a manner that addresses the increased hazards that are inherent in larger structures. The type of construction provisions of Chapter 6 are directly related to the allowable height and area provisions of Chapter 5. Additionally, allowable heights and areas may vary based upon the building's use and occupancy, sprinkler protection, and open frontage. All of these factors may play a role in the determination of acceptable construction types.

Since early in the last century, the fire protection required for the various types of construction has been based on hourly fire-endurance ratings. Prior to this time, fire-resistance requirements were developed by specifying the type and thickness of materials used.

Many of the concepts in previous building codes that have carried over to today were developed from the reports issued by the committee known as the Department of Commerce Building Code Committee, which was appointed by Herbert Hoover, then Secretary of Commerce. The committee was also dubbed the *Little Hoover Commission* and was appointed to investigate building codes. This was an outgrowth of the findings of the Senate Committee on Reconstruction and Production, which was appointed in 1920 to study the various factors entering into the recovery of our economy from the depression of the early 1920s. Although the committee studied a wide-ranging set of those institutions and groups affected by the economy, it was especially interested in construction. During its tenure, the committee held numerous hearings and expressed the following sentiment at their conclusion: "The building codes of this country have not been developed on scientific data, but rather on compromise; they are not uniform in practice and in many instances involve an additional cost of construction without assuring more useful or more durable buildings." Thus, the stage was set for improvement in building regulations, and the timing was especially favorable for the model codes to take advantage of the reports of the Department of Commerce Building Code Committee.

The IBC® classifies construction into five basic categories, listed in a somewhat descending order from the most fire resistant to the least fire resistant. These five types are based on two main groupings, noncombustible (required) construction (Types I and II) and combustible (permitted) construction (Types III, IV, and V). The various types of construction within the five categories are further subdivided based on fire protection and are represented as follows:

1. Noncombustible, protected—Types IA, IB, and IIA
2. Noncombustible, unprotected—Type IIB
3. Combustible and/or noncombustible, protected—Types IIIA, IV-A, IV-B, IV-C, IV-HT, and VA
4. Combustible and/or noncombustible, unprotected—Types IIIB and VB

Although Types III, IV, and V are commonly considered combustible construction, the use of noncombustible materials, either in part or throughout the building, is certainly acceptable. The reference to combustible construction more simply indicates that such construction is acceptable in Types III, IV, and V but not mandated. A perusal of Tables 504.3, 504.4, and 506.2 indicates that the IBC considers Type II, III, and IV-HT buildings to be of comparable protection. For example, Types IIA, IV-HT, and, to some degree, IIIA are permitted the same approximate areas and heights for most occupancy classifications. The same is also true for Types IIB and IIIB. Type IV-A, IV-B, and IV-C mass timber construction height and area limits have been derived from a variety of factors.

Differing from the concept of mixed-occupancy buildings, the code does not permit a building to be considered to have more than one type of construction. In simple terms, classification of a building for construction type is based on the *weakest link* concept. If a building does not fully conform to the provisions of Chapter 6 for type of construction classification, it must be classified into a lower type into which it does conform. Unless specifically permitted elsewhere by the code, the presence of any combustible elements regulated by Table 601 prohibits its classification as Type I or II construction. Similarly, the lack of required fire resistance in any element required by Table 601 to be protected will result in a fully nonrated building.

Table 601 identifies the required fire-resistance ratings of building elements based on the specified type of construction. Reference is made to Section 703.2 for those building elements required to have a fire-resistance rating by Table 601. Section 703.2 establishes the appropriate test procedures for building elements, components, and assemblies that are required to have a fire-resistance rating.

The provisions of Chapter 6 in regard to fire resistance are intended to address the structural integrity of the building elements under fire conditions. Unlike those fire-resistance-rated assemblies, such as fire walls and fire barriers, whose intent is to safeguard against the spread of fire, the protection afforded by the provisions of Chapter 6 is solely that of structural integrity. As such, the protection of door and window openings, ducts, and air transfer openings is not required for building elements required to be fire-resistance rated by Table 601 unless mandated by other provisions of the IBC.

The IBC intends that the provisions of the code are minimum standards. Thus, Section 602.1.1 directs that buildings not be required to conform to the requirements for a type of construction higher than the type that meets the minimum requirements of the IBC based on occupancy. A fairly common case in this regard is where a developer may construct an industrial building that complies in most respects to the requirements of the code for a Type IIIB building, but the occupancy provisions are such that a Type VB building would meet the requirements of the code. In this latter case, it would be clearly inappropriate and, in fact, a violation of the code for the building official to require full compliance with requirements for a Type IIIB building. However, where the building does comply in all respects to Type IIIB, the building official may so classify it.

602.2 Types I and II. Buildings classified as Type I and II are to be constructed of noncombustible materials unless otherwise modified by the code. The various building elements in these noncombustible buildings are regulated by Table 601. Although Type I and II buildings are defined as noncombustible, it is evidenced by Section 603 that combustible materials are permitted in limited quantities. Wood doors and frames, trim, and wall finish are permitted, as well as combustible flooring, insulation, and roofing materials. Where these combustibles are properly controlled, they have proven, over the years, to not add significantly to the fire hazard.

Furthermore, Type I buildings are to be of the highest levels of fire-resistance-rated construction. The fire-resistance ratings required for Type I buildings historically have provided about the same protection over the years and, thus, have proved to be satisfactory for occupancies housed in buildings of considerable height and area, particularly those designated as high-rise buildings. Type IB construction is very similar to Type IA construction except for a reduction of 1 hour in the required ratings for interior and exterior bearing walls, and the structural frame, while providing a ½-hour reduction for roof construction. Thus, and particularly because of the reduction in the fire-resistance rating required for the structural frame, the Type IB building does not enjoy all of the unlimited heights and areas that accrue to the Type IA building. It will be noted from Tables 504.3, 504.4, and 506.2 that Type IB construction typically has height and, to some degree, area limits placed on it.

Buildings of Type II construction, although noncombustible, may be of either protected (Type IIA) or unprotected (Type IIB) construction. The building elements of a Type IIA building are typically required to be protected to a minimum fire-resistance rating of 1 hour. Such elements in a Type IIB structure may be nonrated.

602.3 Type III. The Type III building grew out of the necessity to prevent conflagrations in heavily built-up areas where buildings were erected side by side in congested downtown business districts. After the severe conflagrations of years past in Chicago and Baltimore, it became apparent that some control must be made to prevent the spread of fire from one building to another. As a result, the Type III building was defined. The Type III building is, in essence, a wood-frame building (Type V) with fire-resistance-rated noncombustible exterior walls.

Around the turn of the 20th century, and prior to the promulgation of modern building codes, Type III buildings were known as ordinary construction. They later became known in some circles as ordinary masonry construction. However, as stated previously, the intent behind the creation of this type of construction was to prevent the spread of fire from one combustible building to another. Thus, the early requirements for these buildings were for a certain thickness of masonry walls, such as 13 inches (330 mm) of brick for one-story and 17 inches (432 mm) for two-story buildings of bearing-wall construction. Later, the required fire endurance was specified in hours. Thus, any approved noncombustible construction that would successfully pass the standard fire test for the prescribed number of hours was permitted.

In spite of the requirement for noncombustible exterior walls, Type III buildings are considered combustible structures and are either protected (Type IIIA) or unprotected (Type IIIB). Interior building elements are permitted to be either combustible or noncombustible. There is an allowance for the use of fire-retardant-treated wood as a portion of the exterior wall assembly, provided such wall assemblies have a required fire-resistance rating of 2 hours or less.

602.4 Type IV. Type IV construction is represented by two distinctly different approaches to the use of large wood building elements. A single term, mass timber, represents the various sawn and engineered wood products that make up this unique construction type. Traditional Type IV-HT construction has exterior walls made of noncombustible materials, fire-retardant-treated wood (FRTW), or protected cross-laminated timber (CLT). Interior building elements must be of solid or laminated wood. Columns and roof/floor beams and girders must have minimum nominal dimensions per Table 2304.11. Flooring, roof decking, and partitions must be of a minimum thickness as prescribed in the IBC. Type IV-HT is deemed to meet fire-resistance requirements based on its required minimum dimensions.

In contrast, mass timber used in construction types IV-A, IV-B, and IV-C is required to have a fire-resistance rating. The rating is to be determined in accordance with IBC Section 703.2, which includes calculations in accordance with IBC Section 722. The fire-resistance rating is determined by adding the fire resistance of the unprotected wood element to the protection time provided by any noncombustible protection.

Type IV-HT construction. Although Type IV-HT buildings have traditionally been designated as heavy-timber buildings, they may also have interior building elements of solid wood, laminated wood, or structural composite lumber (SCL). In the eastern United States during the 1800s, a type of construction evolved that was known as mill construction. Mill construction was developed by insurance companies to reduce the significant losses they were facing in the heavy industrialized areas of the Northeast.

Type IV-HT construction has historically been recognized for its slow burning characteristics. Wood under the action of fire loses its surface moisture, and when the surface temperature reaches about 400°F (204°C), flaming and charring begin. Under a continued application of the heat, charring continues, but at an increasingly slower rate, as the charred wood insulates the inner portion of the wood member. There is quite often enough sound wood remaining during and after a fire to prevent sudden structural collapse. In recognition of these characteristics, the insurance interests reasoned that replacement of light-wood framing on the interior of factory buildings with heavy-timber construction would substantially decrease their fire losses.

In the early development of Type IV-HT heavy-timber construction, not only did the heavy-timber members have large cross sections to achieve the slow-burning characteristic, but, furthermore, surfaces were required to be smooth and flat. Sharp projections were to be avoided, as well as concealed and inaccessible spaces. Thus, the intent of the concept was to provide open structural framing without concealed spaces and without sharp projections or rough surfaces, which are more easily ignitable. In this case, flame spread along the surface of heavy-timber members is reduced, and without concealed blind spaces, there is no opportunity for fire to smolder and spread undetected. Current allowances for concealed spaces now recognize that such spaces are necessary to accommodate mechanicals. Therefore, concealment is acceptable where the contents are limited to noncombustible materials other than building elements and compliant electrical, mechanical, fire protection, and plumbing materials. In addition, all concealed spaces must be protected by one of the methods set forth in Section 602.4.4.3.

In accordance with Table 601 and Section 2304.11, current Type IV-HT construction can be a mixture of heavy-timber floor and roof construction and 1-hour fire-resistance-rated bearing walls and partitions. Although Type IV-HT construction is not generally recognized as equivalent to 1-hour fire-resistance-rated construction, the code considers heavy timber to provide somewhat equivalent protection.

In keeping with the concept of slow-burning construction by means of wood members with large cross sections, the IBC specifies minimum nominal dimensions for wood members used in Type IV-HT construction. As the code specifies the size of members as nominal sizes, the actual net surfaced sizes may be used. For example, an 8-inch by 8-inch (203-mm by 203-mm) solid sawn member nominally will actually be a net size of 7¼ inches by 7¼ inches (185 mm by 185 mm). Therefore, even though the code calls for a nominal 8-inch by 8-inch (203-mm by 203-mm) member, the net 7¼-inch by 7¼-inch (185-mm by 185-mm) member meets the intent of the code. As indicated earlier, the minimum sizes for heavy-timber construction are based on experience and the good behavior in fire of Type IV-HT construction.

Wherever framing lumber or sawn timber is specified, structural glued-laminated timber or SCL may also be used, as all have the same inherent fire-resistive capability. However, because solid sawn wood members, glued-laminated timbers, and SCL are manufactured with different methods and procedures, they do not have the same dimensions. Table 2304.11 compares the solid sawn sizes with those of glued-laminated members and SCL to indicate equivalency in regard to compliance with the Type IV-HT construction criteria.

Section 2304.11.2.2 specifies that partitions shall be of either solid-wood construction or 1-hour fire-resistance-rated construction. However, various provisions of the code address the use of fire partitions and fire barriers. In these cases, the fire-resistant-rated fire partitions or fire barriers in Type IV-HT buildings should be constructed as required by the code for the required rating. For example, where there is a requirement for fire-resistance-rated corridors, 1-hour fire-resistance-rated construction must be used rather than solid-wood construction for the fire partitions.

It is highly unusual for any building designed and constructed today to be considered compliant as a Type IV-HT structure. As previously addressed, in order for a building to be properly classified, all portions must be in conformance with the established criteria. Many buildings may have some heavy-timber elements that qualify for use in Type IV-HT buildings; however, the floor construction or some other building element does not fully comply with the prescriptive requirements of Sections 2304.11.3. In such cases, the building cannot be classified as Type IV-HT. Such buildings are most likely Type III or V construction. However, even if the building as a whole is not considered a Type IV-HT structure, the recognition of individual heavy-timber elements is very important. For example, the provisions of Section 705.2.3 recognize heavy-timber projections for use in locations where unprotected combustible construction is not permitted. For this and other reasons, the requirements for Type IV-HT buildings and heavy timber construction must be fully understood.

Types IV-A, IV-B, and IV-C construction. The recent introduction of cross-laminated timber (CLT) in U.S. construction to form solid timber wall, floor, and roof sections has brought about a sea of change in the structural and fire-resistance performance for wood buildings. This necessitated development of three entirely new construction types (IV-A, IV-B, and IV-C) with their own unique material and fire-resistance requirements, and corresponding building size limitations.

The Type IV-A, IV-B, and IV-C construction types are indicative of just how different the structural and fire performance of solid timber walls, floors, and roofs is compared to repetitive-member frame buildings using studs and joists. Fire resistance of solid timber assemblies is very different from assemblies with concealed spaces, and when solid mass timber is protected by gypsum the difference can be dramatic. By way of example, a five-ply CLT wall with one layer of 5/8-inch (15.9-mm) Type X gypsum has achieved 3-hour fire resistance in an ASTM E119 test, whereas a wood stud wall with the same amount of gypsum achieves only a 1-hour fire-resistance rating. The structural performance of CLT walls and floors is also vastly different than their light-frame counterparts.

The greatest challenge in the recognition of Type IV-A, IV-B, and IV-C construction involved developing an approach to compensate for the combustible nature of the material while recognizing its inherent fire resistance and fire performance. The initial step was the development of a protection system that would result in a performance akin to that of existing Type IB construction, then assigning similar fire-resistance requirements to Type IV-B buildings. The same process was used to address Type IV-A and IV-C buildings in turn.

It was determined that fire testing was necessary to validate the established performance objectives. Consequently, five full-scale, multiple-story fire tests were developed to simulate the three construction types (Type IV-A, IV-B, and IV-C). The successful results of those tests, as well as testing for structural performance in accordance with ASTM E119, and additional testing by others, helped establish the basis of the mass timber code provisions.

Performance of all mass timber structural elements, whether protected or exposed, is based on design for a quantified fire resistance (e.g., 2 or 3 hours) consistent with traditional construction types that have performed well for buildings of similar sizes. In the end, the concepts of adding the fire-resistance contribution of noncombustible protection to the fire resistance of the exposed mass timber element and specifying the relative portion of fire resistance that must be provided by the noncombustible protection, were developed specifically for the Type IV-A, Type IV-B, and Type IV-C mass timber construction types.

Because of the close relationship to traditional heavy timber, the established variants are Type IV-A for buildings with structural elements completely protected with noncombustible protection, Type IV-B for buildings with structural elements partially protected with noncombustible protection, and Type IV-C for buildings with most structural elements left unprotected. Even in Type IV-C, noncombustible protection is required for concealed spaces, shaft walls, and the exterior surface of outside walls (see Table 6-1).

For Type IV-B construction, the amount of exposed surface permitted, as well as the required separation between unprotected portions, is intended to limit potential contribution

Table 6-1. Type IV-A, IV-B, and IV-C Construction Comparison

Feature	Type IV-A	Type IV-B	Type IV-C
Description	100% Noncombustible (NC) protection on all mass timber (MT) surfaces	NC protection on all MT surfaces except for partially exposed MT elements	100% exposed MT except: shafts, concealed spaces, and outside of exterior walls
Permitted Materials			
Structural Building Elements, Non-loadbearing Exterior & Interior Walls	MT or NC	MT or NC	MT or NC
Shaft and Exit Enclosures			
High-rise* to 12 stories or 180 ft	NC or MT protected with ≥80 min NC [two layers 5/8" Type X gypsum—or three if 3-hr FRR] each side of shaft or enclosure	NC or MT protected with ≥80 min NC [two layers 5/8" Type X gypsum] each side of enclosure	NC or MT protected with ≥40 min NC [one layer 5/8" Type X gypsum] each side of shaft or enclosure
Above 12 stories or 180 ft	NC	Not Permitted	Not Permitted

*See IBC definition of high-rise

of the structure to an interior fire. The intent of the 15-foot separation distance is to prevent large sections of walls from providing enough radiant energy between them to continue burning. Unprotected portions of mass timber ceilings, including attached beams, are limited to an area equal to not more than 100 percent of the base floor area in any dwelling unit or fire area. Unprotected portions of mass timber walls, including attached columns, are limited to an area of not more than 40 percent of the base floor area. Unprotected portions of both walls and ceilings of mass timber, including attached columns and beams, are limited in area in accordance with an equation for summing two ratios of actual exposed area to allowable exposed area. For example, if 100 percent of the ceiling is unprotected, all portions of all walls must be properly protected. Columns and beams that are not an integral part of a wall or ceiling, such as columns located in the middle of a space or room, are not required to be included in the calculation of ceiling and wall areas, respectively, and are permitted to be unprotected without restriction as to unprotected area or separation from one another. Limitations on unprotected portions of mass timber walls and ceiling are evaluated on a story-by-story basis. Multiple-story floor areas cannot be used to determine the allowable exposed mass timber in walls and ceilings in multi-story dwelling units and fire areas. See Application Example 602-1.

Combustible light-frame materials are specifically and completely excluded from these three new types of construction. Light-frame wood stud or joist assemblies are prohibited, including the use of light-frame wood to fur out concealed spaces. To accommodate mechanicals in Type IV-A, IV-B, and IV-C construction, protected concealed spaces are permitted where protected with noncombustible materials or provided with sprinkler coverage.

Interior exit enclosures and elevator hoistway enclosures in high-rise buildings up to and including 12 stories or 180 feet (54,864 mm) are permitted to be constructed of mass timber having noncombustible protection. For buildings above 12 stories or 180 feet (54,864 mm), interior exit and elevator hoistway enclosures must be constructed of noncombustible materials

In this example, a two-level dwelling unit is 30 feet × 30 feet, having floor areas of 900 square feet each and ceiling heights of 10 feet. Section 602.4.2.2.2, Exception 1.3, permits both exposed wall and ceiling areas within each story of the dwelling unit when the configuration meets the mixed unprotected area requirements of Section 602.4.2.2.3. Equation 6-1 for evaluating the combined area of exposed walls and floors is shown above. Using the graphics below with proposed dimensions for exposed mass timber elements determines compliance.

For story 1:

$$U_{tc} = 15 \text{ ft} \times 30 \text{ ft} = 450 \text{ ft}^2$$
$$U_{ac} = 900 \text{ ft}^2 \times 1.0 = 900 \text{ ft}^2$$
$$U_{tw} = 18 \text{ ft} \times 10 \text{ ft} = 180 \text{ ft}^2$$
$$U_{aw} = 900 \text{ ft}^2 \times 0.40 = 360 \text{ ft}^2$$

Applying these values to Equation 6-1:

$$\left(\frac{450}{900}\right) + \left(\frac{180}{360}\right) = 0.5 + 0.5 = 1.0 \leq 1.0, \text{ which is compliant.}$$

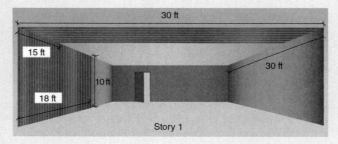

Story 1

For story 2:

$$U_{tc} = 10 \text{ ft} \times 30 \text{ ft} = 300 \text{ ft}^2$$
$$U_{ac} = 900 \text{ ft}^2 \times 1.0 = 900 \text{ ft}^2$$
$$U_{tw} = 24 \text{ ft} \times 10 \text{ ft} = 240 \text{ ft}^2$$
$$U_{aw} = 900 \text{ ft}^2 \times 0.40 = 360 \text{ ft}^2$$

Applying these values to Equation 6-1:

$$\left(\frac{300}{900}\right) + \left(\frac{240}{360}\right) = 0.33 + 0.67 = 1.0 \leq 1.0, \text{ which is also compliant.}$$

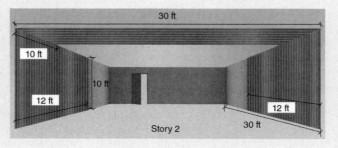

Story 2

Application Example 602-1

for their entire building height. Other shafts in buildings above 12 stories or 180 feet (54,864 mm), like those used for mechanical equipment, are permitted to be constructed of mass timber having noncombustible protection (see Table 6-1).

602.5 Type V. Type V buildings are essentially construction systems that will not fit into any of the other higher types of construction and may be constructed of any materials permitted by the code. The usual example of Type V construction is the light wood-frame building consisting of walls and partitions of 2-inch by 4-inch (51-mm by 102-mm) or 2-inch by 6-inch (51-mm by 152-mm) wood studs. The floor and ceiling framing are usually of light wood joists of 2-inch by 6-inch (51-mm by 152-mm) size or deeper. Roofs may also be framed with light wood rafters of 2-inch by 4-inch (51-mm by 102-mm) size or deeper cross sections or, as is now quite prevalent, framed with pre-engineered wood trusses of light-frame construction. Wood-frame Type V buildings may be constructed with larger framing members than just described, and these members may actually conform to heavy-timber sizes. Such structures sometimes have a significant number of noncombustible building elements. However, unless the building complies in all respects to one of the other four basic types of construction, it is still a Type V building.

Type V construction is divided into two subtypes:

1. Type VA. This is protected construction and required to be of 1-hour fire-resistance-rated construction throughout.
2. Type VB. This type of construction has no general requirements for fire resistance and may be of unprotected construction, except where Section 705.5 requires exterior wall protection because of proximity to a lot line.

Table 601—Fire-Resistance Rating Requirements for Building Elements. This table provides the basic fire-resistance rating requirements for the various types of construction. It also delineates those fire-resistance ratings required to qualify for a particular type of construction. As previously discussed, even though a building may have some features that conform to a higher type of construction, the building shall not be required to conform to that higher type of construction as long as a lower type will meet the minimum requirements of the code based on occupancy. Nevertheless, any building must comply with all the basic fire-resistance requirements in this table if it is indeed the intent to classify it for that particular type of construction. The 12 designations used throughout the code to describe a specific construction type are not specifically defined in the IBC. Rather, their definitions can only be determined through the combination of the provisions of Section 602.2, 602.3, 602.4, or 602.5 along with the minimum fire-resistance ratings set forth in Table 601. For example, the definition of a building identified as being of Type IIA construction is based on Section 602.2 requiring the building to have noncombustible building elements, as well as Table 601 mandating that the primary structural frame, bearing walls, floor construction, and roof construction, as applicable, be provided with minimum 1-hour fire-resistive ratings. It is only through the evaluation of these two provisions that the definition of a Type IIA building can be realized.

Footnote a. Limited in application to buildings of Type I and Type IV construction, the fire-resistance ratings of primary structural frame elements and interior bearing walls supporting only a roof may be reduced by 1 hour. In other words, primary structural-frame members or interior bearing walls providing only roof support shall have a minimum fire-resistance rating of 2 hours in Type IA buildings and 1 hour in Type IB construction. The 1-hour reduction is also permitted for roof supports in Type IV-A, IV-B, and IV-C construction, but it is not permitted for interior bearing walls. Additional provisions addressing the protection of certain primary structural frame members are found in Section 704.

Footnote b. This footnote, an exception to the general rule for roof construction, addresses those situations where the roof and its components are 20 feet (6,096 mm) or more above any floor immediately below. Under these circumstances, the roof and its components, including primary structural frame roof members, roof framing, and decking, may be of unprotected construction. The reduction of the fire-resistance rating would apply to buildings of Type IA, IB, IIA, IIIA, and VA construction. The footnote mandates that all portions of the roof construction must be located at or above the 20-foot (6,096-mm) height requirement. For example, in a sloped roof condition, it is not acceptable to merely protect those portions below the 20-foot (6,096-mm) point and leave the remainder unprotected. See Figure 601-1. It is important to note that the elimination of any required fire-resistance rating is also applicable to elements of the roof construction considered as primary structural frame members.

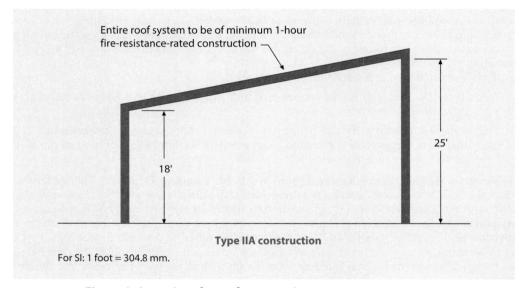

Figure 601-1 **Fire-resistive ratings for roof construction.**

The reduction in rating applies to all occupancies other than Groups F-1, H, M, and S-1, where fire loading is typically higher. In all occupancies other than those just listed, the relaxation of the requirements is based on the fact that where the roof is at least 20 feet (6,096 mm) above the nearest floor or mezzanine, the temperatures at this elevation during most fire incidents are quite low. As a result, fire protection of the roof and its members, including the structural frame, is not necessary. For those occupancies where the fire loading and the consequent potential fire severity is relatively high, such as factory-industrial, hazardous, mercantile, or storage uses, the code does not permit a reduction in roof protection. It is also quite common in these occupancies for combustible or hazardous materials to be located in close proximity to the roof structure, as in the case of high-piled storage.

Footnote c. Applicable to Types IB, IIA, IIB, IIIA, and VA construction, the code permits heavy-timber members complying with Section 2304.11 to be used in the roof construction without any fire-resistance rating as required by the table. It is assumed that roof members sized and constructed in compliance with the details of heavy-timber construction are equivalent to roof construction having a 1-hour fire-resistance rating. In addition, heavy-timber members are permitted to be utilized in the roof construction of an otherwise noncombustible Type IB, IIA, or IIB building. The allowance also includes the use of heavy timber primary structural frame members as a part of the roof construction.

Footnote e. In addition to any required fire-resistance rating based on the type of construction of the building per Table 601, it is also necessary that such rating requirements for exterior walls, both bearing and nonbearing, be in compliance with Table 705.5. The table regulates the hourly fire-resistance ratings for exterior walls based on fire separation distance. This footnote specifically indicates that exterior bearing walls have a fire-resistance rating based on Table 601 or 705.5, whichever provides for the highest hourly rating. Exterior nonbearing walls are totally regulated by the rating requirements found in Table 705.5. The provisions of Section 704.10 must also be consulted where load-bearing structural members are located within the exterior walls or on the outside of the building. See Application Example 705-1 for the appropriate use of these provisions. In addition, applicable provisions in Section 603 regulating combustible material in Type I and Type II construction may apply to exterior walls.

Footnote g. Applicable only to Type IV-HT buildings, heavy timber interior bearing walls supporting more than two floors or a floor and a roof are required to have a minimum 1-hour fire-resistance rating, which is appropriate for vertical elements in mid-rise, multistory buildings. If the designer desires to have an exposed CLT interior bearing wall supporting multiple stories, wall thickness may need to be increased to provide a minimum 1-hour rating. The rating may be calculated in accordance with Chapter 16 of the *National Design Specification for Wood Construction* as allowed in IBC Section 722.1, or a tested wall element can be utilized.

Section 603 *Combustible Material in Type I and II Construction*

Buildings of Type I and II construction are considered noncombustible structures. As such, all of the building elements, including walls, floors, and roofs, are to be constructed of noncombustible materials. There are, however, a variety of exceptions to the general rule that allow a limited amount of combustibles to be used in the building's construction. It has been determined that the level of combustibles permitted by Section 603.1, as well as their control, does not adversely impact the fire-severity potential caused by the materials of construction.

The following listing provides an overview of some of those combustible materials permitted in Type I and II buildings:

1. For other than shaft construction in Group I-2 occupancies and ambulatory care facilities, fire-retardant-treated (FRT) wood may be used in the construction of interior nonbearing partitions where the required fire-resistance rating of the partitions does not exceed 2 hours. In nonbearing exterior walls, FRT wood is permitted provided no fire rating of the exterior walls is mandated. Roofs constructed of FRT wood are also acceptable in most buildings. This would include roof girders, trusses, beams, joists, or decking, as well as blocking, nailers, or similar components that may be a part of the roof system. Wood nailers of other than FRT wood may be used for parapet flashing and roof cants per Item 27. Where the building is classified as other than Type IA construction, the use of FRT wood roof elements is permitted in all cases, regardless of building height. The same allowance is permitted in one- and two-story buildings of Type IA construction. For Type IA buildings exceeding two stories in height, the use of FRT wood in the roof construction is only allowed if the uppermost story has

a height of at least 20 feet (6,096 mm). Logically, the 20-foot measurement would be taken in a manner consistent with that described in Footnote b of Table 601, from the floor to the lowest point of the roof construction above.

The allowances provided in Section 603.1, Item 1.3, do not reduce any required level of fire resistance mandated for wall or roof construction as established by Table 601. Rather, they simply allow the use of FRT wood in the locations listed where noncombustible construction is otherwise required. In reviewing the permitted use of FRT wood, there are two obvious building elements where such materials are not permitted in Type I or II construction. In Type I or II buildings, FRT wood is not permitted to be used in the floor construction and any bearing wall assemblies.

Per Item 13, the use of fire-retardant-treated wood is also permitted in the construction of balconies, porches, decks, and exterior stairways in Type I and II buildings. However, there are two significant conditions for such an allowance. One, the height of the building under consideration cannot be more than three stories above grade plane. And two, the balcony, porch, deck, or exterior stairway must not be considered as a required exit from the building.

2. Combustible insulation used for thermal or acoustical purposes is acceptable, provided the flame-spread index is limited. Additional regulations addressing the use of thermal- and sound-insulating materials within buildings are found in Section 720.

3. Foam plastics installed under the limitations of Chapter 26 are permitted, as are roof coverings having an A, B, or C classification as specified in Section 1505.

4. Wood doors, door frames, window sashes and frames, trim, and other combustible millwork and interior surface finishes are acceptable, as is blocking for handrails, grab bars, cabinets, window and door frames, wall-mounted fixtures, and similar items. Combustible stages and platforms are also permitted when complying with Section 410, and wood-finish flooring may be used when applied directly to the floor slab or installed over wood sleepers and fireblocked in accordance with Section 805.1.

Another allowable use of combustible elements in noncombustible buildings, detailed in Item 11, addresses the situation where nonbearing partitions divide portions of stores, offices, or similar spaces occupied by one tenant only. The key words in this item are "occupied by one tenant only." It is the intent of the IBC that this expression applies to an area or building that is under the complete control of one person, organization, or other occupant. This would be contrasted to multitenant occupancies, where the various tenant spaces in the building would be under the control of two or more individuals, companies, or occupants. In such a multitenant space, the walls common to the public areas and to other tenants would not be regulated under this allowance. However, within each of the tenant spaces, those nonbearing walls and partitions not common with other tenants or public areas could utilize the optional construction methods of Item 11.

Reference is also made under Item 24 to Section 718.5 for the allowance of specific combustible elements within concealed spaces. The allowance for combustible items in concealed spaces is limited because of the increased potential for fire spread. Therefore, the flame spread index and smoke-developed index of the permitted items are often highly regulated. Combustible piping is permitted to be installed within partitions, shaft enclosures, and concealed ceiling spaces of noncombustible buildings. Various combustible materials are also permitted in plenums of Type I and II buildings, including wiring, fire-sprinkler piping, pneumatic tubing, and foam plastic insulation under the limitations imposed by Section 602 of the *International Mechanical Code*® (IMC®).

KEY POINTS

- Buildings are classified in general terms as combustible or noncombustible, as well as protected or unprotected.
- Table 601 identifies the minimum required fire-resistance ratings of building elements based on the specified type of construction.
- Unless a fire wall is utilized, or the special provisions of Section 510 are applied, structures can be classified into only one type of construction.
- The structural frame is regulated in a manner apart from that of walls, floors, and roofs.
- Type I and II buildings are considered noncombustible (required), whereas Type III, IV, and V buildings are viewed as combustible (permitted) construction.
- Mass timber buildings of Types IV-A, IV-B, and IV-C construction vary based upon the percentage of mass timber surfaces that can be unprotected.
- Very few structures fully comply with the provisions for Type IV-HT construction; however, many buildings contain some heavy-timber elements.
- Type V buildings are by far the most common type of construction.
- Various reductions in fire resistance are permitted for nonbearing partitions.
- Combustible materials identified in Section 603 are permitted in otherwise noncombustible construction (Types I and II).

CHAPTER 7

FIRE AND SMOKE PROTECTION FEATURES

Section 702 Multiple-Use Fire Assemblies
Section 703 Fire-Resistance Ratings and Fire Tests
Section 704 Fire-Resistance Rating of Structural Members
Section 705 Exterior Walls
Section 706 Fire Walls
Section 707 Fire Barriers
Section 708 Fire Partitions
Section 709 Smoke Barriers
Section 710 Smoke Partitions
Section 711 Floor and Roof Assemblies
Section 712 Vertical Openings
Section 713 Shaft Enclosures
Section 714 Penetrations
Section 715 Joints and Voids
Section 716 Opening Protectives
Section 717 Ducts and Air Transfer Openings
Section 718 Concealed Spaces
Section 719 Fire-Resistance Requirements for Plaster
Section 720 Thermal- and Sound-Insulating Materials
Section 721 Prescriptive Fire Resistance
Section 722 Calculated Fire Resistance
Key Points

The types of construction and the fire-resistance requirements of the *International Building Code*® (IBC®) are based on the concept of fire endurance. Fire endurance is the length of time during which a fire-resistive construction assembly will confine a fire to a given area, or continue to perform structurally once exposed to fire, or both. In the IBC, the fire endurance of an assembly is usually expressed as a "___-hour fire-resistance-rated assembly." Chapter 7 prescribes test criteria for the determination of the fire-resistance rating of construction assemblies and components, details of construction of many assemblies and components that have already been tested, and other information necessary to secure the intent of the code as far as the fire resistance and the fire endurance of construction assemblies and components are concerned. Additionally, Chapter 7 addresses other construction items that must be incorporated into a building's design in order to safeguard against the spread of fire and smoke.

Section 702 *Multiple-Use Fire Assemblies*

This section serves as a reminder that assemblies can serve multiple purposes and therefore must be capable of meeting all of the applicable requirements imposed due to those differing purposes. For example, a door in a rated corridor generally has only a 20-minute fire-protection rating along with smoke- and draft-control protection. If the door was also serving as a separation between occupancies, it may need a higher fire-protection rating based on the requirements of Tables 508.4 and 716.1(2), but it would still need the smoke- and draft-control protection required by Section 716.2.2.1.1 to protect the opening into the corridor.

Section 703 *Fire-Resistance Ratings and Fire Tests*

It is the intent of the IBC that materials and methods used for fire-resistance purposes are limited to those specified in this chapter. Materials and assemblies tested in accordance with ASTM E119 or UL 263 are considered to be in full compliance with the code, as are building components whose fire-resistance rating has been achieved by one of the analytical methods specified in Section 703.2.2.

703.2 Fire-resistance ratings. This section indicates that building elements are considered to have a fire-resistance rating when tested in accordance with the procedures of ASTM E119 or UL 263. Figures 703-1 through 703-5 depict the fundamental testing requirements of the two standards. The intent of the IBC is that any material or assembly that successfully passes the end-point criteria depicted for the specified time period shall have its fire-endurance rating accepted and the assembly classified in accordance with the time during which the assembly successfully withstood the test. When conducting these fire tests, the standards and the code do not factor in or allow the installation of a sprinkler system as a part of the testing. Thus the ratings of the materials and assembly are established on the basis of its own merits and credit is not given for other features.

Although early fire testing in the United States began as long ago as the 1890s, the standard fire-endurance test procedure using a standard time-temperature curve and specifying fire-endurance ratings in hours was developed in 1918. The significance of 1918 and later standards is the fact that they were and are intended to be reproducible so that the test conducted at one testing facility can be compared with the test of the same assembly conducted at any other testing facility.

Fire-Resistance Ratings and Fire Tests

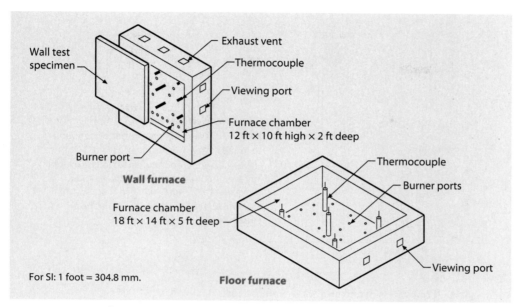

Figure 703-1 **Test furnaces.**

An often-expressed criticism of a standard such as ASTM E119 or UL 263 is that "it does not represent the real world." This is true in many cases, and for that reason it should not be thought of as representing the absolute behavior of a fire-resistance-rated assembly under most actual fires in buildings. There are too many variables that affect the fire endurance of an assembly during an actual fire, such as fuel load, room size, rate of oxygen supply, and restraint, to consider that the

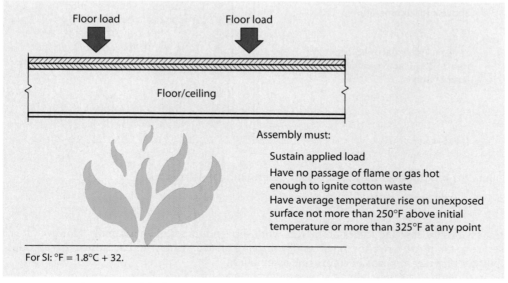

Figure 703-2 **Floor assembly fire test.**

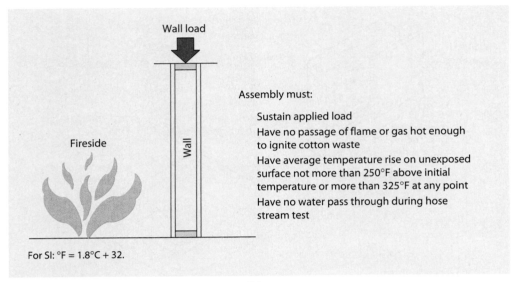

Figure 703-3 **Conditions of acceptance—wall fire test.**

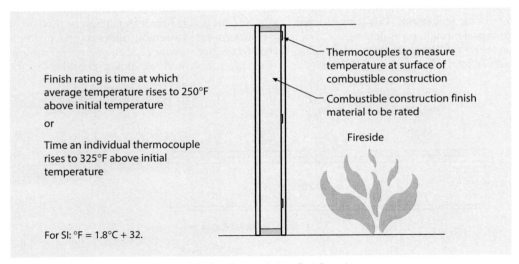

Figure 703-4 **Combustible assembly for determining finish rating.**

test establishes absolute values of the real-world fire endurance of an assembly. However, it is a severe test of the fire-resistive qualities of a material or an assembly, and because of its reproducibility, it provides a means of comparing assemblies.

In addition to the fire-endurance fire ratings obtained from the standard fire tests of ASTM E119 and UL 263, it is also possible to obtain, as expressed in the standards, the protective membrane

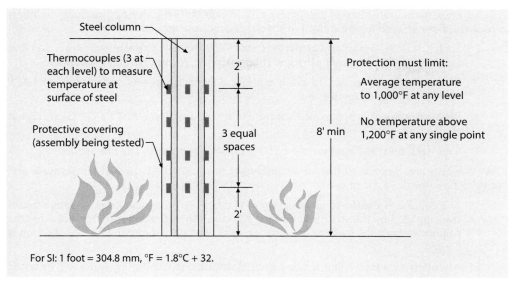

Figure 703-5 **Alternative fire test of steel column protection.**

performance for walls, partitions, and floor or roof assemblies. In the case of combustible walls or floor or roof assemblies, this is also referred to as the finish rating. Although the test standards do not limit the determination of the protective membrane performance to combustible assemblies, its greatest significance is with combustible assemblies.

The end-point criteria for determining the finish rating are that the average temperature at the surface of the protected materials shall not be greater than 250°F (139°C) above the beginning temperature. Furthermore, the maximum temperature at any measured point shall not be greater than 325°F (181°C) above the beginning temperature. These temperatures relate to the lower limit of ignition temperatures for wood. Figure 703-4 illustrates the determination of the finish rating for a wall assembly, which is usually determined during a fire-endurance test of the assembly.

When conducting the standard fire test, the conditions of acceptance, also referred to as failure criteria and end-point criteria, of fire-resistance-rated assemblies are as follows:

1. For load-bearing assemblies, the applied load must be successfully sustained during the time period for which classification is desired.
2. There shall be no passage of flame or gases hot enough to ignite cotton waste on the unexposed surfaces.
3. The average temperature rise on the unexposed surface shall not be more than 250°F (139°C) above the initial temperature during the time period of the test.
4. The maximum temperature on the unexposed surface shall not be more than 325°F (181°C) above the initial temperature during the time period of the test.
5. Walls or partitions shall withstand the hose-stream test without passage of flame or gases hot enough to ignite cotton waste on the unexposed side or the projection of water from the hose stream beyond the unexposed surface.

In addition to the conditions of acceptance just described, load-carrying structural members in roof and floor assemblies are subject to special end-point temperatures for:

1. Structural steel beams and girders—1,100°F (593°C) average at any cross section and 1,300°F (704°C) for any individual thermocouple, for unrestrained assemblies.

2. Reinforcing steel in cast-in-place reinforced concrete beams and girders—1,100°F (593°C) average at any section.

3. Prestressing steel in prestressed concrete beams and girders—800°F (427°C) average at any section.

4. Steel deck floor and roof units—1,100°F (593°C) average on any one span.

As columns are exposed to fire on all surfaces, the standard has special temperature and testing criteria for these members:

1. The column is loaded so as to develop (as nearly as practicable) the working stresses contemplated by the structural design. The condition of acceptance is simply that the column sustains the load for the duration of the test period for which a classification is desired. Or,

2. Alternatively, a steel column may be tested without load, and the column will be tested in the furnace to determine the adequacy of the protection on the steel column. The test and end points are depicted in Figure 703-5.

703.2.1.1 Nonsymmetrical wall construction. At times, an interior wall or partition is constructed nonsymmetrically as far as its fire protection is concerned, with the membrane on one side of the wall differing from that on the opposing side. Where the wall is to be fire-resistance rated, it must be tested from both sides in order to determine the fire-resistance rating to be assigned to the assembly. Based on the two tests, the shortest time period is determined to be the wall's rating. An assembly tested from only one side may be approved by the building official, provided there is adequate evidence furnished to show that the wall was tested with the least fire-resistive side exposed to the furnace. The provisions for exterior walls of nonsymmetrical construction differ somewhat from those addressing interior walls and are regulated by Section 705.5.

703.2.1.3 Restrained classification. A dual classification system is used in ASTM E119 and UL 263 for roof and floor assemblies, including their structural members. This dual classification system involves the use of the terms *restrained* and *unrestrained*. The use of the word *restrained* entails the concept of thermal restraint (restrained against thermal expansion as well as against rotation at the ends of an assembly or structural member).

For example, if a structural beam of a uniform cross section is subjected to heat on its bottom surface, such as would be the case in the standard test furnace, it will attempt to expand in all directions with the longitudinal expansion being the primary component. If the beam is restrained at the ends so that it cannot expand, compressive stresses will build up within the beam, and it will in effect behave in a similar fashion as a prestressed beam. As a result, the thermal restraint will be beneficial in terms of improving the beam's ability to sustain the applied load during the fire test. If the same beam is restrained only for the lower one-half of its cross section, it will tend to deflect upward owing to the conditions of restraint. This upward deflection tendency is also considered to enhance the beam's ability to sustain the applied load during a fire-endurance test.

Conversely, if the end restraint is applied only to the upper half of the beam's cross section, the beam will tend to deflect downward and, in this case, the restraint will be detrimental to the beam's ability to sustain the applied load during the fire-endurance test. As the heat is applied to the bottom surface during a fire, it creates a downward deflection, and the two downward deflections are additive. In an actual building, this could lead to premature failure. It can be seen, then, that thermal restraint may be either beneficial or detrimental to the fire-resistant assembly, depending on its means of application in the building.

General guidance for the building official is provided in ASTM E119 and UL 263 as to what conditions in the constructed building provide restraint. It is generally agreed that an interior panel of a monolithically cast-in-place reinforced-concrete floor slab would be considered to have thermal restraint. Also, Footnote k to Table 721.1(1) provides that "interior spans of continuous slabs, beams and girders may be considered restrained." Conversely, because the restraint present in many construction systems cannot be determined so neatly, the IBC requires that these assemblies be considered unrestrained unless the registered design professional shows by the requisite analysis and details that the system qualifies for a restrained classification. Furthermore, the code requires that any construction assembly that is to be considered restrained be identified as such on the drawings.

703.2.1.5 Exterior bearing walls. This section is intended to modify the acceptance criteria for exterior bearing walls so that the walls will receive a rating based on which of the two following sets of criteria occurs first during the test:

1. Heat transmission or flame and hot gases transmission for nonbearing walls.
2. Structural failure or hose-stream application failure.

The first set of end points measures the wall's ability to prevent the spread of fire from one side to the opposite side. It is considered overly restrictive to require that exterior bearing walls comply with this first set of end points for a longer time than would be required for a nonbearing wall located with the same fire separation distance if it is still structurally capable of carrying the superimposed loads.

703.2.2 Analytical methods. In addition to those assemblies and materials considered fire-resistance-rated construction based on designs documented in approved sources, a number of other methods for determining fire resistance are set forth in this section. Where it can be determined that the fire-resistance rating of a building element is in conformance with one of the listed methods or procedures, such a rating is considered acceptable. The fire exposure and acceptance criteria of ASTM E119 or UL 263 are the basis for applying all of the methods, enabling consistent and compliant application regardless of the method used.

703.2.3 Approved alternate method. This section simply serves as a reminder that the code does accept alternative materials and methods based on Section 104.2.3. Any potential alternate must be "approved" and needs to be shown to be equivalent to the performance that is specified by the code. So while the testing or analytical methods of Sections 703.2.1 and 703.2.2 would generally be used in the vast majority of designs, the code will still permit other options as long as they can be justified and shown to be effective.

703.3 Noncombustibility tests. Throughout the IBC, particularly in Chapter 6, the terms *combustible* and *noncombustible* are used. Under many different conditions, limits are placed on the use of combustible building materials, particularly in buildings of Type I or II construction. This section sets forth the two methods for determining if a material is noncombustible.

For most materials, ASTM E136 is the test standard used to determine if a material is noncombustible. The code will also accept using the ASTM E2652 test standard and equipment provided the ASTM E136 acceptance criteria are used. The ASTM E2652 test is based on the international standard for noncombustibility. The exception will accept a composite of materials such as gypsum board, which have a structural base of noncombustible materials with a surfacing of combustible material. The surfacing material is limited in both thickness and flame spread.

Note that the term *noncombustible* does not apply to surface finish materials or trim.

703.4 Fire-resistance-rated glazing. The use of fire-resistance-rated glazing typically only occurs where the limitations placed on fire-protection-rated glazing make it undesirable or impractical. Fire-resistance-rated glazing is subjected to the ASTM E119 or UL 263 testing criteria, which include stringent limitations on temperature rise through the assembly. Because the glazing is regulated as a wall assembly rather than an opening protective, its use is not limited by any of the provisions of Section 716. It is only regulated under the appropriate code requirements for a fire-resistance-rated wall assembly.

The labeling requirements specific to fire-resistance-rated glazing are set forth in Table 716.1(1). The table indicates that glazing intended to meet the wall assembly criteria be identified with the marking "W-XXX." The "W" indicates that the glazing meets the requirements of ASTM E119 or UL 263, thus qualifying the glazing to be used as a part of a wall assembly. It also indicates that the glazing meets the fire-resistance, hose-stream, and temperature-rise requirements of the test standard. The fire-resistance rating of the glazing (shown in minutes) will then follow the "W" designation.

703.5 Marking and identification. The integrity of fire and/or smoke separation walls is subject to compromise during the life of a building. During maintenance and remodel activities, it is not uncommon for new openings and penetrations to be installed in a fire or smoke separation without the recognition that the integrity of the construction must be maintained or that some type of fire or smoke protective is required. The reduction or elimination of protection that occurs is typically not malicious. Rather, the installation of an inappropriate air opening, or the penetration of the separation without the proper firestopping, is often done due to the lack of information regarding the wall assembly's function and required fire rating.

Through the identification of fire and smoke separation elements, it is possible for tradespeople, maintenance workers, and inspectors to recognize the required level of protection that must be maintained. The requirements apply to all wall assemblies where openings or penetrations are required to be protected. This would include exterior fire-resistance-rated walls as well as fire walls, fire barriers, fire partitions, smoke barriers, and smoke partitions. The identifying markings must be located within 15 feet (4,572 mm) of the ends of the wall and at maximum 30-foot intervals (9,144-mm) to increase the possibility that they would be visible during any work on the wall assemblies. A minimum letter height of 3 inches (76 mm) is also prescribed along with sample language for the marking. See Figure 703-6.

It is intended that the identification marks be located in areas not visible to the general public. Specific locations set forth in the provisions indicate that the identification is to be provided within those concealed spaces that are accessible, such as above suspended ceilings and in attic areas.

703.6 Determination of noncombustible protection time contribution. When dealing with mass timber elements, Section 703.6 provides a performance path to determine the contribution of the noncombustible protection to the overall fire-resistance rating of the assembly or element. This process involves comparative testing of the assembly, both with and without the noncombustible protection to determine the time, in minutes, the noncombustible covering provides. This time contribution will be used in meeting the protection requirements in Section 602.4 and the required fire-resistance ratings needed for Type IV mass timber buildings. While Section 703.6 provides a performance method that requires testing to demonstrate compliance, Section 722.7 provides a prescriptive method that can be used in lieu of this testing procedure.

703.7 Sealing of adjacent mass timber elements. The abutting edges and intersections between mass timber elements are required to be sealed to limit air movement through the gaps. Where mass timber elements are required to fire-resistance rated and they abut other rated elements–either mass timber or other materials—sealants or adhesives are generally

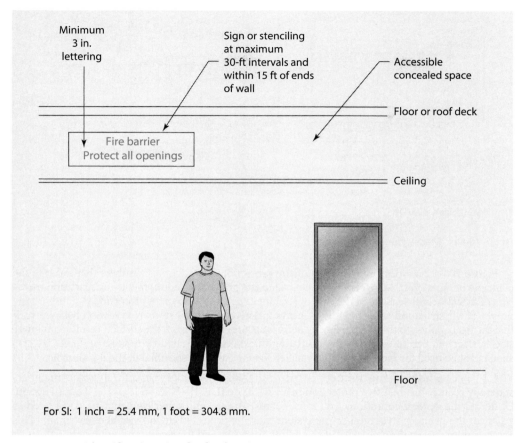

Figure 703-6   Identification sign for fire barrier.

required to resist the passage of air. The importance of this sealing requirement is evident in that it is addressed in Section 1705.20 with a requirement for periodic special inspections.

Section 704 *Fire-Resistance Rating of Structural Members*

Structural frame members such as columns, beams, and girders are regulated for fire resistance based on a building's type of construction. Some types of constructions mandate a higher level of fire endurance for structural members and assemblies on account of the critical nature of their function. Type of construction considerations is based primarily on the potential for building collapse when subjected to fire. Therefore, the structural frame is specifically addressed in Table 601 as to the required fire-resistance ratings. See the discussion in Section 202 on the definitions of primary structural frame members. This section provides further details for the protection of structural members.

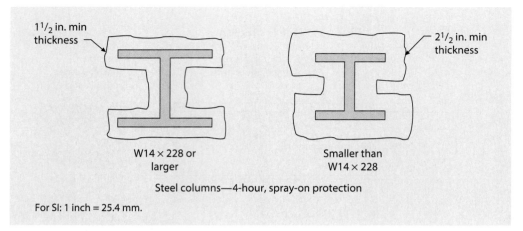

Figure 704-1   **Mass effect.**

Figure 704-1 provides simple details of fire protection of structural members that indicate the principle of *mass effect*. Mass effect is beneficial to the protection requirements for structural members of a heavy cross section. In the case of steel members, the amount of protection depends on the weight of the structural steel member. A heavy, massive structural steel cross section behaves such that the heat applied to the surface during a fire is absorbed away from the surface, resulting in lower steel surface temperatures. Thus, the insulating thicknesses indicated by tests or in Table 721.1(1) should not be used for members with a smaller weight than that specified in the test or table.

704.2 Protection of the primary structural frame.   Primary structural frame members generally require fire-resistive protection in all but Type II-B, III-B, IV-HT, and V-B construction. Under all the general conditions, columns, beams, girders, and other elements are considered as a part of the primary structural frame system must be protected by individual encasement. This protection must occur on all sides of the member and extend for a column's full height.

Where a ceiling is provided, the fire resistance of a column is to be continuous from the top of the foundation or floor/ceiling assembly below through the ceiling space to the top of the column. The fire protection required for the column shall also be provided at the connections between the column and any beams or girders. Where located within a fire-resistance-rated wall assembly as shown in Figure 704-2, the column must be protected through individual encasement unless within the limits of Section 704.3.1. It is not acceptable to simply place an unprotected column within a fire-resistance-rated wall assembly and consider the column as fire-resistant rated.

Primary structural frame members other than columns, such as beams and girders, must also be provided with individual encasement unless Exception 2 is applicable.

Exception 2 recognizes that where structural frame members, except for columns, support no more than two floors or one floor and a roof, nor support a load-bearing wall or nonload-bearing wall more than two stories in height only membrane protection is necessary. In general, the individual encasement is to be protecting all sides of the member, but Exception 1 allows the protection to only be on the "exposed sides."

Where individual encasement is not required, the code allows the use of a floor/ceiling or roof/ceiling assembly to provide protection for the structural member. This is based on the differences in both the testing procedure and the conditions of acceptance that were discussed in Section 703.2. The use of the horizontal assembly's ceiling membrane protection in lieu of individual encasement applies only to horizontal structural members, such as girders, trusses, beams, or lintels. (See Section 704.2 for column protection.)

Examples of various conditions are shown in Figure 704.3.

Fire-Resistance Rating of Structural Members 217

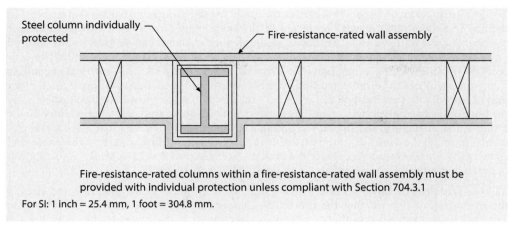

Figure 704-2 **Individual protection of structural columns.**

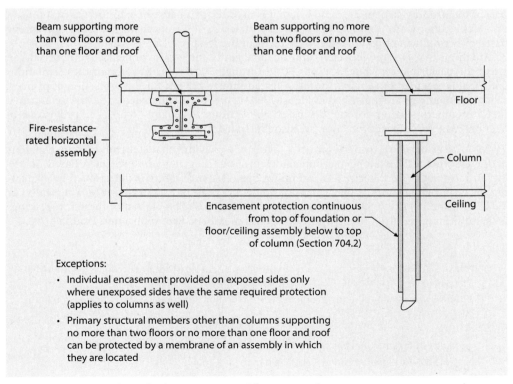

Figure 704-3 **Protection of primary structural frame members.**

704.3 Protection of secondary structural members. Secondary structural members, as defined in Section 202, may be protected in the same manner as primary structural frame members, where a fire-resistance rating is required. Such elements can be individually encased or protected by a membrane of a fire-resistance-rated wall or horizontal assembly. The subsections provide specific exemptions for allowing certain vertical and horizontal secondary structural members to be protected by a membrane as opposed to individual protection. Floor joists and roof joists are examples of secondary members that are permitted to be protected by the horizontal assembly in which they are located. In light-frame wall construction, membrane protection is also permitted for studs, columns, and boundary elements that are integral elements "between the top and bottom plates" of the wall. When dealing with the light-frame wall elements, it is important to recognize the limitations in Section 704.3.1 in order to distinguish between it and the individual encasement required by Section 704.2.

704.4 Truss protection. Trusses can vary greatly from not only the materials and sizes of the members but also whether the protection encapsulates the entire truss or the individual elements. When establishing the appropriate protection, it is the intent of the code that the thickness and details of construction of the fire-resistive protection be based on the results of full-scale tests or of tests on truss components. Approved calculations based on such tests that show that the truss components provide the fire endurance required by the code are also acceptable. One application of this concept is in the use of the encapsulated trusses as dividing partitions between hotel rooms in multistory steel-frame buildings. Because the truss becomes part of the primary structural frame where it is used to span between exterior wall columns, it provides a column-free interior. The fire-resistive design of the encapsulated protection can be based either on tests or on analogies derived from fire tests.

Anything that can reduce, interrupt, or damage the fire-resistive protective material can affect the performance. Thus, Sections 704.5 through 704.8 address some of these situations and what is accepted or required. Some of the additional criteria for the protection of primary structural members are illustrated in Figures 704-4 and 704-5, which depict details for attached metal members and reinforcing discussed in Sections 704.5 and 704.6. The provisions of Section 704.8 for impact protection are also illustrated in Figure 704-6.

704.9 Exterior structural members. The code provides that structural frame elements in the exterior wall or along the outer lines of a building must be protected based on the higher rating of three criteria. The minimum fire-resistance rating is determined by evaluating the requirements for (1) the structural frame per Table 601, (2) exterior bearing walls per Table 601, and (3) fire separation distance per Table 705.5. The highest of these three ratings is the minimum required rating of the structural members. See Application Example 704-1.

GIVEN: An exterior nonbearing wall in a Type IIIB building housing a Group M occupancy. The wall has a fire separation distance of 15 feet to an interior lot line.

DETERMINE: The minimum required fire-resistance rating for structural columns located within the exterior wall.

SOLUTION:

Per Table 601 for structural frame members, a minimum of 0 hours

Per Table 601 for exterior bearing walls, a minimum of 2 hours

Per Table 705.5 for a fire separation distance (FSD) of 15 feet, a minimum of 1 hour

∴ The columns shall have a minimum fire-resistance rating of 2 hours.

Application Example 704-1

Fire-Resistance Rating of Structural Members

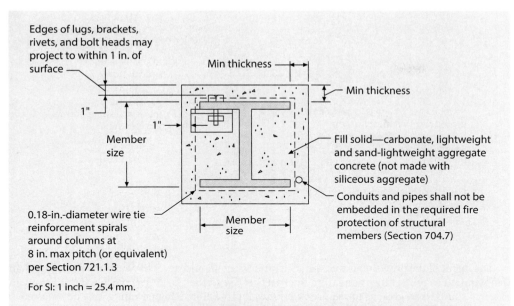

Table 721.1(1)					
Item number	Member size	Minimum thickness (inches)			
		4 hour	3 hour	2 hour	1 hour
1-1.1	6 in. × 6 in. or greater	2½	2	1½	1
1-1.2	8 in. × 8 in. or greater	2	1½	1	1
1-1.3	12 in. × 12 in. or greater	1½	1	1	1

Figure 704-4 **Protection of structural steel column.**

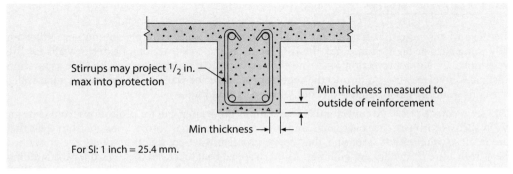

Figure 704-5 **Reinforcing steel in concrete joists.**

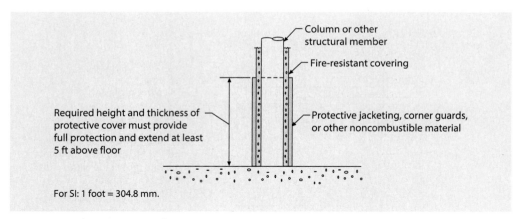

Figure 704-6 **Impact protection.**

The intent of the provision is that the structural frame should never have a lower fire rating than that required to protect the frame from internal fires. Nevertheless, if the exposure hazard from an external source is so great as to require exterior wall protection, a higher rating may be required.

704.10 Bottom flange protection. Exempted from the requirements for fire protection in buildings of fire-resistance-rated construction are the bottom flanges of short-span lintels, and shelf angles or plates that are part of the structural frame. It is assumed by the code that the arching action of the masonry or concrete above the lintel will prevent anything more than just a localized failure. Furthermore, only the bottom flange is permitted to be unprotected and, as a result, the wall supported by the lintel will act as a heat sink to draw heat away from the lintel and thereby increase the length of time until failure that is due to heat. This latter rationale also applies to shelf angles and plates that are not considered as a part of the structural frame. The limitation to spans no greater than 6 feet 4 inches (1,931 mm) is intended to allow such unprotected lintels and angles where a pair of 36-inch (914-mm) doors is installed in the opening.

Section 705 *Exterior Walls*

Because of the potential for radiant heat exposure from one building to another, either on adjoining sites or on the same site, the IBC regulates the construction of exterior walls for fire resistance. Opening protection in such walls may also be required based on the fire separation distances involved. In addition to the regulation of exterior walls and openings in such walls, the code addresses associated projections, parapets, and joints.

705.2 Projections. Architectural considerations quite often call for projections from exterior walls such as cornices, eave overhangs, and balconies. Where these projections are from walls that are in close proximity to a lot line, they create problems that are due to trapping the convected heat from a fire in an adjacent building. As this trapped heat increases the hazard for the building under consideration, the code mandates a minimum distance the leading edge of the projecting element must be separated from the line used to determine fire separation distance. The permitted extent of projections is established by Table 705.2 and based solely on the clear distance between the building's exterior wall and an interior lot line, centerline of a public way, or assumed imaginary line between two buildings on the same lot. Where the distance is less than 2 feet (609 mm), all types of projections are prohibited. As the clear distance increases to 2 feet (609 mm) and beyond, projections are permitted; however, the extent of such projections is regulated.

The exception referencing multiple buildings on the same lot is intended to apply only those projections that extend beyond the opposing exterior walls of the adjacent buildings. For those exterior walls that directly oppose each other, the limits on projecting elements are not applicable where the two buildings are being considered as a single building under Exception 1 to Section 705.3. However, those projections that occur at exterior walls not located in opposition to those exterior walls of an adjacent building are to be regulated by the provisions of Section 705.2. The application of the exception to Section 705.2 is shown in Figure 705-1.

Projections from buildings are further regulated in order to prevent a fire hazard from inappropriate use of combustible materials attached to exterior walls. Thus, the IBC requires that projections from walls of Type I or II buildings be of noncombustible materials. However, it should be noted that certain combustible materials are permitted for balconies and similar projections as well as bay windows and oriel windows in accordance with Sections 705.2.3.1 and 705.2.4.

For buildings that the code considers to be of combustible construction (Type III, IV, or V construction), both combustible and noncombustible materials are permitted in the construction of projections. Where combustible projections are used and extend within a distance of 5 feet (1,524 mm) to the line where fire separation distance is measured (interior lot line, centerline of a public way, or assumed imaginary line between two buildings on the same lot), the code requires that they be of at least 1-hour fire-resistance-rated construction, of heavy-timber construction, constructed of fire-retardant-treated wood (FRTW), or as permitted in Section 705.2.3.1 for balconies and similar projections. This requirement is based on a potential for a severe exposure hazard and, consequently, the code intends that combustible materials be protected or, alternatively, be of heavy-timber construction, which has comparable performance when exposed to fire. An example is shown in Figure 705-2.

Because projections are typically regulated independent of the roof construction, it is entirely possible that their construction methods may be inconsistent. For example, Figure 705-3 shows two situations where the roof construction and resulting projections may differ in their required protection. Figure A relates a Type VA building with a 1-hour fire-resistance-rated roof system but a nonrated projection. On the other hand, Figure B indicates a Type VB building with nonrated roof construction but a minimum 1-hour-protected projection. In each case,

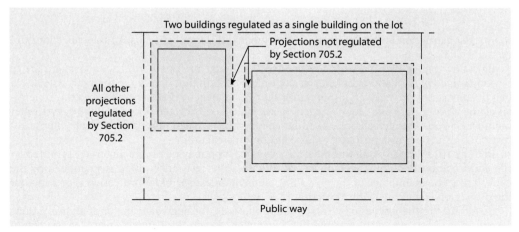

Figure 705-1 **Two buildings regulated as a single building on the lot.**

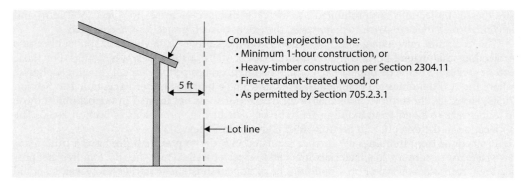

Figure 705-2 **Protection of combustible projections.**

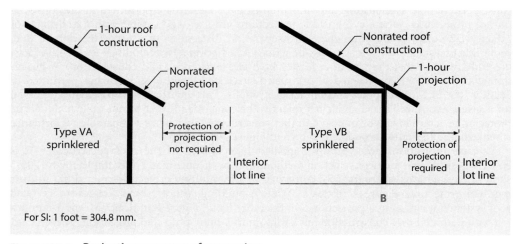

Figure 705-3 **Projection versus roof protection.**

the roof construction and its projection are regulated differently because of the concept of fire resistance being applied.

Balconies and similar projecting elements in buildings of Type I and II construction are to be constructed of noncombustible materials, unless the building is no more than three stories in height above grade plane. Under this condition, FRTW may be used where the balcony or similar element is not used as a required egress path. Unless constructed of complying heavy-timber members, a combustible balcony or similar combustible projection must have a minimum fire-resistance rating equivalent to the required floor construction.

In Type III, IV, and V buildings, balcony construction may be of any material permitted by the code, combustible, or noncombustible. Where a fire-resistance rating is mandated by the code, it must be maintained at the projecting element unless sprinkler protection is provided or the projection is of Type IV construction.

In addition to these types of construction limitations, the aggregate length of all projections cannot exceed 50 percent of the building perimeter at each floor unless sprinkler protection is extended to the balcony areas. In all cases, the use of untreated wood or certain plastic composites is permitted for pickets, rails, and similar guard elements when limited to a height of 42 inches (1,067 mm).

705.3 Buildings on the same lot. The IBC regulates exterior wall construction, opening protection, and projection extent and protection based on the proximity of the exterior walls to lot lines, either real or assumed. This section provides the code requirements for the establishment of imaginary lines between buildings on the same lot. Where two or more buildings are to be erected on the same site, the determination of the code requirements for protection of the exterior walls is based on placing an assumed imaginary line between buildings. Figure 705-4 illustrates an example of two nonsprinklered Type IIIB buildings housing Group S-2 occupancies sharing a 30-foot-wide (9,144-mm) yard, and it is noted that the imaginary line can be located anywhere between the two buildings so that the best advantage can be taken of wall and opening protection, depending on the use and architectural considerations for the exterior walls of the buildings. For example, if unprotected openings amounting to 25 percent of the area of the exterior walls of each nonsprinklered building were desired, the imaginary line would be located so that the distance between it and each building would permit such an amount of unprotected openings. Thus, the code would require that each building be placed at least 15 feet (4,572 mm) from the imaginary line in order to have unprotected openings totaling 25 percent of each opposing wall area. If one of the buildings was to have no openings in the exterior wall, the imaginary line could be placed at the exterior wall of the building without openings. The other building would be located at a distance of 30 feet (9,144 mm) or more from the imaginary line and the other building. In the first case described, the opposing nonbearing exterior walls would both be required to be of minimum 1-hour fire-resistance-rated construction as they are each located less than 30 feet (9,144 mm) from the imaginary line. However, the wall located 30 feet (9,144 mm) from the imaginary line would not require any fire rating. Also, in the last example, Section 705.11 could possibly require that the exterior wall on the assumed lot line be provided with a parapet. See discussion of Section 705.11.

In the case where a new building is to be erected on the same lot as an existing building, the same rationale applies as depicted in Figure 705-4, except that the exterior wall, opening, and projection protection of the existing building determine the location of the assumed imaginary line.

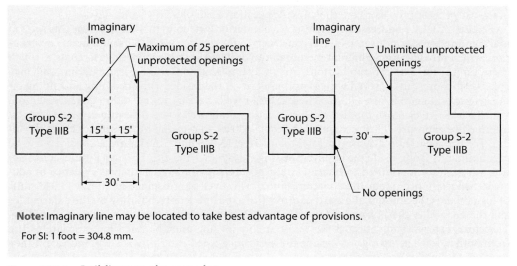

Figure 705-4 **Buildings on the same lot.**

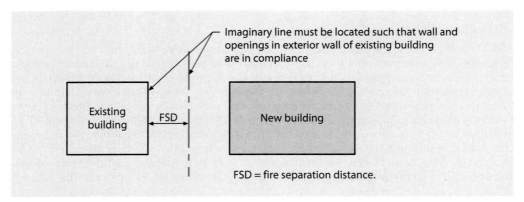

Figure 705-5 **Buildings on the same lot.**

As shown in Figure 705-5, the exterior wall and opening protection of the existing building must remain in compliance with the provisions of the IBC.

As an alternative, where two or more buildings are located on the same lot, they may be considered to be a single building subject to specified limitations. For further discussion of this condition, see the commentary on Section 503.1.2.

705.5 Fire-resistance ratings. The IBC requires that exterior walls conform to the required fire-resistance ratings of Tables 601 and 705.5. Bearing walls must comply with the more restrictive requirements of both tables, whereas nonbearing exterior walls need only comply with Table 705.5. Table 601 is intended to address the fire endurance of bearing walls necessary to prevent building collapse that is due to fire for a designated time period. Table 705.5 is used to determine the required fire-resistance ratings that are due to exterior fire exposure from adjacent buildings, as well as the interior fire exposure that adjacent buildings are exposed to on account of the uses sheltered by the exterior walls. Where structural frame members are located within exterior walls, or on the outside of the building, compliance with Section 704.9 is required. Examples of the use of these provisions are shown in Application Examples 704-1 and 705-1.

Section 703.2.1.1 addresses nonsymmetrical interior wall construction, whereas this section of the code addresses nonsymmetrical construction for exterior walls. This method of construction, which provides for a different membrane on each side of the supporting elements, is much more typical for exterior applications. As an example, a nonsymmetrical exterior wall may consist of wood studs covered with gypsum board on the inside, with sheathing and siding on the exterior side. See Figure 705-6. Where exterior walls have a fire separation distance of more than 10 feet (3,048 mm), the fire-resistance rating is allowed to be determined based only on interior fire exposure. This recognizes the reduced risk that is due to the setback from the lot line. For fire separation distances greater than 10 feet (3,048 mm), the hazard is considered to be predominantly from inside the building. See Figure 705-7. Thus, fire-resistance-rated construction whose tests are limited to interior fire exposure is considered sufficient evidence of adequate fire resistance under these circumstances. However, at a distance of 10 feet (3,048 mm) or less, there is the additional hazard of direct fire exposure from a building on the adjacent lot and the possibility that it may lead to self-ignition at the exterior face of the exposed building. Therefore, exterior walls located very close to any lot line must be rated for exposure to fire from both sides. The listings of various fire-resistance-rated exterior walls will indicate if they were only tested for exposure from the inside, usually by a designation of "FIRE SIDE" or similar terminology. Where so listed, their use is limited to those applications where the wall need

Exterior Walls **225**

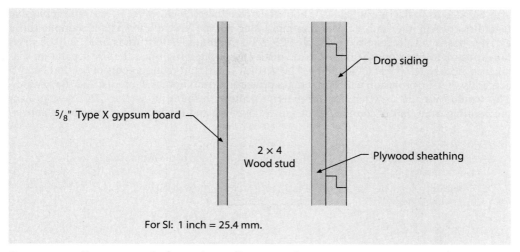

Figure 705-6 **Nonsymmetrical exterior wall construction.**

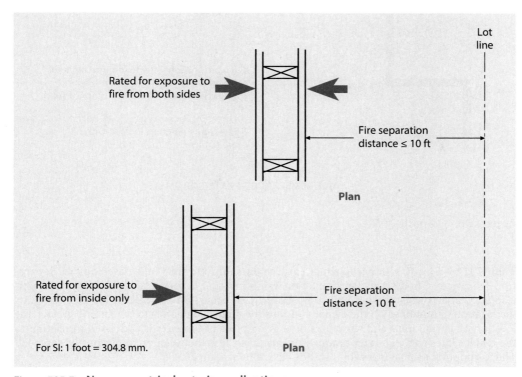

Figure 705-7 **Nonsymmetrical exterior wall ratings.**

only be rated from the interior side. It should be noted that this allowance is generally applicable regardless of why the wall requires a rating. Using a Type VA building as an example, those exterior bearing walls required by Table 601 to be minimum 1-hour walls need only be rated for exposure to fire from the interior side if they are located such that the fire separation distance is more than 10 feet (3,048 mm). The vertical separation requirements of Section 705.9.5 contain a specific provision which mandates protection from both the interior and the exterior. This would override the exclusion for exterior protection found in Section 705.5 and protect the building from flames coming out of a lower opening and impinging on an opening above.

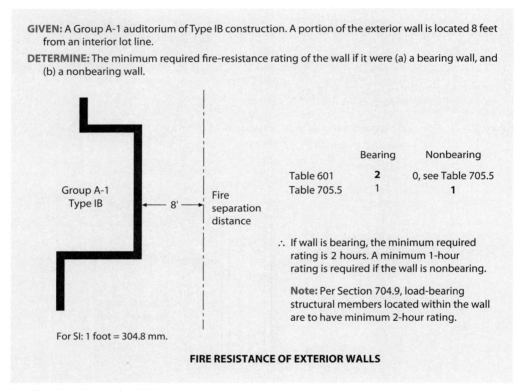

Application Example 705-1

Table 705.5—Fire-Resistance Rating Requirements for Exterior Walls Based on Fire Separation Distances. The IBC, as far as exterior wall protection is concerned, operates on the philosophy that an owner can have no control over what occurs on an adjacent lot and, therefore, the location of buildings on the owner's lot must be regulated relative to the lot line. In fact, the location of all buildings and structures on a given piece of property is addressed in relation to the real lot lines as well as any assumed or imaginary lines between buildings on the same lot. The assumption of imaginary lines was discussed in Section 705.3.

The lot-line concept provides a convenient means of protecting one building from another insofar as exposure is concerned. Exposure is the potential for heat to be transmitted from one building to another under conditions in the exposing building. Radiation is the primary means of heat transfer.

Exterior Walls

The code specifically provides that the fire separation distance be measured to the center line of a street, alley, or public way. As the code refers to public way, this would also be applicable to appropriate open spaces other than streets or alleys that the building official may determine are reasonably likely to remain unobstructed through the years.

The regulations for exterior wall protection based on proximity to the lot line are contained in Table 705.5. The IBC indicates that the distances are measured at right angles to the face of the exterior wall (see definition of "Fire separation distance" in Section 202), which would result in the fire-resistive requirements for exterior walls not applying to walls that are at right angles to the lot line. See Figure 705-8.

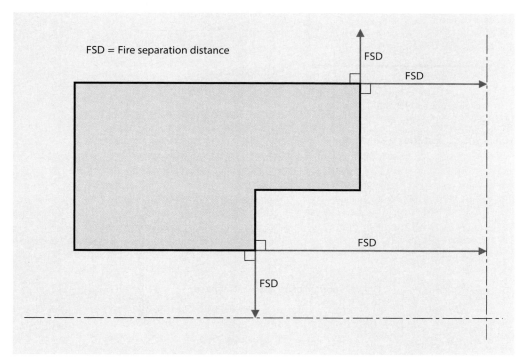

Figure 705-8 **Fire separation distance.**

In order to properly utilize Table 705.5, it is necessary to identify the fire separation distance, the occupancies involved, and the building's type of construction. As the fire separation distance increases, the fire-resistance rating requirements are reduced, based on the occupancy group under consideration. Figure 705-9 illustrates the application of exterior wall protection where the exterior walls of the building are parallel and perpendicular to the lot line. In this case, the illustration assumes that the building is one story of Type VB construction and used for offices (Group B). Referring to Table 705.5, it is noted that exterior walls less than 10 feet (3,048 mm) from the lot line must be of minimum 1-hour fire-resistance-rated construction. Figure 705-10 depicts a similar building located such that one of the exterior walls is not parallel and perpendicular to the lot line, but is at some angle other than 90 degrees (1.57 rad). The regulation of doors, windows, and other openings in exterior walls is addressed in the discussion of Section 705.9.

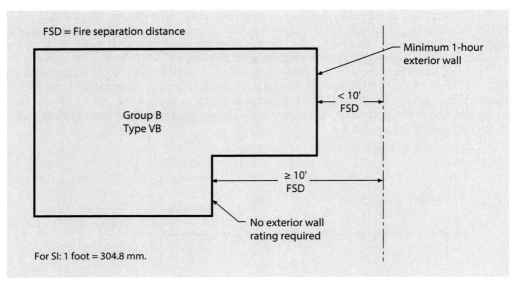

Figure 705-9 **Exterior wall rating.**

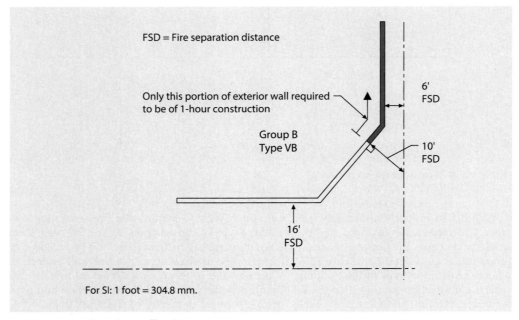

Figure 705-10 **Exterior wall rating.**

Several footnotes to the table address modifications to the general requirements. Footnote a repeats a previous requirement that load-bearing exterior walls must comply with both Tables 601 and 705.5.

Although Table 705.5 requires a Group S-2 occupancy of Type I, II, or IV construction to have a minimum 1-hour exterior wall where the fire separation distance is less than 30 feet (9,144 mm), Footnote c reduces that distance significantly where the building is a complying open parking garage. Under such conditions, a minimum 1-hour fire-resistance-rated exterior wall is required only where the fire separation distance is less than 10 feet. Footnote d indicates that each story of the building is regulated independently for the fire separation distance provisions, as shown in Figure 705-11.

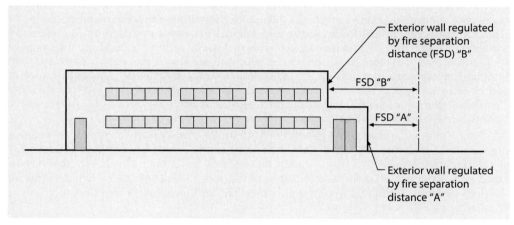

Figure 705-11 **Fire separation distance measurement.**

Table 705.9 allows for an unlimited amount of unprotected openings in exterior walls of a sprinklered building that has a fire separation distance of at least 20 feet (6,096 mm). However, in certain buildings, Table 705.5 requires those same exterior walls to be fire-resistance rated for a minimum of 1 hour. Footnote g recognizes that any nonbearing exterior wall permitted to be entirely open due to the unlimited unprotected opening allowances of Table 705.9 need not be required to have a fire-resistance rating due to fire separation distance.

There are only a small percentage of buildings where this footnote is applicable. It has no effect on:

- Exterior bearing walls
- Group H-1, H-2, and H-3 occupancies
- Nonsprinklered buildings
- Buildings of Type IIB and VB construction, other than Groups H-4 and H-5
- Exterior walls with a fire separation distance of less than 20 feet (6,096 mm)
- Exterior walls with a fire separation distance of 30 feet or more (9,144 mm)

705.6 Continuity. Two methods are available to maintain the continuity or boundaries of fire-resistance-rated exterior walls. The lower boundary of a fire-resistance-rated exterior wall is either the top of the foundation or the top of a floor/ceiling assembly, as applicable.

The upper boundary of a fire-resistance-rated exterior wall is based on one of two options. The first option is the underside of the floor sheathing, roof sheathing, deck, or slab immediately above the lower boundary. The second option for the upper boundary is a floor/ceiling or roof/ceiling assembly with a fire-resistance rating at least equivalent to the exterior wall. Additionally, a fire separation distance greater than 10 feet (3,048 mm) must be provided, recognizing that this second option only addresses fire exposure from the inside of the building. A new provision in Section 705.7.1 (addressed as the following topic in this publication) provides further clarification that in Type III platform construction only the portion of the floor/ceiling or roof/ceiling assembly that supports the exterior wall above is required to have the fire-resistance rating of the exterior wall.

705.7 Structural stability. This section addresses the two conditions regarding elements used to brace the exterior wall. Where such elements are located within the plane of the wall (as part of the wall assembly) or on the outside of the wall, the bracing members are to be regulated for both external fire exposure and internal fire exposure based on the references to Tables 601 and 705.5. Those bracing elements that occur within the building, such as floor joists and/or roof joists that frame into the exterior wall, are also required to be protected from internal fire exposure if mandated by Table 601. However, external fire exposure concerns addressed in Table 705.5 are not specifically addressed. Floor systems, roof systems, and other elements that provide lateral bracing from the interior side of an exterior wall are only required to be fire-resistance rated when required by Table 601.

705.7.1 Floor assemblies in Type III construction. For Type III construction where a nonrated or 1-hour fire-resistance-rated floor intersects and supports a 2-hour exterior wall at each floor level, there are various significant concerns, including maintaining the fire resistance of the exterior wall at the intersection with the floor, as well as material requirements for the floor structure, given that wood framing members and sheathing within the wall itself are required to be FRTW.

Maintaining the fire resistance of supporting construction plays a much more important role in the performance of a fire-resistance-rated wall than the use of FRTW in the supporting floor. There is no demonstrated increase in fire-resistance rating for FRTW when compared to untreated wood. FRTW exhibits reduced flame spread but does not increase the assemblies' fire-resistance rating. In other words, requiring the end of a floor/ceiling or roof/ceiling assembly to be fire-retardant-treated does not increase the fire resistance of the wall. The code does not require elements of the floor to be fire-retardant-treated even if they serve to support the gravity loads from the wall above. However, it does require those supporting floor elements to provide fire resistance equal to that required for the wall.

The first sentence of Section 705.7.1 limits the application of this section to Type III construction. Although it is not stated, the criteria will most commonly apply to platform frame construction where the floor assembly is supported by the top of the wall below and the wall above is supported by that floor assembly. The portion of the floor assembly directly within the gravity load path of the exterior wall is required to provide a 2-hour fire-resistance rating as required by Table 601. This requirement is intended to ensure that the fire-resistance rating required of the exterior wall of the story above will continue through the supporting segment of the floor assembly to the exterior wall of the story below.

The second sentence states that it is permissible to consider the ceiling membrane contribution when assessing the fire-resistance rating of the floor assembly at the exterior wall. A ceiling membrane may or may not be present, but it is an appropriate design assumption to consider the ceiling membrane's contribution when the fire rating of the floor assembly supporting the exterior wall is to be based on fire exposure from the interior of the building.

The last sentence deals with material requirements. Typically, construction material at the perimeter of the floor assembly may include a single rim joist, multiple rim joists, and/or blocking to achieve the required fire-resistance rating, while also maintaining a gravity load path for the duration of the required fire-resistance rating. Requirements for materials within the wall plane that are part of the floor construction are to be consistent with what is required for the interior floor assembly, not the exterior wall. For example, if the exterior wall studs are light-gauge steel to satisfy the noncombustible exterior wall requirement, the perimeter material in the floor assembly which bears on the wall below can be constructed of any material permitted for the interior building elements in Type III construction, provided the required fire-resistance rating, as clarified in this code section, is demonstrated. This detailing can be accomplished by several means, such as providing extra rim board members or blocking, and extra protection for the floor elements at the intersection.

705.8 Unexposed surface temperature. The provisions of this section provide a reduction of the prescriptive fire-resistance requirements for exterior walls under certain conditions. A fire-resistance-rated wall is generally required to meet the conditions of acceptance of ASTM E119 or UL 263 for fire endurance and hose-stream tests on the surface exposed to the test fire, and heat-transmission limits on the unexposed surface. At fire separation distances beyond the point where openings are not permitted, typically 3 feet (914 mm), two more options are available:

1. Where opening protection is required, but the percentage of opening protection is not limited [typically a fire-separation distance of 20 feet (6,096 mm) or more], compliance with the heat-transmission limits of ASTM E119 or UL 263 is not required. This recognizes that, although heat transmission is an important consideration for interior walls, the fire hazard that the limit addresses is substantially reduced once the exterior wall of a building is set back far enough that the fire hazard it presents to (and receives from) a building on an adjacent lot does not warrant a limit on the percentage of opening protection to limit the hazard. It has the effect of compliance with the conditions of acceptance for fire assemblies of the same hourly rating.

    According to NFPA 252 for fire door assemblies and NFPA 257 for fire window assemblies, nearly identical conditions of acceptance for fire endurance and hose-stream tests are required, but without limits on heat transmission. Because an unlimited percentage of opening protection is allowed, the lack of a heat transmission limit for exterior walls is consistent with that for fire door and fire window assemblies. An exterior wall that does not meet the heat transmission limits is considered equivalent to an opening protective of the same hourly rating in its reduced ability to limit heat transmission.

2. Where the percentage of opening protection is limited [typically having a fire separation distance between 3 feet (914 mm) and 20 feet (6,096 mm)], a similar reduction is possible, provided a correction is made according to the formula presented in this section. The formula converts the actual proposed area of protected openings to an increased equivalent area in proportion to the area of exterior wall surface under consideration that lacks adequate control of heat transmission. It places additional limits on the allowable percentage of opening protection.

The formula increases the required percentage of opening protection, whereas Section 705.9 sets limits on the percentage. Relative to the limitations of Section 705.9, this method allows for a smaller percentage of opening protection at the same fire separation distance. Thus, a greater fire separation distance is required to maintain the same percentage of opening protection. The reduction of the heat transmission capacity of the exterior walls is compensated by a reduction in the allowable percentage of opening protection. Without this provision, a fire-resistance-rated exterior wall that does not meet the heat transmission limits would not be allowed.

If actual test results or other substantiating data are available, they may be used in the computations. In their absence, the standard time-temperature curve of ASTM E119 and UL 263 would be used, which results in an equivalent area of protected openings equal to the actual area of protected openings plus the exterior wall area without adequate control of heat transmission per ASTM E119 and UL 263. This is converted into a percentage of opening protection and compared to the limits of Section 705.9. The use of actual test results may reduce this effect. See Section 705.9 for the basic limits prior to modification by this provision.

705.9.1 Allowable area of openings. Openings in an exterior wall typically consist of windows and doors. Occasionally, air openings such as vents are also present. The maximum area of either protected or unprotected openings permitted in each story of an exterior wall is regulated by this section. In addition, both unprotected and protected openings are permitted in the same exterior wall based on a unity formula. The term protected in this section refers to those elements such as fire doors, fire windows, and fire shutters regulated in Section 716. Protected openings have the mandated fire-protection rating necessary to perform their function. Unprotected openings are simply those exterior openings that do not qualify as protected openings. Opening protection presents a higher fire risk than fire-resistance-rated construction insofar as it does not meet the heat transmission limits of ASTM E119 or UL 263, as previously discussed. At increasing distances from where openings are no longer prohibited, the hazard from heat radiation decreases, allowing the percentage of openings, both protected and unprotected, to increase. The high hazard of heat exposure at small fire separation distances justifies the prohibition of openings in order to limit the percentage of wall area without adequate heat transmission limits. As the fire separation distance increases, the percentage of openings is allowed to increase in compensation. At greater distances, the limit on the percentage of opening protection is eliminated. This recognizes that, at greater distances, the lack of adequate control of heat transmission does not pose a significant hazard to adjacent buildings, but containment of the fire to its origin inside the exposed building is still important. The exterior wall and opening protection requirements apply on each individual story of the building based on the fire separation distance. This accounts for the exposure at each separate level and would relax the requirements where the building facade steps back and faces less radiant heat exposure.

There is a distance from lot lines where the hazard is reduced to such a degree that all opening limitations are no longer warranted. At this point, exposure to and from adjacent buildings is not significant and the need for fire resistance at exterior walls is reduced to fire protection of bearing walls and structural members in order to delay building collapse in the event of fire. Arguably the most important provision is Exception 2 to Section 705.9.1. It indicates that if the exterior wall of the building and its primary exterior structural frame are not required by the code to have a fire-resistance rating, then unlimited unprotected openings are permitted. In other words, if the wall does not require a rating, any openings in the wall are unregulated for area and fire protection. An example is shown in Figure 705-12.

Although not stated in the exception, only when Table 705.5 requires a fire-resistance rating does Table 705.9 limit the maximum area of exterior openings. Where some other provision of the code mandates a fire-resistance-rated exterior wall, such as an exterior bearing wall supporting a fire-resistance-rated horizontal assembly, the limitations of Table 705.9 do not apply. It is also not necessary to review Table 601 to apply the exception as the conditions are established in such a manner that Table 705.5 provides all of the necessary information. Directly stated, if Table 705.5 does not mandate a fire-resistance-rated exterior wall, an unlimited amount of unprotected openings are permitted.

The limitation on exterior openings is also not applicable for first-story openings in buildings of other than Group H as indicated in Exception 1 to Section 705.9.1. Limited in application, this exception allows an unlimited amount of unprotected openings at the first story under

Exterior Walls

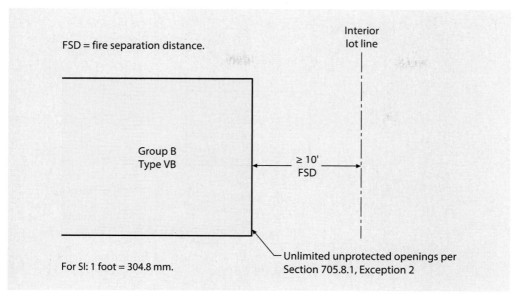

Figure 705-12 Unlimited unprotected openings.

specified circumstances. Exception 1 is basically allowing the first floor of the building to use the entire width of the open space instead of the more limited fire separation distance which is usually a midpoint. Based on the clear space, access to the area, and the ease of protecting a lower story, the code permits these unlimited unprotected openings. The provisions are often used for opposing buildings having storefront systems. See Figure 705-13.

How are exterior openings regulated in fully sprinklered buildings? Table 705.9 also recognizes an increase in the allowable area of unprotected exterior openings for those buildings that are provided with an automatic sprinkler system throughout. For example, the fire-resistance-rated exterior wall of a fully sprinklered building having a fire separation distance of 15 feet (4,572 mm) may have 75 percent of its surface area consisting of unprotected openings. If the building is not sprinklered, the limit on unprotected openings is only 25 percent. In other than higher level Group H occupancies, the maximum permitted area of unprotected openings in an exterior wall is allowed to be the same as the tabulated limitations for protected openings, provided the building is protected throughout with an NFPA 13 automatic sprinkler system. The increased areas permitted due to sprinkler protection are all incorporated directly in Table 705.9. It is important to note that the presence of an automatic sprinkler system does not increase the maximum allowable opening area for protected openings. Whereas the benefits of such an increase would seem justifiable because of the increased level of protection, such an allowance is not addressed in the code. In addition, the unity formula (Equation 7-2) is not applicable to fully sprinklered buildings insofar as the code provides an increased allowance for unprotected openings to the amount permitted for protected openings.

705.9.2 Protected openings. Section 716 is referenced for opening protection and addresses fire windows, fire doors, and fire shutters. The use of sprinklers and water curtains to eliminate the required opening protection is addressed in the exception. It indicates that where the building is sprinklered throughout, those openings protected by an approved water curtain do not need to be fire-protective assemblies. However, the exception has virtually no application when

234   Chapter 7 ■ Fire and Smoke Protection Features

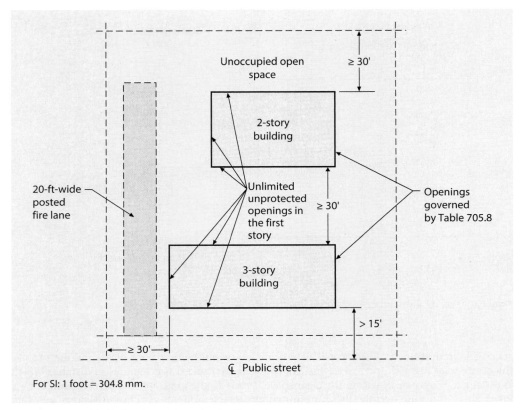

Figure 705-13   **Unlimited openings in the first story exterior wall.**

the provisions of Table 705.9 are implemented, as the table typically allows for the elimination of protected openings in sprinklered buildings (in other than Group H-1, H-2, and H-3 occupancies) without the need for water curtains. There are a number of provisions throughout the IBC where the exception could be used. For example, Section 1029.3 typically mandates ¾-hour fire-protected openings in walls of egress courts less than 10 feet (3,048 mm) in width. In a fully sprinklered building, the use of a complying water curtain would eliminate the need for such openings to have a fire-protection rating. Another example where the exception might be applied includes Section 1023.7 where walls or openings of the building create an exposure hazard to interior exit stairways and ramps.

705.9.4   Mixed openings.   Table 705.9 specifies the maximum allowable percentage of protected and unprotected openings, considered separately and based on fire separation distance alone. The unity formula (Equation 7-2) as set forth in Section 705.9.4 determines the maximum allowable area of protected and unprotected openings where they are proposed together in an exterior wall at an individual story of a nonsprinklered building. It offers a traditional interaction relationship, namely, the sum of the actual divided by the sum of the allowable cannot exceed one. An example of the determination of the maximum area of exterior wall openings, where both protected and unprotected openings are used, is provided in Application Example 705-2. The use of Equation 7-2 is limited to those buildings that are not provided with

Exterior Walls

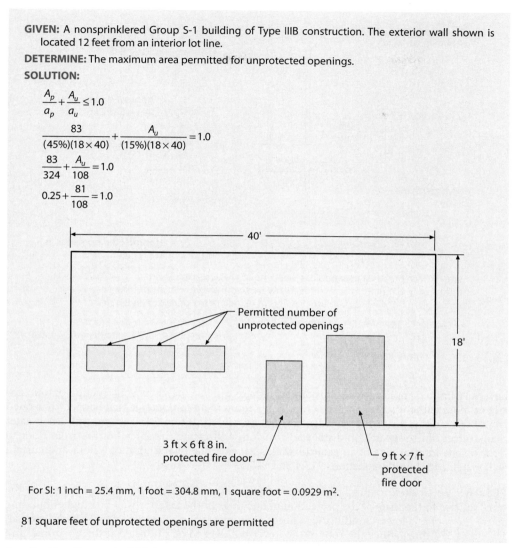

GIVEN: A nonsprinklered Group S-1 building of Type IIIB construction. The exterior wall shown is located 12 feet from an interior lot line.
DETERMINE: The maximum area permitted for unprotected openings.
SOLUTION:

$$\frac{A_p}{a_p} + \frac{A_u}{a_u} \leq 1.0$$

$$\frac{83}{(45\%)(18 \times 40)} + \frac{A_u}{(15\%)(18 \times 40)} = 1.0$$

$$\frac{83}{324} + \frac{A_u}{108} = 1.0$$

$$0.25 + \frac{81}{108} = 1.0$$

For SI: 1 inch = 25.4 mm, 1 foot = 304.8 mm, 1 square foot = 0.0929 m².

81 square feet of unprotected openings are permitted

Application Example 705-2

an NFPA 13 sprinkler system throughout. Where the building is fully sprinklered, the code provides no advantage where protected openings are provided.

705.9.5 Vertical separation of openings. The intent of this section is to limit the vertical spread of fire from floor to floor at the exterior wall of the building. The code requires exterior flame barriers projecting out either from the wall or in line with the wall. These flame barriers are intended to prevent the leap-frogging effect of a fire at the outside of a building. See Figure 705-14. However, there are three exceptions that eliminate the required barriers. The first is for buildings that are three stories or less in height. The second is for fully sprinklered buildings. The third exception is for open parking garages. It is probable that this provision will

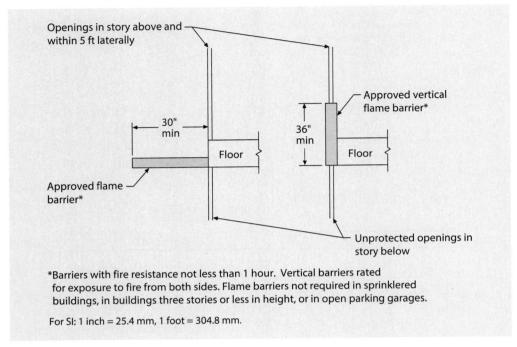

Figure 705-14 Flame barriers.

have very limited application, as it is doubtful there will be much new construction of four stories or more without sprinkler protection. The requirement for exterior wall barrier to be rated from both sides is a specific requirement that overrides Section 705.5 where wall has greater than 10 feet of fire separation distance. Provisions addressing the spread of fire from floor to floor on the interior side of an exterior wall, such as at the intersection of a floor and curtain wall system, are found in Sections 715.4 and 715.5.

705.9.6 Vertical exposure. The scope of this section is limited to buildings located on the same lot and to the issue of the protection of openings in the exterior wall of a higher building above the roof of a lower building. It requires each opening in the exterior wall that is less than 15 feet (4,572 mm) above the roof of the lower building to be protected if the horizontal fire separation distance for each building is less than 15 feet (4,572 mm). See Figure 705-15. There is an exception that applies where the roof construction has at least a 1-hour fire-resistance rating, also illustrated in Figure 705-15. Application of this provision potentially mandates a higher level of protection than that required by the code for two buildings on separate adjoining lots. The presence of a lot line between two buildings institutes the concept of fire separation distance in the regulation of the opposing exterior walls and any openings in such walls. On a single lot with two buildings, the same concept is applied owing to the requirement for the placement of an assumed imaginary line between the buildings. This line is also the basis for regulating exterior wall and opening protection that is due to fire separation distance. The provisions of Section 705.9.6 introduce additional requirements that may not be mandated on account of the fire separation distance concept. In addition, where two buildings are located on the same lot, the provisions of Sections 705.3 and 503.1.2 permit them to be considered a single building if the aggregate area of the buildings is within the limits of Chapter 5 for a single

Exterior Walls

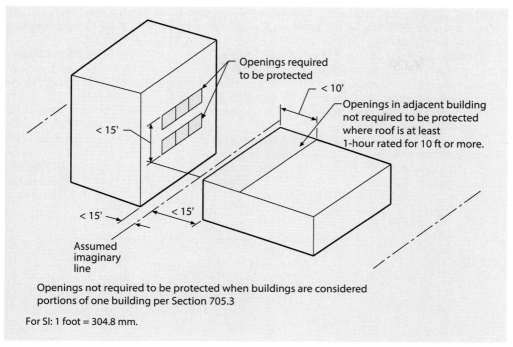

Figure 705-15 **Vertical exposure.**

building. For consistent application of the fire separation distance concept, it would appear that the methodology for buildings on the same lot could be permitted to be used rather than the vertical exposure provisions of Section 705.9.6.

It appears that the provision is intended to mirror the termination requirements and allowances for fire walls where located in stepped buildings as established in Section 706.6.1. It would be logical to assume that if a fire wall is not necessary to obtain code compliance, resulting in no required application of the fire wall termination requirements, the same concept should be considered when applying this provision to buildings on the same lot that can be regulated as a single building.

705.12 Parapets. This section intends that the exterior walls of buildings shall extend a minimum of 30 inches (762 mm) above the roof to form a parapet. There are two reasons for the parapet:

1. To prevent the spread of fire from the roof of the subject building to a nearby adjacent building.

2. To protect the roof of a building from exposure that is due to a fire in an adjacent nearby building.

Most buildings do not have complying parapets, and those that do typically use them to hide the roof slope or roof-top equipment. Therefore, the exceptions to this section tend to become the general rule. Three of the six exceptions listed in the code—1, 3, and 6—involve cases where the parapet would serve no useful purpose. In Exception 2, a concession is made to the small-floor-area building, and in Exceptions 4 and 5, an alternative method for providing equivalent protection is delineated. It is not necessary that all of the exceptions listed apply. Compliance with only one of the exceptions is all that is necessary for the elimination of a complying parapet.

Certainly, walls not required to be of fire-resistance-rated construction would not benefit from a parapet. In the case of walls that terminate at 2-hour fire-resistance-rated roofs or roofs constructed entirely of noncombustible materials, the parapet would be of little benefit, as the construction of the roof would prevent the spread of fire from or into the building. The exception for noncombustible roof construction is not intended to preclude the use of a classified roof covering.

In the case of walls permitted to have unprotected openings in conformance with Exception 6, the code assumes that the exterior wall will be far enough away from either an exposing building or an exposed building so that the protection provided by the parapet will not be necessary. This distance will vary based on the presence of a sprinkler system in the building, as shown in Figure 705-16.

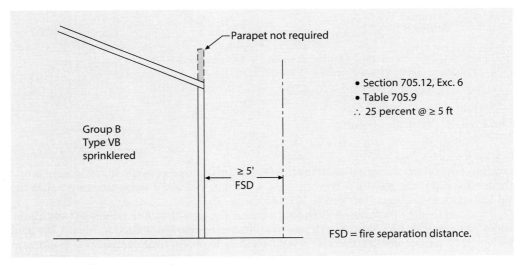

Figure 705-16 **Parapet exception.**

The fourth exception makes a provision for 1-hour fire-resistance-rated exterior walls that are constructed similar to 2-hour fire walls that terminate at the underside of the roof sheathing, deck, or slab. This provides designers with an alternative to the use of parapets while recognizing that these walls provide adequate protection of the structure and its occupants as well as consistency with Section 706.6 for fire walls. See Figure 705-17. Exception 5 applies only to Group R-2 and R-3 occupancies and is intended to protect at the roof line through the use of a noncombustible roof deck, fire-retardant-wood sheathing, or a gypsum-board underlayment.

705.12.1 Parapet construction. The requirements of this section are only applicable to those parapets that are mandated based on Section 705.12. Where a parapet is not required but merely provided for some purpose, such as the screening of rooftop equipment, these construction requirements are not applicable. In addition to having the same degree of fire resistance as required for the wall, the code also requires that the surface of the parapet that faces the roof

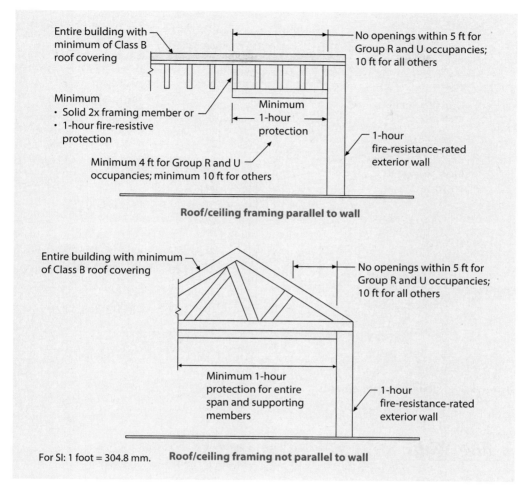

Figure 705-17 Parapet alternative.

be of noncombustible materials for the upper 18 inches (457 mm). Thus, a fire that might be traveling along the roof and reaching the parapet will not be able to continue upward along the face of the parapet and over the top and expose a nearby adjacent building. The requirement only applies to the upper 18 inches (457 mm) of the parapet to allow for extending the roof covering up the base of the parapet so that it can be effectively flashed. The 18-inch (457-mm) figure is based on a parapet height of at least 30 inches (762 mm).

As stated in the code, the 30-inch (762-mm) requirement is measured from the point where the roof surface and wall intersect. Therefore, when a cricket is installed adjacent to the parapet, the 30-inch (762-mm) dimension would be taken from the top of the cricket.

In those cases where the roof slopes upward away from the parapet and slopes greater than 2 units vertical in 12 units horizontal (16.7-percent slope), the parapet is required to extend to the same height as any portion of the roof that is within the distance where protection of openings in the exterior wall would be required. However, in no case shall the height of the parapet be less than 30 inches (762 mm). See Figure 705-18 for an illustration of this requirement.

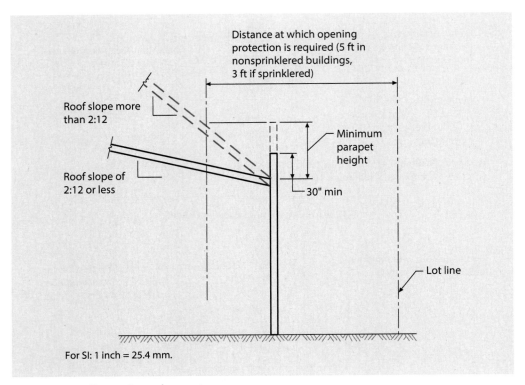

Figure 705-18 **Parapet requirements.**

Section 706 *Fire Walls*

Section 706 provides the technical details of how fire walls are to be constructed. The IBC selectively permits fire walls to be installed within a building, thereby creating one or more smaller area buildings. It further selectively recognizes that fire walls can be used for other purposes, such as where fire areas are created. The concept is based on buildings on adjoining lots having a common party wall or two separate fire-resistance-rated walls located on the lot line. The high level of fire-resistance-rated construction between the two buildings, along with other controls, is deemed adequate for the protection of one building from its neighboring building and thus limiting the size of a potential fire. The use of one or more fire walls within a building is optional, based on a decision by the designer. The code never mandates a fire wall be used, but rather offers it as an alternative to other mandated provisions. The use of one or more fire walls to create separate buildings is seemingly limited by Sections 503.1 and 414.2.3 to only the following applications:

1. Allowable building area. The installation of one or more fire walls reduces the floor area in each of the separated buildings. Smaller floor areas can result in a reduction in the type of construction for one or more of the smaller buildings.

2. Allowable building height. Each building created by the presence of a fire wall can be evaluated independently for height purposes.

Fire Walls

3. **Multiple construction types.** By separating a structure into separate buildings, they each are regulated independently for type of construction. Thus, not all of the structure would need to be classified based on the lowest construction type involved.

4. **Number of control areas.** The allowable number of control areas per building as limited by Table 414.2.2 is permitted to be increased through the formation of "separate buildings" created by the installation of one or more fire walls.

The use of a fire wall for purposes other than creating multiple buildings is selectively established throughout the IBC, including its use as a horizontal exit separation per Section 1026.2 or an increase in residential building height as set forth in Sections 510.5 and 510.6.

Examples of two uses of fire walls are shown in Application Examples 706-1 and 706-2. Another application of the fire wall concept is found in Appendix B of the *International Fire Code®* (IFC®) relating to fire-flow requirements for buildings. Where structures are separated by fire walls without openings, the divided portions may be considered separate fire-flow calculation areas. There are numerous other applications for fire walls, all based on the concept of reducing hazards due to a significant fire separation.

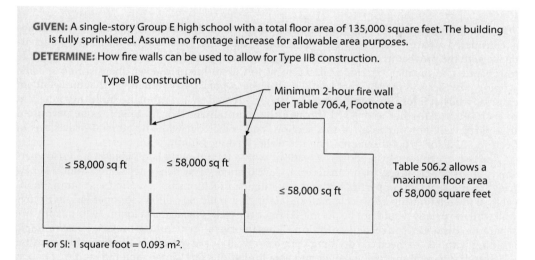

GIVEN: A single-story Group E high school with a total floor area of 135,000 square feet. The building is fully sprinklered. Assume no frontage increase for allowable area purposes.

DETERMINE: How fire walls can be used to allow for Type IIB construction.

For SI: 1 square foot = 0.093 m².

By creating a minimum of three separate buildings under one roof, Type IIB construction is acceptable. Slight increases in allowable area are possible by including available frontage increases; however, each of the three buildings must be evaluated individually for open space.

Application Example 706-1

706.1 General. As previously mentioned, one or more fire walls may be constructed in a manner such that the code considers the portions separated by the fire walls to be separate buildings. Because a fire wall is such a critical element in the prevention of the spread of fire from one separated building to another, it is of great importance that the wall be situated and constructed properly. It must provide a complete separation. It should be noted that when a wall serves both as a fire barrier separating occupancies and a fire wall, the most restrictive requirements of each separation shall apply. The code also prohibits any openings in fire walls that are constructed on lot lines (defined as party walls).

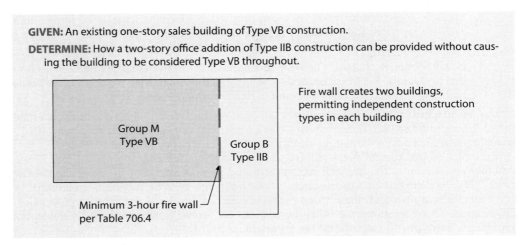

Application Example 706-2

706.1.1 Party walls. A common wall located on the lot line between two adjacent buildings is considered a party wall under this provision of the code. Regulated as a fire wall in accordance with the provisions of Section 706, a party wall can potentially be accepted as an alternate to separate and distinct exterior walls adjacent to the lot line. However, there are no specific scoping provisions in the IBC that would permit the use of a party wall rather that independent exterior walls at a lot line condition. The hazard created by neighboring buildings adjacent to each other is further addressed through the requirement that no openings be permitted in a party wall. For purposes of this section, and consistent with the general provisions of Section 503.1 for structures containing fire walls, separate buildings are created.

The two exceptions help address situations where the property is divided more for ownership purposes (e.g. taxes, loans, insurance, development) and being developed jointly versus truly having separate properties without any intended relationship. The first exception, applicable to the relationship between a mall and its anchor buildings, allows for openings in a party wall which typically would not be permitted since the wall is located at a lot line without any fire separation distance. The second exception allows the overall structure to be reviewed as a single building without the need for dividing with a fire wall or party wall. In this situation, because the overall structure meets the height and area limitations for a single building and that access is for maintenance is provided to both owners to work on either side of the lot line, there is no need for dividing the building.

706.2 Structural stability. The objective of a fire wall is that a complete burnout can occur on one side of the wall without any effects of the fire being felt on the opposite side. Furthermore, the only damage to the wall will be the effects of fire and the shock of hose-stream application on the fire side. The code is very clear that fire walls should remain in place for the expected time period. Therefore, structural failure on either side of the wall shall not cause the collapse of the wall, nor can the required fire-resistance rating be diminished. In addition, structural members (especially members that conduct heat) that penetrate fire walls could limit their effectiveness and do not comply with this provision. Any structural member that passes through a fire wall could also adversely affect the integrity of the required fire-resistance-rated construction.

The intent of this section can be partially traced back to Section 101.3, which states that one of the goals of the code is to provide safety to fire fighters and emergency responders during emergency operations. During a fire, a fire wall provides a safe haven on the non-fire side for fire fighters to stage and fight a fire. It is critical that the fire wall does not pose a threat of collapse to the fire department personnel. This is more easily achieved where the fire wall is a nonbearing wall and is not penetrated by load-bearing elements. However, where a fire wall is proposed as a bearing wall, the building official should ensure that those structural members that frame into the wall will not cause the premature collapse of the fire wall prior to the hourly rating established for the wall. The structural engineer of record should provide evidence to this fact. If all structural elements framing into the fire wall, as well as their supporting members, have the same fire-resistance rating as the fire wall, it is reasonable to assume that the intent of the provision has been met.

As an option to a single fire wall, the code permits the use of a double fire wall if designed and constructed in accordance with NFPA 221. Double fire walls are simply two back-to-back walls, each having an established fire-resistance rating. While acceptable for use in a new structure, double fire walls are most advantageous where an addition is being constructed adjacent to an existing building and the intent is to regulate the addition as a separate building under the fire wall provisions. The exterior wall of the existing building, if compliant, can be used as one wall of the double wall system, with the new wall of the addition providing the second wall.

Fire wall assemblies that comply with the applicable provisions of NFPA 221, *Standard for High Challenge Fire Walls, Fire Walls, and Fire Barrier Walls* are considered acceptable except where in conflict with any provision of the IBC. This standard addresses a number of criteria for double fire walls, including fire-resistance rating, connections, and structural support. In order to meet the minimum fire-resistance rating for a fire wall as set forth in IBC Table 706.4, each individual wall of a double fire wall assembly is permitted to be reduced to 1 hour less than the minimum required rating for a single fire wall. For example, where IBC Table 706.4 requires the use of a minimum 3-hour fire wall, two minimum 2-hour fire-resistance-rated (double) fire walls are required. Similarly, two 3-hour fire walls in a double wall system can be considered as a single 4-hour fire wall, and two 1-hour fire walls used as a double wall qualify as a single 2-hour fire wall.

The intended goal of fire wall construction is to allow collapse of a building on either side of the fire wall while maintaining an acceptable level of fire separation. Therefore, other than the sheathing allowed by the exception, the only connection permitted by NFPA 221 between the two walls that make up the double fire wall is the flashing, if provided. Illustrated in the explanatory material to the standard, the choice of flashing methods must provide for separate flashing sections in order to maintain a complete physical separation between the walls. Each individual wall of the double wall assembly must be supported laterally without any assistance from the adjoining building. In addition, a minimum clear space between the two walls is recommended by NFPA 221 in order to allow for thermal expansion between unprotected structural framework, where applicable, and the wall assemblies that make up the double fire wall. The intent of the exception is to allow the sheathing used for floor or roof diaphragms in the higher seismic zones to continue through the walls. Therefore, allowing the separate buildings to be stronger and react together in a seismic event.

NFPA 221 can also be used as the design standard for the construction of other types of fire walls. Two other high challenge (HC) fire walls are addressed in NFPA 221, cantilevered HC fire walls and tied HC fire walls. Cantilevered HC fire walls must be fully self-supported and nonbearing, with no connections to the buildings on either side other than flashing. Tied HC fire walls are limited to one-story conditions and supported laterally by the building framework with flexible anchors. Centered on a single-column line or between a double-column line, a tied HC fire wall shall be constructed such that the framework on each side of the wall shall be continuous or tied together through the wall.

706.3 Materials. In buildings of other than Type V construction, fire walls shall be constructed of noncombustible materials. The high degree of protection expected from a fire wall mandates that noncombustible construction be used for all but the lowest type of construction.

706.4 Fire-resistance rating. It is obvious that a fire wall performs the very important function of acting as a barrier to fire spread so that a fire on one side of the wall will not be transmitted to the other. On this basis, the fire wall must have a fire-resistance rating commensurate with the occupancy and type of construction of which it is constructed. The IBC provides that fire walls be of 2-hour, 3-hour, or 4-hour fire-resistance-rated construction as specified in Table 706.4. Where the type of construction and/or occupancy group that occurs on one side of a fire wall is inconsistent with that on the other side, the more restrictive fire-resistance rating set forth in Table 706.4 shall apply. See Figure 706-1. Permitted openings in fire walls are addressed in Section 706.8.

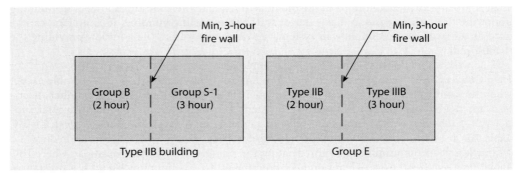

Figure 706-1 **Fire-resistance rating.**

706.5 Horizontal continuity. A fire wall must not only separate the interior portions of the building but must also extend at least 18 inches (457 mm) beyond the exterior surfaces of exterior walls. See Figure 706-2. A number of exceptions permit the fire wall to terminate at

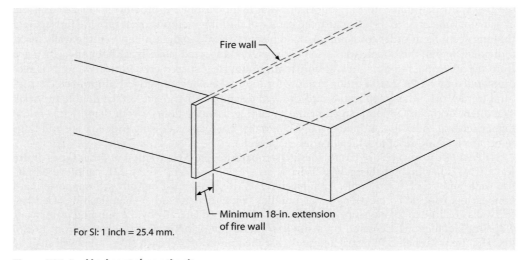

Figure 706-2 **Horizontal continuity.**

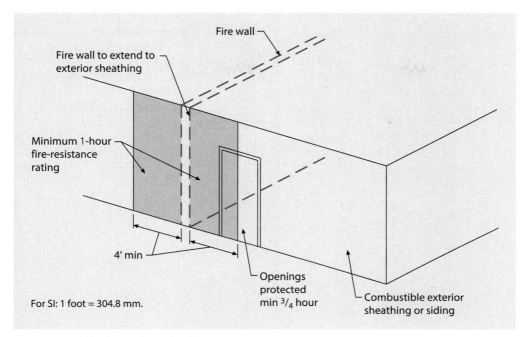

Figure 706-3 **Horizontal continuity.**

the interior surface of the exterior finish material, with Exception 1 illustrated in Figure 706-3. Where combustible sheathing or siding materials are used, the wall must be protected for at least 4 feet (1,220 mm) on both sides of the fire wall by minimum 1-hour construction with any openings protected at least 45 minutes. If the sheathing, siding, or other finish material is noncombustible, such noncombustible materials shall extend at least 4 feet (1,220 mm) on both sides of the fire wall; however, unlike the previous exception, no opening protection is required. As an option, where the separate buildings created by the fire wall are sprinklered, the fire wall may simply terminate at the interior surface of noncombustible exterior sheathing.

706.5.1 Exterior walls. Where a fire wall creating separate buildings intersects with the exterior wall, there is the potential for direct fire exposure between the buildings at the exterior. Unless the intersection of the exterior wall and the fire wall forms an angle of at least 180 degrees (3.14 rad), such as a straight exterior wall with no offsets, a condition occurs similar to that of two buildings located on the same site. The proximity of the two buildings may be such that the distance between them would allow for direct fire or substantial radiant heat to be transferred from one building to the other. This condition is also possible where the two buildings on the lot are portions of a larger structure with fire wall separations.

Where the fire wall intersects the exterior wall to form an angle of less than 180 degrees (3.14 rad), the exterior wall for at least 4 feet (1,220 mm) on both sides of the fire wall shall be of minimum 1-hour fire-resistance-rated construction, and all openings within the 4-foot (1220-mm) portions of the exterior wall are to be protected with 45-minute fire assemblies. See Figure 706-4. As an option, an imaginary lot line may be assumed between the two buildings created by the fire wall and the exterior wall, and opening protection would be based on the fire separation distances to the imaginary lot line. This method is consistent with the provisions of Section 705.3 for addressing two buildings on the same lot. An example is shown in Figure 706-5.

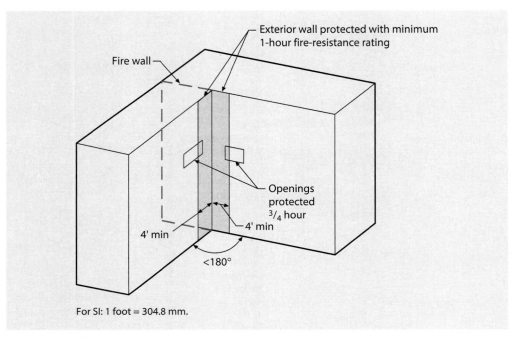

Figure 706-4 **Fire wall intersection with exterior walls.**

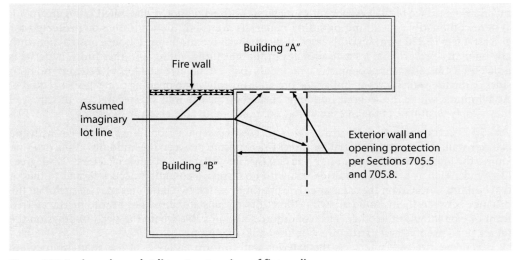

Figure 706-5 **Imaginary lot line at extension of fire wall.**

706.5.2 Horizontal projecting elements. Under the conditions where a horizontal projecting element such as a roof overhang or balcony is located within 4 feet (1,220 mm) of a fire wall, the wall must extend to the outer edge of the projection. This general requirement provides for a complete separation by totally isolating all building elements, including projections, on either side of a fire-resistance-rated wall. However, such a condition is typically not visually pleasing. Therefore, the code indicates the fire wall is not required to extend to the leading edge of the projecting element if constructed in compliance with one of three exceptions. The protection must extend through the projecting element unless the projection has no concealed spaces. Where the projecting element is combustible and has concealed spaces, the fire wall shall extend through the concealed area, whereas in noncombustible construction, the extension need only be 1-hour fire-resistance-rated construction. Under all of the exceptions, the exterior wall behind and below the projecting element is to be of 1-hour fire-resistance-rated construction for a distance not less than the depth of the projecting element on both sides of the wall. All openings within the rated exterior wall are to be protected by fire assemblies having a minimum fire-protection rating of 45 minutes. Figure 706-6 depicts these various conditions addressed in the exceptions.

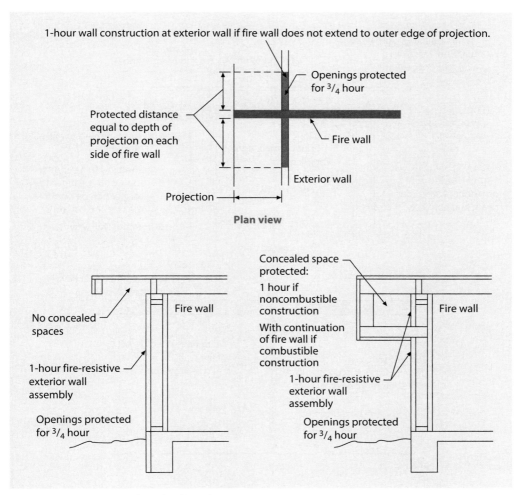

Figure 706-6 **Horizontal projecting elements.**

706.6 Vertical continuity. Having established the intent of the IBC that fire walls prevent the spread of fire around or through the wall to the other side, the IBC further ensures the separate building concept by specifying that the wall shall extend continuously from the foundation to (and through) the roof to a point 30 inches (762 mm) or more above the roof. The 30-inch (762-mm) parapet prevents the spread of fire along the roof surface from the fire side to the other side of the wall.

Several exceptions, some of which are illustrated in Figure 706-7, allow the fire wall to terminate at the underside of the roof sheathing, deck, or slab, rather than terminate in a parapet. The basis for such exceptions includes:

1. Equivalent protection being provided by an alternative construction method;
2. Aesthetic considerations, as parapets disrupt the appearance of the roof; or
3. A combination of the two previous reasons.

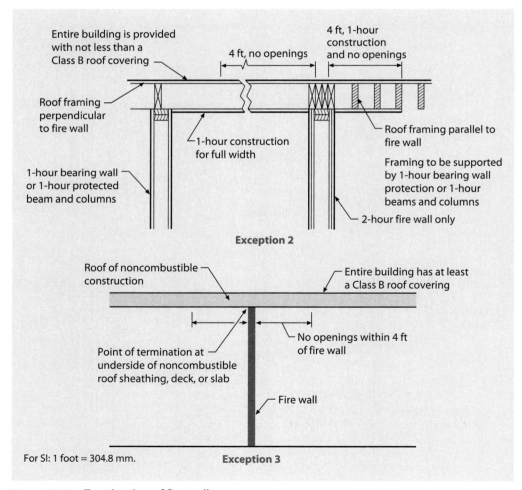

Figure 706-7 **Termination of fire walls.**

It is emphasized that the term *fire wall* also limits its use to vertical walls. Therefore, there can be no horizontal offsets nor can the plan view of the wall change from level to level. See Figure 706-8.

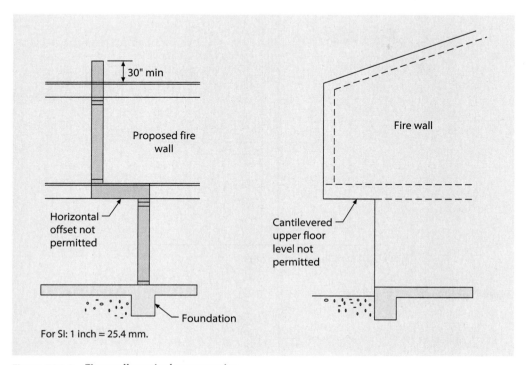

Figure 706-8   **Fire wall vertical community.**

706.6.1 Stepped buildings. Quite often, a fire wall is provided at a point in the building where the roof changes height. Under such conditions, the fire wall must extend above the lower roof for a minimum height of 30 inches (762 mm). In addition, the exterior wall shall be of at least 1-hour fire-resistance-rated construction for a total height of 15 feet (4,572 mm) above the lower roof. The exterior wall shall be of fire-resistance-rated construction from both sides. Any opening that is located in the lower 30 inches (762 mm) of the wall shall be regulated based on the rating of the fire wall. Openings above the 30-inch (762-mm) height, but not located above a height of 15 feet (4,572 mm), shall have a minimum fire-protection rating of 45 minutes. An illustration of this provision is depicted in Figure 706-9.

An alternative is described in the exception that allows the fire wall to terminate at the underside of the roof sheathing, deck, or slab of the lower roof. It is very similar to Exception 2 in Section 706.6. Because the greatest exposure occurs from a fire penetrating the lower roof and exposing the adjacent exterior portion of the fire wall, the code mandates that all protection be applied to the roof assembly of the lower roof. As shown in Figure 706-10, the lower roof assembly within 10 feet (3,048 mm) of the wall shall be of minimum 1-hour fire-resistance-rated construction. In addition, no openings are permitted in the lower roof within 10 feet (3,048 mm) of the fire wall.

250　Chapter 7 ■ Fire and Smoke Protection Features

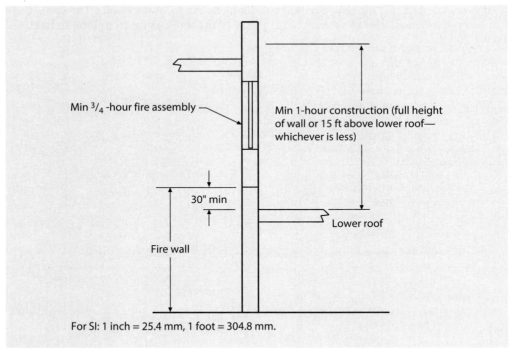

Figure 706-9　**Stepped buildings.**

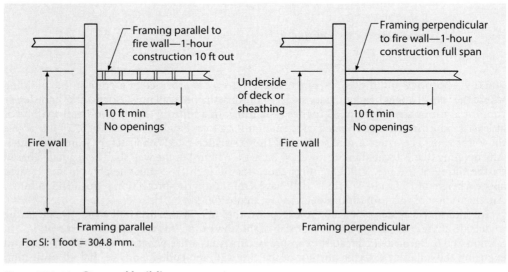

Figure 706-10　**Stepped buildings.**

Section 707 *Fire Barriers*

A common function of a fire barrier is to totally isolate one portion of a floor level from another through the use of fire-resistance-rated walls and opening protectives. Fire-resistance-rated horizontal assemblies are also often used in conjunction with fire barriers in multistory buildings in order to isolate areas vertically. This section identifies the different uses for fire barriers, as well as the method in which fire barriers are to be constructed.

707.3 Fire-resistance rating. A fire barrier shall be used to provide the necessary separation for the following building elements or conditions:

1. Shaft enclosure. The minimum required degree of fire-resistance for fire barriers used to create a shaft enclosure is based primarily on the number of stories connected by the enclosure. A minimum 2-hour fire-resistance rating is mandated where four or more stories are connected, with only a 1-hour rating required where connecting only two or three stories. In all cases, the rating of a shaft enclosure must equal or exceed that of the floor assembly that is penetrated by the enclosure but need not exceed 2 hours.

2. Interior exit stairway construction. The separation between an interior exit stairway (stair enclosure) and the remainder of the building shall be accomplished with fire barriers having either a 1- or 2-hour fire-resistance rating, as required by Section 1023.2. Similar enclosures are required for interior exit ramps.

3. Exit access stairway enclosures. Where exit access stairways are required to be enclosed by Section 1019.3, the enclosure shall include the use of fire barriers.

4. Exit passageway. An exit passageway must be isolated from the remainder of the building by minimum 1-hour fire-resistance-rated fire-barrier walls. Where horizontal enclosure is also required, minimum 1-hour fire-resistance-rated horizontal assemblies must also be used to totally isolate the exit passageway. Where an exit passageway is a continuation of an interior exit stairway, it must, at a minimum, maintain the fire-resistance rating of the stairway enclosure.

5. Horizontal exit. A minimum 2-hour fire-resistance-rated fire barrier may be used to create a horizontal exit when in compliance with all of the other provisions of Section 1026. The fire barrier creates protected compartments where occupants of the building can travel to escape the fire incident.

6. Atrium. Unless a complying glazing system or ¾-hour glass block construction is used, minimum 1-hour fire barriers are required when isolating an atrium from surrounding spaces.

7. Incidental uses. Table 509.1 indicates the required separation or protection required for special hazard areas such as waste and linen collection rooms, laboratories, and furnace rooms. Where a 1- or 2-hour fire-resistance-rated wall is required, it shall be a fire barrier.

8. Control areas. Table 414.2.2 identifies the minimum required fire-resistance rating for fire barriers used to create control areas in buildings housing hazardous materials. A minimum rating of 1 hour is mandated for separating control areas located on the first three floor levels above grade plane, whereas minimum 2-hour fire barriers are required for control area separations on all floor levels above the third level.

9. Separated occupancies. The separation of dissimilar occupancies in the same building is accomplished by fire barriers. Table 508.4 is used to determine the required fire-resistance rating of the required fire barriers, ranging from 1 hour through 4 hours.

10. **Fire areas.** Where a building is divided into fire areas by fire barriers in order to not exceed the limitations of Section 903.2 for requiring an automatic sprinkler system, the minimum required fire-resistance ratings of the fire barriers are set forth in Table 707.3.10. Ranging from a minimum of 1 hour to a maximum of 4 hours, the fire-resistive requirements are based solely on the occupancy classification of the fire areas. The provisions are applicable to both single-occupancy and mixed-occupancy conditions. See the discussion on Section 901.7 for further information.

Note also that fire barriers are required as separation elements in other miscellaneous locations identified by the code, such as stage accessory areas (Section 410.4) and flammable finish spray rooms (Section 416.2). Throughout the code, references are made to fire barriers as the method of providing the appropriate fire-resistance-rated separation intended. In addition, many of the other *International Codes*® also address the use of fire barriers to create protected areas.

707.5 Continuity. Fire barriers must begin at the floor and extend uninterrupted to the floor or roof deck above. Where there is a concealed space above a ceiling, the fire barrier must continue through the above-ceiling space. See Figure 707-1. Fireblocking, required only in combustible construction, must be installed at every floor level if the fire barrier contains hollow vertical spaces. The intent of a fire barrier is to provide a continuous separation so as to completely isolate one area from another. As with many other fire-resistance-rated elements, the supporting construction must be of an equivalent rating to the fire barrier supported. A reduction relates to 1-hour incidental use separations in nonrated construction.

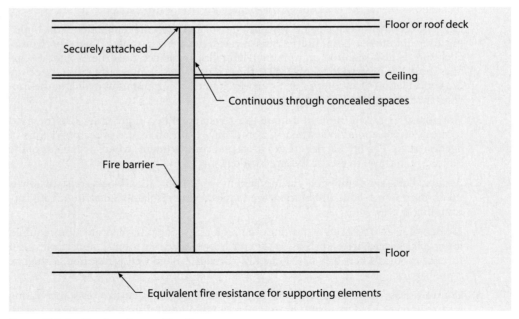

Figure 707-1 **Fire barrier continuity.**

707.6 Openings. The provisions of Section 716 regulate the protection of openings in fire barriers. The fire-protection ratings mandated for fire-barrier openings in Tables 716.1(2) and 716.1(3) vary depending on the fire-resistance rating of the fire barrier as well as its purpose. The required rating may be as little as ¾ hour to as much as 3 hours.

707.9 Voids at intersections. It is not uncommon for a void to be created at the joint between a fire barrier and the floor or roof deck above. Where the joint occurs at a fire-resistance-rated floor or roof deck, Sections 707.8 and 715.3 mandate that the joint be protected by an approved fire-resistant joint system. Protecting the gap or void around an assembly is also important where the joint occurs between a fire barrier and a non-fire-rated floor or roof. Section 707.9 is only intended to address those situations where the roof assembly or exterior wall is not fire-resistance rated. The void need only be filled with an approved material that is securely installed and capable of retarding the passage of fire and hot gases.

Section 708 *Fire Partitions*

This section regulates the design and construction of fire partitions installed in the listed locations. The IBC identifies eight locations where fire partitions are required or used within the code.

708.3 Fire-resistance rating. The minimum fire-resistance rating of fire partitions is to be 1 hour, unless a reduction is permitted by one of two exceptions. Exception 1 refers to Table 1020.2, which identifies the required fire-resistance rating of a corridor based on three factors—the occupancy classification of the area served by the corridor, the occupant load the corridor serves, and whether or not the building is sprinklered. Where conditions warrant, the table indicates that the corridor needs only a ½-hour fire-resistance rating. If no fire-resistance rating is mandated, fire partitions are not required and none of the provisions of Section 708 are applicable. The second exception applies to walls separating dwelling units and sleeping units in buildings of nonrated construction. The presence of an automatic sprinkler system complying with NFPA 13 reduces the required fire-partition rating to 30 minutes. It should be noted that the exception does not permit this reduction where an NFPA 13R system is installed.

708.4 Continuity. Consistent with the required continuity of fire barriers, the general requirement for fire partitions is that they must extend from the floor to the floor or roof deck above. However, unlike the provisions for fire barriers, an alternative construction method is permitted where fire partitions may terminate short of the floor or roof deck under various conditions. Where a fire-resistance-rated floor/ceiling or roof/ceiling with an equivalent fire-resistance rating is provided, a fire partition need only extend to, and be securely attached to, the ceiling membrane. For an example of this provision as it relates to corridor construction, refer to Figure 708-1. Under this condition in combustible construction, fireblocking or draftstopping must be installed at the partition line in the concealed space above the ceiling. In general, any supporting construction is to be at least 1-hour fire-resistance rated, however, this support requirement is waived for many fire partitions in buildings of Type IIB, IIIB, and VB construction.

The following exceptions modify the continuity provisions of this section:

1. Where a crawl space exists below a floor assembly of at least 1-hour fire-resistance-rated construction, the fire partition does not need to extend into the underfloor space. See Figure 708-2.

2. The arrangement shown in Figure 708-3 would meet the code requirement for adequately enclosing a corridor based on applying item 2.1. The corridor walls are protected on the room-side of the occupied use spaces by a fire-resistance-rated membrane extending from the floor to the floor or roof above. In this case, the ceiling over the corridor may be considered as completing the enclosure provided the ceiling membrane is equivalent to the wall membrane. As indicated in item 2.2, the room-side membrane need not extend to the underside of the sheathing above provided the building is sprinklered and sprinkler protection is extended to the concealed space above the ceiling.

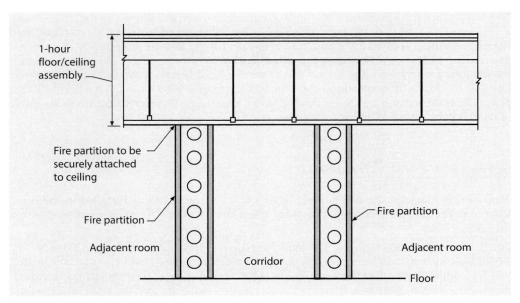

Figure 708-1 **Corridor fire partitions.**

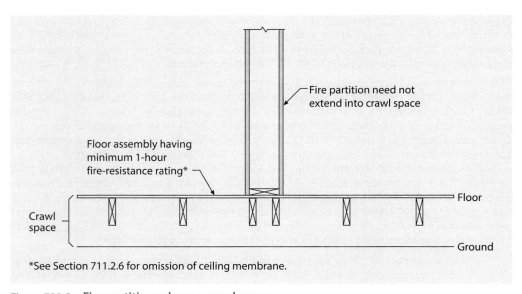

Figure 708-2 **Fire partitions above a crawl space.**

3. The code provides that the corridor ceiling may be of the same construction as permitted for corridor walls as shown in Figure 708-4. In all probability, typical wall construction might not pass the 1-hour test when tested in a horizontal position. However, this arrangement, generally referred to as tunnel construction, is considered to be adequate protection for the corridor separating it from the spaces above.

Fire Partitions   255

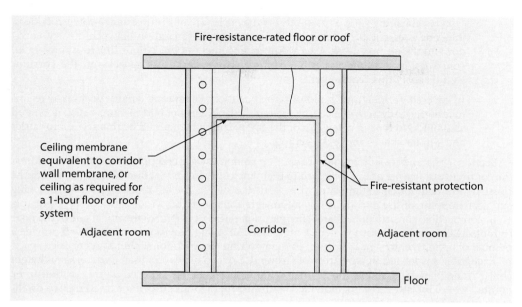

Figure 708-3 **Corridor construction.**

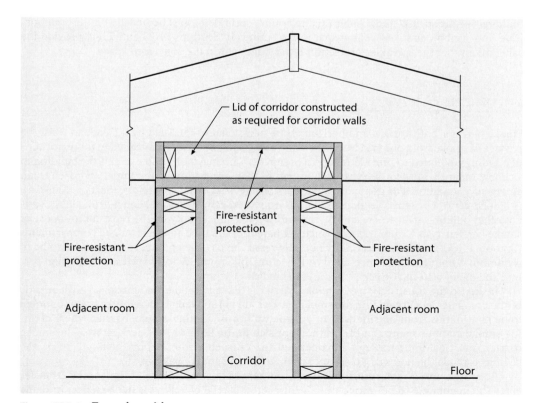

Figure 708-4 **Tunnel corridor.**

By establishing various methods for the enclosure of fire-resistance-rated corridors, the code is essentially attempting to get a minimal separation between the exit corridor and the occupied-use spaces. Any arrangement of the 1-hour fire-resistance-rated construction that effectively intervenes between these use spaces and the corridor would satisfy this requirement.

4. In covered or open mall buildings, fire partitions separating tenant spaces may terminate at the underside of a ceiling, even if the ceiling is not part of a fire-resistance-rated assembly. No type of extension of the fire partition is required by this section for attics and similar spaces above the ceiling.

Section 708.4 and its subsections are split up so that the general provisions of 708.4 address the continuity of fire partitions in regard to their enclosure limits. Section 708.4.1 deals with the construction components supporting fire partitions, while 708.4.2 addresses the fireblocking and draftstopping of fire partitions of combustible construction.

In combustible construction where a fire partition does not extend to the underside of the floor or roof sheathing as addressed in item 1 of Section 708.4, the provisions of 708.4.2 will provide a number of options to prevent a fire from passing over the fire partition within the concealed space.

Exception 4 is for use in attic areas of Group R-2 occupancies less than five stories in height above grade plane and allows the draftstopping or fireblocking required by this section to be omitted where the attic area is subdivided by draftstopping into areas not exceeding two dwelling units or 3,000 square feet (279 m^2), whichever is less.

Exception 1 applies in combustible buildings where the fire partitions stop at the fire-resistance-rated ceiling membrane. Fireblocking or draftstopping is not required at the partition line if the building is fully sprinklered. Under this exception, sprinklers must be installed in the combustible floor/ceiling and roof/ceiling spaces. Exceptions in Sections 718.3 and 718.4 provide the same intent that the sprinkler protection is provided within the concealed space.

Section 709 *Smoke Barriers*

Smoke barriers, both vertical and horizontal, are occasionally mandated by the code to resist the passage of smoke from one area to another. The use of smoke barriers is assigned to those portions of buildings intended to provide refuge to occupants who may not be able to exit the building in a timely manner. In such cases, relocation, rather than evacuation, is the initial approach to an emergency condition. For example, smoke barriers are used in areas of refuge (Section 1009.6.4), in smoke-control systems (Section 909.5), in Group I-3 occupancies (Section 408.6), and in various other building areas where smoke transmission is a concern. By far the most common use of smoke barriers is in Group I-2 occupancies, where they are used to create smoke compartments (Section 407.5). Smoke barriers must not only resist the passage of smoke, they must also be of minimum 1-hour fire-resistance-rated construction. In Group I-3 occupancies, an exception permits the use of 0.10-inch-thick (2.5-mm) steel in lieu of 1-hour construction.

The key to the construction of a smoke barrier is that all avenues for smoke to travel outside of the compartment created by the smoke barrier are eliminated. This requires the membrane to be continuous from outside wall to outside wall or to another smoke barrier where creating smoke compartments, and from the floor slab to the floor or roof deck above. The smoke barriers must continue through all concealed spaces, such as those above ceilings, unless the ceilings provide the necessary resistance against fire and smoke passage.

The most common interpretation of an allowable ceiling membrane to limit smoke transfer is for the membrane to be "monolithic." However, this type of ceiling is not practical in many healthcare settings, because main utility and ductwork lines run above corridor ceilings to minimize their placement in patient care areas. Therefore, the installation of a lay-in ceiling system is permitted in Group I-2 occupancies provided two special conditions are met. The assembly must meet minimum 1 pound per square foot criteria and other code requirements related to combustibility and flame spread. In addition, limiting the HVAC system to ducted systems will preclude the possibility of an open plenum return system. Plenum systems are discouraged due to the required pressure relationships for infection prevention considerations and to maintain more accurate control of the temperature and humidity.

All door openings in smoke barriers are to be protected with assemblies having a minimum fire-protection rating of 20 minutes, per Table 716.1(2). In cross-corridor situations in Group I-2 occupancies, the code mandates a pair of opposite-swinging doors installed without a center mullion. Such doors shall be provided with an approved vision panel; be close fitting; have no louvers or grilles; and undercuts are limited to ¾ inch. Although positive latching is not required, the doors are to have head and jamb stops, astragals, or rabbets at meeting edges, and automatic-closing devices.

Smoke barriers, like smoke partitions regulated by Section 710, are only mandated where specifically identified or referenced by the code. As an example, Section 509.4.2 requires the use of "construction capable of resisting the passage of smoke" as a potential physical separation for incidental uses. Thus, only construction that will perform the intended function is required, and not necessarily a smoke barrier or smoke partition.

Section 710 *Smoke Partitions*

The purpose of a smoke partition is limited to the concerns of smoke movement under fire conditions, with no intent to regulate for the resistance to flame and heat.

Smoke partitions are mandated by the code in limited applications, most commonly in the construction of corridors in Group I-2 occupancies. As such, corridors in hospitals, nursing homes, and similar Group I-2 occupancies are regulated by the provisions of both Sections 710 and 407.3. It requires a comparison of the two sections to determine the requirements for corridor systems, particularly corridor doors and air openings.

710.5 Openings. This section generally prohibits the installation of louvers in doors in smoke partitions. This is consistent with the provisions of Section 407.3.1 mandating an effective barrier against the transfer of smoke. An exception will permit louvers or an undercut to provide exhaust systems with make-up air from non-rated corridors. The provisions of this section also require that doors in smoke partitions be tested in accordance with UL 1784 and be self-closing or automatic closing, but only where required elsewhere in the code. A review of Section 407.3.1 regulating corridor doors in Group I-2 occupancies does not require the UL test, and it specifically states that self-closing or automatic-closing devices are not required. In this case, the provisions in Section 407.3.1 take precedence as a specific requirement (see Section 102.1).

710.8 Ducts and air transfer openings. The provisions of both Sections 717.5.7 and 710.8 mandate the need for smoke dampers in air transfer openings that occur in smoke partitions. Smoke dampers are not required at duct penetrations of smoke partitions, but only at unducted air openings.

Section 711 *Floor and Roof Assemblies*

This section is applicable where floor and roof assemblies are required to have a fire-resistance rating. This will occur where the type of construction mandates protected floor and roof assemblies, such as in Type I, IIA, IIIA, and VA construction, and where the floor assembly is used to separate occupancies or create separate fire areas. For example, in a building of Type IIA construction, Table 601 requires minimum 1-hour fire-resistance-rated floor construction. As another example, where the floor separates a Group A-2 occupancy from a Group B, Table 508.4 addressing separated occupancies mandates a 1- or 2-hour separation.

As referenced in Section 420.3, complementary to the provisions of Section 708 for fire partitions, floor assemblies separating dwelling units or sleeping units are required to be of at least 1-hour fire-resistance-rated construction. The exception to Section 711.2.4.3 reduces the required level of protection for the floor assembly to ½ hour in buildings of Type IIB, IIIB, or VB construction, provided the building is protected by an automatic sprinkler system.

711.2.3 Supporting construction. As a general rule, horizontal assemblies must be supported by structural members or walls having at least the equivalent fire rating as that for the horizontal assembly. For example, in a Type IIA school building of two stories where the floor construction is required to be a 2-hour fire-resistance-rated assembly in order to separate fire areas, any walls or structural members in the first story supporting the second floor would be required to also be of 2-hour fire-resistance-rated construction. This would be the case even though the building generally is required to be only of 1-hour fire-resistance-rated construction. Obviously, if the horizontal assembly is not supported by equivalent fire-resistance-rated construction, the intent and function of the separation are negated if its supports fail prematurely.

711.2.5 Ceiling panels. The protection of a ceiling membrane also includes the adequacy of the panelized ceiling system to withstand forces generated by a fire and other forces that may try to displace the panels. These forces can generate positive pressures in a fire compartment that need to be counteracted. As a result, lay-in ceiling panels that provide a portion of the fire resistance of the floor/ceiling or roof/ceiling assembly should be capable of resisting this upward or positive pressure so that the panels stay in position and continue to maintain the integrity of the system. The code defines the pressure to be resisted as 1 pound per square foot (48 Pa).

711.2.6 Unusable space. Figure 711-1 illustrates how this provision is applied in regard to unusable spaces such as crawl spaces and attics. For 1-hour fire-resistance-rated floor construction over a crawl space, the ceiling membrane is not necessary in the crawl-space area. Similarly, in 1-hour fire-resistance-rated roof construction, the floor membrane is not required in the attic. Note that the elimination of the membranes in the attic and crawl space is only applicable where the required rating of the floor or roof assembly is a maximum of 1 hour.

Section 712 *Vertical Openings*

It is well known that one of the primary means for the spread of fire in multistory buildings, particularly older buildings, has been the transmission of hot gases and fire upward through unprotected or improperly protected vertical openings. The primary cause of death in the hotel portion of the MGM Grand Hotel in Las Vegas, Nevada, as a result of the fire in November 1980, was the upward transmission of smoke through inadequately protected elevator shafts, stair shafts, and heating and ventilating shafts. It is because of this potential for fire spread vertically through buildings that this section requires that vertical openings be appropriately regulated. This section identifies the following applications for addressing vertical openings:

1. The use of a shaft enclosure to protect vertical openings is a time-honored method recognizing that the fire-resistance-rated enclosure of a floor opening provides a near equivalency to a floor with no openings. The shaft enclosure is intended to replace the

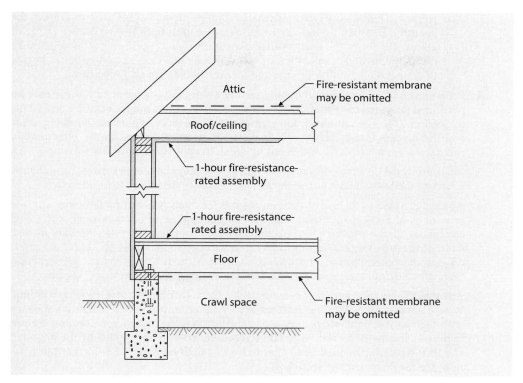

Figure 711-1 **Omission of ceiling or floor membrane.**

floor construction with an equal or better degree of fire resistance. A shaft enclosure is often used for vertical opening protection simply because none of the other protection methods listed in this section are applicable. The specific requirements for shaft enclosures are set forth in Section 713.

2. Within individual dwelling units, unconcealed vertical openings are permitted provided they connect no more than four stories. The allowance for openings is intended to apply to units that are open vertically with unenclosed stairways and floor spaces, such as lofts. Concealed spaces used for the installation of ducts, piping, and conduit between stories do not fall under this allowance.

3. In fully sprinklered buildings, the vertical openings created for an escalator may be protected by one of the following two methods:

    3.1. By limiting the size of the openings, along with the installation of draft curtains and closely spaced sprinklers, the code assumes that the vertical openings in sprinklered buildings do not present an untenable condition. The purpose of the required draft curtain is to trap heat so that the sprinklers will operate and cool the gases that are rising. The curtain is not a fabric; it is constructed of materials consistent with the type of construction of the building. In other than mercantile and business occupancies, the use of this method is limited to openings connecting four or fewer stories.

3.2. Approved power-operated automatic shutters may also be used to cut off the openings between floors. Required to have a minimum fire-resistance rating of at least 1½ hours, the shutters shall close immediately upon activation of a smoke detector. Obviously, operation of an escalator must stop once the shutter begins to close. Other conditions required for compliance are also specified.

4. Penetrations of cables, conduit, tubing, piping, vents, and similar penetrating items are permitted provided the penetrations are protected in conformance with the provisions of Section 714. The allowance for vent penetrations only applies to plumbing vents and those vents that convey products of combustion as defined in the *International Mechanical Code®* (IMC®) and is not intended to apply to exhaust ducts.

5. Joints protected by a fire-resistant joint system, like other building elements protected by approved methods, do not create any additional hazard that needs to be addressed.

6. As an alternative to the use of shaft enclosures for ducts penetrating floor systems, the provisions of Section 717.6 also regulate the penetration of ducts through horizontal assemblies. If the provisions of Section 717.6 do not mandate the installation of a damper, the provisions of Sections 717.1.2 and Sections 714.5 through 714.6.2 are applicable to the penetration of a horizontal assembly without a shaft. Where Section 717.6 mandates a damper in the duct or air transfer opening, then Section 717 applies.

7. Atriums are intended to be open vertically through multiple stories of a building. Where such special building features are designed and constructed in compliance with their own unique provisions, other means of vertical opening protection are not required. The use of an atrium is a design option, voluntarily applied by the designer, and thus typically provided as an alternative to a shaft enclosure. For further information, see the discussion of Section 404.

   It is a widely held belief that the allowance for atriums is also applicable to covered mall buildings that meet the special requirements of Section 402. However, there is no specific provision in Section 712 that specifically addresses multistory covered mall buildings. Vertical openings in the mall portion of the building could be regulated under the provisions of Sections 712.1.7, 712.1.9, and 712.1.12, but a general allowance for the entire building is not available.

   In most occupancies, the use of Section 712.1.7 is limited to those buildings where the floor openings connect three or more stories. In other than Group I-2 and I-3 occupancies, where a floor opening connects only two stories, Sections 712.1.9 and 712.1.12 are applied.

8. A masonry chimney extending through one or more floor levels is permitted where the annular space around the chimney is protected in the manner specified by Section 718.2.5.

9. This provision permits two adjacent stories to intercommunicate with each other without protection of the openings between the two stories, except in the case of Group I-2 and I-3 occupancies. As long as these intercommunicating openings serve only the one adjacent floor, no protection is required. This provision is commonly used where multistory office buildings have a lobby that extends up through the second story so that individuals on the second floor may look down over a guard into the lobby below. It is important that the unprotected floor opening be appropriately separated from floor openings serving other floors.

   In addition, the opening between floor levels cannot be concealed within the construction of a wall or floor-ceiling assembly. The limitation on concealment is intended to prevent unprotected openings that are completely enclosed by walls, partitions, chases, or floor/ceiling assemblies. Where the openings are concealed in this manner, they permit a fire within the concealed space to burn undetected and distribute products of combustion to the upper floor.

10. Parking garages have a number of provisions related to openings in a horizontal assembly. These include:

    10.1 Automobile ramps in both enclosed and open parking garages, when in compliance with the provisions of Sections 406.5 and 406.6, respectively, are permitted. Because the nature of these uses makes it impractical for enclosures, other safeguards are provided by the code in Section 406.

    10.2 The enclosure of elevator hoistways in parking garages is not required for those hoistways that only serve the parking garage.

    10.3 Vertical opening protection is not mandated for the enclosure of mechanical exhaust and supply duct systems in both types of garage facilities. The protection of vertical openings provided to accommodate elevators, as addressed in Section 712.1.10.2, as well as exhaust ducts and supply ducts, is unnecessary since the vehicle ramps of parking garages are permitted to be open at all levels.

11. By definition, a complying mezzanine is intended to be open into the room below. As such, unenclosed floor openings between the mezzanine and the lower floor are permitted.

12. Exit access stairways in compliance with the exceptions listed in Section 1019.3 do not require enclosure. For example, under Exception 1, stairways may be unenclosed provided they connect no more than two adjacent stories and further comply with the conditions of Section 712.1.9.

13. Floor fire door assemblies and ceiling access doors are permitted to protect vertical openings where tested in accordance with the applicable standard.

14. In a Group I-3 occupancy, openings in floors within a housing unit are permitted without a shaft enclosure provided four specific conditions established in Section 408.5 are met.

15. The installation of unprotected skylights and other penetrations through the roof deck of a fire-resistance-rated roof assembly is permitted, provided the structural integrity of the roof construction is maintained.

16. Throughout the code, there may be other allowances for floor penetrations or openings that are adequately regulated as vertical openings. Where permitted, these openings must comply with the specifics of their use.

Section 713 *Shaft Enclosures*

The use of a shaft enclosure has long been an acceptable means to protect vertical openings between stories. The enclosure construction is considered to be equivalent to the floor system and thus is permitted to protect any opening that occurs. Although using a shaft enclosure is just one of many applications listed in Section 712.1 for addressing vertical openings, its use is very common, particularly in buildings with a substantial number of stories. In some buildings, the use of a shaft enclosure is the only viable application available.

713.4 Fire-resistance rating. To provide an acceptable level of protection for vertical openings between floors, this section mandates that all shaft enclosures have a fire-resistance rating at least equivalent to the rating of the floor being penetrated, but never less than 1 hour. Therefore, in Type I construction, or where the shaft enclosure connects four or more stories, a minimum 2-hour enclosure is mandated. A shaft enclosure is never required to have a higher fire-resistance rating than 2 hours.

The exception recognizes the allowance in Section 403.2.1.2 applicable to high-rise buildings no more than 420 feet in height. In such buildings, 1-hour fire-resistance-rated shaft construction is permitted for all shaft enclosures other than enclosures for elevator hoistways and interior exit stairways. Sprinklers must also be installed within the shaft enclosures at the top and at alternate floor levels.

713.5 Continuity. Shaft enclosures are required to be constructed as fire barriers, extending from the top of the floor/ceiling assembly below to the underside of the floor or roof deck above, except as permitted by Sections 713.11 and 713.12. It is important that the walls continue through any concealed spaces such as the area above a ceiling, and that any hollow vertical spaces within the shaft wall be fireblocked at each floor level in buildings of combustible construction. In addition, the supporting elements of any shaft enclosure construction must be of fire-resistance-rated construction equivalent to that of the shaft construction. The enclosure, fire-resistance-wise, should be continuous from the lowest floor opening through to its termination.

713.6 Exterior walls. Unless required to be fire-resistance rated because of the proximity to an exterior exit balcony, interior exit stairway or ramp, or an exterior exit stairway or ramp, the exterior walls of a shaft enclosure need only be protected because of their location on the lot as regulated by Section 705.

713.11 Enclosure at the bottom. Many shafts do not extend to the bottom of the building or structure. Therefore, it is necessary to provide an approved method for maintaining the integrity of the shaft enclosure at its lowest point. This section identifies three methods for enclosing the bottom of a shaft enclosure. First, the shaft can be enclosed with fire-resistance-rated construction equivalent to that for the lowest floor penetrated, with a minimum rating consistent with that of the shaft enclosure. Second, a termination room related to the purpose of the shaft can be considered to be the enclosure at the bottom, provided the room is separated from the remainder of the building with fire-resistance-rated construction and opening protectives equivalent to those of the shaft enclosure. Third, approved horizontal fire dampers can be used to protect openings at the lowest floor level in lieu of the enclosure at the bottom of the shaft enclosure. See Figures 713-1a, b, and c.

The first of the three exceptions eliminates the fire-resistance-rated room separation where there are no other openings into the shaft enclosure other than at the bottom. All portions of the enclosure bottom must be closed off except for the penetrating items, unless the room is provided with an automatic fire-suppression system. An example of this concept would be a vent enclosure. The second exception requires that a shaft enclosure containing a waste or linen chute be used for no other purpose and shall end in a termination room per Section 713.13.4. Exception 3 applies where the shaft enclosure contains no combustible materials and there are no openings to other stories. In this situation, there is no need for either the fire-resistance-rated room separation or protection at the bottom of the enclosure. An example would be a light well that extends through several floor levels to the roof, as illustrated in Figure 713-2. It would be considered an extension of the floor below the level of the floor opening.

713.12 Enclosure at the top. Most shafts extend to or through the roof deck at the exterior, where there is no requirement to maintain the fire-resistance rating of the shaft enclosure construction. However, where the enclosure does not extend to the roof, the top of the shaft must be enclosed. When terminating within the building, the required fire-resistance rating of the shaft lid shall be equivalent to the rating of the topmost floor penetrated by the shaft, but in

Shaft Enclosures   263

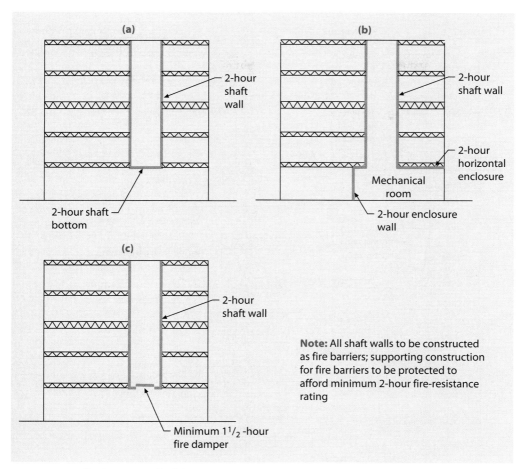

Figure 713-1 Enclosure at shaft bottom.

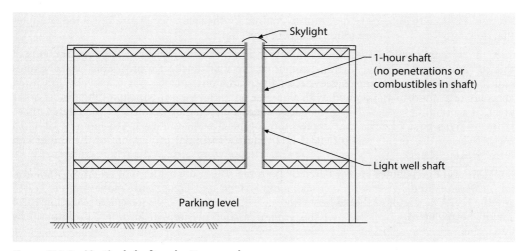

Figure 713-2 Vertical shafts—bottom enclosure.

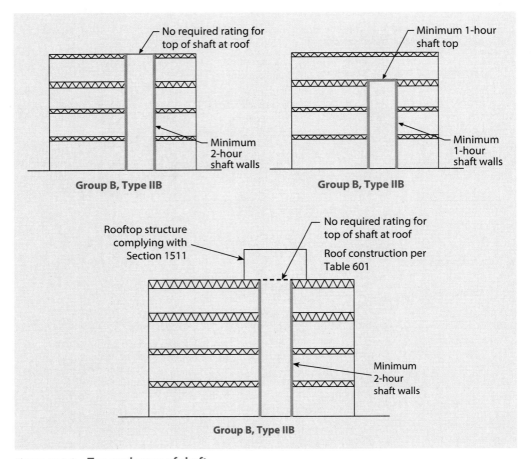

Figure 713-3 **Top enclosure of shaft.**

no case less than the fire-resistance rating required for the shaft. See Figure 713-3 for examples that illustrate the three upper termination options.

713.13 Waste, recycle, and linen chutes and incinerator rooms. The requirements of this section are intended to further strengthen the shaft-enclosure provisions where chutes and discharge rooms for waste, recycle, or linen are constructed. Waste, recycle, and linen areas are often poorly maintained, with a greater potential for a fire incident than most other areas of a building. Coupled with the shaft conditions that are created by the chutes, these types of areas pose hazards that exceed those typically encountered. See Figure 713-4. To further secure the intent, Section 903.2.11.2 requires sprinkler protection for the chutes and termination rooms.

Waste and linen chutes are regulated by both the IBC and Chapter 6 of NFPA 82, *Standard on Incinerators and Waste and Linen Handling Systems and Equipment*. While Section 713.13 and the NFPA 82 standard should be viewed as being the "specific" requirements, Section 102.4 should be reviewed since many of the general shaft provisions of Section 713 will still be applicable.

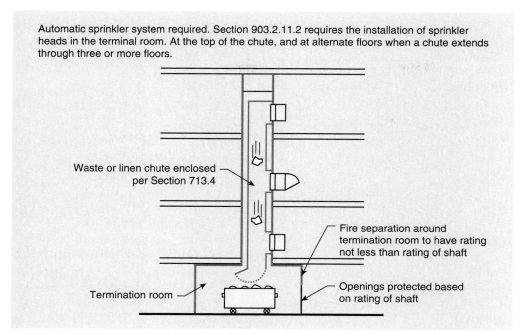

Figure 713-4 Waste and linen chutes.

Section 714 *Penetrations*

The integrity of fire-resistance-rated horizontal and vertical assemblies is jeopardized where penetrations of such assemblies are not properly addressed. Cables, cable trays, conduit, tubing, vents, pipes, and similar items are those types of penetrating items regulated by the code. This section of the IBC identifies the appropriate materials and methods of construction used to protect both membrane penetrations and through penetrations.

714.1.1 Ducts and air transfer openings. Section 717.5 identifies the various conditions under which fire-resistance-rated wall assemblies penetrated by ducts or air transfer openings must be provided with fire and/or smoke dampers. There are a limited number of locations where a damper is not required, such as that permitted by Exception 3 of Section 717.5.2 for fire barrier penetrations and Exception 4 of Section 717.5.4 for fire partitions. See Figure 714-1. In such situations, it is necessary that the penetrations be protected in accordance with the appropriate provisions of Section 714 in order to maintain the integrity of the fire-resistive assembly.

714.3 Sleeves. As illustrated in Figure 714-2, sleeves used in the process of creating a through-penetration of a fire-resistance-rated building element must be properly installed. They must be securely fastened to the assembly that is being penetrated. The installation must comply with the sleeve manufacturer's installation instructions as well as the listing criteria for those systems that are listed. In addition, both the space between the sleeve and the assembly and the space between the sleeve and the penetrating item must be appropriately protected.

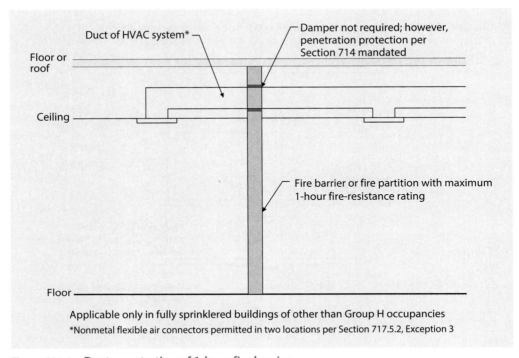

Figure 714-1 **Duct penetration of 1-hour fire barrier.**

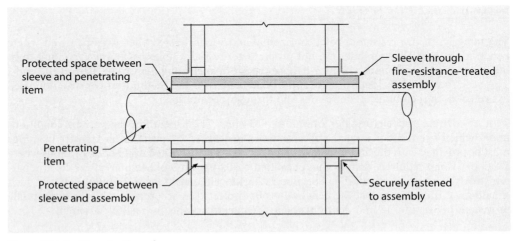

Figure 714-2 **Penetration sleeve.**

714.4 Fire-resistance-rated walls. This section regulates the penetration into or through fire walls, fire barriers, fire partitions, and smoke barrier walls. The protection of penetrations in fire-resistance-rated exterior walls is not addressed; however, where such exterior walls are bearing walls it may be necessary to consider penetration protection in order to maintain the structural integrity of the walls during fire conditions. Fire-resistance-rated interior bearing walls are also not specifically identified as elements regulated for penetrating items; however, any penetrations of such walls should be addressed in order to maintain the necessary structural fire resistance. For the most part, membrane penetrations are addressed in the same manner as through penetrations.

714.4.1 Through-penetrations. As a general rule, through-penetrations (where the penetrating items pass through the entire assembly) are required to be firestopped with approved through-penetration firestop systems when the penetrations pass through fire-resistance-rated walls, unless the approved wall assembly is tested with the penetrations as a part of the assembly. The firestop system is required to have an F rating at least equivalent to that of the fire-resistance rating of the wall penetrated, as shown in Figure 714-3. There is no requirement for a T rating on a wall penetration, justified on the basis that there is no need for such a restrictive temperature rating for the penetration of wall assemblies.

The IBC contains an exception to the general rule for firestopped wall penetrations allowing certain small noncombustible penetrating items no larger than 6-inch (152-mm) nominal diameter to penetrate concrete or masonry walls, provided the full thickness of the wall, or the thickness required to maintain the fire resistance, is filled with concrete, grout, or mortar. The size of the opening is limited to 144 square inches (0.0929 m^2). A second exception that is used extensively will allow the annular space around the same type of noncombustible penetrating item to be filled with a material that prevents the passage of flame or hot gases sufficient to ignite cotton waste when tested under the time-temperature fire conditions of ASTM E119 or UL 263, and under a positive pressure differential of 0.01-inch (0.25-mm) water column. When properly installed around the penetrations of noncombustible items, these materials provide adequate firestopping between the penetrating item and the fire-resistive membrane of the wall. See Figure 714-4.

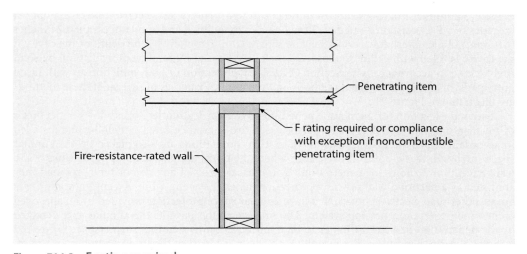

Figure 714-3 **F rating required.**

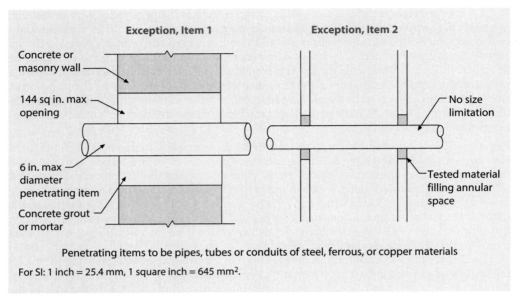

Figure 714-4 Through-penetrations of wall.

714.4.2 Membrane penetrations. This section addresses penetrations through a single membrane of fire-resistance-rated walls. For the most part, a membrane penetration is to be protected by one of the methods established for through-penetrations as previously described. However, there are some membrane penetrations that are allowed without a specific firestopping material in the annular space around such penetrations. Openings for steel electrical boxes are specifically addressed where located in walls with a maximum 2-hour rating, provided that they are no more than 16 square inches (0.0103 m^2) in area and the aggregate area of the boxes does not exceed 100 square inches (0.0645 m^2) for any 100 square feet (9.29 m^2) of wall area. The annular space between the wall membrane and any edge of the electrical box is limited to 1/8 inch (3.2 mm). Also, to prevent an indirect through-penetration, electrical boxes on opposing sides of a fire-resistance-rated wall shall be horizontally separated by no less than 24 inches (610 mm). As an alternative, boxes may be separated horizontally by the depth of the cavity if the cavity is filled with cellulose loose-fill, rockwool, or slag mineral wool insulation; by solid fireblocking in accordance with Section 718.2.1; by protection of both outlet boxes with listed putty pads; or by any other listed methods and materials. Examples of several of these methods are illustrated in Figure 714-5.

A second exception for membrane penetrations of electrical-outlet boxes allows outlet boxes of any material, provided they are tested for use in fire-resistance-rated assemblies and installed in accordance with the instructions for the listing. Limitations are also placed on the annular space surrounding the box and conditions where the boxes are placed on opposite sides of the wall. Exception 3 allows for penetrations by electrical boxes of any size or type provided they are listed as a part of a wall opening protective material system, while Exception 4 addresses boxes, other than electrical boxes, that have annular space protection provided by an approved membrane penetration firestop system. The fifth exception permits the annular space created by the penetration of a fire sprinkler to be unprotected, provided that such a space is covered by a metal escutcheon plate. Because the escutcheon is a part of the listed sprinkler, it is inappropriate to require firestopping at this location. It should be noted that this exception applies to

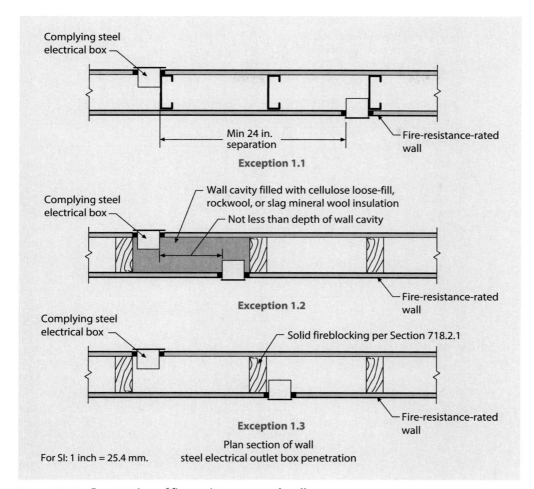

Figure 714-5 Penetration of fire-resistance-rated walls.

the penetration of sprinklers, not simply sprinkler piping or cross mains that might be penetrating fire-resistance-rated construction. See Figure 714-6. Exception 6 addresses steel electrical boxes that exceed 16 square inches (0.0103 m^2) in area as well as those where the aggregate area of the boxes exceed 100 square inches (0.0645 m^2) in any 100 square feet (9.29 m^2) of wall area. Figure 714-7 also illustrates two of the exceptions.

714.4.3 Dissimilar materials. This provision is intended to limit the occasional practice of using a noncombustible penetrating item (such as a short metal coupling) to penetrate a fire-resistance-rated wall, then connect to a combustible item (such as plastic piping or conduit) on the room side of the wall. The building official can accept such a condition where it is demonstrated that the fire-resistive integrity of the wall will be maintained. See Figure 714-7.

714.5 Horizontal assemblies. The shaft enclosure provisions of Section 713 intend to maintain a level of protection that is compromised when one or more openings occur in a floor or floor/ceiling assembly. However, penetrations by pipes, tubes, conduit, wire, cable,

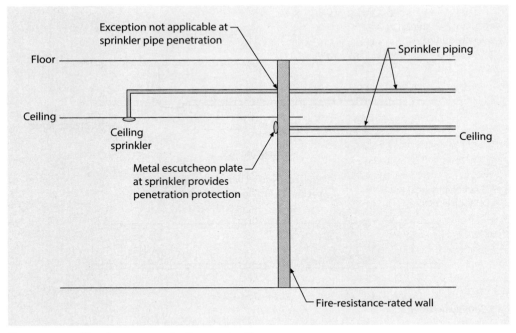

Figure 714-6 **Membrane penetration protection.**

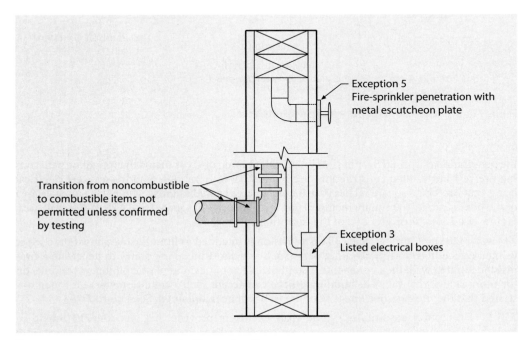

Figure 714-7 **Membrane penetrations of walls.**

and vents are permitted without shaft enclosure protection where in compliance with this section. In addition, this section addresses penetrations that occur in the ceiling of a roof/ceiling assembly. Penetrations occurring in both fire-resistance-rated horizontal assemblies and non-fire-resistance-rated assemblies are addressed.

714.5.1 Through-penetrations. The protection requirements for the through-penetration of fire-resistance-rated horizontal assemblies are very similar to those required for vertical elements. The general provisions state that the penetrations are to be installed as tested in an approved fire-resistance-rated assembly or protected by an approved through-penetration firestop system. Where a firestop system is used, it must have both an F rating and a T rating equivalent to the floor penetrated, but in no case less than 1 hour. Only an F rating is needed if the penetrating item, as it passes through the floor, is contained within a wall cavity above or below the floor. See Figure 714-8. Exception 2 indicates that a T rating is also not required where the floor penetration is a floor drain, tub drain, or shower drain.

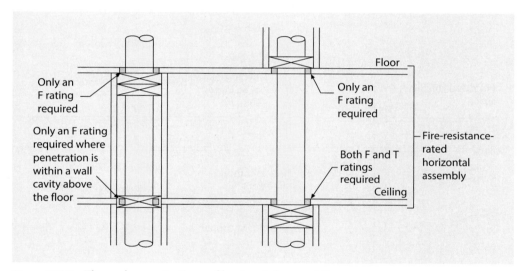

Figure 714-8 **Through-penetrations of horizontal assemblies.**

Noncombustible penetration items are granted exceptions to the general requirements as previously discussed for fire-resistance-rated walls. Where only a single fire-resistance-rated floor is penetrated, the annular space around the noncombustible penetration item need only to be protected with any material that essentially preforms equivalent to an F-rating when tested as stated. There is no limit on the size of the penetrating items, provided they are appropriately protected. Where multiple floor assemblies are penetrated, the size of any penetrating item is limited to 6 inches (152 mm) in nominal diameter. In addition, the area of the opening in the floor is limited to 144 square inches (92,900 mm^2) in any 100 square feet (9.3 m^2) of floor area. Figure 714-9 depicts the use of this exception. Allowances are also provided for noncombustible penetrations of concrete floors, all penetrations of concrete parking garage ramps and floors, and at tested electrical outlet boxes.

714.5.2 Membrane penetrations. Fire-resistance-rated horizontal assemblies must be adequately protected at penetrations of the floor or ceiling membrane. Therefore, they are regulated in much the same manner as through-penetrations addressed in Section 714.5.1.

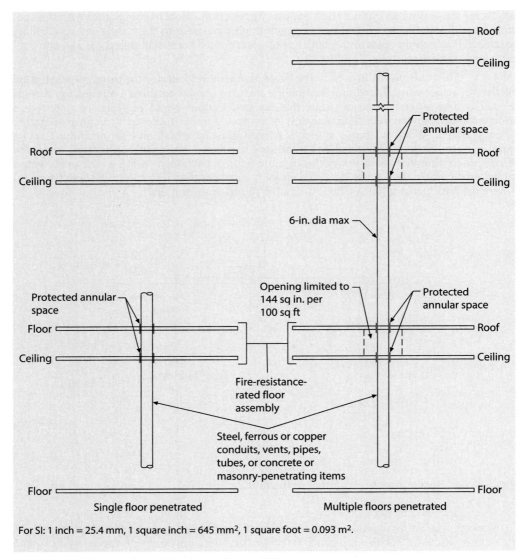

Figure 714-9 **Penetrations of horizontal assemblies.**

The code also specifies that any recessed fixtures that are installed in fire-resistance-rated horizontal assemblies shall not reduce the level of required fire resistance. Exceptions to the general requirement for approved firestop systems apply to noncombustible penetrations, steel electrical boxes, boxes listed as a part of an opening protective material system, listed electrical-outlet boxes, fire sprinklers, noncombustible items cast into concrete building elements or listed luminaires. See Figure 714-10.

Exception 7 allows for the practical application of the code where wood-framed walls extend up and attach directly to the underside of wood floor joists/trusses or roof joists/trusses for structural requirements. However, there are limits to its use. Double wood top plates are allowed

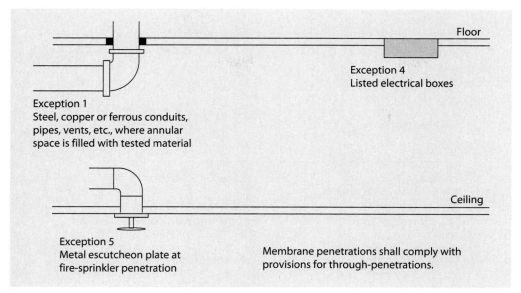

Figure 714-10 **Membrane penetrations of horizontal assemblies.**

to interrupt the gypsum-board membrane of the floor/ceiling or roof/ceiling membrane. The allowance is only permitted where the horizontal assembly has a required fire-resistance rating of 2 hours or less, and the intersecting wall must be sheathed with Type X gypsum wallboard. Piping, conduit, and similar items within the fire-resistance-rated wall must be adequately protected where they penetrate the double-wood top plate. Compliance with the established criteria is deemed to provide for an equivalent degree of fire resistance and protection at the discontinuous portion of the ceiling membrane where the double wood top plates occur. See Figure 714-11.

714.6 Non-fire-resistance-rated assemblies. Figure 714-12 illustrates the provisions for penetrations of those horizontal assemblies not required to have a fire-resistance rating. Section 713 for shaft enclosures will regulate such penetrations where the allowances set forth in this section are not applicable. Where penetrations connect only two stories, the annular space around the penetrating items must simply be protected with a material that resists the free passage of fire and smoke. If the penetrating items are noncombustible, up to five stories may be connected, provided the annular space is filled appropriately.

Section 715 *Joints and Voids*

A fire-resistant joint system is defined in Section 202 as "an assemblage of specific materials or products that are designed, tested, and fire-resistance rated in accordance with either ASTM E1966 or UL 2079 to resist, for a prescribed period of time, the passage of fire through joints made in or between fire-resistance-rated assemblies." The term *joint* is also defined as "the opening in or between adjacent assemblies that is created due to building tolerances, or is designed to allow independent movement of the building, in any plane, caused by thermal, seismic, wind, or any other loading." The approved system should be designed to resist the passage

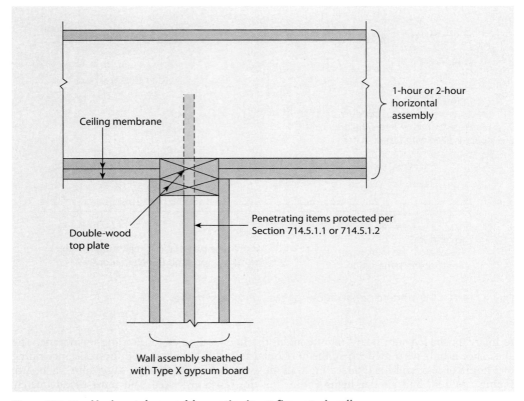

Figure 714-11 **Horizontal assembly continuity at fire-rated wall.**

of fire for a time period not less than the required fire-resistance rating of the floor, roof, or wall in or between which it is installed. See Figure 715-1.

The code lists 10 locations where it is not necessary to provide fire-resistant joint systems. For most of the applications listed, they are also locations where fire assemblies are not required to protect openings in the horizontal or vertical assemblies. Item 9 references maximum ⅝-inch (15.9-mm) control joints when tested in accordance with ASTM E119 or UL 263.

715.4 Exterior curtain wall/fire-resistance-rated floor intersections. Vertical passages without barriers allow fire and hot gases to circumvent the protection for occupants in the floors above. When floors or floor/ceiling assemblies do not extend to the exterior face of a building, this section requires an approved perimeter fire containment system at the intersection at least equal to the fire resistance of the floor or floor/ceiling assembly. See Figure 715-2.

ASTM E2307 is identified as the general specification and testing method used to determine the necessary fire resistance. This test method measures the performance of a perimeter fire-containment system and its ability to maintain a seal to prevent fire spread during the deflection and deformation of the exterior wall assembly and floor assembly expected during a fire condition, while resisting fire exposure from both an interior compartment and the flame plume emitted from a window burner as shown in Figure 715-2.

Joints and Voids **275**

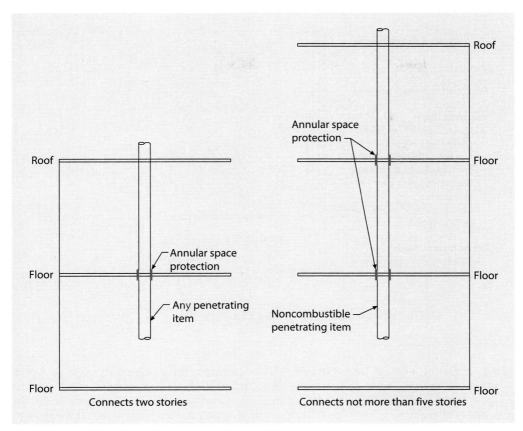

Figure 714-12 **Non-fire-resistance-rated assemblies.**

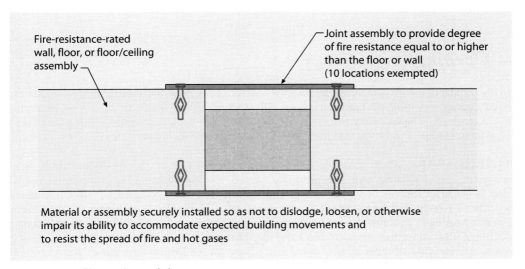

Figure 715-1 **Fire-resistant joint systems.**

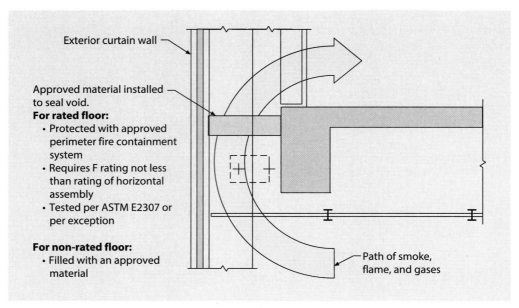

Figure 715-2 **Exterior curtain wall/floor intersection.**

715.5 Exterior curtain wall/nonfire-resistance-rated floor assembly intersections. A minimum level of protection is also mandated at any voids created at the intersection of an exterior curtain wall and a nonrated floor or floor assembly. The required method is consistent with that required in the code for the penetration of ducts and other items through non-fire-resistance-rated floor systems. The void space is to be filled with an "approved" material that retards the free passage of flame and products of combustion between the stories. See Figure 715-2.

Section 716 *Opening Protectives*

In the context of the IBC, an opening protective refers to a fire door, fire shutter, or fire-protection-rated glazing, including the required frames, sills, anchorage, and hardware for its proper operation. Generally, whenever any fire door, fire shutter, or fire-protection-rated glazing is referred to, it is the intent of the code that the entire fire assembly be included.

716.1.2.2.4 Fire-rated glazing that exceeds the code requirements. Both fire-resistance-rated glazing and fire-protection-rated glazing must be appropriately identified for verification of their appropriate application. These markings establish compliance with hose-stream and temperature rise requirements, while also identifying the minimum assembly rating in minutes. It is not unusual for such glazing to be marked indicating a higher degree of protection than mandated by the code. This provision clarifies that the use of glazing marked to indicate a higher level of compliance is permitted for use where such compliance is not required.

716.1.2.3 Fire-resistance-rated glazing. This section is an extension of the provisions in Section 703.4 regarding the use of fire-resistance-rated glazing. Where such glazing is appropriately tested and labeled, its use is permitted in fire doors and fire window assemblies as specified in Tables 716.1(2) and 716.1(3), respectively. Such glazing tested as a part of a wall assembly per ASTM E119 or UL 263 is not required to comply with the provisions of Section 716 where used as a part of the wall.

Table 716.1(1) defines and relates the various test standards for fire-rated glazing to the designations used to mark such glazing. The table reflects the use of the designations "W," "FC," "OH," "D," "T," "H," and "XXX" as markings for fire-rated glazing. Tables 716.1(2) and 716.1(3) set forth the markings required for acceptance in specified applications. The marking of fire-rated glazing does not include the "NH" (not hose-stream tested) and "NT" (not temperature-rise tested) designations, as these designations correspond with test standards, not end uses.

716.2 Fire door assemblies. This section sets forth the test standards and additional criteria necessary for the acceptance of fire door assemblies. In addition, Table 716.1(2) identifies the minimum fire-protection rating for an opening protective based on the type of assembly in which it is installed. For example, a door assembly in a 1-hour fire barrier wall separating hazardous material control areas would need to have a minimum ¾-hour fire-protective rating, whereas a 1-hour fire-resistance-rated interior exit stairway enclosure would require a minimum 1-hour door assembly.

Side-hinged or swinging doors are to be tested for conformance with NFPA 252 or UL 10C. It is important that the NFPA 252 test provides for positive pressure in the furnace as established by this section. See Figure 716-1. For other types of doors, the pressure level at the top of the door need only be maintained as nearly equal to the atmosphere's pressure as possible.

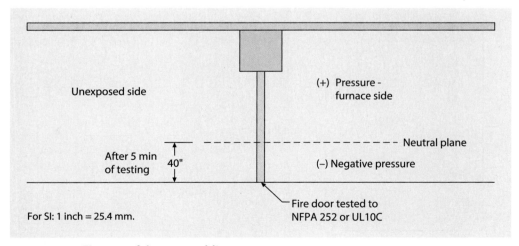

Figure 716-1 **Fire test of door assemblies.**

716.2.2.1 Door assemblies in corridors and smoke barriers. Fire door assemblies located in fire-resistance-rated corridor walls or smoke barrier walls are further regulated where required by Table 716.1(2) to have a 20-minute fire-protection rating. They are commonly referred to as smoke- and draft-control assemblies. Their primary purpose is to minimize smoke leakage around the door and through the opening. For this reason, these doors shall not contain louvers and per Section 716.2.10 must be installed in accordance with NFPA 105.

The protection of fire-rated corridors is intended to protect the corridor from smoke that might be generated by a fire occurring within the adjacent use spaces. However, there may be unique occasions where it is just as important to protect the occupied use spaces from smoke in the corridor. The fire test for corridor and smoke-barrier doors is essentially the same test of the door as for other fire-door assemblies, except that the fire test for the 20-minute assembly does not include the hose-stream test. In addition, Section 716.2.1.4 requires the door assembly to be tested for smoke infiltration through the UL 1784 air leakage test and Section 716.2.2.1.1 identifies the criteria for acceptance. Note that per Section 716.2.1.3, glazing other than in the door itself, such as in sidelights or transoms, must be tested with the hose-stream test as set forth in NFPA 257 or UL 9.

Within Section 716.2.2.1, an exception permits the installation of a viewport through the door for purposes of observation. These viewports must be installed under the limitations of, and in accordance with, the conditions specified in the exception. Corridor door provisions are modified in Section 407.3.1 for Group I-2 occupancies and in multitheater complexes as shown in Figure 716-2. In addition, where horizontal sliding doors are used in smoke barriers of Group I-3 occupancies as specified, the 20-minute fire-protection rating is not required.

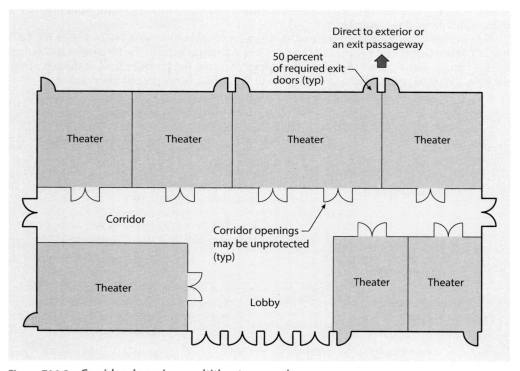

Figure 716-2 **Corridor doors in a multitheater complex.**

716.2.2.3 Doors in interior exit stairways and ramps and exit passageways. In addition to the normal requirement for fire doors, the IBC is concerned that fire door assemblies installed in enclosures for interior exit stairways/ramps and exit passageways shall be capable of limiting the temperature transmission through the door. It specifies that the temperature rise above ambient temperature shall be limited to a maximum of 450°F (232°C) at the end of 30 minutes of the normal fire test. However, in buildings equipped with an automatic sprinkler system, the temperature limitation is not applicable.

The purpose of these highly protected exit elements and their openings is to protect the building occupants while they are exiting the building. It is intended that in a properly enclosed and protected interior exit stairway or ramp enclosure, building occupants from the floors above the fire floor will be able to pass through the fire floor inside the enclosure and eventually pass down and out of the building. The end-point limitation on temperature transmission through the fire door, then, is literally to protect the person inside the enclosure from excessive heat radiation from the fire door as they pass through the fire floor. A similar expectation applies to travel through an exit passageway. In sprinklered buildings, the maximum transmitted temperature end point is not required. It is expected that a sprinkler system will limit the fire growth to the point where such extra care is unnecessary.

716.2.5.1 Size limitations. This section addresses both fire-resistance-rated and the lower performing fire-protection rated glazing. In a door, fire-protection-rated glazing is permitted in wall assemblies rated at 1 hour or less when in compliance with the size limitations of NFPA 80. Where a fire wall or fire barrier requires a rating greater than 1 hour, such glazing is prohibited except for two conditions. First, fire-protection-rated glazing may be used as vision panels in swinging fire door assemblies serving as horizontal exits when limited in size. Owing to the use of a horizontal exit as a required means of egress, it is often beneficial to provide a glass light of limited size so that occupants may view the egress path ahead of them. Second, the maximum size of all types of fire-protection-rated glazing in 1½-hour fire doors is limited to 100 square inches (0.065 m^2) when installed in a fire barrier.

716.2.6.1 Door closing. The code mandates that fire doors be provided with closers to allow them to shut and protect the opening without manual operation. Exception 1 to this broad-based requirement applies to those fire doors located in the common walls between sleeping units of hotels and motels. These doors are so seldom open that it is unreasonable to require door-closing hardware. Other exceptions are specific to fire doors installed for compliance with the storm shelter provisions of ICC 500 and those elevator doors located at the floor level designated for recall.

716.2.6.4 Automatic-closing fire door assemblies. Where automatic-closing devices are used instead of self-closing devices on fire doors, they must also comply with the provisions of NFPA 80 for self-closing action. Although they are generally held in an open position, doors equipped with automatic-closing devices become self-closing when actuated. The use of automatic-closing devices is typically a design decision; however, the code does mandate such devices for doors in four applications. Automatic-closing devices are required by Section 709.5.1 on cross-corridor doors located in Group I-2 occupancies and ambulatory care facilities. They must also be installed on cross-corridor doors located in a horizontal exit as set forth in Section 1026.3. Sections 405.4.3 and 3008.6.3.2 related to protecting doors on elevator lobbies are the other door locations where automatic-closing devices are required. Though not a door, rolling type fire shutters also require the use of automatic-closing devices per Section 716.2.8.

716.2.6.6 Smoke-activated doors. This section identifies eight locations where a smoke detector is to be used to actuate the closing operation for an automatic-closing fire door where such a closing device is provided. The detectors must be installed in accordance with the provisions of Section 907.3 and, furthermore, they must be of an approved type that will release the door in the event of a power failure. Automatic-closing fire door assemblies are often used to increase the reliability of the opening protection. Swinging fire doors with self-closers are all too often propped open with wood blocks or wedges. Although this section regulates the method for activating automatic-closing fire doors, it does not identify where automatic-closing doors are mandated. It is only if automatic-closing doors are installed at these locations that the code then mandates they be provided with smoke-activated closers.

716.2.9 Labeled protective assemblies. Fire doors are required to have an approved label or listing mark permanently affixed at the factory. The label must contain information that identifies the manufacturer, the third-party inspection agency, and the fire-protection rating. Where applicable, the maximum transmitted temperature end point or the smoke- and draft-control designation must be identified.

Listing agencies will typically only label door assemblies that have been tested. However, some door assemblies are too large to be tested in available furnaces. As a result, the code permits the installation of oversized fire doors under the conditions of Section 716.2.9.2. As oversized fire doors are not subjected to the standard fire test, an approved testing agency must provide a certificate of inspection from them certifying that, except for the fact that the doors are oversized, they comply with the requirements for materials, design, and construction for a fire door of the specified fire-endurance rating. An approved agency may also provide a label on the door indicating it is oversized. Where the certificate or label of an approved agency has been provided, there is assurance that the fire door will protect the opening as required by the code.

The letter "S" on a fire door indicates that it is in compliance with UL 1784, the air leakage testing. Through this identifying mark, it is possible to quickly identify the door and frame assembly as appropriate provided listed or labeled gasketing is installed. It can be inferred from this code provision that all smoke- and draft-control door and frame assemblies required by the IBC to be in compliance with UL 1784 be provided with listed or labeled gasketing.

Individual components, such as vision panels, may be installed in labeled fire doors provided such components are listed or classified and labeled for use by a third-party agency.

716.2.9.5 Labeling. Glazing used in fire door assemblies must be identified for verification of its appropriate application. As established in Table 716.1(1), the "D" designation indicates the glazing can be used in a fire door assembly, with the remaining identifiers providing specific information as to the glazing's capability to meet the hose-stream test and temperature limits. See Figure 716-3.

716.3.4 Fire-protection-rated glazing. In many situations, it is necessary to provide glazed openings in fire-resistance-rated walls. The provisions of this section address fire window assemblies installed as opening protectives in fire partitions, smoke barriers, and exterior walls, as well as in some 1-hour fire barriers.

Fire-protection-rated glazing in fire window assemblies must be tested in accordance with NFPA 257 or UL 9. In addition, they must be installed and sized in accordance with NFPA 80. In all cases, a fire-window assembly must include an approved frame, be fixed in position, or be automatic closing.

In interior applications, fire-protection-rated glazing is limited to fire partitions, smoke barriers, and three types of fire barriers (per Table 716.1(3)) having a maximum fire-resistance rating of 1 hour. Per Section 716.3.2.1.2 the total aggregate area of fire windows cannot exceed 25 percent of the area of the common wall between areas, as shown in Figure 716-4. In making this 25-percent calculation, it is permissible to assume the entire area of the common wall even though a portion of that area might be taken up by doors. This gross area is usable in calculating the maximum percentage of area for windows. Where the ceilings are of different heights, the lower ceiling establishes the gross area.

Fire-protection-rated glazing is not permitted to be located in any fire wall or in any fire barrier (with three exceptions); however, Section 716.1.2.3 recognizes that glazing tested as a part of a wall assembly is permitted in all applications and, therefore, not regulated by this section. This type of glazing is referred to as fire-resistance-rated glazing, rather than fire-protection-rated

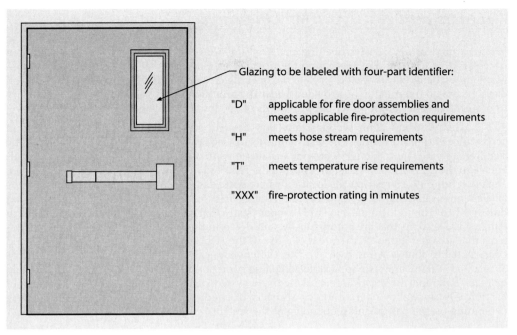

Figure 716-3 **Identification of glazing in fire doors.**

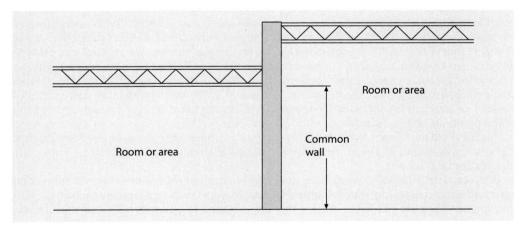

Figure 716-4 **Glazing limitations.**

glazing, and is further addressed in Section 703.4. In addition, the 25-percent area limitation is not applicable. Glazing that has a fire-resistance rating under ASTM E119 or UL 263 that meets the fire-resistance rating of the wall may be used in more applications than fire-protection-rated glazing complying with NFPA 257 or UL 9. Also see the discussion of Section 703.4.

Section 717 *Ducts and Air Transfer Openings*

Where a duct or air transfer opening penetrates a fire-resistance-rated assembly, it is often necessary that a method of protection be provided to maintain the integrity of the assembly. Many times, dampers are used to protect the opening created by the duct or transfer opening. If dampers are not required to be provided under the provisions of this section, it is still necessary to protect the penetration under the provisions of Sections 717.1.2 and 714.1.1.

717.2 Installation. This section states that fire and smoke dampers shall be installed in accordance with their listing. The test standards for each of the types of damper carry specific requirements that manufacturers provide installation and operating instructions, and that a reference to these instructions shall be a part of the required marking information on the damper.

Subsection 717.2.3 restricts the location where static dampers can be used. Only fire dampers labeled for use in dynamic systems shall be installed in systems intended to operate with fans on during a fire. The test standard for fire dampers states that fire dampers are intended for use in either static systems that are automatically shut down in the event of a fire, or in dynamic systems that are operational in the event of a fire. If the HVAC system has not been designed and constructed to shut down in case of a fire, then dynamically listed fire dampers are necessary. Similar provisions are also applicable to ceiling radiation dampers. Special attention should be paid to damper listings when smoke-control systems are installed under the provisions of Section 909 since they operate under a variety of conditions that may be more demanding depending on the design and sequencing of the system.

717.3.1 Damper testing. Dampers must not only be listed but also bear a label indicating that the damper is in compliance with the appropriate standard as identified by this section. For example, for fire dampers the required information on the damper includes the hourly rating; the words "Fire Damper"; whether or not the damper is to be in a dynamic or static system (or both); maximum rated airflow and pressure differential across the closed damper for dampers intended for use in dynamic systems; an arrow showing direction of airflow for dampers intended for use in dynamic systems; the intended mounting position (vertical, horizontal, or both); top of damper; and, of course, the manufacturer's name and model number. UL 555 (which applies to fire dampers) requires that all of this information shall be available on the damper label, which is installed at the factory, and that all labels shall be located on the internal surface of the damper and be readily visible after the damper is installed. UL 555 indicates that fire dampers tested under that standard are intended for use in HVAC duct systems passing through fire-resistive walls, partitions, or floor assemblies.

Just as fire dampers are tested for different hourly ratings, they are also tested for different installation positions. A damper listed for vertical installation cannot arbitrarily be installed in the horizontal position.

Test standard UL 555S is to be used to determine the compliance of smoke dampers. This standard states that leakage-rated dampers (smoke dampers in the IBC) are intended to restrict the spread of smoke in HVAC systems that are designed to automatically shut down in the event of a fire, or to control the movement of smoke within a building when the HVAC system is operational in engineered smoke-control systems.

In addition to fire dampers and smoke dampers, three other types of dampers are referenced in the IBC. Where combination fire/smoke dampers are provided, they must comply with the requirements of both UL 555 for fire dampers and UL 555S for smoke dampers. Figure 717-1 illustrates the installation of an automatic-closing combination fire and smoke damper. Ceiling radiation dampers, intended for installation in air-handling openings penetrating the ceiling membranes of fire-resistance-rated floor/ceiling and roof/ceiling assemblies, have to meet the conditions of UL 555C.

Ducts and Air Transfer Openings 283

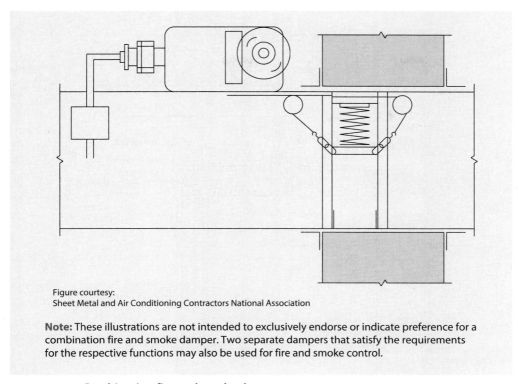

Figure courtesy:
Sheet Metal and Air Conditioning Contractors National Association

Note: These illustrations are not intended to exclusively endorse or indicate preference for a combination fire and smoke damper. Two separate dampers that satisfy the requirements for the respective functions may also be used for fire and smoke control.

Figure 717-1   **Combination fire and smoke dampers.**

Corridor dampers are to be utilized in a specific application, where (1) air ducts penetrate or terminate at horizontal openings, (2) provided such openings occur in the ceiling of a fire-resistance-rated corridor, and (3) if the corridor ceiling is constructed as required for corridor walls as further addressed in Exception 3 of Section 708.4. See Figure 717-2. The listing and testing of a corridor damper is different from the regular combination fire/smoke damper in that the testing is done in a wall assembly installed in the horizontal position.

717.3.2.1   Fire damper ratings.   Test standard UL 555 covers fire dampers ranging from 1½ hours to 3 hours. Because fire dampers carry an hourly rating, plans should reflect the rating required at a particular location if more than one rating is required within a building. Table 717.3.2.1 indicates whether a 1½-hour or 3-hour rating is required for a fire damper, based on the fire-resistance rating of the assembly in which it is installed. In all applications rated at 3 hours or greater, a 3-hour-rated damper is mandated; otherwise, a 1½-hour damper is acceptable.

717.3.2.2   Smoke damper ratings.   A Class I or Class II leakage rating is required for smoke dampers, which also must have an elevated temperature rating of at least 250°F (121°C). The class designation indicates the maximum leakage permitted in cubic feet per minute per square foot (cubic mm per minute per mm^2) for the particular class. The three classes progress from Class I (least leakage or best performance) through Class III (greatest leakage or poorest performance). The IBC requires conformance with Class I or II, so a damper rated as Class III would not be acceptable. These leakage ratings are determined at ambient temperature after exposing the damper to temperature degradation at an elevated temperature, with 250°F (121°C) being the lowest elevated temperature allowed by the code. Dampers can be tested

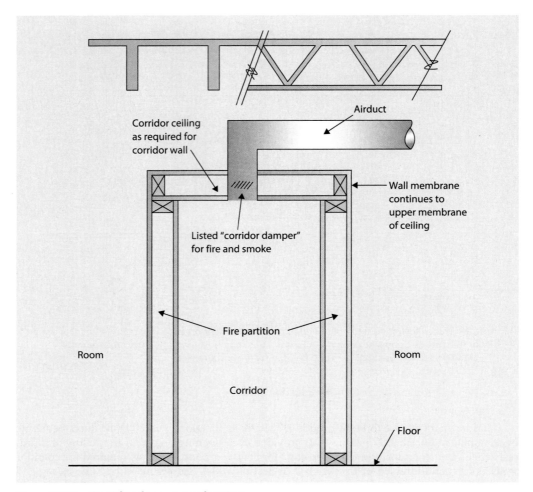

Figure 717-2 **Corridor damper application.**

using higher degradation temperatures [one as high as 850°F (454°C)], but most listed dampers seem to have been tested at either 250°F (121°C) or 350°F (177°C).

The provisions of Section 717.3.3.2 specifically instruct the designer or installer on how to control smoke dampers. Smoke dampers are required to be closed by activation of smoke detectors installed in accordance with Section 907.3 for fire-detection systems and any of the five specified methods of control listed in this provision. These methods of control, each having benefits and drawbacks, were proposed by those individuals involved with damper installation and should provide consistent and logical control methods for the dampers.

717.4 Access and identification. Both fire dampers and smoke dampers shall be installed so that they are accessible for inspection and servicing. Where access is limited or restricted, Section 717.4.1.2 requires that provisions are made for remote inspection. It is important that any access openings in a fire-resistance-rated assembly be adequately protected in order to maintain the integrity of the assembly. This will typically involve the use of an access door having the required fire-protection rating. Permanent identification of the access points to fire-damper and smoke-damper locations is also mandated.

717.5 Where required. This section lists those specific locations where the various dampers are required. Dampers need only be installed in ducts and transfer openings where specifically identified by this section. In some locations, both a fire damper and a smoke damper are required. This means that either two dampers must be installed or a damper listed for both fire and smoke control must be used.

717.5.1 Fire walls. Because of the importance of maintaining the separation provided by fire walls used to divide a structure into two or more separate buildings, the code requires the use of approved fire dampers under all conditions. Such dampers are to be installed at all permitted duct penetrations and air-transfer openings of fire walls. Where the fire wall serves as a party wall as addressed in Section 706.1.1, ducts and air transfer openings are prohibited. There is no requirement for smoke dampers at duct penetrations and air openings through fire walls except for those fire walls serving as horizontal exits.

717.5.2 Fire barriers. Much like fire walls, fire barriers are designed to totally isolate one area of a building from another. Therefore, the general requirement is that all duct penetrations and air transfer openings of fire barriers be protected by complying fire dampers. There are, however, several exceptions that may eliminate the need for dampers. Of special note is the elimination of fire dampers in certain sprinklered buildings. Fire dampers are not required for duct penetrations and air openings in fire barriers where all of the following conditions exist:

1. The penetration consists of a duct that is a portion of a fully ducted HVAC system.
2. The fire-resistance rating of the fire barrier is 1 hour or less.
3. The area is not a Group H occupancy.
4. The building is fully protected by an automatic fire-sprinkler system.

Nonmetal flexible air connectors are acceptable provided they are limited to the following locations:

1. In the mechanical room at the duct connection to the air handling unit.
2. Between an overhead metal duct and a ceiling diffuser located in the same room.

A fire barrier that serves as a horizontal exit must also be provided with a listed smoke damper at each point a duct or air transfer opening penetrates the fire barrier.

717.5.3 Shaft enclosures. Both a fire damper and a smoke damper, or a combination fire/smoke damper, must be installed where a duct or air transfer opening penetrates a shaft enclosure. Five exceptions identify conditions under which fire or smoke dampers are not required. As shown in Figure 717-3, the first exception permits fire dampers to be eliminated where steel exhaust subducts enter an exhaust shaft. The subducts must extend vertically at least 22 inches (559 mm), and there must be continuously powered air flow upward to the outside through the shaft. Smoke dampers are also not required under similar conditions, but the exception is limited to fully sprinklered Group B and R occupancies where the fan providing continuous airflow is on standby power. Exception 5 exempts fire dampers and fire/smoke dampers in kitchen and clothes dryer exhaust systems due to potential obstruction hazards. Smoke dampers are not addressed in this exception, but may be omitted when Exception 2 is applied.

717.5.4 Fire partitions. Fire partitions are not regulated by the code as highly as fire barriers or fire walls, so it is consistent that the requirements for dampers through such partitions are not as restrictive. The general rule is that a fire damper is required in any duct or air transfer opening that penetrates a fire partition. However, where the building is fully sprinklered, ducts penetrating tenant separation walls in covered mall buildings or fire-resistance-rated corridor walls need not be fire dampered. In addition, fire dampers are not necessary for small steel ducts installed above a ceiling, provided the duct does not communicate between a corridor and

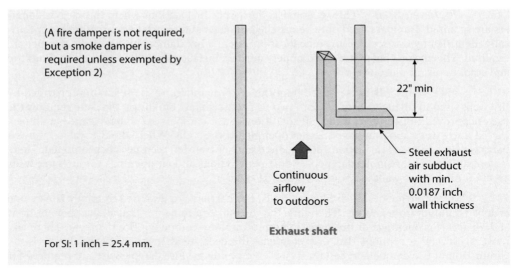

Figure 717-3 **Exhaust subducts penetrating shafts.**

adjacent rooms and does not terminate at a register in a fire-resistance-rated wall. The fourth exception is conceptually consistent with Exception 3 of Section 717.5.2 addressing fire barriers however the allowance for nonmetal flexible air connectors is not provided.

717.5.4.1 Corridors. Because a fire-resistance-rated corridor is intended to be an exit access component providing a limited degree of occupant protection during egress activities, it is logical that air openings into the corridor be addressed. This section mandates, with exceptions, that all corridors required to be protected with smoke- and draft-control doors shall also be provided with smoke dampers where ducts or air transfer openings penetrate the corridor enclosures. Additional protection is required where ducts and air transfer openings penetrate the ceiling of a fire-resistance-rated corridor. Either a corridor damper or a ceiling radiation damper shall be installed depending upon the method of ceiling construction. As illustrated in Figure 717-4, an important exception eliminates the need for smoke dampers where steel ducts pass through, but do not serve, the corridor.

717.5.5 Smoke barriers. Those air openings, both ducts and transfer openings, that penetrate smoke barriers are to be provided with smoke dampers at the points of penetration. Steel ducts are permitted to pass through a smoke barrier without a damper, provided the openings in the ducts are limited to a single smoke compartment. Fire dampers are not required at penetrations of smoke barriers.

717.6 Horizontal assemblies. The code is quite restrictive when it comes to the protection of vertical openings between floor levels, particularly where the floor or floor/ceiling assembly is required to be fire-resistance rated. This section requires the use of a shaft enclosure to address the hazard that is created where a duct or air transfer opening extends through a floor, floor/ceiling assembly, or ceiling membrane of a roof/ceiling assembly. The remainder of the section modifies this general requirement for through-penetrations, membrane penetrations, and non-fire-resistance-rated assemblies.

717.6.1 Through-penetrations. Ducts and air transfer openings that penetrate horizontal assemblies are initially regulated by the provisions of Section 713 for shaft enclosures. However, a shaft enclosure is not required where a duct that connects only two stories is provided

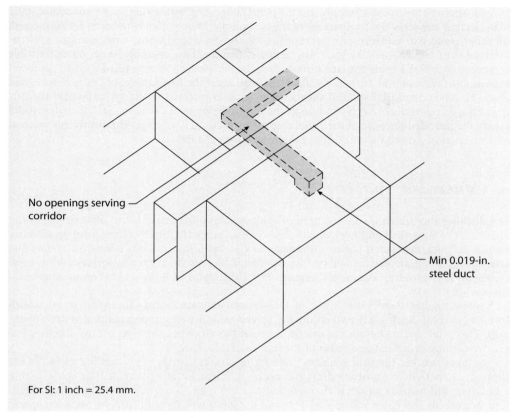

Figure 717-4 **Ducts crossing corridor.**

with a fire damper installed at the floor line of the fire-resistance-rated floor/ceiling assembly that is penetrated. As another option, the duct may be protected in a manner prescribed in Section 714.5 for the penetration of horizontal assemblies. The exception goes on to eliminate the fire damper requirement in all occupancies other than Groups I-2 and I-3, provided the duct connects no more than four stories and all of the applicable conditions are met.

717.6.2 Membrane penetrations. A shaft enclosure need not be provided where an approved ceiling radiation damper is installed at the ceiling line of a fire-resistance-rated floor/ceiling or roof/ceiling assembly penetrated by a duct or air-transfer opening. Designed to protect the construction elements of the floor or roof assembly, the ceiling radiation damper is not required where fire tests have shown that the damper is not necessary to maintain the fire-resistance rating of the assembly. Additionally, ceiling radiation dampers are not required at penetrations of exhaust ducts, provided the penetrations are appropriately protected, the exhaust ducts are contained within wall cavities, and the ducts do not pass through adjacent dwelling units or tenant spaces.

717.6.3 Nonfire-resistance-rated floor assemblies. The elimination of shaft enclosures at vertical openings is also possible where the floor assemblies are not required to be of fire-resistance-rated construction. Two conditions are identified using the filling of the annular space between the assembly and the penetrating duct with an approved noncombustible material that will resist the free passage of fire and smoke. Where only two stories are connected, no other protective measures are necessary. In three-story conditions, a fire damper must be installed at each floor line. The code also mandates that the floor assembly be of noncombustible construction and the annular space surrounding the penetrating duct be filled with an approved noncombustible material, such as a sealant, that will resist the free passage of flame, smoke, and gases. However, the installation of such sealant or other material would typically void the listing of the damper. Under such conditions, the use of the damper's steel mounting angles would satisfy the intended purpose of the annular space protection. An exception permits this method of protection in a dwelling unit without the installation of a fire damper.

Section 718 *Concealed Spaces*

Fireblocking and draftstops are required in combustible construction to cut off concealed draft openings (both vertical and horizontal). The code requires that fireblocking form an effective barrier between floors and between the top story and attic space. The code also requires that attic spaces be subdivided, as will be discussed later, along with concealed spaces within roof/ceiling and floor/ceiling assemblies. Figures 718-1 through 718-4 depict IBC requirements for fireblocking.

Experience has shown that some of the greatest damage occurs to conventional wood-framed buildings during a fire when the fire travels unimpeded through concealed draft openings. This often occurs before the fire department has an opportunity to control the fire, and greater damage is created as a result of the lack of fireblocking.

For these reasons, the code requires fireblocking and draftstops to prevent the spread of fire through concealed combustible draft passageways. Virtually any concealed air space within a building will provide an open channel that can bring air to a fire, or through which high-temperature air and gases will spread. Fire and hot gases will spread through concealed spaces

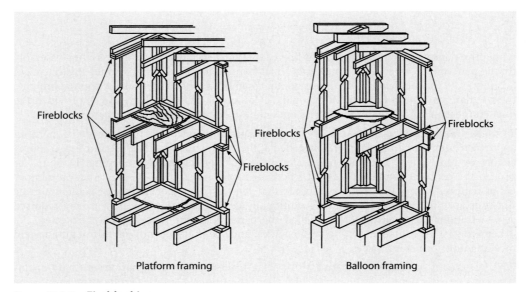

Figure 718-1 **Fireblocking.**

Concealed Spaces   289

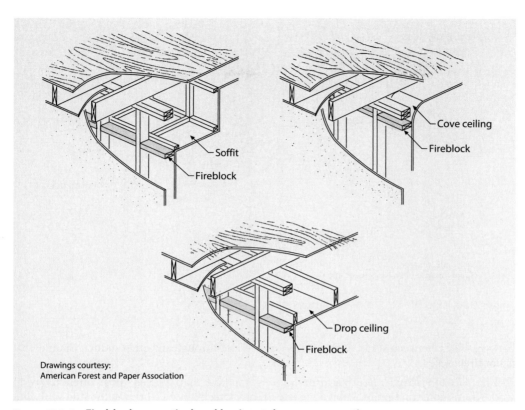

Figure 718-2   **Fireblocks—vertical and horizontal space connections.**

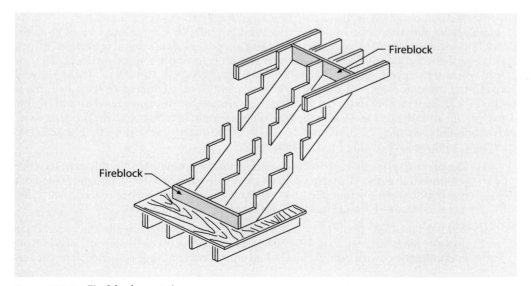

Figure 718-3   **Fireblocks—stairs.**

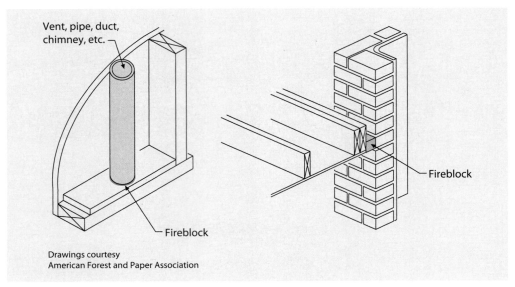

Figure 718-4 **Fireblocks—pipes, chimneys, etc.**

between joists, between studs, within furred spaces, and through any other hidden channel that is not fireblocked.

718.2 Fireblocking. The platform framing method that is used most often today in wood-frame construction provides adequate fireblocking between stories in the stud walls, but care must be exercised to ensure that furred spaces are effectively fireblocked to prevent transmission of fire and hot gases between stories or along a wall. For this reason, the code requires that fireblocking be provided at 10-foot (3,048-mm) intervals horizontally along walls that are either furred out, of double-wall construction, or of staggered-stud construction.

Fireblocking provisions for wood flooring used typically in gymnasiums, bowling alleys, dance floors, and similar uses containing concealed sleeper spaces are found in Section 718.2.7. As long as the wood flooring described in this section is in direct contact with a concrete or masonry fire-resistance-rated floor, there is no significant hazard. However, if there is a void between the wood flooring and the fire-resistance-rated floor, a blind space is created that is enclosed with combustible materials and provides a route for the undetected spread of fire. Therefore, the code requires that where the wood flooring is not in contact with the fire-resistance-rated floor, the space shall be filled with noncombustible material or shall be fireblocked. Two exceptions to these fireblocking requirements are:

1. The first exception exempts slab-on-grade floors of gymnasiums. In this case, the code presumes a low hazard, as gymnasiums are usually only one story in height. If the floor is at or below grade, it is unlikely that any ignition sources would be present to start a fire that would spread through the blind space under the wood flooring.

2. Bowling lanes are exempted from fireblocking except as described in the code, which provides for areas larger than 100 square feet (9.3 m^2) between fire blocks. Fireblocking intermittently down a bowling lane would create problems for a consistent lane surface.

Fireblocking materials are required to consist of lumber, mass timber, or wood (including fire-retardant-treated) structural panels of the thicknesses specified, gypsum board, cement fiber board, mineral wool, glass fiber, or any other approved materials securely fastened in place.

Batts or blankets of mineral wool and glass fiber materials are allowed to be used as fireblocking and work especially well where parallel or staggered-stud walls are used. Loose-fill insulation should not be used as a fireblocking material unless specifically tested for such use. It must also be shown that it will remain in place under fire conditions. Even in the case where it fills an entire cavity, a hole knocked into the membrane enclosure for the cavity could allow the loose-fill insulation material to fall out, negating its function. Therefore, loose-fill insulation material shall not be used as fireblocking unless it has been properly tested to show that it can perform the intended function. The main concern is that the loose-fill material, even though it may perform adequately in a fire test to show sufficient fire-retardant characteristics to meet the intent of this section, would not be adequately evaluated for various applications because of the physical instability of the material in certain orientations.

718.3 Draftstops in floors. Draftstops are often used to subdivide large concealed spaces within floor/ceiling assemblies of combustible construction. Figure 718-5 shows IBC requirements for draftstops in these locations. Gypsum board, wood structural panels, particleboard, mineral wool, or glass fiber batts and blankets, and other approved materials are considered satisfactory for the purpose of subdividing floor/ceiling areas, provided the materials are of adequate thickness, are adequately supported, and their integrity is maintained.

Draftstops are to be installed in floor/ceiling assemblies as follows:

1. Residential occupancies. The code requires that draftstops be installed in line with the fire partitions separating tenants or dwelling units from each other and the remainder of the building, consistent with the provisions of Sections 708.4.3 and 420 for dwelling unit and sleeping unit separations. In this case, a fire originating in a dwelling unit or hotel room will find draftstops in the concealed space blocking the transmission of fire and hot gases into another hotel room or apartment. Where the residential occupancy is fully sprinklered, draftstopping is not required. Where a residential sprinkler system is used, automatic sprinklers must also be installed in the combustible concealed floor areas in order to eliminate the draftstops.

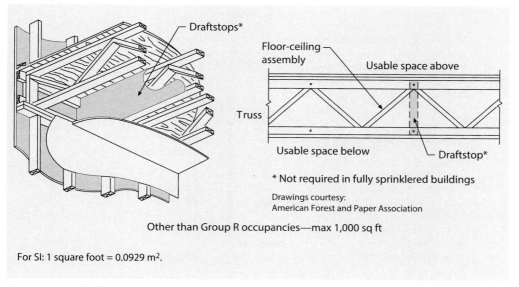

Figure 718-5 **Draftstops—floors.**

292 Chapter 7 ■ Fire and Smoke Protection Features

2. **All other occupancies.** For uses other than residential occupancies, the code intends that the concealed space within the floor/ceiling assembly be separated by draftstops so that the area of any concealed space does not exceed 1,000 square feet (9.3 m²). An exception permits the elimination of draftstops where automatic fire sprinklers are installed throughout the building.

718.4 Draftstops in attics. In attics and concealed roof spaces of combustible construction, the code requires draftstopping under certain circumstances. Consistent with the requirements for fireblocking, draftstopping is not required for spaces constructed entirely of noncombustible materials. Materials used for draftstopping purposes, such as gypsum board, plywood, or particleboard, are to be installed consistent with the provisions of Section 718.3.1 for the draftstopping of floors. The following locations are identified as those requiring draftstopping:

1. **Residential occupancies.** The attic draftstops for residential uses are addressed in Section 708.4.3 due to the reference to that section. Draftstops are to be installed above and in line with the walls separating dwelling units or between walls separating sleeping units. Figure 718-6 explains the intent of Exception 2 in 708.4.3. Exception 4

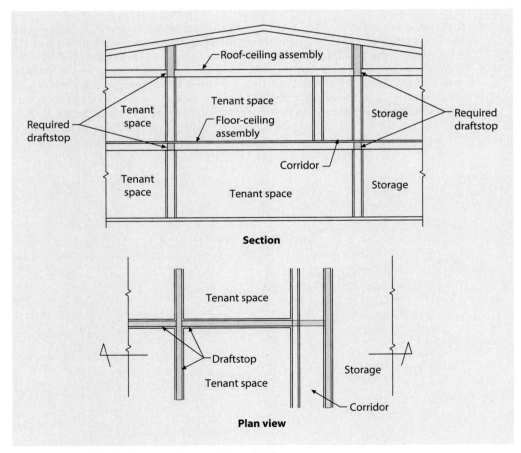

Figure 718-6 **Attic draftstop—Groups R-1 and R-2.**

Concealed Spaces

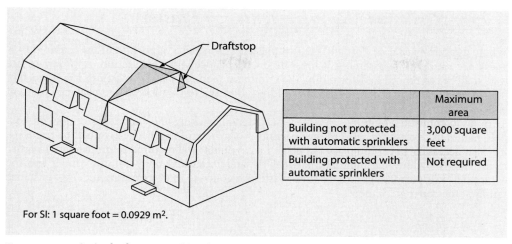

Figure 718-7   Attic draftstop—other than Group R occupancies.

in 708.4.3 applies to Group R-2 occupancies less than five stories in height. In this case, attic areas may be increased by installing draftstops to subdivide the attic into a maximum of 3,000-square-foot (279-m²) spaces, with no area to exceed the inclusion of two dwelling units. Exception 1 eliminates the need for draftstopping in fully sprinklered buildings. Where a residential sprinkler system is used, the system must be extended into the concealed attic space in order to eliminate the draftstop.

2. Other uses. Draftstops are required by the code to be installed in attics and similar concealed roof spaces of buildings other than Group R so that the area between draftstops does not exceed 3,000 square feet (279 m²). As permitted by the exception, draftstopping of the attic space is not required in any building equipped throughout with an automatic sprinkler system. See Figure 718-7.

718.5 Combustibles in concealed spaces in Type I or II construction. Where buildings are intended to be classified as noncombustible, it is intended that combustibles not be permitted, particularly in concealed spaces. The six exceptions to this limitation identify conditions under which a limited amount of combustible materials is acceptable. Exception 1 references Section 603, which identifies 27 applications where combustible materials are permitted in buildings of Type I or Type II construction. It is felt that the low level of combustibles permitted, as well as their control, does not adversely impact the fire-severity potential caused by the combustible materials. Exception 2 permits the use of combustible materials in plenums under the limitations and conditions imposed by IMC Section 602. The third exception allows the concealment of interior finish materials having a flame-spread index of Class A. Exception 4 addresses combustible piping, provided it is located within partitions or enclosed shafts in a complying manner. As an example, the presence of plastic pipe within the wall construction of a Type I or II building does not cause the building to be considered combustible construction. Exception 5 allows for the installation of combustible piping within concealed ceiling areas of Type I and II buildings, while Exception 6 permits combustible insulation on pipe and tubing in all concealed spaces other than plenums.

Section 719 *Fire-Resistance Requirements for Plaster*

Where gypsum plaster or Portland cement plaster is considered a portion of the required fire-resistance rating of an assembly, it must be in compliance with this section. Appropriate fire tests shall be referenced in determining the minimum required plaster thickness. It is important that the material under consideration is addressed in the test, unless the equivalency method of Section 719.2 is used.

In noncombustible buildings, it is necessary that all backing and support be of noncombustible materials. Except for solid plaster partitions or where otherwise determined by fire tests, it is also necessary in certain plaster applications to double the required reinforcement in order to provide for additional bonding, particularly under elevated temperatures. Under specific conditions, it is permissible to substitute plaster for concrete in determining the fire-resistance rating of the concrete element.

Section 720 *Thermal- and Sound-Insulating Materials*

The intent of this section is to establish code requirements for thermal and acoustical insulation located on or within building spaces. This section regulates all insulation except for foam-plastic insulation, which is regulated by Section 2603: duct insulation and insulation in plenums, which must comply with the requirements of the IMC; fiberboard insulation as regulated by Chapter 23; and reflective plastic core insulation, which must comply with Section 2614.

As a general requirement, insulation must have a flame spread index not in excess of 25 and a smoke-developed index not to exceed 450. Facings such as vapor retarders or vapor permeable membranes must be included when testing. Section 720.2.1 waives the flame-spread and smoke-developed limitations for facings on insulation installed in buildings of Type III, IV, and V construction, provided that the facing is installed behind and in substantial contact with the unexposed surface of the ceiling, floor, or wall finish.

Section 721 *Prescriptive Fire Resistance*

In this section, there are many prescriptive details for fire-resistance-rated construction, particularly those materials and assemblies listed in Table 721.1(1) for structural parts, Table 721.1(2) for walls and partitions, and Table 721.1(3) for floor and roof systems. For the most part, the listed items have been tested in accordance with the fire-resistance ratings indicated. In addition, a similar footnote to all of the tables allows the acceptance of generic assemblies that are listed in GA 600, the Gypsum Association's *Fire-Resistance Design Manual*. It is important to review all of the applicable footnotes when using a material or assembly from one of the tables.

Section 721.1.1 intends that the required thickness of insulating material used to provide fire resistance to a structural member cannot be less than the dimension established by Table 721.1(1), except for permitted modifications. An example of the minimum thickness of concrete required for a structural-steel column is shown in Figure 721-1. Note that Figure 704-4 illustrates that the edges of such members are to be adequately reinforced in compliance with the provisions of Section 721.1.3. Figure 721-2 illustrates the provisions of Section 704.6 and the minimum concrete-thickness requirements for protecting reinforcing

Prescriptive Fire Resistance 295

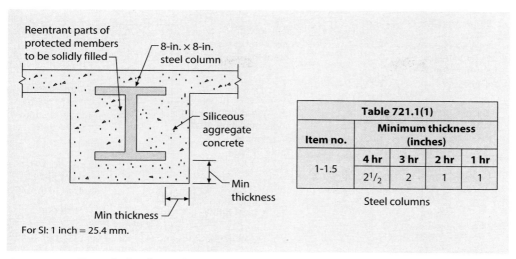

Figure 721-1 **Prescriptive fire resistance.**

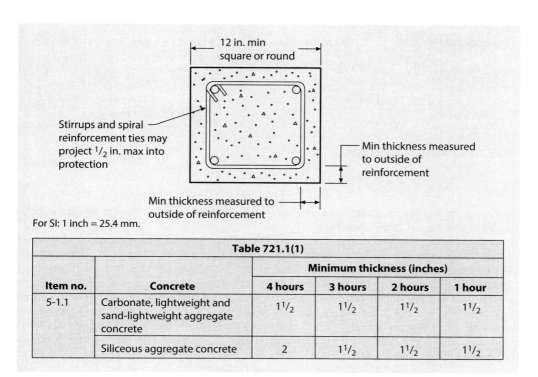

Figure 721-2 **Reinforcing steel in concrete columns, beams, girders, and trusses.**

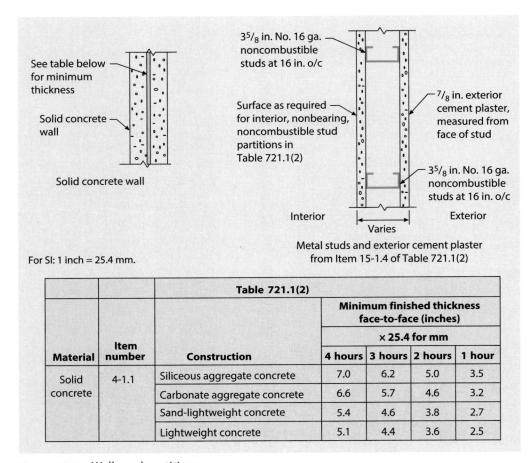

Figure 721-3 **Walls and partitions.**

steel in concrete columns, beams, girders, and trusses. Refer to Section 704 for additional provisions regarding structural members.

As previously mentioned, the fire-resistance ratings for the fire-resistance-rated walls and partitions outlined in Table 721.1(2) are based on actual tests. Figure 721-3 shows two samples from the table. For reinforced concrete walls, it is important to note the type of aggregate as discussed earlier in this chapter. The difference in aggregates is quite significant for a 4-hour fire-resistance-rated wall, as it amounts to a difference in thickness of almost 2 inches (51 mm). For hollow-unit masonry walls, the thickness required for a particular fire-endurance rating is the equivalent thickness as defined in Section 722.3.1 for concrete masonry and Section 722.4.1.1 for clay masonry. Figure 721-4 outlines the manner in which the equivalent thickness is determined.

Table 721.1(3) of the IBC provides fire-resistance ratings for floor/ceiling and roof/ceiling assemblies, and Figure 721-5 depicts the construction of a 1-hour fire-resistance-rated wood floor or roof assembly. Of special note is Footnote n, which exempts unusable space from the flooring and ceiling requirements. See Figure 711-1.

Prescriptive Fire Resistance

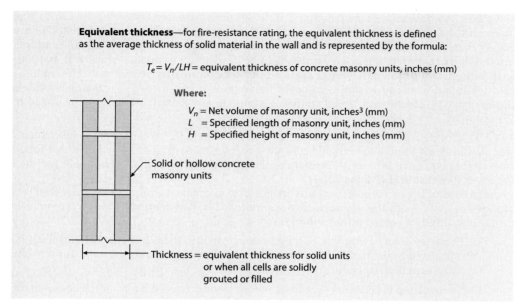

Figure 721-4 Equivalent thickness of clay brick and tile masonry walls.

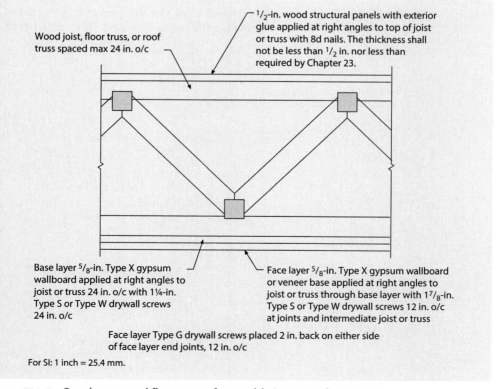

Figure 721-5 One-hour wood floor or roof assembly item number 21-1.1.

Often, materials such as insulation are added to fire-resistance-rated assemblies. It is the intent of the IBC to require substantiating fire test data to show that when the materials are added, they do not reduce the required fire-endurance time period. As an example, adding insulation to a floor/ceiling assembly may change its capacity to dissipate heat and, particularly for noncombustible assemblies, the fire-resistance rating may be changed. Although the primary intent of the provision is to cover those cases where thermal insulation is added, the language is intentionally broad so that it applies to any material that might be added to the assembly.

721.1.5 Bonded prestressed concrete tendons. Figure 721-6 depicts the requirements specified in Items 1 and 2 for variable concrete cover for tendons. It must be noted that for all cases of variable concrete cover, the average concrete cover for the tendons must not be less than the cover specified in IBC Table 721.1(1).

As prestressed concrete members are designed in accordance with their ultimate-moment capacity, as well as with their performance at service loads, Item 3 provides two sets of criteria for variable concrete cover for the multiple tendons:

1. Those tendons having less concrete cover than specified in Table 721.1(1) shall be considered to be furnishing only a reduced portion of the ultimate-moment capacity of the member, depending on the cross-sectional area of the member.

2. No reduction is necessary for those tendons having reduced cover for the design of the member at service loads.

As the ultimate-moment capacity of the member is critical to the behavior of the member under fire conditions, the code requires the reduction for those tendons having cover less than that specified by the code. However, behavior at service loads is less affected by the heat of a fire; therefore, the code permits those tendons with reduced cover to be assumed as fully effective.

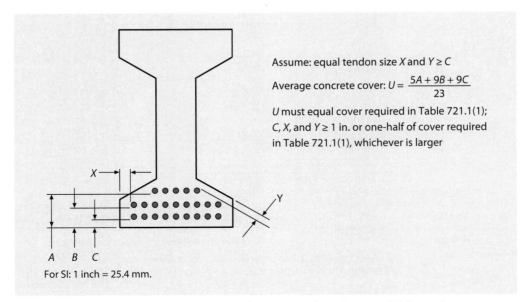

Assume: equal tendon size X and $Y \geq C$

Average concrete cover: $U = \dfrac{5A + 9B + 9C}{23}$

U must equal cover required in Table 721.1(1); C, X, and $Y \geq 1$ in. or one-half of cover required in Table 721.1(1), whichever is larger

For SI: 1 inch = 25.4 mm.

Figure 721-6 **Variable protection of bonded prestressed tendons, multiple tendons.**

Section 722 *Calculated Fire Resistance*

Fire research and the theory of heat transmission have combined to make it possible with the present state-of-the-art technology to calculate the fire endurance for certain materials and assemblies. As a result of this testing and research, this section permits the calculation of the fire-resistance rating for assemblies of structural steel, reinforced concrete, wood, concrete masonry, clay masonry, and mass timber.

While the calculated fire-resistance method is generally more conservative than a tested assembly, it can provide a quick and convenient way of demonstrating code compliance. Therefore, the code users should be aware of the useful information presented in this section, including:

Reference	Subject
Section 722.5.1.2	Attachment of gypsum wallboard around structural-steel columns
Section 722.2.1.3.1	Thickness of ceramic blanket joint material for precast concrete wall panels
Section 722.6.2	Wood wall, floor, and roof assemblies
Section 722.7.2	Gypsum board protection for mass timber

KEY POINTS

- Fire endurance is the basis for the fire-resistance requirements in the IBC.
- Materials and assemblies tested in accordance with ASTM E119 or UL 263 are considered to be in full compliance with the code for fire-resistance purposes.
- Elements required to be fire-resistance rated include structural frame members, walls and partitions, and floor/ceiling and roof/ceiling assemblies.
- The method for protecting fire-resistance-rated elements must be in full compliance with the desired listing.
- In many cases, fire-resistance protection for structural members must be applied directly to each individual structural member.
- Exterior walls of buildings located on the same lot are regulated by the placement of an assumed line between the two buildings.
- Where an exterior wall is located an acceptable fire separation distance from the lot line, the wall's fire-resistance rating is allowed to be determined based only on interior fire exposure.
- Opening protection presents a higher fire risk than fire-resistance-rated construction insofar as it does not need to meet the heat-transmission limits of ASTM E119 or UL 263.
- The maximum area of both protected and unprotected openings permitted in each story of an exterior wall is regulated by the fire separation distance.
- The code intends that each portion of a structure separated by a fire wall be considered a separate building for the purpose of height, area, and type of construction limitations.
- The objective of fire walls is that a complete burnout can occur on one side of the wall without any effects of the fire being felt on the opposite side.
- The purpose of a fire barrier is to totally isolate one portion of a floor from another through the use of fire-resistance-rated walls and opening protectives as well as fire-resistance-rated horizontal assemblies.
- Fire barriers are used as the separating elements for interior exit stairways, exit access stairways, exit passageways, horizontal exits, incidental use separations, occupancy separations, and other areas where a complete separation is required.
- Fire barriers must begin at the floor and extend uninterrupted to the floor or roof deck above.
- The potential for fire spread vertically through buildings mandates that openings through floors be protected with fire-resistance-rated shaft enclosures.
- Various modifications and exemptions for the enclosure of horizontal openings are found in the IBC.
- Fire partitions are used to separate dwelling units, sleeping units, tenant spaces in covered mall buildings, and fire-resistance-rated corridors from adjacent spaces.
- Fire partitions are permitted to extend to the membrane of a fire-resistance-rated floor/ceiling or roof/ceiling assembly.
- Smoke barriers are required in building areas where smoke transmission is a concern.

- The membrane of smoke barriers must be continuous from outside wall to outside wall or to another smoke barrier, and from floor slab to the floor or roof deck above, to eliminate all avenues for smoke to travel outside of the compartment created by the smoke barriers.
- Smoke partitions are intended to solely restrict the passage of smoke.
- Horizontal assemblies are required to have a fire-resistance rating where the type of construction mandates protected floor and roof assemblies, and where the floor assembly is used to separate occupancies or create separate fire areas.
- Penetration firestop systems are approved methods of protecting openings created through fire-resistance-rated walls and floors for piping and conduits.
- A limited level of protection is permitted for penetrations of noncombustible items.
- Both through-penetrations and membrane penetrations are regulated, typically in a similar fashion.
- Joints, such as the division of the building designed for movement during a seismic event, must often be protected if they occur in a fire-resistance-rated vertical or horizontal element.
- An opening protective refers to a fire door, or fire-protection-rated glazing, including the required frames, sills, anchorage, and hardware for its proper operation.
- Table 716.1(2) identifies the minimum fire-protection rating for a fire door assembly based on the type of wall assembly in which it is installed.
- In interior applications, fire-protection-rated glazing is limited to fire partitions and fire barriers having a maximum fire-resistance rating of 1 hour.
- In addition to fire dampers and smoke dampers, ceiling radiation dampers, combination fire/smoke dampers, and corridor dampers are referenced in the IBC.
- Fireblocking and draftstops are required in combustible construction to cut off concealed draft openings.
- Prescriptive methods for fire-resistance-rated construction are detailed for structural parts, walls and partitions, and floor and roof systems.
- The calculation of fire resistance is permitted for structural steel, reinforced concrete, wood, concrete masonry, clay masonry, and mass timber.

CHAPTER 8

INTERIOR FINISHES

Section 802 General
Section 803 Wall and Ceiling Finishes
Section 804 Interior Floor Finish
Section 805 Combustible Materials in
             Types I and II Construction
Key Points

Unfortunately, a number of building code provisions are enacted only after a disaster (usually with a large loss of life) indicates the need to regulate in a specific area. This is true of the interior wall and ceiling finish requirements of Chapter 8. In this case, the 1942 Cocoanut Grove nightclub fire in Boston, with a loss of almost 500 lives, provided the impetus to develop code requirements for the regulation of interior finish. Based on fire statistics, lack of proper control over interior finish (and the consequent rapid spread of fire) is second only to vertical spread of fire through openings in floors as a cause of loss of life during fire in buildings.

The dangers of unregulated interior finish are as follows:

The rapid spread of fire. Rapid spread of fire presents a threat to the occupants of a building by either limiting or denying their use of exitways within and out of the building. This limitation on the use of exits can be created by:

1. The rapid spread of the fire itself so that it blocks the use of exitways.
2. The production of large quantities of dense, black smoke (such as smoke created by certain plastic materials), which obscures the exit path and exit signs.

The contribution of additional fuel to the fire. Unregulated finish materials have the potential for adding fuel to the fire, thereby increasing its intensity and shortening the time available to the occupants to exit safely. However, because ASTM E84 and UL 723 do not require the determination of the amount of fuel contributed, the *International Building Code*® (IBC®) does not regulate interior finish materials on this basis.

Section 802 *General*

It is the intent of the IBC to regulate the interior finish materials on walls and ceilings, as well as coverings applied to the floor. In addition, limitations on the use of trim, decorative materials, and combustible lockers are found in Section 806, with the exception of foam plastics used as trim or finish material. These are addressed in Chapter 26. Combustible materials are permitted as finish materials in buildings of any type of construction, provided the wall, ceiling, or floor finishes are in compliance with this chapter.

As established in Section 803.2, it is not the intent of the IBC to regulate thin materials, such as wallpaper, that are less than 0.036 inch (0.9 mm) thick and are applied directly to a wall or ceiling surface. These thin materials behave essentially as the backing to which they are applied and, as a result, are not regulated. In some cases, however, repeated applications of wallpaper where the original materials are not removed can accumulate to a thickness of such magnitude that they must be regulated. The IBC, as stated in Section 803.3, also does not typically regulate the finish of exposed heavy-timber members complying with Sections 602.4 and 2304.11 insofar as this type of construction is not subject to rapid flame spread. However, such elements are subject to the interior finish requirements where they are in an exposed condition in interior exit stairways, interior exit ramps, and exit passageways.

Section 803 *Wall and Ceiling Finishes*

803.1.1 Interior wall and ceiling finish materials tested in accordance with NFPA 286. All interior wall and ceiling finish materials are permitted to be tested to NFPA 286 *Standard Methods of Fire Tests for Evaluating Contribution of Wall and Ceiling Interior Finish to Room Fire Growth*. Any finish material which complies with the acceptance criteria for NFPA 286 testing is considered as having a Class A flame spread rating and is not required to be tested in accordance with ASTM E84 or UL 723. Since NFPA 286 does not classify materials into flame spread categories, the acceptance criteria for a successful test under NFPA 286 follow as Section 803.1.1.1.

803.1.2 Interior wall and ceiling finish materials tested in accordance with ASTM E84 or UL 723. Section 803.1.2 addresses testing for the determination of flame spread and smoke development characteristics under ASTM E84 and UL 723. This test is commonly known as the Steiner tunnel test. Based on the results of these tests, interior wall and ceiling materials are classified. These classifications are divided into three groups, Class A, Class B, and Class C.

There is a variety of materials and conditions with other characteristics that cannot simply be tested to NFPA 286, ASTM E84, or UL 723; therefore, additional testing criteria are needed. These items are listed in Sections 803.2 through 803.12, where additional testing criteria are provided.

803.5 Textile wall coverings. This section regulates carpet—as well as other textiles that are napped, tufted, looped, woven, or nonwoven—where applied as wall finish materials. Textile wall coverings present a unique hazard because of the potential for extremely rapid fire spread. The code provides three options for the acceptance of textiles used as interior wall finish. One method of testing includes the surface burning characteristics test of ASTM E84 or UL 723. The textile must have a flame-spread index of Class A and be protected by automatic sprinklers. A second option is based on the room corner test for textiles as established in NFPA 265, where the testing must be done in accordance with the Method B protocol. It is important that the testing be done in the same manner as the intended use of the textile materials. A third approach is the use of the ceiling and wall finish room corner test as set forth in NFPA 286. This test is also based on the intended application of the textile material and must include the product-mounting system. Where textiles are intended to be applied as ceiling finish materials, only the methods using ASTM E84, UL 723, and NFPA 286 are to be used.

Where these types of materials are intended to be applied as ceiling coverings, the limitations on their use are similar to those for wall coverings. Although compliance with Section 803.1.1 or 803.5.2 allow for their use, the provisions of Section 803.5.1 utilizing the NFPA 265 test are not to be applied.

803.13 Interior finish requirements based on occupancy. Table 803.13 is divided based on the presence, or lack of, an automatic sprinkler system. Note that an extensive number of notes modify the general provisions of Table 803.13.

As a general rule, interior exit stairways, interior exit ramps, and exit passageways are regulated at the highest level because of their importance as exit components in the means of egress. These types of exits permit unlimited travel distance and are typically single-directional. Corridors and exit access stairways are also highly regulated, but not to the extent of the higher level exit components. Interior finish requirements for rooms and other enclosed areas are not as restrictive as for exitways; however, the wall and ceiling finishes are still regulated to some degree.

When it comes to occupancy groups, the high-hazard, institutional, and assembly occupancies typically have the most restrictive flame-spread classifications. On the other hand, utility occupancies have no restrictions. Using a nonsprinklered office building as an example, the maximum flame spread classification of finish materials, based on occupancy group and location within the building, is shown in Figure 803-1.

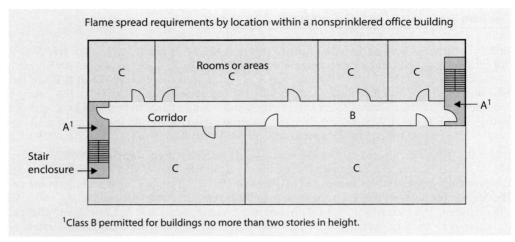

Figure 803-1 **Flame-spread requirements by location within a nonsprinklered office building.**

Based on the cooling that an automatic sprinkler system provides, the code permits a reduction of one classification for many of the occupancies and locations. However, this is not a standard reduction. The table must be referenced to identify those reductions available for sprinklered applications. There is also no allowance for reducing a Class C requirement to a lower classification based on sprinkler protection. It should be noted that the required sprinkler system need only be provided in those exitways or rooms where the classification reduction is taken, and not throughout the entire building.

Figure 803-2 provides flame-spread classifications of woods commonly used in construction and finish work. A glance at the chart will show that most species of wood qualify for a Class C rating. There are few wood species that warrant a Class B rating, and no species is shown that qualifies for a Class A rating. However, there are many paints and varnishes in the market that manufacturers refer to as fire-retardant coatings. Because of intumescence, these paints or coatings bubble up or swell up under the action of flame and heat to provide an insulating coating on the surface of the material treated. Certain intumescent paints and varnishes can reduce the flame spread of combustible finishes to as low as Class A and the smoke density to considerably below 450. These flame-spread-reducing intumescent coatings are particularly useful when correcting an existing nonconforming combustible interior finish.

803.14 Stability. The IBC requires that the method of fastening the finished materials to the interior surfaces be capable of holding the material in place for 30 minutes under a room temperature of 200°F (93°C). If there is any question as to the adequacy of the fastening, appropriate tests should be required to determine compliance with this provision of the code.

803.15 Application of interior finish materials to fire-resistance-rated or noncombustible building elements. This section is applicable only where finish materials are applied on walls, ceilings, or structural elements required to have a fire-resistance rating, or where such building elements must be of noncombustible construction (typically Types I or II construction). The greatest concern is where interior finish is not applied directly to a backing surface, creating concealed spaces that provide the opportunity for fire to originate and spread without detection until the interior finish material has burned through. The installation of furring strips is permitted, provided they are installed directly against the surface of the wall, ceiling, or structural member. In addition, the concealed space created by the furring strips must be

Species of wood	Flame spread	Source*
Birch, yellow	105–110	UL
Cedar, eastern red	110	HUD/FHA
Cedar, Pacific Coast yellow	78	CWC
Cedar, western red	70	HPMA
Cottonwood	115	UL
Cypress	145–150	UL
Fir, Douglas	70–100	UL
Gum, red	140–155	UL
Hemlock, West Coast	60–75	CWC
Lodgepole	93	CWC
Maple flooring	104	CWC
Oak, red or white	100	UL
Pine, eastern white	85	UL
Pine, Idaho white	72	CWC
Pine, northern white	120–215	HPMA
Pine, ponderosa	105–200	UL
Pine, red	142	HUD/FHA
Pine, southern yellow	130–190	CWC
Pine, western white	75	HUD/FHA
Poplar	170–185	UL
Redwood	70	UL
Spruce, northern	65	UL
Spruce, western	100	UL
Spruce, white	65	UL
Walnut	130–140	CWC
Plywoods		UL
Douglas fir, ¼-inch	120	HUD/FHA
Lauan, three-ply urea glue, ¼-inch	110	HUD/FHA
Particleboard, ½-inch	135	HPMA
Redwood, ⅜-inch	95	CRA
Redwood, ⅝-inch	75	CRA
Walnut, ¾-inch	130	HUD/FHA

*Source:
CRA: California Redwood Association
  Data Sheet-2D2-7L (Lumber)
  Data Sheet-2D2-7P (Plywood)
CWC: Canadian Wood Council
  Data File FP-6
HPMA: Hardwood Plywood Manufactures Association
  Test No. 337, Test No. 592, Test No. 596
HUD/FHA: Flame-spread Rating for Various Material
UL: Underwriters Laboratories
  UL 527, May 1971

Figure 803-2 **Flame-spread classification of woods.**

either fireblocked at maximum 8-foot (2,438-mm) intervals, or filled completely with a Class A, noncombustible, or organic material. The maximum depth of the concealed space is limited to 1¾ inches (44 mm). This section is also referenced by Section 803.15.3 for fireblocking in heavy-timber construction.

Where interior finish materials are set out or suspended more than the 1¾ inches (44 mm) specified in Section 803.15.1, the potential exists for the fire to gain access to the space through joints or imperfections and to spread along the back surface as well. In this case, the flame spreads at a much faster rate than on one surface, as the flame front will be able to feed on the material from two sides. Therefore, the provisions of Section 803.15.2 are intended to protect against this type of hazard. In the case where a wall is set out, the wall, including the portion that is set out, is required by the code to be of fire-resistance-rated construction as would be required by the code for the occupancy and type of construction.

It should be noted again that the provisions of Sections 803.15.1 and 803.15.2 are applicable only where the walls and ceiling assemblies are required to be of either fire-resistance-rated or noncombustible construction. Where the walls and ceiling assemblies are of unprotected combustible construction, only the fireblocking provisions of Section 718.2 are applicable.

Section 803.15.4 requires that thin materials—no more than ¼ inch (6.4 mm) thick—other than noncombustible materials, be applied directly against a noncombustible backing unless they are qualified by tests where the material is furred out or suspended from the noncombustible backing. The reason for this requirement is similar to that in Section 803.15.2. There are some buildings where thin paneling, such as luan plywood, is installed on walls and ceilings. When not installed against a noncombustible backing, these materials readily burn through and permit an almost uncontrolled rapid flame spread because the flame proceeds on both surfaces of the material.

Section 804 *Interior Floor Finish*

Floor finishes such as wood, terrazzo, marble, vinyl, and linoleum present little, if any, hazard that is due to the spread of fire along the floor surface. However, other flooring materials such as carpeting are highly regulated by the IBC because of their potential for helping to increase the growth of a fire.

804.2 Classification. For the purpose of regulating floor finishes based on the occupancy designation of the area where the finishes are installed, the code identifies two classes: Class I and Class II. Determined by test standard NFPA 253 or ASTM E648, the classifications are based on the critical radiant flux. The critical radiant flux is determined as that point where the heat flux level will no longer support the spread of fire. Class I is considered to have a critical radiant flux of 0.45 watts per square centimeter or greater, where Class II need only exceed 0.22 watts per square centimeter. Therefore, the Class I material is more resistant to flame spread because of the higher heat-flux-level characteristics.

804.4 Interior floor finish requirements. Interior floor finishes are regulated differently by the code based on two factors: the occupancies in the building and where the finish materials are located in relationship to the means of egress. The IBC selectively requires that interior exit stairways and ramps, exit passageways, and corridors be provided with a floor finish exceeding the critical radiant flux level established by the "pill test" used in DOC (U.S. Department of Commerce) FF-1 and test standard ASTM D2859. In addition, the floor finish materials of all rooms or spaces unseparated from a corridor by full-height partitions are regulated in a like manner, as there is evidence that corridor floor coverings can propagate flame when exposed to a fully developed fire in a room that opens into a corridor. Those rooms that have no direct connection with a corridor are simply regulated for the DOC FF-1 criteria, as are rooms that are separated from a corridor by full-height partitions.

The DOC FF-1 and ASTM D2859 "pill tests" require a minimum radiant flux of 0.04 watts per square centimeter and are used to regulate all carpeting sold in the United States. Fire tests have demonstrated that carpet on the floor that passes the pill test is not likely to become involved in a room fire until the fire has reached or approached flashover. Only those finish materials having a Class I or II classification may be installed in corridors, exit passageways, and exit enclosures of fully sprinklered Group I-1, I-2, and I-3 occupancies. The same limitation holds true for floors of exitways in areas of nonsprinklered buildings housing Group A, B, E, M, or S occupancies. In other occupancies, the interior floor finish in the listed exitways need only comply with DOC FF-1 or ASTM D2859 listing. The commentary above assumes the reduction in the classifications of floor finishes that is permitted where the building is fully sprinklered. Class II floor finish materials are permitted in lieu of Class I materials, whereas materials complying with the DOC FF-1 or ASTM D2859 "pill test" may be used instead of Class II materials. The entire building, and not just the area where the floor finish is located, must be provided with an automatic sprinkler system.

It is important to note that all fibrous floor coverings in those occupancies and locations not required by the IBC to have a Class I or II classification must comply with the "pill test" requirements of DOC FF-1 or ASTM D2859.

Section 805 *Combustible Materials in Types I and II Construction*

Where combustible flooring materials are installed in or on floors in noncombustible buildings, they are regulated by this section. Combustible sleepers may only be installed where the space between the fire-resistance-rated floor deck and the sleepers is completely filled with approved noncombustible materials or is fireblocked in the manner described in Section 718. Finish flooring of wood shall be attached to sleepers, which, if not imbedded, shall be appropriately fireblocked. As long as the wood flooring is in direct contact with a fire-resistance-rated floor, there is no significant hazard. However, if there is a space between the wood flooring and the fire-resistance-rated floor, a concealed space is created that is enclosed with combustible materials and provides a route for the undetected spread of fire. Therefore, the code requires that where wood flooring is not in direct contact with the fire-resistance-rated floor, the space be filled with noncombustible material or be fireblocked. Based on the controls placed on wood sleepers and finish flooring, it is also reasonable that combustible insulating boards be permitted where installed in a similar manner.

KEY POINTS

- Regulation of finish materials by the IBC includes those on walls and ceilings, as well as floor coverings.
- Unregulated interior finish materials contribute to the rapid spread of fire, presenting a threat to the occupants by limiting or denying their use of exitways.
- Interior exit stairways and exit passageways are the most highly regulated building elements for the application of interior finish materials, with corridors and exit access stairways being moderately controlled, and rooms or areas being the least-regulated portions of the building.
- Installation of an automatic sprinkler system often allows a one-class reduction in the requirement for flame-spread classification.
- Textile wall and ceiling coverings are more highly regulated than other finish materials because of the potential for extremely rapid fire spread.
- Carpeting and similar floor covering materials are highly regulated in specific locations of specific occupancies.
- In all occupancies, fibrous floor coverings must, at a minimum, comply with the requirements of the DOC FF-1 "pill test" or with ASTM D2859.
- In certain occupancies, the code regulates curtains, hangings, and other decorative materials as to their flame resistance and limits their use.

CHAPTER 9

FIRE PROTECTION AND LIFE-SAFETY SYSTEMS

Section 901 General
Section 902 Fire Pump and Riser Room Size
Section 903 Automatic Sprinkler Systems
Section 904 Alternative Automatic Fire-Extinguishing Systems
Section 905 Standpipe Systems
Section 907 Fire Alarm and Detection Systems
Section 909 Smoke Control Systems
Section 910 Smoke and Heat Removal
Section 911 Fire Command Center
Section 912 Fire Department Connections
Section 913 Fire Pumps
Section 914 Emergency Responder Safety Features
Section 915 Carbon Monoxide Detection
Section 916 Gas Detection Systems
Section 917 Mass Notification Systems
Key Points

Chapter 9 provides requirements for three distinct fire protection systems considered vital for the creation of a safe building environment. The first of these systems is intended to control and limit fires, protect building occupants, and assist fire fighters with fire-fighting operations. Included are fire-extinguishing systems and standpipe systems. The detection and notification of an emergency condition are addressed by the second system. Manual fire alarms, automatic fire detection, emergency alarm systems, and mass notification are included in this group. The third system is intended to control smoke migration. Included are design and installation standards for smoke control systems, as well as smoke and heat vents and mechanical smoke removal systems. In addition to the provisions for fire protection systems, criteria are provided to increase the efficiency and safety of fire department personnel during emergency operations. Topics addressed include emergency responder safety features, fire command centers, fire department connections, and fire pump rooms.

Section 901 *General*

It is the intent of Chapter 9 to require fire protection systems in those buildings and with those uses that through experience have been shown to present hazards requiring the additional protection provided by fire protection systems. The installation, repair, operation, and maintenance of such systems are based on the provisions of the IBC® and the *International Fire Code®* (IFC®). Furthermore, it is the intent of the code to prescribe standards for those systems that are required. However, there are times when the installation of a fire protection system is not based on a code mandate. Such fire protection and life safety systems may be installed as a requirement of the insurance carrier, a choice of the designer, or a desire of the owner. In such situations, the nonrequired system must still meet the provisions of the code. Once fire protection and life safety systems are provided to some degree, it is expected that the system is properly installed and functional.

An exception to this section permits a system or any portion of that system that is not required by the code to be installed for partial or complete protection, provided that the installation meets the code requirements. As an example, fire sprinkler protection may be provided only in a specific area of a building, based on a request by the owner rather than on a requirement of the code. Although the sprinkler system must be installed in accordance with the proper design standard (in most cases NFPA 13), it is not necessary that the sprinkler system extends into other areas of the building.

More than likely, however, a fire protection system is used to take advantage of modifications to other code requirements. Under these conditions, the fire protection system is considered a required system and is subject to all of the requirements imposed by the IBC and IFC.

901.7 Fire areas. The fire area concept is based on a time-tested approach to limiting the spread of fire in a building. Through the use of fire-resistive elements, compartments can be created that are intended to contain a fire for a prescribed period of time. The floor area that occurs within each such compartment is considered to be the fire area. By definition, a fire area is the aggregate floor area enclosed and bounded by fire walls, fire barriers, exterior walls, and/or horizontal assemblies of a building. See Figure 901-1. In addition, any areas beyond the exterior wall that are covered with a floor or roof above, such as a canopy extending from the building, are considered part of the building for fire area purposes. This approach is consistent with the determination of building area in Chapter 2. An example is shown in Application Example 901-1.

General **313**

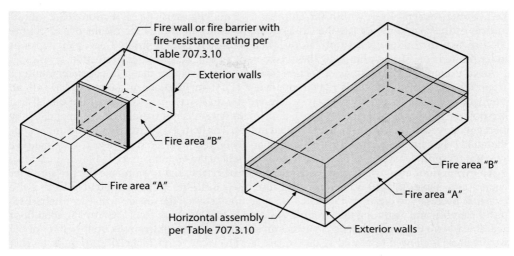

Figure 901-1 **Fire areas.**

Through the isolation of a fire condition to a single fire area through the use of fire separation elements having significant fire-resistance ratings, only that portion of the building within the fire area is considered as risk due to a single fire incident.

The use of fire areas as a fire protection tool is limited almost exclusively to the requirements for automatic sprinkler systems. Other fire protection systems, such as fire alarm systems, are for the most part regulated by methods, such as occupancy classification, that are not based on

GIVEN: A Group M retail sales building that includes a roofed exterior sales area of 3,000 sq. ft.
DETERMINE: If a sprinkler system is required.

SOLUTION: As a single Group M fire area of 14,000 sq. ft a fire sprinkler system is required per Section 903.2.7, Item 1. However, no sprinkler system is required per Section 707.3.10 if a complying 2-hour fire barrier separates the interior and exterior sales areas. Then the largest fire area would be less than 12,000 sq. ft, and the aggregate size of the fire areas would be less than 24,000 sq. ft.

For SI: 1 square foot = 0.093 m^2.

FIRE AREA APPLICATION

Application Example 901-1

the fire area concept. Even within the automatic sprinkler provisions of Section 903.2, only a portion of the requirements use the fire area approach as an alternative means of protection. The fire area methodology set forth in the IBC, applicable only in limited occupancy groups under limited conditions, allows for the omission of automatic sprinkler protection.

As an example, the provisions of Section 903.2.3, Item 1, require that a fire area containing a Group E occupancy that exceeds 12,000 square feet (1,115 m^2) in floor area be provided with an automatic sprinkler system. Conversely, where the fire area size does not exceed the established threshold of 12,000 square feet (1,115 m^2), a sprinkler system is not required unless mandated by another code provision, such as that established in Item 2 or Item 3. Where the building under consideration is limited to a maximum of 12,000 square feet (1,115 m^2), it can be viewed as a single fire area, and no sprinkler system is mandated. However, where the building exceeds 12,000 square feet (1,115 m^2) in floor area, two or more fire areas must be established to eliminate the fire sprinkler requirement. Table 707.3.10 is referenced because it sets forth the minimum required level of fire resistance necessary to create an adequate fire separation between the fire areas that are established. In the example, and assuming the Group E building is 20,000 square feet (1,858 m^2) in total floor area, at least two fire areas must be created as an alternative to sprinkler protection. Neither of the two fire areas is allowed to exceed 12,000 square feet (1,115 m^2), and Table 707.3.10 indicates that the minimum fire separation between the two fire areas must be 2 hours. Therefore, a minimum 2-hour fire-resistance-rated fire wall, fire barrier, or horizontal assembly, or a combination of these elements, would be required. It is important to note that, regardless of the floor area of the Group E fire area, sprinkler protection would still be required if the occupant load of the fire area is 300 or more (Section 903.2.3, #3) or if the fire area is not located solely on the level of exit discharge (Section 903.2.3, #2).

A similar approach is taken in a mixed-occupancy building where the multiple fire areas are of different occupancy classifications. The minimum required fire-resistance rating for the separation between the fire areas would also be based on the requirements of Table 707.3.10. For example, where a building contains a 10,000-square-foot (929-m^2) Group M occupancy and a 6,000-square-foot (558-m^2) Group F-1 occupancy, the minimum fire-resistive separation between the Group M fire area and the Group F-1 fire area would be 3 hours. Although the Group M requirement in Table 707.3.10 only mandates a 2-hour separation, a minimum 3-hour fire separation is required for a Group F-1 occupancy. For further information, see Application Example 901-2 and the discussion of Table 707.3.10.

Section 902 *Fire Pump and Riser Room Size*

The IBC does not require the construction of a fire pump room or fire sprinkler riser room. But where such a room is constructed, specific criteria for its design and construction are provided. The intent of the code requirements is to provide protection for the equipment within these rooms since they are critical to the fire- and life-safety systems in the building. This section would be used in conjunction with Section 913 regarding fire pumps. See the discussion of Section 913 for further information.

902.1 Pump and riser room size.
The equipment in the fire pump room or fire sprinkler riser room must be accessible for maintenance, as well as its use and operation during a fire event. This section requires that adequate working clearance be provided around all stationary equipment. This would include items such as the fire pump and the control panel for the fire pump, the jockey pump and control panel for the jockey pump, the control valves on the sprinkler system risers, and any monitoring or supervision equipment. The area must also be large enough to allow for removal and replacement of the largest component in the room. This provision applies to doorways and hallways that provide the path to the exterior.

Automatic Sprinkler Systems

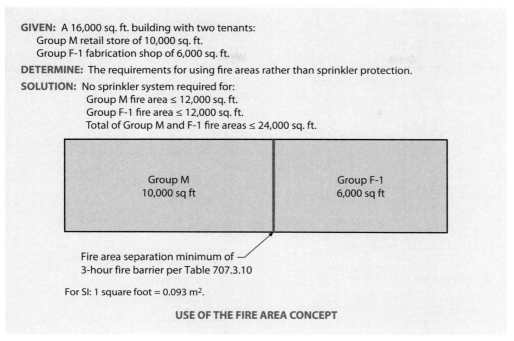

Application Example 901-2

The fire pump room or riser room must be provided with permanent lighting. Permanently installed environmental controls are required to maintain the air temperature above 40°F (4°C). Even though water does not freeze until the temperature is below 32°F (0°C), ice crystals begin to form and the water becomes like a slush as the temperature nears freezing. Maintaining the room temperature at or above 40°F (4°C) provides for some safety margin to protect the water in the fire pump, sprinkler riser, and associated piping.

Section 903 *Automatic Sprinkler Systems*

In general, automatic sprinkler systems are required when certain special features and hazards of specific buildings, areas, and occupancies are such that the additional protection provided by sprinkler systems is warranted.

In addition, there occasionally are inadequate numbers and sizes of openings in the exterior walls from which a fire may be fought from the exterior of the building. The provisions requiring sprinklers in these so-called *windowless buildings* apply to all buildings, regardless of occupancy, except for Group U occupancies.

There are four general situations in which fire sprinkler systems are to be provided within a building. An automatic sprinkler system may be required throughout the building, throughout a fire area, throughout the occupancy, or only in the specific room or space where sprinkler protection is necessary. Examples are depicted in Figure 903-1.

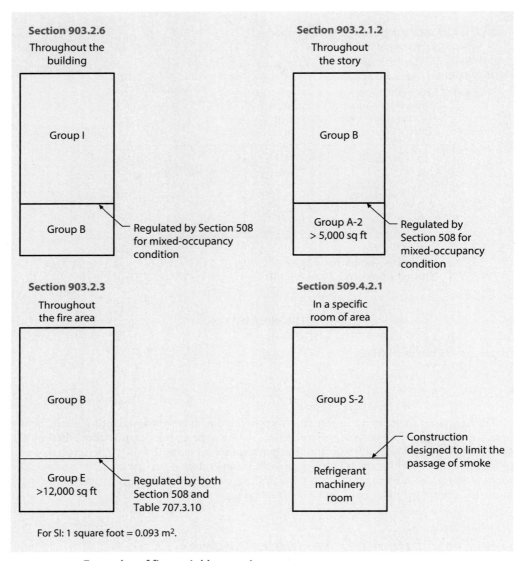

Figure 903-1 **Examples of fire sprinkler requirements.**

1. **Throughout the building.** There are numerous applications of the code that require the entire building to be sprinklered, either mandated because of a code requirement or used as a substitute for other fire- and life-safety features. Examples include the requirement for sprinklers throughout all buildings containing a Group I fire area per Section 903.2.6 and the elimination of corridor fire protection in some occupancies based on Table 1020.2, Footnote c. The extent of required sprinkler protection in Group A occupancies is described in a different manner, but quite often results in a fully sprinklered condition. Initially, the entire story containing a Group A occupancy must

be provided with sprinkler protection where the assembly occupancy exceeds the limits established by Sections 903.2.1.1 through 903.2.1.4. However, as mandated in each of those sections, the fire sprinkler system must also be provided on all stories between the Group A occupancy and the level of exit discharge, including the level of exit discharge. This will commonly result in a requirement that the entire building be provided with an automatic sprinkler system. Additional commentary is provided in the discussion of Section 903.2.1.

2. **Throughout a fire area.** In Section 903.2, a variety of provisions require only those fire areas that exceed a certain size or occupant load, or are located in a specific portion of the structure to be sprinklered. The sprinkler requirements based on fire area include the provisions of Section 903.2.3 for Group E occupancies and Section 903.2.4.1 for Group F-1 woodworking operations.

3. **Throughout an occupancy.** A third variation of sprinkler mandates occurs in those mixed-occupancy buildings containing a Group H-2, H-3, or H-4 occupancy. The code only mandates that the sprinkler system be provided in the Group H portion, not the entire building. Because all other occupancies in a mixed-occupancy building must be appropriately isolated from the Group H occupancy because of the *separated occupancies* provisions of Section 508.4, the result is basically a requirement to sprinkler the Group H compartment.

4. **Specific rooms or areas.** Occasionally, only a specific portion of the building requires the protection provided by a sprinkler system. The sprinkler addresses the particular hazard that occurs at possibly only a single location. For example, the allowance for a reduction in the flame-spread classification of interior finishes from Table 803.13 is based on sprinkler protection in the room, area, or exitway where the finish under consideration is installed. In addition, Table 509.1 addresses incidental uses within a building and for most of the hazards provides an option on mitigating the hazard. For example, a refrigerant machinery room must be protected with either a fire sprinkler system, or it can be separated from the remainder of the building by 1-hour fire-resistance-rated construction. Where the sprinkler option is applied, only the refrigerant machinery room is required to be sprinklered to comply with Table 509.1. Additional commentary is provided in the discussion of Section 509.1.

903.1.1 Alternative protection. Where an automatic sprinkler system is addressed in the IBC, alternative automatic fire-extinguishing systems are acceptable, provided they are installed in accordance with approved standards. These systems, regulated by Section 904, include fire-extinguishing systems required by the IFC and systems installed as a design alternative. This is one of the few provisions in the IBC where approval must come from someone other than the building official. Although the building official is almost always charged with making any decisions regarding the building code, the fire code official is typically better able to evaluate and determine the appropriateness of an alternative fire-extinguishing system. It is important to note that where an automatic sprinkler system is recognized by the code for the purpose of an exception or reduction to a requirement, the use of an alternative fire-extinguishing system will not provide such a benefit. See Section 904.2.1.

903.2 Where required. It is the intent of this section to specify those occupancies and locations where automatic sprinkler systems are required. A fire-extinguishing system is a system that discharges an approved fire-extinguishing agent such as water, dry chemical powder, aqueous foam, or carbon dioxide onto, or in, the area of a fire. A fire sprinkler system is a specialized fire-extinguishing system that discharges water. The code specifies a fire sprinkler system in this section, as it is the intent of the code that water be applied and not one of the other extinguishing agents. Generally, water is the most effective extinguishing agent for fires. Only where water creates problems, such as in magnesium or calcium carbide storage areas, would some other type of extinguishing agent be required. The allowance for the installation of a system other than an automatic sprinkler system is subject to approval by the fire code official. It is important to note that Section 904.2.1 indicates that where an automatic fire-extinguishing system is installed as an alternative to automatic sprinkler system protection, it is not permitted to be used for the purposes of exceptions or reductions allowed by other code provisions. Alternative automatic fire-extinguishing systems are addressed in the discussion of Section 904.

Fire areas. Most of the requirements of this section are based on the concept of fire areas. Where a fire area exceeds a specified size, is located in a certain portion of the building, or exceeds a specified occupant load, the code often requires the installation of an automatic sprinkler system to address the increased hazards and concerns that exist. The provisions for fire areas can be found in various sections of the IBC. The definition of "fire area" is located in Chapter 2. A fire area is "the aggregate floor area enclosed and bounded by fire walls, fire barriers, exterior walls or fire-resistance-rated horizontal assemblies of a building." Complete isolation and separation of a portion of a building from all other building areas is created through the use of fire-resistance-rated construction and opening protectives. The total floor area within the enclosed area, including the floor area of any mezzanines or basements, is considered the size of the fire area. See Application Example 903-1. It is also important to note that "areas of the building not provided with surrounding walls shall be included in the fire area if such areas are included within the horizontal projections of the roof or floor next above."

In addition to fire walls, fire barriers and fire-resistance-rated horizontal assemblies may also be used to create fire areas, provided the fire-resistance-rated construction totally separates one interior area from another. In order to determine the minimum fire-resistance rating of the

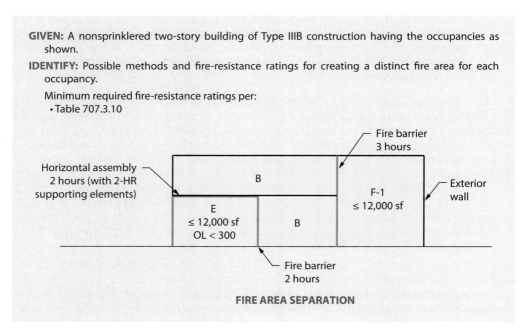

Application Example 903-1

vertical and horizontal elements, the occupancy classifications of the areas being separated must be identified. Table 707.3.10 is then referenced to determine the minimum fire-resistance rating of the separation. The use of this table is applicable to both single-occupancy and mixed-occupancy buildings, as illustrated in Application Example 903-2. Where the fire area separation occurs between two fire areas of the same occupancy, the hourly rating established by Table 707.3.10 for that single occupancy classification is applied. If the fire areas contain different occupancy classifications, the controlling fire-resistance rating of the fire barrier or horizontal assembly separating the occupancies is based on the higher of the ratings as established by Table 707.3.10 for the occupancies involved. For further information, see the discussions of Section 901.7 and Table 707.3.10.

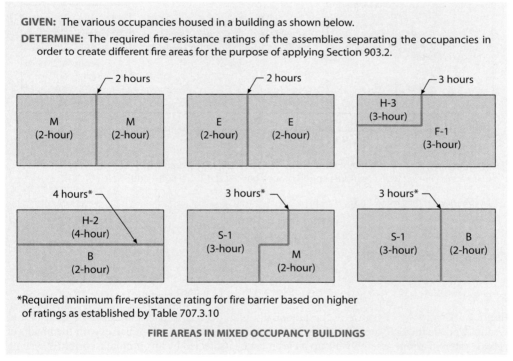

Application Example 903-2

Because the majority of sprinkler provisions are based on the size of the fire area, it is sometimes possible for the designer to eliminate the requirement for sprinklers by reducing the floor area within the surrounding fire-resistance-rated construction. The use of fire walls or fire barriers and fire-resistance-rated horizontal assemblies can subdivide a structure into smaller, less hazardous areas that are of such a size that sprinklers are not necessary. See Application Example 903-3. This concept of compartmentation has been used in building codes for decades as an effective method of reducing the loss of life and property in fires.

The exception to Section 903.2 eliminates the sprinkler requirement in rooms or areas in telecommunications facilities dedicated solely for essential telecommunications and power equipment. Alternative protection is provided through the required installation of an automatic smoke detection system, as well as appropriately rated separation from other areas of the building.

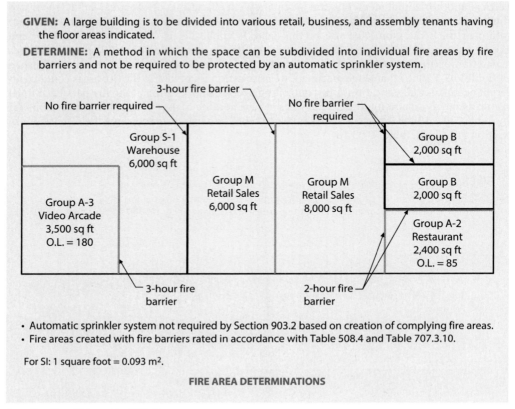

Application Example 903-3

903.2.1 Group A.
Because of the potentially high occupant load and density anticipated in Group A occupancies, coupled with the occupants' probable lack of familiarity with the means of egress system, various assembly occupancies must be protected by an automatic sprinkler system. Where an automatic sprinkler system is required for a Group A-1, A-2, A-3, or A-4 occupancy, the system must be installed throughout the entire story where the Group A occupancy is located. In addition, where the Group A-1, A-2, A-3, or A-4 occupancy is located on a story other than the level of exit discharge, that story and all stories between the Group A occupancy, to and including the level(s) of exit discharge must be sprinklered as well. By expanding the areas of the building required to be protected by an automatic sprinkler system beyond just the assembly areas, the code provides protection adjacent to the Group A areas as well as throughout the means of egress. Figures 903-2 and 903-3 illustrate these fundamental provisions.

It is important to note that where multiple fire areas containing Group A-1, A-2, A-3, or A-4 occupancies exist, additional provisions apply as addressed in the discussion of Section 903.2.1.7.

903.2.1.1 Group A-1.
The combination of highly concentrated occupant loads, high numbers of occupants, reduced lighting levels, and potentially high fuel loads create a level of hazard that justifies the need for sprinkler protection. Therefore, fire areas containing

Automatic Sprinkler Systems

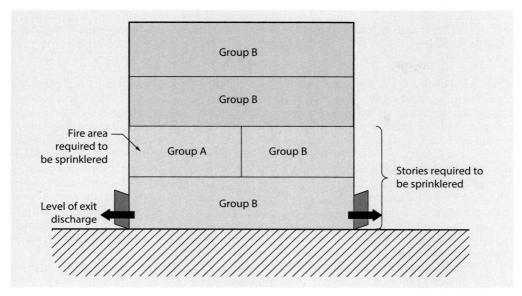

Figure 903-2 Multi-story Group A sprinkler requirements.

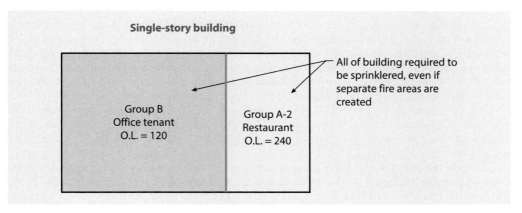

Figure 903-3 Example of single story Group A sprinkler requirements.

theaters and similar assembly uses intended for the viewing of motion pictures or the performing arts shall be provided with an automatic sprinkler system where any one of the following conditions exists:

1. The fire area containing the Group A-1 occupancy exceeds 12,000 square feet (1,115 m²).
2. The occupant load of the fire area exceeds 299.
3. The fire area is located on any floor level other than that of the exit discharge.

It should also be noted that any fire area containing a multitheater complex, defined as two or more theaters served by a common lobby, shall be provided with a sprinkler system throughout the story.

As addressed in the discussion of Section 903.2.1, the sprinkler protection required by Section 903.2.1.1 must extend beyond the Group A fire area to include the entire story of the building, as well as any stories from the Group A occupancy to and including the level(s) of exit discharge.

903.2.1.2 Group A-2. Fire areas housing uses intended for food or drink consumption are regulated for sprinkler protection at a higher level than other enclosed assembly occupancies. Even where the occupant load is not excessive, the hazards associated with such uses warrant the protection provided by an automatic sprinkler system. Oftentimes, the consumption of alcohol beverages by the building's occupants creates an environment more likely to be unsafe. The reduced lighting levels in some uses, along with the probability of loose chairs and tables, also increase the risk for obstructed egress. The record of casualties during fires in buildings housing nightclubs, casino gaming areas, restaurants, and similar types of uses demonstrates the need for the additional protection provided by fire sprinklers or, alternatively, the separation of the use into smaller compartments. The code intends that fire areas exceeding 5,000 square feet (464 m^2) that contain Group A-2 uses be provided with an automatic sprinkler system, as well as such uses having an aggregate occupant load within the fire area of 100 or more, or where the fire area is located on a floor level other than the level of exit discharge.

As addressed in the discussion of Section 903.2.1, the sprinkler protection required by Section 903.2.1.2 must extend beyond the Group A fire area to include the entire story of the building, as well as any stories from the Group A occupancy to and including the level(s) of exit discharge.

903.2.1.3 Group A-3. The sprinkler threshold for a Group A-3 occupancy is identical to that for a Group A-1 occupancy. As such, where any fire area in a Group A-3 occupancy exceeds 12,000 square feet (1,115 m^2), where the fire area has an occupant load greater than 299, or where the assembly occupancy is located on any story other than the exit discharge level, an automatic sprinkler system is required. In applying the provisions of this section, it is important to note that the occupant load threshold is based on the number of people within the entire fire area, not just in each assembly room. See Application Example 903-4.

The discussion of Section 903.2.1 addresses how the sprinkler protection required by Section 903.2.1.3 must extend beyond the Group A fire area to include the entire story of the building, as well as any stories from the Group A occupancy to and including the level(s) of exit discharge.

The code requirements for these types of uses, specifically for exhibition and display rooms, can be strongly attributed to the McCormick Place fire in Chicago on January 16, 1967. McCormick Place was not sprinklered and consisted of three levels, including a main exhibit area of 320,000 square feet (29,728 m^2) on the upper level. Both the upper and lower levels were in the final stages of readiness for a housewares exhibition and were heavily laden with combustibles when the fire broke out. The fire was reported to have originated in the storage area behind an exhibit booth on the upper level. The upper level was almost totally destroyed, and considerable damage occurred to the lower level.

Ordinarily, assembly occupancies are considered to have a very low fire loading; however, the need for built-in fire suppression for assembly uses that house exhibitions or displays was clearly demonstrated by the McCormick Place fire. Display booths are most often constructed with combustible materials, and the storage area behind the booths is a receptacle for combustible materials and packing boxes. Thus, without built-in fire extinguishing systems, the large quantities of combustible materials and large areas combine to create an excessive hazard. Many other assembly occupancies classified as Group A-3 also present significant fire loading such as art galleries, libraries, and museums. Therefore, sprinkler protection is beneficial for all large Group A-3 occupancies.

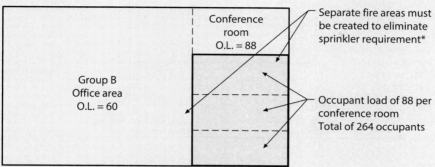

GIVEN: A mixed-occupancy single-story building containing a Group B office area and four Group A-3 conference rooms (each with an occupant load of 88)

DETERMINE: An appropriate method of fire area separation as an alternative to installation of a sprinkler system

*Each Group A fire area to be less than 12,000 sq ft with no more than 299 occupants

Note: Mixed occupancy conditions must also comply with Section 508.

For SI: 1 square foot = 0.093 m².

SOLUTION: A minimum of two fire areas must be created so as not to exceed the 299 occupant load and 12,000 square foot limitations.

AGGREGATE GROUP A OCCUPANT LOADS

Application Example 903-4

903.2.1.4 Group A-4. The fire sprinkler requirements for Group A-4 occupancies (those assembly uses provided with spectator seating for the viewing of indoor activities and sporting events) are identical to the provisions for Group A-1 and A-3 uses. See Section 903.2.1.3 for a discussion of the sprinkler requirements.

903.2.1.5 Group A-5. The fire loading in stadiums and grandstands is typically quite low except for specific accessory areas such as concession stands, storage and equipment rooms, press boxes, and ticket offices. Therefore, assembly occupancies classified as Group A-5 do not require the installation of an automatic sprinkler system except for those enclosed support areas where:

- the area exceeds 1,000 square feet (93 m²) in floor area, or
- the area does not exceed 1,000 square feet (93 m²), but it is not separated from the remainder of the facility by 1-hour fire-resistance-rated construction as required in Section 1030.1.1.1.

The limitation of 1,000 square feet (93 m²) is based on the floor area of each individual area and not on the aggregate area of all such spaces. Where such accessory spaces are of a considerable size, the hazards posed by the potentially large quantities of combustible materials can be reduced where such areas are sprinklered.

903.2.1.6 Assembly occupancies on roofs. Where a large Group A occupancy is located on the roof of a building other than a parking garage of noncombustible Type I or II construction, all floor levels below the occupiable roof must be sprinklered as illustrated in Figure 903-4.

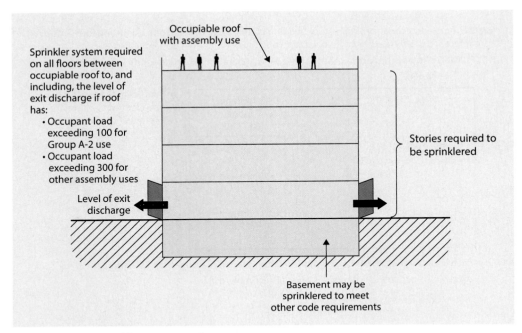

Figure 903-4 **Assembly use on roof.**

The sprinkler system is required where the occupant load exceeds 100 for Group A-2 occupancies, or 300 for all other Group A occupancies. The sprinkler protection shall extend to, and include, the level of exit discharge. Whether building occupants are located on an upper story or on the roof, they are exposed to a similar hazard as they travel down through the building to the discharge level. It should be noted that this provision does not require the roof to be sprinklered or provided with any alternative fire-extinguishing system.

The application of this provision should not be extended to other requirements of the code. For example, the occupiable roof is not considered as *building area*, *fire area*, or a *story*. Therefore, even though an occupiable roof is viewed as being an assembly occupancy, the limitations for allowable area and many other code provisions would not apply. Note that the height limits as regulated by Section 504 for allowable number of stories are regulated in a special manner as described in the discussion of Section 503.1.4.

903.2.1.7 Multiple fire areas. The option of compartmentalizing Group A occupancies into separate fire areas simply to avoid any sprinkler requirement is limited in its application. Where the fire areas share a common egress system and the combined occupant load of the Group A fire areas exceeds an occupant load of 299, sprinkler protection is required in accordance with Section 903.2.1.

Figure 903-5 depicts a condition where a sprinkler system is required even though multiple complying fire areas have been provided. An occupant load of 400 must be assigned to the corridor based upon the full contributing occupant loads of the Group A occupancies. Note that for purposes of this provision, 100 percent of the occupant loads of the Group A-2 and A-3 spaces must be assigned to the corridor because partial occupant loads are not to be considered.

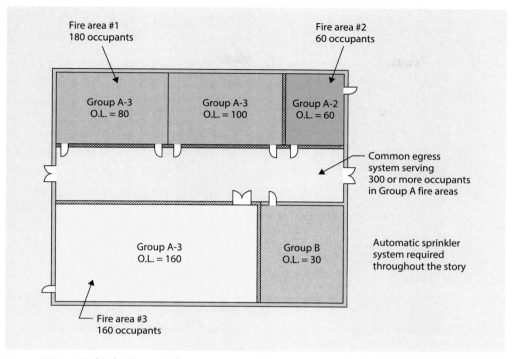

Figure 903-5 **Multiple Group A fire areas example.**

903.2.2 Group B. As a general rule, Group B occupancies do not require a sprinkler system based solely on their occupancy classification. However, Section 903.2.2.1 mandates that a Group B ambulatory care facility be provided with an automatic sprinkler system when either of the following conditions exists at any time:

- Four or more care recipients are incapable of self-preservation, or
- One or more care recipients who are incapable of self-preservation are located at other than the level of exit discharge.

Although such facilities are generally regarded as moderate in hazard level due to their office-like conditions, additional hazards are typically created due to the presence of individuals who are temporarily rendered incapable of self-preservation due to the application of nerve blocks, sedation, or anesthesia. While the occupants may walk in and walk out the same day with a quick recovery time after surgery, there is a period of time where a potentially large number of people could require physical assistance in case of an emergency that would require evacuation or relocation. The installation of an automatic sprinkler provides an important safeguard that enables the moderate-hazard classification of Group B.

The sprinkler system, when required, must extend throughout the entire story on which the ambulatory care facility is located. In addition, in multistory buildings where ambulatory care is provided below the exit discharge level, the sprinkler system must be installed on those stories between the level of ambulatory care and the level of exit discharge, inclusive. Where the ambulatory care facility is located above the first floor, the code stipulates that the automatic sprinkler system be provided on all floors below. This would include a basement even though the level of exit discharge may be above the basement level. See Figure 903-6. Open parking

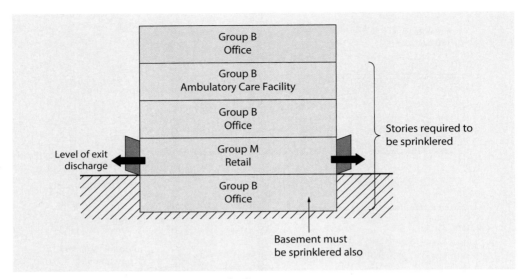

Figure 903-6 **Group B ambulatory care facility.**

garages located below the ambulatory care facility are not required to be sprinklered due to their limited contribution to the fire load.

The other uses classified as Group B where a sprinkler system is required based on occupancy classification include the research, development, and testing of lithium-ion and lithium metal batteries. The required sprinkler protection is intended to address those activities where there is an increased risk of thermal runaway and where, in some cases, thermal runaway is intentionally caused as a part of the battery development process.

903.2.3 Group E. History has shown that educational occupancies perform quite well when it comes to fire- and life-safety concerns. Much of this can be attributed to the continuous control and supervision that takes place within schools, as well as the students' knowledge of egress responsibilities in case of a fire or other emergency. However, because of the potential for moderate to high combustible loading, fire areas in Group E occupancies that exceed 12,000 square feet (1,115 m^2) in floor area must be provided with an automatic sprinkler system. In addition, fire sprinklers are required for Group E fire areas that occur above or below the level of exit discharge, as well as those fire areas that have an occupant load more than 299. Of particular note is that, unlike the requirement for sprinklered Group A occupancies, the sprinkler protection is only required within the applicable Group E fire area and need not extend to other portions of the story or building. It should also be noted that the requirements for automatic sprinkler protection are applicable to day care facilities classified as Group E occupancies.

903.2.4 Group F-1. Without an automatic sprinkler system to limit the size of a fire, the fire can spread very quickly to other portions of the structure. This is particularly true for large floor-area buildings containing combustible materials such as manufacturing facilities, warehouses, and retail sales buildings. The IBC requires an automatic sprinkler system to be installed throughout the entire building if it contains a Group F-1 occupancy and the fire area containing the Group F-1 exceeds 12,000 square feet (1,115 m^2). A fire sprinkler system is also required where the building housing the Group F-1 occupancy is four stories or more in height or has an aggregate of Group F-1 fire areas in the building of more than 24,000 square feet (2,230 m²). The aggregate Group F-1 fire area would also include the floor area of any mezzanines involved. The 24,000-square-foot (2,230-m²) limitation would also be applicable in both single-story and multistory structures. Where lithium-ion or lithium metal batteries are manufactured, the building must be provided with automatic sprinkler production throughout. A fully-sprinklered building is also required where such batteries are part of the manufacturing of vehicles, energy storage systems or equipment.

The subsections of 903.2.4 address the specific hazards of woodworking operations and the manufacture of distilled spirits, upholstered furniture, or mattresses. Due to the increased fire hazard, these uses will either always require the sprinkler protection, or the threshold for the protection is reduced well below the general provisions for a Group F-1 occupancy. Since the sprinkler is addressing a specific hazard, the system only needs to be installed throughout the fire area as opposed to applying throughout the entire building. These provisions allow for additional compartmentation in lieu of sprinkler protection or installing the system only in the area to address the increased hazard. However, once the general limits of Section 903.2.4 are exceeded, the sprinkler systems would be required to extend throughout the building. The 2,500-square-foot (232-m²) threshold for upholstered furniture or mattresses, although arbitrary in nature, represents a reasonable top-end limit where sprinkler protection is not required. It is also consistent with the sprinkler requirements of IFC Table 3206.2 addressing high-piled combustible storage of high-hazard commodities in buildings not typically accessible to the public.

903.2.4.1 Woodworking operations. Because of the special hazards involving combustible dusts and waste created during woodworking operations such as sanding and sawing, this section requires that an automatic fire-sprinkler system be installed in Group F-1 fire areas containing woodworking occupancies where the floor area of such operations exceeds 2,500 square feet (232 m²). Where equipment, machinery, or appliances that generate finely divided combustible waste or that use finely divided combustible materials are a portion of a woodworking operation, the size of the higher hazard operations is strictly limited unless sprinklers are installed. An example of this provision is shown in Figure 903-7. The provision is based on the floor area where the sanding, sawing, and similar operations occur, not necessarily the floor area of the entire woodworking operation. For example, any floor area devoted solely to the assembly of wood cabinetry would not be included in the determination. However, because these types of operations occur quite often as an integral part of the overall woodworking activities, rather than isolated in their own room or area, some means of regulating and controlling the hazard should be provided.

903.2.5 Group H occupancies. Group H occupancies are high-hazard uses, and one special feature is that, in addition to presenting a local hazard within the building, it has a potential for presenting a high level of hazard to the surrounding properties. Therefore, the code requires sprinkler protection for all Group H occupancies. Note that the sprinkler system is not necessarily required throughout the entire building that contains a Group H-2, H-3, or H-4 occupancy. Only the Group H occupancies must be provided with a sprinkler system. In the case of a Group H-1 occupancy, no other occupancies are permitted in the same building. Therefore, a building housing a Group H-1 occupancy must be sprinklered throughout. In addition, buildings containing Group H-5 occupancies require sprinklers throughout the entire building based on the original premise that the primary protection feature of this highly protected use is the automatic sprinkler system. For the purpose of sprinkler-system design, all areas of a Group H-5 are considered Ordinary Hazard Group 2, except for storage rooms with dispensing operations, which are considered Extra Hazard Group 2.

While not always classified as a Group H occupancy, buildings where pyroxylin plastics are manufactured, stored, or used in quantities greater than 100 pounds (45 kg) are required to

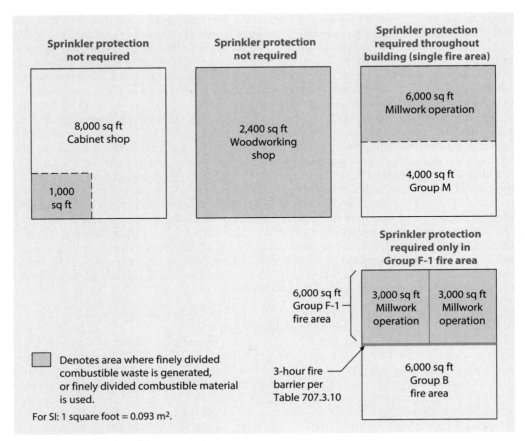

Figure 903-7 **Woodworking operations.**

be protected with an automatic sprinkler system. Pyroxylin plastics present a high hazard and are extremely flammable, as such those areas where these materials occur in larger quantities must be protected with fire sprinklers.

903.2.6 Group I. Because the mobility of the occupants of Group I occupancies is greatly diminished (in the case of hospitals and detention facilities, the self-preservation is essentially nonexistent), the code requires an NFPA 13 automatic sprinkler system throughout any building where a Group I fire area exists. For supervised residential facilities classified as Group I-1, Condition 1 occupancies, allowances are made for the use of an approved NFPA 13R system. The similarities between this Group I use and those uses classified as Group R justify the reduction in sprinkler protection.

Exception 2 provides the only instance in which a building containing a Group I occupancy is not required to be provided with some type of automatic sprinkler system. Day-care facilities classified as Group I-4 occupancies are not required to be sprinklered where the day-care operations only occur at the level of exit discharge and every room where care is provided has an exit door that leads directly to the exterior. This allowance is similar in some respects to the conditions established in Section 308.5.1 where children's day-care operations consistent with a Group I-4 classification may be classified as a Group E occupancy.

903.2.7 Group M. The typical American supermarket evolved during the construction boom that followed World War II. At that time, the typical supermarket consisted of a one-story building of moderately large area, e.g., 15,000 to 25,000 square feet (1,394 to 2,323 m^2). During the 1950s, fire statistics indicated that large-area supermarkets without sprinkler protection were subject to a larger proportion of fires than were usually attributable to this use in the past. As a result, building codes began requiring sprinklers in larger retail sales occupancies. The present requirements, detailed in the discussion of Section 903.2.4, are based on any of three factors: the size of the fire area, the number of stories, or the combined fire area on all stories. In addition, reference to the IFC is made for sprinkler protection in mercantile buildings where merchandise is placed in high-piled or rack storage.

The installation of an automatic sprinkler system is also mandated in any Group M fire area that is used for the display and sale of upholstered furniture or mattresses where the floor area devoted to such goods exceeds 5,000 square feet (464 m^2). Thus it would allow the fire area to exceed 5,000 square feet (464 m^2) without sprinkler protection provided the floor area used for the display and sale of the upholstered furniture and mattresses does not exceed the 5,000 square-foot (464 m^2) threshold. The requirement does not apply to the display and sale of furniture that is not upholstered, such as furniture constructed entirely of wood, plastic, or metal. Similar provisions are established for Group F-1 and S-1 occupancies where upholstered furniture or mattresses are manufactured or stored, however the sprinkler thresholds for Group M and S-1 occupancies are based only on the actual "area used" while the Group F-1 provisions are based on the entire fire area size. See the discussion of Section 903.2.4 for further commentary.

The Group M requirements regulate battery storage within a mercantile occupancy and tie the sprinkler requirements back to either the storage requirements for these batteries addressed in IFC Section 320 or the standard high-piled storage requirements of the IFC. Since high-piled storage applies to commodity storage in any occupancy, it is appropriate to use that trigger here for Group M occupancies. Because lithium-ion and lithium metal batteries would be classified as "high-hazard" commodities under the high-piled storage provisions, a sprinkler would be required within the storage area where the storage exceeds 500 square feet (46.4 m^2) in area and 6 feet (1,829 mm) in height.

903.2.8 Group R. In hotels, apartment buildings, dormitories, and other Group R occupancies, occupants may be asleep at the time of a fire, and may experience delay and disorientation in trying to reach safety. In addition, fire hazards in residential uses are often unknown to most occupants of the building, as they are created within an individual dwelling unit or guestroom. This helps to explain why these occupancies have a poor fire record when it comes to injury and loss of life. Annually, the largest loss of life occurs in Group R occupancies. The National Fire Protection Association (NFPA) reports that in 2022 a total of 2,910 civilians perished as a result of fires in the United States. Of those fatalities, 95 percent occurred in residential occupancies. Therefore, an automatic sprinkler system is required throughout any building containing a Group R occupancy. The sprinkler requirement applies to the entire building and not just the fire area containing the Group R occupancy.

903.2.9 Group S-1. In a manner consistent with that for Group F-1 and M occupancies, buildings containing combustible storage and warehousing uses must be provided with an automatic sprinkler system where the floor area or story level exceeds the specified threshold. The sprinkler requirement is based on the probable presence of large amounts of combustible materials, typically arranged in a highly concentrated manner.

Although the storage of commercial trucks, buses, and similar motor vehicles is typically regulated under the provisions of Section 903.2.10.1, there are situations where the parking of such vehicles occurs in the same area with Group S-1 uses. Therefore, these multipurpose spaces, such as fire station bays and public works facilities, are more appropriately classified as

Group S-1 occupancies. In such cases, a more restrictive threshold of 5,000 square feet (464 m²) is used to require sprinkler protection. By definition, a commercial motor vehicle is used to transport passengers or property and either has a gross vehicle weight rating of 10,000 pounds (4,540 kg) or more, or is designed to transport 16 or more passengers, including the driver.

Specific to the storage of vehicles powered by lithium-ion or lithium metal batteries, sprinkler protection is mandated throughout the building where the storage fire area exceeds 500 square feet (46.4 m²). This provision is not applicable to the parking of such vehicles in Group S-2 open and enclosed parking garages as well as Group U private garages.

The storage of distilled spirits, wine, upholstered furniture or mattresses pose an increased fire load. Therefore additional protection requirements exist. The storage of upholstered furniture and mattresses poses much the same hazard as in buildings where such goods are manufactured or displayed. However, the sprinkler requirements are not totally consistent with those for Group F-1 occupancies where upholstered furniture or mattresses are present. As with Group M occupancies, the sprinkler requirements for Group S-1 storage occupancies are based only on the floor area where upholstered furniture and mattresses are stored, not the size of the entire storage space or fire area.

903.2.9.1 Repair garages. The unique hazards associated with vehicle repair garages may be addressed in part through the installation of an automatic sprinkler system. However, the requirement for sprinklers is limited only to those repair garages that present a high level of concern based on size or location. By locating the repair garage above grade in a building of one or two stories, the size of the fire area containing the garage becomes the controlling factor in the determination of whether or not a sprinkler system is required. Where there is vehicle parking in the basement of a building used for vehicle repair, the building must be sprinklered regardless of fire area size. The sprinkler requirement is applicable even where the repair activity occurs only above the basement level. In buildings where commercial motor vehicles are repaired, the threshold for sprinkler protection is consistent with that established in Section 903.2.10.1 for commercial parking garages.

Where a repair garage provides storage for lithium-ion or lithium metal powered vehicles, the building is to be sprinklered consistent with the requirements for a Group S-1 storage building. If the fire area for such vehicle storage exceeds 500 square feet (46.4 m²), the building containing the repair garage shall be sprinklered throughout.

903.2.10 Group S-2 parking garages. Because the bulk of the uses designated as Group S-2 occupancies present very low fire-load potential, there is generally no requirement for these low-hazard occupancies to be sprinklered. However, where the Group S-2 portion of a building is an parking garage, the hazard level is increased. There is a need to protect other uses housed above an enclosed parking garage; thus, a Group S-2 enclosed parking garage is required to be sprinklered where the garage is located below another occupancy. In fact, in such a situation the entire building must be sprinklered, regardless of the size of the garage itself. There is an exception to the sprinkler requirement where an enclosed parking garage is located beneath a Group R-3 occupancy. Where the enclosed parking garage has no uses above, the required point at which an automatic sprinkler system is required is consistent with the threshold established for other moderate-hazard occupancies. The installation of an automatic fire sprinkler system for enclosed parking garages is required where the fire area containing the garage exceeds 12,000 square feet (1,115 m²) in area. The fire behavior in an enclosed parking garage, although similar to that in an open parking garage, is of greater concern since smoke ventilation will be more difficult due to the lack of sufficient exterior openings. This concern is addressed by the required installation of an automatic sprinkler system once an enclosed parking garage exceeds the 12,000-square-foot (1,115-m²) fire area threshold or an open parking garage fire area exceeds 48,000 square feet (4,460 m²). Mechanical-access enclosed garages must be protected due to the limited access into and through the building.

903.2.10.1 Commercial parking garages. Where commercial motor vehicles are stored within a building, the code mandates stringent floor areas when it comes to the requirement for an automatic sprinkler system. Where a fire area containing commercial parking exceeds 5,000 square feet (464 m^2) in area, the building housing the vehicles must be sprinklered throughout. The provision is intended to address those facilities housing larger vehicles. It is not applicable where pick-up trucks and similar-sized vehicles are being used for business activities.

903.2.11.1 Stories without openings. The provisions of this section make specific the intent of the code to require automatic sprinkler protection in "windowless buildings." A structure having inadequate openings on the exterior wall as determined by this section such that fire department access is insufficient is considered a "windowless buildings." The requirements of this section apply to all occupancies except Group U. The provisions are applicable on a floor-by-floor basis and do not apply to any story above grade plane or basement having a floor area of 1,500 square feet (139.4 m^2) or less:

- **On the basis of each individual story above ground.** Each individual story is analyzed for the size and the number of exterior wall openings. Thus, in a multistory building, it is possible to have a requirement that a sprinkler system be installed in one story and not in another.

The code requires that the openings be:

1. **Installed entirely above the adjoining ground level.** This provision is necessary so that effective fire-fighting and rescue can be accomplished from the exterior of the building.

    Where the openings cannot be located entirely above the adjoining ground level, the code permits the use of exterior stairways or ramps that lead directly to grade.

2. **Of adequate size and spacing.** Although it may be argued that the openings required by the code are not the equivalent of automatic fire-sprinkler protection, the access for fire-fighting provided by the openings has proven satisfactory.

    Although not expressly stated in the code, there is an expectation that a below-grade opening used to satisfy this provision be simply a typical 3-foot by 6-foot, 8-inch (914-mm by 2,032-mm) door leading directly to the exterior stairway or ramp. However, above-grade openings are more specifically addressed. A total of 20 square feet (1.86 m^2) of openings is mandated in each 50 lineal feet (15,240 mm) of exterior wall. It is not necessary to obtain all 20 square feet (1.86 m^2) from a single opening, as long as the minimum dimension requirement of 30 inches (762 mm) is met. Multiple openings, each 30 inches by 30 inches (762 mm by 762 mm) would comply; however, they may not be as effective as a larger single opening.

    The intent of the code is that one or more openings totaling at least 20 square feet (1.86 m^2) be provided in each 50 linear feet (15,240 mm) of exterior wall. It may be better stated that *any* wall section of 50 feet in length be provided with complying openings. Thus, an exterior wall 100 feet (30,480 mm) long with 20-square-foot (1.86-m^2) openings located at third points along the wall would comply, as shown in Figure 903-8. There is no portion of the wall that is 50 feet (15,240 mm) in length that does not contain the necessary openings. However, the same wall with such openings located at each end, as depicted in Figure 903-9, will not comply with the intent of the code insofar as there is a length of wall that exceeds the 50-foot (15,240-mm) dimension without an opening. Certainly, the same wall with only one 40-square-foot (3.72-m^2) opening at one end also would not comply.

3. **Accessible to the fire department from the exterior.** Surely, the openings would be of no value for fire-fighting if the fire-fighting forces could not gain access. The mere

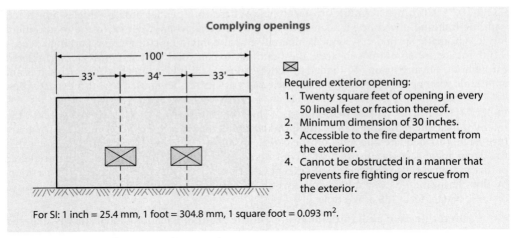

Figure 903-8 **Complying exterior openings.**

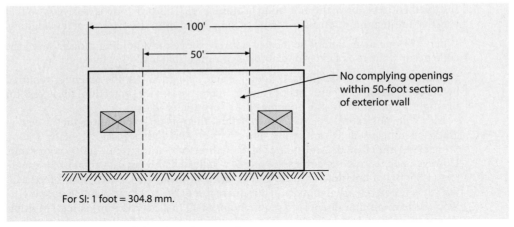

Figure 903-9 **Noncomplying exterior openings.**

fact that the openings may be 30 or 40 feet (9,144 or 12,192 mm) above grade does not mean the openings are inaccessible. However, if, with the resources available to the fire department, access cannot be obtained to the openings, they would be considered inaccessible. The determination of accessibility rests with the building official. However, personnel in a fire department should be consulted for their professional opinions and also for their knowledge of the capabilities of their equipment.

4. **Adequate to allow access for fire-fighting to all portions of the interior of the building.** The code requires that where openings are provided on only one side and the opposite exterior wall is more than 75 feet (22,860 mm) away, sprinklers shall be provided, or, as an alternative, openings shall be provided on at least two sides. The 75-foot

(22,860-mm) distance is a straight-line measurement taken between the two opposing walls. Where complying openings are required in two exterior walls because of the 75-foot (22,860-mm) limitation, the openings are permitted on either two adjacent sides or opposite sides on the assumption that, with two exterior sides having openings, adequate access may be gained to effectively fight the fire. See Figure 903-10A.

In stories above grade plane, the provisions require openings in exterior walls on a maximum of two sides. The code does not dictate specific openings for interior partition arrangements, because the normal openings provided through interior partitions provide adequate access to all interior portions of the building.

5. **Basements.** As previously noted, the provisions of Section 903.2.11.1 apply to every story, including basements, of all buildings where the floor area exceeds 1,500 square feet (139.4 m²). Basements are considered to be somewhat more difficult than stories above grade when it comes to fighting fires from the exterior of the building. Therefore, several additional requirements are imposed in addition to those of Section 903.2.11.1.2. The code provides that when any portion of a basement is located more than 75 feet (22,860 mm) from complying exterior wall openings, the basement is required to be provided with an automatic sprinkler system. The 75-foot (22,860-mm) measurement should be taken in a straight line, resulting in the use of the arc method, as shown in Figure 903-10B. The two methods of providing complying exterior openings set forth in Section 903.2.11.1 are both available for a basement condition. If the openings are available entirely above the adjoining ground level, they are regulated in the same manner as for floor levels above grade. Otherwise, the openings must lead directly

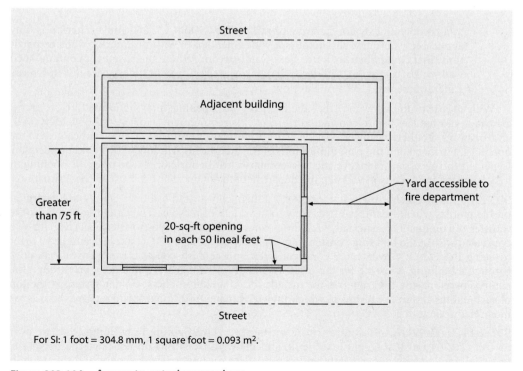

For SI: 1 foot = 304.8 mm, 1 square foot = 0.093 m².

Figure 903-10A **Access to exterior openings.**

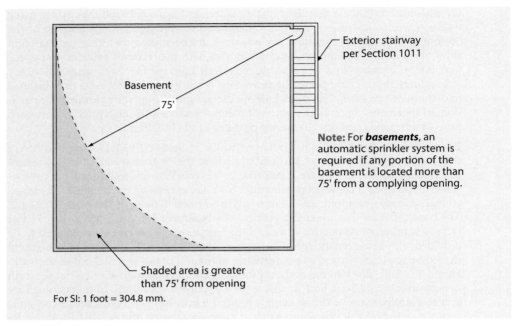

Figure 903-10B **Openings in basements.**

to a complying exterior stairway or ramp. The 75-foot (22,860-mm) criterion is only applicable where the basement is a wide-open space with no interior walls or partitions that could obstruct a fire-hose water stream. Where the basement contains such walls or partitions, a sprinkler system is always required in the basement if it exceeds 1,500 square feet (139.4 m²) in floor area.

With regard to the allowance that the exterior wall openings may be located below grade, areaways and light wells are considered to meet this requirement. However, the window wells and areaways should typically be provided with a stairway or ramp for gaining ready access to the openings. Furthermore, the plan dimensions of the areaway or window well should be adequate to permit the necessary maneuvering to accomplish fire-fighting or rescue from the opening. It may be necessary to consult with the fire department to obtain their expertise in these situations.

903.2.11.2 Rubbish and linen chutes. Linen chutes and rubbish chutes are potential problem areas when it comes to fire safety for a variety of reasons. They are often used for the transfer of combustible materials, including some levels of hazardous materials. They are also concealed within the building construction, possibly allowing a fire to smolder and grow prior to being detected. Of even more concern, linen and rubbish chutes create vertical openings through a building, allowing for the rapid spread of fire, hot gases, and smoke up through the chute. Therefore, the IBC requires the installation of an automatic sprinkler system at the top of such chutes and in the rooms in which they terminate. Additional sprinklers are required as illustrated in Figure 903-11.

903.2.11.3 Buildings 55 feet or more in height. Fire-fighting in buildings that are over 55 feet (16,764 mm) in height is difficult, and many jurisdictions do not have the personnel or equipment to rescue occupants and control fires on upper floors. Therefore, this section requires an automatic fire sprinkler system in those buildings in recognition of this problem. The provision applies to all buildings other than Group F-2 occupancies.

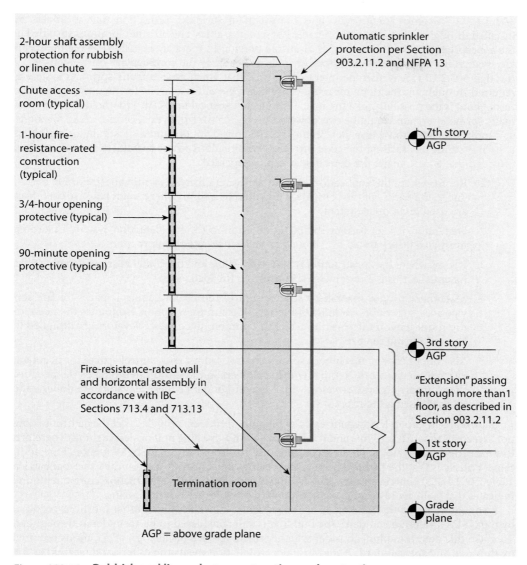

Figure 903-11 **Rubbish and linen chute construction and protection.**

903.3.1.1 NFPA 13 sprinkler systems. Where the code requires the installation of an automatic sprinkler system in a building, it typically is referring to a system designed and installed in accordance with the criteria of NFPA 13. This standard is also applicable for those provisions that permit the installation of a sprinkler system as an alternative to other code requirements. Throughout the code, the use of a sprinkler system "in accordance with Section 903.3.1.1" is referenced. In addition, where an automatic sprinkler system is required by the code with no direct reference to Section 903.3.1.1, the use of an NFPA 13 system is required. The fire sprinkler system in any building can be designed using NFPA 13. However, there are times when the code allows the use of sprinkler systems designed in accordance with NFPA 13R or 13D. The code will make a specific reference to Section 903.3.1.1 when the sprinkler is to be designed to NFPA 13.

903.3.1.1.1 Exempt locations. It is the intent of sprinkler protection that sprinklers be installed throughout the structure, including basements, attics, and all other locations specified in the appropriate standard. However, as discussed previously, there are occasions where the sprinkler system is only required on a story or in a fire area. See the commentary on Sections 903.2 through 903.2.10.1. It is also the intent of the IBC that when an automatic sprinkler system is required throughout, the same meaning is implied. One of the reasons for requiring protection throughout is the possibility of a fire in an unprotected area gaining such a foothold that the automatic sprinkler system would be overpowered. However, over the years, certain areas, locations, or conditions have shown that they require special consideration, and a substitute for sprinklers is permitted. In this section, the code provides five situations where sprinklers are not required provided that an automatic fire detection system is provided.

1. Rooms where the application of sprinkler water creates a significant fire or life hazard due to the nature of the contents. This could be a room where water reactive chemicals are used in an open system.
2. Generator or transformer rooms. In addition to the fire detection system, this room must also be separated by 2-hour fire-resistance-rated construction.
3. Rooms where the construction is noncombustible and the contents are noncombustible. Essentially, this is a room with absolutely no fire load.
4. Machinery rooms, machinery spaces, control rooms, and control spaces for fire service access elevators. Removing the sprinklers in these rooms eliminates the need for the shunt trip on the elevator, which is desirable since these elevators are designed to be used during the fire.
5. Machinery rooms, machinery spaces, control rooms, and control spaces for occupant evacuation elevators. Removing the sprinklers in these rooms eliminates the need for the shunt trip on the elevator, which is desirable since these elevators are designed to be used during the fire.

Bathrooms in Group R occupancies do not require sprinklers provided that the bathroom does not exceed 55 square feet (5 m^2) and it is located within a dwelling unit or sleeping unit. Therefore, this exception would not apply to a common use bathroom in the lobby of a hotel. One additional criteria is that the bathroom walls and ceilings must provide a 15-minute thermal barrier. Table 722.2.1.4(2) contains fire-resistance time ratings for various construction components and indicates that ½-inch (13-mm) gypsum wallboard provides a 15-minute rating.

It should be noted that as long as fire sprinklers are installed in all other locations required by the NFPA 13 design standard, the building is still considered to be sprinklered throughout. To make this determination, consider whether sprinklers are installed in all locations required by the code and the standard. If the sprinkler installation meets the criteria, but sprinklers are not provided in the elevator machine room, the building is still sprinklered throughout since the code stipulates that sprinklers are not required in such rooms in accordance with Item 4 or 5 of Section 903.3.1.1.1.

903.3.1.1.3 Lithium-ion or lithium metal batteries. Because of the potential fire severity and the possibility of thermal runaway in lithium-ion and lithium metal batteries, the design of the automatic sprinkler system must be based on large-scale testing at an approved testing laboratory. The test results will help to ensure that the submitted sprinkler system design can address the potential for a high heat release event.

The NFPA 13–22 *Standard for Installation of Sprinkler Systems* specifically indicates that lithium-ion and lithium metal batteries are not addressed by their general commodity classification provisions, and therefore, the general sprinkler provisions lack the guidance to adequately address the hazard.

Automatic Sprinkler Systems

It should be noted that the special testing requirements within IFC Section 1207.1.7 and Item 2 in Section 1207.5.5, are only applicable to ESS. The inclusion of IBC Section 903.3.1.1.3 as a subsection to 903.3.1.1 reminds code users that these automatic sprinkler systems cannot follow the general NFPA 13 requirements but that special testing must be done to ensure the lithium battery systems are adequately protected. Absent this language requiring evaluation of the specific hazard, it is more likely that sprinkler systems will be incorrectly designed and installed using density requirements for the general occupancy of the area where the batteries are located instead of at a level adequate to address the actual hazard.

903.3.1.2 NFPA 13R sprinkler systems. Although residential sprinkler systems installed in accordance with NFPA 13R may be used to satisfy the requirements of specific institutional and residential occupancies, they are not always recognized as full sprinkler protection for the purposes of exceptions or reductions permitted by other code requirements. However, where specifically mentioned through a reference to this section, such systems may be considered acceptable. Where the code indicates that a benefit can be derived from a sprinkler system installed "in accordance with Section 903.3.1.2," it intends that an NFPA 13R system can be used for the benefit. An important point is that an NFPA 13R sprinkler system is only permitted in residential-type buildings up to four stories in height above grade plane. Of particular note is the requirement that the 4-story limit be measured from grade plane for "podium" buildings as established in Sections 510.2 and 510.4.

Although the height limit of four stories above grade plane is applicable to all buildings protected by an NFPA 13R sprinkler system, a distinction between Group R-2 occupancies and other eligible occupancies exists in the building's maximum allowable height in feet. While residential uses "other than Group R-2 occupancies" are regulated by the 30-foot (9,144-mm) height limit to the floor of the highest story, Group R-2 occupancies are limited to a 45-foot (13,716-mm) height measurement which is coordinated with the three options used in the attic protection provisions of Section 903.3.1.2.3. This approach triggers the need for an NFPA 13 system based on the same three measuring points which addresses concerns of unprotected areas being located beyond the reach of typical fire department equipment. See Figure 903-12.

The code imposes additional sprinkler requirements in the subsections of Section 903.3.1.2. These require the NFPA 13R sprinkler system to protect areas of a building where the standard

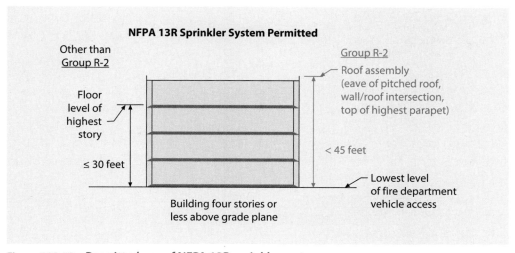

Figure 903-12 Permitted use of NFPA 13R sprinkler system.

would not require them (see Section 102.4). These extra protection requirements address certain balconies and decks, corridors and balconies used for egress, and attic areas.

903.3.1.2.1 Balconies and decks. Experience has shown that numerous fires in apartment buildings have started from grilling or similar activities on the balconies and patios. Because the NFPA 13R sprinkler standard does not mandate sprinklers in such locations, Condition 1 of this section requires such sprinkler protection. The provision is applicable to both dwelling units and sleeping units in buildings of Type VA or VB construction. The automatic sprinkler protection is only required where there is a roof, deck, or balcony directly above a balcony, deck, or patio below. These areas will typically be protected by sidewall fire sprinklers. If there is no horizontal element located directly above an exterior balcony, deck, or ground-floor patio, the additional sprinkler protection is not required. See Figure 903-13. As referenced in Condition 2 of this section, where the balconies are constructed in accordance with Section 705.2.3.1. Exception 3, the balconies are permitted to be constructed of Type V construction without a fire-resistance rating on buildings of Type III, IV, or V construction provided sprinkler protection is extended to protect the balcony areas.

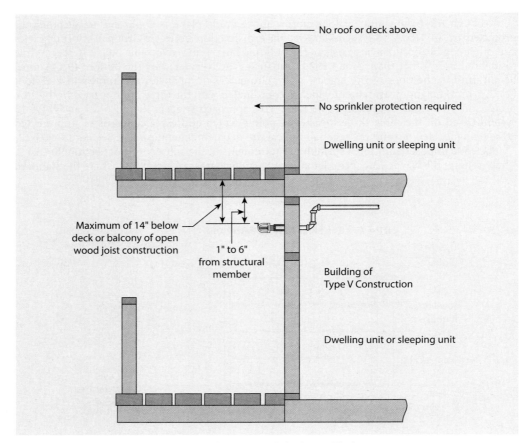

Figure 903-13 **Sprinkler protection of residential decks and balconies.**

903.3.1.2.3 Attic protection. The NFPA 13R sprinkler standard does not require sprinklers in the attic space unless a fuel-fired appliance is located within the attic space, and then only one sprinkler is required, to be located above the equipment. In addition to this condition, the IBC mandates that if the attic is used for, or intended to be used for, living purposes or storage, sprinkler protection is required throughout the attic.

If the building is of Type III, IV, or V construction, and the roof assembly is more than 55 feet (16,764 mm) above the lowest level of fire department vehicle access, attics that are not required to be sprinklered to protect storage or a living space must still be protected by one of four methods:

- Install fire sprinkler protection throughout the attic.
- Construct the attic of noncombustible materials.
- Construct the attic of fire-retardant-treated wood.
- Fill the entire attic space with noncombustible insulation.

In applying this requirement for attic protection, the code specifies how the height of the roof assembly is to be measured. The measurement is from the lowest required fire department vehicle access road to the eave of the highest pitched roof, or to the intersection of the highest roof and the exterior wall, or to the top of the highest parapet, whichever is the greatest distance. Consider the top point of measurement to be where the fire department would rest the top of ladder. See Figures 903-14 and 903-15.

Attics in Group R-4, Condition 2 occupancies must all be protected regardless of the height of the roof assembly. Where such attics are not used or intended for living or storage purposes, they are to be protected with either sprinklers throughout, or by one of four alternate methods:

- Install a heat detection system throughout the attic. The detection system must activate the fire alarm in the facility.
- Construct the attic of noncombustible materials.
- Construct the attic of fire-retardant-treated wood.
- Fill the entire attic space with noncombustible insulation.

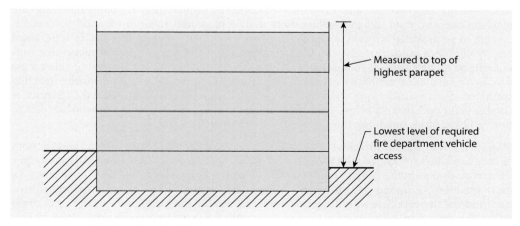

Figure 903-14 **Height of roof with parapets.**

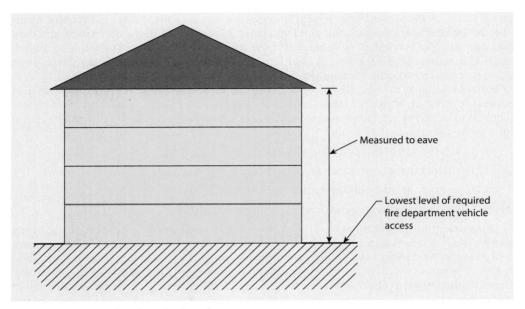

Figure 903-15 **Height of pitched roof.**

903.3.1.3 NFPA 13D sprinkler systems. The NFPA 13D standard is designed for use in one- and two-family dwellings and manufactured homes. The code permits such systems designed to this standard to be installed in one- and two-family dwellings, Group R-3 occupancies, and Group R-4, Condition 1 occupancies. Note that an NFPA 13D system is not allowed to be installed in a Group R-4, Condition 2 facility. Those facilities require a sprinkler system designed to NFPA 13 or 13R.

903.3.2 Quick-response and residential sprinklers. Based on the timely performance of quick-response and residential automatic sprinklers, the code requires that they be installed in those occupancies where response or evacuation may not be immediate because of the condition of the occupants. Therefore, all spaces within a Group I-2 smoke compartment containing care recipient sleeping units and all dwelling units and sleeping units in Group R and I-1 occupancies are to be provided with these types of sprinklers. Similarly, ambulatory care facilities must be provided with quick-response or residential sprinklers in all smoke compartments that contain one or more treatment rooms. Any Group I-2 smoke compartment containing either a gas fireplace appliance or decorative gas appliance must also be protected with the quicker reacting sprinklers. Such sprinklers are also required in light-hazard occupancies, where the quantity or combustibility of contents is low. Light-hazard occupancies included places of worship, education facilities, office buildings, museums, and seating areas of restaurants and theaters.

903.3.8 Limited area sprinkler systems. There are situations where only a specific room or area needs to be protected with fire sprinklers. See the discussion of Section 903. For example, there is an option in the code to only sprinkler certain incidental uses listed in Table 509.1. In these situations, a limited area sprinkler system could be used. These systems are limited to a maximum of six sprinklers within any one fire area, but a building with multiple fire areas could utilize this option in each fire area. The water supply for these fire sprinklers is the domestic water piping in the building, there is no separate sprinkler piping or control valve. NFPA 13 would be consulted to select the appropriate sprinkler, determine the minimum pressures and flows of the sprinklers, and design of the piping based on hydraulic calculations. Because of the limited capability of these systems, they are limited to only Light Hazard or Ordinary Hazard Group 1 occupancies as defined in NFPA 13.

Section 904 *Alternative Automatic Fire-Extinguishing Systems*

The code permits the use of automatic fire-extinguishing systems other than automatic sprinkler systems for those circumstances approved by the fire code official. However, the use of an alternative system in lieu of a sprinkler system does not gain the benefit of exceptions or reductions in code requirements. Only those buildings or areas protected by automatic sprinkler systems can take advantage of the allowances provided throughout the code.

The installation of automatic fire-extinguishing systems shall comply with this section, which for the most part refers to the appropriate design and test standard and listing for each of the various types of systems. Those types addressed include wet-chemical, dry-chemical, foam, carbon dioxide, halon, clean-agent systems, water mist, and aerosol fire-extinguishing systems.

The inspection and testing of the system is emphasized because of the importance of a fully operating system. Specific items are identified for inspection, including the location, identification, and testing of the audible and visible alarm devices.

904.14 Commercial cooking systems. Fire-extinguishing systems are typically required for commercial cooking operations. The requirement for installing a fire-extinguishing system originates in the *International Mechanical Code®* and *International Fire Code*, but this section deals with installation criteria. Several requirements span the various codes, such as the requirement for the fire-extinguishing system to shut down the electrical power and fuel gases to the cooking equipment under the exhaust hood when it operates.

904.15 Domestic cooking facilities. Domestic cooking equipment, such as cooktops and ranges, is permitted to be installed in Group I-1 occupancies, Group I-2 occupancies, and Group R-2 college dormitories. The allowance of these domestic cooking appliances is justified by the requirement to protect the cooking appliances with an automatic fire-extinguishing system, or install cooking appliances listed to prevent ignition of cooking oil when their burners are set on high and unattended for 30 minutes. Essentially, these cooking appliances do not create as much heat allowing for a safer operation.

Section 905 *Standpipe Systems*

A standpipe system is a system of piping, valves, and outlets that is installed exclusively for firefighting activities within a building. Standpipes are not considered a viable substitute for an automatic fire sprinkler system. They are needed in buildings of moderate height and greater, and when used by trained personnel provide an effective means of fighting a fire. Section 202 provides definitions for the three different classes of standpipe systems, as well as the five different types of standpipes. It is important to note that Section 905.3 identifies those conditions under which a standpipe system is required. Section 905.4 establishing the locations where standpipe connections are required is only to be applied where such systems are mandated by Section 905.3.

Table 905-1. **Required Standpipe Systems**

	Class I	Class II	Class III
Buildings where highest floor level is located more than 30 feet above lowest point of fire department vehicle access			X [1,2]
Buildings where lowest floor level is located more than 30 feet below highest point of fire department vehicle access			X [1,2]
Buildings which are 4 or more stories above or below grade plane			X [1,2]
Nonsprinklered Group A buildings with an occupant load >1,000	X		
Covered and open mall buildings	X		
Landscaped and vegetative roofs			See Note 3 below
Underground buildings	X		
Buildings with a heliport or helistop on the roof	See IFC Section 2007.5		
Marinas and boatyards	See IFC Chapter 36		
Groups A-1 and A-2 having more than 1,000 occupants		X [4]	

For SI: 1 foot = 304.8 mm, 1 square foot = 0.093 m².
[1]Exceptions allow for Class I standpipe in sprinklered buildings, parking garages, sprinklered basements, and Group B and E occupancies.
[2]Only a Class I standpipe is required in buildings where the occupant-use hose lines will not be used by trained personnel or the fire department.
[3]Where building with vegetative or landscaped roof is provided with a standpipe system, the system shall extend to such roof level.
[4]Connections to be located on each side of rear of auditorium and on each side of balcony.

905.3 Required installations. This section provides the scoping criteria for when a standpipe system must be provided. Table 905-1 provides the basic requirements for when a standpipe system is required in a building. Building height is the primary consideration for the installation of a standpipe system. The general requirement calls for Class III standpipe systems where the vertical distance between the highest floor level in the building and the lowest level of fire department vehicle access exceeds 30 feet (9,144 mm). A similar condition occurs where a floor level is significantly below grade. See Figure 905-1. For measurement purposes, it is not necessary to consider any level of fire department vehicle access that, because of topographic features, makes access to the building from that point impractical or impossible. Where the building has four or more stories above or below grade, a standpipe system is required regardless of the building's height in feet. Several exceptions allow the use of Class I rather than Class III standpipes. Additional standpipe requirements may apply to assembly occupancies, covered and open mall buildings, landscaped and vegetative roofs, and underground buildings.

905.4 Location of Class I standpipe hose connections. The code intends that Class I standpipes are for the use of the fire department to fight fires within a building. Therefore, once a Class I or Class III standpipe system is required by Section 905.3, the necessary locations for the Class I standpipe hose connections are identified. The code requires that standpipe outlets be in every required interior exit stairway and exterior exit stairway. The connections are to be located at the main floor landing of those stairways required by the code. As an alternative, the hose connections are permitted to be located at the intermediate landing between

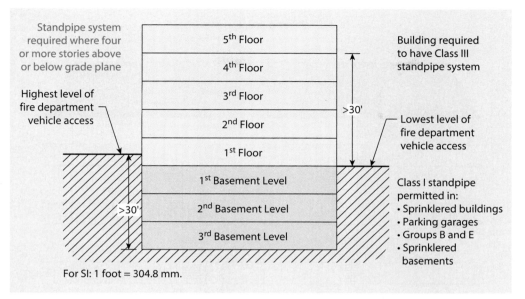

Figure 905-1 Class III standpipe systems.

floor levels, but only where specifically allowed by the fire code official. The installation of hose connections at intermediate landings is often preferred in order to avoid congestion at the stairway door. With these locations for standpipe outlets, the fire department personnel can bring a hose into the stair enclosure and make a hookup to outlets in a relatively protected area. Such protection for firefighters is also provided at exterior exit stairways. Standpipe connections are not required for exit access stairways permitted to be unenclosed by Section 1019.3, as such stairways provide no protection for fire department personnel. A single hose connection can be installed where open breezeways or open corridors are provided between open stairways (exterior exit stairways), and the open stairways are no more than 75 feet (22,860 mm) apart. These stairways are considered as very close together when you consider the fact that the fire department would typically be using a 100-foot (30,480-mm) hose line. Additionally, since the code allows for the open corridors and open stairways, there is no fire-resistance-rated separation between the two stairs so the single connection is allowed to serve both stairs. See Figure 905-2.

Because a horizontal exit provides a barrier having a minimum fire-resistance rating of 2 hours, it is a logical location for Class I standpipe outlets in those buildings required to be provided with a Class I or Class III standpipe system. Such outlets are to be provided on both sides of the horizontal exit wall adjacent to the egress doorways through the horizontal exit wall, regardless of whether or not egress is provided from both directions. An exception permits the omission of the standpipe hose connection at the horizontal exit opening where there is a limited distance between the opening and the stairway hose connection. The elimination of the hose connection is permitted on one side of the horizontal exit, as depicted in Figure 905-3, or on both sides, provided the "100-foot of hose plus 30-foot of hose stream" distance is not exceeded. The application of the exception is most common where the horizontal exit is provided to allow for the termination of a fire-resistance-rated corridor at an intervening room, or where necessary to address an inadequate number of

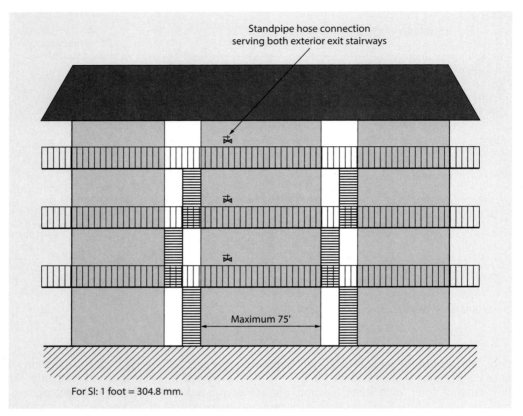

Figure 905-2 **Class I standpipe connections at exterior exit stairways.**

exits or insufficient egress width. In those cases where the horizontal exit is provided because of a problem with travel distance, the omission of the standpipe connections will seldom be permitted. It should again be noted that a standpipe system is not required simply because a horizontal exit occurs within the building. This provision addressing required hose connection locations is only applicable where a horizontal exit is provided in a building described in Section 905.3 as requiring a Class I or Class III standpipe system.

Under certain circumstances, standpipe outlets located in interior exit stairways, exterior exit stairways, or at horizontal exit doorways may not provide adequate coverage for all portions of the story. This occurs when the most remote portion of the story is more than 150 feet (45,720 mm) from the standpipe connection. This distance is increased to 200 feet (60,960 mm) when the entire floor is sprinklered. Under these circumstances, additional standpipe outlets must be provided in approved locations when required by the fire code official.

Exit passageways provide both protection and good access points for fire fighter access. Where the building requires a Class I or Class III standpipe system, a Class I standpipe connection is required at the entrance to the exit passageway from other portions of the building. In other words, as fire fighters use the exit passageway to access the interior of the building, this is the point where the fire fighters would leave the exit passageway to fight the fire. An exception similar to that applicable to horizontal exit conditions eliminates standpipe connections at exit passageway locations where floor areas are reachable from other connections.

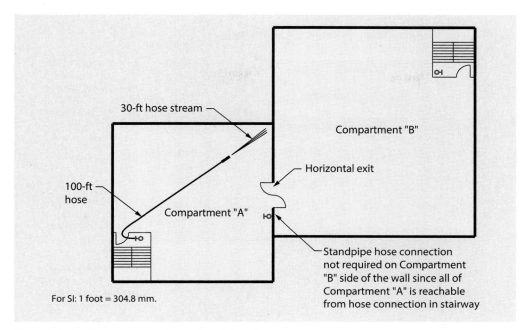

Figure 905-3 Connections at a horizontal exit.

The enclosure for an interior exit stairway also provides protection for the standpipe and piping system. In those cases where the risers and laterals are not within interior exit stairways, the code requires that they be protected by equivalent fire-resistant construction. The exception to this requirement assumes that the automatic fire sprinkler system will keep the laterals cool enough so that they will not be damaged by fire. Figure 905-4 depicts the required locations of Class I standpipe connections.

On those roofs that have a flat enough slope for fire fighters to move about, the code requires at least one roof outlet so that exposure fires can be fought from the roof. The roof slope must be less than 4:12 for this requirement to apply. This connection is to be located at the roof access from the interior exit stairway at the roof level. Frequently, in multi-story buildings, more than one interior exit stairway is provided. However, only one stairway is required to provide access to the roof. This section only requires one connection at the roof level and it should be at the stairway providing roof access.

The interconnection of the standpipe risers at the bottom for multiple standpipe systems is intended to increase reliability and use during firefighting operations.

905.5 Location of Class II standpipe hose connections.   Class II standpipes are designed for occupant use, similar to portable fire extinguishers. It is the intent of the code to require the location of hose cabinets for Class II standpipes at intervals ensuring that all portions of a building will be within 30 feet (9,144 mm) of a nozzle attached to 100 feet (30,480 mm) of hose. In plan review, this would necessitate allowing for pulling the hose down corridors and through rooms such that several right-angle turns may be necessary before the hose stream can be placed on the fire. Therefore, judgment is necessary in the determination of standpipe locations. One method to account for this type of partitioning in a building where the future location of partitions is unknown is to subtract 30 feet (9,144 mm) from the straight-line distance between the hose cabinet and the remote location and then multiply the remainder by 1.4. If the result is more than 100 feet (30,480 mm), an additional standpipe connection will be required. Figure 905-5 illustrates the location of a Class II standpipe

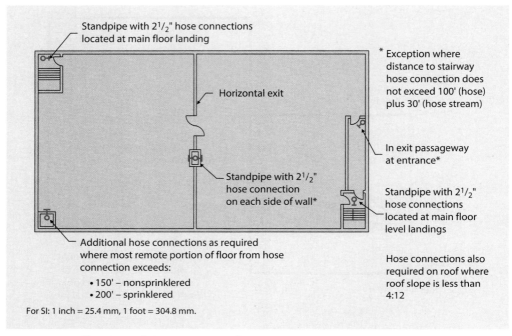

Figure 905-4 **Class I standpipe connection locations.**

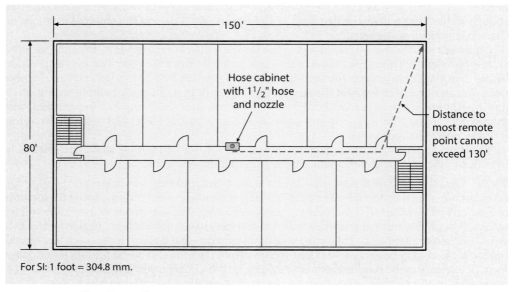

Figure 905-5 **Class II standpipe connections.**

in a building where an office floor has a central corridor with offices on each side. In this particular arrangement, it is obvious from the layout that the one standpipe will suffice.

As there are no scoping provisions in Section 905.3 for a Class II or III standpipe system in a sprinklered Group A occupancy, the provisions of Section 905.5.1 have no application. Therefore, the required locations for Class II standpipe hose connections are limited to those buildings or areas required to have Class III standpipe systems.

Because Class II standpipe systems are designed for occupant use, the code does not require fire-resistive protection. Class II standpipes are designed for first aid type of application, early in the fire event. Therefore, the piping is not expected to survive the fire for any duration.

905.6 Location of Class III standpipe hose connections. Because Class III standpipes are a combination of the benefits of Class I and II systems, containing both 1½-inch (38-mm) outlets for occupant use and 2½-inch (64-mm) outlets for fire department use, it is only logical that they be located so as to serve the building as required for both Class I and II standpipes. Figure 905-6 shows the typical arrangement for a Class III standpipe in a building. Usually, the hose rack for 1½-inch (38-mm) outlets and 2½-inch (64-mm) hose outlets are both located within a stair enclosure. Where the coverage requirements for Class II standpipes are such that

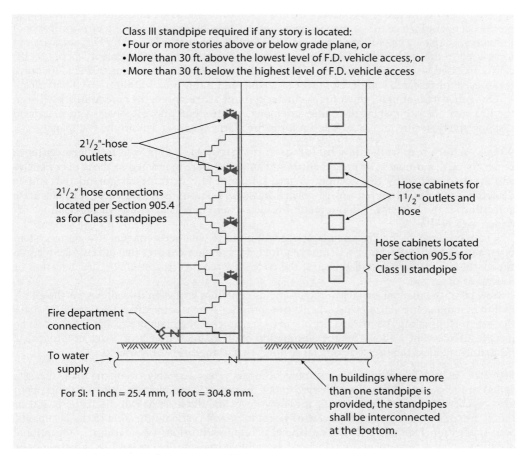

Figure 905-6 **Class III standpipe connections.**

interior exit stairway and other required locations will not cover the entire building, laterals are usually run to other locations in order to provide for the required coverage.

Class III standpipe systems and their risers and laterals are required to have fire-resistive protection the same as for Class I standpipes. In a sprinklered building, the laterals not located within the stair enclosure do not need to be protected as the sprinkler water will provide protection during the fire.

Section 907 *Fire Alarm and Detection Systems*

One of the most effective means of occupant protection in case of a fire incident is the availability of a fire alarm system. An alarm system provides early notification to occupants of the building in the event of a fire, thereby providing a greater opportunity for everyone in the building to evacuate or relocate to a safe area. This section covers all aspects of fire alarm systems and their components.

Unlike most of the provisions of Section 903.2 addressing the required installation of an automatic sprinkler system, those requirements mandating a fire alarm system are typically not applied based on the fire area concept. Manual fire alarm systems are most often required based on the occupant load of the occupancy group under consideration, including any occupants of the same occupancy classification that may be identified in another fire area. Fire areas are not to be used in the application of this section regarding fire alarm systems except where specifically addressed, such as in Sections 907.2.2.1 (ambulatory care facilities) and 907.2.2.2 (laboratories used for research and development or testing of lithium-ion or lithium metal batteries).

One other item of note, when fire sprinkler systems are required, they are to be installed in the fire area, the occupancy, the entire story, or the entire building. However, with fire alarm systems, the typical requirement is to limit the fire alarm system to within the occupancy.

907.2 Where required—new buildings and structures. Approved fire alarm systems, either manual, automatic, or both manual and automatic, are mandated in those occupancies and areas identified by this section. Where automatic fire detectors are mandated, smoke detectors are to be provided unless normal business operations would cause an inaccurate activation of the detector. All automatic fire detection systems are to be installed in accordance with NFPA 72.

The requirements for fire alarm systems will indicate whether a manual fire alarm system or an automatic fire detection system is required. This section requires that at least one manual fire alarm box must be installed even if the code only requires an automatic fire detection or waterflow detection.

Several of the manual fire alarm requirements contain a provision that allows for the elimination of manual fire alarm boxes when a fire sprinkler system is installed. Section 907.2 states that when eliminating the manual fire alarm boxes, at least one manual fire alarm box must be installed. Exception 2 allows for the elimination of all manual fire alarm boxes in Group R-2 occupancies unless the fire code official requires that a single box be installed.

907.2.1 Group A. Where an assembly occupancy has a sizable occupant load, the safe egress of the occupants becomes an even more important consideration. When the occupant load reaches 300, or when the occupant above or below the level of exit discharge is 100 or more, the code mandates a manual alarm system to provide early notification to the occupants. See Application Example 907-1. The IBC also requires portions of a Group E educational occupancy that are occupied for associated assembly purposes, such as a school lunchroom or library, to have alarms as required for the Group E use, rather than based on the less restrictive

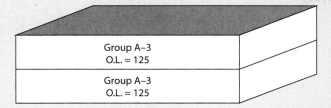

GIVEN: Two-story art gallery with a total occupant load of 250, divided equally between the two floors.

DETERMINE: If a fire alarm system is required.

SOLUTION: When the total occupant load of 250 is considered, a fire alarm system is not required since it is not 300 or greater. When each floor is considered, the second floor has an occupant load of 100 or more. Therefore a manual fire alarm system is required. The fire alarm system is required throughout both floors of the Group A occupancy.

FIRE ALARM SYSTEM IN GROUP A OCCUPANCIES

Application Example 907-1

requirements mandated for Group A occupancies. In addition, the Group E fire alarm provisions are applicable to those areas within a Group E school building that may be classified as Group A occupancies, such as a gymnasium or auditorium. As is the case in many other occupancy classifications, manual alarm boxes are not required in those Group A occupancies where an automatic sprinkler system is installed that will immediately activate the occupant notification appliances (audible and visible devices) upon water flow.

Where there are two or more Group A occupancies within the same building, the occupant loads of all such occupancies shall be used in evaluating whether or not a manual fire alarm system is required unless multiple fire areas are created in accordance with Section 707.3.10. Where the fire area concept is applied, the alarm requirement is based on the occupant load within each individual fire area. Only assembly uses within the fire area having an aggregate Group A occupant load of 300 or more, or a Group A occupant load of 100 or more above or below the level of exit discharge, are required to be provided with a manual fire alarm system.

In assembly occupancies containing much larger occupant loads (1,000 or more people), activation of the required fire alarm system must initiate a prerecorded announcement via an emergency voice/alarm communications system. The emergency voice/alarm communications system, upon approval of the building official, may also be used for live voice emergency announcements originating from a constantly attended location. For more information on emergency voice/alarm communication systems, see the discussion of Section 907.5.2.2.

907.2.2 Group B. Larger business occupancies classified as Group B require the installation of a fire alarm system. Where the total occupant load exceeds 499 persons, or where more than 100 persons occupy Group B spaces above or below the lowest level of exit discharge, a manual fire alarm system shall be installed. See Figure 907-1. Similar to exceptions for other occupancies, the manual fire alarm boxes are not required in a sprinklered building where sprinkler water flow activates the notification devices, other than one box in an approved location.

Ambulatory care facilities must be provided with a manual fire alarm system as well as an electronically supervised automatic smoke detection system. The manual fire alarm system must be provided within the fire area in which the ambulatory care facility is located, while the

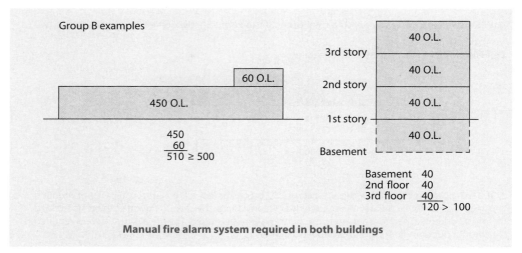

Figure 907-1 **Example of required fire alarm system for Group B.**

smoke detection system must serve the care facility as well as the adjoining public areas. In both cases, the presence of an automatic sprinkler system modifies the requirements where occupant notification appliances will activate upon water flow.

Lithium-ion and lithium metal batteries have a potential for an increased fire hazard due to the possibility of thermal runaway. Once the batteries fail in this manner, it is almost impossible to extinguish them until they completely burn out. One of the best approaches to prevent and reduce this potential hazard is through early detection, a mitigation plan, and cooling and suppression via an automatic sprinkler system. Section 907.2.2.2 requires a fire alarm system which is activated either by an air sampling-type smoke detection system or a radiant energy-sensing detection system in fire areas of Group B occupancies used for research and development or the testing of such batteries. Similar requirements are mandated in other moderate-hazard occupancies where the manufacture, installation, sales, or storage of such batteries occurs.

When lithium-ion and lithium metal batteries begin to fail, they often begin to off-gas certain chemicals or increase in temperature. With early detection and notification, occupants can exit the building, enact a mitigation plan and hopefully prevent the batteries from entering their thermal runaway phase. While mitigation plans will vary based on individual installations, the techniques may be as simple as disconnecting batteries from their power supplies, cooling the batteries or isolating the failing battery. Various provisions throughout Section 907 address the detection aspect of protecting these systems and occupancies. They are also part of a coordinated package including an automatic sprinkler system (Section 903.2), as well as other code requirements including those referenced in IFC Section 320.

907.2.3 Group E. The IBC follows the philosophy of society in general that our children require special protection when they are not under parental control. Therefore, in addition to the other life-safety requirements for educational occupancies, manual fire alarm systems are required whenever the occupant load of any Group E occupancy is more than 50. In addition, when the building is provided with a smoke detection or sprinkler system, such systems shall be connected to the building's fire alarm system. However, the more probable reason for such a low occupant load threshold being established is the exceptional value of such a system in an educational use. Students tend to react quickly and efficiently at the first notification of the alarm system, making safe egress possible. In addition, periodic fire drills reinforce the appropriate egress activity. Where the four conditions are met as listed in Exception 3, manual alarm boxes are not required. Item 3.4 of the exception requires a manual means to alert occupants and notify emergency services in the event a fire is discovered in an area such as a classroom or storage area that is allowed without the required detection from the other items. This requirement for at least one manual station is similar in intent to that of Section 907.2. Exception 4 also exempts manual fire alarm boxes where, in a sprinklered building, the sprinkler water flow activates the notification appliances and, consistent with Section 907.2, requires manual activation from a normally occupied location.

The fire alarm system for a Group E occupancy must include an emergency voice/alarm communications system where the occupant load exceeds 100. This type of system will not only provide the necessary notification of occupants under possible fire conditions, but it is also valuable in ensuring the necessary level of life safety inside of the building during a lockdown situation. Because of concerns of school campus safety serving kindergarten through 12th grade facilities, it is beneficial to provide an effective means of communication between the established central location and each remote secured area.

907.2.4 Group F. In multistory manufacturing occupancies, a fire alarm system is mandated where an aggregate occupant load of 500 or more is housed above and/or below the level of exit discharge. Similar to several other occupancies, the alarm system is necessary where there are very large occupant loads and where those occupants must travel vertically to exit. Application of the provision where occupants are located both above and below the lowest level of exit discharge is similar to that for Group B and M occupancies. In Group F occupancies, a manual fire alarm system is mandated where the aggregate occupant load of those levels, other than the exit discharge level, exceeds 499. Where the building is sprinklered, only a single manual alarm box is required if the water flow of the sprinkler system activates the notification appliances.

In buildings where lithium-ion or lithium metal batteries are manufactured, or where such batteries are installed in vehicles, energy storage systems or equipment, a fire alarm system is required. See the discussion of Section 907.2.2 for further information.

907.2.5 Group H. Only those Group H occupancies associated with semiconductor fabrication or the manufacture of organic coatings need to be provided with a manual alarm system. Note that there is no exception allowing elimination of manual fire alarm boxes. Because of the increased hazard in these occupancies all Group H are required to be sprinklered, and all of the manual alarm boxes must be installed. Areas containing highly toxic gases, organic peroxides, and oxidizers shall be protected by an automatic smoke detection system when required in the *International Fire Code*.

907.2.6 Group I. As many patients, residents, or inmates of Group I occupancies are incapable of self-preservation, Section 907.2.6 provides for early warning of the occupants and staff, thus enhancing life safety. The general provisions require a manual fire alarm system to be installed in all Group I occupancies, including those classified as Group I-4. However, an automatic smoke detection system is only required in those occupancies classified as Group I-1, I-2, or I-3. Where a Group I-1, Condition 1 facility is provided with an automatic sprinkler system designed to NFPA 13, only corridors and waiting areas open to such corridors need to be equipped with the smoke detection system. The corridor smoke detection system required in Group I-2 facilities may be omitted where the patient sleeping rooms have smoke detectors that are connected to the fire alarm system and provide a visual display on the corridor side of each patient room, as well as a visual and audible alarm at the appropriate nursing station. Another exception exempts the requirement for corridor smoke detection where sleeping room doors are equipped with automatic door closers having integral smoke detectors on the room side that performs the required alerting functions.

Because of their special nature, in Group I-3 occupancies the provisions for fire alarm systems are greatly expanded. The manual and automatic fire alarm system is to be designed to alert the facility staff. Actuation of an automatic fire-extinguishing system, an automatic sprinkler system, a manual fire alarm box, or a fire detector must initiate an automatic fire alarm signal, which automatically notifies staff. For obvious reasons, manual fire alarm boxes need only be placed at staff-attended locations having direct supervision over the areas where boxes have been omitted. Under certain conditions, the installation of an approved smoke detection system in resident housing areas is required.

907.2.7 Group M. The threshold at which a manual fire alarm system is required for a Group M occupancy is the same as that for a Group B occupancy. During that portion of time when the building is occupied, the signal from a manual fire alarm box or waterflow switch may be designed to only activate a signal at a constantly attended location, rather than provide the customary visual and audible notification. However, to allow this to occur the fire alarm system must be an emergency voice/alarm communication system. When an alarm occurs, the emergency voice/alarm communications system can be used to notify the customers of the emergency conditions. This provision is helpful in eliminating nuisance alarms that may occur because of the presence of fire alarm boxes.

In rooms where lithium-ion or lithium metal batteries are stored (as required by IFC Section 320) within a Group M occupancy, a fire alarm system is required within the room. See the discussion of Section 907.2.2 for further information.

907.2.8 Group R-1. When asleep, the occupants of residential buildings will usually be unaware of a fire, and it will have an opportunity to spread before being detected. As a result, a majority of fire deaths in residential buildings have occurred because of this delay in detection. It is for this reason that the IBC requires fire alarm systems in addition to smoke detection in certain residential structures.

In hotels and other buildings designated as Group R-1, the general provisions mandate that both a manual fire alarm system and an automatic fire detection system be installed. There is an exception that eliminates the requirement for a manual alarm system for such occupancies less than three stories in height where all guestrooms are completely separated by minimum 1-hour fire partitions and each unit has an exit directly to a yard, egress court, or public way. This exception is based on the compartmentation provided by the separations between units, the requirement that the Group R-1 occupancy is sprinklered, and by the relatively rapid means of exiting available to the occupants. Where guestrooms are limited to egress directly to the exterior, early notification, although important, is not as critical. A second exception requires the alarm system, but does not mandate the installation of fire alarm boxes throughout buildings that are protected throughout by an approved supervised fire-sprinkler system. There is, however, a need for at least one manual fire alarm box installed in a location approved by the building official. In addition, sprinkler flow must activate the notification appliances. The automatic smoke detection system required by this section need only be provided within all corridors that serve guestrooms. An exception eliminates the requirement for the automatic smoke detection system in buildings where egress does not occur through interior corridors or other interior spaces.

907.2.9 Group R-2. Group R-2 buildings such as apartment houses are to be provided with a fire alarm system based on the number of dwelling units and sleeping units, as well as the location of any such units in relationship to the level of exit discharge. Where more than 16 dwelling units or sleeping units are located in a single structure, or where such units are placed at a significant distance vertically from the egress point at ground level, it is beneficial that a detection and notification system be provided. If any one of the three listed conditions exists, a manual fire alarm system is required unless exempted or modified by one of the three exceptions. Exceptions similar to those permitted for Group R-1 occupancies apply to Group R-2 buildings as well.

Provisions addressing the installation of an automatic smoke detection system in Group R-2 college and university buildings are essentially the same as those required for Group R-1 occupancies. The single difference is that the detection system must also be provided in laundry rooms, mechanical equipment rooms, and storage rooms, as well as all common areas located outside of the individual sleeping units or dwelling units. It is intended that these provisions be applicable only to those residential buildings that are operated by a university or college. It is not incumbent that the facility be located on the college or university campus. The requirement is applicable to all facilities operated by a college or university.

907.2.10 Group S. A manual fire alarm system is required in the corridors and common areas of self-storage facilities three or more stories in height. As with many of the other occupancies, the pull stations for the manual alarms can be eliminated when the building is sprinklered and the alarm system is activated by the water flow. While users of these storage facilities may know how to get to their unit, they are typically not familiar with the entire building and the alarm system can notify them of problems within the building. The provisions also establish that the visible alarm notification is only required out in the corridor and common areas and not within the storage units themselves.

Where lithium-ion or lithium metal batteries are stored within a Group S storage facility as required by IFC 320, a fire alarm system is required within the fire area containing the batteries. See the discussion of Section 907.2.2 for further information.

907.2.11 Single- and multiple-station smoke alarms. As indicated in the introduction to the residential fire alarm provisions, residential fire deaths far exceed those of any other building classification. Thus, the IBC requires smoke alarms in all residential buildings and in certain institutional occupancies. In Group R-1 occupancies, single- or multiple-station smoke alarms are to be installed in all sleeping areas, in any room along the path between the sleeping area and the egress door from the sleeping unit, and on each story within the sleeping unit. In all other residential occupancies, the code requires that smoke detectors be located in the sleeping rooms and the corridor or area giving access to the sleeping rooms. In addition, at least one smoke alarm shall be installed on each story of a dwelling unit, including basements. Where split levels occur in guestrooms or dwelling units, a smoke alarm need only be installed on the upper level, provided there is no intervening door between the adjacent levels. See Figure 907-2 for illustrations of these provisions.

In order to notify occupants throughout the dwelling unit or sleeping unit of a potential problem, multiple smoke alarms need to be interconnected. Therefore, when activation of one of the alarm devices takes place, activation of all the alarms must occur. This requirement is also applicable where listed wireless alarms are installed in lieu of physically interconnected smoke alarms. The intent of the code is that the alarms be audible throughout the dwelling, particularly in all sleeping rooms.

As effective as smoke alarms are in detecting and alerting to a fire, they can be a source of nuisance alarms, such as activation when the toast is burnt or reaction to steam from a shower or cooking operation. Therefore, the code specifies minimum separation distances from these potential sources of nuisance alarms. Table 907-1 provides the minimum required separation distances based on various conditions. The separation from the bathroom door is only required if the bathroom contains a bathtub or shower.

Sections 907.2.12 through 907.4.1 contain additional requirements for the installation of fire alarm systems. The previous requirements discussed were based on the occupancy classification; these sections are based on the use or operation within the building without regard for the occupancy classification. Table 907-2 lists these special uses and provides the requirement for the installation of a fire alarm system.

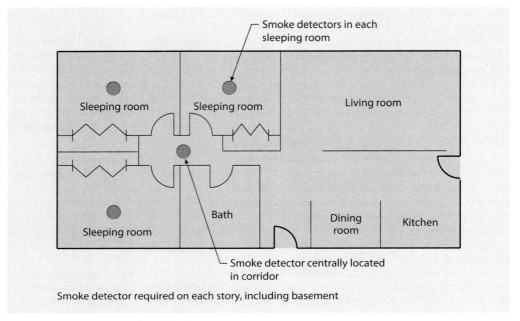

Figure 907-2 **Location of smoke alarms.**

Table 907-1. **Smoke Alarm Separation from Nuisance Alarm Sources**

	Distance from Permanently Installed Cooking Appliance [1]	Distance from Bathroom Door [2]
General requirements	10 feet (3,048 mm)	3 feet (914 mm)
Exception[3]	6 feet (1,829 mm)	3 feet (914 mm)

[1] The separation distance is a horizontal distance, such as measured on plan view.
[2] The separation distance is only required where the bathroom contains a bathtub or shower.
[3] Reduction permitted where necessary to comply with location requirements of Section 907.2.11.1 or 907.2.11.2.

907.4.2 Manual fire alarm boxes. Manual fire alarm boxes are one of the four initiating options for fire alarm systems allowed by Section 907.4. The IBC often requires a manual fire alarm system because of the special occupants or hazards that exist within the building. Manual fire alarm boxes, defined as manually operated devices used to initiate an alarm signal, and often referred to as pull stations, are used in many situations as a means for occupants to notify others of a potential fire emergency. This section identifies the proper locations for the installation of these alarm boxes.

In order that manual fire alarm boxes are readily available and accessible to all occupants of the building, they are to be located in close proximity to the point of entry to each exit. This would include placement within 5 feet (1,524 mm) of exterior exit doors, as well as doors entering interior exit stairways and exit passageways, as well as those doors accessing exterior exit stairways and horizontal exits. By placing the boxes adjacent to the exit doors, they will be available to occupants using any of the available exit paths. Additional boxes may be required in extremely large nonsprinklered structures, as the maximum travel distance to

Table 907-2. Required Fire Alarm Systems for Special Uses

Use or Operation	Required Fire Alarm System		
	Manual Alarm	Fire Detection[1]	Emergency Voice/Alarm Communication
Special amusement building		X	X
High-rise building		X	X
Atriums connecting >2 stories		X	
High-piled combustible storage		X	
Aerosol storage rooms and warehouses	X		
Lumber, wood structural panel, and veneer mills	X		
Underground buildings		X	
Deep underground buildings	X	X	X
Covered and open mall buildings >50,000 sq. ft. (4,645 m²)			X
Residential aircraft hangars		X[2]	
Airport traffic control towers		X	
Energy storage systems		X[3]	
HVAC system >2,000 cfm (0.94 m³/s)		X	
Special locking systems		X	
Elevator emergency operation		X	
Fire alarm control unit		X	

[1] Fire detection systems shall utilize smoke detectors, unless smoke detection is not appropriate for the location.
[2] Single- or multiple-station smoke alarms are acceptable in lieu of a fire detection system.
[3] Smoke detection or radiant-energy detection system permitted.

the nearest alarm box cannot exceed 200 feet (60,960 mm). In sprinklered buildings, travel distance to a fire alarm box is not limited. A manual fire alarm box, required to be red, must be located in a position so that it can be easily identified and accessed. The maximum height of 48 inches (1,372 mm), measured from the floor to the activating lever or handle, is based on the high-end-reach range limited by the accessibility provisions of ICC A117.1. The minimum height of 42 inches (1,067 mm) keeps the activating mechanism in a position readily viewed and providing ease of manipulation. See Figure 907-3.

Unless the fire alarm system is monitored by an approved supervising station, a sign must be installed on or adjacent to each manual fire alarm box advising the occupants to notify the fire department. Where a supervised alarm system is in place, notification becomes automatic and the sign is not necessary.

Often it is necessary to develop a means of reducing the accidental or intentional damage or activation of the alarm-initiating device. Therefore, the code gives the fire code official authority to accept protective covers placed over the listed boxes. The alarm box should remain easily identifiable when covered, with adequate instructions for operation. See Figure 907-4.

907.5 Occupant notification. Once a fire alarm system has been activated by one of the four initiating options listed in Section 907.4, the occupants need to be alerted and notified of the hazard.

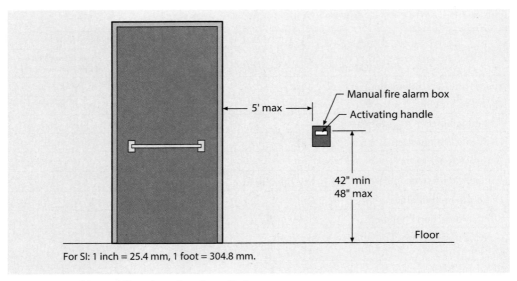

Figure 907-3 **Manual fire alarm box installation.**

Figure 907-4 **Manual fire alarm box with protective cover.**

In most situations, it is intended that all occupants in the building are notified. There are times when this is not the best approach, such as in a hospital or detention facility where the occupants cannot respond on their own. In certain cases, the code allows the notification to be to staff so that appropriate actions can be taken.

Notification to the occupants and staff occurs by either an audible signal or a visible signal, or both. Where the occupants are expected to respond the alarm, such as an office building, an audible alarm must be heard throughout the facility. Section 907.5.2.1 addresses audible alarms and requires that the audibility must be at least 15 decibels above the average ambient sound level. This must be accomplished without any single device exceeding 110 decibels. Sound pressure above 110 decibels can cause damage to the ear.

The type of audible signal used in sleeping rooms of Group I-1, R-1, and R-2 occupancies must be a low-frequency (520 Hz) signal instead of the typical 3 kHz that is normally used for alarms. The low-frequency signal has been shown to improve the waking effectiveness for several higher-risk occupant types (e.g., people over 65 who are hard of hearing, school-age children, and people who are alcohol impaired).

907.5.2.2 Emergency voice/alarm communication systems. As we have seen in certain occupancies or uses, an emergency voice/alarm communication (EV/AC) system is required. This system is a fire alarm system with a different type of notification process. When the fire alarm system is activated, the EV/AC sounds a prerecorded message rather than the beep, beep, beep of traditional fire alarm systems. The system can be designed with a different prerecorded message for different activation scenarios. For example, when a smoke detector activates on the east side of the building, the message can direct occupants to the west exits. The EV/AC can also be zoned so that when a fire occurs on a floor in a high-rise building, the notification is delivered to the fire floor and the floor above and the floor below. The messages are to be approved and would be addressed in the fire safety and evacuation plan required by Section 1002.2.

The speakers placed throughout the building must be grouped into paging zones to allow for different messages for specific locations within the building. At a minimum, a separate paging zone is required for each floor level, elevator groups, areas of refuge, and interior exit stairways.

Another function of the EV/AC system is the ability to provide live announcements. The initial message may have been delivered, but now a different message is needed based on the spread of the fire. For example, using the EV/AC system the fire department can direct occupants to different stairways or advise them to stay in place.

907.5.2.3 Visible alarms. Visible alarms are intended to alert hearing-impaired individuals to a fire emergency. Visible alarms are also required in areas where the ambient noise level exceeds 95 decibels since the audible alarm sound pressure level must be 15 decibels above ambient, but not more than 110 decibels. As with those locations where audible alarm systems are required, the IBC identifies the specific conditions under which visible alarm notification appliances are also mandated. In those portions of the building deemed to be public areas or common areas, visual alarms shall be provided in addition to audible alarms. For example, in an office building, the lobby, public corridors, and public restrooms would be considered public areas, whereas the corridors, toilet rooms, break areas, and conference rooms inside an office suite would be considered common areas. On the other hand, the individual offices of each employee would be considered private use. Although such offices would not require the installation of visible alarm notification appliances, wiring must be in place for future installation of the alarms as necessary. The potential of additional visible alarm notification appliances is taken into account by requiring at least 20 percent spare capacity for the appliance circuits. Keep in mind that visible alarms are only required in those occupancies where an alarm system is first required by Section 907.2.

Table 907.5.2.3.2 is used to determine the number of sleeping units in Group R-1 or I-1 occupancies that must be provided with visible alarm notification appliances. In both occupancies, the appliances are to be activated by both the in-room smoke alarm and the building's fire alarm system. The number of sleeping units required to have both visible and audible alarms is based on the total number of units in the building. Not strictly limited to sleeping units, the visible and audible alarm requirements are also applicable where dwelling units are located within a Group R-1 occupancy. In Group R-2 apartment houses and similar occupancies required by the code to have a fire alarm system, provisions must be made for the future installation of visible alarm notification devices in accordance with ICC A117.1 as they become necessary.

Section 909 *Smoke Control Systems*

The provisions of this section are applicable to the design, construction, testing, and operation of mechanical or passive smoke control systems only when they are required by other provisions of the IBC. Section 909 specifically exempts smoke- and heat-removal requirements that appear in Section 910 and are discussed in the next section of this handbook. Also, this section states that mechanical smoke control systems are not required to comply with Chapter 5 of the *International Mechanical Code* (IMC®) for exhaust systems unless their normal use would otherwise require compliance.

The provisions of this section establish minimum requirements for the design, installation, and acceptance testing of smoke control systems, but nothing within the section itself is intended to imply that a smoke control system is to be installed. Some sections that specifically reference Section 909 are the requirements for atriums (Section 404.5), underground buildings (Section 405.5), and windowless buildings housing Group I-3 occupancies (Section 408.9). Smoke control systems are intended to provide a protective environment in areas outside that of fire origination to allow for the evacuation or relocation of occupants in a safe manner. The provisions are not designed to protect the contents from damage or assist in fire-fighting activities. Much of this section was originally based on the American Society of Heating, Refrigerating, and Air-Conditioning Engineers (ASHRAE) publication, *Design of Smoke Control for Buildings*; NFPA 92-A, Standard for "Smoke-Control Systems Utilizing Barriers and Pressure Differences"; and its companion NFPA 92-B, "Standard for Smoke Management Systems in Malls, Atria, and Large Spaces." These three documents have been replaced and updated since then, and NFPA 92, "Standard for Smoke Control Systems" has replaced the previous NFPA 92-A and 92-B standards and is now referenced.

Although this section covers both passive and active smoke control systems, the majority of the material presented addresses the three mechanical methods—pressurization (Section 909.6), airflow (Section 909.7), and exhaust (Section 909.8)—with other sections addressing related subjects such as the design fire; equipment, including fans, ducts, and dampers; power supply; detection and control systems; and the fire fighter's smoke control panel.

An important segment of Section 909 addresses acceptance testing of the smoke control system. Smoke control system installation requires special inspection testing per Section 909.18.8 to be performed during erection of ductwork and prior to concealment. These inspections are intended for the purpose of testing for leaks, as well as for recording the specific device locations. The latter creates, in effect, as-built drawings for the system. Additional testing and verification prior to occupancy is also mandated. Section 909.18.8.3 requires the work of the special inspector to be documented in a final report. The report shall be reviewed by the responsible registered design professional who is required to certify the work. The final, designer-approved report, together with other information addressed in Section 909.18.9, shall be provided to the fire code official, and a copy shall be maintained on file at the building.

909.20 Smokeproof enclosures. The provisions of this section identify the methods for complying with Section 1023.12 for the construction of smokeproof enclosures. Smokeproof enclosures are required by Sections 403.5.4 for high-rise buildings, 405.7.2 for underground buildings, and 412.2.2.2 for airport traffic control towers. There are two methods for construction of a ventilated smokeproof enclosure, both of which use an enclosed interior exit stairway or ramp. Either an exterior balcony, or a vestibule used in conjunction with pressurization of the stair enclosure and the vestibule, can be used as the buffer between the floor of the building and the exit stairs or ramp. In addition, pressurization of the stair or ramp shaft without an exterior balcony or vestibule is a permitted alternative.

Unless the pressurization provisions of Section 909.20.4 are used where a smokeproof enclosure is required, the exit path to the stair or ramp shall include a vestibule or an open exterior balcony. The minimum size of the balcony or vestibule is established in Section 909.20.1 and illustrated in Figure 909-1. A minimum 2-hour fire-resistance-rated fire barrier separates the smokeproof enclosure from the remainder of the building and also separates the stairway or ramp from the vestibule. Per Section 1023.4, the only openings permitted into the enclosure are the required means of egress doors. Where ventilating equipment is used for either the vestibule, stair or ramp enclosure, it must comply with Section 909.20.6. Construction of an open exterior balcony is based on the required fire-resistance rating for the building's floor construction, which would typically be 2 hours.

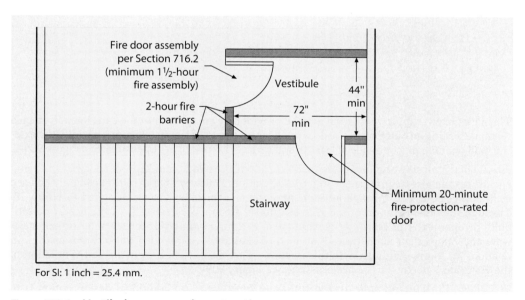

Figure 909-1 **Vestibule access and construction.**

909.20.3 Natural ventilation alternative. In this section, the code provides the details of construction where natural ventilation is used to comply with the concept of a smokeproof enclosure. Where an open exterior balcony is provided, fire doors into the stairway or ramp shall comply with Section 716.2. In a vestibule scenario, a minimum 90-minute fire door assembly is required between the floor and the vestibule. Between the vestibule and the stairway or ramp, the door assembly need only have a 20-minute fire-protection rating.

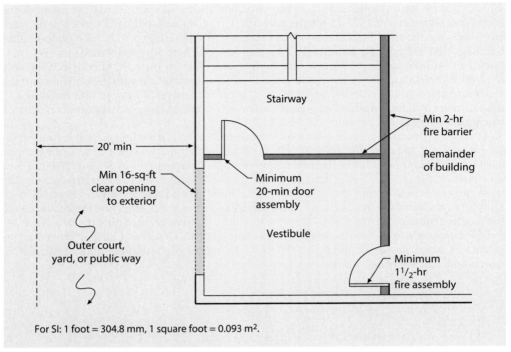

Figure 909-2 **Natural ventilation of vestibule.**

The necessary vestibule ventilation is to be provided by an opening in the exterior wall at each vestibule. Facing an outer court, yard, or public way at least 20 feet (6,096 mm) in width, the exterior wall opening must provide at least 16 square feet (1.5 m²) of net open area. See Figure 909-2.

909.20.4 Stairway and ramp pressurization alternative. Smokeproof stairway and ramp enclosures can also be protected using pressurization. Stairway pressurization can only be used in buildings equipped throughout with an automatic sprinkler system designed to NFPA 13. Through the pressurization of the stair or ramp shaft to a prescribed level, the need for vestibules or open exterior balconies is eliminated. Pressurization levels for the interior exit stairways and ramps shall fall between 0.10 inch (25 Pa) and 0.35 inch (87 Pa) of water column in relationship to the building. With this method of pressurization, a vestibule and ventilation of the vestibule is deemed to be unnecessary. See Figure 909-3.

909.20.5 Pressurized stair and vestibule alternative. An additional design option is to provide a combination vestibule/stairway pressurization system. This method allows for slightly lower pressure differentials than required under Section 909.20.4 due to the presence of the vestibule. Having the vestibule serves as a second line of defense to keep smoke from entering the stairway. Under this option, the smoke would need to overcome the vestibule pressurization, and then the increased enclosure pressure in order to enter the stairway system. See Figure 909-4.

909.21 Elevator hoistway pressurization alternative. Elevator shafts must often be protected to limit the spread of smoke from one story to another. Elevator hoistways can be pressurized, similar to stairway pressurization, in lieu of constructing an elevator lobby or meeting other protection criteria as established in Section 3006.3. A difference between

Smoke Control Systems 361

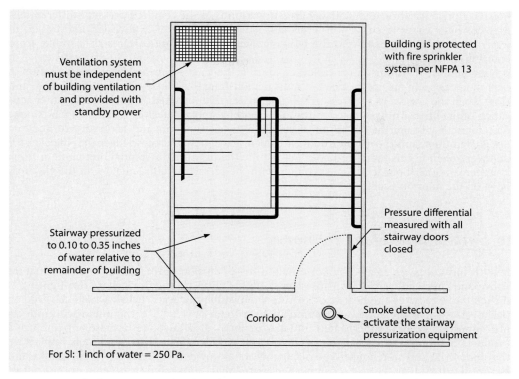

Figure 909-3 **Stairway pressurization alternative.**

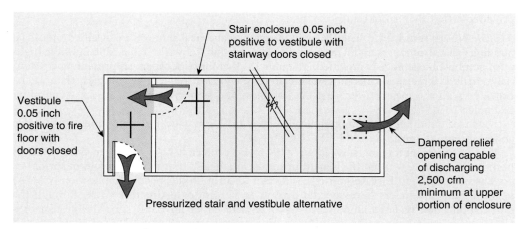

Figure 909-4 **Pressurized stair and vestibule alternative.**
Fan system to have smoke detector to allow for automatic system shut down.

hoistway pressurization and stairway pressurization is in the required pressure differentials. Elevator hoistways pressurization levels fall between 0.10 and 0.25 inch (25 and 67 Pa) of water column in relationship to the building. The increased pressure restricts smoke from entering the hoistway.

One concern is the use of the elevator during the time when hoistway pressurization is operating. As such, the acceptance test must demonstrate that all elevator doors will open and close when the hoistways are pressurized. The pressurization system is activated by either activation of the fire alarm system, or activation of a lobby smoke detector. When the lobby smoke detector activates, per the ASME A17.1 standard the elevator cars are captured, returned to a designated floor, and the elevator doors are open. Exception 4 acknowledges that the elevator doors are open in this mode and does not require the pressure differential on that floor level; however, all other floor levels must comply with the pressure requirements with the elevators open on the floor of recall.

Section 910 *Smoke and Heat Removal*

Smoke and hot gases created by a fire rise to the underside of the ceiling or roof structure above and then build up. This causes reduced visibility to the point where fire-fighting is difficult and radiates the heat back to the combustible materials below which can lead to flashover. Also, as the hot gases accumulate near the roof structure, the unburned products of combustion become superheated, and if a supply of air is introduced, these hot, unburned products of combustion will ignite violently. Thus, it has been found that it is imperative that industrial and warehouse-type occupancies be provided with some type of smoke and heat removal system. This is accomplished with either smoke and heat vents in the roof or a mechanical smoke removal system. Although this section is typically applied to one-story buildings, as well as one-story portions of multistory buildings, smoke and heat removal may also be required in a multistory structure. Where the space requiring smoke and heat removal is not located at the upper story of the building, a mechanical smoke removal system must be provided rather than smoke and heat vents.

910.2 Where required. The IBC requires smoke and heat removal in specified industrial buildings and warehouses and in any occupancy, as required by IFC Table 3206.2, where high-piled combustible storage is provided. The intent is that those occupancies presenting a potential to include large areas of combustible materials, and as a result, large volumes of smoke, be provided with the means to exhaust the hot gases and smoke during a fire. See Application Example 910-1.

For those buildings containing high-piled combustible storage, the requirements for smoke and heat removal are addressed in IFC Chapter 32.

It is not uncommon for such large areas to be protected by a sprinkler system that is designed to operate without the benefit of smoke and heat removal. These types of sprinklers are Early Suppression Fast Response (ESFR) and Control Mode Specific Application (CMSA) systems. Such sprinklers have specific designs to control or extinguish a fire, through the quick application of large amounts of water. Where either an ESFR sprinkler system or a CMSA sprinkler system using sprinklers with a response time index (RTI) of 50 or less and designed to operate with no more than 12 sprinklers, are installed, smoke and heat removal is not required.

In buildings where smoke and heat removal is required, it can be accomplished with either smoke and heat vents or a mechanical smoke removal system. The method chosen is up to the designer; however, there are some limitations. In nonsprinklered buildings, smoke and

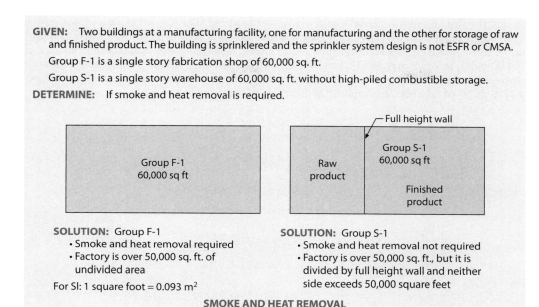

Application Example 910-1

heat vents are the only acceptable method. In sprinklered buildings, either method will suffice provided that the protected area is only one story in height. When it is more than one story, smoke and heat vents are not an option and a mechanical smoke removal system must be used.

910.3 Smoke and heat vents. As smoke and heat vents are intended to release smoke and hot gases from a fire within the building, the code requires that they operate automatically but with the capability to open manually. The smoke and heat released should not put any adjacent building or property into danger, so the smoke and heat vents must be placed a minimum of 20 feet (6096 mm) from fire walls and from any lot line in order to reduce exposure to adjacent property. Smoke and heat vents shall also be located at least 10 feet (3048 mm) from fire barriers. Smoke and heat vents need to be distributed uniformly to provide smoke removal throughout the protected area. Such conditions as roof pitch, curtain location, sprinkler head location, and structural members shall be considered in the location of vents. The fusible link temperature to open the vents is set high enough to ensure that the sprinkler system will activate prior to the vents opening.

Note that the formula to determine the required venting area is different for sprinklered buildings versus nonsprinklered buildings. For sprinklered buildings, the entire volume of the protected area is used and then divided by 9,000 to determine the minimum vent area. In nonsprinklered buildings, the floor area is used rather than the volume. In the nonsprinklered scenario, the floor area is divided by 50 to determine the minimum vent area.

910.4 Mechanical smoke removal systems. A mechanical smoke removal system is only intended to remove the smoke for fire-fighting operations; it is not a life-safety system, and it is not a smoke control system as regulated in Section 909. The system is designed to operate during the fire and remove smoke to provide ventilation similar to the smoke and heat vents. The design criteria are to provide two air changes per hour for the entire volume of the protected area. Ventilation is accomplished with exhaust fans, each with a maximum capacity of 30,000 cubic feet per minute (14.2 m³/s). As with any exhaust system in a building, makeup air needs to be provided at a ratio of 8 square feet per 1,000 cubic feet per minute (0.74 m² per 0.4719 m³/s) of exhaust. See Application Example 910-2.

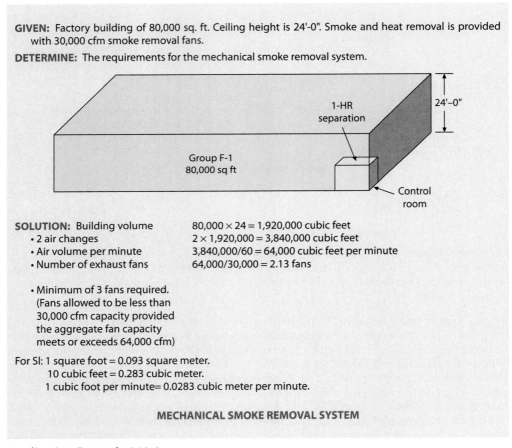

Application Example 910-2

The smoke removal system can be used as part of the normal building ventilation, as long as the exhaust fans shut down automatically upon sprinkler water flow. The fire department can control the exhaust fans manually with the manual controls that override the automatic shut down and allow on/off control of the fans.

The mechanical exhaust fans are expected to operate during the fire occurrence. To allow this operation, the smoke removal system must be wired ahead of the main disconnect and the wiring must be protected against an interior fire exposure of 1,000°F (538°C) for a period of not less than 15 minutes. This can be accomplished with fire-resistance-rated construction or cable listed for this purpose.

Section 911 *Fire Command Center*

The fire command center is the control center for the fire- and life-safety systems in a complex building. The IBC mandates a fire command center be provided in only very special structures such as high-rise buildings, airport traffic control towers, buildings with smoke-protected assembly seating and large Group F-1 or S-1 buildings (see Sections 403.4.6, 412.2.3.2, 909.16, and 911.1, respectively). The purpose of the command center is to provide a central location where fire personnel can operate during an emergency and coordinate fire-fighting operations and building systems. Located as determined by the fire department, the fire command center shall be isolated from the remainder of the building by a minimum 1-hour fire-resistance-rated fire barrier. In general, the size of the fire command center is dependent on the building size, but must be a minimum size of 200 square feet (19 m^2) in area, but a 96 square foot (9 m^2) size is permitted for the F-1 and S-1 buildings if approved by the fire code official. The room must contain all of the system components, controls, display panels, indicators, devices, furnishings, and plans that are listed in Section 911.1.6.

Except for the large Group F-1 and S-1 buildings, the provisions of Section 911 are only applicable where some other provision of the code specifically mandates that a fire command center be provided. As an example, Section 909.16 mandates that the fire fighter's smoke-control panel required for buildings provided with a smoke-control system must be located in a fire command center if the smoke-control system is used to address smoke-protected assembly seating conditions. For other smoke-control applications, a fire command center is not required to be provided.

Section 912 *Fire Department Connections*

Buildings equipped with an automatic sprinkler system designed to NFPA 13 "Standard for the Installation of Sprinkler Systems" or NFPA 13R "Standard for the Installation of Sprinkler Systems in Low-Rise Residential Occupancies" will typically have a fire department connection (FDC). An FDC is not required for sprinkler systems designed to NFPA 13D "Standard for the Installation of Sprinkler Systems in One- and Two-Family Dwellings and Manufactured Homes". Class I and III standpipes will also be equipped with an FDC, while Class II standpipe systems will not. The FDC is used to augment the water supply in a fire sprinkler system, or to provide water in a standpipe system. In a fire sprinkler system, this allows the fixed sprinkler piping to apply more water directly onto the fire, many times before the fire-fighting crews can reach the fire with hose lines. Since the FDC is critical to effectively fighting the fire, it must be available and usable. Section 912 contains criteria intended to keep the FDC in an accessible and usable condition. The hose threads must match the threads on the fire hose carried by the fire department. The FDC is to be located on the street side of the building or on a side facing a fire apparatus access road. The FDC must be immediately accessible and requires a minimum clear space of 36 inches (762 mm) on all sides and 78 inches (1981 mm) vertically. This clear space needs to take into account the building, fences, railings, and landscaping. The location should be chosen so that the hose lines from the closest fire hydrant to the FDC do not obstruct on-site fire apparatus access roads, although this is not always possible. See Figure 912-1.

Figure 912-1 **Fire department connection and fire hydrant on same side of driveway.**

Section 913 *Fire Pumps*

Many taller buildings or buildings with high-piled combustible storage require a fire pump to supply adequate pressures and flows for the automatic sprinkler system or standpipe system. The fire pump can be located within the building, or located outside. Since the fire pump is expected to operate for the duration of the fire in the building, it must be protected. The fire pump room must be separated from the remainder of the building by at least 2-hour fire-resistance-rated construction in high-rise buildings and 1-hour construction in all other buildings if they are fully sprinklered. If the fire pump is located outside, NFPA 20 "Standard for the Installation of Stationary Fire Pumps for Fire Protection" requires that it is located at least 50 feet (15,240 mm) from the building which protects it from an exposure fire. See Figure 913-1.

Additional issues to be addressed for the protection of the fire pump are freezing concerns and power supply. The pump room must be maintained above 40°F (5°C). This is necessary to keep the water from freezing or becoming slush in the small piping of many of the components and controls in the pump room. Even though water freezes at 32°F (0°C), it will start to form ice crystals and become slushy as the temperature approaches freezing. The threshold of 40°F (5°C) is also used when determining whether a fire sprinkler system can be a wet-pipe sprinkler system or dry-pipe sprinkler system. Where located outside of a protected pump room, electrical circuits needed for the survivability and operation of the fire pump must be protected by 1-hour construction, listed cables with a fire-resistance rating of 1-hour or be encased in concrete.

Section 914 *Emergency Responder Safety Features*

Section 914 is intended to provide correlation to the current requirements in the IFC for the identification of shaftway hazards and the location of fire protection systems and controls for utility and mechanical equipment. These requirements are located in Sections 316.2 and 509.1 of the IFC.

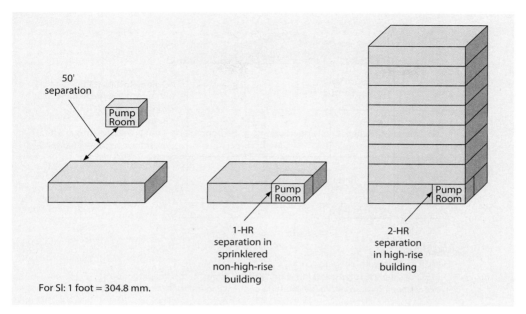

Figure 913-1 Protection of fire pump room from fire exposure.

Section 101.3 of the IBC states that the safety of emergency responders is part of its scope and purpose. This section reinforces that purpose by specifying that interior and exterior shaftway hazards be identified as well as the location of fire protection systems, such as fire alarm control units or automatic sprinkler risers, and rooms housing HVAC and elevator controls.

Section 915 *Carbon Monoxide Detection*

Carbon monoxide (CO) is an odorless and colorless gas, and it can be deadly. CO is produced when the use of a solid, liquid, or gaseous fuel does not result in a complete combustion process. In these instances, CO is produced. Some examples where CO can be produced are a gas or wood fireplace, a gas stove, heating/ventilation/air-conditioning equipment powered by fuel oil or gas, and a vehicle stored in the garage. Non-fire-related CO fatalities in the United States average more than 400 per year. CO detectors, with exceptions, are required in all buildings where one or more of four listed conditions exist.

In residential-type occupancies where CO detectors are required, they are to be installed in locations similar to smoke alarms. Detection is required on each floor with sleeping units and located outside of each sleeping area, but in the immediate vicinity of the sleeping rooms. When more than one CO detector is provided, they must be interconnected. Where the potential CO source is in a sleeping room or bathroom, detection must be installed within the sleeping room. See Figure 915-1. In Group E, detection shall be installed inside each classroom. The primary power supply is the building wiring, and each CO detector must have a battery backup. In all other occupancies, detectors are to be located on the ceiling of any enclosed rooms or areas containing CO-producing devices or served by a CO source forced-air furnace.

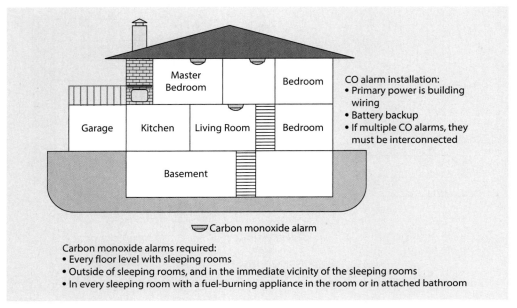

Figure 915-1 **Carbon monoxide alarms in a dwelling unit.**

The definition of *carbon monoxide source* in Chapter 2 helps limit the scope of the provisions. The definition and the code requirements of Section 915 are intended to cover permanent CO sources, including regularly used vehicles (vehicles within garages, propane forklifts, etc.) and recurrent situations, such as floor cleaners that are brought in periodically. Potential inappropriate situations such as people using barbeque grills or gas-powered generators inside their homes during a power outage are also contemplated. This definition would also exclude the provisions from applying to items such as candles or gelled alcohol cans used in chaffing dishes.

Section 915.1 addresses circumstances where CO detection is needed. Section 915.2 specifies locations where CO detection devices are required to provide the best protection. Sections 915.3 through 915.5 then deal with installation, including compliance with NFPA 72 *National Fire Alarm and Signaling Code*, electrical hardwiring and interconnection. Depending on other protection features existing within the building, this may permit (1) the use of standalone CO alarms where there is no fire alarm system present, (2) the addition of CO detectors (with notification) in places where a fire alarm system is already present, or (3) the addition of CO detectors attached to an existing CO detection system—when installed per NFPA 72.

Section 916 *Gas Detection Systems*

Gas detection systems are required to detect leaks in many uses and hazardous operations as regulated by both the IBC and IFC. For example, gas detection systems are required in semiconductor fabrication facilities, hydrogen fuel gas rooms, refrigerated liquid CO_2 beverage dispensing operations, extraction facilities using flammable gases, areas of storage and use of highly toxic or toxic gases, and ozone gas generator rooms, along with other situations where gas detection is used as an option to ventilation. This section addresses the general requirements for gas detection systems.

Each gas detection system must be designed for use with the specific gas being detected. Secondary power is required unless the gas detection system provides a trouble signal at an approved location upon power loss.

The process of detection is intended to be a continuous ongoing process. However, for HPM gases, the detection system must obtain a sample at least every 30 minutes. A sampling interval of 5 minutes is permitted where the gas detection system is provided in lieu of an exhaust treatment system as allowed for toxic gases in accordance with IFC Section 6004.2.2.7. The sample is evaluated, and when the gas concentration exceeds 25 percent of the lower flammable limit (LFL) for flammable gases, or one-half of the IDLH (immediately dangerous to life and health) for nonflammable gases, the system shall activate a visible alarm or audible signal.

Section 917 *Mass Notification Systems*

An evaluation of the need of a mass notification system on college and university campuses is addressed in this section, triggered when there are multiple buildings on the campus and the aggregate occupant load of those buildings is 1,000 or more. A similar requirement is also applicable to Group E educational occupancies where two conditions exist: a fire alarm system is required in the building, and the building has an occupant load of 500 or more. NFPA 72 *National Fire Alarm and Signaling Code* addresses two types of mass notification: in-building mass notification and wide-area mass notification.

> **In-building mass notification** provides real-time information and messages to building occupants by voice communication or visible text, graphics, or signals.
>
> **Wide-area mass notification** can provide real-time information and messages to outdoor areas.

A risk analysis must be completed to determine if mass notification is required. The risk analysis is to evaluate fire and non-fire emergencies, such as geological events, meteorological events, biological events, human-caused intentional events, human-caused accidental events, and technical events. The mass notification system must provide prerecorded messages and have the capability of delivering live voice messages.

If the risk analysis indicates that an emergency communication system is necessary for mass notification, then a mass notification system must be provided. The risk analysis is required prior to building construction, so it should be available at the time of plan review.

KEY POINTS

- Automatic sprinkler systems are typically installed because they are mandated by the code, or because they are to be used as equivalent protection to other code requirements.
- Because of the potentially high occupant load and density anticipated in Group A occupancies, coupled with the occupants' probable lack of familiarity with the means of egress system, large assembly uses must be protected by an automatic sprinkler system.
- Most school buildings must be sprinklered throughout unless complying compartmention is provided.
- Large manufacturing buildings and warehouses, when containing combustible goods or materials, must be sprinklered to limit the size of a fire.
- The IBC requires sprinkler protection for all Group H occupancies owing to local hazards within the building and the potential for presenting a high level of hazard to the surrounding properties.
- Because the mobility of the occupants of Group I occupancies is greatly diminished, the code requires an automatic sprinkler system in such occupancies.
- On account of their fire record, hotels, apartment buildings, assisted-living facilities, and all other residential occupancies must always be sprinklered.
- Adequate openings must be provided in exterior walls for fire department access, or a sprinkler system must be installed.
- Standpipe systems are typically required based upon the building's height or number of stories.
- The required locations of Class I standpipe connections are specifically identified in the code.
- The required locations of Class II standpipe hose cabinets are based on the distances that the fire hose can reach throughout the building.
- One of the most effective means of occupant notification in case of a fire incident is the availability of a fire alarm system.
- Pressurization, airflow, and exhaust are the three methods of mechanical smoke control.
- Three means are established to create a smokeproof enclosure: the natural ventilation alternative, the stairway and ramp pressurization alternative, and the pressurized stair and vestibule alternative.
- Smoke and heat vents or a mechanical smoke removal system is required in large, open areas of manufacturing and warehouse occupancies, as well as retail sales with high-piled stock.
- CO detection is required under specific conditions in all occupancies where a CO source is present.
- A risk analysis is required to determine if a mass notification system is required for larger college and university campuses.

CHAPTER 10

MEANS OF EGRESS

Section 1001 Administration
Section 1002 Maintenance and Plans
Section 1003 General Means of Egress
Section 1004 Occupant Load
Section 1005 Means of Egress Sizing
Section 1006 Number of Exits and Exit Access Doorways
Section 1007 Exit and Exit Access Doorway Configuration
Section 1008 Means of Egress Illumination
Section 1009 Accessible Means of Egress
Section 1010 Doors, Gates, and Turnstiles
Section 1011 Stairways
Section 1012 Ramps
Section 1013 Exit Signs
Section 1014 Handrails
Section 1015 Guards
Section 1016 Exit Access
Section 1017 Exit Access Travel Distance
Section 1018 Aisles
Section 1019 Exit Access Stairways and Ramps
Section 1020 Corridors
Section 1021 Egress Balconies
Section 1022 Exits
Section 1023 Interior Exit Stairways and Ramps
Section 1024 Exit Passageways
Section 1025 Luminous Egress Path Markings
Section 1026 Horizontal Exits
Section 1027 Exterior Exit Stairways and Ramps
Section 1028 Exit Discharge
Section 1029 Egress Courts
Section 1030 Assembly
Section 1031 Emergency Escape and Rescue
Key Points

This chapter establishes the basic approach to determining a safe exiting system for all occupancies. It addresses all portions of the egress system and includes design requirements as well as provisions regulating individual components, which may be used within the egress system. The chapter specifies the methods of calculating the occupant load that are used as the basis of designing the system and, thereafter, discusses the appropriate criteria for the number of exits, location of exits, width or capacity of the egress system, and the arrangement of the system. This arrangement is treated in terms of remoteness and access to the egress system. The issue of access handled both in terms of the system's usability by building occupants and in terms of it being available within a certain maximum distance of travel. After having dealt with general issues that affect the overall system or multiple zones of the system defined as the exit access, exit, and exit discharge, the chapter then establishes the design requirements and components that may be used to meet those requirements for each of the three separate zones.

In interpreting and applying the various provisions of this chapter, it would help to understand the four fundamental concepts on which safe exiting from buildings is based:

1. A safe egress system for all building occupants must be provided.

2. Throughout the system, every component and element that building occupants will encounter in seeking egress from the building must be under the control of the person wishing to exit.

3. Once a building occupant reaches a certain degree or level of safety, as that occupant proceeds through the exiting system, that level of safety is not thereafter reduced until the occupant has arrived at the exit discharge, public way, or eventual safe place.

4. Once the exit system is subject to a certain maximum demand in terms of the number of persons, that system must thereafter (throughout the remainder of the system) be capable of accommodating that maximum number of persons.

Egress for individuals with physical disabilities is to be provided under the provisions of this chapter, primarily through the design of an accessible means of egress system. Because many of the elements composing the egress system (doors, landings, ramps, etc.) may also form part of the accessible routes as required by Chapter 11, such requirements must be referenced where applicable.

The IBC® uses a three-part approach to the "means of egress." The three-part system, or zonal approach as it is now used, was introduced by the National Fire Protection Association (NFPA) in 1956 and was incorporated over the years into all of the legacy model codes. This approach has established terms that are used throughout the design and enforcement communities to deal with the means of egress system. The three parts of the means of egress system are the exit access, the exit, and the exit discharge. For conceptual ease, the exit access is generally considered any location within the building from where you would start your egress travel, and continues until you reach the door of an exit. The exit access would include all the rooms or spaces that you would pass through on your way to the exit. This may be the room you are in; an intervening room; a corridor; an exterior egress balcony; and any doors, ramps, unenclosed stairs, or aisles that you use along that path. An exit is the point where the code considers that you have obtained an adequate level of safety so that travel distance measurements are no longer a concern. Exits will generally consist of fire-resistance-rated construction and opening protection that will separate the occupants from other areas of the building where a fire condition could exist. Elements that are considered exits include exterior exit doors at ground level, interior exit stairways, interior exit ramps, exit passageways, exterior exit stairways, exterior exit ramps, and horizontal exits. Exterior exit doors, exterior exit stairways, and exterior exit ramps will not provide the fire-protection levels that the other elements provide, but insofar as the occupant will be outside the building, they will provide a level of safety by removing the occupants from the hazard area and eliminating the concerns of smoke accumulation. The last of the three parts is the exit discharge. The *International Building Code*® (IBC) will generally view exterior areas at ground level as the exit discharge portion of the exit system. Therefore, the exit access will be the area within the building that gets the occupants to an exit, whereas the exit discharge will be the exterior areas at grade where the occupants go upon leaving the building in order to reach the public way.

Section 1001 *Administration*

This section requires that every building or portion thereof comply with provisions of Chapter 10. In dealing with portions of buildings, it is important to understand that the code intends this chapter to apply to all portions that are occupiable by people at any time. Therefore, areas such as storage rooms and equipment rooms, although often unoccupied, will still be regulated under the provisions of the chapter.

In order to provide an approved means of egress at all times, it is critical that the exiting system be maintained appropriately. Section 1032 of the *International Fire Code*® (IFC®) regulates maintenance of the means of egress for the life span of the building. Should there be alterations or modifications to any portion of the building, Section 1001.2 mandates that the number of existing exits not be reduced, nor the capacity or width of the means of egress be decreased, below that level required by the IBC. Section 1104 of the IFC also provides a limited number of specific provisions addressing the means of egress in existing buildings.

Section 1002 *Maintenance and Plans*

While Chapter 10 of the IBC and IFC are generally identical, the IFC includes a Section 1032 which addresses the maintenance of the means of egress. This information is not included within the IBC due to the code's scoping, and therefore a reference to the IFC is provided.

A reference to the IFC provisions addressing fire safety, evacuation, and lockdown plans recognizes the need to provide consistent and effective fire- and life-safety operations during emergency conditions. Section 404 of the IFC requires the fire-safety and evacuation plans in those occupancies and building types as set forth in IFC Section 403. These plans are required to include or address a number of different types of issues that may affect the egress of occupants from the building. Along with other items, these include the identification of potential hazards, exits, primary and secondary egress routes, and occupant assembly points, as well as establishing procedures for assisted rescue for people who are unable to use the general means of egress unassisted.

Section 1003 *General Means of Egress*

The requirements and topics addressed in this section are used as basic provisions and are to be applied throughout the entire egress path as applicable. Examples of the types of general issues that are found here include ceiling height, protruding objects, floor surface, elevation change, and egress continuity. For consistency purposes, the provisions for ceiling height and protruding objects are identical to the accessibility criteria of ICC A117.1.

1003.2 Ceiling height. In order to provide an exit path that maintains a reasonable amount of headroom clearance for the occupants, this section requires the means of egress to have a minimum ceiling height of 7 feet 6 inches (2,286 mm). The intent of the provision is to address all potential paths of exit travel that can be created based on multiple directions of egress and the layout of the room or space insofar as furniture, equipment, and fixtures are concerned. Any portion of the floor area of the building that can reasonably be considered a possible exit path should be provided with a minimum 7-foot 6-inch (2,286-mm) clear height, unless reduced by exceptions permitted for sloped ceilings, dwelling and sleeping units in residential occupancies, stairway and ramp headroom, door height, and protruding objects. Additional exceptions reduce the minimum required clear height to 7 feet 0 inches (2,134 mm) in parking garage vehicular and pedestrian traffic areas as well as above and below floors considered as mezzanines.

1003.3 Protruding objects. Limitations are placed on the permitted projection of protruding objects for two purposes. First, to maintain an egress path that is essentially free of obstacles. Second, to provide a circulation path that is usable by all occupants, including those individuals with sight-related disabilities. For this reason, provisions regulate the accessibility concerns regarding protruding objects as well as the egress concerns. Note that projections into the required egress width and the minimum clear width of accessible routes are also limited by other provisions of the code.

1003.3.1 Headroom. Consistent with the allowance for stair headroom and doorway height to be reduced below the required egress height of 7 feet 6 inches (2,286 mm), other portions of the egress system may likewise be reduced to a minimum height of 80 inches (2,032 mm). The reduction for signage, sprinklers, decorative features, structural members, and other protruding objects is limited to 50 percent of the ceiling area of the egress path. See Figure 1003-1. Though projections at an 80-inch (2,032-mm) height are not unusual to building occupants, it is necessary to maintain a majority of the egress system at 7 feet 6 inches (2,286 mm) or higher. Passage through a doorway may be further reduced in height to 78 inches (1,981 mm) at the door closer, overhead door stop or frame stop, as well as at any power door operator or electromagnetic door lock. This reduction at doors is also permitted for accessibility purposes by Section 307.4 of ICC A117.1. Where a vertical clearance of 80 inches (2,032 mm) cannot be achieved, the reduced-height portion of such floor area cannot be used as a portion of the means of egress system. It is also necessary to provide some type of barrier that will prohibit the occupant from approaching the area of reduced height. This is of particular importance where the occupant is sight impaired, with no method other than a barrier to identify the presence of an overhead protruding object. The mandated barrier is to be installed so that the leading edge is no more than 27 inches (686 mm) above the walking surface, as shown in Figure 1003-2. By limiting the height of the barrier edge, it will be located in a manner so that a sight-impaired individual using a long cane will detect the presence of an obstruction and maneuver to avoid the hazard.

1003.3.2 Post-mounted objects. Free-standing objects mounted on a post or pylon that are located along or adjacent to the walking surface are potential hazards, particularly to a sight-impaired individual. Objects such as signs, directories, or telephones that are mounted on posts or pylons are, therefore, limited to an overhang of 4 inches (102 mm) maximum if located

General Means of Egress

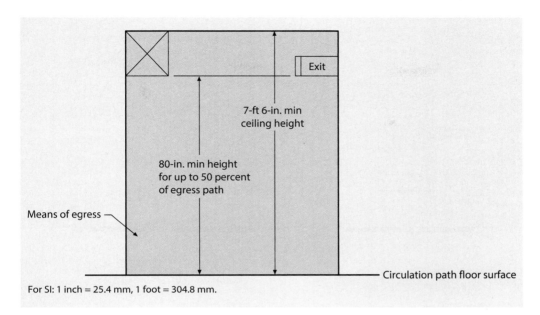

Figure 1003-1 **Means of egress headroom.**

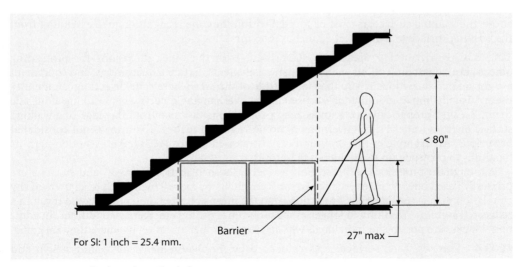

Figure 1003-2 **Reduced vertical clearance.**

more than 27 inches (686 mm), but not more than 80 inches (2,032 mm), above the floor level. By limiting the overhang to 4 inches (102 mm), a cane will hit the post or pylon prior to the individual impacting the mounted object. See Figure 1003-3. Free-standing objects mounted at or below 27 inches (686 mm) will fall within the cane-detection zone, and objects mounted at 80 inches (2,032 mm) or higher are sufficiently above the walking surface. Similar concerns

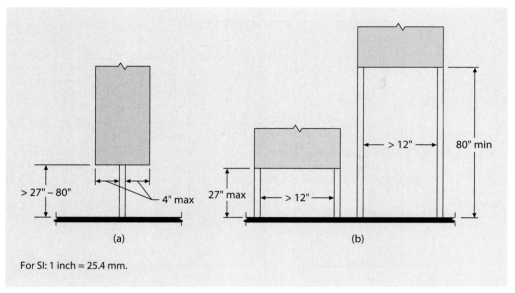

Figure 1003-3 **Post-mounted protruding objects.**

are addressed where the obstruction is mounted between posts located more than 12 inches (305 mm) apart. Unless the lowest edge of the obstruction is at least 80 inches (2,030 mm) above the walking surface, it must be located within the cane recognition area extending from the walking surface to a height of 27 inches (686 mm).

1003.3.3 Horizontal projections. Consistent with the other provisions for protruding objects, horizontal projections such as structural elements, fixtures, furnishings, and equipment are considered hazardous where they fall outside of the area where cane detection can identify them. Visually impaired individuals cannot detect overhanging objects when walking alongside them. Because proper cane techniques keep people some distance from the edge of a walking surface or from walls, a slight overhang of no more than 4 inches (102 mm) is not considered hazardous. An example of this provision is illustrated in Figure 1003-4. An exception permits handrails to protrude up to 4½ inches (114 mm).

Although the provisions of this section, as well as those in Sections 1003.3.1 and 1003.3.2, are primarily based on clearances established for accessibility purposes, their value to all users of any circulation path is considerable. Projections into the means of egress potentially could result in a reduced travel flow, resulting in longer evacuation times during emergency conditions. In addition, injuries are possible to individuals who fail to pay proper attention to where they are going.

1003.4 Slip-resistant surface. As evidenced by the requirements for ceiling height and protruding objects, the potential for exit travel to be impeded by obstructions is addressed throughout Chapter 10. Various provisions attempt to eliminate the opportunity for hazards along the exit path to slow travel. This section recognizes one area that is often taken for granted when it comes to egress—the floor surface of the means of egress. It is typically assumed that a floor surface that provides adequate circulation, and often accessibility, throughout a building will be acceptable for egress purposes as well. Although this is usually true, it is stated in the code that the egress path should have a surface that is slip resistant and securely attached so there is no tripping or slipping hazard that would result in an obstruction of the exiting process. Although the regulation of floor surfacing materials is typically recognized for interior

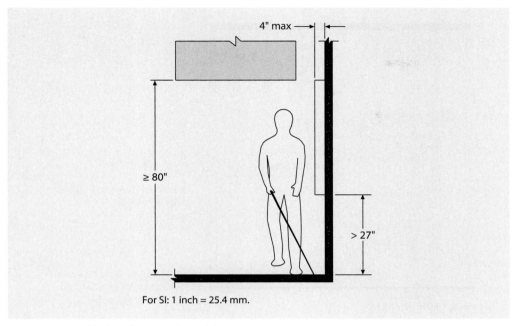

Figure 1003-4 **Limits of protruding objects.**

circulation paths, the provision is also applicable to exterior egress paths. The performance criteria of this code section provide a basis for determining the appropriateness of any questionable exit discharge elements.

1003.5 Elevation change. The code is concerned that along the means of egress there is no change in elevation along the path of exit travel that is not readily apparent to persons seeking to exit under emergency conditions. Therefore, along the means of egress, any change in elevation of less than 12 inches (305 mm) must be accomplished by means of a ramp or other sloping surface. A single riser or a pair of risers is not permitted. See Figure 1003-5. Steps used to achieve minor differences in elevation frequently go unnoticed and as a consequence can cause missteps or accidents.

This limitation on the method for a change of elevation, however, does not apply in certain locations. Where Section 1010.1.4 permits a level change at a door, a level change along the egress

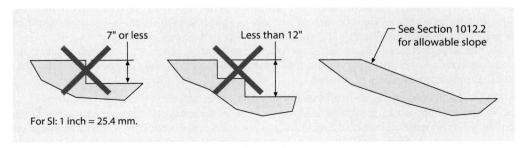

Figure 1003-5 **Longitudinal section through corridor or other exit path.**

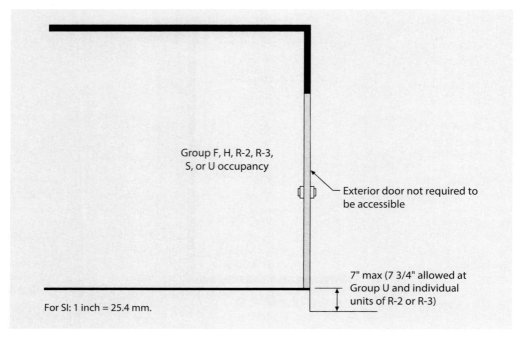

Figure 1003-6 **Single step at exterior door.**

path could occur in certain occupancies and locations. See Figure 1003-6. A second exception allows, under specific conditions, a stair with a single riser or with two risers and a tread at those locations not required to be accessible by Chapter 11. In this case, the risers and treads must comply with Section 1011.5, but the tread depth must be at least 13 inches (330 mm), and a minimum of one complying handrail must be provided within 30 inches (762 mm) of the centerline of the normal path of egress travel on the stair. See Figure 1003-7. A third exception applies to seating areas not required to be accessible. Risers and treads may be used on an aisle serving the seating where a complying handrail is provided.

1003.6 Means of egress continuity. This section emphasizes that wherever the code imposes minimum widths on components in an exiting system, such widths are to be clear, usable, and unobstructed. Nothing may project into these required capacities and minimum widths so as to reduce the usability of the full dimension, unless the code specifically and expressly states that a projection is permitted. Two notable examples of permitted projections are doors, either during the course of their swing or in the fully open position, and handrails. The limitations on the amount of such projections are specified within the appropriate sections of the chapter. Additionally, this section places into code language one of the four basic concepts that were previously discussed—that once the exit system is subject to a certain maximum demand in terms of number of persons, that system must thereafter be capable of accommodating that maximum number of persons.

1003.7 Elevators, escalators, and moving walks. For a variety of reasons, elevators, escalators, and moving walks are not to be used to satisfy any of the means of egress for a building. These building components are intended for circulation purposes and do not conform with the detailed egress requirements found in Chapter 10. The only exception is for elevators used as an accessible means of egress as addressed in Section 1009.4, which truly is more of a means of assisted rescue.

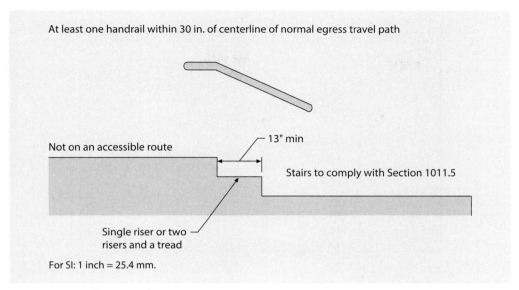

Figure 1003-7   **Elevation change.**

Section 1004 *Occupant Load*

1004.1 Design occupant load. This section prescribes a series of methods for determining the occupant load that will be used as the basis for the design of the egress system. The basic concept is that the building must be provided with a safe exiting system for all persons anticipated in the building. The process for determining an appropriate occupant load is based on the anticipated density of the area under consideration. Because the density factor is already established by the code for the expected use, variations in occupant load are simply a function of the floor area assigned to that use. It is apparent that in many situations the occupant load as calculated is conservative in nature. This is appropriate because of the extent that the means of egress provides for life-safety concerns. The egress system should be designed to accommodate the worst-case scenario, based on a reasonable assumption of the building's use.

1004.2 Cumulative occupant loads. This provision mandates that the occupant loads are to be cumulative as the occupants egress through intervening spaces. Under the conditions of Section 1016.2, the path of travel through the intervening space must be discernable to allow for a continuous and obvious egress path. Egress travel is permitted to pass through complying adjoining rooms provided the design occupant load is increased to account for those potential occupants who are assigned to that specific egress path. See Figure 1004-1. Another common application occurs as users of the means of egress merge at aisles, corridors, or stairways, as shown in Figure 1004-2. Where alternative means of egress are provided, only the number of occupants assigned to each of the egress paths is used in the cumulative occupant load determination. See Figure 1004-3. Where significant occupant loads are anticipated such that a room or space requires at least two means of egress, the provisions of Section 1005.5 must also be considered for the proper distribution of the occupant load capacity.

The practice of accumulating occupants along the means of egress also applies vertically where travel from a mezzanine level leads into a room on an adjacent level, rather than directly

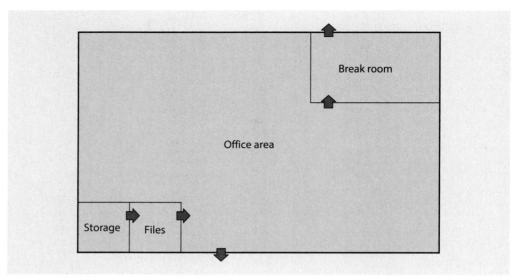

Figure 1004-1 **Combination occupant loads.**

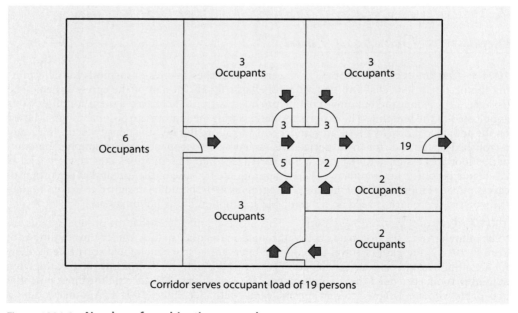

Figure 1004-2 **Number of combination example.**

into an enclosure for an interior exit stairway. Where travel occurs within the story, occupant loads are to be cumulative vertically as well as horizontally. An example is shown in Figure 1004-4.

Where the means of egress occurs on stairways that connect two or more stories, Section 1005.3.1 indicates that the capacity of the exitways be based on the individual occupant loads of each story. In other words, the number of persons for which the capacity of the

Occupant Load **381**

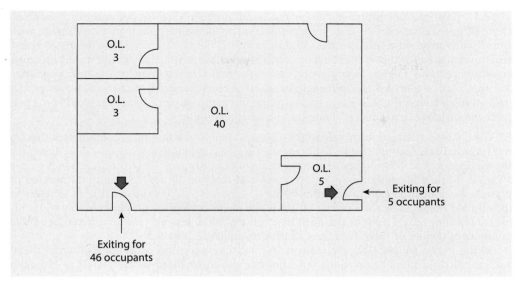

Figure 1004-3   **Assignment of occupant load example.**

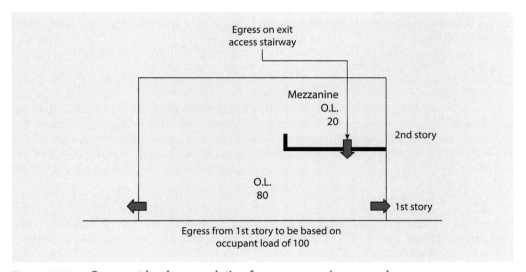

Figure 1004-4   **Occupant load accumulation from a mezzanine example.**

stairway is designed is not based on any cumulative total number of persons, but rather on the required capacity of the exits at each particular story. In no case, however, shall exit capacity decrease along the path of egress travel. A more in-depth discussion of this issue is found in Section 1005.1 and is illustrated in Figure 1005-2. Where exiting occurs from a mezzanine, the provisions of Section 1004.2.2 apply rather than those of Section 1005.3.1. A mezzanine is considered a portion of the story it is located in and therefore Section 1004.2.2 is used for mezzanines instead of the provisions for adjacent stories.

1004.4 Multiple occupancies. In many buildings there are two or more occupancies. Quite often, one or more of the egress paths from an individual occupancy will merge with egress paths from other occupancies. Within each individual occupancy, the means of egress shall be designed for that specific occupancy. However, where portions of the means of egress serve two or more different occupancies, the more restrictive requirements of the occupancies involved shall be met. An example might be where a sizable assembly occupancy shares an exit path with a business use. The more restrictive requirement for panic hardware would be applicable for any doors encountered along the shared egress route.

1004.5 Areas without fixed seating. The vast majority of buildings contain uses that do not include fixed seating. Unlike auditoriums, theaters, and similar spaces, in most instances the maximum probable number of occupants may not be known. Therefore, the code provides a simple formula for determining an occupant load that constitutes the minimum number of persons for which the exiting system must be designed. As a consequence, the code refers to the number obtained by the formula as the design occupant load. Egress systems for all buildings or building spaces must be designed to accommodate at least this minimum number. Basic examples of the use of Table 1004.5 are illustrated in Figure 1004-5.

As the person responsible for interpreting and enforcing the code, the building official will be called on to make decisions regarding the categories listed in Table 1004.5. Although Table 1004.5 contains occupant load factors that will serve the code user under most conditions, there will be occasions when either the table does not have an occupant load factor appropriate for the intended use or the occupant load factor contained in the table will not have a realistic application. In such instances, the building official has the authority to establish an appropriate occupant load factor or an appropriate occupant load for those special circumstances and those special buildings.

It may be meaningful to point out that the first column of Table 1004.5 is headed "Function of Space." The categories listed in the first column are not the specific groups identified in Chapter 3 for the purpose of assigning an occupancy classification but are the basic generic uses of building spaces. It has been pointed out in the discussion of the various occupancy groups that it is possible to have a classroom classified as a Group E occupancy, a Group B occupancy, or, possibly, a Group A occupancy. In terms of occupant density, however, a classroom is a classroom, and it is reasonable to expect the same density of use in a classroom regardless of the occupancy group in which that classroom might be classified. Therefore, the table specifies that when considering classroom use, one must assume there is at least one person present for each 20 square feet (1.86 m^2) of floor area.

In specifying how the occupant load is to be determined, the code intends that it is to be assumed that all portions of a building are fully occupied at the same time. It may be recognized, however, that in limited instances not all portions of the building are, in fact, fully occupied simultaneously. An example of this approach for support uses might include conference rooms in various occupancies or minor assembly areas such as lunch rooms in office buildings or break rooms in factories. It is important to note that the code does not provide for a method to address such conditions; thus, full occupancy should always be assumed. Only under rare and unusual circumstances should the building official ever consider reducing the design occupant load because of the nonsimultaneous use concept. In such situations, they must determine that there are support spaces that ordinarily are used only by persons who at other times occupy the main areas of the building; therefore, it is not necessary to accumulate the occupant load of the separate spaces when calculating the total occupant load of the floor or building. It is always necessary, however, to provide each individual space of the building with egress as if that individual space was fully and completely occupied.

Another type of support area that must be considered in occupant-load calculation includes corridors, closets, toilet rooms, and mechanical rooms. These uses are typical of most buildings

Occupant Load

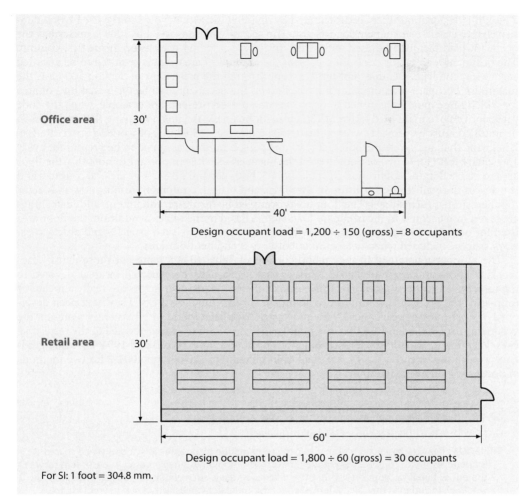

Figure 1004-5 **Design occupant load examples.**

and are to be included by definition in the gross floor area of the building. A quick review of Table 1004.5 will show that most of the uses listed are to be evaluated based on gross floor area, with no reduction for corridors and the like. However, a few of the listings indicate the use of the net floor area in the calculation of the occupant load. An example would be the determination of an occupant load in a school building. The building official should calculate the occupant load in such buildings using only the administrative, classroom, and assembly areas. It is generally assumed that when corridors, restrooms, and other miscellaneous spaces are occupied, they are occupied by the same people who are at other times occupying the primary use spaces.

The occupant load that can be expected in different buildings depends on two primary factors—the nature of the use of the building space and the amount of space devoted to that particular use. Different types of building uses have a variety of characteristics. Of primary importance is the density characteristic. Therefore, in calculating the occupant load of different uses, by means of the formula, the minimum number of persons that must be assumed to occupy a building or portion thereof is determined by dividing the area devoted to the use by that density characteristic or occupant load factor. The second column of Table 1004.5 prescribes the occupant load factor to be used with respective corresponding uses listed in the first column. The occupant load factor does not represent the amount of area that is required to be afforded each occupant. The IBC does not limit, except through the provisions of Section 1004.5.1, the maximum occupant load on an area basis. Rather, the occupant load factor is that unit of area for which there must be assumed to be at least one person present. For example, when the code prescribes an occupant load factor of 150 gross for business use, it is not saying that each person in an office must be provided with at least 150 square feet (13.94 m^2) of working space. Rather, it is saying that, for egress purposes, at least one person must be assumed to be present for every 150 square feet (13.94 m^2) of floor area in the business use. It is important to note that the floor area to be used in the application of Table 1004.5, both net and gross, is to include counters and showcases in retail stores, furniture in dwellings and offices, equipment in hospitals and factories, and similar furnishings. The floor areas occupied by furniture, equipment, and furnishings are taken into account in the occupant load factors listed in the table. In addition, the floor area used for occupant circulation purposes within a room or space, such as aisles and aisle accessways, is to be included when determining both gross and net floor area.

The numbers contained in the second column of Table 1004.5 represent those density factors that approximate the probable densities that can usually be expected in areas devoted to the respective functions listed. For this purpose, the occupant load factors are really a means of estimating the probable maximum density in the varying function areas. They have been developed over a period of years and, for the most part, have been found to consistently represent the occupant/furnishing densities that one might expect in building spaces devoted to the respective uses. Where multiple occupancies or functions occur, Sections 1004.3 and 1004.4 provide guidance to address the egress issues. See Application Examples 1004-1 and 1004-2 for two methods of occupant load determination.

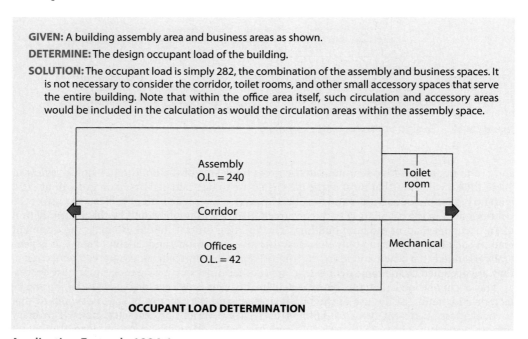

Application Example 1004-1

> **GIVEN:** A 1,600-square-foot conference room in a hotel.
>
> **DETERMINE:** The design occupant load of the room.
>
> **SOLUTION:** Because a variety of assembly activities can occur within the room, the use creating the largest occupant load would be evaluated.
>
>   (1) Dining use with tables and chairs
>
>   1 person per 15 sq ft = 106.67 = 106 occupants
>
>   (2) Conference/seminar use with chairs only (auditorium-style seating)
>
>   1 person per 7 sq ft for seating = 228.57 = 228
>
> Therefore for egress purposes, a design occupant load of 228 shall be used. Note that other potential uses of the room (seminars, receptions, dances, etc.) would also utilize the applicable factors.
>
> **OCCUPANT LOAD DETERMINATION**

Application Example 1004-2

The exception to this section allows for a reduction in the calculated design occupant load on a very limited case-by-case basis. The building official is granted authority for the discretionary approval of lesser design occupant loads than those established by calculation. Although the provision allows the building official to be accommodating by recognizing the merits of the specific project, its use should be limited to very unique situations such as extremely large manufacturing or warehousing operations. See Application Example 1004-3. Where the exception is enacted in order to reduce the occupant load, the building official will typically impose specific conditions to help ensure compliance. It is critical that the reasoning for the occupant load reduction be justified and documented.

1004.5.1 Increased occupant load. The provisions of Section 1004.5 specify the method to be used in determining the anticipated occupant load for areas without fixed seating. The occupant load determined by this method is the minimum number of persons for which the exiting system must be designed. The provisions do not, as previously pointed out, intend that the maximum permitted occupant load be regulated or controlled on a floor-area basis other than in the manner described by this section.

The provisions of this section specify how the maximum permitted occupant load in a building or portion of the building is to be determined. Here, the approach is taken that the occupant load determined as previously provided may be increased where the entire egress system is adequate, in all of its parts, to accommodate the increased number. In no case, however, shall the occupant load be established using an occupant load factor of less than 7 square feet (0.65 m^2) of floor space per person. See Application Example 1004-4.

In order to analyze any increased occupant load, the building official must carefully review all aspects of the arrangement of space as well as the details of the total egress system, not only from the immediate space but continuously through all other building spaces that might intervene. In many cases, a diagram will be required indicating the approved furnishing and equipment layout. It is important to recognize that a significant increase in occupant load over that calculated based on Table 1004.5 will most likely result in noncompliance in the means of egress. Using a dining area as an example, a sizable increase in occupant load will also include a sizable increase in the required number of tables and chairs. In order to provide floor area for the additional furnishings, less floor space can be devoted to the means of egress. In this case, Section 1030.13.1 addressing seating at tables should be reviewed to verify egress compliance.

GIVEN: A 210,000-square-foot industrial building designed for final assembly of commercial aircraft.
DETERMINE: The design occupant load of the building.

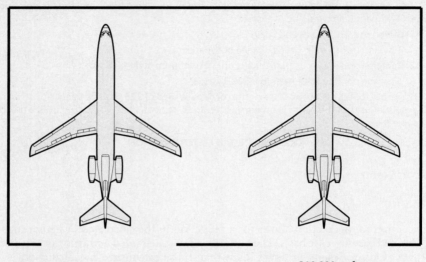

Based on Table 1004.5, the design occupant load would be 2,100.

$$\left(\frac{210{,}000 \text{ sq ft}}{100 \text{ (factor for industrial areas)}} \right)$$

Where approved by the building official, a more realistic design occupant load is permitted based on the actual maximum number of occupants anticipated in the building.

OCCUPANT LOAD DETERMINATION

Application Example 1004-3

GIVEN: A restaurant where the occupant load of the dining area is calculated at 135, based on Table 1004.5 (2,025 sq ft/15). The restaurant's owner would like to establish a higher occupant load.

DETERMINE: The maximum permitted occupant load of the dining area.

SOLUTION: The absolute maximum occupant load per Section 1004.5.1 appears to be 289 (2,025/7). However, it is obviously impossible for such an occupant load to safely occupy the space, even if adequate exit doors were provided. If tables and chairs were provided to seat 289 customers, there would be inadequate aisle accessways (Section 1030.13) and aisles (Section 1030.9). If addition, the potential for egress obstruction would be significant. The appropriate maximum occupant load would be approved by the building official on a case-by-case basis, relying on the specific design of the space, the furniture and/or equipment layout, and the egress patterns created.

MAXIMUM OCCUPANT LOAD

Application Example 1004-4

Although it is critical that the building's means of egress system be designed to accommodate the increased occupant load, all other code requirements that are based on the number of occupants must also be reviewed based on the increased number. For example, if it is intended to increase the calculated occupant load of 258 in a Group A-3 conference facility to 340 occupants, all code requirements shall be applied based on the occupant load of 340. This would include the provisions of Section 903.2.1.3 that require an automatic sprinkler system, those of Section 907.2.1 mandating a manual fire alarm system, and the main exit requirements of Section 1030.2. An additional occupant-load-based provision that must be considered is that for plumbing fixtures as addressed in Section 2902.

1004.6 Fixed seating. The method of calculating occupant load discussed to this point—that is, the formula that divides an appropriate occupant load factor into the amount of space devoted to a specific function—is used when dealing with building spaces without fixed seating. Where fixed seats are installed, the code specifies that the occupant load be determined simply by counting the number of seats. Although the code does not define the term *fixed seats*, it is intended by this term that the seats provided are, in fact, fastened in position, not easily movable, and maintained in those fixed positions on a more or less permanent basis. A primary example of a fixed-seat facility would be a performance theater. In determining the occupant load for this type of facility, only the number of fixed seats is used because the code also requires that the space occupied by aisles may not be used for any purpose other than aisles and, therefore, may not be used for accommodating additional persons. The aisle system within a fixed-seating facility is, in fact, the exiting system for those fixed seats and, as such, must remain unobstructed. Therefore, the code does not assume any occupancy in the areas that make up the aisles.

Under varying circumstances, fixed-seating assembly spaces may include other assembly areas capable of being occupied. Such areas could include wheelchair spaces, waiting areas, and/or standing room. Performance areas and similar spaces would also be evaluated and assigned an appropriate occupant load. The occupant load of all such areas must be added to that established for the fixed seating in the calculation of the total occupant load per Section 1004.3. An example is shown in Figure 1004-6. The inclusion of these additional occupiable areas provides for a more accurate determination of the potential number of persons who could occupy the room or space.

In addition to those fixed-seating arrangements where the seating is provided by a chair-type seat, there will be those that use continuous seating surfaces such as benches and pews. When this type of seating is provided, it is necessary to assume at least one person present for each 18 inches (457 mm) of length of seating surface. Where seating is provided by use of booths, as is frequently done in restaurants, it must be assumed that there is a person present for each 24 inches (610 mm) of booth-seating surface. If the booth seating is curved, the code specifies that the booth length be measured at the backrest of the seating booth. Where seating is provided without dividing arms, such as for benches and booths, it is reasonable to base the occupant load individually to each bench or booth. Similarly, it is appropriate to round the calculated occupant load down to the lower value, as this section only regulates each full 18 inches or 24 inches of width. See Application Example 1004-5.

The method for determining occupant load in a small restaurant is depicted in Application Example 1004-6.

1004.7 Outdoor areas. Occupiable roofs, yards, patios, and courts that are used by occupants of the building must be provided with egress in a manner consistent with indoor areas. This provision is applicable to outdoor areas, including building rooftops, that are occupied for a variety of uses, but is primarily applied to outdoor dining at restaurants and cafés. The building official shall assign an occupant load in accordance with the anticipated use of outdoor areas. If an area's occupants need to pass through the building to exit, the cumulative total of the outdoor area and the building shall be used to determine the exiting requirements. This concept is consistent with the provisions of Section 1004.2.1. See Figure 1004-7.

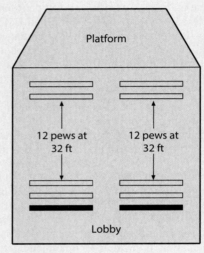

Application Example 1004-5

Another example that is becoming more common is that of secured exterior areas serving nursing homes (Group I-2) or assisted living facilities (Group I-1, R-3, or R-4). Such exterior spaces are often provided to enhance the livability of the facilities by providing outdoor spaces where the patients or residents are free to roam without individual supervision. When evaluating these spaces for egress purposes, there are several issues to consider. If the secured yard is provided with a means of egress independent of the facility, the gates must comply with all of the requirements for egress doors. Electric locking systems, sensor release devices, delayed egress devices, or controlled locking systems installed in accordance with Sections 1010.2.10 through 1010.2.13, as applicable, would be permitted as a means of addressing occupant safety for these areas. Without compliant gates, the means of egress must be designed for travel back through the facility. The facility must also egress independent of the secured yard unless all means of egress from the secured exterior area comply with the code.

The judgment of the building official is very important to the application of these provisions because the building official must determine exactly what occupant load should be

GIVEN: Information as shown in illustration.

DETERMINE: The occupant load for the small restaurant.

SOLUTION: Section 1004.6 states that where booths are used in dining areas, the occupant load shall be based on one person for each 24 inches (610 mm) of booth length. Based on this requirement, each 4-foot 6-inch booth would have an occupant load of four, or a total occupant load for the booth area of 16.

The fixed seats at the counter number eight, which, based on the first paragraph of Section 1004.6, would establish an occupant load of eight.

The open dining area (tables and chairs), having a floor area of 600 square feet (55.7 m²) and using an occupant load factor of 15 square feet (1.39 m²) per occupant, as set forth in Table 1004.5, would have an occupant load of 40.

The cooking area, having a floor area of 200 square feet (18.6 m²) and using an occupant load factor of 200 square feet (18.6 m²) per occupant, as set forth in Table 1004.5, would have an occupant load of one.

The total occupant load is as follows:

Booths 16
Counter 8
Dining area (tables and chairs) 40
Cooking area 1
  Total occupants: 65

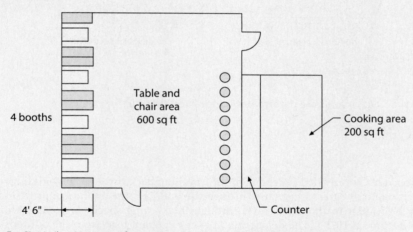

For SI: 1 inch = 25.4 mm, 1 foot = 304.8 mm, 1 square foot = 0.0929 m².

DETERMINATION OF OCCUPANT LOAD

Application Example 1004-6

considered and to what degree the area is accessible and usable by the building occupants in order to establish the egress requirements. Some cases that will require judgment include large spaces that might have a very limited anticipated occupant load such as areas that are primarily for the service of the building. Where a portion of the required means of egress for

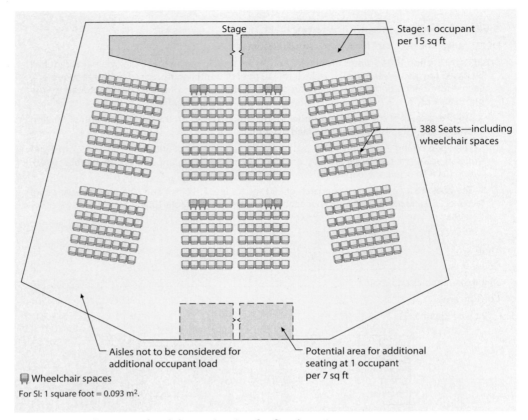

Figure 1004-6 **Occupant load determination for fixed seating.**

the outdoor area is provided independent of travelling back through the building, or where all of such required egress must pass through the building, the applicable provisions would generally be similar to those for travel through intervening spaces or comply with Exception 8 in Section 1010.2.4. The distribution of the occupant load from the outdoor area will depend on how many exits are required and how many means of egress paths are available.

1004.9 Posting of occupant load. Where a room or space is to be used as an assembly occupancy, this section requires the posting of a sign indicating the maximum permitted occupant load. This sign serves as a reminder to the occupants of the space, as well as building employees, that any larger occupant load would create an overcrowded condition. In order to be effective, the sign must be conspicuously located near the main exit or exit access doorway from the room or space, and must be permanently maintained. An example of an occupant load sign is shown in Figure 1004-8. Where multiple uses causing varying occupant loads are anticipated, it is appropriate to designate the maximum occupant load for each use, as shown in Figure 1004-9.

Occupant Load

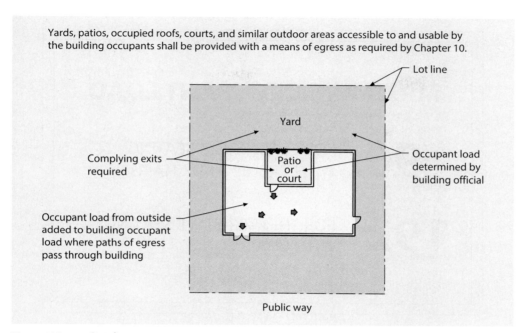

Figure 1004-7 **Outdoor areas.**

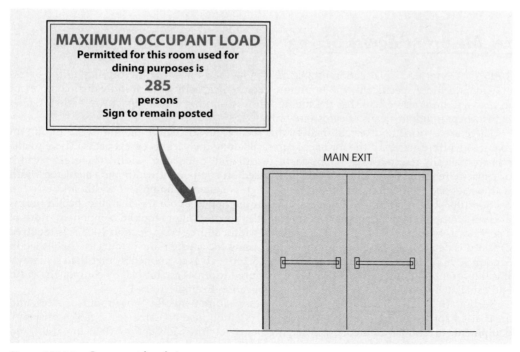

Figure 1004-8 **Occupant load sign.**

```
┌─────────────────────────────────────────────┐
│                                             │
│         MAXIMUM OCCUPANT LOAD               │
│                                             │
│      90      FOR CLASSROOM SEATING          │
│              WITH TABLES AND CHAIRS         │
│                                             │
│     192      FOR THEATER-STYLE SEATING      │
│              WITH CHAIRS ONLY               │
│                                             │
└─────────────────────────────────────────────┘
```

Figure 1004-9   **Posting of occupant load.**

Section 1005 *Means of Egress Sizing*

1005.1 General. This section establishes the method for sizing the capacity of the egress system, and more specifically, the minimum required capacity of each individual component in that system. It also establishes the method for distributing egress capacity to various egress paths where multiple means of egress are provided.

There are two methods established by this section for the determination of the minimum required width/capacity of the means of egress all along the various egress paths. These methods are typically referred to as "component" width and "calculated" width. The greater width or capacity required for each egress element based on component width and calculated width is to be applied in the design of the means of egress system. Component width, addressed in Section 1005.2 as minimum width, is specified throughout the code based on the specific means of egress component under review. As an example, the minimum required component width of a corridor is based on Table 1020.3. Calculated width, addressed in Section 1005.3 as required capacity, is determined based on occupant load using the appropriate formula as established in Section 1005.3.1 for stairways and Section 1005.3.2 for egress components other than stairways. It is important to note again that the greater required width as established by component width and calculated width is to be provided. See Application Example 1005-1.

Section 1030 provides special means of egress requirements for those rooms, spaces, and areas used for assembly purposes. Where an assembly use contains seats, tables, displays, equipment, or similar elements, the requirements set forth in Section 1030 for aisles and aisle accessways are to be applied in addition to the general means of egress requirements of Chapter 10.

GIVEN: Various egress components and the occupant load served by each component.

DETERMINE: The minimum required width for each component in a Group B nonsprinklered occupancy.

Egress Component	Occupant Load Served	Minimum Required Calculated Width	Minimum Required Component Width	Minimum Required Width
Aisle	32	32 (0.2) = 6.4"	36" Sec. 1018.3	36"
Corridor	130	130 (0.2) = 26" 26"/2 = 13"*	44" Sec. 1020.3	44"
Stairway	200	200 (0.3) = 60"	44" Sec. 1011.2	60"
Door	180	180 (0.2) = 36"	32" Sec. 1010.1.1	36"

For SI: 1 inch = 25.4. mm.
*Assuming two means of egress from corridor

Application Example 1005-1

1005.2 Minimum width based on component. Egress components all have a minimum width established by other provisions in the code. It should be noted that where the component width is the appropriate method for determining egress width, in many situations the required component widths may lessen along the egress path. For example, in an office building with an occupant load of 68 persons, a corridor required to be at least 44 inches (1,118 mm) in width by Table 1020.3 may lead to an exit door with a minimum clear width of 32 inches (813 mm) as regulated by Section 1010.1.1. In this example, the minimum component widths for the corridor and the exit door provide for greater widths than the capacity required by Section 1005.3.2.

1005.3 Required capacity based on occupant load. The formula for means of egress capacity based on occupant load is very succinct. It states that the total required capacity of the means of egress shall not be less than that obtained by multiplying the total occupant load served by an egress component by the appropriate factor as set forth in Section 1005.3.1 or 1005.3.2, as applicable. It should be noted that the calculation of egress capacity for aisles and aisle accessways in assembly occupancies is not regulated by this section, but rather is governed by Section 1030.6. Where an assembly space contains seats, tables, displays, equipment, or other fixtures or furnishings, it must comply with the means of egress provisions of Section 1030.

In designing the means of egress system, it is first necessary to determine the occupant load that must be accommodated through each individual portion of the system. The occupant load anticipated to be served by each individual component is the basis for sizing each component. The design occupant load is to be used when determining both the component width and the calculated width. Continuing to use a corridor as an example, Table 1020.3 requires a minimum component width of 44 inches (1,118 mm) where serving an occupant load of 50 or more, but only 36 inches (914 mm) where serving an occupant load less than 50. For calculated width (required capacity), multiplying the occupant load by the appropriate factor in Section 1005.3.1 or 1005.3.2 will result in the minimum required capacity in inches (mm) necessary to accommodate the occupant load. An example is shown in Application Example 1005-2.

> **GIVEN:** A home-improvement center has an occupant load of 1,590. The building is fully sprinklered and has an emergency voice alarm communication (EVAC) system.
>
> **DETERMINE:** The total required egress width from the building at the exit doors.
>
> **SOLUTION:** For a Group M occupancy in a sprinklered building, Exception 1 to Section 1005.3.2 indicates a capacity factor of 0.15 inches per occupant for egress components other than stairways. 1,590 (0.15) = 238.5 inches of clear door width to be distributed among available exit doors.

Application Example 1005-2

It cannot be emphasized too strongly that when the code discusses width in terms of an egress system or component, it is referring to the clear, unobstructed, usable width afforded along the exit path by the individual components. Therefore, if it is determined, for example, that a means of egress must have a width of at least 3 feet (914 mm), it shall be arranged so that it is possible to pass a 36-inch-wide (914-mm) object through that egress path and each of its components. Unless the code specifically states that a projection is permitted into the required width by Section 1005.7, nothing may reduce the width of the component required to provide the necessary exit capacity.

Egress width in assembly spaces. As previously mentioned, where the provisions of Section 1030 are applicable for assembly uses, the egress width and capacity of aisles and aisle accessways shall be determined based on such provisions. The requirements of Section 1030 regulate all assembly spaces containing seats, tables, displays, equipment, or other material. Thus, it is typical that the egress capacity requirements of Section 1030.6 are to be followed rather than those of Section 1005.1. Section 1030.6 addresses egress capacity for seating areas in assembly uses, based primarily on whether or not smoke-protected assembly seating is provided. A further discussion of this subject is provided in the analysis of Section 1030.

1005.3.1 Stairways. This provision establishes the method for determining the required capacity of an egress stairway, based on the occupant load assigned to the stairway. As a basic requirement, the capacity of a stairway is based on the occupant load assigned to the stairway multiplied by 0.3 inches per occupant. Where the building is provided with both an automatic fire sprinkler system and an emergency voice/alarm communication system, a factor of 0.2 inches per occupant is to be used. The presence of these two fire-protection systems allows for a 33 percent reduction in the minimum required calculated egress capacity. Again, the minimum required component width must also be addressed. This reduction to 0.2 inches per occupant is not permitted in Group H and I-2 occupancies.

Provided the occupants are traveling in the same direction, there is no need to combine the loads from adjacent stories. The IBC assumes that in exiting multistory buildings there will be occupants feeding into the exit stairs at various stories. This approach recognizes that using the cumulative occupant load for vertical travel often results in stairway widths of significant size. The reduced sizes established by the floor-by-floor method may result in some increase in occupant evacuation time; however, the slower evacuation rate is deemed acceptable since the travel occurs within a fire-resistance-rated enclosure. This is in contrast to the methodology applied to horizontal travel where cumulative occupant loads must be addressed due to the fact that such travel is typically in unprotected portions of the exit access. The capacity factors are substantially higher for stairways than for horizontal egress components, primarily because the speed of exiting on a stair is substantially less than the speed of exiting on level or nearly level surfaces. On stairways there is a forced reduction in normal stride, as the length of stride on a stair must coincide with the stair's run. A study of this difference shows that a stairway requires an increase in width of approximately 50 percent above that for horizontal travel in order to maintain equivalent flow rates.

At one time, building codes addressed a cascading effect when analyzing and determining the required width of stairs, but this concept is no longer applicable. The required width of a stair is calculated on a story-by-story application of the formula.

In the design of multistory buildings, it is quite common that different stories have different occupant loads. Thus, the occupant load calculated for each story must be considered. As a matter of fact, it is not uncommon for buildings to have assembly uses on the top story. As a consequence of that configuration, it is entirely possible that the top story of a building will have an occupant load greater than any other. Under such a condition, it will be necessary to calculate the required stairway capacity based only on the occupant load of the uppermost story. This required capacity must be maintained through the successive stories until it serves a greater occupant load from a lower story or until the occupants have reached the public way or ultimate safe place. Figure 1005-1 illustrates this requirement.

It should be noted that the same concept of determining stairway width that is discussed above applies where building occupants exit upward through the stairway. This is the situation in buildings with basements and sub-basements. Occupants of those below-grade floors must exit up the

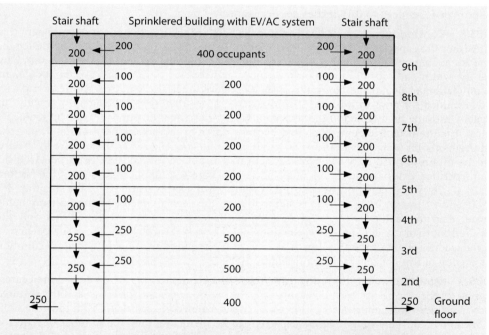

Note: First-story occupants may exit using interior exit stairways or through independent exits. Width required at ground level is the same for each case.

> Minimum stairway width or capacity required
> 4th through 9th stories: 200 × 0.2 = 40 in. (44 in. per Section 1011.2)
> 1st through 3rd stories: 250 × 0.2 = 50 in.
>
> Minimum stair exit door to exterior clear width required based on capacity
> 250 × 0.15 = 37.5 in.

For SI: 1 inch = 25.4 mm.

Figure 1005-1   Width of exits—multistory buildings example.

stairway, onto the landing on the ground floor, and then out through the exterior exit door from that landing. The largest capacity calculated from any floor level will be the controlling factor of the exit at ground level, as well as the exit doorway from the stairway. Of course, it is possible that occupants on the ground floor will exit through an enclosure of an interior exit stairway. If so, the required capacity determined based on that condition may govern. However, the ground floor usually has adequate width of exits independent of any paths through the stairway enclosures. Thus, the occupant load of the ground floor is usually not an issue in the determination of the exterior exit doorway capacity from the stairway enclosure. In any case, the occupant load of the ground floor would not be added to any occupant load of floors above for determining egress capacity from the building. Where upper and lower floors converge at an intermediate level, see the discussion of Section 1005.6 on egress convergence.

1005.3.2   Other egress components.   A means of egress capacity factor of 0.2 inches per occupant applies to those egress components other than stairways, such as doorways, ramps, aisles, and corridors. The factor may be reduced in all occupancies other than Groups H and I-2, provided the building is fully sprinklered and provided with an emergency voice/alarm communication system. Where both of these fire-protection features are provided, the capacity factor may be reduced to 0.15 inches per occupant.

1005.4   Continuity.   As stated earlier, it is the width of the most restrictive component that establishes the capacity of the overall exit system. To ensure that a design does not reduce the capacity at some point throughout the remainder of the egress system, this section stipulates that the width and capacity may not be reduced and that the design must accommodate any accumulation of occupants along the path. Therefore, once the required width or capacity is determined for any story (in fact, from any room or other space), that required width or capacity must be maintained until the occupants have reached the public way or ultimate safe place. It is important to remember that because different factors are used to determine the requirements for stairways than for all other components, it is really the required capacity, not the minimum width, of the egress system that is not permitted to be reduced. An aisle, door, corridor, exit passageway, or other horizontal egress component located at the bottom of any stairway may generally be reduced in width from what is required for the stairway. See Application Example 1005-3. This is simply due to the capacity factor of the stairway being greater than the factor for other egress elements. It must again be mentioned that only the required width or capacity needs to be maintained. The actual width of the means of egress may be reduced throughout the travel path as long as the required width is provided. A common application of this concept is shown in Figure 1005-2.

1005.5   Distribution of minimum width and required capacity.   When the required egress capacity of the system has been determined by the use of one of the formulas, and the required number of exit access doorways or exits has been determined in accordance with the provisions of Section 1006, the required egress capacity can be divided among the number of required means of egress. In fact, where additional complying means of egress are provided above the number required by the code, they too can be used for distribution purposes. The manner of distribution shall be such that the loss of any one means of egress will not reduce the available capacity to less than 50 percent of that required. Thus, after the loss of one means of egress, at least one-half of the required capacity must be available. See Application Example 1005-4. It is the intent of this section that there be reasonable distribution of the egress capacity necessary to serve a given occupant load.

The primary reason for requiring multiple egress paths is the fact that in a fire or other emergency, it could be possible that at least one of the routes will be unavailable or blocked by fire. If the egress capacity was concentrated at one exit point, it could very easily be that path that affords the greatest portion of the egress capacity that would be lost. The resulting limitation on the occupants' ability to exit a building or portion thereof is simply unacceptable. In addition,

Means of Egress Sizing 397

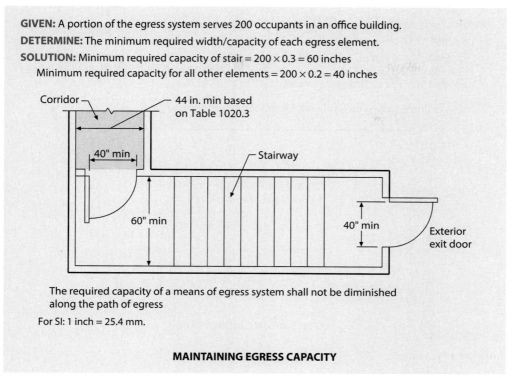

Application Example 1005-3

the presence of two or more means of egress allows for a distribution of occupants, which should provide for more efficient and orderly egress under emergency conditions. A third benefit of egress distribution is the potential reduction in the distance occupants must travel to reach an exit or exit access doorway.

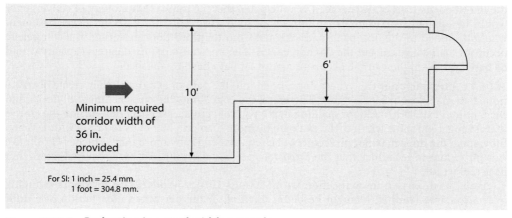

Figure 1005-2 Reduction in actual width example.

GIVEN: A retail store having three exits, with a total required exit capacity of 140 in.

DETERMINE: The manner in which the exit capacity may be distributed.

SOLUTION: Any distribution of egress capacity is acceptable provided that at least 50 percent of the required egress capacity (70 in.) is available after the capacity of the largest exit (66 in.) has been deducted from the total capacity provided (165 in.).

In this example, (165″ − 66″) ≥ 70″, so it is an acceptable solution.

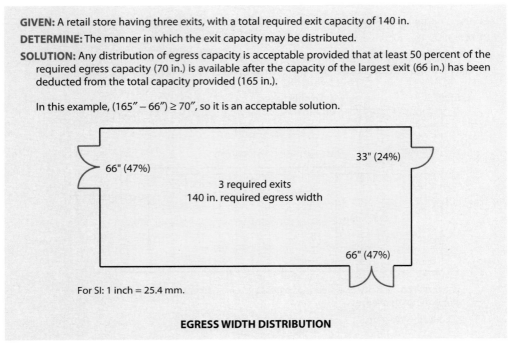

For SI: 1 inch = 25.4 mm.

EGRESS WIDTH DISTRIBUTION

Application Example 1005-4

1005.6 Egress convergence. This section directly addresses those situations where occupants from floors above and below converge at an intermediate level, rather than traveling in the same direction, as discussed in Sections 1005.3.1 and 1004.2.3. The code states that the proper approach for this condition would be to add the occupant loads together—a method that is also used when converging aisles or merging corridors. In these cases, it can be assumed that the occupants arrive at the same point at the same time and, therefore, the capacity of the system must accommodate the sum of these converging floors. See Figure 1005-3. Although the code does not specify the approach to be taken where there are multiple floors both above and below the intermediate level of discharge, it is anticipated that the same methodology of convergence would be applied. For example, the occupant load assigned from a second-level basement would be added to the occupant load assigned from the third floor, resulting in a congregate occupant load converging at the discharge level. The minimum required egress capacity would be based on the highest of the occupant loads that have been established.

1005.7 Encroachment. Where doors open into the path of exit travel, they create obstructions that may slow or block egress. Therefore, the code limits the encroachment of doors into the required exit width. A door opening into a path of egress travel may not, during the course of its swing, reduce the width of the exit path by more than one-half of its required width. When fully open, the door may not project into the required width by more than 7 inches (178 mm). It is important to recognize that the provisions are based on the exitway's required width, not its actual width.

Again, as discussed in connection with Sections 1010.1.5 and 1011.6 for doors swinging over a stairway landing, it might be better to think of the permitted obstruction of a door during the course of its swing from a positive viewpoint. So stated, each door, when swinging

Means of Egress Sizing **399**

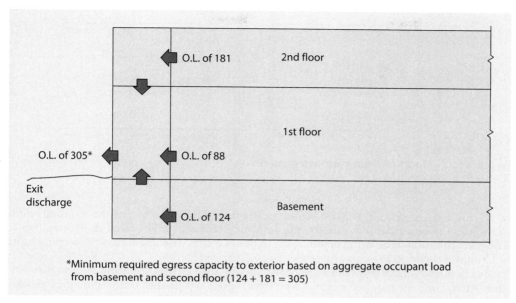

Figure 1005-3 **Egress convergence example.**

into an egress path such as an aisle or a corridor, must leave unobstructed at least one-half of the required width of the path of travel during the entire course of its swing. At least one-half of the required width must always be available for use by the building occupants. When the door is in its fully open position, the required egress width, minus 7 inches (178 mm), must be available. See Figure 1005-4.

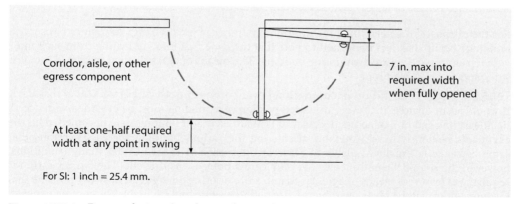

Figure 1005-4 **Egress obstruction due to door swing.**

In applying the requirements for projections, the code imposes these limitations on a door-by-door basis. It is desirable, but not required that doors be arranged so as not to have two doors directly opposing each other on opposite sides of the exit path. Better design would

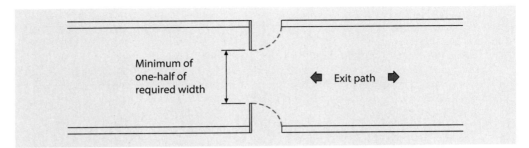

Figure 1005-5 **Doors swinging into egress path.**

avoid this arrangement. The intent of the code is that at least one-half of the required width of the exitway be available for use by the building occupant as illustrated in Figure 1005-5. The restrictions on door swing do not apply to doors within dwelling units and sleeping units of Groups R-2 and R-3.

Where nonstructural projections other than doors, such as trim and similar decorative features, extend into the required width of a means of egress component, the limit on their projection into the required width of egress components is 1½ inches (38 mm) unless such projection is specifically prohibited by the code. In reviewing the provisions for aisles in Section 1018.1, corridors in Section 1020.4, and exit passageways in Section 1024.2, the encroachments established in this section are specifically permitted. Handrail projections are also permitted provided such projections do not extend beyond the limitations established in Section 1014.9. Of specific note is a reference to Section 407.4.3 regarding permissible projections into the required width of nursing home corridors.

Section 1006 *Number of Exits and Exit Access Doorways*

For the purposes of determining the minimum number of means of egress required to serve a building's occupants, it is important to note that the code first looks at exiting from each individual room, space, or area within the building. The means of egress is then regulated for each story and any occupied roof.

1006.2.1 Egress based on occupant load and common path of egress travel distance. It would seem obvious that every occupied portion of a building must be provided with access to at least one exit or exit access doorway. It is assumed that if buildings are occupied, then the occupants obviously have a method of entering the various building spaces. Therefore, that same entrance is usually available to serve as the means of egress. Under many conditions, however, the use of the entrance as the only egress point is insufficient. The basic reason for requiring at least two means of egress is that in a fire or other emergency, it is very possible that the entry door will be obstructed by the fire and, therefore, not be usable for egress purposes. A second exit or exit access doorway can provide an alternative route of travel for occupants of the room or area. However, it is often unreasonable to require multiple egress paths from small spaces or areas with limited occupant loads. It is also seldom beneficial because of the relatively close proximity in which such exits or exit access doorways must be located. Therefore, the code does not require a secondary egress location from all rooms, areas, or spaces where in compliance with Table 1006.2.1.

The IBC establishes two basic criteria for providing adequate egress for occupants from any space within the building. First, at least two exits or exit access doorways must be provided when the occupant load of the space exceeds the values set forth in Table 1006.2.1. Second, two or more egress doorways are required when the common path of egress travel exceeds the limitations also established in Table 1006.2.1. Only where both the design occupant load of the room, space, or mezzanine and the common path of egress travel distance set forth in Table 1006.2.1 are not exceeded, it is permissible to have a single means of egress. See Application Examples 1006-1 and 1006-2.

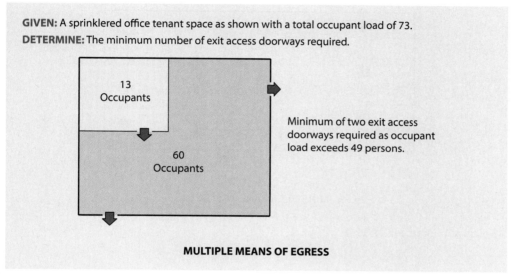

Application Example 1006-1

As seen in Table 1006.2.1, the allowance for a single exit or exit access doorway from a space varies based on the occupancy designation of the space. The variations are caused by conditions associated with the specific uses that occur within the room or area. Factors that contribute to the differences in occupant load include the concentration of occupants, occupant mobility, and the presence of hazardous materials. Exception 1 recognizes that the occupant load does not affect the *number* of exits from a foyer or lobby where spaces egress through them. However, the required capacity of the exit from the foyer or lobby is to be based on the total cumulative occupant load served. See Application Example 1006-3. The second exception references the provisions of Section 407.4.4 for Group I-2 care suites, while the third exception addresses unoccupied mechanical rooms and penthouses.

In addition to the occupant load limitations imposed by Table 1006.2.1 for single exit rooms and spaces, the common path of egress travel distance is also strictly regulated. The definition of a common path of egress travel is found in Section 202. Described as "that portion of exit access travel distance measured from the most remote point of each room, area, or space to that point where the occupants have separate and distinct access to two exits or exit access doorways." Therefore, a common path of egress travel is that portion of the means of egress system where only one path is available to the occupant, and ends only once separate paths are available to at least two exitways. The length of the common path is measured from the most remote point of a room or area to the *nearest* location where multiple exit paths to separate exits are available. See Figures 1006-1 and 1006-2.

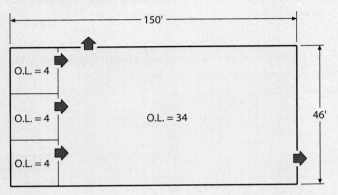

Application Example 1006-2

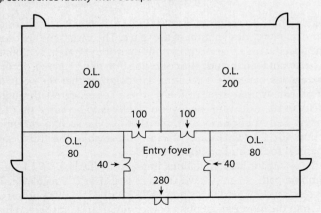

Although entry foyer serves ≥ 50 people, only one exit is required. Required capacity of entry foyer doorway to be 280 occupants

Application Example 1006-3

Number of Exits and Exit Access Doorways

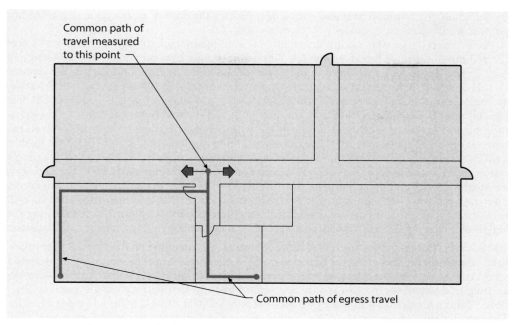

Figure 1006-1   **Common path of egress travel.**

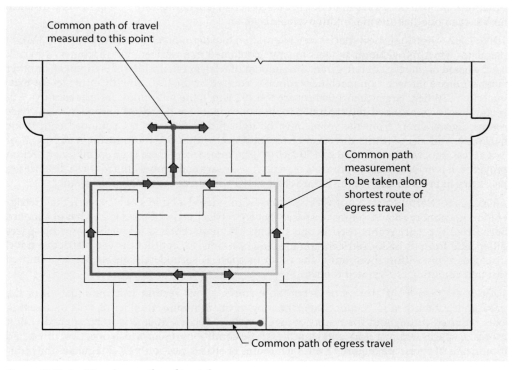

Figure 1006-2   **Merging paths of travel.**

By definition, the common path also ends when reaching the door of an exit because it is solely a portion of *exit access* travel distance.

As a general rule, the maximum length of a common path of egress travel is 75 feet (22,860 mm), as shown in Table 1006.2.1. Because of unique risks potentially encountered in a high-hazard occupancy, the common path of travel is limited to 25 feet (7,620 mm) in Groups H-1, H-2, and H-3. In certain occupancies and under specific conditions set forth in the table, an increase in the common path of travel to a distance of 100 feet (30,480 mm) or 125 feet (38,100 mm) is permitted. Other than the end point, the method of measurement of the common path of egress travel should be consistent with that for measuring exit access travel distance as regulated in Section 1017. See the discussion of Sections 1017.2 and 1017.3 for further guidance. The most obvious example of a "common path" condition is a room with a single exit or exit access doorway. Although there are numerous paths that may lead to the doorway, eventually they all end up at the same point. Where two or more complying exits or exit access doorways are provided, common path conditions do not exist. As shown in Figure 1006-2, two separate paths that merge are considered as a common path condition. In determining the length of the common path, the distance should be measured along the shortest path available.

1006.2.2.1 Boiler, incinerator, and furnace rooms. Depending on the capacity of the fuel-fired equipment located in a large boiler room or similar area, it may be necessary to provide a secondary egress doorway because of the potential hazards. At least two exit access doorways are required where the room exceeds 500 square feet (46 m^2) and any single piece of fuel-fired equipment such as a furnace or boiler exceeds 400,000 Btu (422,000 KJ) input capacity. The requirement is based on the size of a single piece of equipment, not the aggregate total of all fuel-fired equipment in the room. Access to one of the two exit access doorways may be accomplished through the use of a ladder or an alternating stair device. As with other multiple-exit situations, the two doorways must be adequately separated, in this case by a horizontal distance no less than one-half the maximum overall diagonal.

1006.2.2.2 Refrigeration machinery rooms. The *International Mechanical Code*® (IMC®) mandates when refrigeration systems must be contained in a refrigeration machinery room. It can be based on the type of refrigerant, the amount of refrigerant, the type of equipment, or other factors. Once a refrigeration machinery room is required by the IMC, the IBC provides the exiting criteria. Where larger than 1,000 square feet (92.9 m^2), the room must have at least two exit access doorways, accessed and separated in the same manner as described for boiler and furnace rooms. Egress doors from the room must be tight-fitting, self-closing, equipped with panic hardware, and swing in the direction of travel. Travel within the room to the nearest exit or exit access doorway is restricted to 150 feet (45,720 mm) where there is no sprinkler protection provided. It is evident that the presence of multiple exitways and a more limited travel distance are necessary in order to address the hazards associated with areas containing refrigerants.

1006.2.2.3 Refrigerated rooms or spaces. Considered a bit less of a concern than refrigeration machinery rooms, rooms or spaces that are refrigerated still pose somewhat of a hazard because of the refrigerants used in the system. The requirements for such rooms or spaces differ little from those for refrigeration machinery rooms, except that the less restrictive travel distances apply within the room if the room or space is sprinklered and egress is permitted through adjoining refrigerated rooms.

1006.3 Egress from stories or occupiable roofs. This section is a continuation of the provisions established in Section 1006.2 relating to the minimum number of exits or access to exits required throughout the means of egress system. Section 1006.2 is to be applied within rooms or spaces within a story and identifies those conditions under which one, two, three, and four means of egress are required from such rooms or spaces. Section 1006.3 regulates the minimum number of exits or access to exits required from each story of a building. Section 1006.3.4

allowances for a single exit, as well as the requirements for multiple exits, are established for each individual story, as well as for any occupiable roof.

As few as one exit may be permitted, based on occupancy classification, occupant load, and story location within the building. On the other hand, up to four exits may be required based primarily on the occupant load of the story. In all cases, once a minimum required number of exits or exit access stairways has been established by the code, that number cannot be reduced.

1006.3.2 Path of egress travel. It is possible to use a combination of interior/exterior exit stairways and exit access stairways for means of egress purposes. It is permissible that the means of egress from a story of a building consist of all exit access stairways as regulated by Section 1019. However, per Section 1006.3.2, the exit access stairways generally cannot pass through more than one adjacent story before providing access to the required exit. This limits the travel distance and exposure to hazards the occupants may face prior to reaching the safety of an exit.

Vertical travel on an exit access stairway must be considered in the evaluation of travel distance and such travel is usually only available to a single adjacent story. The use of an exit access stairway as a required means of egress component is limited to one story of travel, at which point the occupants must use an interior exit stairway or other exit element. See the example in Figure 1006-3. The exceptions address situations where exit access can pass through more than one adjacent story prior to reaching an exit. These are typically situations where vertical compartmentation/separation between stories is not required or between occupied roofs.

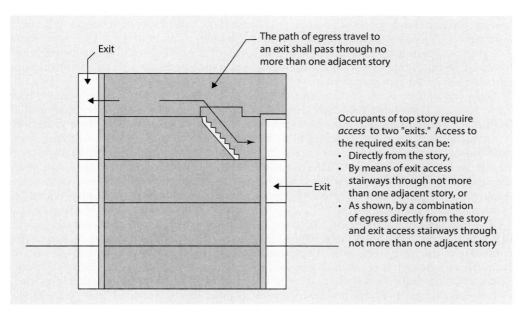

Figure 1006-3 **Access to exits at adjacent levels.**

1006.3.3 Egress based on occupant load. As established in Table 1006.3.3, every story or occupiable roof shall be provided with two exits, or access to at least two exits. A single means of egress is only permitted under the conditions identified in Section 1006.3.4. Table 1006.3.3 indicates that, based solely on occupant load, three, or even four, exits may be required.

Any story of a building that has an occupant load in excess of 500, up to and including 1,000, shall be provided with three exits, or access to at least three exits. Any story having an occupant load in excess of 1,000 must be provided with four exits, or access to not less than four exits. Under no circumstances does the IBC require more than four exits for any building or portion thereof based on the number of persons present. It must be noted, however, that additional exits will sometimes be required to satisfy the other egress requirements of Chapter 10. An additional interior exit stairway may also be required in high-rise buildings over 420 feet (128 m) in height per Section 403.5.2.

1006.3.4 Single exits. As a general requirement expressed in Table 1006.3.3, every story and occupiable roof shall be provided with two exits, or access to at least two exits. However, a single means of egress is permitted where the story or occupiable roof is within the limits of Table 1006.3.4(1) or Table 1006.3.4(2) as permitted by Exception 1, or where one of the other four exceptions to this section specifically permits a single exit.

Although it is desirable that a minimum of two means of egress be provided to building occupants in order to provide for a more reliable and efficient evacuation process, the code recognizes that there are instances where the life-safety risk is so minimal that it is reasonable to permit a single means of egress. Under limited conditions, this allowance extends from a single-level basement to a three-story-above-grade-plane condition. Tables 1006.3.4(1) and 1006.3.4(2) identify those stories where a single exit, or access to a single exit, is permitted. The tables are based on varying criteria, including occupancy classification, number of stories above the grade plane, occupant load, number of dwelling units, and exit access travel distance. Examples illustrating the use of these tables are shown in Figure 1006-4. It should be noted that special consideration is given in Footnote b of Table 1006.3.4(2) for certain Group B, F, and S occupancies. Also illustrated in Figure 1006-4 is a permitted condition for a basement where a single exit is permitted. Exception 4 recognizes that all Group R-3 and R-4 occupancies may have a single exit, while Exception 2 addresses stories at the exit discharge level complying with Section 1006.2.1 for spaces requiring only one means of egress.

Only one exit or access to a single exit is required from a story where permitted by Table 1006.3.4(1) or Table 1006.3.4(2), regardless of the number of means of egress required from other stories in the building. For example, a Group B occupancy on the second story of a multistory building is only required to have one exit, or access to one exit, from the story, provided its occupant load does not exceed 29 and the maximum exit access travel distance to an exit does not exceed 75 feet (22,860 mm). The number of occupants and travel distances on the other stories do not affect the determination of the second story as a single-exit story. See Application Example 1006-4. Where applicable, other stories are also regulated independently as to the number of means of egress.

The two tables referenced by Exception 1 differ only in the occupancy classifications to which they are applicable. Table 1006.3.4(1) is only to be used for Group R-2 occupancies consisting of dwelling units, such as apartment buildings. The allowance for a single exit, or access to a single exit, is regulated based on the story's relationship to grade plane, the number of dwelling units on the story, and the maximum exit access travel distance provided. Complying emergency escape and rescue openings must also be provided in all of the dwelling units. Table 1006.3.4(1) does not apply to dormitories, fraternity, sorority houses, and similar Group R-2 occupancies comprised of sleeping units based on footnote b of the table. Table 1006.3.4(2) applies to all occupancies other than those specific Group R-2 occupancies addressed by Table 1006.3.4(1). A single means of egress from each story of such occupancies is regulated by the occupancy classification, number of occupants per story, maximum exit access travel distance, and the story's relationship to grade plane.

Tables 1006.3.4(1) and 1006.3.4(2) also establish as to when it is permissible to provide a single exit or access to a single exit for an occupiable roof. The scoping of Table 1006.3.4(1) for

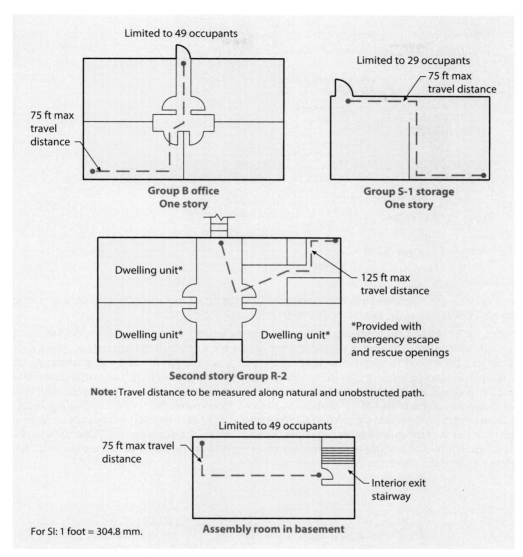

Figure 1006-4 Stories with one exit examples.

egress from occupiable roofs accessed through and serving individual dwelling units in Group R-2 occupancies permits a single means of egress for occupiable roofs that are located over a first-story above grade plane condition or over a second-story above grade plane condition. Conversely, an occupiable roof over dwelling units located on the third story above grade plane would require a minimum of two means of egress. It should be noted than the allowance for a single means of egress is only applicable to those occupiable roofs that are an extension of a single dwelling unit. An allowance is also provided in Table 1006.3.4(2) for all occupancies not addressed by Table 1006.3.4(1). Where an occupiable roof serves these occupancies, a single

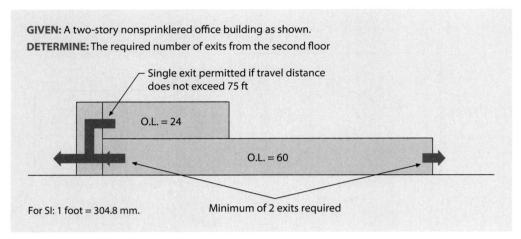

Application Example 1006-4

exit or access to a single exit is only permitted in those situations where the occupiable roof is located directly over the first story above grade plane.

1006.3.4.1 Mixed occupancies. Where multiple tenants or occupancies are located on a specific story, they are to be regulated independently for single-exit determination. The provisions can be applied to specific portions of the story, rather than the story as a whole. As an example, the second story of a building houses two office tenants, each with its own independent means of egress. Each tenant would be permitted a single, but separate, means of egress provided each had an occupant load of less than 30 and an exit access travel distance not exceeding 75 feet (22,860 mm). This portion-by-portion philosophy also applies to a mixed-occupancy condition, provided each of the individual occupancies does not exceed the limitations of Table 1006.3.4(2). See Figure 1006-5. If a single or common means of egress is provided for the mixed occupancies, then a sum of the ratios can determine compliance.

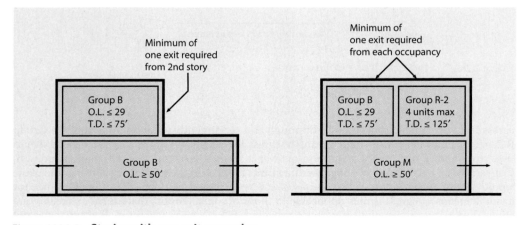

Figure 1006-5 Stories with one exit examples.

Section 1007 *Exit and Exit Access Doorway Configuration*

In addition to providing multiple means of egress, it is imperative that egress paths remain available and usable. To ensure that the required egress is sufficiently remote, the code imposes rather strict requirements relative to the location or arrangement of the different required exits or exit access doorways with respect to each other. The purpose here is to do all that is reasonably possible to ensure that if one means of egress should become obstructed, the others will remain available and will be usable by the building occupants. As a corollary, this approach assumes that because the remaining means of egress are still available, there will be sufficient time for the building occupants to use them to evacuate the building or the building space.

1007.1.1 Two exits or exit access doorways. This remoteness rule in the IBC is sometimes referred to as the one-half diagonal rule. The one-half diagonal rule states that if two exits or exit access doorways are required, they shall be arranged and placed a distance apart equal to or greater than one-half of the maximum overall diagonal of the space, room, story, or building served. Such a minimum distance between the two means of egress, measured in a straight line, shall not be less than one-half of that maximum overall diagonal dimension. See Figure 1007-1 for examples of the application of this rule. It should be noted that, by definition, the term *exit access doorway* includes any point of egress where the occupant has a single access point that must be reached prior to continued travel to the egress door. See Figure 1007-2.

The use of the one-half diagonal rule has been beneficial to code users for many years. It quantifies the code's intent when the code requires that separate means of egress be remote. It does not leave the building official with a vague performance-type statement that can, in many

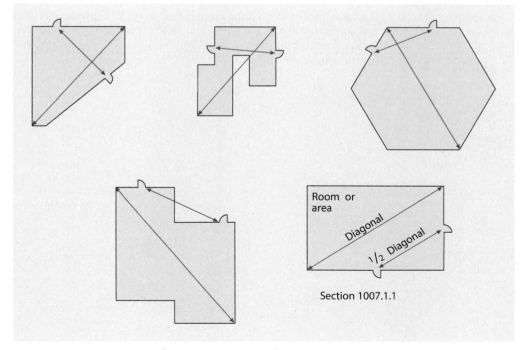

Figure 1007-1 **Separation of exits or exit-access doorways.**

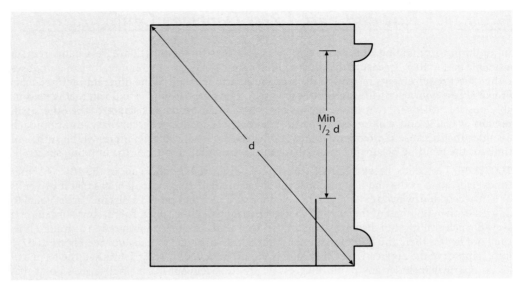

Figure 1007-2 **Egress separation.**

instances, result in a situation where egress separation would be dictated more by the design or desired layout of the building rather than by a consideration for adequate and safe separation of the means of egress.

In applying the one-half diagonal rule to a building constructed around a central court with an egress system consisting of an open balcony that extends around the perimeter of the court, it is important to take the measurement of the diagonal from which the one-half diagonal dimension is derived at the proper locations. Refer to Figure 1007-3 for examples.

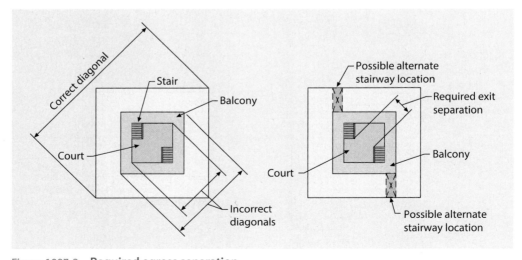

Figure 1007-3 **Required egress separation.**

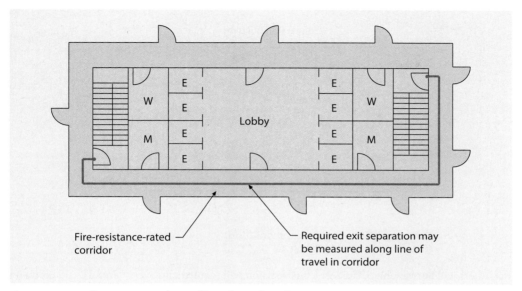

Figure 1007-4 **Core arrangement of interior exit stairways.**

Figure 1007-4 illustrates Exception 1 to the one-half diagonal rule for those buildings, such as core buildings, where the means of egress are sometimes arranged in rather close proximity. The code recognizes the benefits of such a floor arrangement and makes a specific exception in the event there is such a design. If the exits or exit access doorways are connected by a fire-resistance-rated corridor, the distance determined by one-half of the maximum overall diagonal of the space served may be measured along the shortest path of travel inside the corridor between the two exits instead of in a straight line between the two exit doors. It is specific that the connecting corridor is to be of 1-hour fire-resistant construction.

A second exception reduces the minimum length of the overall diagonal dimension between remote exits or exit access doorways in those buildings equipped throughout with automatic sprinkler systems. Because of the presence of sprinkler protection, the separation distance need only be one-third of the length of the overall diagonal dimension. The use of this exception results in a reduction of the required distance between exits or exit access doorways by 33⅓ percent. See Figure 1007-5.

1007.1.1.1 Measurement point. The code specifically indicates the manner in which the straight-line measurement should be taken. Three measurements methods are established to clearly indicate how to measure between doors, stairways, and ramps. Any point within the width of the exit or exit access doorway may be selected by the designer for measurement purposes. This will typically result in measuring between the far edge of one egress doorway opening and the far edge of the other egress doorway opening in order to allow the greatest design flexibility.

1007.1.2 Three or more exits or exit access doorways. When more than two means of egress are required, the remoteness rule takes on more of a performance character. In such an instance, at least two of the required exits or exit access doorways shall be arranged to comply with the one-half diagonal rule (one-third in fully sprinklered buildings). The other means of egress must be arranged at a reasonable distance from the other egress points so that if any one of the required exits or exit access doorways becomes blocked by a fire or any other emergency,

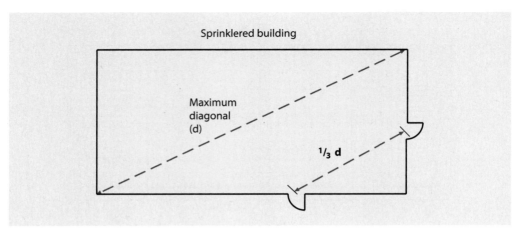

Figure 1007-5 **Exit separation—sprinklered building.**

the others will be available. Obviously, this decision will require some very careful evaluation and judgment on the part of the building official. There may be a sufficient basis for applying that same rule when considering each possible pair of egress components in a multi-exit situation. The code is silent in this particular aspect, and proper code administration does require substantial, careful evaluation and judgment on the part of the building official in ensuring that the number of means of egress required is sufficiently remote so that it is not likely that the use of more than one access to exit will be lost in any fire incident. A similar evaluation should be done where more than two exit access stairways or ramps provide the required means of egress.

1007.1.3 Remoteness of exit access stairways or ramps. Minimum separation distances between exit access stairways and ramps are required to be maintained for the entire length of travel on the stairway or ramp. See Figure 1007-6. This requirement prohibits stairway and ramp runs that meet the separation distance at the first riser or beginning of the slope from converging

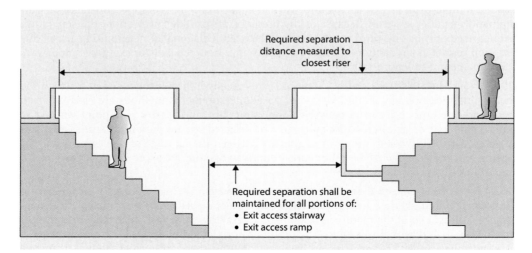

Figure 1007-6 **Measurement points for exit access stairways.**

toward another stair or ramp such that the separation is reduced below the minimum distance as the occupant goes either up or down the stairway or ramp. It is reasonable to expect the egress separation distance to be maintained in order to ensure a fire cannot affect both of the egress paths, whether it is at the beginning of the stair or ramp or at any point until the egress travel is completed.

Section 1008 *Means of Egress Illumination*

1008.2 Illumination required. In order for the exit system to afford a safe path of travel and for the building occupant to be able to negotiate the system, it is necessary that the entire egress system be provided with a certain minimum amount of illumination. Without such lighting, it would be impossible for building occupants to identify and follow the appropriate path of travel. The lack of adequate illumination would also be the cause of various other concerns, such as an increase in evacuation time, a greater potential for injuries during the egress process, and most probably an increased level of panic to those individuals trying to exit the building. Therefore, the code requires that, except in a limited number of occupancies, the egress paths be illuminated throughout their entire length any time the building space served by the means of egress is occupied. The code intends that illumination be provided for those portions of the egress system that serve the parts of the building that are, in fact, occupied. Parts of the exiting system that would not be serving the occupants of the building need not, at that time, be illuminated. For a variety of reasons, there are five exceptions that identify areas where continuous illumination during occupancy is not mandatory. Two exceptions address uses where sleeping is a common activity—dwelling units and sleeping units in Group R-1, R-2, and R-3 occupancies, and sleeping units in Group I occupancies. Another exception addresses utility structures designated as Group U, whereas a fourth exception exempts aisle accessways in Group A assembly uses.

The requirement to provide illumination of the means of egress system also does not apply within any self-service storage unit which is less than 400 square feet (37.2 m^2) in area provided it is accessed directly from the exterior of the building. Typical self-storage units are single-story structures similar to individual garages. Many of these structures do not include any type of electrical lighting system or, if they do, it is often very limited in quantity and size. This approach limits the potential hazards that can be caused by an electrical system while also hopefully discouraging the units from being used for dwelling, manufacturing, or other purposes for which they are not designed. The size of the space is limited to what is generally equivalent to a two-car garage area, and the opening to the exterior recognizes that users entering the space will be aware of the exit location. In addition, adjacent exterior areas are typically already illuminated for both the security and convenience of users.

It is mandated that the entire means of egress serving the occupants be illuminated. Therefore, the exit discharge—that portion of egress travel from the building to the public way—must also be provided with adequate illumination. Although there are often numerous light sources at a building's exterior, such as lighting for landscaping, parking lots, city streets, and adjacent buildings, it is important that the illumination be effective and reliable for use under this provision. It should also be noted that the requirements of this section are simply for general illumination of the entire egress system, and are not the higher level conditions for emergency lighting as mandated in Section 1008.3. The exception within Section 1008.2.3 will allow the illumination to end at a safe dispersal area versus extending all the way to the public way.

1008.2.1 Illumination level under normal power. For other than stairways, such illumination must be capable of producing a light intensity of not less than 1 foot-candle (11 lux) at the walking surface throughout the entire path of travel through the system. One foot-candle (11 lux) of light on a surface is not a great deal of light. It is probably not sufficient light to enable a person to read. However, it is sufficient light to allow a person passing through the exit system to distinguish objects and to identify obstructions in the actual path of travel. The light cast by a full moon on a clear night might approximate the 1 foot-candle (11 lux) light level. When the amount of light intensity is in doubt, it may be necessary to measure it with a light meter. On stairways and their associated landings, the illumination level is required to be not less than 10 footcandles (108 lux) when the stair is in use. This higher illumination level is to ensure that the walking surface is visible and that level changes are more apparent.

An exception recognizes that these illumination levels might interfere with presentations in such places as motion picture theaters and concert halls; therefore, the exception allows a reduction in such building uses to a level of not less than 0.2 foot-candle (2.15 lux). Such a reduced lighting level, however, is permitted only during a performance and would be brought up to the minimum 1 foot-candle (11 lux) level if a fire alarm system were activated. At this point it is important to remember the distinction between a stepped aisle within the assembly seating area and a stairway since the 10 footcandle (108 lux) illumination level only applies to stairways and not to a stepped aisle.

A second method permits the use of self-luminous marking of the steps, landings, and ramps in accordance with Section 1025.

1008.3 Illumination required by an emergency electrical system. Normally, the power for illumination of the egress path is provided by the premises' wiring system. However, where the potential life-safety hazard is sufficiently great, it is considered inadequate to solely provide the illumination of the exit system by such a system. In these cases, it is necessary that emergency power—a completely separate source of power—automatically provide illumination of the exitways. In fundamental terms, separate sources of power are required in all occupancies in which two or more means of egress are required. Therefore, any space, area, room, corridor, exterior egress balcony, or other portion of the egress system requiring access to at least two exits or exit access doors is to be provided with emergency lighting. An example of this application is depicted in Figure 1008-1. Also included are exit stairways, both interior and exterior,

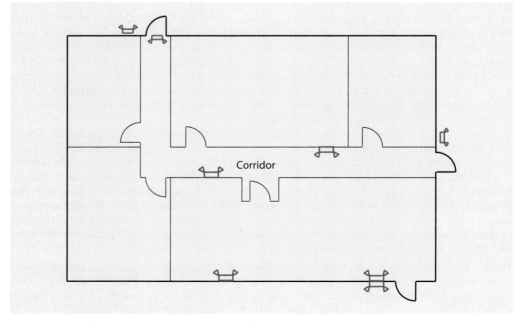

Figure 1008-1 **Emergency power for egress illustration.**

and exterior landings at exit doors in buildings requiring a minimum of two means of egress. Several specific rooms and spaces, also identified in Section 1008.3, must also be provided with emergency power for illumination purposes.

Where emergency power systems are required, they are to be supplied by storage batteries, unit equipment, or an on-site generator. It is the intent that this power source be automatically available even in the event of the total failure of the public utility system. Therefore, a separate, independent source is generally required. Installation of the emergency power system is regulated by referring to the requirements of NFPA 70.

The initial illumination provided by emergency power along the path of egress at floor level shall average at least 1 foot-candle (11 lux), with a minimum level of illumination required to be 0.1 foot-candle (1 lux). Illumination levels are permitted to decline over the required 90-minute duration of the emergency power source to an average of 0.6 foot-candle (6 lux) with a minimum at any point of 0.06 foot-candle (0.6 lux). Recognizing the variation in light levels throughout the exit path, only the average illumination level needs to be determined; however, an absolute minimum level of illumination must be attained. In no case shall the illumination uniformity ratio between the maximum light level and the minimum light level exceed 40 to 1.

Within Group I-2 occupancies, additional redundancy is provided by the requirement that the failure of any single lamp (bulb) in the fixture does not leave the area with inadequate illumination.

Section 1009 *Accessible Means of Egress*

1009.1 Accessible means of egress required. In addition to the access to buildings required by the provisions of Chapter 11, it is important that safe egress for persons with disabilities is provided. Therefore, the code requires that accessible spaces be provided with accessible means of egress consisting of one or more of the following components as set forth in Section 1009.2:

1. Accessible routes complying with Section 1104.
2. Interior exit stairways complying with Sections 1009.3 and 1023.
3. Exit access stairways complying with Sections 1009.3 and 1019.3 or 1019.4 (unless connecting levels in the same story).
4. Exterior exit stairways complying with Sections 1009.3 and 1027 (where serving floor levels other than the level of exit discharge).
5. Elevators complying with Section 1009.4.
6. Platform lifts complying with Section 1009.5.
7. Horizontal exits complying with Section 1026.
8. Ramps complying with Section 1012.
9. Areas of refuge complying with Section 1009.6.
10. Exterior areas for assisted rescue complying with Section 1009.7 (where serving exits at the level of exit discharge).

At least one accessible means of egress must be provided from all accessible spaces. Where more than one means of egress is required from any accessible space, at least two accessible means of egress are required. This ensures that a person with a disability also has an egress option should one path be blocked. An example to illustrate this provision is a large department store requiring multiple exits. Although the number of required exits from the store is addressed in Sections 1006.2 and 1006.3, only two accessible means of egress would be required from the accessible space. Therefore, the store might be required to provide three or more means of egress, but only two accessible means of egress need to be provided.

Two exceptions reduce or eliminate the accessible means of egress requirements. Only one accessible means of egress is required from accessible mezzanines, as well as from sloped-floor or stepped assembly spaces with limited travel to all wheelchair spaces.

1009.2 Continuity and components. As previously mentioned in the discussion of Section 1009.1, the code recognizes various accessible elements as components of an accessible means of egress. The accessible egress travel is required to extend beyond the building itself to the public way, unless an alternative means of protection is provided. If the egress route from the building to the public way is not accessible, it is acceptable to provide a complying exterior area of assisted rescue rather than create an accessible exit discharge path. Addressed further in the discussion of Section 1009.7, the exterior area of assisted rescue performs in much the same manner as an area of refuge inside the building.

1009.2.1 Elevators required. Unlike the general provision found in Section 1003.7 that specifically prohibits considering an elevator as an approved means of egress, this section requires an elevator for assisted rescue purposes under certain conditions. In buildings where a required accessible floor is four or more stories above or below a level of exit discharge, ramps and stairs cannot adequately serve as egress for individuals with a mobility impairment. Therefore, at least one elevator must be provided as an accessible means of egress. The elevator is not required to conform with Section 1009.4 in sprinklered buildings on those floors provided with a conforming ramp or horizontal exit.

In the application of this provision, the second story of a typical building is considered the first story above the level of exit discharge. Accordingly, the building's fifth story, or any accessible occupiable roof above the fourth story, would be the threshold at which an accessible means of egress by elevator would be required. See the discussion of Section 202 for *level of exit discharge* and Figure 1009-1. Under such conditions, a minimum of one accessible means of egress must be a complying elevator.

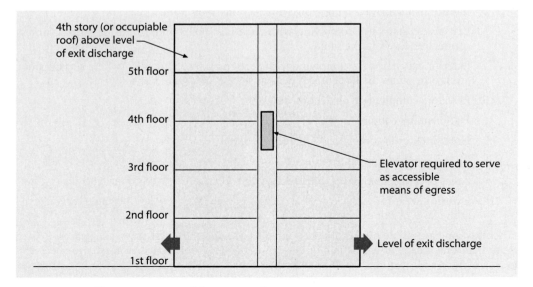

Figure 1009-1 **Elevator as accessible means of egress.**

1009.2.2 Maneuvering clearances at doors. To be accessible, doors along an accessible route must include appropriate maneuvering clearances to enable a user to approach and then position themselves to allow the door to be opened. This general requirement for maneuvering clearances applies to any door which is located along the accessible route. It is sometimes questioned as to whether this maneuvering clearance requirement applies to doors into and out of stairway enclosures. Section 1009.2.2 addresses maneuvering clearances at doorways along accessible routes for accessible means of egress, including doorways at stairways.

While a stairway is not viewed as being an accessible route that someone with mobility impairments can use independently, the stairway is used as a part of an accessible means of egress (AMOE). The key is that the fire department is using the stairways to come into the building, so that is where the people who need assistance and the emergency responders can get together most efficiently. If the stairway includes an area of refuge, maneuvering clearance at the door is required so that people can independently move into that area for assisted evacuation. If a building is sprinklered throughout, an area of refuge is not required in the stairway. Activation of the system automatically notifies the fire department, the sprinkler helps minimize the fire and smoke, and once the fire department arrives, they can use the sprinkler panel to identify which floor they need to move to first. So, per the exception, with no area of refuge in the stairway, a maneuvering clearance is not required at the door to get into the stairway.

At the ground floor, where the exit discharge does not provide an accessible route to the public way, an exterior area of assisted rescue or an interior area of refuge is required. This is typically where the route from the exit door to the public way includes steps, steep slopes or uneven ground. Maneuvering clearances are also required at a door leading to these spaces. If the landing is large enough for someone to turn around and come back in, and the door has hardware on the outside, this section could also require maneuvering clearance at this door on the outside as well as the inside.

The wording of the exception is important to note since it only exempts levels that are "above and below the level of exit discharge." This type of situation is quite common where a corridor on the level of exit discharge passes directly through a stair enclosure to reach the exterior exit door. Under this layout, while the occupants are entering the stair enclosure, they are not required to transit to another level, and thus the egress route should be accessible. Where occupants move through the stairway at the level of exit discharge for either entrance or means of egress, a maneuvering clearance is required in the direction of travel. If the accessible entrance and accessible means of egress are through other doors, maneuvering clearance would not be required.

1009.3 Stairways. Exit stairways are typically used as one or more of the required accessible means of egress in a multistory building. Increasing the width of stairs to 48 inches (1,219 mm) between the handrails allows for the minimum amount of space needed to assist persons with disabilities in the event of a building evacuation. The provisions for an area of refuge or, as specified in Exception 6 of Section 1009.3.3, a horizontal exit address the increased time needed for egress. The limitations of Section 1009.3.1 are intended to address exit access steps in assembly seating or split-level buildings, with a limited change of level, where an individual awaiting assistance or rescue may be required to wait in the middle of a building and not be able to travel toward an exit. There is no intent to prohibit the use of exit access stairways as part of an accessible means of egress from mezzanines and balconies as established the exception in Section 1009.3.1.

The exceptions within the subsections of Section 1009.3 should not be overlooked since they will often be applicable and can dramatically alter the stairway requirements. Two exceptions in Section 1009.3.2 remove the requirement for at least 48 inches (1,219 mm) of clear width between handrails. The 48-inch (1,219-mm) width is not mandated in sprinklered buildings, nor is it required for stairways accessed from a horizontal exit. A number of other exceptions in Section 1009.3.3 eliminate the general requirement for an area of refuge. Areas of refuge are not mandated at exit stairways serving open parking garages. In addition, areas of refuge are not mandated in Group R-2 occupancies and in smoke-protected or open-air assembly seating areas as regulated by Sections 1030.6.2 and 1030.6.3. Exception 1 addresses exit access stairways, eliminating any area of refuge requirement where two-way communication is provided at elevator landings under the provisions of Section 1009.8.

Exception 2 in Section 1009.3.3 is the most commonly applied exception and permits the omission of areas of refuge in buildings protected throughout by an automatic sprinkler system. The purpose of an area of refuge is to provide an area "where persons unable to use stairways can remain temporarily to await instructions or assistance during emergency evacuation." Much of the reasoning in exempting fully sprinklered buildings from the requirement for areas of refuge comes from NISTIR 4770, a report issued by the National Institute of Standards and Technology (NIST) in 1992 titled *Staging Areas for Persons with Mobility Impairments*. The primary conclusion of the report was that the operation of a properly designed sprinkler system eliminates the life threat to all occupants regardless of their individual abilities and can provide superior protection for persons with disabilities as compared to staging areas. The ability of a properly designed and operational automatic sprinkler system to control a fire at its point of origin and to limit production of toxic products to a level that is not life threatening to all occupants of the building, including persons with disabilities, eliminates the need for areas of refuge.

1009.4 Elevators. Although an elevator may be used as an accessible means of egress component in all multilevel facilities, it is only required as such in buildings regulated by Section 1009.2.1. Elevators used as accessible means of egress must comply with the operation and notification criteria of ASME A17.1, Section 2.27. In addition, standby power is required in order to maintain service during emergencies. The general requirement is that any elevator used as an accessible means of egress be accessed from an area of refuge or, as established in Exception 5, a horizontal exit. However, areas of refuge are not required in fully sprinklered buildings, open parking garages, smoke-protected or open-air assembly seating areas, and where elevators are not required to be protected by shaft enclosures. Additional information is provided in the discussion of Section 1009.3. In such cases, the elevator is still considered an accessible means of egress when in compliance with the other criteria of this section.

1009.5 Platform lifts. Except in limited applications, a platform lift is specifically excluded as an acceptable element of a means of egress. The maintenance of the lift as well as the complexity and delay in using a platform lift are considered substantial obstacles in providing acceptable means of egress for persons in wheelchairs. Section 1110.11 specifically sets forth the few instances where platform lifts are permitted for access and, with the exception of Item 10, egress purposes. Where a complying platform lift is used as an accessible means of egress component, it must be provided with standby power, much in the same manner as an elevator used for the same purpose.

1009.6 Areas of refuge. By definition, an area of refuge is an area "where persons unable to use stairways can remain temporarily to await instructions or assistance during emergency evacuation." Unfortunately, the term *temporary* is not defined, so a number of provisions are applied to an area of refuge to increase the level of protection for anyone using it. These provisions include a size large enough to accommodate wheelchairs without reducing exit width, smoke barriers designed to minimize the intrusion of fire and smoke, two-way communications systems, and instructions on the use of the area under emergency conditions. The two-way communications system is intended to allow a user of the area of refuge to identify his or her location and needs to a central control point. Obviously, it is important that someone be available to answer the call for help when a two-way communications system is provided. The system shall have a timed automatic telephone dial-out capability to a monitoring location or 911 that can be used to notify the emergency services when the central control point is not constantly attended. Each area of refuge shall be identified by a sign with the international symbol of accessibility stating that it is an area of refuge.

1009.6.3 Size. Each required area of refuge shall be sized to accommodate at least one wheelchair space not less than 30 inches by 52 inches (762 mm by 1,320 mm). Where the occupant load of the refuge area and the areas served by the refuge area exceeds 200, additional wheelchair spaces must be provided. Because wheelchair spaces are not permitted to reduce the required exit width and should be located so as to not interfere with access to and use of the fire department hose connections and valves, the designer needs to consider access to fire protection equipment and exit width when placing wheelchair spaces in the area of refuge.

1009.6.4 Separation. The primary concern for individuals awaiting assistance in an area of refuge is the intrusion of smoke and toxic gases into the refuge area. Therefore, the code requires a physical separation between an area of refuge and the remainder of the building. The separation is to be a smoke barrier complying with the provisions of Section 709. The smoke barrier is not required where the area of refuge is located within an enclosure for an interior exit stairway due to the inherent protection provided by such an egress enclosure. It is also permissible to create an area of refuge through the use of a horizontal exit.

1009.6.5 Two-way communication. Individuals awaiting assistance in an area of refuge must be provided with a communication means in order to contact a central control point. Where the central control point is not constantly attended, the area of refuge must be provided with a complying two-way communication system with automatic dial-out capability. Both audible and visible signals shall be provided. The requirements for such communication systems are established in Sections 1009.8.1 and 1009.8.2.

1009.7 Exterior area for assisted rescue. Item 10 of Section 1009.2 identifies an exterior area for assisted rescue as an acceptable portion of an accessible means of egress provided it complies with this section. The primary use of an exterior area for assisted rescue is to provide an exterior refuge area for those occupants unable to complete their egress travel due to the lack of an exterior accessible route from the building to the public way. Where exterior steps or other inaccessible site elements are present between the building's discharge level and the public way, it is permissible to use an exterior area for assisted rescue as an alternative to a fully accessible exterior path. Sized in accordance with Section 1009.6.3, the exterior area for assisted rescue in a nonsprinklered building must be separated from the interior of the building in a manner similar to that addressed for egress courts and exterior exit stairways. For nonsprinklered buildings, the exterior wall protecting the landing shall be protected for 10 feet (3,048 mm) horizontally to both sides with a minimum 1-hour fire-resistance-rated wall with any openings protected for at least ¾ hour and from the ground level to a point at least 10 feet (3,048 mm) above the floor level of the exterior area for assisted rescue. Such protection need only extend to the roof line if it is less than 10 feet (3,048 mm) vertically above the floor level of the exterior area for assisted rescue. See Figure 1009-2. In lieu of the 10 foot horizontal protection of the exterior wall, a wing wall which is perpendicular to the exterior wall and extends a minimum of 4 feet is permitted. The required extent of an exterior area for assisted rescue is not described in the code, requiring an individual evaluation where the size of the exterior area for assisted rescue becomes quite large. For an outdoor space to be considered an exterior refuge area, it must be at least 50-percent open so that toxic gases and smoke do not accumulate. Where an exterior stairway serves the exterior area for assisted rescue, an adequate distance between the handrails is mandated. In addition, complying signage is required per Section 1009.9, Item 2, to identify the exterior area as an appropriate refuge location. Where not desired or practical, such as where site restrictions or other conditions do not allow for the installation of an exterior area for assisted rescue, the exception in Section 1009.6.2 does permit the use of an area of refuge within the building on the level of exit discharge.

1009.8 Two-way communications. Unless provided in areas of refuge in multistory buildings, two-way communications systems must be located at the elevator landing of each accessible floor level other than the level of exit discharge. The system is intended to offer a means of communication to individuals who need assistance during an emergency situation. Such a system can

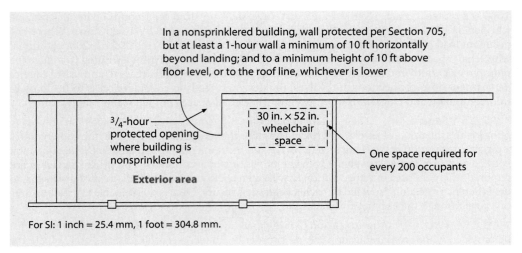

Figure 1009-2 **Exterior area for assisted rescue.**

be useful not only in the event of a fire but also in the case of a natural or technological disaster by providing emergency responders with the location of individuals who will require assistance in being safely evacuated from floor levels above or below the discharge level. See Figure 1009-3.

The first exception exempts the requirement for locating the communication systems at the elevator landings where the building is provided with complying areas of refuge. Since areas of refuge are required by Section 1009.6.5 to be equipped with two-way communication systems,

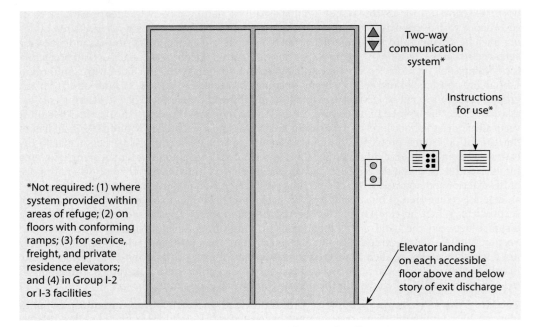

Figure 1009-3 **Two-way communication system at elevator landing.**

there is a limited need to provide such additional systems at the elevator landings. However, where multistory buildings are not provided with areas of refuge, such as is the case with most sprinklered buildings, the installation of communications systems at the elevator landings is important to those individuals unable to negotiate egress stairways during an emergency. As a result, most sprinklered and nonsprinklered multistory buildings must be provided with the means for two-way communications at all accessible floor levels other than the level of exit discharge. A second exemption applies to floor levels that use exit ramps as vertical accessible means of egress elements. Where complying ramps are available for independent evacuation, such as occurs in a sports arena, the two-way communications system is not required at the elevator landings. It should also be noted that multistory buildings without elevators, such as those identified in Section 1104.4, would not be regulated by this section. Thus in other than Group I-2 or I-3 facilities, all multistory buildings, except those exempted by Exception 2 and those without elevators, are required to be provided with two-way communication systems unless the only elevators in the building are freight, service, or private residence elevators.

The arrangement and design of the two-way communication system are specified in Section 1009.8.1. In addition to the required locations specified in Section 1009.6.5 for areas of refuge or Section 1009.8 for elevator landings, a communication device is also required to be located in a building's fire command center or at a central control point whose location is approved by the fire department. The term "central control point" is not a defined term. However, given the intent and function of the two-way communication system, a central control point is a location where an individual answers the call for assistance and either provides aid or requests aid for an impaired person. A central control point could be the lobby of a building constantly staffed by a security officer, a public safety answering point such as a 9-1-1 center, a central supervising station, or possibly any other constantly staffed location. The key functions at the central control point are that an individual is always available to answer the call for assistance and can either provide assistance or is capable of requesting assistance. In addition, the communication system provides visual signals for the hearing impaired and audible signals to assist the vision impaired.

Guidance to the users of the two-way communication system is also specified. Operating instructions for the two-way communication system must be posted and the instructions are to include a means of identifying the physical location of the communication device. If a signal from a two-way communication system terminates at a public safety answering point, such as a fire department communication center, current 9-1-1 telephony technology only reports the address of the location of the emergency—it does not report a floor or area from the address reporting the emergency. The "identification of the location" required by Section 1009.8.2 and posted adjacent to the communication system should ensure that most discrete location information can be provided to the central control point. This will aid emergency responders, especially in high-rise buildings or corporate campuses with multiple multistory structures.

1009.11 Instructions. Instructions on the use of the area of refuge, exterior area for assisted rescue, and elevator landing locations for required two-way communication systems must be provided where applicable. The intent of the instructions is not only to provide directions on the use of the communication equipment, but also to alert the users as to other available means of egress.

Section 1010 *Doors, Gates, and Turnstiles*

1010.1 General. This section applies to doors, gates, turnstiles, or doorways that occur at any location in the means of egress system. The provision found in Section 1022.2 should also be noted insofar as it will require that at least one exterior door that meets the size requirements of Section 1010.1.1 be provided from every building used for human occupancy. As doors pose a potential obstruction to free and clear egress, they are highly regulated.

Additional doors. The IBC establishes criteria for all egress doors, including those that are not required by Chapter 10. Such additional egress doors must comply with all the provisions of Section 1010 for exit doors. Where the doors are installed for egress purposes, whether or not required by the IBC, the building occupant would probably assume that they are a part of the means of egress system. Because the building occupant would then correctly expect the door to provide a safe path from the space, it is imperative that such doors and doorways conform to all applicable code requirements of Section 1010. Two examples are shown in Figures 1010-1 and 1010-2.

Door identification. The primary gist of the provisions on door identification is that egress doors should be installed so that they are readily recognized as egress doors and are not confused with the surrounding construction or finish materials. It is important that they be easily discernible as doors provided for egress purposes. As this provision is performance in nature, each condition would need to be individually evaluated. See Figure 1010-3 for an example of a condition where the means of egress doors may not be easily identifiable. The corollary of this requirement is that exit doors should not be concealed. In other words, they should not be covered with drapes or decorations, nor should they be provided with mirrors or any other material or be arranged in a way that could confuse the building occupants seeking an exit.

1010.1.1 Size of doors. The size of doors is regulated for both their ability to provide clear and efficient egress in emergencies and to function as a part of the building's general circulation system. Therefore, doors are generally to be of such a size as to provide a clear opening width of at least 32 inches (813 mm), as illustrated in Figure 1010-4, with a minimum clear opening door height of 80 inches (2,032 mm). Again, the code requires that the net dimension of clear width be provided by the exit component. Thus, when a swinging door is opened to an angle of 90 degrees (1.57 rad), it must provide a net unobstructed width of not less than 32 inches (813 mm) and permit the passage

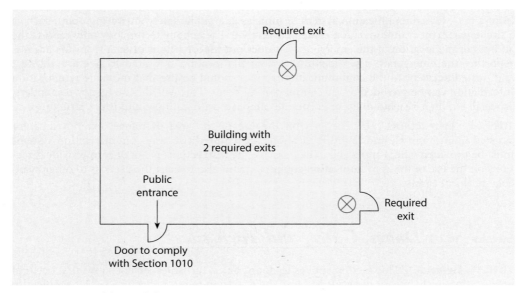

Figure 1010-1 **Additional door.**

Doors, Gates, and Turnstiles

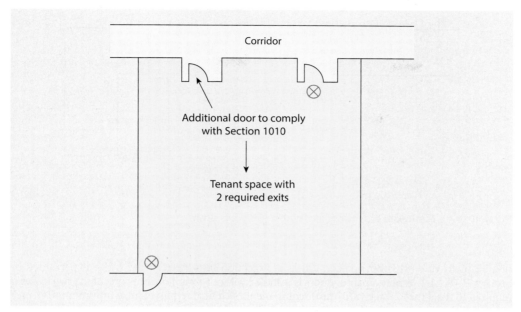

Figure 1010-2 **Additional door.**

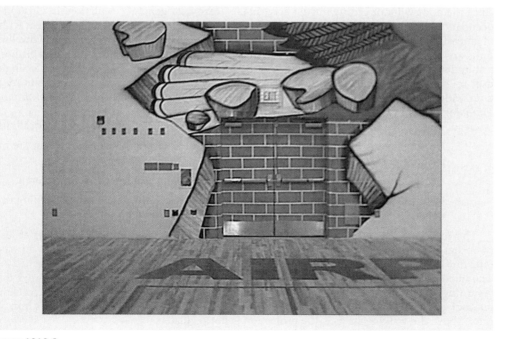

Figure 1010-3

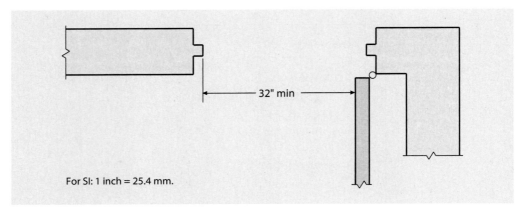

Figure 1010-4 **Minimum clear opening width of egress door.**

of a 32-inch-wide (813-mm) object, unless a projection into the required width is permitted by Section 1010.1.1.1. Where a pair of doors is installed without a mullion, only one of the two leaves is required to meet the 32-inch (813-mm) requirement. As a final requirement, a minimum 41½-inch (1,054-mm) means of egress doorway width to facilitate the movement of beds is mandated for those portions of Group I-2 occupancies where bed movement is likely to occur.

The exceptions address a number of situations where the door size is allowed to be reduced since the door will not greatly impact accessibility, egress, or general circulation. In Group I-3 occupancies, door openings to other than Accessible sleeping units need only have a clear width of at least 28 inches (711 mm). Accessible user passage door openings within Type B dwelling units are permitted a minimum clear width of 31¾ inches (806 mm). Exception 8 establishes a minimum clear opening width of 20 inches (508 mm) for doors serving nonaccessible single-user sauna compartments, toilet stalls or dressing, fitting or changing rooms. The minimum door width for shower compartments is regulated at 22 inches (559 mm) by the *International Plumbing Code*.

In addition, minimum door-opening widths are totally unregulated in the following locations:

1. Door openings in nonaccessible units of Group R-2 and R-3 occupancies that are not part of the required means of egress.
2. Storage closet doors where the closet is less than 10 square feet (0.93 m^2) in area.
3. In I-1, R-2, R-3, and R-4, interior egress doors within a dwelling unit or sleeping unit not required to be an Accessible unit, Type A unit or Type B unit.

The required clear width of door openings shall be maintained up to a height of at least 34 inches (864 mm) above the floor or ground. Projections may then encroach up to 4 inches (102 mm) for a height between 34 inches (864 mm) and 80 inches (2,032 mm). See Figure 1010-5. The maximum 4-inch (102-mm) limitation is based partially on those accessibility provisions regarding protruding objects. Its application allows for the intrusion of panic hardware, or similar door-opening devices, into the required clear width. At a height of 80 inches (2,032 mm) or more above the walking surface, the projection is not regulated.

Although the general requirement for the clear opening door height is a minimum of 80 inches (2,032 mm), the exceptions to Sections 1003.3.1 and 1010.1.1.1 permit door closers, overhead door stops, frame stops, power door operators, and electromagnetic door locks to encroach into this clear height, provided a headroom clearance of at least 78 inches (1,981 mm) is maintained.

Doors, Gates, and Turnstiles   425

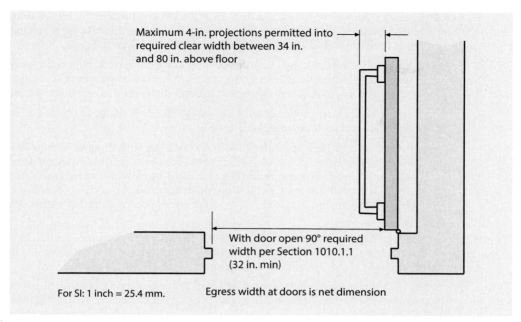

Figure 1010-5 **Egress door width.**

Exceptions 4 and 5 in Section 1010.1.1 permit door openings at least 78 inches (1,981 mm) in height within a dwelling unit or sleeping unit and a minimum height of 76 inches (1,930 mm) for all exterior door openings in nonaccessible dwelling units other than the required exit door. Only Exception 4 along with the previously mentioned exceptions in Sections 1003.3.1 and 1010.1.1.1 are applicable to the height reduction at a required exit door; therefore, required means of egress door openings must have a minimum clear opening height of 80 inches (2,032 mm), but may be 78 inches (1,980 mm) where complying projections exist.

1010.1.2 Egress door types. This section requires that every egress door, with exceptions, be of the side-hinged swinging, pivoted, or balanced door type. In most instances, it is necessary that the egress door encountered be of a type that is familiar to the user and easily operated. Therefore, swinging doors are required under all but the following conditions:

1. Private garages, office, factory and storage areas, and similar spaces where the occupant load of the area served by the doors does not exceed 10. Because of the limitation in occupant load and potential hazard, other types of egress doors are considered acceptable. A common application of this allowance is the use of overhead doors at self-storage facilities.

2. Detention facilities classified as Group I-3 occupancies. The security necessary in this type of use calls for special types of doors.

3. Critical care or intensive care patient room within suites of health-care facilities. In these areas, it is often preferable to use sliding glass doors to allow for visual observation and the efficient movement of equipment.

4. Within or serving an individual dwelling unit in Group R-2 and R-3 occupancies. Because of the limited occupant loads involved, and the familiarity of the occupants with the doors encountered, door types other than swinging doors are permitted.

5. Revolving doors conforming with Section 1010.3.1, where installed in other than Group H occupancies. In other than hazardous occupancies, the use of revolving doors is acceptable subject to the special conditions as set forth in the code.

6. Special purpose horizontal sliding, accordion, and folding door assemblies complying with Section 1010.3.3, where installed in other than Group H occupancies. Conditions for the use of such doors make them equivalent to other doors used in egress situations.

7. Power-operated doors in compliance with Section 1010.3.2. Safeguards provided for power-operated doors create an acceptable level of safety.

8. Bathroom doors within individual dwelling units and sleeping units of Group R-1 occupancies. It is often beneficial to use sliding pocket doors to provide access to hotel bathrooms, mostly due to the minimum 32-inch (813-mm) width requirement for doors in Group R-1 occupancies. Conflicts often occur between door swings and the required clearances for plumbing fixtures or clear floor space required at bathroom doors.

9. The use of a typical horizontal sliding door that is operated manually, such as a "pocket" door or a sliding "patio" door, is deemed acceptable in those instances where the occupant load served by the door is very low.

In addition, any side-hinged swinging, pivoted or balanced door serving an area or room with an occupant load of 50 or more, or those serving any Group H occupancy, shall swing with the flow of egress travel. In 1942, 492 people died in the Cocoanut Grove nightclub fire in Boston. One of the significant contributing factors to that loss of life was the fact that the exterior exit doors swung inward. As a consequence, it was not possible to open the doors because of the press of the crowd attempting to exit the building. This incident was identified as the primary reason for changing building codes to require that, under certain circumstances, exit doors must swing in the direction of exit travel.

1010.1.3 Forces to unlatch and open doors. This section differentiates between the force to unlatch the door by operating the door hardware itself and then the force needed to open the door. The code establishes the maximum force limits for each. Any type of push or pull hardware must operate at a maximum 15-pound (67 N) force while hardware that operates by rotation, like a lever hardware or doorknob, could not require a force exceeding 28 inch-pounds (315 N-cm). To move the door to an open position, interior swinging doors, other than fire doors, must have a maximum opening force of 5 pounds (22 N). All other swinging, sliding, or folding doors as well as fire doors, are to be set in motion with a maximum force of 30 pounds (132 N) but once moving must open with not more than a 15-pound (67 N) force. These forces are applied to the latch side of the door. Most doors are openable with forces less than these maximum limits. However, when in doubt, the actual force required can be easily measured by use of a spring scale.

1010.1.4 Floor elevation. The purpose of this section is to avoid any surprises to the person passing through a door opening, such as a change in floor level. Therefore, it is necessary that a floor or landing be provided on each side of a doorway. It is further intended that such a floor or landing should be at the same elevation on both sides of the door. A variation up to ½ inch (12.7 mm) is permitted because of differences in finish materials. See Figure 1010-6. Landings are required to be level, except exterior landings may have a slope of not more than ¼ inch per foot (6.4 mm per m) for drainage purposes.

Exceptions for individual dwelling units of Group R-2 and R-3 occupancies. An allowance is provided for individual dwelling units where it is permissible to open a door at the top step of an interior flight of stairs, provided the door does not swing out over the top step. The reason for permitting this type of arrangement in dwelling units is that as a building occupant approaches such a door from the nonstairway side, he or she must back away from the door in order to open

Doors, Gates, and Turnstiles

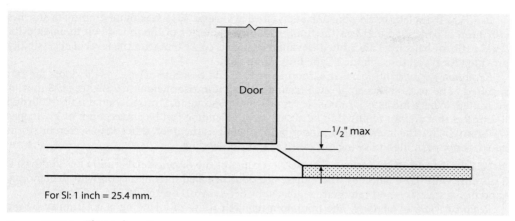

Figure 1010-6 **Floor elevation.**

it. This creates the need for a minimum landing to be traversed before the occupant can proceed to step down onto the stairs. In this situation, with minimal occupant load and familiarity with the unusual condition, the opening may occur at the top of the stairs, but the door must swing out toward the person descending the stairs. In an ascending situation, the stair user should have little difficulty in opening the door while standing on the stair treads, insofar as the door swings in the direction of travel. See Figure 1010-7.

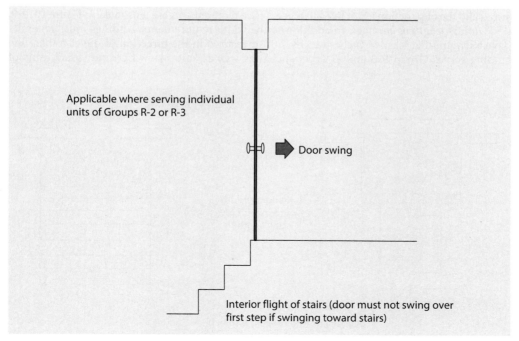

Figure 1010-7 **Floor level at doors.**

In a Type B dwelling or sleeping unit as provided in Chapter 11, a maximum drop-off of 4 inches (102 mm) is permitted between the floor level of the interior of the unit down to an exterior deck, patio, or balcony. This limited elevation change is consistent with the level of accessibility provided for travel throughout a Type B dwelling unit.

Exception for exterior doors. In Group F, H, R-2, and S occupancies exterior doors are not required to be accessible, a single step having a maximum riser height of 7 inches (178 mm) is permitted. Where the exterior door serves a Group U occupancy or individual units of Group R-2 or R-3 that are not required to be accessible, the landing can be a maximum of 7¾ inches (197 mm) below the interior side. Under both exceptions the door, other than screen or storm doors, is not permitted to swing out over the exterior landing.

1010.1.5 Landings at doors. This section contains the dimensional criteria for landings. It deals only with those landings where there is a door installed in conjunction with the landing. Landings at stairways and ramps are regulated by Sections 1011.6 and 1012.6, respectively.

Required width of landings. The minimum required width of a landing is determined by the width of the stairway or the width of the doorway it serves. Figure 1010-8 depicts these relationships. The requirement is that the minimum width of the landing be at least equal to the width of the stair or the width of the door, whichever is greater. The code is concerned that doors opening onto landings should not obstruct the path of travel on the landing. In this regard, the code establishes two limitations. The first states that when doors open onto landings, they shall not project into the required dimension of the landing by more than 7 inches (178 mm) when the door is in the fully open position. Second, whenever the landing serves an occupant load of 50 or more, doors may not reduce the dimension of the landing to less than one-half its required width during the course of their swing. Stated from the positive direction, it requires that doors swinging over landings must leave at least one-half of the required width of the landing unobstructed. Although the obstruction of one-half of the required width of the landing might seem excessive, it must be remembered that when the door is creating such an obstruction, it is in a position where it is free to swing and the obstruction is not fixed in place. These requirements are illustrated in Figure 1010-9.

Required length of landings. In addition to the width requirements, landings must generally have a length of at least 44 inches (1,118 mm) measured in the direction of travel. Where the landing serves Group R-3 and U occupancies, as well as landings within individual units of

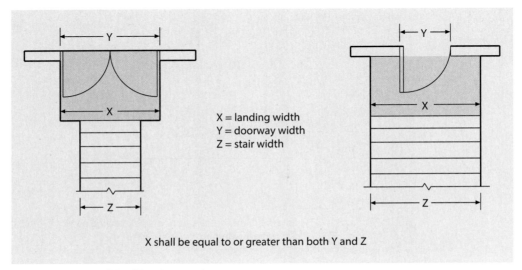

Figure 1010-8 **Width of landing at doors.**

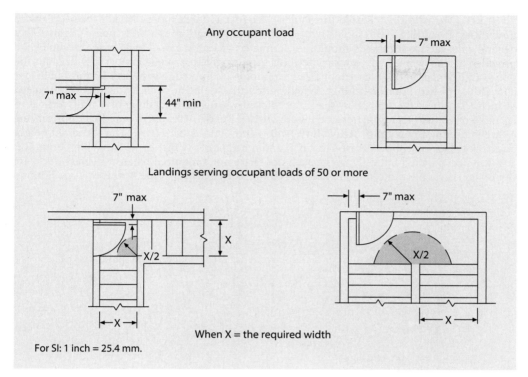

Figure 1010-9 **Doors at landings.**

Group R-2, the length need only be 36 inches (914 mm). These code requirements are illustrated in Figure 1010-10.

It should be noted that these minimum dimensions for landings in both width and length will be modified by the provisions in Chapter 11 where the door or doorway is a portion of the accessible route of travel.

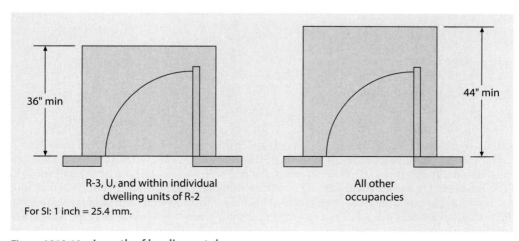

Figure 1010-10 **Length of landings at doors.**

1010.1.6 Thresholds. Raised thresholds make using doors more difficult for people with disabilities. In addition, thresholds with abrupt level changes present a tripping hazard. As a general rule, raised thresholds should be eliminated wherever possible. Where thresholds are provided at doorways, it is necessary to limit their height to provide easy access through the doorway. Changes in floor level and raised thresholds are limited to ½ inch (12.7 mm) in height above the finished floor or landing. Where raised thresholds or changes in floor level exceed ¼ inch (6.4 mm), the transition shall be achieved with a beveled slope of 1 unit vertical to 2 units horizontal (1:2) or flatter. See Figures 1010-11 and 1010-12. For a sliding door serving a dwelling unit, a maximum ¾-inch (19.1-mm) threshold is permitted. The threshold height at exterior doors may be increased to a maximum height of 7¾ inches (197 mm) in Group R-2 and R-3 occupancies, but only where such doors are not a required means of egress door, are not on an accessible route, and are not part of an Accessible unit, Type A unit or Type B unit.

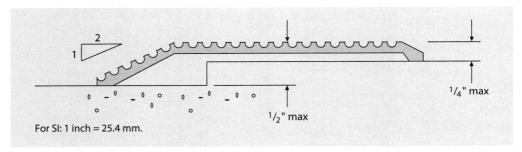

Figure 1010-11 **Threshold height.**

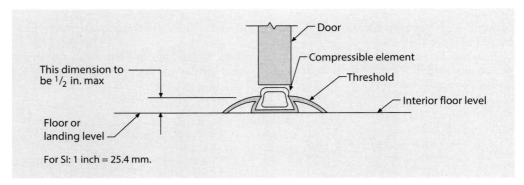

Figure 1010-12 **Threshold height.**

1010.1.7 Door arrangement. Adequate space must be provided between doors in a series to allow for ease of movement through the doorways. In other than dwelling units not considered Type A units, a minimum clear floor space of at least 48 inches (1,219 mm) in length is sized for a wheelchair user to negotiate through the door arrangement. Where a door swings into the space, the clear length shall be increased by the width of the door. As shown in Figure 1010-13, doors in a series must swing in the same direction or swing away from the space between the doors.

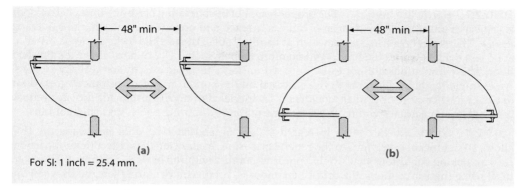

Figure 1010-13 **Two doors in series.**

1010.2 Door operations. This section, along with Section 1010.1.2, is particularly focused on the concept of ensuring that everything in the path of travel through the exit system, particularly doors, shall be under the control of and operable by the person seeking egress. Therefore, as a general statement, this section states that all doors in the egress system are required to be operable from the side from which egress is sought, without the need of a key or any special knowledge or special effort. If a key or special knowledge or effort is required, in all probability the door could not be readily openable by many building occupants. Such devices as combination locks are also prohibited on doors in exiting systems. Essentially, the code intends that the hardware installed be of a type familiar to most users—something that is readily recognizable under any condition of visibility, including darkness, and under conditions of fire or any other emergency.

In addition, the hardware must be readily operable. At times, one will encounter a different type of device such as a thumb turn. The building official must determine if this special type of operating device is acceptable. In many instances, it will be necessary to ensure that the building occupant can, in fact, grip the operating device and operate it. Some thumb turns are so small that they are quite difficult to operate, while others may require multiple twisting operations to achieve unlocking.

Another consideration that needs to be remembered in evaluating the acceptability of operating hardware is the fact that this hardware is going to be in place and in use over a substantial period of time. Unfortunately, doors and their operating hardware do not always get the constant maintenance that they should to keep them in operating order. It is imperative, in accordance with Section 1002.1, that the operation of doors in the egress system be maintained continuously in compliance with this section.

1010.2.1 Unlatching. The installation of multiple devices, or hardware requiring multiple operations, is inappropriate as well. Special effort and special knowledge are often necessary to open a door where more than one operation is required to unlock or unlatch the door. As a result, the multiple operations will typically result in an unacceptable delay in the egress efforts. Three exceptions set forth applications where multiple unlatching or unlocking operations are acceptable.

1010.2.2 Hardware. Where a door is required to be accessible under the provisions of Chapter 11, additional criteria come into play. It is important that all door hardware intended to be encountered by the door user be of a type that does not require tight grasping, pinching, or twisting of the wrist for operational purposes. Individuals with limited hand dexterity must be able to operate the unlatching or unlocking device without any special effort.

1010.2.3 Hardware height. The proper height of the operating hardware on an egress door is critical to ensure that the door user can easily reach and operate the unlatching or opening device. Therefore, operating devices such as handles, pulls, latches, and locks are to be installed at no less than 34 inches (864 mm) and no more than 48 inches (1,219 mm) above the finished floor. These limitations are not to be imposed on locks installed strictly for security reasons. An example might be a locking device installed on the bottom rail of an entrance/egress door of a convenience store or similar retail tenant. A special allowance is also made for latch release mechanisms on gates in required barriers surrounding swimming pools, spas, and hot tubs.

1010.2.4 Locks and latches. In regard to locking and latching door hardware, the IBC allows 10 significant exceptions to the general provision. In allowing these exceptions, it permits certain locking conditions that would appear to conflict with the basic requirement. However, in allowing these exceptions, the code often imposes certain compensating safeguards when the exit doors are to be locked or are provided with noncomplying hardware. It is the intent of the code that if the conditions are satisfied that a balance is struck between security or protecting occupants and the need for egress. These exceptions then essentially afford an equivalent level of safety as would be provided if the door were, in fact, readily openable at all times without the use of a key or any special knowledge or special effort.

1. This exception for locking hardware primarily concerns Group I-3 occupancies such as prisons, jails, correctional facilities, reformatories, and similar uses, where individuals are restrained or secured. By their nature, it is necessary that the occupants in these facilities be limited in their movement. Therefore, the IBC allows alternative locks or safety devices when it is necessary to forcibly restrain the personal liberties of the inmates or patients.

    Where a portion of a building is used for the restraint or security of five or fewer individuals, it is not considered a Group I-3 occupancy. Rather, it is anticipated that the secured area would simply be classified the same as the occupancy to which it is accessory. For example, up to five individuals could be restricted in an area such as a merchandise viewing room for customers in a jewelry store. In such situations, the allowance provided by Item 1 would be applicable for the egress door from the viewing room. A lock or latch that would prevent the expected unlocking or unlatching operation of the egress door would be acceptable. Thus, it is not necessary that the room or space be classified as a Group I-3 occupancy in order to apply this provision.

2. This exception addresses how manual locking systems can be used in assisted living, hospitals, and nursing facilities and still safely address the clinical needs and concerns of security and dementia wandering issues for the care recipients. This is conceptually similar to the provisions for electric locking systems of Section 1010.2.14 and used in similar circumstances. It is important to notice that "all" clinical staff must have the keys or codes to unlock the doors at all times.

3. The provisions of this exception apply to the main door or doors in Group A (having an occupant load of 300 or less), B, F, M, and S occupancies, and in all places of religious worship. It permits the main entrance/exit doors of buildings and spaces in buildings to be equipped with a key-operated locking device if several conditions are satisfied. In the occupancies listed, it is reasonable to assume that if the building or space is occupied, the main entrance/exit will, in all probability, be unlocked. The first condition states that the locking device be readily distinguishable as locked. The use of an indicator integral to the locking device may assist in determining when an unsafe condition is present. A second condition requires that there be a sign that is readily visible, permanently maintained, and located on or adjacent to the door. This sign is required to read THIS DOOR TO REMAIN UNLOCKED WHEN THIS SPACE IS

OCCUPIED. The letters must be at least 1 inch (25 mm) in height and placed on a contrasting background. Both of these requirements are for legibility. Although the language of the sign appears to apply without exception, there will be obvious situations where it should not be taken literally. For example, where an employee is working after hours and may be the lone occupant in the building or space, it is not the intent that the main doors remain in an unlocked position.

Obviously, the sign or the presence of the sign is not going to ensure that the door is unlocked. However, it does advise the occupant that whenever the space is occupied, the law does require, in the interest of reasonable fire safety, that the door be unlocked. In the event the door is not unlocked, the occupant is advised that his or her life may be at risk. The occupant should seek to alleviate that situation. The limitation imposed by this provision is typically applied only when the public is involved. For example, it is not reasonable to assume that the subject door be unlocked during a time period the occupancy is limited to a janitor or other personnel. Note that the use of this exception may be revoked by the building official for due cause.

4. Where egress doors are used in pairs, it is anticipated that each leaf in the pair of doors should be provided with its own operating hardware. This exception permits a special arrangement, however, where one of the door leaves is made inactive. Where installed in accordance with Table 1010.2.4, manual bolts, automatic flush bolts, and constant latching bolts are permitted on the inactive leaf. Additionally, the inactive leaf must not have any operating device such as a lever handle or panic hardware.

5. The fifth exception refers to exit doors from individual dwelling or sleeping units of Group R occupancies. The general requirement of Section 1010.2.1 essentially prohibits the use of dead bolts or other security devices that would be installed in addition to the complying door hardware, as such an arrangement would require multiple operations to unlatch the door. This exception does permit, however, the use of a dead bolt, a security chain, or a night latch when the dwelling or sleeping unit is permitted to have a single means of egress, on the condition that the device be openable from the inside without the use of a key or any special tool. It follows from the basic requirement of this section, however, that the device must not require any undue effort in order to unlatch the door and gain egress.

6. The listed test procedures for a fire door include the disabling of the door operation mechanism. This exception clarifies that once the minimum elevated temperature has disabled the door's unlatching mechanism, the resulting prevention of the door's operation is acceptable.

7. This exception is a simple issue of security and would be similar to allowing a lock on the building's entry door so that people may not enter the building through this door. If the roof is intended to be occupied, then access to the means of egress is required.

8. Exceptions 8, 9, and 10 address various situations where occupied exterior spaces must egress back through the building. These exceptions attempt to strike a balance between security and life safety. As a general concept, the code requires an egress path to always be available and under the control of the user. The code also requires, per Section 1010.2, that the egress doors are "readily openable from the egress side without the use of a key or special knowledge or effort." Where an occupied exterior space must pass back into and through the building, this essentially means the building must be left unlocked to allow entry into the building. These three exceptions provide guidance on where and how exit doors serving the exterior spaces may be locked in order to secure the building when the exterior space is unoccupied or of limited size.

1010.2.6 Stairway doors. The general requirement for interior stairway doors is that they be openable from both sides without the use of a key or special knowledge or effort. Such conditions allow for immediate access from the stairway enclosure to the adjacent floor area for emergency responders. In addition, in the unlikely event that the stairwell becomes untenable during evacuation procedures, occupants may reenter a floor level as an alternative means of egress. However, five exceptions are provided to modify this requirement.

1. Those doors that provide egress from the stair enclosure, discharging directly to the exterior or to an egress component leading to the exterior, are permitted to be locked only on the side opposite the direction of egress travel.

2. In high-rise buildings, stairway doors, other than exit discharge doors, may be locked from the stairway side. Under these conditions, such doors must be capable of being unlocked simultaneously without unlatching upon a signal from the fire command center. Although this exception is not limited to use in high-rise buildings, it is most commonly applied in such situations. When used in buildings that are not considered high-rise, the criteria of Section 403.5.3 may also be applied.

3. Doors are required to be locked from the side opposite the egress side for stairways where emergency personnel have the ability to simultaneously unlock the door. This action must be accomplished upon a signal from a single interior location at the building's main entrance, with the specific location for the actuating device likely approved by the fire code official. Where the building is provided with a fire command center, the signal must be actuated from within the center. The unlocking action shall also be initiated upon activation of a fire alarm signal, as applicable, and upon failure of the locking system's power supply. It is important that the unlocking signal not deactivate the latching devices of the stairway doors. The doors must remain latched in order to maintain their integrity as fire door assemblies.

4. Applicable only to two-story buildings housing a Group B, F, M, or S occupancy, this exception permits the locking of the stairway door from the stairway side provided the only interior access to the tenant space is a single exit stair. Under specified conditions in Section 1006.3.4, a single exit is permitted from the basement or second story of the moderate-hazard occupancies scoped in this exception. Where a single exit stair is acceptable due to a limited occupant load and travel distance, it is permissible to lock the door to the tenant space from the stairway side.

5. This exception is similar in application to Exception 4.

1010.2.7 Locking arrangements in educational occupancies. This section is intended to provide a balance between needed egress availability and security which may be needed to keep intruders in threatening situations from being able to open the door. These provisions may be applied in any educational use whether classified as a Group E, B, or I-4 occupancy. In addition, the provisions are not limited only to the classrooms but can include any other associated occupied rooms or office areas.

1010.2.8 Panic and fire exit hardware. Basically, panic hardware is an unlatching device that will operate even during panic situations, so that the weight of the crowd against the door will cause the device to unlatch. This provision of the code, in harmony with the need to swing the door in the direction of egress travel, is intended to prevent the type of disaster experienced in the Cocoanut Grove fire in Boston in 1942. When a panic hardware device is installed on a swinging door leaf, the press of the crowd, which prevented the opening of the door in the Cocoanut Grove fire, will ensure the automatic opening of the door.

Where panic hardware is provided, it is necessary that the activating member of the device extend for at least one-half of the width of the door leaf. This minimum length ensures that the unlatching operation will take place when one or more individuals impact the door. In addition, the device must be arranged so that a horizontal force not exceeding 15 pounds (66 N), when applied in the direction of exit travel, will unlatch the door.

Where required. Because of the large concentration of people in an assembly occupancy that may reach an exit door at about the same time during an emergency, the Group A occupancy is one of those occupancy groups that requires the installation of panic hardware listed in accordance with UL 305. Such hardware is also required in Group E occupancies for essentially the same reason. Therefore, in these two occupancies, any swinging egress door provided with a latching or locking device that serves a room or area having an occupant load of 50 or more must be provided with panic hardware. In addition, in all Group H occupancies, panic hardware is required on every egress door regardless of the occupant load served. The potential life-safety hazard in these Group H occupancies is such that when it is necessary to evacuate this type of use, exiting from such spaces must be almost immediate. To facilitate the rapid escape from these occupancies, egress doors shall not be provided with a latch or lock unless it is panic hardware. Again, it should be noted that if a door is used in a situation where panic hardware might otherwise be required, it is not necessary to install panic hardware if the door has no means for locking or latching. If the door is free to swing at all times, there is no need to install panic hardware to overcome a lock or latch. Panic hardware is also not mandated on those doors considered as the main exit in a Group A occupancy having an occupant load of 300 or less where the provisions of Section 1010.2.4, Item 3, are applied, permitting the use of key-operated locking devices. Electrically locked doors complying with Section 1010.2.10 are also permitted in lieu of panic hardware. To coordinate with provisions in other sections, Exceptions 3 and 4 allow locking devices on egress doors from exterior spaces that reenter the building and those that serve as the secondary route from courtrooms provided the appropriate limitations are followed. Subsections 1010.2.8.1 and 1010.2.8.2 address the need to immediately escape the hazards found within refrigeration machinery rooms and electrical rooms containing certain types of equipment or exceeding certain ratings.

Panic hardware on balanced doors. Special care must be taken when installing panic hardware on balanced doors. In this instance, push-pad-type panic hardware is to be used, and it must be installed at one-half of the door width nearest the latch side to avoid locating the pad too close to the pivot point.

Section 1010.1.2 requires that exit doors should be side-hinged or pivoted swinging doors. The typical pivoted door has its top and bottom pivot points located near the edge of the door frame opposite the latch side. Balanced doors are nothing more than a specialized type of pivoted door in which the pivot point is located some distance inboard from the door edge, creating a counter-balancing effect. The length of the panic bar is limited for balanced doors because the door cannot be opened if the opening force is applied too close to the pivot point or beyond. Limiting the panic bar length to one-half the door width ensures that those who use the door will apply opening pressure at a distance sufficiently removed from the pivot point to allow the door to open. See Figure 1010-14.

1010.2.9 Monitored or recorded egress and access control systems. Monitored egress is where a device requiring credentials such as a card reader, keypad, or biometrics is used to monitor who is egressing. Access control provides for security at the door location. This section permits these types of systems provided they are incorporated with one of the special locking arrangements listed, or that the door is still capable of being opened without activating the device.

1010.2.10 Door hardware release of electrically locked egress doors. As a general rule, means of egress door hardware must be operable by manual operation to provide for occupant control of the egress system. Locking devices are typically prohibited, as they can interfere or prevent efficient egress through the door during an emergency situation. However, owner concerns that must be considered sometimes require a greater degree of security. In specific occupancies, doors in the means of egress are permitted to be electrically locked if equipped

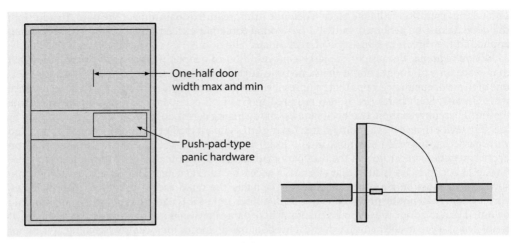

Figure 1010-14   **Panic hardware on balanced doors.**

with listed hardware that incorporates a built-in switch that interrupts the power supply to the electric lock and unlocks the door. The use of this type of locking system provides for a greater degree of security than that offered by other methods addressed in the code, including delayed egress locking systems and sensor released electrically locked egress systems.

In other than Group H, the allowance for electronically locked egress doors is permitted in all occupancies where security can be a major concern. The listed hardware that incorporates a built-in switch is required to comply with UL 294. When the occupant prepares to use the door hardware, the method of operating the hardware must be obvious, even under poor lighting conditions. The operation shall be accomplished through the use of a single hand. This is consistent with the general requirement that the door be readily openable without the use of special knowledge or effort. The unlocking of the door must occur immediately on the operation of the hardware by interrupting the power supply to the electronic lock. As an additional safeguard, the loss of power to the hardware shall automatically unlock the door.

This special type of locking device is permitted to be used in conjunction with panic hardware and fire exit hardware, but only where operation of such hardware also releases the electromagnetic or electromechanical locking device.

1010.2.11 Sensor release of electrically locked egress doors. Security concerns have prompted the use of electrically locked doors to provide controlled access at entrance doors to buildings or tenant spaces. Therefore, this section permits the use of an approved electrically locked egress door at the main entrance of all occupancies except Group H. Sensor-released egress doors will typically be locked from the exterior at all times. In order to ensure that the door is fully operable during a fire incident or other emergency, a number of criteria have been developed. For the most part, when a problem situation is identified, the door operates in a manner like any other egress door. A signal from a sensor, or loss of power to the sensor or to the electric lock, must unlock the electric lock. Activation of the building fire alarm, automatic sprinkler system, or fire detection system will automatically unlock the door. Unlocking must also be possible from a location adjacent to the door, or occur when there is loss of power to the access-control system. It is important to note that emergency lighting must be provided on the egress side of the door and that the security device must be listed in accordance with UL 294, *Access Control System Units*, or UL 1034, *Burglary-Resistant Electric Locking Mechanisms*.

1010.2.12 Delayed egress. The building code provides for a degree of security to egress doors serving Groups B, F, I, M, R, S, and U occupancies, as well as for limited situations for Group E classrooms and Group A-3 or B courtrooms. This section allows the use of a door that has an egress-control device with a built-in time delay under specific conditions. These devices were introduced in the code to resolve the problem of an exit door being illegally blocked by building operators desperate to stop the theft of merchandise through unsupervised, secondary exits. Institutional and residential occupancies are included because it is perceived that they also have security problems that need to be addressed. The devices are sometimes needed in nursing homes or group-care where facility operators must restrict patient egress while still maintaining viable exit systems.

It must be emphasized that under the conditions imposed by this section, and within the reliability of the automatic systems required, there will often be no delay whatsoever at the exit in an actual fire emergency—the door will be immediately openable.

Delayed egress locking conditions. Several conditions should be emphasized at the outset:

1. Such devices may be used only in connection with the specifically listed occupancies.

2. The entire building in which the delayed egress locking system is installed must be completely protected throughout by either an approved automatic sprinkler system or an approved automatic smoke- or heat-detection system.

3. In other than Groups I-1 Condition 2, I-2 and I-3 occupancies, a building occupant shall not be required to pass through more than one door equipped with a delayed egress lock before entering an exit. Groups I-1 Condition 1 and I-4 are also granted this two door/30 second option, but only with specific increased sprinkler requirements.

4. The door shall unlock in compliance with the following criteria: The device must immediately and automatically deactivate on activation of the sprinkler system or fire-detection system, and on the loss of electrical power to the egress lock. There must also be a way of manually deactivating the device by the operation of a signal from the fire command center where provided. Where the operating device is activated, it must initiate an irreversible process that will cause the delayed egress lock to deactivate whenever a manual force of not more than 15 pounds (66 N) is applied for a minimum period of 3 seconds to the operating hardware. The irreversible process must achieve the deactivation of the device within a time period of not more than 15 seconds from the time the operating hardware is originally activated. Where approved by the building official, a delay of not more than 30 seconds is permitted. Upon activation of the operating hardware, an audible signal shall be initiated at the door so that the person attempting to exit the building will be aware that the irreversible process has been started.

A sign must be installed on the door above and within 12 inches (305 mm) of the operating hardware so that the person seeking egress can be informed as to the type and nature of the egress lock. The sign must read PUSH (PULL) UNTIL ALARM SOUNDS. DOOR CAN BE OPENED IN 15 (30) SECONDS. An additional requirement requires emergency lighting to be provided at the door where the delayed egress lock is used. In all cases, the delayed egress locking system must comply with UL 294.

Delayed egress lock reactivation. The code emphasizes that, regardless of the means of deactivation, relocking of the device shall only be by manual means at the door. This requirement ensures that to relock the delayed egress lock, someone must go to the door itself, verify that the emergency no longer exists, and only then relock the door by manual means.

1010.2.13 Controlled egress doors in Groups I-1 and I-2. Group I-1 and I-2 facilities often house dementia and Alzheimer's patients. In order to balance the needs of the facility with the life safety of the occupants, the limitations on locking devices in these types of uses must allow for a safe and secure environment for these patients within the means of egress concepts of the code. Locks are permitted on means of egress doors that serve Group I-1 and I-2 care recipients whose movement is restrained provided a number of conditions are met. Many of the conditions are similar to those set forth in Section 1010.2.12 for delayed egress locks. The building must be fully sprinklered or provided with an approved smoke detection system. The doors must unlock upon actuation of the sprinkler or smoke-detection system, the loss of power to the lock or lock mechanism, or by a signal from the nursing station or other approved location. The staff must also have the means to unlock the doors when necessary. As a further condition, it is mandated that occupants need only pass through one door equipped with this special egress device prior to entering an exit element. In all cases, the door locking system must be in conformance with UL 294 or UL 1034.

Where patients with mental disabilities are housed, it is often necessary that they be restrained or contained for their own safety. In such cases, the level of restraint must be maintained even if the fire protection systems are activated or the power to the lock fails. However, it is still important that the emergency preparedness plan be developed and the clinical staff has the ability to monitor and enable the evacuation. See Figure 1010-15. The provisions are also applicable where the locking system is used in nursery and obstetric areas of Group I-2 hospitals to reduce the risk of child abduction.

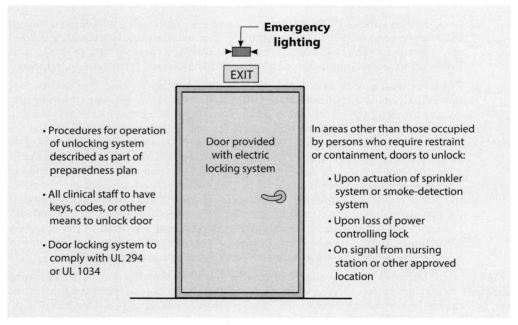

Figure 1010-15 **Locks on egress doors of Group I-1 and I-2 occupancies.**

1010.2.14 Elevator lobby exit access doors. Elevator lobbies are required to have access to at least one means of egress. Egress is prohibited through a space that can be locked, therefore many elevator lobbies must either provide direct access to an exit stairway or they must open into a corridor that then connects to a stairway. Depending on how the story and its exits are arranged, this can either result in the need for an additional stairway or in finding it necessary to divide the tenant spaces in a manner that permits an egress path. As an alternative solution, electrically locked exit access doors providing egress from elevator lobbies are permitted.

This section will allow an elevator lobby to have access to its required exit (Section 3006.4) by passing through an electrically locked exit access door and through a space that would not typically be open from the lobby or allow exiting (Item 4 in Section 1016.2). An example of this type of arrangement is a story where a single tenant space occupies the entire floor level with two exit stairways only being available within the tenant space and an elevator lobby separated from the tenant's reception area. The elevator lobby space could be locked from the tenant space and lobby occupants would not have access to either of the exit stairways. Under these provisions, the lobby doors into the tenant space could be locked, provided the eight listed requirements are followed. These conditions for compliance offer a reasonable solution that (a) provides required egress when needed, (b) provides security by having public areas locked off, and (c) helps minimize disruption to a floor plan that an additional exit or access from a lobby directly to an exit could cause.

Although these exit access doors are set up to unlock and be available as an exit in an emergency (Item 3, 4, or 5), Item 6 requires a two-way communication system within the lobby so that a person does not end up being trapped if there is no emergency. This two-way communication system would allow a person in the lobby area to notify someone if they were unable to get to the exit stairs.

1010.3 Special doors. Based on the provisions in Section 1010.1.2, the code generally requires that doors in exiting systems be of the side-hinged swinging, pivoted or balanced door type. In this section, four different types of doors are identified that may be used under very specific conditions.

1010.3.1 Revolving doors. Revolving doors continue to be used at building entrances. Where once used primarily in cold climates, they are now being installed in all regions, primarily as an energy-conservation measure. The use of revolving doors is specifically permitted as an alternative to enclosed vestibules where such vestibules are required by the *International Energy Conservation Code*® (IECC®) at the entrances to commercial buildings. Exception 5 of Section 1010.1.2 permits the installation and use of revolving doors in all occupancies other than Group H when complying with this section. However, it is not permissible to use revolving doors to supply more than 50 percent of the required egress capacity, nor be assigned a capacity greater than 50 persons. Where used, the door must be an approved revolving door and comply with the specific requirements listed.

All revolving doors must be in full compliance with BHMA A156.27. When revolving doors are installed, they must be of a type where the door leaves will breakout under opposing pressures providing at least 36 inches (914 mm) of aggregate width. Location of the door in relationship to the foot or top of stairs or escalators is regulated, as is the maximum number of revolutions per minute. At least one conforming exit door shall be located in close proximity to the revolving door. In such an arrangement, the adjacent swinging door can be used to satisfy exit capacity requirements. The maximum force levels required to collapse a revolving door vary based on whether or not the door is to be used as an egress component.

1010.3.2 Power-operated doors. Power-operated sliding or swinging doors are often used at the main entry of a building, particularly in mercantile and business occupancies. The same doors are also typically an important aspect in the overall exiting system for the building. There are a number of different types of doors that use a power source to open a door or assist in the manual operation of the door. This may include doors with a photoelectric-actuated mechanism to open the door upon the approach of a person, or doors with power-assisted manual operation. Where such doors are used as a portion of the means of egress, they must be installed in accordance with this section. The main criterion concerns the capability of the door being opened manually in the event of a power failure. Essentially, doors shall have the capability of swinging, and they must be designed and installed to break away from any position in the opening and swing to the fully open position when an opening force not exceeding

50 pounds (222 N) is applied at the normal push-plate location. Power-operated swinging, sliding, and folding doors must comply with BHMA A156.10, whereas BHMA A156.19 applies to swinging doors that are power-assisted and low-energy power-operated. Sliding or folding doors that are low-energy power-operated are regulated by BHMA A156.38.

1010.3.3 Special purpose horizontal sliding, accordion, or folding doors. Used as smoke and/or fire separation elements, these doors are normally in a fully open position and hidden from view. Closing only under specific conditions, they typically are part of an elevator lobby or similar protected area. Eight provisions are identified in the IBC that regulate the use of such doors as a component of means of egress. Fundamental to the use of these special doors is that manual operation of the normally power-operated doors must be possible in the event of a power failure, and no special or complex effort or knowledge should be necessary to open the doors from the egress side.

1010.3.4 Security grilles. Because of the concern of exit doors being obstructed or even completely unusable, the use of security grilles is strictly regulated. By their nature, security grilles are difficult to operate under emergency conditions. Used frequently at the main entrances to retail sales tenants in a covered mall building, such grilles are also permitted in other Group M occupancies as well as Groups B, F, and S. Security grilles, either horizontal sliding or vertical, are only permitted at the main entrance/exit. During periods of time when the space is occupied, including those times where occupied by employees only, the grilles must be openable from the inside without the use of a key or special knowledge. They must be secured in the fully open position during those times where the space is occupied by the general public. Where two or more exits or access to exits are required from the space, a maximum of 50 percent of the exits or exit access doorways are to be equipped with security grilles.

1010.4 Gates. This section serves as a reminder that gates within the means of egress system must comply with all of the requirements for doors.

1010.4.1 Stadiums. Facilities such as stadiums may be enclosed by fencing or similar enclosures. The requirement for panic hardware does not apply to gates through such enclosures, provided that the gates are under constant and immediate supervision while the stadium is occupied. However, there must be a safe dispersal area of a size sufficient to accommodate the occupant load of the stadium based on 3 square feet (0.28 m^2) per person, and located between the stadium and the fence or other enclosure. Such a dispersal area must not be less than 50 feet (15,240 mm) from the stadium it serves.

1010.5 Turnstiles and similar devices. In order to address safety concerns that are due to the use of turnstiles or similar devices that may be placed along the path of exit travel, this section regulates their use when located in a manner to restrict travel to a single direction. The general rule requires that each turnstile be credited with a maximum capacity of 50 occupants, and then only when specific conditions are met. When primary power is lost, the device shall turn freely in the direction of egress travel. Release shall also occur upon manual operation by an employee in the area. When determining the overall egress capacity, turnstiles may only be considered for up to 50 percent of the required capacity. Each device is limited to 39 inches (991 mm) in height and must have a minimum clear width of 16½ inches (419 mm) at and below the height of 39 inches (991 mm). Above the 39-inch (991-mm) height, a clear minimum width of 22 inches (559 mm) is necessary. Obviously, variations in these requirements are necessary where the turnstile is located along an accessible route of travel. Where turnstiles exceed 39 inches (991 mm) in height, they are regulated in a manner consistent with revolving doors.

To address the concern for use of such devices in large occupancies, Section 1010.5.4 requires a side-hinged swinging door for devices other than portable turnstiles. Required at a point where the occupant load served exceeds 300, a swinging door must be located within 50 feet (15,240 mm) of each turnstile. Portable turnstiles are designed to be moved out of the way for large occupancies such as sporting events.

While the previously discussed requirements apply to what may be considered as traditional turnstiles, Section 1010.5.2 deals with a newer style of control device that is generally viewed as providing security access. Where these devices inhibit egress travel they must comply with this section. These barriers have not only different size and capacity requirements, but they can only be used in fully sprinklered buildings and must be set up to automatically open during a number of circumstances.

Section 1011 *Stairways*

In order to apply the requirements of this section on stairways in an appropriate manner, the scope of the provisions must be determined. The definition of a stairway is critical for this determination. Found in Section 202, the definition consists of two parts. First, a stair is considered a change of elevation accomplished by one or more risers. Second, one or more flights of such stairs make up a stairway, along with any landings and platforms that connect to them. Based on these two definitions, a single step would also be considered a stairway under the IBC.

1011.1 General. The scoping provision for this section indicates that the requirements of Section 1011 apply to any stairway serving an occupied portion of a building, eliminating the potential for inappropriate interpretations that view stairways are not required as a means of egress as not regulated by Chapter 10 nor the provisions of Section 1011. Whether stairways are serving as a required portion of the egress system or simply installed in additional numbers beyond the code minimum, as "convenience" stairs, it is appropriate for all stairways to meet the minimum safeguards that the code intends. As a reminder, stairway systems in assembly seating areas are identified as stepped aisles and regulated by Section 1030.

1011.2 Width and capacity. The provisions concerning minimum width and required capacity of stairways are analogous to the provisions relating to corridors discussed in Section 1020.3. If the stair is subject to use by a sufficiently large occupant load, the minimum required width of the stair is determined by using the formula stated in Section 1005.3.1. Otherwise, its required width cannot be less than the width established by this section. In general terms, the minimum required width of any stair must be at least 44 inches (1,118 mm). In the event the stairway serves an occupant load of 49 or less, the required minimum width of the stairway is only 36 inches (914 mm). The entire occupant load of the story served by the stairway is considered, rather than divided by all available stairways. See Application Example 1011-1. Other modifications to the width requirements apply to spiral stairways as addressed in Section 1011.10, and stairways that are provided with an incline platform lift or stairway chairlift. Generally, when the code specifies a required width of a component in the egress system, it intends that width to be the clear, net, usable unobstructed width.

GIVEN: A two-story building with two stairways serving the second floor. The occupant load of the second floor is 68.

DETERMINE: The minimum required width of the stairway.

SOLUTION: Because both of the stairways are required to serve the floor, the minimum required width of each stairway is 44 inches. The occupied load is not divided (as in the case of distributing calculated width) between the two stairways.

For SI: 1 inch = 25.4 mm.

Application Example 1011-1

However, handrails and other projections are permitted to encroach into the required capacity or minimum width of a stairway as established by the provisions of Section 1014.9. It should be noted that "stepped aisles" in assembly spaces must comply with the requirements of Section 1030 and not the general stairway provisions. There are a number of differences such as width, capacity, riser heights, landings, and other items that make this distinction important as to why the specific requirements from Section 1030 should be applied.

1011.3 Headroom. A minimum headroom clearance of 6 feet 8 inches (2,032 mm) is required in connection with every stairway. Such required clearances shall be measured vertically from the leading edge of the treads to the lowest projection of any construction, piping, fixture, or other object above the stairs, and shall be maintained for the full width of the stairway and landing. See Figure 1011-1. This specific height requirement overrides the general means of egress ceiling height requirement found in Section 1003.2 and is modified for spiral stairways by Section 1011.10.

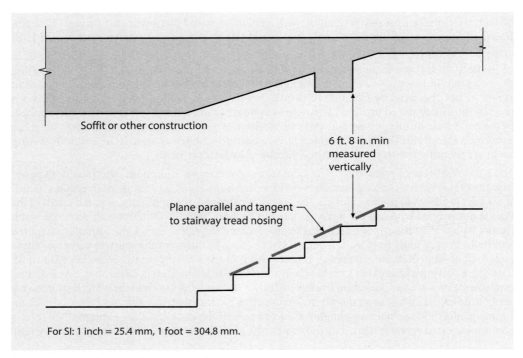

Figure 1011-1 **Stairway headroom clearance.**

1011.5.2 Riser height and tread depth. This section provides for a maximum riser height of 7 inches (178 mm), a minimum riser of 4 inches (102 mm), and a minimum tread run of 11 inches (279 mm) for each step on any stairway. These limiting dimensions are identified in Figure 1011-2. Variations in the requirements for treads and risers apply to alternating tread devices, spiral stairways, and stepped aisles in assembly seating areas. These variations are discussed elsewhere in this chapter. Another exception allows 7¾ inches (197 mm) maximum and 10 inches (254 mm) minimum for rise and run, respectively, for stairways in Group R-3 occupancies, within dwelling units in Group R-2 occupancies, and in Group U occupancies accessory to a Group R-2 or R-3 dwelling unit.

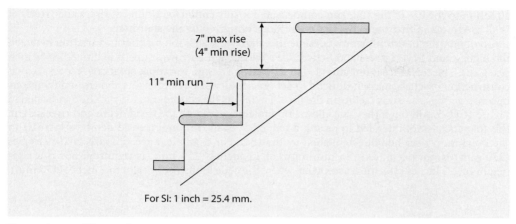

Figure 1011-2 **Rise and run.**

The 7-inch (178-mm) rise and 11-inch (279-mm) run figures for the steps are based primarily on safety in descending the stairs and are the result of much research. Probably at no prior time in the history of codes has the proportionate of stairs enjoyed a better foundation in research.

As one descends a stairway, balance is essential for safety. Therefore, the tread run must be of such a dimension as to permit the user to balance comfortably on the ball of the foot. The appropriate combination of riser height and tread run provides the proper geometry to enable the user to accomplish the necessary balance to descend the stairway with reasonable safety. Consistent with the importance of the tread dimension, the method of measurement of the tread is expressly stated. Specifically, tread depth (or run) is that distance measured horizontally between vertical planes passing through the foremost projections of adjacent treads or landing. As such, the tread dimension is the net gain in the run of the stair. Tread dimension is measured in this manner because any tread surface underneath the overhang of a sloping riser or nosing on the tread above is not available to the person descending the stair. Because descending is the more critical direction, proper dimension of the tread is of paramount importance.

Studies of people traveling on stairways have shown that probably the greatest hazard on a stair is the user. Inattention has been identified as the single factor producing the greatest number of missteps, accidents, and injuries. Inattention frequently results from the user being overly familiar with the stair and its surroundings. It often results from a variety of distractions. It is critical to stair safety that the stair user be attentive to the stair, although attentiveness cannot be codified or dictated. However, stair design and geometry, which usually trigger human error, can be controlled.

Curved stairways, along with spiral stairways, represent somewhat of an exception to what is normally considered a traditional stairway. Where the typical stairway is required to have treads of a consistent and uniform size and shape, these two stairs may have different dimensional characteristics from adjacent treads and vary from one end of the tread to the other. Alternating tread devices and ships ladders are additional types of vertical travel components whose design is inconsistent with that of a typical stairway. The use of these components as a portion of the means of egress system varies; however, other than a straight run stairway, only curved stairways may be used as a part of the means of egress in all occupancies and locations.

1011.5.4 Dimensional uniformity. A significant safety factor relative to stairways is the uniformity of risers and treads in any flight of stairs. The section of a stairway leading from one landing to the next is defined as a flight of stairs. It is very important that any variation that would interfere with the rhythm of the stair user be avoided. Although it is true that adequate attention to the use of the stair can compensate for substantial variations in risers and treads, it is all too frequent that the necessary attention is not given by the stair user.

To obtain the best uniformity possible in a flight of stairs, the maximum variation between the highest and lowest risers and between the widest and narrowest treads is limited to ⅜ inch (9.5 mm). This tolerance is not intended to be used as a design variation, but it does recognize that construction practices make it difficult to get exactly identical riser heights and tread dimensions in constructing a stairway facility in the field. Therefore, the code allows the variation indicated in Figure 1011-3. Although the code allows for a tolerance in both the tread depth and riser height, this tolerance is not intended to permit a reduction in the minimum tread depth or increase in the maximum riser height established by the IBC. For example, a tread depth of 10⅝ inches (270 mm) is not permitted if a minimum of 11 inches (279 mm) is required, nor is a riser height of 7⅜ inches (188 mm) acceptable where the code limits riser height to 7 inches (178 mm).

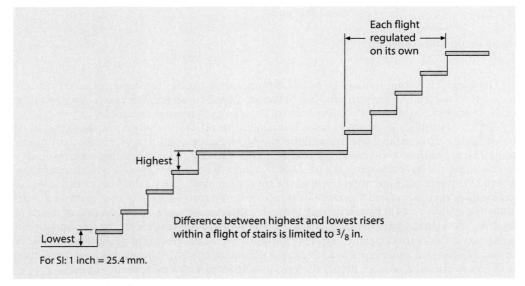

Figure 1011-3 **Stair tolerance.**

Under the provisions of Section 1030.14.2, riser height nonuniformity is permitted for stepped aisles where changes in the gradient of the adjoining assembly seating area are necessitated in order to maintain adequate lines of sight. Another exception permits the transition between a typical straight run stairway and consistently shaped winders under the conditions of Section 1011.5.

1011.5.4.1 Nonuniform height risers. With respect to variation, it is recognized that stairs occasionally descend or rise to areas where the ground or the finished surface is sloping. Where this occurs on private property, the code anticipates that the landing of the stairs be level so that there will not be any variation in the riser height across the width of the stair at that point. However, from time to time, stairs will land on spaces that are not under the control of the property owner, such as a public sidewalk. Therefore, a certain degree of slope across the width of the stair is permitted, resulting in a variation of the height of the riser from one side of the stair to the other. Where this occurs, the height of such a riser may be reduced along the slope to less than 4 inches (102 mm), and the maximum permitted slope shall not exceed 1 unit vertical in 12 units horizontal (8.3-percent slope). Figure 1011-4 shows this condition. It should be clarified that the sloping surface is intended to be an established grade, such as a walkway, public way, or driveway.

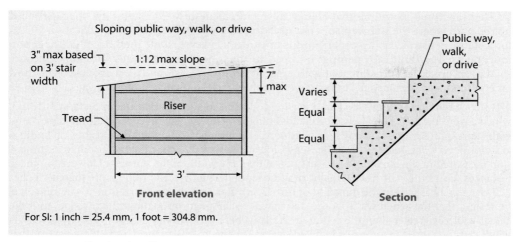

Figure 1011-4 **Sloping landing.**

1011.5.5 Nosing and riser profile. Tread nosings are limited to a maximum radius or beveling of 9/16 inch (14.3 mm) and, where solid risers are required, shall not extend more than 1¼ inches (32 mm) beyond the tread below. A minimum curvature or bevel of 1/16 inch (1.6 mm) is also required. Nosing projections are to be consistent throughout the stair flight, including where the nosing occurs at the floor at the top of a flight. Risers are to be solid and, if sloped, slope no more than 30 degrees (0.52 rad) from the vertical. See Figure 1011-5 for an overview of these provisions.

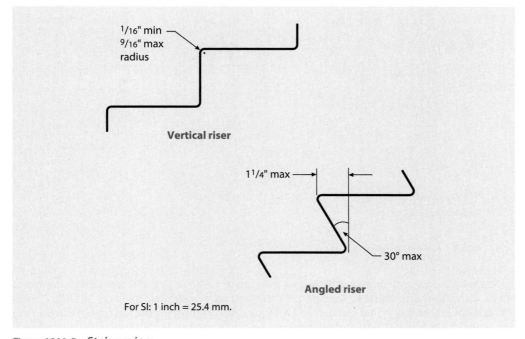

Figure 1011-5 **Stair nosing.**

There are three exceptions that exempt the requirement for solid risers. First, the risers need not be solid where the stairway does not need to comply with the provisions of Section 1009.3 for a stairway used as an accessible means of egress. However, some method of construction must be used to limit the openings between treads to less than 4 inches (102 mm). Second, solid risers are not required in nonpublic areas of Group F, H, I-3, and S occupancies. The third exception indicates that solid risers are not required for complying spiral stairways.

1011.6 Stairway landings. Landings are discussed to some extent under Section 1010.1.5, which covers landings that are used with adjoining doors. This section covers landings associated with stairs in creating a stairway. The basis for determining the required dimensions of landings is simple; every landing must be at least as wide as the stair it serves. It must also have a dimension measured in the direction of travel not less than the width of the stairway. However, in those instances where the stair has a straight run, the landing length need not be more than 48 inches (1,219 mm) measured in the direction of travel. As a stairway changes direction at a landing, it is important that the actual width of the stairway be maintained throughout the travel, even if the actual width is greater than the required width. Where the stairway reaches capacity across its width during egress, a landing of reduced size will create an obstruction to the flow pattern that has been established. Because this condition does not occur in a straight run of stairs, a limitation on length is permitted. The dimensional criteria for stair landings are illustrated in Figure 1011-6. An exception provides that stepped aisles need only comply with Section 1030.

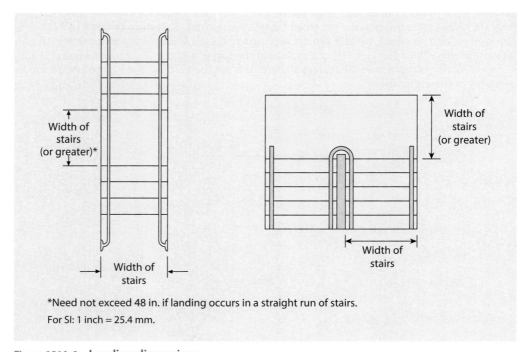

*Need not exceed 48 in. if landing occurs in a straight run of stairs.
For SI: 1 inch = 25.4 mm.

Figure 1011-6 **Landing dimensions.**

Shape of landing. It has generally been viewed that the code permits providing a complete curve with the radius equal to the width of the stairway, as shown in Figure 1011-7. This is accepted by Exception 3 and recognizes that egress travel through the landing maintains the egress width and would not generally extend into the corners. Application of this approach

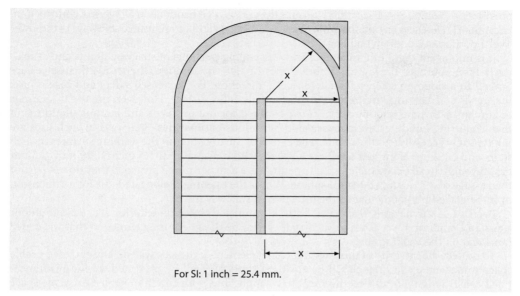

Figure 1011-7 **Alternative shape of landing.**

would provide for the presence of structural columns, standpipes, and similar building components within the nonusable portions of the landing. Exception 2 explains how to measure the landing on a curved stair. This exception is somewhat of a combination between the requirements for a straight run stair and those of Exception 3 and is illustrated in Figure 1011-8.

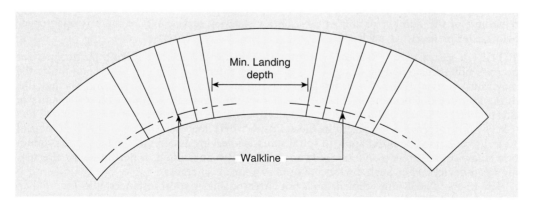

Figure 1011-8 **Landing depth at curved stairway.**

1011.7 Stairway construction. Materials used in the construction of stairways are to be consistent with those types of materials permitted based on the building's type of construction. In other words, stairways in Type I and II buildings are to be constructed of noncombustible materials. In buildings classified as Type III, IV, or V construction, combustible or noncombustible materials may be used for stairway construction. These requirements are applicable to the structural elements of the stairway, including stringers and treads. Finish materials on stairways are regulated under the provisions of Section 804. Exception 1 permits the use of wood handrails in all types of construction. Exception 2 references an option within Section 510.2 that allows combustible stairways in the lower Type I portion of a podium building.

It is important that both ascending and descending portions of stairway flights end at relatively level landings, having maximum slopes of 1:48. In addition, the treads themselves are limited to a slope no steeper than 1:48 in any direction. For increased safety and ease of use, the treads and landings are to have a substantially solid surface. However, the use of gratings or similar open-type walking surfaces is permitted for use in treads and landing platforms in manufacturing buildings, warehouses, and hazardous occupancies. Stairways in such uses are not typically accessible to the public. The extent of the openings in the walking surfaces of such stairs and landings is limited to where a sphere with a diameter of 1⅛ inches (29 mm) cannot pass through. In all occupancies, small openings are also permitted provided they do not permit the passage of a ½-inch (12.7-mm) sphere. Where the opening is elongated, the long dimension must be placed perpendicular to the direction of stairway travel.

Because of the hazards associated with accumulated water on exterior stairways, outdoor stairs and outdoor approaches to such stairs are to be designed in a manner so that water will not stand on the walking surfaces.

To protect the integrity of this very important element of the exit system, any enclosed usable space under either an unenclosed or enclosed stairway shall be protected by fire-resistance-rated construction. For this requirement to be applicable, both conditions must exist: the space under the stairway must be enclosed, and it also must be accessible, such as by a door or access panel. The level of protection of the walls and soffits within the enclosed space shall be equivalent to the fire-resistance rating of the stairway enclosure, but in no case less than 1 hour. In order to better ensure the integrity of the enclosure, access to the usable space cannot occur from within the stair enclosure. Within an individual dwelling unit in Group R-2 or R-3, the mandated fire-resistance-rated construction is not required; however, a minimum of ½-inch (12.7-mm) gypsum board protection must be provided.

Much like interior stairways, enclosed usable space under exterior egress stairways must be protected by fire-resistance-rated construction. The minimum level of fire resistance is 1 hour, regardless of the number of stories the exterior stairway serves. Where there is open space below exterior stairways, such space is not to be used for any purpose.

1011.8 Vertical rise. Negotiating stairs can sometimes become difficult, particularly for persons not accustomed to using stairs, or for the elderly or disabled. The code limits the maximum vertical rise between landings serving stairs to 12 feet (3,658 mm) so that this difficulty does not become excessive. Aside from the physical exertion necessary, stairs of exceptional height can be intimidating. It is necessary at vertical intervals not exceeding 12 feet (3,658 mm) to provide for places in the stairway where the user can rest. The limitation of 12 feet (3,658 mm) does not apply to spiral stairways serving as egress from technical production areas. Up to 20 feet (6,096 mm) of vertical rise between landings is permitted for alternating tread devices when such devices are used as a means of egress.

The 12-foot (3,658-mm) dimension is not unreasonable. In most instances, it will more than accommodate a single-story height so that in most buildings a single flight of stairs could be used, if desired, to negotiate travel from one floor to the next adjacent floor.

Even though a single flight of stairs is permitted by the code, stairs having an intermediate landing between floors are used in the majority of buildings.

1011.9 Curved stairways. Curved stairs are one of the special types of stairways that the code allows as an alternative to the typical straight stair. Figure 1011-9 depicts this type of alternative stair. It is essentially circular in configuration. The basic requirement is that the inside, or least, radius should be at least twice the minimum width or required capacity of the stair. This rule ensures a certain limited degree of curvature deemed acceptable in circular stairs. The only

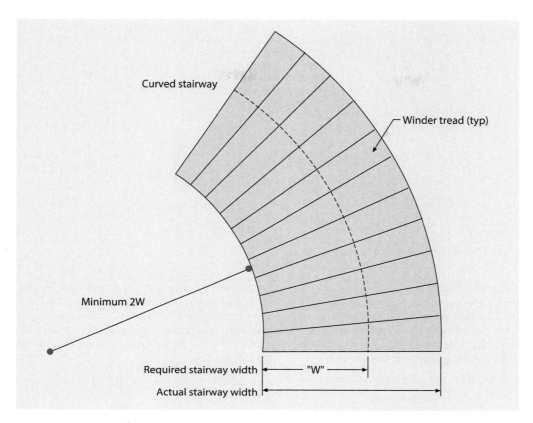

Figure 1011-9 **Curved stairways.**

other criterion specified for circular stairs is that treads comply with the winder tread provisions of Section 1011.5. In designing the curved stair and relating the inside radius to the width of the stair, it is important that only the required width or capacity of the stair need be used. Because of the stair's geometry, this is a fairly comfortable stairway to use. The geometry of the curved stairway for Group R-3 occupancies and within dwelling units of Group R-2 would only be regulated by the winder tread provisions of Section 1011.5.

1011.10 Spiral stairways. Spiral stairways are a special type of stair that the code permits in limited applications. As a means of egress component, spiral stairways may be used within dwelling units, or from spaces having an occupant load of five or less and not exceeding 250 square feet (23 m²) in area. Spiral stairways are also permitted as egress elements from galleries, catwalks, gridirons, and other technical production areas.

A spiral stairway is one where the treads radiate from a central pole. Such a stair must provide a clear width of at least 26 inches (660 mm) at and below the handrail. Each tread must have a minimum dimension of 6¾ inches (171 mm) at a point 12 inches (305 mm) from its narrow end. This tread depth is to be measured in a straight line at the walkline as opposed to being measured perpendicular to the face of the tread. The stair must have at least 6 feet 6 inches (1,981 mm) of headroom measured vertically from the leading edge of the tread. The rise between treads can be as much as, but not more than, 9½ inches (241 mm). The required dimensions of a spiral stairway are depicted in Figure 1011-10.

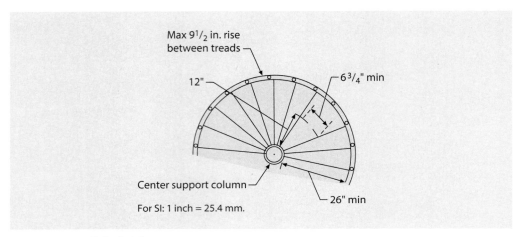

Figure 1011-10 **Spiral stairways.**

1011.11 Handrails. Probably the most important safety device that can be provided in connection with stairways is the handrail. It will never be known how many missteps, accidents, injuries, or even fatalities have been prevented by a properly installed, sturdy handrail. Basically, a handrail should be within relatively easy reach of every stair user. In general, all stairways are required to have handrails on each side. However, the code has the following specific conditions allowing the use of only one handrail:

1. On stairways within dwelling units
2. On spiral stairways

Handrails may be omitted under four conditions. Where a single change in elevation occurs at a deck, patio, or walkway, handrails are not required, provided a complying landing area is present. In Group R-3 occupancies, handrails are not required at a single riser serving an entrance door or egress door. In individual dwelling units and sleeping units of Group R-2 and R-3 occupancies, it is also permissible to provide a change in room elevation of three or fewer risers and not install a handrail. The last exemption is a unique situation where a platform lift is serving as the landing and some type of handhold or gripping surface is provided, such as a wall around the platform lift.

1011.12 Stairway to roof. To provide for easy access to roof surfaces and to facilitate firefighting in buildings four or more stories in height, at least one stairway is required to extend to the roof unless the roof is considered unoccupied. However, even where the roof is considered unoccupiable, access by way of a stairway must be provided where the roof and penthouse contain elevator equipment that must be accessed for maintenance. Where a stairway is not required, access may be achieved through the use of an alternating tread device, a ships ladder, or a permanent ladder. Access is not a requirement on steeper roofs where the slope exceeds 4 units vertical in 12 units horizontal (33-percent slope).

1011.13 Guards. Roof hatches are permitted as a means of access to unoccupiable roofs where such access is required in buildings four or more stories in height. In addition, roof-hatch openings are often provided in low-rise buildings for varied purposes, including access to rooftop equipment. This provision addresses the hazard created where the roof hatch is located very close to the roof edge. See Figure 1011-11. It is necessary that persons accessing the roof by a roof hatch be protected from falling off the roof as a result of a trip or misstep. Occasionally, these

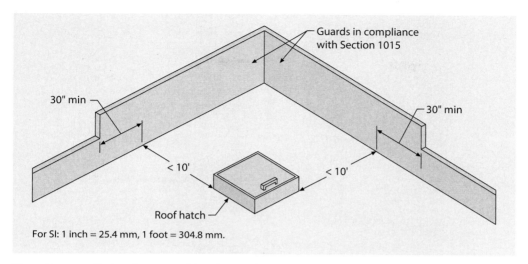

Figure 1011-11 **Protection at roof-hatch openings.**

roof accesses are used during inclement weather, emergency situations, or times of darkness. It is during these conditions when the hazard level is even higher.

1011.14 Alternating tread devices. An alternating tread device is a unique type of stairway that also has some characteristics of a ladder. Because it is considered difficult to use for egress purposes, an alternating tread device may only be used as a means of egress in a limited number of occupancies. In factories, warehouses, and high-hazard occupancies, this device can only be used for egress from a small mezzanine serving a limited number of occupants. In Group I-3 occupancies, an alternating tread device may be the egress path from a small guard tower or observation area. An alternating tread device can also be used for access and egress from an unoccupied roof. See Figure 1011-12.

Alternating tread devices must have complying handrails on both sides. In addition, treads are regulated based on whether or not the alternating tread device is used as a means of egress component. Where it is not used for egress purposes, but rather is for convenience only, the minimum tread depth may be reduced and the maximum riser height may be increased.

Section 1012 *Ramps*

1012.1 Scope. With the exception of ramped aisles in assembly rooms, curb ramps, and vehicle ramps in parking garages used for pedestrian exit access travel, whenever a ramp is used as a component anywhere in a means of egress it is necessary that the ramp comply with the provisions of this section. Note that where ramps are used as a part of the egress system, every ramp must meet these standards, regardless of the occupant load served. Ramped aisles in assembly occupancies need only conform to the provisions of Section 1030.

In addition to those egress provisions, all ramps located within an accessible route of travel shall be made to conform to the requirements found in ICC A117.1.

1012.2 Slope. Whenever any ramp is used in an exit system, the slope of such a ramp shall not be steeper than 1 unit vertical in 12 units horizontal (8.3-percent slope). Those pedestrian ramps not considered a portion of the means of egress may have a greater slope, but may not

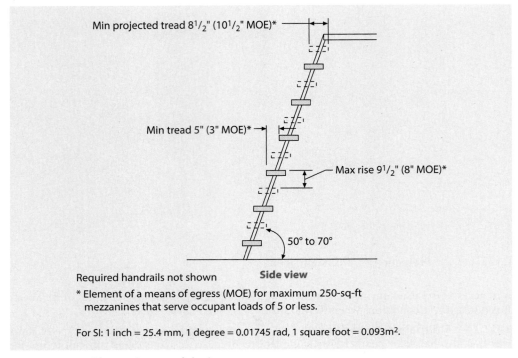

Figure 1011-12 **Alternating tread device.**

be steeper than a slope of 1 unit vertical in 8 units horizontal (12.5-percent slope). For areas of assembly seating, Section 1030.14 allows a maximum 1:8 slope for means of egress ramps not on an accessible route.

1012.5 Minimum dimensions. Because requirements for the widths and capacities of ramps are identical to those requirements for widths and capacities of corridors, see discussion of Section 1020.3. The net clear width between the handrails is to be no less than 36 inches (914 mm), differing from the stairway provisions that allow for handrail projections into the required width. In fact, all projections into the minimum width requirements of the ramp and its landings are prohibited. Where the minimum required width of an egress ramp is established, it shall not be diminished at any point in the direction of travel. Where doors enter onto or swing over ramp landings, the doors may not, during the course of their swing, reduce the minimum dimension of the landing to less than 42 inches (1,067 mm). As depicted in Figure 1012-1, in all cases there must be at least 42 inches (1,067 mm) of width available on the ramp even while the door is swinging.

1012.6 Landings. Any ramp, defined as having a slope steeper than 1 unit vertical in 20 units horizontal (5-percent slope), must have landings at the top and bottom, at any changes of direction, at points of entrance or exiting from the ramp, and at doors. All landings are required to have a dimension measured in the direction of the ramp run not less than 5 feet (1,524 mm). These larger-than-normal dimensions are required to reasonably ensure that disabled persons in wheelchairs will have sufficient space to maneuver on any intermediate landings as well as in the area at the top and bottom adjacent to the ramp. See Figure 1012-2. Exceptions permit

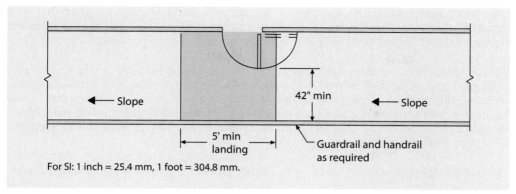

Figure 1012-1 **Intermediate ramp landings.**

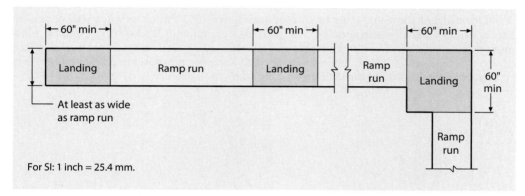

Figure 1012-2 **Intermediate ramp landings.**

a reduction in landing lengths to 36 inches (914 mm) in nonaccessible Group R-2 individual dwelling units and nonaccessible R-3 occupancies, whereas ramps not on an accessible route in other occupancies must only be 48 inches (1,220 mm) in length.

1012.7 Ramp construction. Consistent with the provisions for stairway construction found in Section 1011.7, ramps shall be built of materials consistent with the building's type of construction. In addition, exterior ramps and their approaches should be designed in a manner to avoid water accumulation.

Because a ramp has a sloping surface, the potential for accidents resulting from slips and falls is greatly increased. It is critically important that the surface of the ramp be made slip-resistant to help prevent such accidents. It is imperative that very careful attention be given to the selection of materials for ramp surfaces and the methods of finishing such surfaces. Certainly, slick-finished materials should be avoided on ramps, unless it can be ensured that they have been made slip-resistant by some process.

There are a number of products on the market available for treating ramp surfaces that are not sufficiently slip-resistant. In many instances, the use of these materials has not proven to be completely satisfactory. One method is to install carborundum strips across the width of the ramp at appropriate intervals. Other available products can be installed on the ramp surface with an adhesive. It is essential that slip-resistant treatments and materials be of reasonably permanent nature and securely attached so they can perform the function for which they were installed, while at the same time not becoming a potential tripping hazard. Therefore, the slip resistance should be proven by tests and be an integral part of the ramp surface.

1012.8 Handrails. Whenever any ramp in the means of egress has a rise greater than 6 inches (152 mm), the ramp shall be provided with handrails on both sides. The detailed requirements for the handrails are found in Section 1014. There are no general provisions that allow a handrail on only one side of a ramp, nor is there any requirement for intermediate handrails. It should be remembered that handrails for ramped aisles in assembly occupancies are regulated by Section 1030.16.

1012.10 Edge protection. In order to further protect the user of a ramp, safeguards are required along ramp edges as well as along each side of ramp landings. Two different methods, as shown in Figure 1012-3, are available for providing compliant edge protection at ramp runs and ramp landings. Either method may be used to provide the necessary level of protection. It is acceptable to use a curb with a minimum height of 4 inches (102 mm) to satisfy the first method. A rail located just above the ramp or landing surface will also suffice, as will a wall or similar barrier. As an alternative methodology, the surface of the ramp or landing may extend a minimum distance of 12 inches (305 mm) beyond a vertical plane established from the inside face of a complying handrail. Both methods are deemed adequate for preventing a ramp user from an unacceptable drop-off at the landing or ramp edge.

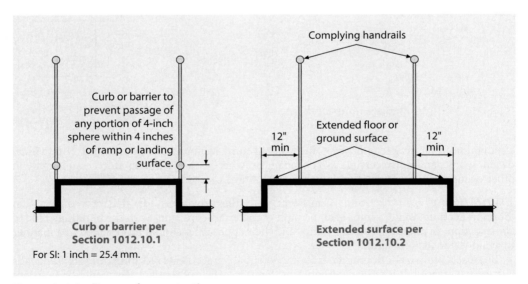

Figure 1012-3 **Ramp edge protection.**

Section 1013 *Exit Signs*

1013.1 Where required. To properly identify the egress path through a building, it is necessary to provide exit signs. Although somewhat vague, these provisions are probably more performance oriented than most requirements. The basic provision is that signs are required so that the exiting path is clearly indicated. By combining this requirement with Exception 1, it will mean that, in general, when two or more means of egress are required from any portion of a building, such as from a room, area, floor, or other space, exit signs must be installed at all required exit and exit access doorways, and at any other location throughout the exiting system where deemed necessary by the building official to clearly identify the path of travel to the exit.

Care must be taken in the placement of the exit signs to ensure they are properly located and properly oriented to the direction of travel in the egress system. They need to be easily read by persons seeking the exit. Particular care should be taken to see that nothing occurs in the means of egress that might tend to obscure or screen the exit sign, or that might cause confusion in identifying the exit sign. It is not uncommon in many building plans to find that exit signs are shown at what appear to be the appropriate locations, only to later find they are actually installed behind ducts, equipment, or other building elements in such a manner that they are not really visible to building occupants, or near or over doors that are not the exit. Additionally, the presence of banners, signs, and other movable elements may obstruct the view of required exit signs. In the placement of exit signs, the use of the space and its impact on clear sight lines to exit signs must be considered.

In certain instances, the exit signs may be present, but they are not oriented to the path of travel and to the approaching building occupant. Therefore, they are not visible and are certainly not legible to the persons seeking the exit. It is strongly advised that, as one of the last points of inspection, and before approving the occupancy of any building, the building inspector or building official carefully walk the egress path to ensure the proper installation and effective location and orientation of exit signs. It is also important that ceiling-suspended exit signs or exit signage extending from a wall do not project below the minimum permitted headroom height of 80 inches (2,032 mm) for projecting elements as required by Section 1003.3.1.

As occupants proceed along the path of travel through a corridor or exit passageway, they typically assume that their direction of travel is taking them to an exit. However, where the corridor or exit passageway length is quite extensive, the users may, at some point, make a determination that they are traveling in the wrong direction. This could cause the occupants to reverse their direction and seek another travel path. For this reason, the provisions require exit sign placement within a corridor or exit passageway at points no more than 200 feet (60,960 mm) apart. This method of placement will ensure that no point within the corridor or exit passageway is more than 100 feet (30,480 mm) from the nearest visible sign.

The use of exit signs to identify paths of egress travel is usually limited to the exit access portions of the building. Once the occupants reach the exits, such signs are typically unnecessary as the paths are often direct and single directional. However, in buildings with more complicated means of egress systems, it is possible that egress travel within the exits may not be immediately apparent to the occupants. For this reason, the mandate for exit signs is extended to those portions within exits, such as exit passageways, where such signs are necessary to provide clear egress direction for the occupants. Evacuees may be hesitant or even confused when traveling within an exit that involves transition from a vertical to a horizontal direction and horizontal extension that includes turns and intervening doors within the path of egress. Where travel direction is not clear within an exit, it creates uncertainty and causes delays in evacuations under threatening conditions. Therefore, the direction of egress travel should be identified by exit signs where such direction may not be clearly understood, and all intervening means of egress doors within the exits must also be provided with complying exit signs. An example of the appropriate use of exit signs within exits is shown in Figure 1013-1.

Omission of exit signs. Although the installation of exit signs is of paramount importance in most instances, there will be cases where the means of egress is abundantly obvious to any building occupant. Where the exit point is clearly identifiable, such as at main exterior entrance/exit doors, the exit is essentially its own identifying sign. Where approved by the building official, it may not be necessary to install an exit sign at such an obvious exit. In addition, exit signs may be omitted in Group U occupancies, and within individual sleeping and dwelling units in

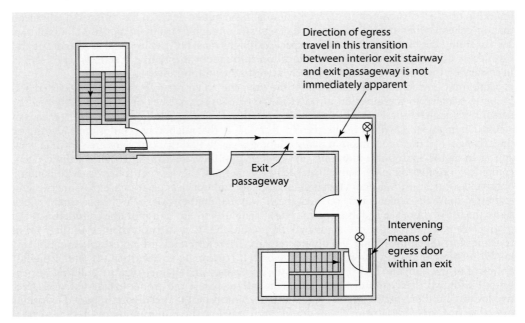

Figure 1013-1 **Exit sign locations.**

Group R-1, R-2, and R-3 occupancies. Because of the residential nature of such uses, it is felt that exit signs are unnecessary. Exit signs may also be omitted in day rooms, sleeping rooms, and dormitories of a Group I-3 occupancy, as requiring exit signs for such an occupancy is considered unnecessary. The first exception, which eliminates the need for exit signs in areas that need access to only one exit or exit access element, has application to a great number of areas within most buildings. It is assumed that the exit path is obvious because of its use as the typical entrance to the area or room. It should be noted that the exception applies even where multiple means of egress are provided, provided only one of them is required. A final exception applies to assembly seating arrangements that occur in Groups A-4 and A-5. Exit signs may be omitted in the seating area, provided that the direction to the concourse area is easily identified and adequately illuminated. Once reaching the concourse, exit signs identifying the exit path and egress doors must be provided.

1013.3 Illumination. Where exit signs are required, they must be illuminated. The source of the illumination is not material, provided the level of illumination at the sign meets the minimum requirements of the code. Tactile exit signs, required by Section 1013.4, are exempted from the illumination requirement for obvious reasons.

1013.5 Internally illuminated exit signs. Internally illuminated signs include all exit signs that generate their own luminosity in some manner. Electrically powered exit signs, including LED, incandescent, fluorescent, and electroluminescent signs, in combination with those signs considered as self-luminous and photoluminescent, represent the full range of product types currently in the market. All such exit signs must be listed and labeled in accordance with the provisions of UL 924, *Standard for Emergency Lighting and Power Equipment.* Consistent with most other installation requirements found in the code, the manufacturer's instructions must be followed. The value of visible exit signs is demonstrated by the requirement that exit signs must be illuminated at all times.

1013.6 Externally illuminated exit signs. Although it is uncommon to provide illumination for an exit sign from an external source, there are occasions where such methods are used. This section regulates the graphics, illumination, and emergency power supply for such signs.

Although no particular color is specified for exit signs, it is required that the color and design of the signs, the lettering, the arrows, and other symbols on the sign provide good contrast so as to increase legibility. The letters of the word *exit* are required to be at least 6 inches (152 mm) in height and have a width of stroke of not less than ¾ inch (19.1 mm). By specifying the letter spacing, the code ensures that the letters are not placed too close together and that the signs will be legible and more effective as a result. Additionally, when a larger sign is provided, it is important the lettering be increased proportionally.

When the illumination of the sign is from an external source, that source must be capable of producing a light intensity of at least 5 foot-candles (54 lux) at all points over the face of the sign. Measurement of the 5 foot-candles (54 lux) from an external source can be made by a light meter. The measurement of the equivalent luminance, however, from an internal source is more difficult. That is why the code relies on a testing laboratory that has examined, certified, and labeled the exit sign. In labeling such a sign, the testing agency is certifying, among other things, that the light level is at least that required by the code.

Externally illuminated exit signs are required to be illuminated at all times and must provide continued operation for a minimum of one- and one-half hours after the loss of the normal power supply. This backup power supply shall be from a complying storage battery system, unit equipment, or other approved on-site, independent source. The exception addresses illumination that is provided independent of external power supplies and, therefore, does not require compliance with the emergency power provisions.

Section 1014 *Handrails*

A handrail is defined in Section 202 as "a horizontal or sloping rail intended for grasping by the hand for guidance or support." The IBC mandates, with limited exceptions, that handrails be provided to assist the users of stairways and ramps during normal travel conditions. In addition, a handrail must be available for support in case of a misstep or other occurrence that might cause the user to stumble and fall. Numerous criteria for the design and installation of effective handrails are set forth in this section.

1014.2 Height. In past building codes, the height of handrails was traditionally established within a range of at least 30 inches (762 mm) and not more than 34 inches (864 mm) vertically above the leading edge of treads of the stairway that the handrails serve. However, research has shown that handrails better serve stair users if located in a range higher than this traditional location. Higher handrails can be more readily reached by adult stair users and, interestingly enough, handrails at the higher elevation are also more usable by very small persons, including toddlers. Where handrails are required, they must be located at least 34 inches (864 mm) and not more than 38 inches (965 mm) vertically, measured from the nosing of the stair treads to the top of the rail. A reduction in the handrail height range is provided for alternating tread devices and ship ladders due to the steepness of travel.

Handrails are not required on the entire stairway but only on the "flight" of the stairway with an additional requirement for extensions at landings. Where a handrail transitions from one flight of stairs to another at a dog-leg or switch-back stair landing, the base code requirement for handrail extensions or for handrail continuity often creates the need for some type of transition. Exception 1 allows for a more gradual variation in the height even though it will allow for portions of the handrail to exceed the normal 38-inch maximum—the belief being that a "continuous" handrail is probably more important than staying within the height limitation. Exception 2 is more comprehensive in that it also addresses the transition from a handrail to a guard; however, its application is limited to residential occupancies.

In many instances, stairways and ramps are constructed in conjunction with landings, balconies, porches, and other building components. A common condition occurs with the typical switchback stairway, where a substantial elevation change can occur from one flight to a lower adjacent flight. When those components or conditions are more than 30 inches (762 mm) above the adjacent ground level or surface below, they must be protected by guards meeting the requirements of Section 1015, which requires guards to be a minimum of 42 inches (1,067 mm) in height. Therefore, compliance with requirements for handrail heights in Section 1014.2 is not considered sufficient by the code for guard protection. The handrail must be supplemented with an additional element at a greater height in order to meet the provisions for guards. See Figure 1014-1. As a note, one item often missed when looking at the provisions is that the glazed side of a stairway must be protected by a guard unless the glazing meets the strength and attachment requirements of Section 1607.9.

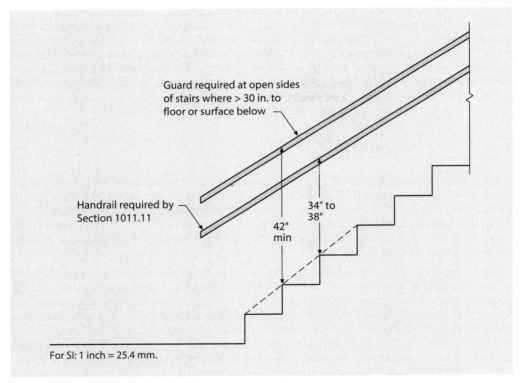

Figure 1014-1   **Guard and handrail.**

1014.3 Lateral location. The location of handrails along a stairway, ramp, ramped aisle, or stepped aisle is important for overall user safety along the path of travel. Having a properly located handrail in a usable position will not only help provide stairway and ramp users with greater stability, but it will also be in an expected location should someone begin to stumble and reach out for the handrail to arrest their fall. To ensure that handrails can easily be reached and usable, the code establishes a maximum horizontal distance the handrail can be placed from the walking surface.

The horizontal location of a handrail is important because it not only will impact the reach of different-sized users (e.g., adults and children) but it also will impact the support and usability the handrail provides for them. As a handrail is located a greater horizontal distance away from the walking surface, the number of users who will be able to simply grasp the handrail, or support themselves while grasping it, begins to be reduced.

To ensure that a handrail is positioned where it is useful for more people, its location is limited to within 6 inches (152.4 mm) of the walking surface of a stairway, ramp, ramped aisle, or stepped aisle as shown in Figure 1014-2. This is very important for occupants that may have limited reach capabilities or strength. This 6-inch (152.4 mm) requirement is based on anthropometric data.

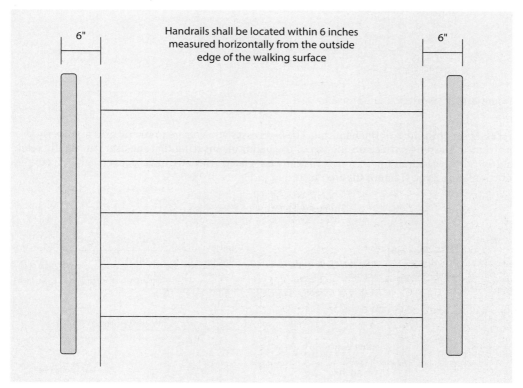

Figure 1014-2

1014.4 Handrail graspability. To be truly effective, handrails must be graspable. The code establishes specific criteria for handrail shape intending to define "graspability" in prescriptive terms. A performance alternative is also provided indicating that the configuration of the handrail can be such that it provides an equivalent, graspable shape. In prescriptive terms, the handgrip portion of the handrail must not be less than 1¼ inches (32 mm) or more than 2 inches (51 mm) in a circular cross-sectional dimension, or where the cross section of the handrail is not circular, an alternative shape is described by the code and illustrated in Figure 1014-3. Such handrails are considered as Type I handrails and are permitted for use in all occupancies. Handrail profiles having a perimeter dimension greater than 6¼ inches (160 mm), identified as Type II handrails, are also acceptable where installed in specified residential applications.

Research has shown that Type II handrails have graspability that is essentially equal to or greater than the graspability of handrails meeting the long-accepted and codified shape and size defined as Type I. The key features of the graspability of Type II handrails are graspable finger

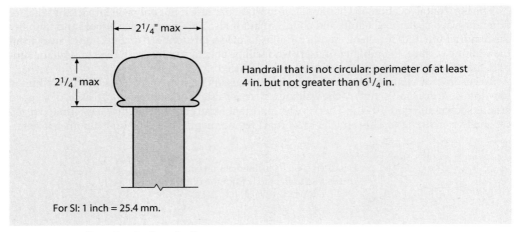

Figure 1014-3 **Noncircular handrail.**

recesses on both sides of the handrail. These recesses allow users to firmly grip a properly proportioned grasping surface on the top of the handrail, ensuring that the user can tightly retain a grip on the handrail for all forces that are associated with attempts to arrest a fall. Examples of complying Type II handrails are illustrated in Figure 1014-4.

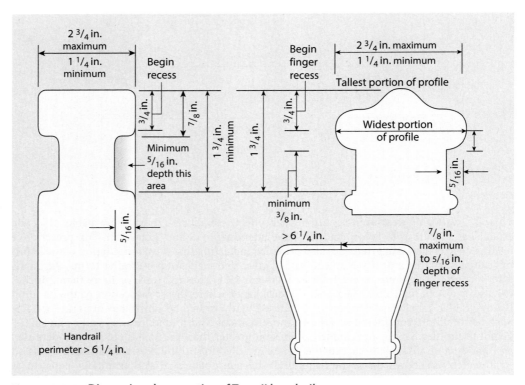

Figure 1014-4 **Dimensional properties of Type II handrails.**

Many persons are incapable of exerting sufficient finger pressure on a plane surface, such as that provided by a rectangular handrail. To get adequate support or to adequately grasp the handrail, it is necessary to provide those persons with one of the shapes that the code specifies. Where a handrail with a complying profile is installed, it is possible for those persons to actually wrap their fingers around that portion of the handrail and, thereby, obtain better support.

1014.5 Continuity. The handrails must be continuous for the full length of the ramp or the flight of stairs, thereby affording the user support throughout the entire flight. Within dwelling units, handrails are permitted to terminate at a starting newel or ramp or stair volute that is located on the first tread. In addition, a newel post may interrupt the handrail continuity at a stair landing provided the stairway is not located in an Accessible unit or a Type A unit. These types of terminations have been found in residences for years without a record of accidents or lawsuits for an unsafe condition.

Handrail brackets or balusters are not considered obstructions, provided they are attached to the bottom surface of the handrail and do not project horizontally beyond the sides of the handrail within 1½ inches (38 mm) of the bottom of the rail, as shown in Figure 1014-5. This provision is used to regulate the method of support so that the handrail is graspable at any point along its length. A lesser vertical clearance at the handrail's bottom surface is permitted where the perimeter of the handrail exceeds 4 inches (102 mm). As the perimeter of the handrail increases, the vertical clearance may be reduced.

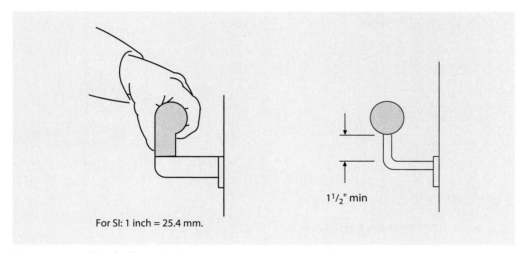

Figure 1014-5   **Handrail continuity.**

1014.7 Handrail extensions. Other than locations where handrails are continuous from one flight of stairs to the next, the handrails must extend horizontally not less than 12 inches (305 mm) beyond the nosing of the top riser. In addition, under such conditions they must also continue to slope beyond the bottom tread nosing for a distance equal to the depth of one tread. See Figure 1014-6. Note that the minimum required extension length must be provided before any change in direction or decrease in clearance occurs. The purpose of the extension of the handrail at the top and bottom of the stairs is to provide minimal additional facility in assisting the stair user. As such, the handrail must extend in the same direction as the travel along the stairway or ramp. This extension is not required for aisle handrails in rooms or spaces used for assembly purposes, for handrails within a dwelling unit that is not required to be accessible,

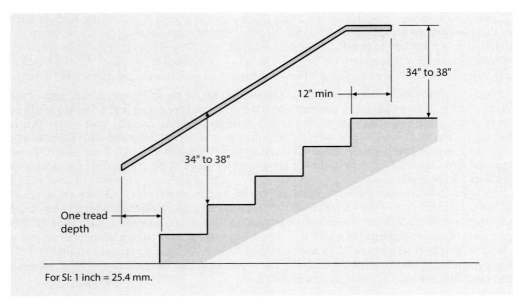

Figure 1014-6 **Stairway handrail extensions.**

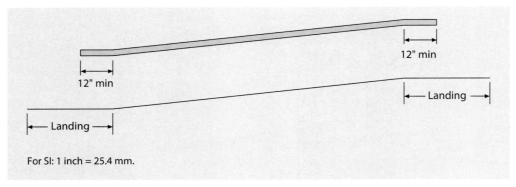

Figure 1014-7 **Ramp handrail extensions.**

as well as handrails for alternating tread devices and ship ladders. Somewhat similar conditions are applicable where ramp handrails do not continue from one ramp run to the next. For such noncontinuous handrails, a minimum extension of 12 inches (305 mm) is required, as shown in Figure 1014-7. Handrails for both stairs and ramps must return to a wall, guard, or the walking surface unless they continue and connect to a handrail of an adjacent stair flight or ramp run. The reason for requiring that the ends of handrails be returned to the wall, floor, or landing, or end at a guard or similar safety terminal, is to avoid the possibility of loose clothing or other articles being caught on the projection of the handrail.

1014.7 Clearance. In view of the desire to make the handrail graspable, and considering the requirement that the handrail be continuous, it is necessary to provide a clear space of at least 1½ inches (38 mm) between the handrail and any abutting construction to avoid injury to fingers. See Figure 1014-8. Exceptions address the curvature or angle of handrail returns as well as mounting flanges at the ends of handrail returns.

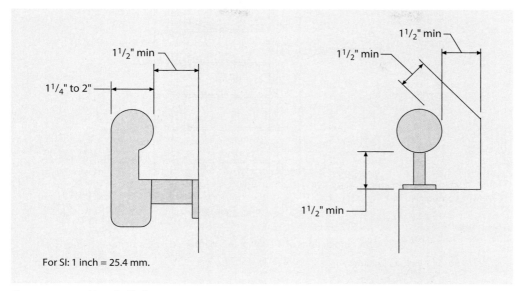

For SI: 1 inch = 25.4 mm.

Figure 1014-8 **Handrail clearance.**

1014.9 Projections. As stated earlier, when the code specifies a required width of a component in the egress system, it generally intends that width to be clear and unobstructed. One permitted projection is for handrails, which may project a maximum distance of 4½ inches (114 mm) from each side of the ramp or stair, as shown in Figure 1014-9. In addition, stringers, trim, and other features are permitted to project into the required width up to 4½ inches (114 mm) at or below the handrail height. Again, unless a projection into the minimum required width of the stairway is expressly allowed, none is permitted.

Unlike the permissible projections of handrails into the required stairway width, such projections into the required ramp width are limited. In no case may the distance between ramp handrails be less than 36 inches (914 mm).

1014.10 Intermediate handrails. Where the occupant load served by a stairway becomes significant, additional handrails may be necessary to assist stair users. The requirement is based on the required capacity of the stair established by Section 1005.3.1, not the actual width, and mandates that at no point shall the required capacity or minimum width be more than 30 inches (762 mm) from a handrail. See Figure 1014-10. It is difficult to determine the exact point at which intermediate handrails are required, as the handrail projection into the required width can vary from one design to the next. It should be noted that the measurement is to be taken in regard to the handrail location, which is permitted to extend a maximum of 4½ inches (114 mm) into the required width. Where the maximum encroachment occurs on each side of the stairway, an intermediate handrail must be provided where the required width exceeds 69 inches (1,752 mm). A lesser required width would apply where the handrails do not extend the full 4½ inches (114 mm) into the minimum required stairway width. As an additional safeguard for wide monumental stairs, the handrails must be located along the anticipated travel path of the stair users.

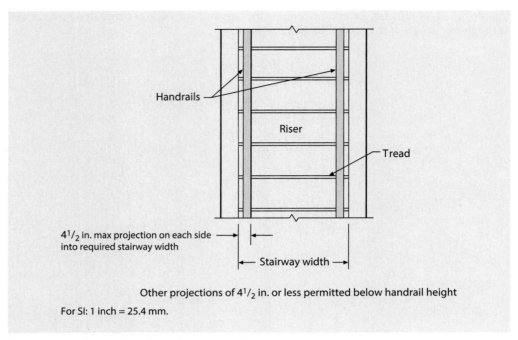

Figure 1014-9 **Projections into stairway width.**

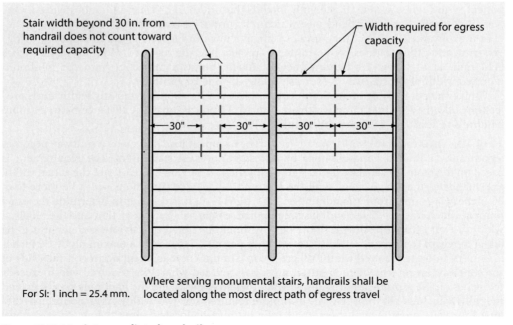

Figure 1014-10 **Intermediate handrails.**

Section 1015 *Guards*

1015.2 Where required. In this section, the code provides for guard protection at open sides along walking surfaces, such as aisles, mezzanines, stairways, ramps, landings, and occupiable roofs that are more than 30 inches (762 mm) above grade or a floor surface below. Also, protection is required for the glazed sides of stairways, ramps, and landings located more than 30 inches (762 mm) above the floor or grade below, unless the glazing complies with the strength and attachment provisions in Section 1607.9 for live loads. The need for guards in these circumstances is evident, although the arbitrary limit of 30 inches (762 mm) is subject to conjecture. Nevertheless, in the case of the IBC, it is assumed that the maximum height differential of 30 inches (762 mm) without guard protection does not create a significant safety hazard.

In the determination of the difference in elevation, an objective method is established for measuring the height of the walking surface above the grade below. Rather than taking this measurement to the ground level or floor directly below the edge of the walking surface, the code requires the height of the walking surface above grade to be based on the lowest point within 36 inches (914 mm) horizontally from the edge of the deck, porch, or other elevated element. This approach recognizes that a sloped site sometimes occurs adjacent to a deck, porch, or similar walking surface, increasing the level of hazard. This method of measurement is illustrated in Figure 1015-1.

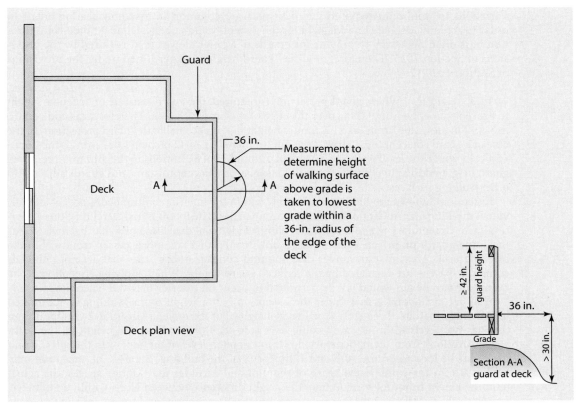

Figure 1015-1 **Determination of guard applicability.**

There are eight exceptions that identify locations or situations where guards complying with Section 1015 are not required. These exceptions fall into essentially three categories where, for obvious reasons, guards would be inappropriate:

1. Commercial and industrial applications. Guards are not required on the loading side of loading docks or piers, nor along vehicle-service pits inaccessible to the public. Such guards would severely restrict the work that takes place in these areas. The guards can also be eliminated at transportation platforms where they would interfere with passenger boarding operations.

2. Stage and platform areas. Because of the nature of activities involving performance stages or platforms, it is impractical to provide guards in various locations. Guards may be omitted on the audience side of stages and raised platforms, at runways or side stages used for presentations, at any vertical openings in the performance area, and at elevated walking surfaces used to access or use lighting or equipment. Guards also may be omitted along any stairs that may lead from the auditorium area to a stage or platform; however, handrails must still be provided under the provisions of Section 1014.

3. Occupiable roofs: An approved barrier as determined by the building official may be installed as a substitute for a guard. It is expected that the barrier provides equivalent fall protection to that provided by a complying guard. Limited elevation changes between roof levels are also exempt from the guard requirement.

Variations in minimum required guard heights are common in assembly seating areas. In order to achieve adequate lines of sight for the assembly audience seated in a tiered configuration, it is often necessary to provide for guards of a height lower than required by the provisions of Section 1015. Therefore, modified guard heights are permitted under the provisions of Section 1030.17.

1015.3 Height. Where guard protection is required, the guard must be of adequate height to prevent someone from falling over the edge of the protected area. Therefore, the code establishes 42 inches (1,067 mm) as a minimum height that is acceptable for guard protection. In the case of a guard adjacent to a stairway, the minimum height of 42 inches (1,067 mm) is measured vertically above the leading edge of the tread. A number of exceptions to the minimum required guard height address two different conditions: guards adjacent to stairs and guards adjacent to other walking surfaces.

Exception 2 allows a reduced guard height "within the interior conditioned space" of an individual dwelling unit in Group R-2 and R-3 occupancies. The intent is to differentiate this exception from Exception 1 where the phrase "within individual dwelling units" has been viewed as including guards on private exterior decks or balconies that are only accessed from within the dwelling unit. Areas that are under dwelling unit control—those areas that are not public or common spaces—are regulated by the reduced guard height. When applying Exception 2, the reduced guard height would not be permitted to occur on exterior decks or balconies.

Another distinction is that under the Exception 2, the height of the building is not limited to buildings that are three stories or less in height. For example, a Group R-2 condominium that has two levels within the unit could have a 36-inch (914 mm) guard height on the walkway or overlook from the upper story down to the main level of the unit even though the unit itself may be located on the 10th and 11th stories of the building. Because the exception only applies within the conditioned space of the unit and excludes the exterior spaces, the actual building height is not relevant. Where the height difference between levels within a multilevel unit exceeds 25 feet (7.62 m), then the exception does not apply, and the 42-inch (1,067 mm) minimum height guard is applicable.

Exception 4 allows the top rail of a stairway handrail in Group R-3 occupancies and within individual dwelling units in Group R-2 occupancies to be considered an adequate guard height, provided the handrail is located between 34 inches (864 mm) and 38 inches (965 mm) measured vertically from the leading edge of the stair tread nosing. Exception 7 is similar, but with additional restrictions for stair handrails that also serve as the guard on stairways in limited Group F areas to coordinate with federal Occupational Safety and Health Administration (OSHA) requirements. Exception 3 coordinates with Exception 4 but in locations where the guard is separate from the handrail. Exception 6 allows alternating tread devices and ship ladders to have guards with top rails serving as handrails located at a reduced height above the tread nosings.

The minimum required guard height is reduced by Exception 1 to 36 inches (914 mm) in certain residential occupancies to provide coordination with the *International Residential Code®* (IRC®). It should be noted that the scope of the exception is consistent with that of the IRC, the building cannot exceed three stories in height and each individual dwelling unit must be provided with a separate means of egress. Exception 5 reminds the code user that modifications are also provided in Section 1030.17 for stadiums, arenas, theaters, and other assembly areas to address the impact guard heights have on the lines of sight in some spectator locations.

It is important to note that the minimum height criteria for guards only apply to those guards that are required by Section 1015.2. The minimum required guard height does not apply to optional barriers that are installed.

1015.4 Opening limitations. Along with a minimum height requirement for guards, the code also requires that for open-type rails, intermediate members or ornamentation be provided so that a sphere 4 inches (102 mm) in diameter cannot pass through any opening: a requirement that prevents individuals, particularly small children, from falling or climbing through the guard assembly. This limitation applies to the lower 36 inches (914 mm) of the guard. At a height above 36 inches (914 mm) to a height of 42 inches (1,067 mm), the opening size may be increased, provided a sphere 4⅜ inches (111 mm) in diameter cannot pass through the opening. See Figure 1015-2.

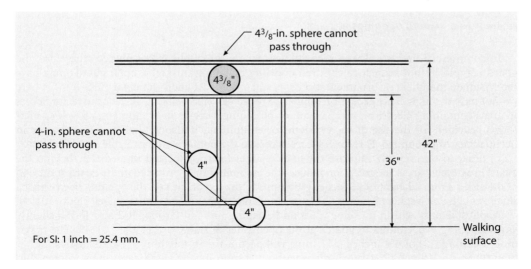

Figure 1015-2 **Guard opening limitations.**

Several exceptions increase the maximum opening size permitted in an open guard. The triangular area formed by the tread, riser, and bottom rail of the guard is limited in size to where a sphere of 6 inches (152 mm) in diameter cannot pass through the opening. Because of the unusual configuration of a stairway and its required guard, as well as the location of the triangular opening, an increased size is deemed reasonable. This configuration is shown in Figure 1015-3.

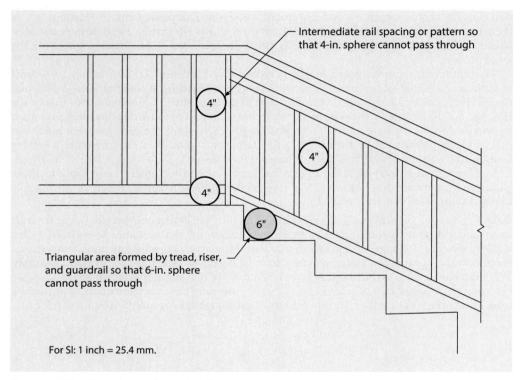

Figure 1015-3 **Guard openings at stairs.**

In commercial, industrial, and security uses where the public is not invited (therefore, the guard is typically not subject to children crawling or falling through), open guards may have intermediate members or ornamentation spaced so that a 21-inch-diameter (533-mm) sphere cannot pass through. This exception applies to those elevated walking surfaces used for access to areas containing electrical, mechanical, or plumbing systems or equipment, as well as platforms provided for the use of such systems or equipment. It also applies to elevated areas in occupancies of Group I-3, F, H, or S where access to the public is not available.

In order to significantly improve the lines of sight for rows located immediately behind the guard in assembly areas, an exception reduces the amount of infill provided in the guard at the end of the aisles. From a height of 26 inches (660 mm) to the top of the rail, the opening need only be small enough that a sphere 8 inches (203 mm) in diameter cannot pass through. See Figure 1015-4.

Applicable only within dwelling units and sleeping units in Groups R-2 and R-3, a slightly larger opening is permitted in guards at the open sides of stairs. See Figure 1015-5. The maximum opening size of 4⅜ inches (111 mm) will permit the installation of two balusters, rather than three, on 10-inch (254-mm) treads without compromising significant safety in regard to infants crawling through the openings. The greater hazard that cannot be addressed in the code is that of infants falling down the stairs.

Guards **469**

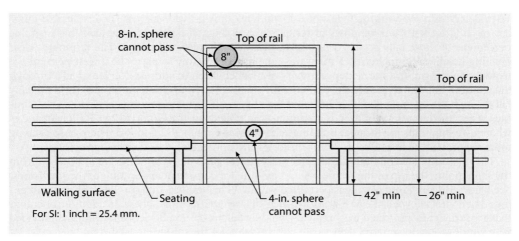

Figure 1015-4 **Assembly seating guards.**

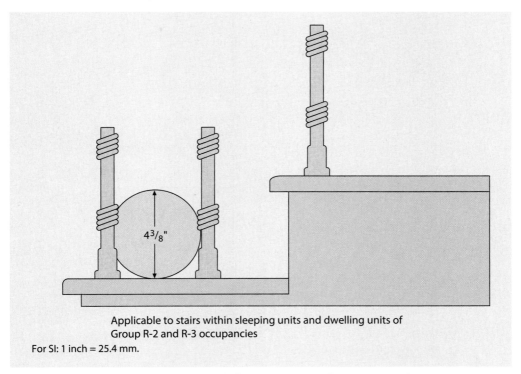

Figure 1015-5 **Guard opening limitations for Group R-2 and R-3 occupancies.**

1015.8 Window openings. Historical data have shown that each year a considerable number of children fall from windows in residential buildings. It has been estimated that a sizable percentage of those falls occurred through windows with a low sill height. The minimum clear opening height of 36 inches (915 mm) was established not only to place the lowest point of the window opening above the center of gravity of most children to address accidental falls through the window, but also to reduce the potential for small children to climb onto the sill and possibly fall through the opening. It should be noted that the requirement is only applicable to those buildings where the hazard to children is common—buildings housing Group R-2 or R-3 occupancies where dwelling units are located. Buildings classified as Groups R-2 and R-3 where only sleeping units are present, such as Group R-2 dormitories, are not regulated under these provisions.

The measurements on both the interior and exterior sides of the building are to be taken from the same point, the top of the window sill, providing for consistent application of the provisions. See Figure 1015-6. Fixed glazing is permitted within 36 inches (915 mm) of the floor, as are openings that do not allow for the passage of a 4-inch-diameter (102-mm) sphere. In addition, openings that are sufficiently protected by complying window fall prevention devices and windows provided with window opening control devices are not regulated for the minimum sill height.

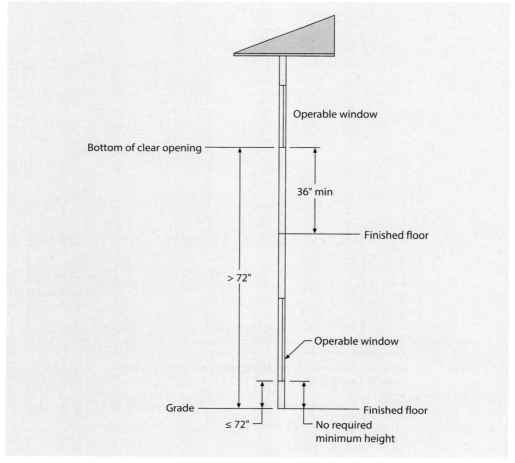

Figure 1015-6 **Minimum window sill height.**

The requirement for window clear opening (sill) height does not affect the application of the IBC provision addressing emergency escape and rescue windows. Where a window regulated by this section occurs in a sleeping room and is designated as the required emergency escape and rescue opening, it will have both a minimum and a maximum required sill height.

Section 1016 *Exit Access*

The exit access is identified as the initial component of the means of egress system—that portion between any occupied point in a building and an exit. Leading to an exterior exit door at the level of exit discharge, an exterior egress stairway or ramp, or the door of an exit passageway, interior exit stairway, interior exit ramp, or horizontal exit, the exit access makes up the vast majority of any building's floor area.

1016.2 Egress through intervening spaces. Basically, the code intends that access to exits should be direct from the room or area under consideration. This section, however, makes some modifications where, under certain circumstances, exit paths may be arranged through adjoining rooms or spaces rather than directly into corridors or exit elements, such as interior exit stairways or exit passageways, or through exterior doors.

Under Item 2, which is applicable to all occupancies other than Group H, it is permissible to provide egress through an adjoining room or space, provided the adjacent rooms or spaces are accessory to each other. In this context, the term *accessory* describes an interrelationship between the adjoining spaces based on their use, and not necessarily their size. Where egress must occur through an intervening space, it is important that such a space be under the same control as the initial space. Egress through such an intervening area to an exit must be direct and obvious so that the occupant is well aware of the exit path. It is assumed that a discernible egress path through an area that essentially is an extension of the area served poses no significant hazard to exiting, provided travel does not enter a hazardous area. See Figure 1016-1.

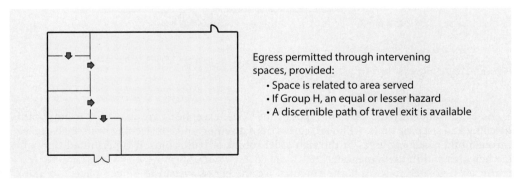

Figure 1016-1 **Egress through intervening space.**

Where the room of origin is a Group H occupancy, Item 3 permits the means of egress to pass through adjoining rooms or spaces designated as equal or lesser hazards. Because the hazard level is not increased along the egress path through the intervening room, the general prohibition does not apply. For example, occupants of a Group H-2 occupancy may travel through a Group H-3 space because of the reduction in hazard level that is anticipated.

This section is also concerned with the arrangement of the exit path in that it puts restrictions on certain spaces that are considered to present an undue probability of obstruction to free egress travel. Therefore, the code prohibits the exit path from passing through kitchens, store rooms, closets, and spaces used for similar purposes where the probability of things obstructing the path of travel is substantially greater. An exception permits travel through a stockroom provided four conditions are met, as illustrated in Figure 1016-2.

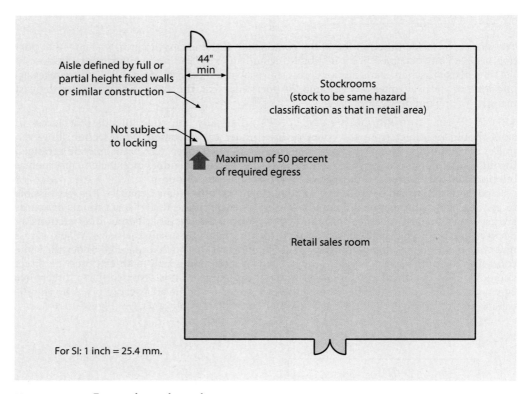

Figure 1016-2 **Egress through stockrooms.**

As is so frequently the case throughout the code, exceptions are made for rooms within dwelling and sleeping units. Although egress from dwelling units or sleeping units shall not lead through other sleeping areas or through toilet rooms or bathrooms, it is permitted through a kitchen area within the same unit.

An even greater concern is the potential for the egress system to be completely unusable. Therefore, the code specifies that exit access cannot pass through any room or area that can be locked to prevent egress. An exception is provided to address elevator lobbies where a complying electrically locked door provides exit access from the lobby. See the discussion of Section 1010.2.14.

1016.2.1 Multiple tenants. Where any floor of a building is occupied by more than one tenant, it is critical that access to all of the required exits be accomplished without passing through any adjacent tenant spaces. This condition also applies to dwelling units and sleeping units. Because it is almost always impossible to have control over what occurs in the tenant spaces of others, it is important that the exit system not be reliant on an egress path through any neighboring tenant spaces. The exception allows relatively small tenant spaces to egress through larger tenant spaces under specified conditions. Independent egress from the smaller tenant spaces is not required where such spaces are limited in size, provided they are classified to an occupancy group similar to that of the main occupancy and their means of egress cannot be locked. An example might be a branch bank or fast-food restaurant tenant that is located within a large retail store.

Section 1017 *Exit Access Travel Distance*

1017.1 General. In this section, the code is concerned that the means of egress be accessible in terms of their arrangement so that the distance of travel from any occupied point in the building to an exit is not excessive. The IBC, therefore, establishes maximum distances to exits from any occupiable point of the building. This distance is referred to as the travel distance. Travel distance is the distance that a building occupant must travel from the most remote, occupiable portion of the building to the door of an interior exit stairway or ramp, an exit passageway, or a horizontal exit; to an exterior egress stairway or exterior egress ramp; or to an exterior exit door located at the level of exit discharge. A travel limit is only imposed to the nearest exit component, not to all required exits from the room, floor, or building.

Each of these exit elements in a means of egress system is considered to represent a sufficient level of safety such that the code is no longer concerned about the distance that the building occupant must travel to reach the eventual safe place. On the top floor of the world's tallest buildings, the distance to exits, referred to as the travel distance, is measured on that floor from the most remote occupiable point to the point where the building occupant enters the interior exit stairway. The fact that the building occupant then has dozens of floors of stairway to traverse before exiting the building is not a consideration in dealing with distance to exits. Indirectly, in establishing a maximum distance of travel to a point of reasonable safety, the code is imposing a time factor on the ability of the building occupant to travel from the point of occupancy to a relatively safe place either outside or within the building.

1017.2 Limitations. Basically, the code states that travel distance to an exit may not exceed the distances found in Table 1017.2. For most moderate-hazard occupancies, the travel limitation is 200 feet (60,960 mm) in nonsprinklered buildings and 250 feet (76,200 mm) in buildings provided throughout with an automatic sprinkler system. However, the travel limitations vary from these distances in occupancies considered as low hazard and high hazard. Fully sprinklered Group B occupancies are permitted travel distances of 300 feet (91,440 mm). In low-hazard factories and warehouses, as well as utility buildings, the maximum travel distance without sprinklers is 300 feet (91,440 mm) and 400 feet (121,920 mm) where the building is fully sprinklered. Travel distance varies for high-hazard occupancies from a low of 75 feet (22,860 mm) in a Group H-1 to a high of 200 feet (60,960 mm) in a Group H-5. Those institutional occupancies classified as Groups I-2, I-3, and I-4 are permitted a maximum travel distance of 200 feet (60,960 mm) in a sprinklered building. In the situations where a Group I-4 is permitted to be without sprinklers, the travel distance is limited to 150 feet (45,720 mm).

To alert code users to the fact that there are also other code sections that affect travel distance requirements, the notes to Table 1017.2 identify several locations where special requirements may be found, including:

1. Section 402.8. Travel distance within mall tenant spaces or within the mall itself. The section limits travel distance within the mall to 200 feet (60,960 mm). The travel distance within individual tenant spaces is also limited to 200 feet (60,960 mm).

2. Section 1030.7. Travel distance within an assembly building having smoke-protected assembly seating or open-air seating. Where smoke-protective assembly seating is provided, the travel distance from every seat to the nearest concourse entrance is limited to 200 feet (60,960 mm). Travel from the concourse entrance to a stair ramp or walk on the exterior of the building is also limited to 200 feet (60,960 mm). Where the seating is open to the exterior, the maximum travel distance from each seat to the building's exterior shall not exceed 400 feet (121,920 mm). In such buildings of Type I or II construction, travel distance is unlimited.

3. Section 3103.4. Travel distance in temporary structures. The maximum exit access travel distance permitted in a temporary structure is 100 feet (30,480 mm).

1017.2.1 Exterior egress balcony increase. As depicted in Figure 1017-1, the travel limitations specified in Table 1017.2 may be increased by an additional 100 feet (30,480 mm) if the increased travel distance is the last portion of the travel distance on an exterior egress balcony. As an example, for a Group B occupancy, either the 200 feet (60,960 mm) in a nonsprinklered building or the 300 feet (91,440 mm) in a sprinklered building may be increased to 300 feet and 400 feet (91,440 mm and 121,920 mm), respectively, if in each instance the last 100 feet (30,480 mm) of travel distance is on an exterior egress balcony. Simply stated, all travel, up to a maximum of 100 feet (30,480 mm), that occurs beyond that permitted by Table 1017.2 must occur on an exterior egress balcony.

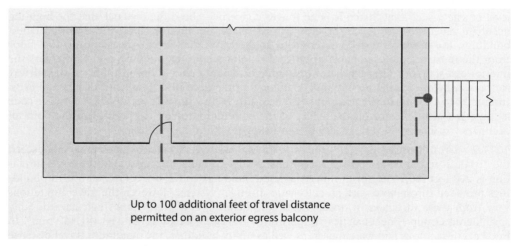

Figure 1017-1 **Egress balcony travel distance increases.**

1017.2.2 Groups F-1 and S-1 increase. The general travel distance limitation of 250 feet (76,200 mm) in fully sprinklered moderate-hazard factory and storage buildings can be increased where two conditions reflecting building height are met. Extended travel to an exit is permitted for Group F-1 and S-1 occupancies where such occupancies are single story and have a minimum floor to deck height of 24 feet (7315 mm). The increased travel distance recognizes information gained through fire modeling and egress times, based in major part on the *Report to the California State Fire Marshal on Exit Access Travel Distance of 400 Feet by Task Group 400, December 20, 2010* and the subsequent *Fire Modeling Analysis Report* dated July 20, 2011. These reports provide the technical basis for increasing the travel distance based on the required conditions.

1017.3 Measurement. Travel distance is one of the most difficult features of the egress system to determine in either the design or the plan review stage. Travel distance is intended to be measured along the natural, unobstructed path available to the building occupant. See Figure 1017-2. That path is often determined by the location of partitions, doors, furniture, equipment, and similar objects. Many of these objects are reasonably portable and, as a consequence, the actual path available is frequently and easily altered. Although it is obvious that travel is to be measured around permanent construction and building elements, how to measure travel in areas with tables, chairs, furnishings, cabinets, and similar temporary or movable fixtures is debatable.

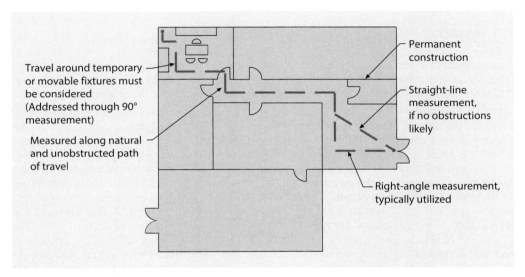

Figure 1017-2 **Measurement of travel distance.**

The preferred approach, conservative in nature, would dictate using the right-angle method for measuring travel distance. This method recognizes that obstacles such as desks, shelving, modular furnishings, and so on would cause travel to negotiate around such objects. The increased distance determined by right-angle measurement would account for such travel. On the other hand, the straight-line method would assume there are no obstructions along the measured travel path. Although such conditions at times exist, it is seldom that a straight-line path of travel will always be available. Although within private offices and similar small areas the straight-line approach would seem to cause no great concern, the method of measurement could be critical in larger spaces such as home improvement centers or department stores. Care should be taken to measure the travel distance in a manner that best represents the actual means of egress travel through the space.

Although the general requirement requires travel distance to be measured until the occupant's path of travel reaches the entrance to an exit, in open parking garages the travel distance need only be measured to the closest riser of an exit access stairway. Coordinated with the provisions of Section 1006.3.2, Exception Item 4, and Section 1019.3, Item 6, this allowance recognizes that stairways in open parking garages need not be enclosed.

1017.3.1 Exit access stairways and ramps. Since travel distance is measured until the entrance of an exit is reached, travel on exit access stairways and ramps must be included in the measurement. Exit access stairways and ramps do not provide the degree of fire resistance required of interior exit stairways and ramps. Therefore, travel on such stairways and ramps must be included in evaluating compliance with the travel distance provisions.

1017.3.2 Atriums. Because of the potential for smoke spread to upper levels within an atrium, the travel distance at upper levels within the atrium is limited to 200 feet (60,960 mm). If the egress path does not pass through the atrium or is located at the bottom of the atrium, the general travel distance requirements of Section 1017.2 are used since there is no increase in hazard.

Section 1018 *Aisles*

As a portion of the exit access, aisles are primarily regulated for minimum width purposes. Aisles typically serve occupant travel from adjoining aisle accessways and often merge into a main aisle. Aisles are commonly found in those buildings that contain furniture, fixtures, or equipment.

1018.1 General. This section requires that aisles and aisle accessways be provided in all occupied portions of the exit access that contain seats, tables, furnishings, displays, and similar fixtures or equipment. Primarily, this would have application to occupied-use areas or rooms where it is necessary to provide a circulation system so that building occupants will have reasonable means for moving around in the occupied spaces, as well as have access to corridors and other components of the egress system. It is customary to think of aisles in such facilities as theaters where an aisle system is installed to serve the fixed-seating areas. However, this section does not apply to such uses. Aisles in assembly areas, including those with seating at tables, such as in restaurants, are regulated solely under the provisions of Section 1030. The provisions of this section apply to circulation systems through open office areas, retail sales rooms, manufacturing areas, warehouse facilities, and other spaces with similar features. As mentioned, aisles serving assembly areas, including grandstands and bleachers, are to comply with Section 1030. Also, all aisles located within an accessible route of travel must also comply with Chapter 11.

The minimum required width of aisles may vary according to the occupancy in which the aisles are located; the nature of the use area that the aisles serve; the occupant load served by the aisles; and even, in some instances, the type of occupant served. In public areas of Group B and M occupancies, such as open offices, retail sales areas, and similar spaces, the minimum required clear width of any aisle is consistent with the width required for corridors based upon the conditions of use. Section 1020.3 is referenced for minimum required width. In nonpublic areas, where the number of employees or other persons served by the aisle is less than 50 and the aisle is not required to be accessible, a minimum required width of only 28 inches (711 mm) is mandated. All aisles are to be unobstructed, except for those permitted projections such as nonstructural trim and doors. In all cases, the minimum width of an aisle cannot be less than the width needed to satisfy the calculated width requirements of Section 1005.3.2.

It is important to realize that the first element of any means of egress system is typically an aisle accessway, which then leads to an aisle. The code regulates aisle accessways to some degree in Group M occupancies, as well as in assembly spaces per Section 1030, but it is silent for all other occupancy classifications. The building official must determine at what point an aisle accessway becomes an aisle and must be regulated for minimum width in accordance with this section.

1018.4 Aisle accessways in Group M. Aisle accessways are regulated within merchandise pads of Group M mercantile occupancies. A merchandise pad is defined as the merchandise display area that contains multiple counters, shelves, racks, and other movable fixtures. Every element within a merchandise pad must adjoin a minimum 30-inch-wide (762-mm) aisle accessway on each side. Travel within a merchandise pad is also limited, with a maximum common path of travel of 30 feet (9,144 mm). This limitation is extended to 75 feet (22,880 mm) in those areas serving a maximum occupant load of 50.

Section 1019 *Exit Access Stairways and Ramps*

By definition, an exit access stairway is "a stairway within the exit access portion of the means of egress system." Any stairway that is not protected by a fire-resistive enclosure as regulated by Section 1023 and used as an "exit" component in the means of egress is to be regulated by this section. Located within the exit access portion of the means of egress, an exit access stairway generally has no inherent fire protection. Travel on an exit access stairway is considered to be unprotected travel. However, where the floor openings created to accommodate exit access stairways and ramps do not meet any of the conditions set forth in Exceptions 1 through 9, enclosure of the exit access stairway is necessary. The required enclosure is intended to restrict the spread of fire, smoke, and gases vertically through a multistory building and is not intended for the protection of occupants during the exiting process. The exceptions provide multiple conditions, which either eliminate the requirement for any type of enclosure or provide some other means of addressing openings between stories. The allowances and exceptions established in this section also apply to exit access ramps.

1019.3 Occupancies other than Groups I-2 and I-3. A variety of exceptions have been established where the enclosure of exit access stairways is not required. However, none of these situations are applicable to Group I-2 and I-3 occupancies. Due to the desire to create smoke compartmentation vertically in such institutional buildings, it is important that all floor openings between stories in Groups I-2 and I-3 be protected with shaft enclosures in order to maintain the mandated fire and smoke separation between smoke compartments.

Exception 1 emphasizes a fundamental concept of the IBC that is applicable under most conditions—two adjacent stories may be open to each other provided such stories do not atmospherically communicate with any other stories. The intent of Exception 1 is that when any two stories are open to one another, neither may be open to yet another story. This is to prevent the formation of an unprotected vertical shaft through more than two stories. This does not mean that the stories under consideration cannot have access to other floors. They can, but complying enclosures must be provided in order to do so. However, this provision does not allow for the enclosure of an interior exit stairway to be interrupted by unenclosed flights.

Exception 4, applicable only in fully sprinklered buildings, addresses limited-size exit access stairway openings that are protected by sprinklers and draft curtains rather than fire-resistance-rated enclosures. By limiting the size of the openings, along with the installation of draft curtains and closely spaced sprinklers, the code assumes that the vertical openings in sprinklered buildings do not present an untenable condition. The purpose of the required draft curtain is to trap heat so that the sprinklers will operate and cool the gases that are rising. The curtain is not a fabric; it is constructed of materials consistent with the type of construction of the building. In other than mercantile and business occupancies, the use of this method is limited to openings connecting four or fewer stories. Although required exit access travel is permitted on stairways meeting Exception 4, such travel is limited to only one adjacent story as set forth in Section 1006.3.2. Therefore, the application of this exception is limited to those stairways that are not a portion of the means of egress system, but rather provided as additional circulation elements (convenience stairways).

Section 1020 *Corridors*

The IBC contains a definition of the term *corridor*. Defined as "an enclosed exit access component that defines and provides a path of egress travel," the determination as to when a corridor exists is essentially left to the building official.

For the purpose of the code, a corridor is typically a space where the building occupant has very limited choices as to paths or directions of travel. The available path is restricted and is usually bordered by occupied-use spaces. As a consequence, it is potentially exposed to fires that might occur in those enclosed spaces. Generally speaking, in a building space of this type, the occupant has only two choices as far as direction of travel through the exiting system is concerned. For that reason, it is sometimes necessary for the building official to evaluate the planned layout of an area and determine whether the space presents a potential fire-hazard exposure to building occupants, as any regular, well-defined corridor might. If the determination is that the fire-exposure potential is the same, the building space should be made to comply with requirements for corridors.

To provide for a greater degree of consistency in the identification or determination of corridors, some jurisdictions have established a set of guidelines that expand on the definition in the IBC. An excellent example addresses four common characteristics of corridors as regulated in the code.

1. It is a space formed by enclosing walls or construction over 6 feet (1,828 mm) in height,
2. It has a length-to-width ratio greater than 3 to 1,
3. Its primary function is for the movement of occupants in the means of egress system, and
4. It has a length greater than that permitted for a dead-end condition.

All four conditions must be present for the element to be considered a corridor. There are many rooms that meet the first, second, and fourth conditions, but their primary use is for something other than the movement of occupants in the egress system. An open office system may have spaces that meet Conditions 2, 3, and 4; however, the walls are not over 6 feet (1,828 mm) in height. In such a space, the egress paths would be regulated as aisles or aisle accessways. It also makes little sense to designate a space as a corridor where it conforms to Conditions 1, 2, and 3 but is very limited in length, as the travel time through the space will be quite short. This approach is just one method for providing uniformity in defining a corridor; there are undoubtedly others that also help in the application of the code's intent. Of course, the definition of a corridor is not as critical where it is not required to be fire-resistance rated.

1020.2 Construction. The thresholds found in Table 1020.2 indicate at what point a fire-resistance-rated corridor must be provided. Where required by the table, walls of corridors shall be considered fire partitions, and as such shall be regulated by the provisions of Section 708. The provisions in Section 708 also address the lid of the corridor in relationship to the continuity of the fire partitions. Examples of appropriate corridor fire-resistance-rated construction are illustrated in the discussion of Section 708.

In a fully sprinklered building, a corridor must be fire-resistance rated only in Groups H, R, I-1, and I-3. In Groups H-1, H-2, H-3, I-1, and I-3, the protection is required regardless of the occupant load served by the corridor. Where a corridor serves an occupant load of more than 10 in a sprinklered Group R occupancy, it must have a minimum ½-hour rating. A minimum 1-hour rating is required for R-3 or R-4 occupancies which use an NFPA 13D sprinkler system. Where a corridor serves an occupant load of more than 30 in a Group H-4 or H-5 occupancy, the walls of the corridors are required to be of not less than 1-hour fire-resistance-rated construction.

In buildings not equipped with an automatic sprinkler system throughout, corridors are required to be fire-resistance rated in all occupancies, based on the occupant load served.

When a corridor serves an occupant load greater than 30 in a Group A, B, E, F, M, S, or U occupancy, or any occupant load in a Group I-4, 1-hour fire-resistance-rated corridor walls are required. Where the occupant load reaches the levels specified, it is appropriate to afford those persons in the means of egress corridor some additional protection from potential fire occurring in the enclosed spaces bordering the corridor. Therefore, a minimum separation for the corridor of 1-hour fire resistance is deemed necessary.

Fire-resistance-rated corridor exceptions. A series of exceptions to the 1-hour corridor requirement addresses a number of special circumstances where it is felt that the fire-resistance-rated separation is not necessary. The first applies to corridors in Group E occupancies. Such corridors need not comply with the 1-hour fire-resistance requirement when:

1. Every classroom that the corridor serves has at least one egress door leading directly to the outside at ground level, and

2. Rooms that are served by the corridor and are used for assembly purposes have at least one-half of their required egress doors leading directly to the outside at ground level.

Exception 2 exempts corridors within dwelling units or sleeping units of Group I-1 and Group R occupancies from the fire-resistance rating. Exceptions 3 and 4 indicate that fire-resistance-rated corridors are not required in open parking garages, nor are they required in spaces of Group B occupancies requiring only a single means of egress. Exception 5 is an extension of the exception to Section 708.5 recognizing that corridor walls that are also exterior walls need to comply only with the wall and opening requirements of Section 705.

1020.3 Width and capacity. The minimum required width and capacity of a corridor is regulated like any other component of the means of egress. The required capacity (calculated width), determined by Section 1005.3.2, must be compared with the minimum (component) width, which in this case is established in Table 1020.3. The greater of the two required widths must be available for egress purposes.

For component width purposes, a corridor is required to be at least 44 inches (1,118 mm) in width. A number of conditions reduce or increase this minimum width to the following dimensions:

1. Twenty-four inches (610 mm) is permitted in those nonpublic areas where access is necessary to service or use electrical, mechanical, or plumbing systems or equipment. These areas are seldom occupied, and then typically only by a single occupant.

2. Thirty-six inches (914 mm) is adequate where the corridor serves as a means of egress for 49 occupants or fewer, or where the corridor is located in a dwelling unit. Small occupant loads can easily egress through corridors of this minimum width. It also allows adequate width for circulation purposes and the movement of furnishings and equipment.

3. Seventy-two inches (1,829 mm) is the minimum width required for school corridors serving at least 100 students and for corridors in ambulatory care facilities where the movement of stretchers is anticipated. The increased width for Group E occupancies provides for better circulation during peak usage, whereas the increase in surgical areas allows the limited movement of nonambulatory patients and equipment.

4. Ninety-six inches (2,438 mm) is required in Group I-2 occupancies for the movement of patients in standard hospital beds. This width is necessary to safely accommodate this kind of use. An exception exempts those areas used as a part of a defend-in-place strategy.

As stated, this section provides for the minimum required widths and capacities of corridors. The code is not totally clear on how the required corridor width is determined when the occupant load is large enough to require corridor widths in excess of these minimums. A common approach to determining the required minimum widths of corridors is that the width should be related to the required capacity of the exit to which the corridor leads. The most logical procedure for determining corridor widths is to determine the required capacity of the exits at the end of the corridor and to size the corridor to provide that same minimum width. See Figure 1020-1. The controlling concept is to mutually equate the required capacity of the corridors and the required capacity of the exits that the corridor serves. However, the minimum required width of 44 inches (1,118 mm) must be maintained for the corridor unless an exception is applicable.

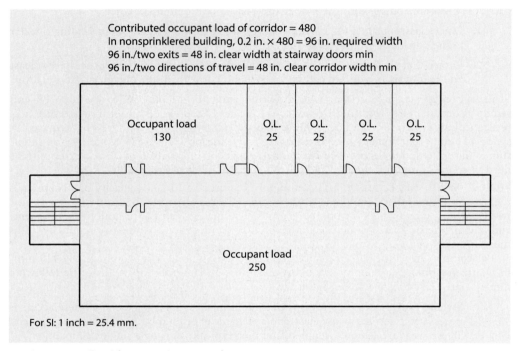

Figure 1020-1 **Corridor capacity example.**

1020.5 Dead ends. This section establishes the limitations on dead-end corridors. Basically, wherever more than one exit or exit access doorway is required, the exit access within a corridor shall be arranged so that occupants can travel in either direction from any point to a separate exit. However, dead ends in corridors are permitted up to a maximum length of 20 feet (6,096 mm), as illustrated in Figure 1020-2. In the event that the conditions of building occupancy permit access to only a single exit, the dead-end limit does not apply. When the basic requirement is for travel in only a single direction to one exit, travel in the opposite direction is not considered to be in violation of the dead-end requirements. However, the maximum distance of travel will be regulated by the common path provisions of Section 1006.2.1.

Two exceptions permit dead ends of a length up to 50 feet (15,240 mm). In Group I-3 occupancies where free movement is allowed to some degree (Condition 2, 3, or 4), and in low-hazard and moderate-hazard occupancies housed in fully sprinklered buildings, the increased length of the dead-end condition has been found to be acceptable. Exception 3 permits a dead end of unlimited length, provided that such length of the dead-end corridor does

Corridors

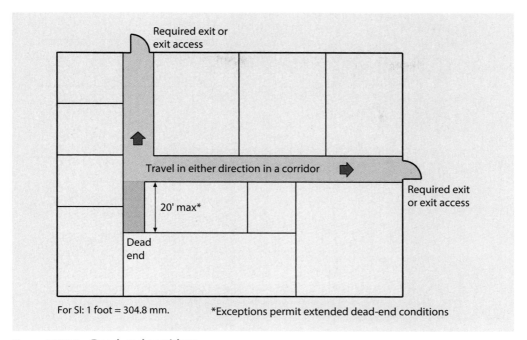

Figure 1020-2 **Dead-end corridors.**

not exceed two- and one-half times the least width of the dead-end portion of the corridor. See Figure 1020-3. The fourth exception allows a 30-foot (9,144 mm) dead end in portions of hospital corridors that do not serve patient rooms or treatment spaces. These areas tend to be similar to business or other moderate-hazard occupancies and therefore a longer dead-end will not impact the patient safety.

The limitation on dead ends is directed toward avoiding portions of the corridor system that could result in entrapment of the building occupant by creating a situation where the building occupant, following a proper path of travel through the means of egress system, might, under fire conditions, take a wrong turn into a portion of the system from which there is no outlet. In such a situation, it is possible that the building occupant will have to proceed all the way to the end of the dead end before learning that there is no way out and, thereafter, will have to retrace steps to arrive at an exit.

1020.6 Air movement in corridors. Corridors are not to be used as a portion of the heating, ventilating, and air-conditioning system by serving as plenums or ducts. They also may not be used for relief or exhaust venting purposes. As corridors are relied on to be an important component of the egress system, it is not appropriate to provide potential avenues for the spread of smoke and gases. It should be noted that this entire air movement provision is applicable to all corridors, not just those that are required to be fire-resistance rated. However, there appears to be inconsistency in the code insofar as adjoining spaces are permitted to be open to a nonrated corridor. Where a lack of physical separation occurs between the corridor and surrounding spaces, how can air be prevented from moving between them? In application, the provision merely is intended to require that those spaces open to the corridor be mechanically designed independent of the corridor (supply, return, exhaust, and make-up air) and not rely on the corridor to move air to or from the space. Exceptions to the general restriction are provided

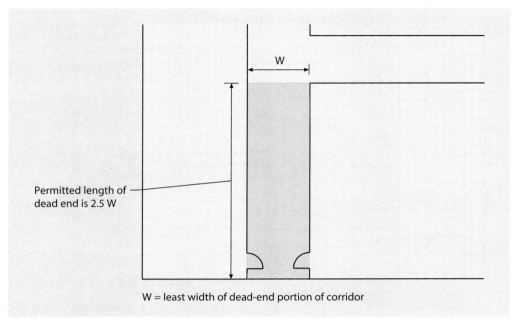

Figure 1020-3 **Dead-end limits.**

where the corridor is a source of make-up air for exhaust systems in adjacent spaces, a corridor in a dwelling unit, and a corridor within a tenant space not exceeding 1,000 square feet (93 m²) in area. The final exception addresses the fact that several locations within a healthcare facility require either a positive or negative pressure relative to other spaces and thus a certain amount of transfer air is needed to maintain the pressure difference.

1020.7 Corridor continuity. It is required that fire-resistance-rated corridors should not be interrupted by intervening rooms, and that they be continuous to the exits. This provision carries out the basic concept, which states that once a building occupant progresses to a certain level of safety—in this case, the safety afforded by a fire-resistance-rated corridor—that level of safety is not thereafter reduced as the building occupant proceeds through the remainder of the means of egress. Therefore, the building occupant, having once reached such a corridor, is not thereafter brought out of the corridor and introduced into another occupied use space of the building; the corridor must be continuous. However, the code emphasizes that it is possible to permit fire-resistance-rated corridors to be conducted through foyers, lobbies, and reception rooms. These are not to be considered intervening spaces as long as they are constructed in accordance with the requirements for the corridor they serve. See Figure 1020-4. The code is silent regarding the maximum size or permitted uses for a lobby or reception room. An extreme example would be the lobby of a large hotel. It is not uncommon to see seating areas, piano bars, breakfast buffets, and similar occupiable areas open to the fire-resistance-rated corridor. The building official must evaluate the appropriateness of all areas open to the extended corridor. It would be desirable for exiting purposes to provide all direct egress from fire-resistance-rated corridors; however, lobbies, foyers, and reception areas are accepted entrance elements that must be considered. Corridor continuity must also be considered where egress within the

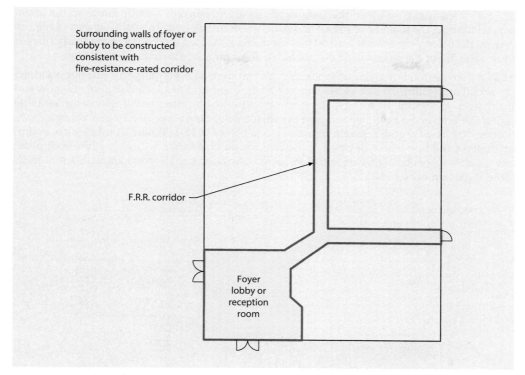

Figure 1020-4 **Corridor continuity.**

fire-resistance-rated corridor system includes vertical travel. A fire-resistance-rated corridor may extend to include two stories where the travel is maintained within a fire-resistance-rated corridor enclosure at both stories and at the exit access stairway connecting the corridors on both stories. The key is that the fire-resistive continuity is maintained continuously throughout the enclosure that extends to include the two interconnected stories.

Section 1021 *Egress Balconies*

Balconies used for egress purposes are similar in many ways to corridors and, as such, are required to meet the corridor provisions addressing width and capacity, headroom, dead ends, and projections.

1021.2 Wall separation. Because there is a potential exposure to a fire condition for occupants using an exterior egress balcony during evacuation of a building, a minimum level of fire protection is mandated. This section requires that the balcony be separated from the interior of the building in a manner consistent with the separation requirements for corridors. However, where two stairs serve the egress balcony, and travel from a dead-end condition to a stair does not occur past an unprotected opening, the separation is not required.

1021.3 Openness. One of the features of any egress balcony is its openness to the atmosphere, limiting the amount of smoke and toxic-gas accumulation. To qualify as part of the exit system, the balcony must be at least 50 percent open on the long side. It is also necessary that the open areas above the guardrail be distributed to allow for adequate natural ventilation.

1021.4 Location. Because of exposure potential from adjacent property, exterior balconies are prohibited within 10 feet (3,048 mm) of an adjacent lot line. Such components must also maintain a minimum 10-foot (3,048-mm) separation from other buildings on the same lot unless the exterior walls and openings of the adjacent building are protected in accordance with Section 705 based on fire separation distance. See Figure 1021-1. Where a building has a court or similar condition where portions of the exterior walls of the same building face each other, any exterior egress balconies on such opposing walls must also be provided with a minimum 10-foot (3048 mm) separation.

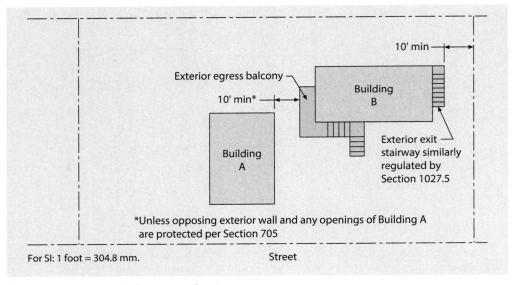

Figure 1021-1 **Egress balconies and stairways.**

Section 1022 *Exits*

Exits constitute those portions of the means of egress where the occupant first achieves a significant level of fire protection. An exit is expected to provide a mandated level of protection, and that level cannot be reduced until arrival at the exit discharge. Travel within exits is not limited and is typically single directional. Therefore, care is taken to ensure that the exit component is available for use by the building occupant during a fire or other emergency. This section makes it clear that the primary function of an exit component is to provide egress, and any other use cannot interfere with that function.

1022.2 Exterior exit doors. Once the provisions of Section 1010 for doors have been reviewed, there are relatively few exterior exit door requirements remaining to address. It should be noted that at least one exterior door that meets the size requirements of Section 1010.1.1 must be provided for every building used for human occupancy.

Section 1023 *Interior Exit Stairways and Ramps*

1023.1 General. The interior exit stairways and interior exit ramps addressed in this section are considered as exit components within a means of egress. As such, they must provide a prescribed degree of fire resistance in order to protect occupants traveling through the system. In addition, interior exit stairways must provide direct travel to the exterior unless interrupted or extended by exit passageways complying with Section 1024. Travel within an exit passageway is considered equivalent to travel within an interior exit stairway or ramp. Due to the importance of an interior exit stairway as an element of the means of egress, it is critical that the stairway enclosure be used for no purpose that would lessen its integrity as an exit component. The presence of equipment, furnishings, stored items, and other obstructions within the stairway system is strictly prohibited.

1023.2 Construction. Consistent with the provisions addressing the enclosure of vertical openings with shaft enclosures as set forth in Section 713, all vertical openings for every interior stairway or ramp must be similarly enclosed within fire-resistance-rated construction. Because vertical openings provide the most readily available paths for fire spreading upward from floor to floor through buildings, it is extremely important that such openings be adequately enclosed. This enclosure is also required in order to protect and separate the vertical exitway from potential fire and products of combustion in other spaces of the building to allow for a usable egress path. Other than where the Group I-3 special provisions of Section 408.3.8, the atrium criteria of Section 404.6, and the podium building allowances of Section 510.2 are applied, there are no exceptions to the fire-resistance-rated enclosure provisions of this section. Unenclosed stairways and ramps are permitted, but only under the limitations established in Section 1019 for exit access stairways and ramps.

The degree of fire resistance required for enclosures for interior exit stairways and ramps is dependent on the number of stories connected by the stairways. Where four or more stories are connected, the enclosing construction must be at least 2-hour fire-resistance-rated construction. In all other instances, required enclosures of interior exit stairways and ramps must be a minimum of 1-hour fire-resistance-rated construction. For the determination of the required fire resistance of the enclosure, the number of stories also includes any basements that may exist within the building, but not mezzanines. The only exception to these general rating requirements is found in Exception 3, which references the specialized construction requirements for podium buildings in Items 3 and 4 of Section 510.2.

The fire-resistance rating of the enclosure for an interior exit stairway or ramp is also regulated in a manner consistent with shaft enclosures in regard to the rating of the floor(s) being penetrated by the enclosure. Where the floor construction penetrated by the enclosure has a fire-resistance rating, the enclosure must have the same minimum rating. For example, an interior exit stairway enclosure that penetrates a 2-hour floor assembly must have a minimum fire-resistance rating of 2 hours, regardless of the number of stories the enclosure connects. The fire-resistance rating of an interior exit stairway or ramp enclosure need never exceed 2 hours. If the floor assembly penetrated requires a minimum 3-hour fire-resistance rating, the enclosure rating is only required to be 2-hour fire-resistance rated.

1023.4 Openings. Because enclosures for interior exit stairways and ramps are so fundamental to the safety of building occupants and their ability to safely exit a multistory building during a fire emergency, the code is careful to protect the integrity of these vertical enclosures in every way possible. In addition to the fire-resistance-rated construction required of the enclosure, the openings that are permitted to penetrate the fire-resistant enclosure are narrowly limited. This section very clearly establishes that only those openings necessary to provide exit facilities for occupants of the building spaces and allowed openings in the exterior walls are permitted.

Because the exterior walls of an interior exit stairway or ramp enclosure are not protecting the exit from other building spaces, openings through the exterior wall into the atmosphere are permitted. In fact, in buildings that are located on the lot so that there would be no requirement for the fire-resistance-rated construction of the exterior wall or for the protection of the openings in such walls, the exterior wall of an enclosure could be eliminated entirely. However, such openings must comply with the proper requirements of the code relating to the location of those openings with respect to lot lines or to other potential exterior fire exposures.

This provision also makes it clear that it is the intent of the code to prohibit openings from typically unoccupied spaces directly into the interior exit stairway or ramp enclosure. Therefore, it is not permitted to provide openings from such spaces as store rooms, toilet rooms, equipment rooms, machinery rooms, electrical rooms, and similar rooms directly into the enclosure. In addition, elevators are specifically prohibited from opening into an enclosure for interior exit stairways and ramps.

Where openings for exit doorways are provided, it is necessary that they be protected with a fire-rated assembly. Fire-door assemblies and other opening protectives permitted under this section are addressed in Section 716. Where an exit passageway is used as an extension to the exterior of the interior exit stairway or ramp enclosure, Section 1023.3.1 indicates that a fire separation with a fire door assembly is required to separate the vertical enclosure from the exit passageway. Such a fire door assembly shall have a fire-protective rating in accordance with Section 716.2.

However, the entire fire separation is not required where there are no openings into the exit passageway extension as illustrated in Figure 1023-1 or where both the stairway enclosure and the exit passageway are pressurized.

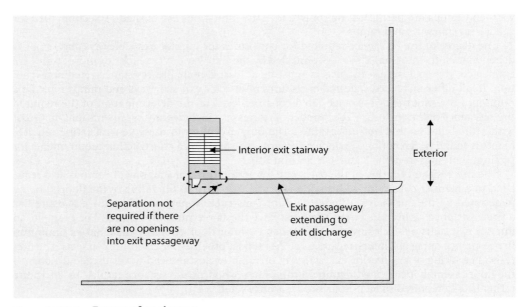

Figure 1023-1 **Extent of enclosure.**

1023.5 Penetrations. Penetrations into an interior exit stairway or ramp are prohibited unless necessary to service or protect the exit component. Acceptable penetrations include fire protection systems, security systems, two-way communication systems, electrical conduits serving the stairway or ramp enclosure and structural elements supporting the stair, the stair enclosure, or the roof of the enclosure. All such penetrations shall be made in a manner that will maintain the structural and fire-resistance integrity of the enclosure. Under no circumstances shall there be communicating openings or penetrations between adjacent interior exit stairways. The exception permits complying membrane penetrations, but only on the outside of the enclosure. There is no limit on the type or purpose of the penetration other than limiting the location to the exterior membrane and requiring the proper protection.

1023.6 Ventilation. Equipment and ductwork necessary for independent pressurization of an enclosure for an interior exit stairway or ramp is permitted under specific conditions. Where ventilation of the enclosure is desired, the ventilation systems shall be independent and isolated from the other building ventilation systems. There are three methods set forth to regulate the installation of the ventilation equipment and ductwork. The equipment and ductwork may be located at the exterior of the building, within the enclosure, or within the building. In all cases, the provisions of Section 713 for shaft enclosures will regulate the separation of the equipment and ductwork from the remainder of the building.

1023.7 Interior exit stairway and ramp exterior walls. Whenever a stairway or ramp is installed as a component in an exiting system, it is important to protect the stair or ramp user from potential exposure to any fire that might occur in the building. Therefore, where exterior walls of interior exit stairway and ramp enclosures are nonrated and openings in such walls are unprotected, the location and protection of adjacent portions of the building must be considered. Only those walls and openings within 10 feet (3,048 mm) horizontally and located at an angle less than 180 degrees (3.14 rad) from the enclosure walls need to be protected. Such walls shall have a minimum fire-resistance rating of 1 hour and any openings shall be protected at least ¾ hour. The extent of the protected construction shall be from the ground to a point at least 10 feet (3,048 mm) above the topmost landing of the stairway or ramp, or to the roof line if it is lower than 10 feet (3,048 mm). An example of this provision is shown in Figure 1023-2.

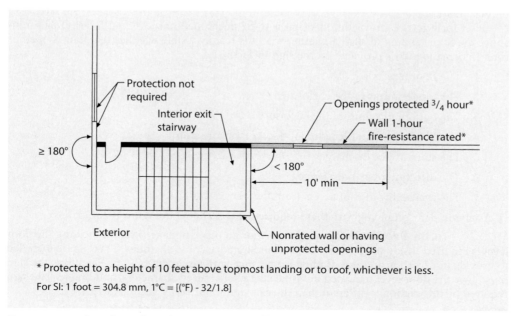

Figure 1023-2 **Interior exit stairway exterior walls.**

In all cases, the exterior walls of the enclosure must comply with the provisions of Section 705 based on fire separation distance.

This provision is based on the enclosure exterior walls being of nonrated construction with any openings left unprotected. Because the main issue is to protect the egress path, any adjacent building construction that would present an exposure hazard to occupants using the stair or ramp enclosure must be fire-resistance-rated construction. On the other hand, the enclosure walls and openings could be fully protected on all sides, including those on the exterior of the building, under the provisions of Section 1023.2 and, therefore, the requirements of this section would not be applicable. Whether the protection is at the enclosure or at the location of the hazard is ultimately immaterial.

Similar protection requirements apply to nonrated exterior walls of an interior exit stairway or ramp where a nonrated roof assembly creates an exposure hazard, based on the recognition that the exposure hazard can come not only from an adjacent wall segment but also from a roof or skylight which is located adjacent to the stairway. The provisions address the vertical exposure hazard that would be caused by fire through the roof or any unprotected skylight or other openings.

1023.8 Barrier at level of exit discharge. The barrier required by this section is intended to prevent persons from accidentally continuing into the basement. The term "barrier" is used rather than "gate" or "door" to indicate this is a performance provision. The design and location details of the barrier must be approved by the building official. Directional exit signs as specified in Section 1013 are also required, in addition to the physical barrier.

1023.9 Stairway identification signs. This section specifies a system whereby any persons, particularly fire fighters, inside an interior exit stairway or ramp enclosure in a building will be provided with information telling them where they are in the building and where the stairway or ramp leads to both above and below that point. This required sign can be critically important for fire-fighting purposes and is frequently useful to other building occupants. Enclosures connecting three stories or fewer are exempt from the identification requirements.

As set forth in this provision, this sign is to be positioned not less than 5 feet (1,524 mm) above the floor landing in such a manner that it is readily visible whether the door is open or shut. Information to be provided on the sign includes:

1. The floor level.
2. The top terminus of the enclosure.
3. The bottom terminus of the enclosure.
4. The identification or location of the stairway or ramp.
5. The story of exit discharge.
6. The direction of exit discharge.
7. The availability of roof access from the stairway.

A sample sign complying with these requirements is shown in Figure 1023-3.

1023.11 Tactile floor-level signs. In addition to other signage, a sign indicating the floor level is required inside the stairway on the landing adjacent to the door. This sign is intended to provide information and way-finding for persons with visual disabilities. Therefore, the sign must have the floor level indicated in visual characters, raised characters, and braille. The sign required by this section is different than the exit sign required by Section 1013.4.

1023.12 Smokeproof enclosures. Under certain conditions of building occupancy, protection of the vertical exitways over and above that required in the more usual situations is

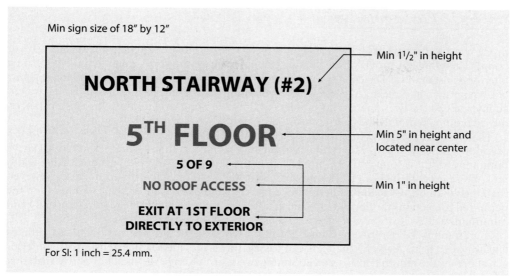

Figure 1023-3   Typical stairway floor number sign.

warranted. In such situations, it is necessary that the vertical exitway be constructed not only with a vertical enclosure, but also with either an outside balcony, by a pressurized stair and vestibule, or by stairway pressurization meeting the requirements of Section 909.20. A smokeproof enclosure is a special arrangement of the vertical exitway to minimize, if not prevent, the infiltration of smoke and other products of combustion into the actual stairway. Therefore, a smokeproof enclosure is one where the building occupant does not enter the stair enclosure directly from the occupied space of the building, unless the stairway enclosure pressurization alternative addressed in Section 909.20.4 is used.

Only those buildings regulated by Section 403 (high-rise buildings), or Section 405 (underground buildings), or Section 412.2 (airport traffic control towers) are required to use smokeproof enclosures in their means of egress. For high-rise buildings, it is required that all exits in such buildings be smokeproof enclosures for each of the exits that serve stories where the floor surface is located more than 75 feet (22,860 mm) above the level of fire-department vehicle access. For underground buildings, they are required where at least one floor is more than 30 feet (9,144 mm) below the level of exit discharge serving such floor levels. Details of the various features required of smokeproof enclosures are found in the discussion of Section 909.20.

Where a smokeproof enclosure reaches the level of exit discharge, it is to exit directly to a public way or to a yard or open space providing access to a public way. As an option, egress is permitted to travel through areas on the level of exit discharge if the conditions of Exception 1 or 2 of Section 1028.2 are met. Where the enclosure is not located at the exterior wall, an exit passageway may be used to maintain the protection of the exitway. The exit passageway must be of 2-hour fire-resistance-rated construction and have no openings that could adversely affect the integrity of the enclosure. As previously mentioned, the access to a stairway within a smokeproof enclosure must be made through a pressurized vestibule or an open exterior balcony. The one exception is the use of a pressurized system complying with the provisions of Section 909.20.4.

Section 1024 *Exit Passageways*

In many ways, the role of an exit passageway is identical to that of an interior exit stairway (or ramp). Therefore, the provisions regulating exit passageways are very similar to those governing interior exit stairways. The code specifies that both components are not to be used for any purpose other than as a means of egress. Once a building occupant is inside an exit passageway or an interior exit stairway, there is no subsequent limitation on travel distance. In most instances, the exit component must be continuous to the exit discharge or a public way.

The width and capacity of an exit passageway are regulated in the same manner as for a corridor. Those passageways serving an occupant load of less than 50 are permitted to have a minimum width of 36 inches (914 mm). Where the occupant load is 50 or more, such a width must be at least 44 inches (1,118 mm) based on the required capacity. In no case, however, shall the width be less than that determined by calculation in Section 1005.3.2. Other than the permitted projections for doors, trim, and similar decorative features, the required capacity and minimum width of exit passageways are to be clear and unobstructed.

A minimum 1-hour fire-resistance rating is required for the walls, floors, and ceilings enclosing exit passageways. Where the exit passageway is provided as an extension of an interior exit stairway, the required fire-resistance rating of the exit passageway shall not be less than that required for the connecting interior exit stairway. For example, an exit passageway serving as a horizontal extension of stairway travel in a 2-hour fire-resistance-rated interior exit stairway must also be 2-hour enclosed.

Openings and penetrations that occur in the enclosure elements of an exit passageway are strictly controlled. In fact, the limitations are consistent with provisions regulating openings into and penetrations through interior exit stairways. See the discussion of Sections 1023.4 and 1023.5 for an analysis of the provisions that are applicable to both exit passageways and interior exit stairways.

Uses of exit passageways. Exit passageways are commonly used in several different exiting situations. In many cases, it is required that interior stairways be enclosed and that the enclosure extend completely to the exterior of the building, including, if necessary, an exit passageway on the floor of the level of exit discharge. When used in this configuration, the exit passageway assumes the same fire-protection requirements as for the stairway enclosure it serves. An exit passageway is also required to maintain the protective continuity of an interior exit stairway at horizontal travel above the discharge level. Where travel on a story is required from the termination point of one interior exit stairway to the entrance to another interior exit stairway in order to provide a continuous exit path, the connecting element must be an exit passageway.

An ongoing use of exit passageways is in connection with covered mall buildings. An example of the use of exit passageways in covered mall buildings results from the fact that travel distance in the mall is limited to a maximum of 200 feet (60,960 mm). It will occasionally be necessary between major exits from the mall to introduce an exit passageway or provide an additional exit to satisfy the requirements limiting travel distance. By the use of such exit passageways, it is possible to locate the main entrance/exit points to the mall building at substantially greater intervals. In addition, the use of an exit passageway can potentially eliminate dead-end conditions that occur where back-of-tenant egress is provided along the same path as mall egress.

A historic use of exit passageways is in buildings that have very large floor areas. In such buildings, it is sometimes not possible to get the building occupants to the exits at the building's perimeter within the limitations of the permitted travel distance. Therefore, an exit passageway is used to literally bring the exit to the interior of the spaces and to the building occupant so that it is possible for any building occupant to reach and enter the exit passageway within the permitted travel distance. This type of exit passageway is frequently accomplished by constructing, in effect, a special type of fire-resistive corridor. It can also be accomplished by constructing either an overhead, fire-resistant, enclosed passageway, or a tunnel. By these latter means, it is possible to avoid manufacturing processes and other functions at the floor level within the building.

Another use of exit passageways occurs when a separation of multiple exit doors is insufficient. By extending the points of egress through the use of one or more exit passageways, it is possible to relocate the exit doors to the point where they comply with Section 1007.1. The exit separation distance would then be measured in a straight line between the exit doors, which could each enter into an exit passageway.

Again, it should be noted that once in an exit passageway, the building occupant is considered to be in a relatively safe location and travel distances within the exit passageway are not limited, just as travel distances within interior exit stairways are not limited.

In most cases, the primary difference between an exit passageway and an exit corridor lie in their respective requirements for opening protection and the fact that the passageway requires a complete enclosure, including ceiling and floor, of at least 1-hour fire-resistance-rated construction. Permitted openings and penetrations in exit passageways are also much more limited than what is allowed into a corridor.

Section 1025 *Luminous Egress Path Markings*

Improving the visibility of stair treads and handrails under normal and emergency conditions is a significant factor in raising the level of occupant safety for individuals negotiating stairs during egress of a high-rise building. A second source of emergency power for exit illumination, exit signs, and stair shaft pressurization systems in smokeproof enclosures is mandated for high-rise buildings. In the event of an emergency that disconnects utility power, the emergency power source should engage, causing the stair shaft to be illuminated and maintained smoke-free by the pressurization system. Unfortunately, such systems can fail under demand conditions. The mandate for luminous egress path markings adds an additional level of safety to the egress activity. The installation of photoluminescent or self-illuminating marking systems that do not require electrical power and its associated wiring and circuits provide an additional means for ensuring that occupants can safely egress a building via exit stairs, even if the emergency power supply and system fails to operate.

The use of photoluminescent or self-illuminating materials to delineate the exit path is required in Group A, B, E, I-1, M, and R-1 occupancies having occupied floors or an occupiable roof more than 75 feet (22,860 mm) above the lowest level of fire-department vehicle access. In such high-rise buildings, the required use of these markings is limited to interior exit stairways and ramps, as well as exit passageways. The selected materials must meet the requirements of UL 1994, *Luminous Egress Path-Marking Systems*, or ASTM E2072, *Standard Specification for Photoluminescent (Phosphorescent) Safety Markings*.

All markings are required to be solid and continuous stripes. A key requirement for marking systems is that their design must be uniform. The placement and dimensions of markings shall be consistent throughout the same stairway or ramp enclosure. By specifying standard marking dimensions, the requirements ensure that the marking is visible during dark conditions and provides consistent and standard application in the design and enforcement of exit path markings. Markings installed on stair steps, perimeter demarcation lines, and handrails must have a minimum width of 1 inch (25 mm). For stair steps and perimeter demarcation lines, their maximum width cannot exceed 2 inches (51 mm). The provisions for stair steps, perimeter demarcation lines, and handrails allow the width of the marking to be reduced to less than 1 inch (25 mm) when marking stripes are listed in accordance with UL 1994.

Markings are required along the entire length of the leading edge of each stair step and along the leading edge of stair landings. Markings are also required along the perimeter of stair landings and other floor areas within the enclosure. These demarcation lines serve to identify the transition from the stair steps to the landing, which is important to minimize the risk of a fall inside of a stairway or ramp enclosure that is not illuminated. In order to discern the transition from the stair to the floor, the demarcation line is located either across the bottom of the door or on the floor in front of the door.

Selected materials used in the construction of the luminous egress path markings must comply with either UL 1994 or ASTM E2072. ASTM E2072 allows the use of paints and coatings, which can be useful because it avoids a potential tripping hazard, especially in locations where the surface substrate may not be even. The luminescence of the selected marking system must provide an illumination of at least 1 foot-candle (11 lux), which is consistent with the requirement in IBC Section 1008.2.1 for the general illumination of walking surfaces. This degree of illumination must be provided for at least 60 minutes.

Analogous to rechargeable batteries, many photoluminescent and self-illuminating egress path markings require exposure to light to perform properly. Thus, photoluminescent egress path markings must be exposed to a minimum 1-foot candle (11 lux) of light energy at the walking surface for at least 60 minutes prior to the building being occupied. The charging rate for photoluminescent egress path markings is based on the wattage of lamps used to provide egress path illumination. Therefore, it is important to verify that the specified lamps have sufficient wattage to meet the required illumination levels.

Section 1026 *Horizontal Exits*

The horizontal exit may well be the least understood and most under-used component in the means of egress. It can be a very effective method for providing adequate required exiting capacity while at the same time realizing some very substantial construction cost and space savings.

A horizontal exit consists essentially of separating a story into portions by dividing it with construction having a fire-resistance rating. The construction of one or more horizontal exit walls divides the floor into multiple compartments. The concept of the horizontal exit is to permit each of these fire compartments to serve as a refuge area for occupants in one or more of the fire compartments in the event of a fire emergency. Building occupants in the compartment of fire origin may then pass through the fire-resistance-rated horizontal exit into the compartment of refuge, thereby gaining sufficient protection and sufficient time for either the extinguishment of the fire and elimination of the fire threat, or the orderly use of the remaining exits from the compartment serving as the refuge area.

The horizontal exit is used effectively in hospitals, as well as in detention and correctional facilities, where total evacuation from the building may present numerous physical and other problems. If in a health-care facility it is necessary to move patients from their rooms in a fire emergency, it is desirable to avoid the need for moving them vertically by stairs. Therefore, if an arrangement can be provided whereby patients would only be subject to horizontal movement, the safety of the building occupant can be far more easily achieved. Although horizontal exits are most frequently used only where the more traditional forms of egress design are not satisfactory, they can often be used quite effectively in many situations. In fact, in those instances where fire walls having a fire-resistance rating of not less than 2 hours are provided, the resulting arrangement is often tailor-made for use as a horizontal exit. However, fire walls are not required to be viewed as creating a horizontal exit. The use of a fire wall as a horizontal exit is a design decision, and compliance with all aspects of Section 1026 would be needed before it is accepted as a compliant exit.

1026.1 General. When properly constructed, a horizontal exit may serve as a required exit. It may, in fact, be substituted on a one-for-one basis for other types of exits. However, in only very limited situations may horizontal exits be used as the only exit from a portion of a building, or to provide more than one-half the total required number, width, and capacity of exits from any building space. See Figure 1026-1 for the general rule. In a Group I-2 occupancy, horizontal exits may comprise up to two-thirds of the required exits from any building or floor area, and Group I-3 occupancies may use horizontal exits for 100 percent of the required egress.

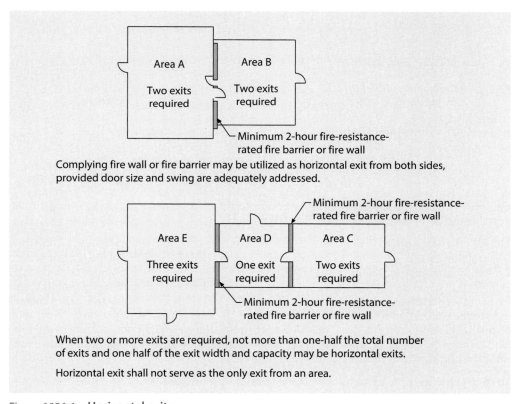

Figure 1026-1   **Horizontal exits.**

In the design of a horizontal exit system, it must be emphasized that it is not necessary to accumulate the total occupant load of the two compartments and then provide exit capacity for that total occupant load from each of the compartments. Were that the case, the horizontal exit would often provide little or no benefit. In designing a horizontal exit arrangement, one simply treats the separate compartments as if they were, in fact, separate buildings. Each compartment is provided with an exit system that is in compliance with all of the various criteria for a total exit system from a building. The main difference is that in this configuration, one of the exits from

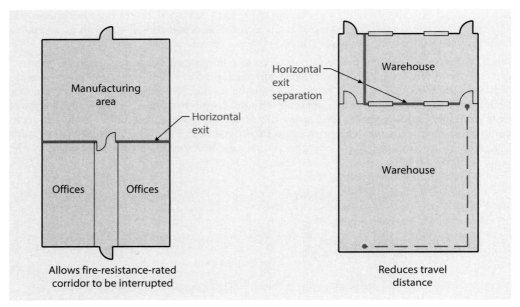

Figure 1026-2 **Horizontal exit uses.**

the separate compartments is a horizontal exit into the compartment of refuge. Figure 1026-2 depicts possible arrangements for horizontal exits.

1026.2 Separation. As previously mentioned, a fire barrier and a fire wall are the only two construction methods acceptable for the separation provided by a horizontal exit. In both cases, a minimum fire-resistance rating of 2 hours is required. A fire wall provides a natural horizontal exit. For a fire barrier, acting as a horizontal exit, to completely divide a floor of a building into two or more separate refuge areas, the fire barrier walls must be continuous from exterior wall to exterior wall. In addition, a fire barrier used as the horizontal exit must extend vertically through all levels of the building, unless 2-hour fire-resistance-rated floor assemblies are provided with no unprotected openings. This method of isolation affords safety in the refuge area from fire and smoke in the area of incident origin.

1026.3 Opening protectives. Openings through a horizontal exit are required to be protected, and thus they must have a fire-protection rating, consistent with the fire-resistance rating of the wall, as required by Section 716. Where installed through a horizontal exit wall, door openings must be self-closing, or smoke-detector-actuated automatic-closing assemblies must be installed in accordance with Section 716.2.6.6. In fact, when a horizontal exit is installed across a corridor, or when a corridor terminates at a horizontal exit, smoke-detector-actuated automatic-closing assemblies must be used. As is the case with any exit door, doors in horizontal exits must swing in the direction of exit travel when serving an occupant load of 50 or more, as provided in Section 1010.1.2.1. When the horizontal exit is to be used for exiting in both directions, it will often be necessary to provide separate exit doors for the separate directions to satisfy this requirement. Most likely, this will require the use of one or more pairs of opposite-swinging doors. See Figure 1026-3.

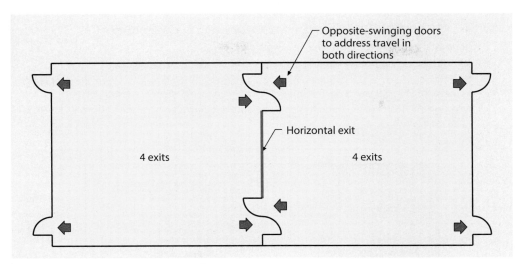

Figure 1026-3 **Opposite-swinging doors.**

1026.4 Refuge area. The area of refuge in a horizontal exit configuration must be sized to provide sufficient space for the original occupant load of the compartment of refuge plus an anticipated occupant load from the fire compartment for a limited period of time. The occupant load assigned from the fire compartment is determined by calculating the capacity of the horizontal exit doors that enter the area of refuge or the fire compartments design occupant load and using the smaller of the two. It is necessary for building occupants to remain in the area of refuge only long enough to permit the extinguishment of the fire and elimination of the fire threat, or to allow the combined occupant load of the two compartments to use the remaining exit facilities from the compartment of refuge.

To reasonably accommodate the combined occupant loads in the compartment of refuge, the code requires that at least 3 square feet (0.28 m^2) of net clear floor area be provided for each occupant. Where the horizontal exit also serves to form a smoke compartment in a Group I-1, I-2, I-3, or a Group B ambulatory care facility, then additional refuge size requirements apply. The references ensure the refuge area is capable of accommodating the increased sizes needed for care recipients in beds, stretchers, or with extra equipment. For example in a Group I-2 occupancy, an area of 30 square feet (2.8 m^2) is needed for each patient in a bed or stretcher, while 6 square feet (0.56 m^2) is needed for ambulatory patients who are not in a bed or stretcher. An example of calculating refuge area capacity is illustrated in Application Example 1026-1.

Although these area figures will permit a rather dense occupancy in the refuge area, it must be remembered that the occupancy of that space is only temporary and that the occupants of the refuge area will continue to evacuate the refuge area by use of the remaining exit or exits. In a fire emergency, such space per person is considered adequate for a short period of time.

It is also important to provide such a required refuge area for occupants in spaces that will, in fact, be available to the occupants of the building as they enter the compartment of refuge. Such spaces can be provided in corridors, lobbies, and other public areas, as long as they are sufficient to accommodate the total occupant load at the appropriate rate of area per person. Spaces in the refuge area occupied by the same tenant as in the area of fire origin are also permitted.

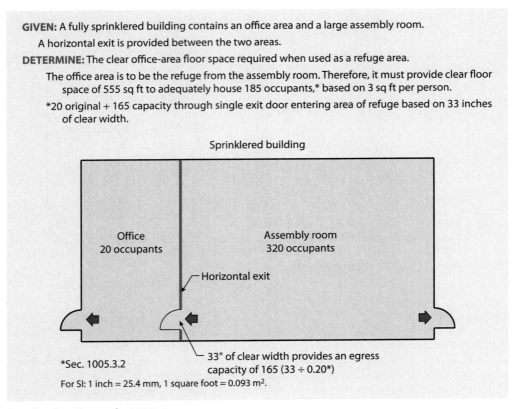

Application Example 1026-1

Section 1027 *Exterior Exit Stairways and Ramps*

1027.2 Use in a means of egress. An exterior exit stairway or ramp may serve as an exit component in the means of egress system in all occupancies other than Group I-2. The use of an exterior exit stairway or ramp as a required means of egress element is limited to non-high-rise buildings with a maximum of six stories above grade plane.

1027.3 Open side. To be classified as an exterior stair, it must be open on at least one side. The open side must then adjoin open areas such as yards, courts, or public ways. In order to qualify as an open side, there must be at least 35 square feet (3.3 m²) of aggregate open area adjacent to each floor level and at the level of each intermediate landing. In addition, the required open area must be at least 42 inches (1,067 mm) above the adjacent floor or landing level. See Figure 1027-1. By limiting the amount of enclosure by the exterior walls of the building, an exterior stair will be sufficiently open to the exterior to prevent accumulation of smoke and toxic gases. Any stairway that does not comply with these criteria is considered an interior stairway.

1027.5 Location. Because of exposure potential from adjacent property, exterior stairways and ramps are prohibited within 10 feet (3,048 mm) of an adjacent lot line for other than individual units of Group R-3. Such components must also maintain a minimum 10-foot (3,048-mm)

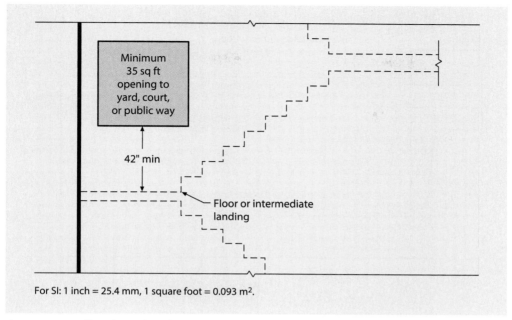

Figure 1027-1 **Exterior exit stairways.**

separation from other buildings on the same lot unless the exterior walls and openings of the adjacent building are protected in accordance with Section 705 based on fire separation distance. See Figure 1021-1.

1027.6 Exterior exit stairway and ramp protection. In order to adequately protect the building occupants as they travel an exterior stairway or ramp during egress, the exterior exit path must be adequately separated from the interior of the building. With exceptions, this section requires that an exterior stairway or ramp be separated from the interior of the building in the same manner and to the same degree as is provided by an interior exit stairway. See the discussion of Section 1023.2. Consistent with these provisions, openings are not permitted unless necessary for egress from normally occupied spaces to the exterior stair. Adjacent exterior walls must also be addressed where they expose the stairway or ramp. The provisions are consistent with those applied to interior exit stairways and ramps as set forth in Section 1023.7. There are four situations where the separation between the exterior exit ramp or stairs and the building's interior is not required:

1. In buildings no more than two stories above grade plane housing other than Group R-1 and R-2 occupancies, separation is not required where the level of exit discharge is at the first story. In such a scenario, only one story of vertical exit travel would be required. This would limit the exposure to a degree that protection is deemed unnecessary and coordinates with the fact an unprotected exit access stairway could be used in an interior condition.

2. If an open exterior balcony provides access to at least two remote exterior stairways or other exits, the separation is not required. See Figure 1027-2. To be considered open, the balcony must be open to the exterior for at least 50 percent of its perimeter.

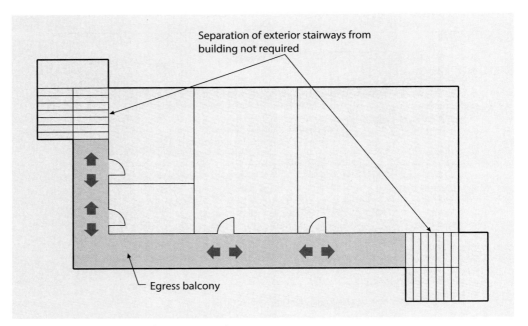

Figure 1027-2 Exterior stairway protection.

This length must then be open vertically at least 50 percent of the wall height, with openings extending at least 7 feet (2,134 mm) above the balcony. Where the building occupant has an alternative choice of exit travel, the occupant is not forced to use the stairway or ramp adjacent to the hazard. Therefore, wall and opening protection is not required.

3. An open-ended corridor is simply a corridor that is open to the outside at the exterior of the building, leading directly to an exterior stairway or ramp at each end with no intervening doors or enclosures at the exterior wall. Where open-ended corridors are used, such as in a breezeway design, the code identifies under what conditions the fire separation is not necessary between the building interior and the exterior stairway or ramp. First, the building, including the open-ended corridor and stairs, must be sprinklered throughout. Second, the corridor shall meet all of the corridor provisions of Section 1020. Third, the exterior stairways at the ends of the open-ended corridor are to comply with the general exterior exit stairway and ramp provisions of Section 1027. Fourth, exterior walls and permitted openings adjacent to the exterior exit stairway or ramp must be protected when required by Section 1023.7. Fifth, where a change in direction of more than 45 degrees (0.79 rad) occurs in the corridor, an exterior stairway, exterior ramp, or openings to the exterior shall be provided. The openings shall be such that the accumulation of smoke and toxic gases will be minimized, but in no case less than 35 square feet (3.25 m^2) in area. See Figure 1027-3.

4. Exterior stairways serving limited height individual Group R-3 dwelling units may also eliminate the separation. This exception coordinates with what is allowed for interior exit access stairways in Section 1019.3. An exterior stairway would present no greater hazard than the interior exit access stairway within the unit.

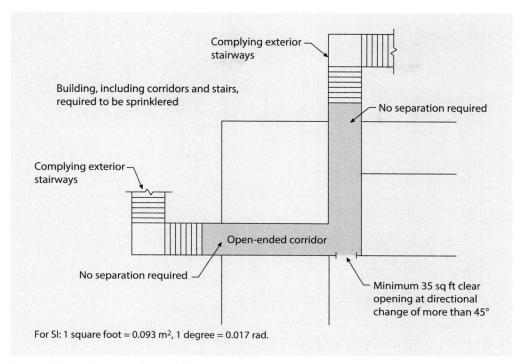

Figure 1027-3 **Open-ended corridor.**

Section 1028 *Exit Discharge*

1028.2 Exit discharge. Exits are intended to discharge directly to the exterior of the building. Three exceptions permit the exit path to include a portion of the building beyond the exit component. An exception to the requirements for the continuity of interior exit stairways (and ramps) is permitted where a maximum of 50 percent of the exits pass through areas on the level of exit discharge. The path of travel to the exterior must be unobstructed and easily recognized. Sprinkler protection is required for the egress path between the termination of the interior exit stairway to the building's exterior, as is fire-resistance-rated construction isolating any areas below the discharge level. See Figure 1028-1. Egress separation is also established where an exit access stairway is used as a part of the means of egress system. Effectively, all portions of the discharge level that provide access to the egress path must be sprinklered as well. A second exception permits egress from an interior exit stairway to enter a vestibule where limited to 50 percent of the number and capacity of the total stairways provided. The vestibule must be separated from the areas below by fire-resistance-rated construction, must be limited in size and shape, cannot be used for purposes other than egress, and must lead directly to the exterior. A minimum degree of construction providing fire and smoke protection, through the use of a fire partition and complying openings, shall separate the vestibule from other portions of the level of exit discharge. See Figure 1028-2. Although it is acceptable to use both of these exceptions in the same building, their combined use cannot exceed 50 percent of the number and capacity of the required exits.

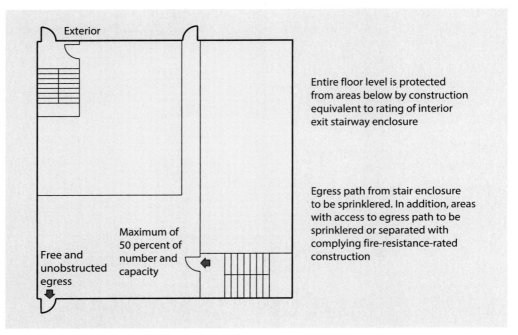

Figure 1028-1　**Exit discharge through building.**

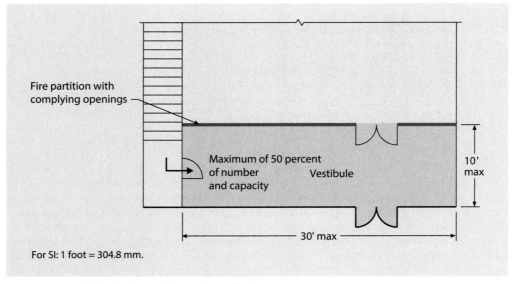

Figure 1028-2　**Exit discharge through vestibule.**

Therefore, this limitation will typically permit only one of the exceptions to be applied. The third exception simply recognizes that horizontal exiting does not provide egress directly to the outside of a building.

Exit discharge is prohibited from reentering a building to ensure that the user does not go back into any portion of the building once he or she has reached the exit discharge. This prohibition can be viewed as excluding the reentry into interior exit stairways and ramps, exit passageways, or any portion that would be considered a component of the exit.

1028.4 Exit discharge components. The general concept of the exit discharge portion of the means of egress is that the components be sufficiently open to the exterior to prevent the accumulation of smoke and toxic gases. As occupants reach the exterior of the building at grade level, they expect to have arrived at a point of relative safety. Where adequate natural cross ventilation is available to disperse any smoke or gases that may be present, one of the major fire- and life-safety concerns is assumed to have been addressed. It is common that roofed areas be provided at the exterior side of exterior exit doors in order to provide weather protection and/or decoration. Such areas are typically considered as a portion of the exit discharge where they are adequately open on the sides to provide for adequate natural ventilation.

Section 1029 *Egress Courts*

An egress court is defined as any court or yard that provides access to a public way for one or more required exits. As such, the code requires that every egress court discharge into a public way. See Figure 1029-1.

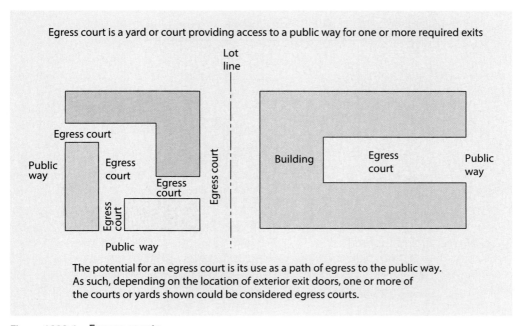

Figure 1029-1 **Egress courts.**

1029.2 Width or capacity. The minimum width and required capacity of an egress court are determined in a similar manner to that of a corridor. An egress court must provide at least 44 inches (1,118 mm) of width in all occupancies except Groups R-3 and U, where the width may be reduced to 36 inches (914 mm). When egress courts are subject to use by a sufficiently large occupant load, the minimum width may be wider than 44 inches (1,118 mm) based upon the required capacity. Such a greater width would be determined in accordance with the applicable provisions of Section 1005.3.2 based on the occupant load served. In no event may the minimum width be less than the required capacity. As is the case for most other egress components, limited encroachments into the required width are permitted for doors and handrails. Whatever the required width of the egress court, there must be at least 7 feet (2,134 mm) of unobstructed headroom.

1029.3 Construction and openings. Because an egress court is a component in a means of egress system, building occupants using that component must be afforded sufficient protection to reasonably ensure that they will reach the safety of the public way. Therefore, in other than a Group R-3 occupancy, any time an egress court serves an occupant load of 10 or more and is less than 10 feet (3,048 mm) in width, the walls of the exit court must be of 1-hour fire-resistance-rated construction for a minimum height of 10 feet (3,048 mm) above the floor of the court. By this means, a fire-resistant separation is maintained between the persons in the court and the occupied-use spaces of the building. Should any openings occur in the portion of the egress court wall required to be fire-resistance-rated construction, those openings must be protected by assemblies having a fire-protection rating of not less than ¾ hour. See Figure 1029-2.

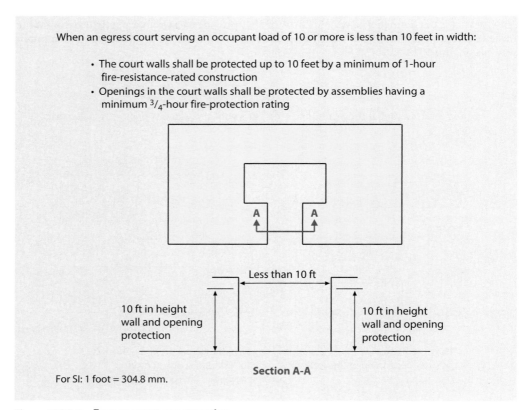

Figure 1029-2 **Egress court construction.**

Exception 3 eliminates this section's mandate for protection of the exterior wall and openings where multiple exit options exist and each of the exit paths is adequately sized to accommodate the anticipated egress requirements. Conceptually, only a single fire incident is contemplated within a building. If one pathway were to be blocked or exposed to the fire, the other path would theoretically be open and available. Increasing egress options reduces risk and therefore justifies a reduced level of protection. A similar concept exists in the exception to Section 1021.2 regarding protection on egress balconies. This exception will not eliminate any protection that may be required based on the building's type of construction or fire separation distance. For example, Table 705.5 requires all buildings with a fire separation distance of less than 10 feet (3,048 mm) to have a minimum 1-hour fire-resistance rating, and Section 705.9 limits unprotected openings that could be used in such a narrow space. While not all situations involving egress courts will be impacted by the fire separation distance requirements, exterior walls facing lot lines or facing an imaginary line between two buildings on the same lot could potentially be regulated.

Section 1030 *Assembly*

Because of the potential for large occupant loads in concentrated areas, assembly uses are regulated for egress a bit differently than other occupancies. This section addresses a variety of issues that are specific to buildings and portions of buildings that are used for assembly purposes, including exit capacity, aisle widths, and smoke-protected assembly seating. Although the provisions of this section do not apply to all assembly-type uses, they are applicable to most. The means of egress for any assembly room or space containing seats, tables, displays, shelving, equipment, fixtures, or similar elements must comply with this section.

Section 1030 is applicable to assembly uses regardless of their occupancy classification. Although the provisions are most often applied to Group A occupancies, they would also address other assembly uses, such as those assembly rooms accessory to Group E occupancies. While the classification of such areas may be based on the educational function of the building, it is important to recognize that from a means-of-egress perspective the large assembly uses still function as assembly spaces. All of the hazards involved with assembly-type functions are present in school assembly rooms, regardless of whether the room is classified as Group A or E. Therefore, the specific means of egress provisions of Section 1030 are to be applied to all assembly rooms or spaces, not just those classified as Group A.

Bleachers, grandstands, folding seating, and telescoping seating are not regulated by Section 1030, but rather by ICC 300, *Bleachers, Folding and Telescopic Seating, and Grandstands*. This standard, developed by the ICC Consensus Committee on Bleacher Safety, addresses the means of egress for such special types of seating arrangements. Simply an extension of the IBC, ICC 300 also addresses structural design and construction features.

1030.1.1.1 Spaces under grandstands and bleachers. In order to protect the assembly seating in bleachers and grandstands from an exposure to hazards, a minimum separation of 1-hour fire-resistance-rated construction is required to protect the seating area from the spaces below. This separation requirement applies where the space beneath the seating is used for any purpose other than restrooms of any size, limited size ticket booths, and other accessory spaces that are sprinklered and not more than 1,000 square feet (92.9 m^2) in floor area. While Section 903.2.1.5 would not generally require the sprinkler system until the space exceeded 1,000 square feet (92.9 m^2), this requirement is specific that the smaller spaces must be sprinklered in order to eliminate the 1-hour separation. Conceptually, this requirement for protection on the underside of the seating is similar to the required protection of enclosed usable space beneath an exterior stairway (see Section 1011.7.4). It is intended that the provision apply to storage rooms, concession stands, locker rooms, and similar areas where there is a concern due to the amount of combustible loading directly beneath an assembly seating area. It should be assumed that the requirement for a protective covering need not be applied where the means of egress passes below the seating area due to the absence of combustible materials.

1030.2 Assembly main exit. This section provides special exiting requirements for assembly buildings, rooms, and spaces. The first specification is that every assembly use having an occupant load greater than 300 must be provided with a main exit. The minimum required width of this main exit is determined by two criteria—calculated width (required capacity) and component width (minimum width). The greater required width for the main exit is the one that would govern the required minimum size.

First, in any assembly use having such a sizable occupant load, the designated main exit must have sufficient capacity to accommodate at least one-half of the total occupant load. Second, its capacity shall not be less than the total required capacity of all means of egress such as aisles, corridors, exit passageways, and stairways that lead to the main exit. The main exit must always be connected by appropriate exit components to ensure that the occupants have continuous and unobstructed access to a public way.

Basically, the requirement that the main exit be adequate to accommodate 50 percent of the total occupant load takes into account a characteristic of human nature. The majority of the occupants are, in all probability, not completely familiar with the facility and its exit system. As a consequence, in the event of an emergency it is typical that people will attempt to reach the exit through which they entered. This natural tendency could put an unduly large load on the main exit if it were not sized according to this requirement.

Distribution of exits is permitted where there is no well-defined main exit or where multiple main exits are provided. In no case, however, may the total capacity of egress be less than 100 percent of the required capacity. Should there be multiple points of entry into the building, it would be unreasonable to size each point as if it were a main exit. However, where a main exit is specifically identified, the 50 percent rule is applicable.

1030.3 Assembly other exits. In addition to requiring that the main exit accommodate 50 percent of the total occupant load, it is also necessary that every assembly use with an occupant load of more than 300 be provided with additional means of egress. They are required to be of sufficient capacity to accommodate at least 50 percent of the total occupant load served by that level. As expected, these additional means of egress must also comply with Section 1007.1 addressing doorway arrangement.

1030.4 Foyers and lobbies. In Group A-1 occupancies, typically theaters and similar uses, persons often occupy a lobby to await a presentation in the major-use area. It is important that while the lobby is occupied, the required capacity and minimum egress width from the major-use area is maintained through the lobby to the exit from the building.

1030.5 Interior balcony and gallery means of egress. Consistent with the basic requirement for the number of exits in most assembly uses as reflected in Table 1006.2.1, every balcony, gallery, and press box that has an occupant load of 50 or more must be provided with a minimum of two remote, separate means of egress. In some assembly uses, more than one such feature is provided. These requirements for exits apply to any and all balconies, galleries, and press boxes in the building. The means of egress may include unenclosed exit access stairways and ramps between the balcony, gallery, or press box and main assembly floor in uses such as theaters, churches, and auditoriums.

1030.6 Capacity of aisle for assembly. The method for calculating assembly aisle capacities has three distinct categories. These are buildings without smoke protected seating, with smoke protected seating, and open-air assembly seating. Thus, the first thing that must be established is the category in which the egress must be analyzed. Section 202 defines smoke-protected assembly seating as "seating served by means of egress that is not subject to smoke accumulation within or under a structure for a specified design time by means of passive design or by mechanical ventilation." Section 1030.6.2 regulates smoke-protected assembly seating in regard to smoke control, roof height, and automatic sprinklers. Section 1030.6.3 will regulate open-air assembly seating which is defined as "seating served by means of egress that is not subject to smoke accumulation within or under a structure and is open to the atmosphere."

If the reduced width requirements of Table 1030.6.2 are applied in a building that has smoke-protected assembly seating, a Life Safety Evaluation complying with NFPA 101 must be conducted. An acceptable evaluation would include such criteria as pedestrian flow rates, movement characteristics of persons using exit systems, the nature of the means of egress system beyond the aisles, location of potential hazards, staff-response capabilities, facility pre-planning and training, and other items relating to occupants' safety. Should it be determined that a building is not smoke protected, or that no life-safety evaluation has been performed, the aisle capacity will need to be calculated using the more stringent non-smoke-protected conditions. Open-air assembly seating is regulated by separate capacity factors that are given in Section 1030.6.3, but through a reference there are permitted to use the smoke-protected capacity factors in larger buildings where Table 1030.6.2 would give a reduced requirement.

1030.6.1 Without smoke protection. For egress on stepped aisles, two modifications to the basic width factor of 0.3 inch (7.6 mm) per occupant are to be considered. The general requirement of 0.3 inch (7.6 mm) is based on stepped aisles having riser heights of no more than 7 inches (178 mm). Where the riser height exceeds 7 inches (178 mm), at least 0.005 inch (0.127 mm) of additional aisle capacity for each occupant shall be provided for each 0.10 inch (2.5 mm) of riser height above 7 inches (178 mm). The following formula represents this method of width increase:

$$W = 0.3 + 10(R - 7.0)(0.005)$$

where:

W = Required capacity in inches per occupant
R = Riser height in inches

Formula 10-1

The second potential modification applies only where egress is accomplished on the stepped aisle in a descending fashion. Under such a condition, an increase of at least 0.075 inch (1.9 mm) of additional width per occupant must be provided where no handrail is provided within a horizontal distance of 30 inches (762 mm). This egress-capacity increase is illustrated by the following formula:

$$W = 0.3 + (0.75) = 0.375$$

where:

W = Required capacity in inches per occupant

Formula 10-2

It is also very possible that both increases will be applied to the same stepped aisles. Where the riser height exceeds 7 inches (178 mm), and a portion of the required stepped aisle capacity is more than 30 inches (762 mm) from a handrail, both modifications to the basic requirement of 0.3 inch (7.6 mm) per occupant are applicable. The increase in required egress capacity can be shown as:

$$W = 0.3 + 10(R - 7.0)(0.005) + (0.075)$$

where:

W = Required capacity in inches per occupant
R = Riser height in inches

Formula 10-3

Where a ramped aisle is used as a portion of the means of egress, the difference in calculated width is based on the slope of the aisle. Ramped aisles having a slope steeper than 1 in 12 (1:12) shall have a required capacity based on 0.22 inch (5.6 mm) per occupant served. Note that this condition would only apply to aisles not regulated by Chapter 11 for accessibility. The required capacity of level or ramped egress paths having a slope of 1 in 12 (1:12) or less is based on 0.20 inch (5.1 mm) for each occupant.

The required increases in egress capacity are all based on one of the primary IBC concepts regarding means of egress: where it is anticipated that users of the egress system may be slowed in their egress travel, the path is required to be widened to compensate for the reduced travel speed. By increasing the width of the exit path, the occupants are expected to continue to travel at a relatively consistent rate through the exit system. For stepped aisle travel, a riser height over the customary maximum of 7 inches (178 mm) creates an uncomfortable condition for most stair users. In such cases, it is natural for persons to slow their travel in order to more safely use the stepped aisle. Occupants also tend to travel more slowly when descending a stepped aisle where a handrail is not within easy reach. Without the ability to easily grasp a handrail in case of a misstep, the stepped aisle user tends to be a bit more cautious. On the other hand, the presence of an adjacent handrail provides a degree of security that encourages faster travel. In ramped aisle travel, a slope steeper than that normally encountered also creates a small degree of hesitancy, requiring a greater width to compensate for the reduction in travel speed.

The increases in required egress width prescribed by this section are only intended to apply where the minimum width is calculated based on the number of occupants served by the egress path (required capacity). The width increases are not applicable to the component widths established in Section 1030.9.1 (minimum width). For example, a minimum width of 36 inches (914 mm) is mandated for stepped aisles with seating on only one side. This condition would typically allow for a single handrail, which would be located more than 30 inches (762 mm) horizontally from a portion of the required aisle width. It is not appropriate to increase the minimum 36-inch (914-mm) required width based on the factors established in Item 3 of this section. Additionally, if the stepped aisle includes risers that exceeded a height of 7 inches (178 mm), the adjustment found in Item 2 would also not be applied to the minimum component width as established in Section 1030.9.1.

1030.6.2 Smoke-protected assembly seating. Calculation of egress capacity in an assembly space located in a smoke-protected environment is based on Table 1030.6.2. As the total number of seats in the smoke-protected assembly occupancy increases, egress-capacity requirements continue to decrease until reaching an end point of 25,000 or more seats. As addressed earlier, a Life Safety Evaluation must be done for a facility using the reduced capacity requirements

GIVEN: A 10,000-seat arena with seating sections as shown.
  Case I—Smoke-protected assembly seating for which a life-safety evaluation has been provided.
  Case II—Nonsmoke-protected assembly seating.
DETERMINE: Required aisle capacity for both cases assuming stepped aisles each serve 400 occupants.
SOLUTION:
  Case I—Use Table 1029.6.2. For a 10,000-seat arena, the capacity equation for stepped aisles is 0.130 inch per seat served, and the required capacity of aisle becomes:

  $$400 \times 0.130 = 52 \text{ Inches or } 4.33 \text{ feet } (1321 \text{ mm})$$

Since 52 inches (1321 mm) exceeds the minimum required width of 48 inches (1219 mm), it becomes the governing width.

Case II—For nonsmoke-protected assembly seating, the capacity is based on 0.3 inch per seat served ($400 \times 0.3 = 120"$). However, since a center handrail will exceed the 30-inch rule, an increase of 0.075 inch per person is required. Thus, the factor is 0.375 inch per seat and the required capacity of aisle becomes:

$$400 \times 0.375 = 150 \text{ inches}$$

Thus, the aisle required for nonsmoke-protected assembly seating is almost three times as wide as the aisle required for smoke-protected assembly seating.

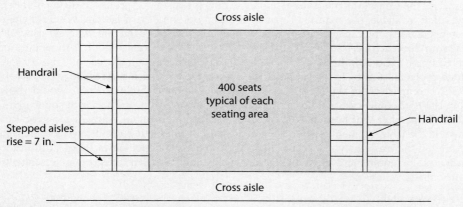

For SI: 1 inch = 25.4 mm, 1 foot = 304.8 mm.

Application Example 1030-1

of Table 1030.6.2. Application Example 1030-1 shows how the calculated width of exit provisions would be applied for smoke-protected seating and seating without smoke protection. The permitted reduction in egress capacity applies to aisles, but only to the extent that the aisles and aisle accessways are smoke protected. Any egress elements that are not provided with complying smoke protection are subject to the greater capacities established for areas that are not smoke protected.

1030.6.2.1 Smoke control. To maintain an essentially smoke-free means of egress system, a smoke-control system complying with Section 909 must be provided. As stated in Section 909.1, a smoke-control system should be designed to provide a tenable environment for the evacuation or relocation of occupants. When it can be satisfactorily demonstrated to the building official, a design incorporating a natural venting system is permitted. The natural or mechanical ventilation must be designed to maintain the smoke level at a point 6 feet (1,829 mm) or more above the floor level of any portion of the means of egress within the smoke-protected assembly seating area.

1030.6.2.2 Roof height. Whenever smoke-protected assembly seating is covered by a roof, a minimum clearance of 15 feet (4,572 mm) is required between the highest aisle or aisle accessway to the lowest portion of the roof. In an outdoor stadium, the roof canopy need only be 80 inches (2,032 mm) or more above the highest aisle or aisle accessway, provided there are no projections or obstructions below the 80-inch (2,032-mm) level. By providing an adequate roof height above the occupiable portion of the building or structure, a smoke-containment area is created. Smoke control or removal would then limit smoke migration into the egress environment.

1030.6.2.3 Automatic sprinklers. Another condition for the use of the liberal egress width provisions for smoke-protected assembly seating areas is the installation of an approved automatic sprinkler system. In general, the sprinkler system is required in all areas enclosed by walls and ceilings in buildings or structures containing smoke-protected assembly seating. However, two exceptions identify locations where sprinklers may be omitted. Where the area on the assembly room floor is used for low fire-hazard uses such as performances, contests, or entertainment, sprinklers may be omitted, provided the roof construction is more than 50 feet (15,240 mm) above the floor level. Small storage facilities and press boxes under 1,000 square feet (92.9 m²) in area are also exempt.

1030.6.3 Open-air assembly seating. Where the facilities are outdoors, such as in a stadium, and the egress system for the assembly seating is considered protected owing to the natural ventilation available, the required capacity is based on one of two factors. Where the egress is by stepped aisles, the capacity factor is 0.08 inch (2.0 mm) per occupant, whereas it is 0.06 inch (1.52 mm) for level aisles and ramped aisles. An example of this calculation is shown in Application Example 1030-2. If, however, the capacity calculated through the use of Table 1030.6.2 for smoke-protected seating is determined to be a lesser capacity, such a lesser capacity is acceptable. A quick comparison of the factors in Section 1030.6.3 and Table 1030.6.2 would show that Table 1030.6.2 could start yielding smaller capacity factors as the total occupant count approaches 20,000. Remember that Section 1030.6.2 does permit interpolation between the specific values shown in the table. As mentioned in Section 1030.6.2.3, automatic sprinkler systems are generally required in these assembly seating facilities, but in the case of open-air facilities a third exception eliminates the requirement where the seating facility and the means of egress are essentially open to the outside.

> **GIVEN:** An open-air assembly seating area having stepped aisles serving an occupant load of 300. The total number of seats in the facility is 2,600.
>
> **DETERMINE:** The required aisle capacity.
>
> CASE 1: Using Section 1030.6.3, the capacity per person is 0.08 inch. 300 persons (0.08 inch) = 24 inches < 48 inches minimum required for seating on both sides by Section 1030.9.1, #1.
>
> CASE 2: Using the exception in 1030.6.3 and Table 1030.6.2, the capacity per person is 0.200 inch. 300 persons (0.20 inch) = 60 inches.
>
> ∴ Minimum required width is 48 inches.

Application Example 1030-2

1030.7 Travel distance. Travel distance in assembly occupancies is regulated under one of three conditions: seating without smoke protection, smoke-protected seating, and open-air seating. The measurement of this distance shall be along the line of travel, including along the aisles and aisle accessways, from each seat to the nearest exit door. It is improper to measure this travel distance over or on the seats. Using the code's general requirements in Section 1017, within a building without the benefit of smoke protection, travel distance is limited to 200 feet (60,960 mm) in nonsprinklered buildings and 250 feet (72,200 mm) in sprinklered buildings. In smoke-protected assembly seating, a maximum travel distance could reach up to 400 feet (122 m). This is limited to a maximum of 200 feet (60,960 mm) being permitted from each seat to the nearest entrance to an egress concourse, and then up to another 200 feet (60,960 mm) of travel is permitted from the egress concourse entrance to an egress stair, ramp, or walk at the building exterior. In open-air seating facilities where all portions of the means of egress are open to the outside, the maximum travel distance is 400 feet (121,920 mm). This distance is measured from each seat to the exterior of the seating facility. When the outdoor seating facilities are of Type I or II noncombustible construction, the travel distance may be unlimited.

1030.8 Common path of egress travel. By providing persons in an assembly occupancy a choice of travel paths, egress opportunities will be enhanced. The distance an occupant must travel prior to reaching a point where two paths are available is more limited than permitted by the general provisions of Section 1006.2.1. In most assembly areas having fixed seating, the common path of travel is limited to 30 feet (9,144 mm). In smoke-protected or open-air areas, up to 50 feet (15,240 mm) of common path travel is permitted. Where the seating area has a limited occupant load, additional travel is also permitted.

1030.8.1 Path through adjacent row. Because the common path of travel is most often limited to 30 feet (9,144 mm), single-access seating areas are often required to egress through rows of the adjoining seating area to reach another aisle in order to comply. Such seating areas are served by an aisle on only one side, and the aisle is single directional (top loading or bottom loading only). Where this condition occurs, the code (1) limits the maximum number of seats in the adjoining row to 24 and (2) increases the minimum required width of the aisle accessway serving the row. The exception allows additional seating in the row and modifies the increase in the aisle access width for smoke-protected or open-air facilities. A similar concept is applied in Section 1030.9.5 for increasing the maximum permitted length of dead-end aisles.

1030.9 Assembly aisles are required. This section requires that aisles be provided in all occupied portions of any assembly occupancy that contains seats, tables, displays, and similar fixtures or equipment. Although the intent of the aisle provisions is to provide safe access to components of the egress system such as exits or exit access doorways, the provisions would also have application to occupied-use areas where it is necessary to provide a circulation system so that building occupants will have reasonable means for moving around in the occupied spaces. Egress travel within restaurants, classrooms, and similar uses with seating at tables is regulated by Section 1030.13.1.

1030.9.1 Minimum aisle width. Where seating is arranged in rows, the clear width shall not be less than the following, while still conforming to the calculated aisle-width (required capacity) provisions:

1. Forty-eight inches (1,219 mm) for stepped aisles where seating is provided on both sides of an aisle. A 36-inch (914-mm) aisle is permitted where the aisle serves less than 50 seats.

2. Thirty-six inches (914 mm) for stepped aisles where seating is provided only on one side of an aisle. Only 23 inches (584 mm) of aisle width is required between a stepped aisle handrail and the nearest seat where an aisle serves no more than five rows on only one side.

3. Twenty-three inches (584 mm) between a stepped aisle handrail and the nearest seat where the rail is within the aisle.

4. Forty-two inches (1,067 mm) for level or ramped aisles where seating is provided on both sides of an aisle. A 36-inch (914-mm) aisle is permitted where the aisle serves less than 50 seats, and a minimum of 30 inches (762 mm) is allowed where serving 14 seats or less and not serving as an accessible route.

5. Thirty-six inches (914 mm) for level or ramped aisles where seating is provided on only one side of an aisle. Only 30 inches (762 mm) of aisle width is required where the aisle serves no more than 14 seats and does not serve as an accessible route.

This section must be used in concert with the other provisions of Section 1030 to completely define and arrange an assembly seating area.

1030.9.2 Aisle catchment area. The occupant load served by an aisle is the determining factor in establishing its minimum required width based on capacity. In this determination, it is assumed that the egress travel is distributed evenly among the adjacent travel paths. The tributary occupant load would be assigned to each aisle proportionally, based on the arrangement of the means of egress.

1030.9.3 Converging aisles. Where aisles converge to form a single aisle, the capacity of that single aisle shall not be less than the combined required capacity of the converging aisles. There is no penalty for providing aisles that are wider than the minimum code requirement.

1030.9.4 Uniform width and capacity. Where egress is possible in two directions, the shape of an aisle cannot be of an hourglass configuration. The minimum width and required capacity must be maintained throughout. A tapered aisle is allowed only for dead-end aisle conditions or where the provided width exceeds the required minimums for that capacity.

1030.9.5 Dead-end aisles. In the arrangement of a seating area, all aisles serving the seating area must end in a cross aisle, foyer, doorway, vomitory, or concourse. In large facilities, it is not uncommon to find a number of aisles leading to a number of cross aisles. The required egress capacity of a cross aisle is the same as that for converging aisles, the combined required capacities of the aisles leading to the cross aisle. Although dead-end aisles are allowed, their length is limited to 20 feet (6,096 mm) except where the seats served by the dead-end aisle are not more than 24 seats from another aisle measured along a row having a minimum clear width of 12 inches plus 0.6 inch (305 mm plus 15 mm) for each additional seat over seven in a row (10 where seats are without backrests). The extended dead-end allowances are permitted for aisles longer than 20 feet (6,096 mm) provided the 24-seat limitation is applied to the seats beyond the 20-foot (6,096 mm) dead-end aisle. Where complying with the common path limitations set forth in Section 1030.8, dead-end aisles serving fewer than 50 seats are also permitted. For smoke-protected assembly seating, up to 21 rows are permitted in a dead-end vertical aisle except where the seats served by the dead-end aisle are not more than 40 seats from another aisle measured along a row having a minimum clear width of 12 inches plus 0.3 inch (305 mm plus 7.6 mm) for each additional seat over seven in a row (10 where seats are without backrests).

1030.10 Transitions. An important aspect of safe travel, for both egress and circulation purposes, is the consistency of the building elements along the path. Where there is a transition from one element to another, it is important to provide a condition such that the user is not subjected to a hazardous condition. Where a transition between a stairway and a stepped aisle does not alter the rise/run configuration that has been established, there is no need to apply any special requirements. However, where the riser and/or tread dimensions differ between the two elements, a means of providing a safe and smooth transition must be provided. Distinctive marking is also required to better identify the transition. Section 1030.12 should be reviewed where vomitories exist and transitions occur.

1030.13 Aisle accessways. The provisions of this section are specific to aisle accessways serving two different conditions: seating at tables and seating in rows.

1030.13.1 Seating at tables. Because it is often difficult to determine the exact dimensions of chairs and other seating that may encroach into an aisle or aisle accessway, a measurement of 19 inches (483 mm) is used to account for any potential obstruction caused by the seating. The 19-inch (483-mm) distance, applicable for movable seating, is measured perpendicular to the edge of the table or counter where the seating is provided. Where fixed seating is provided, the width is to be measured to the back of the seat. Examples of the use of this provision are illustrated in Figure 1030-1. Where other side boundaries are present, the clear width is to be measured to walls, tread edges, seating edges, or similar elements.

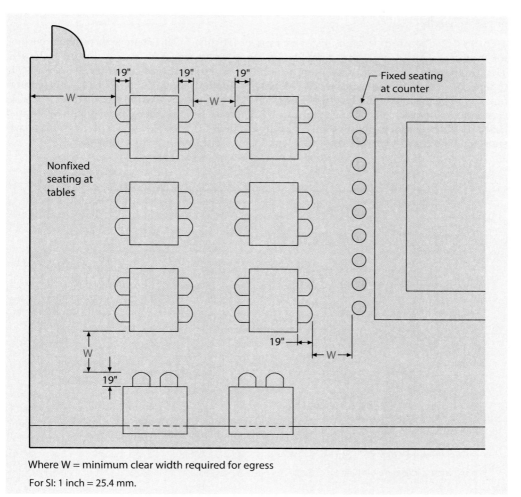

Where W = minimum clear width required for egress

For SI: 1 inch = 25.4 mm.

Figure 1030-1 **Seating at tables and counters.**

1030.13.1.1 Aisle accessway capacity and width for seating at tables. The clear width of aisle accessways serving various arrangements of seating at tables or counters is based on 12 inches (305 mm) plus ½ inch (12.7 mm) of additional width for each 1 foot (305 mm), or fraction thereof, beyond 12 feet (3,658 mm) of aisle accessway length. This length is measured from the center line of the seat farthest from the aisle.

> Minimum clear width = 12 inches + 0.5 inch (x – 12 feet), where:
>
> x = the distance in feet between the aisle and the center line of the most remote seat.

It should be remembered that in no case should the width be less than that based on the capacity requirements of Section 1005.3.2. On the other hand, there is no minimum width required for those portions of aisle accessways serving four or fewer persons and 6 feet (1,829 mm) or less in length.

1030.13.1.2 Seating at table aisle accessway length. Consistent with other provisions for means of egress, single-directional travel along aisle accessways is also limited in length. The distance from any seat to the point where two or more paths of egress travel are available to separate exits is limited to 30 feet (9,144 mm). In application, the 30-foot (9,144-mm) limitation would be the maximum length of a dead-end aisle accessway. Within that distance, the aisle accessway must reach an aisle or other means of egress element where travel would be limited based on the common path provisions of Section 1006.2.1. Various examples of the width provisions for aisle accessways serving seating and tables are shown in Figure 1030-2.

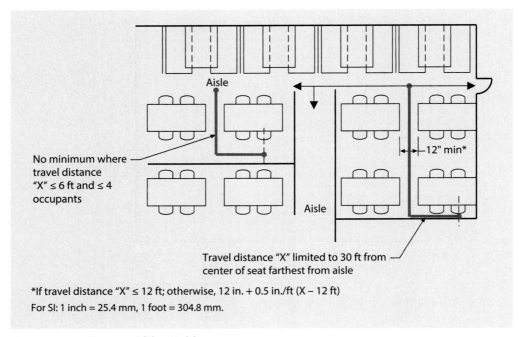

Figure 1030-2 **Egress width at tables.**

1030.13.2 Clear width of aisle accessways serving seating in rows. The minimum clear width between rows of seats is 12 inches (305 mm), measured between the rearmost projection of the seat in the forward row and the foremost projection of any portion of the seat in the row behind, where the rows have 14 or fewer seats. If automatic or self-rising seats are used (such as in movie theaters), the minimum clear width shall be measured with the seat up. If any chair in the row does not have an automatic- or self-rising seat, the measurement must be made with the chair in the down position. See Figure 1030-3. Seats with folding tablet arms are regulated in a special manner. Unless the tablet arm is of a type that automatically returns to the stored position by gravity when raised manually to the vertical position, the row spacing measurement must be taken with the arm in the position in which it is used.

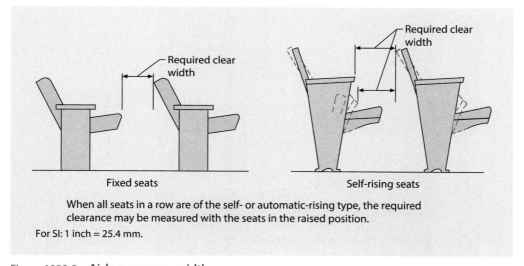

Figure 1030-3 Aisle accessway width.

Such clear width must be increased whenever the number of seats in a row exceeds 14, but in no case shall the number of seats in any row exceed 100. The increased width when seating rows are served by aisles or doorways at both ends of a row is the minimum 12 inches plus 0.3 inch (305 mm plus 7.6 mm) for each additional seat over 14; however, the width need not exceed 22 inches (559 mm). The establishment of the minimum required width is illustrated in the following formula:

Clear width, in inches = 12 inches (305 mm) + 0.3 inch (7.6 mm) $(x - 14$ seats$)$
where:

x = the number of seats in a row.

By interpreting the formula, we find that the maximum required clear width occurs when the number of seats in a row reaches 48. See Figure 1030-4.

In smoke-protected and open-air facilities, the maximum number of seats permitted for a 12-inch-wide (305-mm) aisle accessway with dual access is permitted to exceed the 14-seat limit for seating areas without smoke protection. With a maximum that varies up to 21 seats

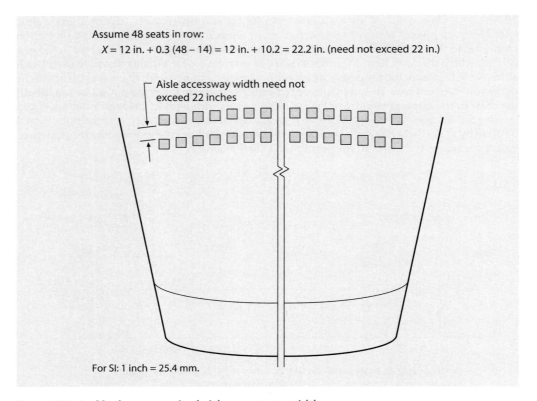

Figure 1030-4 **Maximum required aisle accessway width.**

for seats with backrests, the limitations indicated in Table 1030.13.2.1 are based on the total occupant load devoted to assembly seating. Increases on single-access aisle accessways are also available. A similar version of the previous formula can be used with Table 1030.13.2.1 to determine the minimum required aisle accessway width when the number of seats permitted for a 12-inch-wide (305-mm) aisle accessway is exceeded in a building provided with smoke-protected assembly seating or for an open-air facility. As an example, assume a 16,000-seat arena provided with smoke-protected assembly seating. A dual-access row contains 34 seats. The minimum required aisle accessway (row) width is determined by the following formula:

12 inches + [(0.3 inch)($x - y$)]
where:

x = the number of seats in the row (34 in this example), and
y = the maximum number of seats permitted with a 12-inch aisle accessway (19 in this example)

12 + [(0.3)(34 − 19)] = 12 + 4.5 = 16.5 inches minimum width

For rows of seating served by an aisle or doorway at only one end of a row, the formula for the clear width is the minimum 12 inches plus 0.6 inch (305 mm plus 15 mm) for each seat over seven, but the clear width need not exceed 22 inches (559 mm). This is similar to the previous provision with one major difference: a maximum 30-foot (9,144-mm) path of travel is permitted from the occupant's seat to a point where there is a choice of two directions of travel to an exit. This can be to the adjacent two-way aisle or along a dead-end aisle to a cross-aisle or doorway. See Figure 1030-5. Where one of the two paths of travel is across the aisle through a row of seats to another aisle, the maximum of 24 seats rule described previously is in effect. For smoke-protected assembly seating, single direction travel distance can be increased per Table 1030.13.2.1.

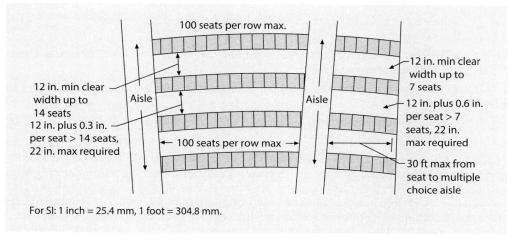

Figure 1030-5 **Row spacing.**

1030.14 Assembly aisle walking surfaces. Where aisles have a slope of 1 unit vertical in 8 units horizontal (12.5-percent slope) or less, steps in the aisles are prohibited because occupants in low-slope aisles have a tendency to not notice steps as readily as they would in the steeper aisles. Continuous surfaces are safer surfaces.

Where an aisle has a slope steeper than 1 unit in 8 units horizontal (12.5-percent slope), a stepped aisle with a series of risers and treads must be used. These risers and treads shall extend the entire width of the aisle, shall in general, have a rise of no more than 8 inches (203 mm) nor less than 4 inches (102 mm), and shall be uniform for the entire flight. The tread shall not be less than 11 inches (279 mm) and shall be uniform throughout the flight. As a general rule, variations in run or height between adjacent treads or risers shall not exceed 3/16 inch (4.8 mm).

One provision that helps the user of an aisle notice a step is the provision requiring a contrasting strip or other approved marking on the leading edge of each tread. Designed to identify the edge of each tread when viewed in descent, this marking strip may be omitted where it can be shown that the location of each tread is readily apparent.

Where changes in riser height are needed to match the changes in seating sightlines, an exception permits variations in riser height to exceed 3/16 inch (4.8 mm) between risers, provided the exact location of such a variation is clearly identified with a marking strip at the nosing or leading edge adjacent to the nonuniform risers. This edge marking strip, having a width between 1 inch and 2 inches (25 mm and 51 mm), shall be distinctively different from the contrasting marking strip required on each tread. In another exception to the riser height provision, riser heights may be increased to 9 inches (229 mm) where it can be demonstrated that lines of sight would otherwise be impaired. The permitted variations between adjacent risers are controlled by Section 1030.14.2.2.1 and depend on whether the risers are intended to be consistent or if they are designed to match the seating sightlines.

1030.15 Seat stability. Because of the potential obstructions to the paths of egress travel caused by loose seating, this section requires seats in assembly uses to be securely fastened to the floor. The following six exceptions identify conditions under which the securing of seating is either impractical or unnecessary:

1. Where 200 or fewer seats are provided on a flat floor surface
2. Where seating is at tables and the floor surface is flat
3. Where more than 200 seats are fastened together in groups of three or more on a flat floor surface
4. Where seating flexibility is critical to the function of the space, and 200 or fewer seats are provided on tiered levels
5. Where level seating is separated by railings or similar barriers into groupings of 14 seats or fewer
6. Where seating is separated by railings or similar barriers and limited to use by musicians or other performers

1030.16 Handrails. All ramped aisles having a slope steeper than 1 unit vertical in 15 units horizontal (6.7-percent slope), and all stepped aisles shall have handrails complying with Section 1014. The handrails can be placed on either side of or down the center of the aisle served and can project into the required width no more than 4½ inches (114 mm). Where aisles are 74 inches (1,880 mm) or greater in width and have seating on only one side, then a minimum of two handrails are required and one must be located within 30 inches (762 mm) of the edge of the stepped aisle. This requirement addressing these wider aisles is intended to help regulate what are often called "social stairs" which are somewhat a hybrid between stairways and assembly seating. Figure 1030-6 illustrates this requirement.

Handrails may be omitted where the slope of the aisles is not greater than 1 unit vertical in 8 units horizontal (12.5-percent slope) with seating on both sides, or where a guard that conforms to the size and shape requirements of a handrail is located at one side. The first exception intends the seating to be a substitute for handrails. The second exception permits a graspable top rail of a guard to be used as the handrail where a drop-off occurs on the one side of the aisle.

Handrails located within the aisle width shall not be continuous, but shall provide gaps at intervals not exceeding five rows. The width of these gaps should not be less than 22 inches (559 mm), nor more than 36 inches (914 mm). This is to provide access to seating on either side of the rail and to facilitate the flow of users on the aisle. A rail located 12 inches (305 mm) below the main handrail is required to prevent users from ducking under the handrail and hindering flow. Also, it provides a handrail for toddlers who may be using the aisle. Where a stepped aisle is required to have two handrails as would be required for wider aisles on social stairways, the mid-aisle handrail is also to be discontinuous. This is illustrated in Figure 1030-6 and places a stepped aisle adjacent to the seating platforms and leaves the remaining portion of the stepped aisle to serve more like a traditional stairway.

1030.17 Assembly guards. The code requires minimum 26-inch-high (660-mm) guards between aisles parallel to seats (cross aisles) and the adjacent floor or grade below where an elevation change of 30 inches (762 mm) or less occurs. An exception exists where the backs of seats on the front of the cross aisle project 24 inches (610 mm) or more above the adjacent floor of the aisle. See Figure 1030-7. Where the elevation change adjacent to a cross aisle exceeds 30 inches (762 mm), the general guard height requirements of Section 1015.3 shall apply.

Assembly 517

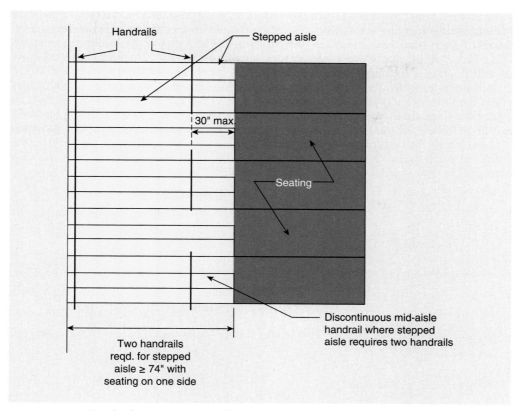

Figure 1030-6 **Handrails requirements of social stairway.**

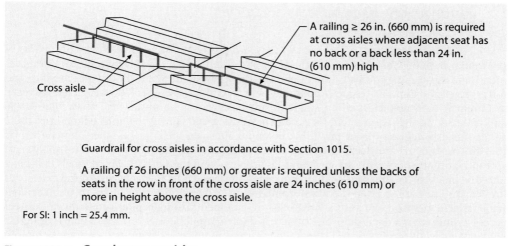

Figure 1030-7 **Guards at cross aisles.**

The intent of this provision is to provide a certain degree of protection from falls that may occur while occupants are using a cross aisle adjacent to a drop-off. Even if the drop is minimal, the conditions of egress from an assembly use, particularly in low light, dictate the need for an increased level of safety. In addition, where the top of the seat backs are less than 24 inches (610 mm) above the aisle floor, an unintentional impact of the seat back could cause a fall over the seats.

In order to provide for proper viewing in auditoriums, theaters, and similar assembly uses where the floor or footboard elevation is more than 30 inches (762 mm) above the floor or grade below, a guard in front of the first row of fixed seats, and that is not at the end of an aisle, may be 26 inches (660 mm) in height. Under such conditions, a guard height of at least 36 inches (914 mm) high shall extend the full width of the aisle at the foot of the aisle. In addition, the top of the guard shall be located at least 42 inches (1,067 mm), measured diagonally, from the nosing of the nearest tread, as depicted in Figure 1030-8.

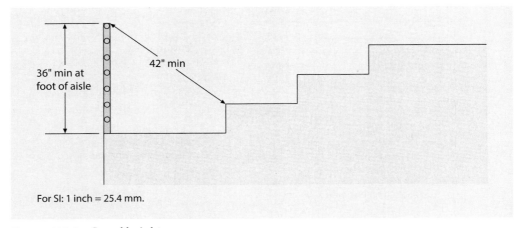

For SI: 1 inch = 25.4 mm.

Figure 1030-8 **Guard heights.**

Section 1031 *Emergency Escape and Rescue*

1031.2 Where required. Because so many fire deaths occur as the result of occupants of residential buildings being asleep at the time of a fire, the IBC selectively requires that basements and all sleeping rooms below the fourth story have windows or doors that may be used for emergency escape or rescue. Applicable only to Groups R-3 and R-4 occupancies, as well as Group R-2 occupancies with a single means of egress as permitted by Tables 1006.3.4(1) and 1006.3.4(2), the requirement for emergency escape and egress openings help ensure these single means of egress spaces provide a potential alternate means to escape. The concern is that when residents are sleeping and unaware of their surroundings, a fire will usually have spread before the occupants are aware of the problem, and the normal exit channels will most likely be blocked. The reason for the requirement in basements is that access to the exterior is limited and they are so often used as sleeping rooms. An exception eliminates the requirement for emergency escape and rescue openings for basements and sleeping rooms having direct access by means of an exit door or exit access door to a public way or a yard, court, or egress balcony that leads to a public way. Emergency escape and rescue openings are also not required in basements with a limited ceiling height or a small floor area, provided no habitable space is provided. Basement sleeping rooms in sprinklered Group R-2 and R-3 dwelling and sleeping units are not required to be provided with an escape and rescue opening provided one of two conditions occurs in the basement, giving occupants a choice of two paths of travel or escape.

The scope of this section is of particular importance as it applies to Group R-2 occupancies. Where at least two exits, or access to at least two exits, are provided on each story of a Group R-2 building, then the provisions of Tables 1006.3.4(1) and 1006.3.4(2) are not applicable. Therefore, the provisions of Section 1031 addressing emergency escape and rescue openings also do not apply. However, where the allowances of Table 1006.3.4(1) or 1006.3.4(2) permitting a single means of egress are used, then the Group R-2 dwelling units must be provided with complying emergency escape and rescue openings. In those situations where, in multistory buildings, one or more stories may have access to two or more means of egress and there are other stories with access to only one exit, the requirements of this section would only be applied to those stories with access to just one exit.

The code intends that the openings required for emergency escape or rescue be located on the exterior of the building so that rescue can be affected from the exterior or, alternatively, so that the occupants may escape from that opening to the exterior of the building without having to travel through the building itself. Therefore, where openings are required, they shall open directly onto a public street, public alley, yard, or court. This provision ensures that continued egress can be accomplished after passing through the emergency escape and rescue opening.

1031.2.1 Operational constraints and opening control devices. As stated in the code, these openings used for emergency escape or rescue must be operational from the inside of the room. Where windows are used, the intent is that they be of the usual double-hung, horizontal sliding, or casement windows operated by the turn of a crank. Although there is no specified limit on the type of window that may be used, Section 1031.3.2 requires that the net clear opening dimensions must be provided through normal operation of the window. The building official should evaluate special types of windows other than those just described based on the difficulty of operating or removing the windows. If no more effort is required than that required for the three types of windows just enumerated, they could be approved as meeting the intent of the code as long as no keys or tools are required. The code does accept compliant fall prevention devices with emergency release mechanisms as the opening control.

1031.3.1 Minimum size. The dimensions prescribed in the code, and as illustrated in Figure 1031-1 for exterior wall openings used for emergency egress and rescue, are based in part on extensive testing by the San Diego Building and Fire Departments to determine the proper relationships of the height and width of window openings to adequately serve for both rescue and escape. The minimum of 20 inches (508 mm) for the width in Section 1031.3.2 was based on two criteria—first, the width necessary to place a ladder within the window opening, and second, the width necessary to admit a fire fighter with full rescue equipment. The minimum 24-inch (610-mm) height dimension was based on the minimum necessary to admit a fire fighter with full rescue equipment. By requiring a minimum net clear opening size of at least 5.7 square feet (0.53 m^2), the code ensures that an opening of adequate dimensions is provided. Where the opening occurs at grade level, the opening need only be 5 square feet (0.46 m^2) because of the increased ease of access from the exterior.

1031.3.3 Maximum height from floor. In order to be relatively accessible from the interior of the sleeping room or basement, the emergency escape and rescue opening cannot be located more than 44 inches (1,118 mm) above the floor. The measurement is to be taken from the floor to the bottom of the clear opening.

1031.4 Emergency escape and rescue doors. Where a door is used for the emergency escape and rescue opening, this code section requires that a swinging door or sliding door be used. Other than that limitation, the door would be expected to meet all the general size and operational constraint requirements from Sections 1031.3 and 1031.2.1, which are applicable to all emergency escape and rescue openings.

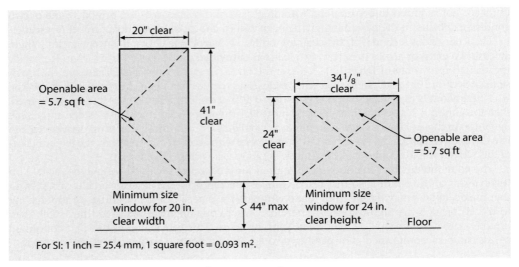

Figure 1031-1 **Emergency escape and rescue opening.**

1031.5 Area wells.   Area wells in front of emergency escape and rescue openings also have minimum size requirements. These provisions address those emergency escape and rescue openings (windows or doors) that occur below grade. Obviously, just providing the standard emergency escape opening criteria to the actual opening will get occupants through them but the area well may actually trap them against the building without providing for their escape from the area well or providing for fire-fighter ingress.

The minimum size requirements in cross section are similar in intent to the emergency escape and opening criteria; that is, to provide a nominal size to allow for the escape of occupants or ingress of fire fighters. See Figure 1031-2. The ladder or steps requirement is the main difference.

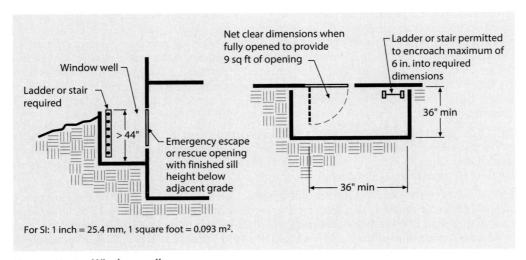

Figure 1031-2 **Window wells.**

Emergency escape openings below the fourth story are not required to have an escape route down to grade; however, those openings below adjacent grade are so required. When the depth of an area well exceeds 44 inches (1,118 mm), a ladder or steps from the window well are required. The details for the ladder or steps are given in Section 1031.5.2 and are shown in Figure 1031-3. Because ladders and steps in window wells are provided for emergency use only, they are not required to comply with the provisions for stairways found in Section 1011.

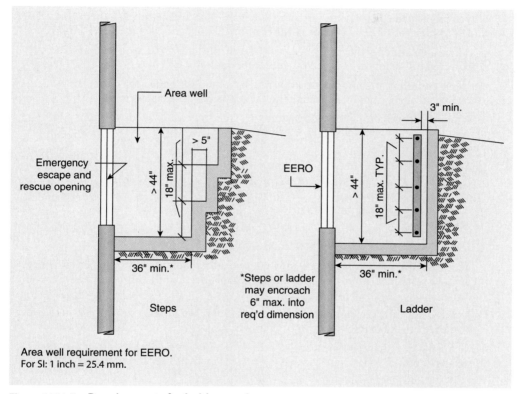

Figure 1031-3 **Requirements for ladders and steps.**

1031.6 Bars, grilles, covers, and screens. The ever-increasing concern for security, particularly in residential buildings, has created a fairly large demand for security devices such as grilles, bars, and steel shutters. Unless properly designed and constructed, the security devices over bedroom windows can completely defeat the purpose of the emergency escape and rescue opening. Therefore, the IBC makes provisions for security devices, provided the release mechanism has been approved and is operable from the inside without the use of a key, tool, or force greater than that which is required for normal operation of the escape and rescue opening. Fire deaths have been attributed to the inability of the individual to escape from the building because the security bars prevented emergency escape.

The very essence of the requirement for emergency escape openings is that a person must be able to effect escape or be rescued in a short period of time because the fire will have spread to the point where all other exit routes are blocked. Thus, time cannot be wasted in figuring out means of opening rescue windows or obtaining egress through them. Therefore, any impediment to escape or rescue caused by security devices, inadequate window size, difficult operating mechanisms, and so on, is not permitted by the code.

KEY POINTS

- The means of egress is an exiting system that begins at any occupiable point in a building and continues until the safety of the public way is reached.
- Three distinct elements compose the means of egress—the exit access, the exit, and the exit discharge.
- Occupant load, the driving force behind the design of an exiting system, must be determined for the expected use or uses of a building.
- Components along the path of egress travel must be sized to accommodate the expected occupant load served by the components.
- Specific minimum component widths, such as those provided for doors, aisles, corridors, and stairways, often dictate the capacity of the means of egress.
- Where multiple complying exitways are provided, the calculated width may be dispersed among the various exits or exit-access doorways.
- With limited exceptions, the means of egress must have a minimum clear height of 7 feet 6 inches (2,286 mm) throughout the travel path.
- The code regulates the means of egress so that there is no change in elevation along the path of exit travel that is not readily apparent to persons seeking to exit under emergency conditions.
- Exit access describes the vast majority of a building's floor area that provides the access necessary to reach a protected area (an exit).
- Access to at least two means of egress is typically required from floor levels above the first story, and rooms or areas having sizable occupant loads or excessive travel distance.
- As a general rule, exit signs are required from rooms or areas requiring access to two or more paths of exit travel.
- Requiring continuous illumination, exit signs must be provided with a secondary source of power.
- In those rooms or areas requiring access to at least two exitways, a second source of power is required for maintaining illumination to the exit path.
- Guards must be designed to reduce the probability of falls from one level to a lower level that exceeds 30 inches (762 mm) in elevation difference.
- Guards are typically required to be at least 42 inches (1,067 mm) in height.
- In all public areas, guards must have limited openings to prevent individuals from falling or climbing through the guard assembly.
- In addition to providing proper access to and through a building, an accessible means of egress must be provided.
- Areas of refuge are mandated for certain buildings where stairs and/or elevators occur along the accessible means of egress.
- Doors are highly regulated in the IBC because of their potential for obstructing the means of egress.
- The use of revolving, overhead, and sliding doors for egress purposes is strictly limited.

- Doors swinging in the same direction as egress travel are mandated for all hazardous uses, as well as areas in other uses having an occupant load of 50 or more.
- Criteria for an acceptable latching or locking device on an egress door are very basic in that no key, special effort, or special knowledge is necessary to open the door.
- Where security issues are as important as those addressing fire and life safety, the IBC permits the installation of delayed egress locks.
- Panic hardware is typically mandated for swinging doors in Group A and E occupancies having an occupant load of 50 or more, as well as in all Group H occupancies.
- Gates located in the means of egress are regulated in a manner similar to doors.
- A stair is considered a change of elevation accomplished by one or more risers, whereas one or more flights of such stairs make up a stairway, along with any landings that connect to them.
- There are two different stairways addressed in the IBC—exit access stairways and interior exit stairways.
- Exit access stairways may be unenclosed where connecting no more than two adjacent stories.
- Treads and risers must be appropriately sized and uniform throughout the stair flight.
- Spiral stairways, curved stairways, and alternating tread devices are limited in their use because of the uniqueness of their configurations.
- Handrail design is regulated for height, size, shape, and continuity.
- Ramps must be designed for egress purposes as well as for accessibility.
- Multiple exit paths must be arranged in order to minimize the risk of a single fire blocking all of the exitways.
- Egress from a room through a nonhazardous accessory area is permitted, provided there is a discernible egress path that is direct and obvious.
- Travel distance is limited within the exit access portion of the means of egress; however, once an exit is reached, travel distance is no longer regulated.
- In rooms where seating is at tables, additional limitations are placed upon the aisles and aisle accessways used for egress purposes.
- Corridors are intended to be used for circulation and egress purposes, and at times must be constructed as a protected element for use as a path for egress travel.
- The exit is the portion of the means of egress that provides a degree of occupant protection from fire, smoke, and gases.
- Horizontal exits, exit passageways, interior exit stairways and ramps, exterior exit doors at the level of exit discharge, exterior exit ramps, and exterior exit stairways are the exit components addressed in the IBC.
- Interior exit stairways are to be constructed of either 1-hour or 2-hour fire-resistance-rated construction with protected openings.
- Openings and penetrations into an interior exit stairway or ramp are strictly limited because of the hazards involved with vertical egress.

- An exit passageway is similar to a corridor but is built to a higher level and limited in much the same manner as a vertical exit enclosure.
- The concept of a horizontal exit is the creation of a refuge area to be used by occupants fleeing the area of fire origin.
- The use of an exterior exit stairway as a required means of egress element is limited to buildings not exceeding six stories or 75 feet (22,860 mm).
- In high-rise buildings, luminous egress path markings are required in exit enclosures and exit passageways of Group A, B, E, I-1, M, and R-1 occupancies.
- Egress travel outside of the building at grade level is considered exit discharge, continuing until the public way is reached.

CHAPTER 11

ACCESSIBILITY

Section 1101 General
Section 1102 Compliance
Section 1103 Scoping Requirements
Section 1104 Accessible Route
Section 1105 Accessible Entrances
Section 1106 Parking and Passenger Loading Facilities
Section 1107 Motor-Vehicle-Related Facilities
Section 1108 Dwelling Units and Sleeping Units
Section 1109 Special Occupancies
Section 1110 Other Features and Facilities
Section 1111 Recreational Facilities
Section 1112 Signage
Key Points

This chapter addresses accessibility and usability of buildings and their elements for persons with physical disabilities. Where a facility is designed and constructed in accordance with this chapter and other related provisions throughout the *International Building Code®* (IBC®), it is considered accessible.

A historical perspective. In 1961, the American National Standards Institute (ANSI) published ANSI Standard A117.1. The President's Committee on Employment of the Handicapped and the National Easter Seal Society were designated as the secretariat for the standard. Since that time, a number of historic events have occurred that have brought accessibility and usability issues to the forefront of not only building code enforcement, but of society as a whole.

In the early 1970s, all three United States legacy model code groups approved code changes to make buildings more accessible and usable for people with disabilities. These independent developments resulted in confusion in the regulatory design and construction community. As a result, the Council of American Building Officials (CABO) requested that the Board for the Coordination of Model Codes (BCMC) review the regulations and suggest provisions to all of the model codes that would result in uniformity. In addition, ANSI requested that CABO become the secretariat for ANSI A117.1.

In October 1987, BCMC began its assignment to provide regulations that set forth when, where, and to what degree access must be provided (commonly referred to as scoping) for persons with disabilities. The ANSI A117.1-1986 standard contained design specifications intended to provide buildings and facilities accessible to and usable by people with disabilities but did not specify scoping provisions. Authorities who chose to employ ANSI A117.1 found it necessary to adopt amendments to establish when, where, and to what degree its provisions applied. During the BCMC work on accessibility, it became apparent that safe egress for people with disabilities was essential if access to buildings was to be increased. Therefore, the final BCMC report addressed both access and egress for people with disabilities. While BCMC was working on scoping provisions for ANSI A117.1-1986, the ANSI A117.1 committee continued to study revisions to their standard for public review.

In 1988, the United States Congress passed the Fair Housing Amendments Act to cover multifamily housing of four units or more on a site. On July 26, 1990, President George Bush signed the Americans with Disabilities Act (ADA), which set forth comprehensive civil rights protection to individuals with disabilities in the areas of employment, public accommodations, state and local government services, and telecommunications. The reasons legislators supported the ADA were the recognized inadequacy, limited application, and nonuniformity of existing protection for individuals with disabilities. One year later, on July 26, 1991, the United States Department of Justice (DOJ) issued its final rules, the Americans with Disabilities Act Accessibility Guidelines (ADAAG), that provided for access and usability for disabled persons in public accommodations and commercial facilities. Both acts were born from the Civil Rights Act of 1964. The ADA set forth statutory deadlines for when certain requirements became effective. One of these requirements was that new facilities designed and constructed for first occupancy after January 26, 1993, must be accessible.

The public review draft of revisions to the ANSI A117.1 standard dated January 24, 1992, was submitted to the DOJ with a request for technical assistance. A staff comparison of ANSI A117.1 and ADAAG yielded only a few areas in which ADAAG was deemed to provide greater accessibility. Generally, the differences found between ADAAG and ANSI A117.1 indicated that the ANSI standard provided for greater overall accessibility. At the BCMC meetings in May and June of 1992, the committee reviewed suggestions that would incorporate ADA guidelines into the ANSI A117.1 standard and other regulations.

At the BCMC meeting on June 8, 1992, the committee finalized its report. From June 9 through 11, 1992, the ANSI A117.1 committee finalized its standard.

The final BCMC report of June 8, 1992, and the final draft of the ANSI A117.1 standard were soon adopted by all of the model code groups as their accessibility requirements. The final BCMC report of June 8, 1992, and the CABO/ANSI A117.1-1992 standard were submitted to the DOJ for a technical review. The resulting letter received in November 1995 described nine general problems, many of which were differences in philosophy. Noncode items such as laboratory equipment, automated teller machines, and telephones were also a concern of the DOJ. During the DOJ review, a joint task force was established to make recommendations for changes to both the CABO/ANSI A117.1, *Accessible and Usable Buildings and Facilities*, and ADAAG. This harmonization effort over many months resulted in suggested revisions to both documents. In an effort to reduce conflict, confusion, and frustration among all users, the results of the harmonization report were accepted in July 1996 by the ADAAG Review Advisory Board. That board presented its final report to the Architectural and Transportation Barriers Compliance Board, which in turn resulted in additional rulemaking by the Access Board.

The CABO/ANSI revisions were finalized and subsequently approved by ANSI's Board of Standards Review on February 13, 1998. Because CABO was incorporated into the International Code Council® (ICC®) in November 1997, the resulting standard was retitled ICC A117.1-1998. The 2009 edition of ICC A117.1 was finalized in late 2009, and its technical provisions and format closely paralleled those technical requirements in what at that time was called the proposed "new" ADA/ABA AG (Americans with Disabilities Act and Architectural Barriers Act Accessibility Guidelines). These new federal regulations were ultimately approved by DOJ and are now known as the *2010 ADA Standards for Accessible Design*.

These updated federal requirements were the first major rewrite of ADAAG since 1990 and were published in the Federal Register on July 23, 2004. The updates resulted in the reconciliation of most differences that occurred between the A117.1 standard and the original ADAAG. Although the coordination effort was not fully completed, it was expected that the efforts of all parties would result in a system that greatly enhanced compliance with the *2010 ADA Standards for Accessible Design*. These updated federal standards became an allowable alternative to the original ADAAG in March 2011 and ultimately became mandatory on March 15, 2012, when they replaced the original ADAAG.

The 2017 edition of the A117.1 standard is adopted by reference into the 2024 IBC. This edition of the standard has continued to move forward and improve the level of access and clarity it provides. Although it has in limited circumstances moved away from being harmonized with the ADA, the latest A117.1 provisions generally exceed those of the ADA and thus will not compromise compliance with the minimum federal requirements.

On another front, the ICC has worked closely with the U.S. Department of Housing and Urban Development (HUD) to ensure that both the IBC scoping requirements and the A117.1 standard's technical provisions are coordinated to either meet or exceed the accessibility requirements of the federal Fair Housing Act (FHA). HUD has recognized that the IBC and the A117.1 provisions do meet or exceed the federal Fair Housing Accessibility Guidelines (FHAG) by granting "safe harbor" status to several of the ICC documents. This status recognizes that these various documents—if followed in their entirety without modification or waiver—will keep residential buildings compliant with the FHA. As of the publication of this handbook, HUD has not completed their review or approval of either the 2017 edition of the A117.1 Standard or the 2021 and 2024 editions of the IBC. Because HUD has granted safe harbor status to every edition of the A117.1 Standard between and including the 1986 and 2009 editions, and every edition of the *International Building Code* between and including the 2000 through 2018 editions, the assumption is that the 2024 IBC along with the 2017 A117.1 Standard will also receive this safe harbor status.

Section 1101 *General*

The term *accessible* takes on a very broad meaning by requiring compliance with Chapter 11. Space requirements must be addressed for all portions of the building to be considered accessible or to provide an accessible route within a building. Space requirements apply to adequate maneuvering space, clearance width for doors and corridors, and height clearances.

Location of controls, switches, and other forms of hardware becomes a function of forward and parallel reach ranges if they are to be considered accessible. To assist persons with limited dexterity, controls, and other forms of hardware need to be operable without tight gripping, grasping, or twisting of the wrist.

People with visual impairments are provided with accessible routes, by the inclusion of provisions for clear and unobstructed routes that are free of protrusions created by benches, overhanging stairways, poles, posts, and low-hanging signs. Many hazards can be eliminated by using different textural surfaces in the accessible route to alert sight-impaired persons. Visually impaired or partially sighted persons can be assisted with proper signage of the correct size, surface, and contrast. Signage is also provided for the hearing impaired. Directional signage should always be clear, concise, and appropriately placed.

To properly apply the code, users must understand the terminology and recognize the limits of defined terms. It is important to note that the defined terms listed in the IBC are specifically for this code and may have different meanings from definitions in other accessibility provisions or regulations, such as the ADA Standards for Accessible Design. The accessibility terms defined specifically in the IBC are all located in Section 202. Additional definitions that apply to these provisions can be found in Section 107 of ICC A117.1.

An accessible route is defined in both the IBC and ICC A117.1, and is described in Section 402 of the latter document. Accessible routes have a number of components, such as walking surfaces with a slope not steeper than 1:20, marked crossings at vehicular ways, clear floor spaces at accessible elements, access aisles, ramps, curb ramps, and elevators. Each component must comply with specific applicable standards and requirements.

Several key elements to review when addressing accessible routes are the requirements related to protruding objects and ramps. An accessible route may contain a number of protruding objects that can affect its use. These objects include ordinary building elements such as railings, telephones, water fountains, signs, directories, and automatic teller machines.

Protruding objects and other such obstructions are regulated by Section 307 of ICC A117.1. Similar provisions governing the means of egress system are located in IBC Section 1003.3. Ramps with proper slopes may serve as acceptable means of egress, as well as part of an accessible route. A sloped surface that is steeper than 1:20 is considered to be a ramp and is required to comply with both the requirements of Section 1012 for egress paths and Section 405 in ICC A117.1. To be considered acceptable, ramps shall have a slope not steeper than 1:12, with a maximum rise for any ramp not to exceed 30 inches (762 mm) and a minimum width of at least 36 inches (914 mm).

Three types of dwelling units and sleeping units are defined and then required by the code: Accessible, Type A, and Type B. The Accessible units are deemed to be fully accessible and must be in compliance with the IBC and Section 1102 of ICC A117.1. Type A units are considered dwelling units and sleeping units that are designed and constructed for accessibility in accordance with Section 1103 of ICC A117.1. Designed and constructed in accordance with Section 1104 of ICC A117.1, a Type B dwelling unit or sleeping unit is intended to be consistent with the technical requirements of HUD's Fair Housing Guidelines. A Type A unit is considered to provide a significant degree of accessibility, whereas a Type B unit provides for only a minimum level.

Several other definitions in Section 202 assist in clarifying the intent of the provisions. An employee work area is limited to those spaces used directly for work activities and does not include those common use areas such as toilet rooms and break rooms. Public use areas are identified as those spaces used by the general public and may be interior or exterior.

Section 1102 *Compliance*

1102.1 Design. This section adopts ICC A117.1, more specifically the 2017 edition, as the adopted design standard to be used to ensure that buildings and facilities are accessible to and usable by persons with disabilities. With this section providing the accessible design and construction standards for buildings, the remaining sections of the chapter provide the scoping provisions that set forth when, where, and to what degree access must be provided.

The importance of this section and the requirements it imposes should not be overlooked. As previously stated, buildings must be designed and constructed to the minimum provisions of Chapter 11, along with the other applicable provisions of the IBC and ICC A117.1, to be considered accessible. Therefore, prior to applying the code provisions for accessibility, it is important for the code user to review the technical requirements found in ICC A117.1. Although these technical requirements are not directly within the IBC, they are still applicable due to being scoped from the IBC and the provisions of IBC Section 102.4 related to referenced standards. Such elements include, but are not limited to, space allowance and reach ranges, accessible route, protruding objects, and ramps.

Space requirements can vary greatly depending on the nature of the disability, the physical functions of the individual, and the skill or ability of the individual in using an assistive device. However, it is generally accepted that spaces designed to accommodate persons using wheelchairs will be functional for most people.

It is important to note that not all portions of ICC A117.1 are referenced by the IBC. For example, the criteria set forth in Section 504 for stairways as well as the stair handrail provisions of Section 505 have no application insofar as accessible stairways are not specifically scoped by IBC Chapter 11. As a result, the provisions of IBC Sections 1011 and 1014 solely regulate all stairways and associated handrails for accessibility purposes. As another example, accessible telephones and automatic teller machines are addressed in ICC A117.1, Sections 704 and 707, respectively. However, these elements are only regulated in the IBC by Appendix E, which must be specifically adopted to be in effect.

Section 1103 *Scoping Requirements*

In general, access to persons with disabilities is required for all buildings and structures, whether temporary or permanent. There are, however, certain conditions under which sites, buildings, facilities, and elements are exempt from the provisions where specifically addressed. Various modifications to the general requirements for accessibility are also found throughout other areas of Chapter 11. For example, an exception to Section 1104.4 eliminates the requirement for an accessible route of travel to levels having relatively small floor areas, applicable in all but a few types of uses. There are also a number of exceptions that apply only within dwelling units.

Where the building under consideration is existing, only the requirements found in *International Existing Building Code*® apply. Intended to include historic buildings, the provisions apply to the maintenance, alteration, or change in use of an existing structure. An exception relating to Type B dwelling units and sleeping units states that they need not be provided in existing buildings unless substantial alterations in conjunction with a change of occupancy take place.

In commercial applications, individual employee workstations are not required to be fully accessible. They must, however, comply with the appropriate provisions for visible alarms, accessible means of egress, and common use circulation paths. In addition, such work areas shall be located on an accessible route in order to provide access to, into, and out of the work area. Modifications to each individual workstation can then be made in order to address the specific needs of the employee. None of these limited provisions are applicable for small, elevated work areas where the area must be elevated because of the nature of the work performed. A number of other employee-use areas are also considered spaces that need not be provided with accessibility. Raised areas used for security or safety purposes, limited access spaces, equipment spaces, and highway tollbooths have been identified as those types of areas where it seems unreasonable, if not impossible, to provide full access.

Observation galleries, prison-guard towers, fire towers, lifeguard stands, and similar elevated observation areas need not be accessible or served by an accessible route. Ladders, catwalks, crawl spaces, freight elevators, very narrow passageways or tunnels, and any other space deemed to be nonoccupiable require no access. Spaces such as elevator pits and penthouses; mechanical, electrical, or communications equipment rooms; equipment catwalks; water or sewer treatment pump rooms and stations; electric substations in transformer vaults; and any other areas accessed solely by personnel for maintenance, repair, or monitoring of equipment are not required to be accessible. In addition, highway tollbooths where access occurs only by passageways elevated above grade or buried below grade are not required to be accessible. Accessibility is also not required to walk-in cooler and freezer equipment provided they are accessed only from employee work areas.

Section 1103.2.5 recognizes that some activities directly associated with construction projects will not be safe for persons with certain physical disabilities. Therefore, structures, sites, and equipment used or associated with construction are not required to be accessible. The limited scope of this provision is important insofar as the accessibility provisions generally do apply to the temporary buildings. Although accessibility would not be required to a construction trailer, it must be provided to sales trailers that are common in new subdivisions or multifamily projects. It is also necessary to know that where pedestrian protection is required by Chapter 33 and Table 3306.1, such walkways are required to be accessible per Section 3306.2.

For the most part, occupancies designated as Group U are exempt from the provisions of Chapter 11. One exception requires agricultural buildings that are associated with the general public to be provided with access to those public areas. An example might be a produce stand set up just inside the entry of the agricultural greenhouse. In addition, all paved work areas for agricultural buildings must be provided with access to such areas. A second exception mandates that where accessible parking is provided in private garages or carports, such parking structures shall be accessible.

Many religious architectural building features are based upon traditions or rituals that are often conducted from either raised or recessed areas within the sanctuary or worship area. Areas such as altars, bimahs, baptisteries, pulpits, minibars, and minarets are examples of the types of areas exempted from accessibility by Section 1103.2.8. The size and elevation differences selected for this exemption are consistent with those for employee work areas. The provision does not indicate that the 300-square-foot limitation is aggregate in its application, so it is permissible to have several elements such as a high altar, baptistery, and pulpit all addressed separately.

A live/work unit, primarily regulated by Section 508.5, is considered to be a dwelling unit or sleeping unit in which a significant portion of the space includes a nonresidential use operated by the tenant. Although the entire unit is classified as a Group R-2 occupancy in the IBC, for accessibility purposes it is viewed more as a mixed-use condition. The residential portion of the unit is regulated differently for accessibility purposes than the nonresidential portion. The floor area of the dwelling unit or sleeping unit that is intended for residential use is regulated under the provisions of Section 1108.6.2 for Group R-2 occupancies. See Sections 1108.6.2.1 and 508.5. The requirements for an Accessible Type A or B unit would be applied based on the specific residential use of the unit and the number of units in the structure. The exceptions for Type A and B units set forth in Section 1108.7 would also exempt such units where applicable. In the nonresidential portion of the unit, full accessibility would be required based on the intended use. For example, if the nonresidential area of the unit is used for hair care services, all elements related to the service activity must be accessible. This would include site parking where provided, site- and building-accessible routes, the public entrance, and applicable service facilities. In essence, this portion of the live/work unit would be regulated in the same manner as a stand-alone commercial occupancy.

Smaller transient lodging occupancies, both Group R-1 and R-3, are not required to comply with the accessibility provisions of Chapter 11 provided two conditions are met, the facility must be proprietor-occupied and the facility cannot contain more than five residential units.

Section 1104 *Accessible Route*

An accessible route is defined as a continuous unobstructed path that complies with the provisions in Chapter 11. This route connects all accessible elements or spaces of the building or facility, including corridors, aisles, doorways, ramps, elevators, lifts, and clear floor space at fixtures. Code users should review the accessible route provisions in Chapter 4 of ICC A117.1. In addition, exterior portions of accessible routes must be evaluated and may include parking access aisles, curb ramps, crosswalks at vehicular ways, walks, ramps, and lifts.

1104.2 Within a site. The clear intent of these provisions is to provide, on sites with single or multiple buildings or facilities, access to each accessible element from all accessible parking areas, as well as from one accessible element to another on the same site. Where the only means of access between accessible facilities on a site is a vehicular way, an accessible route is not required between such facilities. Should a sidewalk, walking path, or similar circulation route be provided connecting the site elements, the route is to be designed and constructed as an accessible route.

1104.3 Connected spaces. Those portions of a building that are required to be accessible must be connected by at least one accessible route of travel. Such route shall connect to all accessible entrances and lead to accessible walkways connecting other accessible site elements and potentially the public way. One exception clarifies that in assembly areas with fixed seating an accessible route must only be provided to the accessible wheelchair seating areas. Other exceptions reference various provisions throughout Chapter 11 where either accessible routes are not required or their extent limited.

1104.3.1 Employee work areas. Although employee work areas are typically exempt from the accessibility provisions of the code, the "common use" circulation paths within such areas that are used by multiple employees are regulated. The paths must be designed and constructed as a complying accessible route unless exempted by one of the three exceptions. In those cases where an exception is applicable, it is still necessary to connect the work area to an accessible route such that persons with disabilities can approach, enter, and exit the area as stated in Section 1103.2.2. For the purpose of applying any of the three exceptions, it is important to note that *common use* is defined as nonpublic areas shared by two or more individuals. An example is shown in Figure 1104-1.

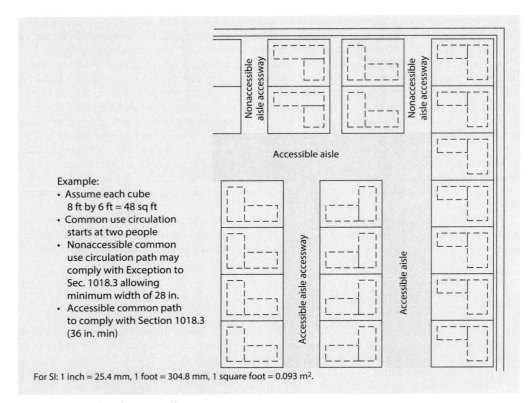

Figure 1104-1 **Employee work areas.**

1104.4 **Multistory buildings and facilities.** In addition to providing an accessible route to each portion of a building, each accessible story and/or mezzanine in a multilevel facility, and an occupiable roof, where applicable, must be connected via at least one accessible route of travel. The first exception waives the requirement for an accessible route to floors above and below any accessible level, provided such inaccessible levels have an aggregate floor area of not more than 3,000 square feet (278.7 m^2). This allowance is not permitted, however, where the vertically inaccessible level contains offices of health-care providers, passenger transportation facilities, multitenant sales facilities, where the building is a government facility, or where the facility contains four or more dwelling units.

This exception eliminates the requirement for an elevator or other means of vertical access; however, it does not reduce or eliminate the obligation of compliance with other provisions for accessibility. As an example, a toilet room on an inaccessible level permitted by the exception would still be required to comply with all of the provisions for accessible toilet rooms. In addition, whereas the floor either above or below the accessible grade-level floor would not be required to meet the applicable accessible route provisions, any facilities located on these floors would also be required to be provided on the accessible floor. For example, if toilet facilities are located on the floor either above or below the accessible floor, then the same facilities are required on the accessible floor and shall be constructed as accessible facilities.

1104.5 **Location.** Where an interior route of travel is provided within a building, any required accessible route shall also be interior. The intent of the provisions is to provide equal means of access. For example, an interior stairway and exterior ramp fails the equality test.

Section 1105 *Accessible Entrances*

In general, at least one public entrance, but not less than 60 percent of all such entrances to a building or individual tenant space, must be accessible.

In Figure 1105-1, two entrances (Doors A and B) are considered to be public entrances to the entire building. As such, both entrances must be accessible. If Doors C and D are public tenant entrances, they must be accessible in addition to the accessible entrances provided into each tenant space from the common lobby. One easy requirement to remember is that if there are only one or two public entrances into a building or tenant, those entrances must be accessible. In addition to the general provisions for public entrances, the code mandates accessible entrances under other conditions, such as between a parking garage and the building served by the garage. Even though the minimum number of accessible entrances has been provided at other locations, the direct access between the parking garage and the building must be accessible. Exceptions are provided for entrances used exclusively for loading and service, as well as entrances to spaces not required to be accessible.

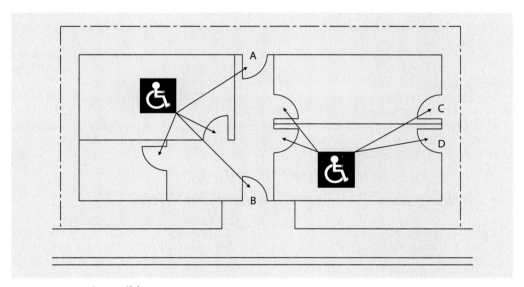

Figure 1105-1 **Accessible entrances.**

An important aspect of accessibility and accessible entrances is obviously the door assembly and its related components, including the threshold, hardware, closers, and opening force. Important requirements that affect doors and their accessibility are in IBC Section 1010 and in Section 404 of ICC A117.1.

To improve the access at public entrances, automatic doors are to be provided for certain larger occupancies. These uses tend to involve people that are more transient and less familiar with the facility and only apply where the occupant loads exceed the thresholds listed in Table 1105.1.1. When dealing with a mixed occupancy building, the "sum of the ratios" method is to be utilized to determine if the power-operated door requirement applies.

Landings on both sides of doorways are also important accessibility features. When access for persons with disabilities is required, a floor or landing shall generally not be more than ½ inch (12.7 mm) lower than the threshold of the doorway according to Section 1010.1.6 of the IBC.

Where the level change or threshold exceeds ¼ inch (6.4 mm), it is required to be beveled at a slope of one unit vertical in two units horizontal (1:2) or less. Similar provisions are found in Sections 303 and 404.2.4 of the A117.1 standard.

Door hardware, including handles, pulls, latches, locks, or any other operating device on accessible doors, is required to be of a shape that is easy to grasp with one hand and does not require a tight pinching, tight grasping, or twisting of the wrist to operate. Many individuals have great difficulty operating door hardware that does not include push-type, U-shaped handles, or lever-operated mechanisms.

Door closers with delayed action capability are also important and allow a person more time to maneuver through a door. These closers are required to be adjusted so that from an open position of 90 degrees (1.57 rad), the time required to move the door to an open position of 12 degrees (0.21 rad) will be a minimum of 5 seconds. Door closers are required to have minimum closing forces in order to close and latch the door; however, for other than fire doors and exterior doors, maximum force levels are set to limit the force levels for pushing or pulling open doors. Opening forces and the methods used to measure them are specified in Section 404.2.8 of ICC A117.1 and IBC Section 1010.1.3.

Minimum maneuvering clearances at doors, other than those that are for automatic doors, are based on a combination of forward- and side-reach limitations, the direction of approach, and minimum clear width required for wheelchairs. They also permit enough space that a slight angle of approach can be gained, which then provides additional leverage or opening force by the user. Without these required clearances, there is a possibility of interference between the edge of the door and the footrest on the wheelchair. This could render the door inaccessible to someone using a wheelchair.

In addition to the traditional provisions for doors, ICC A117.1 contains many detailed provisions in illustrations that are found in Section 404.

Care should be exercised when considering the maneuvering space necessary to make doors accessible. An inadvertent reversal of the latch side to hinge side may render a door inaccessible to an individual in a wheelchair.

Section 1106 *Parking and Passenger Loading Facilities*

The number of accessible parking spaces required on a site varies by the total number of spaces provided. For other than specific residential and medical uses, Table 1106.2 is used as the basis for calculating the required number of spaces. Rehabilitation facilities, as well as those facilities providing outpatient physical therapy, require a larger percentage of accessible spaces than addressed in the table. Obviously, this is due to the much higher probability of individuals with a mobility impairment visiting the facility. On the other hand, the required number of accessible parking spaces for Groups R-2, R-3, and R-4 is typically reduced from Table 1106.2.

For every six accessible parking spaces, at least one of the spaces must be an accessible van space. As an example, consider a parking lot with 23 total parking spaces. According to Table 1106.2, at least one accessible parking space is to be provided, and it must be designed and constructed to be van accessible. As another example, a parking lot with 202 parking spaces must be provided with a minimum of seven accessible spaces. At least two of those seven spaces shall be van-accessible.

An indicator of the critical nature of site development is the requirement for the shortest accessible route of travel. On a site with multiple buildings and a number of requirements for each, accessibility may pose complex site-design problems related to direct routes from parking areas. Where the parking facilities do not necessarily serve a particular building, the provisions require the accessible parking spaces to be located as closely as possible to the accessible pedestrian entrance to the parking facility. Early involvement or early attention to the location of multiple accessible elements during site development may tend to eliminate most, if not all, related problems.

Where multiple distinct parking facilities are provided on a site, such as a parking garage and a surface parking lot, the provisions of Section 1106.2 require that the total minimum required number of accessible parking spaces be determined individually. See Application Example 1106-1. However, Section 1106.7 allows accessible spaces to be relocated from remote lots to locations near accessible building entrances. This addresses a concern that in a large facility, the dispersion of accessible parking spaces into remote lots may result in decreased access for persons with disabilities.

If a passenger loading zone is provided, it shall have an adjacent access aisle that is part of the accessible route to the building. In accordance with ICC A117.1, the space shall have a vertical clearance of at least 114 inches (2,895 mm) at the zone and along the vehicle access route on the site. There are only three conditions under which the code mandates the installation of a passenger loading zone, where a driver would drop-off or retrieve their car, such as at valet parking services or mechanical access parking garages, and at accessible entrances of specified medical facilities. However, in other occupancies the loading zone must be accessible if it is provided.

GIVEN: A 440-space parking garage and 160-space surface parking area.
DETERMINE: The minimum required number of accessible parking spaces in each facility.

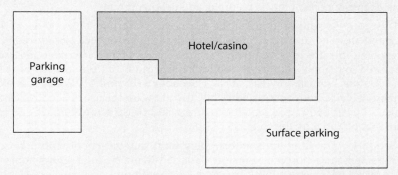

SOLUTION: Per Table 1106.2, the minimum required number of accessible parking spaces is:

	Total spaces	Van spaces
Parking garage	9	2
Surface parking	6	1
	15	3

Of the minimum 15 required accessible parking spaces, at least 3 must be van accessible. They need not be located in the individual facilities as calculated, provided substantially equivalent or greater access is provided in terms of distance from an accessible entrance or entrances, parking fee, and user convenience.

CALCULATION OF ACCESSIBLE PARKING SPACES

Application Example 1106-1

Section 1107 *Motor-Vehicle-Related Facilities*

This section regulates the various services related to motor vehicles including electrical vehicle charging stations (EVCS) and the general fuel-dispensing systems. There are several important factors to notice with the EVCS. First off is the fact that these stations are not required but are regulated if they are provided. Secondly is the fact the provisions regulate that at least one of "each type" of charging system be accessible. This is important since there are a variety of charging methods and connectors currently available and not any type of standardized system. One last aspect to recognize is that while these EVCS spaces will use the size requirements for an accessible van parking space, these are considered as a service and not as being a parking space. Therefore, the signage which would be required by Section 1112 and limit who could use the space is not required, but it may also be helpful to determine some way of marking which spaces do provide the accessible charging elements.

Section 1108 *Dwelling Units and Sleeping Units*

This section is limited to the accessibility provisions related to dwelling units and sleeping units as defined in Chapter 2. It provides guidance as to the conditions under which Accessible units, Type A units, and Type B units are mandated. These scoping provisions generally indicate where some degree of accessibility is required, and to what extent.

1108.2 Design. As addressed under the discussion of Section 1101, there are three types of dwelling and sleeping units that provide varying degrees of accessibility required by the code. Accessible units are provided with the most comprehensive accessibility requirement and are required to comply with those applicable provisions of Section 1102 of ICC A117.1. Type A and B units, regulated by ICC A117.1 Sections 1103 and 1104, respectively, not only provide a reasonable degree of accessibility, but also allow for the use of adaptive features. This section reflects how it is always acceptable to design and construct to a higher degree of accessibility than that required by the code.

1108.5 Group I. Dwelling units and sleeping units in a variety of Group I occupancies, including nursing homes, hospitals, nurseries, assisted-living facilities, group homes, and care facilities, are regulated for accessibility. In Group I-1, Condition 1 occupancies, at least 4 percent of the dwelling and sleeping units shall be Accessible units. Where the Group 1-1 facility is considered as a Condition 2 use, no less than 10 percent of the units must be Accessible units. In the Group I-2 category, however, the percentage of Accessible patient rooms varies based on the type of institutional facility. Hospitals and rehabilitation facilities that specialize in the treatment of conditions affecting mobility are required to have all patient rooms, including the toilet rooms and bathrooms, designed and constructed as Accessible units. General-purpose hospitals, psychiatric facilities, and detoxification facilities are required to have at least 10 percent of their patient rooms be Accessible units for their patient population. It is assumed that, in most cases, not more than 10 percent of the facility's patients would need accessible rooms. In nursing homes and long-term care facilities, at least 50 percent of the dwelling and sleeping units are required to be Accessible units. This increase in the percentage of accessible rooms recognizes that the care recipients of these facilities may be ambulatory on admission, but may become nonambulatory or have further mobility limitations during their stay. In Group I-3 facilities, at least 3 percent of the resident dwelling units and sleeping units must be Accessible units.

While most of the accessibility provisions are intended to allow a person to use the facility independently, the code now recognizes that in certain situations (Groups I-1, I-2 Condition 1, and in I-2 rehabilitation facilities) that this is not always possible and that the patients may be better served and need assistance with toileting and bathing. Therefore, exceptions will allow for changes to be made to the general Accessible unit provisions to allow for this assistance.

1108.6.1 Group R-1. Unless intended to be occupied as a residence, a Group R-1 occupancy is typically regulated for Accessible units only. The minimum required number of Accessible units is easily determined from Table 1108.6.1.1. For example, a hotel with 185 guestrooms must provide a minimum of eight guestrooms that comply as Accessible units. The number of buildings in which the guestrooms, referred to in the code as dwelling units or sleeping units, are located does not impact the result, provided each building does not contain more than 50 units. Where more than one building on the site contains guestrooms, the aggregate number of rooms is used to determine the minimum requirements. For larger buildings, those with more than 50 units, the minimum number of required Accessible units must be provided on an individual building basis by applying Table 1108.6.1.1 to that individual building.

Where multiple types of dwelling or sleeping units are provided, the Accessible units must be represented in each room type. However, it is not necessary to provide for additional Accessible units above and beyond the number required by Table 1108.6.1.1. As an example, a motel requiring two Accessible units and providing three room types (such as a double, king, and king suite) would only require two of the three room types to be Accessible units.

Although all of the Accessible units must be provided with accessible bathing facilities, only a portion are required to have roll-in showers. The table indicates the minimum number of Accessible units that must contain roll-in showers as described in Section 608 of ICC A117.1. The minimum required number of units provided with roll-in showers is complemented by requiring a minimum required number of Accessible units without roll-in showers. The intent is to provide persons with physical disabilities a range of options equivalent to those available to other persons served by the facility. If the standard rooms have bathtubs, then some of the Accessible units should also be provided with bathtubs, and a small percentage of rooms should incorporate roll-in showers. Likewise, if all of the standard rooms have shower compartments, then most of the Accessible units should have transfer showers, with again a small percentage being provided with roll-in showers. Where all of the units contain showers and none contain bathtubs, the total number of required Accessible units may be provided with showers.

1108.6.2 Group R-2. The provisions for Group R-2 occupancies are divided into three general categories, those applicable to: live/work units; typical apartment buildings; and dormitories, fraternity houses, and sorority houses. Live/work units are only required to provide Type B units, and only in specific cases. Apartment houses, along with monasteries and convents, are regulated for Type A and B units. On the other hand, dormitories and similar congregate living facilities do not need to contain Type A units, but do require one or more Accessible units.

The mandate for Type A units in apartment buildings applies where 21 or more dwelling units or sleeping units are contained within the building. If two or more buildings are located on the same site, the aggregate number of units is used to determine the minimum required number of Type A units. Assuming a site contains four apartment buildings, each containing 30 units, the provisions are based on the sum total of 120 units. Using the 2-percent rule, at least three of the dwelling units are required to be designed and constructed as Type A units. Assuming that throughout the four buildings three types of units (studio, one bedroom, and two bedroom) are represented, a minimum of one unit of each type shall be a Type A unit.

All three Type A units are permitted in the same building, and all may be located on the same floor level, which would typically be at grade. Those units that are not required to be Type A must be designed and constructed as Type B units unless exempted by the provisions of Section 1108.7.

In dormitories, sorority houses, fraternity houses, and boarding houses, Accessible units are required in the same manner as Group R-1 occupancies. At least one of the dwelling or sleeping units within the building must be an Accessible unit, with additional Accessible units as required by Table 1108.6.1.1. Where multiple Type A units are provided, at least one must include a bathroom with a complying shower. As with other Group R-2 occupancies, the remaining units must be designed and constructed as Type B units unless allowed to be reduced by Section 1108.7.

1108.7 General exceptions. The required number of Type A and B dwelling units and sleeping units required in Section 1108 may be reduced under the provisions of this section. However, there is no provision allowing the reduction or elimination of Accessible units mandated by Sections 1108.5 and 1108.6. Five exceptions to these requirements are included in this section. The first four exceptions relate to buildings with limited or no elevator service and are relatively self-explanatory as worded in the code. The fifth exception relates to buildings that are required to have raised floors at the primary entrances as measured from the elevations of the vehicular and pedestrian arrival points in order to accommodate the minimum required lowest floor elevation. Where no such arrival points are within 50 feet (15,240 mm) of the primary entrance, the closest arrival points shall be used.

As noted, most of the exceptions are only applicable to buildings with no elevator service. It is expected that once an elevator is installed to access floors above the grade level, the additional measures to make the units more accessible are appropriate. However, the provisions do not mandate the installation of an elevator simply to make the upper floors accessible. Several of the applications of this section are shown in Figures 1108-1 through 1108-3.

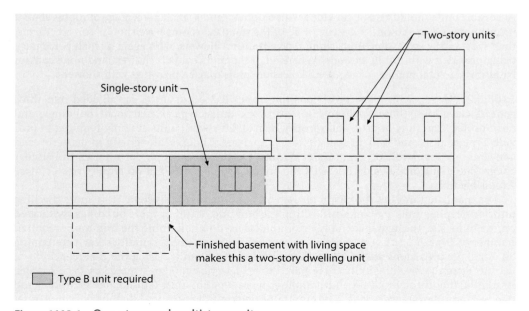

Figure 1108-1 **One-story and multistory units.**

Dwelling Units and Sleeping Units

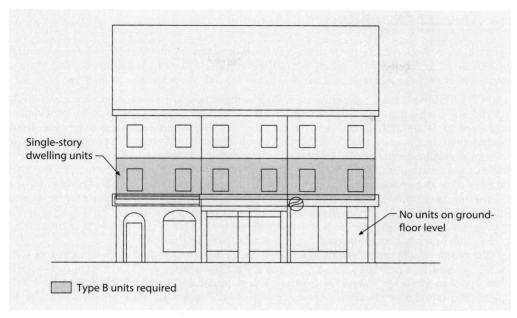

Figure 1108-2 **One story of Type B units required.**

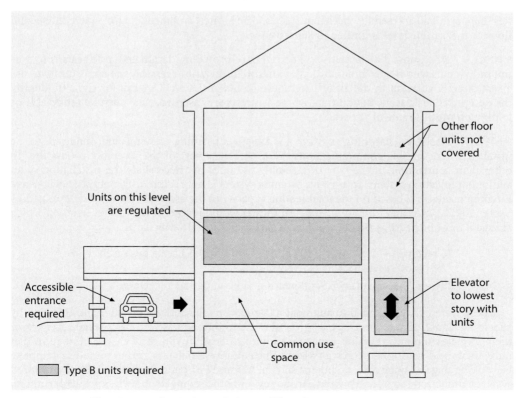

Figure 1108-3 **Elevator service to lowest story with units.**

Section 1109 *Special Occupancies*

In addition to those requirements that apply to all occupancies in a general fashion, a number of provisions are specific to certain occupancy groups. Assembly, institutional, and storage occupancies are uniquely regulated because of the special characteristics of their use.

1109.2 Assembly area seating. In stadiums, theaters, auditoriums, and similar assembly seating areas with fixed seats, the number of accessible wheelchair spaces to be provided is addressed in Table 1109.2.2.1. Unlike parking requirements where each lot is to be considered separately in the calculation of the minimum required number of accessible parking spaces, the minimum required number of accessible wheelchair spaces for grandstands and bleachers serving a single function should be based on the aggregate number of seats. For example, the total number of seats provided in a high school football facility would be the basis for determining the minimum required number of wheelchair spaces even though the bleachers may be located on opposite sides of the playing field. However, the accessible spaces must be dispersed in an appropriate manner to accommodate individuals on both sides of the field. Where the seating is in a suite or a box area, separate accessible seating may be required.

The requirements for accessible wheelchair locations can be found in Section 802 of ICC A117.1, including a provision mandating that at least one seat be provided beside each wheelchair space to allow for companion seating. The required dispersion of accessible wheelchair spaces is comprehensively addressed in the ICC A117.1 standard. Such spaces should be an integral feature of any seating plan to provide individuals with physical disabilities with a choice of admission prices and a line of sight comparable to that provided to the general public. There are two exceptions to the general rule for the location of wheelchair spaces in multilevel facilities that permit all accessible wheelchair spaces to be located on the main level where the second floor or mezzanine level is limited in capacity.

1109.2.5 Designated aisle seats. The intent of providing designated aisle seats is to permit individuals who might find it difficult to negotiate an aisle accessway the opportunity to use a seat directly adjacent to the aisle. The requirements of ICC A117.1 call for signs to identify the designated aisle seats. In addition, where armrests are installed, they must be retractable or folding on the aisle side of the seat.

1109.2.7 Assistive listening systems. Assistive listening systems are required to be installed where audible communications are a necessary part of the assembly room's use. In other than courtrooms, if the type of assembly use does not necessitate the installation of an audio-amplification system, an assistive listening system is not required. The number of assistive listening receivers is based on the total seating capacity of the assembly area and determined by Table 1109.2.7.1. As an example, an 800-seat theater would require at least 30 receivers, of which at least 8 must be hearing-aid compatible, as shown in the following calculation:

20 receivers + 1 receiver [(800 seats − 500 seats)/33 seats per receiver] = 29.1 = 30 receivers.

25% of 30 receivers = 7.5 = 8 hearing-aid compatible receivers.

These systems are intended to augment a standard public address or other audio system by providing signals that are free of background noise to individuals who use special receivers or their own hearing aids. Further provisions found in Section 706 of ICC A117.1 require that individual fixed seating served by an assistive listening system be located to provide a complete view of the stage, playing area, or cinema screen. AM and FM radio-frequency systems, infrared systems, induction loops, hard-wired earphones, and other equivalent devices are all permitted.

1109.2.9 Dining and drinking areas. The general rule for accessibility in dining rooms is that the total floor area be accessible. This requirement includes both interior dining/drinking spaces and those that are outdoors. The second exception to this section eliminates the requirement for an elevator or ramp system to a mezzanine seating area in a dining room under strict conditions. Caution should be exercised when determining the same services mentioned in this exception. It is generally accepted that the same services not only address actual food and drink service but also include decor, views, and ambiance and it is equally important that a specific accessible space or area not be set aside and restricted for use only by individuals with disabilities.

For spaces at accessible fixed tables or counters, at least one table in the facility shall be provided for wheelchairs. When more than one is required by this section, then they shall be equally distributed around the facility so as not to isolate an accessible area from the rest of the establishment.

1109.3 Self-service storage facilities. The number of individual storage spaces that must be accessible at self-service storage facilities is determined from Table 1109.3. Assuming a total of 280 spaces in the facility, at least 12 accessible storage spaces must be provided as shown:

10 spaces + [2% (280 spaces −200 spaces)] = 11.6 = 12 storage spaces.

Where different sizes or classes of storage space are available, the individual accessible spaces shall be appropriately dispersed among the sizes or classes available, but only up to the total number of accessible spaces that are required. All accessible storage spaces complying with this section are permitted to be located in a single building.

1109.4 Judicial facilities. Courtrooms, holding cells, and visitation areas of judicial facilities are required to be accessible to the degree mandated by this section. Courtrooms typically have elements unique to their use, such as witness stands, jury boxes, judge's benches, and similar areas. Provisions are in place to provide for a limited degree of accessibility to such areas. The spectator area of a courtroom is regulated under the provisions for assembly seating. Workstations are regulated differently based on their use. Employee workstations, such as the judge's bench, bailiff's station, and court reporter's station, need to be on an accessible route, but the portion of the route leading to such elevated work areas is not required at the time of construction. It is only necessary to provide adequate space and support such that a complying ramp, platform lift, or elevator can be installed in the future without extensive reconstruction. Workstations for other than employees, such as the litigant's station, counsel station, and lectern, must provide full accessibility.

Section 1110 *Other Features and Facilities*

This section provides scoping provisions to ensure that certain elements or areas within buildings, where specific services are provided or activities are performed, are made accessible. Certain components in ICC A117.1 or other accessibility regulations have not been included in this section. Items such as telephones, automatic-teller machines, and fare machines are referenced in Appendix E, insofar as these are not typically considered items regulated by the building code.

1110.2 Toilet and bathing facilities. As a general rule, all toilet rooms and bathing facilities shall be accessible. The term *accessible* applies to the doors, fixtures, clear floor space, operable parts, towel dispensers, and mirrors, among other elements. Typically, the main components of toilet facilities are the water closet, the toilet stall, and the lavatory; and the main accessibility issues are the clear space, door swing, transfer capability, height of fixtures, grab bars, and controls. In bathing facilities, a number of items related to the shower or bathtub and its location, controls, and grab bars need to be considered. The ICC A117.1 standard contains details and many illustrations that depict these requirements.

A number of exceptions revise, reduce, or eliminate the requirements for toilet rooms and bathing facilities under specific conditions. Within the facility, at least one of each type of fixture, element, control, or dispenser must be accessible. There is an allowance for a nonaccessible urinal where only a single urinal is located within a toilet room.

Exception 9 permits toilet facilities to be designed using the children's size provisions of the A117.1 standard and to still be considered as being accessible. In facilities or portions of buildings that are primarily designed for children's use, the general accessibility requirements may result in the elements not being usable by the major portion of the occupants. This exception provides the "scoping" requirement that will allow designers to use the children's height requirements of the A117.1 standard when determining the accessibility requirements for areas that are primarily for children's use. The A117.1 standard defines *children's use* as "spaces and elements specifically designed for use primarily by people 12 years old and younger." Examples of where the adults are in the minority and the space is "primarily for children's use" include most areas of an elementary school, preschool, or kindergarten; a children's library; or a children's museum. Adult-dimensioned fixtures should be provided in other areas or spaces where there is a mix of all ages or where the space serves staff, parents, older students, or the general public. Therefore, if a restroom is provided in the staff area of an elementary school, that toilet room should meet the adult requirements and not attempt to use the reduced-size children's provisions in that location. So even though the code permits children-sized accessible elements to substitute for accessible adult-sized provisions, the intent is to clearly limit the application to the areas that the children occupy and the fixtures that they would be using.

1110.2.1 Family or assisted-use toilet and bathing rooms. Specific provisions are provided in this section on accessibility requirements for family or assisted-use bathing and toilet rooms. Such bathing and toilet rooms are required to comply with this section and ICC A117.1. The primary issue relative to family or assisted-use toilet/bathing facilities is that some people with disabilities may require the assistance of persons of the opposite sex and, therefore, require a toilet or bathing facility that accommodates both persons.

Bathing facility requirements for recreational facilities require that an accessible family or assisted-use bathing room be provided where separate-sex bathing facilities are provided. If each separate-sex bathing facility has only one shower fixture, family or assisted-use bathing facilities will not be required. Accessible family or assisted-use toilet room requirements for Group A and M occupancies are mandated by this section where an aggregate of six or more male and female water closets are required. These occupancies generally have high occupant loads with a minimum stay of approximately one hour for the occupants; thus, a high probability exists that there will be occupants who will need the use of such facilities. In determining the total number of fixtures required in a building as mandated by Chapter 29, those fixtures located in family or assisted-use facilities may be included in the total fixture count.

Family or assisted-use bathing rooms are to be provided with a water closet and lavatory in addition to the single shower or bathtub fixture. Family or assisted-use toilet rooms are typically limited to a single water closet and/or lavatory. However, it is permissible to include a urinal, a child-height water closet, a child-height lavatory, and an adult changing/bathing station as additional fixtures. A complying family or assisted-use bathing room may be considered the family or assisted-use toilet room. In order to provide the appropriate privacy for a family or assisted-use facility, doors to a family or assisted-use toilet room or bathing room must be capable of being secured from the inside.

Family or assisted-use toilet and bathing rooms shall be located on an accessible route. Family or assisted-use toilet rooms are to be located not more than one story above or below separate-sex toilet facilities. The accessible route from any separate-sex toilet room to a family or assisted-use toilet room must not exceed 500 feet (152,400 mm). Additionally, in passenger transportation facilities and airports, the accessible route from separate-sex toilet facilities to a family or assisted-use toilet room shall not pass through security checkpoints. The restriction regarding crossing through security checkpoints at airports and similar facilities is intended to eliminate any potential delays that may cause missed flights and/or connections.

1110.2.2 Water closets designed for assisted toileting. As was discussed with Section 1108.5, some Accessible units within certain Group I occupancies can be modified from the general Accessible unit provisions to allow for assisted toileting and bathing. Sections 1010.2.2 and 1010.2.3 address these modifications. Because greater clearances are required around the water closet to allow staff to maneuver beside the person they are assisting, changes are needed since the side wall and grab bars would not be located within the typical reach range. Having additional space both beside and in front of the water closet provides the space needed for the staff to assist whether physically standing next to them or using a lift to maneuver them. Because of the greater clearance width and the fact that fixed grab bars could obstruct the assistance, the use of swing-up grab bars is permitted under specific limitations. Typically swing-up grab bars are only allowed in a Type B unit and not in an Accessible unit or any public restroom.

1110.2.3 Standard roll-in type shower compartment designed for assisted bathing. As scoped by Section 1108.5, these showers for assisted bathing will only be used in a certain percentage of Accessible units where the residents are typically incapable of unassisted bathing. The primary change this section makes from the normal requirements in an Accessible unit is the elimination of the requirement for a fixed or folding seat within the shower. In facilities where bathing is assisted, the residents typically would be using a rolling shower chair and not transferring to the seat, and the seat tends to obstruct space needed for the staff person assisting the resident with their bathing. Due to the elimination of the seat, an additional grab bar may be required on what was typically the "seat wall." However, where the shower size exceeds the minimum sizes, then some of the grab bars can be shortened or eliminated.

1110.2.4 Water closet compartment. At least one wheelchair-accessible compartment shall be provided in all toilet rooms or bathing facilities where compartments are installed. Also, an ambulatory-accessible water closet compartment must be provided in addition to the wheelchair-accessible compartment in those toilet rooms having an aggregate total of six or more water closet compartments and urinals. The ambulatory-accessible compartment benefits those individuals who, although not wheelchair users, have physical limitations or impairments that make it difficult to use other types of toilet compartments. Both wheelchair-accessible and ambulatory-accessible compartments are to be in compliance with ICC A117.1. It should be noted that the threshold for this provision is based on a room-by-room evaluation, not the aggregate number of fixtures throughout the facility. In those unusual situations where a toilet room or bathing room contains more than 20 water closet compartments, one or more additional wheelchair accessible compartments will be required based on the 5-percent requirement. Similarly, where the number of water closets and urinals provided exceeds 20, one or more additional ambulatory-accessible compartments are mandated. Proportionate accessibility is provided based on the recognition that it is likely that more than one user within such a large toilet or bathing room will be using a wheelchair, crutches walker, or a scooter. Therefore, it is appropriate to provide additional wheelchair compartments and ambulatory-accessible compartments that allow improved access and support options.

1110.2.5 Lavatories. The provisions specific to lavatories are a subsection to the other provisions for toilet and bathing facilities. The requirement for a minimum of 5 percent of the lavatories to be accessible results in a single mandated accessible lavatory in almost every toilet room and bathing room. Only where more than 20 lavatories are provided additional accessible lavatories must be provided.

Additional accessibility features are required where the total number of lavatories in a toilet room or bathing room exceeds five. Where six or more lavatories are provided, a minimum of one lavatory must be provided with enhanced reach ranges. It is permissible for the lavatory with the enhanced reach range to serve as the required accessible lavatory. The technical requirements for such lavatories with enhanced reach ranges are found in ICC A117.1 Section 606.5. These types of lavatories are usable by individuals who have a limited obstructed reach depth. Such individuals can often only reach faucets and soap dispensers up to a reach depth of 11 inches (279 mm) in lavatories with a height of 34 inches (864 mm). The maximum 11-inch (279-mm) depth is possible by locating the faucet controls to the side of the bowl while leaving the spout toward the back, mounting the faucet on a sidewall, installing the faucet on the side of the bowl, or other potential locations that will provide the necessary access.

1110.4 Adult changing stations. Provisions establish both where adult changing stations are required and where they must be located. They further mandate that changing stations that are voluntarily provided must also comply with the same location and safety features. Because the size and weight of an adult will far exceed the limits of infant and toddler changing tables, the tables must be secure and safe for a user.

Section 1110.4.1 addresses the four situations where adult changing stations are required. Item 1 addresses larger assembly and mercantile occupancies where greater occupant loads are found and where the odds of an adult user needing such facilities is greater. This may include stadiums, theaters, shopping centers, etc. and is consistent with when family and assisted-use toileting and bathing facilities are required.

Items 2 and 3 address educational uses but in a slightly different fashion depending on the occupancy served. Item 2 addresses adult education ("above the 12th grade") and requires that areas serving classrooms and lecture halls with larger occupant loads are covered. This would not include other portions of the college or university such as administrative office areas. By including "lecture halls" it incorporates larger assembly rooms that may be located in a Group B occupancy educational building.

Item 3 covers lower-aged educational occupancies (up "through the 12th grade") but it only regulates larger assembly-type spaces. This item does not factor in the general classroom occupants but just those areas "used for assembly purposes." Since IBC Section 303.1.3 allows assembly spaces "associated with Group E occupancies" to be classified as a part of the Group E use, as opposed to being separately classified as a Group A occupancy, this item uses a similar threshold from item 1 to provide these large assembly areas in the school with changing stations.

Item 4 is somewhat unique and not something that may typically be thought of as being a code issue. This provision is only intended to address rest stops and service plazas that are located along highways. It is not intended to regulate a separate/private business that may be located near a highway that primarily serves people who are traveling. Rather, it is intended to regulate those types of uses that can only be accessed from the highway and not by other means.

Sections 1110.4.2 through 1110.4.4 regulate the location and type of space where adult changing stations are to be provided. These requirements are generally comparable to those for family or assisted-use toilet and bathing rooms which are found in Section 1110.2.1. The primary difference is found within Section 1110.4.4 where the travel distances are increased. This section will allow adult changing stations to be located one additional story and an extended distance beyond what Section 1110.2.1.4 requires for family or assisted-use toilet and bathing rooms. This travel distance increase helps limit the number of these larger adult changing stations that will need to be installed while still ensuring that they are provided within facilities where their need is anticipated.

Introduced and approved as supplement material, Section 613 in the ICC A117.1 standard addresses the technical requirements for adult changing stations.

1110.7 Drinking fountains. This section ensures that all of the drinking fountains provided are made accessible. This is another of the many provisions where it is important to focus on the word *provided* and note that the provision does not require drinking fountains to be installed.

Drinking fountains must be accessible based on use from a wheelchair while still providing access to people with a limited ability to bend or stoop. The technical provisions set forth in Section 602 of ICC A117.1 address drinking fountains designed for wheelchair access, as well as those intended for use by standing persons. Other than the requirement for clear floor space, the provisions of Section 602 are applicable to both heights of fountains. Where a single water fountain with two separate spouts (combination *hi-lo* unit) can comply with the requirement for both user groups, it is permitted to be used in lieu of two individual fountains.

The minimum required number of drinking fountains is established by the *International Plumbing Code®* (IPC®). At times, drinking fountains are installed even though they are not required by the code. It is also not uncommon for the actual number of fountains provided to exceed the minimum required. Under such conditions, one-half of the number provided must comply with the provisions of ICC A117.1 for wheelchair users, and the remainder shall accommodate standing persons. If an odd number of fountains are installed, an additional drinking fountain is not required in order to meet the 50-percent criteria. The remaining water fountain after the 50/50 split is permitted to be of either type. For example, if three drinking fountains are provided on the first floor of a building, at least one, but not more than two, is to be wheelchair accessible.

Exception 2 to Sections 1110.5.1 and 1110.5.2 addresses drinking fountains used primarily by children. The current height of 38 to 43 inches (965 mm to 1,090 mm) specified in the A117.1 for standing-height drinking fountains is too high for small children. The exceptions override the standard and will allow the height to be reduced to 30 inches (762 mm).

Where bottle-filling stations are provided, either separately or in conjunction with drinking fountains, IBC Section 1110.6 will require they be accessible. The technical requirements for the bottle-filling stations are found in A117.1 Section 602 with the drinking fountain provisions.

1110.11 Lifts. The provisions of this section address the limited acceptance of platform (wheelchair) lifts as a portion of an accessible route. To use platform lifts in lieu of an elevator or ramp, they must comply with Section 410 of ICC A117.1 and be installed in one of the 10 listed locations. A companion provision, Section 1009.5, allows platform lifts to be used as an accessible means of egress for the first nine locations established in this section. However, the platform lift cannot be used as a portion of an accessible means of egress where the lift is used along the accessible route because of existing exterior site constraints as described in Item 10. Because of the slow operation of a platform lift, it is considered inappropriate for egress purposes where there is the potential for a large number of lift users.

1110.13 Detectable warnings. Detectable warnings must be provided on edges of passenger transit platforms where a drop-off occurs, unless platform screens or guards are present. Not applicable to bus stops, detectable warnings should be standardized to assist in the universal recognition and reaction of persons using these platforms. Detectable and tactile warning devices also serve guide dogs and should contrast visually with the adjoining surfaces in a light-on-dark or dark-on-light application.

1110.18 Controls, operating mechanisms, and hardware. To be able to operate controls, operating mechanisms, and hardware, a clear floor space and reach ranges for either forward or parallel approaches must be provided. The A117.1 standard addresses windows in Section 506 and then also within the Accessible and Type A unit provisions of Sections 1102 and 1103, respectively.

Section 1111 *Recreational Facilities*

1111.2 Facilities serving Group R-2, R-3, and R-4 occupancies. Recreational facilities serving Accessible units in Group R-2, R-3, and R-4 occupancies shall all be accessible as applicable. Where recreational facilities are provided serving Type A or B units in Group R-2, R-3, or R-4 occupancies, at least 25 percent but not less than one of each type of such facilities shall be accessible. All recreational facilities of each type on a site shall be considered to determine the total number of each type that is required to be accessible. This requirement recognizes that not all recreational facilities need to be accessible, nor would such a requirement be feasible. However, at least one of every four recreational facilities shall be available to someone with a disability. Where multiple residential buildings are on a site, each type of recreational facility, such as a basketball court, a handball court, a weight room, an exercise area, a game room, or a television room serving each building, must be fully accessible. Each type of recreational facility on a site, such as a racquetball court in an apartment complex, is considered to determine the total number required to be accessible. Thus, if there were 12 racquetball courts on the site, at least 3 of the courts would need to be accessible. The distribution of the accessible recreation facilities will also depend on the building and site they serve. Where the accessible dwelling units are in multiple buildings which may each have separate recreation facilities, then at least 25 percent but not less than one of each type serving each building must be accessible.

1111.4 Recreational facilities. The overall intent of these requirements is to provide access to recreational facilities so that persons with mobility impairments can participate to the best of their ability. The extent of the provisions is not intended to change any essential aspects of that specific recreational activity. The requirements are generally intended to coordinate with those of the ADA Standards for Accessible Design.

Section 1112 *Signage*

This section lists those required accessible elements and locations that must be identified with appropriate signage. Elements such as accessible parking spaces and loading zones, accessible areas of refuge, accessible dressing rooms, and toilet and bathing rooms that are accessible where some are not shall be provided with the International Symbol of Accessibility, as shown in Figure 1112-1. Directional signage, including the International Symbol of Accessibility, is to be used to indicate the nearest route to a like accessible element. Directional signage must be provided at inaccessible building entrances, inaccessible public toilets and bathing facilities, elevators not serving an accessible route, separate-sex toilet and bathing facilities (to indicate the location of the nearest family or assisted-use facility), and exits and exit stairways serving a space required to be accessible but not providing an accessible means of egress. Additional signage will be needed where assistive-listening systems are available for assembly occupancies: at doors to egress stairways, exit passageways, and exit discharge; at areas of refuge; at areas of rescue assistance; and at two-way communication systems. Where signage identifying rooms is provided, it shall include visual and raised characters as well as braille unless exempted. Appendix Chapter E contains additional signage provisions such as those for directional signs. The appendix signage provisions are intended to mirror the federal law, but are only enforceable

Figure 1112-1 **International symbol of accessibility.**

under the building code if the appendix chapter is adopted. Similar to the general room designations of Section 1112.6, signs identifying toilet or bathing rooms are required to have visual characters, raised characters and braille. Pictograms may also be used on the sign. The technical details regarding signage can be found in Section 703 of the ICC A117.1 standard.

KEY POINTS

- Chapter 11 of the IBC provides the scoping provisions for accessible spaces and elements within buildings.
- ICC A117.1-2017 is the referenced standard for regulating the facilities and elements of buildings that are required to be accessible by IBC Chapter 11 and is used as the basis for discussions in this chapter of the handbook.
- Virtually all buildings, both temporary and permanent, must be designed and constructed for accessibility and usability.
- Accessible routes must be provided to connect all accessible elements within a building, as well as all accessible elements on the site.
- At least 60 percent of all public building entrances must be accessible.
- The number of accessible parking spaces required on a site varies by the total number of spaces provided.
- All occupancies require some degree of accessibility, with special provisions for assembly, institutional, residential, and utility uses.
- Assembly areas are further regulated for wheelchair spaces and assistive-listening systems.
- Both Type A and B dwelling units may be required in a large apartment building.
- A variety of building elements are regulated as facilities required to be accessible, including toilet rooms, bathing facilities, drinking fountains, elevators, stairs, platform lifts, fixed or built-in seating, storage, customer-service facilities, controls, operating mechanisms, and alarms.

- Under specific conditions, accessible family or assisted-use bathing rooms and toilet rooms must be provided in addition to the accessible separate-sex facilities.
- Varying degrees of access to a variety of recreational facilities are mandated to allow for the expanded use of such facilities.
- Appropriate signage is mandated in order to properly identify the various accessible elements.

CHAPTER 12

INTERIOR ENVIRONMENT

Section 1202 Ventilation
Section 1203 Temperature Control
Section 1204 Lighting
Section 1205 Yards or Courts
Section 1206 Sound Transmission
Section 1207 Enhanced Classroom Acoustics
Section 1208 Interior Space Dimensions
Section 1209 Access to Unoccupied Spaces
Section 1210 Toilet and Bathroom Requirements
Key Points

Chapter 12 ■ Interior Environment

> This chapter is designed to address those issues related to the interior environment aspects of a building's use, such as ventilation, lighting, temperature control, yards and courts, sound transmission, classroom acoustics, and room dimensions.

Section 1202 *Ventilation*

Ventilation in buildings is regulated based on the ventilating method used. This section addresses the use of natural ventilation, whereas the use of mechanical ventilation is regulated by the *International Mechanical Code®* (IMC®).

1202.2.1 Ventilated attics and rafter spaces. During cold weather, condensation is deposited on cold surfaces when, for example, warm, moist air rising from the interior of the building and through the attic comes in contact with the roof deck. This alternative wetting and drying that is due to condensation creates dry rot in the wood, and preventive measures are required. In attic areas of noncombustible construction, it is also important to ventilate the area, particularly in light-gauge steel construction. Therefore, enclosed attics and enclosed rafter spaces formed where ceilings are applied directly to the underside of roof-framing members, such as in cathedral ceiling applications, are to have cross ventilation for each separate space. Ventilation of the attic prevents moisture condensation on the cold surfaces and, therefore, will prevent dry rot on the bottom surfaces of shingles or wood roof decks. Figure 1202-1 provides three compliant examples of attic ventilation. In the areas where the moisture condensation is a particular problem or where the normal requirement for attic ventilation cannot be provided, ventilation of the attic by mechanical exhaust fans may be required. Exhaust fans are particularly beneficial in all cases because of the extra movement of the air provided.

The method and arrangement of providing ventilated openings is an important aspect in the proper ventilation of attic spaces. It is critical that any such openings be protected against the entrance of rain and snow. In addition, blocking and bridging that is installed must be located so as to not interfere with the movement of air. At least 1 inch (25 mm) of air space must be provided between the insulation and the roof sheathing, as shown in Figure 1202-2. The net-free ventilating area must be at least 1/150 of the area of the space ventilated. Two exceptions permit a reduction in the amount of ventilating area to 1/300 of the attic area being ventilated, provided specific criteria are met. Exception 2 is illustrated in Figure 1202-3.

Something often overlooked when sizing attic vents is that the code requires that the area provided be the net free area. The net free area can be as much as 50 percent less than the gross area. For example, one manufacturer's 24-inch (610-mm) square gable vent [gross area equals 576 square inches (0.37 m^2)] is listed in their catalog as having a net free area of 308 square inches (0.20 m^2), which is about 53 percent of the gross area. The manufacturer's literature for the specific vents being used needs to be consulted in order to obtain accurate free area information.

1202.2.2 Openings into attic. Exterior ventilation openings are required by the code to be screened in order to prevent entry of birds, squirrels, rodents, and other similar creatures. A mesh size between 1/16 inch (1.6 mm) and 1/4 inch (6.4 mm) is required to address the problems of both smaller openings being blocked by debris and spider webs, and larger openings permitting access to small rodents. In addition to the use of corrosion-resistant-wire cloth screening, it is also permissible to use hardware cloth, perforated vinyl, or any other similar material that will prevent unwanted entry. A cross-reference is also provided to IMC Chapter 7 to remind users that there are special requirements where combustion air is obtained from the attic area.

Ventilation

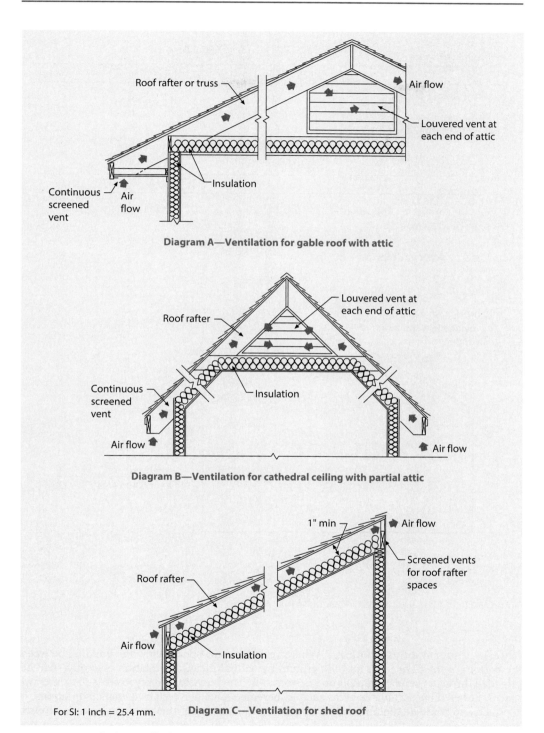

Figure 1202-1 Attic ventilation.

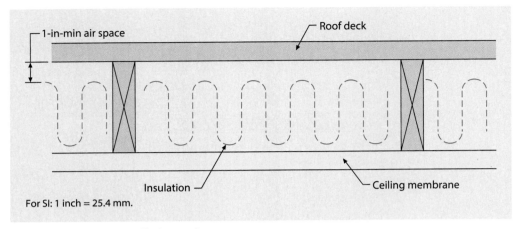

Figure 1202-2 Attic ventilation—air space.

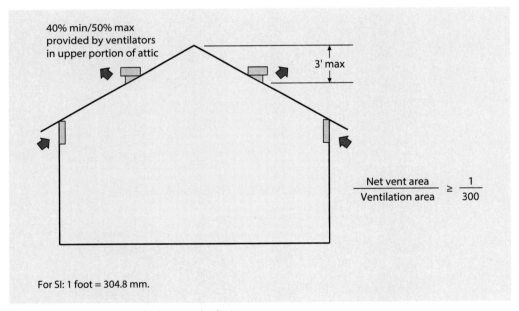

Figure 1202-3 Attic ventilation—calculations.

1202.4 Under-floor ventilation. Where ventilating the space below the building between the bottom of the floor joists and the ground by natural means, ventilation openings shall be provided through foundation walls or exterior walls. The provisions apply to areas such as crawl spaces, rather than occupiable areas such as basements. Under certain climatic conditions, it is possible to ventilate the under-floor space into the interior of the building, or continuously operated mechanical ventilation may be provided in lieu of ventilation openings where the ground surface is covered with an approved vapor retarder.

To determine the minimum net area of ventilation openings required, at least 1 square foot of opening area shall be provided for each 150 square feet (0.67 m² for each 100 m²) of crawl space area where the crawl space has an uncovered earth floor. See Figure 1202-4. A significant reduction in the minimum required ventilating area is permitted where the ground surface is covered with a Class I vapor retarder. In this case, the total area of ventilation openings need not exceed 1/1,500 of the under-floor area. In both cases, it is important that adequate cross-ventilation be provided in the under-floor space.

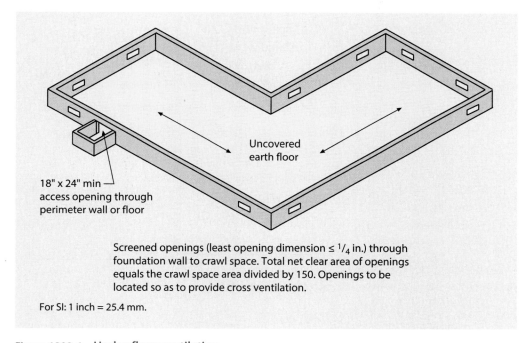

Figure 1202-4   **Under-floor ventilation.**

It is critical that ventilation openings be completely covered with a substantial material to prevent the entrance of insects and animals. Corrosion-resistant wire mesh, with the least dimension not exceeding 1/8 inch (3.2 mm), is one of the six materials identified by the code to address this concern.

Where the under-floor space is conditioned, it is unnecessary to provide ventilation openings. It has been shown that by insulating the perimeter walls, covering the ground surface with a Class I vapor barrier, and conditioning the space in accordance with the *International Energy Conservation Code®* (IECC®), unvented crawl spaces outperform vented under-floor areas.

1202.5 Natural ventilation.   Where buildings are not provided with adequate mechanical ventilation as specified in the IMC, natural ventilation through openings directly to the exterior must be provided. In order to determine the amount of ventilation air required, the minimum required openable area to the outdoors is based on 4 percent of the floor area being ventilated.

In those cases where rooms or spaces do not have direct openings to the exterior, it is still necessary to ventilate the interior space, often through an adjoining room. In this case, the opening to the adjoining room should be unobstructed and have an area not less than 8 percent of the floor area of the interior room or space. In no case should the opening between the rooms be less than 25 square feet (2.3 m²). Where the intervening room is a thermally isolated sunroom or patio cover, the minimum openable area of 8 percent is still applicable; however, the opening need only be at least 20 square feet (1.86 m²). See Figure 1202-5.

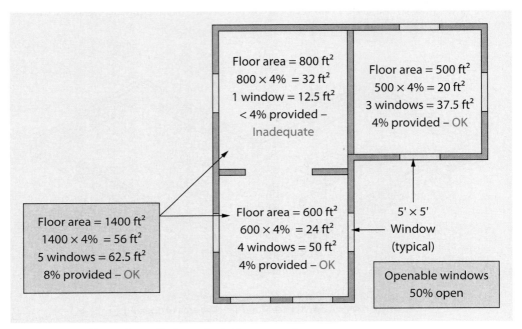

Figure 1202-5 **Natural ventilation calculation example.**

As previously mentioned, in calculating the total openable area to the outdoors, such opening area shall not be less than 4 percent of the total floor area being ventilated. In those conditions where openings that provide the natural ventilation are located below grade, the outside horizontal clear space measured perpendicular to the opening is required to be at least one- and one-half times the depth of the opening.

Where rooms contain bathtubs, showers, spas, and similar bathing fixtures, natural ventilation is not an acceptable method. Because of the common reluctance to open exterior windows, particularly in cold-weather conditions, a mechanical system provides for more consistent ventilation. Therefore, bathrooms and similar spaces must be mechanically ventilated in accordance with the IMC. Where flammable and combustible hazards or other contaminant sources are present within an interior space, ventilation-exhaust systems shall be provided as required by the IMC and the *International Fire Code*® (IFC®).

Section 1203 *Temperature Control*

For those interior spaces where the primary purpose is associated with human comfort, it is important that a minimum indoor temperature of 68°F (20°C) can be maintained, measured at 3 feet (914 mm) above the floor on the design heating day. Although the code does not require that this temperature be constantly provided, it does mandate that such interior spaces be provided with equipment or systems having the capability of maintaining the desired temperature.

The exceptions recognize that there are conditions and occupancies where a mandate for space-heating systems is unnecessary. Environmental comfort requirements for Group F, H, and S occupancies are governed by OSHA regulations. In Group U occupancies, such space heating is typically not necessary, and in Group H occupancies it is often undesired.

Section 1204 *Lighting*

Almost every occupancy requires some level of lighting that is due to its use. Means of egress illumination is also required by Section 1008. In spite of the obvious need for interior light as a necessary part of a building's function, the code mandates that some degree of lighting, whether artificial or natural, be provided to every occupiable space.

1204.2 Natural light. Where glazing to the exterior is used as the method for providing natural light, the exterior openings shall open directly onto a public way, yard, or court in compliance with Section 1205. Exterior wall openings used to provide natural light must have an area that is computed based on the net glazed area for windows and doors, and not the nominal size of the opening. The minimum net glazed area shall be not less than 8 percent of the floor area of the room served. Where a room is not located on an exterior wall, the provisions of this section permit the borrowing of light from an adjoining room. Figure 1204-1 illustrates the requirements for this condition.

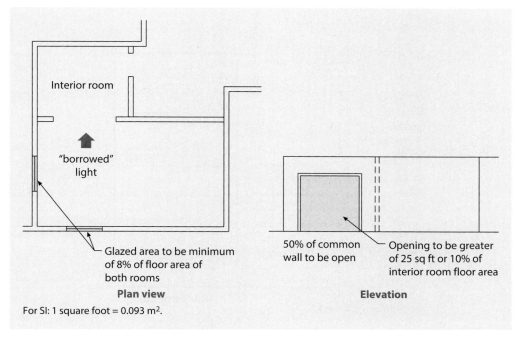

Figure 1204-1 **Borrowing natural light.**

Section 1205 *Yards or Courts*

The *International Building Code®* (IBC®) contains provisions for yards and courts where they are used to provide the required light and ventilation to exterior openings in the building. Most modern-day zoning ordinances also have requirements for yards and courts, and quite often these are more than adequate to gain the lighting and ventilation required by the code. In addition, where the alternatives of artificial light and mechanical ventilation as provided by Sections 1204.3 and 1202.1, respectively, are used in lieu of natural light and ventilation, the provisions of Section 1205 are not applicable.

To be considered providing adequate natural light and ventilation, each yard or court must have a minimum width of 3 feet (914 mm). In addition, these yards and courts must be increased in width, depending on the height of the building. See Figure 1205-1. As a building's height increases, it is important to have an increased yard or court width so that light will be able to reach the lower stories of the building, and this is only possible where the width of the yard or court is in proper relationship to the height of the building. The requirements in the code are an obvious compromise between optimum light and ventilation at the bottom of a building and the need to build as much building area on the lot as possible for economic purposes. For buildings exceeding 14 stories above grade plane, the required dimensions of the yard or court are calculated on the basis of 14 stories above grade plane.

Inner courts that are enclosed by the walls of the buildings, sometimes referred to as light wells, obviously need some means to remove accumulated trash at the bottom, provide for adequate drainage, and provide for circulation of air for ventilation purposes. In keeping with this intent, the code requires that an air intake be provided at the bottom of courts for buildings more than two stories in height, that grading and drainage be addressed, and that all courts be provided with access for cleaning.

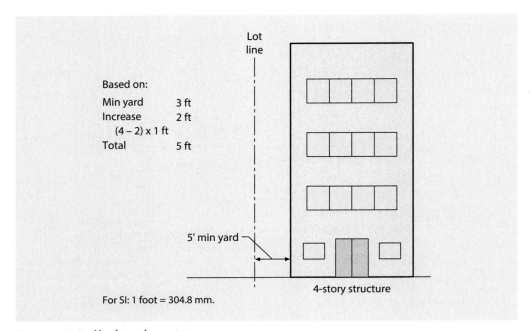

Figure 1205-1 **Yards and courts.**

Section 1206 *Sound Transmission*

Applicable only to buildings containing dwelling units and sleeping units, this section is intended to provide regulations addressing sound transmission control. It pertains to wall and floor/ceiling assemblies separating dwelling units from each other and from public space, such as interior corridors, stairs, or service areas. These separation elements must be provided with airborne-sound insulation for the walls and both airborne- and impact-sound insulation for the floor/ceiling assemblies.

For airborne-sound insulation, the separating walls and floor/ceiling assemblies must be provided with a sound transmission class (STC) of not less than 50 (minimum of 45 NNIC in accordance with ASTM E336 when field tested) as defined by ASTM E90. As an alternative for concrete masonry and clay masonry assemblies, the STC may be calculated per TMS 302. Penetrations or openings through the assemblies must be sealed, lined, insulated, or otherwise treated to maintain the required ratings. Entrance doors only need to be tight-fitting to the frame and sill. Floor/ceiling assemblies between separate dwelling units must also provide a minimum impact insulation class (IIC) of 50 (minimum of 45 NISR in accordance with ASTM E1007 when field tested), as defined in ASTM E492. Both the STC and the IIC are also permitted to be established by engineering analysis by a registered design professional where a comparison is made to an assembly that has been tested.

Although the scope of these provisions has historically always been limited to apartment buildings and other residential structures containing dwelling units, the application of sound transmission controls is also applicable to all buildings containing sleeping units. Although this would include those vertical and/or horizontal elements separating hotel guest rooms and dormitory rooms, it also addresses sound separations between sleeping units in institutional facilities, such as hospitals, nursing homes, and assisted living facilities.

Section 1207 *Enhanced Classroom Acoustics*

Good classroom acoustics are essential to support language acquisition and learning for all children, particularly younger children. There are three primary styles of learning in a classroom: visual, auditory, and kinesthetic or tactile. Visual learning is a teaching and learning style in which ideas, concepts, data, and other information are associated with images. This style highlights the importance of good lighting in classrooms, so students can read or see different images. Auditory learning is a learning style in which a person learns through listening. An auditory learner depends on hearing and speaking. In the instruction of auditory learners, teachers use techniques such as verbal direction, group discussions, verbal reinforcement, group activities, reading aloud, and putting information into a rhythmic pattern, such as a rap, poem, or song. The third style, kinesthetic or tactile learning, is learning while carrying out a physical activity, such as physical manipulation or experiments. Learning modality strengths can occur independently or in combination; they can change over time and they become ingrained with age. It is estimated that auditory learners make up about 30 percent of the population.

For children who have hearing loss and those who use cochlear implants, there is no substitute for a good acoustic environment. Assistive technologies typically only amplify the teacher and do not amplify discussions among children or between the teacher and individual child. Additionally, children with disabilities not related to hearing, such as those with autism and learning disabilities, may be adversely affected by high ambient-noise levels, while students learning English can significantly benefit from clear speech and low background noise.

The enhanced acoustic requirements only apply to those spaces where the room volume doesn't exceed 20,000 cubic feet (566 m³). For example, where a 10-feet (3,048-mm) ceiling height occurs, all such classrooms of up to 2,000 square feet (186 m²) in floor area are required to meet these requirements. The criteria in ICC A117.1 Section 808 are intended to be applicable to standard-sized self-contained classrooms, while not addressing larger spaces used for activities such as band and choir. The criteria are also not intended to apply to ancillary learning spaces, such as individual tutoring areas.

A classroom is limited to a maximum reverberation time for classroom acoustics, determined through the measurement of room performance or by a calculation based on materials on the floor, ceiling, and walls. Performance in a fully furnished, unoccupied classroom should include a maximum reverberation time of 0.6–0.7 seconds, depending upon the size of the classroom, and a maximum background noise of either 35 dBA (A-weighted sound pressure level) or 55 dBC (C-weighted sound pressure level). The ambient sound levels must be measured in both dBA and dBC. The dBA filter measures mid-range frequencies, while the dBC filter measures low and high frequencies.

Reverberation time measures how quickly a sound decays in a room. The volume of the space and the surfaces in the room will affect the reverberation time. The intent of the 0.6–0.7 second reverberation time is to increase the sound level from the teacher throughout the room while maintaining clarity. ASTM E2235, *Standard Test Method for Determination of Decay Rates for Use in Sound Insulation Test Methods*, is used for testing sound decay rates.

The prescriptive calculation requires noise reduction coefficient (NRC) ratings for every surface finish, including finishes on the floor, ceiling and all walls. Ratings range from 0 to 1, with 1 indicating the surface absorbs most of the speech sound energy and 0 indicating the surface reflects most of the speech sound energy. The NRC for each material is multiplied by the square footage of that material. The surface area of these materials are to be subtracted from the area of the mounting surface—for example, if a wall is 10 feet (3,048 mm) in length by 10 feet (3,048 mm) in height, and three 2-feet by 10-feet (610 mm by 3,048 mm) panels are attached to the wall, the calculation would be based on 60 square feet (5.57 m²) of panels and 40 square feet (3.72 m²) of uncovered wall.

The ICC A117.1 criterion also considers other sound sources, including ambient sound sources outside the classrooms. These sources may include playground noises, airplane or traffic sources, and student movement in hallways. There are multiple ways to mitigate these issues. Some outdoor environment sources can be addressed by careful placement of the classrooms on the site or through the choice of the exterior materials for walls and roofs. Building noises can be mitigated by the proper selection of wall systems, location of doors, and placement of mechanical systems.

Section 1208 *Interior Space Dimensions*

Room size, tightness of construction, minimum ceiling height, number of occupants, and ventilation all interact with each other to establish the interior living environment insofar as odors, moisture, and transmission of disease are concerned. Therefore, the IBC regulates room sizes to assist in maintaining a comfortable and safe interior environment, and the minimum room sizes become increasingly important as buildings become even tighter in their construction because of energy-conservation requirements.

1208.1 Minimum room widths. Section 1208.1 requires that habitable spaces, other than kitchens, be at least 7 feet (2134 mm) in any direction. Kitchens require a clearance of at least 3 feet (914 mm) between counter fronts and appliances or walls.

1208.2 Minimum ceiling heights. Section 1208.2 regulates ceiling height, not only to assist in maintaining a comfortable indoor environment, but also to provide safety for the occupants of the building. It is important that tall individuals be able to move about without accidentally striking projections from the ceiling.

The basic requirement is that the ceiling height be not less than 7 feet 6 inches (2,286 mm) for occupiable spaces, habitable spaces, and corridors (see definitions of occupiable space and habitable space in Section 202). Kitchens, bathrooms, toilet rooms, storage rooms and laundry rooms may have a ceiling height less than 7 feet 6 inches (2,286 mm), but under no circumstances may such a height be less than 7 feet (2,134 mm) measured from the finished floor to the lowest projection from the ceiling.

For those ceilings within one- and two-family dwellings having exposed beams that project down from the ceiling surface, the ceiling beam members may project no more than 6 inches (152 mm) below the required ceiling height, provided the beams or girders are spaced at not less than 4 feet (1,219 mm) on center.

For rooms with sloped ceilings, the code requires only that the prescribed ceiling height be maintained in one-half the area of the room. However, no portion of the room that has a ceiling height of less than 5 feet (1,524 mm) shall be used in the computations for floor area. In the case of a room with a furred ceiling, the code requires the prescribed ceiling height in two-thirds of the area and, as in all cases for projections below the ceiling, the furred area may not be less than 7 feet (2,134 mm) above the floor.

1208.3 Dwelling unit size. Section 1208.3 sets the minimum dwelling unit size at 190 square feet (17.7 m^2) of habitable space.

1208.4 Room area. Section 1208.4 requires that every dwelling unit have at least one room with not less than 120 square feet (11.2 m^2) of net floor area. Sleeping units and other habitable rooms of a dwelling unit require a net floor area of at least 70 square feet (6.5 m^2). Kitchens are not required to have a minimum floor area.

1208.5 Efficiency dwelling units. This section of the code provides for a specific type of dwelling unit—a dwelling unit consisting of only one habitable room. Such a unit is commonly referred to as a "studio" apartment or residential condominium. Many of the requirements in this section are redundant, as this chapter already requires many of these provisions. See Table 1208-1. However, there are some requirements that are unique to the efficiency dwelling unit:

1. The single room (which serves as a living room, bedroom, and kitchen) is to have not less than 190 square feet (17.7 m^2) of floor area. It is the intent of the code that this floor area be the total gross floor area, less the area occupied by built-in cabinets and other built-in appliances that are not readily removed and that preclude any other use of the floor space occupied by the built-in cabinets and fixtures.

Table 1208-1. **Comparison of Dwelling Units**

Dwelling Unit	Standard	Efficiency
Primary room—Minimum Size	120 ft^2	190 ft^2
Other rooms—minimum size	70 ft^2	Not applicable
Cooking facilities required	Yes	Yes
Sanitation facilities required	Yes	Yes
Closet required	Not required in primary room	Yes

2. A closet is required.

3. A kitchen sink, cooking appliance, and refrigeration facilities are required, each providing a clear working space of not less than 30 inches (762 mm) in front.

4. A separate bathroom containing a water closet, lavatory, and bathtub or shower is required.

Section 1209 *Access to Unoccupied Spaces*

Access to crawl spaces and attic spaces is regulated by this section. Though typically unoccupied, it is sometimes necessary that these normally concealed areas be accessed for various reasons.

1209.1 Crawl spaces. This section of the code mandates that under-floor areas be accessible by a minimum 18-inch by 24-inch (457-mm by 610-mm) access opening. Where the access opening opens to the exterior of the building, the code intends it to be screened or covered to prevent the entrance of insects and animals. Also, it is the intent of the code that all portions of the under-floor area be provided with access including beneath or around obstructions created by pipes, ducts, and so on.

1209.2 Attic spaces. Because enclosed attics provide an avenue for the undetected spread of fire in a concealed space, the code requires that access openings be provided into the attic so that fire-fighting forces may gain entry to fight the fire. To be of any value, the access openings must be of sufficient size to admit a firefighter with fire-suppression gear and must also have enough headroom so that entry into the attic may be secured. Although not specified, the access should be located in a readily accessible location. A public hallway is typically an excellent location for attic-access openings. Fire department personnel will not then have to open private offices, apartments, or hotel rooms in order to enter the attic. Attic access may be provided through a wall as well as through a ceiling. In split-level buildings with multiple attics, an attic-access opening must be provided to each attic space.

Attic access is only required by the IBC for those attic areas having a clear height greater than 30 inches (762 mm). Where such conditions occur, an attic-access opening of not less than 20 inches by 30 inches (559 mm by 762 mm) shall be provided.

1209.3 Mechanical appliances. If a crawl space or attic contains mechanical equipment, the opening may need to be enlarged to gain compliance with the IMC.

Section 1210 *Toilet and Bathroom Requirements*

The primary emphasis of this section is to provide easily cleanable, sanitary, and water-resistant surfaces in toilet rooms and shower areas.

1210.2.1 Floors and wall bases. Except for dwelling units, the code requires toilet room and bathing room floors to have a smooth, hard, nonabsorbent surface. Finishes such as concrete and ceramic tile are certainly acceptable, as are other approved materials that may also be used. It is the intent of the code that the building official determines the suitability of the proposed floor surface insofar as cleanability and water resistance are concerned.

Although the materials used for the floor covering and the wall base are to all have a smooth, hard, nonabsorbent surface, they can be of different materials. For example, a top set rubber base extending at least 4 inches (102 mm) up the wall is generally viewed as acceptable in conjunction with a floor covering of vinyl composition tiles.

Toilet and bathing room floor-finish requirements apply to all uses and occupancies except for dwelling units. Because motel and hotel rooms are typically not dwelling units, toilet room flooring in these uses must often comply with this section.

1210.2.2 Walls and partitions. Walls and partitions within 2 feet (610 mm) of the front and sides of urinals, water closets, and service sinks are required to have a smooth, hard, non-absorbent finish. Required finishes shall extend to a height of at least 4 feet (1,219 mm) above the floor and shall be of a type not adversely affected by moisture. Materials used as a base for tile or wall panels must satisfy the limitations set forth in Section 2509. Although the code does not specifically state that concrete walls must be sealed, concrete is not, strictly speaking, a non-absorbent material and needs some type of surface treatment. Sealing is particularly important for concrete block walls.

Special wall finishes at water closets and urinals are not required for dwelling units, sleeping units, and toilet rooms not for use by the public that contain only one water closet. Note that the exceptions for floor finishes and wall finishes are not the same. A private toilet room containing one water closet would be required to have flooring that complies with Section 1210.2.1, but there would be no special requirements for wall finishes.

In all occupancies, including dwelling units, penetrations of the water-resistant surfacing of the walls for the installation of accessories such as grab bars and towel bars are required to be sealed to protect the structural elements from moisture. The intent of the code is that because the structural elements are not required to be moisture resistant, the penetrations should be sealed to protect the structural elements. Although sealing is required for all walls in a toilet room, sealing is obviously most critical in areas of water splash such as in showers or behind or adjacent to lavatories.

1210.2.4 Showers. All showers must have floor and wall finishes that are smooth, non-absorbent, and not affected by moisture. Wall finishes must extend not less than 72 inches (1,829 mm) above the drain inlet.

KEY POINTS

- The method and arrangement of providing ventilated openings is an important aspect in the proper ventilation of attic spaces.
- Under-floor ventilation is to be provided through foundation walls or exterior walls in order to adequately ventilate the space below the building.
- Ventilation of interior spaces may be accommodated through exterior openings or by a mechanical system.
- The IBC mandates that some degree of lighting, whether artificial or natural, is to be provided to every occupiable space.
- Wall and floor/ceiling assemblies separating dwelling units and sleeping units from each other or from public or service areas must be provided with sound transmission control.
- Most Group E educational classrooms must be provided with enhanced acoustics as established in ICC A117.1.
- In specific areas of a building, floors and walls are required to have a smooth, hard, nonabsorbent finish.

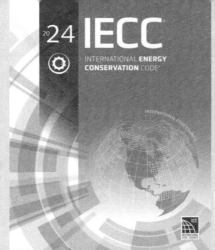

CHAPTER 13

ENERGY EFFICIENCY

This chapter of the *International Building Code®* (IBC®) references the *International Energy Conservation Code®* (IECC®) for provisions regulating the design and construction of buildings for energy efficiency. The IECC sets forth minimum requirements for new and existing buildings by regulating their exterior envelopes. In addition, it addresses the selection of heating, ventilating, and air-conditioning; service water-heating; electrical distribution; and illumination systems and equipment in order to provide for efficient and effective energy usage.

The IECC contains two separate sets of provisions, one for commercial buildings and one for residential buildings. Each set of provisions is applied separately to buildings within their scope. The IECC—Residential Provisions apply to all detached one- and two-family dwellings and townhouses as well as Group R-2, R-3, and R-4 buildings three stories or less in height above grade plane. All other buildings are regulated for energy efficiency under the IECC—Commercial Provisions.

CHAPTER 14

EXTERIOR WALLS

Section 1402 Performance Requirements
Section 1403 Materials
Section 1404 Installation of Wall Coverings
Section 1405 Combustible Materials on the Exterior Side of Exterior Walls
Section 1406 Metal Composite Material (MCM)
Section 1407 Exterior Insulation and Finish Systems
Section 1409 Insulated Metal Panels
Section 1412 Soffits and Fascias at Roof Overhangs
Key Points

This chapter establishes the basic requirement for exterior wall coverings, namely that they shall provide weather protection for the building at its exterior. The other important item necessary for complete weather protection, the roof system, is addressed in the next chapter. Chapter 14 presents general weather-protection criteria and special requirements for veneer. Vapor retarder materials, classifications, and installation options are also established. In addition, exterior wall coverings, balconies, and similar architectural appendages constructed of combustible materials are regulated in Section 1405.

Section 1402 *Performance Requirements*

This section provides specifications for the protection of the interior of the building from weather; therefore, there are requirements for:

1. Protection of the interior wall covering from moisture that penetrates the exterior wall covering.
2. Flashing of openings in, or projections through, exterior walls.
3. Vapor retarders used to resist the transmission of water vapor through the exterior assembly

It is also necessary to provide for the removal of any water that may accumulate behind the exterior veneer. The code requires a means for draining water to the exterior that enters the assembly. The intent of the provision is to merely utilize the felt building paper or other water-resistive barrier, along with flashing, to drain the water to the outside, rather than provide an air space between the siding and the water-resistant barrier. Thus, the practice of placing siding directly over the barrier is acceptable, provided flashing is correctly installed to direct water within the wall assembly to the exterior.

The code exempts the need for a weather-resistant wall envelope over concrete or masonry walls designed in accordance with Chapters 19 and 21, respectively. The *International Building Code*® (IBC®) requirements for weather coverings, flashings, and drainage are also not applicable where the exterior wall assembly has been tested in accordance with ASTM E331 and shown to resist wind-driven rain. In addition, where an exterior insulation and finish system (EIFS) is installed in accordance with Section 1407.4.1, the requirements of Section 1402.2 need not be followed.

Exterior walls of those buildings required to have noncombustible exterior walls (Types I, II, III, and IV) must be tested in accordance with NFPA 285 where they contain a combustible exterior wall covering, combustible insulation, exterior wall veneers manufactured using combustible adhesives, or a combustible water-resistive barrier and are over 40 feet (12,192 mm) in height. The use of this testing method is necessary to address the potential vertical and lateral flame spread that can occur either on or within exterior wall systems that contain combustible materials. An exception is provided for water-resistive barriers where the exterior wall covering is a specified noncombustible material, or where the barrier is limited in peak heat release rate, total heat release, effective heat of combustion, flame spread index, and smoke-developed index.

Section 1403 *Materials*

Using appropriate materials on the exterior of exterior wall assemblies is critical to ensuring weather resistance and assembly performance. The code provides references to industry standards for various materials. Where no standards exist, code provisions may be provided. Where neither a code provision nor referenced standard exists, approval by the building official is always an option. See Table 1403-1.

Table 1403-1. Materials Used for Construction of Exterior Walls

IBC Section	Material	IBC or Referenced Standard
1403.2	Water-resistive barriers	
	No. 15 felt	ASTM D226, Type 1
	Permeable flexible sheets	ASTM E2556, Type I or II
	Foam plastic insulating sheathing	Section 1402.2
	Fenestration and curtain walls	ASTM E331 and Section 1402.2
	Other approved materials	Section 104.2.3
1403.3	Wood	Chapter 23
1403.3.1	Basic hardboard	ANSI A135.4
1403.3.2	Hardboard siding	ANSI A135.6
1403.4	Masonry	Chapter 21
1403.5	Metal	
	Cold-formed or structural steel	Chapter 22
	Aluminum	Chapter 20
1403.5.1	Aluminum siding	AAMA 1402
1403.5.2	Cold-rolled copper	ASTM B370
1403.5.3	Lead-coated copper	ASTM B101
1403.6	Concrete	Chapter 19
1403.7	Glass-unit masonry	Chapter 21
1403.8	Vinyl siding	ASTM D3679
1403.9	Fiber-cement siding	ASTM C1186, Type A (or ISO 8336, Category A)
1403.10	Exterior insulation and finish systems	Section 1407
1403.11	Polypropylene siding	ASTM D7254 and Section 1404.18
1403.11.1	Flame spread index	ASTM E84 or UL 723
1403.11.2	Fire separation distance	Not less than 10 feet (3048 mm)
1403.12	Foam plastic insulation	Chapter 26
1403.13	Fiber-mat reinforced cementitious backer units	ASTM C1325
1403.14	Insulated vinyl siding	ASTM D7793

As most exterior wall coverings are permeable by moisture, the code requires that there should be a water-resistive barrier placed under the exterior wall covering and over the sheathing. The purpose of the barrier is to prevent moisture infiltration to the sheathing and subsequently the interior wall surfaces. The required water-resistive barrier is identified as at least one layer of No. 15 felt complying with ASTM D226 *Specification for Asphalt-Saturated Organic Felt Used in Roofing and Waterproofing* for Type I felt, Type I or II vapor permeable flexible sheets per ASTM E2556 *Standard Specification for Vapor Permeable Flexible Sheet Water-resistive Barriers Intended for Mechanical Attachment*, compliant foam plastic insulating sheathing, fenestration, and curtain wall protection as set forth in ASTM E331 *Standard Test Method for Water Penetration of Exterior Windows, Skylights, Doors, and Curtain Walls by Uniform Static Air Pressure Difference*, or other approved materials installed per the manufacturer's installation instructions. All necessary flashing must be installed as described in Section 1404.4 in order to complete the water-resistant envelope.

The water-resistive barrier material needs to be continuous to the top of walls and terminated at openings and appendages to avoid water accumulation and maintain a path for water to drain to the exterior. No. 15 asphalt felt and flexible sheets per ASTM E2556 must be installed horizontally with a 2-inch lap and a 6-inch lap at vertical joints.

Section 1404 *Installation of Wall Coverings*

Exterior wall coverings, along with soffits and fascia, are to be installed in accordance with the provisions of this section. Of primary concern are the types and thicknesses of the weather protection provided by the wall coverings. Vapor retarders are mandated to protect against condensation in framed exterior wall assemblies of buildings constructed in specified Climate Zones. In addition, it is important that flashing be installed in those areas prone to leakage. Various types of veneer are addressed, including wood, masonry, stone, metal, glass, vinyl, fiber, and cement. The specifics for each veneer type, including fastening requirements, are described in detail.

1404.2 Weather protection. A broad range of exterior wall-covering materials is listed in Table 1404.2 as being satisfactory when utilized as protection against the elements of nature. Covering types addressed include wood and particleboard siding, wood shingles, stucco, anchored and adhered masonry veneer, aluminum and vinyl siding, stone, and structural glass. For most of the 30 types listed, the minimum required thickness for weather protection is mandated. In the case of particleboard and exterior plywood without sheathing, reference is made to Chapter 23.

1404.3 Vapor retarders. In order to protect against condensation in framed exterior wall assemblies of buildings constructed in the specified Climate Zones, complying vapor retarders are mandated. Table C301.1 in the *International Energy Conservation Code®* (IECC®) establishes the appropriate Climate Zones in the United States based on counties. Those counties designated as Zones 5, 6, 7, and 8, as well as those classified as Zone 4 Marine must be provided, with exceptions, with either a Class I or II vapor retarder on the interior side of any exterior framed walls. The use of Class III vapor retarders is also acceptable provided such retarders are set forth in Table 1404.3(3) for the appropriate Climate Zones.

In contrast, low-permeability vapor barriers are prohibited in warmer climate zones. By prohibiting the Class I and II retarders on the interior side of walls in Climate Zones 1 and 2, and the Class I retarders in Climate Zones 3 and 4 (other than Marine 4), the IBC helps

Installation of Wall Coverings

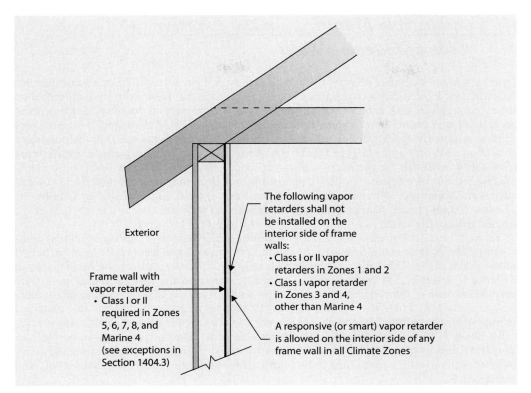

Figure 1404-1 **Vapor retarders.**

to address the problem where warmer moist air from the exterior enters into the wall and then is trapped and condenses against the cooler interior side of the wall. See Figure 1404-1. It should be noted that the provision does not prohibit these particular vapor retarders in the specified climate zones, but that it is simply mandating that they are not placed on the interior side of the wall. If the better vapor barriers are installed on the interior side of the wall in the warmer climates, a reversed vapor retarder is essentially created, which can lead to the moisture being trapped and condensed within the wall.

A responsive (or smart) vapor retarder is made of material complying with Class I or II, but that also has a vapor permeance of 1 perm or greater. A responsive vapor retarder is allowed on the interior side of any frame wall in all Climate Zones.

1404.4 Flashing. The code requires that all points subject to the entry of moisture be appropriately flashed. Roof and wall intersections, as well as parapets, are especially troublesome, as are exterior wall openings exposed to the weather, and particularly, those exposed to wind-driven rain. Even though the code may not cover every potential situation that might occur, it intends that the exterior envelope of the building be weatherproofed so as to protect the interior from the weather. Furthermore, for buildings with human occupancy, the interior must be sanitary and livable. Therefore, whether prescribed in the code or not, any place on the envelope of the building that provides a route for admission of water or moisture into the building is required to be properly protected.

Section 1405 *Combustible Materials on the Exterior Side of Exterior Walls*

Exterior walls of buildings of Types I, II, and III are typically required to be of noncombustible construction. However, it is common for some limited combustible elements to be installed on the exterior side of such exterior walls. Type IV construction permits combustible elements in an exterior wall assembly with some limitations. Since Type IV-A, IV-B, and IV-C have very specific requirements for exterior wall assembly construction in Section 602.4, the provisions of Section 1405 only apply to Type IV-HT construction. Type V buildings are permitted to use combustible materials for the entire wall construction. Where the exterior walls of a building are covered on the exterior side with combustible materials, the materials must be in compliance with this section. The only exception is for plastics, which are regulated by Chapter 26. The installation of such materials in compliance with the code does not negatively impact the desired level of fire safety.

Section 1405.1.1 provides for the use of combustible wall coverings on otherwise noncombustible exterior walls, provided the building is limited to 40 feet (12,192 mm) in height above grade plane. The amount of such materials is also limited where the fire separation distance is 5 feet (1,524 mm) or less. The reductions from the general requirements are granted because of the restricted height and, in some cases, the limited amount of combustible materials. At a height exceeding 40 feet (12,192 mm), the use of combustible materials or supports is prohibited, except where the use of various compliant materials are specifically allowed. Those materials include: metal composite material systems, exterior insulation and finish systems, high-pressure decorative exterior-grade compact laminate systems, exterior wall coverings containing foam plastic insulation, fire-retardant-treated wood, and wood veneer.

Where permitted by Section 1405.1.1, the combustible wall covering must be subjected to the test method of NFPA 268. As the exterior wall is moved farther and farther from the lot line, the tolerance to radiant heat energy need not be as great, allowing the use of alternative materials. Table 1405.1.1.1.2 identifies the tolerable level of incident radiant heat energy based on fire separation distance.

Section 1406 *Metal Composite Material (MCM)*

Metal composite materials (MCM) consist of a thin, extruded plastic core encapsulated within metal facings. Although having some of the same characteristics as foam-plastic insulation, light-transmitting plastic, plastic veneer, and combustible construction, MCM used as exterior wall coverings are specifically regulated by this section. See Figure 1406-1.

Where installed on exterior walls of buildings, MCM systems are regulated for surface-burning characteristics. In buildings of Type V construction, the flame-spread index cannot exceed 75 and the smoke-developed index is limited to 450. In buildings of Type I, II, III, or IV construction, the flame spread index is limited to 25. In such buildings required to have noncombustible exterior walls, the use of a thermal barrier is also necessary to separate the MCM from the interior of the building unless specifically approved by appropriate testing. Where MCM is installed on a Type I, II, III, or IV building to a height of greater than 40 feet (12,192 mm) above grade plane, the MCM must also comply with the acceptance criteria of NFPA 285.

Figure 1406-1   Metal Composite Material system used for an urban building.

Section 1407 *Exterior Insulation and Finish Systems*

Exterior Insulation and Finish Systems (EIFS) are nonload-bearing exterior wall coverings that are used extensively throughout North America, Europe, and the Pacific Rim. The provisions of Section 1407 are primarily intended to reference the applicable ASTM standards that are specific to EIFS. Reference is made to E2273 *Standard Test Method for Determining the Drainage Efficiency of Exterior Insulation and Finish Systems (EIFS) Clad Wall Assemblies,* E2568 *Standard Specification for PB Exterior Insulation and Finish Systems,* and E2570 *Standard Test*

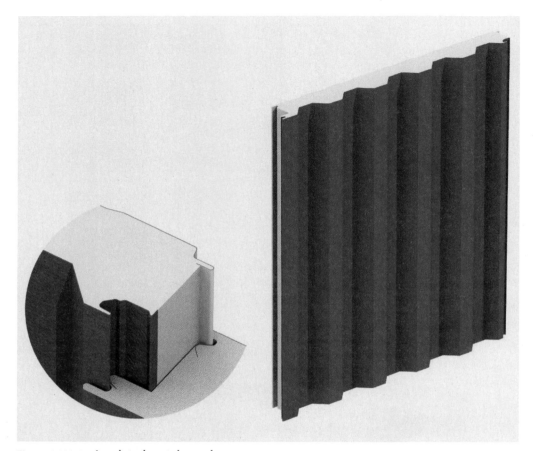

Figure 1409-1 **Insulated metal panels.**

Method for Evaluating Water-Resistive Barrier (WRB) Coatings Used Under Exterior Insulation and Finish Systems (EIFS) or EIFS with Drainage. In addition to the several ASTM standards previously identified, reference is also made to various provisions found elsewhere in the IBC that are applicable to EIFS. Current ICC ES Acceptance Criteria further establish requirements for EIFS and related components, and numerous EIFS manufacturers hold evaluation reports to demonstrate code compliance.

Section 1409 *Insulated Metal Panels*

Insulated metal panel (IMP) systems are factory-manufactured panels composed of an insulation core with metal facers. See Figure 1409-1. Because IMP systems are permitted to include combustible foam plastic insulation as well as other combustible and noncombustible insulating materials, it is important to provide the details of how these products should be regulated. While IMP systems are similar to the metal composite materials (MCM) regulated in Section 1406, MCM systems must by definition be installed over a solid plastic core and therefore are not allowed to be insulated by the materials permitted with IMP requirements. IMP provisions address structural design requirements, durability, fire testing, and weather resistance aspects. While IMPs can be used in a variety of ways including as exterior walls, exterior wall coverings, roof assemblies, or roof coverings, Section 1409 focuses on their application as either an exterior wall or exterior wall covering.

Section 1412 *Soffits and Fascias at Roof Overhangs*

Soffits and fascias at roof overhangs must be designed to resist the component and cladding loads for walls determined per IBC Chapter 16 using an effective wind area of 10 square feet. Vinyl and aluminum soffit panels must be installed using fasteners specified by the manufacturer and fastened at both ends to a supporting component such as a nailing strip, fascia, or subfascia component.

Soffit provisions are applied based on an assumed 30 psf design wind pressure. For most of the country, design wind pressures will be 30 psf or less, even in Exposure Category D areas. Table 1412-1 shows soffit panel material thickness and attachment requirements for this wind pressure limit for various soffit panel materials. At or below this limit, vinyl or aluminum soffit panels must be fastened at each end and an unsupported span cannot exceed 16 inches unless permitted by the manufacturer's product approval (Figure 1412-1).

Table 1412-1. Soffit Panel Material Thickness and Attachment Requirements for 30 psf or Less Wind Pressure

Soffit Panel Material	Minimum Thickness (inches)	Fastener Requirements
Vinyl or aluminum	Per manufacturer	Per manufacturer
Fiber-cement	¼	Per manufacturer
Hardboard	7/16	2½" × 0.113" siding nails ≤ 6" o.c. at panel edges and ≤ 12" o.c. at intermediate supports
Wood structural panels	⅜	2" × 0.099" nails ≤ 6" o.c. at panel edges and ≤ 12" o.c. at intermediate supports

For higher wind regions, enhanced soffit construction is needed where the design wind pressure exceeds 30 psf. Vinyl or aluminum soffit panels must be fastened at each end, and an unsupported span cannot exceed 12 inches unless permitted by the manufacturer's product approval (Figure 1412-1). A prescriptive alternative for wood structural panel soffits that comply with higher wind pressures is also provided in the code.

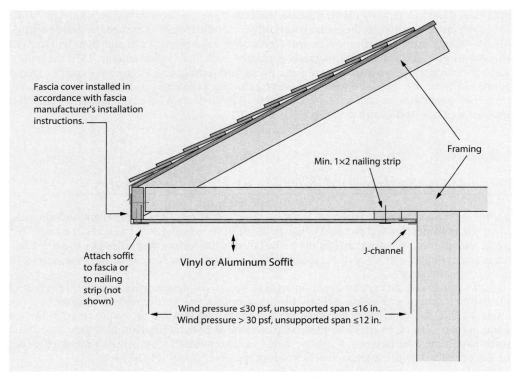

Figure 1412-1 **Typical single-span vinyl or aluminum soffit panel support.**

KEY POINTS

- The interior of a building must be protected with a weather-resistant envelope, including wall coverings, vapor retarders, flashing, and drainage methods.
- The IBC regulates numerous veneer materials for exterior applications.
- A limited amount of combustible wall coverings is permitted on the exterior side of exterior walls of Type I, II, III, and IV buildings.
- Metal composite material (MCM) used as exterior cladding is strictly regulated, particularly when installed at a height of more than 40 feet (12,192 mm) above grade plane.
- The applicable referenced standards are identified for the installation of exterior insulation and finish systems (EIFS).
- Insulated metal panel (IMP) systems are factory-manufactured panels composed of an insulation core with metal facers that are permitted to include combustible foam plastic insulation as well as other combustible and noncombustible insulating materials.
- Soffits and fascias at roof overhangs must be designed and constructed to resist wind loads.

CHAPTER 15

ROOF ASSEMBLIES AND ROOFTOP STRUCTURES

Section 1504 Performance Requirements
Section 1505 Fire Classification
Section 1506 Materials
Section 1507 Requirements for Roof Coverings
Section 1511 Rooftop Structures
Section 1512 Reroofing
Key Points

In addition to the requirements for roof assemblies and roof coverings, this chapter regulates roof insulation and rooftop structures. Rooftop structures include elements such as penthouses, tanks, cooling towers, spires, towers, domes, and cupolas. Roofing materials and components are regulated for quality as well as installation.

The provisions in Chapter 15 for roof construction and roof covering are intended to provide a weather-protective barrier at the roof and, in most circumstances, to provide a fire-resistant barrier to prevent flaming combustible materials such as flying brands from nearby fires from penetrating the roof construction. This chapter is essentially prescriptive in nature and is based on decades of experience with the various traditional roof-covering materials. These prescriptive rules are very important to ensuring the satisfactory performance of the roof covering. The provisions are based on an attempt to prevent observed past unsatisfactory performances of the various roofing materials and components.

Those measures that have been shown by experience to prevent past unsatisfactory performance generally are included in the manufacturer's instructions for application of the various roofing materials. In many cases, the manufacturer's instructions are incorporated in this code by reference. The code intends, then, that they be followed as if they were part of the code.

The overriding safety need of roofs is resistance to external fire factors. In this regard, the necessary regulation is driven by Table 1505.1, as well as the appropriate standards for fire-resistive roof assemblies and roof coverings, including ASTM E108, UL 790, and ASTM D2898. Typically, a roof covering by itself cannot be a listed fire-resistive roof. Therefore, the regulations clearly separate assemblies and coverings to enforce construction of listed roof assemblies to the level at which listed wall and floor/ceiling assemblies are regulated.

Section 1504 *Performance Requirements*

This section of the code regulates the performance of a roof against two concerns: wind and impact. For wind resistance, roofs must comply with this section and Chapter 16. Low-slope roofs must demonstrate that they are resistant to wind and impact damage by complying with the appropriate standards.

The use of aggregate as a roof-covering material is regulated based on specified conditions in an effort to reduce property loss that is due to high winds. Field assessments of damage to buildings caused by high-wind events have shown that aggregate blown from the roofs of buildings has exacerbated damage to other buildings because of breakage of glass.

Table 1504.8 considers aggregate size, roof height, wind speed, and exposure category to determine the minimum required parapet wall height for aggregate-surfaced roofs. Such parapets must be provided on the perimeter of the roof at all exterior sides except where an adjacent wall extends above the roof to a height at least equivalent to that required for the parapet. An engineering and scientific basis for roof design to prevent aggregate blowoff was achieved through the use of wind tunnel tests and subsequent field studies of hurricane damage. The provisions are based on wind speeds for blowoff and only deal with smaller aggregate used for the surfacing of built-up roofs and sprayed polyurethane foam roofs, both of which are different systems than ballasted roofs.

Section 1505 *Fire Classification*

As a minimum, the IBC® generally requires Class B or C roof assemblies for most buildings. These are roof assemblies that provide protection of the roof against moderate and light fire exposures, respectively. The various sizes of brands used for testing are shown in Figure 1505-1. These exposures are external and are generally created by fires in adjoining structures, wildfires (brush fires and forest fires, for example), and fire from the subject building that extends up the exterior and onto the top surface of the roof. Wildfires and some structural fires create flying and flaming brands that can ignite nonclassified roofing.

The roof assembly classifications required by the code, which are related primarily to type of construction, are delineated in Table 1505.1.

Section 1505 defines the following roof assemblies and roof coverings:

1. **Class A roof assemblies.** Roof assemblies recognized as Class A are effective against severe fire-test exposures. A variety of prescriptive materials and assemblies are considered as Class A roof assemblies, as well as any roof assembly or roof covering tested and listed as Class A. Class A roof assemblies are permitted for use on all buildings, regardless of the building's construction type.

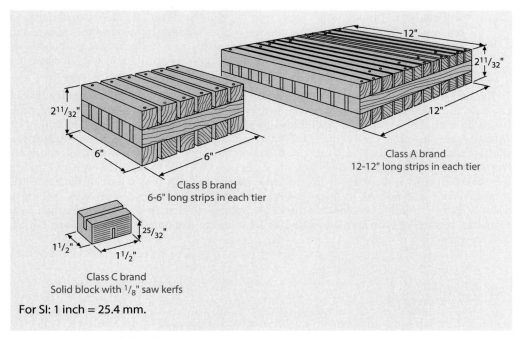

Figure 1505-1 **Brands for Class A, B, and C tests.**

2. **Class B roof assemblies.** Class B roof assemblies are effective against moderate fire-test exposures and are considered appropriate for all types of construction. There are no prescriptive roof assemblies or roof coverings considered as Class B, as they must be listed as such. Consistent with the universal acceptance of Class A assemblies, Class B roof assemblies are also permitted for use on all buildings.

3. **Class C roof assemblies.** Buildings not required to be of fire-resistant construction—Types IIB, IIIB, and VB—are permitted to use Class C roof assemblies. Such assemblies are effective against light fire-test exposures. Class C roof assemblies are also required to be fire tested and listed.

4. **Fire-retardant-treated wood shingles and shakes.** Where fire-retardant-treated wood shakes and shingles are components of Class A, B, or C roof assemblies, they must comply with the criteria of Section 1505.6. AWPA C1, *All Timber Products-Preservative Treatment by Pressure Processes*, is referenced as the standard for regulating the pressure treatment for fire-retardant purposes. In addition to the required markings that identify the shakes or shingles and their manufacturer, a label is mandated to identify the appropriate fire classification.

5. **Nonclassified roofing.** Roof coverings that are considered nonclassified roof coverings are approved for use by the IBC on Group U buildings where the roof is located at least 6 feet (1,829 mm) from all lot lines.

6. **Special-purpose roofs.** These roofs are either of wood shingles or of wood shakes and are applied with a minimum ⅝ inch (15.9 mm) Type X water-resistant gypsum backing board or gypsum sheathing panel. The intent of the provisions for special-purpose roofs is to provide a roof covering that, although it may be ignited by flying brands, will not burn through to the interior of the building. Also, the special underlayment tends to prevent fires from the interior of the building from burning through and igniting the roof covering, which helps prevent flying brands. Special-purpose roofs are permitted in limited applications on buildings of Type IIB, IIIB, and VB construction by Footnote c to Table 1505.1.

Because of the inconsistent use of the terms *roof assembly* and *roof covering* throughout Chapter 15, there is confusion as to the proper use of combustible materials at the roof. Based on Item 4 of Section 603.1, combustible roof coverings that have an A, B, or C classification are permitted on buildings of Type I or II construction. This does not include the structural deck materials, which must be of noncombustible construction or fire-retardant-treated wood in compliance with Item 1.3 of Section 603.1. It would, however, permit the use of wood structural panels or foam plastic insulation boards as a part of the classified roof covering where used in combination with a noncombustible roof deck. Where a Class A, B, or C roofing assembly includes a combustible structural deck, other than fire-retardant-treated wood where permitted, it is limited to use on a building of Type III, IV, or V construction.

Section 1506 *Materials*

Roofing materials must comply with quality standards embodied in the IBC for Chapter 15. Furthermore, identification of the roofing materials is mandatory in order to verify that they comply with quality standards. In addition to bearing the manufacturer's label or identifying mark on the materials, roof-covering materials are required by the code to carry a label of an approved testing agency.

Section 1507 *Requirements for Roof Coverings*

Installation provisions of various roofing materials typically include roof slope limits and adherence to applicable standards and manufacturer's instructions. Certain roof systems covered by the code do not normally require underlayment including built-up roofs, modified bitumen roofing, single-ply roofing, sprayed polyurethane foam roofing, liquid-applied roofing, and vegetative and landscaped roofs.

Other roofing materials, that require underlayment, include asphalt shingles, clay and concrete tile, metal roof shingles, mineral-surfaced roll roofing, slate and slate-type shingles, wood shingles, wood shakes, metal roof panels, and BIPV roof coverings. Critical to roof performance is its ability to withstand high wind loads. Accordingly, the code differentiates the type, application, and fastening of underlayment at a wind threshold of 130 mph in hurricane-prone regions and 140 mph outside of hurricane-prone regions. Underlayment must conform to the applicable ASTM standards. Underlayment materials typically require a label indicating compliance with the standard and, if applicable, type or class indicated in Table 1507.1.1(1). Underlayment must be applied per Table 1507.1.1(2) and provisions typically vary based on roof slope. Underlayment must be attached per Table 1507.1.1(3). See Table 1507-1.

Table 1507-1. Underlayment Type, Application, and Fastening

Roof Covering	Section	V < 130 mph in Hurricane-Prone Regions or V < 140 mph Outside Hurricane-Prone Regions			V ≥ 130 mph in Hurricane-Prone Regions or V ≥ 140 mph Outside Hurricane-Prone Regions		
		Type per Table 1507.1.1(1)	Application per Table 1507.1.1(2)	Fastening per Table 1507.1.1(3)	Type per Table 1507.1.1(1)	Application per Table 1507.1.1(2)	Fastening per Table 1507.1.1(3)
• Asphalt shingles • Clay and concrete tiles • BIPV roof coverings	1507.2 1507.3 1507.16	ASTM	Prescriptive	Sufficient	ASTM	Prescriptive	Prescriptive
• Metal roof panels • Metal roof shingles • Mineral-surfaced roll roofing • Slate shingles • Wood shingles • Wood shakes	1507.4 1507.5 1507.6 1507.7 1507.8 1507.9	ASTM	Manufacturer's instructions	Manufacturer's instructions	ASTM	Prescriptive	Prescriptive

Building-integrated photovoltaic roof panels also require underlayment, but the wind threshold for these roof coverings is different than those shown in Table 1507-1. Installation requirements are provided in Section 1507.17.

Section 1511 *Rooftop Structures*

Penthouses and other roof structures are regulated by the IBC as if they were appurtenances to the building rather than occupiable portions.

In fact, if a penthouse is used for any purpose other than shelter of mechanical or electrical equipment, tanks, elevators, and related machinery, or shelter of stairways, or vertical shaft openings, the code requires that it be considered an additional story of the building.

As intended by the IBC, roof structures are equipment shelters, equipment screens, platforms that support mechanical equipment, water-tank enclosures, and other similar structures generally used to screen, support, or shelter equipment on the roof of the building. This section also regulates towers and spires, which are addressed separately in Section 1511.5.

The IBC regulates penthouses, roof structures, tanks, towers, and spires to prevent hazardous conditions that are due to internal and external fire concerns or structural inadequacy, and to ensure their proper use as equipment shelters.

1511.2 Penthouses. The code does not regulate the height of penthouses and roof structures on Type I buildings. However, for buildings of other construction types, the code limits the height of penthouses and roof structures to 18 feet (5,486 mm) above the height of the roof, except where the penthouse is used to enclose a tank or elevator. In such cases, a maximum height of 28 feet (8,534 mm) is permitted. As the aggregate area of all penthouses and roof structures is limited to one-third the area of the roof, the additional height permitted for penthouses and roof structures above that allowed by Section 504 for building height does not pose any significant fire- and life-safety hazard that is due to other restrictions that this section places on construction. See Figure 1511-1.

Penthouses and other rooftop structures are also regulated by the exception to Section 504.3, where the requirements are based on the effect of a rooftop structure on the allowable height permitted for the entire building. Where applicable, the provisions from both Chapters 5 and 15 are in effect, and where there is a conflict, the most restrictive condition will apply.

As the code has reduced requirements for construction of penthouses and roof structures, it is logical that their use should be limited as specified in this section. Thus, if other uses are made of penthouses or other roof structures, it also seems appropriate that they should be constructed as would be required for an additional story of the building. It is the intent of the code that a penthouse or roof structure complying with this section not be considered to create an additional story above that permitted by Section 504.

If a rooftop structure qualifies as a penthouse, it is intended that the floor of the penthouse is only required to meet the roof provisions of Table 601. As an extension of this recognition, any fire-resistance-rated shaft that extends to or through the penthouse floor does not need to be protected at the penthouse floor line. Under both conditions, the floor of the penthouse is solely regulated as the building's roof. Regarding any required means of egress from the penthouse, the provisions for an occupied floor are not applicable, nor are the requirements of Section 1006.3 for an occupied roof. It would seem appropriate that the access provisions of the *International Mechanical Code*® (IMC®) for rooftop equipment also provide for adequate egress.

1511.2.4 Type of construction. The intent of the code is that penthouses be constructed with the same materials and the same fire resistance as required for the main portion of the building. However, because of the nature of their use, the code does permit exceptions for

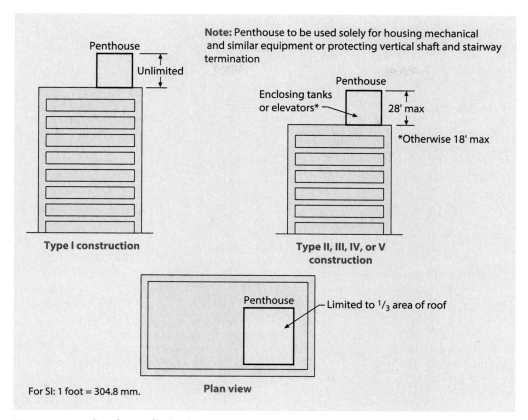

Figure 1511-1 **Penthouse limitations.**

exterior walls, roofs, and interior walls. Where the exterior walls of penthouses are more than 5 feet (1,524 mm) [or in some cases 20 feet (6,096 mm)] from the lot line, reductions in any required fire resistance are typically permitted. See Figure 1511-2. It is important to note that the intent of the exceptions is to reduce any required fire-resistance rating where applicable. Where the building's exterior walls and roof are permitted by other provisions of the code to have ratings less than those established in the exceptions, then those fire-resistance ratings set forth in the exceptions do not apply.

1511.5 Towers, spires, domes, and cupolas. The IBC intends that towers, spires, domes, and cupolas be considered separately from penthouses and other roof structures. The towers contemplated in this section are towers such as radio and television antenna towers, church spires, and other roof elements of similar nature that do not support or enclose any mechanical equipment and that are not occupied. As with penthouses and other roof structures, the code intends to obtain construction and fire resistance consistent with that of the building to which they are attached. Under a variety of conditions, however, towers and similar elements are required to be constructed of noncombustible materials, regardless of the building construction.

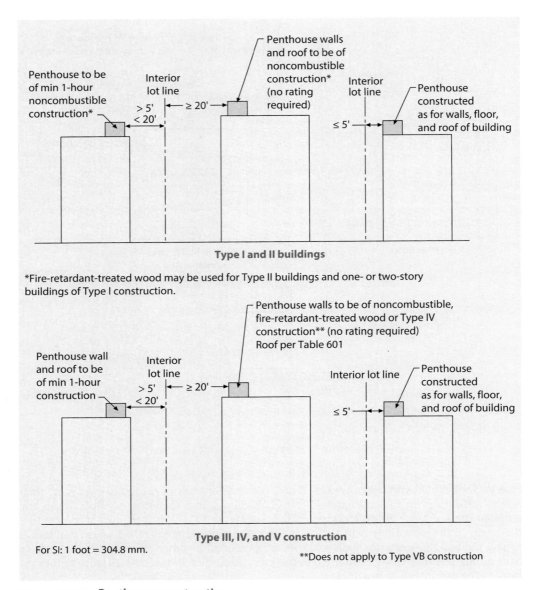

Figure 1511-2 **Penthouse construction.**

Section 1512 *Reroofing*

When a roof is replaced or recovered, it must comply with Chapter 15 for the materials and methods used with exceptions for low-sloped roofs as well as a waiver of the secondary drainage provision when reroofing an existing building where the roof drains properly. Section 1512 does not mandate that the entire roof be replaced but simply that the portion being replaced comply with Chapter 15.

Section 1512.1 General. The exceptions in Section 1512.1 emphasize that during the reroofing process, snow loads need to be evaluated and susceptible bays must be analyzed for ponding instability due to rain loads. The intent is to reduce the likelihood that inadequate drainage and ponding caused by new loading or roofing configurations might lead to either a roof collapse or failure. It is often difficult, if not impossible, to make changes such as to the roof slope or to increase the drainage system when reroofing is being done.

Section 1512.2 Roof replacement. The exceptions in Section 1512.2 are intended to address situations that occur during roof replacement where self-adhered underlayment membranes from the existing roof covering system may remain in place. As a general requirement, all existing layers of roof assembly materials are expected to be removed. Exceptions recognize situations where an existing self-adhered membrane can remain in place, and either a new ice barrier membrane, a compliant underlayment, or a second self-adhered underlayment can be installed over it. Given that many self-adhered products cannot be easily or completely removed from the roof deck without causing damage to the decking, it is appropriate to allow the existing membrane to remain. Where the sheathing is water-soaked or inadequate to serve as the base for the roof covering it should be removed. Note that the requirements reference compliance with the "new" manufacturer's installation requirements. The intent is that a new underlayment or ice barrier will perform properly and not be impacted by interaction with an existing membrane. An example of this situation is where an asphalt system may be interfacing with a butyl system.

Section 1512.3 Roof recover. Installation of a new roof covering over an existing roof covering is permitted in many situations. Exceptions include where the existing roof covering is water-soaked or inadequate to serve as the base for additional roof covering; the existing roof covering is slate, clay, cement, or asbestos-cement tile; or the existing roof already has two or more applications of any type of roof covering.

KEY POINTS

- Roofs and roof coverings are addressed for both weather protection and fire resistance.
- The performance of a roof is regulated against wind and impact.
- The IBC generally requires roof coverings that provide protection for the roof against moderate or light fire-test exposures.
- Roof assemblies are classified as either Class A, B, or C roofing assemblies; nonclassified roofing; fire-retardant-treated wood shingles and shakes; or special-purpose roofs.
- Installation provisions of various roofing materials typically include roof slope limits and adherence to applicable standards and manufacturer's instructions; the code differentiates the type, application and fastening of underlayment at different wind thresholds.
- Penthouses and other roof structures are regulated by the code as if they were appurtenances to the building rather than occupiable portions.
- When a roof is replaced or recovered, it must comply with Chapter 15 for the materials and methods used with exceptions for low-sloped roofs as well as a waiver of the secondary drainage provision when reroofing an existing building where the roof drains properly.

CHAPTER 16

STRUCTURAL DESIGN

Section 1601 General
Section 1602 Notations
Section 1603 Construction Documents
Section 1604 General Design Requirements
Section 1605 Load Combinations
Section 1606 Dead Loads
Section 1607 Live Loads
Section 1608 Snow Loads
Section 1609 Wind Loads
Section 1610 Soil Loads and Hydrostatic Pressure
Section 1611 Rain Loads
Section 1612 Flood Loads
Section 1613 Earthquake Loads
Section 1614 Atmospheric Ice Loads
Section 1615 Tsunami Loads
Section 1616 Structural Integrity
Key Points

> Significant portions of Chapter 16 provisions related to load combinations and the determination of snow, wind, and seismic loads are included only by reference to ASCE 7-22 *Minimum Design Loads and Associated Criteria for Buildings and Other Structures*. Portions of these provisions that remain in the IBC relate to alternate methods and to the local geologic, terrain, or other environmental conditions that many building officials will wish to specify when adopting the model code by local ordinance.
>
> Numerical examples have been included where they serve to illustrate the design requirements.

Section 1601 *General*

Chapter 16, Structural Design, governs the structural design of IBC-regulated buildings, nonbuilding structures, and portions thereof. A building is defined in Section 202 as any structure used or intended for supporting or sheltering any occupancy.

The various types of structural loads specified by Chapter 16 are either gravity, uplift, or lateral loads. Gravity loads specifically addressed are dead loads, live loads, snow loads, and rain loads. Lateral loads specifically dealt with are those due to wind, earthquakes, soil pressure, hydrostatic pressure, flood, or tsunamis. Uplift loads can be caused by wind, earthquakes, or hydrostatic pressures. Loading conditions, such as uniformly distributed and concentrated live loads, impact loads, and most importantly, design load combinations, are also regulated by Chapter 16 provisions. Section 1601 presents the scope of the chapter. Section 1602 lists notations that are used in structural equations that are provided throughout Chapter 16. Section 1603 specifies the minimum information that must be provided on construction documents. Section 1604 gives general design requirements and specifically addresses strength criteria and serviceability criteria, including deflection limitations, structural analysis, risk categories, load tests, anchorage, and wind and seismic detailing. Section 1605 addresses the vital topic of how to apply design load combinations. This section references Chapter 2 of ASCE 7 for general load combination requirements, but also provides an alternate approach to the ASCE 7 provisions if desired by the engineer. Section 1606 specifies design dead loads. Section 1607 provides minimum uniformly distributed live loads and minimum concentrated live loads for various types of uses and occupancies. This section also permits a reduction of design live loads under certain conditions. Section 1608 specifies the design snow loads, largely by reference to Chapter 7 of ASCE 7. Section 1609 contains structural design requirements for wind and tornado loads, largely by reference to Chapters 26 through 30 and 32 of ASCE 7. Sections 1610, 1611, and 1612 address soil lateral and uplift loads, rain loads, and flood loads, respectively. The rain load provisions of Section 1611 include reference to Chapter 8 of ASCE 7. Section 1613 is devoted to the seismic design of buildings and other structures and references Chapters 11 through 18 (excluding Chapters 14 and 16) of ASCE 7. Section 1614 references Chapter 10 of ASCE 7 for atmospheric ice loads. Section 1615 prescribes tsunami load requirements per ASCE 7 Chapter 6 for Risk Category III or IV buildings and structures located within tsunami design zones. Section 1616 includes provisions intended to provide a minimum level of structural integrity in certain types of Risk Category III or IV high-rise buildings.

Section 1602 *Notations*

Both Chapters 16 and 21 include a "Notations" Section at the beginning of the Chapter. The purpose of this Section is to provide a definition for notations used in structural equations that are provided within that Chapter. When trying to understand what an equation is requiring in Chapter 16, it is often helpful to refer back to Section 1602.

Section 1603 *Construction Documents*

This section details the items to be shown on construction documents.

Construction documents are a part of the submittal documents required by Section 107, and the term is defined in Section 202. The indicated loads are required to be equal to or greater than the minimum loads required by the code. This information is useful to the building official in performing plan review and field inspection. It is also an extremely useful piece of information when additions or alterations are made to a structure at a later date. Each of the items indicated in Sections 1603.1.1 through 1603.1.9 is an important parameter in ensuring that appropriate live, snow, wind and tornado, earthquake, soil, flood, rain, and special loads are considered in the building's structural system(s) design. Notably, for snow, wind, tornado, and earthquake design, the risk category of the building must be provided.

The exception to Section 1603.1 simplifies the structural design information required for wood buildings constructed to the conventional light-frame construction provisions of Section 2308. A registered design professional is not required for the design of such buildings. However, many of the requirements of Sections 1603.1.1 through 1603.1.9 clearly would require the services of such a professional in order to provide the specified design data. The requirements in the exception are intended to provide adequate information for the building official to verify the structural design basis of wood-frame buildings built to these conventional construction provisions.

A key provision of Section 1603.1.4 is that designers are required to not only list the design wind pressures for component and cladding materials, but they are now required to specify the applicable zones on the plans and provide dimensions. Adding the zones to the submittal information is intended to improve guidance for locating tighter nailing zones and other changes in construction requirements on a roof or wall surface.

Section 1603.1.7 provides a pointer to Section 1612.4, which is titled "Flood Hazard Documentation." Section 1612.4 requires statements to be included in the construction documents if certain flood-related situations exist.

Section 1603.1.8 requires that special loads that affect the design of the structure be listed. Examples are provided in the text, such as machinery or equipment that weigh more than the specified floor or roof loads. These special loads should be noted on the construction plans and their locations specified. Subsection 1603.1.8.1 specifically requires that the dead load for rooftop-mounted solar panels and their supports be noted on the plans.

Section 1604 *General Design Requirements*

1604.1 General. This section requires buildings and other structures and all portions thereof to be designed and constructed in accordance with the general strength and serviceability requirements contained in Section 1604.

1604.2 Strength. The basic strength requirement is that buildings and structural systems be capable of supporting the factored loads without exceeding the applicable material strength limit. For structural elements designed using nominal rather than factored loads, the actual design stresses are not to exceed the applicable allowable material design stresses.

1604.3 Serviceability. The requirements for serviceability mean that structural systems and members must have adequate stiffness to limit deflection and lateral drift to an appropriate degree, based on the intended use. Specific requirements are given in Sections 1604.3.1 through 1604.3.7. Table 1604.3 contains deflection limits of structural members as a function of span and load type.

Table 1604.3. **Deflection Limits**

	L or L_r	S or W	D + L
Roof members:			
Supporting plaster or stucco ceiling	$l/360$	$l/360$	$l/240$
Supporting nonplaster ceiling	$l/240$	$l/240$	$l/180$
Not supporting ceiling	$l/180$	$l/180$	$l/120$
Floor members	$l/360$	—	$l/240$
Exterior walls:			
With plaster or stucco finishes	—	$l/360$	—
With other brittle finishes	—	$l/240$	—
With flexible finishes	—	$l/120$	—
Interior partitions:			
With plaster or stucco finishes	$l/360$	—	—
With other brittle finishes	$l/240$	—	—
With flexible finishes	$l/120$	—	—
Farm buildings	—	—	$l/180$
Greenhouses	—	—	$l/120$

For SI: 1 foot = 304.8 mm

The general statement provided in the IBC is based on Section 1.3.2 of ASCE 7. ASCE 7 requires that members have adequate stiffness to not only limit deflections and lateral drift, but also to consider vibration or any other deformations that could adversely affect the performance of the building. The IBC excludes reference to "vibration or any other deformations" for the following reasons:

1. The code has no objectively defined standard for structural vibration. Acceptable vibration limits are frequently subjective and highly dependent on the specific requirements of occupants of a building. This information is not necessarily available to the building official.

2. It is impossible for the building official to anticipate everything that can "adversely affect the intended use and performance" of a building. Sections 1604.3.1 through 1604.3.7 provide objectively defined deflection limits that are deemed to suffice for a wide range of structures. Limits more restrictive than these should be a matter of the design professional understanding the client's needs and goals, but they are not typically part of a minimum life-safety building code. For example, there are situations in which sensitive computer, optical, or mechanical equipment require extraordinary measures to limit their movement or vibration. These measures are often very complex and well beyond the life-safety requirements of most structures.

1604.4 Analysis. The first two paragraphs are reproduced with minor modifications from Section 1.3.5 of ASCE 7. The third paragraph is perhaps the most important when performing building plan reviews or inspections. This portion of the code highlights the importance of providing "…a ***complete load path*** capable of transferring loads from their point of origin to the load-resisting elements." There are several load paths that must be considered with relation to any structure. The first is the gravity load path which considers all vertical loads such as dead, live, snow, rain, and ice. Figure 1604-1 shows the gravity load path for a vertical load on a floor slab. Notice that the load is picked up by multiple structural elements until it is finally transferred to the foundation soils through the footings. While this is a good example of a "complete load path," Figure 1604-2 provides an instance where an incomplete load path is provided. Notice that in this figure support is not provided at the end of the floor girder at the wall and subsequently the vertical forces cannot be transferred through the footings and into the surrounding soils.

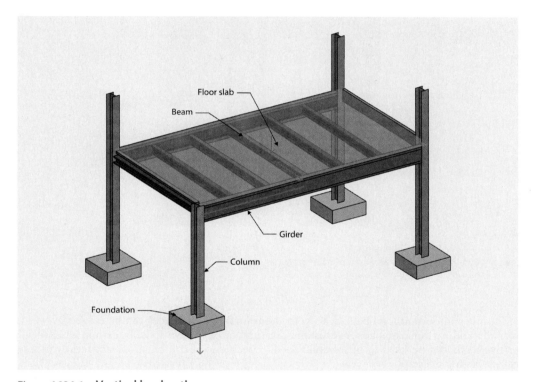

Figure 1604-1 **Vertical load path.**

The second load path that must be addressed in relation to every structure is the lateral load path. As previously discussed, lateral loads are caused by wind, tornadoes, earthquakes, soil and hydrostatic pressure, flooding, and tsunamis. Discussion here will focus on wind and seismic loads. While both wind and seismic forces are applied differently to a building, the lateral load path within the structure is the same for each. Horizontal forces are collected by the floor and roof diaphragms which then transfer these forces to the vertical resisting elements (i.e., shear walls, moment frames, braced frames, and cantilevered columns). These forces are then transferred from the vertical resisting elements down into the foundations and then into the surrounding soils. Figure 1604-3 shows how lateral forces due to wind and seismic loads are applied to, and then move through, a structure.

Figure 1604-2 **Incomplete vertical load path.**

Another important load path deals with uplift loads. Uplift forces can be caused by wind, tornado and earthquake loads, hydrostatic pressure, or expansive soils. In certain cases, dead or live loads can be used to offset uplift loads. Specific load combinations account for these counteracting effects.

One of the most important parts of building plan review and inspections is to ensure that a complete gravity and lateral load path are provided, as referenced in Section 1604.4. Careful review of plans and construction in the field is necessary to ensure that loads are transferred from one portion of the structure to the foundation.

The structural analysis must be based upon "well-established principles of mechanics." These principles are equilibrium, stability, geometric compatibility, and material properties. Although the code in general does not intend to specify the design method used by the engineer, it does intend that the design method be rational and follow standard design principles. Departures from this latter requirement can still be made based on the provisions of Section 104.2.3 when approved by the building official. For example, when the structural adequacy of a building or building element does not lend itself to a rational analysis, full-scale testing may be the only reasonable way to determine its structural behavior. If the testing program shows that a certain

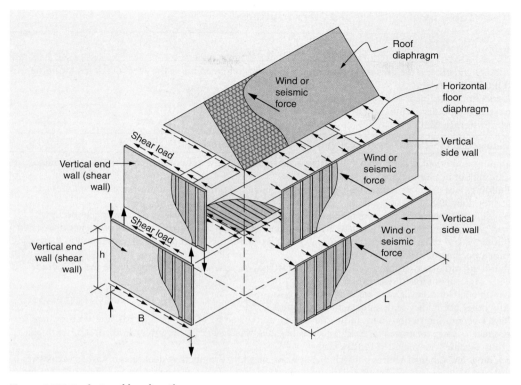

Figure 1604-3 **Lateral load path.**

building or element can safely resist the loads required by the code, the building official may approve the construction based on those test results.

Section 1604.4 highlights two items for lateral loads that must be considered within the rational analysis that is provided. The first clarifies that the design professional must distribute lateral forces to vertical resisting elements based upon their rigidity and that increased forces due to torsion may need to be considered. The second item specifies that the analysis must show that the structure is able to resist all forces described in Chapter 16, and specifically addresses overturning, uplift, and sliding forces.

1604.5 Risk category.   This section and Table 1604.5 make a combined presentation of the risk categories of buildings and other structures. The risk category reflects the relative anticipated seriousness of consequence of failure from lowest hazard to human life (Risk Category I) to highest (Risk Category IV), and is used to relate the criteria for maximum environmental loads specified in the code to the consequence of loads being exceeded for structures and their occupants.

IBC Table 1604.5 provides more comprehensive coverage of buildings assigned to various risk categories. The ASCE 7 Commentary provides background for the disparity. "The reason for this generalization is that the acceptable risk for a building or structure is an issue of public policy rather than purely a technical one.... Individual communities can alter these lists when they adopt local codes based on the model code, and individual owners or operators can elect to

Table 1604.5. **Risk Category of Buildings and Other Structures**

Risk Category	Nature of Occupancy
I	Buildings and other structures that represent a low hazard to human life in the event of failure, including but not limited to: • Agricultural facilities. • Certain temporary facilities. • Minor storage facilities.
II	Buildings and other structures except those listed in Risk Categories I, III, and IV.
III	Buildings and other structures that represent a substantial hazard to human life in the event of failure, including but not limited to: • Buildings and other structures whose primary occupancy is public assembly with an occupant load greater than 300. • Building and other structures containing one or more public assembly spaces, each having an occupant load greater than 300 and a cumulative occupant load of these public assembly spaces of greater than 2,500. • Buildings and other structures containing Group E or Group I-4 occupancies or combination thereof, with an occupant load greater than 250. • Buildings and other structures containing educational occupancies for students above the 12th grade with an occupant load greater than 500. • Group I-3, Condition 1 occupancies. • Any other occupancy with an occupant load greater than 5,000.[a] • Power-generating stations with individual power units rated 75 MW_{AC} (megawatts, alternating current) or greater, water treatment facilities for potable water, wastewater treatment facilities, and other public utility facilities not included in Risk Category IV. • Buildings and other structures not included in Risk Category IV containing quantities of toxic or explosive materials that: Exceed maximum allowable quantities per control area as given in Table 307.1(1) or 307.1(2) or per outdoor control area in accordance with the *International Fire Code*; and Are sufficient to pose a threat to the public if released.[b]

IV	Buildings and other structures designated as essential facilities and buildings where loss of function represents a substantial hazard to occupants or users, including but not limited to:
	• Group I-2 occupancies.
	• Ambulatory care facilities having emergency surgery or emergency treatment facilities.
	• Group I-3 occupancies other than Condition 1.
	• Fire, rescue, ambulance, and police stations and emergency vehicle garages.
	• Designated earthquake, hurricane, or other emergency shelters.
	• Designated emergency preparedness, communications and operations centers, and other facilities required for emergency response.
	• Public utility facilities providing power generation, potable water treatment, or wastewater treatment.
	• Power-generating stations and other public utility facilities required as emergency backup facilities for Risk Category IV structures.
	• Buildings and other structures containing quantities of highly toxic materials that:
	    Exceed maximum allowable quantities per control area as given in Table 307.1(2) or per outdoor control area in accordance with the *International Fire Code*; and
	    Are sufficient to pose a threat to the public if released.[b]
	• Aviation control towers, air traffic control centers, and emergency aircraft hangars.
	• Buildings and other structures having critical national defense functions.
	• Water storage facilities and pump structures required to maintain water pressure for fire suppression.

[a] For purposes of occupant load calculation, occupancies required by Table 1004.5 to use gross floor area calculations shall be permitted to use net floor areas to determine the total occupant load. The floor area for vehicular drive aisles shall be permitted to be excluded in the determination of net floor area in parking garages.

[b] Where approved by the building official, the classification of buildings and other structures as Risk Category III or IV based on their quantities of toxic, highly toxic, or explosive materials is permitted to be reduced to Risk Category II, provided that it can be demonstrated by a hazard assessment in accordance with Section 1.5.3 of ASCE 7 that a release of the toxic, highly toxic, or explosive materials is not sufficient to pose a threat to the public.

design individual buildings to higher occupancy categories based on personal risk management decisions...[this] also provides individual communities and development teams the flexibility to interpret acceptable risk for individual projects."

Risk Category I contains buildings and other structures that represent a low hazard to human life in the event of failure, either because they have a small number of occupants or because they have a limited period of exposure to extreme environmental loading.

Risk Category II contains all occupancies other than those in Risk Category I, III, or IV, and are sometimes referred to as *ordinary* for the purpose of risk exposure.

Risk Category III contains those buildings and other structures that have large numbers of occupants, that are designed for public assembly, or in which physical restraint or other incapacity of occupants hinders their movement or evacuation. Therefore, these structures represent a substantial hazard to human life in the event of failure. Risk Category III also includes important infrastructure such as power-generating stations (≥ 75 MW_{AC}) and water treatment facilities, where a failure may not create an unusual life-safety risk, but can cause large-scale economic impact and/or mass disruption of day-to-day civilian life.

Note that footnote "a" to Table 1604.5 makes a distinction between gross and net floor area for the calculation of occupant load to determine whether a building qualifies for Risk Category III because the occupant load exceeds 5,000. The calculation of the occupant load for the purpose of Table 1604.5 can be based on the net floor area even when Table 1004.1.2 requires gross floor area to be used. The floor area for vehicular drive aisles can be excluded in the determination of net floor area in parking garages. Further, the net floor area, as defined in IBC Chapter 2, does not include corridors, stairways, elevators, closets, accessory areas, and structural walls and columns. This is because the two tables serve different purposes. Table 1004.1.2 provides minimum occupant loads for the purpose of egress design, and egress design is determined on a floor-by-floor basis, assuming that the maximum occupant load can occur on a certain floor at a given time. However, for the purpose of Table 1604.5, occupant loads are calculated for the whole building, and it is very unlikely that all floors of a building would have the maximum occupant load at the same time. This is similar to the reasoning that is behind permitting live load reductions in the code.

Risk Category IV contains buildings and other structures that are designated as essential facilities such as hospitals and fire stations, which are intended to remain operational in the event of extreme loading. Risk category IV also includes facilities where loss of function represents a substantial hazard to users such as public utility power generation, potable water treatment, or wastewater treatment. Also included are structures that are supplementary to Risk Category IV structures and required for the operation of Risk Category IV facilities during an emergency, for example, facilities to maintain water pressure for fire suppression. Structures holding extremely hazardous materials are also included in Risk Category IV because of the potentially adverse effect a release of those materials may cause to the public or the environment.

For further clarity, Table 1604.5, in some cases, defines the risk categories based on the occupancy classifications defined in Chapter 3 of the code. For instance, hospitals and health-care facilities are referred to as Group I-2 occupancies and are subject to the requirements of Risk Category IV. On the other hand, care facilities having Group I-1 classification are not intended to be assigned to Risk Category III or IV. Except for Condition 1, which is assigned to Risk Category III, all Group I-3 facilities such as jails and detention centers, have egress and free movement impeded by locks, rendering occupants incapable of self-preservation. These facilities are therefore assigned to Risk Category IV.

1604.5.1 Multiple occupancies. In cases where there are multiple occupancies in a structure, the highest (or most restrictive) risk category is to be assigned to the structure unless the portions are structurally separated. In other words, when a lower risk category impacts a higher risk category, the higher risk category must either be structurally independent of the other, or the two must be in one structure designed to the requirements of the higher risk category. In cases where the two uses are structurally independent but are functionally dependent, both portions are required to be assigned to the higher risk category. To be a structural separation for seismic applications, there must be a physical separation (not just an expansion joint) that conforms to the requirements of ASCE 7 Section 12.12.2. If portions do not meet the minimum separation distance, then they are not structurally separated and must be designed and interconnected to act as one unit.

An exception is provided to clarify risk category in buildings with storm shelters. As long as the shelter is constructed in accordance with ICC 500 *Standard for the Design and Construction of Storm Shelters*, the code does not require the design professional to consider the shelter occupancy in the risk category analysis, but rather allows the use of that space based upon the normal occupancy of the building to be considered. This exception does not apply if the space is a designated emergency shelter as listed in Table 1604.5.

1604.5.2 Photovoltaic (PV) panel systems. A ground-mounted PV system is defined as an independent system without usable space underneath, installed directly on the ground. An elevated PV system is defined as an independent PV panel support structure designed with usable space underneath, a minimum clear height of 7.5 feet, and intended for secondary use such as providing shade or parking for motor vehicles.

Ground-mounted PV panel systems serving only R-3 buildings meet the description of RC I, as they "represent a low hazard to human life in the event of failure" and are assigned to this category as a low-risk structure. Elevated PV support structures are assigned to Risk Category II as having a typical hazard risk similar to most buildings. Where PV panel systems are mounted on building roofs, whether attached or unattached, they are assigned the same risk category as the building on which they are mounted. Where PV panels plus an energy storage system (ESS) are the only direct source of backup power for an essential services facility (RC IV)—with a transfer switch or other equipment enabling the ESS to operate independently from the grid during a time of grid power outage—the panels and ESS are assigned to RC IV. If PV panels plus ESS are not designed to operate in the event of a grid power outage, then they may be assigned the same risk category as the building to which they supply power. There are cases where elevated PV support structures are installed on the same site as a Risk Category IV building over surface parking spaces that are designated for emergency services vehicles. The elevated PV support structures must be assigned as RC IV as damage to the panels or support structure can take the emergency vehicles out of service. See Table 1604-1.

Table 1604-1. **Risk Categories for Photovoltaic Panel Systems**

	PV Systems and Elevated PV Support	Risk Category (RC)			
		I	II	III	IV
1	Ground mounted for Group R-3 only	x			
2	Ground mounted other than 1 and 5		x		
3	Elevated other than 4, 5, and 6		x		
4	Rooftop and elevated PV on top of buildings	Same as building RC			
5	Paired with ESS and dedicated backup for RC IV building				x
6	Elevated and used for emergency vehicle parking				x

1604.6 In-situ load tests. Whenever there is reasonable doubt as to the stability or load-bearing capacity of a completed building, structure, or portion thereof for the expected loads, an engineering assessment may be required by the building official. The engineering analysis may involve either a structural analysis or an in-situ load test, or both. See IBC Section 1708 for more details.

1604.7 Preconstruction load tests. In evaluating the physical properties of materials and methods of construction that are not capable of being designed by approved engineering analysis, or that do not comply with applicable material design standards listed in Chapter 35, the structural adequacy must be predetermined based on the load test criteria given in Section 1709.

1604.8 Anchorage. This section addresses anchorage of various components of a building to resist uplift and sliding forces that result from application of code-prescribed lateral forces. It intends that all members be tied together or anchored to resist uplift and sliding forces. The section differentiates between uplift and sliding forces to be resisted in general (Section 1604.8.1) and the lateral support required for structural walls (Section 1604.8.2). Anchorage requirement of decks attached to an exterior wall is provided in Section 1604.8.3.

1604.8.2 Structural walls. These requirements pertain to "structural walls," which are essentially defined within this section to include load-bearing walls and shear walls. This section establishes the minimum wall anchorage capacity, based on the structure's seismic design category (SDC). In accordance with provisions of ASCE 7, Section 1.4.4 is referenced for the determination of minimum anchorage forces in structures that are classified as SDC A. The referenced ASCE 7 provision provides minimum connection forces based on the consideration of a minimum level of structural integrity, and in the context of Section 1604.8.2, it means that wall anchorage needs to be adequate to carry a horizontal force equal to 0.2 times the wall weight tributary to the anchor, but not less than 5 psf (0.239 kN/m^2) (see Figure 1604-4). For all other SDCs, the more stringent requirements in Section 12.11 of

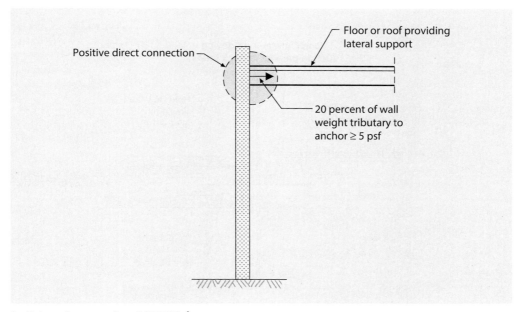

For SI: 1 pound per square foot = 0.0479 kN/m^2

Figure 1604-4 **Minimum anchorage of structural walls in SDC A.**

ASCE 7 are referenced. If the wall being laterally supported is of hollow-unit masonry or a cavity wall, the required anchors must be embedded in a reinforced grouted structural element of the wall.

Properly anchoring structural walls to the floor and roof diaphragms is perhaps the most important mechanism provided to keep a structure from collapsing. In addition to the IBC, the *International Existing Building Code®* (IEBC®) has numerous provisions within it that require an existing building undergoing alteration to have the roof-to-wall connections evaluated. This evaluation can be triggered by reroofing the building or by undergoing a significant remodel. The IEBC focuses specifically on buildings that are in high-wind regions, older concrete and masonry buildings with flexible diaphragms, or unreinforced masonry buildings that are in areas of moderate- to high-seismicity.

1604.8.3 Decks. When a deck is attached to an exterior wall, such attachment needs to be capable of resisting all vertical and lateral loads. As an example, the *International Residential Code®* (IRC®) provides prescriptive requirements for this attachment. For the vertical load attachment, the IRC specifies how the ledger is to be attached to the exterior wall based on deck joist spans. For the lateral connection, the IRC requires that tension devices, such as hold-downs, be used to tie the deck joists to the primary structure's framing members in at least two locations (see Figure 1604-5). Oftentimes, design professionals feel that the ledger attachment provides a sufficient vertical and lateral connection to the exterior walls; however, experience has shown that many decks pull away from the exterior walls as this ledger is typically attached directly to a rim/band joist as shown in Figure 1604-6. This highlights the need for a separate lateral attachment, similar to hold-downs at the ends of a shear wall, to keep the deck from pulling away from the primary structure (see Figure 1604-5).

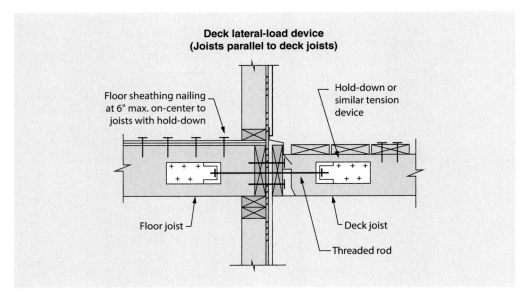

For SI: 1 inch = 25.4 mm

Figure 1604-5 Deck lateral tension device.

General Design Requirements   **597**

Figure 1604-6   **Deck failure.**

The IBC also clarifies that the vertical and lateral attachments must be verified during inspection. If these connections cannot be visually inspected, the deck will need to be capable of supporting itself without any attachment to the exterior wall.

When a deck is supported by an exterior wall at one end as well as by an intermediate support from a framing member (Figure 1604-7) such that a portion of the deck cantilevers beyond

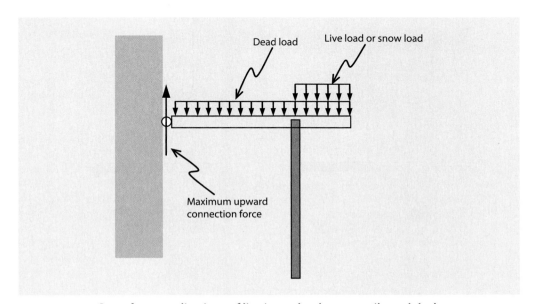

Figure 1604-7   **One of two applications of live/snow load on a cantileverd deck.**

the intermediate support, two load cases need to be considered for the purpose of designing the attachment to the wall. In the first case, in addition to the dead load, the full span of the deck is subjected to the live load or the snow load, whichever is higher. This load case provides the maximum possible downward reaction in the wall attachment.

In the second load case, only the cantilevered portion of the deck is subjected to the live load or the snow load in order to obtain the maximum uplift force in the wall attachment. In both cases, loads need to be combined in accordance with Section 1605 of the IBC.

1604.9 Wind and seismic detailing. The forces induced on a structure by a seismic event are the result of motions in the ground on which it rests. The response (i.e., the magnitude and distribution of forces and displacements) of a structure resulting from such ground motion is influenced by the properties of both the structure and the foundation, as well as the character of the exciting motion.

A simplified picture of the behavior of a building during an earthquake is illustrated in Figure 1604-8. As the ground on which the building rests is displaced, the base of the building moves with it. However, the inertia of the building's mass resists this motion and causes the building to suffer a distortion (greatly exaggerated in the figure). This distortion wave travels along the height of the structure in much the same manner as a stress wave in a bar with a

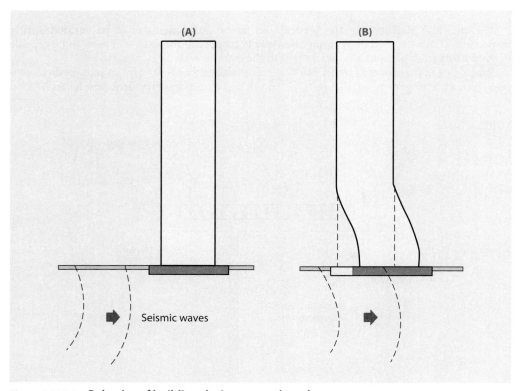

Figure 1604-8 **Behavior of building during an earthquake.**

free end. The continued shaking of the base causes the building to undergo a complex series of oscillations.

It is important to draw a distinction between forces that are due to wind and those produced by earthquakes. Although both wind and earthquake forces are dynamic in character, a basic difference exists in the manner in which they are induced in a structure. Whereas wind loads are external loads applied and, therefore, proportional to the exposed surface of a structure, earthquake forces are essentially inertial forces. The latter results from the distortion produced by both the earthquake motion and inertial resistance of the structure. Their magnitude is a function of the mass of the structure rather than its exposed surface. Also, in contrast to the structural response to essentially static gravity loading or even to wind loads, which can often be validly treated as static loads, the dynamic character of the response to earthquake excitation can seldom be ignored. Thus, although in designing for static loads one would feel greater assurance about the safety of a structure made up of members of heavy section, in the case of earthquake loading, a stiffer and heavier structure does not necessarily represent the safer design.

Experience has shown that structures designed to the level of seismic horizontal forces required by current codes can survive major earthquake shaking. This is because of the ability of well-designed structures to dissipate seismic energy by inelastic deformations in certain localized regions of structural members. This means avoiding all forms of brittle failure and achieving adequate inelastic deformability by the yielding of certain localized regions of certain members (or of connections between members) in flexure, shear, or axial action. This is precisely why the materials chapters (Chapters 19, 21, 22, and 23), reference design standards that provide detailing requirements and other limitations that go hand in hand with the code-prescribed seismic forces. The design earthquake forces (Chapter 16) and the detailing requirements and other restrictions of the materials chapters form an integral package for seismic design.

Wind and earthquake effects are not considered to occur simultaneously on the structure in U.S. design practice. Section 1605 and ASCE 7 contain gravity load combinations, combinations of gravity and wind loads, as well as combinations of gravity loads and earthquake forces. Design of every critical section of every structural member must be done considering all of these load combinations. If the gravity and wind load combinations produce demands that are closer to the design strength of a section than do the combinations of gravity loads and earthquake effects, then wind rather than earthquakes govern the design of that section. If wind loads are largest for every critical section of a structure, then wind governs the design of that entire structure. However, seismic detailing requirements of the materials chapters still need to be applied. Theoretically, it could be argued that if wind effects were larger than unreduced earthquake effects (earthquake effects corresponding to the elastic response of the structure to the design earthquake), then the detailing requirements of the code could be dispensed with. However, even that would not be allowed by the IBC. Totally irrespective of the severity of wind effects, the SDCs defined in Section 1613.2 determine the applicability of the seismic detailing requirements. The SDCs are used in the code, irrespective of the severity of wind effects, to determine permissible structural systems, limitations on height and irregularity, components of the structure that must be designed for seismic resistance, and the type of lateral force analysis to be performed.

1604.10 Loads on storm shelters. Section 423 of the IBC requires that storm shelters be provided for specific occupancies in areas subject to tornado wind speeds of 250 mph (110 m/s). This includes Group E occupancies (>50 occupants), 911 call stations, emergency operations centers, fire stations, and police stations. Structural loads and load combinations for storm shelters shall be in accordance with ICC 500 (Figure 1604-9).

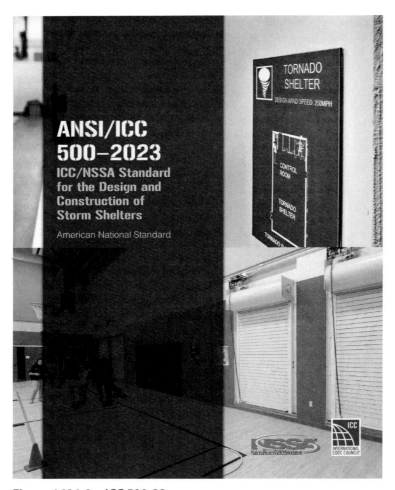

Figure 1604-9   ICC 500-23.

Section 1605 *Load Combinations*

1605.1 General. The importance of load combinations in structural design cannot be over emphasized. IBC Section 1605.1 notes that buildings, structures, and portions thereof must be designed to resist either strength or allowable stress design load combinations. IBC Chapter 16 assists designers in determining appropriate dead, live, snow, wind, tornado, seismic, and other loads that will be applied to a structure, but that is just part of the equation. The designer must also account for the potential simultaneous application of these loads.

Most structures will see most, if not all, of the loads noted above throughout their design life. These loads are applied to the structure at different times and to various magnitudes. As an example, it is not reasonable to require a designer to analyze a structure considering that the worst-case snow, seismic, and flood loads will act concurrently. For that reason, load combinations were developed that require consideration of a "portion" of different design loads acting concurrently on a structure or element. Load factors account for the reduced probability of simultaneous occurrence of transient live loads.

Figure 1605-1 shows the strength design load combinations that are provided in ASCE 7 Section 2.3.1. There are five basic load combinations provided in Section 2.3.1, which blend portions of dead, live, roof live, snow, rain, tornado, and wind loads. A designer must evaluate each load combination with the one that produces the highest load effect governing design of the building, structure, or portion thereof. While ASCE 7 Section 2.3 includes other load combinations for fluid loads, ice loads, lateral earth pressure, ground water pressure, and more, reviewing the basic load combinations in Figure 1605-1 will help clarify the importance of combining loads to find the governing combination that will control design of the structure or element.

#	Load Combination
1.	$1.4D$
2.	$1.2D + 1.6L + (0.5L_r$ or $0.3S$ or $0.5R)$
3.	$1.2D + (1.6L_r$ or $1.0S$ or $1.6R) + (L$ or $0.5W)$
4.	$1.2D + 1.0(W$ or $W_T) + L + (0.5L_r$ or $0.3S$ or $0.5R)$
5.	$0.9D + 1.0(W$ or $W_T)$

Notations
D = dead load
L = live load
L_r = roof live load
S = snow load
R = rain load
W = wind load
W_T = tornado load

Figure 1605-1 ASCE 7 strength design load combinations.

The design professional can choose to use one of three sets of load combinations when performing the analysis of a building, structure, or portion thereof. Those are (1) strength design load combinations, (2) allowable stress design (ASD) load combinations, or (3) alternate ASD load combinations. The strength design and ASD load combinations are referenced in the IBC to ASCE 7 Sections 2.3 and 2.4, respectively. The alternative allowable stress design load combinations are provided in Section 1605.2.

If a designer chooses ASD, the calculated load (i.e., required strength) must be less than the material or component strength (i.e., allowable strength). After calculating the required strength, the designer then analyzes the structure or component and applies a safety factor to obtain the allowable strength. The safety factor that is used is not associated with the load combinations but is based on the specific material being used. The safety factor varies based on the type of force being applied (e.g., axial, shear, bending moment). These safety factors have been established over years of experience, testing, and historical data. In many cases, the safety factor is built into published allowable design values for structural elements.

Strength design, also known as load and resistance factor design (LRFD), follows a slightly different path than ASD. When using LRFD, the designer still needs to show that the required strength is less than the strength of the structure or component. In LRFD, the term "available strength" or "reference resistance" is used instead of ASD's "allowable strength." Rather than utilize a safety factor, LRFD relies on resistance factors to determine the "available strength." These resistance factors vary for different material types as well as the type of load being applied (e.g., bending, tension, compression).

Designers must be careful when analyzing a building or structure and mixing both LRFD and ASD methodologies. As an example, it is quite common for foundations to be designed using ASD, while the steel structure the foundations are supporting is designed using LRFD.

Using different methods for these elements is quite common but requires that the design professional adjust applied loads accordingly. With that said, at no time should a designer mix both LRFD and ASD methodologies indiscriminately while designing a structural frame.

1605.2 Alternative ASD load combinations. The IBC alternative ASD load combinations have existed since the initial 2000 IBC and were included in legacy codes. Through the 2018 IBC, the alternative ASD load combinations permitted the use of a one-third increase in allowable stresses when evaluating load combinations containing short-term transient loads including wind and seismic loads. The basic ASD combinations did not permit the reduction in loads, but applied a factor of 0.75 to transient loads, including live, snow, wind, and seismic loads, when more than one of these loads was considered simultaneously.

The ASCE 7 load combinations have further permitted an increase in allowable stresses when a material, such as wood, has increased available strength under short-term loading, i.e., loads due to snow, rain, wind, or earthquakes. This increase is not intended to be used for the design of masonry, concrete, or steel structures, because the strength of these materials does not vary significantly with the duration of the load.

Unfortunately, the one-third stress increase has routinely been applied incorrectly. To limit misuse of the stress increase, the omega factor, ω, was deleted from the alternative ASD load combinations in the 2021 IBC.

In the 2024 IBC, a factor of 0.7 is applied to snow loads based on the change to strength-based snow loads (see Section 1608). Accordingly, Exception 2 was also revised to reflect a new strength-based threshold of 45 psf for flat roof snow loads when used in combination with seismic loads.

Section 1606 *Dead Loads*

1606.1 General. Dead loads consist of the weight of materials of construction incorporated into a building, including but not limited to walls, floors, roofs, ceilings, stairways, built-in partitions, finishes, cladding and other similarly incorporated architectural and structural items, and fixed service equipment, including the weight of cranes.

To establish uniform practice among designers, the ASCE 7 Commentary Tables C3.1-1 and C3.1-2 provide an extended list of weights for commonly used building materials that can be useful to the designer and building official alike.

1606.2 Weights of materials of construction. Dead loads associated with the actual weight of construction materials are what most people think of when considering dead load. While special cases will inevitably arise, authority is granted to the building official to deal with such cases. The next three sections serve as a reminder to both code officials and designers about dead loads that must be considered in a structural analysis.

1606.3 Weight of fixed service equipment. The weight of fixed service equipment includes both the empty weight of the equipment and the maximum weight of the contents. For example, the weight of liquids is to be included in the dead load of piping and tanks, and the weight of conduit and wiring is to be included in the dead load of cable trays. In addition, as content weight may be variable, it cannot be counted upon to counteract the effects of overturning, sliding, and uplift forces. Exceptions specifically address the calculation of variable loads for liquids and movable equipment.

1606.4 Photovoltaic panel systems. The weight of photovoltaic (PV) panel systems, their support, and ballast should be considered as dead load.

1606.5 Vegetative and landscaped roofs. The weight associated with rooftop landscaping, including the weight of vegetation and saturated soils must be considered as dead loads in the analysis. This section also points out that rooftop hardscaping (i.e., pavers, decomposed granite, etc.) must be calculated as a dead load.

The dead load must be computed considering both fully saturated soil and drainage layer materials and fully dry soil and drainage layer materials to determine the most severe load effects. For example, fully saturated soil and drainage layer material would likely be used to determine the dead load for seismic design while fully dry soil and drainage layer material would be used to offset wind uplift loads.

Section 1607 *Live Loads*

1607.1 General. Live loads are defined as loads produced by the use and occupancy of the building or other structure, and do not include construction or environmental loads such as wind, tornado, snow, rain, earthquake, flood, or dead loads.

1607.2 Loads not specified. For occupancies and uses not specifically included in Table 1607.1, the method of determination of the design live load is subject to the approval of the building official. Extremely valuable information is provided in the commentary to ASCE 7, Tables C4.3-1 and C4.3-2, concerning determination of design live loads for occupancies not listed in Table 1607.1.

1607.3 Uniform live loads. This section charges the designer to use the unit live loads set forth in Table 1607.1 and specifies that these loads must be considered minimum live loads. In other words, floors must be designed for the maximum live loads to which they are likely to be subjected during the life of the building based on its intended use, but in no case should the design loads be less than those given in Table 1607.1. The commentary to ASCE 7 advises that in selecting the occupancy and use for the design of a building, the owner should consider the possibility of later changes of occupancy involving loads heavier than originally contemplated. The lighter loading appropriate to the first occupancy should not necessarily be selected, when an owner has reason to anticipate different uses for the building in the future. The owner's planning should also consider the possibility of temporary changes in the use of a building as in the case of clearing a portion of dormitory space for a dance or other recreational activity. To help prevent floors from being overloaded, Section 106 of the code provides requirements for the posting of live loads. In addition, Section 106.3 makes it unlawful to place a load greater than the code permits on any floor or roof of a building.

1607.3.1 Partial loading of floors. Where loads are uniformly distributed on continuous structural members, they shall be arranged so as to create maximum bending moment in any given critical section. This may require a design to consider so-called skip loading or alternate span loading, as shown in Figure 1607-1.

1607.3.2 Partial loading of roofs. This is nearly identical to Section 1607.12 (see discussion above). However, the provision applies only when the uniform roof live loads are reduced to less than 20 psf (0.96 kN/m^2). ASCE 7 Section 7.5 is referred to for consideration of partial loading of snow.

1607.4 Concentrated live loads. Many uses are susceptible to the movement of equipment, files, machinery, and so on. Therefore, the code requires that floors and roofs, which are listed in Table 1607.1, be designed for the indicated concentrated load placed on a space 2½ feet square (762 mm by 762 mm) whenever this load, on an otherwise unloaded floor or roof, produces stresses greater than those caused by the uniform loads required by the code. As this concentrated

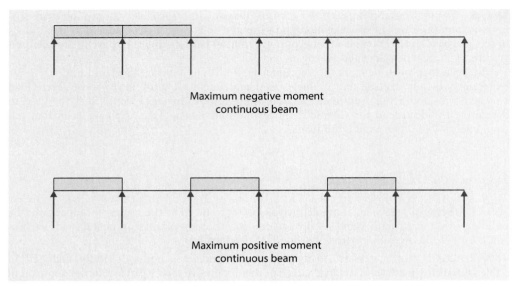

Figure 1607-1   **Alternate span loading for continuous beams.**

load can take many forms in the occupied structure, and as the structural engineer usually does not know in advance what form the load will take and how it will be applied, the best compromise to cover most situations is to consider the concentrated load to be applied through a rigid base 2½ feet square (762 mm by 762 mm). Note that the term "load effects" refers not only to member forces and the resulting stresses, but also to member deformations (see deflection criteria in Section 1604.3).

1607.5 Partition loads. In those uses where partition arrangements are periodically reconfigured such as office buildings and flexible-plan school buildings, the code requires that the floor system be designed to support a partition live load of 15 psf (0.72 kN/m^2). This partition load does not need to be considered where the floor live load is 80 psf (3.83 kN/m^2) or greater. This requirement is irrespective of whether partitions are shown on the construction documents. The 15 psf (0.72 kN/m^2) value was arrived at by assuming 10-foot-high (3,048 mm) partition walls of wood or steel stud wall construction with ½-inch (12.7 mm) gypsum board on each side, and arranged in a square grid of 10-foot (3,048 mm) sides. This assumption was thought to provide a fairly conservative estimate of partition loading. However, the ASCE 7 Commentary also advises designers to consider a larger partition load if a higher density of partitions is anticipated. It should be noted that the uniformly distributed load of 15 psf (0.72 kN/m^2) is considered by the code to be a live load. Thus, it should be included with other live loads in the load combinations. However, because the nature of the distribution of partitions is fundamentally different from office live loads, the partition live load is not reducible in the design for gravity loads. It should be noted that where the 15 psf (0.72 kN/m^2) gravity load is used, ASCE 7 requires a minimum of 10 psf (0.48 kN/m^2) to be added to the effective seismic weight.

1607.6 Helipads. Minimum design loading criteria for helicopter landing surfaces includes:

1. A minimum uniform live load.
2. The weight of the helicopter.
3. The landing impact effect of the helicopter.

A live load of 40 psf (1.92 kN/m²) is applied when the landing area is used by helicopters with a take-off weight not exceeding 3,000 pounds (13.35 kN). For larger helicopters, a 60-psf (2.87 kN/m²) design live load would apply. Landing pads must be labeled to indicate the maximum takeoff weight in the bottom right corner when viewed from the primary approach path.

The takeoff weight limitation is indicated in units of thousands of pounds and placed in a box at least 5 feet in height. See Figure 1607-2.

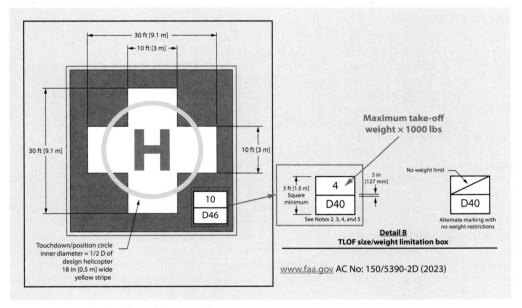

Figure 1607-2 **Helipad marking requirements.**

Helipads are also to be designed for a combination of concentrated live loads to represent the landing impact of the helicopter. These concentrated loads are not required to act concurrently with other uniform or concentrated live loads.

1607.7 Passenger vehicle garages. Garages and other areas where motor vehicles are stored are to be designed for concentrated loads or, alternatively, the uniform loads as specified in Table 1607.1. In the case of garages for storage of private pleasure-type vehicles, Section 1607.7 prescribes a single 3,000-pound (13.35 kN) load over a 4.5-inch by 4.5-inch (114 mm by 114 mm) area. For only mechanical parking structures without slab or deck for the storage of passenger cars, a load of 2,250 pounds (10 kN) per wheel is specified.

1607.8 Heavy vehicle loads. This section provides criteria for addressing heavy vehicle loads—those with gross weights exceeding 10,000 pounds (4,536 kg). Where heavy highway-type vehicles have access onto a structure, this section requires that the structure be designed using the same code and requirements that are applicable to roadways and bridges in that jurisdiction. This loading could be the loading from AASHTO's *LRFD Bridge Design Specifications*, or the loading specified by the jurisdiction for elements such as lids of large detention tanks or utility vaults. The registered design professional should consult with the jurisdiction for design loads that are applicable.

This section also establishes loading criteria for certain categories of heavy vehicle loads such as fire trucks and other similar emergency vehicles—in addition to forklifts and other movable equipment. As a precaution against overloading a structure, the maximum weight of the vehicles that are anticipated and used in the design should be posted by the owner or owner's agent—see also Section 106.1.

1607.9　Loads on handrails, guards, grab bars, and seats. The majority of this section is adapted from ASCE 7. These requirements are intended to provide an adequate degree of structural strength and stability to handrails, guards, grab bars, accessible seats and benches, and their attachments.

1607.9.1　Concentrated load. The basic requirement of this section calls for application of a 200 pound (0.89 kN) concentrated load and Section 1607.9.1.1 requires a 50-plf (0.73 kN/m) uniform load for handrails and guards. The uniform and concentrated loads need not be applied concurrently. The second exception in Section 1607.9.1.1 allows the design load to be reduced from 50 to 20 plf (0.73 to 0.29 kN/m) for guards that are within a Group I-3, F, H, or S occupancy in an area that is not accessible to the general public and that has an occupant load of less than 50. These are the same occupancies listed in Exception 4 to Section 1015.4, which are permitted to have larger openings in the guard.

The first exception of Section 1607.9.1.1 states that while one- or two-family dwellings are exempted from the requirement for a minimum distributed load of 50 plf (0.73 kN/m), the requirement for a single concentrated load of 200 pounds (0.89 kN) specified in Section 1607.9.1 still applies.

The third exception in Section 1607.9.1.1 only requires handrails and guards to be designed for a 200-pound concentrated load for unoccupiable roofs. Unoccupiable rooftops are not factory, industrial, or storage occupancies and therefore do not qualify for a reduced uniform live load; however, unoccupiable roofs have, at most, a few maintenance workers on them at intermittent times and arguably pose less hazard to the public than guard in one- and two-family dwellings and the other occupancies to which this exception currently applies. Unoccupiable rooftop areas meet the two other requirements—namely that they are areas not accessible to the public and serve an occupant load not greater than 50. Therefore, the uniform live load is dropped, and the concentrated live load requirement is maintained as rooftops remain a distinct fall hazard.

Section 1607.9.1.2 requires that guard components such as balusters, panel fillers, and guard infill must be designed to resist a 50 pound (0.22 kN) concentrated load per ASCE 7 Section 4.5.1.2.

1607.9.2　Grab bars, shower seats, and accessible benches. The components listed in this section are those that are typically required in providing accessibility in accordance with Chapter 11. Where accessibility is required, IBC Section 1102.1 mandates conformance with ICC A117 *Accessible and Usable Buildings and Facilities*. ICC A117 determines where elements such as grab bars or shower seats must be provided. The design loading on these elements that is specified herein is also consistent with *ADA Standards for Public Accommodations and Commercial Facilities*.

1607.10　Fixed ladders. Ladders are to be designed to resist a single 300-pound (1.33-kN) concentrated load while each side rail extension should be designed for a single concentrated load of 100 pounds (0.445 kN). This section also highlights that ship's ladders should be designed to resist the loads prescribed for stairs in Table 1607.1.

1607.11　Vehicle barriers. This section requires the application of a 6,000-pound (26.70 kN) design load that accounts for impact. Because of differing vehicle bumper heights, barrier configurations, and anchorage methods, the specified load must be applied over a range of heights so that the maximum load effects are determined for design of the barrier.

1607.12　Impact loads. These provisions are adapted from ASCE 7. Where unusual vibration or impact forces are likely to occur, their effect may be to produce additional stresses and deflections in the structural system. This section requires that the structural design take these effects into account. Impact loads specifically listed in the IBC include elevators, machinery, facade and building maintenance attachments, and fall arrest or lifeline anchorages.

1607.13 Reduction in uniform live loads. The live load reduction provisions of Section 1607.13.1 are based on ASCE 7, while the alternative floor live load reduction provisions are based on legacy model codes. Small portions of a floor are more likely to be subjected to the full uniform load than larger floor areas. Unloaded or lightly loaded areas tend to reduce the total load on the structural members supporting those floors. In recognition that the larger the tributary area of a structural member, the lower the likelihood that the full live load will be realized, the specified uniformly distributed live loads from Table 1607.1 are permitted to be reduced.

The alternative floor live load reduction provisions, based on tributary floor area (Figure 1607-3), is used for reducing floor live loads only. Roof live load reductions are covered in Section 1607.14.

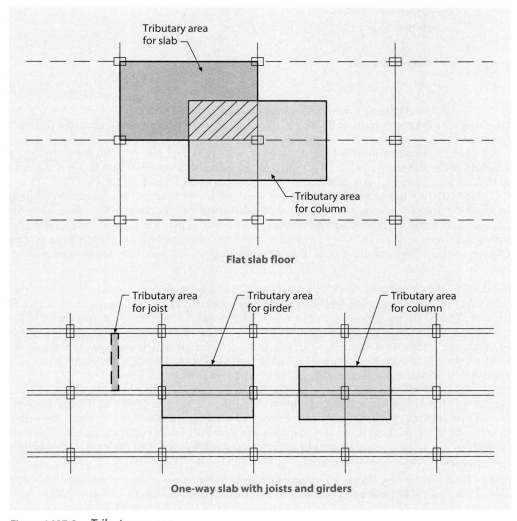

Figure 1607-3 **Tributary areas.**

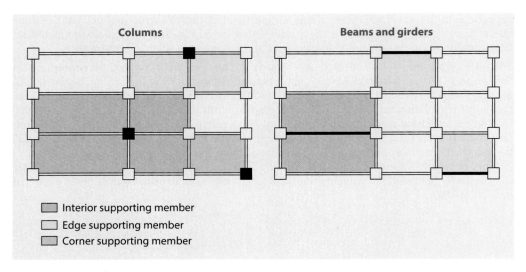

Figure 1607-4 Influence areas.

Figure 1607-4 illustrates the influence area concept. The influence area is considered to be the floor area over which the influence surface for structural effects is significantly different from zero.

In the basic uniform live load approach of Section 1607.13.1, the permitted reduction is a function of the tributary area, A_T, multiplied by the element factor, K_{LL}. This factor is the ratio of the influence area (A_I) of a member to its tributary area (A_T), i.e., $K_{LL} = A_I/A_T$.

Table 1607.13.1 has established K_{LL} values (derived from calculated K_{LL} values) to be used in Equation 16-7 for a variety of structural members and configurations. K_{LL} values vary for column and beam members having adjacent cantilever construction, and Table 1607.13.1 values have been set for these cases to result in live load reductions that are slightly conservative (Figure 1607-5). For unusual shapes, the concept of significant influence for structural effect needs to be applied.

1607.13.1 Basic uniform live load reduction. Reductions in the minimum uniformly distributed live load are permitted, based on an influence area, $K_{LL} A_T$, of 400 square feet (37.16 m²) or more. Essentially, the influence area of a structural element is the total floor area surrounding the element from which it derives any of its loads. The basis for the permitted reduction is that in the design of structural elements with large influence areas, it is highly unlikely that the floors will be fully loaded over their entire area. The reduction of certain live loads is prohibited unless a specific exception applies—see Sections 1607.13.1.2 and 1607.13.1.3.

Live load reduction in excess of 50 percent is not permitted for columns or other structural elements (such as bearing walls) that support the loads of a single floor (see Figure 1607-6). In essence, this means that the influence area may not exceed 3,600 square feet (334.4 m²) in calculating the reduced unit floor live load. For columns or other structural elements that support two or more floors, the sum of the reduced live loads from all floors must not be less than 40 percent of the sum of the unreduced live loads.

1607.13.1.1 One-way slabs. This section allows live load reduction in one-way slabs—see Table 1607.13.1 for K_{LL} value of a one-way slab. Although both one-way and two-way slabs have the same K_{LL} value of 1.0, this section imposes an upper limit on the slab width that can be used

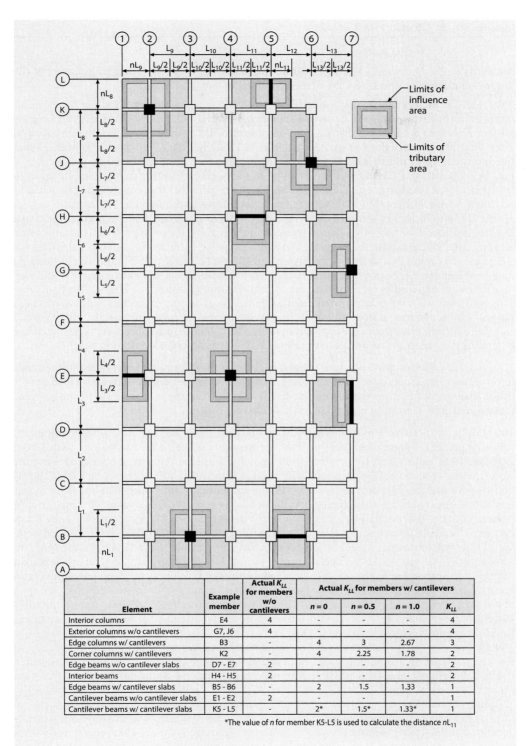

Element	Example member	Actual K_{LL} for members w/o cantilevers	Actual K_{LL} for members w/ cantilevers			K_{LL}
			$n = 0$	$n = 0.5$	$n = 1.0$	
Interior columns	E4	4	-	-	-	4
Exterior columns w/o cantilevers	G7, J6	4	-	-	-	4
Edge columns w/ cantilevers	B3	-	4	3	2.67	3
Corner columns w/ cantilevers	K2	-	4	2.25	1.78	2
Edge beams w/o cantilever slabs	D7 - E7	2	-	-	-	2
Interior beams	H4 - H5	2	-	-	-	2
Edge beams w/ cantilever slabs	B5 - B6	-	2	1.5	1.33	1
Cantilever beams w/o cantilever slabs	E1 - E2	2	-	-	-	1
Cantilever beams w/ cantilever slabs	K5 - L5	-	2*	1.5*	1.33*	1

*The value of n for member K5-L5 is used to calculate the distance nL_{11}

Figure 1607-5 **Typical tributary and influence areas.**

to calculate A_T of a one-way slab as 1.5 times the slab span. This restriction is indicative of the lower redundancy in one-way slabs when compared to two-way slabs.

1607.13.1.2 Heavy live loads. In the case of occupancies requiring relatively heavy minimum uniform live loads, such as storage buildings, several adjacent floor panels can be fully loaded. Field surveys indicate that rarely is any story loaded with an average actual live load of more than 80 percent of the average design live load. Thus, the ASCE 7 committee concluded that the minimum uniform live load should not be reduced for the floor and beam design, but that it may be reduced a flat 20 percent for the design of members supporting more than one floor. In Exception 1, the IBC further qualifies this allowance to require that the reduction be calculated in accordance with Section 1607.13.1, with the maximum reduction limited to 20 percent.

The IBC also includes a second exception to the prohibition on reduction, permitting additional live load reduction for uses other than storage when the registered design professional can provide an acceptable substantiation for doing so. The rationale is that there can be uses other than storage where the maximum design live loads may exceed 100 psf (4.79 kN/m^2), but the average load on members with large tributary areas may be less. For example, floors supporting heavy machinery may have very high uniform loads that are concentrated mostly over a small part of the floor area. This provision will allow the registered design professional to present to the building official a rational load reduction proposal if those scenarios apply. Since there are no specific criteria stated and the reduction is subject to the approval of the building official, this exception is very much like reiterating the concept of an alternative method of design as described in Section 104.2.3.

1607.13.1.3 Passenger vehicle garages. There are no significant variations in the loads imposed on these facilities, which are often fully loaded, thus the prohibition on live load reductions. An exception permits a 20-percent live load reduction for members supporting two or more floors. The reasoning is the same as provided for Section 1607.13.1.2.

1607.13.2 Alternative uniform live load reduction. This section establishes a minimum tributary area of 150 square feet (13.94 m^2) as the threshold for live load reductions computed from Equation 16-8. Figure 1607-6 provides a comparison of live load reductions calculated using the basic live load reduction of IBC Equation 16-7 and the alternative live load reductions of IBC Equation 16-8 for members supporting one floor. For members supporting more than one floor, higher reductions are permitted with both methods. Live loads in excess of 100 psf (4.79 kN/m^2) may not be reduced with two exceptions. First, the design live loads on members supporting two or more floors may be reduced by 20 percent, and second, for usage other than storage, reduction is permitted when found acceptable by a registered design professional through rational analyses, as explained in Section 1607.13.1.2 above. Also, reduction is not permitted for passenger-vehicle garages except for a maximum 20-percent reduction for members supporting two or more floors. An upper limit is also specified for the tributary width of a one-way slab for reduction calculations as 0.5 times the span of the slab, as explained in Section 1607.13.1.1. The maximum live load reduction permitted is 40 percent for members receiving loads from one floor level only and 60 percent for other members (such as columns or transfer girders).

Equation 16-8 was derived so that if a structural member supporting a tributary area of sufficient size to qualify for the maximum reduction allowed by the equation was subjected to the full design live load over the entire area, the overstress would not exceed 30 percent (Figure 1607-6).

It may be noted from Equation 16-9 that the maximum live load reduction is proportional to the ratio of dead load to live load. Therefore, for heavy framing systems, the reduction is permitted to be greater than it would be for lighter framing systems. This reflects the thinking that for a given magnitude of overload on a structural system, the system with the heavier dead load is overstressed proportionately less than one with a lighter dead load. For example, if a floor system weighing 30 pounds per square foot (psf) (1.44 kN/m^2) and designed for a live

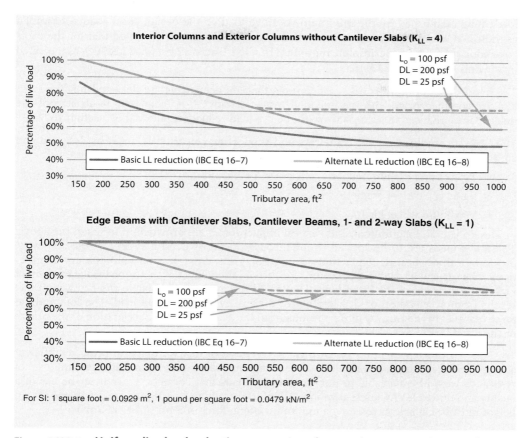

Figure 1607-6 Uniform live load reduction comparison for members supporting one floor.

load of 40 psf (1.92 kN/m²) was subjected to a 20-psf (0.96 kN/m²) overload, the amount by which the structural system would be overloaded is about 30 percent, assuming the system was designed to support just the minimum design live load of 40 psf (1.92 kN/m²). If this floor had a dead load of 60 psf (2.87 kN/m²), the overload would be only 20 percent, again assuming that the system was designed to support just the 40-psf (1.92 kN/m²) live load.

1607.14 Reduction in uniform roof live loads. The reduced load values that are permitted are meant to act vertically upon the projected area and have been selected as minimum roof live loads, even in localities where little or no snowfall occurs. This is because it is considered necessary to provide for occasional loading that is due to the presence of workers and materials during maintenance or repair operations. The live load reduction for roofs is a function not only of tributary area, but also of the slope of the roof. This is because it becomes less probable that the loads on a roof member will reach the maximum levels as the slope of the roof increases. It is also worth noting that no live load reduction is allowed for awnings and canopies consisting of fabric construction supported by a lightweight rigid skeleton structure—see Section 1607.14.1.

Where the design roof snow load exceeds the minimum roof live load in the applicable load combination(s), the snow load is to be used for design of the roof. Only the greater of the roof load value established by the minimum roof live load or the design snow load determined as indicated above is required to be applied to the roof. It should be noted that for the roofs covered in this section the absolute minimum roof live load is 12 psf (0.58 kN/m^2), as indicated by the limits in Equation 16-10.

1607.14.2 Occupiable roofs. Since roofs that are occupiable are designed for live loads that are commensurate with the intended use—see Item 28 of Table 1607.1—these roof live loads may be reduced in accordance with the live load reductions allowed by Section 1607.13.

1607.15 Crane loads. All craneways and supporting construction must be designed and constructed in compliance with ASCE 7 Section 4.9. The crane live loads depend on:

- Type of crane (monorail, bridge (classes A-F), hand-geared)
- Rated capacity of crane
- Maximum wheel loads

Design lateral, longitudinal, and vertical impact forces are provided in terms of the above. These live loads are to be applied simultaneously to the structural system of the craneway, including runway beams, connections, support brackets, cross-bracing, columns, and foundations. The vertical impact force accounts for the vibration effect of the crane bridge movement and the movement of the lifted load. The lateral force (perpendicular to the runway girder) results from acceleration or deceleration of the trolley and the lifted load. The longitudinal force (parallel to the runway girder) results from the acceleration or deceleration of the bridge or the lifted load.

1607.16 Interior walls and partitions. The intent of this section is to provide sufficient strength, stiffness, and durability of wall framing and wall finish, so that a minimum level of resistance would be available to nominal impact loads that commonly occur in the use of a facility and due to HVAC pressurization. All interior walls and partitions greater than 6 feet in height must be capable of resisting a minimum lateral force of 5 psf (0.240 kN/m^2).

1607.16.1 Fabric partitions. Fabric partitions are defined in Section 202 as follows:

FABRIC PARTITIONS. A partition consisting of a finished surface made of fabric, without a continuous rigid backing, that is directly attached to a framing system in which the vertical framing members are spaced greater than 4 feet (1,219 mm) on center.

The definition clearly differentiates them from other more traditional-type partitions, which include partial-height office partitions that contain rigid panels finished with fabric and attached to a rigid frame. In the case of a fabric partition, there is no rigid panel to which the fabric is attached. The fabric simply spans the open space between the rigid frame over which it is stretched and attached. In the definition, it states that the vertical framing members are spaced greater than 4 feet (1,219 mm) on center, again to differentiate this type of partition from a more traditional partition where the vertical framing members are spaced at 4 feet (1,219 mm) on center or less. Typically, these partitions are not intended to be full (ceiling) height, so they would normally not be attached directly to the ceiling. They are usually supported by the floor, except under conditions where, because of the layout and the height of the ceiling, it may be more appropriate to hang the partition from the ceiling grid or use a combination of floor supports and ties to the ceiling grid or a special structural ceiling grid designed to support such partitions. Figure 1607-7 shows some examples of fabric partitions.

The intent of the concentrated load of 40 pounds (0.176 kN) was to ensure that the partition does not tip over if someone were to inadvertently lean up against the frame or the fabric, and that a person inadvertently leaning against the fabric would not cause the fabric to tear and the person to fall abruptly. In the case of the 5-psf (0.24 kN/m^2) horizontal load, it was decided

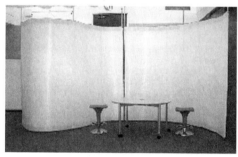

Figure 1607-7 **Examples of fabric partitions.**

that 5 psf (0.24 kN/m²) is to be distributed by calculating the total load based on the area of the fabric and then having that load distributed proportionally over the horizontal and vertical structural framing members of the partition. Thus, the framing system, in effect, will be resisting the total horizontal distributed load of 5 psf (0.24 kN/m²) even though 5 psf (0.24 kN/m²) is not applied over the field of the fabric between the supports.

1607.16.2 Fire walls. IBC Chapter 7 requires fire walls to be structurally stable, which allows for structural collapse on either side without collapse of the wall. The IBC indicates that fire walls are required to resist a 5-psf (0.24 kN/m²) horizontal ASD load.

1607.17 Library stack rooms. While Table 1607.1 prescribes uniform and concentrated live loads for libraries, this section of the code clarifies that those load provisions assume a maximum book stack height of 90 inches (2,286 mm), shelf depth of no more than 12 inches (305 mm), and aisles separating double-faced book stacks having a minimum width of 36 inches (914 mm). Should conditions exceed any of these limitations, the live loads used for the structural analysis should be approved by the building official.

1607.18 Seating for assembly uses. This section refers the designer to the ICC *Standard for Bleachers, Folding and Telescopic Seating and Grandstands* (ICC 300) for the structural design requirements associated with those structures and then prescribes requirements for the horizontal sway loads that must be considered in relation to fixed seats provided in stadiums and arenas. In this instance, stadiums with fixed seats must consider the standard live loads prescribed in Section 1607 in addition to the horizontal sway loads stipulated in Section 1607.18.1.

1607.18.1 Horizontal sway loads. Stadium or arena fixed seats require design to resist 24 pounds per linear foot (plf) (0.35 kN/m) parallel to each row of seats and 10 plf (0.15 kN/m) perpendicular to each row. These loads are not to be applied simultaneously.

1607.19 Sidewalks, vehicular driveways, and yards subject to trucking. This section requires any flatwork that will be subject to truck loading to consider an appropriate uniform live load as well as the concentrated wheel load listed in Table 1607.1 applied over an area of 4.5 inches by 4.5 inches (114 mm by 114 mm).

1607.20 Stair treads. The concentrated stair live load noted in Table 1607.1 should be applied over an area of 2 inches by 2 inches (51 mm by 51 mm) and does not need to be considered to act at the same time as the prescribed uniform live load.

1607.21 Residential attics. This section describes three separate attic live loads that could exist. The first is listed as an "uninhabitable attic without storage," then there is an "uninhabitable attic with storage," and finally "attics served by stairs."

1607.21.1 Uninhabitable attics without storage. Table 1607.1 notes that uninhabitable attics without storage should be designed considering a live load of 10 psf (0.48 kN/m^2). These are attics where the clear height between the rafter and ceiling joists, or the clear height between top and bottom chord of trusses (considering web configurations), is less than 42 inches (1,067 mm).

1607.21.2 Uninhabitable attics with storage. Figure 1607-8 shows the requirements for uninhabitable attics with storage. This occurs when a clear height of 42 inches (1,067 mm) is provided between a joist and a rafter, or if a box having a height of 42 inches (1,067 mm) and width of 24 inches (610 mm) can be accommodated between the webs of two adjacent trusses. In this instance the bottom chord of the truss, or the ceiling joist, must be designed for a uniform live load of 20 psf (0.96 kN/m^2).

1607.21.3 Attics served by stairs. Attic spaces accessible by stairways other than the pull-down type need to be designed to support the habitable attic and sleeping room live load of at least 30 psf (1.44 kN/m^2).

1607.22 Photovoltaic panel systems. Several sections add specific design requirements for roof structures supporting PV solar panels and modules. Applicable uniform and concentrated roof loads need to be considered both with and without PV panel system dead loads. Roofs that support solar panels must be designed for the full panel and ballast dead load, including concentrated loads from support frames in combination with roof live load, and any other applicable loads, including any loads due to snow drifting. Elevated PV support structures with open grid framing and without a roof deck or sheathing must be designed to support uniform and concentrated roof live loads, except that the uniform roof live load can be reduced to 12 psf.

Section 1608 *Snow Loads*

The snow load provisions reference ASCE 7 Chapter 7. Only the provisions regarding determination of ground snow loads in the contiguous United States and Alaska are contained in the code, which many building officials may wish to specify when adopting the model code by local ordinance. Based on the provisions in the IBC and ASCE 7, Figure 1608-1 has been developed to show design snow load determination for new buildings. The variables that must be determined for design snow load calculations include ground snow load, exposure factor, and thermal factor. Other considerations include a rain-on-snow surcharge, partial loading, and ponding instability from melting snow or rain on snow. The discussions about these provisions are presented later.

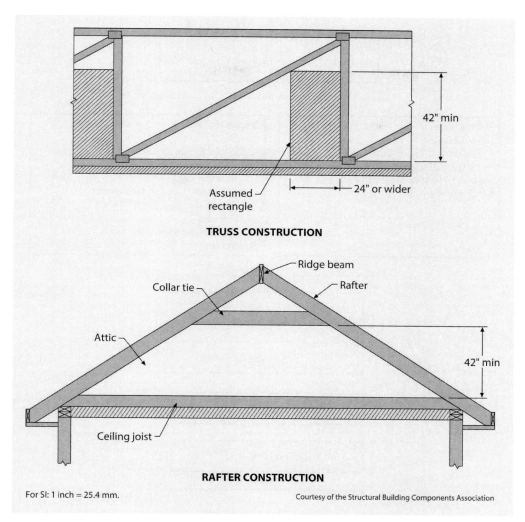

Figure 1607-8 **Uninhabitable attics with storage.**

1608.1 General. The IBC stipulates in this section that design snow loads be determined in accordance with ASCE 7 Chapter 7. Requiring that the design roof load shall not be less than the roof live loads that are determined in Section 1607 merely restates what is required by the load combinations of Section 1605.

1608.2 Ground snow loads. Figures 1608.2(1) through 1608.2(4) for the contiguous United States and Table 1608.2 for Alaska give the ground snow loads to be used in determining design snow loads for roofs. The IBC states: "Snow loads are zero in Hawaii, except in mountainous regions as approved by the building official." The ground snow loads on the maps and in the location-specific Ground Snow Load Geodatabase of the ASCE Hazard Tool (ascehazardtool.org) are based on the risk category of the building or structure. See Figure 1608-2. Each of these maps (and associated datasets) is based on calculations that target the reliability objectives of

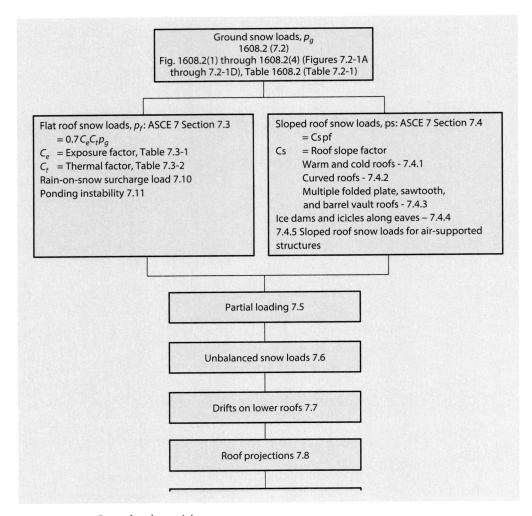

Figure 1608-1 **Snow load provisions.**

Chapter 1 of ASCE 7-22. The adoption of reliability-targeted (strength-based) design ground snow loads is a significant change from ASCE 7-16 and prior editions, which previously used ground snow loads with a 50-year mean recurrence interval (MRI). The mapped values indicate ground snow loads in pounds per square foot (psf).

Regardless of whether one is selecting the ground snow load from the IBC figures, ASCE 7 or a case study, the design professional should always verify the snow load with the local building official prior to initiating their design of the building or structure.

1608.2.1 Ground snow conversion. Given that ground snow load values have been provided as allowable stress loads up to the 2024 IBC, many provisions within the IBC rely on allowable stress design (ASD) values. Therefore, Section 1608.2.1 provides a conversion from the strength-based ground snow load values now provided to an equivalent ASD value, $p_{g(asd)}$. Note that the ASCE Hazard Tool (ascehazardtool.org) also provides ASD ground snow loads.

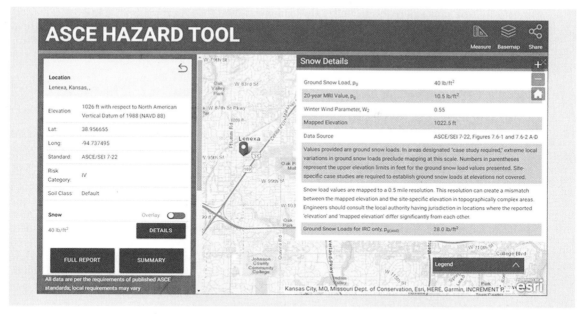

Figure 1608-2   ASCE Hazard Tool web application for snow loads.

ASCE 7 Section 7.3   Flat Roof Snow Loads.   This section converts the ground snow load, p_g, to flat roof snow load, p_f. This calculation utilizes an exposure factor and a thermal factor. The effects of exposure are handled with two factors. First, Equation (7.3-1) contains a basic exposure factor of 0.7. Second, the type of surface roughness and roof exposure are handled by exposure factor, C_e. This two-step approach generates ground-to-roof load reductions that range from 0.49 to 0.84.

ASCE 7 Section 7.3.1   Exposure Factor, C_e.   The roof exposure factor depends on the wind exposure category (or terrain category) as well as the exposure of the roof, as described in the footnotes to Table 7.3-1. The roof snow load is higher for a sheltered site such as a wooded area than it is for an exposed site that is flat and open. Table 7.3-1 footnotes require that the terrain category and roof exposure condition chosen shall be representative of the anticipated conditions during the life of the structure. An exposure is required to be determined for each roof of a structure.

ASCE 7 Section 7.3.2   Thermal Factor, C_t.   The thermal factor accounts for the heat that is transmitted from the interior of the structure, which tends to reduce the snow depth on the roof. Structures that are kept just above freezing, that are unheated, or freezer buildings have a C_t factor of greater than 1.0 per ASCE 7 Table 7.3-2 since snow will stay on the roof of these structures longer than it would on that of a heated structure. For heated structures with unventilated roofs, ASCE 7 Table 7.3-3 provides thermal factors based on a ground snow load ranging from 10 to 70 psf and the insulation value of the roof (R_{roof} or U_{roof}). C_t is larger for buildings in warmer climates with lower p_g values and smaller for buildings in colder climates with higher p_g values.

ASCE 7 Section 7.3.3   Minimum Snow Load for Low-Slope Roofs, p_m.   This section accounts for a number of situations where the ground-to-roof conversion factor of 0.7 as well as the factors C_e and C_t may underestimate the snow load, such as the load resulting from a single storm in an area where the ground snow load is less than shown in ASCE 7 Table 7.3-4. This section also includes the clarification that it is not necessary to combine the minimum snow load with drift load and sliding snow load, as well as unbalanced or partial snow loads. When this minimum snow load is applicable, it is considered a separate snow load case.

ASCE 7 Section 7.4   Sloped Roof Snow Loads, p_s.   The flat roof snow load is used to determine the sloped roof, or balanced, snow load. It does so by applying the roof slope factor C_s. When the value of C_s is 1.0, then the snow load that applies is essentially the flat roof snow load. Roof slope factors are given by Section 7.4.1 for warm and cold roofs, Section 7.4.2 for curved roofs, and Section 7.4.3 for multiple folded plate, sawtooth, and barrel vault roofs. The standard notes, parenthetically, that this sloped roof snow load is the "balanced" snow load. Section 7.6 further clarifies that the balanced and unbalanced snow loads are analyzed separately.

ASCE 7 Section 7.5   Partial Loading.   In many situations, a reduction in snow load on a portion of a roof by wind scour, melting, or snow removal operations may simply reduce the stresses in the supporting members. However, in other cases, removal of snow from an area may induce higher stresses in the roof structure than can occur when the entire roof is loaded. Cantilevered roof joists are a good example; removing half the snow load from the cantilevered portion increases the bending stress and deflection of the adjacent continuous span. In some situations, adverse stress reversals may result. This section requires consideration to be given in design to those adverse situations.

ASCE 7 Section 7.6   Unbalanced Roof Snow Loads.   Snow on the roof of a building rarely accumulates evenly. The code intends that the designer investigate conditions of imbalance by requiring that roof designs consider unbalanced snow loading. One case would be the loading of one slope of a gable roof with snow while the other slope is unloaded (Figure 1608-3). If this creates a less favorable condition in the design of the roof structural members, then it is the loading to be considered. This type of loading covers the case where the snow may have been removed from one side of the roof because of sliding or melting.

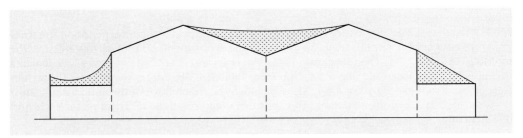

Figure 1608-3   **Potential snow accumulation, gable roofs with sidesheds.**

ASCE 7 Section 7.7   Drifts on Lower Roofs (Aerodynamic Shade).   It is extremely important to consider localized drift loads when designing roofs, because drifts onto lower roofs are a common cause of roof failures after a heavy snow (Figure 1608-3). This section separately addresses windward drift that forms on the windward side of a high-bay wall area, and leeward drift that forms on the leeward side of a high-bay wall area. It provides the formulations for determining drift load intensity and prescribes which portion of the lower roof is subject to drift loading. Provisions are also included for drifting due to adjacent structures and intersecting drifts.

ASCE 7 Section 7.8   Roof Projections and Parapets.   Solar panels, mechanical equipment, parapet walls, and penthouses are examples of roof projections that may cause windward drifts on the roof around them. The drift-load provisions of this section cover most of these situations adequately. Drift loads are not required when the side of the projection is less than 15 feet (4,572 mm) or when the projection has a clear distance of at least 2 feet (610 mm) to the height of the balanced snow load as shown in Figure 1608-4.

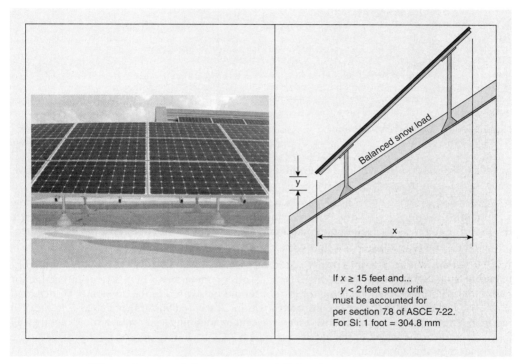

Figure 1608-4   **PV panels potentially cause snow drift loads.**

ASCE 7 Section 7.9   Sliding Snow.   In cold climates where significant snow loads can accumulate on the roof, the design professional must pay special attention to areas where sliding snow can occur. Sliding snow is where a mass of snow from an upper roof falls onto a lower roof or another structure such as a deck. As shown in Figure 1608-5, the sliding snow load from an upper roof should be distributed over an area of 15 feet (4,572 mm) in accordance with Section 7.9 of ASCE 7. This applies for slippery upper roofs with slopes greater than ¼ on 12, and for nonslippery upper roofs with slopes greater than 2 on 12. Sliding snow has been the cause of many structural failures and the potential for sliding snow should be addressed in the design.

ASCE 7 Section 7.10   Rain-on-Snow Surcharge Load.   This load accounts for the increased weight of snow after it rains on the snow. Research has shown that the major portion of the rain-on-snow surcharge load comes from the rainwater flowing along the slope of the roof by percolating through the snow layer immediately above the roof. Thus, the surcharge load increases with larger ridge-to-eave spans (W) and smaller roof slopes. This additional load applies only

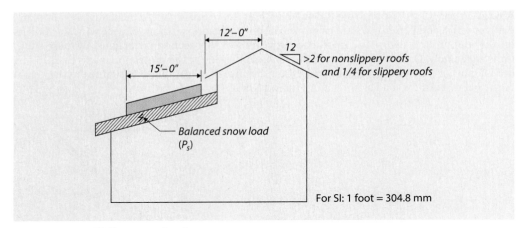

Figure 1608-5 **Sliding snow load.**

to the sloped roof (balanced) load case and need not be used in combination with drift, sliding, unbalanced, minimum, or partial loads.

ASCE 7 Section 7.11   Ponding Instability.   Because of roof deflection caused by snow weight, positive slope to roof drains may be lost, thereby causing ponding. This section requires the roof framing to be stiff enough to maintain positive slope to the roof drains, so as to prevent potential instability from progressive deflection caused by ponding.

ASCE 7 Section 7.12   Existing roofs.   Existing roofs are required to be evaluated for the potential effects of increased snow loads caused by additions or alterations. Owners or agents acting for owners of structures with an existing lower roof should be advised of the possibility of increased snow loads where a higher roof is constructed within 20 feet (6,096 mm) of an existing roof. To reduce or eliminate leeward snow drifts on an adjacent or adjoining lower roof, a snow capture wall (e.g., parapet) can be added to the upper roof. ASCE 7 Commentary Section C7.7 provides more information on this approach.

Section 1609 *Wind Loads*

1609.1   Applications.   The following basic requirement is specified: Wind loads must not be decreased considering the effect of shielding by other structures. As the ASCE 7 Commentary points out, the one possible exception to this rule would be the use of wind tunnel tests in accordance with Chapter 31 of ASCE 7.

1609.1.1   Determination of wind loads.   The IBC does not provide any methods for determining wind loads. Instead, this subsection states that wind loads on every building or structure are to be determined in accordance with ASCE 7 Chapters 26 through 30.

Section 1609.1.1 does allow the designer to utilize either the IBC or ASCE 7 to determine the type of opening protection, the basic wind speed (V), and the exposure category for the site. This subsection then provides exceptions to these design requirements. The exceptions address, with limitations, the use of the following design guides or methods:

- ICC 600 *Standard for Residential Construction in High Wind Regions* used for Group R-2 and R-3 buildings,

- AWC WFCM *Wood Frame Construction Manual for One- and Two-Family Dwellings* used for residential structures,
- AISI S230 *Standard for Cold-Formed Steel Framing—Prescriptive Method for One- and Two-Family Dwellings* used for residential structures,
- NAAMM FP 1001 *Guide Specifications for Design of Metal Flagpoles*,
- TIA-222 *Structural Standard for Antenna Supporting Structures, Antennas and Small Wind Turbine Support Structures*,
- Wind tunnel testing in accordance with ASCE 7
- Temporary structures complying with Section 3103.6.1.2.

The methods used to compute wind loads are found in one of the six ASCE 7 chapters. Chapter 26 covers general wind load design requirements; Chapters 27 and 28 cover the directional and envelope procedures for design of main wind-force-resisting systems (MWFRS); Chapter 29 addresses building appurtenances and other structures using MWFRS (e.g., rooftop equipment, screen walls, signs, etc.); Chapter 30 covers design of components and cladding (C&C); and Chapter 31 covers requirements for the wind tunnel procedure. Chapter 27 (directional procedure) applies to enclosed, partially enclosed, partially open, and open buildings of all heights. Chapter 28 (envelope procedure) applies to low-rise buildings. See Figure 1609-1.

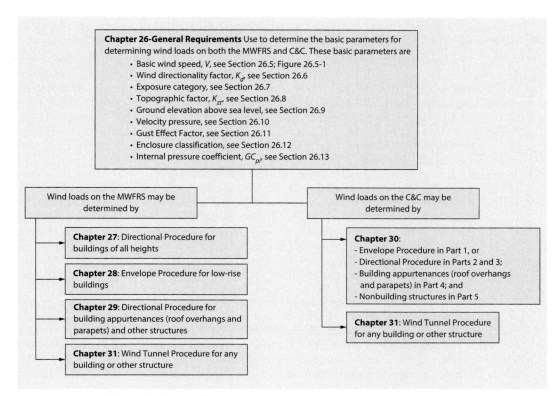

Figure 1609-1 **ASCE 7 Wind procedures.**

An important part of the wind design process is determining which building elements will be designed using MWFRS pressures versus C&C pressures. The MWFRS is a defined term in Section 202: "An assemblage of structural elements assigned to provide support and stability for the overall structure. The system generally receives wind loading from more than one surface." Examples include shear walls, moment frames, and braced frames. See Table 1609-1.

Table 1609-1. **Examples of MWFRS and C&C Building Elements**

Main Wind-Force-Resisting Systems (MWFRS)	Components and Cladding (C&C)
Diaphragms	Lateral framing (studs and connections)
Shear walls	Suction on wall/roof sheathing & cladding (and connections)
Moment frames	Rafters and purlins
Braced frames	Curtain walls
Roof framing uplift (>3:12 slope)	Roof framing uplift (<3:12 slope)

Building envelope elements not qualifying as part of the MWFRS are considered C&C. The C&C elements transfer wind loads to the MWFRS. Examples of C&C are roof and wall sheathing and their connections, rafters and purlins, curtain walls, windows, or doors. See Table 1609-1. Certain members may need to be designed for more than one type of loading. For example, trusses that are part of the MWFRS must be designed for those load effects while individual truss chords must be designed for loads associated with C&C pressures. The ASCE 7 Commentary Section C26.2 provides additional insight into MWFRS versus C&C elements.

1609.2 Protection of openings. In wind-borne debris regions, glazing in buildings must be made impact resistant or be protected with an impact-resistant covering in compliance with the following: (1) glazed openings located within 30 feet (9,144 mm) of grade that meet requirements of the large missile test of ASTM E1996 and (2) glazed openings located more than 30 feet (9,144 mm) above grade that meet the provisions of the small missile test of ASTM E1996. The mandatory protection requirements are meant to maintain building envelope integrity and thereby prevent building damage due to wind and water and the consequent financial and functional losses.

An exception to the above requirement provides the builder and/or owner of buildings with a mean roof height not greater than 33 feet that are classified as Group R-3 or R-4 occupancy with a low-cost alternative to the installation of permanent shutters or laminated glass meeting the requirements of ASTM E1996. This exception requires the builder to provide wood structural panels precut to fit each glazed opening and further requires that corrosion-resistant panel attachment anchors be permanently installed on the building. This is to allow the building owners to attach the protective panels quickly and efficiently while making sure that proper anchorage requirements are met to adequately withstand the required components and cladding loads of hurricane winds.

The second and third exceptions permit any openings in Risk Category I buildings and openings located at least 60 feet (18,288 mm) above ground in Risk Category II, III, or IV buildings to remain unprotected. Additionally, the second part of the third exception recognizes that loose roof aggregate that is not protected by a high parapet can act as a potential source of wind-borne debris. When such a source is present within 1,500 feet (457 m) of a Risk Category II, III, or IV building, an opening needs to be at least 30 feet (9,144 mm) above the source roof to be exempted from the protection requirements. More discussion can be found in ASCE 7 Commentary Section C26.12.

1609.3 Basic wind speed. ASCE 7 uses basic wind speed, V, in mph, for the determination of wind loads. The basic wind speed is determined by Figures 1609.3(1) through 1609.3(4), or the corresponding maps in Chapter 26 of ASCE 7. The basic wind speed for use in the design of buildings and structures is based on Risk Category, which corresponds to specific maps as follows: Risk Category I from Figure 1609.3(1), Risk Category II from Figure 1609.3(2), Risk Category III from Figure 1609.3(3), and Risk Category IV from Figure 1609.3(4).

Wind speeds for the US Virgin Islands, Puerto Rico, and Hawaii are only included in the ASCE Wind Design Geodatabase and are no longer represented by maps in ASCE 7-22. These maps have also been removed from the 2024 IBC and replaced with a pointer to the ASCE Wind Design Geodatabase. Wind speeds in the updated Special Wind Regions also are available for designers and code officials in the ASCE Wind Design Geodatabase. This database of geocoded wind speed design data is available via the ASCE Hazard Tool (ascehazardtool.org).

In nonhurricane-prone regions, when the basic wind speed is estimated from regional climatic data, V is to be determined in accordance with Section 26.5.3 of ASCE 7.

1609.3.1 Wind speed conversion. Where required, such as for use with an older design standard, V must be converted to the allowable stress design wind speed, V_{asd}, using either Table 1609.3.1 or Equation 16-18.

1609.4 Exposure category. For each wind direction considered, an exposure category must be determined based on the building location with respect to its proximity to the surface roughness categories described in Section 1609.4.2.

The IBC exposure category definitions are structured similarly to the ASCE 7 provisions. A Surface Roughness Category is first determined for the building site, and the Exposure Category of the building is then determined based on the Surface Roughness Category prevailing over specified distances in the upwind direction. The definitions for the three types of surface roughness categories (B, C, and D) and the three types of exposure categories (B, C, and D) are identical to those given in ASCE 7.

Once a surface roughness category has been established for a given wind direction, an exposure category is determined from the provisions of Section 1609.4.3. Exposure B applies where Surface Roughness B prevails in the upwind direction for a distance of at least 1,500 feet (457 m) for buildings with mean roof heights less than or equal to 30 feet (9,144 mm). When the mean roof height is greater than 30 feet (9,144 mm), the upwind distance is 2,600 feet (792 m) or 20 times the height of the building, whichever is greater. Exposure D is applicable when Surface Roughness D prevails in the upwind direction for a distance of at least 5,000 feet (1,524 m) or 20 times the building height, whichever is greater. This exposure is to extend into downwind areas of Surface Roughness B or C for a distance of 600 feet (183 m) or 20 times the building height, whichever is greater. Exposure C applies where Exposures B and D do not apply. Commentary Section C26.7 of ASCE 7 contains a method to determine surface roughness categories and exposure categories in cases where a more detailed assessment is required or desired.

In IBC Section 1609.4.1, clear guidelines are provided for determining the governing exposure category whereby two upwind sectors, each enclosing an angle of 45° (0.79 rad) from the selected wind direction (Figure 1609-2), are to be considered separately, and the exposure resulting in the highest wind load governs the design for the selected wind direction. This is particularly helpful in establishing a more objective determination of exposure category for buildings located in a transition zone of two types of exposures. Considerable discussion of determination of exposure category is provided in ASCE 7 Commentary Section C26.7.

1609.5 Tornado loads. Buildings and other structures classified as Risk Category III or IV and located in the tornado-prone region (generally east of the continental divide as shown in Figure 1609-3) are to be designed and constructed to resist the greater of tornado loads or the straight-line wind loads determined in accordance with ASCE 7. Tornado wind speeds

Chapter 16 ■ Structural Design

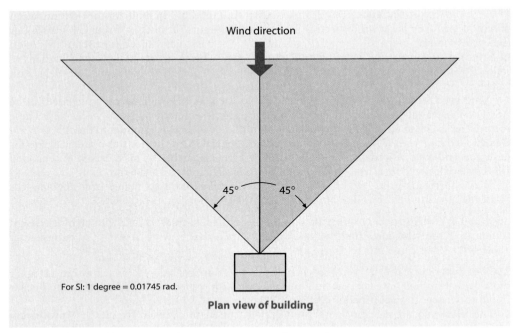

Figure 1609-2 **Upwind sectors for determination of governing exposure category.**

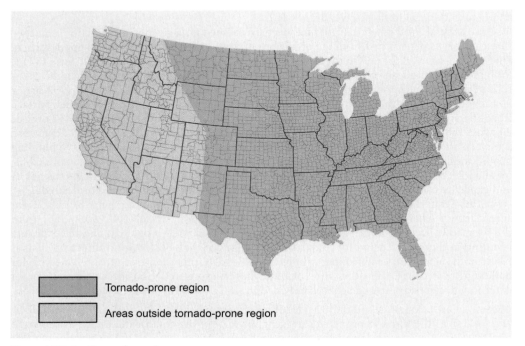

Figure 1609-3 **Tornado regions.**

for RC III and IV structures range from 60 to 138 mph depending on geographic location and effective plan area (which is a function of the building footprint size and shape). This generally corresponds to wind speeds for Enhanced Fujita (EF) Scale EF0-EF2 tornadoes, which are the most common. From 1995 to 2016, over 89 percent of all reported tornadoes were EF0-EF1, and 97 percent were EF0-EF2. These tornado loads are not intended to be used for design of storm shelters which are designed to much higher loads. The ASCE Hazard Tool (ascehazardtool.online) can be used to determine tornado wind speeds.

1609.6 Roof systems. The roof system consists of the roof deck and the roof covering. Provisions for each are discussed below.

1609.6.1 Roof deck. The roof deck is a structural component of the building. Thus, it is required to be designed to resist the greater of wind pressures or tornado pressures determined by the provisions of ASCE 7.

1609.6.2 Roof coverings. If the roof deck is relatively impermeable, the wind or tornado pressures will act through the roof deck to the building framing system. If the roof covering is also relatively impermeable and fastened to the roof deck, the two components will be subject to the same wind or tornado pressures. If the roof covering is air permeable, wind or tornado pressures are able to develop on both the top of the roof covering and underneath the roof covering. This venting action of the roof negates some of the wind or tornado pressures on the roof covering.

1609.6.3 Rigid tile. As explained above, in certain types of installations, the roof covering is not subject to the same wind or tornado pressures as the roof deck. Such installations include concrete and clay tile roof coverings. The gaps occurring at the joints allow some equalization of pressure between the inner and outer faces of the tiles, leading to reduced pressures. The procedure is based on practical measurements on real tiles to determine the effect of air being able to penetrate the roof covering.

1609.7 Elevators, escalators, and other conveying systems. ASME A17.1/CSA B44 *Safety Code for Elevators and Escalators* does not currently address wind provisions. A reference to elevators and escalators in Section 1609.7 points the designer to applicable wind loads for these structures per ASCE 7.

Section 1610 *Soil Loads and Hydrostatic Pressure*

1610.1 Lateral pressures. IBC Table 1610.1 and ASCE 7 Table 3.2-1 list at-rest and active soil pressures for a number of different types of soils. The basis of the soil classification of the various backfill materials listed is ASTM D2487 *Standard Practice for Classification of Soils for Engineering Purposes (Unified Soil Classification System)*. Table 1610-1, in addition to showing soil lateral loads from the IBC and ASCE 7, also shows soil lateral load values using the calculation procedure that formed the basis of the values included in the legacy BOCA *National Building Code*. The calculated soil lateral loads are presented for both moist and saturated conditions. Table 1610-1 shows that for gravels and sands, IBC, ASCE 7, and the calculated soil lateral loads for moist conditions closely agree. For silts and silt–clay mixtures, the IBC values tend to agree more closely with the calculated soil lateral loads for moist conditions, whereas ASCE 7 values are closer to the calculated soil lateral loads for saturated conditions. Footnote a to IBC Table 1610.1 as well as ASCE 7 Table 3.2-1 states that the design lateral soil loads are given for moist soil conditions, which appears to be the case for the IBC soil loads.

According to the exception to Section 1610.1, foundation walls (basement walls) extending not more than 8 feet (2,438 mm) below grade and laterally supported at the top by

Table 1610-1. **Soil Lateral Loads—Calculated versus Code and Standard Design Values**

Description of Backfill Material	Unified Soil Classification	Design Lateral Soil Load Active Pressures (pounds per square foot per foot of depth)		Calculated Lateral Soil Load Active Pressures (pounds per square foot per foot of depth)	
		IBC	ASCE 7	Moist	Saturated
Well-graded, clean gravels, gravel–sand mixes	GW	30	35	34	81
Poorly graded, clean gravels, gravel–sand mixes	GP	30	35	34	80
Silty gravels, poorly graded gravel–sand mixes	GM	40	35	40	84
Clayey gravels, poorly graded gravel-and-clay mixes	GC	45	45	44	86
Well-graded, clean sands, gravelly sand mixes	SW	30	35	32	80
Poorly graded, clean sands, sand–gravel mixes	SP	30	35	32	79
Silty sands, poorly graded sand–silt mixes	SM	45	45	38	82
Sand-silt clay mix with plastic fines	SM-SC	45	85	40	84
Clayey sands, poorly graded sand–clay mixes	SC	60	85	42	85
Inorganic silts and clayey silts	ML	45	85	39	82
Mixture of inorganic silt and clay	ML-CL	60	85	40	83
Inorganic clays of low to medium plasticity	CL	60	100	46	86
Organic silts and silt clays, low plasticity	OL	Unsuitable	Unsuitable	—	—
Inorganic clayey silts, elastic silts	MH	Unsuitable	Unsuitable	—	—
Inorganic clays of high plasticity	CH	Unsuitable	Unsuitable	—	—
Organic clays and silty clays	OH	Unsuitable	Unsuitable	—	—

For SI: 1 pound per square foot per foot of depth = 0.157 kPa/m, 1 foot = 304.8 mm.

flexible diaphragms are permitted to be designed for active pressure. In order to be considered "restrained," the flexible diaphragm must brace the top of the foundation wall. There are two scenarios that must be verified to ensure the diaphragm is bracing the top of the wall. First, the diaphragm itself should be anchored directly to the top of wall. A cripple wall cannot be introduced between the diaphragm and the foundation wall or a hinge is created, and the wall is not properly braced. Second, when floor framing runs parallel to the foundation wall, appropriate blocking and boundary nailing must be provided to stiffen the diaphragm where it connects to the top of wall (see Figure 1610-1).

Expansive soils are found in many regions of the United States. Without special design considerations, expansive soil can cause serious damage to basement walls. Footnote b of Table 1610.1 prohibits the use of expansive soils as backfill because of the potential for very high lateral pressures acting against walls. In this case, it is preferable to excavate expansive soils and backfill with suitable nonexpansive material such as sands and gravels.

Rain Loads

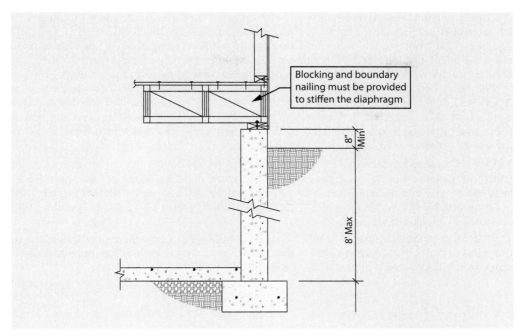

For SI: 1 foot = 304.8 mm

Figure 1610-1 **Restrained foundation wall.**

1610.2 Uplift loads on floor and foundations. Just as expansive soils should be avoided as backfill for foundation walls, they can also cause serious damage to concrete slabs-on-grade. The upward movement of expansive soils is often referred to as soil heave. This new section was added to the 2021 IBC and requires that concrete slabs-on-grade be designed for uplift pressures if they are founded on expansive soils or located in areas that have a shallow water table.

Section 1611 *Rain Loads*

Like many other sections in Chapter 16, Section 1611 is to be used in conjunction with ASCE 7. Chapter 8 of ASCE 7 is used to design primary and secondary drainage systems of a flat roof. The design of the roof drainage systems must be in accordance with Chapter 11 of the *International Plumbing Code*® (IPC). When a primary roof drain becomes blocked by debris, ponding occurs where the roof construction is conducive to retaining water.

1611.1 Design rain loads. In the 2024 IBC, Figures 1611.1(1) through 1611.1(5) were removed because they were 100-year "hourly" rainfall maps, which did not provide rainfall intensities for the required 15-minute duration storms. Furthermore, rainfall rates must now be determined based on the risk category of the building. New Table 1611.1 defines the design storm return period by risk category consistent with the determination of rainfall intensity per ASCE 7-22. Note that return periods are now 200 years and 500 years for Risk Category III and IV structures, respectively. The ASCE Hazard Tool (ascehazardtool.org) can be used to determine rain loads based on location and risk category. Both 15-minute and 60-minute rainfall intensities are provided. The design load is computed using IBC Equation 16-20 (ASCE 7 Equation 8.2-1). Both equations include the addition of the ponding head (d_p) directly into the rain load calculation. In ASCE 7-16 and earlier editions, there was a requirement to perform a ponding analysis, yet limited guidance was provided on how to perform that analysis.

The term "secondary drainage system for structural loading (SDSL)" is consistent with ASCE 7-22. Activation of the SDSL is intended to serve as a warning that the primary drainage system is blocked. Per ASCE 7, the elevation of the SDSL must be at least 2 inches above that of the primary drainage system so that the SDSL is not frequently activated, which would decrease the urgency of the warning and also make the SDSL more susceptible to blockage.

1611.2 Ponding instability. Ponding instability refers to the potential for structural failure or collapse due to progressive deflection from ponding water or full snow loads. Sections 1608.3 and 1611.2 previously referred to the defined term "susceptible bay" for ponding instability evaluation. ASCE 7-22 has dropped this term but still takes ponding into account for snow and rain loads.

1611.3 Controlled drainage. When a roof has hardware to slow the rate of water flow in the primary drain, the secondary system should be placed at an elevation that limits additional water accumulation. Such roofs also need to be checked for ponding instability.

Section 1612 *Flood Loads*

1612.1 General. Section 1612.1 simply states that buildings and structures in flood hazard areas as established in Section 1612.3 must be designed and constructed to resist the effects of flood hazards and flood loads, as prescribed in this section. This section also establishes the requirement of complying with the "substantial improvement" and "substantial damage" requirements. Specifically, a structure located within a special "flood hazard area" that is not currently compliant with the finished floor elevation requirements of the National Flood Insurance Program (NFIP) must be elevated so as to bring it into compliance with the NFIP as found within ASCE 24.

The flood provisions referenced by the IBC are found within ASCE 24, *Flood Resistant Design and Construction.* Much of the impetus for flood-resistant design has come from initiatives of flood insurance and flood-damage mitigation sponsored by the federal government. The NFIP is based on an agreement between the federal government and participating communities that have been identified as being flood prone. The Federal Emergency Management Agency (FEMA), through the Federal Insurance Administration, makes flood insurance available to the residents of communities, provided that the community adopts and enforces adequate floodplain management regulations that meet the minimum requirements. Included in the NFIP requirements, found under Title 44 of the U.S. Code of Federal Regulations, are minimum design and construction standards for buildings located in special flood hazard areas.

Special flood hazard areas are those identified by FEMA's Mitigation Directorate as being subject to inundation during the 100-year flood (flood having a 1-percent chance of being equaled or exceeded in any given year), special flood hazard areas are shown on flood insurance rate maps that are produced for flood-prone communities. Special flood hazard areas are identified on such maps as A Zones and Coastal A Zones (A, AE, A1-30, A99, AR, AO, or AH) or V Zones (V, VE, VO, or V1-30). In the flood hazard areas, communities are encouraged to adopt and enforce NFIP-compliant, flood-damage-resistant design and construction practices.

1612.2 Design and Construction. This section requires the design and construction of buildings and structures located in flood hazard areas, including coastal high-hazard areas and coastal A zones, to be in accordance with Chapter 5 of ASCE 7 and ASCE 24. ASME A17.1/CSA

B44 Section 8.12 specifically directs that elevators and escalators comply with ASCE 24. An exception is provided for public-occupancy temporary structures not needing to be designed for flood loads if controlled occupancy procedures are implemented.

ASCE 7-22 Supplement 2 includes flood load provisions that protect against 500-year flood events—up from 100-year flood provisions. The requirements tie flood hazard mitigation design to risk category, which is consistent with other environmental hazards in ASCE 7. The provisions require Risk Category II structures and above to use the 500-year floodplain to determine flood loads. The most commonly used category, Risk Category II structures include one- and two-family buildings, low- to medium-occupancy businesses, or recreational facilities. Risk Category I structures, including agricultural buildings such as barns and sheds, can still follow 100-year flood criteria.

The provisions of ASCE 24 meet or exceed the building science provisions of the NFIP and present the consensus state-of-the-art approach to flood-resistant construction. ASCE 24 requires that the design of structures within flood hazard areas be governed by the loading provisions of ASCE 7. That standard requires that the structural systems of buildings be designed, constructed, connected, and anchored to resist flotation, collapse, and permanent lateral movement due to the action of wind loads and loads from flooding associated with the design flood. Wind loads and flood loads may act simultaneously at coastlines, particularly during hurricanes and coastal storms. Flood loads are the loads or pressures on the surfaces of buildings and structures caused by the presence of floodwaters. These loads are of two basic types—hydrostatic and hydrodynamic. Wave loads are considered a special type of hydrodynamic load. Additionally, impact loads result from objects transported by floodwaters striking against structures. Minimum design hydrostatic loads, hydrodynamic loads, wave loads, and impact loads are specified in ASCE 7. Wave loads may be determined by using an analytical procedure outlined by more advanced numerical modeling procedures, or by laboratory test procedures or physical modeling.

ASCE 24 requires that the design and construction of structures located in flood hazard areas consider all flood-related hazards, including hydrostatic loads, hydrodynamic loads and wave action, debris impacts, alluvial fan flooding, flood-related erosion, ice flows or ice jams, and mudslides in accordance with the requirements of that standard if specified or, if not specified in that standard, then in accordance with the requirements approved by the authority having jurisdiction. Important to the enforcement of these provisions, design documents are required to identify, and take into account, flood-related and other concurrent loads that will act on the structure. Design documents must also consider, but not be limited to, the applicable conditions listed below:

1. Wave action
2. High-velocity floodwaters
3. Impacts due to debris in the floodwaters
4. Rapid inundation by floodwaters
5. Rapid drawdown of floodwaters
6. Prolonged inundation by floodwaters
7. Wave- and flood-induced erosion and scour
8. Deposition and sedimentation by floodwaters.

1612.3 Establishment of flood hazard areas. Flood hazard areas are established by the local jurisdiction through adoption of a flood-hazard map and supporting data. Minimum requirements imposed on the jurisdiction are spelled out. Two key pieces of information that are not part of the minimum requirement are the design flood elevations and designation of floodways (floodways are areas along riverine bodies that convey the bulk of floodwaters). By definition, the design flood elevation is the elevation of the "design flood," including wave height, relative to the datum specified on the community's legally designated flood hazard map. In areas designated as Zone AO, the design flood elevation shall be the elevation of the highest existing grade of the building's perimeter plus the depth number (in feet) specified on the flood hazard map. In areas designated as Zone AO where a depth number is not specified on the map, the depth number shall be taken as being equal to 2 feet (610 mm). A large percentage of areas that are mapped as special flood hazard areas by the NFIP do not have flood elevations and do not have floodway designations. In such an absence, the building official can require that the design flood elevations be either obtained from any pertinent data available from the federal, state, or other sources, or determined by a registered design professional in accordance with accepted hydrologic and hydraulic engineering practices.

Development in riverine floodplains can increase flood levels and loads on other properties, especially if it occurs in areas known as floodways that must be reserved to convey flood flows. The floodway, by definition, is a river, creek, or other watercourse and the adjacent land areas that "must be reserved in order to discharge the base flood without cumulatively increasing the water surface elevation more than a designated height." For the purpose of this section, the designated height is 1 foot (305 mm).

1612.4 Flood-hazard documentation. A number of documents must be furnished to the building official. The requirements are different for construction in coastal high-hazard areas and coastal A zones compared to other flood hazard areas. A flood emergency plan is required for nonresidential buildings along with provisions for breakaway walls in coastal high hazard areas and coastal A zones.

Section 1613 *Earthquake Loads*

Overview. In the IBC, the seismic load provisions are largely adopted through reference to ASCE 7. The earthquake provisions of ASCE 7 are mainly derived from the Building Seismic Safety Council's *National Earthquake Hazards Reduction Program* (NEHRP) *Recommended Seismic Provisions for New Buildings and Other Structures (Vol. I)*. For information regarding the application of the IBC seismic provisions, one should refer to the NEHRP *Provisions Part 2—Commentary*. In addition, the ASCE 7 Commentary provides some background on the seismic requirements. One other valuable reference is the *2021 IBC SEAOC Structural Seismic Design Manual, Volume 1, Code Application Examples*, which is a compilation of sample calculations developed by the Structural Engineers Association of California.

Using these provisions, restrictions on building height and structural irregularity, choice of analysis procedures that form the basis of seismic design, as well as the level of detailing required for a particular structure are all governed by a structure's Seismic Design Category (SDC), a classification that considers the nature of the occupancy along with the soil-modified ground motion at the site of the structure. In other words, when designing for earthquakes under the IBC, the required detailing as well as other seismic system limitations are affected by the soil characteristics at the site.

Also see Section 1604.9 discussion on building behavior due to earthquake ground motions.

1613.1 Scope. Section 1613 incorporates by reference the seismic load provisions of ASCE 7, except for the items covered in Chapters 14 and 16. After having stated that every structure and portions thereof shall, at a minimum, be designed and constructed to resist the effects of earthquake motions and assigned a SDC in accordance with Section 1613 or ASCE 7, the IBC permits a number of important exceptions:

1. Detached one- and two-family dwellings in SDC A, B, or C are exempt from seismic design requirements.

2. The seismic-force-resisting systems of wood-frame buildings that strictly conform to the provisions of Section 2308 (Conventional Light-Frame Construction) need not be analyzed as required by Section 1613.1. However, conventional wood-frame buildings must meet all of the limitations imposed by Section 2308.2.

3. Agricultural storage structures intended only for incidental human occupancy are exempt from all seismic design requirements. This is because of the very low risk to life that is involved.

4. Certain special structures such as "vehicular bridges, electrical transmission towers, hydraulic structures, buried utility lines and their appurtenances, and nuclear reactors" are placed outside the scope of Section 1613. These structures require special consideration of their response characteristics and environment, and need to be designed in accordance with other regulations.

5. ASCE 7 Chapter 14 as required by the IBC.

6. Temporary structures complying with Section 3103.6.1.4 where seismic loads may be reduced by 25 percent in SDC C through F.

1613.2 Determination of seismic design category. The seismic design category (SDC) classification provides a relative scale of earthquake risk to structures. The SDC considers not only the seismicity of the site in terms of the mapped spectral response accelerations but also the site soil profile and the nature of the structure's risk category.

2024 IBC Section 1613 was simplified by providing SDC maps that users can reference to quickly determine a project's SDC based on default site conditions. These maps are derived based on new multi-period response spectra procedures of ASCE 7-22. IBC Figures 1613.2(1) through 1613.2(7) provide SDC maps for the conterminous United States, Alaska, Hawaii, Guam, the Northern Mariana Islands and American Samoa. The SDC maps are one of two methods provided in the IBC to determine SDC. Users are still allowed to determine the SDC following ASCE 7 provisions, where more refined information such as site-specific soils data can be considered.

These new maps will allow building officials, nonstructural engineers, component manufacturers and others to quickly identify a conservative SDC based on location alone. The ASCE Hazard Tool (ascehazardtool.org) is a free resource for determining seismic design parameters, including SDC, based on location, soil class, and risk category.

Figure 1613-3 shows the IBC SDC maps for the conterminous United States. The key indicates the SDC based on the Risk Category of the structure. The following discussion describes the six design categories that are key to establishing the design requirements for any building based on its occupancy and on the level of expected soil-modified seismic ground motion. The IBC uses SDCs to determine permissible structural systems, limitations on height and irregularity, the type of lateral force analysis that must be performed, the level of detailing for structural members, and joints that are part of the lateral-force-resisting system and for the components that are not.

SDC A applies to structures, irrespective of their occupancy, in regions where anticipated ground motions are minor, even for very long return periods.

SDC B includes Risk Category I, II, or III structures in regions where moderately destructive ground shaking is anticipated.

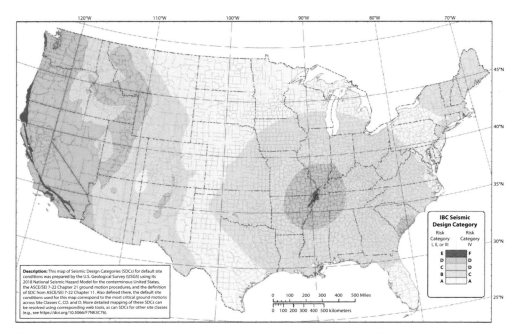

Figure 1613-3 **Seismic design categories for default site conditions for the conterminous United States**

SDC C includes Risk Category IV structures where moderately destructive ground shaking may occur as well as Risk Category I through III structures in regions with somewhat more severe ground-shaking potential.

SDC D includes all Risk Category structures located in regions of severe seismicity and above.

Structures located close to major active faults are directly assigned to SDC E or F. SDC E includes Risk Category I, II, or III structures in regions located close to major active faults, and SDC F includes Risk Category IV structures in those locations (see Figure 1613-3).

The soil site class is a very important parameter when determining the SDC. For projects requiring a geotechnical report, the soil site class can be found in the report while those not having, or requiring, a report should be determined using Table 20.2-1 of ASCE 7 (see Table 1613-1). The ground motions provided in Figures 1613.2(1) through 1613.2(7) correspond to the most critical ground motions across Site Classes C, CD, and D in Table 1613-1. If Site Class DE, E, or F soils are present at the site the SDC shall be determined in accordance with ASCE 7. Site Class DE, E, and F soils consist of loose sand or medium stiff clay, very loose sand or soft clays, peats, and liquefiable soils.

1613.3 Simplified design procedure. The alternate simplified design procedure of ASCE 7 Section 12.14 applies to simple bearing wall or building frame systems. The structure being three stories or less and classified as Risk Category I or II are just two of the twelve conditions. If the building and site meet all conditions of ASCE 7 Section 12.14, the SDC shall be determined in accordance with ASCE 7.

1613.4 Ballasted photovoltaic panel systems. This section includes specific seismic requirements for ballasted, roof-mounted PV solar panels. Rooftop solar panels without positive attachment to the roof structure are limited to low-profile panels with a low height-to-depth

Table 1613-1. Soil Site Classification

Site Class	v_s Calculated Using Measured or Estimated Shear Wave Velocity Profile (ft/s)
A. Hard rock	>5,000
B. Medium hard rock	>3,000 to 5,000
BC. Soft rock	>2,100 to 3,000
C. Very dense sand or hard clay	>1,450 to 2,100
CD. Dense sand or very stiff clay	>1,000 to 1,450
D. Medium dense sand or stiff clay	>700 to 1,000
DE. Loose sand or medium stiff clay	>500 to 700
E. Very loose sand or soft clay	≤500
F. Soils requiring a site response analysis in accordance with ASCE 7 Section 21.1	See ASCE 7 Section 20.2.1

For SI: 1 foot per second = 0.3048 meters per second

ratio that respond by sliding on the roof surface without overturning. ASCE 7 limits unattached PV systems to Risk Category I, II, and III structures six stories or less in height. The roof slope is limited because PV panels on sloped surfaces tend to displace in the downslope direction when subjected to seismic shaking, and the displacement increases with greater roof slope. Displacement-based design of panels includes verifying that there is roof capacity to support the weight of the displaced panel and that wiring to the panel can accommodate the design panel displacement without damage.

1613.5 Elevators, escalators, and other conveying systems. ASME A17.1/CSA B44 and ASCE 7 both outline seismic requirements for elevators and conveying systems. Section 1613.5 is intended to ensure that seismic provisions of both standards are satisfied.

1613.6 Automatic sprinkler systems. Experience has shown that the interaction of nonstructural components and sprinkler drops and sprigs is a significant source of damage and can result in serious consequential damage, as well as compromising the performance of the fire protection system. Clearance for sprinkler drops and sprigs and other nonstructural components needs to be addressed beyond NFPA 13 *Standard for the Installation of Sprinkler Systems*. The minimum clearance value provided in ASCE 7-22, is based on observations of past earthquake damage.

Section 1614 *Atmospheric Ice Loads*

This section provides charging text in the IBC, which references the technical provisions of ASCE 7 for atmospheric ice loads. This section relies on the determination of which structures are ice sensitive in order to decide the need to comply with the applicable provisions of ASCE 7. An "ice-sensitive structure" is defined in Section 202 and provides the technical basis for determining which structures are ice sensitive. This includes, but is not limited to, lattice structures, guyed masts, overhead lines, light suspension and cable-stayed bridges, aerial cable systems (e.g., for ski lifts or logging operations), amusement rides, open catwalks and platforms, flagpoles and signs.

Nominal ice thickness maps are shown in ASCE 7-22 Figures 10.4-2 (lower 48 states), 10.4-3 (Alaska), 10.4-4 (Columbia Gorge region), and 10.4-5 (Lake Superior region) for Risk Categories I, II, III, and IV. Location-specific nominal ice thicknesses may be determined using the ASCE Hazard Tool (ascehazardtool.org). This geodatabase provides nominal ice thickness, concurrent wind and concurrent temperature data.

In the IBC, an exception is included for temporary structures complying with Section 3103.6.1.5. Ice loads on public-occupancy temporary structures are permitted to be determined with a maximum nominal thickness of 0.5 inch (13 mm), for all risk categories. Ice loads need not be considered, provided that controlled occupancy procedures are implemented.

Section 1615 *Tsunami Loads*

1615.1 General. Many coastal areas in the western United States including Alaska, Washington, Oregon, California, and Hawaii are subject to potentially destructive tsunamis. These communities, which need tsunami-resistant design of critical infrastructure and essential facilities, require a standard for designated tsunami vertical evacuation refuge structures as an alternative to nonexistent high ground. This section requires that critical infrastructure and essential facilities designated as Risk Category III or IV be designed in accordance with Chapter 6 of ASCE 7. Location-specific tsunami design zones and related criteria may be determined using the ASCE Tsunami Hazard Tool (asce7tsunami.online).

In the IBC, an exception is included for temporary structures complying with Section 3103.6.1.6. Public-occupancy temporary structures in a tsunami design zone are not required to be designed for tsunami loads. However, controlled occupancy procedures must be implemented.

Section 1616 *Structural Integrity*

This section applies specifically to high-rise buildings that are classified as Risk Category III or IV. The code defines a high-rise building as one that has an occupied floor located more than 75 feet (22,860 mm) above the lowest level of fire department vehicle access. Detailing requirements are provided for concrete- and steel-frame structures, whereas minimum requirements for vertical, longitudinal, transverse, and perimeter ties are provided for bearing-wall structures.

The purpose of this section is to provide a minimum level of structural integrity by establishing requirements for tying together primary structural elements. These provisions have been developed by a broad coalition of industry members involved in steel, concrete, masonry, and wood structures, and apply as appropriate to structures of all materials. Although the structural integrity requirements already included in ACI 318, coupled with common design and construction practices adopted in the United States, are generally adequate to provide a satisfactory level of structural integrity to most concrete structures, this section expands on those provisions in order to include other building materials as well.

KEY POINTS

- This chapter prescribes minimum structural loading requirements for the design of buildings and other structures regulated by the IBC.
- In addition to the general requirements for construction documents given in Section 107, Section 1603 contains specific requirements for construction documents pertaining to structural loads.
- The American Society of Civil Engineer's *Minimum Design Loads and Associated Criteria for Buildings and Other Structures* (ASCE/SEI 7-22) is the referenced standard for structural loads.
- General construction requirements are provided in Section 1604 for strength, serviceability, risk category, load tests, anchorage, and wind or seismic detailing.
- How various structural loads are required to be combined is prescribed in Section 1605.
- Buildings and other structures must resist all prescribed loads and be stable against overturning, sliding, and buoyancy.
- In addition to dead and live loads, buildings and other structures must resist the effects of environmental loads such as snow, wind, tornado, earthquake, soil and hydrostatic pressure, rain, flood, tsunami, and ice loads as required by this chapter and ASCE 7.
- High-rise buildings assigned to Risk Category III or IV must also meet minimum structural integrity requirements of Section 1616 based on whether they are frame structures (gravity loads supported by columns) or bearing-wall structures (gravity loads supported by walls) as defined in Section 202.

CHAPTER 17

SPECIAL INSPECTIONS AND TESTS

Section 1701 General
Section 1702 New Materials
Section 1703 Approvals
Section 1704 Special Inspections and Tests, Contractor Responsibility, and Structural Observation
Section 1705 Required Special Inspections and Tests
Section 1706 Design Strengths of Materials
Section 1707 Alternate Test Procedure
Section 1708 In-Situ Load Tests
Section 1709 Preconstruction Load Tests

Example 17-1 Sample Statement of Special Inspections
Key Points

The primary goal of Chapter 17 of the *International Building Code®* (IBC®) is to improve the quality and workmanship of certain structural systems by requiring structural testing, special inspection, and structural observation. Many of the requirements are specifically intended to improve the quality of the lateral-force-resisting system when buildings are subjected to the design wind or seismic event. Special inspections are also specified for many nonstructural components within the building to ensure the life safety of the occupants. To accomplish these goals, the chapter sets forth provisions for quality of materials, workmanship, testing, and labeling of materials incorporated into the construction of buildings or structures regulated by the code. Generally, all components and materials used in new buildings must conform to the requirements in the code, or the applicable standards referenced by the code. Specific tests and standards are referenced in other parts of the code. Chapter 35 contains a complete alphabetical list of all the testing and material standards referenced in the IBC.

This chapter provides the requirements for special inspection and verification of the construction at various stages, special inspection for wind and seismic resistance, special testing and qualification for seismic resistance, structural observation by the registered design professional, and alternative methods to establish test procedures for products that do not have applicable standards.

All of the administrative requirements related to special inspection, the statement of special inspections, contractor responsibility, and structural observations, are contained in Section 1704. All of the verifications required for various systems, elements, and materials, including special inspection requirements for wind and seismic, are now contained in Section 1705.

Section 1701 *General*

This section sets forth the scope for Chapter 17 and the general requirements for both new and used construction materials. Because Chapter 17 deals with construction documents and submittals, some discussion of the subject is warranted. The term *construction documents* is defined in Section 202 as "written, graphic and pictorial documents prepared or assembled for describing the design, location and physical characteristics of the elements of a project necessary for obtaining a building permit." Construction documents are a part of the submittal documents required by Section 107. It should be noted that there are also specific requirements for construction documents in Section 1603 for structural loads, as well as in the structural material chapters, such as Section 1901.5 for concrete structures. There are also requirements for specific types of submittals, which are considered as a part of the construction documents. For example, Section 1803.6 for geotechnical reports, Section 2207.4 for steel joist drawings, Section 2206.1.3.1 for cold-formed steel truss drawings, and Section 2303.4.1.1 for wood truss drawings.

The following terms have specific definitions in Section 202 and are further defined or discussed in Chapter 17.

APPROVED AGENCY. The definition of this term is needed in order to affect the requirements of Section 1703.1. The word *approved* means "acceptable to the building official" or authority having jurisdiction (Section 202). The basis for approval of an agency for a particular activity by the building official includes the competence or technical capability to perform the work in accordance with Section 1703.

APPROVED FABRICATOR. An approved fabricator is a qualified person, firm, or corporation that is approved by the building official to perform specified work without special inspection because they have approved quality-control procedures and are subject to periodic auditing of fabrication practices by an approved special inspection agency. Approved fabricators issue certificates of compliance for their work product. See discussion of Section 1704.2.5.

CERTIFICATE OF COMPLIANCE. An *approved* fabricator is required to submit a Certificate of Compliance for work performed without special inspection. See Section 1704.2.5 for more detailed discussion of fabricator approval. Figure 1703-3 shows an example of an AITC certificate of compliance for a glued-laminated timber.

DESIGNATED SEISMIC SYSTEM. The designated seismic system consists of those architectural, electrical, and mechanical systems and their components that require design in accordance with Chapter 13 of ASCE 7 for which the component importance factor, I_p, is greater than 1.0, as prescribed in Section 13.1.3 of ASCE 7. Section 13.1.3 of ASCE 7 lists four conditions requiring an importance factor of 1.5, which include components required to function after an earthquake including fire sprinkler systems and egress stairways that are not an integral part of the building structure; components containing hazardous, toxic, or explosive materials; and components in Risk Category IV structures that are necessary for continued operation of the facility. All other components are assigned a component importance factor of 1.0. Risk Category IV structures are described in detail in IBC Table 1604.5. Note that Section 13.1.4 of ASCE 7 lists those nonstructural components that are exempt from the seismic design requirements of Chapter 13 and are therefore not considered to be part of the designated seismic system.

FABRICATED ITEM. Fabricated items are materials assembled prior to installation in a building and are referred to in Section 1704.2.5. The definition is provided to clarify the intent of the code, as the term *fabricated items* could easily be interpreted to mean items for which special inspection is not intended by the code. An example of a fabricated item for which special inspection is required is roof trusses not manufactured in accordance with the in-plant quality control requirements of the TPI 1 standard. The section also describes elements that are not considered fabricated items, such as rolled structural steel, reinforcing bars, masonry units, and wood structural panels that are fabricated under specific quality control standards. Because special inspections are analogous to the quality control (QC) programs required by some standards referenced in the code, items produced in accordance with standards listed in Chapter 35 that have quality control by a third-party quality-control agency are not considered "fabricated items." This provision makes it clear that the code does not intend to impose a redundant special inspection requirement for such items.

MAIN WINDFORCE-RESISTING SYSTEM. The main windforce-resisting system (MWFRS) is one of the structural systems that must be designed to resist wind loads. The other system that is designed to resist wind loads is components and cladding. The MWFRS comprises those structural elements that provide lateral support and stability for the overall structure. In general, these are diaphragms, chords, collectors, vertical lateral-force-resisting elements such as shear walls, braced frames, and foundations. In general, the MWFRS receives wind loading from more than one surface. In contrast, components and cladding are generally on the exterior envelope of the building and receive loading directly from the wind, such as roof and wall covering.

SEISMIC FORCE-RESISTING SYSTEM. The seismic force-resisting system (SFRS) is that part of the structural system that has been considered in the design to provide the required resistance to the prescribed seismic forces. This definition provides for the correct application of special inspections of seismic force-resisting systems. Unless classified as Seismic Design Category (SDC) A, the type of SFRS must be selected to conform to Section 12.2 of ASCE 7.

Providing a level of detailing that is appropriate for the selected system is critical to complying with the code's earthquake provisions.

SPECIAL INSPECTION. That category of field inspection for which special knowledge, special attention, or both are required. For example, inspection of complete penetration welds requires both special knowledge and special attention to ensure that the requirements of the codes and standards are met. Note that the special inspector does not have the same authority as that of the jurisdiction inspector. The role of the special inspector is to report discrepancies between the construction in the field and the approved construction documents to the contractor. If uncorrected, discrepancies should be reported to the design professional in responsible charge and to the building official.

SPECIAL INSPECTION, CONTINUOUS. Continuous full-time inspection is required where compliance of the work or product cannot be determined after incorporation into the building or structure. For example, one cannot determine whether a multipass fillet weld is in compliance with the code requirements unless each pass of the weld is inspected during the welding process. Whether a particular special inspection is continuous or periodic is specified in the various inspection and verification tables in Section 1705.

SPECIAL INSPECTION, PERIODIC. Intermittent or part-time inspection, which may be allowed when the compliance of the work or product can be determined after being incorporated into the structure. For example, compliance with the design nailing requirements of a wood shear wall can be determined after construction of the wall (but before closure); hence, verification by periodic special inspection is adequate in this case. Another good example is placement of reinforcing bars. The special inspector need not be continuously present during placement. Whether a particular special inspection is periodic or continuous is specified in the various inspection and verification tables in Section 1705.

STRUCTURAL OBSERVATION. Structural observation is intended to ensure general conformance with the design intent, and not necessarily specific conformance with the code. For example, in a building under construction in a high seismic area, the registered design professional acting as the structural observer may focus on crucial elements of the seismic-force-resisting system. Like the special inspector, the structural observer does not have the same authority as that of the jurisdiction inspector. The role of the structural observer is to report discrepancies between the construction in the field and the approved design documents to the contractor. If unresolved, discrepancies should be reported to the building official. Note that structural observation by the registered design professional does not replace or substitute for any of the requisite special inspections required by Section 1705 or the jurisdiction inspections required by Section 110.

Section 1702 *New Materials*

Testing is required for all materials that are not specifically provided for in the code or referenced standard. For example, a composite wood material that is not listed in Chapter 23 would be required to follow the procedures set forth in this chapter. A similar provision for acceptance of alternative materials, systems, or methods for which the standards are not adopted in the code is set forth in Section 104.2.3. This section restates that alternative or new materials and methods may be used if it can be established by tests or other means that the performance of the new material or method will equal that required by the code for the replaced product. As noted, International Code Council (ICC) Evaluation Service Reports are often used by building officials as the basis for approving items that are not specifically addressed by the code or referenced standards.

Used materials. Materials may be reused, provided they meet *all* the code requirements for new materials. Note, however, that Section 104.9.1 specifically restricts the use of used equipment and devices unless approved by the building official. One should always exercise caution in approving reuse of materials. The applicable material or design standards should be consulted to determine if reuse of materials is allowed or prohibited. For example, reuse of high-strength A490 structural bolts is prohibited by the Research Council on Structural Connections *Specification for Structural Joints Using High-Strength Bolts (2020)*. Even a piece of used structural steel should be carefully checked for conformance to the design specifications, applicable standards, and code requirements.

Section 1703 *Approvals*

1703.1 Approved agency. The basis for approval of an agency for a particular activity by the code or building official may include the capacity or technical capability and expertise necessary to perform the work in accordance with Section 1703. Special inspection agencies, testing laboratories, and the inspection agencies that provide quality assurance of concrete and steel fabricators should be accredited by a recognized Accreditation Body. For example, testing laboratories should be accredited to ISO/IEC 17025 *General requirements for the competence of testing and calibration laboratories*.

Accreditation through the International Accreditation Service (IAS), a subsidiary of the ICC, is the most straightforward method for the building official to assure that an agency provides quality services and meets all of the requirements of the building code. Quality standards for agency approvals through IAS are defined by the following standards:

- Special inspection agencies should be accredited to IAS AC291.
- Fabricators of reinforced and precast/prestressed concrete should be inspected by an approved special inspection agency, or by an inspection agency, which is accredited to IAS AC157.
- Fabricators of structural steel should be inspected by an approved special inspection agency, or by an inspection agency, which is accredited to IAS AC172.
- Fabricators of metal building systems should be inspected by an agency, which is accredited to IAS AC472.
- Fabricators of cold-formed steel structural and nonstructural components not requiring welding should be inspected by an agency, which is accredited to IAS AC473.

Other accreditation programs are also available that may be approved by the building official, such as the AISC Certification Program for Structural Steel Fabricators, the Plant Certification Program for steel joists by the Steel Joist Institute, or the Plant Certification Program for precast concrete products by the Precast/Prestressed Concrete Institute.

Accreditation helps to ensure that the agency has the necessary equipment and employs qualified persons. For example, if an agency wishes to be approved for the special inspection of structural welds, the agency should submit evidence that its welding inspector is certified in accordance with applicable ICC, American Welding Society (AWS), or American Society of Nondestructive Testing (ASNT) requirements.

1703.1.1 Independence. The agency should have objectivity as well as competence. Objectivity can be measured by the agency's financial and fiduciary independence. The agency should be independent of the contractor responsible for the work and have no financial ties to the organization it inspects. For example, a testing laboratory checking concrete strength should have no financial ties to the contractor, its subcontractors, or the concrete supplier.

1703.1.2 Equipment. If any agency is not accredited as described above, the building official should evaluate the agency to confirm the compliance with the requirements of applicable standards and that it has the appropriate equipment to perform required tests or inspections.

1703.1.3 Personnel. If the agency is not accredited, the building official should evaluate both the experience and qualifications of personnel. An agency may have personnel with the appropriate certifications but not the necessary experience. Supervisory and inspection personnel should have the appropriate certifications as well as the requisite experience and/or education.

If the services being provided by an inspection or testing agency come under the purview of the professional registration laws of the state or jurisdiction, the building official should request evidence that personnel are qualified to perform the work in accordance with the requisite professional registration.

1703.2 Written approval. A written approval by the building official is required for all material, appliances, equipment, or systems incorporated into the work in order to have a documented record of approval and the basis for approval.

1703.3 Record of approval. Records must be kept for all approvals, including conditions and limitations of approval. The approvals must be kept on file and available for public inspection. The records must demonstrate compliance with the applicable code requirements of any material, appliance, equipment, or system incorporated into the structure.

1703.4 Performance. When conformance with the code is predicated on the performance and quality of materials or products, the building official must require the submission of test reports from an approved agency establishing this conformance. In the absence of such reports, the building official should require specific data that show compliance with the intent of the applicable code requirements in accordance with Chapter 16 (see IBC Sections 1604.6 and 1604.7). For example, core tests of in-situ concrete could be used to determine compliance with design strength if the sample cylinders required by ACI 318 were destroyed.

Materials and products must be subjected to various levels of quality control and identification in order to determine that the material complies with the requirements of the code. The degree of quality control and identification to which a material must be subjected is based on its relative importance to the structure's performance and function. See Section 202 for definitions of the terms *mark, certificate of compliance, manufacturer's designation*, and *label*. By use of these terms, the code establishes quality control and identification as follows:

- Level 1: Manufacturer identifies the material or product with the name of the material or product, the manufacturer's name, and the intended usage (see *Mark*). Alternatively, the manufacturer provides a certificate stating that materials and products meet specified standards or that work was done in compliance with *approved construction documents* (see Certificate of compliance).

- Level 2: Manufacturer identifies the material or product as in Level 1 and also self-certifies compliance with a given standard or set of rules (see *Manufacturer's designation*).

- Level 3: Manufacturer identifies the material or product as in Level 1. An approved quality control agency performs periodic audits of the manufacturer's facilities and QA/QC procedures (see *Label*).

The term *mark* is an identification applied on a product by the manufacturer indicating the name of the manufacturer and the function of a product or material. A *certificate of compliance*

Figure 1703-1 **Certificate of compliance might accompany a welded steel railing.**

might be used instead of a *mark* if appearance is important for the product in its intended use (Figures 1703-1 and 1704-1).

The term *manufacturer's designation* refers to an identification applied on a product by the manufacturer indicating that a product or material complies with a specified standard or set of rules. An example of a manufacturer's designation would be the designation on ASTM A706 reinforcing bars where the "W" indicates the bars conform to the ASTM A706 standard for weldable rebar. See Figure 1703-2.

The term *Label* refers to an identification applied to a product by an approved agency containing the name of the manufacturer, the function, and the performance characteristics, and indicates that the product or material has been inspected and evaluated by an approved agency. Examples of a labels are shown in Figure 1703-3.

1703.4.1 Research and investigation. This section is intended to implement the requirements of Section 104.2.3 for use of alternative materials, design, and methods of construction. For example, an innovative prestressed concrete system that does not emulate the performance of cast-in-place concrete could be evaluated in accordance with this section. Note that the section requires that the costs of collecting data and preparing reports are to be borne by the applicant.

1703.4.2 Research reports. This section is similar to Section 104.2.3.6 regarding the use of alternative materials, design, and methods of construction. Evaluation reports prepared by approved agencies, such as those published by the organizations affiliated with the model code groups, may be accepted as part of the data needed by the building official to form the basis of approval of a material or product not specifically covered by the code. ICC Evaluation Service

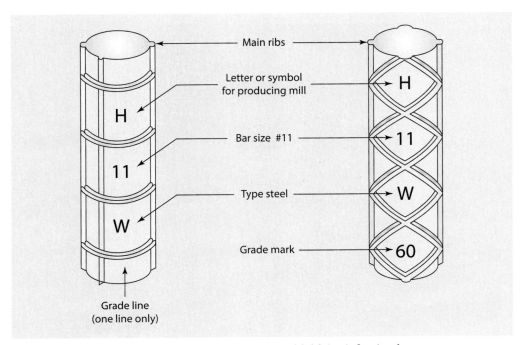

Figure 1703-2 Required marking for ASTM A706 (weldable) reinforcing bars.

Reports are considered by many building officials to be the most straightforward method for the building official to review products and systems that are not specifically covered by the code. Such reports supplement the resources of the building official and eliminate the need for the official to conduct detailed analysis on every new product that is not covered by the code or a referenced standard. Because evaluation reports issued by the model code-affiliated organizations are advisory in nature, the building official is not mandated to accept or approve them. Technically, such products and reports are approved under the alternative materials and methods of construction provisions of Section 104.2.3.

1703.5 Labeling. When materials or assemblies are required to be labeled by the code, such as wood structural panels, lumber, and fire doors, the labeling must be in accordance with the procedures outlined in this section and its subsections. Labeling of materials or assemblies is an indication that the materials or assemblies have been subjected to testing, inspection, and/ or other operations by the labeling agency. The presence of a label does not necessarily indicate compliance with code requirements. For example, use of plywood siding, although labeled, would not comply with the code if Structural I sheathing was specified to resist the design lateral forces. The installation of labeled products must comply with the specific requirements and limitations of the label. For example, the actual span in the field for wood structural panel roof sheathing must meet the panel span rating shown on the label (stamp). See Figure 1703-3 for examples of typical lumber, panel and engineered wood product labels. Another example is a fire-rated door that is labeled with the hardware requirements specified on the label. The building inspector must ensure that the hardware used in the installation of the door meets the labeling requirements to ensure that the door complies with the code. See Figure 1703-4.

See Section 202 for the definition of the terms *label* and *labeled*.

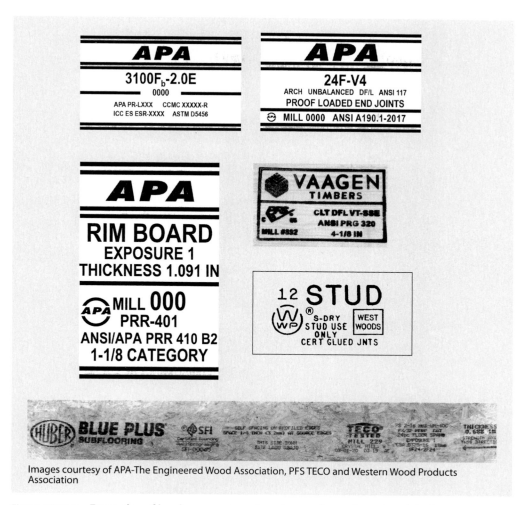

Figure 1703-3 Examples of lumber, panel, and engineered wood product labels.

1703.5.1 Testing. For a material or product to be labeled, the labeling agency is required to perform **specific** testing on representative samples of the material or product in accordance with the applicable standards referenced by the code. For example, quality-control testing for strength and stiffness of machine stress-rated lumber to ensure that the products meet structural requirements in accordance with the American Lumber Standard system. Another example is factory-built fireplace assemblies that must be tested in accordance with the requirements of UL 127 (see Chapter 35).

1703.5.2 Inspection and identification. The approved agency whose label is applied to a material or product must perform periodic inspections of the manufacture of the material or product to determine that the manufacturer is indeed producing the same material or product as tested and labeled. For example, if the labeling agency had tested ½-inch (12.7 mm) C-C plywood sheathing made with five plies but the manufacturer was now making the plywood with only three plies, the agency would need to withdraw the use of its label and listing.

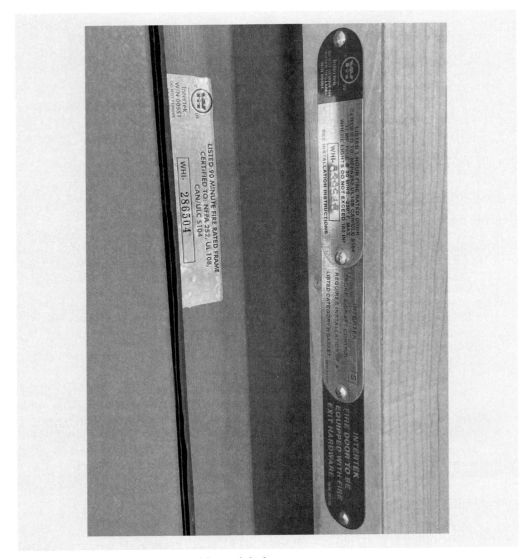

Figure 1703-4 **Fire-rated door and frame labels.**

1703.5.3 Label information. This section specifies the minimum information necessary on a label for the inspector to determine that the installed material conforms to the approved plans. See Figure 1703-3 for examples of typical lumber, panel, and engineered wood product labels. The manufacturer (mill number), the performance characteristics (grade or category), standards or evaluation reports, and approved agency are shown.

1703.5.4 Method of labeling. This section specifies that labels that are required to be permanent must be of a nature that once applied, their removal would cause the label to be destroyed. Examples of labels that are required to be permanent are fire-resistive glazing and opening protectives.

1703.6 Evaluation and follow-up inspection services. This provision applies where a structural component cannot be inspected after completion of a prefabricated assembly. An example might be a prefabricated shear wall consisting of wood structural panel sheathing over a welded steel frame; the welding must be inspected prior to the installation of the sheathing and the entire assembly must be inspected after it is fabricated. The testing and inspection should follow the provisions described in Section 1703.6.

Section 1704 *Special Inspections and Tests, Contractor Responsibility, and Structural Observation*

This section provides administrative requirements for special inspections and tests (1704.2), preparation and submittal of the statement of special inspections (1704.3), the contractor's statement of responsibility (1704.4), submittals to the building official (1704.5), and structural observation by the registered design professional (1704.6).

One of the oldest mechanisms for providing quality assurance in construction is the process known as *special inspection.* The purpose of special inspection is to ensure proper fabrication, installation, and placement of components or materials that require special knowledge or expertise, such as welding of structural steel or placement of grouted masonry. The knowledge and duties of a special inspector differ from that of the jurisdiction building inspector in that the special inspector's expertise is narrower in scope, such as that of a structural steel welding or prestressed concrete special inspector. Another difference is the special inspector is required to be present during certain field operations, whereas the jurisdiction inspector may not be present.

The special inspection requirements in the IBC address three areas:

1. Adequacy or quality of materials, such as concrete strength.
2. Adequacy of fabrication, such as verification of manufacturer's certified test reports.
3. Adequacy of construction techniques, such as proper placement of reinforcing.

Special inspection is that category of inspection requiring special knowledge, special attention, or both. The knowledge required is generally more specialized than that required by a general building inspector. An individual with a high degree of specialized knowledge is generally required; hence, the term *special* inspection.

Most building departments do not have the staff necessary to do detailed inspections on large or complex structures, nor do the permit fees allow the level of inspection necessary for the types of construction where extra care in quality control must be exercised to ensure compliance with the code. In some cases, the special inspection must be continuous during a particular operation, such as pretensioning high-strength bolts or making complete penetration groove welds. Hence, there is a need for a special inspector with specialized expertise who can be there continuously during a particular operation.

ICC offers a publication titled *Special Inspection Manual* that provides the building official with guidance on the administration and implementation of the special inspection requirements of the IBC. The guidance is based on recommended practices and the consensus of building officials and design professionals, as well as inspection and testing agencies. The duties and responsibilities of the building official, special inspector, project owner, engineer or architect of record, contractor, and building official are covered in the guide. Suggested forms are also included that can be easily adapted to the specific needs of the jurisdiction. Examples of checklist documents for special inspection are also included.

Note that in addition to the general special inspection requirements in Section 1705, there are additional special inspection requirements in Section 1705.12 that specifically apply to wind resistance, and in Section 1705.13 that specifically apply to seismic resistance.

1704.1 General. This section outlines the general requirements for special inspections, the statement of special inspections, contractor's statement of responsibility, submittals to the building official, and structural observations. The actual inspections and verifications required are contained in Section 1705.

1704.2 Special inspections and tests. The owner is responsible for the employment of special inspectors meeting the approval of the building official and all costs associated with the employment of special inspectors. Note that the special inspectors must be employed by the owner, or the responsible registered design professional acting as the owner's agent, not by the contractor or builder. This ensures independence of the special inspector and avoids any potential conflict of interest that could occur if the special inspector were employed by the contractor or builder. Note that the special inspections required by Chapter 17 are in addition to, not in lieu of, the jurisdiction inspections required by Section 110.

There are exceptions to the requirement for special inspections for minor work or work not required to be designed or sealed by a registered design professional such as Group U occupancies accessory to a residence, and prescriptive cold-formed steel or wood light-frame construction. Engineered seismic-force-resisting systems are very common in residential structures in Seismic Design Categories D, E, and F. Group R-3 occupancies often have components that require special inspection such as high-load diaphragms, high-strength concrete, structural steel frames, high-strength bolting, complete penetration groove welds, engineered masonry, and deep (pile) foundation systems. The exemption for Group U occupancies accessory to a residential occupancy is for those structures that are typically not required to be designed by a registered design professional or those designed and constructed in accordance with the *International Residential Code*® (IRC®). Exception 1 does not necessarily mean that inspections are not required, only that they are not required to be made by a special inspector. However, the above comments regarding structural systems used in the R-3 occupancies could also apply to U occupancies. It is not inconceivable that a large private garage could have structural components that would require special inspection.

Exception 1 refers to "conditions in the jurisdiction" as a possible exception. The primary conditions envisioned by the code in this case refer either to the jurisdiction having the resources and skill level necessary to perform the requisite special inspection tasks, thus obviating the need for a special inspector hired by the owner, or the work is of a minor nature in the opinion of the building official. Note that this exception for special inspection cannot be invoked by the owner. One purpose of the exception is to allow jurisdictions to perform special inspections if the jurisdiction so desires. Exception 3 waives special inspection for prescriptive light-frame construction of cold-formed steel or conventional wood structures. Section 2206.1.2 applies to prescriptively framed detached one- and two-family dwellings and townhouses less than or equal to three stories constructed of cold-formed steel in accordance with AISI S230. Exception 4 acknowledges that a contractor is allowed to hire the special inspector when the contractor is also the owner of the structure.

1704.2.1 Special inspector qualifications. Special inspectors are required to be qualified and demonstrate competence to the satisfaction of the building official. The registered design professional acting as an approved agency may provide special inspection for work designed by them, provided they demonstrate to the building official that they are qualified to perform the special inspections involved.

1704.2.2 Access for special inspection. This section requiring access to the construction specifically applies to special inspection.

1704.2.3 Statement of special inspections. The permit applicant must submit a detailed statement outlining the required special inspections and designate those who will perform the special inspections. In general, the responsible registered design professional is required to prepare the statement of special inspections, because the special inspections relate directly to the design and construction documents required by Sections 107.1 and 1603. The statement of special inspections must conform to the detailed requirements described in Section 1704.3. The exception in Section 1704.2.3 allows the statement of special inspections to be waived for prescriptive light-frame construction of cold-formed steel or conventional wood structures. The exception in Section 1704.3 allows the statement of special inspections to be prepared by a qualified person instead of a registered design professional where approved by the building official for construction that is not designed by or required to be designed by a registered design professional.

Although a statement of special inspections is not required for cold-formed steel structures constructed in accordance with the provisions of Section 2206.1.2 or conventional wood frame structures constructed in accordance with the prescriptive provisions of Section 2308, it should be noted that Sections 2308.3 and 2308.4 permit portions and elements of an otherwise conventional building to be designed in accordance with the engineering provisions of the code; therefore, these engineered elements and portions could require special inspection. For example, an otherwise conventional wood frame residence that uses a steel moment frame to resist lateral forces at the clear story entrance foyer. Such an element is outside of the conventional construction provisions, requires engineering under Section 2308.4, and could require special inspection under Section 1705.2.

1704.2.4 Report requirement. Records of each inspection must be kept and submitted to the building official so as to document compliance with the code. The records must include all inspections made at approved frequencies, nature and location of tests, compliance with the code requirements, as well as all discrepancies or violations. A final report must show that all required special inspections have been made and that discrepancies have been resolved before a certificate of occupancy can be issued by the building official. It is the responsibility of the special inspector to document and submit inspection records to the building official and to the registered design professional in responsible charge. The final special inspection report documenting resolution of discrepancies must be submitted at a time agreed upon by the permit applicant and the building official prior to the commencement of work.

1704.2.5 Special inspection of fabricated items. Unless structural items are fabricated by an approved fabricator as described in the exception, special inspection is required during the fabrication process. See Section 1704.2.5.1 regarding fabricator approval requirements.

1704.2.5.1 Fabricator approval. Special inspection at a fabrication plant is not required where a fabricator has been specifically approved by the building official. This approval may be based on the building official's own review of the fabricator's written procedures and quality assurance/quality control program. The fabricator should be periodically audited by an independent, approved special inspection agency. This section is intended to apply to fabricators accredited by organizations such as IAS, the AISC Certification Program for Structural Steel Fabricators, the Plant Certification Program for steel joists by the Steel Joist Institute, or the Plant Certification Program for precast concrete products by the Precast/Prestressed Concrete Institute. Some premanufactured low-rise metal building manufacturers meet both the AISC Certification Program for Structural Steel Fabricators and the IAS Fabricator Inspection Program and are recognized by most building officials as having the appropriate personnel, organization, experience, knowledge, and equipment to produce the quality required for structural steel buildings.

This section states that these items "provide a basis for control of materials and workmanship." Having these procedures in place in addition to having periodic third-party audits provides assurance to the building department that the quality of materials and workmanship meet the intent of Chapter 17, therefore allowing approved fabricators to be exempt from the special inspection provisions.

An important item that is often overlooked in relation to approved fabricators is the certificate of compliance. Upon completion of fabrication, the fabricator is to provide the building official with a certificate noting that the item has been fabricated in accordance with the approved plans. Figure 1704-1 is a sample certificate of compliance that is used by the City of San Diego. This certificate provides the building official and the owner with a level of assurance that the item was constructed correctly, similar to the final special inspection report for a project.

1704.3 Statement of special inspections. When special inspections are required by Section 1705, the *registered design professional in responsible charge* is required to prepare a detailed statement of special inspections. There are many areas where it has been simply assumed that this is the engineer of record for the project. However, Chapter 17 requires special inspections for several nonstructural components. As such, this statement of special inspection should actually be developed by the entire design team, with input from the architect, mechanical engineer, electrical engineer, and the structural engineer.

After the statement has been developed, it is the responsibility of the permit applicant to provide the statement to the building official as a condition of the permit issuance (see Section 1704.2.3). In some regions this statement is included on the construction plans while in others it is provided to the building department as a separate document, similar to project specifications.

1704.3.1 Content of statement of special inspections. The statement of special inspections must identify the work requiring special inspection or testing. It should include the type and extent of each special inspection or test and indicate whether the inspections are to be continuous or periodic. Where required, the statement also includes the additional special inspection or testing requirements for wind or seismic resistance as prescribed by Sections 1705.12, 1705.13, and 1705.14. Deferred submittal items that require a supplemental statement of special inspections must also be noted. A sample project-specific statement of special inspections is provided as an example at the end of this chapter.

1704.3.2 Seismic requirements in the statement of special inspections. The statement of special inspections must include elements of the SFRS, the designated seismic system and additional systems, components listed in Section 1705.13, and the seismic testing and qualifications required by Section 1705.14. The terms "seismic force-resisting systems" and "designated seismic systems" are defined in Section 202.

1704.3.3 Wind requirements in the statement of special inspections. The statement of special inspections must identify the elements of the MWFRS and components and cladding (that are subject to special inspection) where the basic wind speed, V, meets the criteria specified in Section 1705.12. See discussion under Section 1609 for descriptions of the basic wind speed and Exposure Categories.

1704.4 Contractor responsibility. The intent of the section is to require the contractor responsible for construction of the MWFRS, the seismic force-resisting system, the designated seismic system, and those wind-resisting components listed in the statement of special inspections to submit a statement of responsibility to the owner and building official. The statement of responsibility is to be submitted prior to commencement of work on a particular structural system or component. This statement of responsibility is intended to improve construction quality for seismic and wind resistance by stressing the contractor's role in providing adequate quality control during the construction process. The contractor must acknowledge awareness of special requirements contained in the statement of special inspections related to the lateral-force-resisting systems for wind and seismic loads.

Special Inspections and Tests, Contractor Responsibility, and Structural Observations 651

City of San Diego
Inspection Services Division
Planning and Development Review
9601 Ridgehaven Court • Suite 220 • MS 1102-B
San Diego, CA 92123
Information (858) 492-5070 • FAX (858) 492-5098

Certificate of Compliance for Off-Site Fabrication

➡ **(MUST BE FILED WITH THIS OFFICE AND APPROVED <u>PRIOR TO</u> ERECTION OF FABRICATED COMPONENTS)** ⬅

FABRICATION CO. NAME _____ TELEPHONE NO. (____) _____

FABRICATION SHOP ADDRESS _____ FAX NO. (____) _____

CITY _____ STATE _____ ZIP CODE _____

BUILDING PERMIT NO./S _____ PLAN FILE NO. _____
(For projects with multiple permit numbers but with the same plan file number, you may list all permit numbers and addresses on a separate sheet.)

JOB SITE ADDRESS _____

DESCRIPTION OF COMPONENTS FABRICATED _____

FABRICATOR IS CURRENTLY CERTIFIED BY (refer to BNL 17-6): ☐ ICBO   ☐ AISC   ☐ ACI   ☐ PCI   ☐ pti
☐ Other _____ Certificate No. _____ Expiration Date _____

SPECIAL INSPECTOR ASSIGNED (BY BUILDING OWNER) _____ Certificate No. _____
(Not required for a fabricator registered and approved by the City of San Diego, unless otherwise noted on the approved permitted plans and specifications.)

FABRICATION DRAWINGS REVIEWED AND ACCEPTED BY DESIGNER OF RECORD?   ☐ YES   ☐ NO

STRUCTURAL FRAMING PLANS, DETAILS & CALCULATIONS REVIEWED AND APPROVED BY THE CITY OF SAN DIEGO BUILDING OFFICIAL?   ☐ YES   ☐ NO

FABRICATION COMMENCEMENT DATE _____ FABRICATION COMPLETION DATE _____

We hereby certify that the components described and listed herein comply with the approved permitted plans, specifications and workmanship provisions of the California Building Code as amended by The City of San Diego and other applicable regulations. We further certify that each fabricated member or component has been inspected and meets all the requirements of the California Building Code as amended by The City of San Diego and other applicable regulations. We further certify that each fabricated member or component listed herein shipped to the project job site with an identifying mark or tag. We understand this certificate shall be submitted to the Inspection Services Division and the architect / engineer of record <u>prior to</u> erection of the fabricated members and components.

NOTE: This certification is limited to fabrication of the components or members described above to be used in the structures identified by the permit numbers shown and is not transferrable to any other fabrication work or other construction sites.

Name (Print): _____ Title: _____

Any person signing this application as an agent of the fabrication company declares under penalty of perjury to be an authorized agent of record of the fabrication company having the authority to execute this document.

Executed on this _____ day of _____ / _____ .
                                      MONTH   YEAR

Signature: _____

This information is available in alternative formats for persons with disabilities.
To request this information in alternative format, call (619) 236-7703 or (800) 735-2929 (TT)

DS-311 (3-00)
Printed on Recycled Paper

Reset Button

Figure 1704-1 **Fabricator certificate of compliance.**

1704.5 Submittals to the building official. This section itemizes all the reports and certifications that must be submitted to the building official. Having all these reports and certifications listed in one section helps clarify the requirements.

1704.6 Structural observations. The purpose of structural observation is to ensure that critical elements of the lateral-force-resisting system are constructed in general conformance with the design as shown in the approved structural drawings and specifications. Because the registered design professional is most familiar with the design and the details of the lateral-force-resisting system, he or she is the most appropriate person to execute the requisite observations.

Structural observation consists of visual observation of specific structural systems by the registered design professional for general conformance with the approved construction documents at significant construction stages and at the completion of the structural system. The structural observer is expected to "visually" observe systems, details, and load paths. Structural observation is separate and distinct from special inspections. Special inspections are very detailed inspections of components and materials within structural systems.

Structural observations are broad, general, visual overviews of a bigger picture. Broad knowledge of structural design issues and specific knowledge of their application to the project are necessary. An observation answers the question, "Is the structural frame of the building being built as directed within the construction documents?" Structural observation does not include or waive inspections performed by the jurisdiction as required by Section 110 or special inspections required by Section 1705. See the definition of "Structural observation" in Section 202.

Prior to performing structural observations, the observer is required to notify the building official in writing of the frequency and extent of structural observations. At the conclusion of work, the structural observer is required to submit a written statement that the site visits have been made and identify any unresolved deficiencies that may exist.

1704.6.1 Structural observations for structures. Structural observations are required for all Risk Category III or IV structures, for high-rise buildings, when specifically required by the registered design professional, and when specifically required by the building official. If the project is located within Seismic Design Category E and is greater than two stories above the grade plane, structural observations are required regardless of the Risk Category of the building.

Risk Category III and IV Occupancies are defined in Table 1604.5, which also provides several examples of essential facilities and buildings or other structures that represent a substantial hazard to human life in the event of failure. Given the relative risk and hazard, it is appropriate to require that a structural engineer conduct site visits to verify general conformance to the design intent for these types of structures. Observations for seismic- or wind-resisting systems and load paths are performed by the same registered design professional that observes gravity systems and load paths. Lateral force–resisting systems and load paths are integrated with gravity force–resisting systems, components, and cladding so there is no differentiation of "wind or seismic resistance observations" from "Structural Observations."

Section 1705 *Required Special Inspections and Tests*

This section applies to special inspections and tests of elements and nonstructural components of buildings and structures.

1705.1.1 Special cases. Special inspection is required for proposed work that is unique or unusual in nature and products or systems that are not specifically addressed in the code or in standards referenced by the code. This section is used to apply special inspection requirements to items that are not specifically covered in the code but are approved under the alternate design and methods of construction provisions in Section 104.2.3. Many ICC Evaluation

Service Reports (ESR) for structural products require special inspection in accordance with this section. For example, ESR 2302 for the Hilti Kwik Bolt 3 requires periodic special inspection during installation.

1705.2 Steel construction. This section sets forth the special inspection requirements for the fabrication and erection of steel structures. The exception eliminates the need for special inspection of steel structures in certain cases. Special inspection is not required if the fabricator does not alter the properties of the parent material by welding, thermal cutting, or heating operations. For example, if the members being fabricated were cut by mechanical means, such as a band saw, and punched or drilled for bolt holes, with no application of heat, special inspection would not be required. But if the same members were cut with an oxy-acetylene torch, special inspection *would* be required. Even if special inspection is not required, the fabricator must provide evidence that tracking procedures are adequate to verify that the material used to fabricate any member meets the required specification, is of the proper grade, and has an associated mill certificate of compliance. Testing may be required if the mill certificate is separated from the materials or is otherwise missing (see Figure 1705-1).

1705.2.1 Structural steel. This section references the quality assurance provisions of AISC 360 *Specification for Structural Steel Buildings* for special inspection of structural steel. AISC 360 Chapter N includes comprehensive quality control and quality assurance requirements for structural steel construction, including requirements pertaining to the structural steel fabricator

Photo courtesy of Smith Emery Laboratories

Figure 1705-1 Bend test on No. 18 rebar per ASTM A370.

and erector, quality assurance requirements pertaining to the owner's inspecting and/or testing agencies and requirements for shop- or field-applied coatings. AISC 360 Chapter N provides the foundation for the quality control and quality assurance requirements for general structural steel construction. The inspection requirements in AISC 360 of the quality assurance inspector are generally equivalent to those specified for the special inspector in IBC Chapter 17.

1705.2.2 Structural stainless steel. Special inspections and nondestructive testing of structural stainless steel elements in buildings must be in accordance with AISC 370 *Specification for Structural Stainless Steel Buildings*. AISC 370 Chapter N provides quality control and quality assurance provisions similar to those outlined in AISC 360 Appendix N with the exception of shop- or field-applied coatings.

1705.2.3 Cold-formed steel deck. The *Standard for Quality Control and Quality Assurance for Installation of Steel Deck* (SDI QA/QC) is referenced for inspection of steel floor and roof decks. The SDI QA/QC standard contains provisions for quality assurance inspection of steel floor and roof deck. The standard clarifies the scope of required inspections and responsibilities of both the installer's quality control personnel and quality assurance inspector, and contains tables of inspection tasks that specifically list inspection requirements for material verification, deck installation, welding, and mechanical fastening.

1705.2.4 Open-web steel joists and joist girders. The structural design and installation of open-web steel joist and joist girder systems warrants special inspection by personnel with sufficient expertise and approval by the building official having the necessary competence to inspect the installation of steel joist systems. Examples of important items that require special inspection are bearing seat attachments, field splices, and bridging attachments. Where steel joist systems are used in roof and floor diaphragms, chords and collectors are important critical elements of the lateral-force-resisting system. Although the standard specifications for open-web steel joists and joist girders (SJI 100) and composite steel joists (SJI 200) by the Steel Joist Institute (SJI) contain provisions for inspections, they are limited to quality control inspections by the manufacturer before shipment to verify compliance and workmanship with the specifications. Quality control and assurance are discussed in Section 2207.1 as shown in IBC Table 1705.2.4. Although sections of the SJI standards noted above are also referenced in Section 4 of the *Code of Standard Practice for Steel Joists and Joist Girders* (SJI COSP) and the *Code of Standard Practice for CJ-Series Composite Steel Joists* (SJI CJ COSP), these SJI codes of standard practice are not directly referenced in the IBC.

Table 1705.2.4 Required Special Inspections of Open-web Steel Joists and Joist Girders. This table specifies what items require special inspection and references the applicable standard listed in Section 2207.1. Note that the inspections are periodic meaning that the special inspector can do the inspection after installation and need not necessarily be present during the operation.

1705.2.5 Cold-formed steel trusses spanning 60 feet or greater. This section requires special inspection for cold-formed steel trusses spanning 60 feet (18,288 mm) or more. Because long-span trusses have significant loads and support reactions, proper installation and bracing is of critical importance. The special inspector must verify that the temporary and permanent truss restraint/bracing are installed in accordance with the approved truss engineering and submittal package. A similar requirement exists for metal-plate-connected wood trusses in Section 1705.5.2.

1705.2.6 Metal building systems. Metal building systems are significantly different from other forms of steel construction, especially regarding the shared design responsibilities between the metal building system manufacturer and registered design professional for the project. The code includes metal building system provisions, a definition, and material

Figure 1705-2 **Metal building system.**

requirements which clarify the design requirements for these systems (see IBC Section 2210 for more information). Metal building systems often contain assemblies made of a variety of components, such as structural steel, cold-formed steel, and steel cables (see Figure 1705-2). While the individual components are often covered by fabricator special inspections and tests found in the subsections of Section 1705.2 as previously discussed, the "systems" used in metal building systems are often unique and are not specifically identified in the steel sections. This section and referenced table include explicit requirements for the special inspection of commonly used systems in metal buildings.

Table 1705.2.6 Special Inspections of Metal Building Systems. Metal building systems are typically highly optimized structures heavily dependent on bracing components to function per the design. Some bracing components consist of materials that are not considered to be "structural steel." This table includes periodic special inspection requirements for rafter, beam, and column flange bracing; purlin and girt bracing and bridging; X-bracing; and purlins and girts including required lapping.

1705.3   Concrete construction.   This section provides the special inspection and verification requirements for concrete construction such as buildings, foundations, and other elements. Detailed requirements are covered in Table 1705.3, which provides the specific verification and inspection requirements, whether the frequency of inspection is to be continuous or periodic, the applicable referenced standard, and/or the IBC code section where appropriate. Exceptions for special inspection are isolated spread footings for buildings three stories or less in height, lightly loaded elements such as slabs on grade and concrete foundations including conventional foundations for light-frame construction and nonstructural concrete flatwork such as sidewalks, patios, and driveways. Where certain criteria are met, Exceptions 1 and 2 exempt conventional foundations, either spread (pad) footings or continuous footings, for buildings three stores or less in height. Under Exception 2, Item 2.3 exempts continuous concrete footings from special inspection where the structural design of the footing is based on a concrete strength of 2,500 psi (17.2 MPa), even if a higher compressive strength is specified in the construction documents for other reasons such as durability. This allows engineers to specify a higher concrete strength as a matter of preference even if a 2,500-psi (17.2 MPa) design strength is used for sizing footings. Note that Section 1808.8.1 requires a specified concrete strength of 3,000 psi (20.7 MPa) for most foundations in Seismic Design Category D, E, or F. This means that foundation concrete in Seismic Design Category D or higher must be designed for a specified concrete strength of 3,000 psi (20.7 MPa) concrete, and therefore special inspection is required. However, Table 1808.8.1 allows a specified concrete strength of 2,500 psi (17.2 MPa) for Group R or U of light-frame construction two stories or less in height for foundations in Seismic Design Category D, E, or F. Therefore, 2,500 psi (17.2 MPa) concrete is permitted for these structures, and special inspection would not be required.

Concrete basement or foundation walls constructed in accordance with the prescriptive provisions in Section 1807.1.6.2 are not required to be engineered and are also exempt from special inspection.

1705.3.1 Welding of Reinforcing Bars. This section clarifies special inspection requirements for welds of reinforcing bars. While the majority of the special inspection requirements for concrete derive from ACI 318, all items related to the welding of reinforcing steel are outlined in AWS D1.4. This includes not only the actual special inspections that are to be performed, but also the qualifications for the special inspectors.

1705.3.2 Material tests. Constituent materials for concrete such as aggregate, cement, admixtures, and water must conform to the requirements set forth in Section 1903 and the standards of Chapter 3 of ACI 318. When sufficient documentation is not available to verify that constituent materials conform to these requirements, the building official should require testing of the materials.

Table 1705.3 Required Special Inspections and Tests of Concrete Construction. Table 1705.3 presents the requirements for special inspection of concrete structures in a concise format along with the inspection frequency, applicable referenced standard, and/or code section. The table summarizes the required inspections and test samples necessary to verify that the in-place concrete meets code requirements. Refer to ACI 318 and the applicable ASTM standards for more details on constituent material tests, sampling of fresh concrete, testing of slump or air content, casting of test specimens, and other test requirements.

Item 1—Reinforcement. Periodic special inspections of reinforcing should verify that tendons and reinforcement are of the correct size and grade, as required by the approved drawings and specifications, and is properly placed prior to placement of concrete. Proper placement of reinforcement has a significant impact on the integrity and strength of reinforced concrete. The reinforcement should be placed within the tolerances set forth in ACI 318 and ACI 117, *Specification for Tolerances for Concrete Construction and Materials*. Additional requirements of ACI 318, such as surface conditions of reinforcement, spacing limitations, and concrete protection for reinforcement or cover, must also be checked.

Item 2—Welding reinforcement. Item 2, inspection of reinforcing steel welding, refers to AWS D1.4 and ACI 318, for requirements pertaining to welding reinforcing steel. Reinforcing steel per ASTM A706 *Standard Specification for Deformed and Plain Low-Alloy Steel Bars for Concrete Reinforcement* is typically specified. Other rebar must be tested and verified by a periodic special inspection before welding begins to determine whether welding is an option or if mechanical splices are required for the reinforcement. There are several requirements for continuous special inspection of reinforcement welding including special moment frames, boundary elements of special structural walls and coupling beams, reinforcement splices and primary tension reinforcement in corbels. Other welds, including single-pass fillet welds up to 5/16 inch (7.9 mm), only require periodic inspections.

Item 3—Cast-in-place anchors. Item 3 requires periodic special inspection when anchor bolts are cast in concrete. Proper placement and embedment of anchor bolts is of extreme importance. If an anchor bolt is set too low for proper thread engagement, there are few satisfactory methods to remedy the situation. The common practice of placing a puddle weld in the nut is of questionable value, as neither the bolt nor the nut may actually be weldable material. The chemistry of nuts is not controlled by the ASTM requirements, and the chemistry allowed for an ordinary A307 bolt is such that it may or may not be weldable. In either case, the weld is not a prequalified weld in accordance with AWS D1.1 *Structural Welding Code—Steel*, and a Procedure Qualification Record should be developed to qualify the welding procedure. Note also that if the anchor bolt is a high-strength bolt such as an ASTM A325 or ASTM A449, the bolt is quenched and tempered, and application of heat by welding may alter the strength of the bolt and make the bolt brittle.

Item 4—Post-installed anchors. Item 4 requires special inspection for all anchors post-installed in hardened concrete. Continuous special inspection is required for post-installed adhesive anchors in hardened concrete where the anchors are installed in horizontally or upwardly inclined orientations and are designed to resist sustained tension loads. All other adhesive and mechanical anchors require periodic special inspection. Footnote b accounts for post-installed anchors approved through the alternate methods of construction provisions of Section 104.2.3, such as anchors installed in accordance with ICC Evaluation Service Reports. It is also intended to distinguish between the requirements for special inspection of anchors designed to comply with the IBC alone versus those qualified by approved research reports in accordance with ACI 355.2, *Post-Installed Mechanical Anchors in Concrete—Qualification Requirements and Commentary*. Typically, items requiring special inspection that are approved under Section 104.2.3 are covered by Section 1705.1.1, Special Cases. Where special inspection requirements are not provided in a research report, the special inspection requirements must be specified by the registered design professional, who would indicate whether inspections are continuous or periodic, and be approved by the building official prior to commencement of the work.

Item 5—Design mix. Item 5 is a particularly important periodic special inspection verification on larger projects that may have many required mix designs with differing strength requirements and aggregate sizes.

Item 6—Concrete testing. Sampling of fresh concrete for making specimens for strength tests is extremely important for proper quality control. Properly sampled and prepared specimens are necessary to determine that the concrete will meet or exceed the design strength. The continuous special inspection frequency of sampling should be in accordance with ACI 318 Section 26.12.2—one set of specimens for each class of concrete not less than once per day, once per 150 cubic yards (115 m^3), or once per 5,000 square feet (465 m^2) of slab or wall. Sampling should be done in accordance with ASTM C172 *Practice for Sampling Freshly Mixed Concrete* to ensure representative samples for determining compressive strength. The tests specified in Item 6 may be supplemented by other tests such as unit weight or air content.

Item 7—Concrete placement. Continuous special inspection of the actual placement is important to determine that the fresh concrete is properly handled so that it does not segregate during placement and that the concrete is properly consolidated by vibration. The mixing, conveying, and depositing requirements (see ACI 318 Section 26.5) should be strictly enforced to ensure proper placement with adequate consolidation and without segregation. Concrete voids or *rock pockets* can adversely affect design strength and are unattractive.

Item 8—Curing. Maintenance of proper cure is essential to obtaining quality concrete that will reach the design strength. Concrete that is not properly cured often will be below design strength and may suffer degradation at the surface from use or from environmental effects much earlier than properly cured concrete. This is a periodic special inspection.

Item 9—Prestressed concrete. The continuous special inspections required by Item 9 are of extreme importance as the strength of a prestressed member is highly dependent on proper prestressing. When checking the application of prestressing force, both the force applied to the tendon and the tendon elongation should be checked simultaneously to ensure that the tendon has not been hung up in the tendon sheath. See ACI 318 Section 26.10.

Item 10—Precast concrete. Criteria for the erection procedures of precast concrete must be provided on the design drawing by the design engineer. The drawings should identify each panel to be cast and should specify: dimensions and thickness of panels, reinforcement grade, size and location, location of inserts, and minimum concrete strength at lifting and in-service. The special inspector should ensure with periodic special inspections that the erection process complies with the approved procedures. See ACI 318 Section 26.9.

Item 11—Precast Concrete Diaphragm Connections and Reinforcement. ACI 318 has provisions for design of precast concrete diaphragms in Section 18.12.11. Such diaphragms are required to comply with the requirements of ACI 550.5. ACI 318 Section 26.13.1.3 and ACI 550.5 have continuous special inspection requirements for precast concrete diaphragm connections or reinforcement at joints classified as medium or high deformability elements (MDE or HDE), installation of embedded parts, completion of the continuity of reinforcement across joints, and completion of connections in the field in structures assigned to SDC C, D, E, and F.

Item 12—Precast Concrete Diaphragm Connection Installation Tolerances. ACI 318 Section 26.13.1.3 requires that installation tolerances of precast concrete diaphragm connections be periodically inspected for compliance with ACI 550.5.

Item 13—Concrete strength. Concrete strength must be verified with periodic special inspections prior to stressing tendons used in post-tensioned concrete. Form supports for prestressed concrete should not be removed until sufficient prestressing has been applied. Forms and shoring for conventionally reinforced concrete members such as beams and structural slabs should not be removed until adequate concrete strength is achieved. See ACI 318 Section 26.11.2 for requirements related to removal of formwork.

Item 14—Formwork. Periodic special inspection of concrete forms for proper shape, dimensions, and location is essential for adequate performance on concrete members such as beams, columns, walls, and structural slabs. See ACI 318 Section 26.11 for general requirements related to formwork.

1705.4 Masonry construction. The Masonry Society (TMS) is responsible for developing the structural masonry standards referenced in the IBC. The two main referenced standards developed by TMS are the *Building Code Requirements for Masonry Structures* (TMS 402) and the *Specification for Masonry Structures* (TMS 602). The level of quality assurance required by Section 3.1 of the TMS 402 is driven by the type of design used and the risk category of the structure. Table 1705-1 reflects the level of quality assurance required based on whether the design is per engineered or prescriptive methods and risk category of the structure.

Table 1705-1. **Quality Assurance Level Required by TMS 402 Section 3.1**

Type of Masonry Structure	Risk Category I, II, III	Risk Category IV
Engineered design methods	Level 2	Level 3
Prescriptive design methods	Level 1	Level 2

There are three exemptions from special inspection for masonry structures. Glass unit masonry or masonry veneer in Risk Category I, II, or III structures are exempt from special inspection when constructed in accordance with Section 2110 or Chapter 14. Note that the TMS 402 requires Level 1 quality assurance for these structures, which requires no specific testing and only consists of verification of certificates of compliance for masonry materials in accordance with TMS 602 Article 1.5. Level 2 quality assurance requires verification of the design strength prior to construction unless exempted by the code, such as for nonstructural veneer and glass block. Level 3 quality assurance requires ongoing verification of materials during the construction of a project. The other two exceptions are for masonry foundation walls constructed in accordance with the prescriptive tables of Section 1807.1.6, which do not require engineering, and masonry fireplaces, heaters, and chimneys constructed in accordance with Section 2111, 2112, or 2113.

1705.4.1 Glass unit masonry and masonry veneer in Risk Category IV. Special inspection for glass unit masonry and masonry veneer in Risk Category IV structures must comply with TMS 602 Level 2 Quality Assurance requirements. See Table 4 of the TMS 602 for specific details. As noted, Table 1705-1 shows the level of quality assurance required based on the type of design and risk category of the structure. When prescriptive design is used for glass unit masonry and masonry veneer, Level 1 quality assurance is permitted for construction in Risk Categories I, II, and III and Level 2 is required for Risk Category IV structures. When engineered design, including Limit Design is used, Level 3 quality assurance is required for Risk Category IV structures and all other risk categories require Level 2. Glass fiber reinforced polymer (GFRP) reinforced masonry requires engineered design and would require Level 2 or 3 quality assurance. Glass unit masonry and masonry veneer are allowed in buildings such as fire houses, emergency operation centers, and hospitals which are classified as Risk Category IV structures.

1705.4.2 Vertical masonry foundation elements. The intent of this section is that the special inspection requirements for vertical masonry foundation elements are as noted in Section 1705.4 and its subsections. In other words, vertical masonry foundation elements may be required to comply with quality assurance and inspection provisions of TMS 602, they may be required to only comply with the Level 2 Quality Assurance requirements prescribed in Section 1705.4.1, or they may be exempt from special inspection in accordance with one of the exceptions. Section 1808.9 references the TMS 402 for design requirements for vertical masonry foundation elements.

1705.5 Wood construction. This section requires special inspection for certain types of wood construction. The requirement is for inspection of prefabricated elements such as wood trusses and refers to Section 1704.2.5 regarding special inspection of fabricators. For portions of wood structures designated as the SFRS, see also the special inspection requirements for wind and seismic resistance discussed in the analysis of Sections 1705.12 and 1705.13. Special inspection is required for site-built assemblies, high-load diaphragms, wood trusses having a span of greater than 60 feet (18,288 mm), and mass timber elements and connections in Type IV-A, IV-B, and IV-C construction. See Table 1705-2.

1705.5.1 High-load diaphragms. When lateral loads become significant, designers often use high-load diaphragms, which have multiple rows of fasteners. Proper construction of high-load diaphragms is critical so this section requires periodic special inspection. Design values for high-load diaphragms fastened with nails are in Table 4.2B of the *Special Design Provisions for Wind and Seismic* (SDPWS), and high-load diaphragms fastened with staples are in Table 2306.2(2). The special inspector is required to inspect the diaphragm for proper sheathing grade and thickness; proper size, species, and grade of framing members; and proper fastener type, size, spacing, and edge distance from sheathing and framing members.

Table 1705-2. **Wood Construction Special Inspections**

Inspection Item	IBC Section	Frequency
Prefabricated wood structural elements and assemblies, where the fabricator **has NOT been approved** to perform work without special inspections in accordance with Section 1704.2.5.1.	1705.5 1704.2.5	Per AHJ and RDP
Prefabricated wood structural elements and assemblies, where the fabricator **has been approved** to perform work without special inspections in accordance with Section 1704.2.5.1.	1705.5 1704.2.5	Submit certificate of compliance
High-load diaphragms designed in accordance with SDPWS Table 4.2B or IBC Section 2306.2. Verify the following are in accordance with approved construction documents: • Wood structural panel sheathing grade and thickness • Nominal framing member size at adjoining panel edges • Nail or staple diameter and length • Number of fastener lines • Spacing between fasteners in each line and at edge margins	1705.5.1	Per AHJ and RDP
Metal-plate-connected wood trusses with clear spans of 60 feet (18,288 mm) or greater. Verify the following are installed in accordance with approved truss submittal package: • Permanent individual truss member restraint/bracing • Temporary restraint/bracing	1705.5.2	Per AHJ and RDP
Mass timber elements and connections in Type IV-A, IV-B, and IV-C construction	1705.5.3 Table 1705.5.3	Per AHJ and RDP

AHJ = Authority Having Jurisdiction; RDP = Registered Design Professional.

1705.5.2 Metal-plate-connected wood trusses spanning 60 feet or greater. Any trusses having spans of 60 feet (18,288 mm) or more require periodic special inspections. Special inspections are required to ensure that temporary and permanent truss member bracing is installed in accordance with the approved truss package.

1705.5.3 Mass timber construction. These provisions address special inspection provisions for mass timber elements and connections in Types IV-A, IV-B, and IV-C construction. This unique type of construction requires a level of inspection consistent with other large buildings and applications. These special inspections are similar to what is required for other prefabricated systems such as precast concrete and structural steel. These provisions are not applicable to heavy timber (IV-HT) construction.

Specific elements requiring periodic special inspection for construction Types IV-A, IV-B, and IV-C include:

1. Connection of mass timber elements to timber deep foundation elements. These connections are critical to transfer loads from the mass timber elements to the piles, particularly for lateral loading. Connections to concrete foundations are addressed in IBC Table 1705.3, Items 3 or 4.

2. Erection of mass timber elements. Similar to pre-cast concrete (IBC Table 1705.3, Item 10), tall wood buildings utilizing prefabricated elements need to have verification that the correct elements are placed in the right location in accordance with the design drawings.

3. Specialized connections between mass timber products that utilize threaded, bolted, or concealed connections are considered periodic in a similar manner that concrete special inspections are required in Table 1705.3. The strength of many connection designs is predicated on specific screw lengths and installation angles. Bolted connections require specific diameters, and for lag screws, specific lengths. Concealed connectors, many of which are proprietary, must be installed correctly for structural performance.

Adhesive anchorage installed in horizontal or upwardly inclined positions resisting tension loads shall be continuously inspected, similar to Table 1705.3, Item 4a. This is required because of issues with creep of adhesives under long-term tension loading. However, consistent with requirements for precast concrete, all other adhesive anchors need only be inspected periodically (ref. Table 1705.3, Item 4b).

1705.6 Soils. The table for verification and inspection of soils clarifies specific inspection tasks that are required for soils. This section covers special inspection requirements for soils such as site preparation, engineered fills, and materials supporting load-bearing foundations. The load-bearing capacity of the site soil and any fill has a significant impact on the structural integrity of a building supported by fill. For example, settlements in improperly compacted fill can cause significant structural distress. Differential settlements of ¼ inch (6.35 mm) in a 20-foot (6,096 mm) grade beam can induce stresses that exceed the yield limits. Hence, fills should be engineered and compaction carefully controlled. The special inspection tasks outlined in Table 1705.6 must be performed to verify compliance with the approved construction documents and the geotechnical report per Section 1803.2. What is required to be included in geotechnical reports is covered in Section 1803.6.

The special inspector is required to (1) verify that materials below footings are adequate to achieve the design bearing capacity; (2) verify that excavations are extended to the proper depth and have reached proper bearing material; (3) perform classification and testing of controlled fill materials; (4) verify use of proper materials, densities, and lift thicknesses during placement and compaction of controlled fill material; and (5) verify that the site and subgrade have been properly prepared prior to placement of controlled fill. All of these special inspections are periodic except Item 4 which requires continuous special inspection. The exception requires that the special inspector verify a minimum of 90-percent compaction.

1705.7 Driven deep foundations. Special inspection is required for installation and testing of driven deep (pile) foundations. The special inspection tasks outlined in Table 1705.7 are required to verify compliance with the approved construction documents and the geotechnical report per Section 1803.2. What is required to be included in geotechnical reports is covered in Section 1803.6. Table 1705.7 requires continuous special inspection to (1) verify that pile materials, sizes, and lengths comply with the requirements; (2) determine capacities of test piles and conduct additional load tests; (3) observe driving operations and maintain complete and accurate records for each pile; and (4) verify placement locations and plumbness, confirm type and size of hammer, record the number of blows per foot of penetration, determine required penetrations to achieve design capacity, record tip and butt elevations, and document any pile damage.

For steel and concrete elements, additional special inspections are to be performed per Sections 1705.2 and 1705.3, respectively. For specialty piles, the special inspector must perform additional inspections in accordance with the registered design professional's recommendations. It should also be noted that when a deep foundation system is required, the geotechnical engineer should specify the actual testing that is required in accordance with Section 1803.5.5.

1705.8 Cast-in-place deep foundations. Chapter 18 has two general types of foundations—shallow foundations and deep foundations. Continuous special inspection is required for installation and testing of cast-in-place deep foundations. The special inspection tasks outlined in Table 1705.8 are required to verify compliance with the approved foundation and geotechnical report per Section 1803.2. The table requires the special inspector to observe drilling operations and maintain complete and accurate records for each pile, and verify placement locations and plumbness, and confirm pile diameters, bell diameters (if applicable), lengths, embedment into bedrock (if applicable), and adequate end-bearing strata capacity. The volume of concrete or grout placed is often the first indicator of potentially significant problems during construction. Item 2 requires the special inspector to record the volumes of concrete or grout placed as a critical diagnostic tool. For concrete cast-in-place deep foundations, Item 3 of the table refers to Table 1705.3 for special inspection tasks related to concrete construction.

Augered piles are drilled and then grouted in place. For each drilled pile, data should include a drilling log showing the types of soils encountered for each foot and the material stratum at the required tip elevation. The inspector should verify that the soil at the required tip elevation is the correct soil. The drilling log should also give information on the duration and cause of any delays, data on the rebar cage and concreting procedures, casing or other procedures necessary to prevent intrusion of ground water, and results of any concrete strength tests.

1705.9 Helical pile foundations. Helical piles are deep foundation systems used for new construction, for foundation repair to mitigate structure settlement, or for tiebacks in shoring applications. Helical piles can be installed with a column of high-strength grout that encases the helical shaft to provide additional compressive and lateral load capacity. The section requires continuous special inspection during installation of helical pile systems. The installation procedure must be in accordance with the recommendations of the registered design professional and the approved geotechnical report.

1705.10 Structural Integrity of Deep Foundation Elements. Significant defects affect the structural strength of deep foundation elements and need to be detected and corrected prior to further construction to prevent foundation failures. When the integrity of a deep foundation element is in doubt (e.g., due to installation records or difficult soil conditions), the deep foundation element should be tested during installation to assure no material defects. Sections 1705.7, 1705.8, and 1705.9 address visual inspections. The ASTM tests specified in this section provide a means to assess portions of deep foundation elements that cannot be visually inspected. Most foundation failures are caused by inadequate geotechnical capacity. This section addresses other possible failure modes—namely lack of structural integrity.

1705.11 Fabricated items. This section references Section 1704.2.5 for specific requirements pertaining to special inspection of fabricated items. See further discussion of fabricator approval under Section 1704.2.5.

1705.12 Special inspection for wind resistance. Unless specifically exempted in the general requirements provided in Section 1704.2, special inspection for wind resistance is required for buildings sited in Exposure B where the basic wind speed (V) is 150 mph (67 m/sec) or more, and Exposure C or D where the basic wind speed (V) is 140 mph (62.6 m/sec) or more. The specific requirements apply to structural wood, cold-formed light-frame steel structures, and exterior cladding. The inspections are continuous or periodic depending on the type of structural elements involved. Table 1705-3 summarizes the requirements in Sections 1705.12.1 (wood), 1705.12.2 (cold-formed steel), and 1705.12.3 (cladding).

Table 1705-3. **Special Inspection for Wind Resistance**

Type of Structural Element	Type of Special Inspection	
	Continuous	Periodic
Wood–Field gluing of MWFRS elements	X	
Wood frame–Nailing, bolting, anchoring, fastening components of MWFRS such as shear walls, diaphragms, braces, collectors, hold-downs[a]		X
Cold-formed steel light frame–Welding of MWFRS elements		X
Cold-formed steel light frame–Screws, bolting, anchoring, fastening components of MWFRS such as shear walls, diaphragms, braces, collectors, hold-downs[a]		X
Roof cladding, roof deck, and roof framing connections		X
Wall cladding fastening and wall connections to roof and floor diaphragms and framing		X

[a]Note that exceptions for both wood frame cold-formed steel do not require special inspection where the fastener spacing of the sheathing is more than 4 inches (102 mm) on center at the panel edges.

1705.12.1 Structural wood. Continuous special inspection is required for any field gluing of the MWFRS (see Section 202 for definition of main windforce-resisting system). Periodic special inspection is required for fastening (nailing, bolting, anchoring) of elements of the MWFRS such as shear walls, diaphragms, chords, collectors (drag struts), braces, and hold-downs. The exception applies where the fastener spacing of the sheathing is more than 4 inches on center at the panel edges.

1705.12.2 Cold-formed steel light-frame construction. Periodic special inspection is required for any welding of the MWFRS. Periodic special inspection is required for fastening (screw attachment, bolting, anchoring) of elements of the MWFRS such as shear walls, braces, diaphragms, collectors (drag struts), and hold-downs. The first exception applies when the sheathing is gypsum board or fiberboard because of their relatively low load capacity. Similar to wood framing, the second exception applies where the fastener spacing of the sheathing is more than 4 inches (102 mm) on center at the panel edges.

1705.12.3 Wind-resisting components. Damage to buildings in high-wind events often begins with failure of the cladding system. Thus, periodic special inspection is required for roof covering, the roof deck, and roof framing connections. Exterior wall covering and connections to the roof or floor diaphragms and framing also require periodic special inspection.

1705.13 Special inspections for seismic resistance. Unless specifically exempted in the general requirements in Section 1704.2, special inspection for seismic resistance is required in addition to the general special inspection requirements covered in Section 1704. The requirements are triggered by the seismic design category of the building. Continuous special inspection is required for the SFRS for steel systems and elements, field gluing operations for structural wood elements, and cold-formed steel (CFS) light-frame construction. Other items such as storage racks, base isolation systems, CFS special bolted moment frames, fastening of wood lateral-resisting systems, designated seismic systems, and architectural, plumbing, mechanical, and electrical components require periodic special inspection. See a summary of special inspection requirements for seismic resistance in Table 1705-4.

Table 1705-4.   Special Inspection for Seismic Resistance

Type of Structural Element	IBC Section	SDC	Continuous	Periodic
Structural Steel—SFRS*	1705.13.1.1	B-F	X	X
Structural Steel Elements—Elements not covered in Section 1705.13.1.1 including struts, collectors, chords and foundation elements*	1705.13.1.2	B-F	X	X
Structural Wood—Field gluing operations of SFRS elements*	1705.13.2	C-F	X	—
Structural Wood—Nailing, bolting, anchoring and other fastening of SFRS elements including wood shear walls, wood diaphragms, drag struts, braces, shear panels and hold-downs*	1705.13.2	C-F	—	X
Cold-formed Steel Light-frame Construction—Welding of SFRS elements; and screw attachment, bolting, anchoring and other fastening of SFRS elements including shear walls, braces, diaphragms, collectors (drag struts), and hold-downs*	1705.13.3	C-F	—	X
Designated Seismic Systems—Verify that labels, anchorage and mounting conform to certificate of compliance for designated systems requiring seismic qualification in accordance with ASCE 7 Section 13.2.2	1705.13.4	C-F	—	X
Architectural Components—Erection and fastening of exterior cladding, interior and exterior nonbearing walls, and interior and exterior veneer*	1705.13.5	D-F	—	X
Access Floors—Anchorage of access floors	1705.13.5.1	D-F	—	X

Plumbing, Mechanical, and Electrical Components
• Anchorage of electrical equipment for emergency and standby power systems
• Anchorage of other electrical equipment (SDC E-F only)
• Installation and anchorage of piping systems designed to carry hazardous materials and associated mechanical units
• Installation and anchorage of ductwork designed to carry hazardous materials
• Installation and anchorage of vibration isolation systems where approved construction documents require nominal clearance of ≤1/4 inch between equipment support frame and restraint
• Installation of mechanical and electrical equipment, including duct work, piping systems and their structural supports, where automatic fire sprinkler systems are installed to verify one of the following:
  o Minimum clearances have been provided as required by ASCE 7 Section 13.2.4
  o Nominal clearance ≥3 inches provided between fire protection sprinkler system drops and sprigs and: structural members not used collectively or independently to support the sprinklers, equipment attached to the building structure, and other systems' piping. Where flexible sprinkler hose fittings are used, special inspection of minimum clearances is not required
	1705.13.6	C-F	—	X

Storage Racks—Anchorage of storage racks that are ≥8 feet in height	1705.13.7	D-F	—	X
Seismic Isolation Systems—Fabrication and installation of isolator units and energy dissipation devices in seismically isolated structures	1705.13.8	B-F	—	X
Cold-formed Steel Special Bolted Moment Frames—Installation of cold-formed steel special bolted moment frames in the SFRS	1705.13.9	D-F	—	X

*See code for exceptions. For SI: 1 inch = 25.4 mm, 1 foot = 304.8 mm.

The SFRS consists of those structural elements and systems that provide lateral stability of the structure and are specifically designed to provide resistance to the anticipated seismic forces. The designated seismic system consists of those architectural, electrical, and mechanical systems and components that require design in accordance with Chapter 13 of ASCE 7 for which the component importance factor, I_p, is greater than 1.0, as prescribed in Section 13.1.3 of ASCE 7 (see Section 202). Section 13.1.3 of ASCE 7 lists four classes of components that have a component importance factor of 1.5. These are components required to function after an earthquake including fire-sprinkler systems and egress stairways that are not an integral part of the building structure, components containing or conveying toxic materials, components in Risk Category IV structures that are deemed to be necessary for continued operation of the facility, and components containing or conveying hazardous materials. Refer to Section 13.1.3 of ASCE 7 for complete descriptions of these systems.

In addition to the general exceptions in Section 1704.2, there are three specific exceptions where special inspection for seismic resistance is not required: Light-frame structures not over 35 feet (10,668 mm) in height and in which S_{DS} does not exceed 0.5; buildings with reinforced masonry or reinforced concrete seismic-force-resisting systems not over 25 feet (7,620 mm) and in which S_{DS} does not exceed 0.5; and detached one- or two-family dwellings not exceeding two stories that do not have the horizontal or vertical structural irregularities specified.

1705.13.1 Structural steel. For special inspection requirements for structural steel in seismic force-resisting systems, IBC Section 1705.13.1.1 references the quality assurance requirements of AISC 341. The first exception exempts special inspection of structural steel seismic-force-resisting systems in Seismic Design Category B or C that have not been specifically detailed for seismic resistance as allowed in Table 12.2-1 of ASCE 7. This exception does not however apply to cantilever column systems. The second exception does not require specific special inspections of the structural steel seismic-force-resisting systems in SDC D, E, or F if the design and detailing is only required to comply with AISC 360, as specified in Table 15.4-1 of ASCE 7. This second exception only applies to nonbuilding structures. IBC Section 1705.13.1.2 includes similar requirements for other structural steel elements, such as struts, collectors, chords and foundation elements not covered in Section 1705.13.1.1. The exceptions are intended to mirror those of Section 1705.13.1.1.

1705.13.2 Structural wood. The seismic special inspections required for wood structures are primarily to ensure continuity of load path within the seismic lateral-force-resisting system. The diaphragms must transfer their inertial loads to the walls, which in turn transmit the inertial loads through the lateral-force-resisting system to the foundation. Periodic inspection is allowed except in the case of field gluing operations, which are to be continuous. The section requires periodic special inspection for fastening such as nailing, bolting, and anchoring components of the SFRS, which includes chords, collectors (struts), braces, and hold-downs. Particular care should be given to the nailing of diaphragms and shear walls. Common nails are often specified in the design, but smaller-diameter sinkers or power-driven (gun) nails are often substituted in the field and the smaller-diameter nails have a lower lateral resistance. For example, when connecting two nominal 2-inch-thick (51 mm) members, the lateral resistance of a 0.131-inch-diameter (3.33 mm) power-driven nail used as a replacement for a 10d common nail in Douglas Fir-Larch is only 97 pounds (see ESR 1539), whereas the value for the 10d common nail is 118 pounds (see NDS). Of additional importance is the connection of collectors to shear walls and the proper installation and tightening of hold-down bolts or screws in shear walls. The exemption from special inspection applies where diaphragm and shear wall construction has a fastener spacing of more than 4 inches (102 mm) on center at the panel edges. The 4-inch (102 mm) spacing for wood construction roughly translates to a minimum capacity of 350 plf (5.11 kN/m) (ASD) for 7/16-inch (11.1 mm) wood structural panel sheathing (see SDPWS). Where the fastener spacing is greater than 4 inches (102 mm) at panel edges, there is lower demand and less potential for splitting, and special inspection is not required.

1705.13.3 Cold-formed steel light-frame construction. Similar to wood framing, the seismic special inspections for cold-formed steel structures are to ensure continuity of load path within the seismic lateral-force-resisting system; that is, the diaphragms must transfer their inertial loads to the walls, which in turn transmit the inertial loads through the lateral-force-resisting system to the foundation. The section requires periodic special inspection of welding and fastening such as screw attachment, bolting, and anchoring components of the seismic-force-resisting system including shear walls, diaphragms, collectors (struts), braces, and hold-downs. Particular care should be given to connections of braces and hold-downs. Wood and cold-formed steel light-frame constructions have similar requirements for their lateral-force-resisting systems. Therefore, the exception for structural wood also applies, with the appropriate adaptation, to cold-formed steel light-frame construction. Thus, where fastener spacing for wood structural panel sheathing or steel sheets is greater than 4 inches (102 mm) at the panel edges, there is lower demand and special inspection is not required. The exception for gypsum board and fiberboard shear walls is based on the fact that the capacity of these materials is below the demand threshold.

1705.13.4 Designated seismic systems. As noted above, the designated seismic system consists of those architectural, electrical, and mechanical systems and components that require design in accordance with Chapter 13 of ASCE 7 for which the component importance factor, I_p, is greater than 1.0, as prescribed in Section 13.1.3 of ASCE 7 (see Section 202). Section 13.1.3 of ASCE 7 lists four classes of components that have a component importance factor of 1.5. Refer to Section 13.1.3 of ASCE 7 for complete descriptions of these systems. The special inspector must verify that elements of the designated seismic system that require certification according to Section 1705.14.3 are properly labeled and anchored or mounted in accordance with the certificate of compliance. See discussion of Section 1705.14.3.

1705.13.5 Architectural components. Exterior cladding and veneers can be a serious safety hazard if they become detached from the structure during seismic ground motion as well as potentially blocking required exit paths. The code requires periodic special inspection for exterior cladding, nonbearing walls and partitions, and veneer in Seismic Design Category D, E, or F. The exemption from special inspection applies to these elements when they are relatively low or lightweight. Thus, special inspection is not required when these elements are 30 feet (9,144 mm) or less above grade or the walking surface, are light-weight cladding and veneer weighing 5 psf (0.24 kN/m^2) or less, or light-weight partitions weighing 15 psf (0.72 kN/m^2) or less.

1705.13.5.1 Access floors. Access floors consist of a system of panels and supports that create a raised floor above the actual structural floor system. The space between the raised floor and the structural floor contains components like wiring for power, voice, and data. The space may also be used for HVAC distribution either as a plenum or with ductwork. Because failure of the floor system can pose a threat to the occupants in high seismic areas, periodic special inspection is required for anchorages of access floors in structures in Seismic Design Category D, E, or F.

1705.13.6 Plumbing, mechanical, and electrical components. Inspection is necessary for components that must function in post-earthquake conditions, such as emergency electrical systems, or for anchorage of mechanical equipment, piping, and ducting using or carrying flammable or hazardous materials. Periodic special inspection is required for the following items in structures in Seismic Design Category C, D, E, or F: (1) during the anchorage of electrical equipment for emergency or standby power systems; (2) during the installation of anchorage or other electrical equipment in Seismic Design Category E or F; (3) during installation of piping systems intended to carry hazardous contents and their associated mechanical units; (4) during the installation of HVAC ductwork that contains hazardous materials; (5) during installation of vibration isolation systems in structures where the design requires ¼ inch (6.35 mm) or less between the equipment support frame and restraint; and (6) to ensure that adequate clearance is provided between mechanical and electrical equipment and automatic fire sprinkler systems. Experience from earthquakes has shown that pounding between sprinkler piping and other nonstructural components has led to connection failures and accidental activations.

1705.13.7 Storage racks. Tall storage racks such as those typically found in large "big box" building supply stores can pose a threat to the public in high seismic areas (Figure 1705-3). Periodic special inspection is required for steel storage racks and steel cantilevered storage racks 8 feet (2,438 mm) or greater in height in Seismic Design Category D, E, or F. Design and installation of storage rack components is based on minimum material thicknesses, yield strengths, anchorage, etc., and compliance with these requirements is imperative during component fabrication and storage rack installation. Storage rack systems can be complex, and installation must comply with permitted drawings on file with the local building department. Table 1705.13.7 outlines the required inspections and applicable standards.

Figure 1705-3 **Steel storage racks.**

1705.13.8 Seismic isolation systems. The performance of seismic isolators is critical to the performance of isolation systems and energy dissipation devices. Periodic special inspection is required during fabrication and installation of these devices. See Section 1705.14.4, which refers to Section 17.8 of ASCE 7 for testing of seismic isolation systems.

1705.13.9 Cold-formed steel special bolted moment frames. This section requires periodic special inspection for cold-formed steel special bolted moment frames (CFS-SBMF) in structures assigned to Seismic Design Category D, E, or F. The CFS-SBMF is a type of cold-formed steel moment frame SFRS designed to withstand anticipated seismic forces by dissipation of energy through controlled inelastic deformation. The system is listed in Table 12.2-1 of ASCE 7 (see Item C.12; $R = 3.5$, $\Omega_o = 3.0$, $C_d = 3.5$). Examples of items that warrant special inspection are critical elements of the system such as installation of the beam-to-column connections and the anchorage to the foundation. The *North American Standard for Seismic Design of Cold-Formed Steel Structural Systems* (AISI S400) contains provisions for inspections and quality control by the fabricator. The AISI S400 standard is used in conjunction with AISI S100, *North American Specification for the Design of Cold-Formed Steel Structural Members*.

1705.14 Testing for seismic resistance. Unless specifically exempted in the general requirements in Section 1704.2, testing for seismic resistance is required in addition to the special inspection requirements for seismic resistance covered in Section 1705.13. The requirements are triggered by the seismic design category of the building. Testing is required for structural steel, nonstructural components, designated seismic systems, and seismic isolation systems.

1705.14.1 Structural steel. Nondestructive testing of structural steel elements in seismic-force-resisting systems, including struts, collectors, chords, and foundation elements of buildings assigned to Seismic Design Category B through F is required in accordance with the quality assurance requirements of AISC 341. Similar to Section 1705.13.1, there are two separate exceptions in each subsection. The first exception for SDC B or C, exempts from nondestructive testing structural steel seismic force–resisting systems and elements that have not been specifically detailed for seismic resistance as allowed in Table 12.2-1 of ASCE 7, although this does not apply to cantilever column systems. The second exception does not require specific nondestructive testing of the steel systems or elements in seismic force–resisting systems in SDC D, E, or F if the design and detailing is only required to comply with AISC 360, as specified in Table 15.4-1 of ASCE 7. This second exception only applies to nonbuilding structures.

1705.14.2 Nonstructural components. For structures assigned to Seismic Design Category B through F, nonstructural components, supports, and attachments must meet the seismic qualification requirements of Item 2 of Section 13.2.1 of ASCE 7. Item 2 requires submittal of the manufacturer's certification that the component is qualified by at least one of the following: (a) analysis, (b) testing in accordance with the alternative found in Section 13.2.6, or (c) experience data in accordance with the alternative prescribed in Section 13.2.7. The registered design professional is required to indicate on the construction documents the requirements for seismic qualification and certificates of compliance documenting that qualification requirements are met and must be submitted to the building official as required by Section 1704.5.

1705.14.3 Designated seismic systems. Structures assigned to Seismic Design Categories C through F with designated seismic systems that are subject to the requirements of Section 13.2.3 of ASCE 7 require certification as follows: (1) Active mechanical and electrical equipment that must remain operable following the design earthquake ground motion is to be certified by the manufacturer as operable. Active parts or energized components must be certified on the basis of approved shake table testing in accordance with Section 13.2.6 or experience data in accordance with Section 13.2.7 unless it can be shown that the component is "inherently rugged" by comparison with similar seismically qualified components. (2) Components assigned a component importance factor of 1.5 with hazardous substances must be certified by the manufacturer that they maintain containment following the design earthquake ground motion by (a) analysis, (b) approved shake table testing in accordance with Section 13.2.6 (see example in Figure 1705-4), or (c) experience data in accordance with Section 13.2.7. As noted above for the

Photo courtesy of Smith Emery Laboratories

Figure 1705-4   Floor diaphragm cyclic test per AISI S907 cantilever test method for cold-formed steel diaphragms.

SFRS, the registered design professional is required to indicate on the construction documents the requirements for seismic qualification and certificates of compliance documenting that the qualification requirements are met and must be submitted to the building official as specified in Section 1704.5.

1705.14.4 Seismic isolation systems.   Seismic isolation systems in seismically isolated structures assigned to Seismic Design Category B through F must be tested in accordance with Section 17.8 of ASCE 7. ASCE 7 Section 17.8 requires the deformation characteristics and damping values of the isolation system used in the design and analysis be based on tests of a selected sample of the components prior to construction. The isolation system components to be tested are required to include wind-restraint systems where these systems are used in the design.

1705.15   Sprayed fire-resistive materials (SFRM).   This section provides requirements for special inspection of SFRMs for floor, roof, and wall assemblies, and structural members. For an SFRM to perform as intended, its application must be within the proper range for certain parameters as determined by the system manufacturer. Requirements are provided for physical and visual testing to demonstrate compliance with the required fire-resistance rating and applicable listing, with specific requirements given for structural member surface conditions, proper application, thickness, density, and bond strength. These parameters must be checked prior to and during installation of the SFRM.

1705.16 Intumescent fire-resistive materials. Where fire-resistant coatings are used on structural members and decks, the special inspector must verify that they are applied in accordance with the Association of the Wall and Ceiling Industry Technical Manual 12-B (AWCI 12-B), *Standard Practice for the Testing and Inspection of Field Applied Thin Film Intumescent Fire-Resistive Materials,* based on the required fire resistance as shown in the approved construction documents.

1705.17 Exterior insulation and finish systems (EIFS). Special inspection is required for EIFS based on the approved research report and manufacturer's installation instructions. Critical areas necessary for adequate EIFS performance are proper installation of the waterproofing membrane and installation of flashings at windows, doors, joints, eaves, corners, and penetrations. The Association of the Wall and Ceiling Industry (AWCI) offers training and certification programs for performing proper special inspection of EIFS. The exceptions from requiring special inspection are for EIFS installed over a water-resistive barrier, or concrete or masonry walls.

1705.17.1 Water-resistive barrier coating. Where a water-resistive barrier coating is applied between the EIFS and the sheathing, special inspection is required for the barrier coating. The coating must be properly applied to provide additional protection to the building from incidental moisture intrusion that may occur through the building envelope.

1705.18 Fire-resistant penetrations and joints. This section requires special inspection of fire-resistant penetrations and joints in high-rise buildings, buildings in Risk Category III or IV, or fire areas containing Group R occupancies with an occupant load greater than 250. Through-penetration and membrane-penetration firestop systems, as well as fire-resistant joint systems and perimeter fire barrier systems, are critical to maintaining the fire-resistive integrity of fire-resistance-rated construction elements, including fire walls, fire barriers, fire partitions, smoke barriers, and horizontal assemblies. The proper selection and installation of such systems must be in compliance with the code or appropriate listing. Where these systems are used in categories of buildings considered "high risk," they must be included in the special inspection program.

Although the proper application of firestop and joint system requirements is very important in all types and sizes of buildings, the requirement for special inspection is limited to specific building types that represent a substantial hazard to human life in the event of a system failure or that are considered to be essential facilities. Inspection conforming to ASTM E2174 *Standard Practice for On-site Inspection of Installed Fire Stops* and ASTM E2393 *Standard Practice for On-Site Inspection of Installed Fire Resistive Joint Systems and Perimeter Fire Barriers* increases the level of quality assurance.

1705.18.1 Penetration firestops. A primary method of addressing a penetration of a fire-resistance-rated wall assembly is through the use of an approved firestop system installed in accordance with ASTM E814 *Standard Test Method for Fire Tests of Penetration Firestop Systems* or UL 1479 *Fire Tests of Penetration Firestops.* The system must have an F rating that is not less than the fire-resistance rating of the wall being penetrated. It is critical that the firestop system be appropriate for the penetration being protected. The choice of firestop systems varies based on the size and material of the penetrating item, as well as the construction materials and fire-resistance rating of the wall being penetrated. Special inspection of the firestop system is intended to verify that the appropriate system has been specified and the installation is in conformance with its listing.

1705.18.2 Fire-resistant joint systems. A joint is defined as a "The opening in or between adjacent assemblies that is created due to building tolerances, or is designed to allow independent movement of the building in any plane caused by thermal, seismic, wind, or any other loading." The joint creates an interruption of the fire-resistant integrity of the wall or floor system, requiring the use of an appropriate fire-resistant joint system. The code mandates general installation criteria for such systems and requires them to be tested in accordance with ASTM E1966 *Standard Test Method for Fire-Resistive Joint Systems* or UL 2079 *Tests for Fire Resistance of Building Joint Systems*. Much like the inspection of penetration firestop systems, the proper choice and installation of fire-resistant joint systems can be verified through a comprehensive special inspection process. Although regulated under the provisions for fire-resistant joint systems, a second type of system is technically not a joint, but rather an extension of protection afforded by a horizontal assembly. The void created at the intersection of an exterior curtain wall assembly and a fire-resistance-rated floor or floor-ceiling assembly must be filled in a manner that maintains the integrity of the horizontal assembly. The system utilized to fill the void must be in compliance with ASTM E2307 *Standard Test Method for Determining Fire Resistance of Perimeter Fire Barriers Using the Intermediate-Scale, Multistory Test Apparatus* and able to resist the passage of flame for a time period equal to that of the floor assembly. Special inspection is necessary to ensure that the appropriate joint system is chosen and properly installed.

1705.19 Testing for smoke control. This section, although related to mechanical systems rather than structural or architectural systems, is required by the code because the mechanical ductwork and signaling devices are likely to be concealed during the building construction, and the ductwork needs to be leakage tested prior to concealment.

1705.19.1 Testing scope. The special inspection applies to leakage testing of the ductwork prior to concealment and overall system performance of the completed system prior to occupancy.

1705.19.2 Qualifications. The special inspection agencies that perform the testing are required to have specific expertise in fire protection or mechanical engineering and air balancing. The Associated Air Balance Council (AABC) is the certifying association for air balance testing. AABC has certifications for technicians and equipment commissioning.

1705.20 Sealing of Mass Timber. Periodic special inspections of sealants or adhesives are required where sealant or adhesive required by IBC 703.7 is applied to mass timber building elements as designated in the approved construction documents.

Some mass timber panels are manufactured under proprietary processes to ensure there are no voids at abutting edges and intersections. Where this proprietary process is incorporated and tested, there is no requirement for sealant or adhesive and an exception is provided for this instance. Where sealant is not required and is not specifically excluded it is still considered good practice covered by this section.

The intent of the sealant or adhesive is to prevent the passage of air. If noncombustible material such as a concrete topping slab is provided on top of the CLT assemblies, it should accomplish this purpose and sealants or adhesives would not be required at abutting edges of the CLT panels. Note however that Type IV-C construction does not require "noncombustible material not less than 1 inch (25 mm) in thickness above the mass timber" as is required for Type IV-A or Type IV-B construction (see IBC Sections 602.4.1.3 and 602.4.2.3). So, if the concrete topping were to be removed at some point in the future (e.g., change of ownership or occupancy), a method to satisfy 703.7 would have to be provided.

Special inspection of mass timber sealing requirements does not apply to "joints" as defined in IBC 202 since joints have their own requirements for the placement and inspection of fire-resistant joint systems in IBC Section 1705.18. Joints are defined as having an opening that is designed to accommodate building tolerances or to allow independent movement. Panels and members that are connected do not meet the definition of a joint since they are rigidly connected and do not have an opening.

Mass timber structural elements manufactured in a fabricator shop are addressed by IBC 1704.2.5 and 1704.2.5.1 regarding fabrication.

Section 1706 *Design Strengths of Materials*

This section requires the strength of structural materials to conform to the applicable design standards, or in the absence of such standards, must conform to accepted engineering practice.

1706.2 New materials. Materials not explicitly covered by the code are allowed, subject to testing that demonstrates adequate performance. Section 1707 provides alternative test procedures, and Section 104.2.3 provides administrative means of accepting alternative methods of design and construction.

Section 1707 *Alternate Test Procedure*

Test reports from approved agencies may be used as the basis for approval of materials not specifically covered by the code or approved standards. New materials not covered in the code are also mentioned in Section 1701.2. This section references Section 104.2.3 in regard to the use of new, innovative, or alternative materials. The building official has the authority to accept such materials based on reports from an approved agency that is independent from the material supplier. ICC Evaluation Service Reports are the most straightforward method used by building officials to review products and systems that are not specifically covered by the code. The cost associated with performing testing and issuing research reports is to be borne by the applicant which is the owner or the owner's agent.

Section 1708 *In-Situ Load Tests*

This section covers structural analysis or load testing for existing structures or portions thereof where the load-bearing capacity or stability is in doubt.

1708.1 General. When there is doubt as to the structural integrity, load-carrying capacity, or stability of an existing structure or portion thereof, the building official has the option of requiring a structural analysis or a load test, or both. If an engineering analysis is required, it should be based on the as-built conditions and actual material properties. If the structural analysis shows that the structure is not capable of safely carrying the code-design loads, the structure must be load tested. If the structure is found to be inadequate or unstable, the structural system must be modified to provide adequate structural integrity.

1708.2 In-situ load tests. In-situ tests fall into two categories: procedures specified and procedures not specified. Load testing must simulate the loading required by Chapter 16 and be performed under the supervision of the registered design professional.

1708.2.1 Load test procedure specified. When the applicable standard has a test procedure and acceptance criteria specified, the requirements of the standard should be applied.

1708.2.2 Load test procedure not specified. When there is no applicable load test procedure in a referenced standard, the structure or portion thereof should be tested with a loading protocol that simulates the actual loads and deformations, both lateral and vertical, that the structure is expected to receive. For the gravity system, the vertical loads should be equal to the factored design loads.

This section clarifies static load test requirements and specifies how to test components that carry dynamic loads. The code language also considers differences influenced by load duration effects when testing wood elements.

Section 1709 *Preconstruction Load Tests*

This section applies to materials or methods of construction that are not capable of being designed by conventional engineering analysis or do not comply with referenced standards.

1709.1 General. When the load-carrying capacity and the physical properties of materials or methods of construction are not amenable to analysis by accepted engineering methods, or the material or method does not comply with applicable standards, the structural load capacity and physical properties must be determined by tests specified in this section. Where tests meet the requirements of the code and approved procedures, the building official is required to accept certified reports of tests conducted by an approved testing agency. This section is applicable to components, assemblies, and elements of structures, for example, windows. This section is not applicable to load testing of existing structures (see discussion of Section 1708).

1709.2 Load test procedures specified. When an applicable standard has a test procedure and acceptance criteria, the standard should be applied.

1709.3 Load test procedures not specified. When there is no applicable load test procedure in a referenced standard, the structure or portion thereof should be tested with a loading protocol that simulates the applicable loading conditions and deformations, both lateral and vertical, that the structure is expected to sustain. For components, assemblies, and elements that are not part of the SFRS, the loading and acceptance criteria are set forth in Section 1709.3.1.

1709.3.1 Test procedure. The loading procedure and acceptance criteria use commonly accepted engineering practices to test the adequacy of the component or assembly to resist structural failure at the design loads. The test load is 2 times the design load for 24 hours, and must recover at least 75 percent of the maximum deflection within 24 hours after the test. The test assembly is then reloaded until failure occurs or 2½ times the load corresponding to the maximum allowable deflection is reached (see Section 1604.3) or 2½ times the design load is reached. This procedure should be used only for loads for which the upper bounds can be established with a reasonable degree of accuracy such as dead and live loads. This procedure should not be used for earthquake loads and should be used with care for wind loads.

1709.3.2 Deflection. Deflection under design load is limited to the allowable limits set forth in Section 1604.3.

1709.4 Wall and partition assemblies. Walls and partitions must be tested for simultaneous vertical and lateral loads, both with and without door and window framing.

1709.5 Exterior window and door assemblies. Door and window assemblies are generally qualified by tests specified in Section 1709.5.1, which requires testing and labeling in accordance with the American Architectural Manufacturers Association/Window and Door Manufacturers Association/Canadian Standards Association (AAMA/WDMA/CSA) standard, or testing in accordance with Section 1709.5.2 and the referenced ASTM and Door & Access Systems Manufacturers Association (DASMA) standards. The exception permits allowable wind pressures for differently-sized units made of identical components, including glass thickness, to have different design values than the tested assembly as determined by engineering analysis or an additional test, provided that the additional test is done on the assembly with the alternative allowable design pressure. Where engineering analysis is used, either an AAMA or WDMA analysis procedure is required.

1709.5.1　Exterior windows and doors.　Exterior windows and sliding doors must be tested and labeled as conforming to the AAMA/WDMA/CSA101/I.S.2/A440 standard. Side-hinged exterior doors can comply with the standard or meet the requirements of Section 1709.5.2, as discussed below. The products tested and labeled in accordance with the AAMA/WDMA/CSA standard need not comply with the analysis and testing requirements of Section 2403.2 and the deflection limitations specified in Section 2403.3.

1709.5.2　Exterior windows and door assemblies not provided for in Section 1709.5.1. Exterior window and door assemblies not covered by Section 1709.5.1 must be tested in accordance with ASTM E330. The test load is equal to 1.5 times the design pressure, as determined per Chapter 16, applied for 10 seconds. Exterior window and door assemblies covered by this section and containing glass are required to conform to the requirements of Section 2403.

1709.5.2.1　Garage doors and rolling doors.　Requirements for structural performance of garage doors are found in ASTM E330 and ANSI/DASMA 108 for sectional garage doors and rolling doors. This section further requires that garage doors have a permanent label that provides a way for building owners, homeowners, and others to be able to determine their performance characteristics after the building has been occupied.

1709.5.3　Windborne debris protection.　In windborne debris regions, glazing in buildings shall be impact resistant or protected with an impact-resistant covering meeting the requirements of an approved impact-resistant standard.

1709.5.3.1　Impact protective systems testing and labeling.　Impact protective systems in windborne debris regions require testing by an approved independent laboratory per ASTM E1886 and ASTM E1996 and design wind pressure per ASTM E330. This section further requires that impact protective systems have a permanent label that provides a way for building owners, homeowners, and others to be able to determine their performance characteristics after the building has been occupied.

1709.6　Skylights and sloped glass.　The section refers to Chapter 24 for sloped unit skylight and tubular daylighting device (TDD) requirements. A TDD is typically field-assembled from a manufactured kit, unlike a unit skylight, which is typically shipped as a factory-assembled unit. The dome of a TDD is not necessarily constructed out of a single panel of glazing material. Thus, a separate definition of a TDD is included in Chapter 2 based on the definition in AAMA/WDMA A440, and TDD's are included in Section 2405.5.

1709.7　Test specimens.　Test specimens should be representative of what is actually used in practice. Tests must be conducted or witnessed by an independent approved agency.

Example 17-1   Sample Statement of Special Inspections

The following "Statement of Special Inspections" (SSI) should address the special inspection and structural testing items required by IBC Chapter 17. Per IBC 1704.3 the SSI should list all items requiring special inspections or structural tests as well as describe the extent and frequency (i.e., periodic or continuous) of the tests and inspections.

Statement of Special Inspections

1. Special inspections and structural testing shall be provided by an independent agency employed by the Owner for the items identified in this section and in other areas of the approved construction plans and specifications, unless waived by the Building Official (see IBC Chapter 17).

2. The names and credentials of the Special Inspectors to be used shall be submitted to the Building Official for approval.

3. Duties of the Special Inspector:

    a. The Special Inspector shall review all work listed below for conformance with the approved construction plans and specifications and the IBC.

    b. The Special Inspector shall furnish special inspection reports to the registered design professional (RDP) in responsible charge, Contractor, Owner and Building Official on a frequency as required by the approved construction documents or Building Official. All items not in compliance shall be brought to the immediate attention of the Contractor for correction, and if uncorrected, to the RDP in responsible charge and the Building Official.

    c. Once corrections have been made by the Contractor, the Special Inspector shall submit a final signed report to the Building Official stating that the work requiring special inspection was, to the best of the Special Inspector's knowledge, in conformance with the approved construction plans and specifications as well as the applicable workmanship provisions of the IBC.

4. Duties and responsibilities of the Contractor:

    a. The Contractor shall submit a written statement of responsibility to the Owner and the Building Official prior to the commencement of work. In accordance with IBC 1704.4, the statement of responsibility shall contain acknowledgement of the special inspection requirements contained within this "Statement of Special Inspections."

    b. The Contractor shall notify the responsible Special Inspector that work is ready for inspection at least one working day (24 hours minimum) before such inspection is required.

    c. All work requiring special inspection shall remain accessible and exposed until it has been observed by the Special Inspector.

5. Please see the "Special Inspection Schedule" for the types, extents and frequency of specific items requiring special inspections and structural tests as part of this project.

SPECIAL INSPECTION SCHEDULE			
	Frequency		
Areas requiring special inspection:	Continuous	Periodic	Comments:
FABRICATORS (IBC 1704.2.5)			
	◆		If fabricator is approved, on-site inspection is not required but a certificate of completion must be provided to the B.O. (IBC 1704.2.5.1)
SOILS (IBC 1705.6)			
Verify adequate materials below footings		◆	Prior to placement of concrete.
Excavation extend to proper depth and materials		◆	Prior to placement of compacted fill or concrete.
Classification and testing of fill materials		◆	Check classification and gradations at each lift, but not less than once for each 10,000 ft^2 of surface area.
Verify proper fill materials, lift thicknesses and in-place densities	◆		
Verify properly prepared site and subgrade		◆	Prior to placement of concrete.
CONCRETE CONSTRUCTION (IBC 1705.3)			
Reinforcing steel placement		◆	Verify size, clearances, splices and proper ties.
Embedded bolts or plates		◆	
Verify required design mix		◆	Verify mix design meets strength and exposure requirements listed on approved plans.
Concrete placement/sampling	◆		Includes sampling for air, slump, strength and temperature tests
Inspect formwork		◆	Verify shape, location and member dimensions.
Post-installed anchors	◆		In accordance with approved ICC-ES Report. Periodic inspections allowed if stated in ES Report.
COLD-FORMED STEEL LIGHT-FRAME CONSTRUCTION (IBC 1705.12.2 and 1705.13.3)			
Components of wind- and seismic-force resisting systems.		◆	Verify proper screw attachment, bolting and anchoring of shear walls, braces and holdowns having a fastener spacing ≤ 4″ o.c. at panel edges.
STEEL CONSTRUCTION (IBC 1705.2, 1705.12, 1705.13)			
Steel Roof & Floor Deck (Appendix 1, ANSI/SDI QA/QC-2022):			
Verify materials and accessories prior to steel deck placement		◆	Identification markings per applicable ASTM standard
Verify that installation conforms to approved plans and manufacturer requirements		◆	Per ANSI/SDI QA/QC.
Verify use of qualified welders; proper environmental conditions for welding; the size, location and type of welds.		◆	Verify that welds conform to AWS D1.3, SDI SD.

SPECIAL INSPECTION SCHEDULE *(continued)*			
Areas requiring special inspection:	Frequency		Comments:
	Continuous	Periodic	
STEEL CONSTRUCTION *(continued)*			
Steel Roof & Floor Deck *(continued)*			
Verify proper tools and mechanical fasteners are used; installed per mfr. Recommendations; proper spacing, type and installation of fasteners.		♦	Per ANSI/SDI QA/QC.
Open-Web Steel Joists and Joist Girders *(IBC Table 1705.2.4):*			
Verify end connections		♦	Per SJI specifications.
Verify proper bridging is installed		♦	Per SJI specifications.
Prior to Welding *(Table N5.4-1, AISC 360-22):*			
Verify welding procedures	♦		
Material identification		♦	Verify type and grade of material.
Welder identification		♦	Verify there is a system in place to identify the welder who has welded a joint or member.
Fit-up groove welds		♦	Verify joint preparation, dimensions, cleanliness, tacking and backing.
Access holes		♦	Verify configuration and finish.
Fit-up fillet welds		♦	Verify alignment, gaps at root, cleanliness of steel surfaces, tack weld quality and location.
During Welding *(Table N5.4-2, AISC 360-22):*			
Control and handling of welding consumables		♦	Verify packaging and exposure control.
Cracked tack welds		♦	Verify welding is not over a cracked tack weld.
Environmental conditions		♦	Verify wind speed is within limits as well as precipitation and temperature.
Welding Procedure Specifications followed		♦	Verify items such as welding equipment settings, travel speed, welding materials, shielding gas type/flow rate, preheat applied, interpass temperature maintained, and proper position.
Welding techniques		♦	Verify interpass and final cleaning, each pass is within profile limitations, and quality of each pass.
After Welding *(Table N5.4-3, AISC 360-22):*			
Welds cleaned		♦	Verify that welds have been properly cleaned.
Size, length, and location of welds	♦		
Welds meet visual acceptance criteria	♦		
Arc strikes	♦		
k-area	♦		
Access holes (flange > 2″)	♦		
Backing & welding tabs removed	♦		

SPECIAL INSPECTION SCHEDULE *(continued)*			
	Frequency		
Areas requiring special inspection:	Continuous	Periodic	Comments:
STEEL CONSTRUCTION *(continued)*			
After Welding (continued):			
Repair activities	♦		
Document acceptance/ rejection of weld	♦		
Nondestructive Testing (Section N5, AISC 360-22):			
CJP welds (Risk Cat. II)		♦	Ultrasonic testing shall be performed on 10% of CJP groove welds in butt, T- and corner joints subject to transversely applied tension loading in materials 5/16-inch thick or greater. Testing rate must be increased to 100% if > 5% of welds have unacceptable defects.
Welded joints subject to fatigue	♦		
Other Steel Inspections (Tables J9.1 and J10.1, AISC 341-22)			
Anchor rods/embeds supporting structural steel		♦	Shall be on the premises during the placement of anchor rods/embedments. Verify diameter, grade, type, and length of element and the extent or depth of embedment prior to placement of concrete.
Reduced beam sections (RBS)		♦	Verify contour and finish as well as dimensional tolerances *(see Table J9.1 of AISC 341).*
Protected zones		♦	Verify that no holes or unapproved attachments are made within the protected zone *(see Table J9.1 of AISC 341).*
MASONRY CONSTRUCTION (IBC 1705.4)			
Minimum Verification (Table 3, TMS-602-20):			
Verification of compliance of submittals.		♦	Prior to construction, per Article 1.5 of TMS 602.
Verification of f'_m and f'_{AAC}.		♦	Prior to construction, per Article 1.4 B of TMS 602.
Verification of Slump Flow and Visual Stability Index (VSI) for self-consolidating grout.		♦	During construction, per Art. 1.5 and 1.6 of TMS 602.
As Construction Begins (Table 4, TMS-602-22):			
Proportions of site-prepared mortar		♦	Verify that mortar is type and color specified on approved plans, it conforms to ASTM C270, and complies with Articles 2.1, 2.6 A & 2.6 C of TMS 602.
Grade, type and sizes of reinforcement, connectors, anchor bolts, and anchorages		♦	Verify reinforcement, joint reinforcement, anchor bolts and veneer anchors comply with approved plans and Articles 3.4 & 3.6 A of TMS 602.

SPECIAL INSPECTION SCHEDULE *(continued)*			
Areas requiring special inspection:	Frequency		Comments:
	Continuous	Periodic	
MASONRY CONSTRUCTION *(continued)*			
Prior to Grouting *(Table 4, TMS-602-22):*			
Grout space		♦	Verify grout space is free of mortar droppings, debris, loose aggregate, and other deleterious materials and that cleanouts are provided per Article 3.2.D and 3.2.F of TMS 602.
Placement of reinforcement, connectors and anchorages.		♦	Verify reinforcement, joint reinforcement, anchor bolts and veneer anchors are installed per approved plans and Articles 3.2.E and 3.4 of TMS 602.
Proportions of site-prepared grout.		♦	Verify grout proportions meet ASTM C476 and a slump between 8 and 11 inches. See also Articles 2.6 B of TMS 602.
During Construction *(Table 4, TMS-602-22):*			
Materials and procedures comply with approved construction documents.		♦	Per Article 1.5 of TMS 602.
Construction of mortar joints		♦	Verify mortar joints placed in accordance with Article 3.3.B of TMS 602.
Size and location of structural elements		♦	Verify locations of structural elements per approved plans and confirm tolerances meet Article 3.3.G of TMS 602.
Type, size and location of anchors, frames, etc.		♦	Verify correct anchorages and connections are provided per approved plans and Chapter 6 of TMS 402.
Placement of grout.	♦		Per Article 3.5 of TMS 602.
Preparation, construction and protection of masonry during cold weather (<40°F) or hot weather (>90°F).		♦	Verify cold-weather construction complies with Article 1.8.C of TMS 602 and hot weather construction per Article 1.8.D of TMS 602.
Observation of grout specimens, mortar specimens, and/or prisms.		♦	Confirm specimens/prisms are performed as required by Article 1.4 of TMS 602.

MASS TIMBER CONSTRUCTION TYPES IV-A, IV-B, AND IV-C (IBC 1705.5.3)			
Areas requiring special inspection:	Frequency		Comments:
	Continuous	Periodic	
Foundation system anchorage		♦	Verify anchorage and connections of mass timber construction to timber deep foundation systems.
Mass timber construction		♦	Verify that installation conforms to approved plans and manufacturer requirements.
Connections			
Threaded fasteners		♦	Verify use of proper installation equipment.
		♦	Verify use of pre-drilled holes where required.
		♦	Inspect screws, including diameter, length, head type, spacing, installation angle, and depth.
Adhesive anchors installed in horizontal or upwardly inclined orientation to resist sustained tension loads	♦		Verify that installation conforms to approved plans and manufacturer requirements.
Adhesive anchors not defined above		♦	Verify that installation conforms to approved plans and manufacturer requirements.
Bolted connections		♦	Verify that installation conforms to approved plans and manufacturer requirements.
Concealed connections		♦	Verify that installation conforms to approved plans and manufacturer requirements.

1 inch = 25.4 mm, 1 sq. ft = 0.0929 m^2

KEY POINTS

- Chapter 17 provides requirements for quality assurance for construction of buildings and other structures regulated by the IBC through special inspection and verification, structural testing, and structural observation.
- Specific provisions for approvals, approved agencies, records, labeling, and testing are provided in Section 1703.
- General requirements for special inspections, contractor responsibility, structural observation, fabricator approval, and the statement of special inspections are provided in Section 1704.
- Specific items that require verification by the special inspector are outlined in Section 1705 including special cases, steel, concrete, masonry, wood, soils, and deep foundations.
- Specific verifications are required for wind resistance based on the basic wind speed at the site and for seismic resistance based on the seismic design category of the building.
- Special testing and qualification for seismic resistance is required for certain structural elements and architectural, mechanical, and electrical components based on the seismic design category of the building.
- Requirements for alternative test procedures, in-situ load testing and preconstruction load testing are provided.

CHAPTER 18

SOILS AND FOUNDATIONS

Section 1801 General
Section 1802 Design Basis
Section 1803 Geotechnical Investigations
Section 1804 Excavation, Grading, and Fill
Section 1805 Dampproofing and Waterproofing
Section 1806 Presumptive Load-Bearing Values of Soils
Section 1807 Foundation Walls, Retaining Walls, and Embedded Posts and Poles
Section 1808 Foundations
Section 1809 Shallow Foundations
Section 1810 Deep Foundations
Key Points

Satisfactory performance of the building structural system is critical in the overall performance of any building during its life cycle. The IBC® structural provisions in Chapters 16, 17, 18, 19, 20, 21, 22, and 23 work together to provide for the overall performance and safety of the structural system. Section 1604.4, Analysis, requires that all structural systems have a continuous load path from the point of origin to the resisting element. The resisting element for buildings and other structures is the foundation that ultimately transfers the loads to the supporting soil. Therefore, the satisfactory performance of the foundation system is critical to the satisfactory performance of the overall structure. Most building structures are designed assuming a fixed unyielding base that is not subjected to large total or differential settlements or displacements. Shallow foundations on firm soils will generally perform satisfactorily if the requirements of these provisions are followed.

Foundation design, however, becomes a significant factor for large structures, embedded structures such as a tall building constructed over a multilevel basement garage, structures on soft soils, structures supporting rotating or reciprocating equipment, and structures sensitive to differential displacements. It is important to have a good knowledge of the behavior of the various foundation types, including their limitations. In addition, for structures subject to high-wind forces or seismic ground motion, special consideration must be given to the lateral load path, and, in the case of deep foundation-supported structures, the ability of the deep foundation to survive the displacements and curvatures imposed on the pile by seismic ground motion.

Sufficient understanding of the behavior and limitations of the various deep and shallow foundation systems is necessary to determine that the foundation and its supported structure will provide the intended serviceability. It is important to determine whether the estimated total and differential foundation settlements are compatible with the selected structure type. For example, a stiff bearing-wall structure with openings may be more sensitive to differential settlements than a more flexible light-frame structure. When considering seismic ground motion, sufficient knowledge of ground-shaking effects on the foundation is important, particularly in soft soils.

Section 1801 *General*

1801.1 Scope. The requirements of this chapter apply to all building and foundation systems of any type and any location. For example, buildings located in beach-front properties that would be subject to wave run up and inundation during hurricanes can still be designed using applicable loads from ASCE 7 *Minimum Design Loads and Associated Criteria for Buildings and Other Structures* and ASCE 24 *Flood Resistant Design and Construction* with proper geotechnical investigation and engineering analysis. ASCE 7 Chapter 5 and ASCE 24 are referenced in Section 1612.2 for the design and construction of buildings and structures in flood hazard areas, including coastal high-hazard areas and coastal A zones.

Section 1802 *Design Basis*

Bearing pressures, stresses, and lateral pressures used in this chapter are allowable pressures or stresses, not strength level values. These allowable foundation pressures are to be used with the allowable stress design load combinations set forth in Section 2.4 of ASCE 7 unless noted otherwise. Material requirements are outlined in Chapters 19, 21, 22, and 23. Site excavations and grading are covered in Section 1804.

Section 1803 *Geotechnical Investigations*

1803.1 General. A geotechnical investigation must be conducted when required by the building official. In general, the investigation should be required unless the foundation is designed and constructed in accordance with the presumptive allowable foundation pressures and lateral-bearing pressures set forth in Section 1806. Some minimal knowledge of soil classification at the bearing elevation of the foundation is required by Section 1806. A registered design professional should be used as required by the professional practice laws of the state in which the jurisdiction is located. The practice of geotechnical or soils engineering is a branch of civil engineering and is generally regulated by the various states.

1803.2 Investigations required. A foundation and soils investigation is required for any of the adverse subsurface conditions listed in Sections 1803.5.2 through 1803.5.12.

In certain cases, an exception allows the building official some flexibility in requiring a foundation and soils investigation when the soil conditions of the site are already known from other soils reports. For example, if the site is located in an area where there are reasonably uniform and horizontal soil strata and a soils report is available for the adjacent parcels, then a new soils report should not be necessary. This exception does not apply to some very specific conditions such as analysis of lateral pressure, liquefaction potential, and ground stabilization techniques in Seismic Design Categories (SDC) D, E, and F. Other conditions where the exception does not apply and a geotechnical investigation is mandated are excavation near existing foundations, compacted fill materials, and controlled low-strength materials (CLSMs).

1803.3 Basis of investigation. The investigation cannot be solely based on theoretical analysis and research; rather, analysis and observation through tests of materials by borings, test pits, or other subsurface explorations in appropriate locations are needed. The number and types of tests, equipment used, type of site inspections, and other issues relevant to the scope of the investigation are the responsibility of and must be under the supervision of a registered design professional experienced in soils exploration. Some states and local municipalities have set criteria on the minimum number of investigations required. Additional studies shall also be made as necessary for conditions such as slope stability, soil strength, position and adequacy of load-bearing soils, the effect of moisture variation on soil-bearing capacity, compressibility, liquefaction, and expansiveness.

1803.4 Qualified representative. Whenever the allowable bearing capacity is in doubt or a geotechnical investigation is necessary, exploratory borings are necessary to determine the soil characteristics and the load-bearing capacity. The investigation procedures must be outlined and identified by the registered design professional in accordance with the accepted engineering practice and acceptable standards. The apparatus and equipment used in the investigation must be calibrated and perform as expected for accurate results. Special inspection requirements for soil-related activities are provided in Chapter 17 of the IBC. Qualified representatives must be present on site during the boring or sampling operations. Qualifications of investigative or inspection agencies and individuals are typically established by their experience, certification, and accreditation through an approved accreditation body such as the International Accreditation Service (IAS).

1803.5 Investigated conditions. Those cases where geotechnical investigations are required have been provided in Sections 1803.5.2 through 1803.5.12. It should be emphasized that the exception discussed in Section 1803.2 that allows the building official to waive the investigation can be used in most of these conditions except those that are critical or are site specific and cannot be determined based on existing geotechnical investigations of surrounding areas (lateral pressure, liquefaction potential, and ground stabilization techniques in SDC D, E, and F, excavation near foundations, compacted fill materials, and CLSMs). Soil materials are required to be classified in accordance with ASTM D2487 *Practice for Classification of Soils for Engineering Purposes (Unified Soil Classification System)*. Rock shall be classified in accordance with ASTM D5878 *Standard Guides for Using Rock-Mass Classification Systems for Engineering Purposes*. IBC Table 1610.1 provides the Unified Soil Classification for some backfill materials. This table establishes the minimum soil lateral loads for the design of foundation walls and retaining walls.

1803.5.2 Questionable soil and rock. Where the classification, strength, moisture sensitivity, or compressibility of the soil or rock is uncertain, or where bearing capacity in excess of the presumptive value is claimed, the code allows the building official to obtain a geotechnical report.

1803.5.3 Expansive soil. Expansive soils are those that shrink and swell appreciably because of changes in soil moisture content. Reference should be made to Section 1808.6 for mitigation methods and design for expansive soils. Frost heave is not considered in this section. Expansive soils are present or prevalent in all 50 states of the United States and in many other countries around the globe. Soils must meet all four of the criteria to be classified as expansive, not just a high Plasticity or Expansion Index.

The section allows two different ways to identify expansive soils. The first option involves meeting four criteria: (1) Plasticity Index (PI) of 15 or greater determined by ASTM D4318 *Test Methods for Liquid Limit, Plastic Limit and Plasticity Index of Soils*; (2) more than 10 percent of the soil particles pass a No. 200 sieve determined by ASTM D6913 *Standard Test Methods for Particle-Size Distribution (Gradation) of Soils Using Sieve Analysis*; (3) more than 10 percent of the soil particles are less than 5 micrometers in size as determined by ASTM D6913; and (4) an Expansion Index greater than 20 according to ASTM D4829 *Test Method for Expansion Index of Soils*. The second option is to determine if the Expansion Index (according to ASTM D4829) is greater than 20. If the Expansion Index is determined, the first three tests need not be conducted. In other words, Expansion Index > 20 is both a necessary and sufficient condition by itself.

1803.5.4 Ground-water. This section may cause significant changes to the conventional approach to foundation design and construction for light-frame buildings with subsurface floors, either a basement or a hillside building with a floor cut into the hillside. A foundation and soils investigation is required to show that the ground-water depth is at least 5 feet (1,524 mm) below the elevation of the lowest floor level and will affect the design and construction of the structure.

1803.5.5 Deep foundations. Section 1803.6 lists the general items that must be included in a geotechnical report. This includes items such as a plot of the locations of soil investigations, soil profiles, the elevation of the water table, and shallow foundation recommendations. When deep foundations are required on a project, the geotechnical report should also list the specific items noted in Section 1803.5.5. This includes the recommended types of deep foundations and their capacities and suitability for the environment, the maximum center-to-center spacing, the driving criteria, installation procedures, special inspection and reporting requirements, load test requirements, the required bearing level of foundation elements, and reductions for group actions.

The initial geotechnical recommendations for a project typically consider shallow foundations, but after reviewing the recommendations the design team or owner may decide to switch to a deep foundation system. It could be that a significant amount of over-excavation and replacement of structural fill is required prior to installation of shallow foundations. After pricing this option, deep foundations become more feasible. Regardless of when the decision to use deep foundations occurs, it is important to obtain the specific recommendations for the deep foundations from the geotechnical engineer as outlined in Section 1803.5.5.

Two items worth highlighting in this section are the types of deep foundations and the load test requirements. There are many types of deep foundation elements and not all are ideal for each site. As an example, helical piers would not be ideal in areas where a significant amount of cobbles exist in the subsurface soils. Likewise, some cast-in-place (CIP) systems would not be ideal where caving soils exist. Many deep foundations require test piles or piers to be installed and load tests to be conducted prior to installing the systems throughout the site. Load test criteria are discussed in detail in Section 1810.3.3 but this information should be specified by the geotechnical engineer as specified in Section 1803.5.5.

1803.5.6 Rock strata. If a rock stratum is being used for bearing, the characteristics of the layer must be known in sufficient detail to classify the rock per Section 1803.5.1.

1803.5.7 Excavation near foundations. The intent of this section is that lateral and subjacent support of any foundation must not be removed unless an investigation is conducted to identify the potential problems that might be created and how to provide the needed lateral support. For example, the area shown in Figure 1803-1 should not be excavated without conducting an investigation to address support for the foundation.

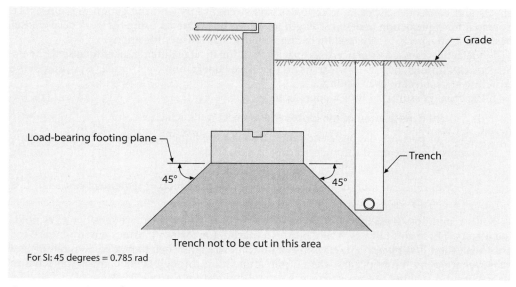

Figure 1803-1   **Lateral support.**

1803.5.8 Compacted fill material. When compacted fill of more than 12 inches (305 mm) in depth is used for shallow foundation support, the geotechnical investigation required in Section 1803 must also contain the seven items listed in this section.

1803.5.9 Controlled low-strength material (CLSM). CLSM is a self-compacted, cementitious material used as backfill instead of compacted backfill. It has also been referred to as "flowable fill" and "lean mix backfill." CLSM has a compressive strength of 1200 psi (8,274 kPa) or less. Most CLSM applications require unconfined compressive strengths of 200 psi (1,379 kPa) or less to allow for future excavation of CLSM. CLSM is composed of water, portland cement, aggregate, and fly ash. It is a fluid material with typical slumps of 10 inches (254 mm) or more and has a consistency similar to a milk shake.

1803.5.10 Alternate setback and clearance. This alternate procedure allows the building official to approve alternate setbacks and clearances from slopes, provided that the intent of Section 1808.7 is met. This section gives the building official the authority to require a geotechnical investigation by a qualified geotechnical engineer to establish that the intent of Section 1808.7 is met and specifies the minimum parameters to be investigated.

1803.5.11 Seismic design categories C through F. For all structures in SDC C through F, a geotechnical investigation is required to evaluate liquefaction, slope stability, total and differential settlement, and surface displacement caused by faulting or lateral spreading. Significant ground motion can occur even in areas with moderate seismic risk. Liquefaction typically occurs at sites with loose sands and high water tables. Surface rupture generally occurs with large-magnitude earthquake events. Lateral spreading often happens adjacent to waterways where a saturated soil has a free edge.

1803.5.12 Seismic design categories D through F. In addition to the investigation required for SDC C through F by Section 1803.5.11, the investigation must evaluate the additional lateral pressures on basement or retaining walls from ground shaking, the detrimental effects and consequences of liquefaction or soil strength loss, and mitigation measures. The potential for liquefaction and soil strength loss must be evaluated using the peak acceleration based on a site-specific study and including the effects of soil amplification.

Liquefaction causes loss of bearing capacity resulting in large differential or total settlements. Structures with high height-to-width ratios that have liquefaction occur under a portion of the structure are subject to overturning.

Mitigation measures for liquefaction include:

1. In-situ densification of the loose sands subject to liquefaction.
2. Use of pile foundations penetrating through the liquefying layers. The pile capacity must be developed through bearing or skin friction in the soils below the liquefying layers.
3. Use of rigid raft foundations that can minimize the effects of settlement caused by loss of bearing capacity.

Soil strength loss is associated with "sensitive" or "quick" clays that are sensitive to remolding effects. These soft clays typically occur in marine (or former marine) environments and are often called *"bay muds."* The clays lose significant strength when remolded. Remolding and subsequent strength loss occurs when a pile foundation through these clays is subjected to strong ground shaking. As a consequence, lateral support for the pile is lost.

Mitigation measures for soils susceptible to strength loss such as "sensitive" or "quick" clays include:

1. Replacement of the soft clay.
2. In-situ consolidation of the soft clays by preloading or water removal.
3. Use of pile foundations penetrating through the soft clay layers. The pile capacity must be developed through bearing or skin friction in the soils below the soft clay layers.

Foundation or retaining walls that support 6-feet (1,829 mm) of backfill must consider seismic earth pressure. Shorter walls are not required to consider this additional earth pressure load in the wall design. If such walls exist on a project the site-specific geotechnical report must list the seismic earth pressures that should be considered in the design. These pressures can significantly alter the required design of the walls in question. Other design pressures that need to be considered include static earth pressures, surcharge loads, hydrostatic loads, sloped backfill, and more.

1803.6 Reporting. The objective of a geotechnical investigation is to produce all the necessary information for design and construction of a structure's foundation. To ensure that the report of the foundation and soils investigation meets this objective and provides enough data to ensure compliance with the code and a safe foundation, the report must contain at a minimum the items enumerated in this section. See also report requirements for deep foundations in Section 1803.5.5, compacted fill in Section 1803.5.8, and for CLSM in Section 1803.5.9. Other data such as results of consolidation tests, compaction curves, sieve analyses, Atterberg Limits tests, and Plasticity and Expansion Index tests should be included in the report where available.

Section 1804 *Excavation, Grading, and Fill*

1804.1 Excavations near foundations. Excavations should not reduce the lateral support provided by any foundation without first protecting the foundation against detrimental movements. See discussion under Section 1803.5.7.

1804.2 Underpinning. This section requires that underpinning activities comply with both Chapter 18 and Chapter 33 on safeguards during construction. Excavations within existing buildings, or near adjacent buildings, can cause significant damage. This section makes it clear that should any excavations be performed near existing foundations specific protocols will need to be followed. It also indicates that the plans should note the "sequence" at which such excavations will occur.

1804.3 Placement of backfill. Backfill should be performed in accordance with an approved soils report. If no soils report is needed, the backfill should be placed in maximum 6-inch (152-mm) layers free from any rocks or cobbles larger than 4 inches and compacted to a minimum 90-percent relative density (Modified Proctor per ASTM D1557 *Test Methods for Laboratory Compaction Characteristics of Soil Using Modified Effort*) using the appropriate compaction equipment. CLSM is an acceptable backfill material that needs not be compacted. See further discussion under Section 1803.5.9.

1804.4 Site grading. The general requirement is that surface water must drain away from foundations. Minimum slope is 5 percent for a distance of 10 feet (3,048 mm). An alternate method is permitted if physical obstructions or lot lines prohibit the 5-percent slope for a minimum of 10 feet (3,048 mm) horizontally. In this case, swales must have a minimum 2-percent slope where located within 10 feet (3,048 mm) of the building foundation. Impervious services within 10 feet of the building can have a minimum 2-percent slope. An exception also allows a 2-percent slope where climatic or soil conditions allow it. Another exception is provided that allows impervious surfaces to be sloped less than 2 percent where the surface serves a door landing or ramp required for egress purposes in accordance with Section 1010.1.4, 1012.3, or 1012.6.1.

1804.5 Grading and fill in flood hazard areas. In general, grading is not permitted in designated flood hazard areas. Changes in the configuration or shape of floodways by grading or fill can divert erosive flows and increase wave energies that could increase forces and adversely affect adjacent buildings and structures. The restrictions in the code are consistent with provisions related to fill in ASCE 24, *Flood Resistant Design and Construction*. The exceptions permit grading in flood hazard areas, provided an engineering analysis demonstrates that the proposed grading will not increase flood levels or otherwise adversely affect the design flow. The last exception allows grading in flood hazard areas that are not designated floodways, provided the overall effect of the encroachment does not increase the design flood elevation by more than 1 foot (305 mm) at any point.

1804.6 Compacted fill material. See discussion under Section 1803.5.8 for geotechnical report requirements. The exception to the geotechnical report requires compacted fill to be no more than 12 inches (305 mm) deep, have specific density requirements, and undergo special inspection in accordance with Section 1705.6.

1804.7 Controlled low-strength material (CLSM). See discussion under Section 1803.5.9.

Section 1805 *Dampproofing and Waterproofing*

This section covers the requirements for waterproofing and dampproofing those parts of substructure construction that need to be provided with moisture protection. Sections 1805.1 through 1805.3 identify the locations where moisture barriers are required and specify the materials to be used and the methods of application. The provisions also deal with subsurface water conditions, drainage systems, and other protection requirements.

Dampproofing requirements are outlined in Section 1805.2, and waterproofing requirements are covered in Section 1805.3. Although both terms are intended to apply to the installation and the use of moisture barriers, dampproofing does not furnish the same degree of moisture protection as does waterproofing.

Dampproofing generally refers to the application of one or more coatings of a compound or other materials that are impervious to water, which are used to prevent the passage of water vapor through walls or other building components, and which restrict the flow of water under slight hydrostatic pressure. Waterproofing, on the other hand, refers to the application of coatings and sealing materials to walls or other building components to prevent moisture from penetrating in either a vapor or liquid form, even under conditions of significant hydrostatic pressure. Hydrostatic pressure is created by the presence of water under pressure. This pressure can occur when the ground-water table rises above the bottom of the foundation wall, or the soil next to the foundation wall becomes saturated with water caused by uncontrolled storm water runoff.

1805.1 General. This section is an overall requirement specifying that waterproofing and dampproofing applications are to be made to horizontal and vertical surfaces of those below-ground spaces where the occupancy would normally be adversely affected by the intrusion of water or moisture. Moisture or water in a floor below grade can cause damage to structural members such as columns, posts, or load-bearing walls, as well as pose a health hazard by promoting growth of bacteria or fungi. Moisture can adversely affect any mechanical and electrical appliances that may be located at that level. It can also cause a great deal of damage to goods that may be located or stored in that lower level. These vertical and horizontal surfaces include foundation walls, retaining walls, underfloor spaces, and floor slabs. Waterproofing and dampproofing are not required in locations other than residential and institutional occupancies where the omission of moisture barriers would not adversely affect the use of the spaces. An example of a location where waterproofing or dampproofing would not be required is in an open parking structure, provided the structural components are individually protected against the effects of water. Waterproofing and dampproofing are not permitted to be omitted from residential and institutional occupancies where people may be sleeping or services are provided on the floor below grade. A person waking in a flooded basement may find themselves in a very hazardous situation particularly if the possibility exists of an electrical charge in the water caused by electrical service at that level.

Section 1805.1.1 addresses the type of problem faced when a portion of a story is above grade, whereas Section 1805.1.2 limits any infiltration of water into crawl spaces so as to protect this area from potential water damage and prevent ponding of water in the crawl space. These sections reference other applicable sections of Chapter 18. Section 1805.1.3 deals with ground-water control systems.

1805.1.1 Story above grade plane. The provisions of this section require that where a basement is deemed to be a story above grade plane, the section of the basement floor that occurs below the exterior ground level and the walls that bound that part of the floor are to be dampproofed in accordance with the requirements of Section 1805.2.

The use of dampproofing, rather than waterproofing, is permitted here because high hydrostatic pressure will not tend to develop against the walls if the basement is a story above grade plane and the ground level adjacent to the basement wall is below the basement floor elevation for not less than 25 percent of the basement perimeter.

Any water pressure that may occur against the walls below ground or under the basement floor would be relieved by the water drainage system required in this section. The drainage system would be installed at the base of the wall construction in accordance with Section 1805.4.2 for a minimum distance along those portions of the wall perimeter where the basement floor is below ground level.

Because of the relationship of grade to the basement floor and the inclusion of foundation drains, the potential for hydrostatic pressure buildup is not significant. Therefore, a ground-water depth investigation, waterproofing, and a basement floor gravel base course are not required.

The objective of this section is to prevent moisture migration in basement spaces. In story-above-grade plane construction that meets the requirements of this section, the basement floor would be only partly below ground level (sometimes a small part) and the need for section-required moisture protection is unnecessary. Dampproofing of the floor slab would be required, however, in accordance with Section 1805.2.1.

1805.1.2 Under-floor space. The requirements of this section are designed to prevent any ponding of water in under-floor areas such as crawl spaces. Crawl spaces are particularly susceptible to ponding of water as they are usually uninhabitable spaces that are infrequently observed. Water can build up in these spaces and remain for an extended period of time without being noticed by the building occupants. This is also to prevent water from ponding under the structure if it is flooded. Under-floor spaces of Group R-3 buildings located in flood hazard areas that meet the requirements of FEMA-TB-11 *Crawlspace Construction for Buildings Located in Special Flood Hazard Areas* need not comply with the requirements in this section.

Stagnant water collected under a building can result in a serious health concern. Water buildup in a crawl space can also damage the structural integrity of the building. Untreated wood exposed to water can deteriorate and rot, and concrete and masonry exposed to water can deteriorate with a loss of strength.

Steel exposed to water or high humidity can eventually rust to the extent that effective structural capability is jeopardized. Water buildup in a crawl space can also damage mechanical or electrical appliances located in the space by causing corrosion of electrical parts or metal skins and deterioration of insulation used to protect heating elements.

Where it is known that the water table can rise to within 6 inches (152 mm) of the outside ground level, or where there is evidence that surface water cannot readily drain from the site, then the finished ground surface in under-floor spaces is to be set at an elevation equal to the outside ground level around the building perimeter unless an approved drainage system is provided.

For the drainage system to be approved, it must be demonstrated that the system will be adequate to prevent water infiltration into the under-floor space. This is done by determining the maximum possible water flow near the foundation wall and footing, and designing the drainage system to remove that flow of water as it occurs, thereby preventing water buildup at the foundation wall.

To prevent water ponding in the under-floor space from a rise in the ground-water table, or from storm water runoff, the finished ground level of an under-floor space is not to be located below the bottom of the foundation footings.

Dampproofing, waterproofing, or providing subsoil drainage is not necessary if the ground level of the under-floor space is as high as the ground level at the outside of the building perimeter, as the foundation walls do not enclose an interior space below grade.

Compliance with Sections 1805.2, 1805.3, and 1805.4 would be required where the finished ground surface of the under-floor space is below the outside ground level and a drainage system is not provided.

1805.1.3 Ground-water control. After completion of building construction, it is necessary to maintain the water table at a level that is at least 6 inches (152 mm) below the bottom of the lower floor to prevent the flow or seepage of water into the basement. Where the site consists of well-draining soil and the highest point of the water table occurs naturally at or lower than the required level, there is no need to provide any kind of a site drainage system specifically designated to control the ground-water level. Where the soil characteristics and the site topography are such that the water table can rise to a level that will produce a hydrostatic pressure against the basement structure, a site drainage system may be installed to reduce the water level. When ground-water control in accordance with this section is provided, waterproofing in accordance with Section 1805.3 is not required.

There are many types of site drainage systems that can be employed to control ground-water levels. The most commonly used systems may involve the installation of drainage ditches or trenches filled with pervious materials, sump pits and discharge pumps, well point systems, drainage wells with deep-well pumps, sand-drain installations, and so on. This section requires that all such systems be designed and constructed using accepted engineering principles and practices based on considerations that include soil permeability, amount and rate at which water enters the system, pump capacity, disposal area capacity, and other such factors that are necessary for the complete design of an effective drainage system.

1805.2 Dampproofing. Where a ground-water depth investigation has established that the high water table will occur at such a level that the building substructure will not be subjected to significant hydrostatic pressure, then dampproofing in accordance with this section and a subsoil drain in accordance with Section 1805.4 are considered sufficient to control moisture in the floor below grade.

Wood foundation systems specified in Section 1807.1.4 are to be dampproofed as required by the American Wood Council *Permanent Wood Foundation Design Specification* (AWC PWF).

1805.2.1 Floors. Floors requiring dampproofing in accordance with Section 1805.2 are to employ materials specified in Section 1805.2.1. The dampproofing materials must be placed between the floor construction and the supporting gravel or stone base, as shown in Figure 1805-1. Even if a floor base is not required, dampproofing should be placed under the slab.

The installation is intended to provide a moisture barrier against water vapor passage or seepage into below-ground spaces.

The dampproofing material most commonly used for underslab installations consists of a polyethylene film not less than 6 mil (0.006 inch; 0.152 mm) in thickness, which is applied over the gravel or stone base required in Section 1805.4.1. Care must be taken in the installation of the material over the rough surface of the base and during the concreting operation so as not to puncture the polyethylene. Joints must be lapped at least 6 inches (152 mm).

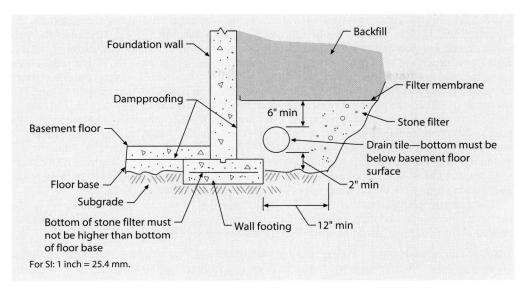

Figure 1805-1   **A foundation drainage system.**

Dampproofing materials can also be applied on top of the base concrete slab if a separate floor is provided above the base slab, because the dampproofing is provided to prevent moisture infiltration of the interior space, and not the concrete slab.

1805.2.2   Walls.   Walls requiring dampproofing in accordance with Section 1805.2 are first to be prepared as required in Section 1805.2.2.1, then coated with bituminous materials or by other approved materials and methods of application. Approved materials are those that will prevent moisture from penetrating the foundation wall.

Coatings are applied to cover prepared exterior wall surfaces extending from the top of the wall footings to slightly above ground level so that the entire wall that contacts the ground is protected. Surfaces are usually primed to provide a bond coat and then dampproofed with a protection coat of asphalt or tar pitch.

Dampproofing materials for walls may be any of the materials specified in Section 1805.3.2 for waterproofing. Surface-bonding mortar complying with ASTM C887 *Specification for Packaged, Dry Combined Materials for Surface Bonding Mortar* may be utilized. This specification covers the materials, properties, and packaging of dry, combined materials for use as surface-bonding mortar with concrete masonry units that have not been prefaced, coated, or painted. Because this specification does not address design or application, the manufacturer's recommendations should be followed. This standard covers proportioning, physical requirements, sampling, and testing. The minimum thickness of the coating is ⅛ inch (3.2 mm).

Acrylic-modified cement coatings may be utilized at the rate of 3 pounds per square yard (16 N/m^2). These types of materials have been used successfully as dampproofing materials for foundation walls. Surface-bonding mortar and acrylic-modified cement are limited in use to dampproofing. The ability of these two types of products to bridge nonstructural cracks, as required in Section 1805.3.2 for waterproofing materials, is not known. Therefore, their use is limited to dampproofing and they are not permitted to be used as waterproofing. Dampproofing may also include other materials and methods of installation acceptable to the building official.

1805.2.2.1 Surface preparation of walls. Before applying dampproofing materials, the concrete must be free of any holes or recesses that could affect the proper sealing of the wall surfaces. Air trapped beneath the dampproofing coating or membranes can cause blistering. Rocks and other sharp objects can puncture membranes. Irregular surfaces can also create uneven layering of coatings, which can result in vulnerable areas of dampproofing. Surface irregularities commonly associated with concrete wall construction can be sealed with bituminous materials or filled with portland cement grout or other approved materials.

Unit masonry walls are usually parged (plastered) with a ½-inch-thick (12.7 mm) layer of portland cement and sand mix (1:2½ by volume) or with a Type M mortar proportioned in accordance with the requirements of ASTM C270 *Specification for Mortar for Unit Masonry* and applied in two ¼-inch-thick (6.35 mm) layers. In no case is parging to result in a final thickness of less than ⅜ inch (9.5 mm). The parging is to be coved at the joint formed by the base of the wall and the top of the wall footing to prevent water accumulation at that location.

The moisture protection of unit masonry walls provided by the parging method may not be required where approved dampproofing materials such as grout coatings, cement-based paints, or bituminous coatings can be applied directly to the masonry surfaces.

1805.3 Waterproofing. Waterproofing installations are intended to provide moisture barriers against water seepage that may be forced into below-ground spaces by hydrostatic pressure.

Where a ground-water depth investigation has established that the high water table will occur at such a level that the building substructure will be subjected to hydrostatic pressure, and where the water table is not lowered by a ground-water control system, as described in the discussion of Section 1805.4.2, all floors and walls below ground level are to be waterproofed in accordance with Sections 1805.3.1 and 1805.3.2.

1805.3.1 Floors. Because floors required to be waterproofed are subjected to hydrostatic uplift pressures, such floors must, for all practical purposes, be made of concrete and designed and constructed to resist the maximum hydrostatic pressures possible. It is particularly important that the floor slab be properly designed, as severe cracking or movement of the concrete would allow water seepage into below-ground spaces. The ability of the waterproofing materials to bridge cracks is limited. Concrete floor construction is to comply with the applicable provisions of Chapter 19.

Materials used for waterproofing below-ground floors are to conform to the requirements of Section 1805.3.1. Below-ground floors subjected to hydrostatic uplift pressures are to be waterproofed with membrane materials placed as underslab or split-slab installations, including such materials as rubberized asphalt, butyl rubber, and neoprene, or with polyvinyl chloride or polybutylene films not less than 6 mil (0.006 inch; 0.152 mm) in thickness, lapped at least 6 inches (152 mm). All membrane joints are to be lapped and sealed in accordance with the manufacturer's instructions to form a continuous, impermeable waterproof barrier. There are many proprietary membrane products available that are specifically made for waterproofing floors and walls (i.e., polyethylene sheets sandwiched between layers of asphalt), which may be used when approved by the building official. Products that have an ICC-ES evaluation report are acceptable in most jurisdictions if the requirements of the evaluation report are followed. ICC-ES reports are intended to address the technical aspects and requirements of new and innovative products that are approved by the building official under the alternative materials and methods of construction provisions of Section 104.2.3. All ICC-ES evaluation reports are posted online at www.icc-es.org.

1805.3.2 Walls. Walls that are required to be waterproofed in accordance with Section 1805.3 must first be prepared as required in this section and then waterproofed with the required membrane-type installations.

The walls must be designed to resist the hydrostatic pressure anticipated at the site, as well as any other lateral loads to which the wall will be subjected, such as soil pressures or seismic loads. As with the floors required to be waterproofed, it is particularly important that the walls required to be waterproofed be properly designed to resist all anticipated loads, as cracking and other damages would allow water seepage into below-ground spaces. Concrete or masonry construction must comply with the applicable provisions of Chapters 19 and 21, respectively.

Waterproofing installations are to extend from the bottom of the wall to a height not less than 12 inches (305 mm) above the maximum elevation of the ground-water table. The remainder of the wall below ground level (if the height is small) may be either waterproofed as a continuation of the installation or must be dampproofed in accordance with the requirements of Section 1805.2.

This section requires that waterproofing consist of two-ply hot-mopped felts. Waterproofing industry practice is to select the number of plies of membrane material based on the hydrostatic head (height of water pressure against the wall). As a general practice, if the head of water is between 1 and 3 feet (305 and 914 mm), two plies of felt or fabric membrane are used; between 4 and 10 feet (1,219 and 3,048 mm), three-ply construction is needed; and between 11 and 25 feet (3,353 and 7,620 mm), four-ply construction is necessary.

Waterproofing installations may also use polyvinyl chloride materials of not less than 6-mil (0.006 inch; 0.152 mm) thick, 40-mil (0.040 inch; 1.02 mm) polymer-modified asphalt, or 6-mil (0.006 inch; 0.152 mm) polyethylene. These materials have been widely recognized for their effectiveness in bridging nonstructural cracks. Other approved materials and methods may be used, provided that the same performance standards are met. All membrane joints must be lapped and sealed in accordance with the manufacturer's instructions.

1805.3.2.1 Surface preparation of walls. Before applying waterproofing materials to concrete or masonry walls, the wall surfaces must be prepared in accordance with the requirements of Section 1805.2.2.1, which requires the sealing of all holes and recesses. Surfaces to be waterproofed must also be free of any projections that might puncture or tear membrane materials that are applied over the surfaces.

1805.3.3 Joints and penetrations. This section requires that all joints occurring in floors and walls and at locations where floors and walls meet, as well as all penetrations in floors and walls, be made watertight by approved methods. Sealing the joints and penetrations in the waterproofing is of primary importance to ensure waterproofing effectiveness. If the joints or penetrations are not sealed properly, they can develop leaks, which become a passageway for water to enter the building. Because the remainder of the foundation is wrapped in waterproofing, moisture can actually become trapped in the foundation walls or floor slab, and serious damage to these structural components can occur.

Methods may involve the use of construction keys between the base of the wall and the top of the footing, or, if there is a hydrostatic pressure, floor and wall joints may require the use of manufactured waterstops made of metal, rubber, plastic, or mastic materials. Floor edges along the walls and floor expansion joints may employ any of a number of preformed expansion joint materials, such as asphalt, polyurethane, sponge rubber, self-expanding cork, and cellular fibers bonded with bituminous materials, which all comply with applicable ASTM standards or other approved specifications. A variety of sealants may be used together with the preformed joint materials. Gaskets made of neoprene and other materials are also available for use in concrete and masonry joints. The National Roofing Contractors Association (NRCA) roofing and waterproofing manuals provide details for the reinforcement of membrane terminations, corners, intersections of slabs and walls, through-wall and slab penetrations, and other locations.

Penetrations in walls and floors may be made watertight with grout or manufactured fill materials and sealants made for the purpose.

1805.4   Subsoil drainage system.   Subsoil drainage systems are required to drain the area adjacent to basement walls to eliminate hydrostatic loads. This section covers subsoil drainage systems in conjunction with dampproofing (see Section 1805.2) to protect below-ground spaces from water seepage. Such systems are not used where basements or other below-ground spaces are subject to hydrostatic pressure, because they would not be effective in disposing of the amount of water anticipated if hydrostatic pressure conditions exist. Ground-water tables may be reduced to acceptable levels by methods described in the discussion of Section 1805.1.3. The details of subsoil drainage systems are covered in the requirements of Sections 1805.4.1 through 1805.4.3.

1805.4.1   Floor base course.   This section requires that floors of basements, except for story-above-grade-plane construction, must be placed on a gravel or stone base not less than 4 inches (102 mm) thick. Not more than 10 percent of the material is to pass a No. 4 (4.75 mm) sieve to provide a porous installation and provide a capillary break. Material that passes a No. 4 (4.75 mm) sieve would be silt or clay that does not permit the free movement of water through the floor base, but allows for upward migration by capillary action.

This requirement serves three purposes. The first is to provide an adjustment to the irregularities of a compacted subgrade so as to produce a level surface upon which to cast a concrete slab. The second purpose is to provide a capillary break so that moisture from the soil below will not rise to the underside of the floor. Finally, where required, the porous base can act as a drainage system to expel underslab water by means of gravity, or the use of a sump pump or other approved method.

The exception allows for the omission of the floor base when the natural soils beneath the floor slab consist of well-draining granular materials such as sand, stone, or mixtures of these materials. Some caution, however, is justified in the use of this exception. If the granular soils contain an excessive percentage of fine materials, the porosity and the ability of the soil to provide a capillary break may be considerably diminished. The exception should be applied only if the natural base is equivalent to the floor base otherwise required by this section.

1805.4.2   Foundation drain.   This section describes in considerable detail the materials and features of construction required for the installation of foundation drain systems.

This type of drain system is suitable where the water table occurs at such elevation that there is minimal hydrostatic pressure exerted against the basement floor and walls and where the amount of seepage from the surrounding soil is so small that the water can be readily discharged by gravity or by mechanical means into sewers or ditches. The objective is to combine the protection afforded by the dampproofing of walls and floors (see Section 1805.2) and that given by the perimeter drains to maintain below-ground spaces in a dry condition.

A foundation drain system usually consists of the installation of drain tiles made of clay or concrete or of drain pipes of corrugated metal or nonmetallic pipes surrounded by crushed stone or gravel and a filter membrane material (filter fabric). The foundation drain is set adjacent to the wall footing and extends around the perimeter of the building. Drain tiles are placed end to end with open joints to permit water to enter the system. Metallic and nonmetallic drains are made with perforations at the invert (bottom) section of the pipe and are installed with connected ends. Where drain tile or perforated drain pipe is used, the invert must not be set higher than the basement floor line so that water conveyed by the drain does not seep into the filter material and then create a hydrostatic pressure condition against the foundation wall and footing. The inverts should not be placed below the bottom of the adjacent wall footings to avoid carrying away fine soil particles whose loss, in time, could possibly undermine the footing and cause settlement of the foundation walls.

Tile joints or pipe perforations should be covered with an approved filter membrane material to prevent them from becoming clogged and to prevent fine particles that may be contained in the surrounding soil from entering the system and being carried away by water.

The gravel or crushed stone fill material around the drain tiles or pipes should contain not more than 10 percent of material that passes a No. 4 (4.75 mm) sieve. The crushed stone or gravel should be selected to prevent the movement of particles from the protected soil surrounding the drain installation into the drain. Crushed stone or gravel fill material is to be placed in the excavation so that it will extend out from the edge of the wall footing a distance of at least 12 inches (305 mm), with the bottom of the fill being no higher than the bottom of the base under the floor (see Section 1805.4.1) and not less than 6 inches (152 mm) above the top of the footing.

Requiring the bottom of the foundation drain to be no higher than the bottom of the floor base is necessary so that if the water table rises into the floor base, it will also be able to rise unobstructed into the foundation drain. The foundation drain will then drain the water away from the building, as required by Section 1805.4.3. The top of the fill material must be covered with an approved filter membrane to allow water to pass through to the perimeter drain tile or pipes without allowing fine soil materials to enter the drainage system.

Drain tiles or pipes are to be installed on at least 2 inches (51 mm) of crushed stone or gravel material and covered with at least 6 inches (152 mm) of the same material to maintain good water flow into the drain tile or pipe. See Figure 1805-1.

1805.4.3 Drainage discharge. This section references the *International Plumbing Code®* (IPC®) for piping system installation requirements for disposal of water from the floor base and the foundation drains. Chapter 11 of the IPC considers the piping materials, applicable standards, and methods of installation of subsurface storm drains to facilitate water discharge either by gravity or by mechanical means.

Where the soil at the site consists of well-drained granular materials such as gravel or sand-gravel mixtures to prevent the occurrence of hydrostatic pressure against the foundation walls and under-floor slab, the use of a dedicated drainage system as prescribed in the IPC is not required, because the site soils would permit natural drainage.

Section 1806 *Presumptive Load-Bearing Values of Soils*

1806.1 Load combinations. The presumptive bearing values in the code are allowable stress values, not strength level values. The allowable foundation pressures are permitted to be increased by one-third with the alternative ASD load combinations of Section 1605.3.2 for combinations including wind or earthquake even though the one-third stress increase has been removed for structural materials (see discussion of Section 1605.2 for more information).

1806.2 Presumptive load-bearing values. The presumptive allowable bearing and lateral pressures must be used unless a geotechnical investigation substantiates higher values. The term "undocumented fill" refers to fill that was not placed and compacted in accordance with an approved soils report.

The classifications in Table 1806.2 are from the Unified Soil Classification System. Most foundation-bearing strata can be classified into one of the classifications in the table. The allowable bearing pressures and lateral-bearing values are based on long-term experience with the behavior of these materials. However, the presumptive value for mud, organic silt and organic clays (OL, OH), and peat (Pt) may not be conservative and should be substantiated. The selection of an allowable bearing pressure should take into account the strength of weaker underlying soil strata so that the pressure in any weaker stratum does not exceed the allowable pressure, particularly in clay soils. Because of this, it is important to know the soil profile and

Table 1806-1. **Presumptive Load-Bearing Values**

Class of Materials	Vertical Foundation Pressure (psf)	Lateral-Bearing Pressure (psf/ft below natural grade)	Lateral Sliding Resistance	
			Coefficient of friction[a]	Cohesion (psf)[b]
1. Crystalline bedrock	12,000	1,200	0.70	—
2. Sedimentary and foliated rock	4,000	400	0.35	—
3. Sandy gravel and gravel (GW and GP)	3,000	200	0.35	—
4. Sand, silty sand, clayey sand, silty gravel, and clayey gravel (SW, SP, SM, SC, GM, and GC)	2,000	150	0.25	—
5. Clay, sandy clay, silty clay, clayey silt, silt, and sandy silt (CL, ML, MH, and CH)	1,500	100	—	130

For SI: 1 pound per square foot = 0.0479 kPa, 1 pound per square foot per foot = 0.157 kPa/m.
a. Coefficient to be multiplied by the dead load.
b. Cohesion value to be multiplied by the contact area, as limited by Section 1806.3.2.

classifications of the different strata. An exception allows the building official to approve a presumptive load-bearing capacity if deemed adequate for the support of lightweight or temporary structures.

1806.3 Lateral load resistance. Sections 1806.3.1 through 1806.3.4 deal with lateral load calculations when using the presumptive load-bearing values of Table 1806.2. The limitation on frictional resistance for silts and clays is intended to provide structural stability and improve serviceability.

The formulae for lateral bearings employing posts and poles (found in IBC Section 1807.3, embedded posts and poles) were originally for outdoor advertising structures. For these structures, deflections of ½ inch (12.7 mm) at the surface do not affect serviceability. Thus, the allowance of two times the tabular value for these structures is permitted.

Section 1807 *Foundation Walls, Retaining Walls, and Embedded Posts and Poles*

1807.1 Foundation walls. Foundation walls of materials such as concrete and masonry, rubble stone, or wood are addressed in Section 1807. Per Section 1807.1.1, lateral soil loads that must be considered in foundation wall design are covered in Section 1610. If a drainage system is not placed behind the wall to drain ground water away from the wall, hydrostatic pressures, which can easily exceed the lateral pressures from the retained soil, will occur. For example, the active lateral pressure from a well-graded granular soil may be in the range of 30 to 35 pounds per square foot (1.44 to 1.68 kPa), whereas the hydrostatic pressure is 62.4 pounds per cubic foot (999.6 kg/m^3). Hence, the equivalent fluid pressures on a wall that was designed as drained, but constructed without an effective drainage system, could be subjected to pressures approximately three times the design pressure.

Unbalanced backfill height and its method of measurement are presented in Section 1807.1.2 and rubble stone foundation walls in Section 1807.1.3. Because rubble masonry uses rough stones of irregular shape and size, a larger thickness, as compared to hollow unit masonry or concrete, is required for adequate bonding of the stone and mortar. Rubble stone foundation walls are not permitted in SDC C, D, E, or F.

1807.1.4 Permanent wood foundation systems. The requirements set forth in AWC PWF must be followed. The wood foundation system is an assembly similar to a fire assembly—no substitution of materials or methods is allowed. All lumber and plywood must be treated in accordance with AWPA U1 (Commodity Specification A, Special Requirement 4.2) and must be identified and labeled in accordance with Section 2303.1.9.1.

All hardware and fasteners must be corrosion resistant. Metals in contact with the preservative salts will corrode at a much faster rate than normal because of the influence of the salts. Hence, only corrosion-resistant fasteners made of silicon bronze, copper, or Type 304 or 316 stainless steel may be used, except that hot-dipped galvanized nails may be used when installed under the specific conditions set forth in the technical report for surface treatment of the nails and moisture protection of the foundation.

1807.1.5 Concrete and masonry foundation walls. Foundation walls must be designed within the applicable provisions of IBC Chapters 19 and 21. However, if the foundation wall is laterally supported at the top and bottom, such as a basement wall laterally supported by a floor diaphragm at the top and a basement floor slab at the base, the wall may be constructed in accordance with the prescriptive provisions of Section 1807.1.6 and its associated tables. These tables allow the use of unreinforced and plain (lightly reinforced) concrete or masonry walls that have been used in low or very low seismic risk areas.

Additionally, if the prescriptive provisions are used, sufficient soil investigation should be done to properly classify the retained soils as indicated in the tables in accordance with the Unified Soil Classification Method (see IBC Section 1803.5.1).

1807.1.6 Prescriptive design of concrete and masonry foundation walls. The requirements and provisions of Sections 1807.1.6.1 through 1807.1.6.3.2 are applicable to the prescriptive design of concrete and masonry foundation walls that are laterally supported at the top and bottom. The minimum wall thicknesses are specified in the appropriate sections, based on the thickness of the supported wall, soil loads, unbalanced backfill height, and overall height of the wall. These minimum thickness provisions are to facilitate support of the wall above. These thickness provisions are empirical and have been used successfully in low or very low seismic risk areas. Additional seismic requirements for concrete and masonry foundation walls are covered in Sections 1807.1.6.2.1 (concrete) and 1807.1.6.3.2 (masonry), based on the SDC of the building.

1807.1.6.2 Concrete foundation walls. This section specifies the material requirements for walls constructed in accordance with the prescriptive tables. Note the effective depth, "d," in Section 1807.1.6.2(3). Placement of reinforcing at the prescribed "d" is critical to develop adequate flexural strength necessary to resist the combined vertical and lateral loads.

The concrete section contains seven specific requirements to prescribe a foundation wall. A conservative maximum unfactored axial load of $1.2tf'_c$ for concrete and $1.2tf'_m$ for masonry are included in the requirements. The maximum unfactored axial load is based on a compressive stress on the outside face of the wall that is due to the axial load and bending moment induced by the backfill that is well below that permitted by ACI 318 or TMS 402. Although this axial load limitation has merit, it requires a calculation to determine actual maximum axial load acting on a given wall. Table 1807-1 shows the maximum unfactored allowable axial load for the typical concrete foundation wall.

Table 1807-1. **Maximum Permissible Axial Load for Concrete Walls Based on $1.2tf'_c$ in Pounds per Foot of Wall**

Wall Thickness, t (inches)	f'_c = 2,500 psi	f'_c = 3,000 psi
7.5	22,500	27,000
9.5	28,000	34,200
11.5	34,500	41,400

For SI: 1 inch = 25.4 mm, 1 pound per square inch = 0.00689 N/mm^2

1807.1.6.2.1 Seismic requirements (concrete). Seismic requirements for concrete foundation walls constructed in accordance with Table 1807.1.6.2 are as follows:

1. SDC A and B—One #5 bar is required at a minimum around window and door openings, which must extend beyond the corners of the openings or be anchored so as to develop f_y in tension at the corner of openings.

2. SDC C, D, E, and F—The prescriptive tables are not allowed to be used except as permitted for plain concrete members in accordance with Section 1905.6.2, which modifies ACI 318, Section 14.1.4. The modification states that structural plain concrete members are not permitted in SDC C, D, E, or F except for structural plain concrete basement, foundation, or other walls below the base in detached one- and two-family dwellings three stories or less in height constructed with stud-bearing walls. Additional restrictions apply to dwellings in SDC D or E, where the walls cannot exceed 8 feet (2,438 mm) in height, cannot be less than 7.5 inches (190 mm) thick, and can retain no more than 4 feet (1,219 mm) of unbalanced fill. A final requirement states that the walls must be reinforced in accordance with ACI 318 Section 14.6.1.

1807.1.6.3 Masonry foundation walls. This section contains requirements for masonry foundation walls, both plain and with reinforcement. Masonry is required to be solid in order to distribute the concentrated force, or hollow units must be solidly grouted as noted in Footnote c of the plain masonry foundation wall Table 1807.1.6.3(1). The masonry section contains 10 specific requirements for a prescriptive masonry foundation wall.

Table 1807-2 shows the maximum unfactored allowable axial load for the typical masonry foundation wall. Also see the discussion under Section 1807.1.6.2, Concrete foundation walls.

Table 1807-2. Maximum Permissible Axial Load for Masonry Walls Based on $1.2tf'_m$ in Pounds per Foot of Wall

Wall Thickness, t (inches)	f'_m = 1,500 psi	f'_m = 2,000 psi
7.625	13,725	18,300
9.625	17,325	23,100
10.625	19,125	25,500
11.625	20,925	27,900

For SI: 1 inch = 25.4 mm, 1 pound per square inch = 0.00689 N/mm²

1807.1.6.3.1 Alternative foundation wall reinforcement. The code permits equivalent cross section of reinforcing, provided the spacing does not exceed 72 inches (1,829 mm) and the bar size does not exceed #11 (M #36). If alternative reinforcement is used, it is preferable to reduce bar size and spacing rather than increase bar size and spacing. In development of the reinforcing, the bar size must be small enough that the reinforcing, and any splices, can be adequately developed. Development refers to the embedment of the reinforcing to adequately develop the bond between the reinforcing and the grout. A good rule of thumb to prevent splitting of concrete masonry is that the bar size number should not exceed $t - 1$, where t is the nominal thickness of the wall in inches.

1807.1.6.3.2 Seismic requirements (masonry). Table 1807-3 summarizes requirements for masonry foundation walls based on SDC and gives applicable sections of the TMS Code.

Table 1807-3. **Seismic Requirements for Masonry Foundation Walls**

Seismic Design Category	TMS 402 Section
C	7.4.3
D	7.4.4
E, F	7.4.5

Seismic requirements for masonry foundation walls constructed in accordance with Tables 1807.1.6.3(1) through 1807.1.6.3(4) are as follows:

1. SDC A and B—No additional requirements apply.
2. SDC C—Additional requirements cover discontinuous members that are part of the lateral-force-resisting system, such as columns, pilasters, and beams that support reactions from walls or frames, but no specific requirements for foundation walls. Refer to Section 7.4.3 of TMS 402 for other requirements.
3. SDC D—Must conform to the requirements of SDC C, as well as Section 7.4.4 of TMS 402.
4. SDC E and F—Must conform to the requirements of SDC C and D, as well as Section 7.4.5 of TMS 402.

1807.2 Retaining walls. The IBC requires retaining walls to be designed to resist overturning, sliding, excessive foundation pressure, and water uplift. The design of walls retaining more than 6 feet of soil and located within SDC D or above are also required to be checked for seismic earth pressure.

The design is required to meet a minimum safety factor of 1.5 in relation to sliding and overturning. If the design includes the seismic earth pressure, the safety factor for sliding and overturning can be reduced to 1.1. Figure 1807-1 shows typical loads that are considered in retaining wall design. Dry-cast concrete units used in the construction of segmental retaining walls shall comply with ASTM C1372 *Standard Specification for Dry-Cast Segmental Retaining Wall Units*.

Where required, provisions for guards on retaining walls are provided consistent with guard requirements elsewhere in the code. The exception exempts conditions where a retaining wall is not accessible to the public.

1807.3 Embedded posts and poles. The design criteria established in 1807.3.1 through 1807.3.3 was originally developed for design of outdoor advertising structures, relate to lateral bearing and apply to a vertical pole considered a column embedded in either earth or in a concrete footing in the earth and used to resist lateral loads. ASABE EP 486.3 *Shallow-post and Pier Foundation Design* contains procedures for determining adequacy of shallow, isolated post and pier foundations resisting applied structural loads. It is limited to post and pier foundations that are vertically installed in relatively level terrain with concentrically-loaded footings.

1807.3.1 Limitations. The limitations imposed by this section are intended for both structural stability and serviceability. The limitations on the types of construction that use the lateral support of poles are based on the brittle nature of the materials. To prevent excessive distortions that would cause the cracking of these brittle materials, the code limits the use of the poles unless some type of rigid cross-bracing is provided to limit the deflections to those that can be tolerated by the materials.

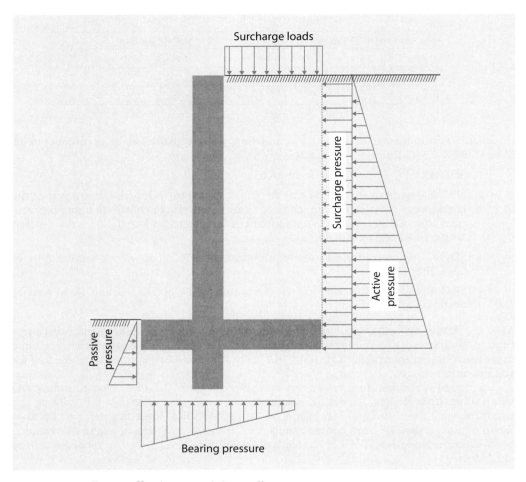

Figure 1807-1 **Forces affecting a retaining wall.**

Wood poles must be treated in accordance with AWPA U1. Sawn timber posts are Commodity Specification A, Use Category 4B, and round timber posts are Commodity Specification B, Use Category 4B.

1807.3.2 Design criteria. The IBC employs allowable lateral-bearing stresses in IBC Table 1806.2, which are considerably lower than those developed for this section. Consequently, Section 1806.3.4 permits a doubling of the lateral-bearing values for isolated poles and poles supporting structures that can safely tolerate the ½-inch (12.7 mm) movement at the ground surface, for example, signs, flagpoles, and light poles.

1807.3.2.1 Nonconstrained. This formula should be used only for minor foundations of moderate size, which will fit within the constraints of the data from which the formula was developed. For large-sized piers, that is, more than 2 feet (610 mm) in diameter, a more appropriate method should be used. Equation 18-1 requires an iterative process to arrive at a satisfactory depth of embedment.

1807.3.2.2 Constrained. A rigid floor or slab-on-ground using reinforced concrete that will form a fulcrum for the column. Columns in flexible pavements such as asphalt concrete must use the formula in Section 1807.3.2.1 for unconstrained conditions.

1807.3.2.3 Vertical load. There is no requirement to consider a combined lateral and vertical load.

1807.3.3 Backfill. In order for the pole to meet the conditions of research that resulted in the code formula, the code requires that the backfill in the annular space around a column that is not embedded in a concrete footing be either of 2,000 psi (13.8 MPa) concrete, CLSM, or of clean sand thoroughly compacted by tamping in layers not more than 8 inches (203 mm) in depth.

Section 1808 *Foundations*

1808.1 General. The provisions of Section 1808 apply to all foundations. The two general types of foundations are shallow foundations and deep foundations. Specific requirements for shallow foundations and deep foundations are located in Sections 1809 and 1810, respectively. Section 202 defines a shallow foundation as an individual or strip footing, a mat foundation, a slab-on-grade foundation, or a similar foundation element. A deep foundation is defined as a foundation that does not meet the definition of a shallow foundation.

1808.2 Design for capacity and settlement. Footings should be designed for approximately equal settlements to minimize differential settlements. For footings on sands, this may require unequal footing pressures to affect equal settlements. Expansive soils are addressed in Section 1808.6.

1808.3 Design loads. Footings are to be designed using full dead load (including overlying fill materials), floor and roof live loads, snow load, wind or seismic forces, and any other loads required by Section 1605 that will produce the most severe loading. Live loads acting at the foundation may be reduced based on the reduced probability of simultaneous occurrence of maximum live loads. This section specifically permits live load reduction as specified in Sections 1607.13 and 1607.14 for the foundation design.

1808.3.1 Seismic overturning. When strength design loads are used to proportion the foundations, the seismic overturning effects are permitted to be reduced in accordance with ASCE 7 Section 12.13.4. This is in recognition that the seismic forces determined in accordance with the ASCE 7 standard are based on strength design, not allowable stress design (ASD). Foundations proportioned in accordance with ASD procedures have historically performed satisfactorily. Because of expected deviation from the results from the equivalent lateral force method, which assumes a fixed base of the building, overturning effects at the foundation are permitted to be reduced 25 percent for structures other than inverted pendulum or cantilever column systems when designed by the equivalent lateral force procedure. Overturning effects at the foundation are permitted to be reduced 10 percent for structures designed by the modal analysis method because of the higher degree of accuracy of the procedure. Note that these reductions cannot be used with the alternative basic ASD load combinations of Section 1605.2.

1808.3.2 Surcharge. This section describes how surcharge loads could affect an adjacent structure. Fill or other surcharge loads are not permitted to be placed adjacent to a building or structure unless it is capable of withstanding the additional loads caused by the surcharge load. Existing footings or foundations that could be affected by an excavation must be protected from detrimental lateral or vertical movement and settlement. The exception allows minor grading for landscaping with limited grading heights using walk-behind equipment which does not induce high forces against an adjacent foundation or wall when approved by the building official.

1808.4   Vibratory loads.   Footings supporting equipment should be designed to minimize the transmission of vibratory loads to the soils. The dynamic interaction of the footing, equipment, and soil mass should be analyzed, and the footing "tuned" to minimize the transmission. As a rough rule of thumb, footings for rotating or reciprocating equipment should have a mass that is at least four times the mass of the equipment.

Vibratory loads from equipment foundations that are transmitted to the soil can cause significant and damaging settlements. The transmitted vibration will cause densification of granular materials, particularly loose or medium dense sands. The reduction in volume can cause large settlements depending on the initial density of the sands. In saturated granular materials, such as loose or medium dense sands with a high water table, the transmitted vibrations can cause a buildup of pore pressure and liquefaction of the sands, with resulting loss of bearing capacity and settlements. In saturated clays, the vibrations can enhance the drainage of water from the pores and increase long-term settlements.

1808.5   Shifting or moving soils.   An example of shifting or moving soils would be loose sands or expansive clay. This section of the IBC clarifies that footings should not be placed on shifting soils but rather placed on suitable soils. There are many means of meeting this requirement such as using deep foundation systems that extend through the shifting soils down to suitable materials or removing the poor soils and replacing with structural fill. Loose soils may also be improved by methods such as deep dynamic compaction, pressure grouting, and stone columns. If foundations will be placed in areas where shifting soils or moving soils exist, a geotechnical report should be provided per Section 1803 recommending what mitigation measures should be followed for the placement of foundations in these areas.

1808.6   Design for expansive soils.   The requirements to mitigate the effects of expansive soils are set forth in this section. In addition to mitigation by foundation design, the effects of expansive soils may also be mitigated by removal of the expansive soils or stabilization by chemical means, pre-saturation, or dewatering. Expansive soils are cohesive soils, typically high plasticity clays, with a high Plasticity Index and a high Swell Index.

1808.6.1   Foundations.   The large volume changes in expansive soils caused by changes in the soils' water content can cause significant differential deflections in a building if not uniform. In a typical building on expansive soils, the soils at the perimeter of the building will have seasonal moisture changes, whereas the soils at the interior of the building will remain at a fairly constant moisture content. The perimeter foundations will rise and fall with the seasonal volume changes in moisture content, whereas the soil at the interior footings or slab will not have any volume changes, because of constant moisture content. The resulting differential displacements between the interior and exterior footings can cause significant structural distress. Hence, the requirements that the foundation be designed to resist the differential volume changes and to minimize racking or differential displacements in the structure.

1808.6.2   Slab-on-ground foundations.   The slab-on-ground or raft foundation design methods cited in this section result in a raft that has sufficient stiffness to bridge differential displacements caused by the volume changes in the supporting soil. Conventionally reinforced (nonprestressed) foundations on expansive soils can be designed in accordance with WRI/CRSI *Design of Slab-on-Ground Foundations*, and post-tensioned foundations on expansive soils can be designed in accordance with PTI *Standard Requirements for Design and Analysis of Shallow Post-Tensioned Concrete Foundations on Expansive and Stable Soils*. Design moments, shears, and deflections can also be determined from each of the aforementioned applicable standards.

The code also permits alternative methods of analysis, provided the methodology is rational and the basis for the analysis and design parameters are available for peer review.

1808.6.3 Removal of expansive soil. Removal of the expansive soil is an acceptable mitigation method and is the preferred method if the stratum of expansive soil is near the surface and reasonably thin. This method may also be the least expensive method if the expansive soil is at the surface.

1808.6.4 Stabilization. Expansive soils may be stabilized so that the moisture content does not change; hence, there will be no volume changes to cause differential displacements. Stabilization can be by chemical methods, by pre-saturating the soils to a maximum swell and capping the expansive layer to keep the moisture content constant, or by dewatering to a minimum shrinkage and providing drainage to keep the moisture content constant.

1808.7 Foundations on or adjacent to slopes. The provisions of this section apply only to buildings placed on or adjacent to slopes steeper than one unit vertical to three units horizontal.

1808.7.1 Building clearance from ascending slopes. This setback requirement is intended to provide protection to the structure from shallow slope failure (sloughing) and protection for erosion and slope drainage. The setback space also provides access around the structure and helps to create a light and open environment. IBC Figure 1808.7.1 depicts the criterion for the clearance as the smaller of H/2 and 15 feet (4,572 mm). Figure 1808-1 also depicts the criteria set forth in this item for determination of the toe of the slope when the slope exceeds 1:1.

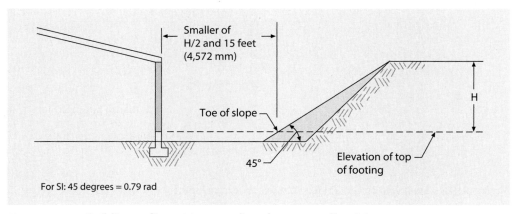

Figure 1808-1 Buildings adjacent to ascending slope exceeding 1:1.

1808.7.2 Foundation setback from descending slope surface. The setback requirement at the top of slopes is intended to provide vertical and lateral support for the foundations and minimize the possibility of shallow bearing failure of the foundation because of lack of lateral support. The setback also provides space for drainage away from the slope without creating too steep a drainage profile, which could cause erosion problems. The setback space also provides access around the structure and helps to create a light and open environment. IBC Figure 1808.7.1 depicts the criterion for the setback or clearance. Figure 1808-2 depicts the criteria set forth in this item for determination of the toe of the slope when the slope exceeds 1:1.

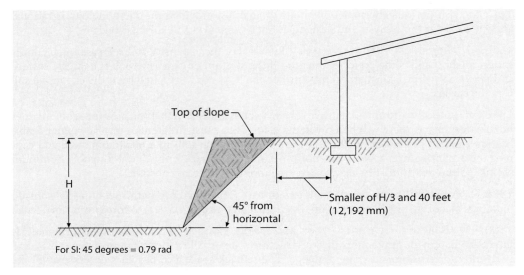

Figure 1808-2 **Buildings adjacent to descending slope exceeding 1:1.**

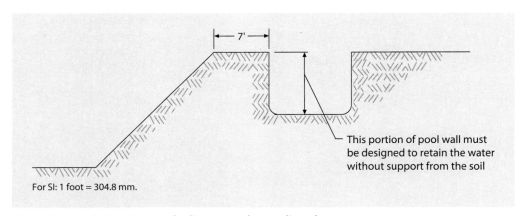

Figure 1808-3 **Swimming pool adjacent to descending slope.**

It is possible to locate a structure closer to the slope than indicated in IBC Figure 1808.7.1. The footing of the structure may be located on the slope itself, provided that the depth of embedment of the footing is such that the face of the footing at the bearing plane is set back from the edge of the slope at least H/3 or 40 feet (12,192 mm), whichever is less.

1808.7.3 Pools. Figure 1808-3 depicts the criteria for the design of swimming pool walls near the top of a descending slope. The pool wall that is within 7 feet (2,134 mm) of the top slope must be sufficient to resist the hydrostatic water pressure without support from the soil to protect against failure of the pool wall should localized minor slope movement or sloughing occur. The pool setback should be established as one-half of the setback required by IBC Figure 1808.7.1 which translates to the smaller of H/6 and 20 feet (6,096 mm).

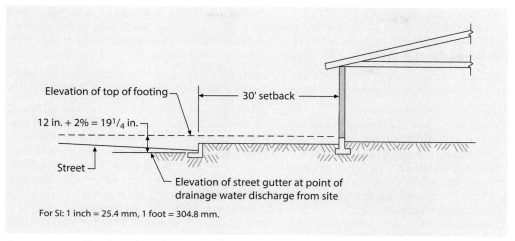

Figure 1808-4 **Footing elevation on graded sites.**

1808.7.4 Foundation elevation. Figure 1808-4 depicts the criteria from this section for the elevation of the exterior foundations relative to the street, gutter, or point of inlet of a drainage device.

The elevation of the street or gutter shown is that point at which drainage from the site reaches the street or gutter. This requirement is intended to protect the structure from water encroachment in the case of heavy or unprecedented rains. This requirement may be modified if the building official finds that positive drainage slopes are provided to drain water away from the building and that the drainage pattern is not subject to temporary flooding from clogged drains, landscaping, or other impediments.

1808.7.5 Alternate setback and clearance. This alternative procedure allows the building official to approve alternative setbacks and clearances from slopes, provided that the intent of Section 1808.7 is met. This section gives the building official the authority to require a geotechnical investigation by a qualified geotechnical engineer. The parameters for such geotechnical investigation are established in Section 1803.5.10 and include consideration of material, height of slope, slope gradient, load intensity, and erosion characteristics of the slope material.

1808.8 Concrete foundations. Footings may be designed, or the requirements of Table 1809.7 may be used, for structures with light-framed walls of conventional construction, where frost heave or expansive soils are not a problem. Table 1809.7 is based on anticipated dead and live loads from the floors and roof and an assumed soil classification of ML, MH, CL, or CH.

1808.8.1 Concrete or grout strength and mix proportioning. Table 1808.8.1 provides the minimum specified compressive strength, f'_c, for concrete or grout to be used in specific foundation types. This table provides values for various SDCs as well as for piles and shafts. A compressive strength of 2,500 psi (17.2 MPa) is the minimum value for most footings of structures in SDC A, B, or C and for the footings of residential light-frame and utility structures, one or two stories in height, in SDC D, E, and F. The minimum specified concrete strength of 2,500 psi (17.2 MPa) is set to provide a material of adequate strength and durability. Concrete of lower strength may not have adequate durability, particularly in freeze-thaw areas.

Slump requirements must be adjusted for specific conditions. For example, concrete placed in uncased drilled holes needs to be in the 6- to 8-inch (152- to 203- mm) range so that the concrete flows readily into the irregularities in the drilled hole. Use of superplasticizers will provide the desired slump while keeping the water to cementitious material ratio low.

1808.8.2 Concrete cover. Cover requirements are set to provide protection and minimize steel corrosion. All concrete cover requirements of Chapter 18 are organized in Table 1808.8.2.

1808.8.3 Placement of concrete. Holes should be free from debris, loose soils, or water. Placement of concrete through water should be avoided because of the increased risk of segregation and dilution of the concrete paste. When concrete is placed under water, by tremie or other approved method, the mix must be different from the standard mix used for ordinary concrete foundations. The mix must be proportioned so that it is plastic with high workability and will flow without segregation. The desired consistency can be obtained by using rounded aggregates, high sand contents, entrained air, and superplasticizers. Higher cement contents are necessary to compensate for the increase in the water to cementitious materials ratio caused by dilution from placement through water. Minimum cement content should be 600 pounds per yard (56 k/m^3). Placement from the top of deep foundations can cause segregation of concrete mix also, and proper measures such as the use of a funnel hopper (elephant trunk) should be taken to avoid the potential for segregation.

When depositing concrete from the top of a deep foundation, the IBC requires concrete to be chuted directly into smooth-sided pipes and tubes or through a centering funnel hopper. The main purpose of the centering funnel for drilled piles is to prevent the concrete from encountering the soil at the perimeter of the hole, which is generally not a problem for pipes and tubes. The term smooth sided is included in the code to prevent possible segregation from the ridges or corrugations if nonsmooth pipes or tubes are used.

1808.8.4 Protection of concrete. Concrete footings should not be placed during rain, sleet, snow, or freezing weather without protection against either freezing or increase in water content at the surface from rain while plastic. If concrete placement is undertaken under such conditions without adequate protection, numerous complications can be expected. Projects may be halted while concrete core samples, testing, analysis, and investigation are performed on the hardened concrete to determine suitability of the deposited concrete and of the structural elements. See ACI 306R *Guide to Cold Weather Concreting* and ACI 306.1 *Standard Specification for Cold Weather Concreting* for cold-weather concrete operations.

1808.8.5 Forming of concrete. The soil should have sufficient strength and cohesion that the shape, dimensions, and vertical sides of the excavation can be maintained without sloughing prior to and during the concreting operations. Excavations in loose granular materials must be formed. Where required, formwork shall be in accordance with ACI 318 Section 26.11.

1808.8.6 Seismic requirements. Specific requirements for foundations in SDC C, D, E, or F are contained in Section 1905, which contains additional seismic requirements to ACI 318. Per Table 1808.8.1, in SDC D, E, or F, concrete must have a specified compressive strength of not less than 3,000 psi (20.7 MPa), except that 2,500 psi (17.2 MPa) concrete strength is permitted in Group R or U occupancies of light-frame construction two stories or less in height.

Buildings in SDC C, D, E, and F are required to comply with the seismic requirements of ACI 318, except for detached one- and two-family dwellings of light-frame construction two stories or less in height. ACI 318 covers foundation requirements in general, and also provides requirements for footings, mat foundations, pile caps, grade beams, and slabs on grade.

Note that plain concrete is either unreinforced or lightly reinforced concrete that contains less reinforcing than required to meet the minimum reinforcement requirements set forth in ACI 318.

1808.9 Vertical masonry foundation elements. Vertical masonry foundation elements that are not foundation piers are to be designed as piers, walls, or columns in accordance with TMS 402.

Section 1809 *Shallow Foundations*

1809.1 General. Foundations are divided into two major categories of shallow foundations and deep foundations. Shallow foundations are individual or strip footings, the bottoms of which are typically close to the surface. Examples include mat foundations, slab on grade, or similar foundation types. Shallow foundations are regulated in Sections 1809.2 through 1809.13. Deep foundations are those that are not classified as shallow foundations.

1809.2 Supporting soils. Because the supporting soil for shallow foundations is close to the surface, in order to minimize differential settlement, shallow foundations must be constructed on undisturbed native soil, compacted fill material, or CLSM. Where constructed on fill, the material must be properly placed and compacted to achieve adequate density in accordance with Section 1804.6. CLSM must be placed and tested in accordance with Section 1804.7.

1809.3 Stepped footings. Footings are required to be stepped when the slope of the bearing surface exceeds 1 in 10. No recommendations or restrictions are provided. Figure 1809-1 schematically represents a satisfactory stepped foundation. The figure shows a recommended horizontal overlap of the top of the foundation wall beyond the step in the foundation to be larger than the vertical step in the foundation wall at that point. This is recommended to keep any crack propagation approximately at a 45-degree (0.79 rad) angle. To keep this cracking to a minimum, it is also recommended that the height of each step not exceed 1 to 2 feet (305 to 610 mm). Other measures to protect against cracking, such as special reinforcing details, may be needed. See Section 2308.10.8.3 for specific requirements pertaining to stepped foundations for conventional construction.

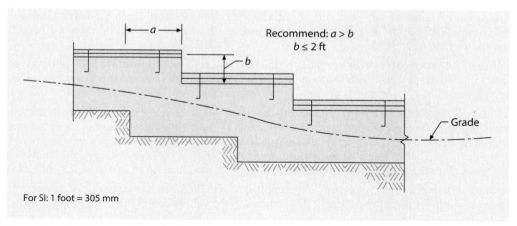

For SI: 1 foot = 305 mm

Figure 1809-1 **Stepped foundations.**

1809.4 Depth and width of footings. Footings should always be placed a minimum of 12-inches (305 mm) deep on either firm, undisturbed earth or properly compacted fill, and be at least 12 inches (305 mm) in width.

1809.5 Frost protection. To prevent frost heave during winter and subsequent settlement upon thawing, foundations and building supports should be placed on a stratum with adequate load-bearing resistance that is below the frost line. Frost heave occurs because of the increased soil volume from the freezing of pore water in the soil. Clay soils, particularly saturated clays, are most susceptible to frost heave. Well-drained sands and gravels will not be susceptible to significant movement. If foundations are built on soils that can freeze, the resulting frost heave, which is rarely uniform, can cause serious damage from differential settlements.

The frost line is defined as the lowest depth below the ground surface to which a temperature of 32°F (0°C) extends. The factors governing the depth of the frost line are air temperature, the length of time the air temperature is below freezing (32°F) (0°C), and the soil's thermal conductivity. Frost lines vary significantly throughout the country from no penetration in southern Florida to 100 inches (2,540 mm) in the northern regions of Michigan and Maine. Data on frost penetration are available from the U.S. Department of Commerce Weather Bureau.

The code offers three options for ensuring adequate frost protection for shallow foundations. The first option—the most common and simplest to accomplish—is to construct the bottom of the footing below the frost line for the particular locality. The second option is to construct the footing in accordance with the referenced standard, ASCE 32, *Design and Construction of Frost-Protected Shallow Foundations*. The third option, which is often encountered in areas where bedrock is prevalent, is to construct the footing on solid rock.

Note the exception where a frost-protected foundation is not required: free-standing buildings classified in Risk Category I (see Section 1604.5), floor area of 600 square feet (56 m^2) or less for light-frame construction or 400 square feet (37 m^2) or less for other than light-frame construction, and eave height of 10 feet or less. Note that the term *light-frame construction* is defined in Section 202 as a system that uses repetitive wood or cold-formed steel framing members.

The code prohibits footings from bearing directly on frozen soil unless the soil is permanently frozen. Permafrost may not meet this condition as permafrost is considered soil that remains in a frozen state for more than two years in a row.

1809.5.1 Frost protection at required exits. Concrete landings that are placed at "required" exits must meet the frost protection requirements. Landings are typically considered nonstructural concrete flat work that is not extended down to frost depth. As frost heave can cause these landings to lift and render exits unusable, landings at all required exits must reach the appropriate frost depth.

1809.6 Location of footings. This restriction is intended to minimize the influence of vertical and lateral loads from footings at a higher elevation on footings at a lower elevation. See Figure 1809-2.

1809.7 Prescriptive footings for light-frame construction. Footings of concrete or masonry units may be designed, or the requirements of Table 1809.7 may be used, for structures with light-frame walls of conventional construction, where frost heave or expansive soils are not a problem. IBC Table 1809.7 shown herein as Table 1809-1 is based on anticipated dead and live loads from the floors and roof and an assumed soil classification of ML, MH, CL, or CH.

1809.8 Plain concrete footings. In compliance with ACI 318, the edge thickness of plain concrete footings in other than light-frame construction must not be less than 8 inches (203 mm). In accordance with ACI 318, the thickness or depth used to compute footing stresses (flexure, combined axial load and flexure, or shear) should be 2 inches (51 mm) less than the actual thickness of the footing for footings cast against soil. This is done to allow for unevenness of excavation and contamination of the concrete adjacent to the soil.

The edge thickness of plain concrete footings can be reduced to 6 inches (152 mm) for Group R-3 occupancies (R-3 occupancies are described in Section 310), provided that the edge distance (projection) of the footing beyond the face of the stem wall does not exceed the

Shallow Foundations

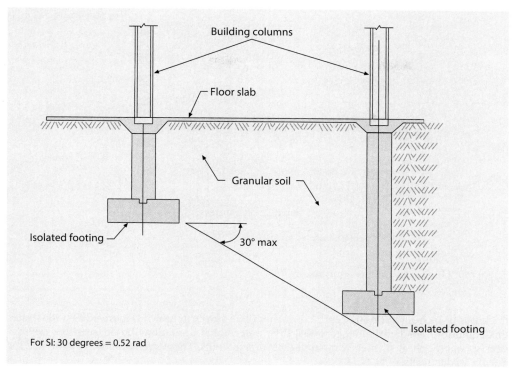

For SI: 30 degrees = 0.52 rad

Figure 1809-2 **Isolated footing foundation.**

Table 1809-1. **Prescriptive Footings Supporting Walls of Light-Frame Construction**[a,b,c,d,e]

Number of Floors Supported by the Footing[f]	Width of Footing (inches)	Thickness of Footing (inches)
1	12	6
2	15	6
3	18	8[g]

For SI: 1 inch = 25.4 mm, 1 foot = 304.8 mm.

a. Depth of footings shall be in accordance with Section 1809.4.
b. The ground under the floor shall be permitted to be excavated to the elevation of the top of the footing.
c. Interior stud-bearing walls shall be permitted to be supported by isolated footings. The footing width and length shall be twice the width shown in this table, and footings shall be spaced not more than 6 feet on center.
d. See Section 1905 for additional requirements for concrete footings of structures assigned to Seismic Design Category C, D, E, or F.
e. For thickness of foundation walls, see Section 1807.1.6.
f. Footings shall be permitted to support a roof in addition to the stipulated number of floors. Footings supporting a roof only shall be as required for supporting one floor.
g. Plain concrete footings for Group R-3 occupancies shall be permitted to be 6 inches thick.

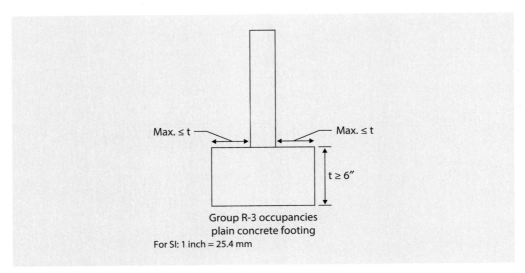

Figure 1809-3 **Plain concrete footing.**

thickness [6 inches (152 mm) depth = 6 inches (152 mm) extension]. Figure 1809-3 illustrates this condition. For lightly loaded walls, this dimensional limitation should keep the flexural stresses and the shear stresses below applicable design limits. These stresses should be checked for heavily loaded walls.

1809.9 Masonry-unit footings. Masonry footings were widely used until the middle of the last century when they were replaced by steel or wood grillage footings, which in turn were replaced by more economical plain or reinforced concrete footings. Masonry footings were often built of stone cut to a specific size, or rubble masonry of random-sized stones bonded with mortar. Although seldom used, masonry footings may be constructed of hard-burned brick set in cement mortar to support lightweight buildings.

1809.9.1 Dimensions. Type M mortar is suitable for unreinforced masonry below grade or with earth contact. Type S should be used for reinforced footings. Projections of the footing beyond a wall or pier should not exceed one-half of the footing depth to keep the shear and flexural stresses in the footing within safe limits. For example, a footing with a 12-inch (305 mm) depth should project no more than 6 inches (152 mm) beyond the face of the wall. The footing width shall be at least 8 inches (203 mm) wider than the supported wall.

1809.9.2 Offsets. The stepping back, or racking, of successive courses of the foundation wall supported by a masonry footing must not exceed 1½ inches (38.1 mm) for a single course or 3 inches (76 mm) for a double course. Where wide footings are necessary for bearing, the wall must be stepped back to keep the footing projection within the limits of Section 1809.9.1. See Figure 1809-4.

1809.10 Pier and curtain wall foundations. Pier foundations may not be used to support structures assigned to SDC D, E, and F because seismic detailing requirements for higher SDCs have not been developed. These foundation elements have been popular particularly in the Eastern United States, and under this section are allowed only for the support of light-frame construction of not more than two stories above grade plane.

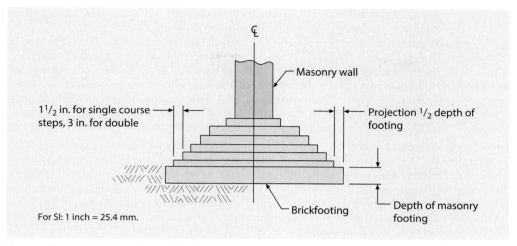

Figure 1809-4 Brick footing wall offsets.

1809.11 Steel grillage footings. The development of reinforced concrete made the use of steel grillage footings obsolete, except for underpinning work. A typical grillage consists of two or more tiers of steel beams, with each tier placed at right angles to the tier below. The beams in each tier are usually held together by a system of bolts and pipe spacers. For construction of new grillage footings, the beams should be clean and unpainted and the entire grillage system filled with and encased in concrete with at least 4 inches of cover. The grillage should be placed on a concrete pad at least 6 inches thick to distribute the load evenly to the soil.

1809.12 Timber footings. Use of timber footings is allowed only for Type V structures. Timber must be pressure treated to American Wood Protection Association (AWPA) U1 (Commodity Specification A, Use Category 4B) standards, except for foundations permanently below the ground-water table. The pressure preservative treatment protects the timber from decay, fungi, and harmful insects. The AWPA Use Category System (UCS) is based on the end use hazard, similar to other international standards for wood treatment. The Use Category System (UCS) is used to specify the wood treatment based on the desired wood species and the environment of the intended end use. Stronger preservatives are necessary to prevent marine borers when timber foundations are used in coastal brackish or marine environments.

Preservative treatment by the pressure process within the limitations of the AWPA standards should not significantly affect the strength of the wood. Part of the process, however, involves the conditioning of the wood prior to treatment by steaming or boiling under vacuum. This conditioning can cause reduction in strength. This strength loss is recognized in the AWC *National Design Specification for Wood Construction* (AWC NDS) by use of the condition treatment factor, C_{ct}, which provides a reduction to certain design values for timber poles and piles. Note that the NDS states that load duration factors greater than 1.6 are not allowed for structural members that are pressure treated with water-borne preservatives. This restriction would apply to impact loads that have a load duration factor of 2.0.

Untreated timber may be used when the footings are completely embedded in soil below the ground-water table. Experience has long shown that timber permanently confined in water will stay sound and durable indefinitely. Wood submerged in fresh water cannot decay, because the necessary air is excluded. Because ground-water levels can sometimes change appreciably, untreated timber should only be used at depths sufficiently below the water table so that small drops in the water level will not expose the timbers to air.

The reference compression design values perpendicular to grain specified in the *NDS Supplement: Design Values* for *Wood Construction* are species group average values associated with a deformation level of 0.04" for a steel plate on wood member loading condition. One method for limiting deformation in special applications where it is critical, is use of a reduced compression design value perpendicular to grain. The NDS uses an equation to calculate the compression design value perpendicular to grain for a reduced deformation level of 0.02" (0.51 mm) where a 0.73 factor is used. This is comparable to the 0.7 factor specified in 1809.12.

1809.13 Footing seismic ties. Interconnection of individual spread footings is required for structures in SDC D, E, and F sited on soils in Site Class E or F. The footings must be interconnected with ties capable of transmitting a force equal to the larger footing load multiplied by the short period response acceleration, S_{DS}, divided by 10 and 25 percent of the smaller footing design gravity load, whichever is smaller. The intent of this requirement is to minimize differential movement or spreading between footings during ground shaking, and have individual footings act as a unit. If slabs on grade, or beams within slabs on grade, are used to meet the tie requirement, the load path from footing to slab or beam/slab and across joints in the slab or beam/slab should be checked for continuity. The slab or beam must be reinforced for the design tension load.

1809.14 Grade beams. The same grade beam provisions contained in the deep foundation section are included in the shallow foundation section. The grade beam must either be ductile or strengthened to resist forces from a maximum considered earthquake by including overstrength in its design. The exception indicates that only the ductile detailing provisions in ACI 318 Section 18.13.3.1 are exempt when grade beams are designed for the overstrength factor. All other provisions of ACI 318, such as durability, reinforcing steel cover, etc., are still applicable. Further, the exception is only permissible when differential settlements are less than one-fourth of the thresholds in ASCE 7 Table 12.13-3.

Section 1810 *Deep Foundations*

1810.1 General. Deep foundations are regulated in Sections 1810.1 through 1810.4 and are those typically referred to as piles, piers, or caissons that transfer load from a superstructure such as a building to the underlying soils or rock that provides support for the superstructure. See Table 1810-1 for an overview of the deep foundation provisions contained in the IBC. Deep foundations are generally used when the loads are too high to be supported by shallow footings or mats. Load resistance is provided by skin friction between soil and the sides of the pile and by end bearing at the tip. Settlement has to be limited within acceptable levels and may control the pile capacity for design purposes.

Deep foundations can be subjected to compression, tension, and lateral loads. The loads can be static and dynamic forces resulting from soil pressure, wind pressure acting on the building, or earthquake load effects from ground motion. In many cases, there are groups of piles, and group interaction must be considered. Closely spaced deep foundation elements have lower capacities.

Deep foundations can be constructed of cased or uncased concrete, precast-prestressed concrete, timber, steel pipe or H sections, or other special types such as micropiles and helical piles. Piles can be either driven in place or drilled in place. The installation method for a particular project depends on a variety of factors such as the geotechnical engineer's experience with local conditions, settlement limitations, acceptable noise levels, and so on.

Table 1810-1. **Deep Foundation Provisions**

IBC Section	Title
1810.1	General
1810.1.1	Geotechnical Investigation
1810.1.2	Use of Existing Deep Foundation Elements
1810.1.3	Deep Foundation Elements Classified as Columns
1810.1.4	Special Types of Deep Foundations
1810.2	Analysis
1810.2.1	Lateral Support
1810.2.2	Stability
1810.2.3	Settlement
1810.2.4	Lateral Loads
1810.2.4.1	Seismic Design Categories D Through F
1810.2.5	Group Effects
1810.3	Design and Detailing
1810.3.1	Design Conditions
1810.3.1.1	Design Methods for Concrete Elements
1810.3.1.2	Composite Elements
1810.3.1.3	Mislocation
1810.3.1.4	Driven Piles
1810.3.1.5	Helical Piles
1810.3.1.6	Casings
1810.3.2	Materials
1810.3.2.1	Concrete
1810.3.2.2	Prestressing Steel
1810.3.2.3	Steel
1810.3.2.4	Timber
1810.3.2.5	Protection of Materials
1810.3.2.6	Allowable Stresses
1810.3.2.7	Increased Allowable Compressive Stress for Cased Mandrel-Driven CIP Elements
1810.3.2.8	Justification of Higher Allowable Stresses
1810.3.3	Determination of Allowable Loads
1810.3.3.1	Allowable Axial Load
1810.3.3.2	Allowable Lateral Load
1810.3.4	Subsiding Soils or Strata
1810.3.5	Dimensions of Deep Foundation Elements
1810.3.5.1	Precast
1810.3.5.2	CIP or Grouted-In-Place
1810.3.5.3	Steel

Table 1810-1. **Deep Foundation Provisions** *(Continued)*

IBC Section	Title
1810.3.6	Splices
1810.3.6.1	Seismic Design Categories C Through F
1810.3.7	Top of Element Detailing at Cutoffs
1810.3.8	Precast Concrete Piles
1810.3.9	CIP Deep Foundations
1810.3.9.1	Design Cracking Moment
1810.3.9.2	Required Reinforcement
1810.3.9.3	Placement of Reinforcement
1810.3.9.4	Seismic Reinforcement
1810.3.9.5	Belled Drilled Shafts
1810.3.9.6	Socketed Drilled Shafts
1810.3.10	Micropiles
1810.3.10.1	Construction
1810.3.10.2	Materials
1810.3.10.3	Reinforcement
1810.3.10.4	Seismic Reinforcement
1810.3.11	Pile Caps
1810.3.11.1	Seismic Design Categories C Through F
1810.3.11.2	Seismic Design Categories D Through F
1810.3.12	Grade Beams
1810.3.13	Seismic Ties
1810.4	Installation
1810.4.1	Structural Integrity
1810.4.1.1	Compressive Strength of Precast Concrete Piles
1810.4.1.2	Shafts in Unstable Soils
1810.4.1.3	Driving Near Uncased Concrete
1810.4.1.4	Driving Near Cased Concrete
1810.4.1.5	Defective Timber Piles
1810.4.2	Identification
1810.4.3	Location Plan
1810.4.4	Preexcavation
1810.4.5	Vibratory Driving
1810.4.6	Heaved Elements
1810.4.7	Enlarged Base CIP Elements
1810.4.8	Hollow-Stem Augered, CIP Elements
1810.4.9	Socketed Drilled Shafts
1810.4.10	Micropiles
1810.4.11	Helical Piles
1810.4.12	Special Inspection

The term *caisson* refers to deep foundations that are required when very heavy loads are supported such as high-rise building towers. Caissons are large-diameter concrete elements installed with a casing that can be left in place. The caisson bottom can be expanded, called a bell bottom, to provide more bearing area or it may include a rock socket. Trump Tower in Chicago is a 92-story building supported on 230 caissons bearing on rock and hardpan. The core caissons have a record-breaking 500-ksf (23.95 MPa) bearing capacity with rock sockets up to 10 feet (3,048 mm) in diameter.

Within Section 1810 provisions are found for seismic design and detailing for various types of deep foundations in SDC C through F. These mostly reflect concerns that significant ground motions can occur in SDC C. Additional requirements are imposed on deep foundation elements in SDC D through F, including a requirement that the upper portion of piles be detailed as special moment-resisting-frame columns, to prevent failure of the piles under severe ground motions. These provisions intend to include the deep foundation element bending and curvatures resulting from horizontal ground movement during an earthquake in the structural design. The reinforcement in the deep element, required to resist the curvature effect, increases ductility of the foundation such that bending or shear failure is precluded.

1810.1.1 Geotechnical investigation. See Section 1803.5.5 for discussion on geotechnical investigations.

1810.1.2 Use of existing deep foundation elements. This section allows the reuse of existing deep foundations where sufficient information is submitted to the building official to demonstrate they are adequate. This introduces flexibility for both the building designer and the building owner to make use of existing materials where it makes sense to do so.

Deep foundations remaining after structures that are demolished should not be used for the support of new loads, unless evidence shows them to be adequate. This is typically because of the lack of soil data or detailed information on the piling material, or unavailability of the pile-driving or pier construction records created during construction of older buildings or structures. As such, the true condition of the deep foundation elements may be unknown and, over time, they may have deteriorated, or their load capacity may have been reduced. Such deep foundation elements may be used, however, if they are load tested or in the case of piles, if they are retracted and redriven to verify their capacities. Only the lowest allowable load capacity as determined by test data or redriving information should be used in the design.

1810.1.3 Deep foundation elements classified as columns. The code requires that deep foundation elements standing unbraced in air, water, or material not capable of providing adequate lateral support shall be designed in accordance with the column formulas of the code. Most soils provide lateral support, although exceptionally loose and unconsolidated fills, liquified sands, and remolded clays are inadequate to provide lateral support. CIP elements can be designed as a "pedestal" under conditions where unsupported height to least horizontal dimension is three or less.

1810.1.4 Special types of deep foundations. Deep foundations such as pile or pier types are basically classified in accordance with the structural material used, such as concrete, steel, or wood. They can also be categorized in accordance with the method of construction or installation. There are many variations of foundation types used in the construction of deep foundations, including some special or proprietary types beyond the scope of the code. Section 1810 includes only those basic pile types commonly used in current construction practice.

Special deep foundation types that are not specifically included in the provisions of the code are not precluded from use, provided that adequate information covering test data, calculations, structural properties, load capacity, and installation procedures is submitted and accepted by the building official.

1810.2 Analysis. Sections 1810.2.1 through 1810.2.5 are related to the analysis methodology for deep foundations and apply to all deep foundations in all locations except where a specific analysis is called for in higher seismic active areas. The discussion of lateral support, stability, settlement, lateral loads, SDC D through F, and group effects discussed below is related to analysis procedures and requirements.

1810.2.1 Lateral support. This section provides needed guidance to the designer and building official on what constitutes adequate lateral support for deep foundations. It specifies that any soil other than fluid soil is allowed to be considered to provide lateral support to prevent buckling of deep foundation elements. Liquefaction causes loss of lateral-bearing capacity with resulting loss of support for deep foundation elements. Loss of lateral support can also occur from the soil strength loss associated with sensitive or quick clays that are sensitive to remolding effects. These clays lose significant strength when remolded, as might occur when a pile foundation is moved through muds by seismic-induced displacements.

Portions of deep foundation elements standing unbraced in air, water, or material not capable of providing adequate lateral support are permitted to be considered laterally supported at a depth of 5 feet (1,524 mm) in stiff soils and at a depth of 10 feet (3,048 mm) in soft soils, as determined by the geotechnical investigation.

1810.2.2 Stability. A group of deep foundation elements such as piles designed to support a common load or to resist horizontal forces must be braced or rigidly tied together to act as a single structural unit that will provide lateral stability in all directions. Deep foundation elements connected by a rigid, reinforced concrete pile cap are deemed to be sufficiently braced to meet the intent of this provision.

This section clarifies that for pile or pier groups to be considered to provide lateral stability, they must meet the radial spacing requirements defined herein. Three or more deep foundation elements are generally used to support a building column load or other isolated, concentrated load. In a three-element group, lateral stability is assured by requiring that the elements are located such that they will not be less than 60 degrees (1.05 radians) apart as measured from the centroid of the group in a radial direction.

For stability of deep foundation elements in a group supporting a wall structure, the elements are braced by a continuous, rigid footing and are alternately staggered in two lines at least 1 foot (305 mm) apart and symmetrically located on each side of the center of gravity of the wall. Other approved deep foundation element arrangements may be used to support walls, provided the elements are adequately braced and lateral stability of the foundation construction is assured.

Exception 1 allows isolated CIP deep foundation elements without lateral bracing where the minimum dimension is 2 feet (610 mm), and analysis demonstrates that the element can support the required loads, including mislocations required by Section 1810.3.1.3, without harmful distortion or instability.

Exception 2 allows one- and two-family dwellings of lightweight construction, such as R-3 buildings, not exceeding two stories above grade plane or 35 feet (10,668 mm) in height, a single row of piles located within the width of the wall.

1810.2.3 Settlement. The purpose of a settlement analysis is to provide the data needed to design a deep foundation system that will maintain the stability and structural integrity of the supported building or structure. The load-bearing stratum of every soil must support the loads transferred through the deep foundation system, as well as the weight of all soil above. The capability of the strata underlying the deep foundation element to support additional loads without detrimental settlement can often be determined by analytical procedures. For example, serious settlements in a pile foundation system, particularly differential settlements, can cause great structural damage to the supported structure and the foundation itself.

Although the settlement analysis of an individual deep foundation element is complex, the analysis of a group of elements is significantly more complicated because of the overlapping soil stresses caused by closely spaced elements. Analytical procedures vary with the type of deep foundation element and especially with the soil conditions. Settlement analysis would generally include cases involving point-bearing piles on rock, and in granular soils and hard clay. It would also involve friction piles in sand and gravel soils, and in clay materials.

Load tests are often used to aid in the analysis. In the case of pile foundations in clay soils, however, there are no practical ways to determine long-term settlement from load tests, and therefore only approximations of settlement may be derived from laboratory tests.

1810.2.4 Lateral loads. Deep foundation moments, shears, and deflections must be based on nonlinear soil-deep foundation element interaction. If using deep foundations in soils with lenses of soft clays, lenses subject to liquefaction, or soils susceptible to strength loss from remolding, the effects of these layers on the curvature and, hence, moment demands on the deep foundations should be investigated. Often these soil-induced curvatures place a much higher moment demand on the deep foundations than would be determined from conventional lateral force analysis. If the ratio of the depth of embedment of the deep foundation element to the element diameter or width is less than or equal to six, the deep foundation element may be assumed to be rigid.

1810.2.4.1 Seismic design categories D through F. In addition to the requirements for SDC C, moments, shears, and deflections must be based on nonlinear soil–pile interaction. If using pile foundations in soils with layers of soft to medium stiff clays, layers subject to liquefaction, or soils susceptible to strength loss from remolding (see Site Class E and F), the effects of these layers on the curvature and, hence, moment demands on the pile must be investigated. Often these soil-induced curvatures place a much higher moment demand on the piles than would be determined from conventional lateral force analysis. In addition, at the interfaces of the layers described above and stiffer layers, plastic hinging may occur. Hence, confinement reinforcement per the concrete special-moment frame provisions must be provided at these interfaces and at the pile-to-cap connection for concrete piers and piles (see ASCE 7 Chapter 12). The curvature capacity requirements are considered to have been met without the analysis outlined in this section under two conditions. One condition is for precast-prestressed concrete piles where the transverse reinforcement detailing complies with ACI 318 Chapter 18. These provisions were developed specifically for precast-prestressed piles to meet the curvature requirements. The other condition is for CIP concrete deep foundation elements that meet a minimum longitudinal reinforcement ratio of 0.005 for the full length of the element and are detailed in accordance with ACI 318 Chapter 18.

1810.2.5 Group effect. Where deep foundation elements are spaced far apart, they are considered as individual elements and analyzed and designed accordingly based on various requirements of the code. As spacing between the elements is reduced, the loads and stresses from one element may affect surrounding deep foundation elements. This is the spacing at which group action must also be considered. The spacing for group action consideration in analysis of lateral loads is a center-to-center spacing of less than eight times the least horizontal dimension of an element, and for axial loads is a center-to-center spacing of less than three times the least horizontal dimension of a deep element. See Figure 1810-1.

The last sentence of this section clarifies that group effects on uplift of deep foundations should be evaluated where element spacing is less than three times their least horizontal dimension. Section 1810.3.3.1.6 is not specific in regard to the spacing that necessitates evaluation of group effects for uplift. Cross referencing Section 1810.3.3.1.6, uplift capacity of grouped deep foundation elements, helps further clarify the intent.

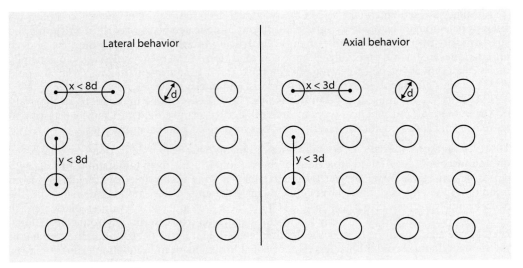

Figure 1810-1 **Group effect.**

Determination of the proper spacing of a pile group in relation to the type of pile foundation employed and the soil conditions encountered is a matter of design. Pile spacing must be such that the loads transferred to the load-bearing strata do not exceed the safe load-bearing values of the supporting strata as determined by test borings, field load tests, or other approved methods.

1810.3 Design and detailing. Sections 1810.3.1 through 1810.3.13 are related to the design and detailing of deep foundations and address general requirements as well as special seismic requirements where applicable. The subsections of Section 1810.3 are listed in Table 1810-1.

Driven piles and helical piles are required to be designed and manufactured in accordance with accepted engineering practice to consider handling, driving, and service conditions and are addressed briefly in Sections 1810.3.1.4 and 1810.3.1.5.

1810.3.1.3 Mislocation. Because of subsurface obstructions or other reasons, it is sometimes necessary to offset deep foundation elements a small distance from their intended locations so that they are not driven out of position. In such cases, the load distribution in a group of deep elements may be changed from the design requirements and cause some of them to be overloaded. To prevent major problems from displacements, the foundation system and the superstructure are required to be designed to resist the effects of deep foundation mislocation of at least 3 inches (76 mm). This section requires that the maximum compressive load on any pile caused by mislocation not exceed 110 percent of the allowable design load. Deep elements such as piles exceeding this limitation must be extracted and redriven in the proper location or other approved remedies applied, such as installing additional piles to balance the group.

1810.3.1.4 Driven piles. Except for steel H-piles, driven piles covered in Chapter 18 are the displacement type. That is, as the pile is driven, a volume of soil is displaced by the pile volume, resulting in compaction of the surrounding soils. Piles must be designed to resist driving and handling stresses in addition to anticipated service loads. In long piles, tensile stresses resulting from driving may govern the design. In shorter piles, the handling loads may dominate. Driven uncased piles are displacement piles and are constructed by driving a temporary casing, removing the soils from the casing, and placing concrete in the hole as the casing is removed.

The casing is driven with a closed end, thereby displacing and compacting adjacent soil during driving. The casing is kept closed either by a detachable tip, which is left in place when the casing is withdrawn, or by a mandrel that closes off the casing tip during driving.

1810.3.1.5 Helical piles. "Helical pile" is a manufactured steel deep foundation element consisting of a central shaft and one or more helical bearing plates. These are piles that are installed by rotating them into the ground very much like screwing in a rotating plate. Additional provisions for helical piles are found in IBC Sections 202 (definition), 1810.3.3.1.9 (allowable axial load), 1810.3.5.3.5 (dimensions), 1810.4.11 (installation), and 1705.9 (special inspection).

1810.3.1.6 Casings. Casings are used in many deep foundation systems. The steel-cased pile is the most widely used type of CIP concrete pile. This pile type is characterized by a thin steel shell and is a displacement pile. This pile type consists of a closed-end light-gauge steel shell or a thin-walled pipe driven into the soil and left permanently in place, reinforced when required for uplift, lateral bending, or seismic-induced curvatures, and filled with concrete. The shell or pipe is usually driven with a removable mandrel. The shell is either a constant section or a tapered shape. Steel-encased piles are generally friction piles. The steel shell must have sufficient strength to remain watertight and not collapse from ground pressure when the mandrel is removed.

1810.3.2 Materials. Deep foundations of various materials such as concrete, steel, and timber are covered in this section including seismic hook requirements. Other relevant issues such as protection of materials and allowable stresses is also discussed.

1810.3.2.1 Concrete. Concrete used for the bulb type of pile must have a zero slump to be stiff enough to be compacted by the drop weight. The maximum sized aggregate allowed is ¾ inch to allow proper compaction and prevent segregation. To prevent the pull-out of the hoops, spirals, and ties in higher seismic areas of SDC C through F, seismic hoops as defined in ACI 318 should be used at the end of such hoops, spirals, and ties.

1810.3.2.2 Prestressing steel. Prestressing steel used in deep foundations must comply with ASTM A416 *Standard Specification for Low-Relaxation, Seven-Wire Steel Strand, for Prestressed Concrete*. Table 1808.8.1 lists when higher strength concrete is used in prestressed piles to reduce volume changes, which reduce prestress losses, and to provide a more dense concrete to reduce cover requirements.

1810.3.2.3 Steel. Structural steel H-piles, sheet piling, steel pipe piles, and fully welded steel piles shall conform to the appropriate and applicable ASTM standard referenced in this section of the code. H-piles are typically available in ASTM A36 and A572 steel. Pipe piles are fabricated from ASTM A252 or A283 plate. ASTM A252 is a specification specifically for welded pipe piles. Piles using ASTM A690 and A992 are used in common practice.

1810.3.2.4 Timber. Timber piles, although not having the high load capacities of steel or concrete piles, are the most commonly used type of pile, mainly because of their availability and ease of handling. Timber piles are shaped from tree trunks and are tapered because of the natural taper of the trunk. Round timber piles are generally made from Southern Pine in lengths up to 80 feet (24.38 m) or Pacific Coast Douglas Fir in lengths up to 120 feet (36.58 m). Other species that are used are red oak and red pine in lengths up to 60 feet (18.29 m).

Untreated timber piles that are embedded permanently below the ground-water level (fresh water only, not brackish or marine conditions) may last indefinitely. If embedded above the water table, the piles are subject to decay, and if above ground are also subject to insect attack. Hence, piles should be preservative treated.

Timber piles may be either end-bearing or friction piles. ASTM D25 *Specification for Round Timber Piles* sets forth the minimum circumference at the butt and the tip based on pile taper, as well as quality of the wood and tolerances on straightness, knots, twist of grain, and other requirements.

Piles must be preservative-treated to prevent decay and insect attack. The IBC references AWPA U1 (Commodity Specification E, Use Category 4C) for round timber piles and AWPA U1 (Commodity Specification A, Use Category 4B) for sawn timber piles. See also discussion in Section 1809.12 for timber footings.

1810.3.2.5 Protection of materials. Unless properly protected, deep foundation elements may deteriorate because of biological, chemical, or physical actions caused by particular conditions that exist or that may later develop at the site. The durability of deep foundation elements will be long lasting if care is taken in the selection and protection of element materials.

Some of the problems associated with deep foundation element durability are:

- Untreated timber piles may be successfully used if they are entirely embedded in earth and their butts (cutoffs) are below the lowest ground-water level or are submerged in fresh water. The risk, however, is in situations where unexpected lowering of the water table occurs and exposes the upper parts of piles to decay and insect attack. Such conditions may occur where the water table is significantly lowered by pumping or deep drainage. There is also the remote possibility that wood piles will be damaged by the percolation of ground water heavily charged with alkali or acids.

- Wood piles extending above the water table or exposed to air or saltwater are subject to decay and attack by insects and marine borers. The piles need to be treated with preservatives conforming to the requirements of AWPA U1.

- Steel piles that are driven and embedded entirely in undisturbed soil are generally not significantly affected by corrosion caused by oxidation, regardless of soil types or soil properties. The reason for this is that undisturbed soil is so deficient in oxygen at levels only a few feet below the ground line or below the water table that progressive corrosion is inhibited. However, where upper portions of steel piles protrude above ground into the air, where piles are placed in corrosive soils, or where ground water contains deleterious substances from sources such as coal piles, alkali soils, active cinder fills, chemical wastes from manufacturing operations, or other sources of pollutants, the steel may be subject to corrosive action. Under such conditions, steel piles may be protected by a concrete encasement or a suitable coating, extending from a level slightly above ground to a depth below the layer of disturbed earth. For piles above ground level that are exposed to air and subject to rusting, the steel should be protected by being painted, as any other type of structural steel construction would be protected.

- Steel piles installed in saltwater or exposed to a saltwater environment are subject to corrosion, and therefore should be protected with approved coatings or encased in concrete.

- Concrete deep elements, plain or reinforced, that are entirely embedded in undisturbed earth are generally considered permanent installations. The level of the water table does not normally affect the durability of concrete deep elements. Ground water that readily flows through either granular materials or disturbed soil and contains deleterious substances can have deteriorating effects. Concrete piles embedded in impervious clay materials will not generally suffer from ground water containing harmful substances. The primary deleterious substances that attack concrete are acids and sulfates. In the case of acids, it is best to use an alternative pile material if the acid attack is potentially destructive as coating piles may be ineffective because of soil abrasion during driving. Concrete can be attacked, however, by exposure to soils with high sulfate content. In the case of high alkaline soils with sulfate salts, Type V portland cement may be used. Where the exposures are only moderate, a Type II portland cement will usually be adequate. If the piles are in a marine environment, Type II or V portland cement is also indicated to provide the necessary sulfate resistance.

Deep foundation elements installed in saltwater, such as for buildings or other structures in waterfront construction, are subject to chemical action on concrete materials coming from polluted waters, frost action on porous concrete, spalling of concrete, and rusting of steel reinforcement. Spalling may become particularly serious under tidal conditions where alternate wetting and drying occurs coupled with cycles of freezing and thawing. Spalling can be minimized or prevented by providing additional cover over the reinforcement; by the use of rich, dense concrete; by air entrainment and suitable concrete admixtures; and, in the case of precast piles, by careful handling to minimize stresses and avoid cracking during placement. See discussion of Chapter 19 and ACI 318 for more detailed information on concrete quality and concrete materials.

1810.3.2.6 Allowable stresses. Allowable stresses for various types of deep foundations have been summarized in Table 1810.3.2.6. The allowable stress limits are set to provide an adequate margin of safety.

For precast-prestressed concrete, the term $0.33 f_c'$ is the same as conventionally reinforced piles. Because prestressing places additional compressive stresses on the pile, this stress must be subtracted from the allowable compressive stress; hence, the subtractive term $-0.27 f_{pc}$. The term f_{pc} is the effective prestress on the gross area, which is the prestressing force remaining after losses have occurred.

For H-piles, the stresses allowed consider stability. Tests have shown that for H-piles driven to refusal in rock through soils that provide full lateral support, the stresses at failure can approach the yield stress of the material. Hence, the higher tabulated stresses are allowed if a soil investigation and load tests are performed as indicated in Section 1810.3.2.8.

For concrete CIP without a permanent steel casing, pile capacity must be based on concrete strength alone without consideration of soil capacity, and the area of the internal cross section of the pile, that is, casing inside diameter. The allowable stresses for drilled uncased piles are the same as CIP piles for which holes are formed by machine drilling with auger or bucket-type drills, with or without temporary casing. Concrete is placed by conventional methods including tremies or funnel hoppers. In augered piles, concrete is injected through the hollow stem auger as the auger is withdrawn. Reinforcement, if placed without lateral ties, is also placed through ducts in the hollow stem auger. Concrete drilled or augered CIP uncased piles have an allowable stress limit of 30 percent of the 28-day specified compressive strength (f_c'), and an allowable compressive stress in the reinforcement of 40 percent of the yield strength of the steel [≤ 30,000 psi (206 MPa)]. These limits are consistent with ASCE *Standard Guidelines for the Design and Installation of Pile Foundations* (ASCE 20). For steel-cased deep concrete foundation elements, the allowable stress is $0.33 f_c'$ as for other concrete piles because the steel shell is too thin to act as a composite pile but does act as confinement reinforcement. The allowable stress is $0.40 f_c'$ if the shell meets the confinement conditions in Section 1810.3.2.7 or Section 1810.3.5.3.4.

1810.3.2.7 Increased allowable compressive stress for cased mandrell-driven CIP elements. The conditions discussed in Section 1810.3.2.7 and Table 1810.3.2.6 where the allowable concrete compressive stress is allowed to be increased consist of six conditions. Note that the shell thickness is not considered load carrying, the shell must be seamless or spirally welded, the yield strength of steel normally used for the casings is 30 ksi (206.8 MPa), and the maximum diameter is set so that the volumetric ratio of shell to concrete is sufficient to provide confinement for the concrete.

1810.3.2.8 Justification of higher allowable stresses. In those sections of Chapter 18 that specifically deal with the types of elements most commonly used in the construction of deep foundations, there are limitations placed on the stresses that can be used in the deep foundation element design. In most cases, the allowable stresses are stated as a percentage of some limiting strength property of the deep foundations material. For example, in the case of piles made of steel materials, the allowable stresses are prescribed as a percentage of the yield strength of the several grades of steel that can be used for pile construction. For concrete, the allowable stress is stated as a percentage of compressive strength of the material. The allowable stresses for timber piles are based on tabulations of already reduced stresses. The reduced stresses are based on the strength values of different species of wood and reductions in strength caused by preservative treatment.

The allowable design stresses designated in Chapter 18 for each of the different types of deep foundations are intended to provide a factor of safety against the dynamic forces of deep foundation element driving that may cause damage to the element, and to avoid overstresses in the element under the design loads and other loads that may be induced by subsoil conditions.

This section allows the use of higher tabulated allowable stresses shown in Table 1810.3.2.6 when evidence supporting the values is submitted and approved by the building official. The data submitted to the building official should include analytical evaluations and findings from a foundation investigation as specified in Section 1803, and the results of load tests performed in accordance with the requirements of Section 1810.3.3.1.2. The technical data and the recommendation for the use of higher stress values must come from a registered design professional that is knowledgeable in soil mechanics and experienced in the design of deep foundations. This registered design professional must supervise the deep foundation design work and witness the installation of the deep foundation so as to certify to the building official that the construction satisfies the design criteria. In any case, the use of greater design stresses should not result in design loads that are larger than one-half of the ultimate axial load capacity (see Section 1810.3.3.1.2).

1810.3.3 Determination of allowable loads. The IBC specifies that the determination of allowable loads shall be based on one of three methods:

1. An approved driving formula
2. Load tests
3. Foundation analysis

In most cases, the allowable loads will be determined by a combination of Items 2 and 3. However, there may be circumstances where the soil conditions, such as granular soils, and the type of deep foundation selected are such that the use of an approved dynamic deep foundation driving formula can be an aid to a qualified practitioner in establishing reasonable but safe allowable loads for the foundation system.

There are two general considerations for determining capacity as required for the design and installation of deep foundations. The first consideration involves the determination of the underlying soil or rock characteristics. The second is the application of approved driving formulas, load tests, or accepted methods of analysis to determine the pile capacities required to resist the applied axial and lateral loads, as well as to provide the basis for the proper selection of pile-driving equipment.

Allowable axial load determination is further addressed in Sections 1810.3.3.1.1 through 1810.3.3.1.9.

1810.3.3.1.1 Driving criteria. Deep foundation elements must be of a size, strength, and stiffness capable of resisting without damage:

- Crushing caused by impact forces during driving.
- Bending stresses during handling.
- Tension from uplift forces or from rebound during driving.

- Bending stresses caused by horizontal forces during driving.
- Bending stresses caused by deep element curvatures.

Additionally, the deep element must be capable of transmitting dynamic driving forces to mobilize the required ultimate deep element capacity within the soil without severe elastic energy losses. Driveability depends on the deep element stiffness, which is a function of deep element length, cross-sectional area, and modulus of elasticity. Yield strength does not affect stiffness. Thus, caution should be observed in the use of high-yield-strength steels for high loads on smaller cross sections requiring high dynamic driving energy. For allowable loads greater than 40 tons (356 kN), a wave equation method of analysis reflects deep element stiffness or driveability. The selection of deep foundation types and dimensional requirements for driveability is a function of soil characteristics.

This section limits the allowable compression load on a deep foundation element as established by an approved driving formula to a maximum of 40 tons (356 kN). Generally, the use of pile-driving formulas to determine pile capacity should be avoided except, perhaps, in cases involving small jobs where the piles are to be driven in well-drained granular soils and load testing cannot be economically justified.

1810.3.3.1.2 Load tests. This section specifies the standards to be used to load test deep foundation elements, where higher compressive loads than allowed in other sections of the code are exceeded or where CIP deep foundation elements have an enlarged base formed either by compacting concrete or by driving a precast base. See the discussion under Section 1810.3.3.1.3 for test evaluation methods.

Many standards, including ASTM D4945 *Test Method for High-Strain Dynamic Testing of Deep Foundations*, indicate that a dynamic test may not be sufficient without a static test (ASTM D1143 *Standard Test Methods for Deep Foundation Elements Under Static Axial Compressive Load*) to calibrate the results, but leave it up to the registered design professional to decide if the dynamic test is sufficient. Other standards require a static load test to calibrate the dynamic test.

The safest method for determining deep foundation element capacity is by load test. The load-bearing capacity of enlarged base piles is specifically required to be determined from load tests. A load test should be conducted wherever feasible and used where the deep foundation element capacity is intended to exceed 40 tons (356 kN) per element (see Section 1810.3.3.1.1). Deep elements for testing are to be of the same type and size as intended for use in the permanent foundation and installed with the same equipment, by the same procedure, and in the same soils intended or specified for the work.

Load tests are to be conducted in accordance with the requirements of ASTM D1143, which covers procedures for testing vertical or batter foundation piles, individually or in groups, to determine the ultimate pile load (pile capacity) and whether the pile or pile group is capable of supporting the loads without excessive or continuous settlement. Recognition, however, must be given to the fact that load-settlement characteristics and pile capacity determinations are based on data derived at the time and under the conditions of the test. The long-term performance of a pile or group of piles supporting actual loads may produce behaviors that are different than those indicated by load test results. Judgment based on experience must be used to predict pile capacity and expected behavior.

The load-bearing capacity of all deep foundation elements, except those seated on rock, does not reach the ultimate load until after a period of rest. The results of load tests cannot be deemed accurate or reliable unless there is an allowance for a period of adjustment. For piles driven in permeable soils such as coarse-grained sand and gravel, the waiting period may be as little as two or three days. For test piles driven in silt, clay, or fine sand, the waiting period may be 30 days or longer. The waiting period may be determined by testing (i.e., by redriving piles) or from previous experience.

This section also requires that at least one deep foundation element be tested in each area of uniform subsoil conditions. The statement should not be misconstrued to mean that the area of test is to have only one uniform stratum of subsurface material, but rather that the soil profile, which may consist of several layers (strata) of different materials, must represent a substantially unchanging cross section in each area to be tested.

The allowable deep foundation element load to be used for design purposes is not to be more than one-half of the ultimate deep element capacity. The rate of penetration of permanent deep foundation elements must be equal to or less than that of the test element(s).

All production deep foundation elements should be of the same type, size, and approximate length as the prototype test elements, as well as installed with the same or comparable equipment and methods. They should also be installed in soils similar to those of the test element.

1810.3.3.1.3 Load test evaluation methods. Three specific methods are given that are acceptable for evaluating deep foundation load tests. Other methods are permitted at the discretion of the building official.

1810.3.3.1.4 Allowable shaft resistance. Resistance that is due to skin friction is limited to one-sixth of the bearing value of the soil material at minimum depth as set forth in Table 1806.2, up to a maximum of 500 psf (24 kPa) unless a greater value is permitted by the building official based on recommendations of an approved geotechnical investigation or a greater value is substantiated by load test methods described in Section 1810.3.3.1.2.

1810.3.3.1.5 Uplift capacity of a single deep foundation element. This section gives both the designer and building official needed guidance on criteria to use for design of a single deep foundation element for uplift. The IBC requires an approved method of analysis with a safety factor of 3 or a test in accordance with ASTM D3689 *Test Methods for Deep Foundations under Static Axial Tensile Load*. The maximum allowable uplift load cannot exceed the ultimate load capacity determined by the methods described in Section 1810.3.3.1.2 divided by a factor of safety of two.

When deep foundation elements are designed to withstand uplift forces, they act in tension and are actually friction elements. The amount of tension that can be developed not only depends on the strength properties of the element, but also on the frictional or cohesive properties of the soil. The uplift or tensile resistance of a deep element is not necessarily a function of its load-bearing capacity under compressive load. For example, the tensile resistance of a friction pile in clay will usually be about the same value as its load-bearing capacity, as the skin friction developed in such soils is very large. In contrast, a friction pile in sand or in other granular materials will develop a tensile resistance considerably less than its load-bearing capacity.

Where the properties of the soil are known, the ultimate uplift resistance value of a pile can be determined by approved analytical methods. This section requires that where the ultimate tensile value is determined by analysis, a safety factor of 3 must be applied to establish the allowable uplift load of the deep foundation element.

The best way to determine the response of a vertical or batter pile to a static tensile load (uplift force) applied axially to the pile is by applying an extraction test in accordance with the requirements of ASTM D3689. The maximum allowable uplift load is not to be more than one-half of the total test load. This section of the code gives a limitation on the upward movement of the pile in compliance with the provisions of the ASTM D3689 test method. The measurements of pile movement in the standard test procedure, however, are time-dependent incremental measurements and should be adhered to in determining allowable pile load.

To be effective in resisting uplift forces as tension members of a foundation system, deep foundation elements must be well anchored into the cap by adequate connection devices. In turn, the cap must be designed for the uplift stresses. Deep foundation elements must also be designed to take the tensile stresses imposed by the uplift forces. For example, concrete piles must be reinforced with longitudinal steel to take the full net uplift. Special consideration needs to be given in the design of pile splices that are intended to act in tension. When design uplift is due to wind or seismic loading, the factor of safety for the analytical method requires a factor of safety of 2 while the load test method requires a factor of safety of 1.5.

1810.3.3.1.6 Allowable uplift load of grouped deep foundation elements. The allowable uplift load on a group of deep foundation elements may need to be reduced from the value obtained on a single element as described in Section 1810.3.3.1.5 of the code and in compliance with comprehensive analytical methods. The capacity of deep foundation groups is limited to two-thirds of the weight of the group and the soil contained in the group plus two-thirds of the ultimate shear resistance along the soil block. This is consistent with requirements in other sections in the code on uplift and overturning, where the dead load resistance is limited to two-thirds of the weight. The IBC allows the use of two-thirds of the effective weight of the pile group, two-thirds of the weight of the soil contained within a block defined by the perimeter of the group and the length of the piles, plus two-thirds of the ultimate shear resistance along the soil block. Where the center-to-center spacing of deep foundation elements is at least three times the least horizontal dimension of the largest single element, the allowable working uplift load for the group must be calculated by an approved method of analysis.

1810.3.3.1.7 Load-bearing capacity. The load-bearing capacity of a deep foundation element is determined as a deep element–soil system. For example, the load-bearing capacity of a single pile is the function of either the structural strength of the pile or the supporting strength of the soil. The load-bearing capacity of the deep foundation element is controlled by the smaller value obtained in the two considerations. The load-bearing capacity of a deep element group may be greater than, equal to, or less than the capacity of a single element multiplied by the number of elements in the group, depending on deep element spacing and soil conditions.

Because the supporting strength of the soil generally controls the load-bearing capacity of a deep foundation element, this section requires that the ultimate load-bearing capacity of an individual element or a group of elements be at least twice the design load capacity of the supporting load-bearing strata.

Sometimes, weaker layers of soil underlie the soil load-bearing strata supporting a pile foundation and may cause damaging settlements. Under such subsurface conditions, it must be determined by an approved method of analysis that the safety factor has not been reduced to less than 2. Otherwise, the piles are to be driven to deeper load-bearing soils to obtain adequate and safe support, or the design capacity is to be reduced and the number of piles increased.

1810.3.3.1.8 Bent deep foundation elements. Deep foundation elements that are discovered to have sharp or sweeping bends because of obstructions encountered during the driving operations or for any other cause are to be analyzed by an approved method, or a representative deep element is to be load tested to determine its load-carrying capacity. Otherwise, the deep foundation elements could be used at some reduced capacity as determined by test or analysis; or, if necessary, they can be abandoned and replaced.

1810.3.3.1.9 Helical piles. Helical piles are comprised of a steel shaft and steel bearing plates, also known as helices or cutters that are driven into the ground. The IBC requires that six specific items be checked in order to determine the allowable capacity of the helical piles. The designer must first determine the "base capacity" of the helical pile which considers the bearing capacity of the combined bearing plates and the skin friction against the steel shaft. Figure 1810-2 depicts how the applied load on a helical pile is resisted by the skin friction on the shaft in addition to the combined bearing capacity of the helical plates.

In addition to the base capacity of the pile, the ultimate capacity must be checked in relation to load test data and historical data based on installation torque. The ultimate capacity of the steel shaft, shaft couplings, and sum of the steel bearing plates must also be considered.

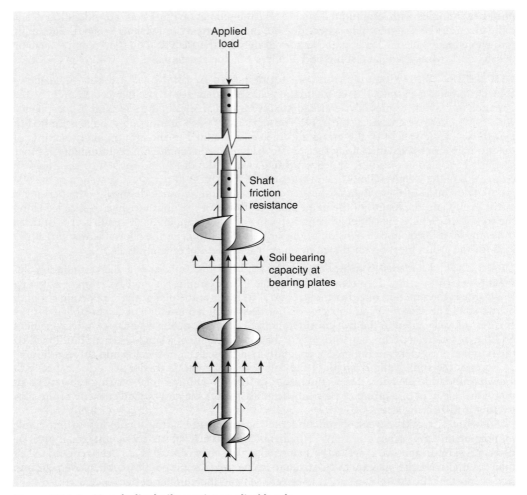

Figure 1810-2 **How helical piles resist applied loads.**

In the end, the least value controls the ultimate axial capacity of the pile and the allowable load to the piles must not exceed one-half of the ultimate capacity.

1810.3.3.2 Allowable lateral load. Because of wind loads, unbalanced building loads, earth pressures, and seismic loads, it is inevitable that individual deep foundation elements or groups of vertical elements supporting buildings or other structures will be subjected to lateral forces. The distribution of these lateral forces to the deep elements largely depends on how the loads are carried down through the structural framing system and transferred through the supporting foundation to the deep foundation elements. The amount of lateral load that can be taken by the deep foundation element is a function of (1) the type of deep foundation element used; (2) the soil characteristics, particularly in the upper 10 to 30 feet (3,048 to 9,144 mm) of the deep foundation element; (3) the embedment of the deep foundation element head (fixity); (4) the magnitude of the axial compressive load on the deep foundation element;

(5) the nature of the lateral forces; and (6) the amount of horizontal deep foundation element movement deemed acceptable.

The degree of fixity of the deep foundation element head is an important design consideration under very high lateral loading unless some other method, such as the use of batter piles, is employed to resist lateral loads. The fixing of the deep foundation element head against rotation reduces the lateral deflection. In general, pile butts are embedded 3 to 4 inches (76 to 102 mm) into the pile cap (see Section 1810.3.11) with no ties to the cap. These pile heads are neither fixed nor free, but somewhere in between. Such construction is satisfactory for many loading conditions, but not for high seismic loads.

The magnitude of friction developed between the surfaces of two structural elements in contact with each other is a function of the weight or load applied. The larger the weight, the greater the frictional resistance developed. In the design of deep foundation elements, frictional resistance between the soil and the bottom of the deep element caps (footings) should not be relied on to provide lateral restraint, because the vertical loads are transmitted through the deep foundation elements to the supporting soil below and to the ground immediately under the deep element caps. Only the weights of the caps can supply some frictional resistance insofar as such footings are constructed by placing fresh concrete on the soil, thus providing a positive contact. The weight of the caps in comparison to the magnitude of loads and lateral forces transmitted to the deep foundation elements is nominal and not significant from a structural design standpoint. Also, in rare occurrences, soils have been known to settle under caps, leaving open spaces and thus eliminating the development of any frictional restraint.

Where vertical deep elements are subjected to lateral forces exceeding acceptable limitations, the use of batter piles may be required. Lateral forces on many structures are also resisted by the embedded foundation walls and the sides of the deep foundation element caps.

The allowable lateral-load capacity of a single deep foundation element or group of such elements is to be determined either by approved analytical methods or by load tests. Group effects are to be evaluated where required by Section 1810.2.5. Load tests are to be conducted to produce lateral forces that are twice the proposed design load; however, in no case is the allowable deep foundation element load to exceed one-half of the test load, which produces a gross lateral element movement of 1 inch (25.4 mm) as measured at the ground surface or the top of foundation element, whichever is lower. This criterion can be exceeded if it can be shown that the predicted lateral movement will not cause any harmful distortion of or instability in the structure and that no element will be loaded beyond its capacity.

1810.3.4 Subsiding soils or strata. Where deep foundation elements are driven through subsiding soils or strata and derive their support from underlying firmer materials, the subsiding soils or strata cause an additional load to the deep foundation elements through so-called negative friction. This negative friction is actually a downward friction force on the deep foundation elements, which increases the axial load on such elements. The code permits an increase in the allowable stress on the deep foundation elements if an analysis of the geotechnical investigation indicates that the increase is justified.

1810.3.5 Dimensions of deep foundation elements. Deep foundation elements must have minimum dimensions as described in Sections 1810.3.5.1 through 1810.3.5.3 for precast, CIP cased, and CIP uncased deep foundations.

1810.3.5.1 Precast. Eight inches (203 mm) is the minimum practical lateral dimension to accommodate reinforcement.

1810.3.5.2 Cast-in-place or grouted-in-place. Eight inches (203 mm) is the minimum practical nominal outside diameter to accommodate reinforcement for cased CIP deep foundation elements. For uncased CIP deep foundation elements, the minimum 12-inch (305 mm) diameter is for inspection purposes. The length-to-diameter ratio limit of 30 is based on construction and stability considerations but may be exceeded if a registered design professional supervises design and installation.

1810.3.5.2.3 Micropiles. A micropile is defined as a bored, grouted-in-place deep foundation element that develops its load-carrying capacity by means of a bond zone in soil, bedrock, or a combination of soil and bedrock. The maximum outside diameter of a micropile is 12 inches (305 mm).

1810.3.5.3 Steel. Steel deep foundation elements are generally H-piles, piles fabricated from welded plates, sheet piling, steel pipes and tubes, and helical piles.

1810.3.5.3.1 Structural steel H-piles. H-piles are usually used as deep end-bearing piles because they are essentially nondisplacement-type piles that can readily penetrate solid strata to reach rock or other suitable hard-bearing strata such as dense gravels. Ideally, steel H-piles are driven to hard or medium hard rock.

H-piles are proportioned to withstand the impact stresses from hard driving. The flange and web thicknesses are usually equal. The flange widths are proportioned such that the section modulus, S_y, in the weak axis is approximately one-third of the strong axis section modulus, S_x. H-piles in SDC D-F must be designed and detailed in accordance with AISC 341.

1810.3.5.3.2 Fully welded steel piles fabricated from plates. Although they are in a separate section, the requirements for steel piles fabricated from welded plates are the same as for steel H-piles except for the high seismic detailing requirements.

1810.3.5.3.3 Structural steel sheet piling. Provisions for structural steel sheet piling require that the profiles conform to manufacturer's specifications and the general requirements in ASTM A6 *Standard Specification for General Requirements for Rolled Structural Steel Bars, Plates, Shapes and Sheet Piling*.

1810.3.5.3.4 Steel pipes and tubes. Driven pipe piles are displacement-type piles if driven closed, and are nondisplacement-type piles if driven open. Pipe piles are made of seamless or welded pipes and are frequently filled with concrete after driving. Pipe piles conforming to ASTM A252 *Specification for Welded and Seamless Steel Pipe Piles* are used in both friction and end-bearing applications. Open-ended pipe piles are generally used when the geotechnical investigation shows rock or a suitable end-bearing stratum close to the ground surface, especially if the loads to be supported are large. The pipe is driven to bearing, the soils forced into the pipe during driving are cleaned out, and the pipe is filled with concrete. Closed-end piles are generally used as friction piles when a suitable bearing stratum is not available at suitable depths. There are several proprietary closed-end pipe piles available.

When steel pipes are driven open ended, minimum thickness to diameter is related to hammer energy by requiring a minimum area per kip-foot of energy. Open-ended pipes require a minimum area of 0.34 square inch (219 mm^2) to resist each 1,000 ft-lb (1,356 Nm). This requirement equates to a wall thickness of 0.27 inch (6.86 mm) for a 10-inch (254 mm) pipe driven with a hammer energy of 25 kip-feet (33.9 kN-m). Note that if the wall thickness is less than 0.179 inch (4.55 mm), a driving shoe is required to prevent local buckling at the tip from hard driving, regardless of diameter or hammer energy. The 0.179-inch (4.55 mm) thickness is consistent with the most common minimum thickness for closed-end pipe piles.

Concrete-filled steel pipes or tubes in structures assigned to SDC C, D, E, or F shall have a wall thickness of not less than 3/16 inch (4.76 mm). The pipe or tube casing for socketed drilled shafts shall have a nominal outside diameter of not less than 18 inches (457 mm) and a wall thickness of not less than ⅜ inch (9.5 mm). The pipe is a welded or seamless pipe conforming to ASTM A252. ASCE 20 as well as the recommendations of the Driven Pile Committee of the Deep Foundations Institute list 0.179 inches (4.55 mm) as the minimum wall thickness.

Concrete-filled steel piles are either seamless or welded pipe, or closed-end tubular piles with either straight or tapered sections that are driven into the soil. The piles may be installed as either friction or end-bearing piles. This pile type is characterized by a steel shell that is thicker than the thin shell used in some steel-cased piles, hence both the concrete and steel shell are assumed to carry load compositely. If driven open ended, the earth core is removed from the shell prior to concreting. The shell may be driven with an internal mandrel in which case the minimum wall thickness is 1/10 inch (2.54 mm).

1810.3.5.3.5 Helical piles. See Section 1810.3.3.1.9.

1810.3.6 Splices. This section specifies the requirements for splicing of deep foundation elements. The 50-percent requirement provides more strength where the bending moments are low, insofar as it is based on the capacity of the deep foundation element, not the design loads. Splices of deep foundation elements of different materials or types are required to develop the full compressive strength and not less than 50 percent of the tension and bending strength of the weaker section. An exception for buildings in SDC A or B is provided, noting that splices do not need to comply with the 50 percent provision where justified by supporting data.

Although it is physically and economically better to drive piles in one piece, site conditions sometimes necessitate that piles be driven in spliced sections. For example, when the soil or rock-bearing stratum is located so deep below the ground that the leads on the driving equipment will not receive full-length piles, it becomes necessary to install the piles in sections or, where possible, to take up the extra length by setting the tip in a preexcavated hole (see discussion, Section 1810.4.4). When piles are installed in areas such as existing buildings with restricted headroom, they are also required to be placed in spliced sections. There are a number of other reasons for field-splicing piles, such as restrictions on shipping lengths or the use of composite piles.

This provision requires that splices be constructed to provide and maintain true alignment and position of the deep foundation element sections during installation. Splices must be of sufficient strength to transmit the vertical and lateral loads on the deep foundation elements, as well as to resist the bending stresses that may occur at splice locations during the driving operations and under long-term service loads. Splices are to develop at least 50 percent of the value of the deep foundation element in bending. Consideration should be given to the design of splices at locations where the deep foundation elements may be subject to tension. Splices that occur in the upper 10 feet (3,048 mm) of pile embedment are to be designed to resist the bending moments and shears at the allowable stress levels of the pile material, based on an assumed pile load eccentricity of 3 inches (76 mm), unless the pile is properly braced. Proper bracing of a spliced pile is deemed to exist if stability of the pile group is furnished in accordance with the provisions of Section 18.10.2.2, provided that other piles in the group do not have splices in the upper 10 feet (3,048 mm) of their embedded length.

There are different methods employed in splicing deep foundation elements depending on the materials used in the deep element construction. For example, timber piles are spliced by one of two commonly used methods. The first method uses a pipe sleeve with a length of about four to five times the diameter of the pile. The butting ends of the pile are sawn square for full contact of the two pile sections, and the spliced portions of the timber pile are trimmed smoothly around their periphery to fit tightly into the pipe sleeve. The second splicing method involves the use of steel straps and bolts. The butting ends of the pile sections are sawn square for full contact and proper alignment, and the four sides are planed flat to receive the splicing straps. This type of splicing can resist some uplift forces.

Splicing of precast concrete piles usually occurs at the head portions of the piles. After the piles are driven to their required depth, pile heads are cut off or spliced to the desired elevation for proper embedment in the concrete pile caps. Any portion of the pile that is cracked or shattered by the driving operations or cutting off of pile heads should be removed and spliced with fresh concrete. To cut off a precast concrete pile section, a deep groove is chiseled around the pile exposing the reinforcing bars, which are then cut off (by torch) to desired heights or extensions. The pile section above the groove is snapped off (by crane) and a new pile section is freshly cast to tie in with the precast pile.

Steel H-piles are spliced in the same manner as steel columns, usually by welding the sections together. Welded splices may be welded-plate or bar splices, butt-welded splices, special welded splice fittings, or a combination of these. Spliced materials should be kept on the inner faces of the H-pile sections to avoid forcing a hole in the ground larger than the pile, causing at least a temporary loss in frictional value and lateral support that might result in excessive bending stresses.

Steel pipe piles may be spliced by butt welding, sometimes using straps to guide the sections and to provide more strength to the welded joint. Another method is to use inside sleeves having a driving fit, with a flange extending between the pipe sections. By applying bituminous cement or compound on the outside of the ring before driving, a water-tight joint is obtained.

1810.3.6.1 Seismic design categories C through F. Splices of deep foundation elements in SDC C through F are addressed in this section and require that the splices develop the lesser of two criteria: (1) the nominal strength of the deep foundation element and (2) the axial and shear forces and moments from the seismic load combinations including overstrength factor of ASCE 7, Sections 2.3.6 or 2.4.5. For a discussion of overstrength factor, see Section 1810.3.11.2.

1810.3.7 Top of element detailing at cutoffs. The requirements of this section are intended to account for conditions where a deep foundation element encounters refusal at a shallower depth than anticipated and a portion of the deep element is cut off. It is imperative that the required reinforcement be provided at the top of the deep foundation element when the excess deep element length is cut off.

1810.3.8 Precast concrete piles. Precast concrete piles are manufactured as conventionally reinforced concrete or as prestressed concrete. Both types can be formed by bed casting, centrifugal casting, slipforming, or extrusion methods. Piles are usually square, octagonal, or round, and either solid or hollow. Precast piles, which may be either friction or end-bearing piles, are of the displacement type and are driven into place.

Design and detailing requirements for precast piles are referenced to ACI 318 with two exceptions. These exceptions allow the use of the ASCE 7 overstrength factor by applying the factor to the load equations with seismic loads included to replace a design following ACI 318. These provisions were not included in ACI 318-19 and thus remain in the IBC. Exceptions 1 and 2 recognize that the volumetric ratio of spiral reinforcement need not be greater than that required for driving and handling stresses when a pile foundation system is designed for load combinations including overstrength. The minimum spiral reinforcement required per ACI 318, Section 13.4.5.6 for driving and handling stresses is the minimum spiral reinforcement required in SDC A and B. Increased axial forces, shear forces, and bending moments provide a large factor of safety against nonlinear pile behavior when the design includes overstrength effects.

1810.3.9 Cast-in-place deep foundations. CIP deep foundations are covered in Sections 1810.3.9.1 through 1810.3.9.6. CIP concrete piles are installed by placing concrete into holes preformed by drilling or by driving a temporary or permanent casing to the required bearing depth. Drilled or augered piles are also known as cast-in-drill-hole or CIDH piles. Drilled or augured uncased piles are nondisplacement piles and are installed by drilling or augering a hole and filling the uncased hole with concrete, either during or after withdrawing the auger. CIP piles may be either cased or uncased. Uncased piles are difficult to construct when below the ground-water table. Except for enlarged base piles, the concrete in CIP piles is not subjected to driving forces, only the forces imposed by the service loads and down-drag from settlement. One advantage of drilled CIP piles is that the tip elevation can easily be adjusted to have the tip on the correct bearing stratum. Reinforcement is installed during the concreting operation. CIP-drilled piles are of the nondisplacement type, that is, the soil is not displaced or compacted by the drilling operation. CIP piles constructed by first driving a closed-end shell are displacement piles, where the soil surrounding the shell is displaced and compacted during the driving operation. CIP piles constructed by driving an open-ended casing without a mandrel or temporary tip closure are nominally a nondisplacement pile, although some compaction around the shell may occur in cohesive soils, and densification may occur from driving in granular soils.

The steel-cased pile is the most widely used type of CIP concrete pile. This pile type is characterized by a thin steel shell and is a displacement pile. This pile type consists of a closed-end light-gauge steel shell or a thin-walled pipe driven into the soil and left permanently in place, reinforced when required for uplift, lateral bending, or seismic-induced curvatures, and filled with concrete. The shell or pipe is usually driven with a removable mandrel. The shell is either a constant section or a tapered shape. Steel-encased piles are generally friction piles.

1810.3.9.1 Design cracking moment. The design cracking moment that is established as nominal moment capacity multiplied by the factor is determined from the equation

$$\varphi M_n = 3\sqrt{f'_c} S_m$$

The equation and the reinforcement requirements of Section 1810.3.9.2 make the IBC consistent with ACI 318 requirements. For both uncased and cased CIP deep foundation elements (but not concrete-filled pipes and tubes), reinforcement must be provided where moments exceed a reasonable lower bound for the capacity of the plain concrete section, known as the cracking moment.

1810.3.9.2 Required reinforcement. See Section 1810.3.9.1.

1810.3.9.3 Placement of reinforcement. With three exceptions, reinforcing steel must be assembled into a cage and placed in the hole or casing prior to concreting, not stabbed after concreting. One exception is for dowels less than 5 feet (1,524 mm) in length, and the other exception is for auger-injected piles, which are placed by injecting the concrete through a hollow stem auger. The third exception is for smaller residential and utility buildings (Group R-3 and U occupancies) where the method of reinforcement placement after concrete placement must be approved by the building official.

1810.3.9.4 Seismic reinforcement. There are four specific cases where prescriptive provisions are provided to comply with the seismic reinforcement requirements in SDC C through F. Other than these four exceptions, CIP deep foundations must comply with Section 1810.3.9.4.1 for seismic reinforcement in SDC C and Section 1810.3.9.4.2 for seismic reinforcement in SDC D, E, and F.

1810.3.9.4.1 Seismic reinforcement in seismic design category C. Minimum steel requirements are established to provide some ductility. The minimum reinforcement must be continued throughout the flexural length of the pile. Minimum reinforced length is not a defined term in the IBC; rather, in this section there are four criteria to determine the length.

1810.3.9.4.2 Seismic reinforcement in seismic design categories D through F. The requirements in SDC D, E, or F are similar to those in SDC C, except that the minimum reinforcement ratio is higher and extends over a longer length to improve ductility. In addition, closed ties or spirals are required to provide confinement in regions of plastic hinging, that is, at the pile-cap interface, at the interface of soft to stiff layers, and in liquefaction zones. Confinement reinforcing should also be used for bay muds and sensitive clays (see Sections 1803.5.11 and 1803.5.12).

1810.3.9.4.2.1 Site classes A through DE. Transverse confinement reinforcement requirements for stiffer soil and rock sites are in this section, which references applicable provisions in ACI 318.

1810.3.9.4.2.2 Site classes E and F. Transverse confinement reinforcement requirements for soft or sensitive soil sites are in this section, which references applicable provisions in ACI 318.

1810.3.9.5 Belled drilled shafts. Bells are designed to increase the bearing surface on which the loads will be transferred. Bells are typically designed in more cohesive soils so that there will not be any collapsing of the bell walls or roof.

1810.3.9.6 Socketed drilled shafts. Socketed drilled shaft deep foundations are also called caisson piles or are more commonly known as drilled-in caissons. They are installed as a special type of high-load-capacity pile and are characterized by a structural steel core, an upper-cased section extending to bedrock, and a lower uncased tip that is socketed into rock.

The socketed drilled shaft deep foundation element is a cased CIP concrete pile that is formed by (1) driving a heavy-wall open-ended pipe down to bedrock, (2) cleaning out the soil materials within the pipe, (3) drilling an uncased socket into the bedrock, (4) inserting a structural steel core into the pipe, and (5) filling the entire pipe and drilled socket with concrete.

The core material is usually made of hot-rolled structural steel wide-flange or I-beam sections, or steel rails. This section specifies that the steel core is to extend full length from the base of the drilled socket to the top of the steel pipe or, as an alternative and depending on design requirements, the steel core may extend halfway up the pipe or as a stub core to a distance in the pipe at least equal to the depth of the socket. The strength of the deep foundation element is developed in combined friction and end-bearing of the rock socket.

1810.3.10 Micropiles. Micropiles are bored, grouted-in-place deep foundation elements that develop their load-carrying capacity by means of a bond zone in soil, bedrock, or a combination of soil and bedrock. Provisions are included for construction, materials, reinforcement, and seismic design.

1810.3.11 Pile caps. Pile caps are to be of reinforced concrete and designed in accordance with the requirements of ACI 318. For footings (pile caps) on piles, computations for moments and shears may be based on the assumption that the load reaction from any pile is concentrated at the pile center. See ACI 318 for loads and reactions of footings on piles.

The soil immediately under the pile cap should not be considered to provide any support for vertical loads. For a more detailed explanation of this requirement, see the discussion of Section 1810.3.3.2, allowable lateral load.

The heads of all piles are to be embedded not less than 3 inches (76 mm) into pile caps, and the edges of the pile caps are to extend at least 4 inches (102 mm) beyond the closest sides of all piles. The degree of fixity between a pile head and the concrete cap depends on the method of connection required to satisfy design considerations.

1810.3.11.1 Seismic design categories C through F. Deep foundation elements in SDC C through F must have a positive connection to the pile cap for sliding or uplift purposes. This is achieved by connecting the deep foundation element to the pile cap either by embedding the deep element reinforcement in the pile cap or by field-placed dowels anchored into the element and extended into the pile cap in accordance with ACI 318.

Piles in structures subject to seismic ground shaking are likely to be subjected to uplift (tension) forces, either by design or because of insufficient resistance to overturning forces by gravity loads. Hence, concrete piles and concrete-filled steel pipe piles must be able to develop the strength of the pile (in tension) in the connection to the cap. This is accomplished by the requirement that the pile be embedded in the cap at least two times the cap embedment length, but not less than the reinforcement tension development length.

Similarly, the various types of steel piles are required to develop the strength in tension and to transmit this strength to the cap by positive means other than bond to the bare steel; for example, welded studs or welded reinforcement must be used.

Splices must develop the full strength of the pile, both tension and compression, for all pile types.

1810.3.11.2 Seismic design categories D through F. Anchorage of piles or piers into pile caps must consider the combined effects of uplift and pile fixity. The anchorage must develop at least 25 percent of the strength of the pile in tension. For piles subject to uplift or required to provide rotational restraint, the anchorage must develop the lesser of the nominal tensile strength of the longitudinal reinforcement in a concrete element, the nominal tensile strength of a steel element, and 1.3 times the frictional force developed between the element and the soil. Because of the large variability in soils, it would be prudent to design these piles for the full tensile capacity rather than 1.3 times the uplift capacity (frictional force). Exceptions allow the use of ASCE 7 Section 2.3.6 or 2.4.5 for design of the anchorage to resist axial tension forces or, in the case of rotational restraint, the design of the anchorage to resist axial and shear forces, and moments.

Steel piles used in higher SDCs are expected to yield just under the pile cap or foundation because of combined bending and axial loads. AISC 341 design and detailing requirements for piles are intended to produce a stable plastic hinge formation in steel piles. ACI 318 provisions are similarly intended to produce stable plastic hinges in concrete piles. Because piles can be subjected to tension caused by an overturning moment, mechanical means to transfer such tension must be designed for the required tension force, but not less than 10 percent of the pile compression capacity.

The load combinations with overstrength only apply where specifically required by the seismic provisions. They constitute an additional requirement that must be considered in the design of specific structural elements to account for the maximum earthquake load effect, E_m, which considers "system overstrength." This system characteristic is accounted for by multiplying the effects of the lateral earthquake load by the overstrength factor, Ω_0, for the seismic-force-resisting system involved. It represents the upper bound system strength for purposes of designing nonyielding elements for the maximum expected load. Under the design earthquake ground motions, the forces generated in the seismic-force-resisting system can be much greater than the prescribed seismic design forces. If not accounted for, the system overstrength effect can cause failures of structural elements that are subjected to these forces. Because system overstrength is unavoidable, design for the maximum earthquake force that can be developed is warranted for certain elements. The intent is to provide key elements with sufficient overstrength so that inelastic (ductile) response/behavior appropriately occurs within the vertical resisting elements. It should be noted that these load combinations are only to be applied where specified in the earthquake load provisions or in other structural chapters of the code. The requirement for using load combinations with overstrength in Section 1810.3.11.2 is such a case.

Batter piles have performed poorly in past earthquakes. This is because the batter piles are laterally stiff relative to vertical piles and resist most of the seismic-induced inertial forces. The piles are not usually designed to resist the actual forces, but are designed to resist an inertial force reduced by an assumed ductility. However, the batter piles are axially stiff and generally not detailed for ductility; hence the failures. To preclude this type of failure in batter piles, the piles and their connections must be designed to resist the anticipated maximum earthquake forces from the load combinations with overstrength factor in Section 2.3.6 or 2.4.5 of ASCE 7.

1810.3.12   Grade beams.   In high seismic regions, grade beams must be designed as ductile in accordance with the provisions of ACI 318 unless the beam is not subject to differential settlement exceeding one-fourth of the thresholds specified in ASCE 7 Table 12.13-3 and is designed to resist the anticipated maximum earthquake force as set forth in the load combination with overstrength factor of Section 2.3.6 or 2.4.5 of ASCE 7. That is, grade beams must be either strong or ductile.

1810.3.13   Seismic ties.   Interconnection of piles and socketed drilled shaft deep foundations is necessary to prevent differential movement of the components of the foundation during an earthquake. It is well known that a building must be thoroughly tied together if it is to successfully resist earthquake ground motion.

Individual piles, piers, or pile caps required for structures must be interconnected with foundation seismic ties as prescribed in Section 18.13.4 of ACI 318. The intent of this requirement is to minimize differential movement or spreading between the footings during ground shaking. If slabs on grade, or beams within slabs on grade, are used to meet the tie requirement, the load path from footing to slab or beam/slab and across joints in the slab or beam/slab should be checked for continuity. The slab or beam must be reinforced for the design tension load.

Subject to approval of the building official, an exception for Group R-3 and U occupancies of light-frame construction allows deep foundation elements supporting foundation walls, isolated interior posts detailed for vertical loads only, or exterior decks and patios to not be subject to interconnection where the soils are of adequate stiffness.

1810.4   Installation.   Care must be used during installation to prevent damage during handling and driving. The proper cushion must be used at the driving end. Precast concrete pile recommendations for design, manufacture, and installation are given in ACI 543R *Guide To Design, Manufacture, and Installation of Concrete Piles*. Damage to piles can be classified into four types:

1. Spalling at the butt or head (driving end) caused by high or irregular compressive stress concentrations. The spalling may be caused by insufficient cushioning, pile butt not square with the pile longitudinal axis, hammer and pile not aligned, reinforcing steel not flush or below the top of the pile allowing the hammer force to be transmitted through the steel, or insufficient transverse reinforcement.

2. Spalling at the tip, which is usually caused by an extremely high driving resistance such as when the tip is bearing on a rock.

3. Breaking or transverse cracking. This is caused by the rarefaction wave reflected from the tip. When the hammer strikes the cushion or head, a compression wave is produced that travels down the pile. The wave can be reflected from the tip as a rarefaction (tension) wave or a compression wave depending on the soil stiffness. Rarefaction waves usually occur when the soil at the tip is soft with very little resistance to penetration, causing tension waves that can cause significant tensile damage. This phenomenon usually occurs only in long piles exceeding 50 feet (15,240 mm). Prestressed piles have more resistance to rarefaction damage than do conventionally reinforced piles. The hammer energy should be reduced when driving long piles through soft soils.

4. Spiral or transverse cracking may be caused by a combination of torsional stress and rarefaction stress. Torsion is usually caused by excessive restraint in the leads.

In the case of precast-prestressed piles, because of the precompression, less care is needed in the handling and driving than for conventionally reinforced piles, and prestressed piles are, in general, more durable than conventionally reinforced precast concrete piles.

If a deep foundation element consists of two or more sections of various materials or different types of deep foundation elements spliced together, each section is required to satisfy the applicable installation requirements of this section. Generally referred to as "composite piles," these refer to deep foundation elements placed in series, such as a CIP concrete deep foundation element placed over a submerged wood pile.

1810.4.1 Structural integrity. Deep foundation elements can be exposed to damage or could potentially cause damage to surrounding areas, especially piles that are generally installed by either driving, vibration, jacking, jetting, direct weight, or a combination of such methods. Most types of piles are exposed to some degree of damage during placement. However, with knowledge of soil conditions and the proper selection of equipment, installation methods, and techniques, damage may be prevented or minimized.

Due care must be exercised during pile installation to avoid interference with adjacent piles or other structures so as to leave their strength and load capacity unimpaired. If any pile is damaged during installation so as to affect its structural integrity, the damage must be satisfactorily repaired or the pile rejected.

Displacement piles have their own special issues. As displacement piles are driven within a group, progressive compaction of the surrounding soil occurs, particularly where it involves closely spaced piles. This can cause piles to be deflected off-line because of the buildup of unequal soil pressures around the piles. Soil compaction during driving operations can cause extreme variations in pile lengths within a group, with some piles failing to reach specified load-bearing material. Ground heave is another effect of soil compaction (see discussion, Section 1810.4.6).

To prevent or significantly reduce the problems associated with soil compaction, the driving sequence of pile installations becomes an important consideration. For example, if the outer piles of a group are driven first so that the inner piles, because of soil compaction, fetch up to specified sets (hammer blows) at much higher elevations than the outer piles, the total load-bearing value of the group will be adversely affected. As another example, starting pile driving at the edge of a group makes the piles progressively more difficult to drive and results in a one-sided bearing group. The general driving practice is to work from the center of a group outward. For large groups consisting of rows of widely spaced piles, driving can be done progressively from one side to the other.

The provisions within Section 1810.4 are to ensure pile structural integrity by adhering to proper installation procedures. The code cannot cover all possibilities, however, thus establishing the need for the general nature of this section.

1810.4.1.1 Compressive strength of precast concrete piles. Handling and driving forces will likely govern pile strength requirements. Through the use of steam cure and Type III cements, 75 percent of specified strength, which is required prior to driving, can be achieved relatively quickly.

1810.4.1.2 Shafts in unstable soils. During construction of drilled piles, potential problems include caving, ground water, and other issues. Most of these problems generally relate to soil conditions, including soil or rock debris accumulating at the base of the pile or occurring in the pile shaft, reductions in the shaft cross section caused by the necking of soil walls because of soft materials or earth pressures, discontinuities in the deep foundation shaft, hollows on the surface of the shaft, and other problems related to the drilling operations.

1810.4.1.3 Driving near uncased concrete. These requirements are intended to prevent damage to uncured concrete in adjacent deep foundation elements from the soil displacements caused by driving adjacent elements. The spacing requirements should not be construed to mean that the center-to-center spacing must not be closer than six average diameters of a cased element in granular soils nor within one-half the pile depth in cohesive soils; just that deep elements cannot be driven at that spacing within 48 hours of placement of concrete.

1810.4.1.4 Driving near cased concrete. The restrictions on driving within four and one-half average diameters of a cased element filled with concrete less than 24 hours old are primarily to avoid damage to uncured concrete in adjacent deep foundation elements.

1810.4.1.5 Defective timber piles. Damage to the pile, including breakage, should be suspected when there is a sudden drop in penetration resistance while driving that cannot be explained by the soil profile. The pile should be withdrawn for examination. If penetration resistance should suddenly increase, driving should be stopped to avoid possible damage. A significant problem encountered during installation of timber piles is damage from overdriving. Overdriving can cause failure by bending, brooming of the tip, crushing, brooming at the butt end, or splitting or breaking along the pile section. See Figure 1810-3.

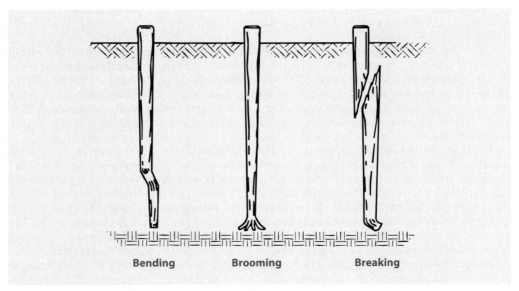

Figure 1810-3 **Effect of overdriving timber piles.**

1810.4.2 Identification. All deep foundation materials must be identified for conformity to the code requirements and construction specifications. Information such as strength (species and grade for timber piles), dimensions, and other pertinent information is required. Such identifications must be provided for all deep foundation elements, whether they are taken from manufacturers' stock or made for a particular project. Identifications are to be maintained from the point of manufacture through the shipment, on-site handling, storage, and installation of the piles. Manufacturers, upon request, usually furnish certificates of compliance with construction specifications. In the absence of adequate data, piles must be tested to demonstrate conformity to the specified grade.

In addition to mill certificates (steel piles), identification is made through plant manufacturing or inspection reports (precast concrete and timber piles) and delivery tickets (concrete). Timber piles are stamped (labeled) with information such as producer, species, treatment, and length.

Identification is essential when high-yield-strength steel is specified. Frequently, pile cutoff lengths are reused and pile material may come from a jobber, a contractor's yard, or a material supplier. In such cases, mill certificates are not available and the steel should be tested to see if it complies with the code requirements and the project specifications.

1810.4.3 Location plan. A plan clearly showing the designation of all deep foundation elements on a project by an identification system is to be filed with the building official before the installation is started. The inspector (see discussion in Chapter 17 on special inspection) must keep piling logs and other records and submit written reports based on this identification system. The use of such a system becomes particularly important at sites where the variations in soil profiles are so extensive that it becomes necessary to manufacture piles of different lengths to satisfy bearing conditions.

The building official should also be furnished copies of all modifications to the original deep foundation location plan that may be necessary as the work proceeds (as-built drawings). This would show elements added, eliminated, or relocated. Such records would facilitate the use of existing deep foundations in the future (see Section 1810.1.2) if the structure is altered or another structure is built on the site.

1810.4.4 Preexcavation. There are several important reasons for the use of preexcavation to facilitate the installation of foundation piles. Some of these purposes are:

- To install piles through upper strata of hard soil.
- To penetrate through subsurface obstructions, such as timbers, boulders, rip rap, thin stone strata, and the like.
- To reduce or eliminate the possibility of ground heave that could lift adjacent structures or piles already driven.
- To reduce ground pressures resulting from soil displacement during driving and to prevent the lateral movement of adjacent piles or structures.
- To reduce the amount of driving required to seat the piles in their proper load-bearing strata.
- To reduce the possibility of damaging vibrations or jarring of adjacent structures, as well as reduce the amount of noise, all of which are associated with pile-driving operations.
- To accommodate the placement of piles that may be somewhat longer than the leads of the pile-driving equipment.

The two most common methods employed in preexcavation operations are prejetting and predrilling. Jetting is usually effective in most types of soils, except very coarse and loose gravel and highly cohesive soils. Jetting is most effective in granular materials. Generally, jetting in cohesive soils is not very practical or especially useful and should be avoided in soils containing very coarse gravel, cobbles, or small boulders. These stones cannot be removed by the jet and tend to collect at the bottom of the hole, preventing pile penetration below that depth.

Jetting operations must be carefully controlled to avoid excessive loss of soil, which could affect the load-bearing capacity of piles already installed or the stability of adjacent structures.

Piles should be driven below the depth of the jetted hole until the required resistance or penetration is obtained. Before this preexcavation method is used, consideration should be given to the possibility that jetting, unless strictly controlled, can adversely affect load transfer, particularly as it involves the placement of nontapered piles.

Predrilling or coring before driving is effective in most types of soils and is a more controllable method of preexcavation than jetting. The risk of adversely affecting the structural integrity of adjacent piles or structures or the frictional capacity of piles is considerably less than jetting.

Predrilling can be performed as a dry operation or as a wet rotary process. Dry drilling can be done by using a continuous-flight auger or a short-flight auger attached to the end of a drill stem or kelly bar. Wet drilling requires a hollow-stem continuous-flight auger or a hollow drill stem employing the use of spade bits. When the wet rotary process of predrilling is used, bentonite slurry or plain water is circulated to keep the hole open. As in the case of jetting, piles should be driven with tips below the predrilled hole. This is necessary to prevent any voids or very loose or soft soils from occurring below the pile tip.

1810.4.5 Vibratory driving. The use of vibratory drivers for the installation of piles is not applicable to all types of soil conditions. They are effective in granular soils with the use of nondisplacement piles, such as steel H-piles and pipe piles driven open ended. Vibratory drivers are also used for extracting piles or temporary casings employed in the construction of CIP concrete deep foundation elements.

Vibratory drivers, either low or high frequency, cause the pile to penetrate the soil by longitudinal vibrations. Although this type of pile driver can produce good results in the installation of nondisplacement piles under favorable soil conditions, the greatest difficulty is the lack of a reliable method of estimating the load-bearing capacity. After the pile has been installed with a vibratory driver, pile capacity can be determined by using an impact-type hammer to set the pile in its final position. Load tests are not required if the pile is to be used for lateral resistance only.

One method to determine pile capacity is to calibrate the power consumption in relation to the rate of penetration. Nonetheless, the use of a vibratory driver is only permitted where the pile load capacity is established by load tests in accordance with the requirements of Section 1810.3.3.1.2.

1810.4.6 Heaved elements. Piles that are driven into saturated plastic clay materials can often displace a volume of soil equal to that of the piles themselves. When this happens, the soil displacement sometimes occurs as ground heave and may lift adjacent deep foundation elements already driven. Under such conditions, heaved deep foundation elements may no longer be properly seated and a loss of pile capacity occurs. Heaved piles must be redriven to firm bearing to again develop the required capacity and penetration. If heaved piles are not redriven, their capacity must be verified by load tests made in accordance with the requirements of Section 1810.3.3.1.2.

This section applies only to piles that can be safely redriven after installation. Heaved uncased CIP concrete deep foundation elements or sectional piles with joints that cannot take tension should be abandoned and replaced. When redriving heaved piles, a comparable driving system or the same as that of the initial driving should be employed. It should be noted that in redriving concrete-filled pipe piles, the driving characteristics of the pile have been altered and the pile is substantially stiffer than when the empty pipe was initially driven. In such cases, the required driving resistance would be less than originally required.

One method used to prevent or reduce objectionable soil displacement is to remove some of the soil in the spaces to be occupied by the piles. This is done by predrilling the pile holes (see discussion, Section 1810.4.4).

1810.4.7 Enlarged base cast-in-place elements. Enlarged base elements are intended to be end-bearing-type deep foundation elements that spread the bearing load over a larger area than a prismatic element, thereby increasing capacity. Enlarged base deep elements may be either cased or uncased. Enlarged base elements are used only in granular soils, which, because of the voids between soil particles, allow densification of the soils around the deep element tip without creating excessive pressures. One type is the compacted base type, which consists of a bulb-shaped footing formed after driving the shaft casing to its final depth. Another type is the concrete-pedestal type, in which a truncated cone- or pyramid-shaped precast concrete tip larger than the steel casing diameter is driven into the soil with the casing. See Figure 1810-4.

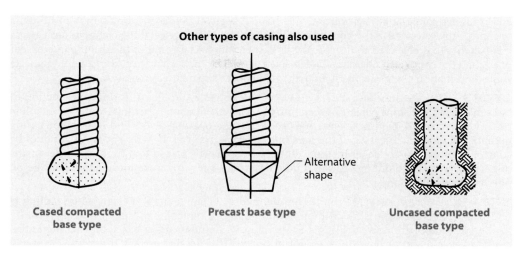

Figure 1810-4 Enlarged base pile cased or uncased shafts.

Installation must employ the same methods used to install the load test piles. The compacted base pile is usually installed by driving a steel casing. A zero slump concrete plug is placed at the tip of the casing and impacted with a heavy drop weight, thereby driving the casing and plug. Sometimes a gravel plug is used rather than zero slump concrete. When the casing and plug have been driven to the required depth, the plug is driven out and the bulb is formed by progressively compacting additional layers of zero slump concrete. A welded reinforcing steel cage is added where required. If the deep foundation element is to be uncased, the shaft is formed by compacting zero slump concrete in small lifts as the casing is withdrawn. If the deep foundation element is to be cased, as would be required for piles through peats or other organic soils, the shaft is formed by inserting a steel shell inside the drive casing after forming the bulb, withdrawing the drive casing, and filling the shell with conventional concrete. The precast base-type pile is installed with the precast base placed at the tip of a mandrel-driven steel shell.

A problem that occurs with the precast base type and the cased bulb-type piles is that an annular space between the casing and the soil remains. Either the pile must be designed as a slender reinforced concrete column governed by buckling, or the annular space must be filled to provide the requisite lateral support. The usual practice is to fill the annular space by pumping grout.

1810.4.8 Hollow-stem augered, CIP elements. In drilling the pile hole, the hollow-stem auger should be rotated and advanced continuously until the required tip elevation is reached. Concrete is then pumped through the hollow stem, filling the hole from the bottom up as the auger is withdrawn. The volume of placed concrete must be monitored and should equal or exceed the theoretical volume of the augered hole. An excess concrete volume of 10 percent is not uncommon, but even larger excesses can be expected where soils are more porous. If the amount of concrete pumped into the hole is considerably more than the anticipated volume, then the cause should be investigated.

If the pile-concreting operation is interrupted, a pressure drop occurs while pumping the concrete, the auger is improperly handled and is raised too fast, or anything else happens that could damage the pile or cause a reduction in the required pile section size, then the hollow-stem auger must be redrilled to the required tip elevation and the pile reformed from the bottom up.

Unless approved by the building official, no new pile holes are to be drilled or injected with concrete if they are located within a center-to-center distance of six pile diameters from other adjacent piles containing fresh concrete that has not been allowed to set for a period of 12 hours or more. This is to reduce the possibility of damage to adjacent piles. If the concrete surface of a completed pile is observed to drop below its cast elevation as a result of the drilling of adjacent piles, then the completed pile should be rejected and replaced. Such an occurrence could indicate some deformation along the pile shaft that could affect its structural integrity.

1810.4.9 Socketed drilled shafts. One of the potential problems during construction of caisson piles is that some water can seep into the bottom of the pile opening through rock joints or socket fissures. Sometimes, water may seep into the opening at the end of the pipe casing because of incomplete seating of the pipe into the rock material. If water in the hole cannot be controlled by ejection or other methods, the concrete may be placed by a tremie or other methods approved by the building official. A brief description of the installation sequence of caisson piles is given in Section 1810.3.9.6.

1810.4.10 Micropiles. Micropiles are defined in Chapter 2 and design and detailing is outlined in Section 1810.3.10. Micropile boreholes are typically advanced by either rotary drilling or rotary percussive drilling. Installation requirements differ based on whether a steel casing is permanent or temporary or not provided (IBC Section 1810.4.10 Items 6, 1, and 2, respectively).

1810.4.11 Helical piles. Helical piles are defined in Chapter 2. See Sections 1810.3.1.5 and 1810.3.5.3.5.

1810.4.12 Special inspection. See analysis and discussions in Chapter 17 regarding special inspection requirements for deep foundations.

KEY POINTS

- Chapter 18 provides requirements for design and construction of foundation systems for buildings and other structures regulated by the IBC.
- Geotechnical investigations are required based on the SDC of the building or where certain site conditions exist.
- The requirement for a geotechnical investigation may be waived by the building official in some cases for buildings in SDC A and B.
- Geotechnical investigations are required for buildings in SDC C through F and where other conditions exist such as compacted fill more than 12 inches (305 mm) deep, foundations bearing on CLSM, and excavations that may compromise lateral support of existing foundations.
- Section 1803.6 specifies the information required to be included in geotechnical reports.
- Requirements for site grading related to foundations are provided in Section 1804. General requirements for site grading are contained in Appendix J.
- Foundations must be dampproof and waterproof where required in accordance with Section 1805.
- Where a geotechnical investigation and report does not specify load-bearing values for soils, presumptive values for vertical bearing, lateral bearing, and lateral sliding friction resistance are provided in the code.
- Prescriptive requirements for concrete and masonry foundation walls are provided based on the properties of the wall and unbalanced backfill.
- Requirements for the design of retaining walls and embedded poles are provided.
- Foundations are divided into two distinct groups: shallow foundations (footings) and deep foundations (piles), with general requirements for both and specific requirements for each.
- Section 1808 contains general requirements for all foundations, Section 1809 covers shallow foundations, and Section 1810 covers deep foundations.
- Where a specific design is not provided, prescriptive requirements for constructing footings supporting light-frame construction are given in Table 1809.7.
- The IBC contains three tables that provide minimum compressive strength for concrete and grout, minimum concrete cover for reinforcement used in foundations, and allowable stresses for materials used in deep foundation elements.

CHAPTER 19

CONCRETE

Introduction
Section 1901 General
Section 1902 Coordination of Terminology
Section 1903 Specifications for Tests and Materials
Section 1904 Durability Requirements
Section 1905 Seismic Requirements
Section 1906 Footings for Light-Frame Construction
Section 1907 Slabs-on-Ground
Section 1908 Shotcrete
Key Points

> Chapter 19 provides minimum accepted practices for the design and construction of buildings and structural components using concrete—both plain and reinforced. Chapter 19 relies primarily on the reference to American Concrete Institute (ACI) 318-19, *Building Code Requirements for Structural Concrete*. Structural concrete must be designed and constructed to comply with this code and all listed standards. There are also specific provisions addressing slabs-on-ground and shotcrete. This chapter was extensively reorganized for the 2024 IBC.

Section 1901 *General*

1901.1 Scope. Chapter 19 provides minimum requirements governing the materials, quality control, design, and construction of structural concrete elements of any structure erected under the requirements of the IBC. The concrete chapter, like other chapters within the IBC, incorporates design requirements by reference and points the user to industry standards.

1901.2 Plain and reinforced concrete. This section requires that structural concrete be designed and constructed in accordance with the requirements of Chapter 19 and ACI 318 as supplemented in Section 1905. Chapter 35 specifies which edition of ACI 318 must be used.

1901.2.1 Structural concrete with GFRP reinforcement. ACI 440.11 *Building Code Requirements for Structural Concrete Reinforced with Glass Fiber-Reinforced Polymer (GFRP) Bars* allows for the use of GFRP in Seismic Design Category (SDC) A for the structural frame and in SDC B and C for concrete members that are not part of the seismic force-resisting system. ASTM D7957 *Standard Specification for Solid Round Glass Fiber Reinforced Polymer Bars for Concrete Reinforcement* is also referenced. These standards allow the design and construction of cast-in-place reinforced concrete using nonmetallic reinforcement.

GFRP-reinforced concrete is especially suitable in highly corrosive environments, such as reinforced concrete exposed to salt water, salt air or deicing salts. It is primarily used to reinforce highway bridge decks where deicing salts are used on the roads and cause severe corrosion to conventional steel reinforcement. Other applications where GFRP-reinforced concrete is considered include marine and coastal structures, parking garages, water tanks, and structures supporting magnetic resonance imaging (MRI) equipment.

Section 1901.2.1 and ACI 440.11 prohibit the use of internally GFRP-reinforced concrete for applications where fire-resistance ratings are required. For a building where reinforced concrete elements require fire-resistance ratings (e.g., Tables 601 and 705.5), the fire-resistance-rated members and assemblies will continue to need steel reinforcement.

1901.3 Anchoring to concrete. Section 1901.3 references ACI 318 for anchorage to concrete, and specifies the types of anchors covered such as cast-in-place bolts and studs and post-installed expansion, undercut, screw, and adhesive anchors. The supplemental provisions in Section 1905.7 are exceptions pertaining to out-of-plane anchorage of walls and in-plane anchorage of light-frame shear walls using wood sills or cold-formed steel tracks.

1901.4 Composite structural steel and concrete structures. Section 2202 is referenced for design of composite structures involving structural steel and reinforced concrete. For other than seismic design, Section 2202 references AISC 360. The seismic design of structural steel systems acting compositely with reinforced concrete using an *R* factor from Table 12.2-1 of ASCE 7 must be designed and detailed in accordance with AISC 341.

1901.5 Construction documents. This section provides a list of eleven items that are required to be shown on the design plans and construction documents. The intent behind this list is to address the basic elements needed for clarity and to properly convey design intent when selecting concrete as the structural frame material. See Sections 107 and 1603 for additional structural-related requirements pertaining to construction documents.

1901.6 Special inspections and tests. This section provides a cross-reference to the special inspection requirements of IBC Chapter 17. Section 1705.3 contains special inspection requirements for concrete construction. Chapter 26 of ACI 318 also contains inspection requirements for structural concrete however, it defers to the building code where adopted.

1901.7 Tolerances for structural concrete. If the design team does not specify concrete tolerances on the construction documents, Section 1901.7.1 refers to ACI 117, *Specification for Tolerances for Concrete Construction and Materials*, for tolerances in relation to cast-in-place concrete elements, and Section 1901.7.2 refers to ACI ITG-7, *Specification for Tolerances for Precast Concrete*, for tolerances associated with precast concrete elements. ACI 117 describes tolerances for materials, foundations, buildings, tilt-up construction, and more. ACI ITG-7 provides both dimensional and erection tolerances for structural and nonstructural precast elements.

Section 1902 *Coordination of Terminology*

This section of the code coordinates the definition for a term used in the IBC, ACI 318, and ASCE 7. Section 1902.1.1 clarifies that the term "design displacement" refers to the Design Earthquake Displacement defined in ASCE 7 Section 12.8.6.3. For diaphragms that can be idealized as rigid in accordance with ASCE 7 Section 12.3.1.2, displacement due to diaphragm deformation corresponding to the design earthquake is permitted to be taken as zero.

Section 1903 *Specifications for Tests and Materials*

1903.1 General. This section specifies that materials required to produce concrete and the testing of such materials must be in compliance with the provisions of ACI 318. The referenced provisions include those regarding cement, aggregates, water, steel reinforcement, admixtures, and storage of materials. Requirements for concrete including materials, proportioning, production, and construction are covered in Sections 26.4 and 26.5 of ACI 318. Requirements for reinforcement including materials and construction are covered in Section 26.6.

1903.2 Glass fiber-reinforced concrete. This section in the IBC, which is not found in ACI 318, requires that glass fiber-reinforced concrete (GFRC) and the materials used in such concrete conform to PCI 128, *Specification for Glass-Fiber-Reinforced Concrete Panels*. This publication contains the latest information on the planning, design, manufacture, and installation of GFRC panels.

1903.3 Flat wall insulating concrete form (ICF) systems. This section references ASTM E2634 *Standard Specification for Flat Wall Insulating Concrete Form (ICF) Systems* for the material used to make forms for ICF systems using molded expanded polystyrene (EPS) insulation panels. ICFs are rigid plastic foam forms used in the construction of cast-in-place concrete structural members. The ICFs remain in place as permanent building insulation for energy-efficient, cast-in-place, reinforced concrete walls, floors, roofs, beams, and columns. The forms are interlocking modular units that are dry-stacked (without mortar) and filled with concrete. ICFs are generally manufactured from polystyrene or polyurethane. Reinforcing steel is placed in the forms and then concrete is pumped into the cavity to form the structural portion of the walls. The forms provide thermal and acoustic insulation, space to run electrical conduit and plumbing, and backing for interior and exterior finishes.

ASTM E 2634 applies to ICF systems that consist of molded EPS insulation panels that are connected by cross ties and act as permanent formwork for cast-in-place reinforced concrete structural members such as reinforced concrete beams, lintels, exterior and interior bearing and nonbearing walls, foundations, and retaining walls. ASTM E 2634 lists test methods appropriate for establishing ICF system performance as a permanent concrete forming system.

Section 1904 *Durability Requirements*

1904.1 Structural concrete. Durability requirements for structural concrete and reinforcing steel may be found within Chapters 19 and 20 of ACI 318 and have been adopted through reference, with certain exceptions. These provisions emphasize the importance of considering durability requirements before selecting a desired material strength (f'_c) and depth of cover for reinforcing steel. They include exposure categories and classes with applicable durability requirements for concrete in a unified format.

Exposure categories and classes. Concrete is required to be assigned to exposure classes in accordance with ACI 318 Section 19.3.1, based on:

1. Exposure to moisture and cycles of freezing and thawing, with or without deicing chemicals (Class F);
2. Exposure to sulfates in water or soil (Class S);
3. Exposure to water (Class W); and
4. Exposure to chlorides from chemicals, salt, saltwater, brackish water, seawater, or spray from these sources, where the concrete has steel reinforcement (Class C).

Concrete properties. Concrete mixtures must conform to the most restrictive maximum water-cementitious materials (w/cm) ratios and maximum specified compressive strength (f'_c) requirements of ACI 318 Section 19.3.2, based on exposure classes assigned in Section 19.3.1.

Since it is difficult to accurately determine the water-cementitious materials ratio of concrete during production, the f'_c specified should be reasonably consistent with the water-cementitious materials ratio required for durability. This will help ensure that the required water-cementitious materials ratio is actually obtained in the field. Because the usual emphasis in inspection is on strength, test results substantially higher than the specified strength may lead to a lack of concern for quality and production of concrete that exceeds the maximum water-cementitious materials ratio. Thus, a concrete strength of 3,000 psi (20.7 MPa) and a maximum water-cementitious materials ratio of 0.45 should not be specified for a parking structure, if the structure will be exposed to deicing salts.

IBC Section 1904.1 includes an exception to the ACI 318 provisions for buildings less than four stories in height that house Group R (residential) occupancies. The exception allows normal-weight concrete subject to weathering (freezing and thawing) to have a minimum specified concrete strength of 3,000 psi (20.7 MPa). The exception is consistent with the *International Residential Code*® (IRC®).

1904.2 Nonstructural concrete. This section prescribes performance requirements for nonstructural concrete. It requires the design professional to assign a freeze-thaw exposure Class (F) based on the anticipated exposure of nonstructural concrete. Nonstructural concrete must have a minimum specified strength of 2,500 psi (17.2 MPa) for Class F0, 3,000 psi (20.7 MPa) for Class F1, and 3,500 psi (24.1 MPa) for Classes F2 and F3. Nonstructural concrete is defined in Chapter 2 as any element made of plain or reinforced concrete that is not used to transfer either gravity or lateral loads to the ground.

Section 1905 *Seismic Requirements*

1905.1 General. A summary of the seismic provisions supplemental to ACI 318 can be found in Table 1905-1. A more detailed discussion of each modification is provided below.

Table 1905-1. IBC Seismic Provisions Supplemental to ACI 318

Item	IBC Section	Subject	ACI 318 Section Modified
1	1905.2	Modification of definitions and addition of terms	Section 2.3
2	1905.3	Supplemental provisions for intermediate precast structural walls	Section 18.5
3	1905.4	Supplemental provisions for foundations designed to resist earthquake forces	Section 18.13
4	1905.5	Addition of provisions for detailed plain concrete structural walls	Section 14.6
5	1905.6	Replacement of the plain concrete provisions in SDC C, D, E, and F	Section 14.1.4
6	1905.7	Addition of anchorage exceptions	Section 17.10.5.2 Section 17.10.6.2

1905.2 ACI 318 Section 2.3. Modifications are made to existing definitions and new definitions supplement those in ACI 318 Section 2.3. ACI definitions of *Detailed Plain Concrete Structural Wall, Ordinary Precast Structural Wall, Ordinary Reinforced Concrete Structural Wall,* and *Ordinary Plain Concrete Structural Wall* are modified to include useful references that are not mentioned in the definitions within ACI 318. Two definitions are added based on ASCE 7 Chapter 14, which is not adopted in the IBC: *Cast-in-place Concrete Equivalent Diaphragm* and *Precast Concrete Diaphragm.*

1905.3 Intermediate precast structural walls. This section modifies ACI 318 Section 18.5 which specifies a minimum ductility requirement for a steel element used in a connection between wall panels or between wall panels and the foundation, so that yielding can be allowed in that element. This section requires that a yielding steel connection element retain at least 80 percent of its design strength at the deformation level corresponding to the design displacement of the structure or use Type 2 mechanical splices.

1905.4 Foundations designed to resist earthquake forces. This section modifies ACI 18.13 by adding the requirement that the seismic design of foundations comply with relevant provisions in ACI 318, unless they are modified by provisions within the IBC

Chapter 18, *Soils and Foundations*. In general, the provisions of the IBC Chapter 18 are more stringent than those found in ACI 318 Section 18.13.

1905.5 Detailed plain concrete structural walls. This section adds text to ACI 318 Section 14.6 to specify the reinforcement requirements for detailed plain concrete structural walls in addition to those required for ordinary plain concrete structural walls.

1905.6 Structural plain concrete. This modification replaces ACI 318 Section 14.1.4 in its entirety, while serving essentially the same purpose, that is, to prohibit the use of plain concrete elements in structures assigned to Seismic Design Category C, D, E, or F, with several specific exceptions that permit the use of plain concrete in foundations. This modification describes these exceptions in more detail than ACI 318 Section 14.1.4 and includes Seismic Design Category C. Section 1905.6.1 further permits structures in Seismic Design Category A or B, detached one- and two-family dwellings less than four stories constructed with stud-bearing walls to have plain concrete footings without longitudinal reinforcement.

1905.7 Design requirements for anchors. This modification to ACI 318 provides flexibility in designing anchors resisting earthquake forces by providing exceptions to the ACI 318 provisions. The three numbered exceptions to ACI 318 Section 17.10.6.2 apply to in-plane shear strength of anchors in light-frame shear walls. Exception 1 applies to the anchorage of wood sill plates to concrete foundations in light-frame construction. Based on light-frame shear wall testing, the wood sill plate controls the ductile behavior of the anchorage assembly. If the anchor meets the requirements of the exception, then the anchor need not meet the requirements of ACI 318. Allowable in-plane shear capacity of the anchor bolt is determined in accordance with NDS Table 12E rather than ACI 318.

Exception 2 applies to the anchorage of cold-formed steel (CFS) track to concrete foundations in light-frame construction. Based on light-frame shear wall testing, the CFS track controls the ductile behavior of the anchorage assembly. If the anchor meets the requirements of the exception, then the anchor need not meet the requirements of ACI 318. Allowable in-plane shear capacity of the anchor bolt is determined from AISI S100 Section J3.3.1 rather than ACI 318.

There are multiple criteria to meet in order to use Exception 1 or 2 for the design of anchor bolts connecting wood sill plates or CFS tracks of shear walls to the foundation in light-frame structures. The maximum nominal diameter of the anchor is ⅝ inches (16 mm); anchors must be embedded a minimum of 7 inches (178 mm) into the footing and located a minimum of 1¾ inches (45 mm) from the edge of the concrete parallel to the wood sill plate or steel track, and a minimum of 15 diameters from the edge of the concrete perpendicular to the wood sill plate or steel track; the wood sill plate must be a nominal 2x or 3x, and cold-formed steel track must be within a thickness range of 33 mil to 68 mil (0.84 mm to 1.73 mm).

For light-frame construction, Exception 3 allows use of Section 17.7.2.1(c) rather than Section 17.10.6.3. Section 202 defines light-frame construction as a system where vertical and horizontal structural elements are primarily formed by repetitive wood or cold-formed steel framing members.

Section 1906 *Footings for Light-Frame Construction*

In general, design and construction of structural concrete members must comply with ACI 318. Section 1906 provides an exception for plain concrete footings that support Group R-3 occupancies or other occupancies not more than a single story above grade plane. Plain concrete footings having a thickness of 6 inches (152 mm) are allowed as long as the footings do not extend more than 4 inches (102 mm) to either side of the supported wall.

Section 1907 *Slabs-on-Ground*

1907.1 Structural slabs-on-ground. Structural concrete slabs-on-ground are considered structural concrete where required by ACI 318 or where they are designed to transmit vertical loads or lateral forces from other parts of the structure to the soil or from other parts of the structure to foundations.

1907.2 Nonstructural slabs-on-ground. Nonstructural slabs-on-ground are only required to comply with the durability requirements of Section 1904.2, the minimum thickness specified in Section 1907.3, and the vapor retarder requirements of Section 1907.4. Portions of nonstructural slabs-on-ground used to resist uplift or overturning loads must be designed throughout the entire portion designated as dead load to resist those forces.

1907.3 Thickness. This section specifies a minimum slab thickness of 3½ inches (89 mm).

1907.4 Vapor retarder. This section requires that a vapor retarder be installed immediately below the slab to retard vapor transmission. There are, however, a number of exceptions where a vapor retarder is not required.

Although good-quality concrete is practically impermeable to the passage of water (that is not under significant pressure), concrete is not impervious to the passage of water vapors. If the surface of the slab is not sealed, water vapor will pass through the slab. If a floor finish such as linoleum, vinyl tile, wood flooring or any type of covering is placed on top of the slab, the moisture is trapped in the slab. Any floor finish adhering to the concrete may eventually loosen or buckle or blister.

Many of the moisture problems associated with enclosed slabs on ground can be prevented or minimized by installing vapor retarders, such as polyethylene sheeting or other approved materials, between the slab and the ground. Such retarders are needed under slabs in habitable spaces. Where garages, utility areas, and similar spaces are not to be heated or occupied, vapor barriers are not required but may be applied. If moisture migration is not expected to be a problem based on the occupancy of a building, Exception 3 permits the vapor retarder to be omitted.

Section 1908 *Shotcrete*

Shotcrete is concrete conveyed through a hose and pneumatically projected at high velocity onto a surface. Originally the term "shotcrete" referred to wet-mixed ejected concrete and the term "gunite" referred to a proprietary method using dry-mixed concrete that was hydrated at the nozzle. The term "shotcrete" is now generally used for both wet-mix and dry-mix methods of ejecting concrete onto a surface. Section 1908 simply refers to ACI 318 for all provisions related to shotcrete design and construction.

KEY POINTS

- Chapter 19 contains requirements for the design and construction of concrete buildings and other concrete structures regulated by the IBC.
- The 2024 IBC references ACI 318-19, *Building Code Requirements for Structural Concrete*, and in most cases, refers to specific sections in the ACI Standard for detailed design and construction requirements.
- In addition to the general requirements for construction documents given in Section 107, Section 1901.5 contains specific requirements for construction documents pertaining to concrete structures.
- Durability requirements for concrete are covered in Section 1904 and ACI 318 Chapter 19 based on exposure categories F (freezing and thawing), S (sulfate), W (water), and C (corrosion protection for reinforcement), with various classes within each category.
- An exception for Group R occupancies less than four stories above grade plane permits normal-weight aggregate concrete to have a minimum specified concrete strength of 3,000 psi (20.7 MPa) in lieu of the durability requirements of ACI 318.
- Supplemental seismic provisions to ACI 318 are found in Section 1905.
   - Section 1905.6 contains important provisions for structural plain concrete foundation walls, isolated footings, and footings supporting light-frame walls.
   - Section 1905.7 contains exceptions from seismic requirements of ACI 318 for (1) anchors designed to resist tension forces and (2) shear design of anchor bolts attaching wood sill plates or cold-formed steel tracks of bearing or nonbearing walls of light-frame structures to foundations or foundation stem walls, provided several conditions are met.
- Section 1906 provides for 6-inch-thick (152 mm) plain concrete footings that support Group R-3 occupancies or other occupancies not more than a single story above grade plane.
- Requirements for slabs-on-ground, which are not specifically covered by ACI 318 are provided in Section 1907.

CHAPTER 20

ALUMINUM

Section 2002 Materials
Key Points

> This chapter covers the requirements for quality, design, fabrication, and erection of aluminum structures.

Section 2002 *Materials*

The IBC® references two aluminum industry design standards: *Aluminum Design Manual* (ADM—2020) and *Aluminum Sheet Metal Work in Building Construction,* Fourth Edition (ASM 35—00).

2002.1 General. The ADM includes design provisions for aluminum structural members and their connections in addition to providing minimum strengths for aluminum products.

ASM 35 applies to the proper use of aluminum in roofing, flashing, and other sheet metal fabrications used in building construction. ASM 35 does not address proprietary or preformed sheet metal products, such as siding or curtain wall systems, but is intended for site-fabricated installations.

The nominal loads used to design aluminum structures are as required by Chapter 16, which references ASCE 7 for many loads, most notably environmental loads.

KEY POINTS

- Chapter 20 contains requirements for the design and construction of aluminum structures regulated by the IBC.
- The 2024 IBC references ADM—2020 *Aluminum Design Manual* and ASM 35—00 *Aluminum Sheet Metal Work in Building Construction,* Fourth Edition.
- Design loads and load combinations are to be determined in accordance with Chapter 16 or ASCE 7 as required.

CHAPTER 21

MASONRY

Section 2101 General
Section 2102 Notations
Section 2103 Masonry Construction Materials
Section 2104 Construction
Section 2105 Quality Assurance
Section 2106 Seismic Design
Section 2107 Allowable Stress Design
Section 2108 Strength Design of Masonry
Section 2109 Empirical Design of Adobe Masonry
Section 2110 Glass Unit Masonry
Section 2111 Masonry Fireplaces
Section 2112 Masonry Heaters
Section 2113 Masonry Chimneys
Section 2114 Dry-Stack Masonry
Key Points

> This chapter covers the materials, design, construction, and quality of masonry. TMS 402, *Building Code Requirements for Masonry Structures*, TMS 403, *Direct Design Handbook for Masonry Structures*, TMS 404 *Standard for the Design of Architectural Cast Stone*, and TMS 602, *Specification for Masonry Structures* are the referenced standards for masonry. These standards are developed and published by The Masonry Society (TMS).

Section 2101 *General*

The masonry provisions govern the materials, design, construction, and quality for masonry designed in accordance with allowable stress design, strength design, or empirical design, and for glass masonry and masonry fireplaces and chimneys. Masonry veneer is covered in Chapter 14 of the IBC® and Chapter 13 of TMS 402. TMS 402 Chapter 1 lists specific masonry items that need to be included on the construction documents.

2101.2 Design methods. This section references TMS 402 for all design methods except the direct design method and architectural cast stone which are described in TMS 403 and TMS 404, respectively. Prestressed masonry must be designed in accordance with Chapters 1 through 7 and 10 of TMS 402 and the seismic requirements in Section 2106.

A simplified design method for single-story, concrete masonry buildings is found in the referenced standard TMS 403. The methodology is based on strength design provisions and factored load combinations for dead, roof live, wind, seismic, snow, and rain loads in accordance with ASCE 7. The direct design procedure is a table-based structural design method that permits the user to follow a series of steps to design and specify relatively simple, single-story, concrete masonry bearing-wall structures. The method is simple to implement compared to conventional design approaches, but it limits the design to only those configurations addressed by the standard. It introduces slightly more conservatism compared to conventional design procedures as a result of the conditions and assumptions inherent to the design method. There are limitations based on ground snow load, wind speed, mapped seismic spectral acceleration, the type of masonry walls and roof systems used, and minimum reinforcing requirements. TMS 403 is intended to capture many of the commonly designed simple load-bearing masonry structures.

It should be emphasized that irrespective of the design method used, masonry structures must also comply with applicable seismic design requirements prescribed in Section 2106 based on the seismic design category of the building or structure. The seismic design category of a structure is determined in accordance with Section 1613.2. Section 2106 references Chapter 7 of TMS 402 without amendments for seismic design of masonry structures.

2101.2.1 Masonry veneer. Masonry veneer is required to comply with the requirements in Chapter 14 of the code, specifically Section 1404.

2101.3 Special inspection. Section 2101.3 refers to Chapter 17 for special inspection requirements, which in turn references the quality assurance requirements specified in TMS 602. See discussion under Section 1705.4.

Section 2102 *Notations*

This section defines notations that are made throughout the chapter.

Section 2103 *Masonry Construction Materials*

This section contains minimum requirements for various masonry construction materials and references TMS 602 in many cases. The IBC references hundreds of ASTM Standards and Specifications, many of which apply to masonry construction materials.

2103.1 Masonry units. Specifications for various types of masonry units are given in Article 2.3 of TMS 602. Table 2103-1 shows the tables in TMS 602 Commentary that contain requirements for various types of masonry materials. Note that structural clay tile used for nonstructural purposes in fireproofing structural members and wall furring need not meet the specified compressive strength but must meet the fire-resistance rating required by Table 705.5 determined in accordance with ASTM E119 or UL 263.

Table 2103-1. **Concrete Masonry Unit Specifications**

Type of Masonry Unit	Commentary Table
Concrete masonry units	SC-3
Clay brick and tile	SC-4
Stone	SC-5

2103.1.1 Second-hand units. Some masonry materials may be reused, provided that they meet all the code requirements for new materials. Note, however, that Section 104.9.1 restricts the use of used materials unless specifically approved by the building official. One should exercise caution in approving reuse of materials. The applicable material or design standards should be consulted to determine if reuse of used materials should be permitted. For example, glass block units cannot be reused, as indicated in TMS 602 Section 2.3D.

This section allows the use of salvaged or *used* brick. Used bricks are often salvaged from demolition of old unreinforced masonry buildings. Masonry units manufactured in the past may not have the same quality as masonry made to current standards. Caution should also be practiced when specifying used brick as a structural material. Even though the brick may appear clean, the pores in the bedding faces may be filled with cement paste, lime, and other deleterious microscopic particles that may reduce the absorption properties of the brick, thereby adversely affecting the mortar and masonry bond and reducing mortar strength. Testing for the absorption rate of used masonry units in accordance with ASTM C 67 *Test Methods of Sampling and Testing Brick and Structural Clay Tile* will give an indication of the bonding qualities of used units with mortar. Used masonry units are best used as veneers, where the units are not relied upon for structural strength.

2103.2 Mortar. Masonry mortar, surface bonding mortar, mortar for ceramic tile, and mortar for adhered masonry veneer are covered in Section 2103.2.1, 2103.2.2, 2103.2.3, or 2103.2.4.

2103.2.1 Masonry mortar. Mortar for use in masonry construction must conform to ASTM C 270 *Specification for Mortar for Unit Masonry* in accordance with Article 2.1 of TMS 602. Mortars proportioned in accordance with Table 1 of ASTM C270 should have a cube strength in excess of that required by ASTM C 270 for the various mortar types. Type M mortar is high-strength mortar having a minimum average 28-day compressive strength of 2,500 psi (17.2 MPa). Type M mortar is suitable for general use, and is recommended where maximum compressive strength is required such as in unreinforced masonry below grade. Type S mortar is recommended where a high lateral strength is required and is specifically recommended for reinforced masonry. The minimum average compressive strength for Type S mortar is 1,800 psi (12.4 MPa). Type N mortar is a medium-strength mortar having a minimum average compressive strength of 750 psi (5.2 MPa). Type N mortar may be used where high compressive or lateral strength is not required. Type N is generally used in exposed masonry above grade where exposed to weather. Type O mortar is a low-strength mortar having a minimum average compressive strength of 350 psi (2.4 MPa). Type O mortar should only be used for nonbearing walls not exposed to severe weathering.

Table 2 in ASTM C270 allows for the proportioning of mortar to meet the minimum requirements for 28-day compressive strength, minimum water retention, and maximum air content.

2103.2.2 Surface-bonding mortar. The specifications for premixed mortar, masonry units, and other materials used in constructing dry-stack, surface-bonded masonry, as well as the construction requirements, are referenced in this section.

2103.2.3 Mortars for ceramic wall and floor tile. The code references ANSI A108.1 *Installation of Ceramic Tile in the Wet-set Method, with Portland Cement Mortar* for Portland cement mortars used to lay ceramic tile. Table 2103.2.3 gives compositions of hydrated lime, cement, and sand for different applications. Specific provisions are provided for dry-set Portland cement mortars, latex-modified Portland cement mortar, epoxy mortar, furan mortar and grout, modified epoxy-emulsion mortar and grout, organic adhesives, and Portland cement grouts.

Mortar requirements for autoclaved aerated concrete (AAC) masonry are covered in TMS 602. The requirements for thin-bed mortar and mortar for leveling courses are in Articles 2.1 D.1 and D.2 of TMS 602. Thin-bed mortar is used in AAC masonry construction with joints 0.06 inch or less. The mortar used for the leveling courses of AAC masonry must be Type M or S.

2103.2.4 Mortar for adhered masonry veneer. TMS 402 Section 13.3 provides criteria for mortar used with adhered masonry veneer.

2103.3 Grout. Grout proportion requirements are given in Article 2.2 of TMS 602, which references ASTM C476 *Standard Specification for Grout for Masonry*. The grout proportioning by volume is in ASTM C476 Table 1. ASTM C476 permits grout to comply with the minimum strength requirement of 2,000 psi or to comply with the proportions given in Table 1. Grout batched in accordance with the proportions in the table will have strength in excess of 2,000 psi.

2103.4 Metal reinforcement and accessories. Requirements for metal reinforcement and accessories used in masonry construction reflect national consensus standards. The various ASTM standards are listed in Table 2103-2. For a complete list of specifications for reinforcement, prestressing tendons, and metal accessories, see Article 2.4 of TMS 602.

Coatings and corrosion protection requirements for mill galvanized, hot-dipped galvanized, and epoxy coating methods are provided in Section 2.4L of TMS 602. Corrosion protection requirements for tendons are provided in Section 2.4M of TMS 602.

Section 2104 *Construction*

The provisions in IBC Section 2104, together with the requirements in TMS 602, provide minimum requirements to ensure the masonry is properly constructed, consistent with the design methods used. The IBC maintains requirements not duplicated in TMS 402 and TMS 602, including provisions for masonry supported by wood members (Section 2104.1.1) and pertaining to molded cornices (Section 2104.1.2). More stringent construction requirements and

Table 2103-2. **ASTM Standards for Metal Reinforcement and Accessories**

Reinforcement Type	ASTM Standard
Deformed reinforcing bars	A615 (Billet steel) A706 (Weldable) A767 (Zinc coated) A775 (Epoxy coated) A996 (Rail and axle steel)
Masonry joint reinforcement	A1064
Deformed reinforcing wire	A1064
Wire fabric	A1064 (Welded deformed steel wire fabric)
Anchors, ties, and accessories	A36 (Plate and bent-bar anchors) A1008 (Sheet metal anchors and ties) A1064 (Welded wire reinforcement) A307 Grade A (Header anchor bolts)
Prestressing tendons	A421 (Wire and low-relaxation wire) A421 (Low-relaxation wire) A416 (Strand and low-relaxation strand) A722 (Bar)

referenced specifications than those contained in the code may be needed to satisfy aesthetic and architectural design criteria. Such requirements should be included in the construction documents.

Section 2105 *Quality Assurance*

The best way to improve the quality of construction in the field is through a quality-assurance program that ensures proper materials and construction methods are used. From a model code perspective, quality assurance during construction is achieved by the special inspection and testing requirements in Chapter 17.

2105.1 General. Section 2105 requires a quality-assurance program to ensure that masonry structures are constructed in accordance with the construction documents. Historically, the quality-assurance program is achieved through the special inspection and testing requirements covered in Chapter 17 of the IBC. However, the specific inspection and verifications for masonry structures are by reference to the quality-assurance provisions in TMS 602. Section 1.6 of TMS 602 has three levels of quality assurance that are based on the risk category. See the discussion of Section 1705.4 in regard to special inspection for masonry structures and Table 1705-1, which reflects the level of quality assurance required based on the type of masonry structure and risk category.

Acceptance relative to strength requirements. Compliance with minimum specified compressive strength is determined by the unit strength method or the prism test method covered in Article 1.4B of TMS 602. Where these methods are not used, masonry prisms saw cut from constructed masonry can be used to establish compliance in accordance with Article 1.4 B.4.

Testing prisms from constructed masonry. When approved by the building official, masonry is permitted to be based on tests of prisms saw cut from the masonry construction. A set of three masonry prisms at least 28 days old are saw cut for each 5,000 square feet (464 m^2) of wall area and not less than one set of three masonry prisms for a given project. The testing and transportation of the prisms must comply with the requirements of ASTM C1532 *Standard Practice for Selection, Removal, and Shipment of Manufactured Masonry Units and Masonry Specimens from Existing Construction* and ASTM C1314 *Standard Test Method for Compressive Strength of Masonry Prisms*.

Section 2106 *Seismic Design*

2106.1 Seismic design requirements for masonry. As with other structural materials, seismic load effects are determined from the IBC, which in turn references ASCE 7, and seismic resistance is determined from the specific material design provisions in TMS 402. Seismic design of reinforced masonry structures is based on the inelastic ductility response of the structure, which reduces the seismic design force level. This is reflected by the *R*-factor for the particular seismic-force-resisting system. Section 1604.9 and ASCE 7 Section 11.1.1 require that structures meet all applicable seismic detailing requirements, even though other loads such as wind may be greater than the seismic load effects.

All masonry walls, unless they are isolated on three edges from the in-plane motion of the basic structural system, must be considered part of the seismic-force-resisting system and designed accordingly. Nonload-bearing walls that are connected to the lateral-force-resisting system could be subject to seismic forces. Per TMS 402, elements are classified as participating or nonparticipating in the seismic-force-resisting system. Participating elements must be designed and detailed to resist seismic forces such as shear walls, columns, beams, and coupling elements. Nonparticipating elements can be any masonry element but are not required to be designed and detailed to resist seismic forces. Element classifications are described in Section 7.3 of TMS 402.

Basic seismic-force-resisting systems. Section 7.3.2 of TMS 402 lists five distinct types of conventionally reinforced shear wall systems based on the expected performance of the walls with various construction techniques. The five types of masonry shear wall systems are ordinary plain (unreinforced) masonry, ordinary reinforced masonry, detailed plain (unreinforced) masonry, intermediate reinforced masonry, and special reinforced masonry shear walls.

In addition to the five types of conventionally reinforced shear wall systems, three prestressed masonry shear wall systems are defined: ordinary plain (unreinforced) prestressed masonry, intermediate reinforced prestressed masonry, and special reinforced prestressed masonry shear walls. The various types of masonry shear wall systems are defined in Section 2.2 of TMS 402.

The various masonry shear wall types are assigned seismic design coefficients found in ASCE 7, such as response modification coefficient, R, based on the expected performance and ductility of the particular shear wall system used. Certain shear wall types are required in some seismic regions, and unreinforced shear wall types are not permitted in intermediate and high seismic-risk regions, based on the seismic design category of the building. See Table 12.2-1 of ASCE 7 for seismic design coefficients for the various types of masonry shear wall systems. ASCE 7 contains only one prestressed masonry shear wall system. See item A13 of ASCE 7 Table 12.2-1 for bearing-wall buildings with prestressed masonry shear wall systems, and item B22 of Table 12.2-1 for building-frame buildings with prestressed masonry shear wall systems. Note that unreinforced prestressed masonry shear wall systems are not permitted in Seismic Design Categories C, D, E, and F.

In addition to the eight shear wall systems described above, TMS 402 has an empirically designed shear wall and three types of AAC masonry shear walls for a total of 12 different types of masonry shear wall systems. The requirements for the different shear walls are given in Commentary Table CC-7.3.2-1 of TMS 402. Table 2106-1 summarizes the requirements for each of the different masonry shear wall systems described above. It should be emphasized, however, that the limitations and restrictions imposed on the different types of masonry shear wall systems are given in Tables 12.2-1 and 12.14-1 of ASCE 7.

Table 2106-1. Shear Wall Types and Requirements[a]

Shear Wall System	TMS 402 Design Methods	Reinforcement Requirements	Permitted in Seismic Design Category
Ordinary plain (unreinforced) masonry shear walls	Section 8.2 or 9.2	None	A, B
Detailed plain (unreinforced) masonry shear walls	Section 8.2 or 9.2	Section 7.3.2.2.1	A, B
Ordinary reinforced masonry shear walls	Section 8.3 or 9.3	Section 7.3.2.2.1	A, B, C
Intermediate reinforced masonry shear walls[b]	Section 8.3 or 9.3	Section 7.3.2.4	A, B, C
Special reinforced masonry shear walls	Section 8.3 or 9.3	Section 7.3.2.5	A, B, C, D, E, F
Ordinary plain (unreinforced) AAC masonry shear walls	Section 11.2	Section 7.3.2.6.1	A, B
Detailed plain (unreinforced) AAC masonry shear walls	Section 11.2	Section 7.3.2.7.1	A, B
Ordinary reinforced masonry AAC shear walls	Section 11.3	Section 7.3.2.8	A, B, C, D, E, F
Ordinary plain (unreinforced) prestressed masonry shear walls[c]	Chapter 10	None	A, B
Intermediate reinforced prestressed masonry shear walls[c]	Chapter 10	Section 7.3.2.10	A, B, C
Special reinforced prestressed masonry shear walls[c]	Chapter 10	Section 7.3.2.11	A, B, C, D, E, F

[a]See ASCE 7 Table 12.2-1.
[b]The maximum spacing for vertical reinforcement in an intermediate reinforced masonry shear wall is 4′-0″ o.c.
[c]ASCE 7 Table 12.2-1 does not allow prestressed masonry shear walls in SDC B-F.

Anchorage of masonry walls. Section 7.2.3 of TMS 402 references the "legally adopted building code" or ASCE 7 for anchorage of masonry walls. Because of the significant out-of-plane seismic forces that develop in concrete and masonry walls during seismic ground motion, the code requires them to be anchored to roofs and floors that provide lateral support for the wall. The anchorage must provide a positive and direct connection capable of resisting the design seismic forces. Although the term *positive and direct* is not explicitly defined in the code, the intent is to transfer the lateral loads as directly as possible and not by indirect circuitous load paths that are less reliable and more prone to failure during an earthquake.

IBC Section 1604.8.2 references Section 1.4.4 of ASCE 7 for walls in structures in Seismic Design Category A and Section 12.11 for walls in structures in other seismic design categories. Section 1.4.4 of ASCE 7 requires the anchorage to develop a design strength equal to 0.2 times the weight of the wall tributary to the anchor but not less than 5 psf (0.239 kN/m^2). Where the anchors are used in masonry walls constructed of hollow units or cavity wall systems, the code requires them to be embedded in a reinforced grouted structural element within the wall. See the discussion under Section 1604.8.2.

For buildings in Seismic Design Categories B through F, anchorage of concrete and masonry walls must be designed in accordance with Section 12.11 of ASCE 7, or Section 12.14.7.5 for the simplified seismic design procedure. Section 12.11 requires the anchorage to be designed for the greater of several seismic forces, with specified minimum forces. Good resources for wall anchorage design are the *2021 IBC SEAOC Structural/Seismic Design Manual Volume 1: Code Application Examples, 2021 IBC SEAOC Structural/Seismic Design Manual Volume 2: Examples for Light-frame, Tilt-up and Masonry Buildings,* and *Structural Load Determination: 2024 IBC and ASCE/SEI 7-22* by David Fanella.

Requirements based on Seismic Design Category. Seismic design is not optional, but the permissible systems and the level of detailing required depend on the seismic design category of the building. Note that the seismic design requirements are cumulative, which means the requirements in Seismic Design Category D include the requirements for Seismic Design Category C, and so on.

Seismic Design Category A. The requirements for Seismic Design Category A are given in TMS Section 7.4.1. The interaction of structural and nonstructural elements that may affect the seismic response of the structure must be considered. Structural elements must be classified as nonparticipating and participating elements. Nonparticipating elements must be isolated from the seismic-force-resisting system. Participating elements are elements of the seismic-force-resisting system and must be designed and detailed accordingly. Masonry elements in the seismic load path to the foundation must be capable of resisting the design seismic forces. TMS 402 does not prescribe anchorage and connection forces, because these are design loads. Thus, TMS 402 references the legally adopted building code or ASCE 7 for determination of anchorage design forces. Although drift limits must be met in Seismic Design Category A, it is permitted to assume that some of the shear wall types are deemed to meet the drift limits required by ASCE 7. See TMS Section 7.2.4. Note that all of the seismic load requirements for Seismic Design Category A are entirely self-contained in Section 11.7 of ASCE 7 which refers to the minimal structural integrity requirements in Section 1.4.

Seismic Design Category B. Masonry partition walls, screen walls, and other masonry elements that are not specifically designed to resist external loads other than the loads produced by their own weight are to be isolated from the rest of the structure so that the forces from the structure are not imparted to these elements. Any joints or connections between the structure and these isolated elements must be designed to accommodate the design seismic story drift. These provisions are the nonparticipating element design requirements of TMS 402 Section 7.4.1.1 and apply to buildings in both Seismic Design Categories A and B. All masonry shear wall types are permitted in Seismic Design Category B.

Seismic Design Category C. Structures in Seismic Design Category C must conform to the requirements for Seismic Design Category B, as well as the additional requirements in Section 7.4.3 of TMS 402. Horizontal or vertical reinforcing is required for nonparticipating elements described in Section 7.4.3.1. The vertical reinforcing requirement for nonparticipating elements is one No. 4 bar spaced not more than 10 feet on center and within 16 inches of the end of masonry walls. Past earthquakes such as the 1971 San Fernando earthquake have demonstrated that connections to columns are particularly vulnerable. Thus, anchor bolts used to connect horizontal members to columns require lateral ties. Anchorages used to transfer seismic forces from roof and floor diaphragms for AAC masonry shear walls must be embedded in grout. Structural clay load-bearing wall tiles are not permitted in the seismic-force-resisting system. Although masonry buildings may be designed to use shear walls to resist seismic forces, columns may be incorporated to support gravity loads. TMS 402 permits piers and columns to support lateral loads where the R-factor is not greater than 1.5. In this case, TMS 402 stipulates that the lateral stiffness at each story level and at each line of lateral resistance be not less than 80 percent of the lateral stiffness provided by the shear walls. Columns that resist loads from discontinuous walls, and beams that resist loads from discontinuous walls or frames, require minimum transverse reinforcement with a ratio of 0.0015. These requirements for strength and toughness are intended to prevent local failure or collapse in elements supporting discontinuous portions of the lateral-force-resisting system. Inherent in the code philosophy is the assumption that the inelastic demands on the structure will be reasonably distributed throughout the lateral-force-resisting system. The value of the response modification factor, R, is based on this assumption, as well as the assumption of sufficient ductility and overstrength to meet the maximum anticipated seismic demands. Elements used to redistribute or transfer the effect of overturning forces and shears from stiff discontinuous elements are susceptible to increased localized or concentrated inelastic demands, which violate the above assumption, and may not achieve the required ductility. The requirement for a minimum amount of transverse reinforcement at a maximum spacing increases the maximum usable masonry strain and ductility in the element so that the element will meet the maximum seismic demand and distribute the overturning forces and shears without failure.

Seismic Design Category D. Structures in Seismic Design Category D must conform to the requirements for Seismic Design Categories B and C and the additional requirements of Section 7.4.4 of TMS 402. The vertical reinforcing requirement for nonparticipating elements is one No. 4 bar spaced not more than 48 inches (1,219 mm) on center and within 16 inches (406 mm) of the end of masonry walls. Lateral ties embedded in grout are required for masonry columns. Lateral ties must be anchored with standard 135° (2.36 rad) or 180° (3.14 rad) hooks. Type N mortar and masonry cement is not permitted to be used in participating elements of the seismic-force-resisting system.

Seismic Design Category E or F. Structures in Seismic Design Category E or F must conform to the requirements specified for Seismic Design Categories B, C, and D and the additional requirements in Section 7.4.5 of TMS 402. Stack bond used in nonparticipating elements is required to have a horizontal reinforcing ratio of at least 0.0015 with a maximum spacing of 24 inches (610 mm) on center and must be constructed of hollow open-end units grouted solid or two wythes of solid units fully grouted.

Special reinforced masonry shear walls. Special reinforced masonry walls are permitted in any seismic design category and are the only type of shear wall allowed in Seismic Design Category D and above. Refer to Tables 12.2-1 and 12.14-1 of ASCE 7 for the seismic design coefficients, limitations, and restrictions pertaining to special reinforced masonry shear walls. The special reinforced shear wall has the most favorable (highest) R-factor of the shear wall types, and as such requires specific reinforcement to improve and ensure ductility.

The special reinforced shear wall reinforcement requirements are given in Section 7.3.2.5 of TMS 402.

Special reinforced shear walls designed by the allowable stress design method. Special reinforced shear walls designed by the allowable stress design method are to be designed to resist 2.0 times the code-prescribed in-plane shear force determined from ASCE 7.

However, the 2.0 multiplier does not apply to the overturning moment or out-of-plane forces. See Section 7.3.2.5.1.1. The intent of this requirement is to ensure that flexural failure dominates in order to ensure ductile performance.

Section 2107 *Allowable Stress Design*

2107.1 General. Section 2107 references the requirements in Chapters 1 through 8 of TMS 402, with additional modifications in Sections 2107.2 and 2107.3. The allowable stress design provisions in the standard are based on the use of full design stresses only, assuming that all engineered structural masonry will have some level of special inspection. These minimum levels of special inspection have accordingly been incorporated into Section 1705.4, which references the quality-assurance provision of TMS 402. It is important to note that the one-third increase in allowable stresses that was permitted for load combinations that include earthquake and wind forces is no longer permitted in TMS 402. From an IBC perspective, masonry structures designed by the allowable stress design provisions would be designed by either the allowable stress design load combinations of ASCE 7 or the alternative load combinations prescribed in Section 1605.2 of the IBC.

2107.2 TMS 402, Section 6.1.7.1, lap splices. This modification allows the reinforcing-bar lap splice lengths to be consistent for all masonry design methods (allowable stress design and strength design). It is a permissible alternative to the lap splice length in TMS 402. The splice lengths in TMS 402 are based on developing the allowable stress in the bar by means of an assumed simplified bond mechanism, which is not conservative for large bar sizes.

Both allowable stress and strength design methods assume that the reinforcing will have sufficient strength to resist the imposed forces. Splices of reinforcing bars must meet this same strength test; the same holds true for development length. In seismic regions, the force in any particular bar is indeterminate but may be at the yield strength of the bar. Required development and splice lengths are based on developing the yield strength of the reinforcing bar, including any apparent overstrength, without distress in the masonry, as is done in the strength design method.

The splice length or development length required by the IBC equation, including the f_s factor, will develop 125 percent of the specified yield strength of the reinforcing bar. This allows for the likely overstrength of reinforcing bars and matches the requirements for welded or mechanical splices, which also must develop 125 percent of the specified yield strength of the bar. The IBC also increases lap splices by 50 percent in areas of high tensile demand (greater than 80 percent of the allowable steel tension stress, F_s).

The amendment in Section 2107.2 is only for lap splices. Although the same length is also needed to develop the reinforcing bar strength, the requirement for development length in TMS 402 is not amended in the IBC.

2107.3 TMS 402, Section 6.1.7, splices of reinforcement. The amendment to the standard is the requirement that bars larger than No. 9 (M #29) be spliced using mechanical connectors. The reason for the amendment is that the allowable stress design method of the standard allows reinforcing bar sizes up to No. 11 (M #36). These large bar sizes are very difficult to lap splice without splitting the masonry. Using mechanical connectors can mitigate the splitting problem that occurs in ordinary lap splices. Hence, the code requires mechanical connectors for bar sizes larger than No. 9 (M #29). TMS 402 Section 6.1.7.2 requires mechanical splices to develop 1.25 times the bar yield strength. Welded splices must be ASTM A706 steel reinforcement.

Section 2108 *Strength Design of Masonry*

2108.1 General. The IBC references general design requirements in Chapters 1 through 7 and the strength design procedure in Chapter 9 of TMS 402 with two modifications. The exception for strength design of AAC masonry indicates that AAC masonry must comply with the design requirements in Chapters 1 through 7 and 11 of TMS 402.

2108.2 TMS 402, Section 6.1.6, development. The required development length of reinforcement is determined by Equation 6-1 of the standard but cannot be less than 8 inches (203 mm). The development length shall be increased by 50 percent for epoxy-coated bars. The modification indicates that the required development length of reinforcement need not exceed 72 bar diameters.

2108.3 TMS 402, Section 6.1.7, splices. Welded splices are not permitted in plastic hinge zones of intermediate or special reinforced masonry shear walls. Wherever used, welded splices must be of ASTM A706 (weldable) steel reinforcement. To achieve adequate performance, splices in reinforcing used in the seismic-force-resisting system that are subjected to high seismic strains must be capable of developing the full strength of the reinforcing steel. In order to be welded properly, the chemistry of the steel must have a limited carbon content as well as other elements such as sulfur and phosphorus. If the chemistry of the steel is not carefully controlled, the welding procedures in AWS D1.4, which are based on the steel chemistry, must be carefully adhered to in order to produce welds that develop the strength of the steel. If the carbon equivalent or the sulfur or phosphorus content is too high, the steel may not be weldable. Because the chemistry of reinforcing steel conforming to ASTM A615, A616, and A617 is not controlled and is often unknown, a quality weld that can adequately develop the strength of the steel is not always possible. In contrast, ASTM A706 steel has controlled chemistry and is always weldable. Welded splices are required to be able to develop only 125 percent of the specified yield strength of the spliced bars. However, because A615, A616, A617, as well as A706 bars can have actual yield strengths in excess of 125 percent of the specified yield strength, a code-conforming welded splice may fail before the spliced bars can yield. This would compromise the inelastic deformability of a structural member. Therefore, welded splices are prohibited within the potential plastic hinge region of members in structural systems in buildings assigned to the higher seismic design categories.

Type 1 mechanical splices are not permitted to be used within a plastic hinge zone or within a beam-column joint of intermediate or special reinforced masonry shear walls, because Type 1 splices may not be able to resist the stress levels that develop within the yielding region. However, Type 2 mechanical splices are permitted in any location within a member because Type 2 splices are required to develop the specified tensile strength of the spliced bars. ACI recommends that good detailing preclude the use of splices within regions subjected to potential yielding. However, if splices cannot be avoided, the designer should investigate and document the force-deformation characteristics of the spliced bar and the ability of the splice to meet the expected inelastic demand. See ACI 318 for discussion of the types of mechanical splices.

Section 2109 *Empirical Design of Adobe Masonry*

The empirical design procedure for adobe masonry is a prescriptive method of sizing and proportioning masonry structures using rules and formulas that were developed over many years. The procedure is based on experience and predates the engineering design methods.

The empirical method was developed for use in smaller buildings with more interior walls and stiffer floor systems than are commonly built today. Gravity loads are assumed to be approximately centered on bearing walls and foundation piers, and the effects of reinforcement are neglected.

The option for empirically designed masonry has been removed from the 2022 edition of TMS 402. In recent years, TMS 402 has curtailed the application of empirical design for contemporary masonry materials, construction methods, and building types because these modern buildings and materials no longer rely on the smaller number and size of openings, more frequent cross walls, and shorter walls assumed in the empirical design approach.

While TMS 402 Appendix A: Empirical Design of Masonry will no longer be included in the 2022 and future editions, retaining reference in the IBC to the previous edition (TMS 402-16) will allow adobe masonry provisions to remain available until a standard specific to adobe construction can be created and approved as an IBC-referenced standard.

2109.1 General. Adobe construction is defined in Section 202 and is designed and constructed according to the empirical method in Appendix A of TMS 402-16 and the specific requirements given in Section 2109.2. From a fire- and life-safety standpoint, adobe construction must meet the requirements for Type V construction specified in Chapter 6.

2109.1.1 Limitations. The limitations are given in Section A.1.2 of TMS 402-16. Empirical design is not permitted in buildings in Seismic Design Category D, E, or F and cannot be used for the seismic-force-resisting system in Seismic Design Category B or C. Where empirically designed masonry walls are used in the lateral-load-resisting system (wind or seismic), the height of the building is limited to 35 feet (10,688 mm). See Table A.1.1 of TMS 402-16 for limitations based on building height and wind speed. The empirical design method cannot be used to design or construct masonry buildings or parts of masonry buildings located in areas where the allowable stress design wind speed (V_{asd}) determined in accordance with Section 1609.3.1 exceeds 110 mph (49 m/s). If a building structure exceeds the limitations prescribed in this section or the standard, then the building or structure must be designed in accordance with the engineering provisions covered in Section 2101.2. Masonry foundation walls may be constructed in accordance with the prescriptive masonry foundation wall provisions covered in Section 1807.1.6.

TMS 402-16 Section A.1.1.2 prohibits the use of empirical design for AAC masonry, which must be designed in accordance with the strength design procedure in TMS 402, by lack of a reference to the material for design. Concrete, clay, and stone masonry may be used with empirical design.

Section 2110 *Glass Unit Masonry*

This section covers the empirical requirements for nonload-bearing glass unit masonry elements used in exterior or interior walls. Glass unit masonry is not permitted to be used in various types of fire-resistive walls, smoke barriers, or load-bearing walls. See further discussion in the *International Building Code Commentary*. Glass unit masonry is designed and constructed according to Chapter 14 of TMS 402 and the specific requirements in Section 2110.

Section 2111 *Masonry Fireplaces*

This section covers requirements for masonry fireplaces and their foundations constructed of concrete or masonry. Note that Section 2111 covers requirements for masonry fireplaces, not masonry chimneys. Specific requirements for masonry chimneys are covered in Section 2113.

The IBC requires concrete and masonry fireplaces to have both seismic reinforcement and seismic anchorage depending on the seismic design category of the structure. In Seismic Design Category A or B, no seismic reinforcement is required. In Seismic Design Category C or D, prescriptive reinforcement and anchorage is provided in Sections 2111.4.1, 2111.4.2, and 2111.5. In Seismic Design Category E or F, the reinforcement and anchorage must be designed in accordance with the engineering requirements of Sections 2101 through 2108, which generally references provisions in TMS 402.

Section 2112 *Masonry Heaters*

This section covers requirements for masonry heaters as defined in Section 2112.1. This is possibly the only definition that still remains in the body of the code and not in Chapter 2. Seismic anchorage and reinforcing is only required in masonry heaters in Seismic Design Categories D, E, and F. The section references Section 2113 for anchorage and reinforcing.

Section 2113 *Masonry Chimneys*

This section applies to masonry chimneys constructed of concrete or masonry. The section covers seismic anchorage and reinforcing, footing support, and general construction requirements for masonry chimneys. The IBC requires concrete and masonry chimneys to have both seismic reinforcement and seismic anchorage depending on the seismic design category of the structure. In Seismic Design Category A or B, no seismic reinforcement is required. In Seismic Design Category C or D, prescriptive reinforcement and anchorage is provided in Sections 2113.3.1, 2113.3.2, and 2113.4. In Seismic Design Category E or F, the reinforcement must be designed in accordance with the engineering requirements of Sections 2101 through 2108, which generally references provisions in TMS 402. Anchorage must be designed in accordance with Section 2113.4.

Section 2114 *Dry-Stack Masonry*

This section limits the use of dry-stack masonry to Risk Category I through III structures and requires the use of concrete masonry units. The design must comply with TMS 402 but shall not exceed the maximum allowable stresses provided in Table 2114.4.

KEY POINTS

- Chapter 21 contains requirements for the design and construction of masonry buildings and other masonry structures regulated by the IBC.
- The 2024 IBC references the 2022 edition of TMS 402 *Building Code Requirements for Masonry Structures,* and TMS 602 *Specification for Masonry Structures,* and in many cases, the IBC refers to specific chapters in the standard for detailed design and construction requirements.
- Masonry structures may be designed by one of the procedures prescribed in Section 2101.2 for allowable stress design, strength design, prestressed masonry design, or the direct design procedure in accordance with TMS 403.
- Allowable stress designed, strength designed, AAC, and prestressed masonry must also meet the applicable seismic design requirements of Section 2106, which references Chapter 7 of TMS 402. Seismic requirements are based on the seismic design category of the structure.
- Glass unit masonry may be designed by provisions in Section 2110 of the code and Chapter 14 of TMS 402.
- Masonry veneer must comply with provisions in Chapter 14 of the code and Chapter 13 of TMS 402.
- TMS 402 contains specific requirements for construction documents pertaining to masonry structures.
- Masonry materials must comply with the requirements and applicable ASTM and other standards referenced in Section 2103.
- Masonry construction methods must comply with the requirements in Section 2104, which references various provisions in TMS 602 or TMS 604.
- A quality-assurance program in accordance with Section 2105 and Chapter 17 is required for masonry structures to ensure that constructed masonry is in conformance with the construction documents.
- Chapter 17 references the quality-assurance provisions in TMS 602, which are based on the risk category of the structure.
- Requirements for masonry fireplaces and chimneys including specific seismic reinforcing and anchorage based on seismic design category are provided in Sections 2111 and 2113, respectively.
- Reinforcement and anchorage of masonry fireplaces and chimneys in structures in Seismic Design Categories E and F must be designed in accordance with the engineering provisions of Chapter 21 and TMS 402.

CHAPTER 22

STEEL

Introduction
Section 2201 General
Section 2202 Structural Steel and Composite Structural Steel and Concrete
Section 2203 Structural Stainless Steel
Section 2204 Cold-Formed Steel
Section 2205 Cold-Formed Stainless Steel
Section 2206 Cold-Formed Steel Light-Frame Construction
Section 2207 Steel Joists
Section 2208 Steel Deck
Section 2209 Steel Storage Racks
Section 2210 Metal Building Systems
Section 2211 Industrial Boltless Steel Shelving
Section 2212 Industrial Steel Work Platforms
Section 2213 Stairs, Ladders and Guarding for Steel Storage Racks and Industrial Steel Work Platforms
Section 2214 Steel Cable Structures
Key Points

> IBC Chapter 22 covers requirements for the quality, design, fabrication, and construction of steel structures, including structural steel, composite structural steel and concrete structures cold-formed steel, steel bar joists, steel cable structures, steel storage racks, structural stainless steel, cold-formed profiled steel diaphragm panels, steel decks, industrial boltless steel shelving, steel work platforms, and metal building systems along with applicable reference standards.

Section 2201 *General*

2201.1 Scope. IBC Chapter 22 covers requirements for the quality, design, fabrication, and construction of steel structures. The 2024 IBC includes an editorial reorganization of Chapter 22 for better flow, usability, and clarification of steel provisions in the building code. New provisions for structural stainless steel, cold-formed profiled steel diaphragm panels, steel decks, industrial boltless steel shelving, steel work platforms, and metal building systems have been added. Based on the extensive reformatting of Chapter 22, Table 2201-1 is provided as a summary of revised section numbers and content.

Table 2201-1. **Reorganization of IBC Chapter 22**

2024 IBC Section	Summary of Section and Content Changes
2201 General	Content moved to 2201.2 Identification, 2201.3 Protection, 2201.4 Connections, and 2201.5 Anchor Rods from previous sections 2202, 2203, 2204, and 2204.3, respectively.
2202 Structural Steel and Composite Structural Steel and Concrete Structures	Renumbered from 2205 and renamed to capture composite structural steel and concrete
2203 Structural Stainless Steel	New section
2204 Cold-Formed Steel	Renumbered from 2210
2205 Cold-Formed Stainless Steel	New section
2206 Cold-Formed Steel Light-Frame Construction	Renumbered from 2211
2207 Steel Joists	No section number change
2208 Steel Deck	New section with content moved from 2210
2209 Steel Storage Racks	No section number change
2210 Metal Building Systems	New section
2211 Industrial Boltless Steel Shelving	New section
2212 Industrial Steel Work Platforms	New section
2213 Stairs, Ladders and Guarding for Steel Storage Racks and Industrial Steel Work Platforms	New section
2214 Steel Cable Structures	Renumbered from 2208

2201.2 Identification. All structural steel used for load-carrying purposes must be properly identified in order to determine conformance with the appropriate specification and grade. Structural steel must be identified in accordance with the American Institute of Steel Construction's AISC 360, *Specification for Structural Steel Buildings*. Chapter N of AISC 360 addresses minimum requirements for quality control, quality assurance, and nondestructive testing for structural steel systems and steel elements of composite members for buildings and other structures.

The identification of cold-formed steel members should be in accordance with the requirements of American Iron and Steel Institute's AISI S100, *North American Specification for the Design of Cold-Formed Steel Structural Members*. Section A of AISI S100 addresses the identification of cold-formed steel. Cold-formed steel used in light-frame construction should be identified in accordance with the additional requirements of AISI S240, *North American Standard for Cold-Formed Steel Structural Framing* for structural members or S220, *North American Standard for Cold-Formed Steel Nonstructural Framing* for nonstructural members. Other types of steel must be identified in accordance with the applicable ASTM standard or other standard referenced in this chapter. The code requires unidentified steel to be tested for conformity to the applicable standard.

2201.3 Protection. Protection of structural steel and cold-formed steel structural members or panels from corrosion is required. Protection may be by painting or galvanizing. Painting of structural steel is required to comply with the requirements of AISC 360. Painting steel joists must comply with the appropriate Steel Joist Institute (SJI) standard for the particular type of joist or girder. Protection of cold-formed steel members should be in accordance with the general requirements of AISI S100, Section B, which addresses the protection of cold-formed steel from corrosion. Cold-formed steel in light-frame construction should be protected in accordance with the additional requirements of AISI S240 or AISI S220, as applicable.

2201.4 Connections. The design and installation of steel connections must be in accordance with the applicable reference standards in this chapter. For special inspection of welding or installation of high-strength bolts, see the discussion of Section 1705.2.

2201.5 Anchor rods. Anchor rods must be set in accordance with the approved construction documents. The code specifies the bolt protrusion such that all of the threads of the nut are fully engaged but the end of the threaded portion does extend beyond the surface of the connected part so the nut does not run out of threads. The most common specification for anchor rods is ASTM F1554 *Standard Specification for Anchor Bolts, Steel, 36, 55, and 105-ksi Yield Strength*. The standard covers straight and bent, headed and headless anchor bolts (also known as anchor rods) made of carbon, carbon boron, alloy, or high-strength low-alloy steel, and having specified yield strengths of 36, 55, and 105 ksi (248, 379 and 724 MPa). The anchor bolts are furnished in two thread classes, and in various sizes, and are intended for anchoring structural supports to concrete foundations. AISC 360 requires threads on anchor rods to conform to ASME B 18.2.6 *Fasteners for Use in Structural Applications*.

Section 2202 *Structural Steel and Composite Structural Steel and Concrete*

Section 202 defines structural steel elements as any steel structural member of a building or structure consisting of a rolled steel structural shape other than cold-formed steel, or steel joist members.

2202.1 General. Structural steel elements and composite structural steel and concrete elements in buildings must be designed, fabricated, and constructed in accordance with the referenced standard, ANSI/AISC 360. AISC 360 is an ANSI-approved standard that provides requirements for the design and construction of structural steel buildings and other structures, and incorporates *both* allowable strength design, and load and resistance factor design methods. The dual-units format provides both U.S. customary and SI units. In addition to AISC 360, AISC publishes the *Steel Construction Manual*. The manual contains the following: AISC 360 *Specification for Structural Steel Buildings*, RCSC *Specification for Structural Joints Using High-Strength Bolts*, and AISC *Code of Standard Practice for Steel Buildings and Bridges*.

2202.2 Seismic design. This section contains provisions for structural steel seismic force-resisting systems and composite structural steel and concrete seismic force-resisting systems and structural steel elements. The IBC references ANSI/AISC 341, *Seismic Provisions for Structural Steel Buildings*. AISC 341 contains an extensive commentary and bibliography. The ANSI-approved specification is a companion to ANSI/AISC 360 that extends coverage to the connection detailing and member design requirements for structural steel and composite systems in high-seismic applications.

In addition to the seismic design standard, AISC publishes the *Seismic Design Manual*. The AISC *Seismic Design Manual* includes AISC 341 and AISC 358, *Prequalified Connections for Special and Intermediate Steel Moment Frames for Seismic Applications*. AISC 358 covers prequalified steel moment connections in intermediate and special moment frames and is also a referenced standard in the IBC.

Requirements for design of structural steel structures resisting seismic forces are based on the seismic design category of the building as determined from Section 1613.2. Requirements for Seismic Design Categories B and C are given in Section 2202.2.1.1, and requirements for Seismic Design Categories D, E, and F are given in Section 2202.2.1.2. For a detailed discussion on determination of seismic design category, see the discussion of Section 1613.

2202.2.1.1 Seismic design category B or C. All structures in Seismic Design Category A need only comply with the general structural integrity requirements prescribed in Section 1.4 of ASCE 7, which is referenced from ASCE 7 Section 11.7. Steel structures assigned to Seismic Design Category B or C may be of any construction permitted in Section 2202. Where a structure is designed using a Response Modification Factor, *R*, from Table 12.2-1 of ASCE 7, the structure must be designed and detailed in accordance with AISC 341. The exception allows structures assigned to Seismic Design Category B or C to be designed as "steel systems not specifically detailed for seismic resistance, excluding cantilever column systems" using an *R*-factor of 3 per AISC 360 and not AISC 341. See Item H of Table 12.2-1 of ASCE 7. It should be noted that this category does not include cantilever column systems, which are specifically covered under Item G in ASCE 7 Table 12.2-1. Although not required, structures in Seismic Design Category B or C may be designed and detailed in accordance with AISC 341 using the appropriate *R*-factor from ASCE 7 Table 12.2-1 or Table 12.14-1 for the type of building system involved. In a sense, this is trading ductile detailing for design force level. Using systems with higher *R* values produces lower lateral forces but requires more ductile detailing to ensure ductile response characteristics. In general, detailing requirements can have a significant impact on construction cost of the seismic-force-resisting system; therefore, providing the required detailing in order to design for lower lateral forces may not be economical in lower seismic design categories.

The IBC clarifies that beam-to-column moment connections that are part of a "special" or "intermediate" moment frame are required to be prequalified, or qualified by testing, in accordance with AISC 341 or to be prequalified as prescribed in AISC 358. This provision also applies to seismic design categories D through F.

2202.2.1.2 Seismic design category D, E, or F. Unlike structural steel systems in Seismic Design Category A, B, or C, steel structures in Seismic Design Category D, E, or F are required to be designed and detailed in accordance with AISC 341 using the appropriate *R*-factor from ASCE 7 Table 12.2-1 or Table 12.14-1 for the building type and seismic-force-resisting system involved. Note that ASCE 7 Table 15.4-1 contains specific detailing for nonbuilding structures similar to buildings. See Section 2202.2.1.1 for discussion of beam-to-column moment connections for special and intermediate moment frames.

2202.2.2 Structural steel elements. This section covers design and construction of steel elements in seismic-force-resisting systems, such as chords, collectors, and foundation elements. These elements must be designed in accordance with AISC 341 where: (1) the structure is in Seismic Design Category D, E, or F, (2) part of nonbuilding structures similar to buildings designed in accordance with ASCE 7 Table 15.4-1, or (3) the structure is in Seismic Design Category B or C and an R-factor greater than 3 is used in the design.

Section 2203 *Structural Stainless Steel*

The 2024 IBC introduced a new section on structural stainless steel with reference to the new AISC 370, *Specification for Structural Stainless Steel Buildings*, standard. This standard provides a uniform practice for the design of structural stainless-steel-framed buildings and other structures.

Section 2204 *Cold-Formed Steel*

2204.1 General. The term "cold-formed steel construction" is defined in Section 202. This section references the *North American Specification for the Design of Cold-Formed Steel Structural Members* (AISI S100-16) with Supplement 2 (2020). AISI S100 is also known as the "North American Specification" and harmonizes cold-formed steel design across the United States, Canada, and Mexico. Chapters A through M and Appendices 1 and 2 comprise the main portion of the document, while Appendix A is specific to the United States and Mexico and Appendix B is specific to Canada.

Sections 2204 and 2208 on cold-formed steel members and decks include a new reference to AISI S310, *North American Standard for the Design of Profiled Steel Diaphragm Panels*, with newly developed Supplement 1. This standard, first published in 2013, includes design provisions for diaphragms consisting of profiled steel decks or panels that include fluted profiles and cellular deck profiles. The diaphragm may be installed with or without insulation between the panels and supports and may be supported by materials made of steel, wood, or concrete. Supplement 1 clarifies the design requirements for diaphragms filled with structural concrete used in the construction of floors. It also provides direction for calculating the resistance of perimeter attachment fasteners, including headed studs, welds, screws, and proprietary fasteners.

2204.2 Seismic design. Where required by ASCE 7, the seismic design of cold-formed steel seismic force-resisting systems is to be in accordance with AISI S100. Section 2204.2.1 references AISI S400 *North American Standard for Seismic Design of Cold-formed Steel Structural Systems, 2020 Edition* which includes design provisions for cold-formed steel special-bolted moment frames (CFS-SBMF). CFS-SBMF systems experience substantial inelastic deformation during a design seismic event, with most of the inelastic deformation occurring at the bolted connections due to slip and bearing (See Figure 2204-1). Cyclic testing has shown that

Figure 2204-1 **Cold-formed steel special-bolted moment frames.**

the system has large ductility capacity and significant hardening. To develop the expected mechanism, requirements based on capacity design principles are provided in the standard for the design of beams, columns, and their connections. Table 12.2-1 of ASCE 7 includes seismic design parameters for the CFS-SBMF system of $R = 3.5$, $\Omega_0 = 3.0$, and $C_d = 3.5$. AISI S400 also includes specific quality assurance and quality control procedures.

Section 2205 *Cold-Formed Stainless Steel*

Cold-formed stainless-steel provisions are included in Section 2205 with reference to standard ASCE 8 *Specification for the Design of Cold-Formed Stainless Steel Structural Members* for design.

Section 2206 *Cold-Formed Steel Light-Frame Construction*

This section lays out the design and installation requirements for structural and nonstructural members used in cold-formed steel light-frame construction where the specified minimum base steel thickness is not greater than 0.118 inch (3.00 mm).

The IBC references several AISI standards including AISI S240, *North American Standard for Cold-Formed Steel Structural Framing*; AISI S400, *North American Standard for Seismic Design of Cold-Formed Steel Structural Systems*; AISI S230, *Standard for Cold-Formed Steel Framing—Prescriptive Method for One- and Two-Family Dwellings*; and AISI S202, *Code of Standard Practice for Cold-Formed Steel Structural Framing*. The AISI S240 standard addresses requirements for construction with cold-formed steel structural framing that are common to prescriptive and engineered light-frame construction.

2206.1 Structural framing. Cold-formed steel light-frame construction must comply with the AISI S240 standard as well as Sections 2206.1.1 through 2206.1.3. The code also clarifies that these provisions apply to floor and roof systems, structural walls, lateral force-resisting systems, and trusses.

2206.1.1 Seismic requirements for cold-formed steel structural systems. Design requirements for cold-formed light-frame construction for Seismic Design Categories B and C are addressed separately from the requirements for Seismic Design Categories D through F. In both instances the design is required to conform to the AISI S400 standard, although for Seismic Design Categories B and C an exception is provided allowing the design to simply conform to AISI S240 if the seismic response modification factor (R=3) selected in Table 12.2-1 Line H of ASCE 7 is not specifically detailed for seismic resistance.

2206.1.2 Prescriptive framing. AISI S230 is referenced for the design and construction of cold-formed steel framing in detached one- and two-family dwellings and townhouses up to three stories in height. This standard incorporates provisions such as the L-header and an efficient design procedure for built-up headers.

2206.1.3 Truss design. AISI S202, specifically Section I1, addresses the responsibilities of the designer, manufacturer, and contractor for cold-formed steel trusses. This section requires that truss drawings be provided and that permanent truss bracing is installed in addition to specifying quality assurance requirements during fabrication and special inspection of cold-formed steel trusses spanning 60 feet (18,288 mm) or more.

2206.2 Nonstructural members. Nonstructural members are to be designed in accordance with AISI S220. In Section 2206.1, structural members are defined as floor and roof systems, structural walls, lateral force-resisting systems, and trusses. Nonstructural members could consist of nonbearing walls, ceiling framing, bracing, or blocking members, etc.

2206.3 Cutting and notching. Cutting holes and notching cold-formed steel framing members must be in accordance with AISI S240 for structural members and AISI S220 for nonstructural members.

Section 2207 *Steel Joists*

Steel joists are required to be designed, manufactured, and tested in accordance with the various specifications published by the SJI and referenced in Section 2207.1.

Joists are used primarily as gravity load-carrying members. When joists are incorporated into the lateral force-resisting system as collectors, chords, and diaphragm ties, care must be taken to ensure that the joists are specifically designed and detailed to properly function as intended. Normally, joists are designed only for gravity and wind uplift loads. A continuous load path is necessary for chords, collectors, and diaphragm ties, and their connections. The typical seat on a steel joist is eccentric to the chord and is not ordinarily designed to carry the axial force and resulting moment through the seat when functioning as a collector or chord, which are in tension and compression. Generally, special seats must be designed and manufactured, or the load path must be specifically directed through the chords by direct chord-to-chord connection.

2207.1 General. The design, manufacture, and installation of open web steel joists and joist girders are required to conform to either SJI 200, *Standard Specification for CJ-Series Composite Steel Joists* or SJI 100, *Standard Specification for K-Series, LH-Series, DLH-Series Open Web Steel Joists, and for Joist Girders.* Additionally, seismic design must be in accordance with the provisions of Section 2202 or Section 2206 for light-frame cold-formed steel construction.

2207.2 Design. The registered design professional must indicate on the construction documents the steel joist or joist girder designations used as well as layout scheme, end support, anchorages, bridging, bridging termination connections, and bearing connections that resist uplift and lateral loads. The construction documents must also include any special loads or conditions that are listed in the code section. The construction documents are part of the submittal documents required by Section 107. See Section 202 for the definition of construction documents.

2207.3 Calculations. The steel joist/girder manufacturer is required to design the steel joists and/or steel joist girders to support the anticipated loads and load combinations prescribed by Chapter 16 in accordance with the applicable SJI specifications. The registered design professional responsible for the design of the building may require joist/girder calculations to be prepared along with a cover letter bearing the seal and signature of the joist manufacturer's registered design professional. In addition to the standard steel joist or joist girder calculations, other items such as non-SJI standard bridging details, nonstandard connection details, field splices, and joist headers are required to be submitted.

2207.4 Steel joist drawings. Steel joist placement plans that indicate the steel joist products specified on the construction documents are required to be submitted to the building department for review and approval. Joist placement plans are primarily used to install the joist system in the field. This code section is very specific as to what the joist placement plans are required to include. Note that the section specifically states that steel joist placement plans do not require the seal and signature of the joist manufacturer's registered design professional.

2207.5 Certification. The steel joist manufacturer is required to submit a certificate of compliance stating that work was performed in accordance with approved construction documents and the applicable SJI standard specifications. See also Section 1704.2.5 regarding exemption of special inspection where work is done by an approved fabricator. A certificate of compliance from the fabricator is required to be submitted to the building official stating that the work was performed in accordance with approved construction documents and with SJI standard specifications.

Section 2208 *Steel Deck*

The section covers design and construction of cold-formed steel floor and roof decks and composite slabs of concrete and steel deck. The newly referenced standard in the 2024 IBC is the Steel Deck Institute's SDI SD *Standard for Steel Deck.* The previously referenced SDI RD (steel roof deck), NC (noncomposite steel floor), and C (composite steel floor) standards were removed and replaced with the combined steel deck standard.

Design of cold-formed steel diaphragms is in accordance with additional provisions of AISI S310. See the discussion in Section 2204 regarding AISI S310.

Section 2209 *Steel Storage Racks*

2209.1 General. In relation to steel storage racks, there are two standards that are referenced in the IBC. These standards developed by the Rack Manufacturer's Institute are ANSI/MH 16.1, *Specification for the Design, Testing and Utilization of Industrial Steel Storage Racks* and ANSI/MH 16.3, *Specification for the Design, Testing and Utilization of Industrial Steel Cantilevered Storage Racks.*

The RMI ANSI/MH 16.1 standard applies specifically to industrial pallet racks, movable shelf racks, and stacker racks made of cold-formed or hot-rolled steel structural members. This standard does not apply to cantilever racks, portable racks, drive-in or drive-through racks, or racks made of materials other than structural steel.

RMI ANSI/MH 16.3 provides design, construction, use, and behavior of cantilevered storage racks with supporting commentary. The behavior of these types of racking systems is significantly different from standard pallet storage racks. Figures 2209-1 and 2209-2 provide a visual comparison of a standard pallet rack and a cantilevered storage rack.

Figure 2209-1 **Steel storage racks.**

Figure 2209-2 **Cantilevered storage racks.**

2209.2 Seismic design. Racking systems that do not comply with the specific limitations of the RMI standards must be designed in accordance with the provisions of Chapter 15 in ASCE 7. ASCE 7 also includes revisions to the ANSI/MH 16.1 standard in relation to base plate design and the use of shims and to the ANSI/MH 16.3 standard for anchor bolt design.

2209.3 Certification. This section requires that racks whose top load level is at 8 feet or more in Seismic Design Category D, E, or F be certified. The installer must provide the owner, or the owner's authorized agent, with a certificate of compliance once installation has been completed. This certificate of compliance must note that the racking provided, and installation, conform to the approved construction documents.

Section 2210 *Metal Building Systems*

Metal building systems are significantly different from other forms of steel construction, especially regarding the shared design responsibilities between the metal building system manufacturer and registered design professional for the project. Metal building systems often contain assemblies made of a variety of components, such as structural steel, cold-formed steel, and steel cables. These systems are typically highly optimized structures that are heavily dependent on bracing components to function as designed. Some bracing components consist of materials that are not considered to be "structural steel." These new provisions and accompanying definition in the 2024 IBC clarify the design requirements for metal building systems. By including the design requirements for different metal building system parts, the special inspection requirements are improved. Section 1705.2.6 provides new special inspection provisions for these systems (see Section 1705.2.6 in this document for more information).

Section 2211 *Industrial Boltless Steel Shelving*

Referenced standard ANSI/MH 28.2 *Design, Testing and Utilization of Industrial Boltless Steel Shelving* applies to boltless shelving placed on mobile carriages; multilevel boltless shelving systems, such as pick modules, catwalks, and deckovers; and boltless shelving used in conjunction with an automated storage and retrieval system. The structural framing components for these systems are made of cold-formed or hot-rolled steel structural members. ANSI/MH 28.2 does not apply to industrial steel pallet racks (addressed by ANSI/MH 16.1), industrial cantilever racks (addressed by ANSI/MH 16.3), boltless shelving structures not fabricated from steel, industrial steel bin shelving or shelving systems built with slotted metal angles. Boltless shelving is typically a prefabricated, free-standing, nonbuilding structure that utilizes a designed framing system. It is usually found within an industrial or warehouse environment that does not allow access by the general public.

ANSI/MH 28.2 applies to boltless shelving structures installed within a building and potentially subjected to seismic loads. Loads from other environmental exposures, such as snow, wind, or rain loads, are not addressed in this standard. ANSI/MH 28.2 does not cover any design requirements that need to be addressed for supported equipment that would subject a shelving system to significant dynamic loading or harmonic vibration that has the potential to cause structural damage or metal fatigue. For applications beyond the scope of ANSI/MH 28.2, evaluation by a qualified design professional is required.

Section 2212 *Industrial Steel Work Platforms*

Referenced standard ANSI/MH 28.3 *Design, Testing and Utilization of Industrial Steel Work Platforms* applies to industrial steel work platforms which are typically prefabricated free-standing nonbuilding structures with an elevated surface that utilizes a predesigned framing system and is located within an industrial or similarly restricted environment. Flooring may include other structural or nonstructural elements, such as concrete, steel, and engineered wood products.

ANSI/MH 28.3 does not apply to platforms whose structural framing components are not made from steel. This standard is based on the work platform being restricted from general public use.

ANSI/MH 28.3 applies to work platforms potentially subjected to seismic loads. Loads from other environmental exposures such as snow, wind or rain loads are not addressed in this standard. For applications beyond the scope of ANSI/MH 28.3, evaluation by a qualified design professional is required.

Section 2213 *Stairs, Ladders and Guarding for Steel Storage Racks and Industrial Steel Work Platforms*

Referenced standard ANSI/MH 32.1 *Stairs, Ladders and Open-Edge Guards for Use with Material Handling Structures* applies to fixed stairways and ladders along with guards for elevated platforms used in material handling structures. The stairways and ladders are attached to structures such as racking pick modules, shelving pick modules and decked-over platforms, and free-standing work platforms. Standard ANSI/MH 28.3 applies to the design, testing, and utilization of industrial steel work platforms (see Section 2212 in this document for background).

Not all stairs, ladders, and guards for steel-framed platforms and work areas are governed by ANSI/MH 32.1. Application is limited to a very specialized subset of stairs, ladders, and guards for "industrial steel work platforms used in material handling structures" within its scope. For example, consider a steel-framed platform used to service HVAC equipment in a factory. The guards on the HVAC access platform must comply with IBC Section 1607.9, which covers loads on handrails and guards. The stairs or ladder used to access the HVAC platform also must comply with the structural and architectural requirements in the IBC, not ANSI/MHI 32.1. Extending the stair, ladder, and guard provisions of the ANSI/MH 32.1 standard to all "industrial steel work platforms" is not appropriate.

Section 2214 *Steel Cable Structures*

2214.1 General. The IBC references ASCE 19, *Structural Applications of Steel Cables for Buildings*, for design and construction of steel cables used in buildings. See the commentary in ASCE 19 for a detailed explanation of the structural requirements for cable structures.

KEY POINTS

- This chapter contains requirements for the design and construction of structural steel buildings and other steel structures regulated by the IBC.
- The 2024 IBC references the 2022 editions of AISC 360 *Specification for Structural Steel Buildings* and AISC 341 *Seismic Provisions for Structural Steel Buildings*, and various AISI standards for cold-formed steel structures.
- Structural steel members must be properly identified in accordance with AISC 360, and cold-formed steel structural members must be identified in accordance with AISI S100.
- Steel members used for structural purposes other than structural steel and cold-formed steel must be properly identified in accordance with applicable ASTM standards.
- Welded and bolted connections must conform to the applicable referenced specification (standard) that governs their specific application.
- Structural steel structures in Seismic Design Category A need not have a designated seismic force-resisting system and need only comply with the basic structural integrity requirements of Section 1.4 of ASCE 7.
- Structural steel structures in Seismic Design Category B or C may be designed in accordance with AISC 360 using an *R*-factor of 3 and need not conform to the seismic detailing requirements of AISC 341, or may be designed to conform to the detailing requirements of AISC 341 where the applicable *R*-factor from ASCE 7 is used.
- Structural steel structures in Seismic Design Category D, E, or F are required to conform to the seismic detailing requirements of AISC 341 using the applicable *R*-factor from ASCE 7.
- Steel joist systems must comply with Section 2207, which includes requirements for design, calculations, joist drawings, and certification.
- Section 2208 requires steel deck design in accordance with Steel Deck Institute or AISI standards.
- Steel storage racks must be designed in accordance with ANSI/MH 16.1 for steel storage racks or ANSI/MH 16.3 for steel cantilevered storage racks, and the applicable seismic design requirements of Section 15.5.3 of ASCE 7.
- Section 2204 references various AISI standards for the design and construction of cold-formed steel structures.
- Section 2206 references various AISI standards for the design and construction of light-frame cold-formed steel structures.
- Sections 2203 and 2205 include provisions for structural stainless steel and cold-formed stainless steel, respectively.
- Section 2210 includes provisions for metal building systems which typically include structural steel, cold-formed steel, steel joists, and steel cable.
- Sections 2211 through 2213 include provisions for industrial boltless steel shelves, industrial steel work platforms and stairs, ladders, and guards for steel work platforms. Standards for these structures are developed by the Material Handling Institute.

CHAPTER 23

WOOD

Section 2301 General
Section 202 Definitions
Section 2302 Design Requirements
Section 2303 Minimum Standards and Quality
Section 2304 General Construction Requirements
Section 2305 General Design Requirements
              for Lateral Force-Resisting Systems
Section 2306 Allowable Stress Design
Section 2307 Load and Resistance Factor Design
Section 2308 Conventional Light-Frame Construction
Section 2309 Wood Frame Construction Manual
Key Points

> Chapter 23 provides minimum requirements for the design of buildings and structures that use wood and wood-based products. The chapter is organized around three design methodologies: allowable stress design (ASD), load and resistance factor design (LRFD), and conventional light-frame construction. In addition, it allows the use of the American Wood Council *Wood Frame Construction Manual* (WFCM) for a limited range of structures. Included in the chapter are references to design and manufacturing standards for various wood and wood-based products; general construction requirements; design criteria for lateral force-resisting systems and specific requirements for the application of the three design methods.

Section 2301 *General*

2301.1 Scope. Chapter 23 covers materials, design, construction, and quality of wood buildings and structures. The structural wood products addressed in this chapter include sawn lumber, glued-laminated timber, cross-laminated timber, wood structural panels, particleboard, fiberboard, hardboard, prefabricated wood I-joists, structural composite lumber, structural logs, round timber poles and piles, and engineered wood rim joists. Provisions for preservative-treated and fire-retardant-treated wood are also included. The chapter additionally contains material specifications, quality requirements, and design provisions, as well as empirical and prescriptive provisions for conventional wood-frame construction.

2301.2 Dimensions. Nominal lumber sizes, for example, 2 inches by 4 inches, are the sizes usually referred to in this chapter. However, actual or net dimensions, smaller than nominal dimensions, are used in structural calculations to determine member section properties, actual stresses, and strength properties. The nominal and actual sizes of dimension lumber are established by Department of Commerce (DOC) Voluntary Product Standard PS 20, *American Softwood Lumber Standard*. The PS 20 standard is available from the National Institute of Standards and Technology (NIST). The nominal size, dressed size, and section properties for sawn lumber and glued-laminated timber are given in Tables 1A, 1B, 1C, and 1D of the *National Design Specification (NDS) for Wood Construction Supplement: Design Values for Wood Construction*. See Section 202 for the definition of nominal size lumber.

Where dimensions of cross-laminated timber (CLT) thickness are specified, they are actual dimensions. CLT is manufactured and identified in accordance with ANSI/APA PRG 320, *Standard for Performance-Rated Cross-Laminated Timber*.

Section 202 *Definitions*

The specific terms related to wood construction are defined in Section 202 to clarify their meaning. Many of the terms clarify elements of the lateral force-resisting systems.

Some specific terms are discussed below.

COLLECTOR. A collector accumulates shear from the (floor or roof) diaphragm and delivers it to vertical lateral force-resisting elements such as shear walls. The term *drag strut* is a colloquial expression that means *collector*. Collectors can also be used to transfer forces within a diaphragm. The IBC recognizes both horizontal and sloped (or nearly horizontal) diaphragms and as such the definition of collector applies to both horizontal and sloped diaphragms. The word *horizontal* was used to differentiate diaphragms from vertical lateral force-resisting elements such as shear walls.

DIAPHRAGM. A diaphragm is a horizontal or nearly horizontal (sloped) structural element that transmits lateral forces to vertical elements (shear walls or frames) of the lateral force-resisting system. An unblocked diaphragm has edge nailing at supported edges only. In an unblocked diaphragm, panel edges are unblocked between supporting structural members. In a blocked diaphragm, all sheathing panel edges are supported by framing members or solid blocking members. Blocked diaphragms have continuity between the sheathing panel edges and therefore have significantly less deflection than unblocked diaphragms because of the stiffness developed by the continuity at blocked panel edges.

It is important to note that the definition of diaphragm also includes horizontal bracing systems.

In terms of distribution of seismic or wind forces, the stiffness of the diaphragm relative to the stiffness of the vertical lateral force-resisting elements (shear walls or frames) is the important parameter by which to classify a diaphragm. Diaphragms can be flexible, rigid, or semi-rigid. Refer to the discussion of ASCE 7 *Minimum Design Loads and Associated Criteria for Buildings and Other Structures* Section 12.3 for additional information on diaphragm classifications.

NATURALLY DURABLE WOOD. Naturally durable wood is the heartwood of a durable listed species. Naturally durable wood includes decay-resistant woods, which are redwood, cedar, black locust, and black walnut, and termite-resistant woods, including redwood, Alaska yellow cedar, Eastern red cedar, and Western red cedar. Note that the heartwood of all listed species are both decay and termite-resistant except black locust and black walnut. The latter are not termite-resistant.

SHEAR WALL. A shear wall is a vertical lateral force-resisting element that is designed to resist lateral seismic, and wind, or soil forces parallel to the plane of a wall. A perforated shear wall is a wood structural panel sheathed wall with openings that has not been specifically designed and detailed for force transfer around the openings. Perforated shear wall segments are sections of the shear wall with full-height sheathing that meet the height-to-width ratio limits specified in *Special Design Provisions for Wind and Seismic* (SDPWS).

Section 2302 *Design Requirements*

Section 2302 lists sections of Chapter 23 applicable to each type of engineered design and prescriptive provisions. Reference is also made to the *Wood Frame Construction Manual* and ICC-400, *Design and Construction of Log Structures*.

Section 2303 *Minimum Standards and Quality*

2303.1 General. Structural lumber, wood structural panels, and other wood products vary in strength and other mechanical properties. The code requires that these materials (defined in the first paragraph of this section) conform to their respective standards and grading rules specified in the code. Furthermore, the code requires that they be identified by a label or be accompanied by a certificate of compliance issued by an approved agency. The label is also required to be placed on the material by an approved agency (see labeling—Chapter 17). The proper use of a wood structural member cannot be determined unless it has been properly identified. Counterfeit grade stamps, for example, do occasionally appear on lumber in the field, and it is important that designers and enforcement personnel be familiar with the approved labels. Examples of labels are shown in Figure 2303-1.

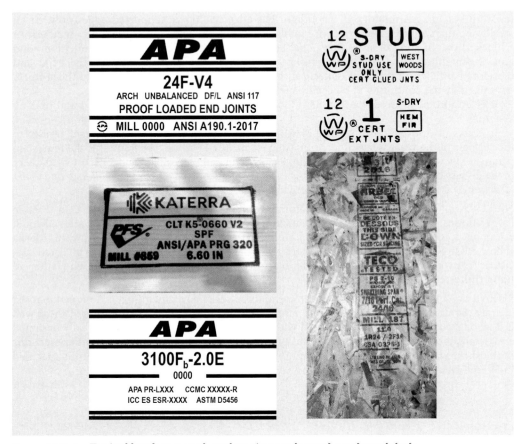

Figure 2303-1 **Typical lumber, panel, and engineered wood products labels.**

2303.1.1 Sawn lumber. This section references the voluntary product standard PS 20 for lumber including end-jointed (finger jointed) or edge-glued lumber, machine stress rated (MSR), or machine evaluated lumber (MEL).

2303.1.1.1 Certificate of inspection. Certification is an acceptable alternative to a grade mark from both U.S. and Canadian grading agencies certified by the American Lumber Standard Committee (ALSC). The code allows certain types of structural lumber to have a certificate of inspection instead of a grade mark. A certificate of inspection is acceptable for precut, remanufactured, or rough-sawn lumber and for sizes larger than 3 inches (76 mm) nominal in thickness. It is industry practice to place only one label (grade mark) on a piece of lumber, which may be removed on precut and remanufactured lumber. Each piece of lumber is graded after it has been cut to a standard size. The grade of the piece is determined based on its size, number, and location of strength-reducing characteristics. Therefore, one log of timber may produce lumber of two or more different grades. It is also industry practice not to label lumber having a nominal thickness larger than 3 inches (76 mm), or rough-sawn material where the label may be illegible. A certificate of inspection from an approved agency is acceptable instead of the label for these types of lumber. The certificate should be filed with the permanent records of the building or structure. If defects exceeding those permitted for the grade installed are suspected, then a certified grader would be needed to determine if the wood is on-grade. To determine if the wood in question is definitely of a suitable grade, the grader must inspect all four faces of the piece.

2303.1.1.2 End-jointed lumber. Approved end-jointed or edge-glued lumber is presumed to be equivalent to solid-sawn lumber of the same species and grade. The NDS permits the use of end-jointed lumber of the same species and grade. When finger-jointed lumber is marked "STUD USE ONLY" or "VERTICAL USE ONLY," such lumber is limited to use where bending or tension stresses are of short duration. The use of the term *approved* is intended to convey the need for quality control during the production of these glued products, and also to establish the qualification tests for the type of end joint used. Joints are tested for strength and for durability, and adhesive manufacturers test their products for durability. Adhesives used in finger-jointed lumber are of two basic types, depending on whether they are to be used for members with long-duration bending loads like floor joists or short-duration bending and tension loads like wall studs. To add elevated-temperature performance requirements for end-jointed lumber adhesives intended for use in fire-resistance-rated assemblies, end-jointed lumber manufactured with adhesives must be designated as "Heat Resistant Adhesive" or "HRA" on the grade stamp.

An example of end-jointed lumber is shown in Figure 2303-2, which illustrates a finger-jointed end joint.

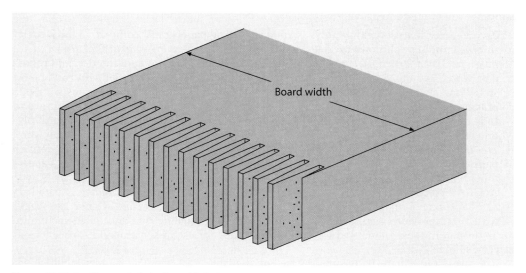

Figure 2303-2 **Finger-jointed end joint.**

2303.1.2 Prefabricated wood I-joists. Wood I-joists are structural members manufactured of sawn or structural composite lumber flanges and wood structural panel webs bonded together with exterior exposure adhesives in the form of an "I" cross-sectional shape. The shear, moment, and stiffness capacities of prefabricated wood I-joists must be established in accordance with ASTM D5055 *Specification for Establishing and Monitoring Structural Capacities of Prefabricated Wood I-Joists*. This standard also specifies that application details, such as bearing length and web openings, are to be considered in determining the structural capacity. Wood I-joists are structural members typically used in floor and roof construction. The standard requires I-joist manufacturers to employ an independent inspection agency to monitor the procedures for quality assurance. The standard specifies that proper installation instructions accompany the product to the job site. The instructions are required to include weather protection, handling requirements and, where required, web reinforcement, connection details, lateral support, bearing details, web hole-cutting limitations, and any special conditions.

2303.1.3 Structural glued-laminated timber. Glued-laminated timbers are manufactured in accordance with ANSI/APA A 190.1 *Product Standard for Structural Glued Laminated Timber*, which references several other standards. ASTM D3737 *Practice for Establishing Allowable Properties for Structural Glued Laminated Timber (Glulam)* is the standard method for establishing allowable stresses for glued-laminated timber.

2303.1.4 Cross-laminated timber. The North American product manufacturing standard, ANSI/APA PRG 320, *Standard for Performance-Rated Cross-Laminated Timber*, provides requirements and test methods for qualification and quality assurance for performance-rated cross-laminated timber. CLT products are manufactured from solid-sawn lumber or structural composite lumber.

Adhesive bonding of wood laminations allows for the manufacture of large wood members. Adhesives used for this purpose must meet specified performance criteria at elevated temperatures. PRG 320 requires such adhesives to pass a bond line shear test at elevated temperatures as a demonstration of this performance. In addition, two test protocols are included in PRG 320 to ensure adhesives used in CLT will not result in significant delamination and fire regrowth. CLT adhesives are required to pass all these tests in order to be qualified under PRG 320.

2303.1.5 Wood structural panels. Wood structural panels must conform to the Department of Commerce voluntary product standards PS 1 *Structural Plywood* or PS 2 *Performance Standard for Wood Structural Panels*. PS 1 is the product standard for construction and industrial plywood, and PS 2 is for wood-based structural-use panels such as oriented strand board (OSB). Wood structural panels include all-veneer plywood; composite panels consisting of a combination of veneer and wood-based material; and mat-formed panels that contain wood strands only, such as OSB and waferboard. The primary distinction between OSB and plywood is that the wood strands in OSB are generally oriented in a manner similar to plywood where "mats" of parallel wood fiber are cross-laminated prior to pressing. Plywood is defined as panels made by cross laminating three or more wood veneers and joining the veneers together with structural adhesives.

ANSI/APA PRP 210 *Standard for Performance-Rated Engineered Wood Siding* is based on APA's PRP 108 *Performance Standards and Qualification Policy for Structural-Use Panels*. ANSI/APA PRP 210 provides requirements and test methods for qualification and quality assurance for performance-rated engineered wood siding intended for use in construction applications as exterior siding.

The term "performance category" is based on the DOC PS 1 and PS 2 standards, which use terminologies of bond classification to reference glue type and performance categories to provide the thicknesses tolerance consistent with the nominal panel thicknesses in the IBC. The performance category value is the "nominal panel thickness" or "panel thickness" both of which are used interchangeably throughout the IBC.

Wood structural panels manufactured in accordance with DOC PS 1 and DOC PS 2 are inspected and labeled to certify compliance by an approved agency. The label identifies the grade and span ratings of the product.

Wood structural panels are also classified by exposure type:

Exterior panels are suitable for applications subject to repeated wetting and redrying or long-term exposure to weather or other conditions of similar severity.

Exposure 1 panels are suitable for uses that are not permanently exposed to the weather, but where panels must resist effects of moisture due to construction delays, high humidity, or other conditions of similar severity. Exposure 1 panels have a fully waterproof bond and are made with the same exterior adhesives used in Exterior-rated panels.

Labels applied to certified structural-use panels must include the exposure durability classification.

Plywood is manufactured from more than 70 species of wood, which are divided into five groups in accordance with their stiffness and strength characteristics. OSB is manufactured from several species of wood. Table 2303-1 provides an overview of wood structural panel use including grade and bond classification, performance category and panel construction (e.g., OSB and plywood).

2303.1.6 Fiberboard. Fiberboard is a smooth textured panel made up of natural fibers such as wood or cane. Fiberboard is used primarily as an insulating board and for decorative purposes, but may also be used as wall or roof sheathing under the provisions of this section. Fiberboard is used in most locations where panels are necessary, including wall sheathing, insulation of walls and roofs, roof decking, doors, and interior finish. Although fiberboard sheathing board may be used for shear walls [see Table 2306.3(2)], fiberboard may not be used for diaphragms. The material is generally labeled to indicate an intended use, strength values, and flame resistance where applicable.

Fiberboard is permitted without any fire-resistance treatment in the walls of all types of construction. When used in fire walls and fire separation walls, the fiberboard must be attached directly to a noncombustible base and protected by a tight-fitting, noncombustible veneer that is fastened through the fiberboard to the base. This will prevent the fiberboard from contributing to the spread of fire.

Fiberboard used in building construction must comply with ASTM C 208, *Specification for Cellulosic Fiber Insulating Board*. When used as structural sheathing, fiberboard must be identified by an approved agency. See SDPWS Table 4.3A for nominal unit capacities for wood-based panel shear walls, including structural fiberboard sheathing.

2303.1.7 Hardboard. Hardboard is used as exterior siding and in interior locations for paneling and underlayment. The code references three Composite Panel Association standards, ANSI A135.6 *Engineered Wood Siding*, ANSI A135.5 *Prefinished Hardboard Paneling*, and ANSI A135.4 *Basic Hardboard*. Manufacturer recommendations for Hardboard products are also referenced for installation requirements.

2303.1.8 Particleboard. Particleboard is a generic term for construction panels and products manufactured from cellulosic materials, usually wood, in the form of discreet pieces and particles as distinguished from fibers. The particles are combined with synthetic resins and other binders and bonded together under heat and pressure.

Particleboard is used as underlayment, siding, and for shear walls. Particleboard used structurally for siding or shear walls must be stamped (labeled) M-S Exterior or M-2 Exterior. The "M" stands for medium density; the "2" designates the strength grade (grades range from 1 to 3); and "S" designates "special grade." Particleboard designated M-S is medium density and has physical properties between an M-1 and M-2 designation. Both must be made with exterior glue. See SDPWS Table 4.3A for nominal unit capacities for wood-based panel shear walls, including particleboard.

Table 2303-1. **Guide to Wood Structural Panel Use**

Panel Grade and Bond Classification	Description and Use	Common Performance Category (in.)	Panel Construction	
			OSB	Plywood Minimum Veneer Grade
Sheathing EXP 1	Unsanded sheathing grade for wall, roof, subflooring, and industrial applications such as pallets and for engineering design with proper capacities.	5/16, 3/8, 15/32, 1/2, 19/32, 5/8, 23/32, 3/4	Yes	Yes
Structural I Sheathing EXP 1	Panel grades to use where shear and cross-panel strength properties are of maximum importance. Plywood Structural I is made from all Group 1 woods.	3/8, 7/16, 15/32, 1/2, 19/32, 5/8, 23/32, 3/4	Yes	Yes
Single Floor EXP 1	Combination subfloor-underlayment. Provides smooth surface for application of carpet and pad. Possesses high concentrated and impact load resistance during construction and occupancy. Touch-sanded. Available with tongue-and-groove edges.	19/32, 5/8, 23/32, 3/4, 7/8, 1, 1-3/32, 1-1/8	Yes	Yes
Underlayment EXP 1 or EXT	For underlayment under carpet and pad. Available with exterior glue. Touch-sanded or sanded. Panels with performance category of 19/32 or greater may be available with tongue-and-groove edges.	1/4, 11/32, 3/8, 15/32, 1/2, 19/32, 5/8, 23/32, 3/4	No	Yes, face C-Plugged, back D, inner D
C-D-Plugged EXP 1	For built-ins, wall and ceiling tile backing. Not for underlayment. Touch-sanded.	1/2, 19/32, 5/8, 23/32, 3/4	No	Yes, face C-Plugged, back D, inner D
Sanded Grades EXP 1 or EXT	Generally applied where a high-quality surface is required. Includes A-A, A-C, A-D, B-B, B-C, and B-D grades.	1/4, 11/32, 3/8, 15/32, 1/2, 19/32, 5/8, 23/32, 3/4	No	Yes, face B or better, back D or better, inner C & D
Marine EXT	Superior Exterior-type plywood made only with Douglas-fir or western larch. Special solid-core construction. Available with medium density overlay (MDO) or high density overlay (HDO) face. Ideal for boat hull construction.	1/4, 11/32, 3/8, 15/32, 1/2, 19/32, 5/8, 23/32, 3/4	No	Yes, face A or face B, back A or inner B

Courtesy American Wood Council. For SI: 1 inch = 25.4 mm

Although similar in characteristics to medium-density Grade 1 particleboard, the particleboard intended for use as floor underlayment is designated "PBU". Particleboard underlayment is often applied over a structural subfloor to provide a smooth surface for resilient-finish floor coverings or textile floor coverings. The minimum ¼-inch thickness is suitable for use over panel-type subflooring. Particleboard underlayment installed over board or deck subflooring that has multiple joints should have a thickness of ⅜ inch. Joints in the underlayment should not be located over the joints in the subflooring.

Fastener lengths are designed to penetrate just through the subfloor and not substantially into the floor joists to minimize nail popping caused by shrinkage of the joists. See Table 2304.10.2.

2303.1.9 Preservative-treated wood. The applicable American Wood Protection Association (AWPA) standards are cited. Different preservative treatments are used depending on whether the wood is above ground or in contact with the ground.

2303.1.9.1 Identification. All wood required to be preservative treated by Section 2304.12 must be stamped (labeled) with the information listed in the section. There are no exceptions.

2303.1.9.2 Moisture content. The intent of this provision is to ensure that any additional moisture driven into the wood from the preservative-treatment process is at 19 percent or less prior to enclosure.

2303.1.10 Structural composite lumber. The term structural composite lumber (SCL) refers to either laminated veneer lumber (LVL), parallel strand lumber (PSL), laminated strand lumber (LSL), or oriented strand lumber (OSL). These materials are structural members bonded with an exterior adhesive. SCL structural properties are set forth in manufacturers' literature and evaluation reports such as those developed by ICC Evaluation Service (ICC-ES). Definitions and provisions for manufacturing and design value development for SCL are based on ASTM D5456 *Standard Specification for Evaluation of Structural Composite Lumber Products*.

LVL is produced by bonding thin wood veneers together. The grain of the veneers is parallel to the long direction of the member. LVL members have enhanced mechanical properties and dimensional stability and offer a broad range of product widths, depths, and lengths. LVL is used for a variety of applications such as headers and beams, hip and valley rafters, rim boards, scaffold planking, studs, flange material for prefabricated wood I-joists, and truss chords.

PSL members consist of long veneer strands in parallel formation and bonded together with adhesive to form the finished structural section. PSL members are commonly used for long-span beams, heavily loaded columns, and beam and header applications where high bending strength is needed.

LSL and OSL are similar to PSL, but are manufactured from flaked wood strands that have a high length-to-thickness ratio. The primary difference between OSL and LSL is the length of strand used to fabricate them. OSL strands are shorter (up to 6 inches) than LSL strands (approximately 12 inches). The strands are combined with an exterior adhesive, and are oriented and formed into a large mat or billet and pressed. Although their strength and stiffness properties are somewhat lower than LVL and PSL members, they have good fastener-holding strength and mechanical connector performance. LSL and OSL members are used in a variety of applications, such as beams, headers, studs, rim boards, and millwork.

Reports prepared by approved agencies or evaluation reports published by ICC-ES may be accepted as part of the evidence and data needed by the building official to form the basis of approval of a material or product not specifically covered in the code. Note that evaluation reports are approved under the alternative materials, design, and methods of construction provisions of Section 104.2.3.

2303.1.11 Structural log members. This section provides structural capacity and grading requirements for logs used as structural members. In general, all log structures require engineering. Establishing structural capacities of logs is based on ASTM D3957 *Standard Practices* *for Establishing Stress Grades for Structural Members Used in Log Buildings* which specifies the requirement for a grading stamp or alternative means of identification. The ICC *Standard on the Design and Construction of Log Structures* (ICC 400) also is available. The goal of the standard is to provide technical design and performance criteria that will facilitate and promote the design, construction, and installation of safe, reliable structures constructed of log timbers. It is intended to be used by design professionals, manufacturers, constructors, and building and other government officials, and continue as a referenced standard in future building codes. Because ICC 400 gives base values and references the NDS for design, either ASD or LRFD can be used.

2303.1.12 Round timber poles and piles. Similar to other sections in the IBC, this section simply directs users to the following referenced standards regarding round timber poles and piles: ASTM D3200 *Standard Specification and Test Method for Establishing Recommended Design Stresses for Round Timber Construction Poles* and ASTM D25 *Specification for Round Timber Piles*.

2303.1.13 Engineered wood rim board. Engineered rim board is an important structural element in many engineered wood applications where both structural load path through the perimeter member and dimensional change compatibility are important design considerations. This section references two standards for products intended for engineered wood rim board applications: ANSI/APA PRR 410 *Standard for Performance-Rated Engineered Wood Rim Boards* and ASTM D7672 *Standard Specification for Evaluating Structural Capacities of Rim Board Products and Assemblies*, which address the fundamental requirements for testing and evaluation of engineered wood rim boards. ASTM D7672 is applicable to determination of product-specific rim board performance (i.e., structural capacities) for engineered wood products that may be recognized in manufacturer's product evaluation reports. The PRR 410 standard also includes performance categories for engineered wood products used in engineered rim board applications. Under PRR 410, products are assigned a grade based on performance category (e.g., categories based on structural capacity) and bear a mark in accordance with the grade.

2303.2 Fire-retardant-treated wood. Fire-retardant-treated wood (FRTW) is defined as any wood product that has been impregnated with chemicals to improve its flame-spread characteristics beyond that of untreated wood. The principal objective of impregnating wood with fire-retardant chemicals is to produce a chemical reaction at specific temperature ranges. This chemical reaction reduces the release of certain intermediate products that contribute to the flaming of wood, and also results in the increased formation of charcoal and water. Some chemicals are also effective in reducing the oxidation rate for the charcoal residue. Fire-retardant chemicals also reduce the heat release rate of the FRTW when burning over a wide range of temperatures. This section gives provisions for the treatment and use of FRTW. However, the fire-retardant chemicals are generally quite corrosive and corrosion-resistant fasteners may be required with FRTW. See Section 2304.10.6.

2303.2.1 Alternate fire testing. ASTM E2768 *Standard Test Method for Extended Duration Surface Burning Characteristics of Building Materials (30 min Tunnel Test)* was developed specifically for code use. It is a modified version of ASTM E84 *Standard Test Methods for Surface Burning Characteristics of Building Materials* with a test extension from 10 minutes to 30 minutes. It measures the flame spread index and flame front progression beyond the centerline of the burners. ASTM E2768 is referenced in the *International Wildland Urban Interface Code* (IWUIC), and the revised IBC provisions are consistent with IWUIC.

2303.2.2 Pressure process. Treatment using a pressure process requires a minimum pressure of 50 psi.

2303.2.3 Other means during manufacture. This section requires treatment using other means during manufacture to be an integral part of the manufacturing process.

2303.2.4 Fire testing of wood structural panels. This section requires wood structural panels to be tested with a ripped or cut longitudinal gap of 1/8 inch.

2303.2.5 Labeling. Each piece of FRTW must be stamped (labeled). The labeling must show the performance of the material. The labeling must also state the strength adjustments, and conformance to the requirements for interior or exterior application.

The FRTW label must be distinct from the grading label to avoid confusion between the two. The grading label gives information about the properties of the wood before it is fire-retardant treated. The FRTW label gives properties of the wood after FRTW treatment. It is imperative that the FRTW label be presented in such a manner that it complements the grading label and does not create confusion over which label takes precedence.

2303.2.6 Design values. Several factors can significantly affect the physical properties of FRTW. These factors are the pressure treatment and redrying processes used, and the extremes of temperature and humidity that the FRTW will be subjected to once installed. The design values for all FRTW must be adjusted for the effects of the treatment and environmental conditions, such as high temperature in attic installations and humidity. The design adjustment values must be based on an investigation procedure, which includes subjecting the FRTW to similar temperatures and humidities. The procedure must be approved by the building official.

Three subsections prescribe specific strength adjustment requirements for treated wood structural panels, lumber, and LVL.

2303.2.6.1 Fire-retardant-treated plywood. This section references the test standard developed to evaluate the flexural properties of fire-retardant-treated plywood that is exposed to high temperatures and high humidity. The referenced standard is limited to softwood plywood. Therefore, judgment is required in determining the effects of elevated temperature and humidity on other types of wood structural panels. The section references ASTM D6305 *Practice for Calculating Bending Strength Design Adjustment Factors for Fire-Retardant-Treated Plywood Roof Sheathing* to be used to evaluate ASTM D5516 *Test Method of Evaluating the Flexural Properties of Fire-Retardant Treated Softwood Plywood Exposed to Elevated Temperatures* test data. This section also requires the manufacturer to publish allowable maximum loads and spans for service as floor and roof sheathing and the modification factors for roof framing for its particular treatment process based on adjusted design values and location.

2303.2.6.2 Fire-retardant-treated lumber. This section references the test standard developed to determine the necessary adjustments to design values for lumber that has been fire-retardant treated including the effects of elevated temperature and humidity. The section references ASTM D6841 *Standard Practice for Calculating Design Value Treatment Adjustment Factors for Fire-Retardant-Treated Lumber* to be used to evaluate ASTM D5664 *Test Methods for Evaluating the Effects of Fire-Retardant Treatments and Elevated Temperatures on Strength Properties of Fire-Retardant-Treated Lumber* test data. This section also requires the manufacturer to publish treatment adjustment factors for roof framing for its particular treatment process based on location.

2303.2.6.3 Fire-retardant-treated laminated veneer lumber. This section requires fire-retardant-treated LVL to be developed per ASTM D8223 *Standard Practice for Evaluation of Fire-Retardant Treated Laminated Veneer Lumber* with provisions for design values and adjustments for treatment effects. The provision requiring that each manufacturer publish reference design values and treatment-based design value adjustment factors is consistent with similar provisions for fire-retardant-treated lumber and plywood.

2303.2.7 Exposure to weather, damp, or wet locations. Some fire-retardant treatments are soluble when exposed to the weather or used under high-humidity conditions. When an FRTW product is to be exposed to weather conditions, it must be further tested in accordance with ASTM D2898 *Standard Practice for Accelerated Weathering of Fire-Retardant-Treated Wood for Fire Testing*. The material is then subjected to the ASTM weathering test and retested after drying.

2303.2.8 Interior applications. When an FRTW product is intended for use under high-humidity conditions, it must be further tested in accordance with ASTM D3201 *Test Method for Hygroscopic Properties of Fire-Retardant Wood and Wood-Base Products*. The material must demonstrate that when tested at 92-percent relative humidity, the moisture content of the FRTW does not increase to more than 28 percent.

2303.2.9 Moisture content. The intent of this provision is to ensure that any additional moisture driven into lumber from the treatment process is at 19 percent or less and 15 percent or less for plywood prior to enclosure. FRTW chemicals are quite corrosive to metal fasteners. Where the moisture content of the treated wood is too high, the corrosivity of the treated wood is even higher and contributes to greater corrosion of metal fasteners.

For wood that is kiln dried after treatment (KDAT), the kiln temperatures cannot exceed that used to dry the lumber and plywood that was submitted for the tests required by Section 2303.2.6.1 for plywood and Section 2303.2.6.2 for lumber.

2303.2.10 Types I and II construction applications. Use of FRTW in Type I or II construction is permitted within the limitations of Section 603.1.

2303.3 Hardwood and plywood. Hardwood plywood and decorative plywood are not used for structural purposes. This section references ANSI/HPVA HP-1 *American National Standard for Hardwood and Decorative Plywood*.

2303.4 Trusses. All types of trusses are included within these provisions, including those with parallel, pitched, or curved chords; bowstring, sawtooth or camelback profiles; heavy-bolted timber or light-gage metal-plate-connected elements. Other trusses that are somewhat common in the industry include those with wood chords and metal tubular webs connected with bolts or pins. Some agricultural structures or other Group U occupancy buildings use plywood gusset plates to connect sawn lumber chords and webs.

This section includes general design requirements; specific and detailed requirements for truss design drawings; requirements for truss member permanent bracing; provisions for truss placement diagrams; specific requirements for the truss submittal package; and specific truss anchorage requirements. The truss submittal package is part of the construction documents, which are part of the submittal documents.

2303.4.1 Design. Wood trusses must be designed in accordance with accepted engineering practice, which is generally governed by state laws that regulate professional engineering. Truss members are permitted to be joined by any acceptable method.

For metal-plate-connected wood trusses, which are most commonly specified, the truss designer should furnish complete calculations substantiating the size of all members and connector plate sizes. The truss calculations should indicate the combined stress index for members subjected to combined stresses from bending and axial compression and tension. The combined stress index should be less than one. The calculations are generally performed with a computer program; therefore, documentation may be required by the building official that substantiates the basis of the computer program used.

2303.4.1.1 Truss design drawings. The section prescribes what information is to be provided on truss drawings. Truss design drawings are to be approved by the building official prior to installation, and the design drawings are to be on the job site.

2303.4.1.2 Permanent individual truss member restraint (PITMR) and permanent individual truss member diagonal bracing (PITMDB). Where permanent diagonal bracing or restraint is required it should be indicated on the truss design drawings and meet the methods prescribed in the section (see Figure 2303-3).

Permanent lateral bracing requirements (e.g., restraint and diagonal bracing) should be specified by the truss designer. Temporary bracing should be left in place until permanent bracing is installed. All lateral bracing must be installed per the truss design so that the truss will have the same structural capacity for which it was designed. In any case, the individual truss member continuous lateral and diagonal bracing locations are to be shown on the truss design drawings.

Permanent bracing must be installed in compliance with the truss industry's permanent bracing standard details that follow sound engineering practice. These details are found in the Building Component Safety Information (BCSI) *Guide to Good Practice for the Handling, Installing, Restraining & Bracing of Metal-Plate-Connected Wood Trusses*, a document produced by the Structural Building Components Association (SBCA). It is the truss industry's guide for job-site safety and truss performance.

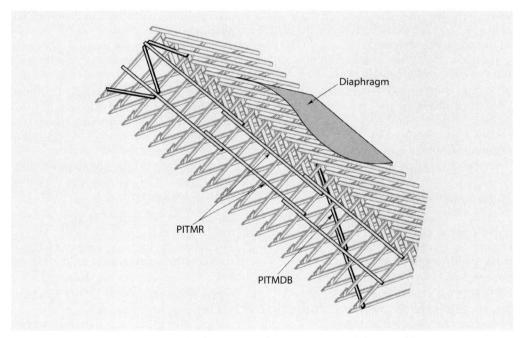

Figure 2303-3 **Permanent individual truss member restraint and diagonal bracing.**

2303.4.1.2.1 Trusses installed without a diaphragm. Trusses installed without a diaphragm on the top or bottom chord require a project-specific PITMR and PITMDB design prepared by a registered design professional (RDP). Group U buildings are exempt from this requirement.

2303.4.1.3 Trusses spanning 60 feet or greater. Trusses spanning 60 feet or more require an RDP to design the temporary and permanent bracing. Note that a similar requirement in Section 2206.1.3 applies to cold-formed steel trusses spanning 60 feet or more.

2303.4.1.4 Truss designer. In general, the truss designer is a specialty engineer that works for the truss manufacturer and is not the same as the RDP responsible for the overall building design. Section 2303.4.1.4.1 specifies what documents need to be stamped and signed by the truss engineer.

2303.4.2 Truss placement diagram. The truss placement diagram is used in the field to facilitate proper installation of the trusses. Note, truss placement diagrams that deviate from the design drawings' layout are required to be stamped and signed by the truss engineer.

2303.4.3 Truss submittal package. The section describes what documents are required to be included in the truss submittal package.

2303.4.4 Anchorage. The requirement in this section is an essential ingredient in providing a complete and continuous load path as required by Section 1604.4. The design of the anchorage required to transfer loads (gravity and lateral) from each truss to the supporting structure is the responsibility of the RDP. This is most critical for lateral loads where forces must be transferred from the roof diaphragm to the collectors and supporting shear walls, but it is also important for gravity loads. To illustrate a simple case, if the trusses are spaced at 24 inches (610 mm) on center and the supporting studs are spaced at 16 inches (406 mm) on center, some trusses will bear on the top plates between studs. Depending on the roof load, such as where high snow loads occur, the double top plate may not be adequate to span between studs and support the truss reaction. In this case, one solution might be for the RDP to require a stud under each truss. Another example is multi-ply girder trusses with very high reactions. In this case, one solution might be for the RDP to require a stud under each ply in the girder truss.

2303.4.5 Alterations to trusses. Truss members should not be cut, notched, drilled, spliced, or altered in any way without approval of the truss engineer. Any addition of loads including but not limited to HVAC equipment, piping, and additional roofing, should be reviewed and approved by the engineer to ensure that the trusses are capable of supporting additional loading.

2303.4.6 TPI 1 specifications. The design, manufacture, and quality assurance of metal-plate-connected wood trusses are to be in accordance with the IBC-referenced standard, TPI 1 *National Design Standard for Metal-Plate-Connected Wood Truss Construction.*

2303.4.7 Truss quality assurance. The TPI 1 standard includes quality-assurance procedures for truss manufacturers. Trusses not manufactured in accordance with the TPI 1 standard or another approved standard that provides quality control under the supervision of a third-party quality control agency require special inspection unless the manufacturer is an approved fabricator in accordance with Section 1704.2.5.

2303.5 Test standard for joist hangers. The section references ASTM D7147 *Specification for Testing and Establishing Allowable Loads of Joist Hangers*, which includes sampling and evaluation criteria as well as further refinements regarding quality of test materials, adjustments for variation in test materials, and limits on design values with materials other than those tested.

2303.6 Nails and staples. This section references ASTM F1667 *Specification for Driven Fasteners: Nails, Spikes and Staples*, including Supplement 1, for driven fasteners such as nails, spikes, and staples. Bending yield strength requirements for nails are provided. The bending yield strength requirements are those used in the NDS lateral strength tables, the IBC and IRC fastener schedules, as well as code evaluation reports. These strengths are standardized within the nail industry for engineered fasteners and are set forth in ASTM F1667.

2303.7 Shrinkage. Because wood shrinkage is a concern, two sections address shrinkage, Sections 2303.7 and 2304.3.3. Section 2303.7 requires the designer to consider the effects of cross-grain dimensional changes (shrinkage) in the vertical direction that can occur in lumber that was fabricated green. According to DOC PS 20, dry lumber is lumber that has been seasoned or dried to a maximum moisture content of 19 percent. Green lumber is lumber that has a moisture content in excess of 19 percent. The *NDS Supplement* provides nominal and minimum dressed sizes for dimension lumber [less than 5-inch (127 mm) nominal dimensions] and timbers [5-inch (127 mm) nominal and greater dimensions]. A practical "rule of thumb" for consideration of cross-grain dimensional change is to anticipate a 1 percent change in dimension for every 4 percent change in moisture content. More precise equations are available for designers seeking more detailed calculations. Shrinkage in the longitudinal direction (along the length) of lumber is approximately one-tenth of the cross-grain directions, so shrinkage along the length of a stud, for example, can typically be ignored.

Section 2304 *General Construction Requirements*

2304.1 General. The requirements of Section 2304 are general and apply to all of the design methods specified in Section 2302.1, which include: Allowable Stress Design, Load and Resistance Factor Design, Conventional Light-Frame Construction, AWC WFCM and, for log structures, ICC 400.

2304.2 Size of structural members. Computations to determine the size of the member required by the design are to be based on the actual sizes or dimensions and not the "nominal" sizes. For example, when computing the cross-sectional area of a 2×4, the actual dimensions of 1.5 inches (38 mm) by 3.5 inches (89 mm) are to be used. The NDS Supplement Table 1A provides nominal and minimum dressed (actual) sizes for dimension lumber and timbers; Tables 1B, 1C, and 1D provide section properties based on actual dimensions for dressed sawn lumber and structural glued-laminated timber.

2304.3 Wall framing. This section addresses minimum prescriptive requirements for wood-framed walls. These provisions include the framing of bottom plates, the framing over openings within the wall, and shrinkage. For all of these provisions, an exception is made when a "specific design is furnished."

2304.3.1 Bottom plates. These prescriptive provisions require sill plates to be at least as wide as the vertical framing members they are supporting and a minimum of 2-inch (51 mm) nominal or 1.5-inch (38 mm) actual thickness.

2304.3.2 Framing over openings. Windows, doors, air-conditioning units, and other service equipment require openings to be provided in wood-stud walls and partitions. Loads imposed above these openings must be transferred by a structural element above the opening to supports on both sides of the opening and then to a foundation, load-bearing wall or partition. Header sizes and spans may be determined per Section 2308.9.5. All headers not selected from span tables must be engineered in accordance with Section 2302.1. In all cases, headers and their supports must be adequate to support the imposed loads (see Figure 2304-1). See discussion in 2308.9.5 regarding prescriptive solutions for openings in walls and partitions.

2304.3.3 Shrinkage. This section imposes requirements for considering wood shrinkage. Wood-framed walls and bearing partitions cannot support more than two floors and a roof unless an analysis shows that there will be no adverse effects on the structure or its systems resulting from anticipated shrinkage or differential movement caused by shrinkage. Systems that typically experience adverse effects include plumbing and mechanical systems. There have been cases reported in multistory wood-frame projects where shrinkage was not considered in the building design, and shrinkage of the framing caused plumbing breaks in the stud walls.

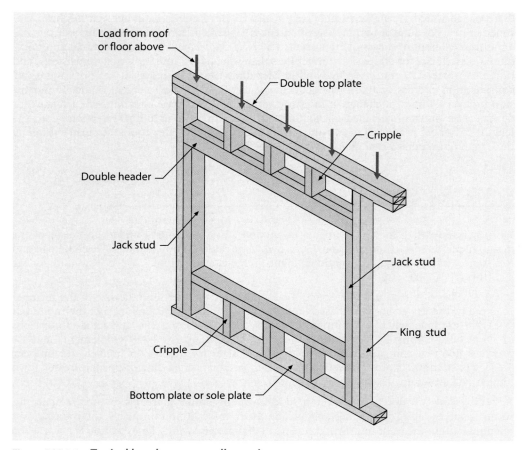

Figure 2304-1 **Typical headers over wall openings.**

With the proliferation of mid-rise (four stories and higher) wood-frame construction, in many areas of the country, designers and code officials need to be aware of proper design and detailing associated with multistory wood-frame buildings.

2304.4 Floor and roof framing. Similar to wall framing, this section addresses minimum prescriptive requirements which apply to wood-framed floors and roofs. These requirements are found in Section 2308. As with wall framing, all of the provisions in this section include an exception when a "specific design is furnished."

2304.5 Framing around flues and chimneys. This section sets the minimum distance combustible framing materials must be from flues, chimneys, fireplaces, and flue openings. Specific wording in the code requires wood framing to be a minimum of 2 inches (51 mm) clear from flues, chimneys, and fireplaces, and 6 inches (152 mm) clear from flue openings. Additional distance requirements are found within the *International Mechanical Code*® (IMC®). The separations specified in this section and in the IMC are intended to prevent the possibility of the wood being subjected to low heat for extended periods of time. Note that 2 inches (51 mm) of masonry will not provide the same level of thermal resistance as 2 inches

(51 mm) of air space. Therefore, adding 2 inches (51 mm) to the required thickness of a fireplace or chimney wall will *not* serve as an acceptable alternative to the 2-inch (51-mm) clear air space requirement.

2304.6 Exterior wall sheathing. This section specifies that wall sheathing on the exterior face of all exterior walls along with the connection of the sheathing to the supporting framing is to be of sufficient strength to span between the studs or other structural members, and capable of resisting wind pressures. The minimum strength or thickness of most of the exterior wall-covering materials listed in Table 2304.6.1 is based on a stud spacing of 16 (406 mm) or 24 inches (610 mm) on center.

2304.6.1 Wood structural panel sheathing. Wood structural panel sheathing used as an exterior wall finish must have an exterior type durability designation. Wood structural panels used under an exterior finish such as roof or wall covering must have an Exterior or Exposure 1 designation. Exposure 1 panels are allowed to be exposed on the underside of eaves.

Table 2304.6.1 provides usage requirements and limitations for wood structural panels used as exterior sheathing to resist wind pressures. For a given wind speed and exposure category, the table gives the minimum nail size, wood structural panel span rating, panel thickness, stud spacing, and nailing schedule. The limitations of the table are: the building must be enclosed (ASCE 7 Section 26.12); the mean roof height must be not greater than 30 feet (9,144 mm) (ASCE 7 Section 26.10), and topographic factor must be equal to 1.0 (ASCE 7 Section 26.8). Nail head pull through and nail shank withdrawal capacity is considered in addition to panel stiffness and bending strength. Panel-fastener capacity is based on tributary area to a single critical nail. Buildings that do not meet the limitations placed on this table are required to use ASCE 7 to engineer the exterior sheathing to resist component and cladding wind pressures.

2304.7 Interior paneling. Softwood wood structural panels used for interior paneling are to be installed in accordance with Table 2304.10.2 and comply with DOC PS 1 or PS 2 or APA PRP 210. Softwood panels installed in accordance with this section are considered interior wall finishes and cannot be used to resist lateral forces. Prefinished hardboard paneling must conform to the requirements of ANSI A135.5, and hardwood plywood must conform to HPVA HP-1.

2304.8 Floor and roof sheathing. This section specifies the wood structural panel sheathing requirements for floors in Section 2304.8.1 and for roofs in Section 2304.8.2.

2304.8.1 Structural floor sheathing. Structural floor sheathing serves three purposes:

1. Provides support for and distributes superimposed loads to floor framing members.
2. Provides lateral support for the top of the floor framing members.
3. Serves as part of the shear-resisting element when the floor acts as a diaphragm to distribute lateral wind or seismic loads to the shear walls or foundation.

Floor sheathing may either be engineered to meet the loading requirements, or be selected from Tables 2304.8(1), 2304.8(2), 2304.8(3), or 2304.8(4). The tables are prescriptive and therefore deemed to comply with the minimum code requirements.

The maximum span for wood structural panels is limited by both strength and deflection. When using the tables, *all* the conditions set forth in the table must be met regarding loads, type, grade, species group, plies, blocking, direction of span, and other requirements. Sheathing systems that do not meet *all* the conditions must be engineered.

Wood structural panels in Table 2304.8(3) are designated by the panel span rating instead of thickness. The rating consists of two numbers such as 32/16. The first number is the allowable span for use as roof sheathing, and the second is the allowable span for use as floor sheathing. Panels intended only for single floor grades are designated by one span rating. The single floor grades have higher-grade faces than Structural I panels and may be preferable for aesthetic reasons as roof sheathing where visible from the underside.

Several thicknesses of structural panel are included under a single panel span rating. This is because different grades of interior and face veneers or strands may be used in different combinations. For example, a five-ply plywood panel will have a higher load rating than a four-ply panel with the same thickness. Conversely, the five-ply panel may be thinner than the four-ply panel, yet have the same panel span rating. The ⅜-inch (9.5 mm) wood structural panel is currently the minimum thickness in the table.

2304.8.2   Structural roof sheathing.   Roof sheathing must be designed or conform to the provisions of Table 2304.8(1), 2304.8(2), 2304.8(3), or 2304.8(5). Wood structural panel roof sheathing must be bonded by exterior glue.

2304.9   Lumber decking.   These provisions cover all decking, including mechanically laminated and solid-sawn decking. The general provisions, Section 2304.9.1, specify the requirements for square ends, beveled ends, and the orientation of tongue-and-groove decking on roofs.

Section 2304.9.2 provides specific requirements for the various layup patterns described in the subsections. Sections 2304.9.3 through 2304.9.5 cover specific requirements for mechanically laminated decking, 2-inch (51 mm) sawn tongue-and-groove decking, and 3- and 4-inch (76 mm and 102 mm) sawn tongue-and-groove decking, respectively.

It should be noted that the capacity of lumber decking is determined using the various layup patterns described in Section 2304.9.2. Depending upon the layup pattern, Section 2306.1.5 gives the design capacity of lumber decking for flexure and deflection formulas within Table 2306.1.5.

2304.10   Connectors and fasteners.   Connectors and fasteners must comply with the applicable provisions of Sections 2304.10.1 through 2304.10.8. The section covers fasteners used to connect wood members, sheathing fasteners, fasteners for joist hangers, and framing hardware, and specific requirements for fasteners used in treated wood. Other fastening methods such as clips, staples, and glue are allowed where approved.

2304.10.1   Fire protection of connections.   Fire protection for connections used in fire-resistance-rated members and assemblies in Type IV-A, IV-B, or IV-C construction are to be determined either by testing in accordance with Section 703.2 where the connection is part of a fire-resistance test or by engineering analysis. The provisions are not applicable to connections in heavy timber (IV-HT) construction, because heavy timber structural members do not have a prescribed fire-resistance rating.

Section 704.2 requires connections of primary structural members to be protected with materials that have the required fire-resistance rating. However, Section 704.2 does not require connections that join elements of the structural frame to be tested in accordance with ASTM E119 *Standard Test Methods for Fire Tests of Building Construction and Materials*. The requirement is for the connections to only be protected with material having the fire-resistance rating required for the structural members which they connect. It is neither practical nor possible to test connections in a standard fire test furnace since there is no capability to test large connections used to transfer gravity loads. In addition, ASTM E119 does not include any provisions for how to test connections and assess their performance. While testing of large gravity load connections is not contemplated by ASTM E119, the fasteners that are used to connect elements of an assembly are considered part of the test and critical to the assembly's fire performance. For example, the fastening system used to connect multiple pieces of abutting nail-laminated timber which forms a floor assembly is considered part of the tested assembly. Proper detailing of the system used to connect abutting segments is a primary factor in meeting the thermal pass/fail criteria established in ASTM E119. The connection system has little bearing on the structural performance but is necessary to prevent the passage of hot gases at abutting edges.

An engineering analysis per Option 2 of Section 2304.10.1 would need to demonstrate that the temperature rise at any portion of the connection is limited to an average temperature rise of 250°F (139°C), and a maximum temperature rise of 325°F (181°C), for a time corresponding to the required fire-resistance rating of the structural element being connected. For the purposes of this analysis, the connection includes connectors, fasteners, and portions of wood members included in the structural design of the connection. Potential options for engineering analysis include the following:

- IBC Section 722 permits structural fire-resistance ratings of wood members to be determined using Chapter 16 of the AWC *National Design Specification for Wood Construction* (NDS). Where a wood connection is required to be fire-resistance-rated, NDS Section 16.5 requires all components of the wood connection, including the steel connector, the connection fasteners, and the wood needed in the connection's structural design, to be protected for the minimum required fire-resistance time in accordance with the AWC *Fire Design Specification* (FDS) *for Wood Construction*. The connection is permitted to be protected by wood, gypsum board, or other approved materials.

- Analysis procedures have been developed that allow protection of these connections to be designed based on test results of ASTM E119 fire tests from protection configurations using the wood member outside of the connection, additional wood cover, and/or gypsum board. AWC *Technical Report 10: Calculating the Fire Resistance of Wood Members and Assemblies* (TR10), which is referenced in the NDS Chapter 16 Commentary, provides examples of connection designs meeting the requirements of IBC Section 704 and NDS Section 16.5 (Figure 2304-2).

It should be noted that Section 110.3.5 establishes inspection requirements where wood cover is specified to provide fire-resistance protection of connections in Types IV-A, IV-B, and IV-C construction.

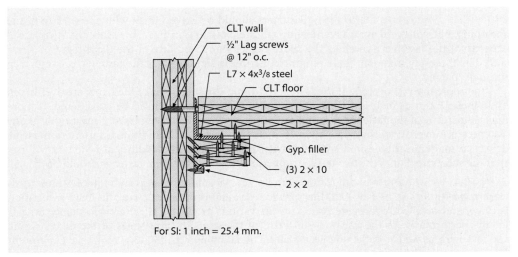

Figure 2304-2   **Mass timber fastener protection example (from AWC TR10, courtesy American Wood Council).**

2304.10.2 Fasteners requirements. While this section states that connectors are to be designed in accordance with Section 2302.1, the minimum number and size of fasteners are found in Table 2304.10.2. The size designations in the table are common to all fastener manufacturers.

This table accommodates the builder of non-designed (conventional) construction, but also provides minimum fastening requirements for designed construction. Details such as end and edge distances and nail penetration are required to be in accordance with the applicable provisions of the NDS. Where required, corrosion-resistant fasteners must be either hot-dipped zinc-coated galvanized steel, stainless steel, silicon bronze, or copper.

Because the pennyweight system of specifying nail sizes is not universally understood, code users sometimes focus on pennyweight (8d = 8 penny, 16d = 16 penny, etc.) and do not pay sufficient attention to the specific type of nail such as common, box, cooler, sinker, finish, and so on. A typical example is substitution of box nails for common nails of the same pennyweight. The specific type of nail is critical because there can be significant differences in strength properties of connections fastened with nails of the same pennyweight but of different nail type. Because dimensions are less subject to misinterpretation, the shank length and diameter in parentheses will help prevent confusion and misapplication of the various types of nails used in wood connections.

Roof sheathing nailing criteria in Table 2304.10.2 is based on ASCE 7 and AWC's WFCM. Wind uplift nailing requirements for common species of roof framing are the basis of nail spacing requirements in Table 2304.10.2 to meet wind uplift loading requirements of ASCE 7 without being overly complex in specification of wood structural panel roof sheathing nailing. The basic roof sheathing nailing schedule is 6 inches (152 mm) on-center at panel edges and 6 inches (152 mm) on-center at intermediate supports in the panel field. For the common case of roof framing spaced at 24 inches (610 mm) on-center, nailing at intermediate supports in the roof's interior portions is 6 inches (152 mm) on-center for wind speeds within the scope of IBC Section 2308. Six inches (152 mm) on-center spacing is also appropriate for edge zones except where basic wind speeds equal or exceed 130 mph (58 m/s) in Exposure B and 110 mph (49 m/s) in Exposure C where 4 inches (102 mm) on-center nail spacing is required. These special cases are addressed by footnote "e." Table 2304.10.2 and IRC Table R602.3(1) are consistent when addressing the structural connection of Items 1 through 39.

2304.10.3 Sheathing fasteners. Fasteners should be driven flush with the surface of the sheathing, but not overdriven. Overdriving of fasteners can significantly reduce the withdrawal capacity of the fastener as well as the shear capacity and ductility of the diaphragm or shear wall. For three-ply material, the strength reduction is significant if the fastener is overdriven through one ply.

If no more than 20 percent of the fasteners around the perimeter of the panel are overdriven by 1/8 inch (3.2 mm) or less, then no reduction in shear capacity need be considered. If more than 20 percent of the fasteners around the perimeter are overdriven by any amount, or if any fasteners are overdriven by more than 1/8 inch (3.2 mm), then additional fasteners must be driven to maintain the desired shear capacity, provided that the additional fasteners will not split the substrate. For every two fasteners overdriven, one additional fastener should be driven.

2304.10.4 Joist hangers and framing anchors. Most joist hangers and other framing hardware have reports issued by building product evaluation services, such as ICC Evaluation Service, and detailed reports are issued for the various products. There are acceptance criteria for the performance of the hardware in terms of the strength limit states of the metal, wood, and fasteners, as well as deflection limit states. For example, ICC AC 13, *Acceptance Criteria for Joist Hangers and Similar Devices*, is used to evaluate devices used to support or attach wood members, such as joists, rafters, purlins, beams, girders, plates, posts, studs, and headers to wood, metal, concrete, or masonry where the attachment is by means of mechanical fastenings such as nails, spikes, lag screws, wood screws, bolts, and so on. The term "device" refers to one or more pieces or units so arranged as to transfer load vertically or laterally, within safe limits, from the end of a supported member (such as a joist) to a supporting member (such as a header, beam, or girder).

Test requirements for joist hangers are found in ASTM D7147. The ASTM D7147 standard includes sampling and evaluation criteria as well as further refinements regarding quality of test materials, adjustments for variation in test materials, and limits on design values with materials other than those tested.

2304.10.5 Other fasteners. Fasteners not specifically cited in the code may be used but are subject to building official approval. See Section 1703.4 for requirements related to research reports and acceptance of evaluation reports.

2304.10.6 Fasteners and connectors in contact with preservative-treated and fire-retardant-treated wood. Some chemicals used in preservative-treated and fire-retardant-treated wood are corrosive. Fasteners, washers, and nuts must be corrosion resistant when used with these materials. Corrosion-resistant fasteners, washers, and nuts are made of type 304 or 316 stainless steel, silicon bronze, copper, or steel that has been hot-dipped or mechanically deposited zinc-coated galvanized with a zinc coating of not less than 1.0 ounce per square foot.

Electro-galvanized steel fasteners do not qualify as corrosion resistant; the zinc coating typically is about 0.1 ounce per square foot (30.5 g/m^2). Electro-galvanized nails are suitable only for occasionally wet locations such as for nailing composition shingles.

The section is subdivided into four categories of fasteners and connectors in treated wood: fasteners in preservative-treated wood, fastening for wood foundations, fasteners in fire-retardant-treated wood in exterior, wet or damp locations, and fasteners in fire-retardant-treated wood in interior locations.

Additionally, there is an exception for fasteners used in preservative-treated wood in an interior dry environment. There is no documented evidence of any detrimental fastener corrosion when plain steel fasteners are used in SBX/DOT (sodium borate) or zinc borate preservative-treated wood in interior, dry environments. In this case, the exception permits plain carbon steel fasteners, including nuts and washers, to be used.

2304.10.7 Load path. The code requires the load path to be continuous from the load's point of origin to the resisting element, which generally means from the roof to the foundation. A continuous load path for both gravity and lateral loads is necessary for adequate performance of the structure. This is especially critical in the case of high wind and load effects from earthquake ground motion. For example, visualize what happens to the wind suction load on a low-slope roof. The upward force that is not offset by the dead load of the roof elements must be resisted by dead load elsewhere. Similarly, where the structure is subjected to lateral load effects from earthquake ground motion, the inertial forces from the floor and roof diaphragms must be effectively transferred to the vertical lateral-force-resisting elements. A positive, properly detailed and constructed continuous load path is essential to ensure the transfer of all gravity, uplift, and lateral loads from the roof and floors through the structural system down to the foundation and supporting soil.

The required minimum thickness of steel straps used to splice discontinuous framing members is consistent with the standard thickness in the AISI Product Data Standard, S201. AISI S201 provides criteria, material, and product requirements for structural and nonstructural members utilized in cold-formed steel framing applications where the specified minimum base steel thickness is between 18 mils (0.0179 inch or 0.455 mm) and 11 mils (0.1180 inch or 3 mm). The minimum thickness specified in the code represented galvanized 20 gage steel, but the term "gage" is no longer a steel thickness designation and is now designated as 0.0329 inch (0.84 mm), which is the minimum base metal thickness of 0.0329 inch (0.84 mm) for steel straps used to splice discontinuous framing members.

2304.10.8 Framing requirements. Columns must be provided with full end bearing to transfer their full compressive load. If not providing full bearing contact, the column's load carrying capacity will be reduced. Connections must also be able to resist lateral and net uplift loads.

2304.11 Heavy timber construction. Table 2304.11 includes a description of heavy timber structural elements based on whether the element supports roof loads and floor loads or only roof loads. The nonsize-related detailing provisions for framing members and connections are included in Section 2304.11.1. Sections 2304.11.2 through 2304.11.4 contain pertinent thickness and detailing requirements for walls, roof, and floor deck construction. Appropriate distinctions between nominal, net finished, and actual dimensions for heavy timber, glulam, SCL, and CLT are provided.

Table 2304.11. **Minimum Dimensions of Heavy Timber Structural Members**

Supporting	Heavy Timber Structural Elements	Minimum Nominal Solid Sawn Size		Minimum Glued-Laminated Net Size		Minimum Structural Composite Lumber Net Size	
		Width, inch	Depth, inch	Width, inch	Depth, inch	Width, inch	Depth, inch
Floor loads only or combined floor and roof loads	Columns; Framed sawn or glued-laminated timber arches that spring from the floor line; Framed timber trusses	8	8	6¾	8¼	7	7½
	Wood beams and girders	6	10	5	10½	5¼	9½
Roof loads only	Columns (roof and ceiling loads); Lower half of: wood-frame or glued-laminated arches that spring from the floor line or from grade	6	8	5	8¼	5¼	7½
	Upper half of: wood-frame or glued-laminated arches that spring from the floor line or from grade	6	6	5	6	5¼	5½
	Framed timber trusses and other roof framing;[a] Framed or glued-laminated arches that spring from the top of walls or wall abutments	4[b]	6	3[b]	6⅞	3½[b]	5½

For SI: 1 inch = 25.4 mm.

[a] Spaced members shall be permitted to be composed of two or more pieces not less than 3 inches nominal in thickness where blocked solidly throughout their intervening spaces or where spaces are tightly closed by a continuous wood cover plate of not less than 2 inches nominal in thickness secured to the underside of the members. Splice plates shall be not less than 3 inches nominal in thickness.

[b] Where protected by approved automatic sprinklers under the roof deck, framing members shall be not less than 3 inches nominal in width.

2304.11.1 Details of heavy timber structural members. This section represents the consolidation of non-size related detailing provisions for heavy timber framing members and connections. This Section is subdivided to address columns in 2304.11.1.1, floor framing in 2304.11.1.2, and roof framing in 2304.11.1.3.

2304.11.1.1 Columns. Columns in heavy timber construction must be continuous or aligned vertically throughout all stories, capable of supporting beams and columns above, and able to transfer vertical and horizontal loads through the joints and connections. Girders and beams at column connections in heavy timber construction are required to be fitted around columns, and adjoining ends must be adequately tied to each other to transfer horizontal loads across the joints.

2304.11.1.2 Floor framing. Wall pockets or hangers are required where wood beams, girders, or trusses in heavy timber construction are supported by masonry or concrete walls. Beams supporting floors are required to bear on girders or be supported by ledgers, blocks, or hangers.

2304.11.1.3 Roof framing. Roof girders and every alternate roof beam in heavy timber construction are to be anchored to supporting members with steel or iron bolts designed to resist vertical and lateral loads of the roof.

2304.11.2 Partitions and walls. This section addresses exterior walls, interior walls, and partitions. The first subsection states that cross-laminated timber at least 4-inch thick, meeting the requirements of Section 2303.1.4, can be used for exterior walls. The second subsection allows a minimum of two layers of 1-inch (25.4 mm) nominal matched boards or laminated construction at least four-inches (102 mm) thick. The latter provides for the use of CLT. Partitions can also be one-hour fire-resistance-rated construction such as a stud and gypsum board FRR assembly.

2304.11.3 Floors. This section has two subsections addressing cross-laminated timber floors and sawn or glued-laminated plank floors: Sections 2304.11.3.1 and 2304.11.3.2, respectively. Items to note include floor decks and floor covering in heavy timber construction cannot extend closer than ½ inch to walls with the space covered by a molding fastened to the wall.

Also notable is an allowance for concealed spaces in traditional Type IV-HT construction complying with Section 602.4.4.3. Protection options include:

- In fully sprinklered buildings, installing automatic sprinkler protection in the concealed space.
- Filling the space with noncombustible insulation.
- Protecting combustible surfaces with not less than 5/8-inch (15.9 mm) Type X gypsum board.

The option of having protected concealed spaces in Type IV-HT buildings is important for encouraging adaptive re-use of existing heavy timber buildings as well as providing for installation of mechanicals (MEP) in Type IV-HT cross-laminated timber construction.

2304.11.4 Roof decks. Subsections 2304.11.4.1 Cross-laminated timber roofs and 2304.11.4.2 Sawn, wood structural panel, or glued-laminated plank roofs address roof deck construction. See preceding discussion in Section 2304.11.3 "Floors" for allowance for concealed spaces in traditional Type IV-HT construction complying with Section 602.4.4.3.

2304.12 Protection against decay and termites. The provisions of this section are intended to protect wood against decay caused by water or termite infestation. The provisions are based on the extensive material on biodeterioration of wood in the *Wood Handbook* published by the U.S. Forest Products Laboratory.

2304.12.1 Locations requiring water borne preservatives or naturally durable wood. Wood used in direct contact with the ground, in areas close to the ground, or exposed directly to weather (Sections 2304.12.1.1 through 2304.12.1.5, 2304.12.2.6, and 2304.12.2.8), must be naturally durable or *preservative-treated* with water borne preservatives. Wood used above ground, if preservative treated, is usually treated with a water borne preservative such as Alkaline Copper

Quaternary (ACQ) or Copper Azole (CA) in accordance with American Wood Protection Association (AWPA) U1 *USE CATEGORY SYSTEM: User Specification for Treated Wood Except Commodity Specification H*. The retention rates are lower than those required for ground contact.

2304.12.1.1 Joists, girders, and subfloor. The code requires a minimum of 18 inches (457 mm) of clearance to joists and 12 inches (305 mm) clearance to wood girders from exposed ground if they are not of naturally durable or preservative-treated wood. See Figure 2304-3.

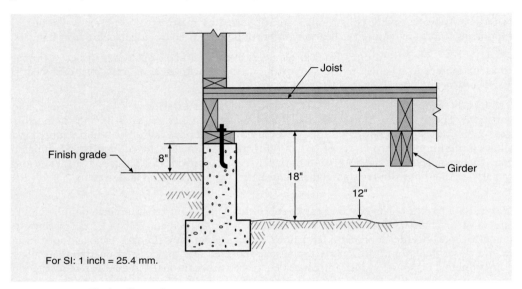

Figure 2304-3 **Under-floor clearances.**

2304.12.1.2 Wood supported by exterior foundation walls. Framing, including sheathing (not siding), must have a minimum of 8 inches (203 mm) of clearance from exposed earth if it is not naturally durable or preservative-treated wood. See Figure 2304-4. Note that Section 2304.12.1.5 allows a 6-inch (152 mm) clearance for wood siding.

2304.12.1.3 Exterior walls below grade. Untreated wood framing attached to the interior side of exterior concrete or masonry foundation walls may draw moisture leading to decay.

2304.12.1.4 Sleepers and sills. Concrete and masonry slabs that are in direct contact with the earth are very susceptible to moisture because of absorption of ground water. Similar to exterior walls below grade, this can occur on interior slabs as well as at the perimeter. This section is intended to prevent the use of untreated wood that may decay under such conditions. Concrete that is fully separated from the ground by a vapor barrier is not considered to be in direct contact with earth.

2304.12.1.5 Wood siding. Wood siding must have a minimum of 6 inches (152 mm) of clearance between the siding and earth unless made of naturally durable or preservative-treated wood. Note that the clearance for wood sheathing under the siding is 8 inches (203 mm) as required by Section 2304.12.1.2. In other words, siding can extend 2 inches (51 mm) below the foundation plate, or framing, whereas the sheathing must be terminated at the sill plate. See Figure 2304-4.

The IBC requires a minimum 2-inch (51 mm) clearance between wood siding and a concrete slab or steps.

General Construction Requirements   **805**

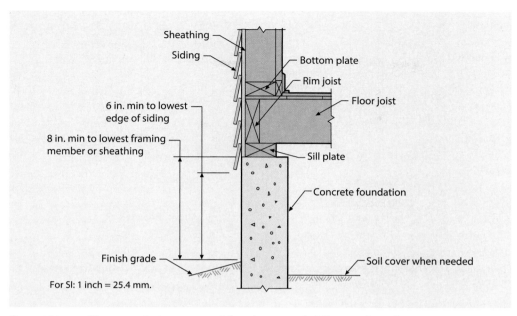

Figure 2304-4 **Clearance between wood framing, wood siding, and earth.**

2304.12.2 Other locations. When preservative-treated wood is used in interior locations, two coats of a protective finish are required to be applied after the wood is dried unless waterborne preservatives are used. The code further specifies that the protective finish is to be urethane, shellac, latex epoxy, or varnish.

2304.12.2.1 Girder ends. A ½-inch (12.7 mm) airspace is required around the top, sides, and ends of wood girders embedded in concrete or masonry exterior walls to reduce moisture takeup that can contribute to decay of an untreated wood member.

2304.12.2.2 Posts or columns. Posts and columns are required to be naturally durable or preservative-treated wood unless they meet the minimum clearances illustrated in Figure 2304-5.

2304.12.2.3 Supporting member for permanent appurtenances. Balconies, porches, and other appurtenances that are directly exposed to weather conditions can collect water in the joints and on the surfaces, creating alternating cycles of wetting and drying conducive to deterioration and decay. In this case, the wood-framed structural supports must be of naturally durable or preservative-treated wood. The exception can be applied to buildings located in areas where climatic conditions are favorable enough to preclude the use of naturally durable or preservative-treated wood. For example, very dry desert areas such as Death Valley, California, and Yuma Valley, Arizona, have extremely low annual precipitation where the exception could be used with the approval of the building official.

2304.12.2.4 Supporting members for permeable floors and roofs. Where wood framing is used to support floors and roofs that are moisture permeable, such as a concrete slab over a patio or a patio slab over a garage, the framing must be of preservative-treated wood or approved naturally durable species, unless the slab is separated from the wood framing by a waterproof membrane that is capable of providing positive drainage. Although these wood-framing

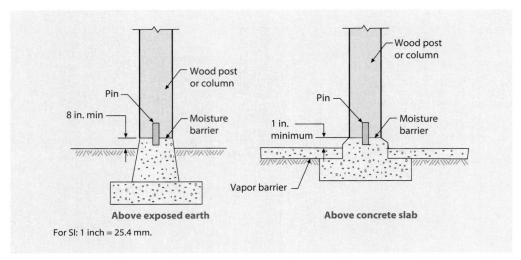

Figure 2304-5 **Wood posts and columns not exposed to weather or protected from weather by roof, eave, or overhang.**

members are not necessarily in direct contact with the ground, their exposure to moisture is similar to that of wood in direct contact with the ground. Therefore, the wood framing must be of naturally durable wood, or it must be preservative treated in accordance with applicable AWPA standards.

2304.12.2.5 Ventilation beneath balconies or elevated walking surfaces. These provisions address required ventilation openings to provide cross-ventilation to enclosed framing supporting exterior balconies and elevated walking surfaces exposed to weather.

2304.12.2.6 Wood in contact with the ground or fresh water. Note specifically the limiting adjective *fresh*, which means this section only applies to wood in contact with the ground or fresh water. The waterborne preservatives used for fresh water are not suitable for brackish or salt water, where attack can also come from marine borers. The first paragraph of this code section allows wood in direct contact with the earth to be naturally durable. However, this only applies to wood in contact with the ground, not posts or columns embedded in concrete. Posts and columns supporting permanent structures are required by Section 2304.12.2.6.1 to be preservative treated. It is also worth noting that an exception for wood entirely submerged in fresh water is provided. This exception acknowledges that the rate of decay is significantly reduced in the absence of free oxygen.

2304.12.2.6.1 Posts or columns. Posts or columns embedded in concrete, such as columns in a post-frame building, where moisture tends to get trapped around the wood member, are subject to decay. Hence, they must be of preservative-treated wood.

2304.12.2.7 Termite protection. Where termites are a significant hazard such as in some southern states, floor framing must be preservative treated, naturally durable, or have some other approved method of termite protection. Section 2603.8 has specific restrictions on the use of foam plastics in areas where the probability of termite infestation is very heavy, based on the termite infestation probability map shown in IBC Figure 2603.8. (See Figure 2304-6.) In areas where the probability of termite infestation is very heavy, naturally durable termite-resistant wood or preservative-treated wood must be used. This map is often updated due to the increasing spread of and threat from termites.

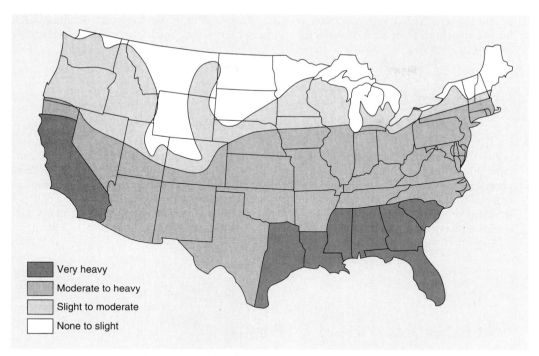

Figure 2304-6 **Termite infestation probability map.**

2304.12.2.8 Wood used in retaining walls and cribs. Wood used in retaining walls, crib walls, bulkheads, and other walls that retain or support the earth must be of preservative-treated wood specified as ground-contact-treated wood. AWPA lists three categories for ground contact use: General Use (UC4A), Heavy Duty (UC4B), and Extreme Duty (UC4C). Heavy Duty protection would be suitable for most circumstances. Extreme Duty would be appropriate in such locations as freshwater lake or pond side walls. AWPA has separate categories for use in marine (saltwater) environments.

2304.12.3 Attic ventilation. Refer to the discussion under Section 1202.2 with regard to attic ventilation.

2304.12.4 Under-floor ventilation (crawl space). Refer to the discussion under Section 1202.4 with regard to under-floor ventilation.

2304.13 Long-term loading. Wood structural members can exhibit long-term creep, which increases deflection, particularly where a high dead load is present. Additional deflection can cause severe cracking in the masonry or concrete. Under sustained loading, wood members exhibit additional time-dependent deformation (creep), which generally develops over long periods of time. The tabulated modulus of elasticity design values, E, in the NDS are intended to be used to estimate the immediate deformation under load. Where dead loads or sustained live loads represent a relatively high percentage of total design load, creep is an appropriate design consideration, which is addressed within the NDS. The total deflection under long-term loading is estimated by increasing the initial deflection associated with the long-term load component by 1.5 for seasoned lumber or 2.0 for unseasoned or wet lumber or glued-laminated timber. In either case, the total deflection, including the effects of long-term loading, are not to exceed those specified within Section 1604.3.1. An exception to the long-term loading check for nonstructural floor or roof surfacing no more than 4-inch thick, such as a gypcrete floor topping, is provided.

Section 2305 *General Design Requirements for Lateral Force-Resisting Systems*

Section 2305 has three subsections: Section 2305.1 is general requirements; Section 2305.2 contains provisions for calculating deflection of stapled diaphragms; and Section 2305.3 contains provisions for calculating deflection of stapled shear walls. All other requirements pertaining to lateral design of wood shear walls and wood diaphragms are in the AWC SDPWS. The SDPWS standard provides complete requirements for design and construction of wood members, fasteners, and assemblies that resist lateral forces from wind and seismic loads (Figure 2305-1).

Figure 2305-1   2021 *Special Design Provisions for Wind and Seismic* standard.

2305.1 General. The section references the AWC SDPWS for design and construction of wood shear walls or wood diaphragms to resist wind, seismic, or other lateral loads. Provisions in Sections 2305, 2306, and 2307 are also required to the extent that they apply.

2305.1.1 Openings in shear panels. This provision predates the perforated shear wall design method. Although the section requires that openings in shear walls that materially affect their strength be detailed and have their edges reinforced, the meaning of the phrase *materially affect their strength* is not defined in the code. It should also be noted that the perforated shear wall design method in SDPWS does permit openings in shear walls without force transfer design, provided certain requirements, adjustments, and restrictions are met.

2305.1.2 Permanent load duration. Reference to the SDPWS is appropriate for the design of wood shear walls and diaphragms to resist wind and seismic loads, but for resistance to permanent lateral loads, such as soil loads in foundation design, the nominal unit shear capacities in SDPWS need further reduction to account for long-term effects. Permanent loads are associated with a permanent load duration factor (allowable stress design, ASD) or time-effect factor (load and resistance factor design, LRFD), as defined by the NDS. Section 2305.1.2 requires the use of a 0.2 factor for ASD and a 0.3 factor for LFRD.

2305.2 Diaphragm deflection. The SDPWS standard does not address deflection calculations for stapled diaphragms. Equation 23-1 includes the parameters necessary to calculate deflection of diaphragms fastened with staples. Stiffness properties (apparent shear stiffness, kip/inch) for diaphragms constructed of wood structural panels and lumber are given in SDPWS for purposes of complying with diaphragm classification, drift, and stiffness compatibility requirements specified in ASCE 7. See the SDPWS Commentary for detailed discussion of calculating diaphragm deflection.

2305.3 Shear wall deflection. As noted earlier, the SDPWS standard does not address deflection of stapled shear walls. Equation 23-2 includes the parameters necessary to calculate deflection of shear walls fastened with staples. Stiffness properties (apparent shear stiffness, kip/inch) for shear walls constructed of wood structural panels and other materials are given in SDPWS for purposes of complying with drift and stiffness compatibility requirements specified in ASCE 7.

Section 2306 *Allowable Stress Design*

2306.1 Allowable stress design. Table 2306.1 includes a list of standards for design and construction of wood elements in structures using ASD. The ANSI/AWC *National Design Specification (NDS) for Wood Construction* is adopted by reference without any amendments (see Figure 2306-1). The NDS includes design information for sawn lumber, structural glued laminated timber, structural-use panels, poles and piles, I-joists, structural composite lumber, cross-laminated timber, and structural connections (nails, bolts, screws). The IBC also references the AWC SDPWS in Sections 2305, 2306, and 2307.

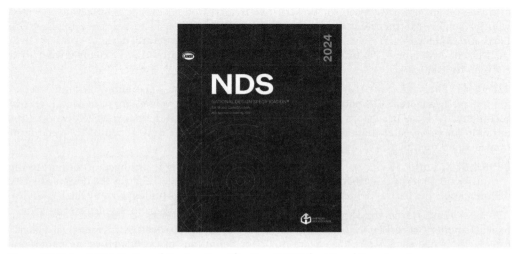

Figure 2306-1   2024 *National Design Specification (NDS) for Wood Construction.*

ASABE EP486.3 *Shallow-post and Pier Foundation Design*, provides a design procedure for shallow post foundations that resist moments and lateral and vertical forces. The standard includes definitions, material requirements, and design equations for designing post foundations. Additional referenced ASABE standards deal with diaphragm design of metal-clad post-frame buildings and mechanically laminated columns used in post-frame construction.

APA—The Engineered Wood Association develops several standards for design of structural glued-laminated timber and related recommendations for connection detailing, field notching and drilling and other application guidelines. APA also promulgates several standards and supplements for wood structural panel design.

The Truss Plate Institute develops TPI 1, *National Design Standard for Metal Plate Connected Wood Trusses*. TPI 1 is the standard for the design, fabrication, quality control, and installation of light-frame metal-plate-connected wood trusses.

AITC standards 104, 110, 113, 119, 120, and 200 provide typical construction details, standard appearance grades and dimensions, hardwood specifications, and quality control systems for structural glued-laminated timber. These standards are all managed by the Pacific Lumber Inspection Bureau (PLIB).

2306.1.1 Joists and rafters. Although the section indicates that rafters may be designed using the AWC *Span Tables for Joists and Rafters* (STJR), the intent is that both joists and rafters are permitted to be designed by the AWC STJR. While several span tables for joists and rafters are included in Section 2308 for conventional light-frame construction, those tables only contain four of the most commonly available softwood lumber species. STJR along with its accompanying *Design Values for Joists and Rafters*, contains span tables for all commercially available species and grades in the U.S.

2306.1.2 Plank and beam flooring. Plank and beam flooring may be designed in accordance with the AWC *Wood Construction Data No. 4* (WCD No. 4), *Plank and Beam Framing for Residential Buildings*. The WCD4 contains information pertaining to principles of design, advantages and limitations, construction details, and structural requirements for the plank-and-beam method of framing, including span and load tables.

2306.1.3 Preservative-treated wood allowable stresses. The allowable unit stresses for preservative-treated wood conforming to AWPA standard U1 as discussed in Section 2303.1.9 do not require adjustment for the effects of the treatment. In other words, design values for untreated wood are to be used for preservative-treated wood. However, preservative-treated wood is subject to other adjustments such as the load duration factors defined in the NDS. Load duration factors greater than 1.6 shall not be used in the structural design of preservative-treated wood members. An example of an application for a load duration factor greater than 1.6 is an impact load.

2306.1.4 Fire-retardant-treated wood allowable stresses. The allowable unit stresses for fire-retardant-treated wood, including connection design values, must be developed in accordance with the provisions of Section 2303.2.6. Similar to preservative-treated wood, load duration factors greater than 1.6 shall not be used in the structural design of fire-retardant-treated wood members.

2306.1.5 Lumber decking. The capacity of lumber decking is arranged according to the various layup patterns described in Section 2304.9.2. Section 2306.1.5 gives the design capacity of lumber decking for moment and deflection according to the formulas given in Table 2306.1.5.

2306.2 Wood-frame diaphragms. The term diaphragm refers to horizontal or sloping panel elements that resist and distribute lateral forces. (Vertical panel elements that resist lateral forces are termed shear walls.) See Section 202 for definitions of various types of diaphragms and diaphragm components. This section references the SDPWS for their design.

Tables 2306.2(1) and 2306.2(2) contain allowable stress design (ASD) values for diaphragms fastened with staples because the SDPWS does not address stapled diaphragms. The design values given in the tables are based on Douglas fir-larch or southern pine framing. When the framing lumber is of a species other than Douglas fir-larch or southern pine, the tabular values must be adjusted to account for the fastener behavior in framing members of other lumber species. The adjustments must be made in accordance with Footnote "a" of each table.

High-load diaphragms are required to have special inspection in accordance with Section 1705.5.1, as indicated in Footnote "g" of Table 2306.2(2). The figures accompanying Table 2306.2(2) show the various multi-row fastening patterns, required spacing, and staggering. These fastening diagrams are intended to clarify proper application of the table and are essential to develop the diaphragm shear values listed.

Footnotes in the diaphragm and shear wall tables [Table 2306.2(1), Table 2306.2(2), and Table 2306.3(1)] provide the necessary factors to convert table values for shear loads of normal or permanent load duration as defined by the NDS. The factors reflect the fact that the values in the tables have a built-in load duration factor of 1.6. As such, to convert to a normal load duration of 1.00, the tabular value must be multiplied by 0.63 ($1.6 \times 0.63 = 1.00$). To convert to a permanent load duration of 0.90, a factor of 0.56 must be used ($1.6 \times 0.56 = 0.90$).

Shear capacity modifications for wind loads. Section 2306.2 permits allowable shear capacities for wood diaphragms fastened with staples to be increased 40 percent for wind design. The 40-percent increase is permitted in Section 2306.2 for diaphragms and Section 2306.3 for wood-framed shear walls. The wind loads determined by ASCE 7 have changed over time, and research has provided a better understanding of how structures are affected by wind, and code-prescribed wind loads have been refined since the diaphragm and shear wall tables were originally developed. The design values in the tables include a 2.8 factor of safety. Because wind loads are essentially monotonic and bounded, a safety factor of 2 is deemed to be sufficient for wind loading. This results in a 40-percent increase in the tabulated values ($2.8 - 2.0/2.0 = 0.4$). This approach is consistent with SDPWS shear wall and diaphragm capacities for assemblies using nails to attach sheathing to framing.

Diagonally sheathed lumber diaphragms. The detailed requirements for diagonally sheathed lumber diaphragms are covered in SDPWS including nominal unit shear capacities for lumber sheathed diaphragms of single horizontal, single diagonal, and double diagonal sheathing.

2306.2.1 Gypsum board diaphragm ceilings. Section 2508.6 allows the use of gypsum board ceiling diaphragms provided certain requirements and restrictions are met. For example, gypsum board ceiling diaphragms cannot be used to resist lateral forces resulting from concrete or masonry. The maximum diaphragm aspect ratio is 1½:1, and cantilevered diaphragms and diaphragm rotation are not permitted.

2306.3 Wood-frame shear walls. Where shear wall panels are fastened with staples, the IBC contains tables with allowable shear values for wood structural panels [Table 2306.3(1)], fiberboard [Table 2306.3(2)], and Portland cement plaster and gypsum materials [Table 2306.3(3)]. Requirements and limitations of SDPWS must also be met for shear walls where panels are fastened to framing members with staples. As noted in Footnote "e" of Table 2306.3(2), fiberboard shear walls fastened with staples are not permitted to be used in Seismic Design Categories D, E, and F. The SDPWS restricts the use of particleboard and structural fiberboard shear walls to Seismic Design Categories A, B, and C and Portland cement plaster (stucco) and gypsum shear walls to Seismic Design Categories A, B, C, and D. Thus, only wood structural panel shear walls are permitted to be used in Seismic Design Categories E and F.

Maximum shear wall aspect ratios (h/b) are prescribed by the SDPWS. Blocked wood structural panel shear walls are limited to 3.5:1 with unblocked set at 2:1. See later discussion on aspect ratio capacity adjustments.

The SDPWS has an exception allowing horizontal blocking to be omitted if specific conditions are met and shear capacity is reduced by the unblocked shear wall adjustment factor.

As with diaphragms, the allowable shear wall capacities where wood structural panels are fastened to framing with staples are 40 percent higher than capacities used for seismic design. See earlier discussion on shear capacity modifications for wind loads.

As previously noted, the IBC references the SDPWS, which covers design and construction requirements for wood members, fasteners, and assemblies designed to resist wind and seismic forces. Table 4.3A of the SDPWS gives nominal unit shear capacities for wood-frame shear walls sheathed with wood-based panels. Tabulated nominal unit shear capacities for wood-frame shear walls are applicable for both wind and seismic design for a given combination of sheathing, fastening schedule, and framing (in prior editions, separate nominal values were tabulated for wind design and seismic design). Coupled with this tabulation of a single nominal unit shear capacity for wind and seismic are ASD reduction factors and LRFD resistance factors for wind and seismic design.

The IBC tables only contain design values for staples since SDPWS does not contain shear wall capacities for panels attached to framing with staples. The design values given in the IBC tables are based on Douglas fir-larch or southern pine framing. When the framing lumber is of a species other than Douglas fir-larch or southern pine, the tabular values must be adjusted to account for the fastener behavior in framing members of other lumber species. The adjustments must be made in accordance with Footnotes "a" and "b" of the wood structural panel and structural fiberboard tables, respectively.

For wood structural panel shear walls utilizing staples, Footnotes "d" and "f" require 3-inch (76 mm) nominal or wider framing at adjoining panel edges where fastener spacing at panel edges is 2 inches (51 mm) on center or less, and where panels are applied to both faces, panel edge fastener spacing is less than 6 inches (152 mm) on center and panel joints are not offset from each other.

Table 2306.3(1) Footnote "g" applies to wood structural panel shear walls in Seismic Design Categories D, E, and F. All framing members receiving edge fastening from abutting panels must be 3-inch (76 mm) nominal when the shear design value exceeds 350 pounds per linear foot (5.1 kN/m) (ASD). Note that two 2-inch (51 mm) nominal members fastened together to transfer shear between framing members in accordance with the NDS is an acceptable alternative. Both panel joint and sill plate fasteners must be staggered to prevent splitting. SDPWS requires 3-inch (76 mm) square by 0.229-inch-thick (5.8 mm) square plate washers with some exceptions. A standard cut washer can be used where all of the following are met: Anchor bolts are designed to resist shear only; the full-height segment design method is used; hold-downs are designed without considering any dead load to resist overturning; the shear wall aspect ratio does not exceed 2:1; and the nominal shear capacity of the shear wall does not exceed 1370 plf (20.0 kN/m) (which translates to an ASD capacity of 490 plf (7.2 kN/m) for seismic and 685 plf (10.0 kN/m) for wind).

Lumber sheathed shear walls. Provisions covering lumber sheathed shear walls are located in the SDPWS including construction requirements for single and double layer diagonally sheathed shear walls, shear walls horizontally sheathed with a single layer of lumber and shear walls sheathed with vertical board siding. The SDPWS only permits diagonally sheathed lumber shear walls in Seismic Design Categories A, B, C, and D. Shear walls horizontally sheathed with a single layer of lumber and shear walls sheathed with vertical board siding are limited to SDC A, B, and C. All lumber sheathed shear walls may be used for wind design.

Particleboard shear walls. Particleboard shear walls are covered in the SDPWS and are permitted to resist wind loads. However, the SDPWS only permits particleboard shear walls in buildings in Seismic Design Categories A, B, and C. Particleboard used for shear walls must conform to ANSI A208.1 *Particleboard*. Note that particleboard shear walls have a maximum aspect ratio of 2:1.

Structural fiberboard shear walls. Structural fiberboard shear walls are addressed in the SDPWS and are also permitted to resist wind loads. However, the SDPWS only permits structural fiberboard shear walls in buildings in Seismic Design Categories A, B, and C. Fiberboard used for shear walls must conform to ASTM C208 *Specification for Cellulosic Fiber Insulating Board*. Comparable to wood structural panel shear walls, SDPWS allows fiberboard shear walls to have an aspect ratio of up to 3.5:1.

Shear wall capacity adjustments. The SDPWS includes adjustments where the shear capacity is reduced by the Aspect Ratio Factor for wind and seismic design. The Aspect Ratio Factor applies for structural fiberboard when the aspect ratio exceeds 1:1. It applies to wood structural panels when the aspect ratio exceeds 2:1. The Aspect Ratio Factors for wood structural panels and structural fiberboard are shown in Table 2306-1 for an 8-foot (2,838 mm) tall shear wall.

Table 2306-1. **Shear Wall Aspect Ratio Factors for 8-Foot Shear Wall Height**

Shear Wall Width, b (feet)	Aspect Ratio, h/b	Aspect Ratio Factor	
		Wood Structural Panel	Structural Fiberboard
2.28	3.5	0.81	0.77
3.00	2.67	0.92	0.85
4.00	2.00	1.00	0.91
5.00	1.60	-	0.95
6.00	1.33	-	0.97
7.00	1.14	-	0.99
8.00	1.00	-	1.00

For SI: 1 foot = 305 mm

Shear walls sheathed with other materials. Shear walls composed of cement plaster and gypsum materials are covered in the SDPWS. The detailed construction requirements for gypsum board, gypsum base for veneer plaster, water-resistant gypsum backing board, gypsum sheathing board, gypsum lath and plaster, and Portland cement plaster shear walls are in SDPWS and IBC Chapter 25. Applicable ASTM standards for the various types of materials are given in SDPWS. The SDPWS permits cement plaster and gypsum shear walls to be used for wind loads and in Seismic Design Categories A, B, C, and D. Shear walls using these materials are considered brittle (nonductile), and as such the strength and stiffness of both plaster and gypsum shear walls decrease significantly under a limited number of cycles when subjected to seismic demands.

Cross-laminated timber shear wall systems. SDPWS contains design and detailing provisions for cross-laminated timber (CLT) shear wall systems that can be used to resist wind and seismic loads. There are two main categories of CLT-based lateral resisting systems that have been actively investigated by researchers around the world since the introduction of this mass timber material. The first type of system is suited for platform-style construction, utilizing CLT panels as shear walls. System strength is provided through nailed connectors at CLT panel edges. Yielding of the nails and metal connectors, and combined rocking and sliding behavior of individual panels, produces the system's ductile response. The CLT wall panel is designed to resist overturning induced compression forces and uses either continuous hold-down rods or conventional hold-downs at shear wall ends to resist overturning induced tension forces. This system is codified in SDPWS (see Table 2306-2 for wind and seismic applications). The second type of CLT-based lateral resisting system is a rocking wall system that can achieve self-centering and energy dissipation through a variety of designed components. Using CLT components in seismic force-resisting systems is an area of considerable ongoing research. Seismic design coefficients for CLT shear wall systems are provided in ASCE 7-22.

Table 2306-2. **Cross-Laminated Timber Shear Wall Capacities**

2021 SDPWS CLT Shear Wall Capacities* (ASD)						
			Seismic		Wind	
Panel Length (ft)	Panel Height (ft)	No. of Angles	Shear (plf)	Hold-down (lbs)	Shear (plf)	Hold-down (lbs)
4	8	2	465	7443	651	7815
4	8	3	698	11164	977	11723
4	8	4	930	14886	1303	15630
4	8	5	1163	18607	1628	19538

*Capacity derived from 2021 SDPWS based on angles along bottom of one panel face and G = 0.42.

Seismic considerations for wood shear wall systems. Technical provisions for seismic design are covered in the ASCE 7 standard. The IBC only contains the design criteria for environmental loads such as seismic and wind. Shear walls resisting seismic loads are subject to the limitations in Section 12.2.1 and Table 12.2-1 of ASCE 7. Similar requirements applicable to the simplified seismic provisions are found in Section 12.14 and Table 12.14-1. Tables 12.2-1 and 12.14-1 contain two types of wood-frame shear walls: light-frame (wood) walls sheathed with wood structural panels rated for shear resistance, and light-frame walls with shear panels of all other materials. See Items 16 and 18 of ASCE 7 Table 12.2-1 under bearing wall systems.

For light-frame shear walls sheathed with materials other than wood structural panels, Table 12.2-1 assigns a response modification factor of 2 and restricts the height of the building to 35 feet (10,668 mm) in Seismic Design Category D. Also, these shear walls are not permitted to be used to resist seismic forces in Seismic Design Category E or F. Further, since ASCE 7 references SDPWS for detailing requirements, and since SDPWS does not permit certain shear wall materials such as particle board and structural fiberboard to be used in Seismic Design Category D, that would limit use of those shear wall materials to Seismic Design Categories A through C. The ASCE 7 standard also limits light-frame wood walls sheathed with wood structural panels to a building height of 65 feet (19,812 mm) in Seismic Design Categories D, E, and F.

ASCE 7 requirements for cross-laminated timber (CLT) shear walls and associated seismic design coefficients are based on research that demonstrates adequate adjusted collapse margin ratios using the FEMA P-695 *Quantification of Seismic Performance Factors* methodology.

Two variants of the CLT shear wall system are addressed in ASCE 7 Table 12.2-1: (a) CLT shear wall systems with a ratio of wall panel height to individual wall panel length of between 2 and 4; and (b) CLT shear wall systems with shear resistance provided by high aspect ratio panels only, where high aspect ratio is defined as wall panel height to individual wall panel length ratio of 4. The response modification coefficient of R = 3 for CLT shear walls and R = 4 for CLT shear walls with shear resistance provided by high aspect ratio panels only recognizes improved displacement capacity associated with use of high aspect ratio CLT panels and has been validated by incremental dynamic analysis results indicating collapse probabilities of less than 10%. Table 2306-3 provides a summary of these requirements.

Table 2306-3. **Summary of Different Types of Wood Shear Wall System Limitations Based on Seismic Design Category**

Shear Wall Type	Permissible Seismic Design Categories per ASCE 7 and SDPWS
Particleboard	A, B, C
Structural Fiberboard	A, B, C
Horizontally Sheathed Single-Layer Lumber	A, B, C
Vertical Board Siding	A, B, C
Gypsum Board, Gypsum Base for Veneer Plaster, Water-Resistant Gypsum Backing Board, Gypsum Sheathing Board, Gypsum Lath and Plaster, Portland Cement Plaster	A, B, C, D[a]
Diagonally Sheathed Lumber - Single Layer Diagonally Sheathed Lumber - Double Layer	A, B, C, D[a]
Wood Structural Panels Over Gypsum Wallboard or Gypsum Sheathing[b]	A, B, C, D, E, F
Wood Structural Panel[b]	A, B, C, D, E, F
Cross-laminated Timber Shear Wall Systems[c]	A, B, C, D, E, F

[a]Limited to 35 feet building height in Seismic Design Category D.
[b]Limited to 65 feet building height Seismic Design Categories D, E, and F.
[c]Limited to 65 feet building height in Seismic Design Categories B through F.

Section 2307 *Load and Resistance Factor Design*

2307.1 General. The section references the NDS and SDPWS for load and resistance factor design (LRFD) of wood structures. Both are dual format standards that permit ASD and LRFD procedures. The NDS approach is to tabulate reference (ASD) capacities for wood members and connections. The following discussion, adapted from the *NDS Commentary*, provides an overview of the LRFD design approach.

The approach of adjusting a reference value in accordance with the applicable adjustment factors to arrive at either an ASD- or LRFD-adjusted design value is illustrated in the following example for sawn lumber bending stress, F_b:

ASD: $F'_b = F_b(C_D)(C_M)(C_t)(C_L)(C_F)(C_{fu})(C_i)(C_r)$

LRFD: $F'_b = F_b(C_M)(C_t)(C_L)(C_F)(C_{fu})(C_i)(C_r)(2.54)(0.85)(\lambda)$

where:

F_b = Reference bending design value, psi
C_D = Load duration factor applicable for ASD only
$(C_M)(C_t)(C_L)(C_F)(C_{fu})(C_i)(C_r)$ = Adjustment factors from NDS
2.54 = Format conversion factor for F_b, K_F, applicable for LRFD only
0.85 = Resistance factor, ϕ_b, applicable for LRFD only
λ = Time effect factor, applicable for LRFD only

Format conversion factors convert reference design values (allowable stress design values based on normal load duration) to LRFD reference resistances. Resistance factors are assigned to various wood properties with only one factor assigned to each stress mode (i.e., bending, shear, compression, tension, and stability). In general, the magnitude of the resistance factor is considered to, in part, reflect relative variability of wood product properties. Actual differences in product variability are accounted for in the derivation of reference design values. Both format conversion factors and resistance factors are determined per ASTM D5457 *Standard Specification for Computing Reference Resistance of Wood-Based Materials and Structural Connections for Load and Resistance Factor Design.*

The time effect factor, λ (LRFD counterpart to the ASD load duration factor, C_D), varies by load combination and is intended to establish a consistent target reliability index for load scenarios represented by applicable load combinations. Except for the load combination for dead load only, each load combination can be viewed as addressing load scenarios involving peak values of one or more "primary" loads in combination with other transient loads. Specific time effect factors for various ASCE 7 load combinations are largely dependent on the magnitude, duration, and variation of the primary load in each combination.

The reduced probability of the simultaneous occurrence of combinations of various loads on a structure, such as dead, live, wind, snow, and earthquake, is recognized for both ASD and LRFD in the IBC and ASCE 7. Specific load reductions for ASD or load combinations for LRFD apply when multiple transient loads act simultaneously.

All modifications for load combinations are entirely separate from adjustments for load duration, C_D, or time effect, λ, which are directly applicable to wood design values. It should be emphasized that reduction of design loads to account for the probability of simultaneous occurrence of loads and the adjustment of wood resistances to account for the effect of the duration of the applied loads are independent of each other and both adjustments are applicable in the design calculation.

The SDPWS approach for shear walls and diaphragms is to tabulate "nominal strength" values. As discussed earlier in Sections 2306.2, "Wood-frame diaphragms," and 2306.3, "Wood-frame shear walls," nominal strength (or nominal capacity) is used to provide a common reference point from which to derive ASD or LRFD reference design values. LRFD resistance factors for sheathing in-plane shear, sheathing out-of-plane bending, and connections are determined per ASTM D5457. Since time effect factors, λ, for load combinations including wind and seismic loads are set to a short-term (10-minute) baseline, a value of unity is assumed.

Section 2308 *Conventional Light-Frame Construction*

2308.1 General. A key feature of the conventional construction provisions is in this statement: "Other construction methods are permitted to be used, provided that a satisfactory design is submitted showing compliance with other provisions of this code." In other words, one need not conform to the restrictions, limitations, and requirements of the conventional light-frame construction provisions if a design is submitted to the jurisdiction that conforms to the engineering requirements of the code.

The provisions of Section 2308 provide prescriptive construction details and methods for light-frame wood construction. Light-frame wood construction consists of 2× construction with walls of 2-inch-nominal-thickness (51 mm) studs typically spaced at 16 or 24 inches (406 or 610 mm) on center, and roofs and floors framed of 2-inch-nominal-thickness (51 mm) rafters or joists spaced at 12, 16, or 24 inches (305, 406, or 610 mm) on center. Construction that meets the restrictions, limitations, and prescriptive provisions of Section 2308 is deemed to comply with the intent of the code without requiring engineering. Perhaps most important is Section 2308.2, which contains the specific restrictions and limitations for the conventional construction provisions.

Detached one- and two-family dwellings and townhouses not more than three stories in height with a separate means of egress are specifically within the scope of the IRC. Only buildings that are within the scope of the IBC and Section 2308.2 are permitted to use the conventional construction provisions of Section 2308. However, the IRC requires structural elements that do not conform to the limits of the IRC to be designed in accordance with accepted engineering practice under the engineering provisions of the IBC. Section 2309 also permits compliance with the AWC *Wood Frame Construction Manual* (WFCM) as an alternative. Tables 2308-1 and 2308-2 compare scoping provisions of the IRC, IBC Section 2308 and the WFCM, all of which contain prescriptive light-frame wood construction provisions. While each of these three documents permit or even provide engineered solutions for light-frame wood construction, the discussion here is limited to prescriptive provisions. Consult each of the codes and the WFCM for exceptions and additional details. Prescriptive wood provisions are applicable to all use groups in the IBC in accordance with the scoping limitations of IBC Section 2308 and the WFCM.

Table 2308-1. **Applications for Prescriptive Light-Frame Wood Construction Provisions**

Representative Building Use and Occupancy	IRC	IBC 2308	IBC 2309 (WFCM)
Detached one- and two-family dwellings and townhouses ≤ 3 stories with separate egress and accessory buildings	X		X
Two-family dwellings and townhouses, without separate egress, or not detached from other occupancies or buildings (R-3)		X	X
All other occupancies as applicable (see Table 2308-2)		X	X

2308.2 Limitations. One of the most important aspects of most prescriptive methods is meeting restrictions and limitations required to use the method. The structures for which conventional construction is applicable are described in this section. The limitations are broken into categories as follows: (1) maximum number of stories, (2) maximum floor-to-floor height, (3) maximum dead, live, and snow loads, (4) maximum basic wind speed, (5) maximum roof span, (6) risk category and highest permissible seismic design category.

Table 2308-2. Scoping Provisions for the IRC, IBC 2308, and IBC 2309

	IRC	IBC 2308	IBC 2309 (WFCM)
Occupancy	One- and two-family dwellings, and townhouses as defined	All buildings except those within the scope of the IRC	All buildings, subject to the limitations of WFCM Section 1.1.3
Risk category	N/A	I, II, III; IV if in SDC A	I and II
Maximum number of stories	3	3	3
Maximum mean roof height	ND	ND	33'
Maximum building length/width	ND except 36'×60' for trusses	ND	80'
Maximum floor/wall height			
(Story) floor-to-floor height	11'-7"	11'-7"	NL
Bearing wall and interior braced wall stud height	10'	10'	10'
Nonload-bearing walls	20'	20'	20'
Roof			
Rafter or roof span	26' lumber 36' truss roof span	40' roof span	26' lumber/I-joists 60' truss roof span
Slope	Flat – 12:12	Flat – 12:12	1.5:12 – 12:12
Gravity load limits			
Floor dead load	20 psf; 10 psf seismic	15 psf (avg)	20 psf
Roof dead load	25 psf	15 psf (avg)	25 psf
Floor live load	40 psf	40 psf, with an exception for concrete slab-on-ground floors not exceeding 125 psf (Risk Categories I and II only)	40 psf
Ground snow load	70 psf	50 psf	70 psf
Basic wind speed (3-sec gust) Exposures B, C, and D	V < 130 mph; or V < 140 mph in non-hurricane-prone regions	V ≤ 130 mph; or V ≤ 140 mph Exp B in non-hurricane-prone regions	90 ≤ V ≤ 195 mph
Seismic			
Seismic design category (SDC)	SDC A-B NL; SDC C-D_2 Townhouses; SDC D_0-D_2 detached one- and two-family	SDC A-E, except Risk Category IV buildings are limited to SDC A only	SDC A-D_2
Story limits	Three stories	SDC A-B ≤ 3 stories; SDC C ≤ 2 stories; SDC D-E = 1 story	Three stories

NL—Not limited.
ND—Not defined.
See the codes and standards for exceptions.
For SI: 1 inch = 25.4 mm, 1 foot = 305 mm

2308.2.1 Stories. Table 2308.2.1 restricts the building height to three stories above grade plane in Seismic Design Category A and B. The table further restricts the building height to two stories above grade plane in Seismic Design Category C, and one story above grade plane in Seismic Design Categories D and E. In Seismic Design Category D or E, cripple walls exceeding 14 inches (356 mm) are considered a story for the purposes of determining the number of stories. Note that cripple walls exceeding 14 inches (356 mm) high in structures classified in any seismic design category are considered an additional story for purposes of bracing as indicated in Section 2308.10.6. Typical conventional raised-floor construction generally uses post and girder construction to support the floor joists within the interior of the structure. Because interior braced wall lines may not necessarily extend down to the foundation, lateral loads must be transferred through exterior cripple walls to the foundation. Because cripple walls are at the lowest level of the structure, they are the most heavily loaded, relative to the braced panel walls. Thus, cripple walls are very important to satisfactory performance of the structure when subjected to lateral loads from wind pressure or earthquake ground motion. By classifying cripple walls exceeding the 14-inch (356 mm) limit as a story, higher bracing strength criteria may be required to resist the higher lateral loads.

2308.2.2 Allowable floor-to-floor height. There are two conditions: a maximum floor-to-floor height of 11 foot 7 inches (3,531 mm) and a maximum bearing wall stud height of 10 feet (3,048 mm). See Figure 2308-1.

2308.2.3 Allowable loads. Section 2308.2.3 limits the average dead load to 15 psf (718 N/m^2) for roofs and exterior walls, floors, and partitions combined. There are two exceptions: (1) stone or masonry veneer up to the lesser of 5 inches (127 mm) thick or 50 psf (2,395 N/m^2) is permitted to a height of 30 feet (9,144 mm) above a noncombustible foundation with an additional 8 feet (2,438 mm) permitted for gable ends, and (2) concrete or masonry fireplaces, heaters, and chimneys are permitted. Live loads are limited to 40 psf (1,916 N/m^2) for floors and in Risk Category I and II the limit is 125 psf (5,985 N/m^2) for slab-on-ground. Allowable stress design ground snow load is limited to 50 psf (2,395 N/m^2). Where required, tornado loads shall not exceed corresponding wind loads on the same elements.

2308.2.4 Basic wind speed. The basic wind speed is limited to 130 mph (57 m/s). Where the basic wind speed, V, exceeds 130 mph (57 m/s), special details are necessary to resist the higher wind pressures and ensure adequate load path continuity. Exceptions allow use of the conventional light-frame construction provisions for buildings in Exposure Category B areas with wind speeds up to 140 mph (63 m/s) if the area is not a hurricane-prone regions. See Section 202 for definition of hurricane-prone region. Additionally, buildings in areas with wind speeds exceeding 130 mph (57 m/s) may use AWC's *Wood Frame Construction Manual* or ICC 600, *Standard for Residential Construction in High-wind Regions*.

2308.2.5 Allowable roof span. The provisions indicate that a ridge board is not to be misconstrued as being a point of vertical support. Conventionally framed roofs with ridge board, rafters, and ceiling joists behave similar to trusses, spanning from wall to wall, with the rafter in compression and the ceiling joist in tension. A structural ridge beam, on the other hand, is designed to support vertical loads (reactions) from the rafters and as such is considered a point of vertical support. See Figure 2308-2.

2308.2.6 Risk category limitation. The conventional construction provisions are not permitted to be used for structures in Risk Category IV assigned to seismic design category B, C, D, or F. Risk categories are determined from Table 1604.5 based on the use of the structure. The seismic design category of a structure is determined based on the risk category (Section 1613.2).

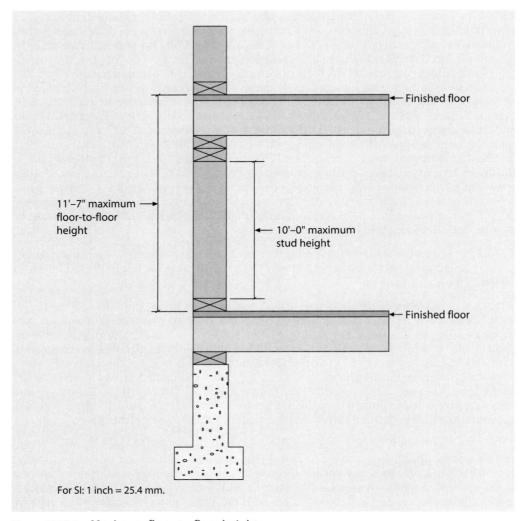

Figure 2308-1 **Maximum floor-to-floor height.**

2308.2.7 Hillside light-frame construction. For light-frame wood dwellings on steep hillsides, the typical assumption of floor loads transferring to braced wall panels based on the tributary area of a flexible wood floor may not provide adequate seismic performance. This building configuration was found to be vulnerable in the 1994 Northridge, California earthquake. Whether the earthquake motion occurs across the slope or perpendicular to the hill, seismic forces follow the stiffest load path to the uphill foundation, rather than distributing evenly to all braced wall panels as assumed in prescriptive IBC seismic wall bracing provisions. These load path provisions provide a correlation between the prescriptive requirements of IBC Section 2308 and IRC Section R301.2.2.6, Item 8. A related provision is also included in ASCE 7 to offer additional guidance to engineers designing these types of structures. See Figure 2308-3.

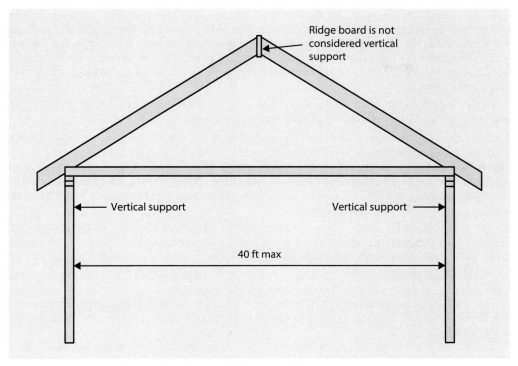

Figure 2308-2 **Typical rafter framing with ridge board.**

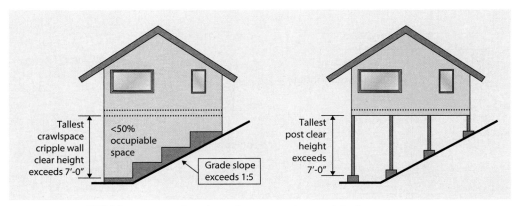

Figure 2308-3 **Hillside construction that triggers engineered design.**

2308.3 Portions or elements exceeding limitations of conventional light-frame construction. In this case, portions of a building as well as the supporting load path must be designed in accordance with accepted engineering practice and provisions of the code. In this section, the term *portions* does not refer to structural elements but to parts of the building that contain volume and area such as a room or a series of rooms.

2308.4 Structural elements or systems not described herein. When a conventionally constructed building contains structural elements or systems that are not specifically covered by Section 2308, the elements and their supporting load path must be designed in accordance with the engineering provisions of the code. Some examples of such elements are engineered floor and roof trusses, I-joists, glued-laminated timber or SCL beams and columns, and cross-laminated timber.

2308.5 Connections and fasteners. This section refers back to Section 2304.10 for requirements for fasteners used in conventionally constructed wood-frame buildings.

2308.6 Cutting, notching and boring of dimensional wood framing. Figure 2308-4 illustrates limits on notches or bored holes in floor joists and ceiling joists. Similar limits apply to rafters. Studs are often notched to accommodate wiring, plumbing, etc. Figure 2308-5 illustrates acceptable cutting or notching of studs. Studs often have drilled holes to accommodate wiring, plumbing, etc. Figure 2308-6 illustrates acceptable dimensions and location of bored holes in studs.

2308.7 Foundations and footings. This section references Chapter 18 for foundation design, which includes prescriptive footings for light-frame construction in Section 1809.7 and prescriptive foundation walls in Section 1807.1.6.

2308.7.1 Foundation plates or sills. Prescriptive requirements for foundation plates and anchorage are given in this section, although these provisions apply only to conventional construction. The requirements in Section 2304.3.1 apply to both engineered and conventional construction. Footing requirements for light-frame construction are contained in Section 1809.7.

Typical sill anchorage is ½-inch-diameter (12.7 mm) bolts spaced not more than 6 feet (1,829 mm) on center, or 4 feet (1,219 mm) on center for structures over two stories above grade. The phrase "or approved anchors spaced to provide equivalent anchorage as the steel bolts" is based on cyclic testing of walls using foundation anchor straps that showed they perform well under cyclic loading.

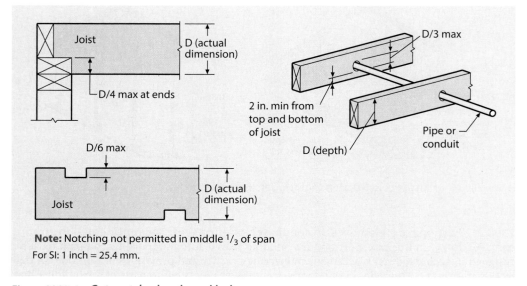

Figure 2308-4   Cut, notched, or bored holes.

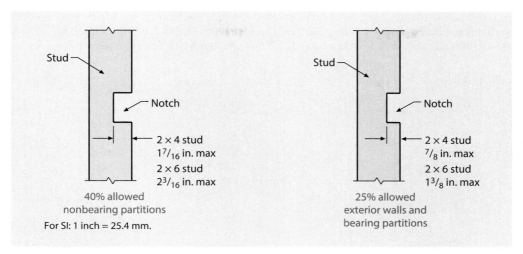

Figure 2308-5 **Cutting and notching of studs.**

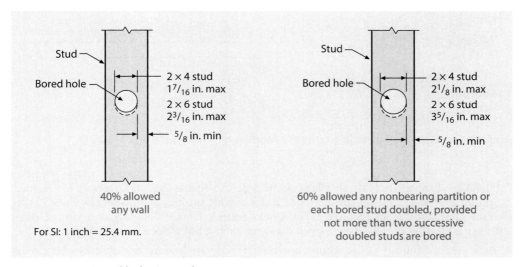

Figure 2308-6 **Bored holes in studs.**

2308.7.1.1 Braced wall line sill plate anchorage in Seismic Design Category D. The requirement for plate washers originated in response to the extensive splitting of wood sills during the Northridge earthquake. The splitting was caused by a variety of factors including cross-grain tension and oversized holes for anchor bolts, in addition to splitting from insufficient edge distance in heavily loaded walls. Subsequent cyclic testing of wood shear wall assemblies showed that use of 2-inch by 2-inch by 3/16-inch, or larger, steel plate washers at the anchor bolts helps prevent the sill from splitting.

The IBC requires 3 × 3 × 0.229 inch (76.2 × 76.2 × 5.8 mm) plate washers. The hole in the plate washer is permitted to be diagonally slotted with a width up to 3/16-inch (81.0 mm) wider than the bolt diameter and a slot length up to 1¾ inches (44.5 mm), provided a standard cut washer is used between the square plate washer and the nut. The IBC, within this section, allows the use of not less than ½-inch diameter (12.7 mm) anchor bolts. See Figure 2308-7.

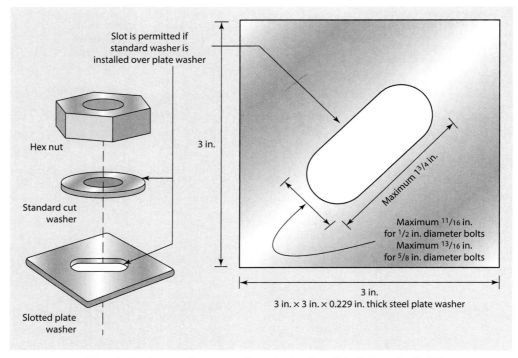

Figure 2308-7 Wood sill plate anchorage to foundation for all buildings in SDC D.

2308.7.1.2 Braced wall line sill plate anchorage in Seismic Design Category E. Similar to Section 2308.7.1.1, this section calls for the use of plate washers and diagonally slotted holes with the same dimensions and tolerances. Unlike Seismic Design Category D, Seismic Design Category E requires not less than ⅝-inch diameter anchor bolts. See Figure 2308-7.

2308.8 Floor framing. The section provides prescriptive framing requirements for girders and floor joists.

2308.8.1 Girders. The section allows the use of a 4 × 6 (102 mm × 152 mm) girder for spans not more than 6 feet (1,829 mm) and spacing not more than 8 feet (2,438 mm). Girders require a minimum of 1½ inches (38 mm) of bearing on wood or metal and 3 inches (76 mm) of bearing on concrete or masonry. The design of girders to support floor loads involves so many variables that the tables in the code are only applicable to the most simple cases. In many cases, girders should be engineered because of unusual configurations or loading conditions. Figures 2308-8 and 2308-9 show details deemed to comply with the splice and tie requirements for posts and girders.

2308.8.2 Floor joists. The floor joist span tables in the IBC are excerpted from the AWC span tables for joists and rafters. Joists selected from these tables are deemed to comply with code requirements for both strength and deflection. Table 2308.8.2.1(1) is for a floor live load

Conventional Light-Frame Construction 825

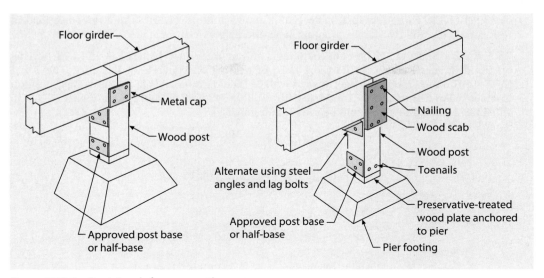

Figure 2308-8 **Post-to-girder connection.**

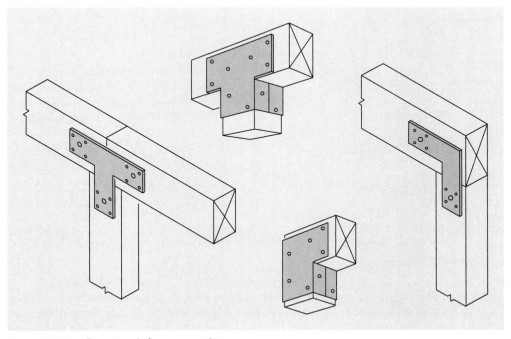

Figure 2308-9 **Post-to-girder connection.**

of 30 psf (1,436 N/m²) because Table 1607.1 allows a 30 psf (1,436 N/m²) floor live load in residential one- and two-family sleeping areas and habitable attics.

2308.8.2.2 Bearing. The code requires a minimum of 1½ inches (38 mm) of bearing on wood or metal and 3 inches (76.2 mm) of bearing on concrete or masonry. The 1 × 4 (25.4 × 101.6 mm) ribbon (ledger) nailed with 3–16d common nails to the studs was common in older balloon-framed multistory structures but is very uncommon today. Figures 2308-10 and 2308-11 illustrate minimum bearing requirements.

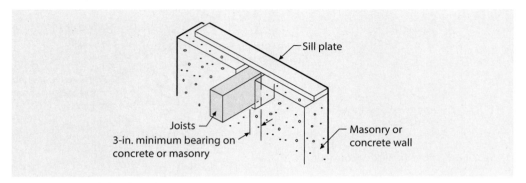

Figure 2308-10 **Joists bearing on concrete or masonry.**

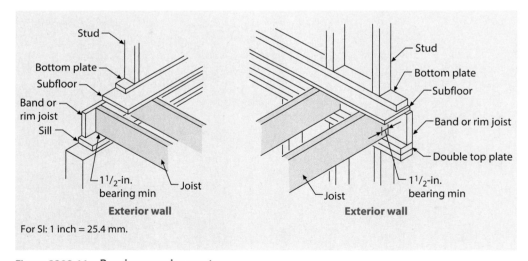

Figure 2308-11 **Bearing requirements.**

2308.8.2.3 Framing details. This section contains basic requirements for framing floor joist systems to provide lateral support at joist ends. Figure 2308-12 depicts the provisions for support by solid blocking or other means. Figure 2308-13 depicts the requirements for lapping and tying joists over a beam or partition.

2308.8.3 Engineered wood products. This section specifically prohibits cuts, notches, and holes bored in trusses, structural composite lumber, structural glued-laminated timber, or I-joists, unless specifically permitted by the manufacturer or where the registered design

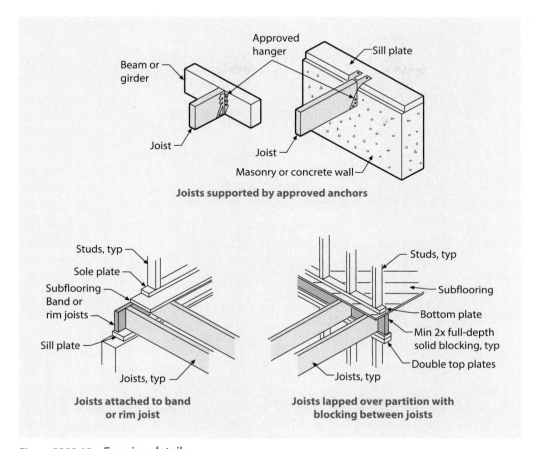

Figure 2308-12 **Framing details.**

professional has considered their effects on the stiffness and capacity of the member and verified the modification.

2308.8.4 Framing around openings. Figures 2308-14 through 2308-16 illustrate typical framing details for floor openings. Figure 2308-14 depicts acceptable framing details for openings not greater than 4 feet (1,219 mm). Figure 2308-15 depicts acceptable framing for openings greater than 4 feet (1,219 mm) but not greater than 6 feet (1,829 mm). Figure 2308-16 shows framing and hangers for openings greater than 6 feet (1,829 mm).

2308.8.4.1 Openings in floor diaphragms in Seismic Design Categories B, C, D, and E. This limitation attempts to place a practical limit on the size of openings in floor diaphragms. See Section 2308.11.5 for detailing requirements for permitted openings in roof diaphragms. In Seismic Design Categories other than A, openings in floor diaphragms with a dimension greater than 4 feet require metal ties and blocking as shown in Figure 2308.8.4.1(1). The section prescribes specific requirements for straps, blocking, and nailing to reinforce the boundaries of the opening. In Seismic Design Categories D and E, the dimensions of openings in floor diaphragms are limited to 50 percent of the distance between braced wall lines or an area greater than 25 percent of the area between orthogonal pairs of braced wall lines as shown in Figure 2308.8.4.1(2). If the

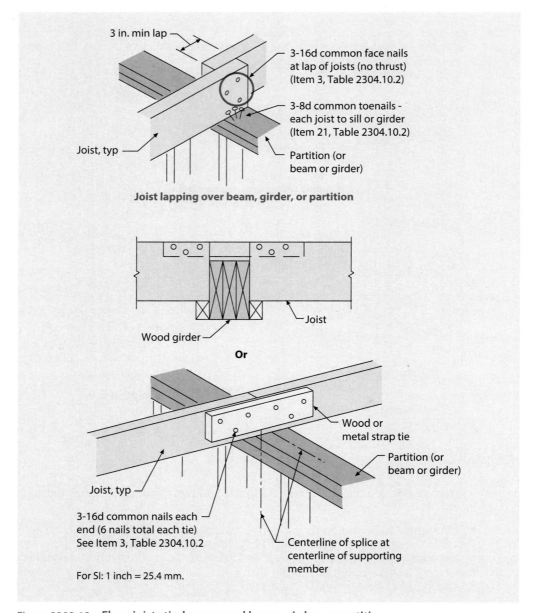

Figure 2308-13 **Floor joists tied over wood beam, girder, or partition.**

restrictions are not met, then the diaphragm and opening reinforcement must be designed to resist lateral forces in accordance with accepted engineering practice.

2308.8.4.2 Vertical offsets in floor diaphragms in Seismic Design Categories D and E. This limitation results from observation of damage that is somewhat unique to split-level wood-frame construction. If floors on either side of an offset move in opposite directions because

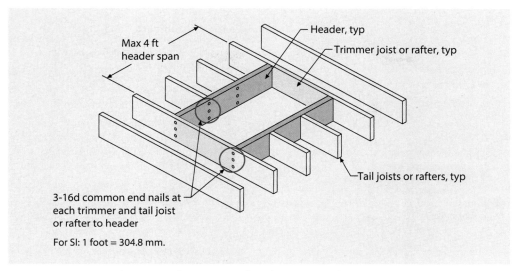

Figure 2308-14   Framing around openings—header span ≤ 4 feet.

of earthquake ground motion or wind loading, the short bearing wall in the middle becomes unstable and vertical support for the upper joists can be lost, resulting in a collapse. If the vertical offset is limited to a dimension equal to or less than the joist depth, and a simple strap tie directly connecting joists on different levels can be provided, then the instability is eliminated. In Seismic Design Categories D and E, portions of floor levels cannot be vertically offset such that the framing members on either side cannot be adequately tied together as shown in Figure 2308.8.4.2. The exception permits the framing to be supported directly by foundations in lieu of being lapped and tied directly together. If the restrictions cannot be met, then the diaphragm and offset irregularity must be designed to resist lateral forces in accordance with accepted engineering practice.

2308.8.5   Joists supporting bearing partitions.   Figure 2308-17 depicts the requirement for joists supporting bearing partitions when partitions are parallel and perpendicular to supporting joists.

2308.8.6   Lateral support.   Rectangular members such as floor and ceiling joists require lateral support to avoid buckling in the weak direction. The requirements are based on the *nominal* depth-to-thickness (d/b) rather than the actual dimensions of the member. Thus, a 2 × 12 has a d/b ratio equal to 12/2 = 6. Figure 2308-18 depicts the lateral support requirements for joists deeper than 2 × 12 unless both edges of the member are held in line by sheathing.

2308.8.7   Structural floor sheathing.   This section references Section 2304.8.1 which references Tables 2304.8(1) through 2304.8(4) for installation requirements for floor sheathing. Flooring systems may consist of a subfloor on which may be placed an underlayment for the floor covering or a combined subfloor and underlayment system, upon which a finished floor-surfacing material is applied. The finished floor-surfacing material may be wood-strip flooring, tongue-and-groove flooring, or various types of resilient floor coverings such as vinyl, tile or carpet. For the noncombined subfloor and underlayment system, underlayment is required to provide a smooth, even surface to which the finished flooring will be attached. However, in some cases, the underlayment is required to add strength to the subflooring.

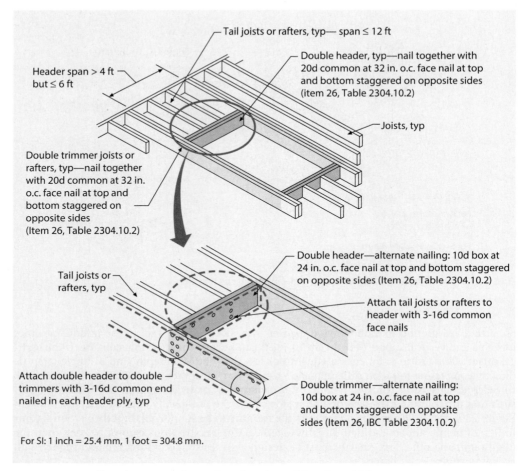

Figure 2308-15 **Framing around openings header > 4 feet but ≤ 6 feet.**

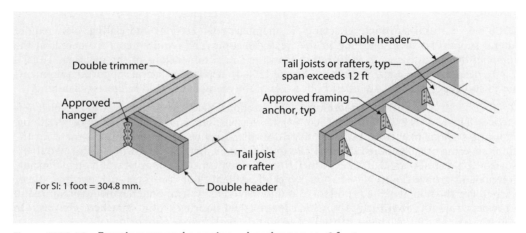

Figure 2308-16 **Framing around opening—header span > 6 feet.**

Conventional Light-Frame Construction   **831**

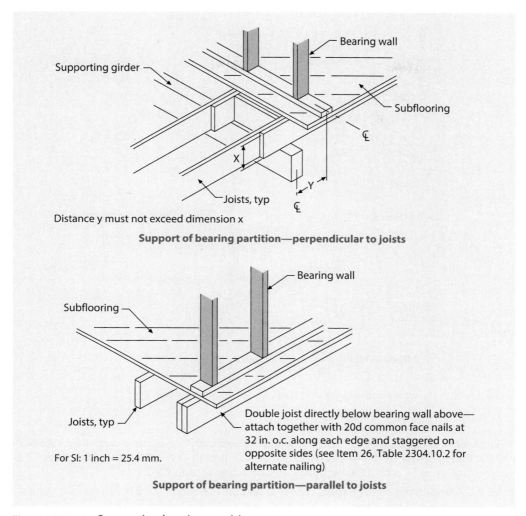

Figure 2308-17 **Supporting bearing partitions.**

Allowable spans and minimum grade requirements for lumber subflooring are in Tables 2304.8(1) and 2304.8(2). The span tables are based on the thickness of the floor sheathing, the orientation of the sheathing with respect to the joists (either perpendicular or diagonal), and the board grade of the lumber being used (see Figure 2308-19).

Wood structural panels may be manufactured for use as either structural subflooring or combined subfloor-underlayment. See Tables 2304.8(3) and 2304.8(4). The allowable spans for structural subflooring and combined subfloor-underlayment are based on the wood structural panel's face grain (strength axis) being continuous over two or more spans with the face grain perpendicular to the supports (see Figure 2308-19). To create a stiffer floor and prevent or minimize squeaking of the floor system after the building has been in use, the subfloor may be glued to the joists. This gluing prevents the relative movement between the panel and the joist that takes place when loads are placed on the floor, and provides additional stiffness.

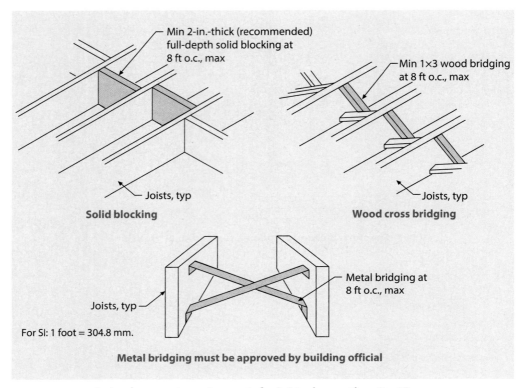

Figure 2308-18 Lateral support requirements for joists deeper than 2 × 12.

Particleboard can be used as underlayment, structural subflooring, or as combined subfloor-underlayment. Where used as underlayment, the code permits Type PBU particleboard not less than ¼-inch thick in accordance with ANSI A208.1. See Section 2303.1.8.

2308.8.8 Under-floor ventilation. See discussion of Section 1202.4 in regard to under-floor ventilation requirements.

2308.8.9 Floor framing supporting braced wall panels. This section references Section 2308.10.7, Connections of braced wall panels, for construction requirements for braced wall panels that are supported by floor framing such as cantilevered floors or set backs from the floor joist support.

2308.8.10 Anchorage of exterior means of egress components in Seismic Design Categories D and E. To prevent possible collapse of means of egress components such as exterior egress balconies and exterior exit stairways, positive anchorage to the main structure is required by this section. Anchors must be spaced at no more than 8 feet (2,438 mm) on center. If anchorage to the main structure is not provided, then the egress component itself must be designed to resist seismic load effects. Toenails and nails in withdrawal are not permitted because past experience has shown that they do not perform well when subjected to cyclic loads from seismic demands.

2308.9 Wall construction. This section contains typical framing requirements for conventionally framed stud walls.

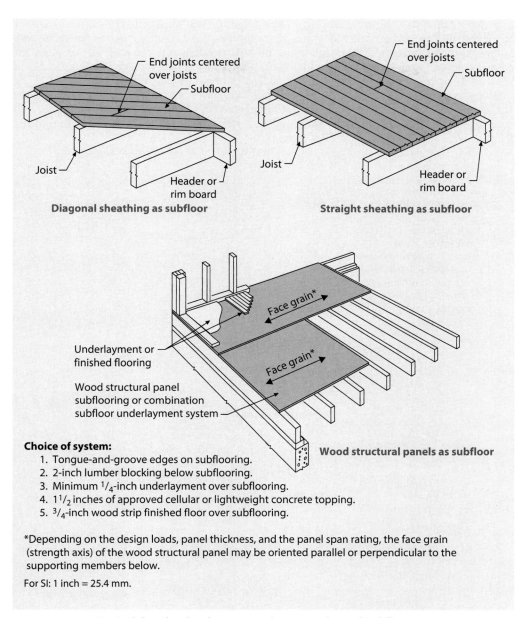

Figure 2308-19   Typical details—lumber or wood structural panel subfloor.

2308.9.1 Stud size, height, and spacing. Figure 2308-20 depicts typical stud requirements for bearing walls. Wall studs must be continuous from a support at the sole plate to a support at the top plate to resist out-of-plane loads perpendicular to the wall. The requirement does not apply to trimmers (king studs), jack studs and cripple studs at openings in walls. Where scissor trusses are used to create a vaulted ceiling, a scissor truss is often used at the gable end wall, and the top plate is not laterally supported because the ceiling follows the profile of the bottom chord of the truss. See Figure 2308-21. Where scissor trusses are used to create a vaulted ceiling profile, the gable end wall studs should be balloon framed up to the

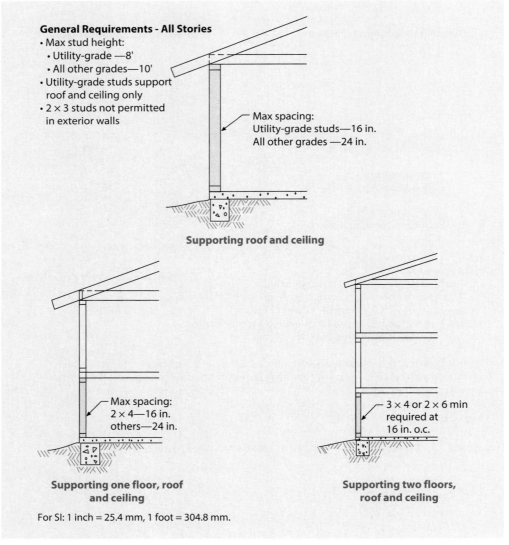

Figure 2308-20 Stud requirements for bearing walls.

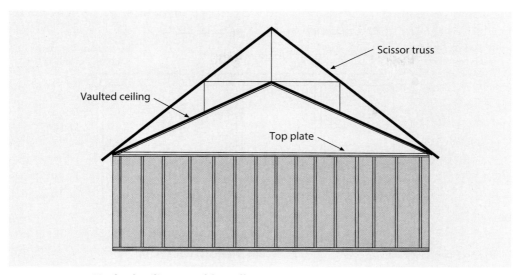

Figure 2308-21 **Vaulted ceiling at gable wall.**

truss bottom chord or to the roof sheathing. If the studs are not supported at the top by a ceiling or roof diaphragm in accordance with this requirement, then an engineered design should be provided for the gable end wall studs to resist out-of-plane (component and cladding) wind loads perpendicular to the wall.

2308.9.2 Framing details. Studs are required to be placed with their wide dimension perpendicular to the plane of the wall for maximum out-of-plane strength and stiffness. Not less than three studs are installed at each corner of exterior walls, although the exception allows two stud corners under specific conditions.

2308.9.3 Plates and sills. The section prescribes requirements for bottom plates and sills, and top plates in conventional stud framed walls.

2308.9.3.1 Bottom plate or sill. Studs are required to have full bearing on a 2-inch (51 mm) nominal bottom plate or sill at least as wide as the supported stud.

2308.9.3.2 Top plates. Top plate splices require face nails on each side of the 4-foot (1,219 mm) lapped joint. The 16d common nails or other equivalent fasteners are required by Item 13 of Table 2304.10.2. Single plate butt-joint splices must have at least twelve 8d box nails on each side of the splice with a 3-inch-wide by 12-inch-long by 0.036-inch-thick (76 mm by 305 mm by 0.91 mm) steel strap. Corners and intersecting walls require only half the number of nails needed for a butt-joint splice. Figure 2308-22 depicts the double top plate splice requirement; Figure 2308-23 depicts the recommended single top plate splice.

Top plates for studs spaced at 24 inches. Figure 2308-24 depicts the requirements for locating the joists or trusses where bearing wall studs are spaced 24 inches (610 mm) on center.

2308.9.4 Nonload-bearing walls and partitions. This section allows for increased stud spacing on nonload-bearing walls, and studs may be turned with the long dimension parallel to the wall for construction of plumbing chases. See Section 202 for definition of nonload-bearing wall. (For wood-frame construction, a nonload-bearing wall is a stud wall that supports 100 plf [1,459 N/m] or less superimposed vertical load.) Note that sheathing materials might require

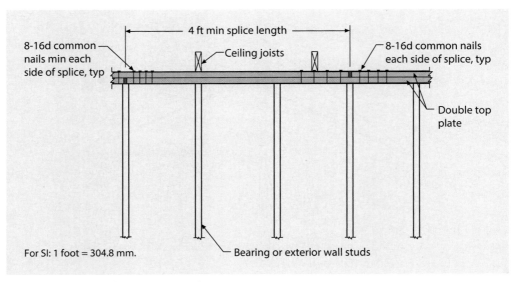

Figure 2308-22 **Double top plate splice.**

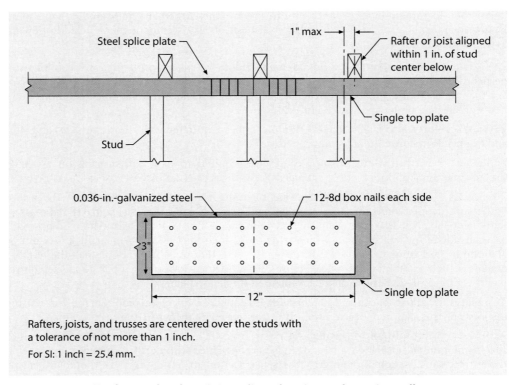

Figure 2308-23 **Single top plate butt-joint splice—bearing and exterior walls.**

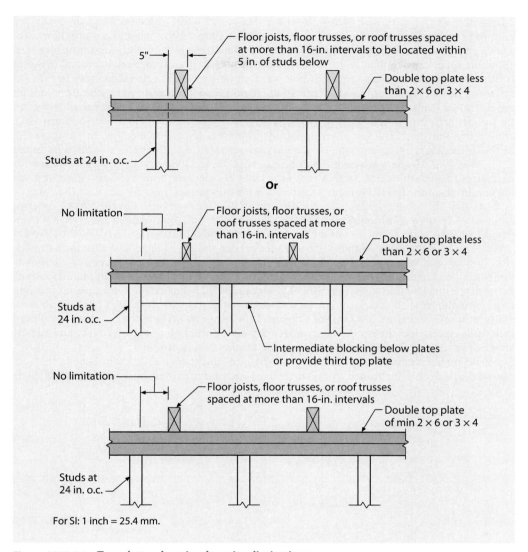

Figure 2308-24 **Top plate—bearing location limitations.**

increased thickness when wider stud spacing is used. Partitions must be capped with at least a single top plate to provide overlapping at corners and at intersections with other walls and partitions. Single top plates need to be continuously tied at joints by 16-inch-long (406 mm) solid blocking equal to the plate size, or by ½-inch by 1½-inch (12.7 mm by 38 mm) metal ties fastened with two 16d nails on each side of the joint.

2308.9.5 Openings in walls and partitions. The section provides prescriptive requirements for conventionally framed openings in exterior and interior walls and partitions.

2308.9.5.1 Openings in exterior bearing walls. Headers over openings in walls are required to carry loads from the wall, floors, and roof above. Allowable header spans for exterior bearing walls are given in Table 2308.8.1.1(1). The table gives the minimum header size, allowable span, and number of jack studs required for various ground snow loads and building widths. The 30-psf (1.44 kN/m^2) allowable stress design ground snow load is also to be used for sizing headers with roof live loads of 20 psf (0.96 kN/m^2) or less or allowable stress design ground snow loads less than 30 psf (1.44 kN/m^2) (see Footnote "e"). The permissible species for lumber are given in the heading on the table (Douglas-Fir Larch, Hem-Fir, Southern Pine, and Spruce-Pine-Fir), and the minimum grade of lumber (No. 2) is specified in Footnote "b." Multiple members must be fastened together in accordance with the prescriptive fastening Table 2304.10.2. Headers selected in accordance with Table 2308.8.1.1(1) may only be used for structures that fit within the parameters of the table. Headers for other configurations or loading conditions must be designed in accordance with the engineering provisions of the code.

Note that the table only includes 2× dimension lumber members. On the West Coast, 4× and 6× headers are common. Span tables that include 4× and 6× members are available from wood industry sources such as the Western Wood Products Association (WWPA). AWC's *Wood Frame Construction Manual* also includes header and girder span tables for both dimension lumber and structural glued-laminated timber members. Use of structural composite lumber in header applications is also prevalent and span tables are available in manufacturers' literature and evaluation reports by ICC-ES. Figure 2308-25 illustrates typical header framing requirements. Headers supported by jack studs as required by Table 2308.8.1.1(1) must have a minimum of 1½ inches (38 mm) of bearing at the supports. Where the number of required jack studs equals one, the header is permitted to be supported by an approved framing anchor attached to the full-height wall stud and to the header (see Footnote "d").

Single-member headers of nominal 2-inch (51 mm) thickness are also addressed in this section stating they shall be framed with a single flat 2-inch-nominal (51 mm) member or wall plate

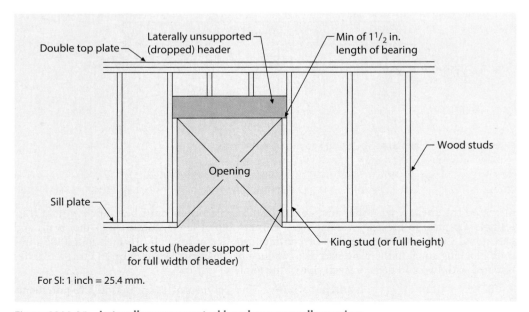

Figure 2308-25 **Laterally unsupported header over wall opening.**

not less in width than the wall studs on the top and bottom of the header. Figures 2308.9.5.1(1) and 2308.9.5.1(2) illustrate these requirements.

Spans are calculated assuming the top of the header or girder is laterally braced by perpendicular framing. Where the top of the header or girder is not laterally braced (for example, cripple studs bearing on the header), Table 2308.8.1.1(1) note "f" indicates that tabulated spans for headers consisting of 2 × 8, 2 × 10, or 2 × 12 sizes shall be multiplied by 0.70 or the header or girder shall be designed. AWC's *Wood Frame Construction Manual* contains header and girder tables for "dropped" and "raised" conditions that correspond to laterally unbraced and laterally braced headers, respectively. See Figure 2308-26 for examples of dropped versus raised headers.

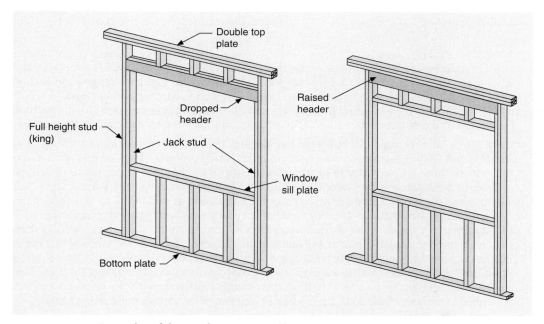

Figure 2308-26   Examples of dropped versus raised headers (courtesy American Wood Council).

2308.9.5.2 Openings in interior-bearing partitions. Allowable header spans for interior-bearing walls supporting floor loads are given in Table 2308.8.1.1(2). The table gives the minimum header size, allowable span, and number of jack studs required. Headers selected in accordance with Table 2308.8.1.1(2) may be used only for structures that fit within the parameters of the table. Headers for other configurations or loading conditions must be designed in accordance with the engineering provisions of the code. Headers supported by jack studs as required by Table 2308.8.1.1(1) or Table 2308.8.1.1(2) must have a minimum of 1½ inches of bearing at the supports.

2308.9.5.3 Openings in interior-nonbearing partitions. Interior-nonbearing walls can have single studs at each end of headers with a minimum of 1½ inches of bearing at each support.

2308.9.6 Cripple walls. Foundation cripple walls are required to be framed of studs not less than the size of the studs above. Exterior cripple wall studs are to be not less than 14 inches (356 mm) in length or be framed of solid blocking. When cripple wall studs exceed 4 feet (1,219 mm) in height, they must be framed of studs having the size required for an additional story. Specific requirements for cripple wall bracing are found in Section 2308.10.6.

2308.9.7 Bridging. Where stud partitions do not have adequate sheathing to brace the studs laterally in their weak (smaller) dimension, and the studs have a height-to-least-thickness (length-to-depth) ratio exceeding 50, the studs are required to have bridging or solid blocking with a minimum nominal thickness of 2 inches (51 mm) and a width the same as the studs. This blocking should be installed at intervals that reduce the height-to-least-thickness ratio below 50. For 2× studs, this translates to 75 inches (1,905 mm) and for 3× studs, 125 inches (3,175 mm).

2308.9.8 Pipes in walls. This section provides requirements for framing walls with pipes such as plumbing walls. Figure 2308-27 illustrates the requirements of this section for pipes in a wood-framed wall.

2308.9.9 Exterior wall sheathing. Exterior wall sheathing other than cement plaster (stucco) must be sheathed with one of the materials specified in Table 2308.9.9. Type, size, and spacing of fasteners must be in accordance with the prescriptive requirements given in Section 2304.10 and Table 2304.10.2 or the fasteners must designed to resist applicable loads in accordance with accepted engineering practice. Other wall sheathing materials and wood structural panels installed and fastened in accordance with Section 2304.6 are a permissible alternative.

2308.10 Wall bracing. All buildings must be able to resist both vertical (gravity) loads and lateral (wind and seismic) loads. Resistance to lateral loads produced by wind pressure or seismic ground motion is provided in part by wall bracing. Buildings that do not require a seismic analysis in accordance with Section 1613.1 and are located in regions where hurricane winds are not anticipated (see Section 2308.2.4) should be braced in accordance with this section.

Conventionally constructed buildings use braced wall panels and braced wall lines to provide lateral support for the structure. Braced wall panels in prescriptive design are analogous to shear walls in engineered design. A braced wall line consists of a series of straight lines through the building plan in both directions (building length and width). The difference between a shear wall and a braced wall panel is a shear wall is designed and detailed to resist a specific portion of the lateral load along a shear wall line, and a braced wall panel is determined from prescriptive rules and methods prescribed in the code. Table 2308-3 provides a description of the various wall bracing methods.

Table 2308-3. **IBC Braced Wall Panel Methods**

Wall Bracing Method	Description
LIB	Let-in brace
DWB	Diagonal wood board
WSP	Wood structural panel
SFB	Structural fiberboard
PBS	Particleboard sheathing
GB	Gypsum board
PCP	Portland cement plaster
HPS	Hardboard panel siding
ABW	Alternate braced wall panel
PFH	Portal frame with hold-downs

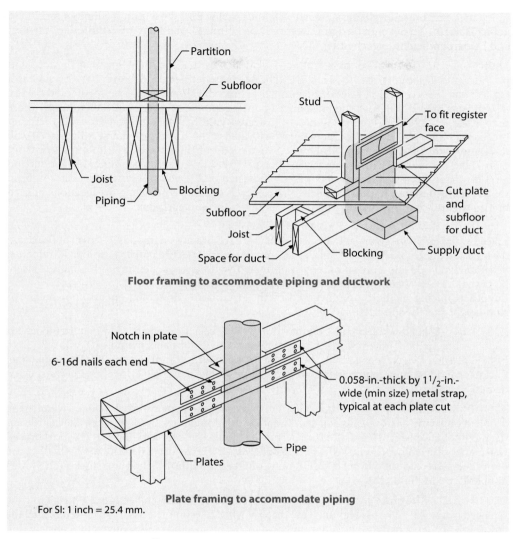

Figure 2308-27 **Pipes in walls.**

Because IBC wall bracing requirements are in some ways similar to the IRC, a good reference and resource for wall bracing in general is *Guide to the 2018 IRC Wood Wall Bracing Provisions* with its *2021 Supplement* jointly published by APA and ICC.

2308.10.1 Braced wall lines. Braced wall lines (BWLs) consist of a series of braced wall panels (BWPs) that are considered to be in a line of lateral resistance as shown in Figure 2308.10.1. The term is defined in Section 202. For determining the amount and location of bracing required along each story level of a building, braced wall lines are considered straight lines in both the longitudinal and transverse direction, and placed in accordance with Table 2308.10.1. IBC Figure 2308.10.1 illustrates the basic components of lateral bracing system consisting of braced wall panels and braced wall lines. Braced wall line spacing must be within the limits specified in Table 2308.10.1. In Seismic Design Category D or E, braced wall lines are required to intersect and be perpendicular to each other.

2308.10.2 Braced wall panels. Braced wall panels are the elements that "brace" and provide lateral support to the structure. The term is defined in Section 202. As noted, a braced wall line consists of a series of complying braced wall panels. To be considered part of a particular braced wall line, the braced wall panels cannot be offset from each other by more than 4 feet (1,219 mm). Figure 2308.10.1 illustrates typical layouts for braced wall panels and braced wall lines. A braced wall panel must be located at the corners of intersecting braced wall lines and within the maximum distance from the end of the braced wall line in accordance with Table 2308.10.1. From a plan review perspective, it is important to note that the section requires that braced wall panels to be clearly indicated on the plans. A good way to accomplish this is for the designer to provide a schematic of each story level, showing the locations of each braced wall line and the braced wall panels assumed to be in each line.

2308.10.3 Braced wall panel methods. Acceptable methods of constructing complying braced wall panels are given in Table 2308.10.3(1). Each bracing method is given an abbreviated designation such as LIB, DWB, and WSP, and each has specific construction requirements. See Table 2308-3. The minimum length of braced wall panels depends on the method used. Note that the ABW and PFH have maximum specified heights of 10 feet (3,048 mm), whereas the other bracing methods are governed by the maximum allowable floor-to-floor height or stud height specified in Section 2308.2.2.

2308.10.4 Braced wall panel construction. The specific construction requirements for each bracing method are given in this section. For example, for panel methods DWB, WSP, SFB, PBS, PCP, and HPS, each panel must be at least 48 inches (1,219 mm) in length. Minimum panel thickness and fastening requirements are provided in Table 2308.10.3(1). See section for specific details.

2308.10.5 Alternative bracing. Alternative braced wall panels were developed to solve the problem of having narrow walls adjacent to garage door openings. In some cases, alternate bracing methods such as ABW or a PFH are permitted to substitute for a 48-inch (1,219 mm) braced wall panel of Method DWB, WSP, SFB, PBS, PCP, or HPS. Also, Method ABW or PFH are permitted to replace the 96-inch (2,438 mm) one-faced Method GB or a 48-inch (1,219 mm) double-faced Method GB.

2308.10.5.1 Alternate braced wall (ABW). The braced panel in Section 2308.10.4 may be replaced by the ABW of this section. Note that the ABW is a complete assembly; therefore, all the requirements described in the section must be met. The specific requirements for constructing alternative braced panels are illustrated in Figure 2308-28. Alternative braced wall panels can only be used on a one-story building or on the first story of a two-story building. They cannot be used on the second story, because the code does not address the specific details and load path connections required for in-plane shear transfer and uplift anchorage to resist overturning forces at a second-story condition. The minimum length of an ABW is 32 inches (813 mm). This allows the alternative braced wall panel to fit in tight conditions in comparison to the 48-inch-long (1,219 mm) braced wall panel.

2308.10.5.2 Portal frame with hold-downs (PFH). The portal frame with hold-downs (PFH) allows a reduction of the width of the full-height segment of ABW panels to 16 inches (406 mm) in width for a one-story building, and 24 inches (610 mm) in width for the first story of a two-story building. Because it is a prescriptive alternative bracing method, no engineering is required if constructed strictly in accordance with all of the requirements of the section, as shown in Figure 2308-29.

Conventional Light-Frame Construction   843

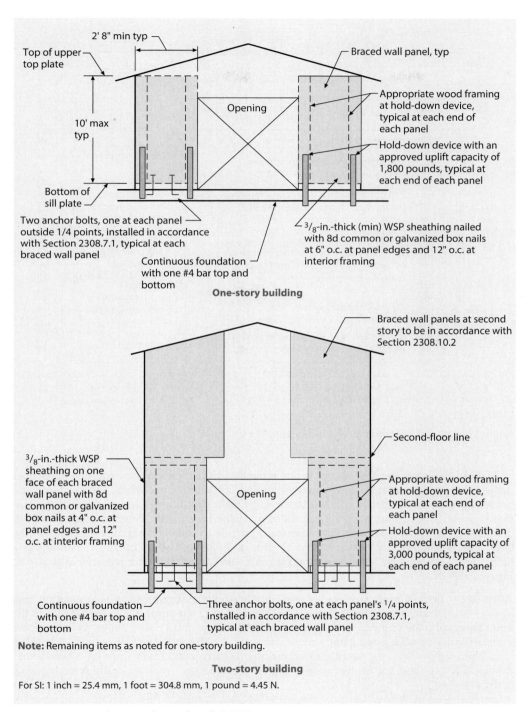

Figure 2308-28 **Alternate braced wall (ABW).**

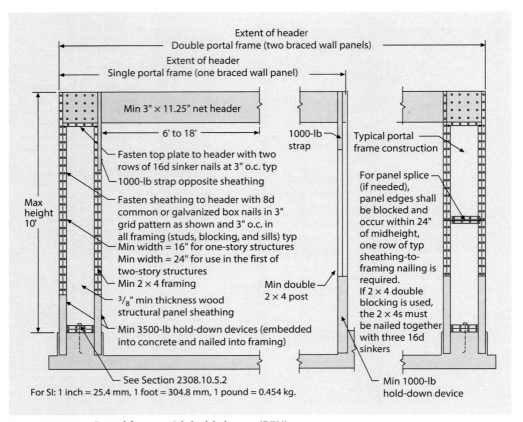

Figure 2308-29 Portal frame with hold-downs (PFH).

Method PFH uses the header over the opening to resist lateral forces by running the header the length of the sheathed bracing panel to the first full-length stud. Where the sheathing and header overlap, the header is fastened to the full-height sheathing with a grid nailing pattern that provides some nominal rigidity at the top. At the base of the sheathed section, embedded hold-down anchors nailed to the edge studs provide a moment-resistant connection at the base. The hold-down anchors are sized to provide uplift and shear capacity, resulting in a reduction in anchor bolt requirements at the plate. Additional straps are required as shown in Figure 2308-29.

2308.10.6 Cripple wall bracing. Cripple walls (sometimes referred to as foundation stud walls or knee walls) are stud walls usually less than 8 feet (2,438 mm) in height that rest on the foundation sill plate and support the first immediate floor and/or wall above. Where cripple walls are exterior walls supporting one or more stories, the wall must be braced with either solid blocking or sheathing based on wall bracing requirements. These walls have been found to rack sideways and fail in moderate and large earthquakes. Adding sheathing or blocking stiffens the wall.

For buildings in Seismic Design Categories (SDC) A, B, and C, where a cripple wall is part of an interior wall line, whether below an interior braced wall line or simply supporting the floor above, there is no requirement for bracing the wall line by blocking or sheathing. These walls are inside a much stiffer exterior foundation wall of concrete or CMU block and will not move independently of the floor and exterior walls during an earthquake.

Cripple wall bracing in SDC D and E is limited to 14 inches (356 mm) in height and must be solidly blocked along both interior and exterior walls. Therefore, buildings may only be one-story with a slab on grade foundation or a crawlspace consisting of studs 14 inches (356 mm) or less in height with solid-blocked cripple walls per Table 2308.2.1. Because cripple walls over 14 inches (356 mm) in height are considered an additional story, a one-story building over cripple walls greater than 14 inches in height is considered a two-story building and prohibited in SDC D and E. The extent of cripple wall stud solid blocking to allow for ventilation and access openings is stipulated.

2308.10.7 Connections of braced wall panels. The general intent of the provisions in this section is to provide prescriptive requirements for achieving a continuous load path for lateral loads from the roof and floor system to the braced wall lines, braced wall panels, and supporting foundation. Section 2308.10.7.1 contains requirements for the bottom plate connections, Section 2308.10.7.2 contains requirements for the top plate connections, Section 2308.10.7.3 includes sill anchorage provisions and 2308.10.7.4 covers anchorage to all-wood foundations. Figures 2308-30 through 2308-33, which originated with the 1997 NEHRP Commentary

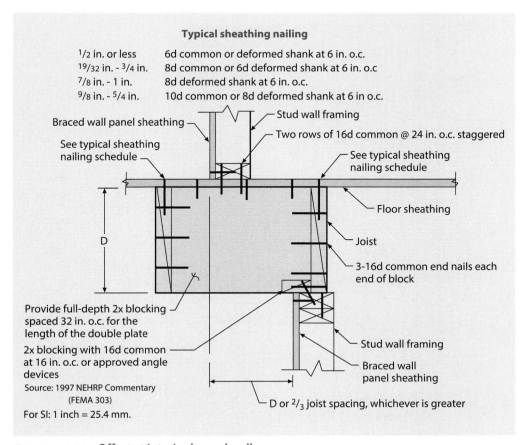

Figure 2308-30   **Offset at interior braced wall.**

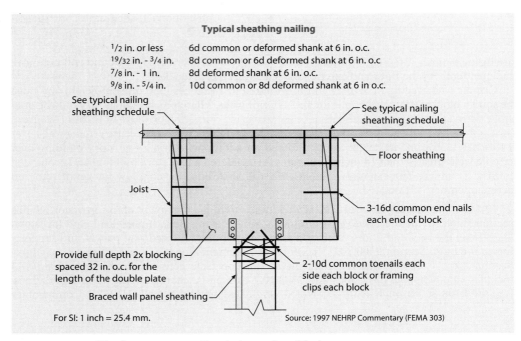

Figure 2308-31 Diaphragm connection to braced wall below.

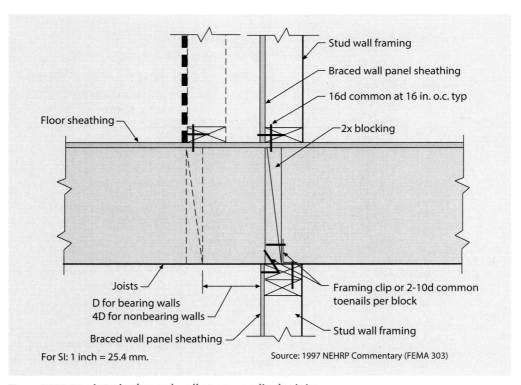

Figure 2308-32 Interior braced wall at perpendicular joist.

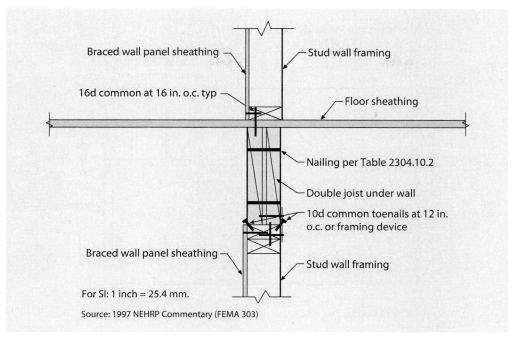

Figure 2308-33 **Interior braced wall at parallel joist.**

(FEMA 303), illustrate *recommended* connection details for various framing conditions. Figures 2308-34 through 2308-37 illustrate methods of transferring lateral loads from the roof system to braced wall panels. These connection details are designed to transfer shears of approximately 100–200 pounds per foot (1.46–2.92 kN/m), which is consistent with the allowable shear resistance of the prescriptive braced wall panel systems.

The purpose of the connections is to transfer wind and seismic lateral forces. The first sentence of Sections 2308.10.7.1 and 2308.10.7.2 clearly states that the connections apply to *braced wall lines* and not the braced wall panel portion of a braced wall line. For example, the connection between blocking and floor framing at a braced wall line top plate does not occur only at the braced wall panel, nor does the connection of the wall bottom plate to the foundation occur only at the braced wall panel. These connections are required along the entire braced wall line.

2308.10.7.1 Bottom plate connection. The bottom plate must be connected to the supporting elements below such as joists, blocking, or the foundation sill plate. The section refers to Table 2304.10.2, which requires two 16d common nails or equivalent spaced at 16 inches (406 mm) on center.

2308.10.7.2 Top plate connection. This section requires braced wall line top plates to be connected over their entire length to the framing above and refers to various fastening options in Table 2304.10.2. The wording of the second paragraph refers to exterior gable end walls because that is the condition where braced wall panels can be practically extended and attached to the roof framing. The phrase "braced wall panel sheathing in the top story shall be extended" provides a clear description of the requirement. In the third paragraph, the language "providing equivalent lateral force transfer" gives the building official a straightforward basis for accepting

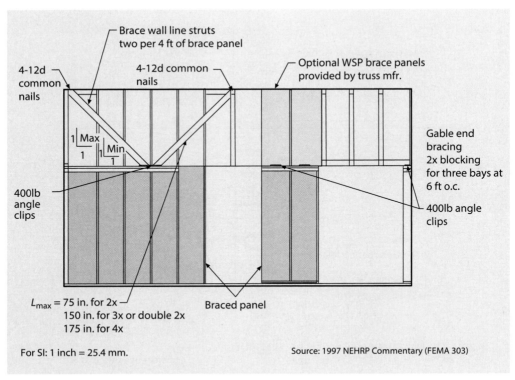

Figure 2308-34 Suggested method for transferring roof diaphragm loads to braced wall panels.

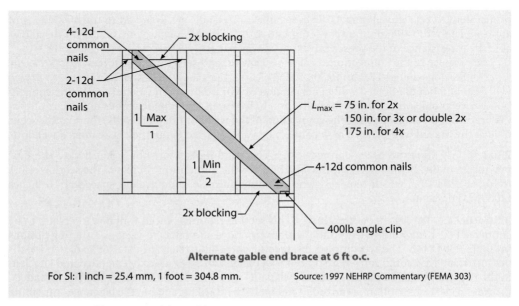

Figure 2308-35 Alternate gable end brace.

Conventional Light-Frame Construction 849

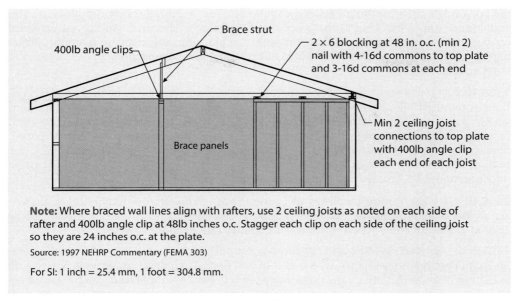

Figure 2308-36 **Wall parallel to truss bracing detail.**

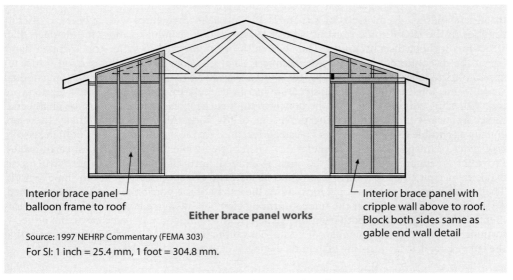

Figure 2308-37 **Wall parallel to truss alternate bracing detail.**

methods other than blocking between ends of trusses. Required minimum blocking size and connection of blocking to the braced wall top plate is specified. Additional guidance on permissible notching or drilling through the blocking is located in Section 2308.6. Figures 2308.10.7.2(1) and 2308.10.7.2(2) illustrate top plate connections where engineered trusses are used.

2308.10.7.3 Sill anchorage. Where braced wall lines are required to be supported by continuous footings, sills must be anchored to the foundation with ½-inch-diameter (12.7 mm) anchor bolts with a maximum spacing of 6 feet (1,829 mm) on center, or 4 feet (1,219 mm) on center for structures over two stories in height. Buildings in Seismic Design Category E require ⅝-inch-diameter (15.9 mm) anchor bolts in accordance with Section 2308.7.1. See Section 2308.10.8 for braced wall line and diaphragm support requirements, and discussion under Section 2308.7 for anchorage requirements.

2308.10.7.4 Anchorage to all-wood foundations. This section requires that anchorage to wood foundations be designed to have at least as much capacity as is required by Section 2308.7 for conventional concrete or masonry foundations.

2308.10.8 Braced wall line and diaphragm support. The section provides requirements for support of braced wall lines as well as roof and floor diaphragms. In general, braced wall lines must be supported by continuous footings, except that interior braced wall lines need not be supported by continuous footings if the maximum plan dimension of the structure does not exceed 50 feet (15,240 mm). In this case, continuous footings are required at exterior walls only and lateral loads on interior braced wall lines are assumed to be transferred to the exterior foundation by the floor system or slab on grade.

Several types of structural irregularities are not permitted in conventionally constructed wood-frame buildings in Seismic Design Category D or E. Figures 2308.8.4.1(1), 2308.8.4.1(2), 2308.8.4.2, 2308.10.8.2(1), and 2308.10.8.2(2) illustrate various conditions to aid the code user in understanding several types of structural irregularities. Structures with geometric discontinuities in the lateral-force-resisting system sustain more earthquake and wind damage than structures without discontinuities. They have also been observed to suffer concentrated damage at the discontinuity location. For Seismic Design Category D or E, this section translates applicable irregularities from ASCE 7 Tables 12.3-1 and 12.3-2 into limitations on conventional wood-frame construction. When a structure has an irregularity and none of the exceptions are met, either the entire structure or the nonconventional (irregular) portions must be engineered in accordance with the engineering provisions of the code. Although conceptually these are equally applicable to all seismic design categories, they are most critical in areas of high seismic risk, where damage caused by irregularities has repeatedly been observed in past earthquakes. Providing an engineered design of the nonconventional portions in lieu of the entire structure is explicitly permitted by the code (see Section 2308.3) and is a common practice in some regions. The registered design professional must judge the extent of the portion required to be designed. This often involves design of the nonconforming element, adequate force transfer into the element, and a complete load path from the element to the foundation. In some cases, the nonconforming portion may have enough of an impact on the behavior of a structure to warrant that the entire lateral-force-resisting system be engineered.

2308.10.8.1 Foundation requirements. In Seismic Design Categories D and E, exterior braced wall panels are required to be in the same vertical plane as the foundation unless one of the exceptions listed in the section are met. If exterior braced wall panels are not in the same vertical plane as the foundation and none of the exceptions are met, then that portion of the structure containing the offset must be designed in accordance with accepted engineering practice as prescribed in Section 2308.3.

Exception 1. This limitation applies when braced wall panels are offset out-of-plane from floor to floor. In-plane offsets are discussed in Exception 2. Ideally, braced wall panels should always stack above each other from floor to floor with the length stepping down at upper floors where less length of bracing panel is required. Because cantilevers and setbacks are very often incorporated into residential construction, the exception offers rules by which limited cantilevers and setbacks can be considered conventional. Exterior braced wall panels are permitted to be offset no more than 4 feet (1,219 mm) from the plane of the foundation, provided the specific requirements given in Items 1.1 through 1.6 are met. Cantilevers or setbacks cannot exceed four times the nominal depth of the floor joists. Floor joists are limited to 2 by 10 [actually 1½ inches by 9¼ inches (38 mm by 235 mm)] or larger, spaced 16 inches (406 mm) on center and doubled at braced wall panel ends to accommodate the vertical overturning reactions at the end of braced wall panels. The ratio of the back span to the cantilever must not be less than 2 to 1. In addition, the ends of the cantilever are attached to a common rim joist to allow for redistribution of load. For rim joists that cannot run the entire length of the cantilever, the metal tie is intended to transfer vertical shear as well as provide a nominal tension tie. Limitations are placed on gravity loads to be carried by the cantilever or setback floor joists so that the joist strength will not be exceeded. The roof loads discussed are based on the use of solid-sawn members where allowable spans limit the possible loads. Where engineered framing members such as trusses are used, gravity load capacity of the cantilevered or setback floor joists should be carefully evaluated.

Exception 2. This limitation applies when braced wall panels are offset in-plane or out-of-plane. Ends of braced wall panels supported on window or door headers below can transfer large vertical reactions to headers that may not be of adequate size to resist these reactions. The exception permits the braced wall panel to extend a maximum of 1 foot (305 mm) over an opening no more than 8 feet (2,438 mm) in width, provided a nominal 4 by 12 or larger header is used. The intent is that the vertical reaction will not result in critical shear or flexure. Other header conditions that do not meet the exception require an engineered design. Wall segments that are not considered braced panels, that is, with minimal lateral resistance, are allowed over openings.

2308.10.8.2 Floor and roof diaphragm support in Seismic Design Categories D and E. This limitation applies to open-front structures or portions of structures. The conventional construction bracing concept is based on using braced wall lines to divide a structure into a series of boxes of limited dimension, with the seismic force to each box being limited by the size. The intent is that each box be supported by braced wall lines on all four sides, limiting the amount of torsion that can occur. IBC Figure 2308.10.8.2(1) illustrates a portion of a roof that is not supported by a braced wall line below. The exception, which permits portions of roofs or floors to extend up to 6 feet (1,829 mm) past the braced wall line, is intended to permit construction such as porch roofs and bay windows, as illustrated in IBC Figure 2308.10.8.2(2). The framing members must be connected to the braced wall line below in accordance with the applicable requirements of Section 2308.10.7.

2308.10.8.3 Stepped footings in Seismic Design Categories B, C, D, and E. IBC Figure 2308.10.8.3 illustrates the requirements of this section. Where a stepped foundation creates braced wall panels that vary in height by more than 4 feet (1,219 mm), this section has specific detailing requirements. The sill must be anchored to the foundation as required by Section 2308.7. This is somewhat redundant because this is a general requirement that applies to all sills anchored to foundations. If there is a segment at least 8-feet long (2,438 mm) where the floor framing bears directly on the sill bolted to the foundation, then the braced wall line is considered sufficiently braced, and the double plate of the cripple wall beyond this segment of footing must be spliced to the sill plate with metal ties as shown in the figure. Item 2 describes the requirements for the splice strap and minimum nailing. Item 3 requires cripple walls to meet the bracing requirements for a story, which is essentially the same as is required by Section 2308.10.6.

2308.10.9 Attachment of sheathing. Wall sheathing must be fastened as specified in Table 2308.10.1 or the general fastening Table 2304.10.2 without substitution. Adhesives are not permitted to be used to fasten wall sheathing due to limited ductility. Nails or staples yield when the braced panels are subjected to high seismic demands from earthquake ground motion and the panels undergo ductile behavior. Panels attached by adhesives are not permitted because they can behave in a brittle manner when subjected to cyclic loading imposed by high seismic demand.

2308.10.10 Limitations of concrete or masonry veneer. Concrete or masonry veneer must comply with Chapter 14 and the requirements of this section.

2308.10.10.1 Limitations of concrete or masonry veneer in Seismic Design Category B or C. In Seismic Design Category B, stone and masonry veneer is permitted to be used in three stories above grade if the lowest story is constructed of concrete or masonry walls. Other wall bracing must be constructed of wood structural panels with a length that is increased to at least one- and one-half times the length required by Table 2308.10.1. Stone and masonry veneer is permitted to be used in two stories above grade, provided the lowest story is constructed of concrete or masonry walls. Where the lowest story is not constructed of concrete or masonry walls, stone and masonry veneer is permitted to be used in two stories above grade, provided the following criteria are met: (1) The bracing required by Section 2308.10.1 must be wood structural panels with an allowable shear capacity of 350 plf (16.8 kN/m^2) minimum; (2) the bracing of the top story must be located at each end, at least every 25 feet (7,620 mm) on center and not less than 25 percent of the braced wall line. The bracing of the first story must be located at each end, at least every 25 feet (7,620 mm) on center and not less than 45 percent of the braced wall line; (3) 2,000-pound (8.9 kN) capacity hold-down devices are required at the ends of braced wall panel segments from the second floor to the first floor wall, and 3,900-pound (17.3 kN) capacity hold-down devices are required from the first floor to the foundation; (4) cripple walls are not permitted.

2308.10.10.2 Limitations of concrete or masonry in Seismic Design Categories D and E. Concrete or masonry walls and stone or masonry veneer above the foundation (basement) level is prohibited entirely in Seismic Design Categories D and E. An exception allows stone and masonry veneer to be used in the first story above grade in Seismic Design Category D, provided the following conditions are met: (1) Wall bracing must be wood structural panels with a minimum allowable shear capacity of 350 plf (16.8 kN/m^2); (2) the bracing on the first story must be located at each end and at least every 25 feet (7,620 mm) on center and not less than 45 percent of the braced wall line; (3) 2,100-pound (9.3 kN) capacity hold-down devices are required at the ends of braced walls from the first floor to the foundation; and (4) cripple walls are not permitted.

2308.11 Roof and ceiling framing. These prescriptive roof framing provisions apply only to roofs with a minimum slope of 3:12 or greater. Where roofs have slopes less than 3:12, the horizontal thrust created by truss action with the ridge board and ceiling joists or rafter ties becomes excessive. Thus, for roof slopes less than 3:12, the members supporting the rafters and ceiling joists such as the ridge beam, hip, and valley rafters must be designed to resist the gravity loads as beams instead of behaving like a truss. See Figures 2308-38 and 2308-39.

2308.11.1 Ceiling joist spans. Allowable spans for ceiling joists may be determined in accordance with the tables in the code. The tables cover live load of 10 psf (0.48 kN/m^2) plus dead load of 5 psf (0.24 kN/m^2), live load of 20 psf (0.96 kN/m^2) plus dead load of 10 psf (0.48 kN/m^2), and a

Conventional Light-Frame Construction 853

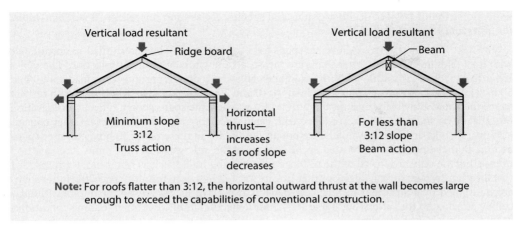

Figure 2308-38 **Roof framing.**

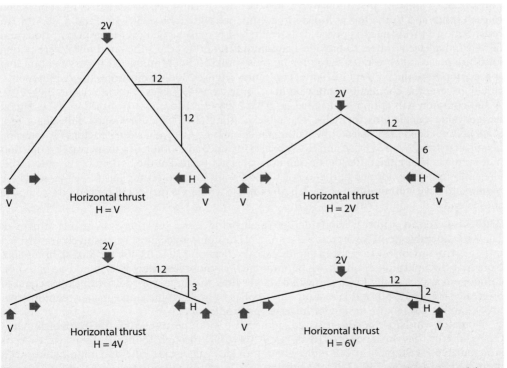

Figure 2308-39 **Roof framing thrusts—truss action.**

deflection limit of $L/240$. For other grades and species, the section references AWC *Span Tables for Joists and Rafters* (AWC STJR).

2308.11.2 Rafter spans. Allowable spans for roof rafters may be determined in accordance with the tables in the code. Rafter spans are measured along the horizontal projection. The tables cover live loads of 20 psf (0.96 kN/m^2) and allowable stress design ground snow loads of 30 and 50 psf (1.44 and 2.39 kN/m^2) plus dead loads of 10 and 20 psf (0.47 and 0.96 kN/m^2). With respect to the indicated allowable stress design ground snow loads, the spans shown reflect consideration of unbalanced snow loads in accordance with established roof snow load criteria. Other considerations, such as drifting snow loads, are not accounted for in these tables. Tabulated rafter spans assume that ceiling joists are located at the bottom of the attic space or that some other method of resisting the outward thrust of the rafters on the bearing walls, such as rafter ties, is provided at that location. Where ceiling joists or rafter ties are located higher in the attic space, the rafter spans are multiplied by adjustment factors. The deflection limit of $L/180$ is used for rafters without a ceiling attached to the rafters. See Table 1604.3 for deflection limits. For other loading conditions, grades and species, the section references AWC STJR.

2308.11.3 Ceiling joist and rafter framing. As noted in the discussion above, the roof rafters in conjunction with the ceiling joists or rafter ties form a simple truss. The ceiling joists or rafter ties resist the horizontal thrusts, as shown in Figure 2308-40, in tension.

2308.11.3.1 Ceiling joist and rafter connections. Ceiling joists and rafters must be nailed to each other and to the top wall plate in accordance with Tables 2304.10.2 and 2308.11.4. To resist outward thrust due to gravity loads, ceiling joists must be tied over interior partitions and fastened to adjacent rafters to provide a continuous rafter tie across the building. Where ceiling joists are not parallel to rafters, a rafter tie must provide a continuous tie across the building spaced not more than 4 feet (1,219 mm) on center. Figure 2308-40 illustrates the requirements. In either case, the connections must be in accordance with Tables 2308.11.3.1 and 2304.10.2. A 10d common nail option is included in Table 2308.11.3.1 footnote "a" based on NDS lateral design value calculations for nails. The table heading clarifies a 10 pounds per square foot (0.48 kN/m^2) dead load basis of the nailing requirements. Also, adjustment factors for rafter tie height, consistent with WFCM and IRC, are also included in footnote "h" to increase connection requirements where the rafter tie, or ceiling joist, is not located at the bottom of the attic space. Generally, rafter ties located at the top of support walls are also ceiling joists. Ceiling joists are required to have a minimum bearing length of not less than 1½ inches (38 mm) on the top plate at each end.

2308.11.4 Wind uplift. Wind suction (main wind force–resisting system pressures) can cause considerable uplift forces on roof framing. The uplift loads must be positively transferred into the structure below to resist the uplift. See also Section 2304.10.7 for additional discussion. Note that the uplift forces on low slope roofs such as those less than 3:12 should be based on component and cladding loads and will be larger than the forces for roofs with slopes equal to or greater than 3:12. Since this section only applies to roofs with a slope greater than or equal to 3:12, uplift ties require design for lower-sloped roofs.

Tabulated wind uplift loads are based on the basic wind speeds used in ASCE 7 which match the required basic wind speeds of IBC Figures 1609.3(1) through 1609.3(4). Basic wind speeds are tabulated for 90 mph to 140 mph, which provides values up to the maximum wind speed permitted in Section 2308.2.4 for conventional light-frame construction. Uplift loads for Exposures C and D with an assumed mean roof height (MRH) of 33 feet are also tabulated. A higher MRH will require engineering analysis for uplift connectors. The uplift loads are in pounds and are based on roof truss or rafter spacing of 24 inches (610 mm) on center. Other spacings must be adjusted. The table values have a built-in allowance for 10-psf (0.48 kN/m^2) roof dead load. The uplift loads are based on 24-inch (610 mm) overhangs. The uplift loads

Conventional Light-Frame Construction   855

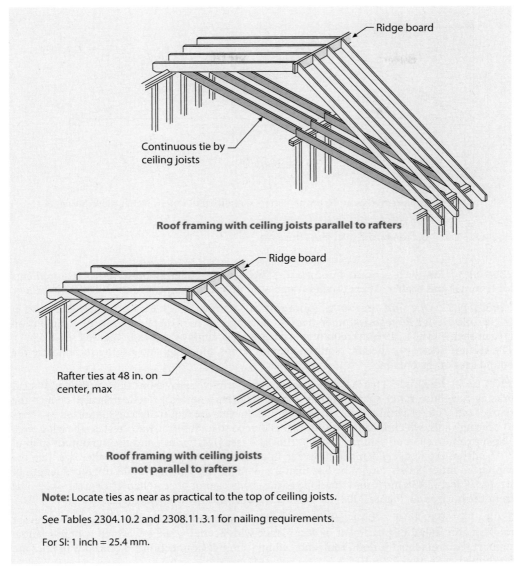

Note: Locate ties as near as practical to the top of ceiling joists.

See Tables 2304.10.2 and 2308.11.3.1 for nailing requirements.

For SI: 1 inch = 25.4 mm.

Figure 2308-40 **Ceiling joist and rafter framing.**

are based on end zone loads from ASCE 7 and allow reductions for connections away from the corners. The uplift loads are permitted to be reduced by 100 pounds (0.44 kN) based on the dead load for each full-height wall above.

An exception allows truss-to-wall connections to be designed using either the loads on the truss design drawings or the construction documents. This is consistent with IRC Section R802.11.1 for truss uplift resistance.

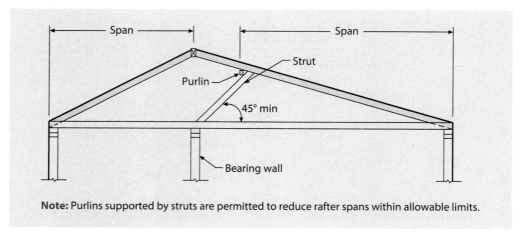

Figure 2308-41 Rafter, purlin and strut framing.

2308.11.5 Framing around openings. This section contains prescriptive requirements for trimmer and header rafters used to frame around openings in conventionally framed roofs.

2308.11.5.1 Openings in roof diaphragms in Seismic Design Categories B, C, D, and E. In Seismic Design Categories other than A, openings in roof diaphragms with a dimension greater than 4 feet (1,219 mm) require metal ties and blocking as shown in Figure 2308.8.4.1(1). The section prescribes specific requirements for straps, blocking, and nailing to reinforce the boundaries of the opening.

2308.11.6 Purlins. This section contains prescriptive provisions for purlin and strut bracing to reduce rafter spans. Purlins and struts are permitted to be installed to reduce the span of rafters to be within allowable span limits. Purlins are required to be supported by struts to bearing walls, which means a purlin that is braced to an interior wall creates a bearing wall. The maximum span of nominal 2 × 4 purlins is 4 feet (1,219 mm), and the maximum span of 2 × 6 purlins is 6 feet (1,829 mm). However, in no case can the purlin be of smaller size than the supported rafter. Struts cannot be less than 2 × 4 members. The maximum unbraced length of struts is 8 feet (2,438 mm), and the slope of the struts cannot be less than 45 degrees (0.79 rad) from the horizontal. Figure 2308-41 illustrates purlin and strut requirements.

2308.11.7 Blocking. Roof rafters and ceiling joists must be supported laterally to prevent rotation and lateral displacement in accordance with Section 2308.8.6, which contains lateral support requirements for floor, roof, and ceiling framing. Connections to braced wall lines are prescribed in Section 2308.10.7.2.

2308.11.8 Engineered wood products. Engineered wood products such as wood I-joists, glued-laminated timber, and structural composite lumber cannot be notched or drilled except where specifically permitted by the product manufacturer or where the effects of notches or drilled holes are specifically considered by the registered design professional.

2308.11.9 Roof sheathing. This section references the appropriate tables for wood structural panel and lumber roof sheathing. Wood structural panels should be labeled Exterior or Exposure 1. See Sections 2304.6.1 and 2304.8.2. Wood structural panels permanently exposed to outdoor conditions should be Exterior type, except that wood structural panels used on the exposed underside of roofs or roof overhangs are permitted to be Exposure 1 type. See Section 2303.1.5.

2308.11.10 Joints. Joints in lumber sheathing must occur over supports unless end-matched (tongue-and-groove) lumber is used. Where end-matched lumber is used, each individual piece must bear on at least two supports.

2308.11.11 Roof planking. Where 2-inch (51 mm) tongue-and-groove wood planking is used, it must be installed in accordance with Table 2308.11.11 or be designed in accordance with the general engineering provisions of the code.

2308.11.12 Wood trusses. Wood trusses are required to be designed in accordance with the engineering provisions of the code and conform to the detailed requirements prescribed in Section 2303.4. As noted in the discussion of Section 2303.4.6, pre-engineered metal-plate–connected wood trusses are considered structural elements that must be designed in accordance with the general engineering provisions of the code and the design, manufacture, and quality assurance provisions of TPI 1.

2308.11.13 Attic ventilation. Refer to the discussion under Section 1202.2.1 regarding attic ventilation.

Section 2309 *Wood Frame Construction Manual*

2309.1 Wood Frame Construction Manual. The IBC references the ANSI/AWC *Wood Frame Construction Manual* (WFCM) *for One- and Two-Family Dwellings* (see Figure 2309-1). The AWC WFCM is a permissible alternative for structural design of wood-frame dwellings in Risk Category I and II subject to the load restrictions in WFCM Section 1.1.2 and the building dimension limitations in WFCM Section 1.1.3. For a summary of the limitations, see Table 1 of the WFCM. Also see Section 2308 and Tables 2308-1 and 2308-2 for a comparison of IRC, IBC Section 2308, and WFCM provisions for conventional light-frame construction. Structures that do not comply with the limitations must be designed in accordance with accepted engineering practice.

Figure 2309-1 The 2024 *Wood Frame Construction Manual (WFCM)*.

The WFCM contains both engineering criteria and engineered prescriptive provisions for wood-frame buildings that may be used for the design of wood-frame structures within its scope. Although WFCM provisions are intended primarily for detached one- and two-family dwellings due to the floor live load assumption associated with those occupancies, many of the provisions for specific geographic wind, seismic, and snow loads are applicable to other types of buildings. For example, wind provisions for sizing of roof sheathing, wall sheathing, fastening schedules, uplift straps, shear anchorage, shear wall lengths, and wall studs for out-of-plane wind loads are included in WFCM and are applicable to other occupancies within the load limitations of the WFCM tables. Similarly, roof rafter size and spacing for heavy snow, and shear wall lengths and anchorage for seismic are applicable within the load limitations of the WFCM tables.

> **KEY POINTS**
> - Chapter 23 contains requirements for the design and construction of wood-frame buildings and other wood structures regulated by the IBC.
> - The IBC references AWC NDS—*National Design Specification (NDS) for Wood Construction* with *NDS Supplement: Design Values for Wood Construction* for general design of wood structures and AWC SDPWS—*Special Design Provisions for Wind and Seismic* (SDPWS) for lateral design of wood structures.
> - In many cases, the IBC refers to specific sections in the referenced standards for detailed design and construction requirements.
> - Wood structures must be designed by allowable stress design (ASD) in accordance with Section 2306 or 2309 where applicable, load and resistance factor design (LRFD) in accordance with Section 2307, or the prescriptive conventional construction provisions of Section 2308.
> - Regardless of the design method used, all wood structures must conform to the minimum quality standards in Section 2303 and the general construction requirements of Section 2304.
> - Log structures designed in accordance with ICC 400 and other applicable provisions of Chapter 23 are deemed to comply as prescribed in Section 2302.1.
> - The quality of wood elements must conform to the requirements and standards referenced in Section 2303.
> - Wood trusses must comply with Section 2303.4, which includes requirements for design, truss design drawings, truss bracing, restraint and anchorage, truss placement diagrams, alterations, and truss quality assurance.
> - Lateral force–resisting systems must be designed and constructed in accordance with Section 2305, which references AWC SDPWS and applicable provisions of 2306 for ASD or 2307 for LRFD.
> - Sections 2306 and 2307 both reference the AWC NDS, which is a dual-format standard allowing both ASD and LRFD procedures.
> - Buildings that meet the scope, restrictions, and limitations of Section 2308.2 may be designed using the prescriptive conventional construction provisions without engineering.

- Portions and elements in otherwise conventionally constructed buildings may be engineered without requiring the entire building to be engineered as prescribed in Sections 2308.3 and 2308.4.
- Wood-framed buildings of conventional construction in Seismic Design Category B, C, D, or E must meet additional seismic specific requirements as prescribed in various applicable sections.
- Buildings designed using the *Wood Frame Construction Manual (WFCM)* are deemed to meet the requirements of Chapter 23 as prescribed in Sections 2302.1 and 2309.

CHAPTER 24

GLASS AND GLAZING

Section 2402 Glazing Replacement
Section 2403 General Requirements for Glass
Section 2404 Wind, Snow, Seismic, and Dead Loads on Glass
Section 2405 Sloped Glazing and Skylights
Section 2406 Safety Glazing
Section 2407 Glass in Handrails and Guards
Section 2408 Glazing in Athletic Facilities
Section 2409 Glass in Walkways, Elevator Hoistways, and Elevator Cars
Key Points

Chapter 24 regulates glass and glazing for essentially two important reasons:
1. To protect against breakage that is due to building distortion under wind and earthquake loads along the wall line or perpendicular to the wall and surface of the glass.
2. To protect against breakage that is due to accidental impact by individuals adjacent to the glazing or debris during wind events.

Glass, light-transmitting ceramic panels, and light-transmitting plastic panels are regulated by Chapter 24 for their use in both vertical and sloped applications.

Section 2402 *Glazing Replacement*

An often overlooked provision, replacement windows and other glazing materials must comply with the requirements for new installations. The building official must be able to ensure that appropriate glazing is used in hazardous locations, in areas where fire-resistant glazing is required, and several other scenarios which are outlined in the code. This requirement is also highlighted throughout the *International Existing Building Code*® (IEBC®) when performing repairs, alterations, changes of occupancy, or when dealing with historic structures.

Section 2403 *General Requirements for Glass*

2403.1 Identification. Because glass is usually incorporated into assemblies that are not manufactured at the building site, the code requires that the glass and glazing be identified as to type and thickness so that it is possible to determine compliance with this chapter as far as permitted areas of glass are concerned. The identifying information furnished on each light may be removable or permanent. Where approved by the building official, an affidavit furnished by the glazing contractor is acceptable in lieu of the manufacturer's mark located on each pane of glass or glazing material. The affidavit must certify that each light is glazed in accordance with approved construction documents, as well as the provisions of Chapter 24. Where an identification mark is utilized, it shall designate both the type and the thickness of the material.

Unless used in a spandrel application, tempered glass is required to be permanently identified by the manufacturer. The identifying mark is to be acid etched, sandblasted, ceramic fired, embossed, or of any other type that is unable to be removed without being destroyed. A removable paper marking provided by the manufacturer is permitted, but only for tempered spandrel glass.

2403.2 Glass supports. Glass firmly supported on all edges is considerably stronger than a glass light with one or more free edges. As a result, where the glass does not have firm support on all edges, the code requires that a design for the glass be submitted to the building official for approval. In this latter case, the design would be based on the number and location of any free edges.

2403.3 Glass framing. The amount of deflection that is permitted to occur within any individual pane of glazing is limited, ensuring the safety of the building occupants as well as those in close proximity to the building exterior. Section 1604.3.7 notes that the load used to determine the glazing deflection shall equal 0.6 times the component and cladding wind loads for the building. Sections 2403.3 and 1604.3.7 each require that any panes of glazing having maximum dimensions no more than 13 feet 6 inches (4,115 mm) shall not deflect more than the length of the pane divided by 175. As an example, consider a 4030 window [4 feet wide by 3 feet high (1,219 mm by 914 mm)]. The maximum pane dimension is 48 inches (1,219 mm) and this is to be divided by 175, resulting in a maximum allowable deflection of approximately ¼ inch (6.4 mm)].

If the glass panes exceed 13 feet 6 inches (4,115 mm), the deflection limit should be measured as the maximum pane dimension divided by 240, plus ¼ inch (6.4 mm). Consider a curtain wall system having a maximum pane dimensions of 14 feet (4,267 mm). The first step is to divide 168 inches (4,267 mm) by 240 and then add ¼ inch (6.4 mm). In this scenario a maximum deflection of 0.95 inch (24 mm) is allowed for a 14-foot (4,267 mm) glass pane. The fact that these requirements are located in both Chapters 16 and 24 of the code highlight their importance.

2403.4 Interior glazed areas. Where interior glazing is installed adjacent to a walking surface, the differential deflection of two adjacent unsupported edges must not be not greater than the panel thickness under a 50 pound per linear foot (plf) (730 N/m) horizontal load applied to one panel at any point 42 inches (1,067 mm) or less above the walking surface.

2403.5 Louvered windows or jalousies. The requirements for louvered windows are based on there being no edge support on the longitudinal edges. See Figure 2403-1. Moreover, for safety purposes, the code requires that the exposed edge be smooth. For the same reason, wired glass, when used in jalousie or louvered windows, shall have no exposed wire projecting from the longitudinal edges. Where a louvered window complies with the provisions of this section, such an application is exempt from the requirements for safety glazing for use in hazardous locations.

Figure 2403-1 **Louvered window.**

Section 2404 *Wind, Snow, Seismic, and Dead Loads on Glass*

Exterior glass and glazing are subject to the same loads as the exterior cladding of the building; therefore, the code requires that glass and glazing in a vertical or near-vertical position should be designed for the same wind loads as specified in Section 1609 for components and cladding.

For seismic considerations, ASCE 7 is referenced for the design of glass in glazed curtain walls, glazed storefronts, and glazed partitions. This design will also include the increased pressures on local areas at discontinuities. As the slope of the glass increases, it may be necessary to address other loads as well, such as the dead load and any snow load. See Figure 2404-1.

Figure 2404-1 **Snow load on glazing.**

Glass sloped 15 degrees (0.26 rad) or less from vertical is defined as vertical glass. Glass sloped more than 15 degrees (0.26 rad) is defined as sloped glass (see Figure 2405-2). Table 2404-1 outlines the applicable code sections for the design of various types and orientations of glass.

Table 2404-1 **IBC Design Provisions for Vertical and Sloped Glass**

Type	Orientation	IBC Section
Glass	Vertical	2404.1
	Sloped	2404.2
Wired	Vertical	2404.3.1
	Sloped	2404.3.2
Patterned	Vertical	2404.3.3
	Sloped	2404.3.4
Sandblasted	Vertical	2404.3.5
Other	Vertical or Sloped	2404.4

Section 2405 *Sloped Glazing and Skylights*

By application, sloped glazing and skylights consist of glazing installed in roofs or walls that are on a slope of more than 15 degrees (0.26 rad) from the vertical. See Figures 2405-1 through 2405-3. The provisions of the *International Building Code®* (IBC®) are intended to protect these glazed openings by providing adequate strength to carry the loads normally attributed to roofs. The provisions are also intended to protect the occupants of a building from the possibility of falling glazing materials.

2405.2 Allowable glazing materials and limitations. Glazing materials and protective measures for sloped glazing and skylights are outlined in this section. The materials and their characteristics and limitations are as follows:

Laminated glass. Laminated glass is usually furnished with an inner layer of polyvinyl butyral, which has a minimum thickness of 30 mil (0.76 mm). Such glass is highly resistant to impact and, as a result, requires no further protection below.

Wired glass. Wired glass is resistant to impact and, when used as a single-layer glazing, requires no additional protection below.

Tempered glass. Tempered glass is glass that has been specifically heat-treated or chemically treated to provide high strength. When broken, the entire piece of glass immediately breaks into numerous small granular pieces. Because of its high strength and manner of breakage, tempered glass has been considered in the past to be a desirable glazing material for skylights without any protective screens. However, as a result of studies that show that tempered glass is subject to spontaneous breakage such that large chunks of glass may fall under this condition, the IBC requires screen protection below tempered glass unless specifically exempted.

Approved light-transmitting plastic materials. See discussion under Section 2606.

Heat-strengthened glass. Heat-strengthened glass is glass that has been reheated to just below its melting point and then cooled. This process forms a compression surface on the outer face and increases the glass strength. However, heat-strengthened glass has the unsatisfactory characteristic

Figure 2405-1   **Sloped glazing as roof.**

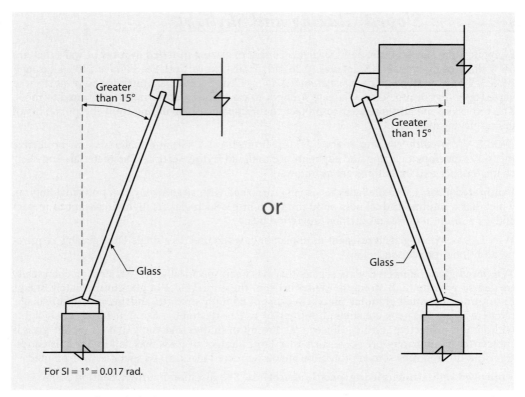

Figure 2405-2 **Sloped glazing.**

of breaking into shards, as does annealed glass. Thus, as a general rule, heat-strengthened glass requires screen protection below the skylight to protect the occupants from falling shards.

Annealed glass. Annealed glass is subject to breakage created by impact, has very low strength, and is unsatisfactory as a glazing material in skylights and sloped glazing. Annealed glass has a further unsatisfactory characteristic for use as a skylight because it breaks up under impact into large sharp shards which, when they fall, are hazardous to occupants of a building. Annealed glass is permitted to a limited degree as sloped glazing and skylights when no walking surfaces are below or when used in some greenhouses, as discussed under Section 2405.3.

2405.3 Screening. Broken glass retention screens, where required, shall be of a noncombustible material at least No. 12 B&S gauge (0.0808 inch) mesh with 1 inch by 1 inch (25 mm × 25 mm) or smaller openings. The screen must be installed within 4 inches (102 mm) of the glass surface and capable of supporting twice the weight of the glass. In corrosive atmospheres, structurally equivalent noncorrosive screen materials must be utilized.

It is also critical that the screen is installed in a manner that will adequately support the weight of the glass. In utilizing a safety factor of 2, the screen and its fastenings must be capable of supporting twice the weight of the glazing. In order to accomplish this, the screen is to be fastened firmly to the framing members.

Monolithic heat-strengthened glass and fully tempered glass must have screens installed below the full area of the glazing material. Multiple-layer glazing systems are also now used quite often

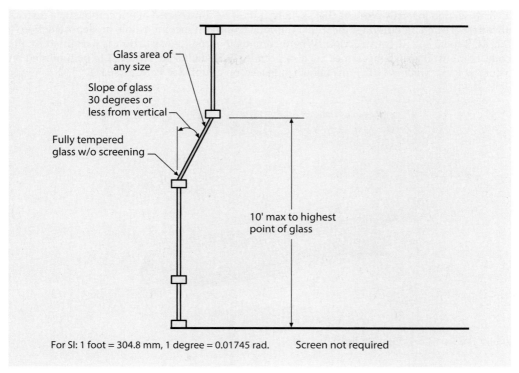

Figure 2405-3 **Sloped glazing without screen.**

for skylights and sloped glazing because of energy-conservation requirements. Where heat-strengthened, fully tempered and wired glass is used as the layer facing the interior, screen protection is required below the skylight.

Five exceptions are provided in Section 2405.3.3 to eliminate the need for protective screens in both monolithic and multilayer sloped-glazing systems. The first exception, shown in Figure 2405-3, permits fully tempered glass in near vertical wall sections, based on the low height plus the very low probability of breakage. Annealed glass may be used only (and is permitted without screening) under the following two circumstances:

1. Where the accessible area below is permanently protected from falling glass.

2. In greenhouses under the limitations outlined in Exception 3.

The fourth and fifth exception allow use of fully tempered glass and laminated glass as single glazing or as both panes in an insulating glass unit within individual dwelling units in Groups R-2, R-3, and R-4, provided each pane of the glass is limited to 16 square feet (1.5 m^2) or less in area, and the highest point of the glass is no more than 12 feet (3,658 mm) above a walking surface or other accessible area. For laminated glass, a 15-mil (0.38-mm) polyvinyl butyral inner layer is required. Tempered glass must be 3/16 (4.8 mm) inch or less in thickness.

2405.4 Framing.   This is an omnibus section containing requirements related to:

1. Combustibility of materials and

2. Leakage protection of the skylight juncture with the roof.

This section logically requires that skylight frames be constructed of noncombustible materials where erected on buildings for which the code requires noncombustible roof construction—that is, Type I and II construction. Where combustible roof construction is permitted by the code, combustible skylight frames are also permitted. Skylights are required to be mounted on a curb at least 4 inches (102 mm) tall when placed on a roof. See Figure 2405-4.

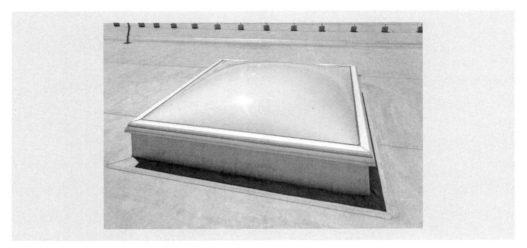

Figure 2405-4 **Curb for a skylight on a roof.**

The provision requiring a curb at least 4 inches (102 mm) above the plane of the roof for mounting skylights is intended to provide a means for flashing between the skylight and the roofing to prevent leaks around the margin of the skylight. The 4-inch (102-mm) curb then provides a vertical surface up to which the flashing can be extended and to which the counterflashing may be attached to cover the flashing.

2405.5 Unit skylights and tubular daylighting devices. This section establishes design provisions for unit skylights (single-panel, factory-assembled units) and tubular daylighting devices that are consistent with those provided for sloped glazing in Section 2404.2, but permit these items to be designed separately for maximum positive and negative pressures. The most critical load on a skylight is determined by the climate in which it is installed. In a colder climate with heavier snow loads and moderate design wind speeds, the positive load on a skylight from the combined snow and dead loads will be more critical than the negative load from wind uplift. The opposite will be the case in warmer, coastal climates with higher design wind speeds and little or no snow load. This method of rating skylights and tubular daylighting devices is addressed in the referenced standard AAMA/WMDA/CSA 101/I.S.2/A440, *Specification for Windows, Doors and Unit Skylights*. The separate rating system for positive and negative pressures on skylights allows the manufacturer to design and fabricate products that are best suited for the climate in which they will be used.

Section 2406 *Safety Glazing*

2406.1 Human impact loads. In areas where it is likely that persons will impact glass or other glazing, the IBC mandates that specific glazing materials be installed. Such specific areas, as identified in Section 2406.4, are considered by the code as "hazardous locations."

Unless exempted, all glazing located in hazardous locations must pass the test requirements established in Section 2406.2 for impact resistance. Special criteria are placed on plastic glazing, glass block, louvered windows, and jalousies.

2406.2 Impact test. Glazing installed in areas subject to human impact as specifically identified by Section 2406.4 is generally required to comply with the CPSC 16 CFR, Part 1201 criteria for Category I or II glazing materials as established in Table 2406.2(1). CPSC 16 CFR, Part 1201, *Safety Standard for Architectural Glazing Materials*, is a federally mandated safety regulation of the U.S. Consumer Products Safety Commission. The exception also allows the installation of glazing materials that have been tested to a different standard, ANSI Z97.1 *Safety Glazing Materials Used in Buildings—Safety Performance Specifications and Methods of Test*, in limited applications. Glazing tested under the ANSI Z97.1 standard must meet the test criteria for Class A or B as set forth in Table 2406.2(2).

For the most part, the differences between the CPSC's 16 CFR Part 1201 standard and the ANSI Z97.1 standard relate to their scope and function. The CPSC standard is not only a test method and a procedure for determining the safety performance of architectural glazing, but also a federal standard that mandates where and when safety glazing materials must be used. It preempts any nonidentical state or local code. In contrast, ANSI Z97.1 is only a voluntary safety performance specification and test method. It does not indicate where and when safety glazing materials must be used.

Glazing in compliance with the appropriate test criteria of CPSC 16 CFR Part 1201 may be used in all hazardous locations. It is also acceptable to utilize safety glazing materials complying with ANSI Z97.1, but only to the extent of those applications other than storm doors, combination doors, entrance-exit doors, sliding patio doors, and closet doors. In addition, such glazing is not permitted in doors and enclosures for hot tubs, whirlpools, saunas, steam rooms, bathtubs, and showers. In all other areas subject to human impact (hazardous locations) as specified in Section 2406.4, required safety glazing is permitted to comply with the applicable requirements of either CPSC 16 CFR 1201 or ANSI Z97.1. See Figure 2406-1.

Where glazing that is tested in accordance with CPSC 16 CFR Part 1201 is utilized, the base requirement is that Category II glazing be used. Only where permitted by Table 2406.2(1) is the use of Category I glazing permitted. A similar approach is provided for those locations where glazing complying with the ANSI Z97.1 criteria can be applied. [See the discussion of Tables 2406.2(1) and 2406.2(2) for information on glazing categories.] As an example, glazing adjacent to stairs as regulated by Section 2406.4.6 is not addressed in Table 2406.2(1). Therefore, glazing tested in accordance with CPSC 16 CFR 1201 must meet the test criteria for Category II. If tested in accordance with ANSI Z97.1, Category A glazing is required. Where Tables 2406.2(1) and 2406.2(2) are applicable, such as for glazed panels adjacent to doors as regulated by Section 2406.4.2, the minimum required category classification is provided based on the size of the glass lite.

Table 2406.2(1)—Minimum Category Classification of Glazing Using CPSC 16 CFR 1201. Glazing tested in accordance with CPSC 16 CFR 1201 is classified as either Category I or II, based on the specifics of the test. The Category II classification is more difficult to achieve and is thus required in those areas where the highest degree of protection is required. Although Category II material is acceptable in all hazardous locations, Category I glazing may only be installed as relatively small lites in doors and areas adjacent to doors. Otherwise, only Category II glazing is acceptable. See Figure 2406-2. As addressed in the discussion of Section 2406.2, Category II glazing is required in those hazardous locations not addressed in Table 2406.2(1).

Table 2406.2(2)—Minimum Category Classification of Glazing Using ANSI Z97.1. The minimum required category classification of glazing tested to ANSI Z97.1, established in Table 2406.2(2), is also based on the glazing location and size of the lite. This approach is similar

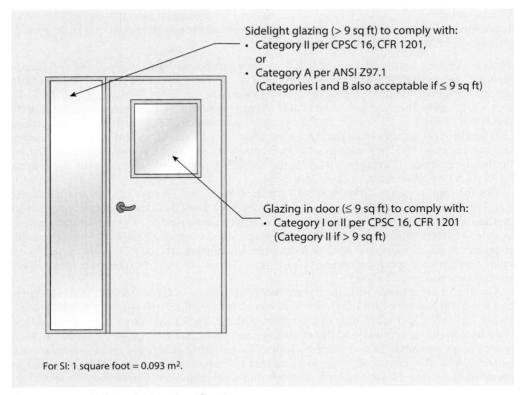

Figure 2406-1 **Safety glazing classification.**

to that for determining the minimum required category classification of glazing tested to CPSC 16 CFR 1201 as set forth in Table 2406.2(1). ANSI Z97.1 addresses three separate impact categories or classes, based on impact performance. ANSI Z97.1 Class A glazing materials are comparable to the CSCS Category II glazing materials and ANSI Z97.1 Class B glazing materials are comparable to the CPSC Category I glazing materials. For those hazardous locations not addressed in Table 2406.2(2), the use of Category A glazing is required. Although there is also a Class C category recognized by ANSI Z97.1, applicable only for fire-resistant glazing materials, it is not viewed by the code as an acceptable safety glazing material. Only Class A and B ANSI Z97.1 glazing materials are recognized in the table.

2406.3 Identification of safety glazing. The code requires the identification of safety glazing for the same reasons as for ordinary annealed glass not subject to human impact. However, in the case of safety glazing, the requirement carries more detail insofar as improperly installed annealed glass in areas of human impact can create a serious hazard. Therefore, it is doubly important that the glazing material be further identified to ensure that the proper glazing material is in place. Not only does proper marking assist in the inspection process, it could also help identify a location where safety glazing must be installed should future replacement be required. The code specifically requires the use of identification marks for glazing installed in hazardous locations.

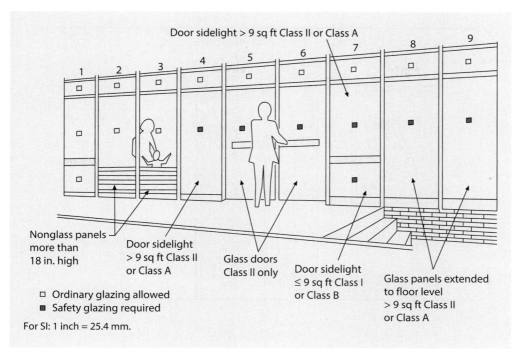

Figure 2406-2 **Hazardous locations.**

Each pane of safety glazing must be individually identified with a manufacturer's designation. It shall specify who applied the designation, the manufacturer or the installer, and the safety glazing standard with which the glazing complies, as well as the type and thickness of the glass or glazing material. Acceptable methods for the designation include acid etching, sand blasting, ceramic firing, and embossing. Additionally, the designation is permitted to be of a type that cannot be removed without being destroyed. This limitation ensures that a designation cannot be transferred to any other glazing materials. As another option, the building official may be willing to accept a certification or affidavit from the supplier and/or installer indicating that the appropriate glazing was provided and installed. Under such a situation, it is critical that the building official be completely satisfied that the appropriate material is utilized for the installation location. The acceptance of an affidavit or similar document is appropriate for all safety glazing materials other than tempered glass.

The only variation to the general requirement for individual identification is for multipane glazed assemblies such as French doors. Where the individual panes do not exceed 1 square foot (0.0929 m^2) in exposed area, only one pane in the assembly is required to be marked as described above. However, all other panes in the assembly must still be identified with "CPSC 16 CFR, Part 1201." See Figure 2406-3. Where multiple panes occur in assemblies other than doors or enclosures regulated by Item 5 of Section 2406.4, glazing tested and labeled as "ANSI Z97.1" is also acceptable.

2406.4 Hazardous locations. This section lists those specific hazardous locations where safety glazing is required. Some of these locations are shown in Figures 2406-5 through 2406-14. In addition to the hazardous locations shown in the various illustrations, safety glazing

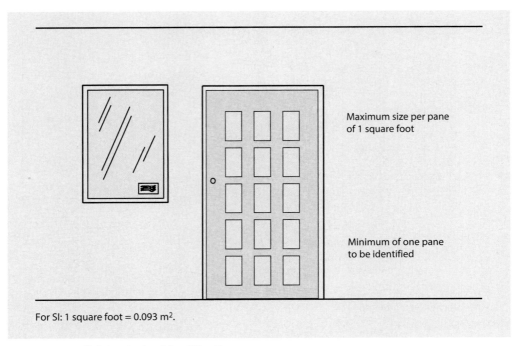

For SI: 1 square foot = 0.093 m².

Figure 2406-3 **Safety glazing identification.**

is also required for a number of other conditions, including fixed and sliding panels of sliding door assemblies, storm doors, and glass railings.

Figure 2406-2 also illustrates several locations where safety glazing may or may not be required. To facilitate discussion, each panel has been numbered. Panels 1, 2, 3, 8, and 9 are addressed under Section 2406.4.3. Under this provision, all four stated conditions must occur before safety glazing is required. These conditions are as follows:

1. The area of an individual pane must be greater than 9 square feet (0.84 m²);
2. The bottom edge must be less than 18 inches (457 mm) above the floor or adjacent walking surface;
3. The top edge must be more than 36 inches (914 mm) above the floor or adjacent walking surface; and
4. One or more walking surfaces must be within 36 inches (914 mm), measured horizontally, of the glazed panel.

Panel 1 is not required to have safety glazing, because a protective bar has been installed in compliance with Exception 2 to Section 2406.4.3, the requirements of which are illustrated in Figure 2406-4. Panels 2 and 3 do not require safety glazing, because their bottom edges are not less than 18 inches (457 mm) from the adjacent walking surface. If Panels 8 and 9 have a walking surface within 36 inches (914 mm) of the interior, safety glazing would be required. This would be true even though the bottom of the panel appears to be greater than 18 inches (457 mm) above the exterior walking surface. However, the exterior condition, because it is adjacent to a stairway,

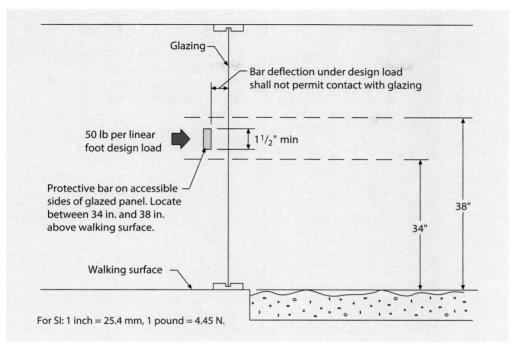

Figure 2406-4 **Protective bar alternative.**

would be regulated by Section 2406.4.6 and/or Section 2406.4.7 for glazing adjacent to stairways. Therefore, multiple conditions mandate the need for safety glazing in Panels 8 and 9. Panels 4 and 7 require safety glazing because they are door sidelights. The exception using a horizontal rail mentioned above does not apply to panels adjacent to a door; therefore, although Panel 7 has been provided with a protective bar, safety glazing is still required.

Another exception to 2406.4.3 deals with outboard panes in insulating glass units or multiple glazing where the bottom exposed edge of the glass is 8 feet (2,438 mm) or more above any grade or walking surface adjacent to the glass exterior. This requirement is impartial regarding the concern of falling out of a building, falling into a building or falling through a window that does not result in a person entering or leaving a building (e.g., an interior window with floor walking surfaces on both sides). It recognizes the need to prevent situations where people on the building's exterior (whether on a roof or adjacent walking surface) could fall through a window into a building. While this is a less likely occurrence than people falling through windows out of a building, it is a concern where windows are located adjacent to the roof level (e.g., a clerestory). Where the vertical distance from the bottom exposed edge of the glass to grade or to a walking surface adjacent to the window exterior is less than 8 feet, safety glazing is required on the outboard pane of insulating glass units (IGU) or multiple glazing. This provision would also apply along a walkway or deck that may be higher than the immediately adjacent walking surface on the other side of the window. For example, if a corridor is adjacent to an atrium and there are multiple panes of glazing, only the pane adjacent to the corridor requires safety glazing. The pane adjacent to the atrium is higher than the 8 foot threshold and would not require safety glazing. See Figure 2406-5.

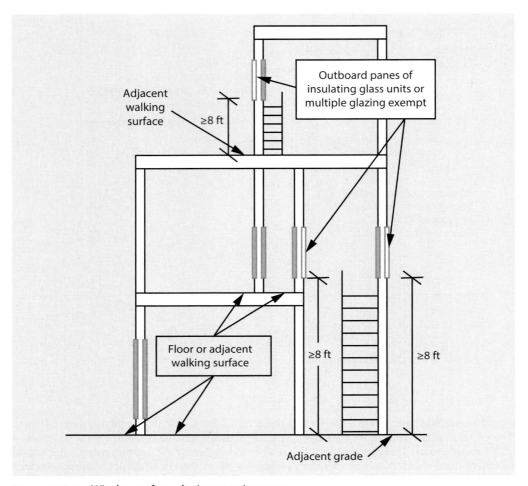

Figure 2406-5 **Window safety glazing requirements.**

Panels 5 and 6 in Figure 2406-2 are glass doors, which require safety glazing based on the provisions of Section 2406.4.1. All ingress and egress doors, unframed swinging doors, and glazing in storm doors require safety glazing, as well as any other swinging, sliding, and bifold doors with fixed or operable glazed panels. See Figure 2406-6. There are several exceptions. If openings in a door will not allow passage of a 3-inch-diameter (76-mm) sphere, the glazing is exempt, as are assemblies of leaded, faceted, or carved glass used for decorative purposes. The latter exception not only applies to doors, but also to sidelights and other glazed panels covered by Sections 2406.4.2 and 2406.4.3.

Figures 2406-7 and 2406-8 illustrate when safety glazing is required for panels adjacent to a door. This requirement applies to both fixed and operable panels. Where there is an intervening wall or permanent barrier as shown in Figure 2406-9, safety glazing would not be required. If the door serves only a shallow storage room or closet, adjacent glazing need not be safety glazing, as depicted in Figure 2406-10. Figure 2406-11 illustrates an exception applicable only to one- and two-family dwellings and within dwelling units in Group R-2 uses.

Safety Glazing **875**

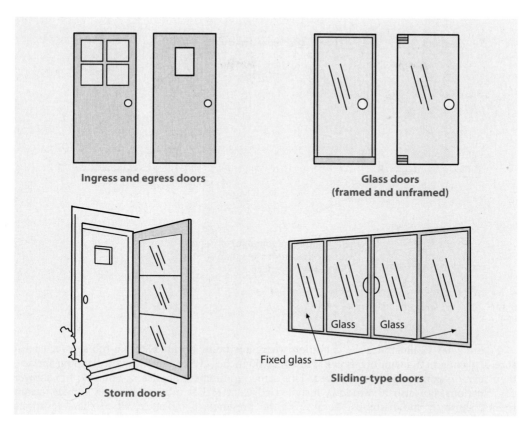

Figure 2406-6 **Glazing in doors.**

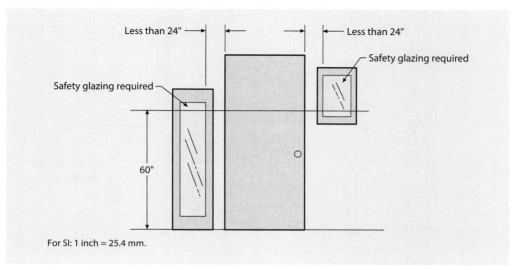

For SI: 1 inch = 25.4 mm.

Figure 2406-7 **Glass in sidelights—elevation.**

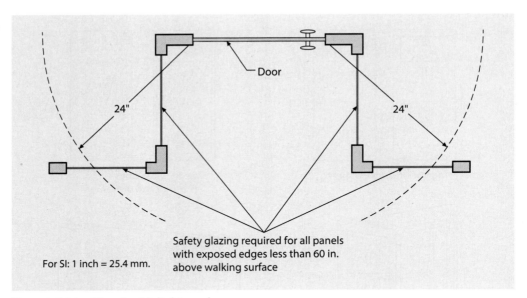

Figure 2406-8 **Glass in sidelights—plan.**

Figure 2406-12 illustrates the condition where a window occurs within a tub/shower enclosure. If glazing in the window shown is less than 60 inches (1,524 mm) above a standing surface, then safety glazing would be required. This same requirement applies not only to tub/shower combinations, but also to windows installed adjacent to hot tubs, whirlpools, saunas, steam rooms, showers, and bathtubs. Because of the presence of moisture, all of these locations

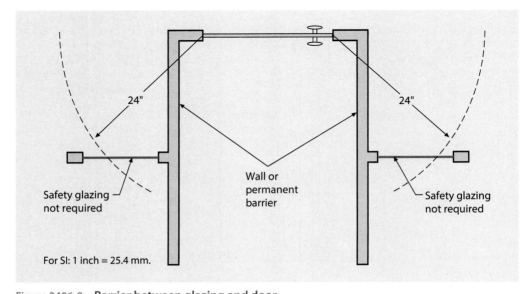

Figure 2406-9 **Barrier between glazing and door.**

Safety Glazing

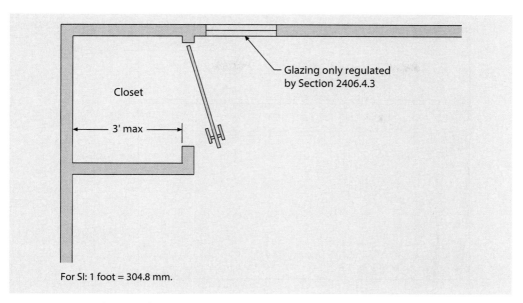

Figure 2406-10 **Glazing adjacent to closet door.**

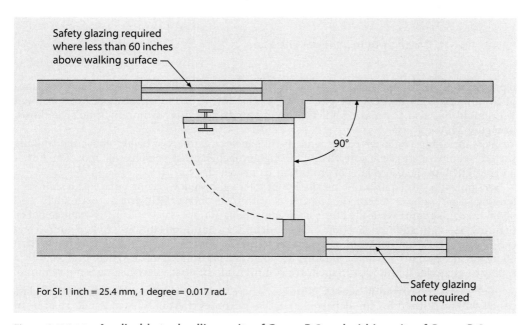

Figure 2406-11 **Applicable to dwelling units of Group R-3 and within units of Group R-2.**

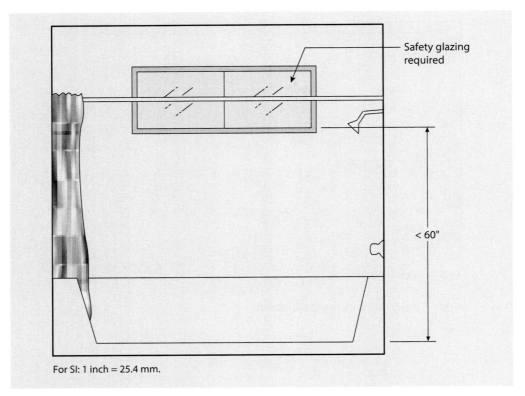

Figure 2406-12 **Glazing within a shower enclosure.**

represent slip hazards and need safety glazing to prevent injury in case of a fall. The provisions of Section 2406.4.5 address not only glazing adjacent to bathtubs, showers, hot tubs, and whirlpools, but also any glazing adjacent to indoor and outdoor swimming pools, as shown in Figure 2406-13.

Glass in railings, baluster panels, and in-fill panels, regardless of height above the walking surface, require safety glazing. Because of the high probability that persons will impact guards, it is critical that an increased level of protection be provided.

Section 2406.4.6 requires the installation of safety glazing for glazing adjacent to stairways, landings, and ramps. Where the glazing is within 36 inches (914 mm) horizontally and 60 inches (1,524 mm) vertically of the adjacent walking surface, safety glazing is mandated. See Figure 2406-14. In addition, Section 2406.4.7 identifies a hazardous location to be within 5 feet (1,524 mm) of the bottom tread of a stairway when the bottom edge of the glazing is less than 60 inches (1,524 mm) above the landing. See Figures 2406-15(a) and (b). Where protected by a railing or guard located at least 18 inches (457 mm) from the glass, safety glazing is not required.

2406.5 Fire department access panels. There may be conditions under which glass panels will be utilized for fire department access purposes. Certainly, such panels shall be of the type that will provide safe conditions under which fire department personnel can enter a building or a specific area of a building. For this reason, glass access panels must be of tempered glass. Where the glazed unit is a multipanel glass assembly, all panes must be tempered. Because access must be provided completely through the panel, it is necessary that all panes be of this specific safety glazing material.

Safety Glazing

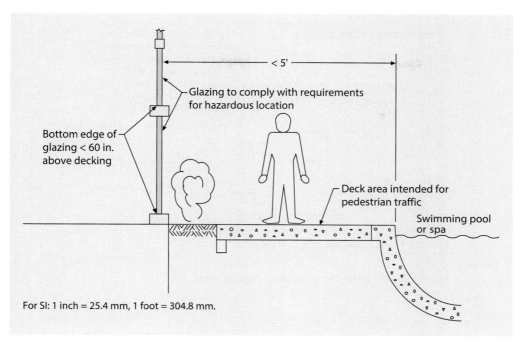

Figure 2406-13 **Glazing adjacent to swimming pool or spa.**

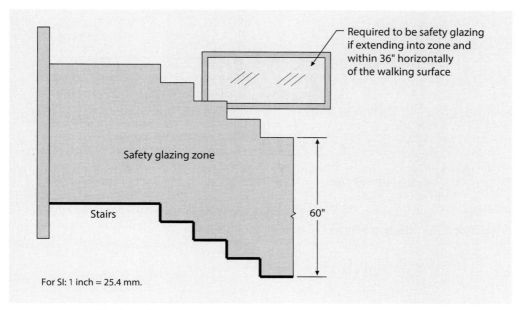

Figure 2406-14 **Glazing adjacent to stairways.**

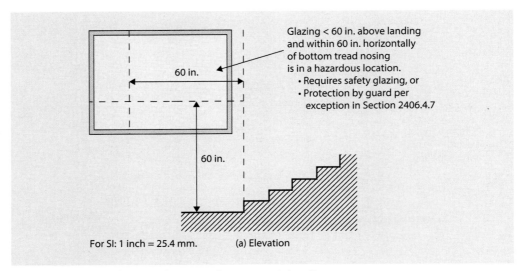

Figure 2406-15(a) **Glazing adjacent to bottom stair landing.**

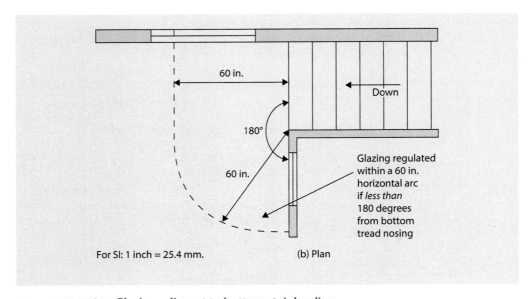

Figure 2406-15(b) **Glazing adjacent to bottom stair landing.**

Section 2407 *Glass in Handrails and Guards*

The desired use of glass in handrail assemblies and guard sections prompted the inclusion of these provisions in the IBC. See Figure 2407-1. These provisions provide uniform regulations identifying the specific types of safety glazing that may be used structurally. Laminated tempered and laminated heat-strengthened glass are the only types considered by the code to be structurally adequate under all conditions of installation. Single tempered glass that is not a laminated product is acceptable under limited applications where there is no walking surface below the glass, or where the walking surface below is permanently protected from the risk of falling glass.

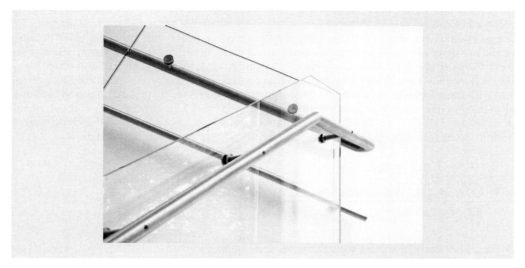

Figure 2407-1 **Glass balusters with handrail.**

Only glazing conforming to the provisions of Section 2406.1.1 is permitted. This would limit glazing in a handrail or guard to materials that have passed the applicable test requirements of CPSC 16 CFR 1201 or ANSI Z97.1. Regardless of the type of glazing, the minimum nominal thickness of the structural balustrade panels in railings is ¼ inch (6.4 mm). Glass handrails and guards and their supporting system must be designed to withstand the loads as specified in Section 1607.9. Calculated stresses in glass elements of handrails and guards due to these loads are limited to a maximum of 3,000 psi (20.7 MPa) for heat strengthened glass and 6,000 psi (41.4 MPa) for fully tempered glass. Guards having glass balusters are not permitted to be installed without an attached top rail or handrail supported by at least three structural glass balusters. The purpose for requiring at least three balusters is that, should one fail, the remaining two balusters will continue to support the handrail or guard section. As an alternative to having three balusters supporting the handrail, the code allows the design team to provide a means of ensuring the handrail will remain in place should a glass baluster fail.

Provisions for glazing in parking garages and windborne debris regions are also covered in this section. Glazing materials are not to be installed in handrails or guards in parking garages except for pedestrian areas where glazing is not exposed to vehicle impact. Glazing installed in exterior handrails or guards in windborne debris regions must be laminated glass complying with Category II of CPSC 16 CFR 1201 or Class A of ANSI Z97.1. Where a top rail is supported by glass, the assembly must be tested according to Section 1609.2 impact requirements and the top rail must remain in place after impact.

Section 2408 *Glazing in Athletic Facilities*

Where glazing forms entire or partial wall sections, or is used as a door or as part of a door, in racquetball courts, squash courts, gymnasiums, basketball courts, and similar athletic facilities subject to impact loading, it shall comply with Section 2408. In racquetball and squash courts, glass walls and glass doors must pass specific test criteria above and beyond those typically required of safety glazing materials. Such special test criteria are necessary to address those glazed areas where impact with the glass is not merely accidental, but rather expected because of the nature of the physical activities involved. Special conditions for compliance are also set forth for glazing subject to human impact in gymnasiums, basketball courts, and other high-intensity activity areas where it is expected that contact with the glazing will occur more often, and with more force, than in most hazardous locations addressed by the code. In such facilities, all glazing in hazardous locations identified by the code must meet the Category II requirement of CPSC 16 CFR 1201 or the Class A requirements of ANSI Z97.1. This would include glazing both in doors (only CPSC 16 CFR 1201 glazing permitted) and adjacent to doors.

Section 2409 *Glass in Walkways, Elevator Hoistways, and Elevator Cars*

Section 2409 deals with three separate building elements, as noted in the title. First, it addresses requirements for elevated glass walkways, mandating that glazing in such walkways consist of laminated glass which complies with ASTM E2751 *Practice for Design and Performance of Supported Laminated Glass Walkways*. This ASTM standard addresses design and support requirements for elevated glass walkways. Section 2409.1 also mandates that the glazing meet the fire-resistance-rating requirements for the applicable floor/ceiling assembly.

Glass within elevator hoistways must comply with the hazardous location requirements in addition to meeting the fire-resistance rating established in Section 716 for hoistways. Provisions are also provided for glass used in hoistway doors. Glass used within elevator cars must consist of laminated glass conforming to Class A of ANSI Z97.1 or Category II of CPSC 16 CFR Part 1201. This includes all glazing within the car, such as doors, walls, ceilings, and lining materials. An exception permits the use of tempered glazing under specified conditions.

KEY POINTS

- Glass and glazing must resist vertical and lateral loads in a manner consistent with other building components.
- Skylights and other sloped glazing are regulated as to the type of glazing material and the need for protective screening below the skylight.
- Safety glazing materials are to be installed in those areas subject to human impact, referred to as hazardous locations.
- The minimum required classification category of safety glazing materials is based on the size and location of the glazing material.
- Common safety glazing materials include tempered or laminated glass, as well as approved plastic.
- In order to verify compliance, the code specifically requires the use of identification marks for glazing installed in hazardous locations.
- Some of the most common locations identified as hazardous include those glazed areas in and around doors.
- Glazing adjacent to showers and bathtubs, where located in a position where impact is likely, must be safety glazed.
- Glass in handrails and guards is regulated for both structural adequacy and human impact.
- Special requirements are mandated for glazing subject to human impact in athletic facilities.
- Special structural and fire-resistant requirements for glazing in walkways, elevator hoistways, and elevator cars are provided.

CHAPTER 25

GYPSUM PANEL PRODUCTS AND PLASTER

Section 2501 General
Section 2508 Gypsum Construction
Section 2510 Lathing and Furring for
　　　　　　 Cement Plaster (Stucco)
Section 2511 Interior Plaster
Section 2512 Exterior Plaster
Key Points

This chapter regulates covering materials for walls and ceilings with requirements for:

1. Providing weather protection for the exterior of the building.
2. Securing the material to the wall and ceiling framing to remain in place during the expected life of the building.

Where these materials are used or required for fire-resistance-rated construction, the code requires that they also comply with the provisions of Chapter 7.

Section 2501 *General*

This chapter of the *International Building Code®* (IBC®) covers the material, design, construction, and quality requirements for wall- and ceiling-covering materials, including their method of fastening and, in the case of plaster, the permitted materials for lath, plaster, and aggregate.

Although plaster has many uses, including ornamental and decorative work, its use in the IBC is regulated purely as a wall- and ceiling-covering material.

The IBC regulates the installation of wall- and ceiling-covering materials as well as quality standards for the materials themselves. The primary wall- and ceiling-covering material in use today is gypsum wallboard; however, other gypsum panel products, as well as lath, plaster, and wood paneling, are sometimes utilized. As wood paneling is covered in Chapter 23, it follows that Chapter 25 only regulates gypsum wallboard, gypsum panel products, lath, and plaster. However, in this section the code permits the installation of other wall- and ceiling-covering materials, provided the materials have been approved. On this basis, the manufacturer's recommendations and conditions of approval should be consulted.

The terms *gypsum panel product, gypsum sheathing, gypsum board*, and *gypsum wallboard* are all specifically defined in Section 202. *Gypsum panel product* is the general name for a family of sheet products consisting essentially of gypsum complying with the standards specified in Tables 2506.2 and 2507.2 and Chapter 35. *Gypsum sheathing* is specifically manufactured with enhanced water resistance for use as a substrate for exterior surface materials. *Gypsum board* consists of a noncombustible core primarily of gypsum with paper surfacing. *Gypsum wallboard* is a *gypsum board* used primarily as an interior surfacing material.

A key aspect of regulation of the use of gypsum panel products and plaster relates to where such materials are installed. Surfaces of walls, ceilings, floors, roofs, soffits, and similar elements, where exposed to the weather, are typically considered weather-exposed surfaces or exterior surfaces. There are three exceptions defined in Section 202 to the general criteria that define weather-exposed surfaces. Exterior conditions considered other than weather-exposed surfaces are illustrated in Figure 2501-1.

Section 2508 *Gypsum Construction*

This section addresses the installation of gypsum materials; primarily that of gypsum wallboard installed as wall and ceiling membranes. As gypsum board is a construction material utilized in almost every construction project, it is important that its application is in compliance with the IBC and the appropriate referenced standards.

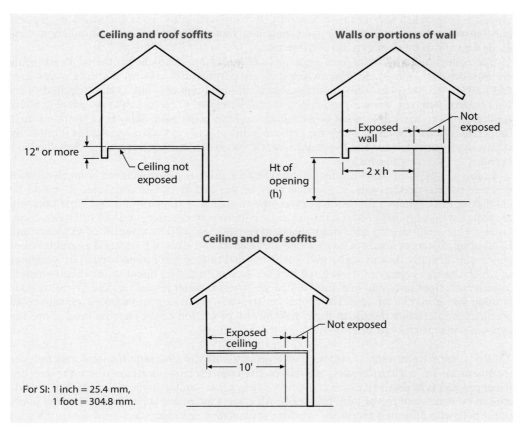

Figure 2501-1 Weather-exposed surfaces.

2508.1 General. The primary installation criteria for various gypsum materials are in referenced standards identified in Table 2508.1. The application of gypsum panel products varies based on panel thickness, wall or ceiling installation, orientation to the framing members, and maximum spacing of the framing members. Based on the specific conditions encountered, the maximum fastener spacing and size of fasteners are identified.

2508.2 Limitations. Because gypsum plaster and gypsum wallboard are subject to deterioration from moisture, the code restricts their use to interior locations and weather-protected surfaces. The definition for weather-exposed surfaces is found in Section 202 and illustrated in Figure 2501-1. Even where installed in interior locations, it is important that gypsum wallboard not be installed in areas of continuous high humidity or wet locations. The IBC further requires that interior gypsum wallboard, gypsum plaster, and gypsum lath shall not be installed until the installation has been weather-protected.

2508.3 Single-ply application. The application of gypsum panel products is specified in this chapter for locations where fire-resistance-rated construction is not provided or for construction where shear resistance or diaphragm action is not required. Chapter 7 and fire-test reports will establish the means of fastening and supporting the ends and edges of gypsum panel products for fire-resistance-rated assemblies. Table 2508.1 provides the installation standards for various types of gypsum construction.

The code requires that the fit of gypsum wallboard sheets be such that the edges and ends are in moderate contact. However, wider gaps are permitted in concealed spaces where fire-resistance-rated construction, shear resistance, or diaphragm action is not required. This requirement is based primarily on appearance. Therefore, where the gypsum panel product application is concealed, it is not objectionable to have wider gaps than those resulting from moderate contact. However, where the gypsum panel product surface is exposed as it normally is, moderate contact is required so that there will be no objectionable cracking when the joint between the sheets is finished.

Unless the gypsum panel product is considered a shear-resisting element or an element of a fire-resistance-rated assembly, fasteners may be omitted at certain locations. It should be emphasized that where a fire-test report or other installation standard indicates that fasteners are required on supports or edges, the fastening pattern may not be modified. Otherwise, those fasteners located at the top and bottom plates of vertical assemblies are permitted to be omitted. In addition, fasteners need not be provided at the edges and ends of horizontal assemblies perpendicular to supports and at the wall line. Note that fasteners are to be applied in a manner in which the face paper is not fractured by the fastener head. The intent of the requirement is to provide a tight fastening but not damage the gypsum panel product to the extent its nail-holding power may be affected. The proper construction procedure for fastening gypsum panel products is to create a dimple in the panel with no projection of the fastener head above the gypsum panel product surface.

2508.5 Joint treatment. Although as a general rule the IBC requires joint and fastener treatment for fire-resistance-rated assemblies, the code exempts locations where the gypsum panel product is to receive a decorative finish or any other similar application, which is considered to be equivalent to the joint treatment. Also, joint treatment is not required where joints occur over wood framing members, or where square-edge or tongue-and-groove edge gypsum panel product is used. In addition, joint treatment is not required for assemblies tested without joint treatment. In general, joint treatment does not materially increase the fire rating, and many partitions have passed the fire test without joint treatment. As indicated earlier in this section, joint treatment is primarily used for aesthetic reasons. One further exception addresses the condition where a multilayer system is constructed. Where two or more layers of gypsum panel products are utilized in the assembly, joint and fastener treatment is not required where the joints of adjacent layers are offset from each other.

2508.6 Horizontal gypsum panel product diaphragm ceilings. Gypsum panel products can be used on wood joists to create a horizontal diaphragm ceiling per Table 2508.6. The maximum aspect ratio of the ceiling diaphragm is 1.5 to 1 between shear-resisting elements. Gypsum panel products at least ½-inch (12.7 mm) thick are to be installed perpendicular to the ceiling framing with end joints staggered. Perimeter edges must be blocked with minimum nominal 2×6 framing members installed flat over the top plate to provide at least 2 inches (51 mm) of nailing surface for the gypsum panel products. Fastener sizes are defined in Table 2508.6 and ceiling diaphragm capacity is based on the spacing of the ceiling framing. Fasteners must be spaced not more than 7 inches (178 mm) on center at all supports, including perimeter blocking, and not more than ⅜ inch (9.5 mm) from the edges and ends of the gypsum panel product. Gypsum panel products are not to be used in diaphragm ceilings to resist lateral forces imposed by masonry or concrete construction. See Figure 2508-1.

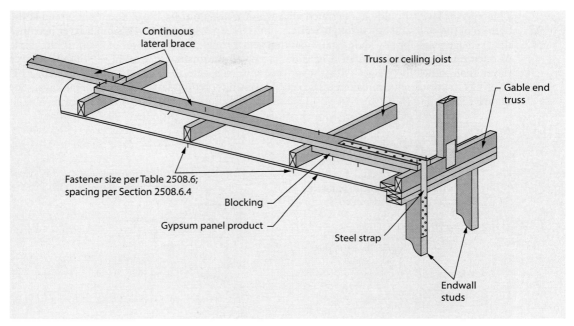

Figure 2508-1   Example of gypsum panel product diaphragm ceiling assembly.

Section 2510 *Lathing and Furring for Cement Plaster (Stucco)*

Cement plaster (stucco) used in both exterior and interior locations must be installed in accordance with this section. In order to ensure reliability and consistency, the materials of construction must comply with the appropriate standards listed in Table 2507.2. Of particular importance is the installation method of the water-resistive barrier (WRB). Section 1403.2 outlines the requirements for WRB's and notes that they shall be continuous and describes acceptable products that can be used. The exception to Section 2510.6 allows for the WRB to be excluded where accumulation, condensation, or freezing of moisture will not damage the materials.

WRB provisions in Section 2510.6 are divided into two categories: WRB requirements for plaster in "dry climates," and provisions for "moist or marine climates." In both scenarios two options are provided. For dry climates, Option #1 is to use two layers of 10-minute Grade D paper. Grade D paper is classified as a Type I WRB in accordance with ASTM E2556 *Standard Specification for Vapor Permeable Flexible Sheet Water-resistive Barriers Intended for Mechanical Attachment.* The greatest benefits of using two layers of WRB materials can only be realized if the method and manner of the installation establish a continuous drainage plane, separated from the stucco. In a two-layer system, each layer provides a separate and distinct function. The primary function of the inboard layer is to resist water penetration into the building cavity.

This interior layer should be integrated with window and door flashings, the weep screed at the bottom of the wall, and any through-wall flashings or expansion joints. This inner layer becomes the drainage plane for any incidental water that gets through the outer layer or at one of the joints or openings or where the outer layer is damaged. The primary function of the outboard layer (layer that comes in contact with the stucco) is to separate the stucco from the WRB. This layer has historically been called a sacrificial layer, intervening layer, or bond break layer. See Figure 2510-1.

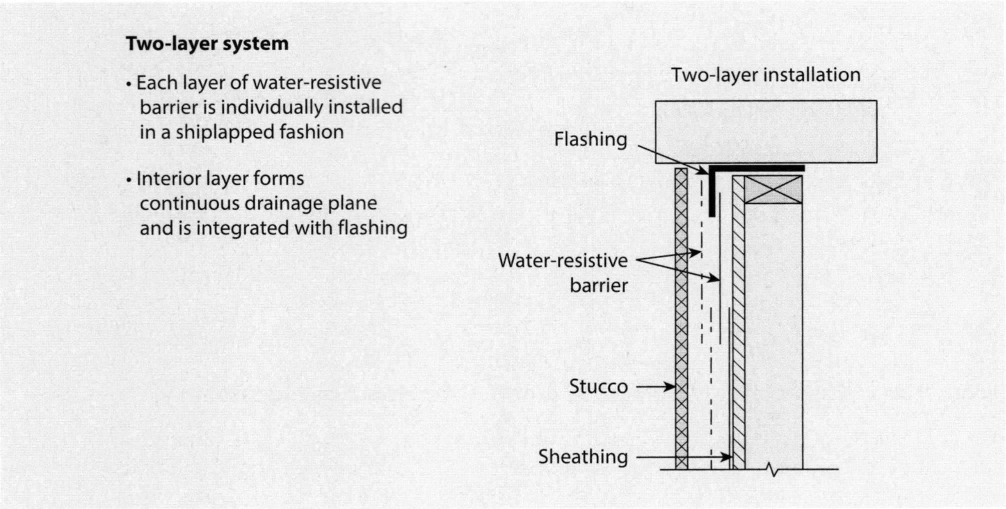

Two-layer system
- Each layer of water-resistive barrier is individually installed in a shiplapped fashion
- Interior layer forms continuous drainage plane and is integrated with flashing

Two-layer installation: Flashing, Water-resistive barrier, Stucco, Sheathing

Figure 2510-1 Installation of water-resistive barrier.

The second option for "dry climates" is to apply a single layer of 60-minute Grade D paper. This is considered a Type II WRB in accordance with ASTM E2556 and must be separate from the stucco by foam plastic sheathing, a non-absorbing layer, or a drainage space.

The options available for "moist or marine climates" incorporate the two options for "dry climates" with some additional provisions. For Option #1 the user can follow either Option #1 or #2 for "dry climates" so long as a space or drainage material having a depth of at least 3/16 inch (4.8 mm) is provided on the exterior side of the WRB. Option #2 requires that option #2 for "dry climates" be followed in addition to confirming that the drainage efficiency on the exterior side of the WRB can be measured at a minimum of 90 percent. Obviously, in wet climates it becomes imperative that adequate drainage at the exterior side of the WRB is provided to afford an additional safety factor to limit moisture from entering the wall cavity.

Section 2511 *Interior Plaster*

Multicoat plastering has been the standard in the western world for over 100 years. It is generally the consensus of the industry that, particularly where plaster application is by hand, multicoat work is necessary for control of plaster thickness and density. Most of the materials used for plaster densify under hand application because of the pressure applied to the trowels, and it is believed that this change in density is more controllable and will be of a more uniform nature where the plaster is applied in thin, successive layers. For these reasons, the IBC requires three-coat plastering over metal or wire lath and two-coat work applied over other plaster bases approved for use by the code. Reducing the requirement for plaster bases other than metal or wire lath is based on the rigidity of the plaster base itself. More rigid plaster bases are not as susceptible to variations in thickness and flatness of the surface. In fact, it may be considered that the first coat applied in three-coat work on a flexible base, such as wire lath, is used to stiffen that base to provide the rigidity necessary to attain uniform thickness and surface flatness.

Fiber insulation board does not have the qualities necessary for a good performing plaster base. It absorbs excessive moisture from the plaster mix, creating problems of workmanship, and it does not have the stability and rigidity required for a proper functioning plaster base. Also, in the colder and damper climates, the fiberboard insulation retains the moisture absorbed from the plaster for a relatively longer period of time than other bases, causing premature failure of the plaster. For these reasons, the code prohibits its use as a plaster base.

Because cement plaster does not bond properly to gypsum plaster bases, the code prohibits its use over gypsum plaster bases. However, the code permits exterior plaster to be applied on horizontal surfaces, soffits, and so on, over gypsum lath and gypsum board when used as a backing for metal lath.

Plaster grounds are utilized to establish the thickness of plaster and usually are wood or metal strips attached to the plaster base. The intent is that plaster grounds are used as a guide for the straightedge in determining the thickness. In many cases, door and window frames are used as plaster grounds.

In plaster work, a base coat is any coat beneath the finish coat. This is true whether the plaster is of two-coat or three-coat application. In three-coat work, the first coat is usually referred to in the trade as the *scratch* coat. It is usually applied over flexible bases, such as metal or wire fabric lath, and is intended to stiffen the base and provide a mechanical bond to the base. Also, as its name implies, the first coat is scratched with a scarifying tool, which provides horizontal ridges or scratches that are intended to provide mechanical keys for the application of the second coat (or brown coat). The brown coat usually constitutes the major bulk of the plaster and, consequently, materially affects the membrane strength. As a result, proportioning and workability are critical, and the mix should have high plasticity for proper application. The term *brown coat* is utilized by the trade to differentiate the relative color of the second coat to the finish coat, which is usually much lighter in color and is sometimes white, depending on the constituents.

The base coats in plaster work provide the strength for the plaster membrane but generally do not provide a proper surface texture for a finished surface. Therefore, a thin, almost veneer, coat of plaster is applied to the base coats as a finish coat. The finish coat may be applied in such a manner as to provide an ornamental or decorative finish, or it may be applied as a smooth surface to act as a flat base over which paint and wallpaper may be applied.

Section 2512 *Exterior Plaster*

Cement plaster is the only material approved by the code for exterior plaster. Gypsum plaster deteriorates under conditions of weather and moisture, which are prevalent on the exterior surfaces of buildings. Exterior cement plaster is required by the code to be applied in not less than three coats when applied over metal lath, wire-fabric lath, or gypsum board backing for the same reasons as discussed for interior plaster. When the cement plaster is applied over other approved plaster bases, the code requires only two-coat work. Additionally, where the plaster surface is to be completely concealed; it is not necessary to provide a finish coat.

The code requires that the exterior plaster be installed to completely cover, but not extend below, the lath and paper on wood or metal-studded exterior wall construction supported by an on-grade concrete floor slab. This requirement, combined with the requirement in Section 2512.1.2 for a weep screed, is intended to prevent the entrapment of free moisture and the subsequent channeling of the moisture to the interior of the building. See Figure 2512-1.

Figure 2512-1 **Exterior stucco construction—wall with wire mesh, paper, and scratch coat.**

2512.1.2 Weep screeds. Water can penetrate exterior plaster walls for a variety of reasons. Once it penetrates the plaster, the water will run down the exterior face of the water-resistive barrier until it reaches the sill plate or mudsill. At this point, the water will seek exit from the wall, and if the exterior plaster is not applied to allow the water to escape, it will exit through the inside of the wall and leak into the building. Thus, the IBC requires a weep screed that, when constructed as shown in Figure 2512-2, will permit the escape of the water to the exterior of the building. In addition, where weep screeds are not provided for plaster exterior walls constructed in cold-climate areas, it is possible that the trapped moisture will freeze and cause a premature failure of the exterior plaster. The water-resistive barrier required by the code must lap the weep screed's vertical flange. Although this section does not specify the amount of overlap, at least 2 inches (51 mm) should be adequate in keeping with the typical weather-resistive-barrier lap requirements.

2512.2 Plasticity agents. Admixtures such as plasticizers should not be added to Portland cement or blended cement unless approved by the building official. Some admixtures can have deleterious effects that more than offset the desired improvement in plasticity. It is preferable that plasticizers be added during the manufacture of the cement in order to ensure product uniformity and proper proportions. When plastic cement is used, the code does not permit any further additions of plasticizers as it is assumed that the amount added during the manufacturing process is adequate and is the maximum permitted. Hydrated lime and lime putty are time-tested plasticizers used with cement plaster, and their use is permitted by the code in the amounts set forth in ASTM C926 *Specification for Application of Portland Cement-Based Plaster*.

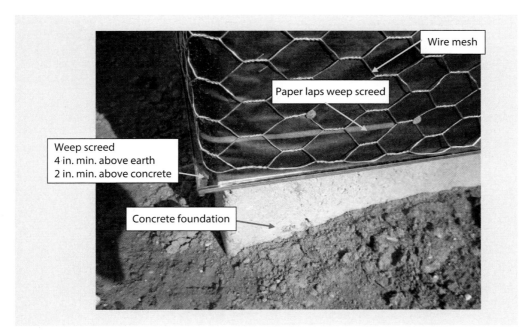

Figure 2512-2 **Weep screed.**

2512.4 Cement plaster. When cement plaster is applied during freezing weather, it loses a high proportion of its strength and, therefore, does not meet the intent of the code. In addition to protecting the plaster coats from freezing for at least 24 hours after set has occurred, application of the plaster should only be done when the ambient temperature is higher than 40°F (4°C). Plaster may be applied in colder temperatures where provisions are made to keep the cement plaster work above 40°F (4°C) during application and for at least 48 hours thereafter.

It is also important that the plaster not be applied to frozen bases or those covered with frost, which will not only weaken the bond of the plaster to its base but will also freeze the layer of plaster adjacent to the frozen base. In those cases where cement plaster is mixed with frozen ingredients or applied to a frozen base, it loses a high percentage of its strength.

KEY POINTS

- Limits on provisions for wall and ceiling coverings are detailed for weather-exposed surfaces.
- Because gypsum plaster, gypsum lath, and gypsum panel products are subject to deterioration from moisture, the code limits their use to interior locations and weather-protected surfaces.
- For exterior plaster walls, the IBC requires a weep screed that will permit the escape of water to the exterior of the building.

CHAPTER 26

PLASTIC

Section 2603 Foam Plastic Insulation
Section 2605 Plastic Veneer
Section 2606 Light-Transmitting Plastics
Section 2608 Light-Transmitting Plastic Glazing
Section 2609 Light-Transmitting Plastic Roof Panels
Section 2610 Light-Transmitting Plastic Skylight Glazing
Key Points

> This chapter covers the use and installation of various types of plastic materials. Included are foam plastic insulation, light-transmitting plastics, plastic veneer, plastic composites, interior plastic trim, and reflective plastic core insulation.

Section 2603 *Foam Plastic Insulation*

Foam plastic insulation code provisions developed as a result of extensive research are centered on two basic concepts:

1. An index limitation of the flame spread and smoke developed to 75 and 450, respectively.
2. Separation of the foam plastic insulation from the interior of the building by an approved thermal barrier. The adequacy of the thermal barrier is related to the time during which the thermal barrier is expected to remain in place under fire conditions.

2603.1.1 Spray-applied foam plastic. Reference is made to ICC 1100 *Standard for Spray-applied Foam Plastic Insulation* for single- or multi-component spray-applied foam plastic insulation. ICC 1100 establishes minimum physical properties and performance requirements for spray-applied products. The standard also addresses specific installation requirements.

2603.1.2 Insulating Sheathing. Foam plastic materials used as insulating sheathing must comply with the provisions of Section 2603 and applicable material standards. The material standard for expanded polystyrene (EPS) and extruded polystyrene (XPS) is ASTM C578 *Standard Specification for Rigid, Cellular Polystyrene Thermal Insulation* and the standard for polyisocyanurate is ASTM C1289 *Standard Specification for Faced Rigid Cellular Polyisocyanurate Thermal Insulation Board*.

2603.2 Labeling and identification. In addition to the flame-spread and smoke-developed criteria, the code also requires that the containers of foam plastic and foam plastic ingredients be labeled by an approved agency to show that the material is compliant. There are many foam plastic products on the market that do not comply with the code and that were not intended for use in construction. The labeling requirement is intended to prevent the misapplication of products not designed for this use.

2603.3 Surface-burning characteristics. It is important that any foam plastic insulation or foam plastic core material found in manufactured assemblies be limited in flame spread and smoke development. In this section, the code limits such foam plastic materials to a flame-spread index of 75 and a smoke-developed index of 450 where tested at the maximum intended thickness of use. Various exceptions are provided for interior trim, cold-storage buildings and similar facilities, interior signs in covered mall buildings, listed roof assemblies, and special approvals.

2603.4 Thermal barrier. Because of the potential hazards involved, foam plastic must typically be separated from the interior of a building by an approved thermal barrier. Gypsum wallboard at least ½ inch (12.7 mm) in thickness, mass timber, or heavy timber satisfy this requirement, as does any equivalent thermal barrier material complying with the criteria of this section. It must be demonstrated by approved testing that the thermal barrier will remain in place for the required 15-minute time period.

The thermal barrier as described is not required when the conditions described in Sections 2603.4.1.1 through 2603.1.15 are met; however, some form of protective membrane is typically mandated.

Masonry or concrete construction. When foam plastics are encapsulated within concrete or masonry walls, or floor or roof systems, the code does not require a thermal barrier as long as the foam plastic is covered by a minimum 1-inch (25-mm) thickness of masonry or concrete. See Figure 2603-1.

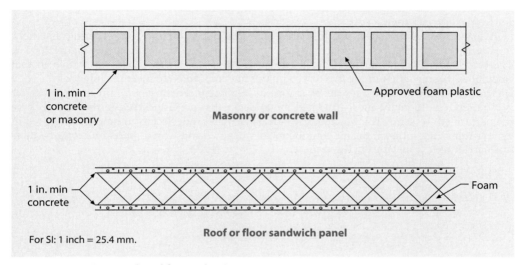

Figure 2603-1 **Encapsulated foam plastic.**

Cooler and freezer walls. Cold storage uses provide a unique condition for foam plastic insulation in that thicknesses are generally required to be greater than 4 inches (102 mm) for proper thermal insulation, although 4 inches (102 mm) is about the maximum that can be tested. The code, in all other cases, places limits on the thickness of the foam plastic insulation to that which was tested. However, because of the nature of the use in cold-storage facilities, the ignition hazards are not great. Therefore, the code permits foam plastic insulation in greater thicknesses, up to a maximum of 10 inches (254 mm), even though tested at a thickness of 4 inches (102 mm). The intent is that the foam plastic will be provided with a complying protective thermal barrier. In the case of interior rooms within a building, the foam plastic is required to be protected on both sides with a complying thermal barrier.

Provisions are included to permit cooler and freezer walls without a thermal barrier if multiple conditions are met including a flame spread index of no more than 25. Again, the code presumes that with a low-hazard use, such as a cold storage and freezer box, the metal covering will prevent the actual impingement of any flames on the foam plastic, and the sprinkler system will provide the cooling necessary to maintain proper low temperatures to prevent ignition of the foam plastic.

Exterior walls—one-story buildings. For one-story buildings, insulated metal panels (IMPs) with foam plastic cores with thicknesses up to 4 inches (102 mm) are permitted to be installed without a thermal barrier, provided the metal cladding complies with the provisions outlined in Section 2603.4.1.4, including a flame spread index of 25 for the insulation core and that the building is protected with automatic fire sprinklers. In this case, the code assumes that the protection and cooling effect provided by automatic sprinklers is a reasonable alternative to the thermal barrier.

Roofing. A thermal barrier is not required for foam plastic insulation that is a part of a Class A, B, or C roof-covering assembly if the assembly is tested or constructed according to code-prescribed prescriptive provisions. Foam plastic may be used when the foam is separated from the interior of the building by minimum 0.47-inch (12-mm) wood structural-panel sheathing bonded with exterior glue. The edges of the wood structural-panel sheathing must be supported by blocking or be of tongue-and-groove construction or of any other approved type of edge support. In this case, the thermal barrier is waived, as well as the smoke-developed index.

If foam plastic insulation is considered an integral part of the roof-covering assembly. A nationally recognized test standard for insulated roof decks is to be utilized. See Figure 2603-2. The test standards for insulated roof-deck construction are adequately conservative so that assemblies passing either of the two test standards are considered to meet the intent of the code without any limit on flame spread or smoke development. Furthermore, no thermal barrier is required. Additionally, most insulated metal decks that are listed require that the deck be nonperforated—essentially a nonacoustical deck. Acoustical decks are commonly proposed in gymnasiums and auditoriums for sound control purposes and would therefore require a thermal barrier unless specifically listed under UL 1256 or NFPA 276.

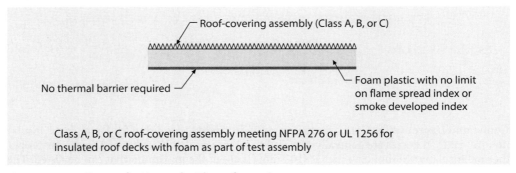

Figure 2603-2 **Foam plastic used with roof covering.**

Attics and crawl spaces. This item describes specific methods used to protect foam plastics located within attics and crawl spaces (in lieu of a complying thermal barrier) where entry is provided only for service of utilities. See Figure 2603-3. The phrase "where entry is provided only for service of utilities" is intended to restrict these reduced requirements for a thermal barrier to those unused areas where there are no heat-producing appliances. In addition, drop lights or portable service lights are often utilized when serving equipment in such concealed spaces, and such lighting devices pose an ignition threat to the foam plastic. Thus, the reduced provisions are intended to provide a barrier whose only purpose is to prevent the direct impingement of flame on the foam plastic.

The reduced level of protection is also applicable in those situations where the service of utilities is not an issue. If there are no utilities within the attic space or crawl space that require service, the minimum described degree of separation between the foam plastic and the enclosed space must still be provided where a complying thermal barrier is not installed. Where the attic or crawl space provides a suitable area that exists for a purpose other than the access to utilities, such as storage, a thermal barrier complying with Section 2603.4 is required.

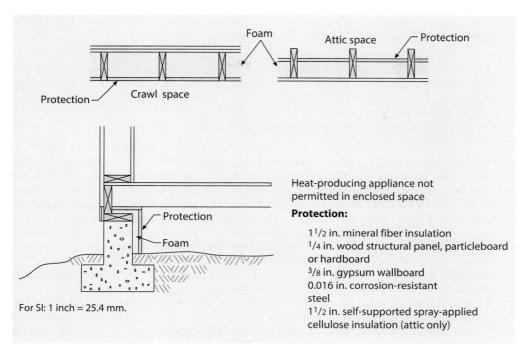

Figure 2603-3 **Foam plastic in attic and crawl spaces.**

Doors not required to have a fire-protection rating. Pivoted or side-hinged doors not required to have a fire protection rating are permitted to be installed without the thermal barrier, provided the door facings are of sheet metal of the thicknesses prescribed in this section. The rationale behind the waiver of the thermal barrier is that the foam plastic is completely encapsulated within the sheet-metal facings, and the quantity of foam plastic in protected doors is quite small.

Garage doors. Garage doors, other than those in garages accessory to dwellings, that contain foam plastic are allowed, provided the door does not require a fire-resistance rating and is faced with materials prescribed by this section. If the garage door containing the foam plastic does not have an aluminum, steel, or wood facing of the minimum thickness prescribed, the door must be tested in accordance with DASMA 107 *Room Fire Test Standard for Garage Doors Using Foam Plastic Insulation.* This provision is intended to regulate the commercial applications of overhead, sectional, and tilt-up types of doors.

Siding backer board. Where it is desired to insulate exterior walls under exterior siding, the code permits foam plastic to be used as a backer board for the siding, provided the insulation has a potential heat of not more than 2,000 Btu per square foot (22.7 mJ/m²). The thermal barrier is not required under these circumstances as long as the siding backer board has a maximum thickness of ½ inch (12.7 mm) and is separated from the interior of the building by at least 2 inches (51 mm) of mineral-fiber insulation or equivalent.

The code also permits the siding backer board without a thermal barrier when the siding is applied as re-siding over existing wall construction. This is reasonable considering the separation provided by the existing construction and limitations on the potential heat imposed by the code.

Type V construction. The use of spray-applied foam plastic has become common in wood-frame construction for sill plates, joist headers, and rim joists. Such a limited amount of foam plastic insulation is considered acceptable without the protection afforded by a thermal barrier. Testing has been conducted to evaluate the behavior of foam plastic having the density, thickness, flame-spread, and smoke-developed indices stipulated. The results indicated no substantial performance difference between a foam plastic-insulated wood floor system and a wood only floor system.

Floors. Section 2603.4.1.14 provides a viable means to protect foam plastic insulation when it is installed within a floor system. The thermal barrier required to separate foam plastic insulation installed beneath a walking surface must not only be an adequate barrier to protect the foam plastic, but also be durable enough to withstand the load and wear-and-tear that is needed for the floor. With society's focus on energy efficiency and conservation, many new types of products are being used that incorporate foam plastic insulation for energy reasons. One example is the use of structural-insulated panels where the foam plastic is laminated between two structural wood facings. These types of panels are often used in floor systems.

Although a ½-inch (12.7-mm) wood structural panel (i.e., plywood or oriented strand board) is not by itself considered as a complying "thermal barrier" as required by Section 2603.4, it will fulfill the dual need for structural strength and thermal protection of the foam insulation. In the case of a floor, the use of such panels will provide sufficient protection because, in the event of an interior fire, the floor faces a reduced exposure and is typically the last building element to be significantly exposed to the fire. See Figure 2603-4.

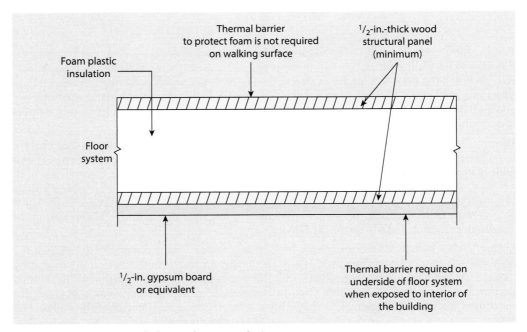

Figure 2603-4   **Floors with foam plastic insulation.**

It is important to note that the protection provided by wood structural panels is only accepted on the walking surface side of the floor. If the floor is used in multistory construction, then the underside of the floor system (ceiling of the room below) must be covered by the typically required thermal barrier. The thermal barrier protection on the bottom side of the assembly does not get to take advantage of the reduction because it will not be used as a walking surface, and it will face a more severe exposure to an interior fire.

The exception is intended to address items such as carpet padding, and other interior floor finishes, that do not need to be covered by a thermal barrier.

Separately controlled climate structures and walk-in coolers. Sections 2603.4.1.3 and 2603.4.1.15 have nearly identical requirements for walk-in coolers and separately controlled climate structures, respectively. In nonsprinklered buildings, foam plastic having a thickness no more than 4 inches (102 mm) and a flame spread index of 75 or less is permitted in separately controlled climate structures and walk-in coolers where the aggregate floor area does not exceed 400 square feet (37 m^2) and the foam plastic is covered by a metal facing at least 0.032-inch (0.81 mm)-thick aluminum or corrosion-resistant steel having a minimum base metal thickness of 0.016 inch (0.41 mm). A thickness of up to 10 inches (254 mm) is allowed if protected by a thermal barrier. The provisions for separately controlled climate structures only apply to Group U buildings.

2603.5 Exterior walls of buildings of any height. The provisions for foam plastic insulation also allow such material in the exterior walls of buildings required to have noncombustible exterior wall construction (Types I, II, III, and IV). Applicable to such buildings of any height, an important provision of this section requires that the wall be tested in accordance with NFPA 285 *Standard Fire Test Method for Evaluation of Fire Propagation Characteristics of Exterior Wall Assemblies Containing Combustible Components*. This test provides a method of evaluating the wall's flammability characteristics that are due to the combustible foam plastic materials within the wall. Wall assemblies need not be tested where they can comply with the provisions of Section 2603.4.1.4; however, this allowance is only applicable to fully sprinklered, one-story buildings. Section 2603.5.1 also requires that test data be provided to show that if a fire-resistance rating is required, the rating of the wall containing the foam maintains the required rating. Moreover, the foam plastic insulation must:

1. Be separated from the interior of the building with a thermal barrier meeting the provisions of Section 2603.4.

2. Not have a potential heat content exceeding that of the foam plastic insulation contained in the wall assembly as tested.

3. Have a maximum flame-spread index of 25 and smoke-developed index of 450. Exterior coatings and facings, tested individually, must also comply with these flame-spread and smoke-development limitations.

4. Be labeled by an approved agency.

5. Comply with the ignition limitations imposed by Section 2603.5.7.

2603.6 Roofing. As previously addressed, complying foam plastic insulation may be utilized as a portion of a roof-covering assembly, provided the assembly with the foam plastic insulation has been tested in accordance with ASTM E108 *Standard Test Methods for Fire Tests of Roof Coverings* or UL 790 *Standard Test Methods for Fire Tests of Roof Coverings*, and has been listed as a Class A, B, or C roofing assembly.

2603.9 Special approval. In this section, the code provides for those cases where foam plastic products and protective coverings do not comply with the specific requirements of Section 2603.4 or 2603.6. The code refers to a number of test standards for determining specific approvals and, in addition, there are others that utilize some variation of the room test and are designed for testing exterior wall applications.

Section 2605 *Plastic Veneer*

Because it is a combustible material, plastic veneer used in the interior of a building is required by the code to comply with the interior finish requirements of Chapter 8.

Where plastic veneer is used on the exterior of a building, the code requires that the veneer be of approved plastic materials as limited by Section 2606.4. This places severe restrictions on the combustibility and smoke development of the plastic materials. Because plastic materials are combustible, the code limits their attachment on any exterior wall to a height no greater than 50 feet (15,240 mm) above grade. Furthermore, the IBC limits the area of plastic veneer to 300 square feet (27.9 m^2) in any one section and requires each section to be separated vertically by a minimum of 4 feet (1,219 mm). The 4-foot (1,219-mm) separation helps control the rapid vertical spread of fire. The code anticipates that local fire-fighting forces can effectively fight a fire that involves plastic veneer up to a height of about 50 feet (15,240 mm). Also, if the plastic veneer involves too large an area, it is conceivable that a fire could overtax local fire-fighting forces. The exception applies to Type VB buildings where the walls are not required to have a fire-resistance rating. In this case, the plastic materials do not present a greatly different hazard than the unprotected wood construction.

Plastic siding used on the exterior of a building is regulated separately from plastic veneer. The provisions of Section 2605 are not appropriate for exterior plastic siding, as the requirements for exterior wall coverings established in Sections 1403 and 1404 are to be applied.

Section 2606 *Light-Transmitting Plastics*

Light-transmitting plastics are those plastics used in the building envelope or with interior lighting to transmit light to the interior of the building. Light-transmitting plastics are regulated because they are combustible materials. The unregulated use of combustible materials in the roof structure and for the exterior walls can possibly defeat the intent of the provisions of the code relating to types of construction. Thus, Sections 2606 through 2611 regulate these materials so that they do not materially affect other code requirements regarding types of construction.

Any use of light-transmitting plastic materials must be approved by the building official and be based on technical data submitted to substantiate their use.

The criteria of Section 2606.4 establishes the combustibility classifications of approved plastic materials, determined to be either Class CC1 or CC2 in accordance with ASTM D635 *Test Method for Rate of Burning and/or Extent and Time of Burning of Plastics in a Horizontal Position*, and serves as the approval basis.

Materials of light-transmitting plastic, such as lenses, panels, grids, or baffles, located below independent light sources are thought of as creating a light-diffusing system. Light-diffusing systems are specifically regulated in Section 2606.7. Regulated as to occupancy, location, support, installation, and size, light-diffusing systems pose potential hazards that are due to their combustibility.

Section 2608 *Light-Transmitting Plastic Glazing*

Because plastic glazing materials are combustible, their use is limited to openings not required to be fire protected. In the case of building construction other than Type VB, their use is further restricted. The glazing of openings not required to be fire protected in Type VB construction is essentially unlimited as to the area, height, percentage, and separation requirements applicable to the individual glazed openings.

For plastic-glazed openings in buildings other than Type VB, restrictions are placed on the area, height, percentage, and separation requirements for the individual glazed openings because plastic glazing materials are combustible. In other types of construction, unprotected combustible materials must be limited in accordance with their real extent and separation. Because of the combustibility of plastic glazing, the code requires flame barriers at each floor level for nonsprinklered multistory buildings to prevent the transmission of flame from one story to another by way of combustible openings.

As with other provisions of the code limiting the height of combustible materials above grade, this section also limits the height of plastic materials above grade to 75 feet (22,860 mm) unless the building is sprinklered throughout.

Section 2609 *Light-Transmitting Plastic Roof Panels*

Plastic roof panels are regulated on the basis of three conditions, of which only one needs to be met in order to utilize light-transmitting plastic panels in roofs of all occupancies other than Groups H, I-2, and I-3. Light-transmitting plastic roof panels may be installed in buildings equipped throughout with an automatic sprinkler system, in buildings where the roof construction is not required to have a fire-resistance rating, or where the roof panels meet the requirements for roof coverings in accordance with Chapter 15.

Because plastic roof panels constitute unprotected openings in the roof, the code requires that they be separated from each other by 4 feet (1,219 mm) horizontally. The minimum 4-foot (1,219-mm) separation is not mandated for fully sprinklered buildings, nor is it required in buildings housing low-hazard occupancies as limited by Exception 2 or 3 to Section 2609.4.

Section 2610 *Light-Transmitting Plastic Skylight Glazing*

In this section, the requirements for skylights are more detailed than those for roof panels in Section 2609 because there is no limit on the type of construction or fire-protection requirements for the roof assembly. Furthermore, skylights have unique requirements, such as those for flashing and resistance to burning brands. Also, as plastic-glazed skylights provide an unprotected combustible assembly in the roof, limitations must be placed on the area, percentage, and separation of each unit. Each unit's location on the roof relative to lot lines is regulated in a manner consistent with that for plastic roof panels.

Two of the primary concerns of plastic-glazed skylights are related to flashing at the intersection with the roof and their ability to resist the effects of flying, burning brands. Therefore, with one exception, the code requires that they be mounted on a curb at least 4 inches (102 mm) above the plane of the roof so that proper flashing may be accomplished. The exception involves skylights on roofs that have a minimum slope of 3 units vertical in 12 units horizontal (25-percent slope) and applies only to Group R-3 occupancies and on buildings having unclassified roof coverings. This slope should provide adequate roof drainage to accommodate skylights. The slope requirements for flat or corrugated plastic-glazed skylights and the rise requirement for dome-shaped skylights are based on the skylights' ability to shed flying brands. However, when the glazing material in the skylights can pass the Class B burning brands test specified in ASTM E108 or UL 790, there is no limitation on slope, either of flat or corrugated glazed skylights, or on rise in the case of dome-shaped skylights.

The requirement for the protection of edges of plastic-glazed skylights or domes is to prevent the rapid spread of fire along the roof, as the edges of the plastic glazing material ignite more readily than the interior portions. Under those conditions where unclassified roof coverings are permitted, the metal or noncombustible edge material is not required.

As with roof panels, the various limitations on area, percentage, and separation of skylights are somewhat arbitrary and, as with roof panels, are based on a consensus among knowledgeable experts on what is reasonable.

KEY POINTS

- Foam plastic is regulated for flame spread and smoke development.
- Separation with a thermal barrier must be provided between foam plastic insulation and the interior of the building.
- Containers of foam plastic and foam plastic ingredients must be labeled to prevent the misapplication of products not designed for their use.
- Foam plastics used in several applications, such as masonry or concrete construction, cooler and freezer walls, roofing, attics and crawl spaces, and doors not required to have a fire-protection rating, may be installed without a thermal barrier under specified conditions.
- When properly tested, foam plastic insulation is permitted in exterior walls of buildings required to have noncombustible exterior wall construction.
- Light-transmitting plastics are regulated in part because they are combustible materials.

CHAPTER 27

ELECTRICAL

Section 2702 Emergency and Standby Power Systems
Section 2703 Lightning Protection Systems
Key Points

> All electrical components, equipment, and systems are to be designed and constructed in accordance with the provisions of NFPA 70: *National Electrical Code®* (NEC®). The *International Building Code®* (IBC®) specifically references the NEC for its technical provisions. The only electrical issues addressed within this chapter of the IBC are those emergency and standby power systems and lightning protection systems that are addressed in other provisions of the code.

Section 2702 *Emergency and Standby Power Systems*

In addition to the provisions of this section, NFPA Standards *70: National Electrical Code, 110: Standard for Emergency and Standby Power Systems*, and *111: Standard on Stored Electrical Energy Emergency and Standby Power Systems* regulate the installation of emergency and standby power systems. Stationary engine generators, where used to provide emergency and standby power, must comply with the requirements of UL 2200 *Stationary Engine Generator Assemblies*. This UL standard provides a benchmark for the evaluation of the safety and reliability of such generators. Section 2702 provides several general provisions for emergency and standby power systems and also identifies 19 specific situations where such systems must be provided.

Emergency power shall be provided as follows for:

1. All required externally illuminated exit signs as required by Section 1013.6.3.
2. Means of egress illumination in occupancies as required by Section 1008.3.
3. Semiconductor fabrication facilities per Section 415.11.11.
4. Power-operated sliding doors or power-operated locks for swinging doors in Group I-3 occupancies, unless a remote mechanical operating release is provided.

Standby power is required under the following conditions:

1. Smoke-control systems as required by Sections 404.7 and 909.
2. Elevators and platform lifts that are a portion of accessible means of egress.
3. Special-purpose horizontal sliding, accordion, and folding doors utilized as a component of a means of egress.
4. Auxiliary-inflation systems in membrane structures exceeding 1,500 square feet (140 m^2) in area.
5. Emergency responder communication coverage systems as required by Section 918.
6. Common exhaust systems for domestic kitchens and clothes dryers located in multistory structures as required by the *International Mechanical Code®* or *International Fuel Gas Code®*.
7. Emergency voice/alarm communication systems.
8. Hydrogen fuel gas rooms as required by the *International Fire Code®*.

Both emergency power and standby power shall be provided in the following situations for:

1. High-rise buildings (with exceptions), defined as those structures having occupied floors more than 75 feet (22,860 mm) above the lowest level of fire-department vehicle access.

2. Underground buildings (with exceptions), defined as those building spaces having a floor level used for human occupancy more than 30 feet (9,144 mm) below the lowest level of exit discharge.

Either emergency or standby power is required for:

1. Gas detection systems as required by the *International Fire Code*.
2. Occupancies having hazardous materials when required by the *International Fire Code*.
3. Higher education laboratory suites located above the sixth story above grade plane or located on a story below grade plane.

In addition to the conditions noted above, critical electrical circuits within Group I-2 occupancies as well as ambulatory care facilities are to comply with both Section 2702.3 and NFPA 99 *Health Care Facilities Code*. Section 2702.3 notes protection requirements for the circuits while Chapter 6 of NFPA 99 requires emergency power for such circuits.

Section 2703 *Lightning Protection Systems*

Lightning protection systems are to be installed and interconnected in accordance with NFPA 780 *Standard for the Installation of Lightning Protection Systems* or UL 96A *Standard for Installation Requirements for Lightning Protection Systems*. UL 96A is not to be utilized for buildings used for the production, handling, or storage of ammunition, explosives, flammable liquids, flammable gases, or other explosive ingredients including dust. Where lightning protection systems are installed, surge protective devices must also be installed in accordance with NFPA 70 and either NFPA 780 or UL 96A.

KEY POINTS

- All electrical components, equipment, and systems are to be designed and constructed in accordance with NFPA 70.
- NFPA Standards 70, 110, and 111 regulate the installation of emergency and standby power systems.
- Stationary engine generators used to provide emergency and standby power must comply with UL 2200.
- Lightning protection systems are to be installed and interconnected in accordance with NFPA 780 or UL 96A and surge protective devices must be installed in accordance with NFPA 70 and either NFPA 780 or UL 96A.
- Only electrical issues regarding emergency and standby power systems and lightning protection systems that are addressed in other IBC provisions are addressed in Chapter 27.

CHAPTER 28

MECHANICAL

This chapter references the *International Mechanical Code®* (IMC®), the *International Fuel Gas Code®* (IFGC®), and Chapter 21 of the *International Building Code®* (IBC®) for various requirements relating to the construction, installation, and maintenance of mechanical equipment and systems. The appropriate code shall be utilized to address heating; air conditioning; refrigeration; mechanical and natural ventilation; plenums; and factory-built chimneys, fireplaces, and barbecues. Reference is made to the IMC and Chapter 21 of the IBC for the regulation of masonry chimneys, fireplaces, and barbecues. Other referenced codes include the *International Fire Code®* and the *International Property Maintenance Code®* regarding use and maintenance activities, as well as the *International Existing Building Code®* for alterations, repairs, replacements, and additions.

CHAPTER 29

PLUMBING

Section 2902 Minimum Plumbing Facilities
Section 2903 Installation of Fixtures
Key Points

Chapter 29 ■ Plumbing

> The intent of Chapter 29 is to reference the *International Plumbing Code®* (IPC®) for the construction, installation, and maintenance of plumbing systems and equipment, and the *International Private Sewage Disposal Code®* (IPSDC®) for the regulation of private sewage disposal systems. In addition, the provisions of this chapter provide for the determination of plumbing fixture counts based on occupancy classification and occupant loads.

Section 2902 *Minimum Plumbing Facilities*

2902.1 Minimum number of fixtures. This section establishes the minimum number of plumbing fixtures that must be provided for various occupancies based on Table 2902.1. The table is based on the general use groupings as classified in Chapter 3 and the corresponding occupant loads calculated for the building. Those fixtures required in most occupancies include water closets, lavatories, drinking fountains, and service sinks. In addition, bathtubs or showers, automatic clothes-washer connections, and kitchen sinks are mandated in some residential occupancies. Bathtubs and/or showers are also required in specified institutional occupancies. It is assumed that at least one fixture of each type as required by Table 2902.1 will be provided in a building designed for human occupancy.

When determining the proper occupant load to be utilized in calculating plumbing fixture count, there is no specific methodology referenced. However, because the only provisions in the IBC® addressing occupant load calculation are found in Chapter 10, it is typically assumed that the approach established in Section 1004 for the means of egress should also be used for plumbing fixture count. The basis for most occupant load determinations, other than areas with fixed seating, is the density factor established in Table 1004.5. An occupant load is determined by dividing the floor area under consideration by the appropriate density factor.

It should be noted that the building official also has the authority as established in the exception to Section 1004.5 to base the occupant load on the actual number of occupants anticipated, rather than the calculated number. Although the use of this exception is typically inappropriate for egress and fire-safety purposes, it is commonly applied for plumbing fixture count. The occupant load utilized for egress and fire protection requirements is purposely conservative by most counts because of the life-safety concerns. The occupant load to be used in calculating the minimum plumbing fixture count should be based on more of a convenience concern, recognizing the need to satisfy any sanitation issues. Therefore, the occupant load used in the plumbing fixture count could differ from that used as the basis for the design of the means of egress system. The building official should rely on all available information that will assist in the appropriate determination of occupant load for fixture count purposes.

2902.1.1 Fixture calculations. Once the appropriate occupant load is determined, the minimum required number of fixtures is calculated by using Table 2902.1. The provisions of Section 2902.1.1 address the method in which the fixtures are to be distributed between the sexes, and also note how gender-neutral provisions can be incorporated.

For the determination of required plumbing fixtures, the total occupant load to be served by the plumbing facilities must first be identified. Unless modified by special conditions, the total occupant load is halved, assuming 50 percent of the occupants to be male, 50 percent to be female. The resulting occupant loads are then used when applying the table, with fixtures calculated individually for each of the sexes. Where the required number of fixtures contains a fraction, an additional fixture is required. See Application Examples 2902-1 and 2902-2.

> **GIVEN:** An exhibition hall classified as an Assembly use. The hall's occupant load is determined to be 8,680.
>
> **DETERMINE:** The minimum required number of (a) water closets for the male occupants, (b) water closets for the female occupants, (c) lavatories for the male occupants, and (d) lavatories for the female occupants.
>
> 1. Assume occupants as 50% male, 50% female per Section 2902.1.1:
>
>   4,340 males
>   4,340 females
>
> 2. (a) $\dfrac{4,340}{125} = 34.72 = 35$ water closets* for males
>
>   *IPC Section 424.2 would require a minimum of 12 or more water closets, with the remainder urinals, to make up the required 35 fixtures.
>
>   (b) $\dfrac{4,340}{65} = 66.77 = 67$ water closets for females
>
> 3. $\dfrac{4,340}{200} = 21.7 =$
>
>   (c) 22 lavatories for males
>   (d) 22 lavatories for females
>
> **MINIMUM REQUIRED PLUMBING FIXTURES**

Application Example 2902-1

Note that the provisions of Section 424.2 of the IPC allow urinals as a substitution for water closets on a 1-to-1 basis, except that assembly and educational occupancies are allowed to replace two-thirds of the required water closets for males with urinals. For example, if nine water closets are required in a men's toilet room in a large nightclub, it is acceptable to provide six urinals and only three water closets. Where the occupancy classification is other than

> **GIVEN:** A manufacturing facility classified as a Factory use. The occupant load is determined to be 684.
>
> **DETERMINE:** The minimum required number of water closets and lavatories.
>
> 1. Assuming a 50:50 split, assign 342 male and 342 female occupants.
>
> 2. $\dfrac{342}{100} = 3.42 = 4$ water closets minimum for each sex
>
>   4 water closets* for males
>   4 water closets for females
>
> Same calculation for lavatories, a minimum of 4 for each sex.
>
> *If urinals are substituted for water closets, a minimum of 2 water closets must be provided, with the remainder urinals, to make up the 4 fixtures.
>
> **MINIMUM REQUIRED PLUMBING FIXTURES**

Application Example 2902-2

Group A or Group E, only one-half of the required number of water closets may be substituted with urinals.

The determination of the minimum plumbing fixture count becomes a bit more complex where the building contains a number of different occupancies. If toilet room facilities are provided independently for each of the occupancies in the building, the basic method of calculating the number of fixtures should be satisfactory. However, if common toilet facilities are designed to serve the occupants from multiple occupancy groups, a different approach is more appropriately warranted. In such a determination, the number of required fixtures for each sex would be calculated for each occupancy, then without rounding, added together to arrive at the minimum fixtures that must be provided. An example of this methodology is shown in Application Example 2902-3. A similar approach could be utilized when substituting urinals for water closets.

GIVEN: A mixed-occupancy building containing a Mercantile use with an occupant load of 368, a Business use with an occupant load of 56, and a Storage use with an occupant load of 78. All of the plumbing fixtures will be located at a single toilet room location.

DETERMINE: The minimum number of water closets that would be required in each of two toilet rooms.

SOLUTION:

Mercantile: 184/500 @ 1 per 500 occupants
Business: 28/25 @ 1 per 25 for first 50 occupants
Storage: 39/100 @ 1 per 100 occupants

184/500 + 28/25 + 39/100 =
0.37 + 1.12 + 0.39 = 1.88

Minimum of two water closets required in each toilet room

MINIMUM REQUIRED PLUMBING FIXTURES

Application Example 2902-3

Exception 2 establishes the concept of multiple-user toilet facilities designed to serve all genders. Additional requirements for this type of facility are limited by Exception 6 of Section 2902.2. The general provisions continue to be applicable for the determination of the minimum required number of plumbing fixtures. A variety of toilet room arrangements are possible, with the regulation of urinal locations as the only privacy feature unique to all-gender facilities beyond what is already required in any multiple-user toilet room intended for use by the same sex. The calculation of fixture count at 100 percent is an exception to the use of the total occupant load divided in half (assuming a 50/50 male/female distribution). Application Example 2902-4 shows how the minimum required number of fixtures is to be determined for a multiple-user facility available to all genders. Fixture count calculations are straightforward where the male/female ratios in Table 2902.1 are identical; however, some adjustment may be necessary where different table ratios are indicated (such as for theaters and similar assembly uses).

2902.1.2 Fixtures in single-user toilet facilities and bathing rooms. Single-user and multiple-user facilities that serve all genders, including family or assisted-use toilet rooms and bathing rooms, are permitted to count toward the minimum required plumbing fixtures. The provisions recognize that the required number of fixtures can be provided based on either

> **GIVEN:** An office building containing a Business use with an occupant load of 60. Toilet facility design chosen assumes one multiple-user gender-neutral facility.
>
> **DETERMINE:** The minimum number of water closets and lavatories that are required.
>
> **SOLUTION:**
>
> Water Closet Ratio: 1 per 25 for the first 50 and 1 per 50 for the remainder > 50
>
> Lavatory Ratio: 1 per 40 for the first 80 and 1 per 80 for the remainder > 80
>
> Water Closets = 50/25 + (60 − 50)/50 = 2.2
>
> Lavatories = 60/40 = 1.5
>
> A minimum of 3 water closets and 2 lavatories are required.
>
> **MINIMUM REQUIRED PLUMBING FIXTURES**

Application Example 2902-4

the required number of separate facilities or the aggregate of any combination of both single-user and separate facilities. An important requirement indicates that single-user, as well as family or assisted-use, toilet rooms and bathing rooms are to be identified to indicate the availability for use by all individuals with no regard as to their sex.

Once the appropriate occupant load is determined, the minimum required number of fixtures is calculated by using Table 2902.1. The provisions of Section 2902.1.1 address the method in which the fixtures are to be distributed. The total occupant load is typically halved, assuming 50 percent male and 50 percent female occupants. However, in many assembly uses where the fixture ratios for males and females are different. A methodology establishes as to how fixtures in single-user and family/assisted use toilet facilities should be calculated toward fulfilling the total requirement for fixtures in a building or tenant space. Application Example 2902-5 illustrates that the reductions, due to the presence of fixtures in single-user and/or family or assisted-use toilet facilities, in the number of fixtures in multiple-user toilet facilities must be made proportionally.

2902.2 Separate facilities. In most buildings where plumbing fixtures are required, separate facilities must be available for each sex. Simply, at a minimum, one women's toilet room and one men's toilet room must be provided. There are conditions, however, where only a single toilet room is mandated. Separate-sex facilities need not be provided within dwelling units and sleeping units. In addition, common facilities are permitted where the number of people (both customers and employees) does not exceed 15. In mercantile occupancies with an occupant load of 100 or less, such as a small retail sales tenant, a single toilet room is also permitted. Another condition where separate sex facilities are not required is where the occupant load of a business occupancy does not exceed 25.

Exception 5 clarifies that separate facilities need not be designated by sex where single-user toilet rooms are provided. Reference is made to Section 2902.1.2 to reinforce the requirement that such single-user facilities are to be identified as being available for use by all persons regardless of their sex.

Separate facilities are also exempted by Exception 6 where a multi-user toilet room is designed to be used by all individuals, regardless of sex, and the water closets and lavatories are available to everyone. It is mandated that water closet privacy be provided with separate compartments, enclosed by partitions and a door. Urinals, where installed, are to occupy a separate area and also provide privacy. Complying partitions shall be placed between each urinal stall.

> **GIVEN:** A theater having 2,000 occupants, the male water closet (WC) ratio is 1:125 and the female ratio is 1:65.
>
> **Design layout:** Four single-user toilet facilities with the remainder of fixtures in multiple-user toilet facilities (separate sex)
>
> **DETERMINE:** What is the minimum number of water closets required in the multiple-user toilet facilities?
>
> **Step 1:** Determine minimum required number of male water closets, all located in a multiple-user toilet facility:
>
> $$1,000/125 = 8$$
>
> Determine minimum required number of female water closets, all located in a multiple-user toilet facility:
>
> $$1,000/65 = 15.4$$
>
> Total = 23.4 WCs
>
> **Step 2:** Proportionally reduce the number of water closets in the multiple-user toilet facilities (because of the presence of water closets located in single-user toilet facilities). In this example, each water closet in a single-user toilet facility allows for a reduction of $8/23.4 = 0.34$ male water closets and $15.4/23.4 = 0.66$ female water closets in the multiple-user toilet facilities.
>
> **Result:** Therefore, four single-user toilet facilities, $(4 \times 0.34) = 1.36$ male WC reduction and $(4 \times 0.66) = 2.64$ female WC reduction. Thus, the multiple- user toilet facilities require a minimum of:
>
> Male: $8 - 1.36 = 6.64 = 7$ water closets
>
> Female: $15.4 - 2.64 = 12.76 = 13$ water closets

Application Example 2902-5

2902.3 Employee and public toilet facilities. Only specific occupancies and uses are required to have public toilet facilities for use by customers, patrons, and visitors of the building. Restaurants, nightclubs, places of assembly, business uses, mercantile occupancies, and similar buildings and tenant spaces intended for public use must be provided with customer toilet facilities located within one story above or below the area under consideration, and with a path of travel not to exceed 500 feet (152 m). Those uses where public use is not expected, such as warehouses, factories, and similar buildings, only require employee toilet facilities.

Two types of public-use buildings are exempted from public toilet facilities, open and enclosed parking garages having no parking attendants, and small quick-transaction tenant spaces. Pizza carryouts, dry cleaners, ATM facilities, shoe repair shops, newspaper stands, and many other similar spaces are unique in that patrons spend a short period of time completing a transaction and then depart. It should be noted that it is the public-access area and not the entire business that is regulated by the 300-square-foot (28 m²) area limitation. This allowance is illustrated in Figure 2902-1.

It is not uncommon for multiple-occupant toilet rooms to have a locking door to facilitate closure of the toilet facility for maintenance or security purposes as necessary. The locking device is generally required by Section 2902.3.6 to be of a type that cannot be engaged from the inside of the toilet room such that persons cannot lock themselves in the room. The requirement is intended to prevent multiple-user toilet rooms from becoming safe havens for illicit activities. However, the toilet room door is permitted to be lockable from the interior side provided three specified conditions are met. Multiple-user toilet facilities can provide for a safe area of refuge in the event of an emergency such as an active shooter situation. Although a simple thumb turn

Installation of Fixtures   917

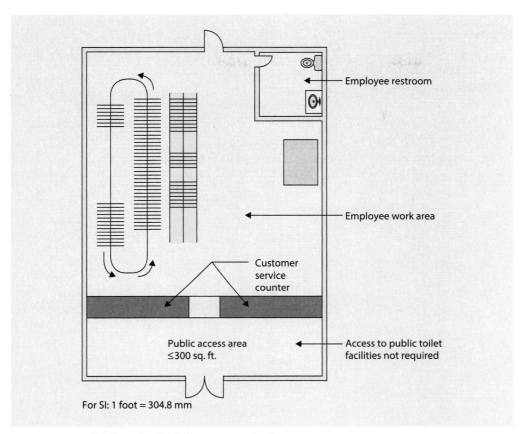

Figure 2902-1   **Public toilet facilities not required.**

latch on the inside of the door would offer the intended protection, such a device defeats the purpose of not having the door lockable from the inside in the first place. The conditions of use allow for a specific design of locking hardware that offers control over who can lock the door from the inside while offering full egress for occupants of the toilet room. Conditions of use limit the ability to lock the egress door from the inside to only those authorized individuals who are provided with a key or other means of unlocking the door. The provisions of Section 1010.2 regulating door operations must be fully met. In addition, the locking device must be capable of being unlocked from outside the toilet room. Only when all three conditions are met is a toilet room egress door locking device permitted to be lockable from inside the room.

Section 2903 *Installation of Fixtures*

Installation requirements are provided for water closets, urinals, lavatories, bidets, water closet compartments, and urinal partitions. These items are to be installed at proper level and in proper alignment. In addition, clearance and dimensional requirements set forth in IBC Section 1210, as well as in the IPC, are replicated to ensure they are not overlooked.

KEY POINTS

- The minimum required number of plumbing fixtures is based on the use of the building and the anticipated number of occupants.
- Single-user, family, and assisted-use toilet rooms are required to be identified as being available for all persons regardless of their sex.
- Multiple-user toilet rooms cannot be lockable from the inside of the room except under specific conditions.
- Installation parameters are provided for standard fixtures.

CHAPTER 30

ELEVATORS AND CONVEYING SYSTEMS

Section 3001 General
Section 3002 Hoistway Enclosures
Section 3003 Emergency Operations
Section 3006 Elevator Lobbies and
    Hoistway Door Protection
Section 3007 Fire Service Access Elevator
Section 3008 Occupant Evacuation Elevators
Section 3009 Private Residence Elevators
Key Points

> Elevators and other types of conveying systems are regulated under the provisions of this chapter. For the most part, the American Society of Mechanical Engineers (ASME) standards are utilized to address the specifics of elevator safety. ICC A117.1 *Accessible and Usable Buildings and Facilities* must also be referenced for all elevators required to be accessible by Chapter 11 of the *International Building Code®* (IBC®).

Section 3001 *General*

3001.2 Elevator emergency communication systems. In multistory buildings, passenger elevators are typically utilized as vertical accessible routes. In some cases, the elevators may also be considered a part of the accessible means of egress. The code requires that an emergency two-way communication system be provided for all elevators. The system must fully comply with the provisions in ASME A17.1/CSA B44, the *Safety Code for Elevators and Escalators* as published by the American Society of Mechanical Engineers.

The elevator code's requirements are intended to serve the broadest number of people who may not be able to communicate verbally. In addition, the system is accessible to people who may speak a different language or who cannot speak due to a medical condition. The system must provide both a text mode and an audible mode that, during operation, allows for live interactive communication between those individuals within the elevator and emergency personnel. The system shall be operational during all times and it shall be designed to allow the selection of the appropriate communication mode by the elevator occupants. Such occupants may select either the text-based or audible mode based on their specific needs. Along with the necessary visual indicators and button indicators, the text-based mode would allow the use of a keyboard by those entrapped individuals who are deaf or hard-of-hearing.

In addition, specific functions are identified where the system must be able to assess whether someone is in the elevator car and that they are entrapped. The provisions also mandate a capability to determine if assistance is needed and to communicate when such assistance is in transit and arrival of the assistance.

3001.3 Referenced standards. Depending on the type of conveying system, Table 3001.3 lists the appropriate standard that must be followed for the design, construction, alteration, installation, and maintenance of the system. As an example, elevators must comply with ASME A17.1/CSA B44 while automotive lifts are required to conform to ALI ALCTV.

3001.6 Structural Design. All interior and exterior elevators, escalators, and other conveying systems and their components must comply with all applicable design loading criteria in Chapter 16, including wind, flood, and seismic loads.

Section 3002 *Hoistway Enclosures*

3002.1 Hoistway protection. This section is essentially a cross reference to Section 713 of the code, which contains the specific requirements for the enclosure of shafts in buildings. Elevator hoistways are to be located in fire-resistance-rated enclosures, unless exempted by Section 712.1.9 addressing two-story-opening conditions or another applicable provision addressing unprotected vertical openings. If required to be fire-resistance-rated, the shaft enclosure must have a 1-hour or 2-hour rating based on the number of stories connected by the shaft enclosure, as well as the required fire-resistance rating of the floor construction. Opening protectives for hoistway enclosures are also regulated by Chapter 7.

3002.2 Number of elevator cars in a hoistway. The basis for limiting the number of cars in a single hoistway is to provide a reasonable level of assurance that a multilevel building served by several elevators would not have all of its elevator cars located in the same hoistway. This could result in a single emergency disabling all elevators within the building. For example, if all elevator cars were allowed to be located in the same hoistway, smoke that entered the enclosure during a fire would render all elevator cars unusable. The code provisions will increase the chance that some elevators within a major building would remain operational during a fire or other emergency. See Figure 3002-1.

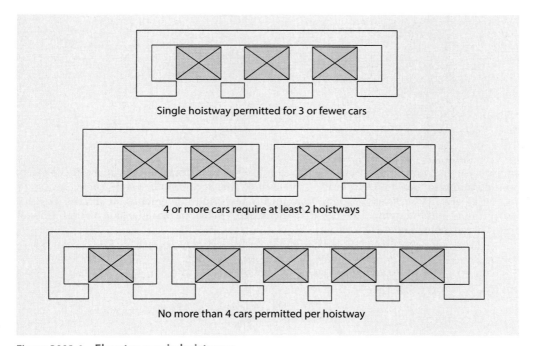

Figure 3002-1   **Elevator cars in hoistways.**

3002.3 Emergency signs. In order to alert occupants to the fact that an elevator is not to be used for egress purposes during a fire incident, this section mandates the placement of a sign adjacent to each elevator call station on each floor of the building. The standardized pictorial sign (an example is illustrated in Figure 3002-2) advises occupants to use the exit stairways rather than the elevators in case of a fire. Where the elevator is a component of the *accessible means of egress*, a sign complying with Section 1009.11 must be provided. The emergency sign does not need to be installed at occupant self-evacuation elevators as described in Section 3008.

3002.4 Elevator car to accommodate ambulance stretcher. Where elevators are provided in buildings of four stories or more, this section of the code requires that at least one elevator serving all floors accommodate an ambulance stretcher. The ability to

Figure 3002-2 **Emergency signs.**

transport an individual on a stretcher in an elevator in a multistory building is a basic life-safety consideration. Immediate identification of elevators that accommodate stretchers is necessary so that emergency-services personnel can quickly respond to emergency conditions. For this reason, an identifying symbol as shown in Figure 3002-3 shall be placed inside on both sides of the hoistway door frame.

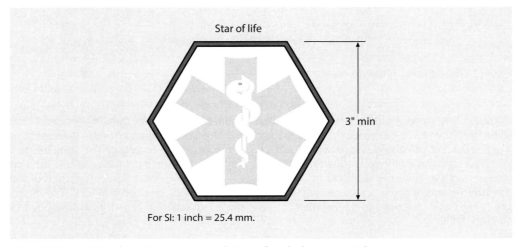

Figure 3002-3 **Sign denoting accommodation of ambulance stretcher.**

Minimum elevator car size requirements have been established to ensure that the typical ambulance stretcher can be accommodated. The minimum 24-inch by 84-inch (610-mm by 2,134-mm) size is further described to address stretchers with rounded or chamfered corners with a minimum 5-inch (127-mm) radius. See Figure 3002-4. By specifically identifying the minimum size requirements, flexibility is provided to the elevator industry in its efforts to provide appropriately sized cars. It is also beneficial to stretcher manufacturers by providing direction to aid in the standardization of their products.

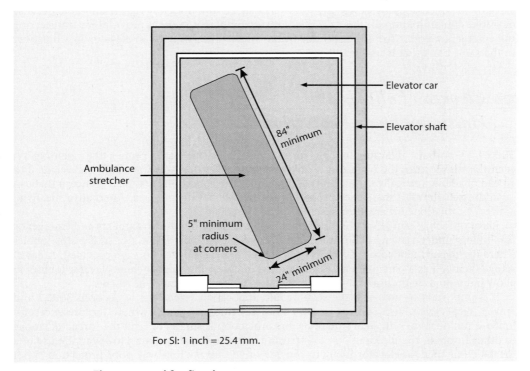

Figure 3002-4   **Elevator used for fire department emergency access.**

3002.6 Prohibited doors or other devices.   The prohibition against installing additional doors or other devices at the point of access to an elevator car is to minimize the possibility of a person becoming trapped between doors, or the possibility of the elevator car being rendered unusable because of blocked access at a particular floor where such an additional door or device has been closed and locked. The only condition under which such doors are permitted occurs where the doors or other devices are readily openable from inside the car without the use of a key, a tool, or any special effort or knowledge.

3002.7 Common enclosure with stairway.   To better ensure that a single fire incident will not restrict or eliminate the use of multiple access or egress components within a building, it is important that elevators not be located in the same shaft enclosure as a stairway. Since there is no requirement for a shaft enclosure for an elevator or exit stairway serving an open parking garage, the isolation of the elevator from the stairway is not required in open parking garages.

Section 3003 *Emergency Operations*

Standby power requirements in this section are applicable for those buildings or structures where standby power is furnished to, or required to, operate an elevator. For example, the provisions of Section 1009.4.1 mandate standby power for any elevator used as an accessible means of egress. The requirement that standby power be manually transferable to all elevators in each bank is necessary to improve their reliability during emergency conditions. As an example, the elevator to which standby power is connected may be in a hoistway that is unusable because of smoke contamination. In this case, the transferability requirement provides for transferring the emergency power from the elevator in the affected hoistway to an elevator in a hoistway in the same bank that is usable.

Section 3006 *Elevator Lobbies and Hoistway Door Protection*

3006.1 General. Elevator lobbies are presented as a means to protect the openings on each floor level from the intrusion of smoke into the elevator hoistway shaft enclosures. The spread of smoke vertically through a building under fire conditions is a major concern that can be addressed through the use of properly constructed elevator lobbies. Alternative means of addressing this concern are also presented in Section 3006.

Elevator lobbies can also be an important refuge area for building occupants and emergency responders under varying conditions. Specific provisions include the use of elevator lobbies where fire department access elevators or occupant evacuation elevators are provided. Areas of refuge included as a part of an accessible means of egress may also include elevator lobbies to allow individuals with disabilities a protected area to await assistance or rescue.

It is important to note that the scoping and application provisions of Sections 3006.2 and 3006.3, respectively, also address elevator hoistway door protection where the presence of a lobby is not necessary. In such situations, the protection is simply to limit the spread of smoke to other floors in the building due to intrusion into the elevator shaft. However, the use of a means other than an elevator lobby to provide the necessary hoistway door protection is not allowed in those conditions addressed in Items 2 through 5.

3006.2 Elevator hoistway door protection required. To reduce the potential for smoke to travel from the floor of fire origin to any other floor of the building by way of an elevator shaft enclosure, some means of elevator hoistway door protection is necessary. Typical elevator hoistway doors, although fire rated, cannot provide the necessary barrier required to keep smoke from passing from floor to floor through an elevator shaft enclosure. To restrict smoke passage from one floor level to other floor levels within the building, there must be some point where the floor level can be adequately separated from the elevator shaft. The acceptable methods of door protection are established in Section 3006.3, and include the use of elevator lobbies. However, hoistway door protection is not required in all multistory buildings, as the requirement is not applicable where the elevator shaft connects only two or three stories. The threshold of four stories is consistent with a number of other code provisions that increase the level of protection where four or more stories are involved.

Where an elevator hoistway required to be enclosed connects four or more stories, and at least one of six listed conditions applies, hoistway door protection must be provided. Buildings containing a Group I-1, Condition 2, Group I-2, or Group I-3 institutional occupancy are addressed. For other occupancies, the presence of a sprinkler system is the deciding factor.

Where an automatic sprinkler system is not provided, the four-story threshold is applicable. However, where sprinkler protection is provided, hoistway door protection is not required until the building is considered a high-rise. In high-rise buildings, door protection is only required for those elevator hoistways that travel more than 75 feet (22,860 mm), measured floor level to floor level. Condition 6 addresses situations where an elevator hoistway door is located in the wall of a fire-resistance-rated corridor.

The three exceptions to hoistway door protection recognize that there are times where such protection is either impractical or unnecessary. The first exception eliminates the door protection for elevators that serve only open parking garages. This exception should also extend to enclosed parking garages since Section 712.1.10.2 indicates that vertical opening protection is not required in both open and enclosed parking garages for elevator hoistways serving only the parking garage. Since hoistway door protection is simply a means to complete the elevator shaft enclosure protection, such door protection is not necessary if a shaft enclosure is not required. Exception 2 exempts the level(s) of exit discharge, provided the discharge level(s) is protected with an automatic sprinkler system. A third exception recognizes that hoistway doors to the exterior have no need to be protected.

3006.2.1 Rated corridors. Where a corridor is identified within a building, it may be necessary to introduce fire-resistance-rated construction and protected openings in order to isolate the means of egress path as defined by the corridor from surrounding spaces. Fire-resistance-rated corridors are intended to provide a protected path of travel within the exit access that will allow occupants to evacuate or relocate during a fire condition that occurs in spaces adjacent to the corridor. Such a corridor must also provide a significant degree of resistance to the movement of smoke into the egress environment. Where an elevator opens into a corridor required to be fire-resistance-rated, the elevator hoistway door can typically achieve the required degree of opening fire protection. However, it cannot comply with the air leakage limits mandated for a smoke- and draft-control assembly. An elevator hoistway door opening into a fire-resistance-rated corridor must be protected in accordance with one of the methods established in Section 3006.3. The protection provided at the elevator hoistway doors restricts the movement of smoke from the fire floor to other floors in the building by way of the elevator shaft enclosure. See Figure 3006-1.

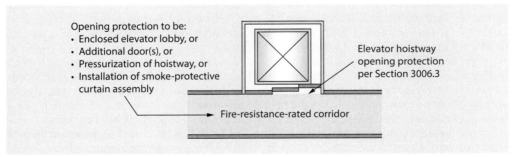

Figure 3006-1 **Elevator hoistway opening protection into a fire-resistance-rated corridor.**

3006.3 Elevator hoistway door protection. While the scoping provisions of Section 3006.2 identify those conditions under which elevator hoistway door protection is required, this section recognizes the methods by which such protection can be provided. Five specific methods are established, two of which utilize elevator lobbies. Method 1 describes elevator lobby protection through the use of fire partitions. Openings and penetrations are to be protected in accordance with Section 716 and Section 717, respectively. Method 2 differs from Method 1 by modifying the means for constructing the elevator lobby. This approach allows for the use of smoke partitions in lieu of fire partitions. Described in Section 710, smoke partitions are not required to have a fire-resistance rating but must only be constructed to limit the transfer of smoke. In Method 3, additional doors or other devices may be used in lieu of an elevator lobby, provided they are installed in accordance with Section 3002.6. In this scenario, the lobby is not necessary, because the smoke-infiltration problem is addressed by the additional door or other device. The fourth method permits pressurization of the elevator shaft enclosure in accordance with Section 909.21. The fifth method allows smoke-protective curtain assemblies for hoistways which are recognized and regulated in Chapter 9 of NFPA 105, *Standard for Smoke Door Assemblies and Other Opening Protectives*.

3006.4 Means of Egress. Because elevator lobbies are required to have access to at least one means of egress and prohibit exiting through a space that can be locked, many elevator lobbies must either provide direct access to an exit stairway or they must open into a corridor that then connects to a stairway. Depending on how the floor and exits are arranged, this can either result in the need for an additional stairway or in needing to divide the tenant spaces in a manner that permits an egress path. As an alternative solution, electrically locked exit access doors providing egress from elevator lobbies are permitted.

Section 3007 *Fire Service Access Elevator*

To facilitate the rapid deployment of fire fighters, Section 403.6.1 contains provisions for fire service access elevators in high-rise buildings that have at least one floor level more than 120 feet (36,576 mm) above the lowest level of fire department vehicle access. Usable by fire fighters and other emergency responders, the specific requirements for the elevators are set forth in Section 3007.

A fire service access elevator has a number of key features that will allow fire fighters to use the elevator for safely accessing an area of a building that may be involved in a fire or for facilitating rescue of building occupants. The elevator is required to be protected by a shaft enclosure that complies with Section 713. For a building four or more stories in height, Section 713.4 requires that the shaft have a minimum 2-hour fire-resistance rating. In the same manner that elevators which can accommodate a stretcher are required to be identified by the "Star of Life" symbol shown in Figure 3002-3, fire service access elevators must include a symbol as illustrated in IBC Figure 3007.6.5 to allow for quick identification.

A fire service access elevator lobby is mandated as a transitional element between the fire service access elevator and the remainder of the floor. In those cases where a fire service access elevator hoistway extends to an occupiable roof, the lobby is not required. Its use as a staging area for the fire department to access the floor(s) above serves no purpose as there are no floor levels above an occupiable roof. Therefore, the requirement of a lobby at each fire service access elevator hoistway opening is unnecessary at the roof level. It is important to note that the new exception only exempts the creation of an elevator lobby. A reasonable degree of fire service access to the occupiable roof must still be provided. See Figure 3007-1.

Because the fire service access elevator lobby will be the location that fire fighters will use as the point of departure to the floor or area of fire origin, it must be constructed to limit the entrance of fire or smoke. The lobby must be separated from each floor by a smoke barrier with a minimum fire-resistance rating of 1 hour, and the lobby door is required to have a minimum fire-resistance rating of 45 minutes. The lobby for the fire service access elevator must be designed such that it has direct or protected access to an enclosure for an interior exit stairway. By providing a direct connection between the elevator lobby and a stair enclosure, efficient access to and

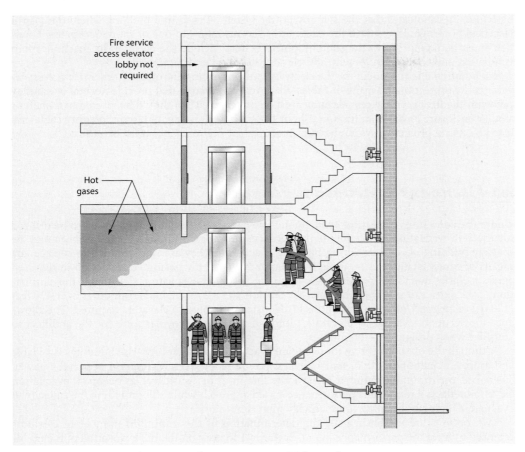

Figure 3007-1 **Fire service access elevator to occupiable roof.**

from other floors is increased. In addition, the required standpipe hose valves located within the enclosure for the interior exit stairway will be directly accessible from the fire service access elevator lobby, providing a protected location for fire fighters to prepare for manual fire-fighting operations and to assess interior fire and smoke behavior near the area of fire origin.

A unique requirement for fire service access elevators is that they must be designed so their status can be continuously monitored in the fire command center. The elevator is to be monitored by a standard emergency services interface meeting the requirements of NFPA 72, *National Fire Alarm Code.* The requirements stipulate that such an interface must be designed and arranged in accordance with the requirements of the organization that will use the device. In the case of a fire service access elevator, the fire department or fire service provider will need to be involved in the arrangement of the interface.

A fire service access elevator requires normal utility power and connection to a standby power system. Provisions further stipulate that the transfer switch for the standby power system operate within 60 seconds of utility power failure (Type 60), and that the power source is designed to operate for at least 2 hours under its design load (Class 2) and meet the requirements of NFPA 110 *Standard for Emergency and Standby Power Systems* for Level 1 service. Level 1 service, as defined by NFPA 110, indicates that the standby power system is used in a building where the loss of electrical power could result in the death of or serious injury to one or more occupants. Loads that must be connected to the standby power system include the elevator, its machine room ventilation and cooling equipment, and elevator hoistway and car lighting.

An additional requirement for the electrical power system serving the fire service access elevator is the protection of wiring or cables. Electrical conductors that provide normal or standby power to the fire service access elevator must be protected. They should be located by a shaft or similar enclosure having a minimum 2-hour fire-resistance rating; or circuit integrity cable or a listed electrical protective system having an equivalent fire resistance must be utilized.

Section 3008 *Occupant Evacuation Elevators*

Under the conditions of Section 3008, public-use passenger elevators are allowed to be utilized for the self-evacuation of occupants in high-rise buildings. Although such elevators may be used by building occupants during building evacuation, they are not intended to replace any means of egress facilities as required by Chapter 10. The only permitted reduction in required egress facilities due to the presence of complying occupant evacuation elevators is the elimination of the extra exit stairway mandated by Section 403.5.2 for high-rise buildings over 420 feet (128 m) in height. Under no conditions is the installation of such elevators required. The allowance for occupant evacuation elevators, although voluntary, provides tools for the architect to consider when designing tall buildings.

Although the use of elevators for occupant evacuation in super high-rise buildings is a viable and more efficient option than stairways, it can require an excessive amount of standby power. Therefore, the minimum number of elevators that must be considered as occupant evacuation elevators reflects a reasonable performance-based approach while still providing the capacity to evacuate a high-rise building more quickly than stairs alone.

Two options are established for the determination of the minimum number of elevators required to meet the performance intent. Scenario 1 focuses on the full evacuation of the building's occupants within a 1-hour time period. This approach does not mandate full building evacuation, but rather is to be considered as a benchmark for the analysis. Scenario 2 is generally based on a more typical phased evacuation process. The 15-minute limit criterion addresses the removal of occupants from the area to which the fire department will respond. The use of five consecutive floors in the analysis provides for a safety factor for the minimum required number of occupant evacuation elevators.

Additional criteria mandate that at least one elevator in every elevator bank be provided for occupant evacuation. Where multiple passenger elevators open into a lobby, a minimum of two occupant evacuation elevators are required. These conditions ensure that as individuals reach an elevator lobby location there will always be one or more elevators available to them for evacuation purposes. In order to better identify the specific elevators that are intended for evacuation activities, signage must be provided indicating their availability.

Section 3009 *Private Residence Elevators*

The design, construction, and installation of elevators, including hoistway enclosures and hoistway opening protection, installed within a residential dwelling unit or installed to provide access to one individual residential dwelling unit must be in accordance with ASME 17.1/CSA B44.

KEY POINTS

- Elevator shafts and elevator machine rooms are typically required to be fire-resistance-rated enclosures, with the rating of either 1 hour or 2 hours based on the number of stories in the building and required fire-resistance rating of the building's floor construction.
- Standby power requirements apply to buildings or structures where standby power is provided or mandated.
- Additional doors are prohibited at the point of access to an elevator car unless they are readily openable from the car side without the use of a key, a tool, or any special effort or special knowledge.
- Fire service access elevators help facilitate the rapid deployment of fire fighters in applicable high-rise buildings.
- Although not intended to replace any required means of egress facilities, public-use passenger elevators are allowed to be utilized for the self-evacuation of occupants in high-rise buildings.
- Private residence elevators must be designed, constructed, and installed in accordance with ASME 17.1/CSA B44.

CHAPTER 31

SPECIAL CONSTRUCTION

Section 3102 Membrane Structures
Section 3103 Temporary Structures
Section 3104 Pedestrian Walkways and Tunnels
Section 3106 Marquees
Section 3111 Solar Energy Systems
Section 3113 Relocatable Buildings
Section 3114 Intermodal Shipping Containers
Key Points

> Those special types of elements or structures that are not conveniently addressed in other portions of the *International Building Code*® (IBC®) are found in this chapter. By special construction, the code is referring to pedestrian walkways and tunnels, awnings, canopies, marquees, telecommunication and broadcast towers, vehicular gates, solar energy systems, and similar building features that are unregulated elsewhere. Unique types of structures are also addressed, including relocatable buildings, membrane structures, temporary structures, shipping container structures, and greenhouses.

Section 3102 *Membrane Structures*

3102.1 General. Because membrane structures have several unique characteristics that set them apart from other buildings, they are regulated in Chapter 31 under the provisions for special construction. The regulations cover all such structures, including air-supported, air-inflated, cable-supported, frame-supported, and tensile membrane structures. The intent of the provisions is that, except for the unique features of membrane structures, they otherwise comply with the code as far as occupancy requirements, allowable area, and other regulations are concerned. Membrane structures are typically limited to one story in height insofar as there is insufficient experience to justify multilevel structures enclosed with a membrane. There is an allowance for use in multistory buildings where the membrane structure is of noncombustible construction and serves as the building's roof.

The membrane structures regulated by the IBC are deemed to be permanent in nature, erected for a period of at least 180 days. Membrane structures in place for shorter periods of time, such as temporary tents, are to be regulated by the *International Fire Code*® (IFC®). Where a membrane structure is erected as a part of a permanent structure, such as a covering for a building, balcony, or deck, it must comply with the provisions of Section 3102 for any time period.

Because of the limited hazards present in structures not used for human occupancy, such as water-storage facilities, water clarifiers, water treatment plants, sewage treatment plants, and greenhouses, only a few provisions are applicable where membrane structures cover these types of facilities. Limitations on the membrane and interior liner material, as well as the structural design, are the only criteria in the IBC that apply to membrane structures covering facilities not typically used for human occupancy.

3102.3 Type of construction. In general, membrane structures are considered to be of Type V construction, except where the membrane structure is shown to be noncombustible. In this case, the membrane structure should be classified as Type IIB construction. Membrane structures supported by heavy-timber framing members are to be considered Type IV-HT construction. The code permits the use of nonflame-resistant plastic material for the membrane of a greenhouse structure that is not available to the general public.

3102.6 Mixed construction. This section permits the use of a noncombustible membrane on any construction type where the membrane is used exclusively as a roof or skylight and is located at least 20 feet (6,096 mm) above any floor, balcony, or gallery. This is similar to Footnote b of Table 601. This will permit nonrated noncombustible membranes to be constructed as roof systems for sports stadiums and similar buildings as well as for atriums. In Type IIB, III, IV-HT, and V construction under the same conditions, the membrane need only be flame resistant in accordance with NFPA 701 *Standard Methods of Fire Tests for Flame Propagation of Textiles and Films*.

3102.8 Inflation systems. Where membrane structures are air-supported or air-inflated, this section addresses the regulations for equipment, standby power, and support. The primary inflation system shall consist of one or more blowers, designed in such a manner that over pressurization is prevented. Air-supported or air-inflated structures exceeding 1,500 square feet (140 m²) in floor area must also be provided with an auxiliary inflation system. This backup system, connected to an approved standby power-generating system, shall operate automatically to maintain the inflation of the structure if the primary system fails.

Additional support for the membrane must also be provided where covering structures that have occupant loads of at least 50 and in those cases where covering swimming pools.

Section 3103 *Temporary Structures*

3103.1 General. Special requirements that apply to temporary structures are included in Section 3103. For applying the applicable code provisions, "temporary" is considered as being erected for a period of less than 180 days. The *International Fire Code* (IFC) also has a significant number of provisions applicable to such structures. Additionally, the IBC identifies key criteria that must be applied to temporary structures, including conformity with structural strength, fire safety, means of egress, accessibility, light, ventilation, and sanitary requirements. Permits are typically required, as are construction documents. The scope of Section 3103 includes special event structures, tents, umbrella structures, and other membrane structures. See Figure 3103-1.

Figure 3103-1 Special event structures at music festival.

Four definitions apply: "temporary structures," "public-occupancy temporary structures," "service life," and "temporary event." Public-occupancy temporary structures are new buildings or structures that are used by the public, or that support public events, where the public expects similar levels of reliability and safety as offered by permanent construction. Public-occupancy temporary structures are often assembled with reusable components and designed for a particular purpose and limited period, which is defined as a temporary event when the period is less than 1 year. Public-occupancy temporary structures in service for a period that exceeds 1 year must meet IBC requirements for new buildings.

While temporary structures are developed for use up to 180 days, many of these structures are used repeatedly at different locations. Thus, their actual service life may be on the order of 5 to 10 years. Such structures are typically subjected to repeated assembly and dismantling with associated wear and tear. Therefore, service life for temporary structures is defined to provide a consistent basis of reliability relative to that of new buildings, and a service life of 10 years is assumed for public-occupancy temporary structures when determining snow and wind load reductions in Section 3103.6.

3103.1.1 Extended period of service time. This section deals with the extension of a public-occupancy temporary structure's service time. The owner and their "qualified person" are to perform any additional inspections required by the building official, and ongoing inspections every 6 months. For the ongoing inspections, the qualified person must keep records should the building official wish to review them. These provisions are comparable to the process for special inspections in Chapter 17, where a building official relies on a special inspector or agency for many of the details of construction. Relocation of a public-occupancy temporary structure (not all temporary structures) requires the owner to apply for a new permit. A request for an extension must be submitted to the building official, and reports resulting from inspections by the qualified person and the registered design professional's examination must be submitted along with the request.

3103.6 Structural Requirements. This section differentiates between temporary structures and its subcategory public-occupancy temporary structures. The latter has specific provisions for structural loads, foundations, installation and maintenance inspection, durability and maintenance of reusable components, serviceability, and controlled occupancy procedures.

Temporary nonbuilding structures (such as stands supporting lights, audio equipment, cameras, etc.) that are associated with either public assemblies or special event structures need to be assigned a risk category consistent with the risk category of the public assembly. Applicable occupant load determination per Section 1004.5 or 1004.6 is to be applied over the assembly area within a radius equal to 1.5 times the height of the temporary nonbuilding structure. Determination of the risk category for each temporary nonbuilding structure in a public assembly area is to be done for each structure individually, not cumulatively, as it is unlikely that simultaneous failure of multiple structures will occur during an event. See Figure 3103-2 for an example.

3103.6.1 Structural loads. Public-occupancy temporary structures should not pose more risk to occupants than permanent structures, but because the code's design-level environmental loads are far less likely during a temporary event, these provisions reduce certain structural load requirements for a consistent level of risk. Minimum uniformly distributed live loads and concentrated loads can be reduced where a registered design professional calls for such reductions based on a rational approach.

3103.6.1.1 Snow loads. Recognizing the relatively short service life of temporary structures, the ground snow load can be reduced to reflect the comparatively low probability that the ASCE 7 *Minimum Design Loads and Associated Criteria for Buildings and Other Structures*

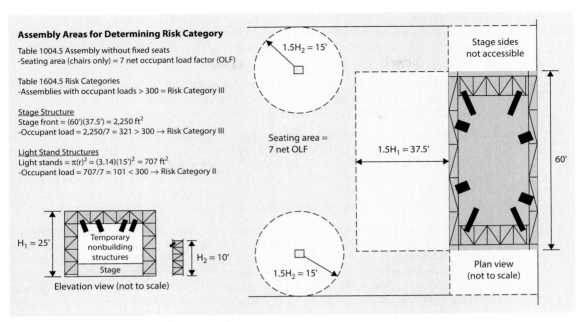

Figure 3103-2 **Risk category determination example.**

ground snow loads will occur during the shorter service life of a temporary structure. If controlled occupancy procedures are implemented, an even larger reduction of ground snow loads is permitted. If the service life of the temporary structure will not occur during months when snow is to be expected, snow loads need not be considered.

3103.6.1.2 Wind loads. In hurricane-prone regions outside of hurricane season, the basic wind speed can be prescriptively set at 115 mph for RC II, 120 mph for RC III, and 125 mph for RC IV structures. For nonhurricane areas, the wind load reduction factors shown in Table 3103.6.1.2 are believed to provide a reasonable level of reliability, given the ability to evacuate or modify temporary structures for strong wind events. If controlled occupancy procedures are implemented, an even larger reduction of wind loads is permitted.

3103.6.1.3 Flood loads. Public-occupancy temporary structures need not be designed for flood loads specified in Section 1612. However, controlled occupancy procedures per Section 3103.8 must be implemented.

3103.6.1.4 Seismic loads. Seismic loads on public-occupancy temporary structures assigned to Seismic Design Categories C through F can be taken as 75 percent of those required by Section 1613, which results in reduced seismic performance relative to permanent structures. This is consistent with the reduction for the evaluation and upgrade of existing buildings and will result in a similar seismic risk to occupants.

3103.6.1.5 Ice loads. Ice loads on public-occupancy temporary structures can be determined with a maximum nominal thickness of 0.5 inch (13 mm), for all risk categories. Where the public-occupancy temporary structure is not subject to ice loads or not constructed and occupied when ice is expected, ice loads need not be considered. However, when the public-occupancy temporary structure is in service when ice is expected, the design must be reviewed and modified to account for ice loads, or controlled occupancy procedures per Section 3103.8 must be implemented.

3103.6.1.6 Tsunami loads. Public-occupancy temporary structures in a tsunami design zone do not have to be designed for tsunami loads specified in Section 1615. However, controlled occupancy procedures per Section 3103.8 shall be implemented.

3103.6.2 Foundations. Public-occupancy temporary structures can be supported on the ground with temporary foundations if approved by the building official. Consideration is needed for the impacts of differential settlement. Presumptive load-bearing values for structures supported on pavement, slab on grade, beach sand, or grass cannot be assumed to exceed 1,000 pounds per square foot unless otherwise determined by a registered design professional. Table 1806.2 soil bearing values are applicable for other typical supporting soil conditions.

3103.6.3 Installation and maintenance inspections. Temporary structures are designed to be assembled and disassembled and transported to different locations as components or as modules. They may be in service during varying weather conditions. Components may be damaged during transportation or installation. Components also may have been manufactured more than a decade prior to the latest use. Therefore, unlike a new structure that is typically constructed with new building materials and components that were not previously used, components for temporary structures need to be inspected regularly, and suitability for reuse needs to be assessed. This is typically done by the installation crews, and this is similar to bleachers regulated by ICC 300, *Standard on Bleachers, Folding and Telescopic Seating, and Grandstands*. The qualified person assigned to perform the regular inspections must be identified by the owner.

3103.8 Controlled occupancy procedures. Provisions for controlled occupancy address cases where an environmental loading hazard cannot be reasonably mitigated and allow for actions based on a preapproved action plan that a building official may use to permit installations that cannot resist code-prescribed loads. For example, public-occupancy temporary structures subject to high wind loads may be evacuated and sections of the structure can be removed to reduce the wind load and prevent components from becoming projectiles. Where such plans are required, they must comply with ANSI ES1.7 *Event Safety Requirements—Weather Preparedness* and an operations management plan per ANSI E1.21 *Entertainment Technology—Temporary Structures Used for Technical Production of Outdoor Entertainment Events* shall be submitted to the building official for approval as a part of the permit documents. ANSI E1.21 includes requirements for monitoring weather forecasts for high winds, tornadoes, thunderstorms, lightning, and other severe conditions, as well as a requirement for mitigating actions for ice and snow to be specified in the operations management plan.

Section 3104 *Pedestrian Walkways and Tunnels*

This section regulates connecting elements between buildings, such as tunnels or pedestrian walkways, that are utilized for occupant circulation. The provisions of this section are only applicable to such tunnels and walkways designed primarily as circulation elements, typically for weather-protection purposes. A covered walkway or bridge connecting two buildings is the most common example of a pedestrian walkway. These elements may be located below, at, or above grade. When in compliance with this section, pedestrian walkways and tunnels are not considered to contribute to the floor area or height of the connected buildings. In addition, those buildings connected by the pedestrian walkway or tunnel are permitted to be considered separate structures. These allowances establish the primary reason for the use of this section. Where multiple structures are attached, they would generally be considered by the code to be a single building and regulated as such. These provisions not only allow each building to be regulated independently, but also limit the requirements of the tunnel or pedestrian walkway to those provisions of this section. A common use of pedestrian walkways is the connection of buildings that are on separate lots, including situations where the buildings are on opposite sides of a public way. Often referenced as *skyways* in northern climates, pedestrian walkways allow for a method to connect buildings across lot lines without the normally mandated fire-resistance-rated exterior walls and opening prohibition associated with a fire separation distance of zero. The pedestrian walkway is treated as a *nonbuilding* and therefore is not regulated where it crosses the lot line. Instead, this approach allows for a level of fire protection at the connection between the buildings and the pedestrian walkway.

It is important to understand that this section is essentially voluntary in application, and is only utilized where the design professional chooses not to consider the walkway and connected buildings a single structure. There is always the option of regulating the entire structure as a single building, in which case this section would have no application. See Figure 3104-1.

This section establishes specific requirements for the protection of walls and openings between the connected buildings and the pedestrian walkway or tunnel, specifies minimum and maximum widths, and limits the length of exit access travel within a pedestrian walkway or tunnel. Section 3104.3 further requires that pedestrian walkways be constructed of noncombustible materials or of fire-retardant-treated wood unless all connected buildings are of combustible construction.

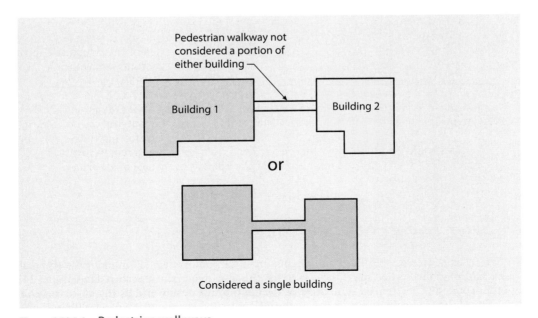

Figure 3104-1   **Pedestrian walkways.**

Section 3106 *Marquees*

A marquee is defined in Section 202 as a canopy with a flat, or relatively flat, roof surface located in close proximity to operable building openings. A canopy is defined as a permanent roof structure or architectural projection that is structurally independent or attached to and supported by a building. A classic example of a marquee is the theater marquee or the entrance marquee at a hotel. The restrictions placed on the construction, projection, and clearances for marquees are intended to prevent interference with:

1. The free movement of pedestrians.
2. Trucks and other tall vehicles using the public street.
3. The fire department in its fire-fighting operations at a building.
4. Utilities.

Thus, the height, construction, and location of the marquee are regulated to meet the intent of the code. Figure 3106-1 depicts the permissible projection and clearances required for marquees.

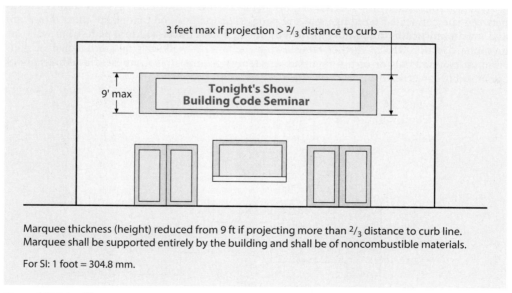

Figure 3106-1 **Marquees.**

Section 3111 *Solar Energy Systems*

Section 3111 addresses specific requirements for both photovoltaic (PV) and solar thermal systems. Most of this section refers the user to the structural requirements covered in Chapter 16, or other codes and referenced standards for the appropriate design and listing of solar PV or solar thermal systems. In addition, this section clarifies that roof-mounted solar PV systems and building-integrated PV systems must meet the fire classification requirements of

Section 1505 and that appropriate access and pathways are to be provided in accordance with Section 1205 of the IFC.

Separate provisions are included for ground-mounted versus elevated PV systems in order to establish appropriate fire testing and listing criteria for overhead PV support structures that could have people or vehicles in the space beneath them. Often referred to as "solar shade structures," overhead PV support structures are most commonly constructed over vehicle parking spaces of surface parking lots and on the uppermost level of parking garages. In addition, such structures are built in a variety of other locations with or without vehicles parked beneath. An associated definition indicates that these elevated PV support structures are to be provided with a clear height of 7 feet 6 inches.

Section 3113 *Relocatable Buildings*

Section 202 defines a relocatable building as "*a partially or completely assembled building constructed and designed to be reused multiple times and transported to different building sites.*" Construction trailers, modular school buildings, and many other prefabricated buildings would likely be classified as relocatable buildings. The only exception applies to a manufactured housing unit that is used as a dwelling. All relocatable buildings must comply with this section as well as Chapter 14 of the *International Existing Building Code®* (IEBC®).

When proposing to install a relocatable building for the first time, or when moving a relocatable building to a new site, a permit must be obtained from the building official. When doing so the applicant is required to submit an application in accordance with Section 105, provide construction documents per Section 107, and provide additional information as outlined in Section 3113.2. Key elements of this supplemental information include the manufacturer's design drawings and applicable structural design loads considered in the design. In addition, Section 3113.3 requires that the manufacturer's data plate be attached on, or near, the electrical panel and that the information on this data plate coincides with the information submitted in the permit application.

Section 3114 *Intermodal Shipping Containers*

The re-use of intermodal shipping containers as buildings or structures has become quite common. Section 3114 addresses the minimum safety requirements that should be considered in relation to these containers without duplicating existing code provisions. This section clarifies additional dimensional and structural requirements that need to be included with the construction documents. In addition, it requires that a third-party inspection be performed prior to re-purposing the container. This third-party inspection will verify that a data plate is attached to the container and that all appropriate information is known prior to retrofitting the container for use within a building or structure. Several other requirements are outlined, including specific structural requirements for shipping containers that conform to ISO 1496-1 *Series 1 Freight Containers—Specification and Testing - Part 1: General Cargo Containers for General Purpose.*

KEY POINTS

- Membrane structures include those buildings that are air-supported, air-inflated, cable- or frame-supported, and tensile-supported.
- Noncombustible membrane structures are classified as Type IIB construction, whereas other membrane structures are considered Type V, except heavy timber frame-supported structures which are classified as Type IV-HT.
- Air-supported or air-inflated membrane structures are regulated for equipment, standby power, and support.
- Special requirements apply to temporary structures which can remain in service for a period of less than 180 days. Public-occupancy temporary structures have similar levels of reliability and safety as offered by permanent construction and if in service for a period that exceeds 1 year must meet IBC requirements for new buildings.
- Complying pedestrian walkways and tunnels are not considered to contribute to the floor area or height of the connected buildings.
- Marquees are to be of noncombustible construction and have dimensional and location limitations.
- Relocatable buildings are required to comply with Section 3113 as well as IEBC Chapter 14.
- Intermodal shipping containers used for, or within, buildings and structures require special documentation in addition to a third-party inspection confirming the data plate information prior to re-use.

CHAPTER

32

ENCROACHMENTS INTO THE PUBLIC RIGHT-OF-WAY

Section 3201 General
Section 3202 Encroachments
Key Points

> By using the language "encroachments into the public right-of-way," the code refers to the projections of appendages and other elements from the building that are permitted to project beyond the lot line of the building site and into the public right-of-way. The intent of the code is that projections from a building into the public right-of-way shall not interfere with its free use by vehicular and pedestrian traffic, and shall not interfere with public services such as fire protection and utilities.

Section 3201 *General*

The *International Building Code*® (IBC®) requires that any construction that projects into or across the public right-of-way be constructed as required by the code for buildings on private property. Furthermore, the intent is that the provisions of this chapter are not considered so as to permit a violation of other laws or ordinances regulating the use and occupancy of public property. Many jurisdictions have ordinances regulating the use and occupancy of public property that go beyond the provisions of this chapter, many of which may be more restrictive and not permit the projections permitted by the IBC. Where there are no other ordinances that prohibit such construction, it is the intent of this chapter to permit the construction of the connecting structure between buildings either over or under the public way.

These types of permanent uses of the public way are often permitted by the jurisdiction under a special license issued by the public works agency. Where these uses create no obstruction to the normal use of the public way, the jurisdiction, by authorizing connecting structures between buildings on either side of the public way, can derive revenue from a licensing agreement, which can be of assistance in maintaining the public right-of-way.

The code prohibits roof drainage water from a building to flow over a public walking surface, as well as any water collected from an awning, canopy, or marquee, and condensate from mechanical equipment. The intent is not necessarily to prevent drainage water from flowing over public property in general, but rather to prevent drainage water from flowing over a sidewalk or pedestrian walkway that is between the building and the public street or thoroughfare.

There are at least two problems that arise when drainage water from a building or structure is allowed to flow over a public sidewalk:

1. Under proper conditions of light and temperature, algae will form where the water flows across the sidewalk and create a hazardous, slippery walking surface.

2. During heavy rain storms, the velocity and force of the water emitting from the drain can create hazardous walking conditions for pedestrians.

Therefore, the usual procedure is to carry the drain lines from the roof or other building elements inside the building through the wall of the building and under the sidewalk through a curb opening into the gutter.

Section 3202 *Encroachments*

The projection of any structure or appendage is generally regulated in regard to its relationship to grade. Those encroachments found below grade, such as structural supports, vaults, and areaways, are governed in one fashion. Those encroachments located above grade, but below 8 feet (2,438 mm) in height, are more severely controlled because of their immediate location adjacent to public areas. Although projections that are located 8 feet (2,438 mm) or more above grade tend to have fewer implications, such encroachments are also regulated.

3202.1 Encroachments below grade. Footings, foundations, piles, and similar structural elements that support the building and are erected below grade are not permitted to project beyond the lot lines. There is an allowance for footings that support exterior walls, provided the projection does not extend more than 12 inches (305 mm) beyond the street lot line, and that the footings or supports are located at least 8 feet (2,438 mm) below grade. In this exception, the code assumes that there will be no utilities or other public facilities below this depth adjacent to the building and, therefore, permits footings to project a limited amount where they are below the 8-foot (2,438-mm) depth. It should be noted that this encroachment is permitted for street walls only and does not apply to interior lot lines. See Figure 3202-1.

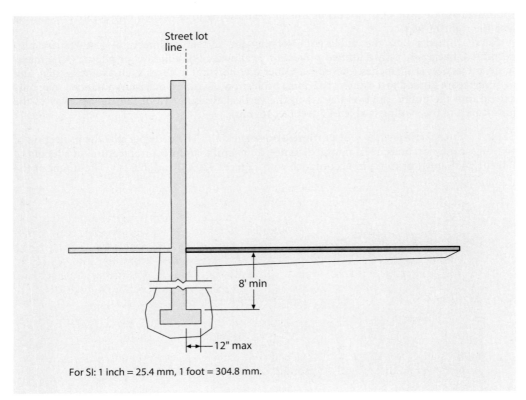

Figure 3202-1 **Encroachments below grade.**

Where vaults, basements, and other enclosed spaces are located below grade in the public right-of-way, they shall be further regulated by the ordinances and conditions of the applicable governing authority. In many cases, the space below the sidewalk or other area adjacent to the building in the public right-of-way is not used for public purposes such as utilities, sewers, and storm drains (except for catch basins); therefore, where local ordinances do not prohibit the use of space beneath the sidewalk or other public area, the code permits its use by adjoining property owners. Usually the basement or other enclosed space of an adjoining building is extended beneath the public right-of-way and is used for any of the uses permitted by the code. This is a revocable permission, as the jurisdiction may find at a later date that there is a public need for the public use of the space beneath grade.

3202.2 Encroachments above grade and below 8 feet in height. Building components that may arbitrarily project into the public right-of-way pose a considerable hazard to pedestrians or other users of the right-of-way. Therefore, doors and windows are prohibited from opening into such an area. However, certain features of the building that are permanent in nature are permitted to encroach into the public right-of-way to a limited degree. Where steps are permitted by other provisions of the IBC, they may project into the right-of-way up to 12 inches (305 mm), provided they are protected by guards at least 3 feet (914 mm) high or similar protective elements such as columns or pilasters. The intent of the provision is that the encroachment be small enough so as not to obstruct pedestrian or vehicular traffic and yet provide some protection for occupants of the building to look each way before proceeding into the right-of-way.

Encroachments into the public right-of-way for certain architectural features are also permitted, but again, only a limited projection is allowed. For columns or pilasters, the maximum projection is 12 inches (305 mm). Other architectural features such as lintels, sills, and pediments are limited to a 4-inch (102-mm) projection. Awnings, including valances, may only extend into the public right-of-way when the vertical clearance from the right-of-way to the lowest part of the awning is at least 7 feet (2,134 mm).

3202.3 Encroachments 8 feet or more above grade. The code permits the projection of awnings, canopies, marquees, signs, balconies, and similar architectural features at a height of 8 feet (2,438 mm) or more in accordance with Figures 3202-2 and 3202-3. The intent of the

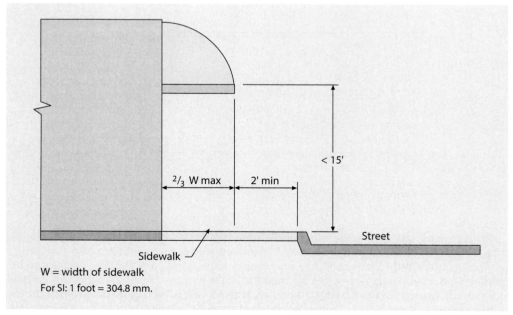

Figure 3202-2 **Awning, canopy, marquee, and sign projections.**

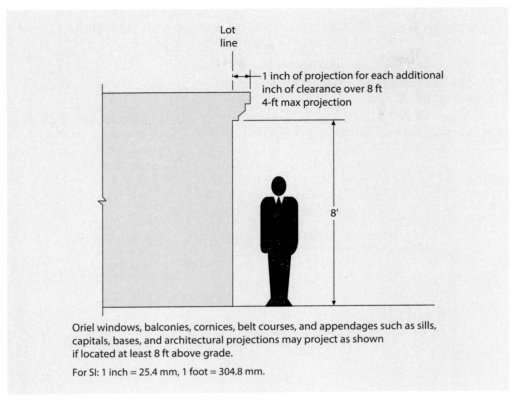

Figure 3202-3 **Balcony and appendage projections.**

code is that no projection should be permitted near the level of the public right-of-way and up to 8 feet (2,438 mm) in height so that the free passage of pedestrians along the sidewalk or other walking surface will not be inhibited. At the 8-foot (2,438-mm) height and above, projections are permitted as long as they do not interfere with public utilities. It is generally assumed that utility lines for telephones or power will not occupy this zone except for the service entrances to the buildings. There are jurisdictions that have high-voltage power lines running along the sidewalk, and the regulations of the agency that regulates the power companies generally require certain clearances from these lines. Therefore, in addition to the requirements shown in Figures 3202-2 and 3202-3, power-line clearances should also be checked, and the requirements of the *National Electrical Code*® should be reviewed when it is adopted by the jurisdiction. Where such awnings, canopies, balconies, and similar building elements are located 15 feet (4,572 mm) or more above grade, their encroachment is unlimited.

It is not uncommon to find a pedestrian walkway or a similar circulation element to be constructed over a public right-of-way. Such a condition is permitted subject to the approval of the local authority having jurisdiction. In no case, however, should the vertical clearance between the right-of-way and the lowest part of the walkway be less than 15 feet (4,572 mm). A clearance of at least 15 feet (4,572 mm) allows not only for unobstructed pedestrian travel but for vehicular access as well.

KEY POINTS

- Encroachments into an adjoining public right-of-way are permitted both above and below grade under specified conditions.
- Where located below grade, structural supports, vaults, and other enclosed spaces are regulated.
- Where located above grade, the types of encroachments addressed include steps, awnings, canopies, marquees, signs, architectural features, and mechanical equipment.

CHAPTER 33

SAFEGUARDS DURING CONSTRUCTION

Section 3301 General
Section 3302 Owner's Responsibility for Fire Protection
Section 3303 Demolition
Section 3304 Site Work
Section 3306 Protection of Pedestrians
Section 3307 Protection of Adjacent Property
Section 3308 Temporary Use of Streets, Alleys, and Public Property
Section 3309 Fire Extinguishers
Section 3310 Means of Egress
Section 3311 Standpipes
Section 3313 Water Supply for Fire Protection
Section 3314 Fire Watch During Construction
Key Points

> This chapter deals with safety practices during the construction process, as well as the protection of property, both public and private, adjacent to the construction site. Included in the provisions are requirements addressing the owner's responsibility for fire protection; demolition; site work; protection of pedestrians; and temporary use of streets, alleys, and public property. Special issues relating to fire extinguishers, exits, standpipes, automatic sprinkler systems, water supply for fire protection and fire watches are also addressed.

Section 3301 *General*

Where an existing building continues to be occupied during the process of remodeling or constructing an addition, it is very important that the life- and fire-safety features of the building continue to be in place. Required exits shall be maintained, the strength of structural elements shall not be diminished, fire-protection devices are to remain in working condition, and sanitary facilities shall be fully functional. These features must be maintained at all times during any remodeling, alterations, repairs, or additions to the building under consideration. Obviously, there will be times when the specific protective feature is the element or device being altered or repaired. In this case, an equivalent method shall be provided to safeguard the occupants. Chapter 33 of the *International Fire Code*® (IFC®) is referenced for fire safety during construction.

Section 3302 *Owner's Responsibility for Fire Protection*

The owner or the owner's authorized agent is responsible for maintaining on-site safety and is required to conduct daily fire safety inspections to enhance site fire safety. The person obligated to perform the daily inspections is the site safety director. The duties and responsibilities of the site safety director include daily inspection of common fire safety hazards and ensuring that contractors working on-site are trained and following operational safety requirements.

These provisions are also found in the *International Fire Code* (IFC) and *International Existing Building Code* (IEBC). Provisions specify that the daily inspections are documented and available for review until the certificate of occupancy is issued. By requiring daily inspections and documentation, any fire or building official can request to see documentation of daily inspections when at the site for any reason, and a clear enforcement path is specified when noncompliance is encountered.

Section 3303 *Demolition*

As the work of demolishing a building is subject to many variations and hazards, the *International Building Code*® (IBC®) authorizes the building official to require the submission of plans and a complete schedule of the demolition. Under certain circumstances, pedestrian protection may be required during the demolition process. Such protection shall be provided in compliance with other provisions of Chapter 33. For some multilevel buildings and for certain types of demolition operations, it may also be necessary to temporarily close the street. Once the structure has been demolished, any resulting excavation is to be filled consistent with the existing grade, or in any other manner in conformance with the ordinances of the jurisdiction. As with construction, Chapter 33 of the IFC is referenced for fire safety during demolition.

Section 3304 *Site Work*

The provisions in Section 3304.1 for excavation and fill are intended to apply to the specific area where the building or structure will be located. This section is not intended to address massive grading on a site. For those requirements, one would turn to Section 1804, as well as any local grading ordinance that might be in effect. To prevent decay and to eliminate an avenue of entrance for termites and other insects, the code requires that the area of the site occupied by the building be free of all stumps, roots, and any loose or casual lumber. Typically applicable in the construction of wood-framed structures, the requirements also address wood forms that have been used in placing concrete, as well as any loose or casual wood that might be in direct contact with the ground under the building.

The IBC limits the slopes for permanent fills or cut slopes for permanent excavations for structures to two units horizontal in one unit vertical (50-percent slope). Although steeper-cut slopes are permitted where substantiating data are submitted, the limitation on filled slopes of two units horizontal in one unit vertical (50-percent slope) is an absolute limitation, and filled slopes may not be steeper. Although cut slopes into natural soils may be excavated with a slope steeper than 50 percent, depending on the nature of the soil materials and the density, the code reasons that fill slopes must not be steeper than 50 percent to cover unprecedented circumstances such as heavy rains creating overly saturated soils or, in the case of seismic activity, vibration of the soils during an earthquake causing failures of steep-filled slopes.

Understandably, the code requires that fill or surcharge loads should not be placed adjacent to an existing building or structure unless the existing building or structure is structurally capable of resisting the additional loads caused by the fills or surcharge. See Figure 3304-1. Alternatively, the existing building can be strengthened in order to resist the additional loading.

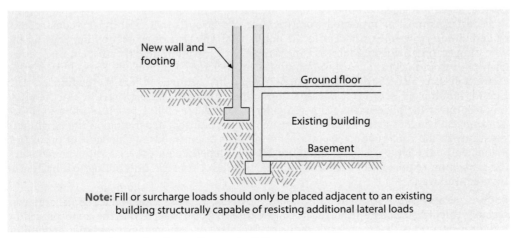

Figure 3304-1 **Surcharge load on adjacent building.**

The code requires that existing footings or foundations that may be affected by an excavation be adequately underpinned or otherwise protected. For obvious reasons, it is the intent of the code that excavations in close proximity to an existing footing shall not be made unless proper protective measures are taken for the existing building. These measures may involve underpinning of the existing foundations, or shoring and bracing of the excavations, so that the existing building foundations will not settle or lose lateral support.

Where the excavation is for a new building foundation and the new footing is at an elevation below, but within reasonably close proximity to, an existing foundation, underpinning is the usual procedure to protect the existing foundation. If the existing footing is for the structure that creates a horizontal thrust, the means of providing lateral support may take the form of a buttressed retaining wall designed to resist the lateral thrust of the existing foundation.

Where buildings are to be supported by fill, the code requires that the fill be placed in accordance with accepted engineering practice. Thus, fill utilized for the support of buildings must be designed by a geotechnical engineer (soils engineer) utilizing the principles of geotechnical engineering so as to provide a proper and adequate foundation for the structure above, and one that will limit settlements to tolerable levels. In order to verify the adequacy of the geotechnical design of the fill, the IBC permits the building official to ask for a soil investigation report outlining the geotechnical design of the fill materials, as well as a report of the satisfactory placement of the fill.

Section 3306 *Protection of Pedestrians*

Both Section 3306 and Chapter 32 regulate the use of public streets and sidewalks. Section 3306 provides those general regulations and criteria for the temporary use of public property, which are generally found to apply to most jurisdictions.

As the nature and philosophy of each jurisdiction varies, so do the regulations each promulgates regarding the use of public property and, in particular, the use of streets and sidewalks. Adjacent property owners have rights of access to their property. The public street also provides access for public services such as fire and police protection, street sweeping and maintenance, street lighting, trash pickup, and other services provided by the jurisdiction or its contractors. Utilities serving adjacent property also use the public street for access.

Pedestrians must be protected from the potential hazards that exist during construction or demolition operations adjacent to the public way. The type of protection depends on the type of operation being conducted. For example, an excavation directly adjacent to a pedestrian path would necessitate a minimum of a construction railing along the side facing the excavation. It is also possible that a barrier would be required. Where the pedestrian path extends into the public street, a construction railing is required on the street side of the walkway to protect the pedestrians from vehicular traffic. Table 3306.1 provides the criteria for determining whether or not additional protection is required, depending on the height and proximity of the construction operations to the pedestrian walkway. The level of protection mandated by the table is shown in Figure 3306-1. Depending on these various parameters, the protection required will vary from none or merely a construction railing to a barrier or covered walkway.

3306.2 Walkways. A public jurisdiction provides sidewalks and streets for the free passage of vehicular and pedestrian traffic. In the case of pedestrian traffic, the usual procedure is to provide a sidewalk on each side of the street. Therefore, when construction or demolition operations are conducted on property adjacent to the sidewalk, the code requires a walkway at

Protection of Pedestrians 951

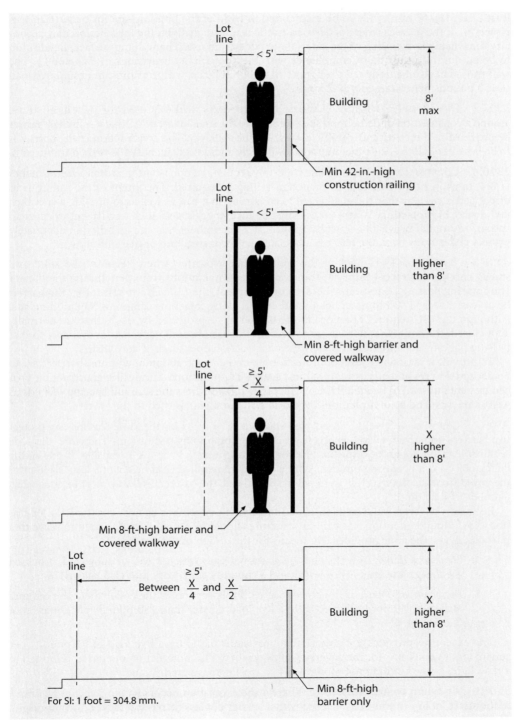

Figure 3306-1 **Pedestrian protection.**

least 4 feet (1,219 mm) wide to be maintained in front of the building site for pedestrian use. However, in those cases where pedestrian traffic is unusually light, the jurisdiction may authorize the fencing and closing of the sidewalk to pedestrian use. The walking surface, in addition to being durable, must be in compliance with the accessibility provisions of Chapter 11. The walkway shall also be designed to support all imposed loads, with a minimum design live load of 150 pounds per square foot (7.2 kN/m^2).

3306.3 Directional barricades. Where the temporary walkway used for pedestrian travel around a construction or demolition site extends into the street, it is critical that a sufficient barrier be provided to direct vehicular traffic away from the walkway. The construction of the barrier is to be large enough to ensure that motorists will easily and quickly identify the revised traffic path.

3306.4 Construction railings. The code only requires an open-type guardrail 3 feet, 6 inches (1,067 mm) in height where a construction railing is required. The intent of the railing is to direct pedestrians as they travel adjacent to construction areas, as well as to provide a very limited degree of protection. Where access must be further regulated, such as adjacent to an excavation, additional provisions must be applied. It is incumbent that the jurisdiction adequately protect pedestrians from the hazards associated with construction or demolition work.

3306.5 Barriers. The intent of the code is that a barrier, where required, be solid and sturdy enough to prevent impacts from construction operations from penetrating the barrier and injuring passing pedestrians. Because of this intent, the code also requires that the barrier be at least 8 feet (2,438 mm) in height and extend along the entire length of the building site. Although the IBC requires openings in the barrier to be protected by doors that are normally kept closed, the code does not intend to prevent the use of small viewports at eye level so that passing pedestrians may stop and view the construction operations if they wish.

To provide a reasonable level of protection from the construction operations, barriers are to be designed to resist loads as required in Chapter 16. As an option, specific construction procedures are outlined in Section 3306.6. The procedures address the size and spacing of studs as well as the size and span limitations for wood structural panels used in the barrier.

3306.7 Covered walkways. A covered walkway is required for the same reasons as a protective barrier—to prevent falling objects from endangering passing pedestrians. Therefore, the code requires that the design be such that the roof and supporting structure is capable of preventing falling objects from breaking through. With a minimum required clear height of 8 feet (2,438 mm) measured from the floor surface to the canopy overhead, the covered walkway must be adequately illuminated at all times.

Because a covered walkway is intended to protect pedestrians against construction operations, including demolition, the canopy structure should be structurally designed to serve that purpose. Thus, the code provides two means for the design of the canopy:

1. *Structural design* to withstand the actual loads to which it will be subjected, provided the design live load is not less than 150 pounds per square foot (7.2 kN/m^2).

2. *Prescriptive provisions* in accordance with this section of the code where the covered walkway will not be subjected to a live load greater than 150 pounds per square foot (7.2 kN/m^2).

The exception provides for prescriptive provisions based on a live load of 75 pounds per square foot (3.6 kN/m^2) for the covered walkway where the construction operation is limited to a new, light-frame building not more than two stories above grade plane.

3306.9 Adjacent to excavations. Where a site excavation occurs within 5 feet (1,524 mm) of the street lot line, it must be enclosed with a barrier not less than 6 feet (1,829 mm) in height. The barrier must be constructed to resist the wind pressures specified in Chapter 16, but need not meet the general barrier requirements of Section 3306.6. The building official also has the authority to require a barrier around excavations in those cases where the excavation is more than 5 feet (1,524 mm) from the street lot line.

Section 3307 Protection of Adjacent Property

It is critically important that during construction, remodeling, or demolition work, any adjacent public or private property be protected from damage. Essentially, all portions of the neighboring structure must be protected from damage, including footings, foundations, exterior or party walls, chimneys, skylights, roofs, and other building elements. In addition, water run-off must be controlled to prevent erosion. Although there are no specific requirements laid out in the IBC for the level of protection required, the intent of the protection is based on common-law precepts, such as that of lateral support where it comes to footings and foundations. Over the years, common law and, more recently, statute law have established the requirements for lateral support. That is, the owner of a piece of real property shall be entitled to the lateral support of the property by adjacent property. Responsible practice would also dictate a satisfactory level of protection for other portions of the adjacent site and building during the construction process.

Where excavation work is to be performed, the person responsible for such work shall notify the owners of any adjacent property or buildings. The written notice, to be delivered no less than 10 days prior to the initiation of the excavation work, shall indicate that excavation work is to take place and that protection of the adjacent property should be considered. Appropriate action should be taken where access to an adjacent property is desired in order to provide the proper protective measures. If shoring or similar excavation retention systems are required, they shall be designed by a registered design professional and monitoring shall be provided to ensure that adverse movement is not occurring within the retention system or adjacent property or structures.

Section 3308 Temporary Use of Streets, Alleys, and Public Property

It is important that access to essential facilities be maintained where construction materials and equipment are temporarily stored on public property. Obstructions shall not block access to fire protection features such as standpipes, fire hydrants, or alarm boxes, as well as any catch basins or manholes that might be present. In regard to maintaining safe and effective vehicular traffic in the area where construction materials and equipment are stored, such materials or equipment must be located at least 20 feet (6,096 mm) from the street intersection, or otherwise located where sight lines are not obstructed.

Section 3309 Fire Extinguishers

It is quite common for combustible debris and waste materials to accumulate in and around a building under construction. Therefore, at least one approved portable fire extinguisher sized for at least ordinary hazard and complying with Section 906 is to be provided at every stairway on all floor levels where combustible material has accumulated. In addition to those extinguishers required within the building, every storage shed and construction shed shall also contain an approved fire extinguisher. Additional extinguishers may be required by the building official where any special or unique hazards exist, such as the presence of flammable or combustible liquids.

Section 3310 *Means of Egress*

Once a building under construction reaches a height of 40 feet (12,192 mm) at least one stairway must be available and usable for egress purposes. The temporary stairway shall be adequately illuminated during those times where there are occupants who may need to utilize the stairway. A temporary stairway need not be provided where there is at least one permanent stairway that is maintained and usable as the construction progresses.

Section 3311 *Standpipes*

For those buildings required to have a permanent standpipe system per Section 905.3.1, during construction operations a standpipe system must also be installed before the construction height exceeds 40 feet (12,192 mm) above the lowest level of fire department access. As construction continues to proceed upward, the standpipes shall also be extended in a timely manner. A standpipe shall always be available within one floor of the highest point of construction having secured decking or flooring. During construction operations, the amount of combustible materials from concrete forms, scaffolding, plastic and canvas tarpaulins, and other materials may be prevalent not only throughout the building, but throughout the construction site itself. Thus, in many cases, the standpipe system provides the only source of water for fire-fighting operations.

A minimum of one standpipe is required, with hose connections to be located in accessible areas adjacent to usable stairs. The standpipes may be either temporary or permanent, but in all cases shall meet the minimum requirements of Section 905 for capacity, outlets, and materials. The water supply may also be either temporary or permanent, as long as it is available at the first sign of combustible material accumulation.

Section 3313 *Water Supply for Fire Protection*

An "approved" water supply is to be provided at the site prior to the arrival of combustible materials. However, it is quite common that the fire protection water infrastructure is not in place by the time vertical construction of the building commences. Section 3313 highlights the need for ensuring that either a permanent or temporary fire protection water source is provided for projects that utilize combustible building materials, with a minimum fire protection water supply of 500 gallons per minute available once combustible materials are brought to the site.

Section 3314 *Fire Watch During Construction*

Some of the most hazardous conditions related to buildings often are present during the construction process. Fires that have occurred at construction sites during times of no activity have demonstrated the need for early notification that can only be provided by fire watch personnel. The lack of fire alarm and detection devices during the construction process requires an alternative approach to identifying and communicating the presence of a fire event. In order to protect adjacent properties from fire in a building under construction, provisions give authority to the code official to require a fire watch during those hours where no construction work is being done.

Fires in sizable buildings under construction have the potential for significant heat release due to the fire loading created by building components and other materials used in the building's construction. For this reason, a fire watch is to be provided where any one of three conditions exist: (1) the height of construction exceeds 40 feet (12,192 mm) above the lowest adjacent grade, (2) multistory construction has more than 50,000 square feet (4,645 m^2) per story, or (3) as required by the fire code official.

It is expected that the requirement applies only to new construction. It is not intended for the provisions to be applied to alterations and other types of minor construction activity. Existing buildings would be regulated by a comprehensive fire-safety plan. Although the potential for a sizable fire load requiring implementation of a fire watch program would be more probable for a building of combustible construction, there are no conditions based on the building's construction type.

KEY POINTS

- Where an existing building continues to be occupied during the process of remodeling or constructing an addition, it is very important that the life- and fire-safety features of the building continue to be in place.
- The owner or the owner's authorized agent is responsible for maintaining on-site safety and is required to conduct daily fire safety inspections to enhance site fire safety.
- Once the structure has been demolished, any resulting excavation is to be filled, consistent with the existing grade.
- The code requires that fill or surcharge loads not be placed adjacent to an existing building unless the existing building is structurally capable of resisting the additional loads caused by the fill or surcharge.
- Pedestrians must be protected from the potential hazards that exist during construction or demolition operations adjacent to the public way.
- Depending on various conditions, required pedestrian protection varies from none, to a construction railing, to a barrier, to a covered walkway.
- During construction, remodeling, or demolition work, any adjacent public or private property must be protected from damage.
- Fire extinguishers, means of egress, standpipes, water supply and fire watch are all important requirements for maintaining fire safety during construction.

CHAPTER 34

RESERVED

Chapter 34 ■ Reserved

This chapter is designated as "Reserved" for future use. The chapter previously contained provisions for the regulation of existing buildings. Section 101.4.7 was established to indicate that the *International Existing Building Code*® is to be applied to all matters governing the repair, alteration, change of occupancy, and addition to and relocation of existing buildings. Chapter 34 is now considered as a placeholder for any future new chapter that may be added to the IBC®.

CHAPTER 35

REFERENCED STANDARDS

The *International Building Code*® (IBC®) is, for the most part, a performance-based code. Thus, the IBC contains numerous references to standards that are intended to assist in the application of the code. Where standards are referenced in the body of the IBC, they are considered as a part of the code. Thus, when a jurisdiction adopts the IBC as its building code, it also adopts the standards identified in this chapter.

Standards can be divided into several different categories—structural engineering standards, materials standards, installation standards, and testing standards. The standards referenced in the IBC are all identified in this chapter. Organized by the promulgating agency of the standard, the chapter provides the identification, effective date, title, and the code section or sections where the standard is referenced in the IBC.

Where one of the listed standards is referenced in the IBC, only the applicable portions of the standard relating to the specific code provision are in force. For example, Chapter 11 references ICC A117.1 for the technical requirements relating to accessibility. The provisions of ICC A117.1 address accessible telephones; however, Chapter 11 does not require telephones to be accessible. Therefore, that portion of the standard is not considered as a part of the IBC, and accessible telephones are not required. Accessible telephones are regulated in Appendix E, so if the jurisdiction adopts the appendix chapter, the technical provisions of ICC A117.1 for telephones will be in force. As a reminder, Section 102.4.1 also indicates that where the provisions of the code and a standard are in conflict, the provisions of the code apply.

APPENDICES

Appendix A   Employee Qualifications
Appendix B   Board of Appeals
Appendix C   Group U Agricultural Buildings
Appendix D   Fire Districts
Appendix E   Supplementary Accessibility Requirements
Appendix F   Rodentproofing
Appendix G   Flood-Resistant Construction
Appendix H   Signs
Appendix I   Patio Covers
Appendix J   Grading
Appendix K   Administrative Provisions
Appendix L   Earthquake Recording Instrumentation
Appendix M   Tsunami-Generated Flood Hazards
Appendix N   Replicable Buildings
Appendix O   Performance-Based Application
Appendix P   Sleeping Lofts

> The appendix chapters to the *International Building Code*® (IBC®) contain subjects that have been determined to be an optional part of the code rather than mandatory, with each jurisdiction adopting all, parts of, or none of the appendix chapters—depending on its needs for enforcement in any given area. The provisions of IBC Section 101.2.1 indicate that the requirements contained in the appendices are only applicable where specifically adopted by the jurisdiction.

Appendix A *Employee Qualifications*

The provisions of this appendix are intended to assist the jurisdiction in qualifying individuals to be employed in key roles in a code compliance agency. Employee qualifications are provided for the position of building official as well as those for chief inspectors, inspectors, and plans examiners. An overview of the qualifications is shown in Table A-1.

In addition to the education and experience criteria, an important consideration is the professional qualification obtained through certification. A comprehensive certification program is available through the International Code Council® (ICC®) that recognizes individuals for their knowledge relating to code enforcement.

Position	Experience			Certification
	Total Years	Supervisory Years	Type	
Building official	10	5	Architect, engineer, inspector, contractor, superintendent of construction	Certified building official
Chief inspector				Certified inspector for appropriate trade: building, plumbing, mechanical, electrical, combination
Inspector	5	—	Types listed above plus supervisor or mechanic in charge of construction	
Plans examiner				Certified plans examiner for appropriate trade

Table A-1 **Employee qualifications.**

Appendix B *Board of Appeals*

This appendix expands on the provisions of Section 113 relating to the board of appeals. Issues dealing with the filing of an appeal, the board membership, meeting notices, and board decisions are also addressed.

This appendix specifies that the board consists of five individuals with appropriate experience and knowledge to make determinations in relation to the code. In addition, two alternate members should be appointed to serve in the absence or disqualification of a regular member. Board members shall not be employed by the jurisdiction or have an apparent conflict of interest in relation to the subject being appealed. As appropriate experience or training is needed to qualify for the board, members are often architects, engineers, and contractors.

It is important that all hearings before the board are considered open meetings and are available to all interested parties. The appellant, the appellant's representative or counsel, the building official, and all other persons who have an affected interest in the decision shall be permitted to address the board. In order to overturn or modify the decision of the building official, at least three members of the board must vote accordingly. Unless acceptable to the appellant, all five members must be present for the board to act on an appeal.

Appendix C *Group U Agricultural Buildings*

The provisions of Appendix C were developed to address the needs of those jurisdictions (primarily unincorporated county territory) whose primary development is agricultural. In these cases, agricultural property usually consists of large tracts of land on which agricultural buildings are placed, usually with large open spaces and with essentially no congestion. Therefore, the provisions for agricultural buildings classify the structures as Group U occupancies and include barns, shade structures, grain silos, stables, and horticultural structures, as well as buildings used for livestock and poultry shelters, equipment and machinery storage, and milking operations.

Because of the generally large open spaces that usually surround the buildings and the relatively low occupant load, the limitations imposed on construction, height, area, mixed uses, and exiting are generally more liberal than the requirements in the body of the code for Group F or S occupancies that would generally otherwise apply.

Appendix D *Fire Districts*

This appendix is available for adoption by jurisdictions wishing to establish fire districts. The use of fire districts provides a method to address fire hazards that are created by a variety of conditions, with the primary concerns based on occupancy and structure density. The provisions apply to new buildings built within the fire district, as well as those buildings undergoing alterations or a change of occupancy.

It is necessary to first establish the territory that is to be included within the fire district. The code identifies three basic types of areas that are of importance in the regulation of fire districts. These include adjoining blocks, buffer zones, and developed blocks. The specifics of each of these areas are illustrated in Figure D-1.

Those buildings already existing when a fire district is established may not be increased in height or area unless in compliance with the code for new buildings. Any new construction must also be of a type permitted within the fire district. Alterations may not increase the level of fire hazard in the building, nor may the occupancy be changed to a classification that is not permitted in the fire district. Any buildings located partially in the fire district are to be regulated as if they are in the district, provided at least 50 percent of the structure lies within the district, or the building extends more than 10 feet (3,048 mm) inside the fire district's boundaries.

Section D105 identifies a number of types of uses and structures permitted within a fire district that might otherwise be excluded. The listed uses are all relatively minor in nature and do not pose a significant hazard to the fire-safety level that is mandated. Such uses include small private garages and sheds, fences, tanks and towers, small greenhouses, and wood decks. This section also permits a limited amount of alteration on dwellings of Type V construction.

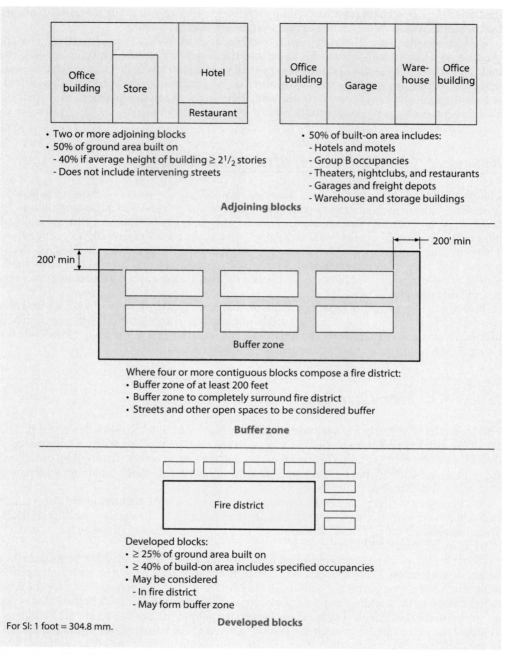

Figure D-1 Fire districts.

Appendix E *Supplementary Accessibility Requirements*

This appendix addresses the design and construction of those accessible facilities not typically addressed by a building code. Although it is important that all reasonable efforts are made to make buildings accessible and usable, the body of the code in Chapter 11 contains only those requirements that directly relate to structures. Therefore, additional criteria are provided in this appendix to expand on the other accessibility features of the built environment.

Many of the facilities regulated by this chapter involve furnishings or equipment. In Section E103.1, the code requires that an accessible route be provided to a raised platform used as a head table or speaker's lectern. Section E104.2 regulates communication features in Group I-3 occupancies and transient lodging facilities.

Portable toilet and bathing rooms, gaming machines, vending machines, mailboxes, automatic-teller machines, and fare machines are other possible features of a building that, through the regulation of this section, can be made more usable for individuals with physical disabilities. Telephones are fully addressed in Section E106, including provisions for wheelchair access, volume controls, and TTYs. Section E107.2 addresses directional and informational signs, as well as other special types of signage.

Three of the other sections of this appendix regulate specific types of uses or buildings. Bus stops and bus shelters must be designed and constructed in a manner that makes them accessible. Fixed transportation facilities, such as stations for rapid rail, light rail, commuter rail, high-speed rail, and other fixed-guideway systems, must selectively have station entrances, signage, fare machines, platforms, TTYs, track crossings, public-address systems, and clocks that are accessible or usable. Some of these same features are regulated for airports as well.

Appendix F *Rodentproofing*

In an effort to reduce the possibility of rodents entering a building, this appendix sets forth construction methods to seal those potential entry points. The provisions apply not only to habitable and occupiable rooms but also to any spaces containing feed, food, or foodstuffs. The obvious intent is to prevent unsanitary conditions and the potential spread of disease that may follow.

All openings in the foundation walls are to be covered or sealed in a prescribed manner to prevent the passage of rodents. Doors and windows are regulated when located adjacent to ground level. It is also the intent of this appendix that an apron or similar protective barrier be installed where the foundation wall is not continuous. The intent of this appendix is an attempt to eliminate all potential avenues for rodent entry that occur around the exterior of a building.

Appendix G *Flood-Resistant Construction*

Most jurisdictions in the United States have specific areas that are subject to flood conditions. This appendix is designed to reduce those losses, both public and private, that occur because of flooding. Administrative procedures and land-use limitations are set forth, intending to meet or exceed the regulations of the National Flood Insurance Program (NFIP). The importance of designating a floodplain administrator to enforce the floodplain management program is highlighted and several tasks for this administrator are laid out in the Appendix.

The building-sciences provisions for flood-resistant design and construction are located in Section 1612. In conjunction with the provisions of this appendix, the regulations are consistent with the NFIP regulations.

Appendix H *Signs*

The design and installation of outdoor signs is regulated by this appendix. This appendix classifies signs based on their location such as ground signs, roof signs, wall signs, projecting signs, marquee signs, and portable signs. The types of signs, including internally illuminated signs, combustible signs, and animated devices, are also addressed.

This appendix identifies the areas of concern when signs are placed on structures. It is important that any exit signs, fire escapes, or egress openings remain unobstructed. Required natural ventilation openings must also remain available. Signs must be able to withstand all imposed loads, including any wind or seismic loads that may be encountered. The combustibility of signs is also regulated, with specific provisions for plastic materials.

Appendix I *Patio Covers*

Applicable only to structures accessory to dwelling units, patio covers regulated by Appendix I are limited to one-story structures not exceeding 12 feet (3,657 mm) in height. Enclosure walls may have any configuration, provided the open area of the longer wall and one additional wall is equal to at least 65 percent of the area below a minimum of 6 feet, 8 inches (2,032 mm) of each wall, measured from the floor. Openings may be enclosed with insect screening, translucent or transparent plastic conforming to Sections 2606 through 2610, or glass conforming to Chapter 24.

Patio covers may be detached or attached to the dwelling units, and shall be used only for recreational, outdoor living purposes and not as carports, garages, storage rooms, or habitable rooms. Any patio cover that falls outside the established scoping limits must be designed and constructed to the standards set forth in the body of the IBC.

Exterior openings required for light or ventilation may open into a patio structure conforming to this section. Where emergency egress or rescue openings from sleeping rooms lead to a patio structure, the structure shall be unenclosed. Where an exit from the dwelling unit passes through the patio structure, the structure shall be unenclosed or exits shall be provided in conformance with Chapter 10.

Patio covers shall be designed and constructed to sustain the applicable snow loads or all dead loads plus a vertical live load of 10 pounds per square foot (0.48 kN/m^2), whichever is greater. The minimum wind and seismic loads shall also be considered in the design.

A patio cover may be supported on concrete slab on grade without footings, provided the slab is not less than 3½ inches (89 mm) thick, and further provided that the columns do not support live and dead loads in excess of 750 pounds (3.34 kN) per column.

Appendix J *Grading*

Not every jurisdiction is located in an area where the topography of the terrain requires extensive grading operations on private property. In those areas where developers need to grade private property, Appendix J provides appropriate administrative and technical regulations to assure the jurisdiction of reasonable safety against slope failure, landslides, and other soil failure hazards.

Appendix K *Administrative Provisions*

Appendix K is provided to allow those communities who adopt NFPA 70, the *National Electrical Code®* (NEC®), to include administrative provisions that will assist in their implementation and enforcement. These provisions assist in the administration of the NEC by providing administrative language that correlates with that of the *International Codes®*. In addition, the provisions established in Section K111 address technical issues that are additions or modifications to the requirements of the NEC.

Appendix L *Earthquake Recording Instrumentation*

Earthquake recording instrumentation measurements provide fundamental information needed to cost-effectively improve understanding of the seismic response and performance of buildings subjected to earthquake ground motions. The requirement only applies to newly constructed buildings of a specified size and located where the 1-second spectral response acceleration, S_1, is greater than 0.40. Because the provision is in an appendix chapter, it is not mandatory unless specifically adopted by the jurisdiction.

Appendix M *Tsunami-Generated Flood Hazards*

The areas designated on State or ASCE Tsunami Design Zone Maps are most likely to suffer significant damage during a design tsunami event. Given the potentially serious life-safety risk presented to structures within these areas, the intent of the provisions in Appendix M is to limit the presence of high-hazard and high-occupancy structures (Risk Categories III and IV) within the designated Tsunami Design Zone. Buildings within the designated hazard zone are only permitted under certain conditions. A vertical evacuation tsunami refuge is permitted when constructed in accordance with FEMA P646 *Guidelines for Design of Structures for Vertical Evacuation from Tsunamis* or where critical facilities are located within the hazard zone to fulfill their function, and they incorporate adequate structural and emergency evacuation features. Vertical evacuation is a central part of the National Tsunami Hazard Mitigation Program, driven by the fact that there are coastal communities along the West Coast of the United States that are vulnerable to tsunamis that could be generated within minutes of an earthquake on the Cascadia Subduction Zone. Vertical evacuation structures provide a means to create areas of refuge for communities in which evacuation out of the inundation zone is not feasible. The referenced FEMA guide includes information to assist in the planning and design of tsunami vertical evacuation structures. Because the provision is in an appendix chapter, it is not mandatory unless specifically adopted by the jurisdiction.

Appendix N *Replicable Buildings*

The concept of replicable building design applies where a given prototype is to be built in a variety of locations while maintaining consistent overall design parameters. The resulting replicable building is reviewed and deemed code compliant by a designated expert, allowing local jurisdictions to accept the structure as substantially code compliant. Such buildings could be approved for construction without a complete plan review, with the construction documents examined by the local authority for compliance with local amendments and conditions only. Appendix N establishes guidelines for replicable buildings that can give jurisdictions a tool they can adopt to help streamline the plan review process in regard to code compliance.

The ICC *Guideline for Replicable Buildings*, ICC G1-2010, provides jurisdictions with guidance regarding the process of expedited plan review for replicable buildings. The Appendix N provisions, based on the ICC G1 guideline, are intended to benefit both owners and local jurisdictions by:

- Enhancing public safety through a more uniform review process,
- Conserving local resources through the elimination of repetitive reviews of transportable plans, and
- Reducing the time between permit submittal and construction mobilization.

One of the major benefits of the *International Codes* (I-Codes®) is the establishment of consistent code provisions for states and local jurisdictions to enforce. Consistent enforcement is very important to national corporations and franchise chains that wish to construct similar or identical buildings across the country. The review process by an approved agency can greatly enhance the consistency of code application for replicable buildings and support consistent enforcement of the I-Codes while maintaining local jurisdictional authority.

Implementing a building document review process to examine and verify the replicable construction documents comply with the current applicable I-Codes can save considerable state and local resources and time by eliminating repetitive code-compliance reviews. Local jurisdictions can then utilize their resources to focus on reviews of complex and high-risk projects. By coupling a global review of replicable documents with a local review of unique jurisdictional requirements that differ from the applicable I-Code, replicable buildings that utilize this optional regulatory review process can be constructed more cost effectively and with greater consistency. Allowing owners, architects, builders, and engineers to submit a design, with its various facades and options, can also allow for more thorough plan examinations, reduced construction times, fewer in-field change orders, more meaningful inspections, and quicker occupancy of the finished building.

The application of a replicable design is to be consistent in all of the fundamental code areas, such as fire- and life-safety, accessibility, structural stability, and energy efficiency, as well as plumbing, electrical, and mechanical systems. The proposed design criteria as submitted must be reviewed by an approved individual or agency for every code discipline reviewed. Site-specific issues must be identified and approved, including variations from the replicable design, local requirements, and foundation design.

Appendix O *Performance-Based Application*

Performance-based design approaches are becoming more common in new and existing construction. Chapter 13 of the *International Existing Building Code®* (IEBC®) provides a performance-based approach that can be used for existing buildings while the ICC *Performance Code® for Buildings and Facilities* includes provisions for new and existing construction. Many building officials have found it difficult to determine what should be provided for review and whether the performance approach provided meets the intent of the IBC. This appendix outlines administrative provisions that can assist building officials in prescribing what is expected for permit applications that incorporate performance-based designs. These administrative provisions were extracted from the ICC *Performance Code* and have been provided in an easy-to-follow format.

Appendix P *Sleeping Lofts*

Appendix P sets forth the scoping limitations and technical criteria for sleeping lofts that are provided within Group R dwelling units and sleeping units. The presence of sleeping lofts in residential units recognizes several important goals. This design concept provides for a balance between flexibility in design and maintenance of a minimum level of safety. As housing affordability has become increasingly important, the use of sleeping lofts allows for densification of multiple-family residential housing, allowing for additional sleeping space within the same building footprint. Such lofts are particularly beneficial for increasing usable space in very small units, which have become more popular based on sustainability and living more simply.

The technical provisions of Appendix P are based in large part on Appendix AQ of the *International Residential Code* regulating tiny houses. Appendix P provides allowances and limitations on designed spaces specifically identified as a sleeping loft, while differentiating these small spaces from mezzanines and other habitable space. Small sleeping lofts are exempt from the provisions of Appendix P provided they are less than 3 feet (914 mm) in depth, less than 35 square feet (3.3 m^2) in floor area, or lack a permanent means of egress. Key areas of regulation include means of egress, guards, and smoke alarms. It is important to note that sleeping lofts must comply with the base provisions of the IBC unless such provisions are modified by the appendix.

METRIC CONVERSION TABLE

Metric Units, System International (SI)
Soft Metrication
Hard Metrication

Metric Units, System International (SI)

The most widely used system of units and measures around the world is the Systeme International d'Unites (SI), which is the modern form of the metric system. The other system of measurement is the U.S. Customary System Units, also known as "English Units," consisting of the mile, foot, inch, gallon, second, and pound. Although the English system is gradually being replaced by the metric system in some sectors of U.S. industry, the full conversion of the U.S. to the metric system is still incomplete.

Using metric units or providing metric equivalents is important because of the following reasons:

1. Use of metric units facilitates understanding and communication in technical areas such as engineering, architecture, building codes, and other scientific areas at a global level.
2. Use of metric units is simpler because variations from smaller to larger units or vice versa are in multiples of 10, but in English units the multipliers could vary with unit and with subject. For example, smaller units of an inch are ½, ¼, ⅛, or 1/16 of an inch, each a multiplier of 2 larger than the other. Conversion of inches to feet is at a multiplier of 12 and from feet to yards at a multiplier of 3. For metric, the smaller unit of a centimeter is a millimeter, which is a centimeter divided by 10. Conversion of centimeters to decimeters is by multiplying a centimeter by 10, and conversion of decimeters to meters is by multiplying by 10, and so on. Accordingly, computations and problem solving are prone to less error.

The conversion to metric units can take two forms—soft metrication and hard metrication.

Soft Metrication

Soft metrication is the use of metric units in specifying measurements, sizes, and other dimensions without changing product sizes and without changing the everyday practice of using English units. For example, a wood-stud member commonly used is a 2 × 4, which is actually 1.5 inches × 3.5 inches. To report or to specify the actual size of this member in metric, a soft conversion of 38 mm × 89 mm is used (rounded from actual 38.1 and 88.9). Another example could be the load-bearing pressure of clay soils of 1,500 pounds per square foot being reported as 72 kPa (kilopascals).

Hard Metrication

Hard metrication goes beyond soft metrication and converts production and manufacturing based on metric sizes. For example, instead of manufacturing 2 × 4 wood studs of 1.5 inches × 3.5 inches (38.1 mm × 88.9 mm, actual dimensions), wood studs of 40 mm by 90 mm might be manufactured. Another example is ½-inch-diameter (12.7-mm) U.S. size automotive bolts versus 13-mm metric bolts, which are manufactured with a diameter of 13 mm. The production of other structural or nonstructural members such as structural steel, plywood, nails, pipes, ducts, insulation panels, and all other such elements would also be done in metric rather than manufacturing in English units and reporting metric equivalents.

More information on the SI system in the United States is available from the National Institute of Standards and Technology (NIST), an agency of the U.S. Department of Commerce: http://physics.nist.gov/cuu/Units/index.html or http://physics.nist.gov/Pubs/SP811/contents.html.

Unit Conversion Tables
SI Symbols and Prefixes

BASE UNIT		
Quantity	Unit	Symbol
Length	meter	m
Mass	kilogram	kg
Time	second	s
Electric current	ampere	A
Thermodynamic temperature	kelvin	K
Amount of substance	mole	mol
Luminous intensity	candela	cd

SI SUPPLEMENTARY UNITS		
Quantity	Unit	Symbol
Plane angle	radian	rad
Solid angle	steradian	sr

SI PREFIXES		
Multiplication Factor	Prefix	Symbol
$1\,000\,000\,000\,000\,000\,000 = 10^{18}$	exa	E
$1\,000\,000\,000\,000\,000 = 10^{16}$	peta	P
$1\,000\,000\,000\,000 = 10^{12}$	tera	T
$1\,000\,000\,000 = 10^{9}$	giga	G
$1\,000\,000 = 10^{6}$	mega	M
$1\,000 = 10^{3}$	kilo	k
$100 = 10^{2}$	hecto	h
$10 = 10^{1}$	deka	da
$0.1 = 10^{-1}$	deci	d
$0.01 = 10^{-2}$	centi	c
$0.001 = 10^{-3}$	milli	m
$0.000\,001 = 10^{-6}$	micro	µ
$0.000\,000\,001 = 10^{-9}$	nano	n
$0.000\,000\,000\,001 = 10^{-12}$	pico	p
$0.000\,000\,000\,000\,001 = 10^{-15}$	femto	f
$0.000\,000\,000\,000\,000\,001 = 10^{-18}$	atto	a

SI DERIVED UNITS WITH SPECIAL NAMES

Quantity	Unit	Symbol	Formula
Frequency (of a periodic phenomenon)	hertz	Hz	1/s
Force	newton	N	kg•m/s^2
Pressure, stress	pascal	Pa	N/m^2
Energy, work, quantity of heat	joule	J	N-m
Power, radiant flux	watt	W	J/s
Quantity of electricity, electric charge	coulomb	C	A-s
Electric potential, potential difference, electromotive force	volt	V	W/A
Capacitance	farad	F	C/V
Electric resistance	ohm	Ω	V/A
Conductance	siemens	S	A/V
Magnetic flux	weber	Wb	V-s
Magnetic flux density	tesla	T	Wb/m^2
Inductance	henry	H	Wb/A
Luminous flux	lumen	lm	cd•sr
Luminance	lux	lx	lm/m^2
Activity (of radionuclides)	becquerel	Bq	1/s
Absorbed dose	gray	Gy	J/kg

Conversion Factors

To convert	to	multiply by
LENGTH		
1 mile (U.S. statute)	km	1.609 344
1 yd	m	0.9144
1 ft	m	0.3048
	mm	304.8
1 in	mm	25.4
AREA		
1 mile^2 (U.S. statute)	km^2	2.589 998
1 acre (U.S. survey)	ha	0.404 6873
	m^2	4046.873
1 yd^2	m^2	0.836 1274
1 ft^2	m^2	0.092 903 04
1 in^2	mm^2	645.16
VOLUME, MODULUS OF SECTION		
1 acre ft	m^3	1233.489
1 yd^3	m^3	0.764549
100 board ft	m^3	0.235 9737
1 ft^3	m^3	0.028316 85
	L (dm^3)	28.3168
1 in^3	mm^3	16 387.06
	mL (cm^3)	16.3871
1 barrel (42 U.S. gallons)	m^3	0.158 9873
(FLUID) CAPACITY		
1 gal (U.S. liquid)*	L**	3.785 412
1 quart (U.S. liquid)	mL	946.3529
1 pint (U.S. liquid)	mL	473.1765
1 fl oz (U.S.)	mL	29.5735
1 gal (U.S. liquid)	m^3	0.003 785 412
*1 gallon (UK) approx. 1.2 gal (U.S.)	**1 liter approx. 0.001 cubic meter	
SECOND MOMENT OF AREA		
1 in^4	mm^4	416 231 4
	m^4	416 231 4 × 10^{-7}
PLANE ANGLE		
1° (degree)	rad	0.017 453 29
	mrad	17.453 29
1′ (minute)	μrad	290.8882
1″ (second)	μrad	4.8481 37
VELOCITY, SPEED		
1 ft/s	m/s	0.3048
1 mile/h	km/h	1.609 344
	m/s	0.447 04

To convert	to	multiply by
VOLUME RATE OF FLOW		
1 ft³/s	m³/s	0.028 316 85
1 ft³/min	L/s	0.471 9474
1 gal/min	L/s	0.063 0902
1 gal/min	m³/min	0.0038
1 gal/h	mL/s	1.051 50
1 million gal/d	L/s	43.8126
1 acre ft/s	m³/s	1233.49
TEMPERATURE INTERVAL		
1°F	°C or K	0.555 556 5/9 °C = 5/9 K
EQUIVALENT TEMPERATURE ($t_{°C} = t_K - 273.15$)		
$t_{°F}$	$t_{°C}$	$t_{°F} = 9/5\, t_{°C} + 32$
MASS		
1 ton (short***)	metric ton	0.907 185
	kg	907.1847
1 lb	kg	0.453 5924
1 oz	g	28.349 52
***1 long ton (2,240 lb)	kg	1016.047
MASS PER UNIT AREA		
1 lb/ft²	kg/m²	4.882 428
1 oz/yd²	g/m²	33.905 75
1 oz/ft²	g/m²	305.1517
DENSITY (MASS PER UNIT VOLUME)		
1 lb/ft³	kg/m³	16.01846
1 lb/yd³	kg/m³	0.593 2764
1 ton/yd³	t/m³	1.186 553
FORCE		
1 tonf (ton-force)	kN	8.896 46
1 kip (1,000 lbf)	kN	4.448 22
1 lbf (pound-force)	N	4.448 22
MOMENT OF FORCE, TORQUE		
1 lbf•ft	N•m	1.355 808
1 lbf•in	N•m	0.112 9848
1 tonf•ft	kN•m	2.711 64
1 kip•ft	kN•m	1.355 82

To convert	to	multiply by
FORCE PER UNIT LENGTH		
1 lbf/ft	N/m	14.5939
1 lbf/in	N/m	175.1268
1 ton/ft	kN/m	29.1878
PRESSURE, STRESS, MODULUS OF ELASTICITY (FORCE PER UNIT AREA) (1 Pa = 1 N/m^2)		
1 tonf/in^2	MPa	13.7895
1 tonf/ft^2	kPa	95.7605
1 kip/in^2	MPa	6.894 757
1 lbf/in^2	kPa	6.894 757
1 lbf/ft^2	Pa	47.8803
Atmosphere	kPa	101.3250
1 inch mercury	kPa	3.376 85
1 foot (water column at 32°F)	kPa	2.988 98
WORK, ENERGY, HEAT (1 J = 1 N • m = 1 W • Ws)		
1 kWh (550 ft•lbf/s)	MJ	3.6
1 Btu (Int. Table)	kJ	1.055 056
	J	1055.056
1 ft•lbf	J	1.355 818
COEFFICIENT OF HEAT TRANSFER		
1 Btu/(ft^2•h•°F)	W/(m•K)	5.678 263
THERMAL CONDUCTIVITY		
1 Btu/(ft•h•°F)	W/(m^2•K)	1.730 735
ILLUMINANCE		
1 lm/ft^2 (footcandle)	lx (lux)	10.763 91
LUMINANCE		
1 cd/ft^2	cd/m^2	10.7639
1 foot lambert	cd/m^2	3.426 259
1 lambert	kcd/m^2	3.183 099

S. K. Ghosh Associates LLC (SKGA)
Seismic and Building Code Consulting

Your technical resource on structural codes and standards

Our technical support includes:

- **Publications** – various code support materials to aid in the learning and comprehension of specific codes.
- **CodeMasters** – a condensed explanation of difficult code requirements on a particular structural topic in a durable reference tool.
- **Webinars** – continuing education (with PDH).
- **Webinar Recordings** – most live webinars available as a recording.
- **Computer Programs** – tools that make code-compliance quick and simple with intuitive interfaces, easy-to-use workflow, and detailed output reports.
- **Seismic and Building Code Consultancy** – on code compliance and other aspects of structural design.

Learn more about our robust list of services & support material
skghoshassociates.com | 847-991-2700